Remote Sensing of Soils
Mapping, Monitoring, and Measurement

Earth Observation Series

Remote Sensing of Soils

Mapping, Monitoring, and Measurement

Edited by

S. Dharumarajan
ICAR-National Bureau of Soil Survey and Land Use Planning, Regional Centre, Bangalore, Karnataka, India

S. Kaliraj
National Centre for Earth Science Studies (NCESS), Ministry of Earth Sciences, Govt. of India, Thiruvananthapuram, Kerala, India

K. Adhikari
USDA-ARS, Grassland, Soil and Water Research Laboratory, Temple, TX, United States

M. Lalitha
ICAR-National Bureau of Soil Survey and Land Use Planning, Regional Centre, Bangalore, Karnataka, India

N. Kumar
ICAR-National Bureau of Soil Survey and Land Use Planning, Nagpur, Maharashtra, India

ELSEVIER

Elsevier
Radarweg 29, PO Box 211, 1000 AE Amsterdam, Netherlands
The Boulevard, Langford Lane, Kidlington, Oxford OX5 1GB, United Kingdom
50 Hampshire Street, 5th Floor, Cambridge, MA 02139, United States

ISBN: 978-0-443-18773-5

For Information on all Elsevier publications
visit our website at https://www.elsevier.com/books-and-journals

Publisher: Candice G. Janco
Acquisitions Editor: Peter Llewellyn
Editorial Project Manager: Joshua Mearns
Production Project Manager: R. Vijay Bharath
Cover Designer: Miles Hitchen

Typeset by MPS Limited, Chennai, India

Contents

Section 2

Section 3

List of contributors

Mohamed A.E. AbdelRahman Division of Environmental Studies and Land Use, National Authority for Remote Sensing and Space Sciences (NARSS), Cairo, Egypt

S. Abinash Environmental Resources Research Centre (ERRC), Thiruvananthapuram, Kerala, India

K. Adhikari USDA-ARS, Grassland, Soil and Water Research Laboratory, Temple, TX, United States

Mrinaljyoti Adhyapok Civil Engineering Department, Jorhat Engineering College, Jorhat, Assam, India

K.S. Anil Kumar ICAR-National Bureau of Soil Survey and Land Use Planning, Regional Centre, Bangalore, Karnataka, India

S. Anirudhan Environmental Resources Research Centre (ERRC), Thiruvananthapuram, Kerala, India

K.V. Archana ICAR-National Bureau of Soil Survey and Land Use Planning, Regional Centre, Bangalore, Karnataka, India

R.S. Ashikraj Environmental Resources Research Centre (ERRC), Thiruvananthapuram, Kerala, India

S. Bhagyashree Department of Geology, Central University of Tamil Nadu, Thiruvarur, Tamil Nadu, India

S.A. Bhoomika Department of Soil Science and Agricultural Chemistry, University of Agricultural Sciences, Bangalore, Karnataka, India

Zhenxing Bian College of Land and Environment, Shenyang Agricultural University, Shenyang, Liaoning Province, China

Amritee Bora Department of Geography, North Eastern Hill University, Shillong, Meghalaya, India

S. Brundha ICAR-National Bureau of Soil Survey and Land Use Planning, Regional Centre, Bangalore, Karnataka, India

Harwinder Singh Buttar Department of Plant Pathology, Punjab Agricultural University, Ludhiana, Punjab, India

Debashis Chakraborty Division of Agricultural Physics, ICAR-Indian Agricultural Research Institute, New Delhi, India

Ranabir Chakraborty ICAR-National Bureau of Soil Survey and Land Use Planning, Regional Centre, Bangalore, Karnataka, India

K. Chandramohan ICSSR-Post Doctoral Fellow, Tribal Research Centre, Tamil University, Thanjavur, Tamil Nadu, India

N. Chandrasekar Centre for GeoTechnology, Manonmaniam Sundaranar University, Tirunelveli, Tamil Nadu, India

B.M. Chittapur Department of Soil and Water Conservation Engineering, College of Agricultural Engineering, UAS Campus, Raichur, Karnataka, India

M. Ciba National Centre for Earth Science Studies (NCESS), Thiruvananthapuram, Kerala, India

Pankaj Dahiya Department of Agricultural Meteorology, CCS Haryana Agricultural University, Hisar, Haryana, India

Mallikarjuna Dandu Department of Soil and Water Conservation Engineering, College of Agricultural Engineering, UAS Campus, Raichur, Karnataka, India

Bappa Das ICAR-Central Coastal Agricultural Research Institute, Old Goa, Goa, India

Parinita Das Department of Agriculture Bioinfromatics, ICAR-Indian Agricultural Statistics Research Institute, New Delhi, India

Saurav Das Department of Agronomy and Horticulture, University of Nebraska, Lincoln, NE, United States

Prava Kiran Dash Department of Soil Science, College of Agriculture, Odisha University of Agriculture and Technology, Bhubaneswar, Odisha, India; Regional Research and Technology Transfer Station, Mahisapat, Dhenkanal, Odisha, India

Anu David Raj Agriculture and Soil Department, Indian Institute of Remote Sensing (IIRS), Indian Space Research Organization (ISRO), Dehradun, Uttarakhand, India

B.K. Desai Department of Soil and Water Conservation Engineering, College of Agricultural Engineering, UAS Campus, Raichur, Karnataka, India

A.G. Devendra Prasad ICAR-National Bureau of Soil Survey and Land Use Planning, Regional Centre, Bangalore, Karnataka, India

Arvind Dhaloiya Department of Soil and Water Engineering, Punjab Agricultural University, Ludhiana, Punjab, India

S. Dharumarajan ICAR-National Bureau of Soil Survey and Land Use Planning, Regional Centre, Bangalore, Karnataka, India

Mrinal Kumar Dutta Civil Engineering Department, Jorhat Engineering College, Jorhat, Assam, India

Pujarini Dutta Agriculture and Soil Department, Indian Institute of Remote Sensing (IIRS), Indian Space Research Organization (ISRO), Dehradun, Uttarakhand, India

P. Elayapillai Department of Literature, Tamil University, Thanjavur, Tamil Nadu, India

Merina Englengpi Civil Engineering Department, Jorhat Engineering College, Jorhat, Assam, India

Chongwei Gan College of Land and Environment, Shenyang Agricultural University, Shenyang, Liaoning Province, China

Deepak Ghimire Department of Agronomy and Horticulture, University of Nebraska, Lincoln, NE, United States

Dibakar Ghosh ICAR-Indian Institute of Water Management, Bhubaneswar, Odisha, India

Dipanwita Haldar Agriculture and Soil Department, Indian Institute of Remote Sensing (IIRS), Indian Space Research Organization (ISRO), Dehradun, Uttarakhand, India

Chunlan Han College of Land and Environment, Shenyang Agricultural University, Shenyang, Liaoning Province, China

G.K. Harikaran Department Soil Science and Agricultural Chemistry, University of Agricultural Sciences, Bangalore, India; ICAR-National Bureau of Soil Survey and Land Use Planning, Regional Centre, Bangalore, Karnataka, India

Prarthana P. Hazarika Department of Soil Science, Assam Agricultural University, Jorhat, Assam, India

Rajendra Hegde ICAR-National Bureau of Soil Survey and Land Use Planning, Regional Centre, Bangalore, Karnataka, India

P. Hittanagi ICAR-National Bureau of Soil Survey and Land Use Planning, Regional Centre, Bangalore, Karnataka, India

Roomesh Kumar Jena ICAR-Indian Institute of Water Management, Bhubaneswar, Odisha, India

R.K. Jhorar Department of Soil and Water Engineering, College of Agricultural Engineering and Technology, Chaudhary Charan Singh Haryana Agricultural University, Hisar, Haryana, India

Xinxin Jin College of Land and Environment, Shenyang Agricultural University, Shenyang, Liaoning Province, China

B. Kalaiselvi ICAR-National Bureau of Soil Survey and Land Use Planning, Regional Centre, Bangalore, Karnataka, India

Justin George Kalambukattu Agriculture and Soil Department, Indian Institute of Remote Sensing (IIRS), Indian Space Research Organization (ISRO), Dehradun, Uttarakhand, India

Ibrahim Kaleel Water Technology, Tocklai Tea Research Institute, Jorhat, Assam, India

S. Kaliraj National Centre for Earth Science Studies (NCESS), Ministry of Earth Sciences, Govt. of India, Thiruvananthapuram, Kerala, India

K.S. Karthika ICAR-National Bureau of Soil Survey and Land Use Planning, Regional Centre, Bangalore, Karnataka, India

K. Karunya Lakshmi ICAR-National Bureau of Soil Survey and Land Use Planning, Regional Centre, Bangalore, Karnataka, India

M.L. Khichar Department of Agricultural Meteorology, CCS Haryana Agricultural University, Hisar, Haryana, India

M. Khysar Pasha Department of General Studies, Jubail Industrial College, Jubail Industrial City, Kingdom of Saudi Arabia

N. Kiruthika Department of Remote Sensing, Bharathidasan University, Tiruchirappalli, Tamil Nadu, India

A.N. Kondareddy Department of Soil Science, Assam Agricultural University, Jorhat, Assam, India

K. Anoop Krishnan National Centre for Earth Science Studies (NCESS), Thiruvananthapuram, Kerala, India

Ashish Kumar Department of Soil and Water Engineering, College of Agricultural Engineering and Technology, Chaudhary Charan Singh Haryana Agricultural University, Hisar, Haryana, India

N. Kumar ICAR-National Bureau of Soil Survey and Land Use Planning, Nagpur, Maharashtra, India

Sachin Kumar Department of Agronomy, CCS Haryana Agricultural University, Hisar, Haryana, India

Sanjay Kumar Department of Soil and Water Engineering, College of Agricultural Engineering and Technology, Chaudhary Charan Singh Haryana Agricultural University, Hisar, Haryana, India

Satender Kumar Department of Soil Science, CCS Haryana Agricultural University, Hisar, Haryana, India

Suresh Kumar Agriculture and Soil Department, Indian Institute of Remote Sensing (IIRS), Indian Space Research Organization (ISRO), Dehradun, Uttarakhand, India

Sweta Kumari ICAR-National Bureau of Soil Survey and Land Use Planning, Regional Centre, Bangalore, Karnataka, India

C.G. Kusuma Department of Soil Science and Agricultural Chemistry, University of Agricultural Sciences, Bangalore, Karnataka, India

C. Lakshumanan Department of Remote Sensing, Bharathidasan University, Tiruchirappalli, Tamil Nadu, India

Narayan Lal ICAR-Indian Institute of Soil Science, Bhopal, Madhya Pradesh, India

M. Lalitha ICAR-National Bureau of Soil Survey and Land Use Planning, Regional Centre, Bangalore, Karnataka, India

N. Maddileti ICAR-National Bureau of Soil Survey and Land Use Planning, Regional Centre, Bangalore, Karnataka, India

B. Mahanta Department of Geology, Central University of Tamil Nadu, Thiruvarur, Tamil Nadu, India

Mallikarjuna Department of Soil and Water Engineering UAS, Raichur, Karnataka, India

B.K. Medhi Department of Soil Science, Assam Agricultural University, Jorhat, Assam, India

Mukesh Kumar Mehla Department of Soil and Water Engineering, College of Technology and Engineering, Maharana Pratap University of Agriculture and Technology, Udaipur, Rajasthan, India; Department of Soil and Water Engineering, College of Agricultural Engineering and Technology, Chaudhary Charan Singh Haryana Agricultural University, Hisar, Haryana, India

B.S. Mipun Department of Geography, North Eastern Hill University, Shillong, Meghalaya, India

Hisham Mohamed Department of Remote Sensing, Bharathidasan University, Tiruchirappalli, Tamil Nadu, India

Pravash Chandra Moharana ICAR-National Bureau of Soil Survey and Land Use Planning, Nagpur, Maharashtra, India

Bhabani Prasad Mondal School of Agriculture, SR University, Warangal, Telangana, India; Division of Agricultural Physics, ICAR-Indian Agricultural Research Institute, New Delhi, India; Department of Agriculture, Brainware University, Kolkata, West Bengal, India

L.G.K. Naidu ICAR-National Bureau of Soil Survey and Land Use Planning, Regional Centre, Bangalore, Karnataka, India

Ravindra Naitam ICAR-National Bureau of Soil Survey and Land Use Planning, Nagpur, Maharashtra, India

A. Natarajan ICAR-National Bureau of Soil Survey and Land Use Planning, Regional Centre, Bangalore, Karnataka, India

K.V. Niranjana ICAR-National Bureau of Soil Survey and Land Use Planning, Regional Centre, Bangalore, Karnataka, India

Asna Nizar National Centre for Earth Science Studies (NCESS), Thiruvananthapuram, Kerala, India

Sheik Mujabar P. Department of General Studies, Jubail Industrial College, Jubail Industrial City, Kingdom of Saudi Arabia

K. Palanivel Department of Remote Sensing, Bharathidasan University, Tiruchirappalli, Tamil Nadu, India

Dinesh Panday Rodale Institute–Pocono Organic Center, Long Pond, PA, United States

Manish Parmar Space Applications Centre (SAC-ISRO), Ahmedabad, Gujarat, India

S. Parvathy ICAR-National Bureau of Soil Survey and Land Use Planning, Regional Centre, Bangalore, Karnataka, India

Ashok Kumar Patra ICAR-Indian Institute of Soil Science, Bhopal, Madhya Pradesh, India

C.M. Pradeep ICAR-National Bureau of Soil Survey and Land Use Planning, Regional Centre, Bangalore, Karnataka, India

Upendra Kumar Pradhan ICAR-Indian Agricultural Statistics Research Institute, New Delhi, India

B.H. Prakash ICAR-National Bureau of Soil Survey and Land Use Planning, Regional Centre, Bangalore, Karnataka, India

Fengkui Qian College of Land and Environment, Shenyang Agricultural University, Shenyang, Liaoning Province, China

G Bhupal Raj School of Agriculture, SR University, Warangal, Telangana, India

Ellur Rajath Agriculture and Soil Department, Indian Institute of Remote Sensing (IIRS), Indian Space Research Organization (ISRO), Dehradun, Uttarakhand, India

N.L. Rajesh Department of Soil and Water Conservation Engineering, College of Agricultural Engineering, UAS Campus, Raichur, Karnataka, India

V. Ramamurthy ICAR-National Bureau of Soil Survey and Land Use Planning, Regional Centre, Bangalore, Karnataka, India

M. Rameshan IHRML Project, Landscape Level Project Management Unit, Munnar, Kerala, India

I. Rashmi ICAR-Indian Institute of Soil and Water Conservation, Regional Centre, Kota, Rajasthan, India

Prasenjit Ray ICAR-Indian Agricultural Research Institute, New Delhi, India

Mallikarjun Reddy Department of Soil and Water Engineering UAS, Raichur, Karnataka, India

R.G. Rejith Environmental Resources Research Centre (ERRC), Thiruvananthapuram, Kerala, India

Partha Deb Roy ICAR-Indian Institute of Water Management, Bhubaneswar, Odisha, India

Udayan Rudra Bhowmick Department of Agronomy, Bidhan Chandra Krishi Viswavidyalaya, Mohanpur, Nadia, West Bengal, India

Jayant Kumar Saha ICAR-Indian Institute of Soil Science, Bhopal, Madhya Pradesh, India

Kaushik Saha Department of Soil Science and Agriculture Chemistry, University of Agricultural Sciences, GKVK, Bangalore, Karnataka, India

Rabi Narayan Sahoo Division of Agricultural Physics, ICAR-Indian Agricultural Research Institute, New Delhi, India

Asha Sahu ICAR-Indian Institute of Soil Science, Bhopal, Madhya Pradesh, India

Nisha Sahu ICAR-Indian Institute of Soil Science, Bhopal, Madhya Pradesh, India

U. Satish Kumar Department of Soil and Water Engineering UAS, Raichur, Karnataka, India; Department of Soil and Water Conservation Engineering, College of Agricultural Engineering, UAS Campus, Raichur, Karnataka, India

K.V. Seema ICAR-National Bureau of Soil Survey and Land Use Planning, Regional Centre, Bangalore, Karnataka, India

P. Senthilvalavan Annamalai University, Annamalai Nagar, Chidambaram, Tamil Nadu, India

Gulshan Kumar Sharma ICAR-Indian Institute of Soil & Water Conservation, Regional Centre, Kota, Rajasthan, India

B.N. Shashikumar ICAR-National Bureau of Soil Survey and Land Use Planning, Regional Centre, Bangalore, Karnataka, India

S.K. Singh ICAR-National Bureau of Soil Survey and Land Use Planning, Regional Centre, Bangalore, Karnataka, India

N.K. Sinha ICAR-Indian Institute of Soil Science, Bhopal, Madhya Pradesh, India

G. Sreenivasulu Department of Geology, Yogi Vemana University, Kadapa, Andhra Pradesh, India

Reji Srinivas National Centre for Earth Science Studies (NCESS), Ministry of Earth Sciences (MoES), Govt. of India, Thiruvananthapuram, Kerala, India

S. Srinivas ICAR-National Bureau of Soil Survey and Land Use Planning, Regional Centre, Bangalore, Karnataka, India

R. Srinivasan ICAR-National Bureau of Soil Survey and Land Use Planning, Regional Centre, Bangalore, Karnataka, India

Amar Suputhra ICAR-National Bureau of Soil Survey and Land Use Planning, Regional Centre, Bangalore, Karnataka, India

R.K. Thakuria Horticultural Research Station, Assam Agricultural University, Jorhat, Assam, India

Pribin Mathew Thomas Environmental Resources Research Centre (ERRC), Thiruvananthapuram, Kerala, India

B. Upendra National Centre for Earth Science Studies (NCESS), Thiruvananthapuram, Kerala, India; Department of Applied Chemistry, Cochin University of Science and Technology (CUSAT), Kochi, Kerala, India

G. Ushakiran ICAR-National Bureau of Soil Survey and Land Use Planning, Regional Centre, Bangalore, Karnataka, India

E. Vairaveni Department of Remote Sensing, Bharathidasan University, Tiruchirappalli, Tamil Nadu, India

R. Vasundhara ICAR-National Bureau of Soil Survey and Land Use Planning, Regional Centre, Bangalore, Karnataka, India

D.H. Venkatesh ICAR-National Bureau of Soil Survey and Land Use Planning, Regional Centre, Bangalore, Karnataka, India

K.N. Vidya ICAR-National Bureau of Soil Survey and Land Use Planning, Regional Centre, Bangalore, Karnataka, India

Vidyashree Department of Soil and Water Conservation Engineering, College of Agricultural Engineering, UAS Campus, Raichur, Karnataka, India

G. Vijayalakshmi Bharath Niketan Engineering College, Theni, Tamil Nadu, India

Shuai Wang College of Land and Environment, Shenyang Agricultural University, Shenyang, Liaoning Province, China; Institute of Bio- and Geosciences, Agrosphere (IBG-3), Forschungszentrum Jülich GmbH, Juelich, Germany; Key Laboratory of Ecosystem Network Observation and Modeling, Institute of Geographic Sciences and Natural Resources Research, Chinese Academy of Sciences, Beijing, China

Yan Wang College of Tourism and Geography, Jiujiang University, Jiujiang, China

Zicheng Wang College of Land and Environment, Shenyang Agricultural University, Shenyang, Liaoning Province, China

Brijesh Yadav ICAR-National Bureau of Soil Survey and Land Use Planning, Regional Centre, Udaipur, Rajasthan, India

Xingyu Zhang College of Land and Environment, Shenyang Agricultural University, Shenyang, Liaoning Province, China

Section 1

Chapter 1

Introduction to soils: soil formation, composition, and its spatial distribution

K. Adhikari[1], M. Lalitha[2], S. Dharumarajan[2], S. Kaliraj[3], Ranabir Chakraborty[2] and N. Kumar[4]
[1]USDA-ARS, Grassland, Soil and Water Research Laboratory, Temple, TX, United States, [2]ICAR-National Bureau of Soil Survey and Land Use Planning, Regional Centre, Bangalore, Karnataka, India, [3]National Centre for Earth Science Studies (NCESS), Ministry of Earth Sciences, Govt. of India, Thiruvananthapuram, Kerala, India, [4]ICAR-National Bureau of Soil Survey and Land Use Planning, Nagpur, Maharashtra, India

1.1 Soil perspective and definition

People have different thoughts and perceptions about soil. From a geologist's point of view, the soil is an epidermal unit of a geological body of material through which all material must pass in the erosion cycle from rock to sediments. Ecologists look at soil as a part of the environment conditioned by organisms and in turn influences the organisms. Soil is one of several records as per anthropologists. To the philosopher, soil is a blend of the forces of life and death. Chemists consider soil as a medium into which mineral matter is placed for plant growth. For farmers, soil is what their livelihood depends on. They cultivate it for food and biomass, they grow, they pray, laugh, fight, and ultimately die with it. To the pedologist, soil is a natural three-dimensional body that has been formed at the earth's surface through the actions and interactions of climate (*cl*) and living organism (*o*) on parent material (*p*) over time (*t*) conditioned by topography (*r*). Its formation can be explained by a fundamental equation: $s = f(cl, o, r, p, t)$, where s is soil, and the variables in the equations are soil-forming factors (Jenny, 1941).

1.2 Composition of soil

Earth was initially in a state of molten magma. Cooling and crystallization of molten magma led to the formation of igneous rocks, which are a source of sedimentary (formed from sediments deposited by water, wind, and glaciers) and metamorphic rocks (transformation of existing rocks by metamorphic processes). The composition of the rocks depends upon the type of minerals present in it. Weathering of the rocks results in the formation of soil parent material, which constitutes about 45% volume of soil solid. Therefore, there is a similarity between the chemical compositions of the soil to that of the earth's crust. Soils are composed of mineral matter, organic materials in various stages of decomposition, and water and gases within soil pores (Fig. 1.1). The amounts of each component and their arrangement into a particular soil depends on soil morphology; and the relative proportion of these four components greatly influences the behavior and properties of soils at a given location.

1.3 Soil genesis

Soil genesis deals about soils with respect to geological perspective, soil-forming factors (pedogenic factors), and soil-forming process (pedogenic process), and as a medium to delivering ecosystem services. The parent material is a state of the soil system at time zero of soil formation. Soil formation proceeds further by the combined interaction of soil-forming factors and soil-forming processes. Soil-forming factors set the stage for the operation of processes of soil formation and have an influence on the properties of the soils. The soil-forming processes are the reorganization and rearrangement of mineral and organic soil constituents resulting in contrasting soil layers at different stages of development leading to soil development, and the distinct soil layers are called soil horizons (Wakatsuki and Rasyidin, 1992). The development of soil

Remote Sensing of Soils. DOI: https://doi.org/10.1016/B978-0-443-18773-5.00015-6

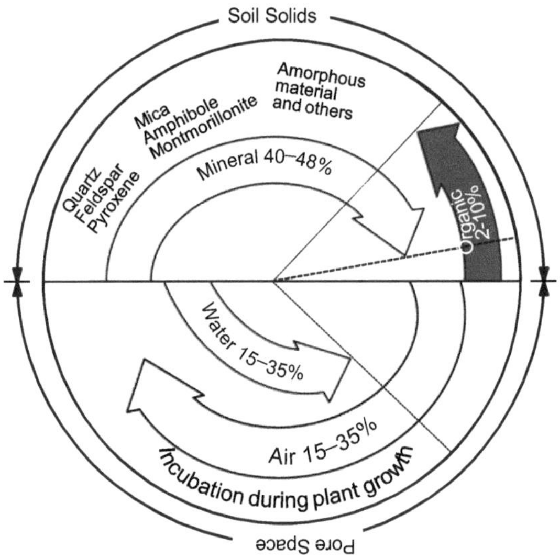

FIGURE 1.1 Volumetric composition of soil under normal conditions.

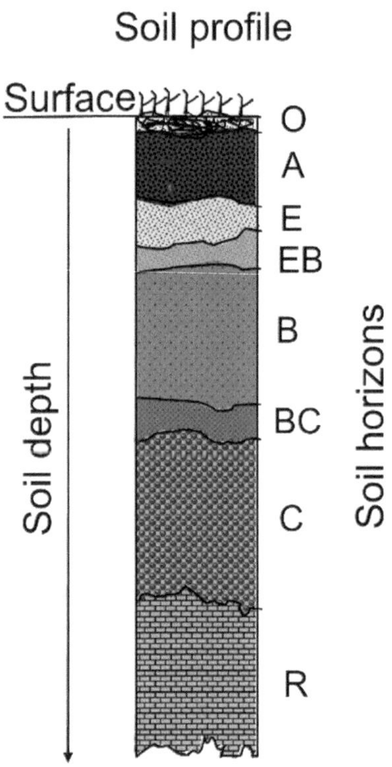

FIGURE 1.2 Schematic representation of a soil profile and the development of soil horizons.

horizons (horizonation) in the upper regolith is a unique characteristic of soil that sets it apart from the deeper consolidated materials. Based on the properties and nature of horizonation, they are designated by horizon symbols, which consist of one or two capital letters for the master horizon and lowercase letter suffixes for subordinate distinctions within master horizons. The master horizons include the capital letters H (saturated and organic), O (unsaturated and organic), A (mineral horizon below O horizon), E (bleached or eluviated horizon), B (illuviated subsurface horizon), C (unconsolidated materials), and R (hard bedrock) (Fig. 1.2). Sometimes two symbols are combined (e.g., EB) to describe a mixed or a transitional horizon. The individual soil unit generally used in field investigations is called a "pedon," which is an imaginary three-dimensional smallest soil sampling unit that displays a full range of properties and characteristics of a particular soil. A group of similar pedon is called a "polypedon," which characterizes a soil individual in a landscape (Brady and Weil, 2008).

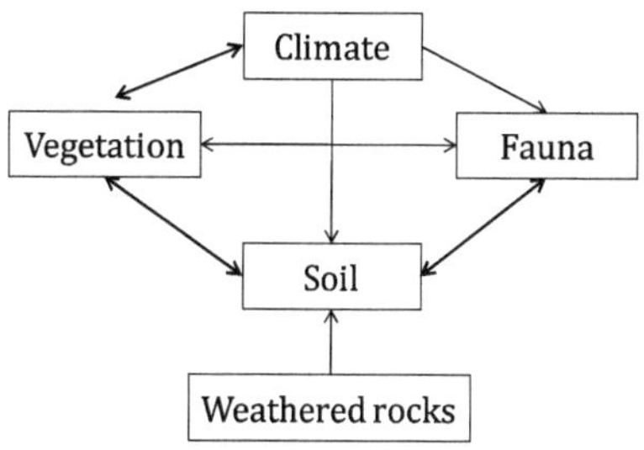

FIGURE 1.3 Interactions of pedogenic factors in soil formation.

There are many concepts of soil genesis and a few are listed here: (1) Pedogenic processes operative over time have varying degrees of expression over space; (2) Present soil at a place is a result of the balance between progressive and regressive pedogenesis; (3) Distinct combination of pedogenic factor and pedogenesis produce distinctive soil; (4) Climate, organisms, relief, parent material, and time are the five significant external factors that influence the internal pedogenic processes; (5) Present-day soils may have imprints of pedogenic factors which was active in the geological past, and (6) Knowledge of soil genesis is important for classification, mapping, management, and use of soil (Fig. 1.3).

1.4 Soil-forming factors

Soil is a complex system and is a dynamic mixture of minerals, organic matter, water, air, and living organisms that form through the interaction of physical, chemical, and biological processes over time. Soil formation is influenced by various factors, including parent material, climate, organisms, relief, and time.

Climate is probably the most significant factor that influences soil formation. Temperature and precipitation patterns affect the rate and type of weathering, erosion, and organic matter decomposition that occurs in the soil. The rate of weathering and other chemical reactions that occur in the soil is largely determined by temperature. Large temperature differences significantly impact the physical weathering of rocks due to uneven expansion of mineral grains, leading to chipping off of the surface layer or exfoliation (Sehgal, 2018). Precipitation is another essential factor that affects soil formation. In climates characterized by excess precipitation over evapotranspiration, there will be a net downward movement of water through the soil profile, and a downward movement of solutes and suspended particles will prevail (Breemen and Buurman, 2003). The amount of precipitation also affects the amount and type of clay minerals in the soil (Tan, 2010).

Living organisms, such as plants, animals, and microorganisms, play a crucial role in soil formation and development. Vegetation protects the soil from rainfall impact, develops stable soil aggregates (Jobbágy and Jackson, 2000), provides organic matter through root exudation and root biomass (Lal, 2005), and accumulates Ca^{+2} from deeper layer (Dijkstra, 2000). Soil microflora, which includes mainly bacteria, fungi, and actinomycetes, play a crucial role in nutrient cycling, organic matter decomposition, and humus formation (López-González et al., 2015; Gierlinger et al., 2004). Soil macrofauna plays a role in creating soil structure by burrowing and tunneling through the soil and improving water and air penetration (Lavelle and Spain, 2001).

Parent material refers to the mineral and organic materials from which soils develop—it is the reference from which soil formation is measured (Breemen and Buurman, 2003). Parent material also influences chemical weathering and the quantity and nature of clay minerals in the profile. The size and shape of the mineral particles influence soil texture, which affects its water-holding capacity, aeration, and nutrient availability (Sposito, 2008). Relief refers to topography or the physical shape of the land surface, and it affects soil formation by influencing water and nutrient flow, erosion, and deposition processes (Jenny, 1941). Topography also influences soil depth and is the main factor of "catena" development. Catena refers to a group of soils that are developed from a similar parent material but occupy distinct topographic positions along the slope in a sequence (top to bottom). Soil-forming processes take time, and the estimation of the relative age or degree of maturity of soils is universally based on the level of weathering and horizon differentiation. In practice, it is generally maintained that the larger the number of horizons and the greater their thickness and intensity, the more mature the soil is (Jenny, 1941) (Fig. 1.4).

1.5 Soil-forming process

The soil-forming process or pedogenesis is extremely complex and dynamic, involving several chemical and biological reactions and processes or a combination of different processes operating under the influence of soil-forming factors.

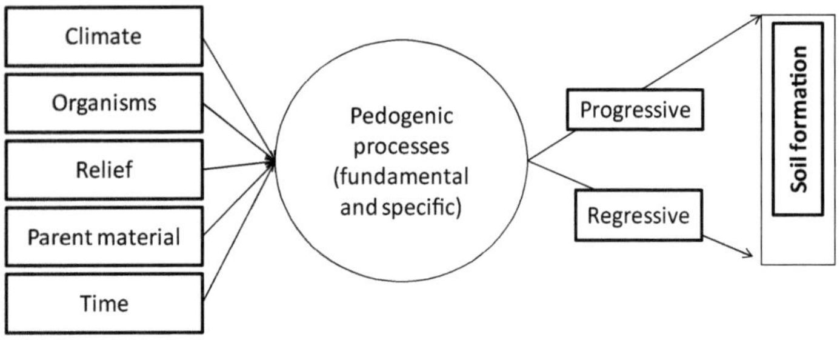

FIGURE 1.4 Factors and processes of soil formation.

TABLE 1.1 List of soil-forming processes and their brief description.

Groups	Processes	Description
Zonal soil formation processes are mainly affected by the prevalent climate and vegetation of a large area.	Calcification	The process by which calcium carbonate accumulates in the soil
	Decalcification	The process by which calcium carbonate is removed from the soil
	Podzolization	The process by which a soil becomes acidic due to the accumulation of organic acids and formation of an iron-rich horizon
	Laterization	The process by which nutrients are leached out and the iron and aluminum oxides accumulates
Intrazonal soil formation processes are mainly affected by local soil conditions.	Salinization	Accumulation of salts in soil profile due to high evapotranspiration demand leading to upward movement of soil water
	Sodization	Partial saturation of the exchange complex with sodium caused by sodium rich groundwater or weathering of a sodic rock in arid climate
	Gleization	Gleization is observed in hydromorphic soils and characterized by a combination of gray and dark (red, brown, black) mottles or concretions of iron and manganese oxides.
	Pedoturbation	Pedoturbation is the process by which soil structure is disrupted through the action of organisms, such as earthworms and termite

Eluviation, illuviation, and humification are the three fundamental soil-forming processes that influence the formation of soil horizons and soil profile development. Eluviation is the process by which minerals and organic matter are removed from the upper soil horizons through the action of water, while illuviation is the process by which these materials are transported and deposited in lower soil horizons. Humification is the process by which organic matter is transformed into humus, a stable organic compound that contributes to soil fertility. It is an important soil-forming process that influences nutrient cycling and soil fertility (Haynes and Swift, 1990). As organic matter decomposes, it is transformed into humus, which is a stable organic compound that can persist in the soil for many years and is instrumental for good soil structure (Sollins et al., 1996). Apart from these fundamental soil formation processes which operate in all soils, there are some specific processes that operate in a particular condition. These processes are divided into two groups: zonal and intrazonal process, and are briefly described in Table 1.1.

Zonal soil formation processes are mainly affected by the prevalent climate and vegetation of a large area. The major zonal processes are calcification, decalcification, podzolization, and laterization. Both calcification and decalcification are important soil-forming processes in arid and semiarid regions. It is closely related to alternate wetting and drying of the profile, separated by long intervals of time or seasonal (Duchaufour, 2020). Podzolization is a dominant process in humid, temperate regions with siliceous parent material and coniferous vegetation. There are three theories of podzolization—the fulvate theory, allophane theory, and low molecular weight organic acid theory—describing the process (Breemen and Buurman, 2003). Laterization is the process by which the soil develops redness and becomes acidic due to the leaching of nutrients and the accumulation of iron and aluminum oxides (Brady, and Weil, 2008; Sehgal, 2018). Intrazonal soil formation processes are mainly affected by local soil conditions. The major processes are salinization, sodification/alkalization, gleization, and pedoturbation.

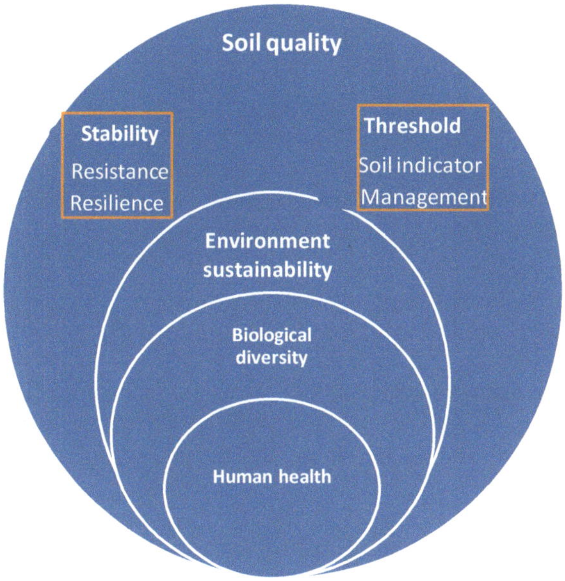

FIGURE 1.5 Concern about soil quality and its challenges. *Modified from Schjonning et al. (2004).*

1.6 Soil health/quality

Soil provides ecosystem services to benefit humankind, and it is linked to different aspects of life on Earth (Adhikari and Hartemink, 2016; Daily et al., 1997; Omuto et al., 2012). Soil health or soil quality is the nexus of environment, human, and societal health through performing a number of ecosystem services. Both soil quality and soil health are interchangeably used in the literature, but according to Lal (2016), soil quality is related to soil functions, whereas soil health considers soil as a living entity that impacts plant health. Soil health status is influenced by physicochemical and biological soil properties (Kibblewhite et al., 2008), which are known as soil health indicators, and The Soil Health Institute (2018) listed 19 major indicators for soil health monitoring and assessment.

Soil quality or soil health is judged either by evaluating the critical soil functions within the ecosystem being performed adequately or by evaluating the number of ecosystem services for which specific soil resources could be used. Despite this, such an evaluation requires assessment and constant monitoring of soil's inherent and dynamic properties. Evaluating inherent soil properties focuses on identifying and grouping soils with similar morphology, properties, or functional characteristics that behave similarly for a given management practice and evaluating each soil group for various uses. The dynamic soil quality evaluation focuses more on the surface soil (0−30 cm) and attempts to describe the status of a specific soil due to relatively recent land use or management decisions (Karlen, 2004). Assessment tools, test kits, minimum datasets, and weightage scores have been developed for the inherent and dynamic properties to identify quality indicators. However, identifying a universal set of indicators may not be appropriate since it changes with each agro-ecological region.

Soil quality monitoring systems are another approach to monitoring soil quality at regular time intervals that involves developing a set of monitoring sites at each agroecological region, land use, and land management. The systems deliver soil quality parameters and changes over time for a given land use and management. Soil quality monitoring sites established within the coffee, rubber, and forest areas of the humid tropics of southern India facilitated the assessment of the soil quality changes upon conversion of forest to coffee and rubber (Nair et al., 2016; Kumar Anil et al., 2016). Evaluating and setting threshold values for soil quality indicators related to crop yield and ecosystem services by considering the nutrient interactions, landscape positions, climate, and land management are crucial for ecosystem sustainability. However, nutrient interactions and land management are yet again dynamic factors. Setting a threshold for land management is also vital in identifying critical soil functions (Fig. 1.5).

1.7 Soil survey, classification, and mapping

Soil is a highly heterogeneous body; thus, it is always difficult to be grouped into different categories. Soil survey provides information on the spatial distribution of different kinds of soils and their characteristics across the landscape.

Though soil survey plays an essential role in increasing basic knowledge on soil and its variability, it is mostly for adopting soil conservation measures, land reclamations, designing irrigation schemes, and transferring agriculture technology, among others. Accordingly, it is classified as either a general purpose or special purpose survey. General-purpose surveys are a systematic inventory of the soil resources of an area, which provides information that can be used directly or interpreted for a wide range of purposes. Special purpose surveys are intended to serve specific objectives, such as designing an irrigation or land reclamation scheme or assessing the suitability of land for growing a particular crop. In either case, soil properties or attributes that are relevant for the objective may be considered. Actual and potential users of soil survey information include agricultural advisory officers, farmers, researchers, foresters, planning agencies, ecosystem conservationists, land development organizations, engineers, and private investors. Soils vary both gradually and abruptly across the landscape since the five soil-forming factors and pedogenic processes vary across the same landscape. Early concepts of pedogenesis were based on soils as products of rock weathering (Simonson, 1986) and the key to the soil classification system was geology. Soil classification is the categorization of soils into groups at varying levels of generalization according to the morphological properties and/or genesis that are important and specific to the purpose of classification.

Very often, soil survey, classification, and mapping are used together to refer to the systematic examination, description, classification, or naming of soil resources, and the production of reports with first-hand soil maps by a pedologist or a soil surveyor. It involves all the necessary processes starting from site description, landscape characterization, soil profile observation, soil sampling, and laboratory analyses, and ultimately the production of reports and soil maps at a given scale.

There are a number of soil classification systems in practice in different parts of the world. One of the oldest systems, the Russian system, which is mainly based on the factors of soil formation, is still widely used in Russia and other countries of the former Soviet Union, and in a few other European countries. Other countries such as China, Belgium, France, Canada, etc. have developed their own system of soil classification which is very specific to the country or region in which the soils are developed. Such national or regional classification systems are of limited use for the rest of the world because they lack a common language for communication and possess different perceptions or understanding of soils and their development, which in many aspects are specific to the local environment. However, a comprehensive and advanced soil classification system—Soil Taxonomy developed by the Soil Survey Staff (Soil Survey Staff, 2010)—has been widely used in the United States, and in more than 55 other countries in the world (Brady and Weil, 2008). Another common soil classification system is the World Reference Base for Soil Resources (IUSS Working Group WRB, 2015), which is the international standard for soil classification systems endorsed by the International Union of Soil Sciences.

Further, in order to have a common system to understand the soil resources of the entire globe, an action toward the preparation of a soil map of the world was started by UN-FAO together with FAO/UNESCO. This task developed a global soil nomenclature system named as FAO Soil Legend based on which the first-ever soil map of the world was published in 1974 at a scale of 1:5,000,000 (FAO-UNESCO, 1974). Later, in 1990, the legend was further revised and an updated Soil Map of the World was published (FAO-UNESCO, 1990). Following this initiative, many countries in the world, including Denmark, have adopted this new soil legend to create national soil maps of their countries at varying scales.

1.8 Limitations

Mapping soils conventionally follows a tacit model of a pedologist or a soil surveyor who explores the relationship of soil to its natural surroundings and delineates different soil types or soil map units. Therefore, soil maps generated may have some limitations, and the process is also time-consuming, less efficient, and expensive. Moreover, soils possess a high spatial variability that generally takes in two forms: (1) continuous, natural variations due to variability in soil-forming factors and (2) discontinuous variation often manifested as inclusions, both in vertical and in horizontal dimensions. The latter form of variation is stochastic, in that measured soil properties cannot be precisely predicted from values of neighboring locations without a large number of samples (Campbell, 1979). Therefore, the uncertainty associated with the misrepresentation of soil boundaries and the variation in subsoil characteristics may or may not be manifested on the surface (Lagacherie et al., 1996). Nevertheless, despite the limitations, it is rather difficult to replace conventional soil surveys with any mechanistic models. However, as an analogy to conventional mapping, the relationship of soil to its natural surrounding can be quantified empirically, and soil maps are generated digitally (e.g., Adhikari et al., 2014). The accuracy of the digital methods depends on the number of profile observations, environmental variables used as predictors, and how well the model quantifies the soil–landscape relationship, among others.

TABLE 1.2 Potential list of environmental variables that can be used as predictors of soil class or soil properties.

Predictor	*Scorpan* factor
Elevation	Relief
Slope	Relief
Aspect	Relief
Topographic position index	Relief
Topographic wetness index	Relief
Plan curvature	Relief
Profile curvature	Relief
Multiresolution of valley bottom flatness index	Relief
Multiresolution ridge top flatness index	Relief
Normalized difference vegetation index	Organism
Enhanced vegetation index	Organism
Soil adjusted vegetation index	Organism
Landsat data (band 1–band 9)	Organism
Land use land cover	Organism
Precipitation	Climate
Average maximum temperature	Climate
Average minimum temperature	Climate
Isothermality	Climate
Temperature seasonality	Climate
Precipitation of wettest month	Climate
Precipitation seasonality	Climate
Lithology or surface geology	Parent material
Gamma radiometric data (potassium, thorium, uranium)	Parent material

1.9 Remote sensing applications in soil survey and mapping

With technological advancement in earth observation, multispectral remote sensing has emerged as an important tool for the collection, mapping, and monitoring of soil resources (Soil Survey Staff, 2017). In landscapes with high soil variability over large areas, soil assessment, and mapping by means of remote sensing data is an important alternative and it has many advantages in terms of coverage and spatial-temporal resolution. For example, Landsat's remotely sensed spectral data serve as useful covariates for soil mapping and are being widely used (Boettinger et al., 2008). These data could be used as a proxy for soil-forming factors to model and map soil distribution using the "scorpan" model following digital soil mapping techniques (McBratney et al., 2003). This "scorpan" model helps in the assessment of quantitative relationships between environmental covariates and soil properties or soil classes to be predicted (Boettinger et al., 2008) and also in the estimation of the error or uncertainty of the prediction. Remote sensing has great potential in digital soil mapping as it provides high-resolution multi-temporal spatial data to benefit soil mapping and predictions. It also offers possibilities to update and improve current soil databases (Mulder et al., 2011). Environmental variables commonly used in predicting soil properties or classes are given in Table 1.2.

$$S_{c/p} = f(s, c, o, r, p, a, n) + e$$

Where $S_{c/p}$ represents soil class or property to be predicted, f is an empirical model or function, s denotes soil, c is the climate, o denotes organisms, r denotes relief, p is the parent material, a denotes age, n is the spatial location, and e is a spatially correlated error.

1.10 Conclusion

Soil is a complex system, and its spatial variability is a natural phenomenon, which makes it rather difficult to predict and map soils accurately. There are different soil classification systems, but Soil Taxonomy and World Reference Base for Soil Classification are the ones that are widely used across nations. Conventional soil mapping is time-consuming and costly, and it can be supplemented with state-of-the-art digital soil mapping techniques. Several environmental variables can be collected efficiently using remote sensing, and the data can be used as a proxy for soil-forming factors for soil predictions and mapping. Terrain attributes derived from the digital elevation model, and spectral bands from Landsat images are some of the examples commonly used in digital soil mapping. Remote sensing has great potential in digital soil mapping.

References

Adhikari, K., Hartemink, A.E., 2016. Linking soils to ecosystem services—a global review. Geoderma 262, 101–111.

Adhikari, K., Minasny, B., Greve, M.B., Greve, M.H., 2014. Constructing a soil class map of Denmark based on the FAO legend using digital techniques. Geoderma 214–215, 101–113.

Boettinger, J.L., Ramsey, R.D., Bodily, J.M., Cole, N.J., Kienast-Brown, S., Nield, S.J., et al., 2008. Landsat spectral data for digital soil mapping. In: Hartemink, A.E., McBratney, A., Mendonca-Santos, M.L. (Eds.), Digital Soil Mapping with Limited Data. Springer-Verlag, Dordrecht, the Netherlands, pp. 193–202.

Brady, N.C., Weil, R.R., 2008. The Nature and Properties of Soils, fourteenth ed. Pearson Education, Inc.

Breemen, V.N., Buurman, P., 2003. Soil Formation. Springer, Dordrecht. Available from: https://doi.org/10.1007/0-306-48163-4_1.

Campbell, J.B., 1979. Spatial variability of soils. Ann Assoc Am Geogr 69 (4), 544–556. Available from: https://doi.org/10.1111/j.1467-8306.1979.tb01281.x.

Daily, G.C., Matson, P.A., Vitousek, P.M., 1997. Ecosystem services supplied by soil. Nature's Services: Societal Dependence on Natural Ecosystems. Island Press, Washington DC, pp. 113–132.

Dijkstra, F., 2000. Effect of tree species on soil properties in a forest of the northeastern United States. PhD Thesis. Wageningen University.

Duchaufour, P., 2020. Handbook of Pedology: Soils, Vegetation, Environment. CRC Press.

FAO-UNESCO, 1974. Soil Map of the World, Legend. FAO, Rome, Italy.

FAO-UNESCO, 1990. Soil Map of the World, Revised Legend. FAO, Rome, Italy.

Gierlinger, N., Jacques, D., Schwanninger, M., Wimmer, R., Pâques, L.E., 2004. Heartwood extractives and lignin content of different larch species (*Larix* sp.) and relationships to brown-rot decay-resistance. Trees 18, 230–236.

Haynes, R.J., Swift, R.S., 1990. Stability of soil aggregates in relation to organic constituents and soil water content. J. Soil. Sci. 41 (1), 73–83.

IUSS Working Group WRB, 2015. World reference base for soil resources 2014, update 2015 international soil classification system for naming soils and creating legends for soil maps. World Soil Resources Reports No. 106. FAO, Rome.

Jenny, H., 1941. Factors of Soil Formation: A System of Quantitative Pedology. McGraw-Hill.

Jobbágy, E.G., Jackson, R.B., 2000. The vertical distribution of soil organic carbon and its relation to climate and vegetation. Ecol. Appl. 10 (2), 423–436.

Karlen, D.L., 2004. Soil quality as an indicator of sustainable tillage practices. Soil Tillage Res 78 (2), 129–130. Available from: https://doi.org/10.1016/j.still.2004.02.001.

Kibblewhite, M.G., Ritz, K., Swift, M.J., 2008. Soil health in agricultural systems. Philos. Trans. R. Soc. B: Biol. Sci. 363, 685–701.

Kumar Anil, K.S., Nair, K.M., Ramesh Kumar, S.C., Srinivas, S., Ramamurthy, V., Maske, S., Jessy, M.D., Hegde, R., Jacob, J., Singh, S.K., 2016. Soil quality monitoring sites for traditional rubber growing areas of southern India. In: NBSS Publ, 1096. National Bureau of Soil Survey and Land Use Planning, Nagpur, India.

Lagacherie, P., Andrieux, P., Bouzigues, R., 1996. Fuzziness and uncertainty of soil boundaries: from reality to coding in GIS. In: Bourrough, P.A., Frank, A.U. (Eds.), Geographic Objects with Indeterminate Boundaries. Taylor & Francis, London, pp. 275–286.

Lal, R., 2005. Forest soils and carbon sequestration. For. Ecol. Manag. 220 (1–3), 242–258.

Lal, R. 2016. Soil health and carbon management, 5, pp. 212–222.

Lavelle, P., Spain, A.V., 2001. Soil organisms. Soil Ecology. Springer, pp. 201–356.

López-González, J.A., Suárez-Estrella, F., Vargas-García, M.C., López, M.J., Jurado, M.M., Moreno, J., 2015. Dynamics of bacterial microbiota during lignocellulosic waste composting: studies upon its structure, functionality and biodiversity. Bioresour. Technol. 175, 406–416.

McBratney, A.B., Mendonça-Santos, M.L., Minasny, B., 2003. On digital soil mapping. Geoderma 117, 3–52.

Mulder, V.L., de Bruin, S., Schaepman, M.E., Mayr, T.R., 2011. The use of remote sensing in soil and terrain mapping — A review. Geoderma 162 (1–2), 1–19.

Nair, K.M., Anil Kumar, Srinivas, S., Nagaraj, J.S., Violet D'Souza, M., Raghuramulu, Y., Hegde, R., Singh, S.K., 2016. Soil Quality Monitoring Sites for traditional coffee-growing areas of India. NBSS Publ.No.1095, National Bureau of Soil Survey and Land Use Planning, Nagpur, India.

Omuto, C., Nachtergaele, F., Rojas, R.V., 2012. State of the Art Report on Global and Regional Soil Information: Where Are We? Where To Go? Food and Agriculture Organization, Rome, Italy.

Schjonning, P., Elmholt, S., Christensen, B.T., 2004. Soil quality management - concepts and terms. CABI Books, CABI. Available from: https://doi.org/10.1079/9780851996714.0001.

Sehgal, J., 2018. Textbook of Pedology: Concepts and Applications. Kalyani Publishers.

Soil Health Institute, 2018. Soil Health Institute, 2018. North American Project to Evaluate Soil Health Measurements.

Soil Survey Staff. 2010. Keys to Soil Taxonomy, eleventh ed. US Department of Agriculture, Natural Resources Conservation Service: Washington, DC.

Simonson, R.W., 1986. Historical aspects of soil survey and soil classification. Part II. 1911–1920. Soil Surv Horiz 27, 3–9.

Sollins, P., Homann, P., Caldwell, B.A., 1996. Stabilization and destabilization of soil organic matter: mechanisms and controls. Geoderma 74 (1–2), 65–105.

Sposito, G., 2008. The Chemistry of Soils, second ed. Oxford University Press.

Tan, K.H., 2010. Principles of Soil Chemistry. CRC Press.

Wakatsuki, T., Rasyidin, A., 1992. Rates of weathering and soil formation. Geoderma 52, 251–263.

Chapter 2

Remote sensing of soils: spectral signatures and spectral indices

S. Dharumarajan[1], M. Lalitha[1], B. Kalaiselvi[1], S. Kaliraj[2], K. Adhikari[3], R. Vasundhara[1], K.V. Niranjana[1], Rajendra Hegde[1], C.M. Pradeep[1], P. Hittanagi[1] and V. Ramamurthy[1]

[1]*ICAR-National Bureau of Soil Survey and Land Use Planning, Regional Centre, Bangalore, Karnataka, India,* [2]*National Centre for Earth Science Studies (NCESS), Ministry of Earth Sciences, Govt. of India, Thiruvananthapuram, Kerala, India,* [3]*USDA-ARS, Grassland, Soil and Water Research Laboratory, Temple, TX, United States*

2.1 Introduction

Soils are mixtures of several mineral and organic constituents and have complex spectral signatures unlike those of vegetation. The most important constituents controlling the shape of soil spectral signatures are iron and organic matter content, as well as moisture and salt content. Soil spectral reflectance is used for quantitative and qualitative analyses of soil properties including soil health indicators, soil fertility, and productivity. Spectral observation results are easy and efficient to obtain, and are reliable with acceptable accuracy. Compared to lab methods which comparatively take more time and are expensive, spectral reflectance based analysis are cost-effective and quick. These methods analyzing electromagnetic wave interaction with the material under study in the ultraviolet, visible (VIS), and infrared (IR) wavelengths. Spectral data of the earth objects indicates their unique characteritics, and they are coupled with multivariate data analysis, then produced the quantitative and qualittative measurements of soil and land for preparation various developmental activities, particularly in soil science.

2.1.1 Spectral characteristics of soils

The use of spectroscopy for the estimation of soil properties has been a reliable alternative to traditional laboratory soil analysis methods. Soil spectral reflectance is the comprehensive representation of soil physical and chemical parameters and their spectral signature can be used for fast and nondestructive estimation of soil properties. Soil chemical bonds of different molecules vibrate at characteristic frequencies when exposed to electromagnetic energy. Energy absorbed, reflected, and scattered in the process may, therefore, be related to specific wavelengths. Such specificity (reflectance at characteristic wavelength) may be treated as a unique spectral feature in unique correspondence to the composition of the end-member. In particular, the specificity allows for the assessment of different soil attributes once spectral reflectance is known and a relationship between the spectral feature and soil attribute is known a priori. Thus, spectral signatures are often considered as inherent soil properties that vary across different soils. Nowadays, severeal remote sensing data products are widely used for estimation of soil properties in local and regional scale, some of widely used data products include Landsat MSS, TM, ETM+, OLI, IRS – LISS 3, LISS 4, AWiFS, Sentinel 2-MSI, IKONS, SPOT, MODIS, Hyperion, ASTER, and AVIRIS-NG, etc.

2.1.2 Spectral sensitivity and soil properties

The factors influencing soil reflectance are soil moisture, organic carbon (OC), clay mineralogy, iron content, particle size, and others (Dharumarajan et al., 2023). *Soil moisture content*: The Oxygen (O) and Hydrogen (H) bond present in water is the strongest absorber of NIR radiation. Fundamental absorption of water in the liquid phase occurs at 3106, 6079, and 2903 nm wavelength, and is associated with the symmetric O–H stretching, H–O–H bending, and

Remote Sensing of Soils. DOI: https://doi.org/10.1016/B978-0-443-18773-5.00033-8

asymmetric O—H stretching, respectively. The overtones of O—H stretching occur near 1400 nm and a combination of the H—O—H bending and O—H stretching occurs around 1900 nm. A combination of metal—OH bending and O—H stretching occurs near 2200—2300 nm. Soil moisture has a dramatic influence on soil albedo and may influence other chromophores. Therefore, an inverse relationship exists between soil moisture content and spectral reflectance. Spectral reflectance decreases as moisture content increases. *OC*: Soil organic matter and its products after decomposition appear to have a noticeable influence on the spectral reflectance of the mineral soils. If OC is removed from the soil, it leads to increase in spectral reflectance. The absorption features associated with organic matter in the very near—IR wavelength are often weak and may not be readily perceptible to the naked eye. But a broad and clear distinction in the overall reflectance with organic matter content is generally seen in the VIS region. *Clay minerals*: Clay minerals have a major influence on the VIS and NIR portion of spectral reflectance. The oxygen and hydrogen functional group is associated with the mineral structure, mostly in 2:1 type octahedral sheets or with water molecules adsorbed to the mineral surface. Thus, all the absorption features related to oxygen hydrogen may also be linked to explain the influence of clay minerals on the spectral reflectance of the soil. Iron content: Iron content has a significant role in defining the shape of the reflectance spectra due to the redox potential of Fe^{2+} and Fe^{3+} present in the soil as iron oxides or impurities. *Particle size*: Spectral reflectance decreases with the increase in particle size. The spectral reflectance of the soil increases exponentially with the decrease in particle size, and the increment was found to be rapid below 0.4 mm diameter. It is observed that the soil surface becomes smooth as the particle size decreases and thus, they related the soil reflectance as a function of its surface roughness. As the soil surface roughness increases, the major portion of the incident energy gets trapped in the inter-aggregate spaces, resulting in lower reflectance.

2.2 Spectral characteristics of different soils

2.2.1 Eroded soils

The spectral evaluation of soil erosion depends on many factors and there is no uniform standard to evaluate the degree of soil erosion. The spectral response of eroded soils shows that the overall level of reflectance spectra increased with an increase in the degree of soil erosion, and both the derivative transformation and waveband reorganization could enhance the spectral information of soil erosion. The eroded soil shows a high correlation between the soil degradation index and spectral parameter. Fig. 2.1 shows spectral signatures of eroded and non-eroded red soils in Northern Karnataka Plateau and, Fig. 2.2 is a representative picture of eroded soils in Karnataka.

2.2.2 Saline and alkaline soils

Soil salinity and alkalinity affects negatively on agricultural productivity. Spectral reflectance can be used for the determination of soil salinity and alkalinity. The vegetation has strong absorption in blue and red light, small reflection peaks in the green light band, and high reflection in the near-IR band, and the spectral curve has obvious peaks and valleys,

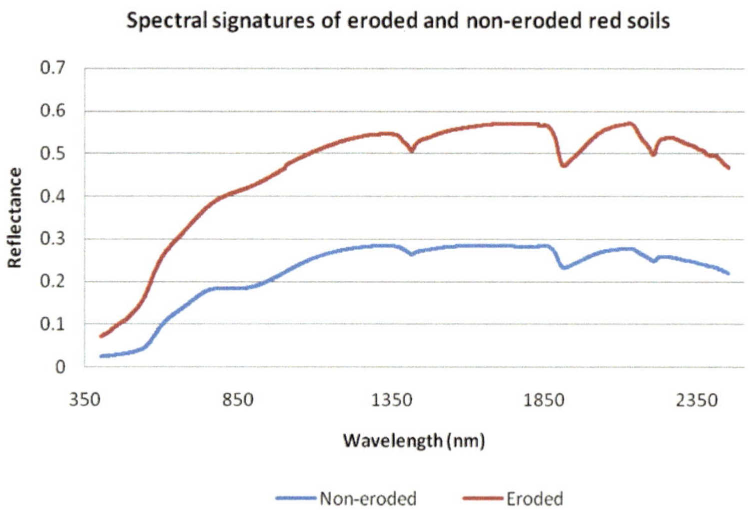

FIGURE 2.1 Spectral signatures of eroded and non-eroded soils of Northern Karnataka.

FIGURE 2.2 Eroded red soils in Chickmagalur district of Karnataka.

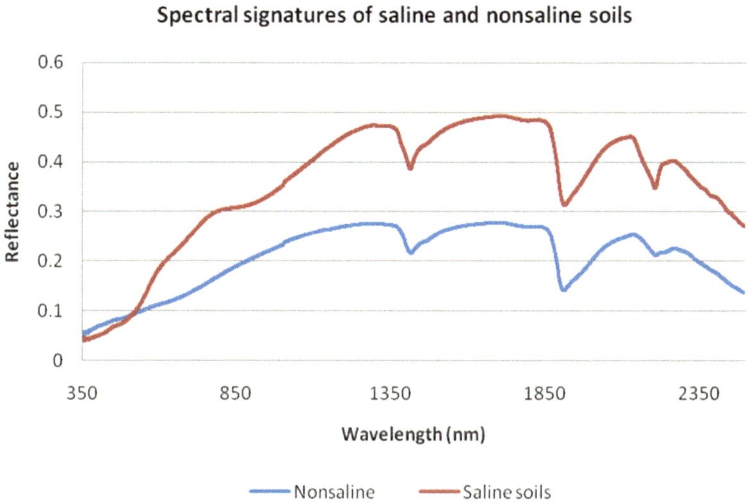

FIGURE 2.3 Spectral signatures of saline and nonsaline soils of Northern Karnataka.

and the whole salt is different. The spectral characteristics of the degree of vegetation tend to be regular; but as the degree of salinization and alkalization increases, the vegetation spectrum increases in the visible light band and decreases in the near-IR band, but does not show regular changes. The spectral reflectance curve of the overlying vegetation canopy on the heavily salinized and alkalized soil is the most gradual. The spectral response of problematic soils shows that the overall level of reflectance spectra increased with increase in the degree of soil salinity and alkalinity. Figs. 2.3 and 2.4 show spectral signatures of saline and alkaline soils in the Northern Karnataka Plateau, and Figs. 2.5 and 2.6 show a representative picture of saline and sodic soils in Bellary district, Karnataka.

2.2.3 Black soil

The soil parent material determines the basic characteristics of reflectance of the black soil, which is a mixture of montmorillonite and illite. Organic matter is the main factor determining the curve shape of black soil reflectance in the region shorter than 1000 nm, and indirectly influencing the reflectance in the region longer than 1000 nm because of the correlation with soil moisture and mechanical composition. The soil moisture mainly changes the reflectance value, but not the curve shape. Black soil reflectance is not influenced significantly by Fe, whereas roughness mainly impacts on the soil reflectance value. Fig. 2.7 shows spectral signatures of red and black soils and, Fig. 2.8 is the corresponding field photos in Northern Karnataka Plateau. Fig. 2.9 shows the spectral signatures of low (OC of $<0.2\%$) and high organic soils (OC of $>1\%$) of Northern Karnataka.

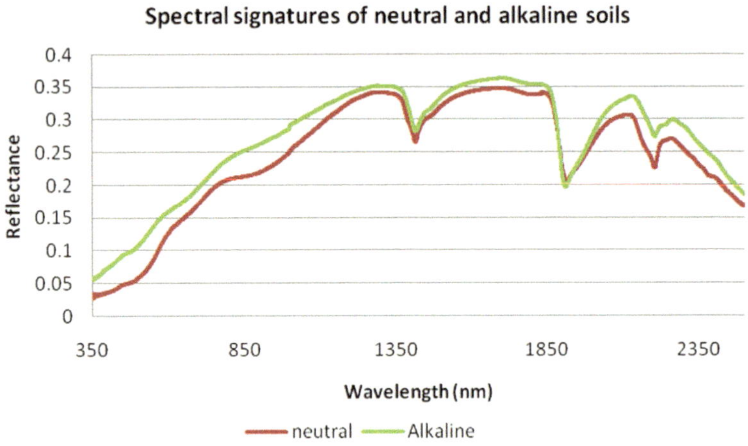

FIGURE 2.4 Spectral signatures of neutral and alkaline soils of Northern Karnataka.

FIGURE 2.5 Barren lands of Karnataka due to high salinity.

FIGURE 2.6 Sodic soils of Bellary district, Karnataka.

2.2.4 Clay minerals

Soil minerals absorb light in the ultraviolet, visible, near-IR, and mid-IR range. Yitagesuab et al. (2011) studied on the spectral characteristics of clay minerals in 2.5–14 μm wavelength region. They found that spectral contrast was high in the 3–5 μm wavelength region, but overall low in the 8–14 μm. The spectra of montmorillonite, illite, and kaolinite were well resolved in the 2.5–14 μm wavelength region. All spectra exhibited strong, broad absorption bands in the

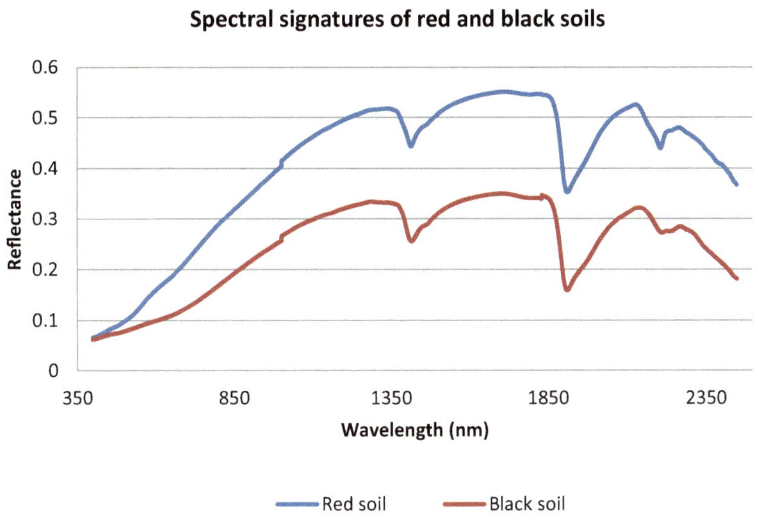

FIGURE 2.7 Spectral signatures of red and black soils of Northern Karnataka.

FIGURE 2.8 Black and red soils in Gadag district, Northern Karnataka.

2.5–3.7 μm wavelength region, with absorption minima at ~2.75 and ~3.1 μm for montmorillonite and illite, being generally broader in the illite spectrum. The kaolinite spectrum showed asymmetric and sharply defined absorption minimum centered at ~2.75 μm. The clay minerals were characterized by strong, diagnostic absorption bands. Thus, spectroscopy in the 2.5–14 μm wavelength region is a useful technique for characterizing clay minerals.

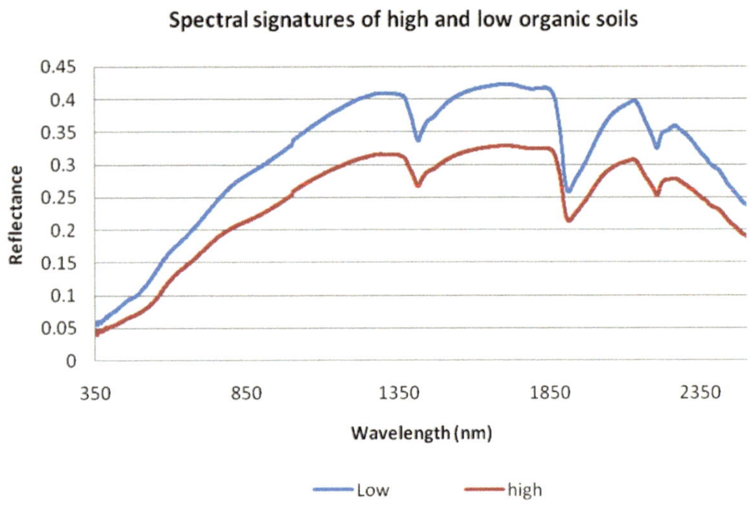

FIGURE 2.9 Spectral signatures of low and high organic soils of Northern Karnataka.

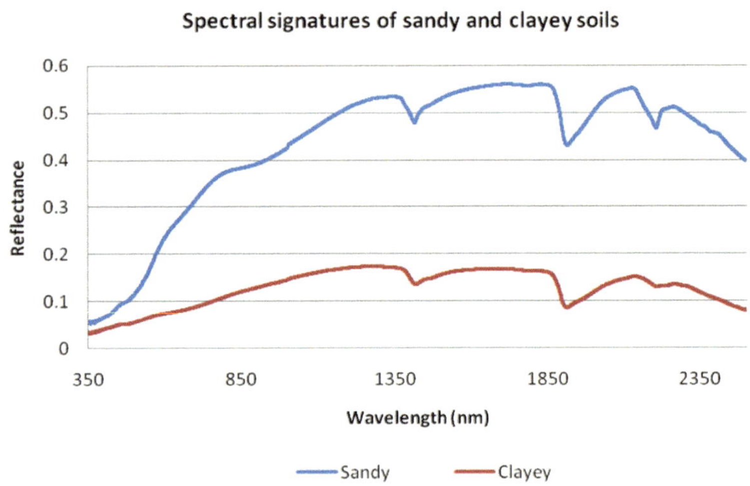

FIGURE 2.10 Spectral signatures of sandy and clayey soils of Northern Karnataka.

2.2.5 Soil texture

Among soil properties, the soil texture is one of the most commonly studied by spectral data, and it has a strong relationship with many other soil properties and functions. The clay particles have direct physical interaction with electromagnetic radiation mainly due to minerals such as iron oxides in the visible portion and aluminosilicates and carbonates in the shortwave-IR region (>2000 nm) (Chabrillat et al., 2002). Among soil textural classes, sandy soil texture has high reflectance compared to loamy and clayey soils. Fig. 2.10 shows the spectral signatures of sandy and clayey soils, and Fig. 2.11 shows the spectral signatures of different textural classes in Northern Karnataka.

2.2.6 Soil orders

The spectral features of soil orders were related to soil horizons and soil properties. Alfisols and Entisols had broad absorption features between 850 and 1100 nm related to the specific absorption shoulder of goethite and hematite. These particular iron oxide absorption bands were not observed in the reflectance spectra of other soil orders because iron oxides are underdeveloped in Inceptisols and Vertisols (Dharumarajan et al., 2022, 2023). The Mollisols, Ultisols, Aridisols, and Inceptisols have similar spectral signatures. Similarly, Alfisols, Aridisols, and Spodosols were fairly well classified from the spectral signatures. Figs. 2.12–2.15 show spectral signatures of different soil orders and corresponding soil profile in Northern Karnataka.

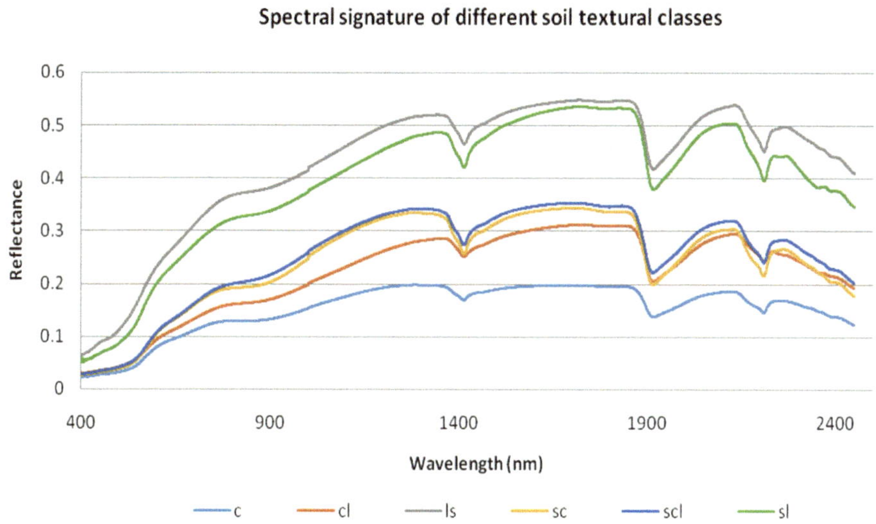

FIGURE 2.11 Spectral signatures of different soil textural classes of Northern Karnataka. c, clay, cl, clay loam, ls, loamysand, sc, sandy clay, scl, sandy clay loam, and sl, sandy loam.

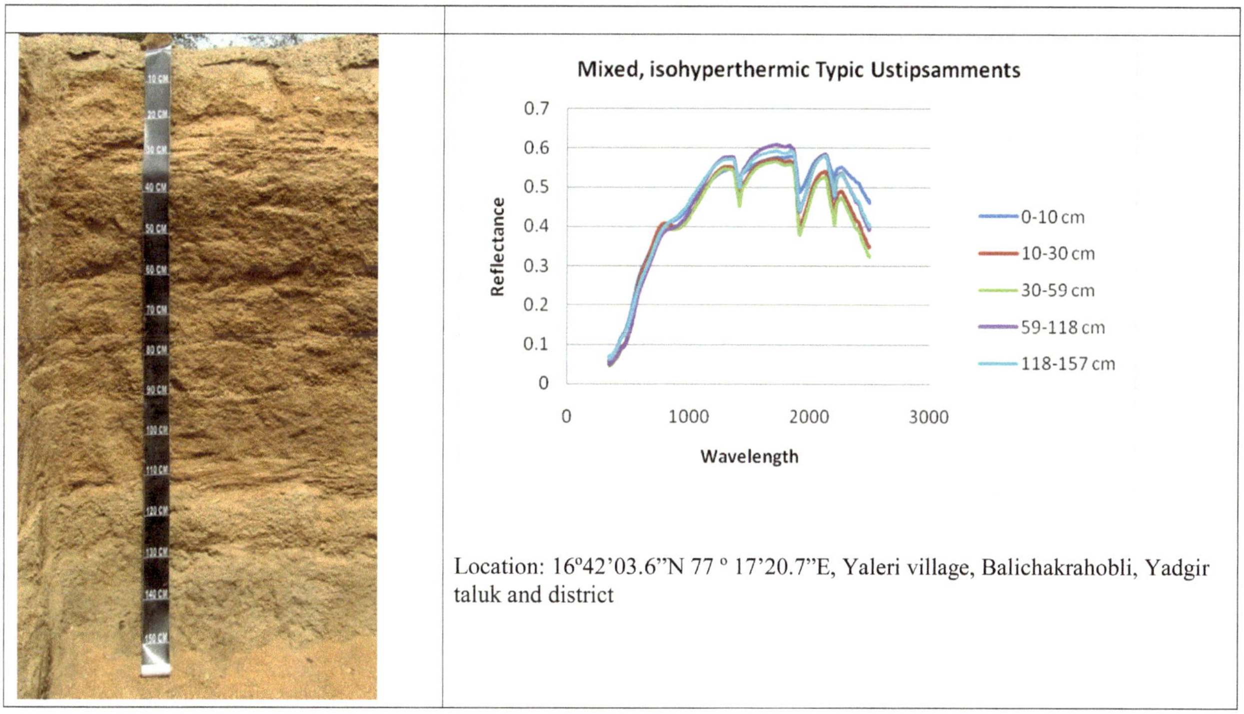

FIGURE 2.12 Spectral signature of Entisol soil profile.

2.3 Spectral indices

The spectral index library offers a simple method for using spectral indices to visualize imagery. The pixel values from two or more spectral bands in a multispectral image are combined to form spectral indices. Spectral indices are intended to draw attention to pixels in an image that demonstrate the relative abundance or absence of a certain type of land cover. Some of the available spectral indices for soil study are structural vegetation indexes, chlorophyll indexes, vegetation water indexes, and spectral absorption metrics. The details of the spectral indices are presented in Table 2.1.

FIGURE 2.13 Spectral signature of Vertisol soil profile.

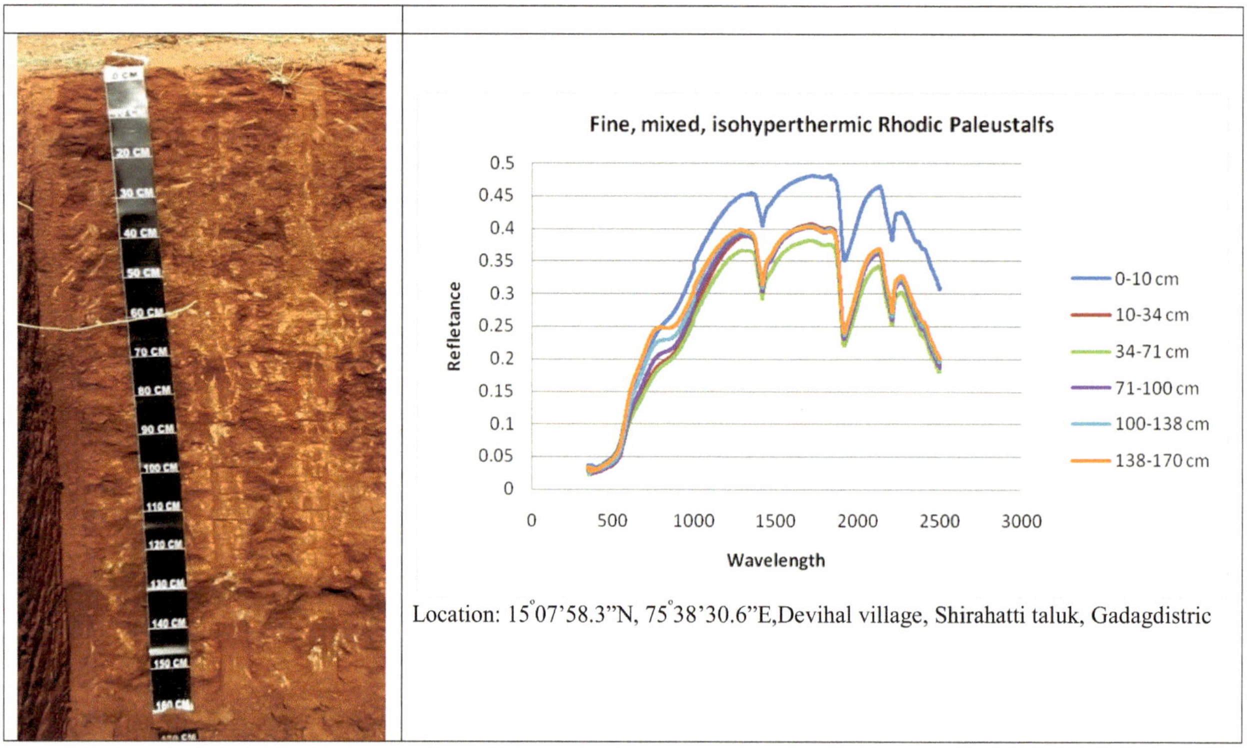

FIGURE 2.14 Spectral signature of Alfisol soil profile.

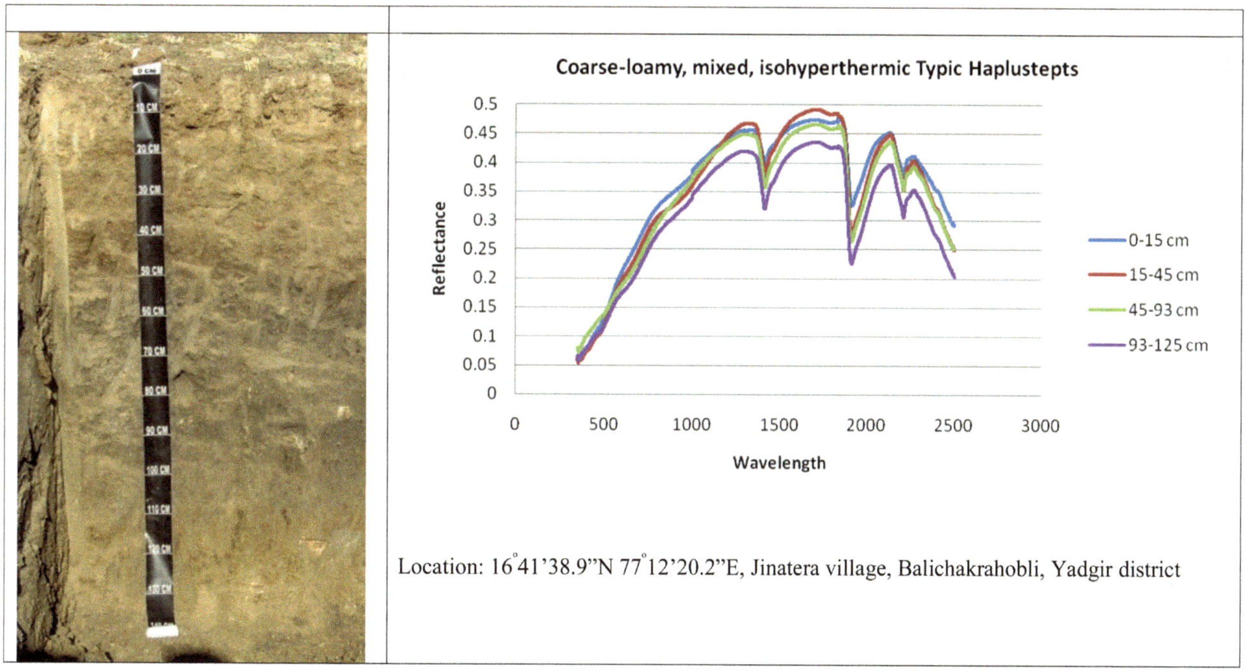

FIGURE 2.15 Spectral signature of Inceptisol soil profile.

TABLE 2.1 Common spectral indices used in soil studies and their references.

Sl. no.	Type of indices	Equation	References
1.0	**Structural vegetation indexes**		
1.1	Normalized difference vegetation index (NDVI)	$NDVI = \frac{(R_{NIR} - R_{red})}{(R_{NIR} + R_{red})}$	Rouse et al. (1974)
1.2	Modified triangular vegetation index (MTVI1)	$MTVI1 = 1.2 \times [1.2 \times (R_{800} - R_{550}) - 2.5 \times (R_{670} - R_{550})]$	Haboudanea et al. (2004)
1.3	Modified triangular vegetation index (MTVI2)	$MTVI2 = \frac{1.5 \times [1.2 \times (R_{800} - R_{550}) - 2.5 \times (R_{670} - R_{550})]}{\sqrt{(2*R_{800}+1)^2 - (6 \times R_{800} - 5 \times \sqrt{R_{670}}) - 0.5}}$	Haboudanea et al. (2004)
1.4	Renormalized difference vegetation index (RDVI)	$RDVI = \frac{(R_{800} - R_{670})}{\sqrt{(R_{800} + R_{670})}}$	Roujean and Breon (1995)
1.5	Simple ratio index (SRI)	$SR = \frac{R_{NIR}}{R_{red}}$	Rouse et al. (1974)
1.6	Modified simple ratio (MSR)	$MSR = \frac{R_{NIR}/R_{red} - 1}{(R_{NIR}/R_{red})^{0.5} + 1}$	Chen (1996)
1.7	Modified chlorophyll absorption in reflectance index (MCARI1)	$MCARI1 = 1.2 \times [2.5 \times (R_{800} - R_{670}) - 1.3 * (R_{800} - R_{550})]$	Daughtry et al. (2000)
1.8	Modified chlorophyll absorption in reflectance index (MCARI2)	$MCARI2 = \frac{1.5 \times [2.5 \times (R_{800} - R_{670}) - 1.3 \times (R_{800} - R_{550})]}{\sqrt{(2*R_{800}+1)^2 - (6*R_{800} - 5 \times \sqrt{R_{670}}) - 0.5}}$	Daughtry et al. (2000)
1.9	Soil-adjusted vegetation index (SAVI)	$SAVI = (1 + L) \times (R_{800} - R_{670})/(R_{800} + R_{670} + L)[L \in (0, 1)]$	Huete (1988)
1.10	Transformed soil-adjusted vegetation index (TSAVI)	$TSAVI = \frac{a(NIR - aRED - b)}{(NIR + aRED - b)}$	Baret et al. (1989)
1.11	Modified SAVI with self-adjustment factor L (MSAVI)	$MSAVI = \frac{1}{2}\left[2 \times R_{800} + 1 - \sqrt{(2 \times R_{800} + 1)^2 - 8 \times (R_{800} - R_{670})}\right]$	Qi et al. (1994)
1.12	Optimized soil-adjusted vegetation index (OSAVI)	$OSAVI = (1 + 0.16) \times (R_{800} - R_{670})/(R_{800} + R_{670} + 0.16)$	Rondeaux et al. (1996)

(*Continued*)

TABLE 2.1 (Continued)

Sl. no.	Type of indices	Equation	References
2.0	**Chlorophyll indexes**		
2.1	Transformed CARI (TCARI)	$TCARI = 3 \times [(R_{700} - R_{670}) - 0.2 \times (R_{700} - R_{550}) \times (R_{700}/R_{670})]$	Haboudane et al. (2002)
2.2	Triangular vegetation index (TVI)	$TVI = 0.5 \times [120 \times (R_{750} - R_{550}) - 200 \times (R_{670} - R_{550})]$	Broge and Leblanc (2000)
2.3	Photochemical reflectance index (PRI)	$PRI = (R_{531} - R_{570})/(R_{531} + R_{570})$	Gamon et al. (1992)
2.4	Cellulose adsorption index (CAI)	$CAI = 0.5 \times (R_{2020} + R_{2200}) - R_{2100}$	Daughtry et al. (1995)
3.0	**Vegetation water indexes**		
3.1	Normalized difference water index (NDWI)	$NDWI = (R_{860} - R_{1240})/(R_{860} + R_{1240})$	Gao (1996)
3.2	Simple Ratio water index (SRWI)	$SRWI = R_{858}/R_{1240}$	Zarco-Tejada et al. (2003)
3.3	Shortwave infrared water stress index (SIWSI)	$SIWSI = \rho 6 - \rho 2/\rho 6 + \rho 2 \rho 2 = 0.841 - 0.876; \rho 6 = 1.628 - 1.652$	Fensholt and Sandholt (2003)
3.4	Plant water index (PWI)	$PWI = R_{970}/R_{900}$	Peñuelas et al. (1997)
4.0	**Spectral absorption metrics**		
4.1	σ = Shape parameter as defined by the inverted-Gaussian curve-fit model	$g(\lambda) = R_{\lambda_i} + (R_{\min 500-600} - R_{\max 700730}) \times \exp\left[\frac{-(\lambda_i - \lambda_{\min 500-600})^2}{2\sigma^2}\right]$	Miller et al. (1990), Bonham-Carter (1988)
4.2	Soil moisture Gaussian model (SMGM)	$g(\lambda) = R_{\lambda_i} + (R_{2800} - R_{\max 1200-1800})^{\times} \exp\left[\frac{-(\lambda_i - \lambda_{2800})^2}{2\sigma^2}\right]$	Whiting et al. (2004)
4.3	Continuum removal band depth	$D_B = \frac{R_c - R_b}{R_c}$	Clark and Roush (1984)
5.0	**Soil spectral index**		
5.1	Bare soil index	$BSI = \frac{(SWIR_2 + R) - (NIR + B)}{(SWIR_2 + R) + (NIR + B)}$	Diek et al. (2017)
5.2	Soil salinity index	$SSI = Red + NIR + SWIR_2$	Fourati et al. (2015)

References

Baret, F., Guyot, G. Major, D.J. 1989. TSAVI: a vegetation index which minimizes soil brightness effects on LAI and APAR estimation. geoscience and remote sensing symposium 1989. In: Canadian Symposium on Remote Sensing 1989, 1355–1358.

Bonham-Carter, G.F., 1988. Numerical procedures and computer program for fitting an inverted gaussian model to vegetation reflectance data. Comput. Geosci. 14, 339–356.

Broge, N. Leblanc, E. 2000. Comparing prediction power and stability of broadband and hyperspectral vegetation indices for estimation of green leaf area index and canopy chlorophyll density. 76(2), 156–172.

Chabrillat, S, Goetz, A.F., Krosley, L., Olsen, H.W., 2002. Use of hyperspectral images in the identification and mapping of expansive clay soils and the role of spatial resolution. Remote Sens. Enviro. 82, 431–445. Available from: https://doi.org/10.1016/S0034-4257(02)00060-3.

Chen, J.M., 1996. Evaluation of vegetation indices and a modified simple ratio for boreal applications. Can. J. Remote. Sens. 22 (3), 229–242.

Clark, R.N., Roush, T.L., 1984. Reflectance spectroscopy: quantitative analysis techniques for remote sensing applications. J. Geophys. Res. 89, 6329–6340.

Daughtry, C.S.T., McMurtrey, J.E., Chappelle, E.W., Dulaney, W.P., Irons, J.R., Satterwhite, M.B., 1995. Potential for discriminating crop residues from soil by reflectance and fluorescence. Agron. J. 87, 165–171.

Daughtry, C.S.T., Walthall, C.L., Kim, M.S., De Colstoun, E.B., Mcmurtrey, J.E., 2000. Estimating corn leaf chlorophyll concentration from leaf and canopy reflectance. Remote. Sens. Environ. 74, 229–239.

Dharumarajan, S., Lalitha, M., Gomez, C., Vasundhara, R., Kalaiselvi, B., Hegde, R., 2022. Prediction of soil hydraulic properties using VIS-NIR spectral data in semi- arid region of Northern Karnataka Plateau. Geoderma Reg. 28, e00475. Available from: https://doi.org/10.1016/j.geodrs.2021.e00475.

Dharumarajan, S., Gomez, C., Lalitha, M., Kalaiselvi, B., Vasundhara, R., Hegde, R., 2023. Soil order knowledge as a driver in soil properties estimation from Vis-NIR spectral data—case study from northern Karnataka (India). Geoderma Reg. 32, e00596. Available from: https://doi.org/10.1016/j.geodrs.2022.e00596.

Diek, S., Fornallaz, F., Schaepman, M.E., de Jong, R., 2017. Barest Pixel Composite for agricultural areas using landsat time series. Remote Sens. 1245. Available from: https://doi.org/10.3390/rs9121245.

Fensholt, F., Sandholt, I., 2003. Derivation of a shortwave infrared water stress index from MODIS near- and shortwave infrared data in a semiarid environment. Remote. Sens. Environ. 87, 111–121.

Fourati, H.T, Bouaziz, M., Benzina, M., Bouaziz, S., 2015. Modeling of soil salinity within a semi-arid region using spectral analysis. Arabian J. Geosci 11175–11182. Available from: https://doi.org/10.1007/s12517-015-2004-3.

Gamon, J.A., Penuelas, J., Field, C.B., 1992. A narrow waveband spectral index thattracks diurnal changes in photosynthetic efficiency. Remote. Sens. Environ. 41, 35–44.

Gao, B.-C., 1996. NDWI—a normalized difference water index for remote sensing of vegetation liquid water from space. Remote. Sens. Environ. 58, 257–266.

Haboudane, D., Miller, J.R., Tremblay, N., Zarco-Tejada, P.J., 2002. Integrated narrow-band vegetation indices for prediction of crop chlorophyll content for application to precision agriculture. Remote. Sens. Environ. 81 (2), 416–426.

Haboudanea, D., Miller, J.R., Patteyc, E., Zarco-Tejadad, P.J., Strachane, I.B., 2004. Hyperspectral vegetation indices and novel algorithms for predicting green LAI of crop canopies: modeling and validation in the context of precision agriculture. Remote. Sens. Environ. 90, 337–352.

Huete, A.R., 1988. A soil adjusted vegetation index (SAVI). Remote. Sens. Environ. 25, 295–309.

Miller, J.R., Hare, E.W., Wu, J., 1990. Quantitative characterization of the vegetation red edge reflectance I. An inverted-Gaussian reflectance model. Int. J. Remote. Sens. 11, 121–127.

Peñuelas, J., Isla, R., Filella, I., Araus, J.L., 1997. Visible and near infrared reflectance assessment of salinity effects on barley. Crop. Sci. 37, 198–202.

Qi, J., Chehbouni, A., Huete, A.R., Kerr, Y.H., 1994. A modified soil adjusted vegetation index. Remote. Sens. Environ. 48 (2), 119–126.

Rondeaux, G., Steven, M., Baret, F., 1996. Optimization of soil-adjusted vegetation indices. Remote. Sens. Environ. 55 (2), 95–107.

Roujean, J.-L., Breon, F.-M., 1995. Estimating PAR absorbed by vegetation from bi-directional reflectance measurements. Remote. Sens. Environ. 51, 375–384.

Rouse, J.W., Haas, R.H., Schell, J.A. Deering, D.W. 1974. Monitoring vegetation systems in the great plains with ERTS. In: Third ERTS-1 Symposium NASA, NASA SP-351, Washington DC, pp. 309–317.

Whiting, M.L., Li, L., Ustin, S.L., 2004. Predicting water content using Gaussian model on soil spectra. Remote. Sens. Environ. 89 (4), 535–552.

Yitagesuab, F.A., Meera, F., Werffa, F., Hecker, C., 2011. Spectral characteristics of clay minerals in the 2.5–14 μm wavelength region. Appl. Clay Sci. 53 (4), 581–591.

Zarco-Tejada, P.J., Pushnik, J.C., Dobrowski, S., Ustin, S.L., 2003. Steady-state chlorophyll a fluorescence detection from canopy derivative reflectance and double-peak red-edge effects. Remote. Sens. Environ. 84 (2), 283–294. Available from: https://doi.org/10.1016/s0034-4257(02)00113-x.

Chapter 3

Remote sensing and geographic information system applications in mapping and assessment of soil resources

S. Kaliraj[1], K. Adhikari[2], S. Dharumarajan[3], M. Lalitha[3] and N. Kumar[4]

[1]*National Centre for Earth Science Studies (NCESS), Ministry of Earth Sciences, Govt. of India, Thiruvananthapuram, Kerala, India,* [2]*USDA-ARS, Grassland, Soil and Water Research Laboratory, Temple, TX, United States,* [3]*ICAR-National Bureau of Soil Survey and Land Use Planning, Regional Centre, Bangalore, Karnataka, India,* [4]*ICAR-National Bureau of Soil Survey and Land Use Planning, Nagpur, Maharashtra, India*

3.1 Introduction

Soil—the upper thin layer of the earth's surface—is a mixture of organic matter, minerals, gases, water, and organisms that together support plant growth. Soil minerals form the basis of soil that originates from the weathering of rocks (parent material) by the processes of erosion, water, wind, temperature change, chemical interaction, pressure fluctuations, and living organisms, and all factors break down parent materials into soil particles like sand, silt, and clay particles (Young, 1975; Osman, 2012; Pereira et al., 2017). Soil's minerals and organic materials vary from place to place depending on parent materials and other contributing soil constituents, which combine to form three-dimensional soil bodies on the surface of the earth. As soils develop over time, they are formed as layers or horizons in their soil profile with two main layers, that is topsoil and subsoil. Furthermore, they exhibit three main horizons such as (1) O—organic or humus layer—which comprises organic matter of decomposed materials and humus and does not occur in all soils and all places, (2) A—surface or topsoil—which is found with humus and enriched nutrients and organic matters, (3) E—eluviation layer or zone of leaching—which occurs with leached minerals and organic matter, wherein, the eluviation process influences soil composition, (4) B—subsoil—which is formed with clay-rich contents in the subsoil, (5) C—parent material—which is the weathered rocks and outcrops involving the formation of A and B horizons, and (6) R—bedrock layer—which is a mass of rocks under subsurface like granite, basalt, quartzite, limestone or sandstone that forms the parent material for some soils (Peverill et al., 1999; Rathinasamy and Saliha, 2017). Soil properties that occur in different horizons determine the spatial pattern of plant diversity. Remote sensing images are widely used for the estimation of soil properties on the multitemporal scale based on the soil's unique spectral reflectance from different wavelengths across the electromagnetic spectrum (Mulders, 1987; Dwivedi, 2017a,b; Baghdadi, and Zribi, 2017; Rakshit et al., 2022). Soil spectral properties are widely used for the estimation of mineral content, soil texture, and organic matter content. Soil spectra in the visible/near-infrared (VNIR bandwidth of 0.4−1.1 μm) and shortwave infrared (SWIR bandwidth of 1.1−2.5 μm) are mainly used for remote sensing−based soil properties measurement. Some optical remote sensing satellite images including Landsat MSS, TM, ETM + , OLI, IRS LISS-III, LISS-IV, AWiFS, Sentinel 2-MSI, IKONS, SPOT, MODIS, Hyperion, ASTER, and AVIRIS-NG, etc. are widely used for estimation of soil and land properties. While thermal infrared (TIR) bands are used in the estimation of temperature, soil, and land surface temperature, soil−water content, and other relations of soil−land−vegetation based on their emissivity characteristics.

3.2 Remote sensing of soils: principles and spectral properties

Remote sensing is the art, science, and technology of obtaining information about the Earth's objects using orbiting payloads (earth-orbiting satellites, aircraft, drones or UAVs, and other aerial vehicles) without physical contact. Fig. 3.1

Remote Sensing of Soils. DOI: https://doi.org/10.1016/B978-0-443-18773-5.00014-4

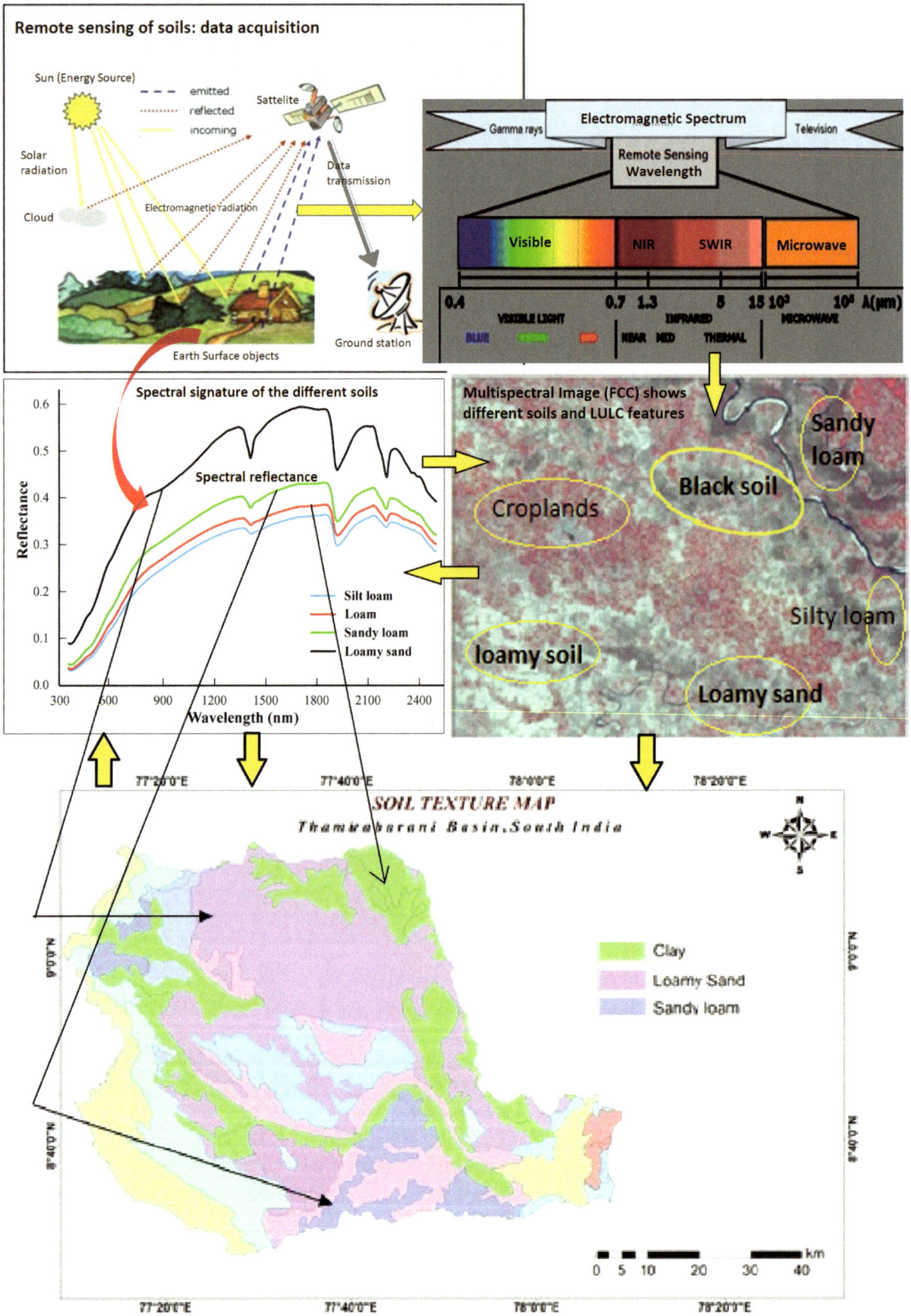

FIGURE 3.1 Remote sensing—operational workflow of earth's object observation and soil spectral characteristics.

shows the remote sensing operational workflow of earth's object observation and soil's spectral characteristics. The sensors onboard on payloads detect and measure the earth's objects by recording the unique spectral reflectance of the electromagnetic spectrum reflected, scattered, observed, and emitted by objects. Satellite images composed of multispectral or hyperspectral bands are recorded in the unique spectra of objects (i.e., soils and land features) to represent their site-specific characteristics and dynamics at a regional and local scale. Remote sensing has played a significant role in global soil mapping (GSM), providing a wealth of data that can be used to identify and map soil properties and characteristics. Soil properties are directly correlated to spectral reflectance of the remotely sensed image that is used for quantitative and quantitative measurements of soils such as surface texture, organic matter, salinity and moisture, total nitrogen, organic carbon (OC), clayey content, ferrous oxides, and other physicochemical properties combined with in situ samplings and LULC features (Baumgardner et al., 1985; Mulder et al., 2011; Babaeian et al., 2019; Fellipe et al., 2021). Some of the remote sensing products that have been used in GSM include (1) Optical remote sensing images, which include multispectral or hyperspectral images used to extract information about soil color, vegetation cover, and other parameters that can be used to map soil properties; (2) Synthetic Aperture Radar (SAR) data which is used for estimation of soil moisture, soil roughness, and other soil properties; (3) DEM data: DEMs are a digital representation of the earth's surface elevation, and they can be used to derive important topographic attributes such as slope, aspect, and terrain roughness; (4) LiDAR (Light Detection and Ranging) data—it can be used to derive soil surface characteristics such as topography, roughness, and vegetation height, which can be used to estimate soil properties; and (5) Field-Spectral Reflectance instrument: Spectral reflectance data collected from field-spectroradiometer can be used to estimate variety of soil properties such as organic matter content, soil texture, and mineral composition in large scale with high accuracy. Each soil has its unique spectral signature in the multispectral image spectrums, such as visible (0.4−0.7 μm), near-infrared (0.78−3 μm), and shortwave infrared (0.9−1.7 μm), and these unique spectra are acting as primary keys for estimation of soil properties. The TIR (10.6−12.51 μm) has a unique emissivity from different soils, and it is used for the estimation of soil and land temperature and its related parameters. For example, visible red and NIR bands are used for normalized differential vegetation index (NDVI) estimation that shows vegetation health and nonvegetative areas like bare soils, build-ups, water, rocks, etc. Similarly, the soil spectra in the bands of visible, NIR, and SWIR, are used to highlight various soil properties like minerals, clay content, iron content, organic matter, salt components, etc., based on strong spectral variability in these wavelengths (Velmurugan and Carlos, 2009; Bhatta, 2011; Yinghai Ke et al., 2015; Kaliraj et al., 2017; Zaitunah et al., 2018). Overall, the remote sensing images provide valuable insights into the spectral properties of soils, which can be used for a wide range of applications including land−soil mapping, soil properties assessment, soil classification, soil moisture estimation, soil fertility monitoring, and land use and crop yield estimation at the regional and local scale. Incorporating remote sensing products with soil sampling data, the geographic information system (GIS) techniques provide a potential database for real-time monitoring and assessment of soil resources. The researchers can develop a comprehensive soil map with more accuracy and reliability by combining remote sensing images−derived soil products with other attributes like geology, land use, and climate factors for sustainable soil−land resource management.

3.3 Digital image processing and soil spectral characteristics

A remotely sensed image is a two-dimensional array of pixels that shows a numeric translation of the original radiance (contains a digital number or simply DN value) received by the sensors that represent the behaviors of earth objects in the form of an image with intensity or gray level at that point. Image has its bandwidth (spectral length), pixel size, and grayscale based on its sensor's specification, for example, Landsat 9's OLI/TIRS sensors produced an image with 11 bands [1 PAN (15 m), 8 VNIR−SWIR (30 m), and 2 TIRS (100 m)]. Digital image processing (DIP) techniques are developed to extract spatial information about the earth's objects from remotely sensed images using computer programming, software tools, numerical algorithms, and statistical models (Lillesand et al., 2011; Kaliraj et al., 2012; Dwivedi, 2017a,b; Liang and Wang, 2019). The different DIP techniques like clustering, layer-stacking, band combination, edge detecting, image histogram, image classification, principal component analysis, spectral indices, etc., are used for extracting soil and land features using multispectral images. The DIP techniques encompass the following steps: (1) image preprocessing—to correct the image from atmospheric and radiometric errors, (2) image enhancement—to enhance the digital image in its spectral and spatial scale for improving the object patterns, (3) image classification—to classify or identify the objects (types, shape, size, area, and distribution) using classifier algorithms based on their spectral signatures, and (4) database preparation—to integrate multiple sets of data or information in the form of maps, layers, table, and graphs, etc., and this can be used for decision-making (Schowengerdt, 2006; Ravisankar and Sreenivas, 2011; Escribano et al., 2017; Dwivedi, 2017a,b; USGS, 2022; Kanwar et al., 2022). Using DIP analysis, multispectral images like Landsat, IRS, Sentinel, and other many data are used for extracting soil and LULC features using the following methods: (1) Visual image interpretation—the process of visually extracting features from the image based on interpretation elements like tone, shape, size, pattern, texture, shadow, and association, etc.; (2) Digital Image Classification (Supervised/Unsupervised)—extracting features using mathematical algorithms by classifying pixels based on spectral values (DN value) of the image; and (3) Spectral indices—used for estimating soil–land–vegetation characteristics using statistical indices by calculating spectral ratio or combining two or more spectral bands. For example, normalized difference vegetation index (NDVI), normalized difference moisture index (NDMI), soil water index, soil moisture index, clay mineral index, soil iron oxide index, and soil salinity index, etc. are used for extracting soil characteristics from multispectral images (Kaliraj et al., 2016; Baghdadi and Zribi, 2017; Asfaw et al., 2018; Li et al., 2018; Zaitunah et al., 2018; Babaeian et al., 2019; Notesco et al., 2020; Taloor et al., 2021). Recently, the DIP used advanced machine learning algorithms for extracting soil-related data from multispectral images (e.g., Landsat 9's OLI, IRS LISS-III, EOS1-Hyperion, AVIRIS-NG) and some widely used techniques such as maximum likelihood classifier, kd-tree, k-means, artificial neural network (ANN), convolution neural network (CNN, or ConvNet), support vector machine, and random forest (RF), and so on. Integrated GIS and remote sensing provide a variety of applications for soil and land resources studies, such as assessment of soil and land fertility, modeling of soil and land degradation vulnerability, estimation of soil properties, soil erosion, soil salinity, and demarcating the deforestation, quarrying, shifting cultivation, human encroachments, and other geo-hazards activities at spatiotemporal scale.

3.4 Remote sensing data products for soil and land resources assessment

Remote sensing image products, including multispectral, hyperspectral, and SAR, can be used for digital soil mapping that is tailored toward detailed large-scale mapping and monitoring of soil properties. Considerably, the integrated remote sensing and GIS methodologies have the potential to facilitate soil and land mapping in support of three components: (1) remote sensing data is used for soil type classification and mapping combined with in situ sampling, (2) remote sensing image's spectral signatures allow to estimate soil properties, its quantitative and quantitative measurements, using GIS-based empirical methods, and (3) remote sensing data supports spatial interpolation of soil properties even under limited data or sparsely sampled data (Lagacherie et al., 2006; Yitagesuab et al., 2011; Srivastava et al., 2016; Bhunia et al., 2017; Liang and Wang, 2019; Manickam et al., 2021). Recently, global digital soil mapping (GDSM) (GlobalSoilMap.Net) has been one of the hot topics among international researchers for agriculture and food security purposes. Worldwide, several remote sensing–based global missions are initiated for soil and land resources mapping and assessment, such as soil moisture active/passive mission (SMAP), GSM, GDSM, global soil partnership, and environmental mapping and analysis program (EnMAP). The advanced development of optical (multispectral), hyperspectral, and microwave (SAR) remote sensing has widely extended a wide range of applications, methodologies, and spatiotemporal modeling in soil studies that act as a key for continuous measurement and monitoring of soil properties, soil conditions, and soil characteristics at the global and regional scale. Generally, soil maps at the scale of 1:250,000 can be prepared from coarser resolution data from IRSLISS-I, AWiFS, and LANDSAT-MSS sensors. To map soils on a 1:50,000 scale, medium-resolution data from LANDSAT-TM, IRS LISS-II, and SPOT-MLA can be used. Satellite data acquired from IRS-P6 (LISS-IV sensor), Cartosat-1 and Cartosat-2, and IKONOS are now being employed for detailed mapping of soils on a 1:10,000 scale or larger. Hyperspectral remote sensing (HRS) can measure the reflectance of light at hundreds of narrow wavelength bands. Some of HRS image products, including EO-1-Hyperion (NASA), PROBA-1 (ESA), PRISMA (Italy), EnMAP (Germany), HISUI (Japan), AVIRIS-NG, and HyspIRI (US), are available for measurement of a verity of soil properties such as mineralogy, texture, and organic matter content combining with in situ sampling data. In addition to that, the digital elevation model (DEM) data products, including SRTM-DEM, ASTER GDEM, CartoDEM, and ALOS3D DEM, etc., available in 30 m spatial resolution are widely used to estimate the bare earth topographic surface like relief, altitude, slope, and bare soil zones, etc.

Table 3.1 shows the list of optical remote sensing products with their specifications used for various soil studies.

Microwave remote sensing or active remote sensing provides significant contributions to soil studies. Synthetic aperture radar (SAR) imaging sensors operate in the microwave portion (1 mm–1 m) of the electromagnetic spectrum backscattered by the sun (passive sensors) or from its laser pulses (active sensors). SAR bands are categorized to Ka, K, Ku, X, C, S, L, and P, with regard to the increase in wavelength, in which, the X and C bands are frequently used for SLR imaging. The SAR data is obtained by transmitting a radar signal toward the earth's surface and measuring the reflected signal, and it is independent of energy source, hence able to collect the data during day and night and even under extreme weather conditions. SAR data have longer wavelengths (i.e., $>$ 6 cm) with base and amplitude properties, and are used for detecting soil moisture content, soil texture, and soil minerals (Lillesand et al., 2011; Liang and Wang,

TABLE 3.1 Optical Remote sensing satellite data products are available for the assessment of soil and land resource resource resource resource resources.

Optical remote sensing data products	Satellite	Sensor	Spectral bands	Wavelength (μm)	Applications in soil ad land resources
	Sentinel 2A & Sentinel 2B	MSI	Visible NIR SWIR	0.43–0.79 0.78–0.9 1.36–2.28	
	Landsat 4–9	MSS TM ETM + OLI TIRS	Visible NIR $SWIR_1$ $SWIR_2$ TIRS	0.43–0.67 0.85–0.88 1.57–1.65 2.11–2.29 10.6–12.51	• Soil classification • Soil type mapping • Soil properties • Soil salinity • Soil erosion zone • Soil alkalinity • Soil minerals • Soil fertility • Land use capacity • Crop suitability • LULC mapping • Soil mineral • Soil texture • NDVI • NDMI • Soil OC • Land degradation • Biomass estimation • Land productivity
	IRS 1 (A to E) Resourcesat 1–2, and 2A Cartosat 1–3 HysIS	LISS-III LISS-IV AWiFS PAN Hyperspectral	Visible NIR SWIR Hyperspectral VNIR Hyperspectral SWIR	0.52–0.86 0.77–0.86 1.55–1.70 0.4–0.95 0.85–2.4	
	Terra MODIS	MODIS	VNIR TIR	0.4–14.4 3.7–14.2	
	SPOT 1 to 5	HRV HRVIR	Visible NIR SWIR	0.50–0.68 0.79–0.89 1.53–1.75	
	EO-1	Hyperion ALI	Hyperspectral (VNIR-SWIR)	0.35–2.57	
	EOS Terra	MISR ASTER	VNIR VNIR SWIR TIR	0.44–0.86 0.52–0.86 1.60–2.43 8.12–11.65	
	IKONOS	MSS	VNIR	0.44–0.85	
	QuickBird Worldview	QuickBird Worldview	VNIR	0.45–0.90	

2019; Srivastava et al., 2022). SAR data products are widely used for the estimation of soil properties; some of the available products are ALOS-PALSAR, ASAR-ENVISAT, ERS-SAR, RADARSAT, SMAP, SMOS-MIRAS, TerraSAR-X, and so on., and these products are effectively used for measuring the soil parameters including moisture, roughness, and texture. Light detection and ranging (LiDAR) data is obtained by emitting laser pulses toward the earth's surface and measuring the time taken for the pulses to be reflected. LiDAR data can be used to create high-resolution digital elevation models, which can produce topographic structures like slope, aspect, and contours involved in digital soil mapping. Overall, integrated remote sensing and GIS provide an effective platform for mapping and monitoring soil and land resources.

Table 3.2 shows the list of microwave remote sensing products with their specifications used for various soil studies.

3.5 Geographic information system—components and concepts

The GIS is a computer-based system for collecting, storing, analyzing, retrieving, and displaying data or objects related to positions on the earth's surface. The GIS represents the input data (earth's objects) with shape, size, structures, relations, and dynamic change in vector and raster format as it looks at the earth's surface within the defined scale and projection system (Chang, 2010; Kaliraj et al., 2016; Pereira et al., 2017; Campbell et al., 2018). The GIS components include hardware (computer, workstation, and cloud data center), software (GIS and image processing tools), data

TABLE 3.2 Microwave Remote sensing (SAR) data products available for soil studies.

Microwave remote sensing data products	Satellite/payload	Type sensor	Bands	Applications in soil studies
	RADARSAT 1 & 2	SAR	C-band	Soil moisture Soil classification Soil texture Surface roughness Soil characterization Land surface temperature Rainfall/precipitation Atmospheric temperature Plant healthiness Crop yield estimation
	ALOS	PALSAR	L-band	
	ENVISAT	ASAR	C-band	
	ERS 1 & 2	ERS-SAR	C-band	
	PALSAR polarimetric	Active	L-band	
	TerraSAR-X	TerraSAR-X	X-band	
	MIRAS	SMOS	L-band	
	Sentinel-1	SAR-C	C-band	
	JERS-1—SAR	JERS-1	L-band	
	SMAP	SMAP	L-band	
	SAOCOM 1A and 1B	SAR	L-band	
	SSM/I	DMSP- SSM/I	V-band	
	METOP-ASCAT	Active	C-band	

sources (i.e., maps, remotely sensed images, tabulated data, in situ data, statistical methods, and numerical algorithms), and brain ware (skilled persons, researchers) and stakeholders or users. Fig. 3.2 shows the GIS components and functionalities used for the assessment of soil properties. The GIS is designed to incorporate the earth's features with attributes to analyze real-world problems and used to investigate the earth's objects as what it is, where it is, when it is, and how it is, based on the following components: (1) location—represented by its coordinates (longitude and attitude), (2) dimension—point, line, and polygon geometry that represent location (sites), shape, size, length, and area of the objects, and (3) spatiotemporal—referring the object's place and time series that shows changing periods and amount of changes occurred on them. The GIS intakes the data formats like (1) spatial data and nonspatial data—raster, vector, and tabular format of the objects with their geometry characteristics and (2) geographic coordinates—nodes and vertices that construct the objects as points (single node or vertex), lines (two or more nodes/vertices, but starting and ending nodes are not joined), and polygon, that is, area and perimeter (three or more vertices and starting node and ending node are the same). GIS technology can be used to display the object's spatial relationships using (1) spatial data—satellite image, aerial photos, topographical maps, atlas, geology maps, and other administrative maps and (2) nonspatial (attribute) data—tabular data and alphanumerical data about metadata, name, place, quantity, and spatial characters. Moreover, integrated GIS and remote sensing technologies are widely being used for various applications in soil and land studies, including (1) soil mapping—high-resolution images are used to produce soil maps at large scale with more attributes; (2) estimation of soil properties like OC, nitrogen, and clay minerals using multispectral images, and (3) soil and land resources assessment—multiple geo-environmental parameters are analyzed to assess soil salinity, soil fertility, soil, and land degradation, etc., using GIS-based empirical models, (4) vulnerability and risk monitoring—statistical approaches and machine learning techniques are used to predict vulnerability at spatiotemporal scale, including soil erosion, landslides, slope failure, deforestation, mining and quarrying, shifting cultivation, human encroachments, and other geo-hazards activities (Pereira et al., 2017; Babaeian et al., 2019; Bonfatti et al., 2020). Satellite images and aerial photographs were used for mapping soil and land resources. The availability of high-resolution satellite images made digital soil mapping more efficient; especially, hyperspectral sensors like AVIRIS-NG data can be used for soil characterization. However, Resourcesat-2—AWiFS data (56 m) with a 5-day repeat cycle, though it has a coarse spatial resolution, can be used for mapping soil and land properties with excellent temporal coverage. Integrated remote sensing and GIS techniques are widely used for mapping and measuring soil and land resources with their variety of attributes. Remote sensing provides multitemporal data products used for the assessment of soil properties like organic matter content, soil moisture, and minerals based on their spectral reflectance characteristics with quantitative measurements.

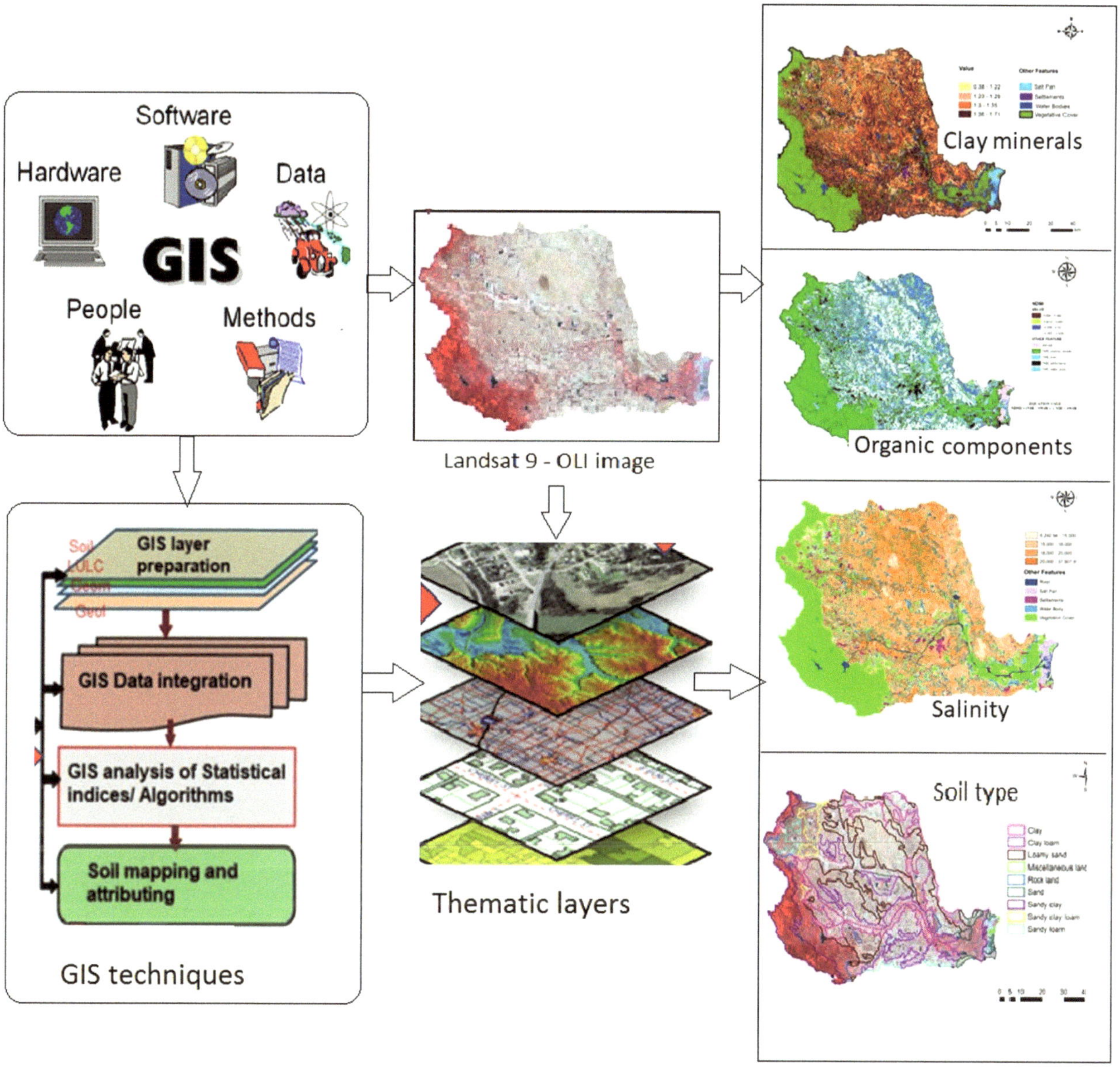

FIGURE 3.2 GIS components and functionalities used for assessment of soil properties.

3.6 Geo-based spatial interpolation techniques for soil properties mapping

The GIS-based spatial interpolation techniques are widely used for soil nutrient assessment and mapping. Spatial interpolation techniques are commonly used in many studies to create a map of the spatial distribution of discrete geo-environmental variables based on their set of sample points, such as soil properties, temperature, precipitation, etc. Mathematical algorithms are used to interpolate the known data of known geographical points or area units to estimate or predict the values at other unknown points or area units in raster layer formats (Velmurugan and Carlos, 2009; Bhatta, 2011; Mohana et al., 2018; Dharumarajan et al., 2023). For this analysis, a variety of statistical algorithms were used for mapping the spatial distribution of soil properties using limited sampling data. For example, spatial variability (distribution) of soil nitrogen is estimated across the basin using a few sampling points. Likewise, mapping of soil fertility, soil nutrients, soil moisture, rainfall, land surface temperature, etc. can be performed using the different spatial interpolation techniques. Some of the spatial interpolation techniques that are widely used are: (1) semi-variogram—which is used to calculate the spatial correlation of variables and to measure the average variance between data as a function of their separation distance, (2) kriging—which produces linear relation among the variables with unbiased estimation at each location to interpolate the value of a random field at an unobserved location from observations of its value at nearby locations, as it is unbiased and has a minimal variance of the prediction error; (3) inverse distance weighted (IDW)—which determines spatial relation by a linearly weighted combination (i.e., the function of inverse distance) of a set of sample points, wherein, the surface being interpolated should be that of a locationally dependent variable, for example, soil macro-elements and rainfall, etc.; and (4) stochastic simulation—which produces multiple equiprobable points (pixels in raster) of the variable using the value of known variables.

Fig. 3.3 shows the GIS-based spatial interpolation techniques used for mapping the soil properties' spatial variability within the interpolated boundary using minimum sampling data. Soil fertility assessment and mapping can be prepared using geo-statistical interpolation techniques that are effectively used for and is vital for fertilizers usage and crop suitability. Soil sampling provides analytical data on soil properties at discrete points, representing their characteristics, but the advantage of the GIS-based geo-statistical interpolation techniques used for mapping and measurement of spatial variability of soil properties and soil nutrients including, sand, silt, and clay content, pH, exchangeable Na, Ca, K, and Mg content, cation-exchange capacity (CEC), base saturation, total available P, and organic matter, etc. Using the different spatial interpolation methods, soil properties and their spatial distribution (also, spatial variability) can be mapped with reliable accuracy using a minimum sampling data collected from selected sites, and these derived outcomes are widely used for decision-making and developmental planning activities.

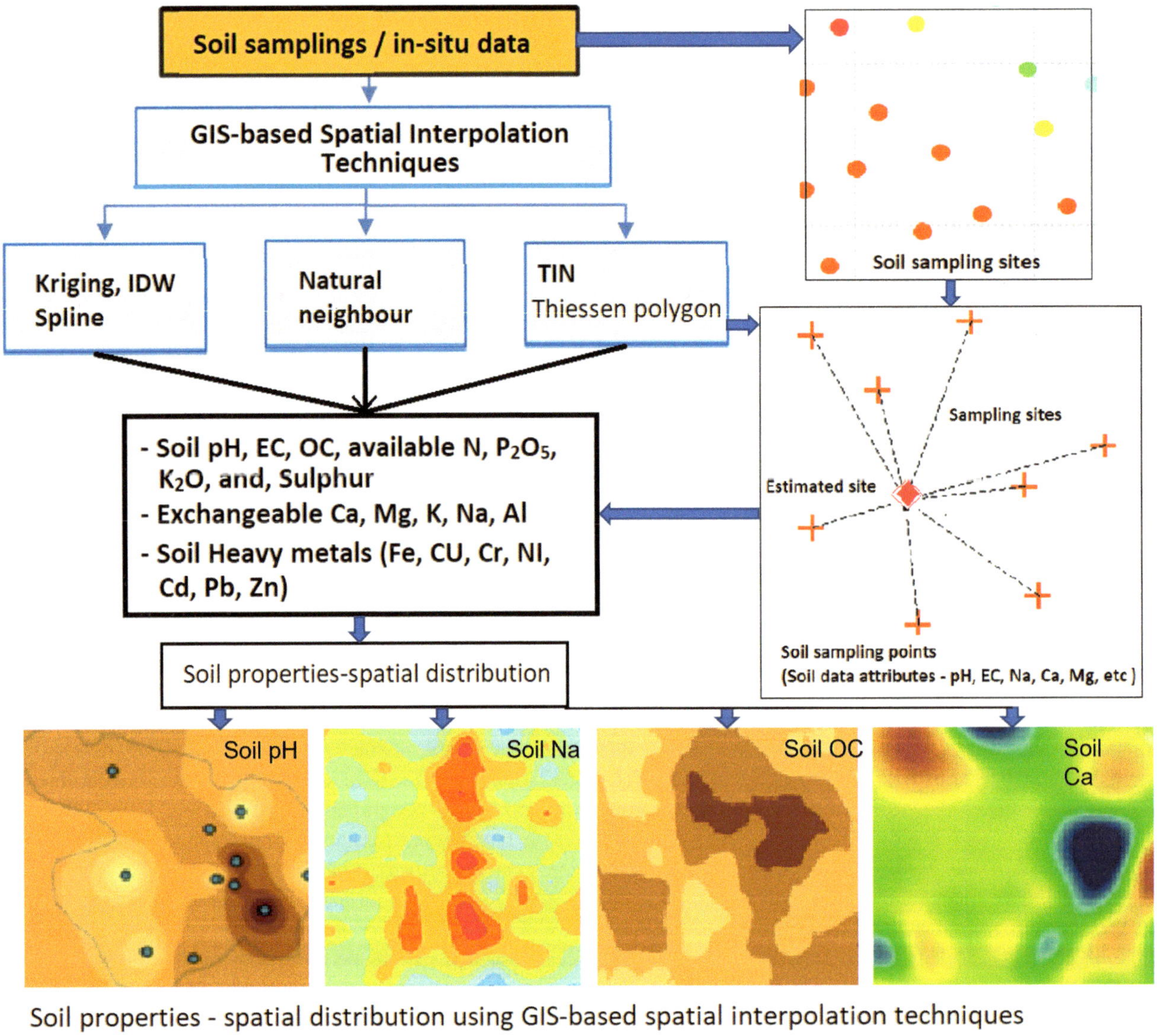

FIGURE 3.3 Schematic diagram of GIS-based spatial interpolation techniques used for soil properties mapping.

3.7 GIS techniques for assessment of soil characteristics and slope stability

Soil characteristics are a primary factor that determines slope stability in the highland areas, especially the Western Ghats and the Himalayas in India. Altering of soil properties by natural and anthropogenic activities induces slope failure followed by a landslide, debris flow, mass wasting, avalanches, and other catastrophic hazards. Assessment of soil characteristics regarding slope stability is a vital process for developmental activities like buildings, highways, railroads, canals, dams, reservoirs, surface mining, and many other excavations activities. The GIS-based slope stability index model is used to identify the zone of slope stability and or possible sites for slope failure. Slope stability at site-specific zones is demarcated by analyzing various soil characteristics including soil texture, soil thickness, soil infiltration rate, soil permeability, soil moisture, friction, coefficient, soil unit weight, and soil−water unit weight, in addition to slope gradients, geologic structure, geomorphic landforms, LULC features, fluctuations in the groundwater table (Horpibulsuk et al., 2013; Kang et al., 2020). All these parameters were derived from Landsat 8's OLI images, SRTM−DEM (30 m) data, GSI published geology and geomorphology maps, CGWB published groundwater table and borehole data, NBSS derived soil maps and in situ soil sampling data. A case study is executed in the Chaliyar River basin of the Western Ghats, India. The basin consists of various soil types such as loam (42.74%), followed by clay (28.66%), clay loam (24.18%), and sandy loam (4.42%). Loamy soils are mainly seen in the lowlands along the alluvial plains, valleys, and coastal plains. Sandy loam occurs in mid-land plateaus, impounded within loamy and clayey soils, and is associated with laterite formations. Clay soil is mainly distributed along the stream flow path in the highlands and river course in the middle parts. Clay loam is exclusively found in the Western Ghats' scarps and foothills and is well-drained as it is formed by higher content sand to sandy clay soils. Fig. 3.4 shows a GIS-

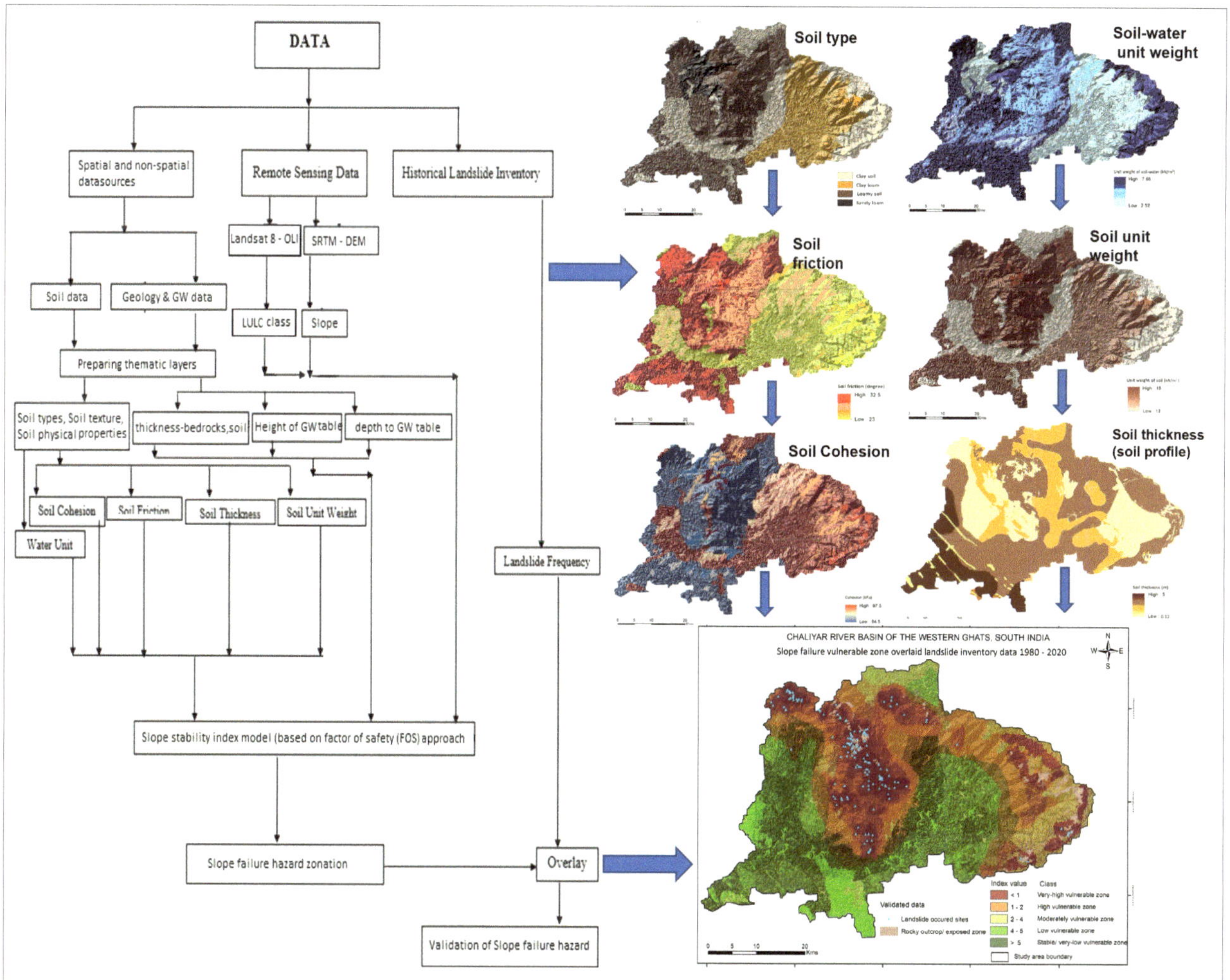

FIGURE 3.4 GIS-based slope stability analysis of soil properties for mapping the slope stability zone—a case study of the Chaliyar river basin of the Western Ghats, India.

based slope stability analysis of soil properties for mapping slope stability zones in the Chaliyar river basin of the Western Ghats, India. The resulting map shows the slope stability zones into five classes, such as (1) stable zone (no risk to slope failure), (2) low-risk zone (high resistance to slope failure), (3) moderately risk zone (medium resistance to slope failure), (4) high vulnerability zone, and (5) very high vulnerability to slope failure. The area of 25.36% and 16.28% fall under the highly vulnerable zone and extremely vulnerable zone respectively, and the areal extent of these classes cover about 996.38 km^2. The uplands consisting of steep valleys, barren lands, and slopes with overburdened materials and weathered surfaces are highly susceptible to severe slope failure hazards while altering soil's physical properties by external factors. The 358.64 km^2 (14.99%) area has been noted as a moderate vulnerability to slope failure. Moreover, the area of 18.16% (434.6 km^2) and 25.18% (602.38 km^2) have demarcated as low vulnerability and stable zones, and these are found in low-lying land of plains, mid-land of plateaus, and gentle slope surface of the basin. The GIS-based comparative technique is applied to validate the resulting map using landslide inventory data (1980–2020) and it is proved that the slope instability zones fall under frequent landslide occurred sites; hence, the output is considered acceptable, and it is very much helpful for decision-making in hazard management purposes.

3.8 GIS modeling of soil erosion assessment and mapping

Soil erosion results from processes of denudation or wearing away of the topsoil particles (upper layer of soil) from an area that is transported downslopes by water, winds, and biological agents, which induce soil infertility, soil and land degradation, loss of crop productivity, and other agriculture-related adverse impacts. Soil erosion assessment is the integrated approach of studying various geo-environmental factors like LULC, geology, geomorphic landforms, topography (slope), relief structures, agricultural practices, rainfall, and other factors. Soil erosion is one of the most serious environmental problems worldwide, and it affects soil quality, agriculture productivity, and environmental ecosystems. GIS and Remote Sensing have been used extensively for mapping soil erosion risk. Integrated GIS and remote sensing techniques play a great role in the measurement and mapping of soil erosion or land degradation in many landscapes. The universal soil loss equation developed by USDA is widely used in many countries, and it is further modified as revised universal soil loss equation (RUSLE) for the measurement of potential soil erosion rate in a basin or catchment scale (USDA, 2002; Lu et al., 2004; Biswas and Pani, 2015; Ganasri and Ramesh, 2016; Mohapatra, 2022). Soil erosion involves the detachment of soil from the soil surface by external factors like water, runoff, wind, etc. Surface runoff from weathered upland ranges transports sediments, nutrients, fertilizers, and pollutants downslope. Multiple geo-environmental data sources are collected from precipitation (IMD rainfall data), slope gradients from DEM data (SRTM DEM, ASTER DEM, Carto DEM, and PolSAR DEM, etc.), soil types and soil depth (soil map and soil samplings), and LULC (extracted from Landsat/ Sentinel/ IRS images). Integrated remote sensing and GIS technique are widely used for soil erosion assessment and mapping. The GIS-based RUSLE model is used for mapping and assessment of the potential rate of soil erosion in the basin area. The RUSLE equation is expressed as $A = R \times K \times LS \times C \times P$. Where A is the estimated soil erosion rate (spatial average of soil loss) over a period in the unit of (t/ha/year). The GIS techniques are used for preparing thematic layers of these RUSLE factors. The GIS-based Raster tool is executed to multiply the raster layers of five factors, which are rainfall erosivity factor (R), soil erodibility factor (K), topographic factor (LS), crop management factor (C), and cultivation practice factor (P). Fig. 3.5 shows the workflow of GIS-based RUSLE modeling of soil erosion assessment in the Thamirabarani river basin, South India. The basin comprises 12 major tributaries that originate from the Western Ghats and confluence into the Bay of Bengal Sea. The basin length is 126 km, which covers an area of about 5717.08 km^2; the land mass of the basin lies in two districts of the Tamil Nadu state, Southern India. The GIS-based RUSLE model is estimated for the soil erosion rate and its risk zones, which shows the various parts that fall under potential soil erosion, and it is estimated at the range of 4.2–152.4 t/ha/year, except water bodies and build-ups. Very high soil erosion is estimated in 10.13 km^2 (0.18%) of the area, mainly in the sloped landscapes, barrens, and fallows due to gully erosion and surface runoff. A higher soil loss rate (18.4–48.6 t/h/year) is noted in the foothills, bajada, and cultivable lands. Moderate soil erosion is found around 726.22 km^2 (12.93%) in some parts of uplands and cultivable lands due to seasonal climatic conditions and human-induced activities. Significantly, an area of 4795.82 km^2 (85.39%) has been noted as a lower soil erosion risk zone in various parts, that is, forest covers, catchments, irrigated lands, and unirrigated lands. In the basin, soil loss variability in different parts is mainly due to impacts of climate change (poor rainfall, monsoonal variability, increasing land surface temperature, and evapotranspiration) and other anthropogenic activities. Overall, soil erosion mainly occurred in sloped landscapes, barrens, and fallows due to activities and gully erosion, surface run-off, and other human-induced activities. Fortunately, the larger area of the basin has been noted under a lower soil erosion risk zone due to proper cultivation practices.

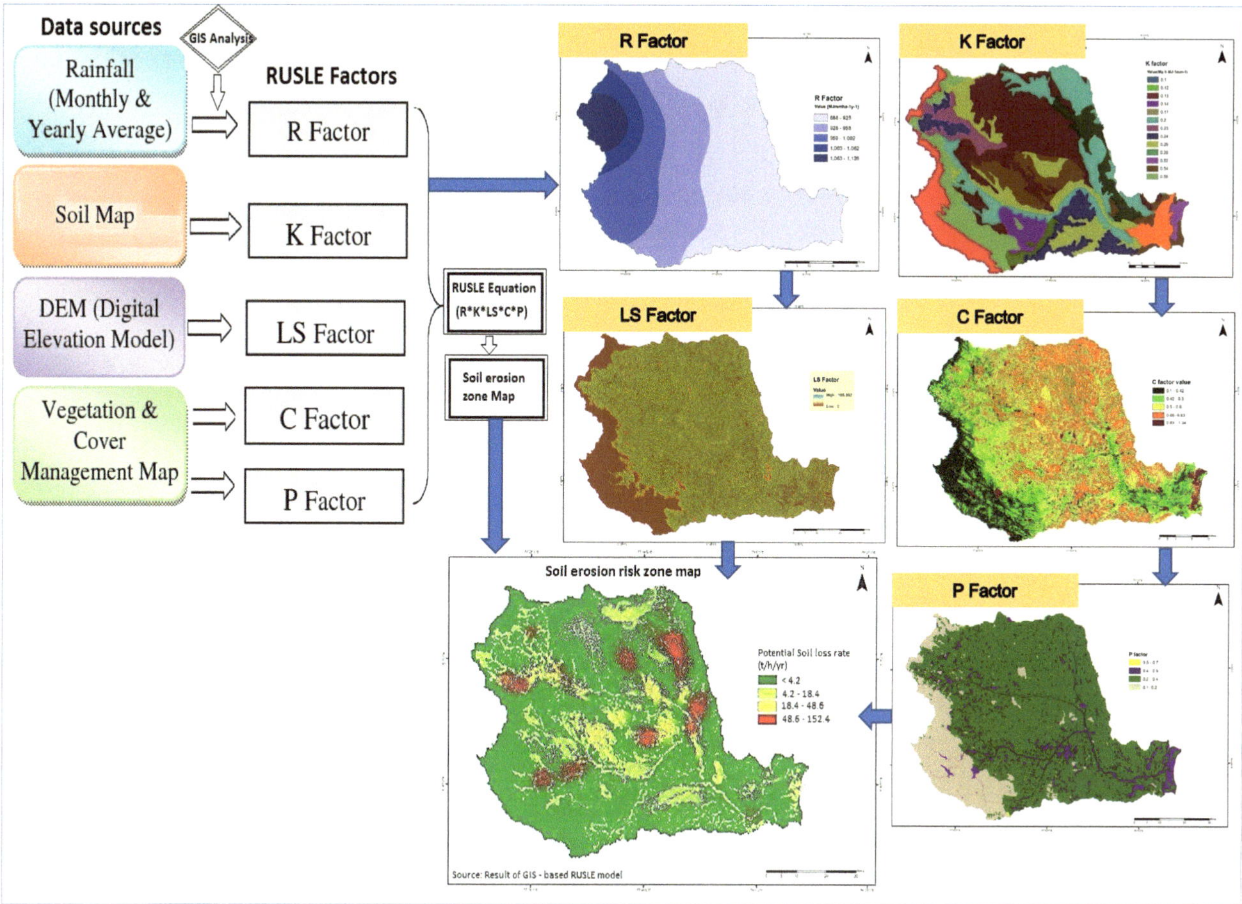

FIGURE 3.5 GIS-based RUSLE modeling of soil erosion assessment in the Thamirabarani river basin, South India.

3.9 GIS techniques for assessment of salt-affected soils

Increasing salt matter in soils is a main factor for soil and land degradation that severely affects crop productivity, particularly in arid and semiarid landscapes. According to FAO (2008), about 6% of the global land mass (of which 2% in India) falls under (SAS) or is affected by soil salinity, and it is likely to be increasing more by 2050. Generally, remote sensing approaches help to estimate the soil salinity and SAS in bare soil areas based on spectral reflectance characteristics. Assessment of SAS is essential for understanding soil and land degradation processes in arid and semiarid environments. Mapping the SAS in various landscapes, especially in arid and semiarid environments, indicates the zone of salt encrustation on topsoil, and it is assessed by analyzing various influencing factors. Multispectral images like Landsat 9 OLI, Sentinel 2-MSI, IRS LISS-III, and all hyperspectral images can be used for the estimation of pH, EC, OC, Na, etc. Fig. 3.6 shows the assessment of SAS through spectral indices in the inland plains of semiarid agroecological regions using remote sensing and GIS techniques. A case study is executed in the semiarid agroecological area of the Villupuram district in Tamil Nadu, India, with an area extent of 28,516 hectares. Major soil types of the area show red loamy, clay loam, and black clay soils, and they are used for cultivating paddy, groundnut, black gram, and sugarcane depending on dug well irrigation and monsoon rainfall. In this analysis, the Resourcesat-1 (IRS-P6) LISS-IV with 5.8 m spatial resolution was used for the delineation of SAS. For mapping, the SAS zones. the IRS LISS-IV data is analyzed by different remote sensing indices like the normalized difference salinity index (NDSI), vegetation soil salinity index (VSSI), NDVI, and salinity index (SI), and the derived outputs are used for mapping and assessment of soil salinity and salt-affected soil zones. The soil properties like pH, EC, OC, calcium carbonate ($CaCO_3$) equivalent (%), Exchangeable sodium percentage (ESP), and exchangeable ammonium acetate were correlated with its corresponding raster value of indices-derived products using linear regression model for validation purposes (Dehni and Lounis, 2012; Chen et al.,

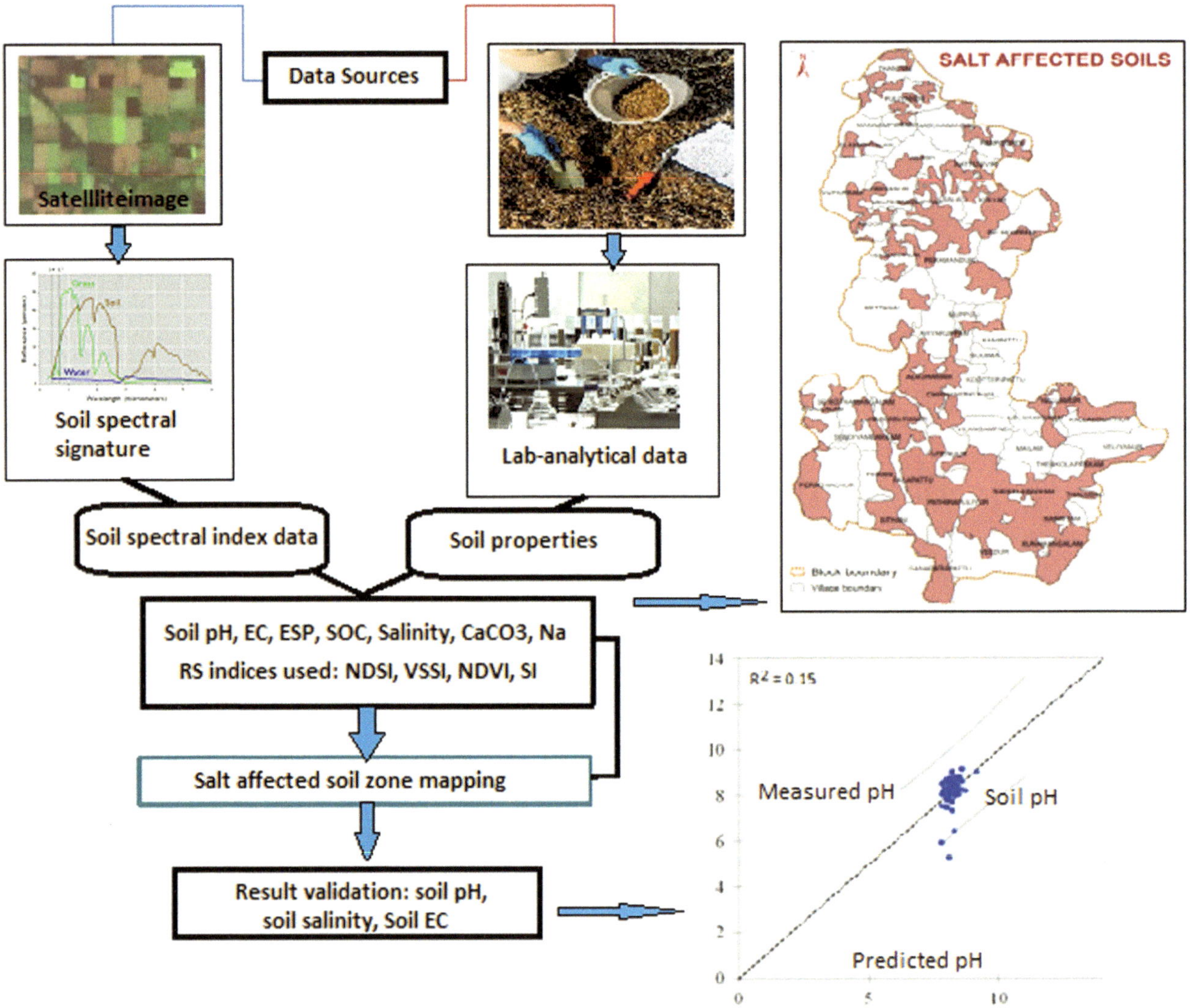

FIGURE 3.6 Assessment of SAS in the semiarid agro-ecological area of the Villupuram district in Tamil Nadu, India, using remote sensing and GIS techniques.

2017; Asfaw et al., 2018; Nguyen et al., 2020; Manickam et al., 2021). The output shows that 39% of the total area falls under SAS, wherein the estimated soil pH varies from 7.8 to 9.7 with a mean value of 8.52, and it is proved that salt matter occurs at a higher rate in these areas. Likewise, the soil EC has been noted at the range of 0.1–2.13 (dS m^{-1}) with a mean value of 0.53. Based on the exchangeable sodium percentage, the SAS were further classified into moderately sodic soil (15.3%), strongly sodic (59.3%), and very strongly sodic soils (25.4%) with the values of pH >8.5, EC <4 dSm^{-1}, and ESP >15%. However, soil EC is highly variable depending on its locations, and the other properties like OC, CEC, ESP, and $CaCO_3$ Equivalent are moderately variable, but soil pH found at least variability in SAS zones. Significantly, the topographic surface [derived from SRTM-DEM (30 m)] indicates that the SAS mainly spread over the landscapes within the elevation range of 40–60 m above MSL, and these mostly cover the lowlands of pediplains. Soil with higher ESP affects soil's structural stability due to heavy soil texture renders the soil drainage, so that salt matters remain in the subsoil and is exposed to the top layer through capillary forces while exceeding potential evapotranspiration than precipitation rate. Moreover, the SAS increase in exchangeable sodium percent, while facing a significant deficit in available phosphorus and zinc contents. Remote sensing spectral indices like VSSI, NDSI, BI, and SI comparatively produce reliable results for soil salinity assessment and mapping in bare soil environments, especially in the arid and semiarid regions; however, it is limited to the image's spectral bands and spectral indices–associated land use and land cover features.

3.10 Remote sensing and GIS techniques for land degradation assessment

Land and soil degradation is one of the significant global issues that affect food production, agriculture, environmental ecosystems, and socioeconomic development. Integrated remote sensing and GIS techniques are effectively used for mapping and assessment of land degradation by **analyzing** multiple geo-environmental parameters (Ravisankar and Sreenivas, 2011; Kaliraj et al., 2015; Baghdadi and Zribi, 2017; Dharumarajan et al., 2018; Zhao et al., 2019). Fig. 3.7 shows the assessment of land degradation vulnerability zones using the GIS-based land degradation vulnerability index (LDVI) model. A case study is executed in two different agro-climatic areas, namely Kasargod district in Kerala (humid tropics) and Virudhunagar district in Tamil Nadu (semiarid). The GIS-based LDVI model is used to analyze multivariate statistical indices, namely climate index (CI), soil index (SI), vegetation index (VI), land use index (LUI), and socioeconomic index (SEI) for demarcating the land degradation zones. In this study, the CI is calculated using global reference evapotranspiration (Global-ET0) data and global aridity index (Global-Aridity_ET0) data. The VI is estimated from Landsat 9's OLI image−derived NDVI products using the equation $VI = (\text{Erosion protection} \times \text{drought resistance} \times \text{fire risk} \times \text{plant cover percentage})^{1/4}$. The SI or (SQI) is calculated using the equation and is expressed as $SI = (\text{Soil texture} \times \text{parent material} \times \text{slope})^{1/3}$ wherein, the soil map, geology map, LULC map, and SRTM-DEM (30 m) are used. The SEI is calculated using the equation $SEI = (\text{Population pressure} \times \text{Unemployment} \times \text{Illiteracy} \times \text{Poverty})^{1/4}$, wherein the census data of 2011 is used. Moreover, the LDVI is calculated by multiplying these indices using the equation $LDVI = (CI \times VI \times SI \times LUI \times SEI)^{1/5}$ for mapping the land degradation zones with their spatial characteristics. A case study of LDVI shows that 8.3% of the area found no significant vulnerability in the Kasargod district, Kerala, while an area of 91.4% shows lower vulnerability and 0.23% of the area falls under moderate vulnerability in specific sites. However, the semiarid area of Virudhunagar district in Tamil Nadu shows an area of 1.4% under higher land degradation, due to severe soil erosion, soil salinization, and other human-induced impacts. However, 65.4% of the area falls under low vulnerability and 33.2% falls under the moderately vulnerable zone. Overall observation reveals that the land and soil resources in various sites are gradually undergoing degradation status that causing adverse impacts on environmental ecosystems. Integrated remote sensing and GIS techniques provide an effective platform for sustainable land resources management on a long-term scale.

3.11 GIS-based decision support system for soil resource management

The GIS-based decision support system (DSS) facilitates developmental planning for soil and land resources management scientifically. Integrated GIS and remote sensing provide an effective platform for mapping and assessment of soil and land resources using remotely sensed images, existing maps, and field-surveyed data. The GIS-based database is used to collect, store, analysis, and retrieve soil information using field measurement and laboratory analytical data coupled with spatial and nonspatial soil inference systems in a GIS environment. The GIS-DSS framework is designed to analyze multiple data sources such as satellite images, aerial photographs, and geological maps applied with more systematic techniques that are used for studying soil and land resources in multiple aspects. Fig. 3.8 shows the schematic diagram of the GIS-DSS architecture and workflow for soil−crop suitability analysis. In this system, GIS tools and functions are systematically arranged to run the analytical models, and it consists of three components such as (1) database management system—it contains the geographical data, (2) software programming interface—made up of software, tools, and functions for analyzing spatial and nonspatial data as user's requirement with the computer system, and (3) library of predefined numerical models or empirical models used to produce the possible outcomes for decision-making (Kothyari and Jain, 1997; NBSS & LUP, 2005; Lagacherie et al., 2006; Dogan, 2008; Mulder et al., 2011; Kaliraj et al., 2012; Trinh, 2016; Bonfatti et al., 2020; Kanwar et al., 2022). The framework of GIS-DSS is designed to execute systematic processes like data collection, storing, analysis, and retrieving for finding solutions, and decision-making beyond the human ideas. Unlike the conventional approach of soil mapping, the GIS-DSS incorporates advanced techniques for the development of a digital soil information system. The system is updated with advanced technologies to make use of GPS surveys, field spectral scanners, remote sensing images, drone/ UAV data, and other several technological advances related to soil studies. Moreover, the system is frequently updated with GIS-based computational methods including map algebra, raster calculator, geostatistical interpolation (i.e., IDW and Kriging,) inference algorithms, digital elevation model, and data mining, and all are included in the mapping of soils and its attribute database. The GIS-DSS was developed to assist the decision-makers in finding solutions to various earth-environmental issues, like soil erosion, soil salinity, soil infertility, and land degradation, etc. Nowadays, the GIS-DSS is constructed with advanced technologies like big-data analysis and cloud computing for performing database management, analytical models, and graphics, to improve the decision-making process. The WebGIS-based DSS is an interlinking global database to facilitate the better management of soil and land resources in a long-term scenario.

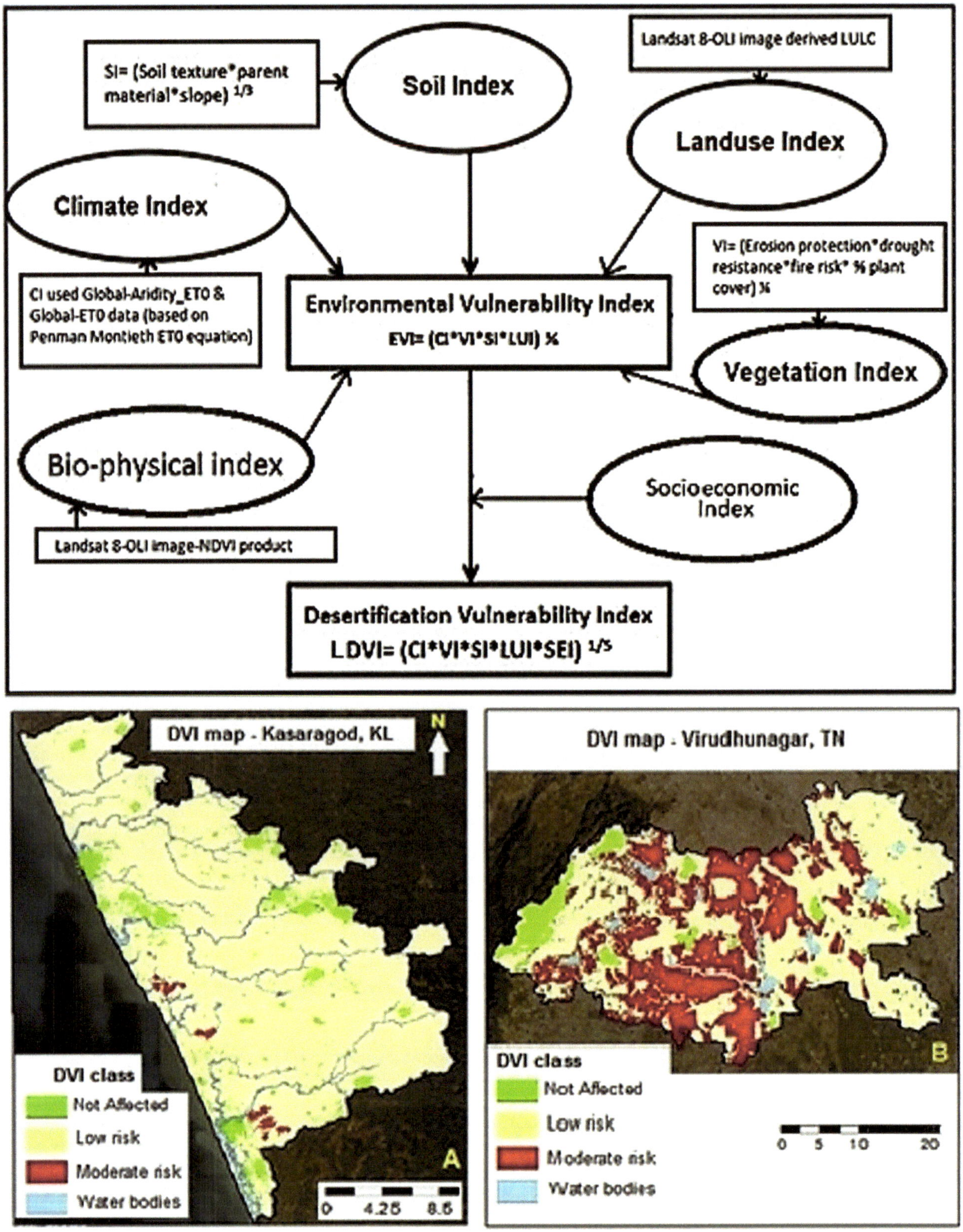

FIGURE 3.7 Assessment of land degradation vulnerability zones using the GIS-based LDVI model in Kasargod district in Kerala (humid tropics) and Virudhunagar district in Tamil Nadu (semiarid).

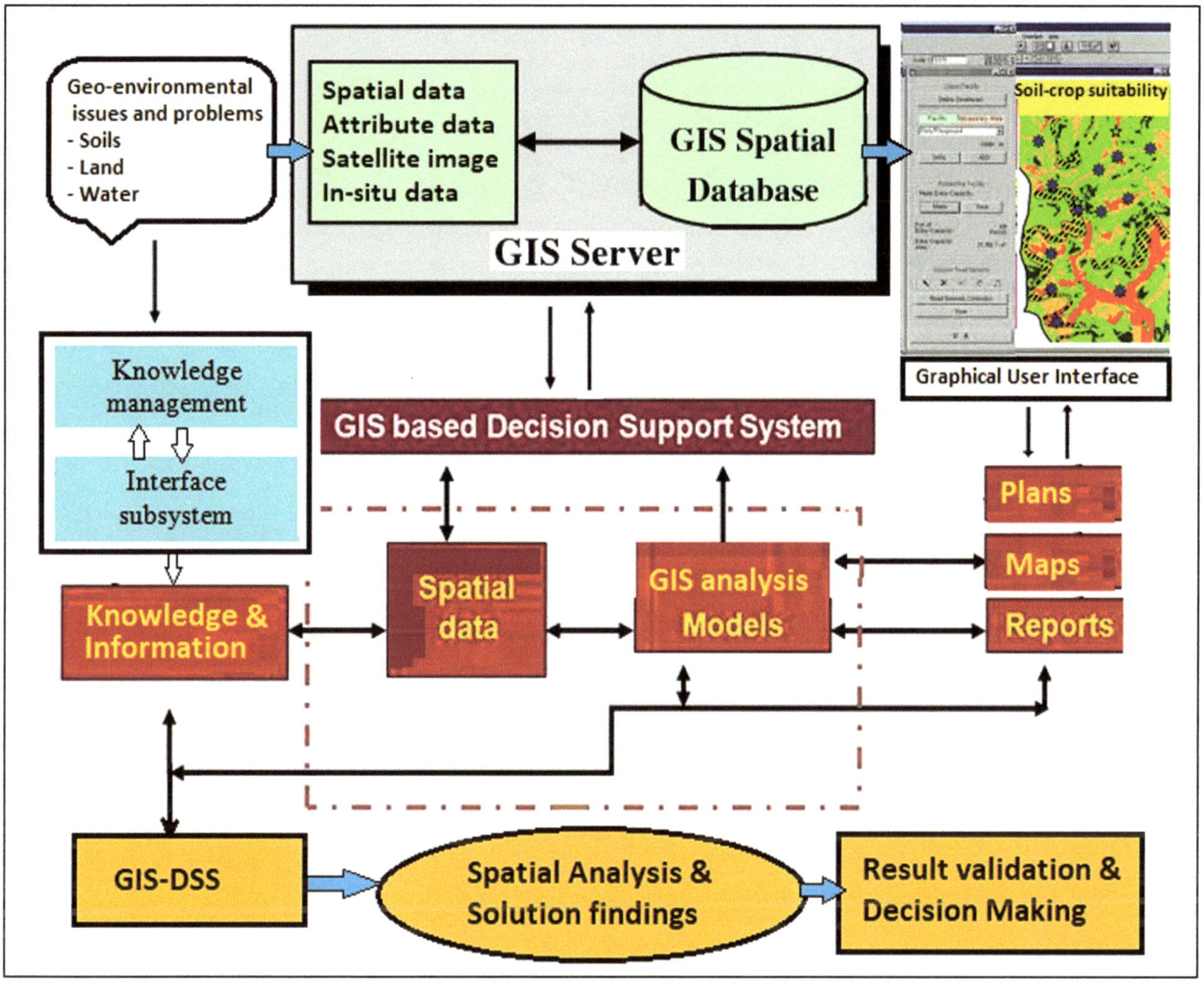

FIGURE 3.8 Schematic diagram of the GIS-DSS architecture and functionalities for soil–crop suitability analysis.

3.12 Conclusions

Integrated remote sensing and GIS technologies are widely used for various applications in soil and land resources studies. Multispectral images are widely used for the estimation of soil properties like salinity, ferrous matter, organic corban, erosion rate, etc., using spectral indices and empirical algorithms. The GIS techniques systematically analyzed multiple geo-environmental parameters for the assessment of soil and land vulnerability and risk zones using multivariate statistical approaches like weighted overlay, spatial interpolation, analytical hierarchical process, the weight of evidence, RF, neural network models (i.e., CNN and ANN), and other machine-learning techniques. GIS-based spatial interpolation techniques, like kriging, IDW, spline, dendrogram, cluster, etc., are used for mapping soil properties with minimum sampling data. Nowadays, the GIS-DSS is developed with machine learning and big-data analysis capabilities that are interlining global databases for better decision-making in soil and land resource conservation and management in a long-term scenario. This study proved that remote sensing and GIS techniques contribute more to soil and land resources mapping, monitoring, and measurement.

Acknowledgments

The corresponding author is thankful to the Director, National Centre for Earth Science Studies (NCESS), Ministry of Earth Sciences (MoES), Government of India, for providing continuous support and encouragement. Authors extend our gratitude to USGS—Earth Explorer, ESA- Copernicus Open Access Hub, IMD, SOI, and NBSS & LUP, for providing necessary data sources.

References

Asfaw, E., Suryabhagavan, K.V., Argaw, M., 2018. Soil salinity modeling and mapping using remote sensing and GIS: the case of Wonji sugar cane irrigation farm, Ethiopia. J. Saudi Soc. Agri. Sci. 17 (3), 250–258. Available from: https://doi.org/10.1016/j.jssas.2016.05.003.

Babaeian, et al., 2019. A new optical remote sensing technique for high-resolution mapping of soil moisture. Front. Big Data 2, 37. Available from: https://doi.org/10.3389/fdata.2019.00037.

Baghdadi, N., Zribi, M., 2017. Land Surface Remote Sensing in Agriculture and Forest. ISTE Press—Elsevier. Available from: https://doi.org/10.1016/C2015-0-01224-2.

Baumgardner, M.F., Silva, M., Biehl, L., Stoner, E.R., 1985. Reflectance properties of soils. Adv. Argon 38, 1–44.

Bhatta, B., 2011. Remote Sensing and GIS, second ed Oxford University Press, New Delhi, India.

Bhunia, G.S., Shit, P.K., Pourghasemi, H.R., 2017. Soil organic carbon mapping using remote sensing techniques and multivariate regression model. Geocarto Int. 34 (2), 215–226. Available from: https://doi.org/10.1080/10106049.2017.1381179.

Biswas, S.S., Pani, P., 2015. Estimation of soil erosion using RUSLE and GIS techniques: a case study of Barakar River basin, Jharkhand, India. Model. Earth Syst. Environ. 1 (4), 1–13.

Bonfatti, et al., 2020. Digital mapping of soil parent material in a heterogeneous tropical area. Geomorphology 367, 107305. Available from: https://doi.org/10.1016/j.geomorph.2020.107305.

Campbell, J., Burkhart, N., Shin, M.E., 2018. Essentials of Geographic Information Systems. FlatWorld, United States.

Chang, K.T., 2010. Introduction to Geographic Information System, fourth ed.

Chen, Z., Zhang, Y., Liu, X., 2017. Mapping and analyzing soil salinity using remote sensing and geostatistics in an arid area of China. J. Arid Land 9 (5), 639–652.

Dehni, A., Lounis, M. 2012. Remote Sensing Techniques for Salt Affected Soil Mapping: Application to the Oran Region of Algeria. Procedia Eng., 33, 188–198.

Dharumarajan, et al., 2018. Status of desertification in South India—assessment, mapping and change detection analysis. Curr. Sci. 115. Available from: https://doi.org/10.18520/cs/v115/i2/331-338.

Dharumarajan, et al., 2023. Soil order knowledge as a driver in soil properties estimation from Vis-NIR spectral data—case study from northern Karnataka (India. Geoderma Reg. 32, e00596. Available from: https://doi.org/10.1016/j.geodrs.2022.e00596.

Dogan, H.M., 2008. Applications of remote sensing and Geographic Information Systems to assess ferrous minerals and iron oxide of Tokat province in Turkey. Int. J. Remote. Sens. 29 (1), 221–233. Available from: https://doi.org/10.1080/01431160701269010.

Dwivedi, R.S., 2017a. Spectral reflectance of soils. Remote Sensing of Soils. Springer, Berlin, Heidelberg. Available from: https://doi.org/10.1007/978-3-662-53740-4_6.

Dwivedi, R.S., 2017b. An introduction to remote sensing. Remote Sensing of Soils. Springer, Berlin, Heidelberg. Available from: https://doi.org/10.1007/978-3-662-53740-4_1.

Escribano, et al., 2017. Chapter 4—Optical remote sensing for soil mapping and monitoring. In: Paulo, P., et al., (Eds.), Soil Mapping and Process Modeling for Sustainable Land Use Management. Elsevier, pp. 87–125. Available from: https://doi.org/10.1016/B978-0-12-805200-6.00004-9.

FAO, 2008. Harmonized World Soil Database (version 1.0). FAO, Rome, Italy and IIASA, Laxenburg, Austria.

Fellipe, et al., 2021. Soil parent material prediction through satellite multispectral analysis on a regional scale at the Western Paulista Plateau, Brazil. Geoderma Reg. 26. Available from: https://doi.org/10.1016/j.geodrs.2021.e00412.

Ganasri, B.P., Ramesh, H., 2016. Assessment of soil erosion by RUSLE model using remote sensing and GIS—a case study of Nethravathi Basin. Geosci. Front. 7 (6), 953–961.

Horpibulsuk, S., Suddeepong, A., Chamket, P., Chinkulkijniwat, A., 2013. Compaction behavior of fine-grained soils, lateritic soils and crushed rocks. Soils Found. 53 (1), 166–172. Available from: https://doi.org/10.1016/j.sandf.2012.12.012.

Kaliraj, S., Chandrasekar, N., Magesh, N.S., 2015. Morphometric analysis of the River Thamirabarani sub-basin in Kanyakumari District, South west coast of Tamil Nadu, India, using remote sensing and GIS. Environ. Earth Sci. 73, 7375–7401. Available from: https://doi.org/10.1007/s12665-014-3914-1.

Kaliraj, S., Chandrasekar, N., Ramachandran, K.K., 2016. Mapping of coastal landforms and volumetric change analysis in the south west coast of Kanyakumari, South India using remote sensing and GIS techniques. Egypt. J. Remote. Sens. Space Sci. 20 (2), 265–282. Available from: https://doi.org/10.1016/j.ejrs.2016.12.006.

Kaliraj, S., Chandrasekar, N., Ramachandran, K.K., Srinivas, Y., Saravanan, S., 2017. Coastal landuse and land cover change and transformations of Kanyakumari coast, India using remote sensing and GIS. Egypt. J. Remote. Sens. Space Sci. 20 (2), 169–185. Available from: https://doi.org/10.1016/j.ejrs.2017.04.003.

Kaliraj, S., Meenakshi, M.S., et al., 2012. Application of remote sensing in detection of forest cover changes using geo-statistical change detection matrices—a case study of Devanampatti reserve forest, Tamil Nadu, India. Nat. Environ. Pollut. Technol. 11, 261–269.

Kang, S., Lee, S.R., Cho, S.E., 2020. Slope stability analysis of unsaturated soil slopes based on the site-specific characteristics: a case study of hwangryeong mountain, Busan, Korea. Sustainability 12 (7), 2839. Available from: https://www.mdpi.com/2071-1050/12/7/2839.

Kanwar, N., Rai, S., Kuniyal, J.C., 2022. Open-source satellite repository and geographic information system (GIS) for soil resource mapping. In: Shit, P.K., Adhikary, P.P., Bhunia, G.S., Sengupta, D. (Eds.), Soil Health and Environmental Sustainability. Environmental Science and Engineering. Springer, Cham. Available from: https://doi.org/10.1007/978-3-031-09270-1_1.

Kothyari, U.C., Jain, S.K., 1997. Sediment yield estimation using GIS. Hydrol. Sci. J. 42 (6), 833–843. Available from: https://doi.org/10.1080/02626669709492082.

Lagacherie, P., McBratney, A., Voltz, M., 2006. Digital Soil Mapping: An Introductory Perspective. Elsevier, London.

Li, P., Xiao, C., Feng, Z., 2018. Mapping rice planted area using a new normalized EVI and SAVI (NVI) derived from Landsat-8 OLI. IEEE Geosci. Remote. Sens. Lett. 15 (12), 1822–1826. Available from: https://doi.org/10.1109/LGRS.2018.2865516.

Liang, S., Wang, J., 2019. Advanced remote sensing. Terrestrial Information Extraction and Applications. Elsevier. Available from: https://doi.org/10.1016/C2017-0-03489-4.

Lillesand, T.M., Kiefer, R.W., Chipman, A., 2011. Remote Sensing and Image Interpretation, sixth ed Wiley, New York.

Lu, et al., 2004. Mapping soil erosion risk in Rondonia, Brazilian Amazonia: using RUSLE, remote sensing and GIS. Land. Degrad. Dev. 15 (5), 499–512.

Manickam, L., Subramanian, D., Khandal, S., et al., 2021. Modeling and mapping of salt-affected soils through spectral indices in inland plains of semi-arid agro-ecological region. J. Indian Soc. Remote. Sens. 49, 1475–1481. Available from: https://doi.org/10.1007/s12524-021-01321-w.

Mohana, P., et al., 2018. Environmental impact assessment of Thamirabarani River Basin, Tamil Nadu using remote sensing and GIS techniques. Indian J. Sci. Technol. 11 (22), 1–7. Available from: https://doi.org/10.17485/ijst/2018/v11i22/106246.

Mohapatra, R., 2022. Application of revised universal soil loss equation model for assessment of soil erosion and prioritization of ravine infested sub basins of a semi-arid river system in India. Model. Earth Syst. Environ. 8, 4883–4896. Available from: https://doi.org/10.1007/s40808-022-01388-5.

Mulder, et al., 2011. The use of remote sensing in soil and terrain mapping—a review. Geoderma 162 (1–2), 1–19. Available from: https://doi.org/10.1016/j.geoderma.2010.12.018.

Mulders, M.A., 1987. Remote Sensing in Soil Science (Developments in Soil Science. Elsevier. Available from: https://doi.org/10.1016/S0166-2481(08)70027-8.

NBSS & LUP, 2005. Reflectance libraries for development of soil sensor for periodic assessment of state of soil resources. Report No. 835, National Bureau of Soil Survey and Land Use Planning, Nagpur, 40.

Nguyen, K.A., Liou, Y.A., Tran, H.P., et al., 2020. Soil salinity assessment by using near-infrared channel and Vegetation Soil Salinity Index derived from Landsat 8 OLI data: a case study in the Tra Vinh Province, Mekong Delta, Vietnam. Prog. Earth Planet. Sci. 7 (1). Available from: https://doi.org/10.1186/s40645-019-0311-0.

Notesco, et al., 2020. Application of hyperspectral remote sensing in the longwave infrared region to assess the influence of dust from the desert on soil surface mineralogy. Remote. Sens. 12 (9), 1388. Available from: https://doi.org/10.3390/rs12091388.

Osman, K.T., 2012. Soils: Principles, Properties and Management. Springer Netherlands, Netherlands.

Pereira, et al., 2017. Soil Mapping and Process Modeling for Sustainable Land Use Management. Elsevier. Available from: https://doi.org/10.1016/C2015-0-04597-X.

Peverill, K.I., Sparrow, L.A., Reuter, D.J., 1999. Soil Analysis: An Interpretation Manual. CSIRO Publishing, doi. 10.1071/9780643101357.

Rakshit, A., et al., 2022. Soil Science: Fundamentals to Recent Advances. Springer. Available from: https://doi.org/10.1007/978-981-16-0917-6.

Rathinasamy, A., Saliha, B.B., 2017. Fundamentals of Soil Science. Scientific Publishers, India.

Ravisankar, T., Sreenivas, K., 2011. In: Roy, P.S., Dwivedi, R.S., Vijayan, D. (Eds.), Soils and Land Degradation in Remote Sensing Applications. NRSC Publication, NRSC, Hyderabad.

Schowengerdt, R., 2006. Remote Sensing: Models and Methods for Image Processing, third ed. Academic Press.

Srivastava, P., Gupta, D., Islam, T., Han, D., Prasad, R., 2022. Radar Remote Sensing: Applications and Challenges. Elsevier. Available from: https://doi.org/10.1016/C2019-0-03955-6.

Srivastava, P., Petropoulos, G., Kerr, Y.H., 2016. Satellite Soil Moisture Retrieval: Techniques and Applications. Elsevier. Available from: https://doi.org/10.1016/C2014-0-03396-5.

Taloor, A.K., et al., 2021. Retrieval of land surface temperature, normalized difference moisture index, normalized difference water index of the Ravi basin using Landsat data. Appl. Comput. Geosci. 9, 100051. Available from: https://doi.org/10.1016/j.acags.2020.100051.

Trinh, L.H., 2016. Application of remote sensing technique to detect and map iron oxide, clay minerals, and ferrous minerals in Thai Nguyen Province (Vietnam. Min. Sci. Technol. 1, 60–65. Available from: https://doi.org/10.17073/2500-0632-2016-1-60-65.

USDA, 2002. NRCS: State Oce of Michigan. Technical guide to RUSLE use in Michigan.

USGS, 2022. Landsat 9 (collection 2) data products information. <https://www.usgs.gov/landsat-missions/landsat-collection-2>.

Velmurugan, A., Carlos, G.G., 2009. Soil resource assessment and mapping using remote sensing and GIS. J. Indian. Soc. Remote. Sens. 37, 511–525. Available from: https://doi.org/10.1007/s12524.

Yinghai Ke, et al., 2015. Characteristics of Landsat 8 OLI-derived NDVI by comparison with multiple satellite sensors and in-situ observations. Remote. Sens. Environ. 164, 298–313. Available from: https://doi.org/10.1016/j.rse.2015.04.004.

Yitagesuab, F.A., Meera, F., Werffa, F., Hecker, C., 2011. Spectral characteristics of clay minerals in the 2.5–14 μm wavelength region. Appl. Clay Sci. 53 (4), 581–591.

Young, R., 1975. Soil Properties and Behaviour. Elsevier.

Zaitunah, A., et al., 2018. Normalized difference vegetation index (ndvi) analysis for land cover types using landsat 8 oli in besitang watershed, Indonesia. IOP Conf. Ser.: Earth Environ. Sci. 126, 012112. Available from: https://doi.org/10.1088/1755-1315/126/1/012112.

Zhao, H., et al., 2019. Exploring an efficient sandy barren index for rapid mapping of sandy barren land from Landsat TM/OLI images. Int. J. Appl. Earth Obs. Geoinf 80, 38–46. Available from: https://doi.org/10.1016/j.jag.2019.04.001.

Section 2

Chapter 4

Remote sensing indices based soil properties measurement – a case study of the Thamirabarani River Basin, South India

S. Kaliraj[1], Reji Srinivas[1], N. Kiruthika[2], E. Vairaveni[2], Hisham Mohamed[2], K. Palanivel[2], C. Lakshumanan[2] and N. Chandrasekar[3]

[1]*National Centre for Earth Science Studies (NCESS), Ministry of Earth Sciences, Govt. of India, Thiruvananthapuram, Kerala, India,* [2]*Department of Remote Sensing, Bharathidasan University, Tiruchirappalli, Tamil Nadu, India,* [3]*Centre for GeoTechnology, Manonmaniam Sundaranar University, Tirunelveli, Tamil Nadu, India*

4.1 Introduction

Soil is the top layer of the earth's surface composed of minerals, organic matter, water, and organisms that together support life. It is interrelating the diverse bio-geo environmental contents and their cyclic processes. Soil formation is related to weathering processes of rocks (parent materials) and decomposed organic biomass. The basic components of soils include minerals, organic matter, biota (plants and microorganisms), soil–water, and air, and relatively their proportions vary from place-to-place depending on environment and climate factors (Young, 1975; Rathinasamy and Saliha, 2017). Soil's spectral properties vary between 300 and 2500 nm wavelength depending on the composition of minerals and organic materials and their environmental factors. Moreover, soil's spectral signatures are key for the measurement of soil properties using remote sensing images. Soil spectra are used to estimate various properties like minerals (mostly clay, iron oxides, and calcium carbonate), organic matter, soil moisture, etc., through remote sensing indices and numerical algorithms (NBSS & LUP, 2005; Mehrez Zribi et al., 2011; Osman, 2012; Dwivedi, 2017). Remote sensing indices can be estimated for soil–land–vegetation characteristics using multispectral and hyperspectral data. Understanding soil–land properties using conventional sampling methods and laboratory analyses may be time–cost consuming process and not efficiently provide information for a larger area. Significantly, each soil type can produce unique spectral signatures in a particular wavelength, and it is a prime factor to measure soil properties. Estimation of soil properties is possible by selecting suitable spectral bands, spectral indices, and algorithms at the pixel scale level (Schowengerdt, 2006; Mulder et al., 2011; Srivastava et al., 2016; Pereira et al., 2017). Remote sensing sensors orbiting the earth are continuously acquiring images of soil layers based on their reflected spectra. Satellite image products of visible and near-infrared (VNIR) bands, short-wave infrared (SWIR), synthetic aperture radar (SAR), and UAV/drone photographs are widely used for the assessment of land and soil properties. So many studies were successfully performed on multispectral images like Landsat TM, ETM + , operational land imager (OLI), Sentinel 2—MSI, IKONOS, SPOT, and ASTER, etc., for measurement of soil components like clay content, organic carbon, ferrous, and iron oxide, etc., by aiding field-surveyed data and in situ samplings (Clark and Roush, 1984; Baumgardner et al., 1985; Dogan, 2008; Ravisankar and Sreenivas, 2011; Kaliraj et al., 2012; Trinh, 2016; Bhunia et al., 2017; Asfaw et al., 2018; Babaeian et al., 2019; Bonfatti et al., 2020; Arisanty et al., 2021; Abay et al., 2022; Kanwar et al., 2022; USGS, 2022). Microwave remote sensing data products include (1) SAR—provides image products used for estimating moisture, roughness, and texture, some example SAR data products, including ALOS-PALSAR, ASAR-ENVISAT, ERS-SAR, RADARSAT, SMAP, SMOS-MIRAS, and TerraSAR-X, etc. and (2) Scatterometer data—provides nonimaging

Remote Sensing of Soils. DOI: https://doi.org/10.1016/B978-0-443-18773-5.00030-2

measurements used for estimations of soil parameters (Yitagesuab et al., 2011; Lillesand et al., 2011; Srivastava et al., 2016; Liang and Wang, 2019; Nguyen et al., 2020; Srivastava et al., 2022)

Remote sensing indices have been developed for qualitative and quantitative measurement of soil and land properties based on their spectral behaviors. Some indices like normalized differential vegetation index (NDVI), soil-adjusted vegetation index (SAVI), TWI, SSQI, Soil Salinity, NDWI, SMI, and LST are widely used to measure soil characteristics like texture, minerals, clay contents, organic matter, moisture, salinity, water index, vegetative cover, and land surface temperature, etc. For example, NDVI is used to estimate the vegetative cover using Red and near-infrared (NIR) bands, while the saline soil is mapped using NIR and SWIR bands. In general, remote sensing indices are developed by mathematical algorithm (or) equations to correlate two or more spectral bands for highlighting particular soil properties. Quantitative data of those index-derived products are widely used for the assessment of various soil–land properties like soil fertility, land use productivity, biomass (chlorophyll) contents, vegetation water content, soil erosion sensitivity, land degradation vulnerability, etc. For example, (1) SAVI—used as a soil brightness correction factor to minimize soil brightness influences on the vegetative cover, (2) soil organic matter index (SOC)—gives the carbon content of the soil after partial decomposition of any material, (3) bare soil index (BSI)—is a numerical indicator of soil variations, (4) Soil salinity index (SSI)—is used to detect salt mineral in soils, (5) normalized difference moisture index (NDMI)—is an indicator of moisture (water) content of soil vegetation. It is used to monitor soil–land interactions like soil infertility, soil–land moisture stress, land vegetation water stress, droughts, land degradation, etc. (Baumgardner et al., 1985; Velmurugan and Carlos, 2009; Dehni and Lounis, 2012; Baghdadi and Zribi, 2017; Escribano et al., 2017; Bonfatti et al., 2020).

In this study, the soil and land characteristics of the Thamirabarani river basin in South India are estimated using Landsat 9—OLI/TIRS image based on the different remote sensing indices and other related data sources, and the outputs can be used for sustainable soil and land resources management purposes.

4.2 The geographical location of the study area

Soil and land properties of the Thamirabarani basin are studied using Landsat 9—OLI image with remote sensing and GIS techniques. The Thamirabarani is only a perennial river flowing eastward from the Western Ghats with a length of 126 km. The basin's geographical extend covers the longitude of 77° 9′ 4.909″ E–78° 9′ 23.213″ E and the latitude of 8° 27′ 714″N–9° 11′ 39.355″ N with an aerial extend of 5565 km^2. Topographically, the basin shows that 83.09% of the landscape covers the gentle slope along the pediplains of the middle parts, 1.72% of uplands consist of steeply sloped landforms, and 5.49% of the landscape showed a moderate slope in various parts (Kaliraj et al., 2015). The basin covers the land mass of the two districts of Tirunelveli and Thoothukudi in Tamil Nadu. Fig. 4.1A shows the geographical location of the study area. Physiographically, the basin consists of three settings like upland hills (elevation range of three hundred, 1725 m), the middle part of pediplains (elevation range of 10–300 m), and the lower part of the estuary and fluvio-marine wetlands with an elevation of less than 10 m. Fig. 4.1B shows the geological formations of the study area. The geological formations are made up of complex crystalline rocks like biotite gneiss, granites, charnockite in the western uplands, and tertiary sedimentary formations consisting of calcareous sandstones and limestones. Quaternary formations of kankar, laterite, shell limestone, alluvium, sand, and silt also occurred in the various parts. Sediment deposits of coarse-grained sands with rock fragments and flood plains formed in the eastern coastal area. About 90% of the southern hill range is formed by quartzite and charnockite of the Archean series. Morphologically, the basin comprises different landforms including the upland hill range of structural hills, rocky outcrops, dissected hills, valley fills, foothill pediments, etc. The middle parts have been predominantly covered with pediments and pediplains complex with river channels and flood plains (Mohana et al., 2018). The lower part of the coastal plains is found with dunes, saltpans, wetlands, and an estuary that confluences with the Bay of Bengal Sea. The basin prevails subtropical climate with an annual average rainfall range of 30.5–650 mm and a temperature range of 27°C–39°C. Major parts of the basin consist of build-ups (urban and rural settlements), agriculture lands, fallows, bare soils, barrens, water bodies (lakes and ponds), upland forest covers, land with shrubs, limestone mining, and saltpans, coastal wetlands, etc. Numerous populations are spread over the middle parts of pediplains, and coastal plains, and their primary occupation is agriculture and its related activities.

4.2.1 Soil type and land use/land cover

The basin consists of different soil types including loamy, clayey (black cotton soil), clay loam, loamy sand, alluvium, sandy loam, and sandy clay loam, etc. Fig. 4.2A shows the soil types of the study area. Major soil types are red loamy

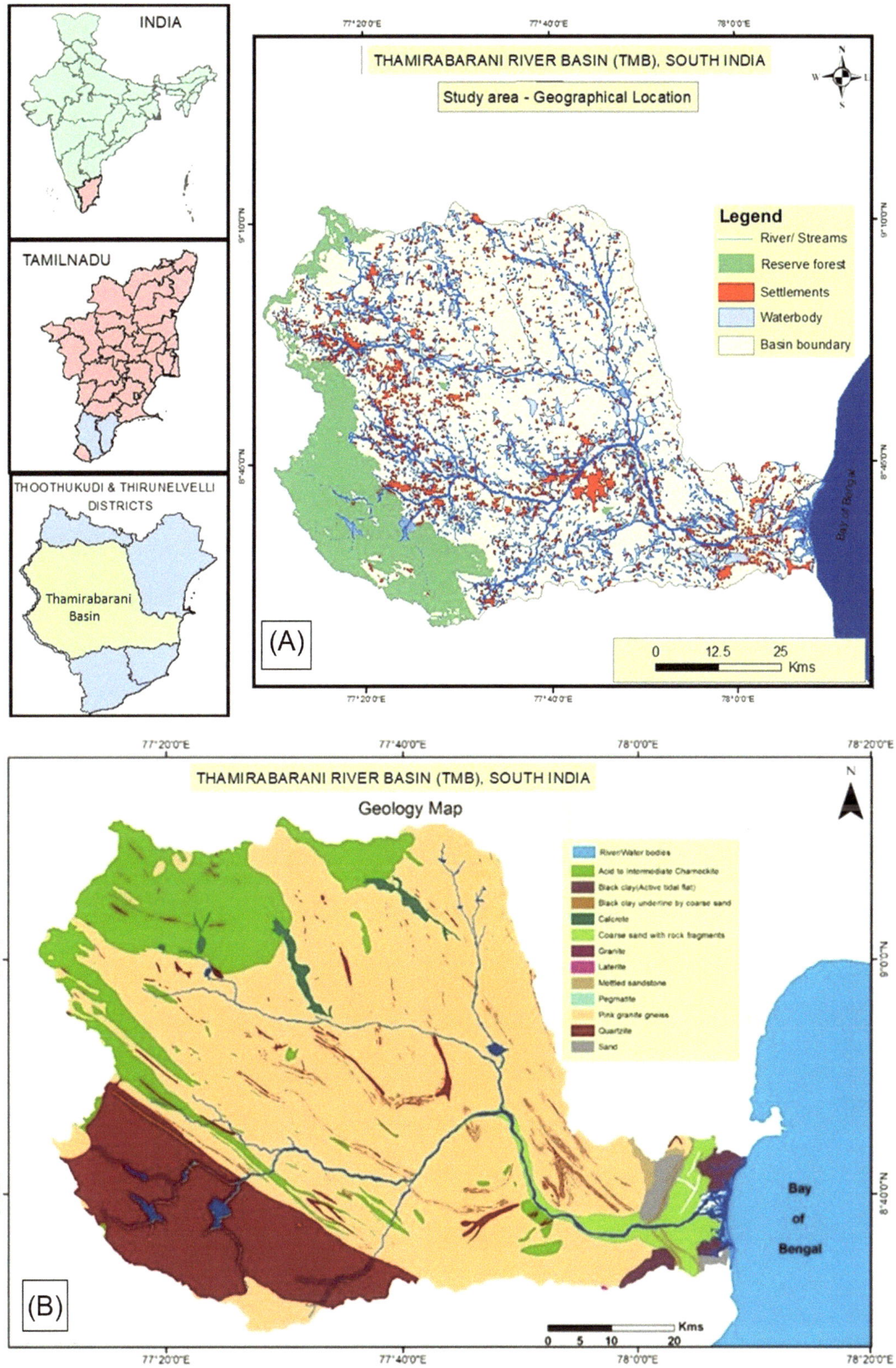

FIGURE 4.1 Location map of the study area.

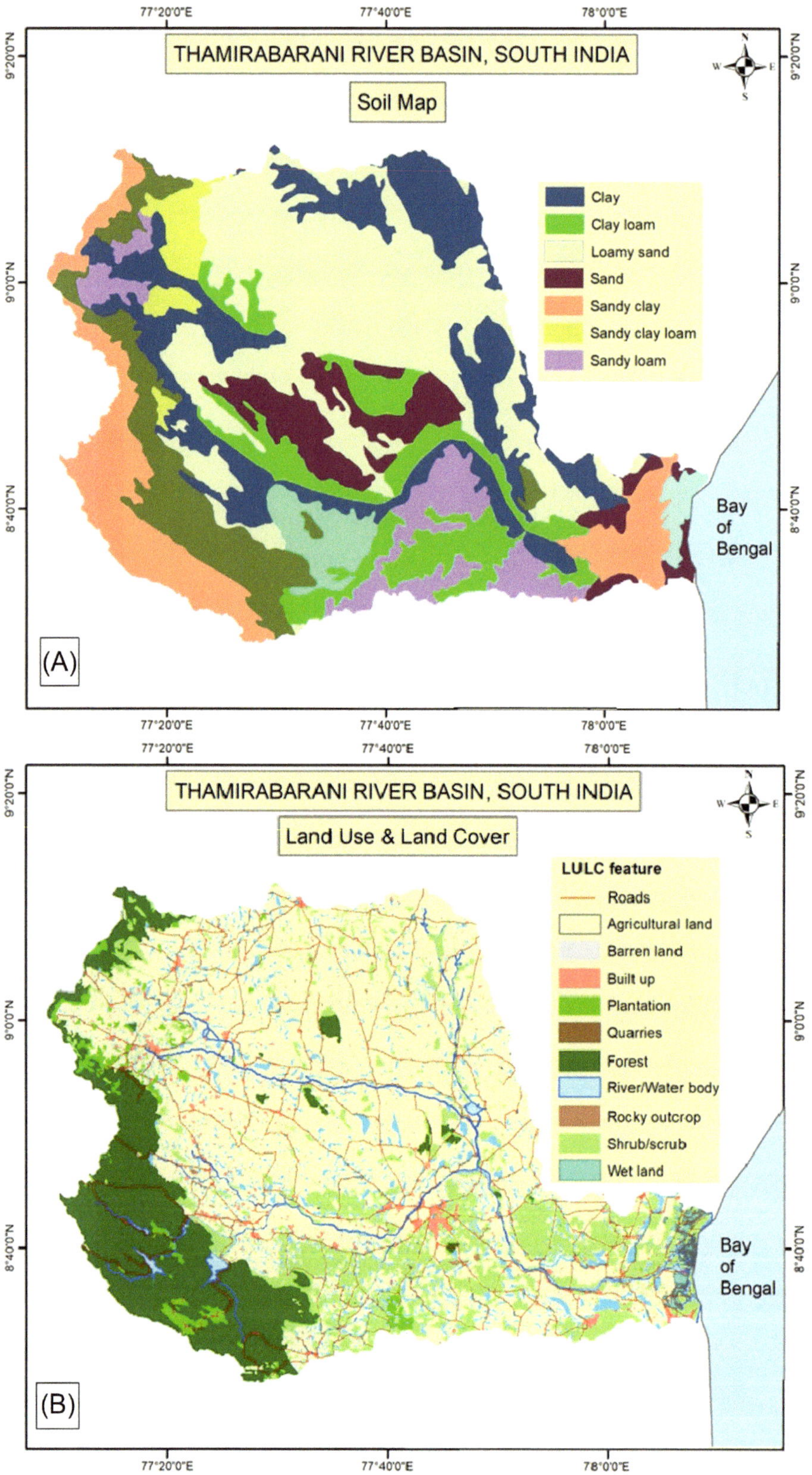

FIGURE 4.2 Thamirabarani river basin; Soil type (A) and land use/land cover (B).

sand and clayey, which spread over larger parts of the buried pediments and pediplains in the middle parts, and these areas found intensive agriculture activities like paddy, banana, vegetables, and sugarcane, etc. Alluvial soils of loamy sand and clay loam are mainly distributed along the riverbed side (Kaliraj et al., 2015; Mohana et al., 2018). The isolated red soil patches are found in various parts, especially in the high ground of northern parts, and it is used for cultivating crops, pulses, and millets during monsoons. Red loamy soils are found in many patches within pediplains, especially in northwestern parts. Black clay soils are found in the middle and northeastern parts, and Sandy alluvium and red sandy soils are found in the southeastern parts. The pediplains of the middle parts are noted, with isolated patches of red loamy and clay soils in the northern parts. Coastal alluvial soils with clay and sandy loam are found within the proxy of flood plains and coastal stretches in the eastern parts. Alkaline and saline soils are also noticed there (Magesh and Chandrasekar, 2014; Kaliraj et al., 2016). The red sandy soil is found in the southeastern parts and is suitable for cultivating groundnut, millets, pulses, etc., The clay soils are found in isolated patches surrounded by red loamy soils in various parts. Black clay soil has enriched clay contents with the calcareous and ferrous matter, and it is highly suitable for cultivating cotton and pulses. Soils of the basin, almost in all parts, are used for crop cultivation during northeast monsoon periods (October–December) and they are found to be fallows, bare soils, and barrens during summer, except the upland forest covers and irrigated croplands.

Fig. 4.2B shows the spatial pattern of land use and land cover (LULC) features in the study area. The LULC features of the basin were extracted from the Landsat 9—OLI image using GIS techniques (Mulder et al., 2011; Kaliraj et al., 2017). The LULC features consist of upland forest cover, plantations, agriculture lands (irrigated and unirrigated lands), fallows, land with shrubs, barren lands, mining and quarries, water bodies (rivers, channels, lakes, ponds, and pools), settlements/built-up (urban and rural), coastal wetlands (estuary and saltpans), etc. About 57% of the area covers agricultural lands, including irrigated croplands mainly on the reverbed side and unirrigated croplands and fallows found in all parts of the pediments and pediplains. About 15% of the forest cover is found in the Western Ghats and other upland areas. Land with shrubs and scrub is noted in many places as isolated batches up to an area of 14.81% and this is mainly associated with barren lands and limestone quarries. Water bodies including rivers, reservoirs, channels, and ponds found in all parts of the basins, and they cover an area of 6.97%. Settlements and built-up cover the area of 3.23%, which includes urban and rural settlements in many locations, mainly nearby waterbodies. Significantly, the agricultural lands are predominantly distributed in all parts, but many of them are noticed as unirrigated croplands, mainly used for one-time cultivation throughout the year depending on monsoonal rainfall.

4.3 Materials and methods

Using GIS techniques, the study area layer is prepared from the map of the River Basin Atlas of India published by the Central Water Commission and ISRO in the year 2012. Fig. 4.3 shows the methodology flowchart used in this study. The bare soil of the basin is extracted by masking forest cover, shrub vegetation, plantations, water bodies, and settlements (urban and rural) areas using the GIS masking tool. The Landsat 9—OLI image (level 2 surface reflectance products of Collection 2 scenes) is used to assess the spatial characteristics of soils and parent materials in the Thamirabarani river basin in South India. Landsat 9—OLI provides multispectral images with four visible spectral bands (0.43–0.67 μm), one NIR band (0.85–0.88 μm), and three shortwave-infrared bands (1.57–2.29 μm) at 30 m spatial resolution, and these multispectral images can be used for the monitoring and assessment of soil and land properties (Nguyen et al., 2020; USGS, 2022; Abay et al., 2022). The remote sensing index is used as an appropriate algorithm to estimate soil property based on the standard protocol. In this analysis, the remote sensing indices namely NDVI, SAVI, BSI, NDMI, soil organic carbon index (SOCI), clay mineral index, ferrous mineral index (FMR), carbonate index, iron oxide index, and SSI are executed on OLI spectral bands (VNIR) to estimate soil properties and its spatial characteristics using GIS techniques (Huete, 1988; Schowengerdt, 2006; Dogan, 2008; Dehni and Lounis, 2012; Baghdadi and Zribi, 2017; Asfaw et al., 2018; Babaeian et al., 2019; Bonfatti et al., 2020; Kanwar et al., 2022).

4.3.1 Normalized differential vegetation index

The NDVI is a vegetative indicator calculated from NIR (band 5) and red (band 4) spectrums of the Landsat 9—OLI image using the equation

$$\mathrm{NDVI} = (\mathrm{NIR} - \mathrm{Red})/(\mathrm{NIR} + \mathrm{Red}) \tag{4.1}$$

For Landsat 9—OLI image, the NDVI equation is expressed as NDVI = (Band 5 − Band 4)/(Band 5 + Band 4). The OLI image of level 2 surface reflectance products (Collection 2 scene) is used. The NDVI calculates the difference

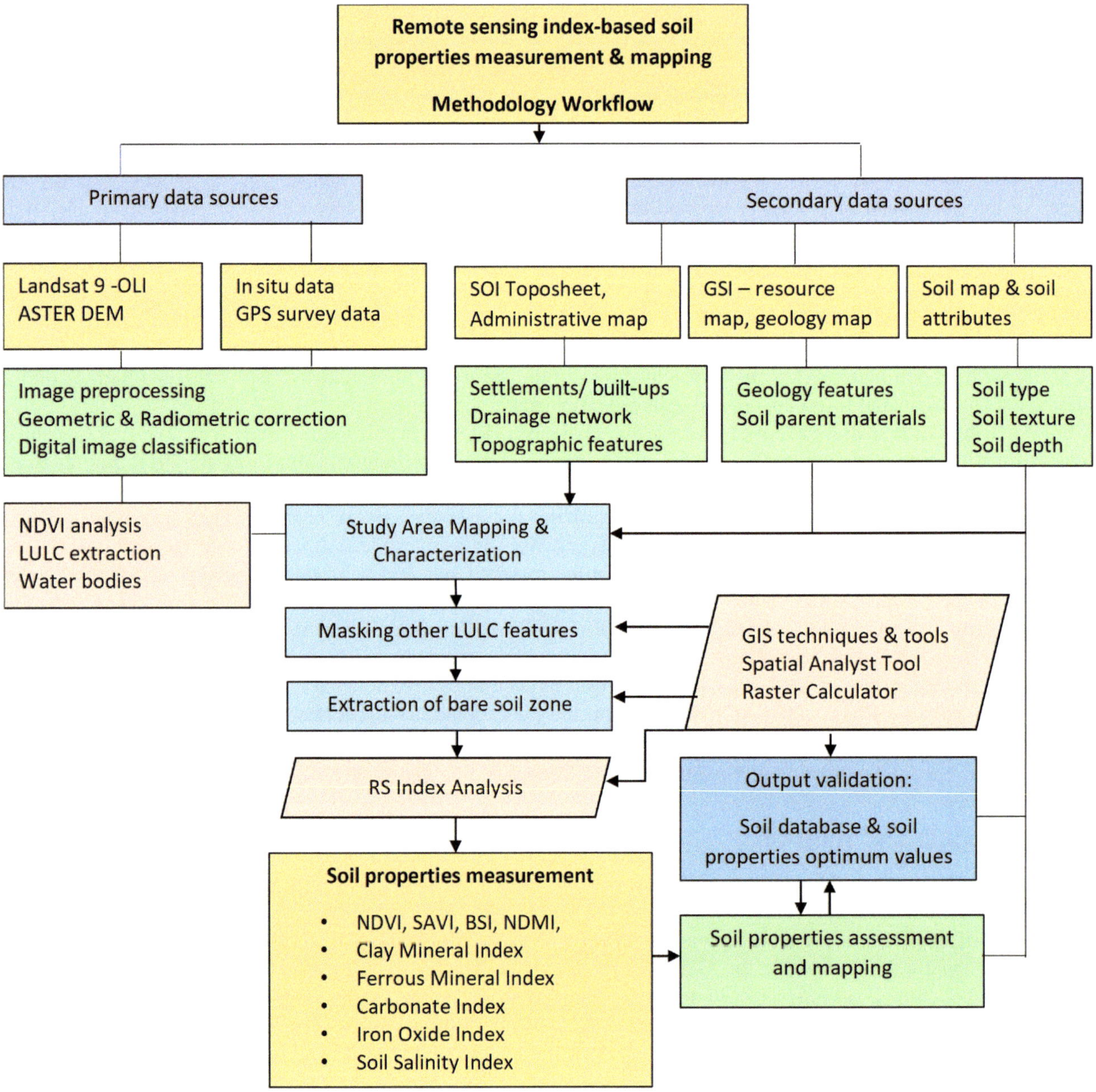

FIGURE 4.3 Flowchart of the methodology used in this study.

between the reflectance of NIR and red to estimate vegetation greenness and nonvegetative features on a patch of land (Bhatta, 2011; Ke et al., 2015; Zaitunah et al., 2018).

4.3.2 Soil-adjusted vegetation index

The SAVI is a vegetation index used to correct the NDVI for the influence of soil brightness in arid or semiarid areas (i.e., areas having lower vegetative cover). This is used to minimize soil brightness influences in NDVI using a soil brightness correction factor (L) using the equation (Huete, 1988; Baghdadi and Zribi, 2017)

$$\text{SAVI} = \big((\text{NIR} - \text{Red})/(\text{NIR} + \text{Red} + \text{L})\big) \times (1 + \text{L}) \tag{4.2}$$

Where *NIR*, near-infrared band; *Red*, red band; and *L*, soil brightness correction factor (i.e., the amount of green vegetation cover); in this, L = 1 for bare soils/nonvegetative cover, L = 0.5 for moderate vegetative cover, and L = 0 for dense vegetation, based on the result of NDVI (Schowengerdt, 2006; Lillesand et al., 2011; Dwivedi, 2017; Li et al., 2018; Fellipe et al., 2021; USGS., 2022). For Landsat 9—OLI image, the SAVI is calculated using the equation SAVI = ((Band 5 − Band 4)/(Band 5 + Band 4 + 0.5)) × (1.5). The OLI image of level 2 surface reflectance products (collection 2 scene) is used for the estimation of NDVI in the area.

4.3.3 Bare soil index

The BSI is a numerical indicator of bare soils (Zhao et al., 2019; Liu et al., 2022), and it is calculated by combining blue (B), red (R), NIR, and SWIR spectral bands of the Landsat 9—OLI image to capture soil variations using the equation (Eq. 4.3), which is expressed as

$$\mathrm{BSI} = ((\mathrm{SWIR} + \mathrm{R}) - (\mathrm{NIR} + \mathrm{B}))/((\mathrm{SWIR} + \mathrm{R}) + (\mathrm{NIR} + \mathrm{B})) \tag{4.3}$$

Where these spectral bands are used in a normalized manner.

4.3.4 Normalized difference moisture index

The NDMI is used to estimate vegetation water content by calculating the ratio of spectral reflectance between NIR and SWIR bands (Taloor et al., 2021), using the equation (Eq. 4.4), which is expressed as

$$\mathrm{NDMI} = (\mathrm{NIR} - \mathrm{SWIR})/(\mathrm{NIR} + \mathrm{SWIR}) \tag{4.4}$$

For Landsat 9—OLI image, the NDMI is calculated as NDMI = (Band 5 − Band 6)/(Band 5 + Band 6) using the level 2 surface reflectance products of NIR and SWIR in the OLI image.

4.3.5 Soil organic carbon index

SOCI index is used for the assessment of available SOC using the Landsat 9—OLI image. SOC values of different soil types were estimated from a combination of the remote sensing indices of NDVI ((NIR − Red)/(NIR + Red)) and Bare Soil Index ((SWIR + R) − (NIR + B))/((SWIR + R) + (NIR + B)) with its coefficient values derived from in situ sampling measurement (Bhunia et al., 2017; Gholizadeh et al., 2018). The LANDSAT 9—OLI images were used to calculate NDVI and BSI. Using multivariate regression analysis, the relationship between indices products and in situ SOC values is estimated to measure the spatial distribution of SOC using the equation, which is expressed as

$$\mathrm{SOC} = 4.24 + 11.90 \times \mathrm{NDVI} - 0.021 \times \mathrm{BSI} \tag{4.5}$$

Where the numerical values refer to the coefficient derived from two variables such as in situ SOC and NDVI and BSI data using the Pearson correlation coefficient method. The SOC obtained from indices-based covariates indicates SOC distribution . In this analysis, remote sensing indices (NDVI and BSI) is used to estimate how the SOC relates with its field SOC data using multivariable regression, and the result is used for finding the optimum SOC in the different soil type in the area.

4.3.6 Clay mineral index

The clay mineral index is used to estimate the soil's clay content based on the ratio of the $SWIR_1$ and $SWIR_2$ bands, and the equation calculates the ratio of SWIR bands that leverages the hydrous minerals such as the clay within the bandwidth (Yitagesuab et al., 2011; Trinh, 2016; Bishop et al., 2017). In general, the clay mineral index is used for ASTER data (15 bands with the spectral range of 520−11650 nm); the equation is expressed as

$$\text{Clay Mineral Index} = \mathrm{SWIR}_1/\mathrm{SWIR}_2 \tag{4.6}$$

In this analysis, Landsat 9—OLI image is analyzed to estimate the clay mineral contents using the equation, which is expressed as clay mineral index = $SWIR_{\text{band } 6}/SWIR_{\text{band } 7}$. The ratio of these two bands is used to demarcate the clay mineral contents in an area based on their high exchangeable ions and high adsorption capacity within the SWIR bands.

4.3.7 Ferrous mineral index or ferrous mineral ratio index

The FMR is used to determine the level of iron concentration in the bare soil by calculating the ratio of SWIR and NIR bands (Dogan, 2008; Trinh, 2016; Abay et al., 2022), and the equation is expressed as

$$\text{Ferrous Mineral Index} = \text{SWIR}/\text{NIR} \tag{4.7}$$

For Landsat 9—OLI image analysis, the equation is expressed as lay mineral index = SWIR $_{(\text{band } 6)}$/NIR $_{(\text{band } 5)}$. Wherein, the ratio of SWIR and NIR bands is used for identifying the iron-bearing minerals (ferrous minerals) in rocks, bare soils, fallows, barren lands, etc.

4.3.8 Soil carbonate index

The carbonate content in soil-bearing materials has significant spectral behavior in B and NIR bands due to their surface reflectance characteristics (Xie et al., 2015; Notesco et al., 2020). The carbonate content is estimated using the equation, which is expressed as

$$\text{Carbonate index} = (\text{Blue} - \text{NIR})/(\text{Blue} + \text{NIR}) \tag{4.8}$$

Using Landsat 9—OLI image, the amount of carbonate content is estimated using the equation, which is expressed as Carbonate index = (Blue $_{\text{band } 2}$ − NIR $_{\text{band } 5}$)/(Blue $_{\text{band } 2}$ + NIR $_{\text{band } 5}$), in which the surface reflectance of B (band 2) and NIR (band 5) of the Landsat 9—OLI image is used to calculate the ratio and it indicates the presence of carbonate-bearing minerals in bare soil areas.

4.3.9 Soil iron oxide index (or) iron oxide ratio index

Soil-bearing iron material is strongly correlated with the spectral reflectance of R and B bands, wherein, higher reflection in the R band is observed, as opposed to strong absorption in the B band (Trinh, 2016; Segal, 2018), and this ratio is significantly highlighted concerning iron fraction minerals in bare soils using the following iron oxide ratio index, which is expressed as

$$\text{Iron Oxide Ratio} = \text{Red}/\text{Blue} \tag{4.9}$$

Using Landsat 9—OLI image, the amount of iron content is estimated using the equation, which is expressed as Iron Oxide Ratio = $\text{Red}_{\text{band } 4}/\text{Blue}_{\text{band } 2}$, in which the ratio between the spectral bands of R (band 4) and B (band 2) in the Landsat 9—OLI image is calculated for estimating the iron-bearing minerals in bare soil areas.

4.3.10 Soil salinity index

Soil salinity assessment in bare soil areas is correlated with spectral properties of NIR and SWIR bands in Landsat 9—OLI image. To identify the soil salinity, the index is principally calculating the ratio between NIR and SWIR spectral bands, and this ratio is used to detect salt matter in soils based on the different spectral responses (Chen et al., 2017; Asfaw et al., 2018; Nguyen et al., 2020). The SSI is developed based on the spectral difference between saline soil and nonsaline soil, and the soil salinity is estimated using the equation, which is expressed as

$$\text{NIR-SWIR Salinity Index(NSI)} = \text{SWIR}_1 \times \text{SWIR}_2/\text{SWIR}_1 \times \text{NIR} \tag{4.10}$$

Wherein, saline soils have higher reflectance values within the visible and NIR bands compared to nonsaline soils, while saline soils' reflectance is extremely low in SWIR compared to nonsaline soils due to the strong absorption of carbonate matters present in saline soils.

4.4 Results and discussion

Spatiotemporal assessment and mapping of soil properties are widely done by using these satellite sensors, namely Landsat—TM, ETM + , MSS, OLI/TIRS, Sentinel 2—MSI, IRS—LISS, IRS-AWiFS, Spot—HRV-XS, MODIS—Terra, and IKONOS-Ms, etc. (Osman, 2012; Dehni and Lounis, 2012; Pereira et al., 2017; Babaeian et al., 2019; Bonfatti et al., 2020; Fellipe et al., 2021). In this study, the Landsat 9—OLI image is analyzed using selected remote sensing indices for assessment of soil characteristics in the Thamirabarani river basin area, and the derived outputs can be used for soil and land resources management purposes. Some of the remote sensing indices used in this study are described as follows.

4.4.1 Assessment of NDVI and its spatial characteristics

The NDVI is used to capture the vegetation's greenness with its health status and to segment the nonvegetative features (Ke et al., 2015; Zaitunah et al., 2018). NDVI value is always estimated within the range from -1 to $+1$. Fig. 4.4 shows the NDVI map of the basin estimated using NIR and R bands of Landsat 9—OLI image. The result shows that the NDVI > 0.6 indicates the presence of healthy vegetation (dense green leaves) in the upland hill range and along the riverbed; NDVI < -0.1 (negative value) indicates the location of water bodies and urbanized areas, and the NDVI value (> 0.1 and < 0.3) indicates the nonvegetative features like soils, fallows, barrens, and rocky outcrops, etc. Higher NDVI denotes the occurrence of healthy vegetation (chlorophyll) that reflects more NIR and green light compared to other wavelengths. NDVI layer is used to extract the nonvegetative zone like soils, fallows, and barren lands from the vegetative cover.

4.4.2 Assessment of SAVI and its spatial characteristics

The SAVI is used to improve the NDVI by minimizing the influences of soil brightness using a soil-brightness correction factor (i.e., amount of vegetation greenness). The derived output can be used to assume soil color, soil moisture, and saturation effects from density vegetation. SAVI is calculated as calculated from the ratio between the surface reflectance of NIR and R bands of Landsat 9—OLI image with a soil brightness correction factor (L), generally defined as 0.5 to accommodate most land cover types (Li et al., 2018; USGS, 2022). However, the L value varies depending on the amount of green vegetative cover. Generally, $L = 1$ in areas with no green vegetation cover, $L = 0.5$ in areas of moderate green vegetative cover, and $L = 0$ in areas with very high vegetation cover (based on the NDVI output). Like NDVI, the SAVI is always estimated within the range from -1 to $+1$. Fig. 4.5 shows the SAVI map of the basin,

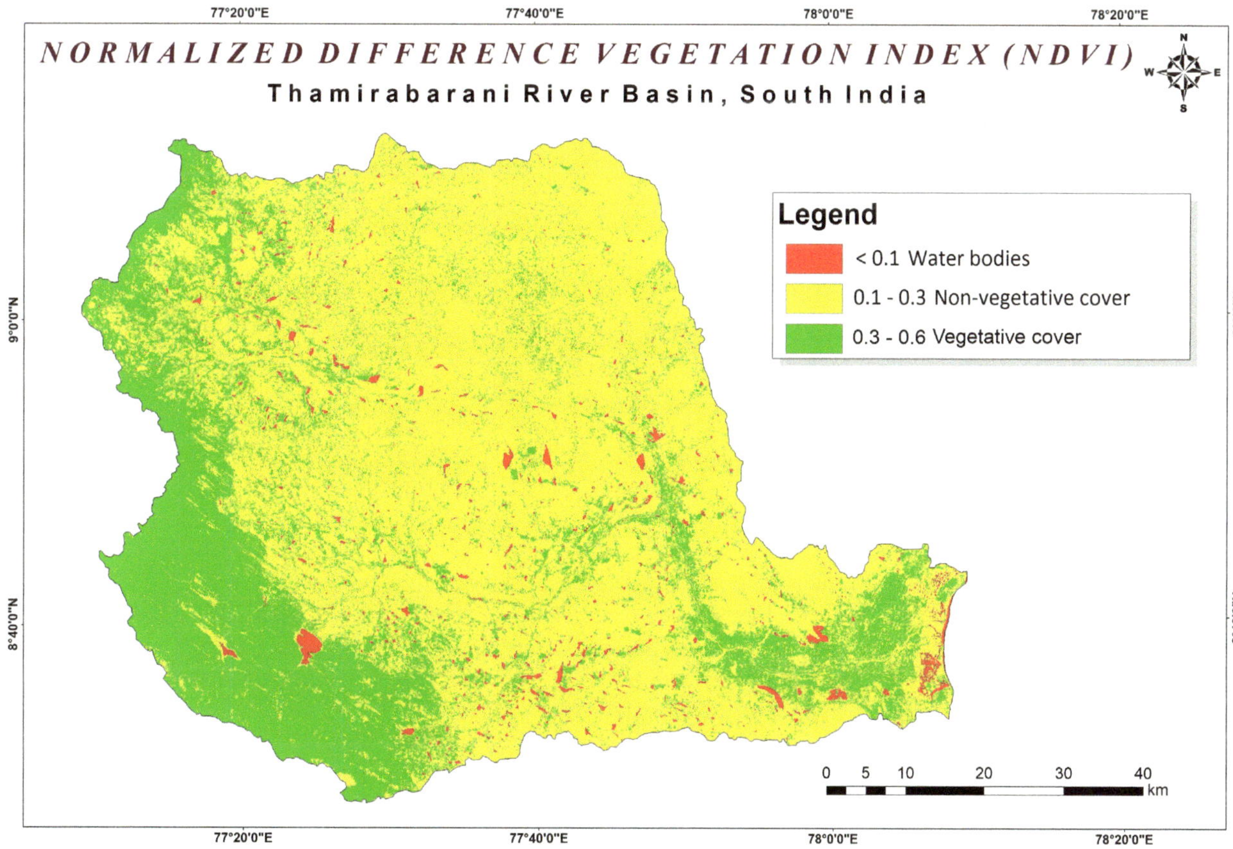

FIGURE 4.4 The NDVI of the basin showing vegetation and nonvegetation zones.

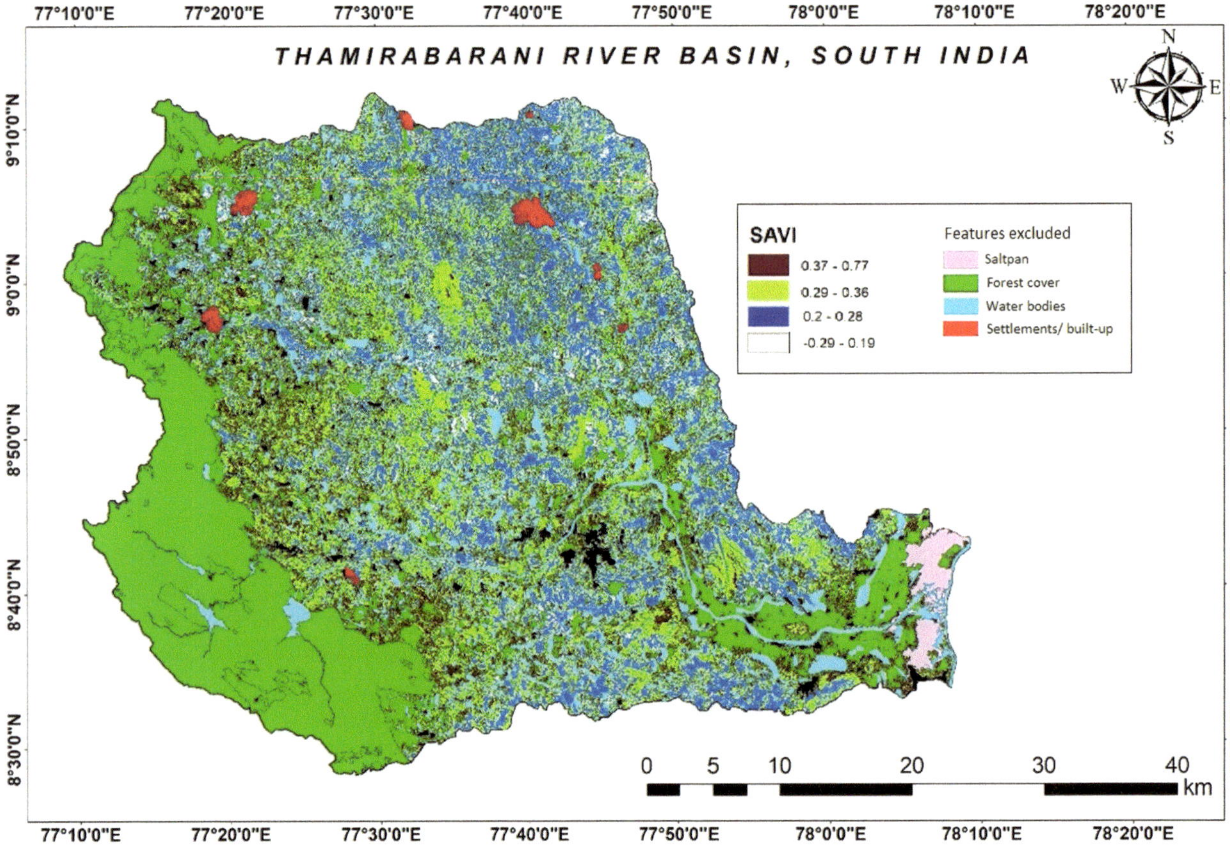

FIGURE 4.5 The SAVI map of the basin.

estimated using Landsat 9—OLI image. The result shows the SAVI is at a high value (0.37–0.77) indicating the denser (and) healthy vegetation, especially plantations and irrigated croplands (herewith, forest cover put under excluded features). SAVI value of 0.29–0.36 indicates unirrigated croplands and land with shrubs. Significantly, the SAVI value at 0.2–0.28 denotes bare soil areas and fallows, and the least SAVI value of 0.29–0.19 shows the nonvegetative features like eroded soils, rocky outcrops, etc. The SAVI can be better used in segmenting vegetation and nonvegetation than NDVI because of its saturating value of > 0.7, but SAVI has this point at only 0.3 and it saturates slower. The SAVI value < 0.2 indicates no vegetation, just bare soil and fallows. However, for the estimation of SAVI, the L parameter needs to be correctly assigned based on the LULC features.

4.4.3 Assessment of BSI and its spatial characteristics

BSI uses the numerical index to capture bare soil zones and their soil variations (Zhao et al., 2019; Kanwar et al., 2022). Bare soil is used to demarcate the area of soil or sand that is not covered by vegetation, grass/pasture, natural turfs, rocks, gravel, and other artificial turfs. In this study, the LULC features, including forest cover, settlements, water bodies, and salt pans, are excluded from BSI analysis using the masking tool in GIS software. Fig. 4.6 shows the BSI map of the study area, derived from the Landsat 9—OLI image. A lower BSI value in the range −0.29 to −0.06 indicates the barren areas of rocks, limestone mining, quarries, and other nonvegetative covers. BSI value in the range −0.061 to −0.009 denotes bare soils like fallows and unirrigated croplands with the least moisture conditions. The BSI is at the moderate value of −0.008–0.0.033, showing the fallows and unirrigated croplands with red loamy and clay soils in semiarid conditions. The high rate of BSI value is estimated at 0.034–0.342 and occurs mainly in irrigated crops, humid bare soils, and sandy soils with significant moisture contents; these lands are found in the proximity of water bodies. The comparative study reveals that the BSI showing bare soils tend to generate rather small NDVI values in the range of 0.1–0.2 due to a lack of vegetation greenness. The BSI ranges of bare soils are relatively distinct in dry and humid bare soils, whereas dry bare soils (arid and semiarid areas) have reflected more in SWIR than that of humid bare soils (irrigated croplands and wetlands) due to higher absorption by soil moisture content. The BSI map can be

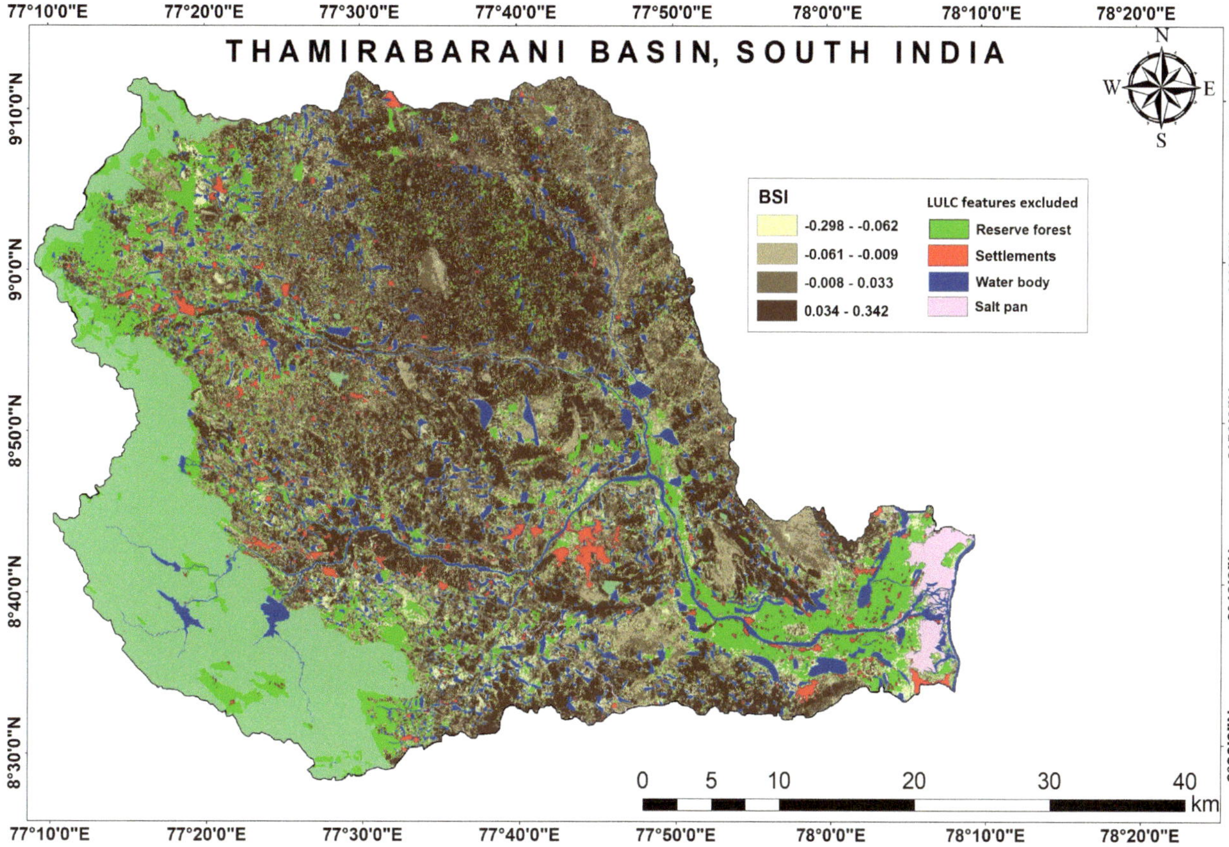

FIGURE 4.6 The BSI map showing bare soil zones of the basin.

used in numerous remote sensing applications, like soil mapping, estimation of SOC, and crop identification in combination with NDVI and in situ data.

4.4.4 Assessment of NDMI and its spatial characteristics

The NDMI is used to detect the vegetation water content (i.e., the moisture level in vegetation) based on the ratio between NIR and SWIR spectral bands (Babaeian et al., 2019; Taloor et al., 2021). The NDMI is a reliable indicator of water stress in crops. In this study, the NDMI analysis is performed on Landsat 9—OLI image to estimate the vegetation water content (moisture level). Generally, the NDMI range is estimated at -1 to $+1$, where the lower values (< -1) indicate the vegetation water content is at a lower rate and the higher values denote the higher vegetation water content (i.e., higher moisture rate). Fig. 4.7 shows the NDMI characteristics in the basin area. The NDMI value is in the lower range ($-0.455$to -0.039), indicating that the vegetation, soils, and lands fall under water stress (i.e., least moisture) conditions, mainly in barren rocks, eroded lands, rocky outcrops, and bare soils. Likewise, the NDMI value ranges from -0.038 to 0.021, indicating that vegetation with lower water content or water stress and lower moisture rate is found in bare soils, fallows, and unirrigated croplands. The NDMI value range of 0.02–0.093 indicate the moderate vegetation water content in the land with shrubs and unirrigated croplands. Moreover, the higher NDMI value range of 0.094–0.484 indicates that the vegetation water content is at a higher rate with good moisture condition, found in dense vegetative covers, plantations, and irrigated croplands within the proxy of water bodies. Wherein, the areas noted the lower NDMI value (< -0.038), mainly in bare soils, fallows, and barren lands, clearly shows the water stress condition across the fields. However, the NDMI values vary in an area depending on the plant's growing seasons and phenological stage.

4.4.5 Assessment of SOC

The SOC is an essential nutrient for crop production activities. It is noted that the amount of carbon present in the soil varies in different soils and places depending on soil organic substances, soil genesis type, and climatic conditions

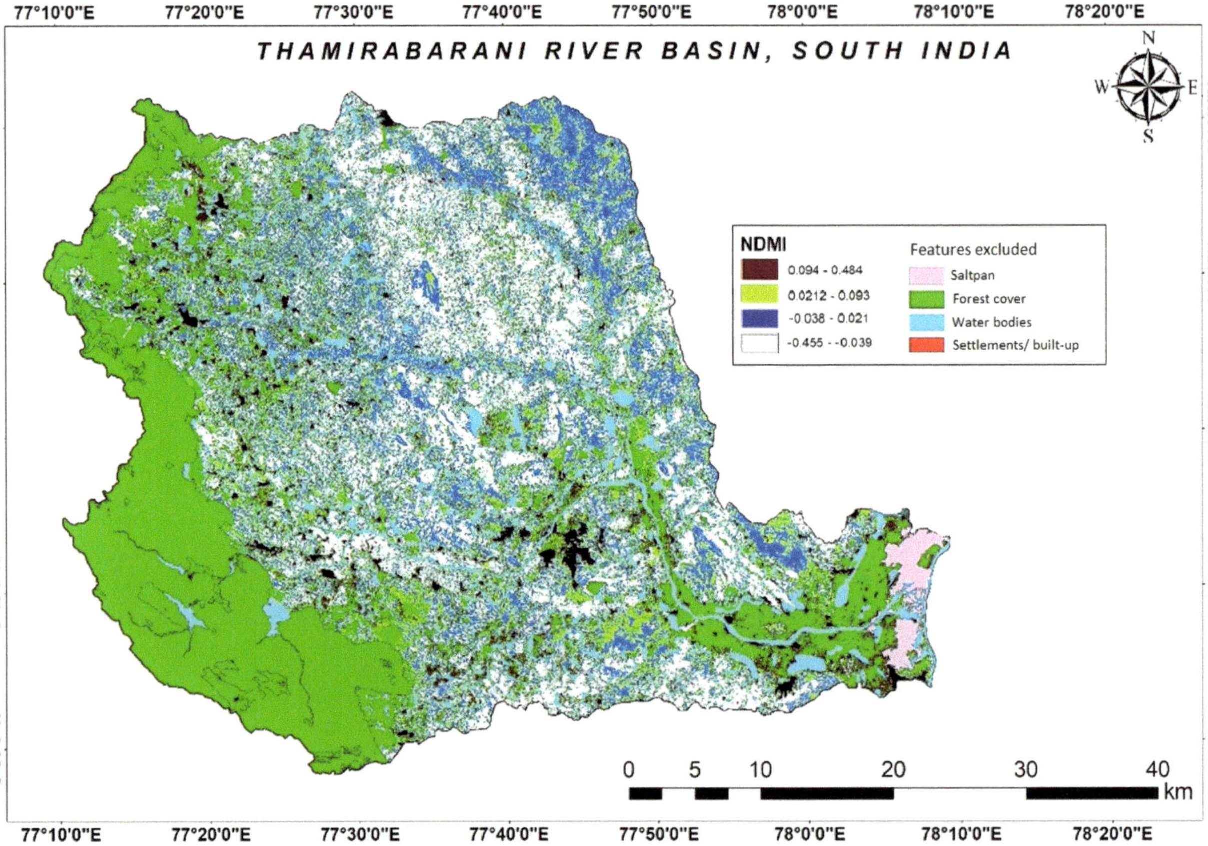

FIGURE 4.7 The NDMI map showing vegetation water content (moisture level) of the basin.

(Mehrez Zribi et al., 2011; Escribano et al., 2017; Bhunia et al., 2017; Gholizadeh et al., 2018). The SOC is estimated using Landsat 9—OLI image based on the indices correlation of NDVI and BSI with in situ SOC-measured data using multivariate regression analysis. Fig. 4.8 shows the optimum SOC estimated for different land use of the basin. The result shows that the estimation of SOC is excluded from the LULC features, including forest cover, plantations, water bodies, and built-up areas. The optimum SOC value estimated is in the range of 0.06%–0.48%, for different soil types, especially in bare soils areas, in which the SOC is at a high rate (0.26%–0.48%) in the irrigated croplands and land with shrubs and it indicates that the availability of organic matters in soils. The SOC is estimated at the range of 0.20%–0.26% in a major part of unirrigated lands of pediplains, including red loamy and black clay soils. Whereas the SOC is at a lower rate (0.06%–0.20%) found in the eroded lands and the lands within the proxy of water bodies due to the removal of topsoil. It is noted that the SOC matters that occurred in soils determine the physical, chemical, and biological function of soils in the basin area.

4.4.6 Assessment of clay mineral content

In general, clay soil is formed by the combination of clay, silt particles, and a small amount of humus and sand particles originating from weathering rocks, boulders, and sediment flows from the upland range. The clay soil is composed of a mixture of mineral contents like silica, alumina, magnesia, or both, and appreciable quantities of potassium, sodium, calcium, etc. The ratio of $SWIR_1$ and $SWIR_2$ bands is used to estimate the clay contents in soils because hydrous minerals like clay absorb more in this spectral bandwidth (Yitagesuab et al., 2011; Bishop et al., 2017). In the basin, clay minerals are often distributed in clay and clay loamy soils across the middle, north, and northeastern pediplains as discrete patches and it is sometimes extended a few feet in thickness within these soil layers. Fig. 4.9 shows the clay minerals exposed in the study area derived from Landsat 9—OLI image. The index value 0.38–1.22 represents low clay mineral regions which are mostly found in fallows of red loamy soils. The index value of 1.23–1.29 and 1.3–1.35 mostly covers the regions of barren land, land without scrubs. The high value of 1.36–1.71 is mostly found on exposed outcrops or near water bodies. The clay mineral content in croplands is also observed at moderate to high quantity.

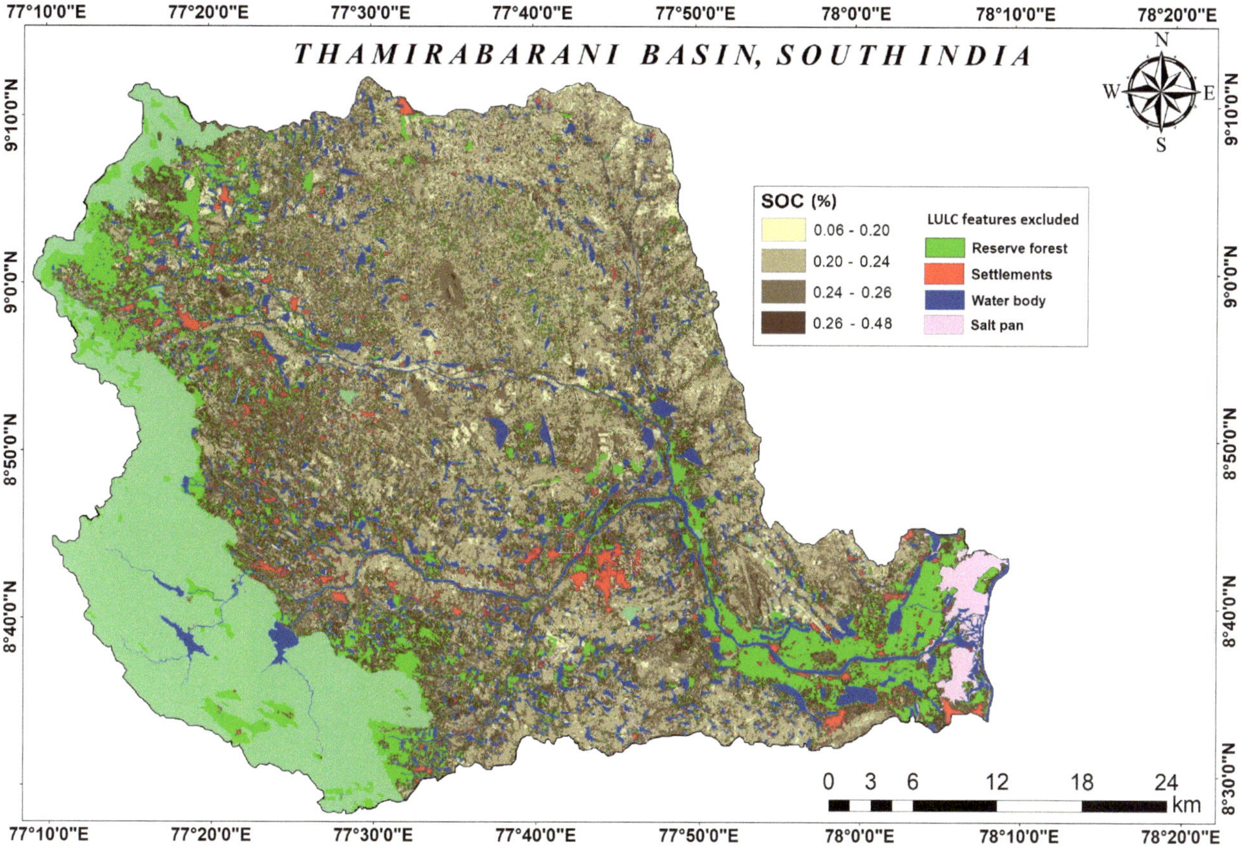

FIGURE 4.8 Spatial distribution of SOC in the basin.

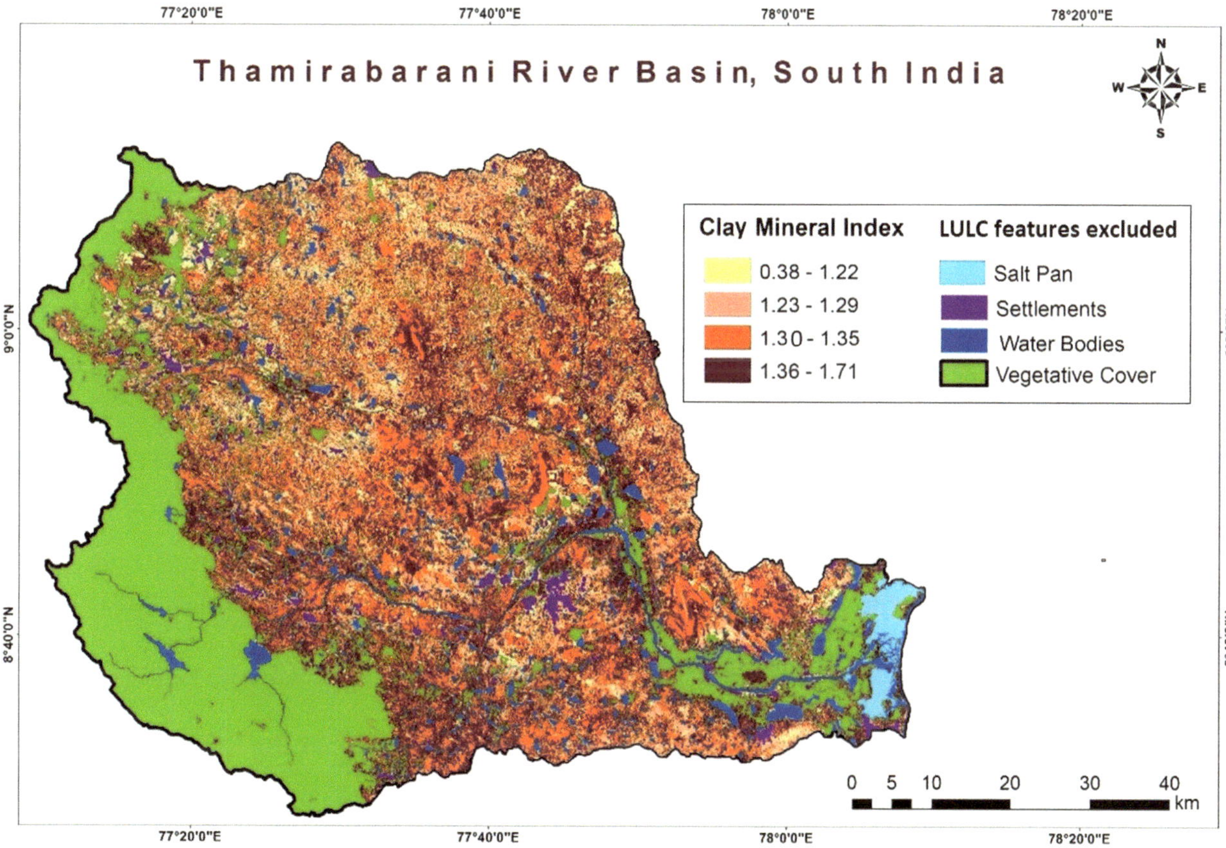

FIGURE 4.9 Spatial distribution of clay mineral content in different soils of the basin.

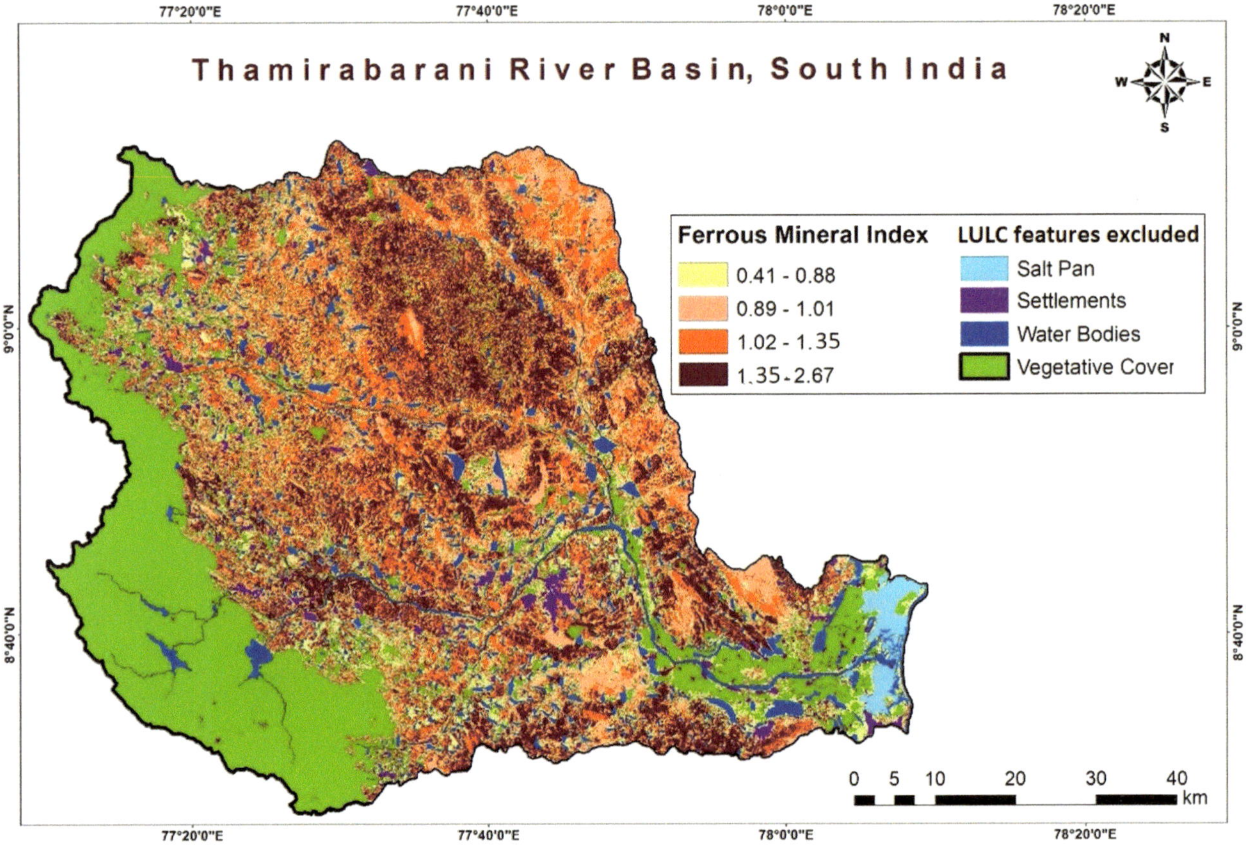

FIGURE 4.10 Spatial distribution of ferrous content in different soils of the basin.

4.4.7 Assessment of ferrous mineral content

The FMR is used to estimate the level of iron concentration (ferrous minerals) in the rocks and bare soil areas. Using the Landsat 9—OLI image, the spectral ratio between SWIR and NIR bands highlights iron-bearing minerals in soils because of their strong influences within these bands (Trinh, 2016; Abay et al., 2022). Fig. 4.10 shows the spatial variability of ferrous mineral contents in the study area. The resultant map highlights the ferric iron-bearing zones in the study area, and the estimated index values range from 0.41 to 2.67; it is classified as 0.41–0.88 (i.e., ferrous minerals at a lower concentration), 0.891.01 (moderate), 1.02–1.1 (High), and 1.11–2.67 (remarkably high). The resulting map shows the spatial distribution of iron-bearing minerals in different soils of the study area. The lower rate (0.41–0.88) of ferrous minerals found in the areas nearby foothills, riverbeds, and coastal wetlands is due to the subsequent removal of topsoil layers. The moderate value 0.89–1.01 is observed in fallows and barrens. The high value of 1.02–1.35 is covered by exposed rocky outcrops, unirrigated croplands, and land with shrubs. The very high value of 1.35–2.67 is denoted by the prominent level of iron concentration (ferrous minerals) in a majority of pediplains covered with loamy and black clay soils under unirrigated croplands across the basin.

4.4.8 Assessment of soil carbonate content

Soil carbonate contents may be composed of calcium carbonate or sodium carbonate, and it is derived from rock weathering or settling of their residual matter by the action of water and wind. Using Landsat 9—OLI image, the spectral bands of B (band 2) and NIR (band 5) are analyzed to estimate subpixel amounts of soil-bearing carbonate content in bare soil areas. Fig. 4.11 shows the carbonate index result derived by bands B and NIR of Landsat 9—OLI image. The resulting map shows the spatial distribution of carbonate-bearing minerals, and it is estimated at the range of 0.11%–0.47%. It is noted that soil-bearing carbonates vary in spectral behavior depending upon the soil parent materials like calcium, magnesium, sodium, etc., which influences their spectral reflectance in visible and NIR bands (Xie et al., 2015; Notesco et al., 2020). The carbonate index is at a lower rate (0.11%–0.19%) in most foothills areas and

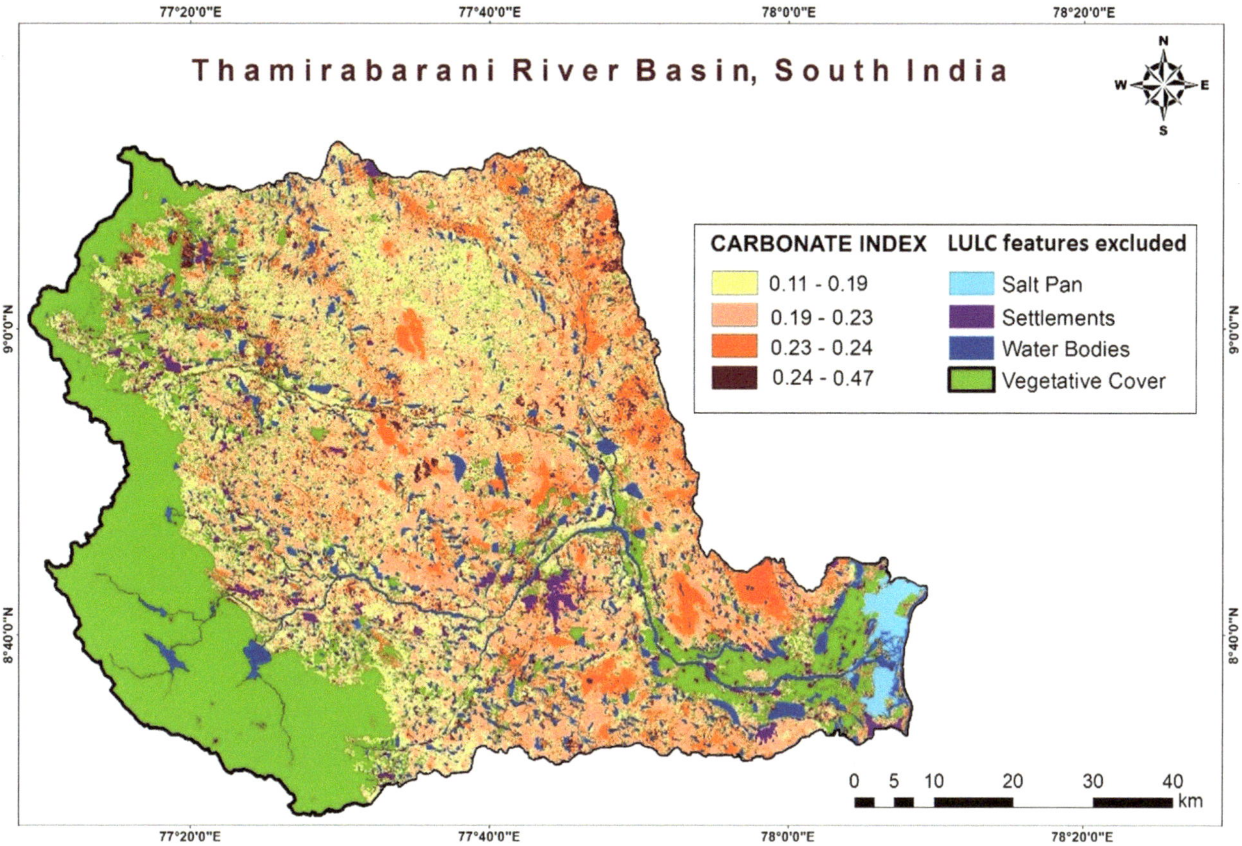

FIGURE 4.11 Spatial distribution of soil carbonate content in the basin.

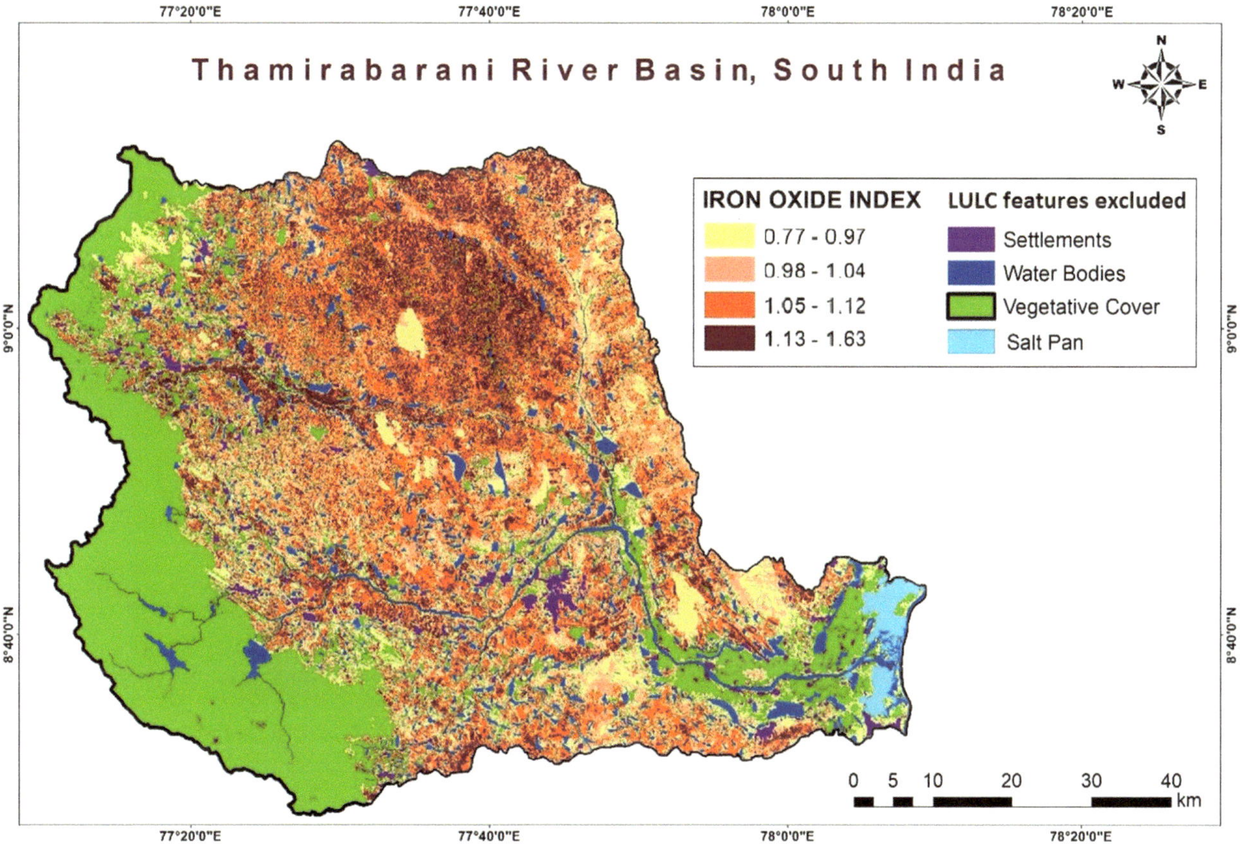

FIGURE 4.12 Spatial distribution of iron oxide content in different soil types of the basin.

pediplains that covers irrigated croplands and fallow land with red loamy and clay soils. The moderate value of 0.19%–0.24% is noted in rocky outcrops, bare soil areas, barrens, and land with shrubs, especially in the northeastern parts. The higher rate of carbonate content (0.23%–0.47%) is found in some parts of barrens lands associated with limestone quarries, discrete patches of salt-affected soils in northeastern parts and coastal wetlands due to the presence of carbonate mixture with residual calcium and sodium matters. In the basin, the high carbonate content seems to be sparely distributed in certain areas resulting from site-specific activities.

4.4.9 Assessment of iron oxide content

In general, soil-bearing iron oxide matter varies in different places based on pedogenic processes that determine the spectral behaviors, that is, higher reflectance in R bands and absolute absorption in B bands (Dogan, 2008; Trinh, 2016; Segal, 2018; Arisanty et al., 2021; Abay et al., 2022). Soil-bearing iron content is formed by the combination of iron cations Fe^{3+} and Fe^{2+}. Fig. 4.12 shows the spatial variability of iron-bearing contents in the study area. The basin comprises iron-bearing contents in various parts of soils, and it is estimated at the range of 0.77%–1.63%. Soils with lower iron content (0.77–0.97) are found in the areas like rocky outcrops and barren lands. The moderate value of 0.98–1.04 is estimated in larger parts of pediplains covered fallows and croplands. The soil-bearing iron content is estimated at a higher rate (1.05%–1.63%) in major parts of black clay soils found in fallows and unirrigated croplands, especially in northern pediplains areas.

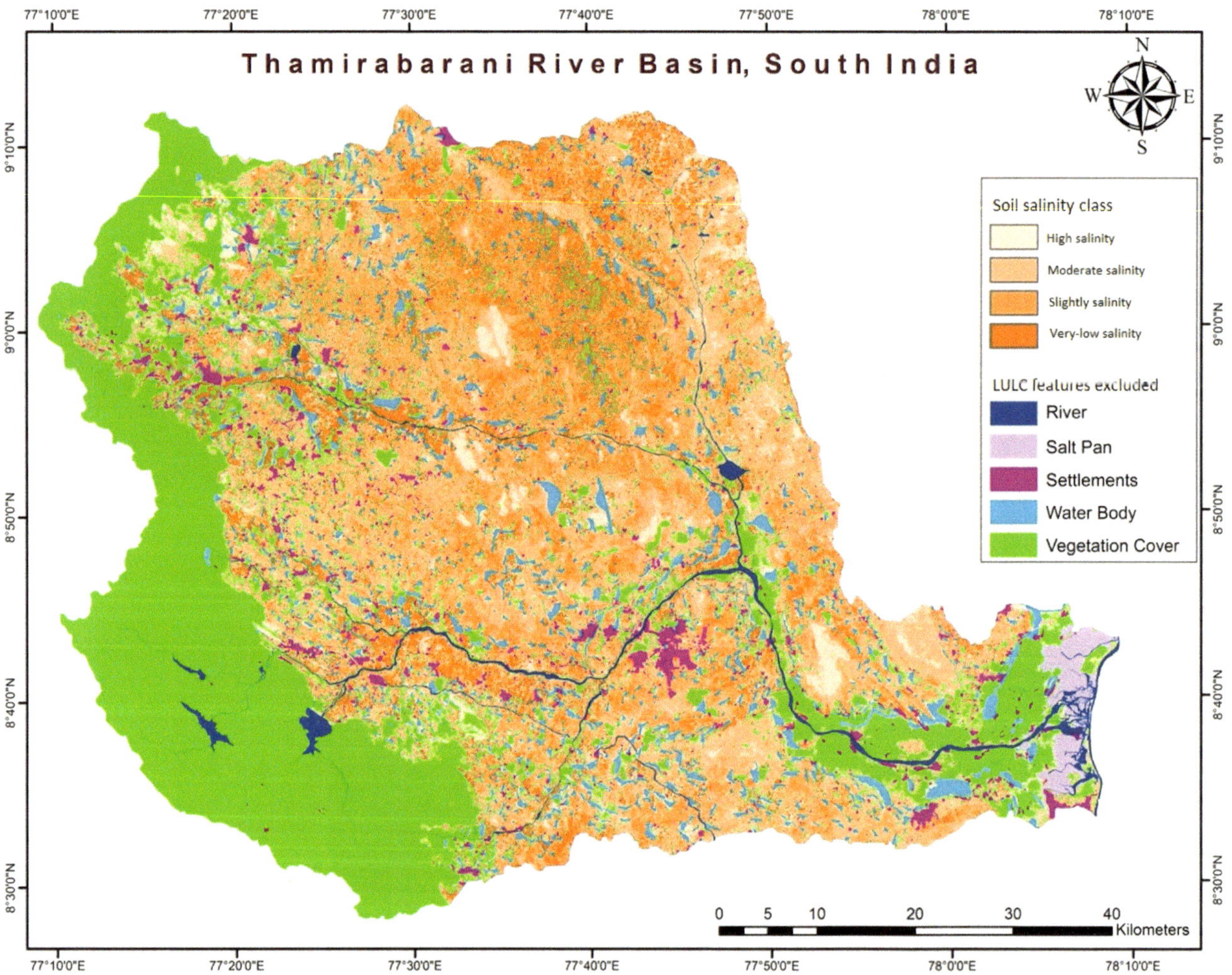

FIGURE 4.13 Soil salinity characteristics of the basin.

4.4.10 Assessment of soil salinity (or) soil salt matter

Soil salinity assessment in bare soil areas is correlated with spectral properties of NIR and SWIR bands in Landsat 9—OLI image. Saline soils have higher reflectance values within visible and NIR bands compared to nonsaline soils, while they have extremely low reflectance in SWIR compared to nonsaline soils due to the strong absorption of carbonate matter present in saline soils (Dehni and Lounis, 2012; Chen et al., 2017; Asfaw et al., 2018; Nguyen et al., 2020). Therefore, the SSI has calculated the ratio between NIR and SWIR bands to detect salt matter in soils, while both saline and nonsaline soils have strong variability spectral reflectance within these bands. Soil salinity indices are principally adjusted to detect salt minerals in soils based on the different responses of salty soils to various spectral bands. Fig. 4.13 shows the spatial characteristics of soil salinity in the study area. The resulting map indicates that soil salinity is at a high level, in areas around the water bodies and some parts of foothills due to the settling of calcareous matters by the action of runoff. The soil salinity is at a moderate level in eroded lands and fallows, especially in the northeastern parts and middle parts of pediplains. Moreover, the soil salinity is at a lower rate in the majority of pediplains that cover croplands, fallows, and barrens across the study area. In the basin, highly saline soil is spatially distributed in various parts of low-lying lands, especially along the channels and eroded lands, due to the settlement of residual salt matters from various sources resulting from soil erosion and surface runoff processes. SSI is key to measuring salt matter based on the degree of spectral reflectance between saline and nonsaline soils within the NIR and SWIR bands, whereas saline soils have extremely low reflectance in the SWIR band than nonsaline soils that indicates carbonate salt matters in soils.

4.5 Conclusions

Assessment and mapping of soil properties and their spatial characteristics are essential for the development of agriculture and food production at the global scale. Integrated remote sensing and GIS techniques make way for better understanding of soils and their spatial characteristics over the area and periods. Soil spectral reflectance plays an important role in determining physical and chemical characteristics at the spatiotemporal scale. Remote sensing indices use multivariate equations and algorithms for the estimation of different properties of soil–land–vegetation from multispectral images like Landsat 9—OLI, Sentinel 2—MSI, IRS—LISS3, and so on. Using Landsat 9—OLI image, the soil and land properties were estimated in the Thamirabarani river basin using different remote sensing indices, and these products were used for further studies related to soil and land resources, agriculture and food production, and environmental ecosystem conservation and management. Remote sensing indices–based products of soil properties provide the required database for studying soil resources for optimal utilization of natural resources on a sustainable basis. Finally, Multitemporal remote data facilitate the mapping and measurement of soils and land properties, even for inaccessible areas, and they provide possibilities for improving the existing soil survey data for digital soil mapping on local and regional scales.

Acknowledgments

The corresponding author is thankful to the Director, National Centre for Earth Science Studies (NCESS), Ministry of Earth Sciences (MoES), Government of India, for providing continuous support and encouragement. Authors extend our gratitude to USGS—Earth Explorer, ESA-Copernicus Open Access Hub, IMD, SOI, and NBSS & LUP, for providing necessary data sources.

References

Abay, et al., 2022. Mapping of ferric (Fe3 +) and ferrous (Fe2 +) iron oxides distribution using ASTER and landsat 8 OLI data, in negash lateritic iron deposit, Northern Ethiopia. Geol. Ecol. Landsc. 1 (18). Available from: https://doi.org/10.1080/24749508.2022.2130556.

Arisanty, et al., 2021. The estimation of iron oxide content in soil based on landsat 8 OLI TIRS imagery in wetland areas. Pertanika J. Sci. Tech. 29. Available from: https://doi.org/10.47836/pjst.29.4.32.

Asfaw, E., Suryabhagavan, K.V., Argaw, M., 2018. Soil salinity modeling and mapping using remote sensing and GIS: the case of Wonji sugar cane irrigation farm, Ethiopia. J. Saudi Soc. Agri. Sci. 17 (3), 250–258. Available from: https://doi.org/10.1016/j.jssas.2016.05.003.

Babaeian, et al., 2019. A new optical remote sensing technique for high-resolution mapping of soil moisture. Front. Big Data 2, 37. Available from: https://doi.org/10.3389/data.2019.00037.

Baghdadi, N., Zribi, M., 2017. Land Surface Remote Sensing in Agriculture and Forest. ISTE Press—Elsevier. Available from: https://doi.org/10.1016/C2015-0-01224-2.

Baumgardner, M.F., Silva, M., Biehl, L., Stoner, E.R., 1985. Reflectance properties of soils. Adv. Argon. 38, 1–44.

Bhatta, B., 2011. Remote Sensing and GIS, second ed. Oxford University Press, New Delhi.
Bhunia, G.S., Shit, P.K., Pourghasemi, H.R., 2017. Soil organic carbon mapping using remote sensing techniques and multivariate regression model. Geocarto Int. 34 (2), 215–226. Available from: https://doi.org/10.1080/10106049.2017.1381179.
Bishop, J.L., Michalski, J.R., Carter, J., 2017. Chapter 14—Remote detection of clay minerals. In: Gates, W.P., et al., (Eds.), Developments in Clay Science, vol. 8. Elsevier, pp. 482–514. Available from: https://doi.org/10.1016/B978-0-08-100355-8.00014-X.
Bonfatti, et al., 2020. Digital mapping of soil parent material in a heterogeneous tropical area. Geomorphology 367, 107305. Available from: https://doi.org/10.1016/j.geomorph.2020.107305.
Chen, Z., Zhang, Y., Liu, X., 2017. Mapping and analyzing soil salinity using remote sensing and geostatistics in an arid area of China. J. Arid. Land. 9 (5), 639–652.
Clark, R.N., Roush, T.L., 1984. Reflectance spectroscopy: quantitative analysis techniques for remote sensing applications. J. Geophys. Res. 89, 6329–6340.
Dehni, A., Lounis, M., 2012. Remote sensing techniques for salt affected soil mapping: application to the Oran region of Algeria. Procedia Eng. 33, 188–198.
Dogan, H.M., 2008. Applications of remote sensing and Geographic Information Systems to assess ferrous minerals and iron oxide of Tokat province in Turkey. Int. J. Remote. Sens. 29 (1), 221–233. Available from: https://doi.org/10.1080/01431160701269010.
Dwivedi, R.S., 2017. Spectral reflectance of soils. Remote Sensing of Soils. Springer, Berlin, Heidelberg. Available from: https://doi.org/10.1007/978-3-662-53740-4_6.
Escribano, et al., 2017. Chapter 4—Optical remote sensing for soil mapping and monitoring. In: Paulo Pereira, et al., (Eds.), Soil Mapping and Process Modeling for Sustainable Land Use Management. Elsevier, pp. 87–125. Available from: https://doi.org/10.1016/B978-0-12-805200-6.00004-9.
Fellipe, et al., 2021. Soil parent material prediction through satellite multispectral analysis on a regional scale at the Western Paulista Plateau, Brazil. Geoderma Reg. 26. Available from: https://doi.org/10.1016/j.geodrs.2021.e00412.
Gholizadeh, A., et al., 2018. Soil organic carbon and texture retrieving and mapping using proximal, airborne and Sentinel-2 spectral imaging. Remote. Sens. Environ. 218, 89–103. Available from: https://doi.org/10.1016/j.rse.2018.09.015.
Huete, A.R., 1988. A soil-adjusted vegetation index (SAVI). Remote. Sens. Environ. 25 (3), 295–309. Available from: https://doi.org/10.1016/0034-4257(88)90106-X.
Kaliraj, S., Meenakshi, M.S., et al., 2012. Application of remote sensing in detection of forest cover changes using geo-statistical change detection matrices—a case study of Devanampatti reserve forest, Tamil Nadu, India. Nat. Environ. Pollut. Technol. 11, 261–269.
Kaliraj, S., Chandrasekar, N., Magesh, N.S., 2015. Morphometric analysis of the River Thamirabarani sub-basin in Kanyakumari District, South west coast of Tamil Nadu, India, using remote sensing and GIS. Environ. Earth Sci. 73, 7375–7401. Available from: https://doi.org/10.1007/s12665-014-3914-1.
Kaliraj, S., Chandrasekar, N., Ramachandran, K.K., 2016. Mapping of coastal landforms and volumetric change analysis in the south west coast of Kanyakumari, South India using remote sensing and GIS techniques. Egypt. J. Remote. Sens. Space Sci. 20 (2), 265–282. Available from: https://doi.org/10.1016/j.ejrs.2016.12.006.
Kaliraj, S., Chandrasekar, N., Ramachandran, K.K., Srinivas, Y., Saravanan, S., 2017. Coastal landuse and land cover change and transformations of Kanyakumari coast, India using remote sensing and GIS. Egypt. J. Remote. Sens. Space Sci. 20 (2), 169–185. Available from: https://doi.org/10.1016/j.ejrs.2017.04.003.
Kanwar, N., Rai, S., Kuniyal, J.C., 2022. Open-source satellite repository and geographic information system (GIS) for soil resource mapping. In: Shit, P.K., Adhikary, P.P., Bhunia, G.S., Sengupta, D. (Eds.), Soil Health and Environmental Sustainability. Environmental Science and Engineering. Springer, Cham. Available from: https://doi.org/10.1007/978-3-031-09270-1_1.
Ke, Y., et al., 2015. Characteristics of Landsat 8 OLI-derived NDVI by comparison with multiple satellite sensors and in-situ observations. Remote. Sens. Environ. 164, 298–313. Available from: https://doi.org/10.1016/j.rse.2015.04.004.
Li, P., Xiao, C., Feng, Z., 2018. Mapping rice planted area using a new normalized EVI and SAVI (NVI) derived from Landsat-8 OLI. IEEE Geosci. Remote. Sens. Lett. 15 (12), 1822–1826. Available from: https://doi.org/10.1109/LGRS.2018.2865516.
Liang, S., Wang, J., 2019. Advanced Remote Sensing: Terrestrial Information Extraction and Applications. Elsevier. Available from: https://doi.org/10.1016/C2017-0-03489-4.
Lillesand, T.M., Kiefer, R.W., Chipman, A., 2011. Remote Sensing and Image Interpretation, sixth ed. Wiley, New York.
Liu, Y., et al., 2022. NDBSI: a normalized difference bare soil index for remote sensing to improve bare soil mapping accuracy in urban and rural areas. CATENA 214, 106265.
Magesh, N.S., Chandrasekar, N., 2014. GIS model-based morphometric evaluation of Tamiraparani subbasin, Tirunelveli district, Tamil Nadu, India. Arab. J. Geosci. 7, 131–141. Available from: https://doi.org/10.1007/s12517-012-0742-z.
Mehrez Zribi, M., Baghdadi, N., Nolin, M., 2011. Remote sensing of soil (editorial note). Appl. Environ. Soil. Sci. 2011 (Special Issue), 1687–7667. Available from: https://doi.org/10.1155/2011/904561.
Mohana, P., et al., 2018. Environmental impact assessment of Thamirabarani River Basin, Tamil Nadu using remote sensing and GIS techniques. Indian. J. Sci. Technol. 11 (22), 1–7. Available from: https://doi.org/10.17485/ijst/2018/v11i22/106246.
Mulder, et al., 2011. The use of remote sensing in soil and terrain mapping—a review. Geoderma 162 (1–2), 1–19. Available from: https://doi.org/10.1016/j.geoderma.2010.12.018.

NBSS & LUP, 2005. Reflectance libraries for development of soil sensor for periodic assessment of state of soil resources, Report No. 835, National Bureau of Soil Survey and Land Use Planning, Nagpur, 40 pp.

Nguyen, K.A., Liou, Y.A., Tran, H.P., et al., 2020. Soil salinity assessment by using near-infrared channel and vegetation soil salinity index derived from Landsat 8 OLI data: a case study in the tra vinh province, Mekong Delta, Vietnam. Prog. Earth Planet. Sci. 7 (1). Available from: https://doi.org/10.1186/s40645-019-0311-0.

Notesco, et al., 2020. Application of hyperspectral remote sensing in the longwave infrared region to assess the influence of dust from the desert on soil surface mineralogy. Remote Sens. 12 (9), 1388. Available from: https://doi.org/10.3390/rs12091388.

Osman, K.T., 2012. Soils: Principles, Properties and Management. Springer Netherlands, Netherlands.

Pereira, P., et al., 2017. Soil Mapping and Process Modeling for Sustainable Land Use Management. Elsevier. Available from: https://doi.org/10.1016/C2015-0-04597-X.

Rathinasamy, A., Saliha, B.B., 2017. Fundamentals of Soil Science. Scientific Publishers, India.

Ravisankar, T., Sreenivas, K., 2011. In: Roy, P.S., Dwivedi, R.S., Vijayan, D. (Eds.), Soils and Land Degradation in Remote Sensing Applications. NRSC Publication, NRSC, Hyderabad.

Schowengerdt, R., 2006. Remote Sensing: Models and Methods for Image Processing, third ed. Academic Press.

Segal, D., 2018. Theoretical basis for differentiation of ferric-iron bearing minerals, using landsat MSS data. In: Proceedings of Symposium for Remote Sensing of Environment, 2nd Thematic Conference on Remote Sensing for Exploratory Geology, Fort Worth, TX (1982), pp. 949–951.

Srivastava, P., Petropoulos, G., Kerr, Y.H., 2016. Satellite Soil Moisture Retrieval: Techniques and Applications. Elsevier. Available from: https://doi.org/10.1016/C2014-0-03396-5.

Srivastava, P., Gupta, D., Islam, T., Han, D., Prasad, R., 2022. Radar Remote Sensing: Applications and Challenges. Elsevier. Available from: https://doi.org/10.1016/C2019-0-03955-6.

Taloor, A.K., et al., 2021. Retrieval of land surface temperature, normalized difference moisture index, normalized difference water index of the Ravi basin using Landsat data. Appl. Comput. Geosci. 9, 100051. Available from: https://doi.org/10.1016/j.acags.2020.100051.

Trinh, L.H., 2016. Application of remote sensing technique to detect and map iron oxide, clay minerals, and ferrous minerals in Thai Nguyen Province (Vietnam). Min. Sci. Technol. 1, 60–65. Available from: https://doi.org/10.17073/2500-0632-2016-1-60-65.

USGS, 2022. Landsat 9 (Collection 2) Data Products Information, https://www.usgs.gov/landsat-missions/landsat-collection-2.

Velmurugan, A., Carlos, G.G., 2009. Soil resource assessment and mapping using remote sensing and GIS. J. Indian. Soc. Remote. Sens. 37, 511–525. Available from: https://doi.org/10.1007/s12524.

Xie, X., et al., 2015. Spectral indices for estimating exposed carbonate rock fraction in Karst Areas of Southwest China. IEEE Geosci. Remote. Sens. Lett. 12 (9), 1988–1992. Available from: https://doi.org/10.1109/LGRS.2015.2441962.

Yitagesuab, F.A., Meera, F., Werffa, F., Hecker, C., 2011. Spectral characteristics of clay minerals in the 2.5–14 μm wavelength region. Appl. Clay Sci. 53 (4), 581–591.

Young, R., 1975. Soil Properties and Behaviour. Elsevier.

Zaitunah, A., et al., 2018. Normalized difference vegetation index (ndvi) analysis for land cover types using landsat 8 oli in besitang watershed, Indonesia. IOP Conf. Ser.: Earth Environ. Sci. 126, 012112. Available from: https://doi.org/10.1088/1755-1315/126/1/012112.

Zhao, H., et al., 2019. Exploring an efficient sandy barren index for rapid mapping of sandy barren land from Landsat TM/OLI images. Int. J. Appl. Earth Obs. Geoinf. 80, 38–46. Available from: https://doi.org/10.1016/j.jag.2019.04.001.

Chapter 5

Soil fertility mapping and applications for site-specific nutrient management: a case study

Roomesh Kumar Jena[1], Pravash Chandra Moharana[2], Upendra Kumar Pradhan[3], Gulshan Kumar Sharma[4], Prasenjit Ray[5], Partha Deb Roy[1] and Dibakar Ghosh[1]

[1]ICAR-Indian Institute of Water Management, Bhubaneswar, Odisha, India, [2]ICAR-National Bureau of Soil Survey and Land Use Planning, Nagpur, Maharashtra, India, [3]ICAR-Indian Agricultural Statistics Research Institute, New Delhi, India, [4]ICAR-Indian Institute of Soil & Water Conservation, Regional Centre, Kota, Rajasthan, India, [5]ICAR-Indian Agricultural Research Institute, New Delhi, India

5.1 Introduction

Soil mapping is an important domain of the soil science discipline. The vast history of soil mapping has been spanned over space and time, which makes it a subdiscipline of soil science. The study in this direction drags attention of many soil scientists with the advancements of spatial technology, algorithm techniques, and data availability. Among several soil thematic maps, soil fertility mapping has its own reservation in the set of soil maps, as demands for such maps are increasing day by day for planning, monitoring, and resource management. The beauty of fertility mapping is to capture the variability which exists in all the directions of a field as this variability poses many constraints for nutrient management in agriculture. Soil nutrient content and the supplying capacity are the two important inherent characteristics which affect the process of crop production. This spatial variability of soil nutrients may be due to changing soil type, landform, vegetation, climate, and human activity. Because of which, it is always a challenging task to capture the dynamic property over space. But as a prerequisite of soil health management, mapping has to be done by considering a particular time frame.

There have been a lot of attempts made by many soil scientists across the globe over a period of time. The methodology starts from traditional polygon-based approach to use of advanced algorithm which will generate continuous fertility maps. A wide range of environmental conditions are prevalent in hilly and mountainous regions, which have an impact on the spatial variability of soils. In addition, the transfer of fertilizer from agriculture to waterbodies result in eutrophication and other associated environmental hazards. Imbalanced fertilizer application in highly variable soils has become a severe danger to sustainable agricultural production (Almasri and Kaluarchchi, 2004; Chen et al., 2010). Yet, in all agricultural soils and crops, nutritional deficits are more prevalent. When nutrients are managed uniformly throughout fields, inputs are frequently applied excessively in regions with sufficient nutrient levels and insufficiently in areas with deficit nutrient levels (Ferguson et al., 2002; Shukla et al., 2014). Excessive fertilizer usage should be avoided in order to maximize returns with the least amount of environmental damage because it does not result in significantly increased crop yields or crop nutrient absorption. Due to the substantial spatial variability of nutrients within individual fields, application rates should ideally be adjusted based on estimations of the needs for optimum output at each area (Page et al., 2005; Ruffo et al., 2005). Earlier, polygon-based mapping techniques failed to address this location-specific fertilizer application, which is presently a domain under precision agriculture. Therefore, effective methods should be used to precisely quantify differences in soil qualities within fields, which is crucial for homogenous management zones (MZs) (Peralta and Costa, 2013). It is not unexpected that there can be significant differences in the composition, availability, and distribution of soil nutrients between fields and within soils (Shukla et al., 2014). The identification of areas (MZs) within the region defined by field boundaries is encouraged by consistent nutrient management, a type of precision agriculture. These subfield regions are parts of the field that share traits including topography,

Remote Sensing of Soils. DOI: https://doi.org/10.1016/B978-0-443-18773-5.00025-9

texture, and nutrient content. The complicated interconnections of all the elements that might impact crop productivity make it challenging to define the MZs precisely. Determining the management by subfields may also be highly challenging since components interact dynamically.

Enhancing site-specific nutrient management in a sustainable way is made possible by the use of geospatial tools and techniques in the definition of MZs. Currently, site-specific analyses of soil fertility are performed at the regional level utilizing cutting-edge methods and equipment including remote sensing, geographic information systems, geostatistics, and kriging (Reza et al., 2016). Additionally, understanding the soil fertility condition may aid in defining the right MZs for certain places in order to effectively manage nutrients, produce crops, and safeguard the environment. For identifying MZs, a number of conventional techniques have been utilized, including soil surveys and topographical maps (Reyniers et al., 2006), the crop yield-based MZ methodology (Flowers et al., 2005; Hornung et al., 2006), the nutrient index technique (Jena et al., 2015; Bandyopadhyay et al., 2018; Padua et al., 2018), and geostatistics-based nutrient MZ methodology (Moharana et al., 2020; Jena et al., 2022). These techniques rely on a broad understanding of crop production to identify significant yield-limiting nutrients that should be considered in the analysis, although the outcomes are frequently only applicable to certain years (Jaynes et al., 2003). In order to split a field into prospective rice crop MZs, Aaron et al. (2004) employed yield data; whereas, Tripathi et al. (2015) used fuzzy clustering of coupled soil parameters. Similar to that, principal component analysis (PCA) is frequently utilized to define MZs (Yan et al., 2007). By combining the main component analysis and fuzzy cluster method with soil fertility parameters, several authors defined the MZs (Davatgar et al., 2012; Tripathi et al., 2015; Shukla et al., 2017).

Due to its strategic location, India's northeastern Himalayan region (NEHR), which makes up 8% of the nation's total geographical area, is one of the world's most vulnerable ecosystems. Because of the intermountain valleys and steep ridges that make up the mountainous topography, crop production is threatened by soil nutrient losses (Bhattacharyya et al., 2015). Additionally, the region is experiencing a massive loss of phytomass, which is the source of soil organic matter, as a result of rapid land use land cover change brought on by large-scale deforestation, which is the conversion of natural forests to temporary agriculture through slash and burn (shifting cultivation) practices, plantations, and other forms (SOM). The primary source of nourishment for the NEHR's rainfed agricultural system, which relies on minimal input and low productivity, is the SOM pool (Ghosh et al., 2009). The NEHR, which is abundant in natural resources, is where the state of Nagaland is situated in the Purvanchal Hills. About 90% of people do agriculture for a living, yet the state cannot feed itself owing to inadequate agricultural management (Tiwari et al., 2010). For agricultural regions, a standard prescription for fertilizer management is now being followed. Due to this, nutrients are applied excessively or insufficiently, regardless of their real nutritious levels. Given the aforementioned facts, the present study was done to (1) characterize the spatial variability of the soil attributes affecting crop productivity using the geostatistical analysis, (2) delineate potential MZs using PCA and fuzzy c-means algorithm based on the variability of the selected soil attributes, and (3) evaluate the potential of the defined MZs for site-specific management in northeastern hilly region of India.

5.2 Materials and methods

5.2.1 Site description

The study was carried out in the NEHR of India, i.e., in the Mokokchung district, Nagaland state. It has a total area of 1615 km^2 and is located between latitudes of 26°11′33.01″ and 26°45′47.18″N and longitudes of 94°17′36.74″ and 94°45′34.45″E (Fig. 5.1). Longleng and Tuensang districts in the east, Wokha district in the west, Zunheboto district in the south, and Assam state in the north share its border. It is distinguished by its mountainous landscape, which consists of alternately spaced high hills and narrow intermountain valleys that are subject to various forms of land degradation. Depending on the elevation changes, which vary from 110 to 1835 m above mean sea level, the climate varies from hot to mild subtropical to warm subtemperate type. The potential evapotranspiration is 1219 mm, while the mean annual rainfall is 1733 mm. Depending on elevation, the mean annual air temperature ranged from 15°C in the winter to around 23°C in the summer. Three broad soil subgroups in the study area are *Umbric Dystruchrepts*, *Typic Dystruchrepts*, and *Typic Udorthents* (Soil Survey Staff, 1999). The land use and land covers are comprised of mixed dense forest, mixed open forest, grasslands, shifting cultivation, and agricultural crop lands.

5.2.2 Sampling and analysis of soil

The schema of grid sampling was followed in the region, and representative surface soil samples (0−25 cm depth) were taken. The samples were collected in a rectangular fashion because of difficult hilly terrain (1 km interval in north to

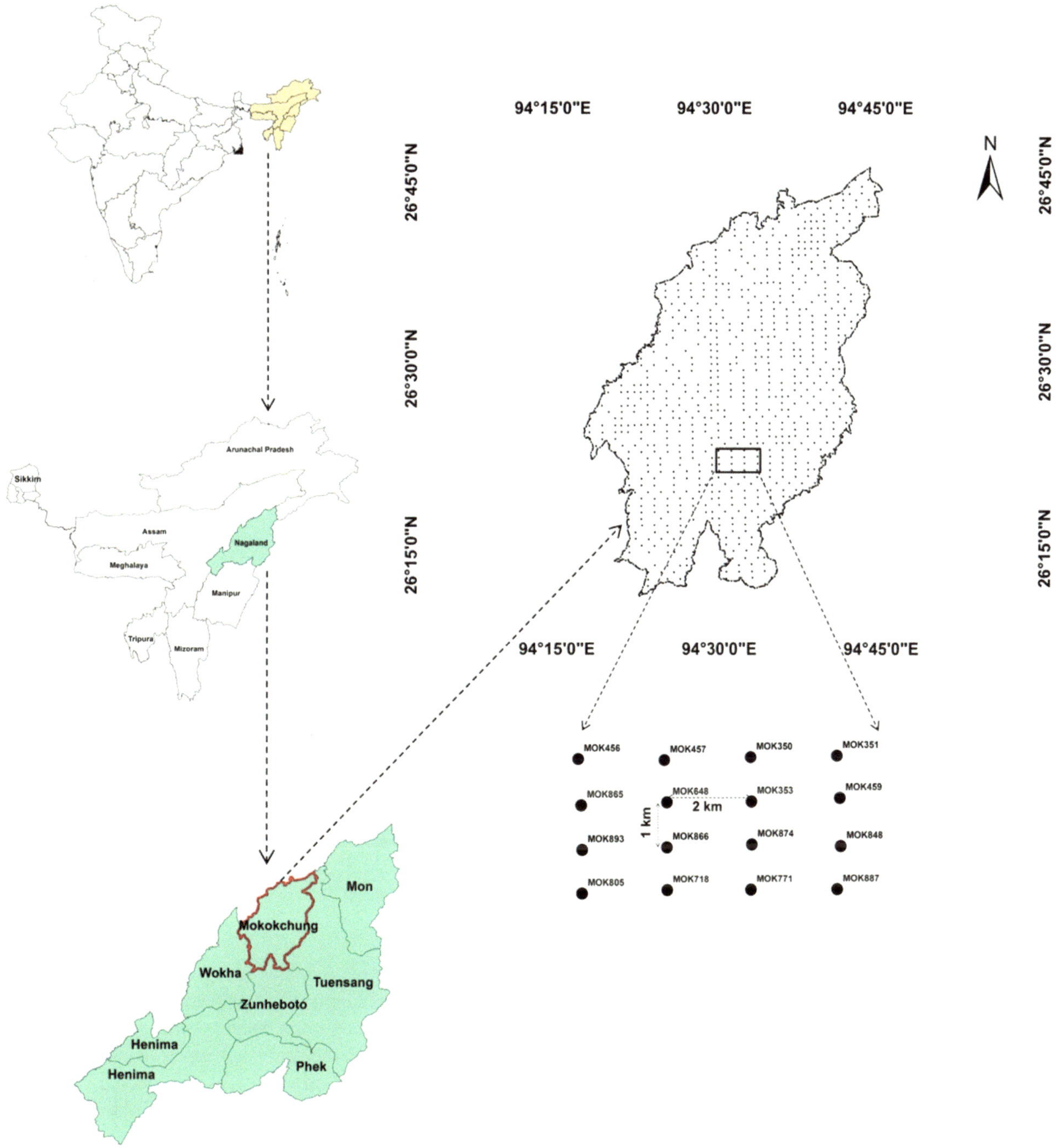

FIGURE 5.1 Study area located in Nagaland state, India, and adopted grid sampling scheme.

south direction and at 2 km interval in east to west direction). A total of 896 soil samples were collected in a rectangular fashion to cover the entire study area. The collected samples were air dried and stones and debris were removed. Then, the samples were grinded to pass through a 2 mm sieve and stored in polythene bottle for analysis of soil properties and available nutrients. The soil samples from grid points were analyzed for pH, organic carbon, available macronutrients (N, P, and K), and micronutrients (Fe, Mn, Zn, and Cu) following standard procedures. Soil pH was measured in 1:2.5 (w/v) soil–water suspension as per the procedure described by Jackson (1973). The soil organic carbon (SOC) was analyzed by Walkley and Black method (Walkley and Black, 1934), available nitrogen (AN) was determined by alkaline $KMnO_4$ method (Subbiah and Asija, 1956), available phosphorus (AP) by Bray P-1 method (Bray and Kurtz, 1945), and available potassium (AK) was determined by neutral normal NH_4OAc method (Hanway and Heidel, 1952) and then measured by flame photometer (ELICO CL 361). The micronutrients (Fe, Mn Zn, and Cu) were extracted by diethylene triamine penta-acetic acid (Lindsay and Norvell, 1978) and measured by using atomic absorption spectrophotometer (Simadzu model No. AA-6300).

5.2.3 Computation of geospatial datasets

The descriptive statistics like minimum, maximum, mean, median, standard deviation, coefficient of variation (CV), skewness, kurtosis, and standard error for each soil property were calculated in R (R-Core-Team, 2013). By performing Pearson's correlation coefficient analysis, the degree of association between all the soil parameters was investigated. The geostatistical analyzer of ArcGIS 10.3.1 software was used to construct semivariogram parameters to define the spatial distribution of accessible nutrients (Webster and Oliver, 2007). Prior to geostatistical analysis, skewed soil parameters were converted using a natural logarithm to a nearly normal distribution. After that, a weighted back transformation approach was used to convert the data back. Semivariogram (h), which gauges the average dissimilarity of data separated by a vector h, is a geostatistical tool that expresses the geographical variability of nutrients that are readily available (Goovaerts, 1998). It was calculated as half of the square root of the average difference between the individual data pair components:

$$\hat{\gamma}(h) = \frac{1}{2N(h)} \sum_{(i=1)}^{N(h)} [z(x_i) - z(x_i + h)]^2$$

Where $z(x_i)$ is the value of the variable at the location x_i and $z(x_i + h)$ is the value of the variable at a lag of h from the position x_i; $N(h)$ is the number of data pairs inside a particular class of distance and direction. Three standard spatial variations parameters were calculated: nugget (C0), sill (C + C0), and range. Experimental semivariograms $[\hat{\gamma}(h)]$ as derived from the above equation were fitted in a standard model using weighted least square approach (a). For a specific lag, weight was allocated as being inversely proportional to the number of pairings. To reduce the border effect, the maximum lag distance used in the semivariogram computation was set at half the minimum extent of the sample region. Since no discernible directional trend was seen, omnidirectional semivariograms were generated for each soil characteristic in this investigation. For each soil attribute, the best-fit model with the lowest residual sum of squares was chosen. To find the semivariogram model that best fits each variable's spatial structure, evaluations of the circular, exponential, tetraspherical, pentaspherical, and rational quadratic semivariogram models were conducted. Then, using a standard kriging technique, the fitted models were utilized to estimate various attributes at unmeasured places as interpolated values for mapping (Krig, 1981). The cross-validation analysis was done to assess the bias and accuracy of the kriging interpolation, which estimates each point measured in a spatial domain by individually removing each point from the domain and treating it as if it had never existed (Tripathi et al., 2015; Shukla et al., 2017).

5.2.4 Principal component analysis and fuzzy cluster algorithm

PCA is a multivariate analytic approach that draws components from a set of variables using complex mathematical concepts. It is based on data transformation, where the orthogonal coordinates are rotated to load as much variation as feasible on the initial components and progressively less variance on succeeding components. Instead of a covariance matrix, a correlation matrix including a subset of soil attributes was employed as the analysis's input, leading to normalized PCA. The study comprised a large number of principal components (PCs) variables. It was expected that among the various PCs, those with high eigenvalues would be the best representations of soil qualities (Schepers et al., 2004). According to Jolliffe (1986) and Shukla et al. (2017), PCs with eigenvalues 1 or more were chosen for the current investigation in order to build the MZs. The influence of altitude on soil nutrient content was shown using a bi-plot, which also looked at the relationship between altitude and soil pH, SOC, and accessible nutrients.

The goal of cluster analysis is to divide a collection of "N" items into "C" clusters so that objects within each cluster are similar to one another and objects in different clusters are distinct from one another. For the compact representation of the dataset into homogenous subsets, clustering may be utilized to quantize the given data and extract a collection of cluster prototypes (Brown, 1998). The fuzzy c-means (FCM) clustering technique, created by Dunn (1973) and modified by Bezdek (1981), is one of the most used clustering algorithms. The goal of the FCM algorithm, which is very similar to the k-means algorithm, is to minimize the objective function, which is defined as follows:

$$\sum_{j=1}^{k} \sum_{x_i \epsilon C_j} u_{ij}^{m} \left(x_i - \mu_j\right)^2$$

Where u_{ij} is the degree to which an observation x_i belongs to a cluster C_j. μ_j is the center of the cluster j and m is the fuzzifier. The variable u_{ij}^m is defined as follow:

$$u_{ij}^m = \frac{1}{\sum_{i=1}^{k} \left(\frac{|x_i - c_j|}{|x_i - c_k|}\right)^{\frac{2}{m-1}}}$$

The degree of belonging, u_{ij}, is linked inversely to the distance from x to the cluster center. The parameter m is a real number greater than 1 ($1 < m < \infty$) and it defines the level of cluster fuzziness. Note that, a value of m close to 1 gives a cluster solution which becomes increasingly similar to the solution of hard clustering such as k-means; whereas a value of m close to infinite leads to complete fuzziness. In fuzzy clustering, the centroid of a cluster is the mean of all points, weighted by their degree of belonging to the cluster:

The relationship between the degree of membership, u_{ij}, and the separation between x and the cluster center is inverse. The parameter m determines the degree of cluster fuzziness and is a real value larger than 1 ($1 < m < \infty$). Be aware that a m value near to 1 results in a cluster solution that becomes closer and closer to the solution of hard clustering, such as k-means, while a m number close to infinity results in total fuzziness. The centroid of a cluster in fuzzy clustering is the mean of all points, weighted by how much they belong to the cluster:

$$C_j = \frac{\sum_{x\epsilon C_j} u_{ij}^m x}{\sum_{x\epsilon C_j} u_{ij}^m}$$

Where C_j is the cluster's centroid. The degree to which an observation (x_i) fits into a cluster (C_j) is indicated by u_{ij}. The nonunique division of the data into a number of clusters is the main concept behind fuzzy clustering. The membership values for each of the clusters are assigned to the data points, and the FCMs clustering technique enables the clusters to take on their natural forms (Pal et al., 2005). Using the R program, the field was divided into 2–7 cluster classes using the FCMs classification process (Packages: Cluster, e1071 and factoextra). We believed there could only be seven feasible MZ clusters (Xin-Zhong et al., 2009; Davatgar et al., 2012; Tripathi et al., 2015). An iterative procedure using a selection of random cluster means was used to identify which samples belonged to each cluster. The closest of these means was used to classify each observation. Based on the separation between the observation and the cluster mean, new means were generated for each cluster. According to the outcome of equal variance and statistical independence, the distance between data points and cluster center points was calculated using the Euclidean distance.

According to Fridgen et al. (2004) and Reyniers et al. (2006), the settings for the R program were as follows: maximum number of iterations = 500, halting criterion = 0.0001, minimum number of zones = 2, maximum number of zones = 7, and fuzziness exponent = 1.5. As indications of the ideal cluster number, Dunn's partition coefficient [F(k), where k is the number of clusters] and silhouette width were utilized as two cluster validity functions. The total squared membership coefficients divided by the total number of observations equals the $F(k)$. Its value ranges from $1/k$ to 1. The coefficient's normalized form, which has a range of 0 to 1, is alternatively written as $(F(k) - 1/k)/(1 - 1/k)$. A low Dunn's coefficient value denotes an extremely fuzzy clustering, whereas a number approaching 1 denotes a clustering that is almost crisp.

5.3 Results and discussion

5.3.1 Descriptive statistics of available nutrients

Table 5.1 summarizes descriptive statistics for a wide range of soil characteristics. The soils range in a pH from 3.60 to 6.50 and are extremely acidic to slightly acidic. With a very high coefficient of variation, the SOC changed greatly. The dense plant cover in Nagaland state caused the SOC content to be high in the soils (Maji et al., 2004). Greater SOC in this region is also attributed to higher altitude and lower mean temperature (compared to the national average of 0.2%–0.5% in India). The quality and amount of litter entry into the soils are altered by the mean annual temperature, which also affects soil physical and chemical features and controls SOC stability (Tian et al., 2016). With values of 427, 44.7, and 306 kg ha^{-1}, respectively, the mean AN (ranging from 51 to 997 kg ha^{-1}), AP (ranging from 21.8 to 118 kg ha^{-1}), and AK (ranging from 67 to 996 kg ha^{-1}) levels are medium to high. The concentration of these micronutrients varied greatly, with mean values of 61.7, 27.5, 1.40, and 0.90 mg kg^{-1} for Fe, Mn, Zn, and Cu in this area, respectively. The various landforms, parent materials, prevalent meteorological conditions, and dynamic land use patterns are blamed for the vast range in nutrient content in the studied region (Meliyo et al., 2015; Shukla et al., 2017; Li et al., 2017). All soil parameters CV values showed low to very high variability. The pH of the soil had the lowest

TABLE 5.1 Descriptive statistics of soil properties of the study area.

Soil properties	Minimum	Maximum	Mean	Median	CV (%)	SD	SE	Skewness	Kurtosis
pH	3.60	6.50	4.56	4.5	8.39	0.38	0.01	0.93	2.05
SOC (%)	0.19	5.12	2.19	2.04	42.7	0.93	0.03	0.31	− 0.53
AN (kg ha^{-1})	50.8	997	427	410	36.6	156	5.22	1.21	4.13
AP (kg ha^{-1})	21.8	118	44.7	39.7	38.9	17.4	0.58	5.00	61.4
AK (kg ha^{-1})	67.0	996	306	271	51.0	156	5.22	1.43	2.70
Fe (mg kg^{-1})	0.51	143	61.6	62.4	38.8	23.9	0.80	0.15	0.05
Mn (mg kg^{-1})	0.03	96.9	27.5	22.2	71.2	19.5	0.65	0.80	− 0.09
Zn (mg kg^{-1})	0.01	7.70	1.40	1.03	83.2	1.16	0.04	2.30	6.43
Cu (mg kg^{-1})	0.07	16.6	0.90	0.75	84.0	0.75	0.03	10.81	212

AK, Available potassium; *AN*, available nitrogen; *AP*, available phosphorus; *CV*, coefficient of variation; *SD*, standard deviation; *SE*, standard error; *SOC*, soil organic carbon; Fe, Mn, Zn, and Cu represent DTPA-extractable iron, manganese, zinc, and copper in soil, respectively.

TABLE 5.2 Correlation matrix for soil properties of the study area.

	pH	SOC	AN	AP	AK	Fe	Mn	Zn	Cu	Altitude
pH	1.00[a]									
SOC	−0.01	1.00[a]								
AN	−0.05	0.53[a]	1.00[a]							
AP	0.33[a]	0.03	0.12[a]	1.00[a]						
AK	0.41[a]	0.28[a]	0.23[a]	0.38[a]	1.00[a]					
Fe	−0.08[a]	0.25[a]	0.22[a]	0.14[a]	0.03	1.00[a]				
Mn	0.30[a]	0.18[a]	0.19[a]	0.24[a]	0.41[a]	0.15[a]	1.00[a]			
Zn	0.18[a]	0.15[a]	0.20[a]	0.25[a]	0.35[a]	0.31[a]	0.49[a]	1.00[a]		
Cu	0.03[b]	0.04	0.09[a]	−0.01	0.12[a]	0.12[a]	0.17[a]	0.33[a]	1.00[a]	
Altitude	0.09[a]	0.45[a]	0.34	0.06[a]	0.21	0.01[a]	0.18[b]	0.07	0.04[a]	1.00[a]

AK, Available potassium; *AN*, available nitrogen; *AP*, available phosphorus; *SOC*, soil organic carbon; Fe, Mn, Zn, and Cu represent DTPA-extractable iron, manganese, zinc, and copper in soil, respectively.
[a]*Correlation is significant at the 0.01 level.*
[b]*Correlation is significant at the 0.05 level.*

CV value of all the soil parameters. Several publications reported modest and moderate variations in soil pH and SOC concentration in acid soils of India, which are consistent with our findings (Tripathi et al., 2015). The macronutrient CV values were modest, ranging from 37% for AP to 51% for AK. With the exception of Fe (39%), the CV values of micronutrients were higher than those of macronutrients (71% for Mn to 84% for Cu). The two most significant variables for variation in soil micronutrient content are changes in parent materials and pedogenic processes, which may account for the considerable spatial variability in soil micronutrient availability (White and Zasoski, 1999). The research area's high CV for soil characteristics demonstrated significant regional variability and suggested that site-specific fertilizer management would be useful to increase crop output.

The correlation matrix revealed that, with the exception of AN, practically all soil chemical parameters were strongly associated with one another (Table 5.2). While Cu, Zn, Mn, AK, and AP were positively impacted by soil pH, the AN, SOC, and Fe had a negative correlation with pH. This shows that variations in soil pH have an impact on the distribution and solubility of micronutrients (Lindsay and Norvell, 1978). Our findings show that a rise in soil pH in this range (3.6–6.5) led to a drop in AN and Fe concentration and an increase in AK, AP, Mn, Zn, and Cu concentration in these soils. In the soils used to grow rice in Odisha, Eastern India, Tripathi et al. (2015) also found a favorable association between soil pH and the amount of Zn and Mn present. In addition to these, SOC and the micronutrients had positive and substantial correlations with each other as well as with the available Fe, Mn, Zn, and Cu. Our findings supported previous researchers' conclusions (Sharma et al., 2000; Wei et al., 2006).

5.3.2 Spatial variability of available nutrients

Fig. 5.2 displays the semivariograms for pH, SOC, available N, P, and K, and micronutrient availability. Table 5.3 lists the critical semivariogram fitting parameters. Exponential semivariogram models were shown to be the best match in this study for soil pH, Mn, and Zn, whereas rational quadratic models were found to be the best fit for SOC, AK, and Cu. Tetraspherical, pentaspherical, and circular models best suit the data for the AN, AP, and Fe, respectively. The distance (h) at which the semi-variance reached its highest value (sill), which roughly equated to the sample variance, was the semi- variogram's "range" (Reza et al., 2012). The diameter of the zone of influence, which reflected the average maximum distance across which a soil attribute of two samples was connected, might be understood as the range stated as distance. Within its geographical range, the measured soil attribute at two sampling stations became comparable as their distance from one another decreased. The pH and SOC have a respective spatial range of 4569 and 8400 m. Reza et al. (2016) observed a similar regional pH range in alluvial soils of India. Fathi et al. (2014) reported a lesser range of pH and organic matter variability. For AN, AP, and AK, the corresponding zones of influence were 8400, 5564, and 8153 m. Reza et al. (2012) showed similar and greater geographical ranges for available N (2.0 km) and K (2.3 km) for soil fertility mapping at the district level. These contradicting results of the geographical distribution of soil attributes might be attributed to the effect of random variables and the gridded soil data's resolution. Positive nuggets were found for all soil characteristics, which may be attributed to sample error, short-range variability, randomness, and intrinsic variability. The degree of the geographic structure that displayed high, moderate, and mild spatial autocorrelation, respectively, could be described by the relative percentage (i.e., nugget/sill) of 25%, 25%–75%, and > 75%. Typically, human activities, intrinsic variables such as soil characteristics and mineralogy, and combined intrinsic and extrinsic causes may be blamed for significant, moderate, and mild spatial dependency of soil properties, respectively (Cambardella et al., 1994; Liu et al., 2006). The nugget to sill ratio (i.e., 57%–94%) indicated that all soil parameters experienced moderate spatial autocorrelations due to a combination of structural and random variables. In the current study, soil variables such as pH, SOC, AK, Fe, Mn, and Cu shows a moderate degree of spatial dependency; whereas, AN, AP, and Zn displayed a poor degree of spatial dependence. The effect of landforms, current meteorological conditions, soil types, and changing land use patterns such as shifting farming are attributed to this. Approximately 101 km^2 of the research region was subject to intermittent shifting agriculture. In soils from the Indian states of Uttarakhand and Madhya Pradesh, respectively, Shukla et al. (2017) and Tagore et al. (2014) observed weak spatial dependency of Zn.

5.3.3 Surface map of soil properties

Distribution maps of soil attributes were created using standard kriging to better understand the geographical distribution of soil properties in the region (Fig. 5.3). In the district, there was a significant difference in the distribution of soil characteristics. The whole region was within a narrow acidic pH range, according to soil pH map. Most areas of the district revealed increased micronutrient availability, which may be attributable to the micronutrient concentration of parent materials such conglomerates, shales, sandstones, and limestones in the area as well as pedogenic processes (Singh, 1971). Additionally, it was observed that soil micronutrient levels are influenced by parent material, geochemical processes, and pedogenic regimes (White and Zasoski, 1999). Additionally, owing to nutrient recycling and biotic and abiotic interactions on a small scale, which are fairly evident in the main forest vegetation in perhumid to subhumid climatic conditions, which makes up about 82% of the research area, AN, AP, and AK content were generally in the moderate range. The maps provide quantitative data on the location of nutrient content dispersion. If the extension worker and farmer in the region become aware with the features connected to the assessed soil properties, the value of such maps for site specific nutrient management in such steep terrain may be further strengthened.

5.3.4 Principal component analysis

There was a strong correlation between the nine soil variables and the one topographical variable (altitude) included in this investigation. In order to aggregate and summarize the variability in the 10 variables while maintaining their major

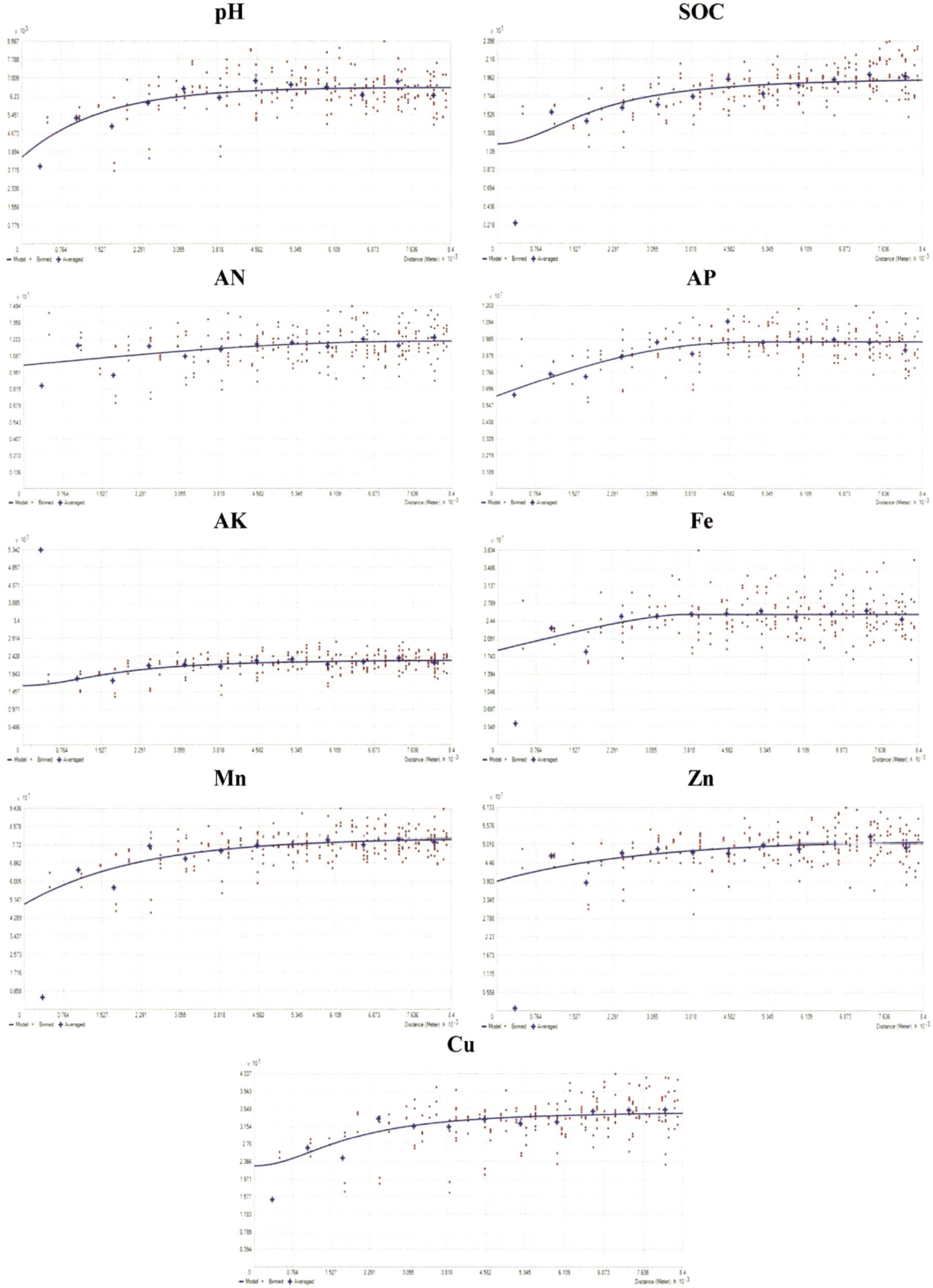

FIGURE 5.2 Calculated semivariograms of soil properties with the lines indicating selected best fit model based on RMSE and r^2 values. *AK*, available potassium; *AN*, available nitrogen; *AP*, available phosphorus; *SOC*, soil organic carbon; Fe, Mn, Zn, and Cu represent DTPA-extractable iron, manganese, zinc, and copper in soil, respectively.

TABLE 5.3 Semivariogram parameters of soil properties of the study area.

Variable	Model	Nugget	Sill	Nugget/sill	Nugget/Sill (%)	Range (m)	RMS	Spatial dependence
pH	Exponential	0.004	0.007	0.57	57	4569	0.35	Moderate
SOC	Rational quadratic	0.118	0.198	0.60	60	8400	0.72	Moderate
AN	Tetraspherical	0.101	0.121	0.83	83	8400	135.46	Weak
AP	Pentaspherical	0.601	0.637	0.94	94	5564	16.41	Weak
AK	Rational quadratic	0.162	0.234	0.69	69	8153	144.6	Moderate
Fe	Circular	0.186	0.257	0.72	72	3642	22.28	Moderate
Mn	Exponential	0.493	0.803	0.61	61	6655	19.13	Moderate
Zn	Exponential	0.391	0.514	0.76	76	8400	1.10	Weak
Cu	Rational quadratic	0.227	0.352	0.64	64	8400	0.72	Moderate

AK, available potassium; *AN*, available nitrogen; *AP*, available phosphorus; SOC, soil organic carbon; Fe, Mn, Zn, and Cu represent DTPA-extractable iron, manganese, zinc, and copper in soil, respectively.

components, PCA was used. Only the top 3 PCs, which accounted for more than 57% of the total variance and had eigen values greater than 1 (Table 5.4), qualified for the final analysis. In Fig. 5.4, three PC maps are displayed. The first PC (PC1), which was dominated by the variables AK and Mn, explained more than 28% of the total variance. As a result, the kriged maps of PC1 and AK and Mn were comparable. Additional 16.39% and 12.93% of the overall variation were explained by the second and third PCs (PC2 and PC3), respectively. The dominant elements in PC2 and PC3 were pH, SOC, AN, Cu, Fe, and Zn, respectively. In order to account for a significant portion of the geographical variability in these variables, the PCA grouped 10 variables into 3 PCs. According to a bi-plot analysis, the research area's pH, SOC, AN, AP, AK, Fe, Mn, Zn, and Cu were all positively linked with height (Fig. 5.5). Our results are consistent with those of Shukla et al. (2017), who observed a positive association between SOC, Zn, Fe, Mn, and Cu with altitude in high-altitude soils in the Himalayan area, with the exception of soil pH.

5.3.5 Clustering analysis for delineation of MZs

The three PCs were divided into MZs using clustering analysis, and the PC scores were then input into the R program to run the FCMs cluster technique. To find the ideal number of MZs, two cluster validity functions partition Dunn's coefficient and silhouette width were utilized. Table 5.5 displays the value of the silhouette width and Dunn's partition coefficient in relation to the number of classes. When Dunn's partition coefficient was at its lowest and the silhouette width was at its greatest, the ideal cluster size was identified. According to the data, the value of Dunn's partition coefficient for three clusters was least (1/k) and the silhouette width was highest (Table 5.5). Therefore, we chose three clusters as the best clusters. The observations were divided into three clusters using the R program (Packages: Cluster, e1071, and Facto extra). It was determined where each observation fit into each cluster. A total of 308 observations were included in the first MZ (MZ 1), while 395 and 193 observations, respectively, were included in the second (MZ 2) and third MZs (MZ 3), respectively. The FCMs method produced the cluster values for each grid. Three fertility MZs were shown on a map created in ArcGIS 10.3.1 using zonal statistics (Fig. 5.6). The efficiency of utilizing a combination of PCA and fuzzy cluster algorithm was evaluated using the Tukey's multiple comparisons test to characterize the geographical variability of soil parameters and the availability of nutrients for identifying MZs. This study demonstrated that the three MZs produced differed noticeably from one another, as other researchers had already noted (Xin-Zhong et al., 2009; Davatgar et al., 2012; Tripathi et al., 2015). A substantial statistical difference ($P < .01$) was also found among the three MZs for all attributes, according to the analysis of variance. (Table 5.6).

MZ 3 is the largest of the three MZs developed, covering 45.9% of the research area. MZ 2 and MZ 1 encompass 31.1% and 23.0% of the study area, respectively. The analysis of variance revealed substantial variations in the

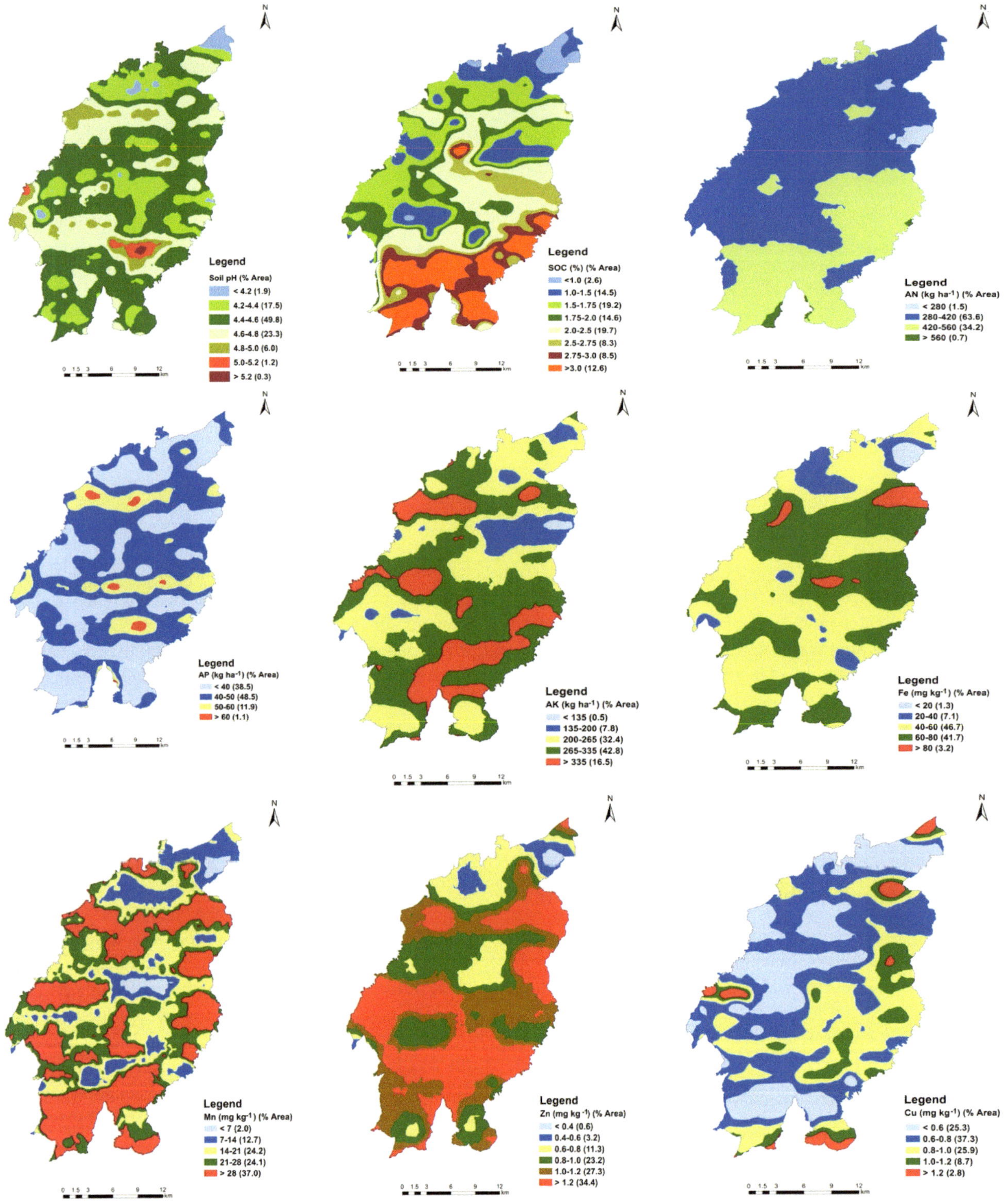

FIGURE 5.3 Spatial distribution map of soil properties in the study area. *AK*, Available potassium; *AN*, available nitrogen; *AP*, available phosphorus; *SOC*, soil organic carbon; Fe, Mn, Zn, and Cu represent DTPA-extractable iron, manganese, zinc, and copper in soil, respectively; and values in parenthesis in legends represent the % area for that particular legend class.

TABLE 5.4 PCA of soil properties and loading coefficient for the first three principal components.

Principal component	Eigen values	Component loading (%)	Cumulative loading (%)							
PC1	2.83	28.35	28.35							
PC2	1.64	16.39	44.74							
PC3	1.30	12.93	57.67							
PC4	0.98	9.8	67.47							
PC5	0.72	7.18	74.65							
PC6	0.63	6.3	80.95							
PC7	0.59	5.96	86.91							
PC8	0.49	4.95	91.86							
PC9	0.43	4.35	96.21							
PC10	0.38	3.79	100							
PC loading for each variable										
	pH	**SOC**	**AN**	**AP**	**AK**	**Fe**	**Mn**	**Zn**	**Cu**	**Altitude**
PC1	0.245	0.330	0.328	0.287	0.418	0.213	0.401	0.393	0.184	0.267
PC2	−0.456	0.487	0.437	−0.317	−0.215	0.199	−0.192	−0.157	−0.033	0.339
PC3	0.319	0.165	0.064	0.132	0.231	−0.487	−0.085	−0.411	−0.488	0.380

AK, available potassium; *AN*, available nitrogen; *AP*, available phosphorus; *SOC*, soil organic carbon; Fe, Mn, Zn and Cu represent DTPA-extractable iron, manganese, zinc and copper in soil, respectively.

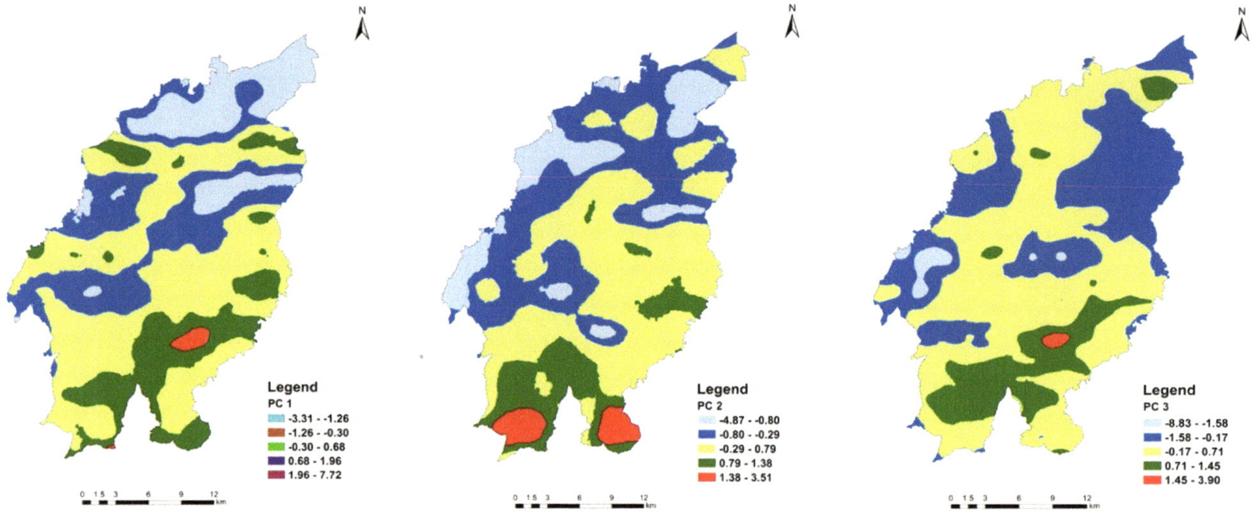

FIGURE 5.4 Kriged maps for the three principal components.

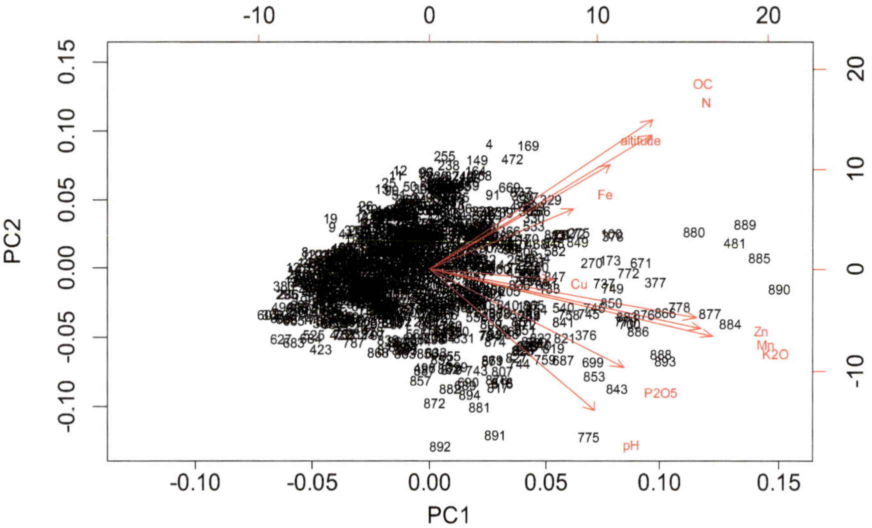

FIGURE 5.5 PCA bi-plot (PC1 vs PC2) of altitude and soil properties.

TABLE 5.5 Dunn's partition coefficient and silhouette width calculated for identifying the optimum clusters for the study area.

Cluster	Dunn's partition coefficient	Normalized Dunn's partition coefficient	Silhouette width
2	0.64	0.27	0.28
3	0.42	0.20	0.29
4	0.52	0.28	0.20
5	0.34	0.18	0.22
6	Not converge		0.21
7	0.27	0.15	0.20

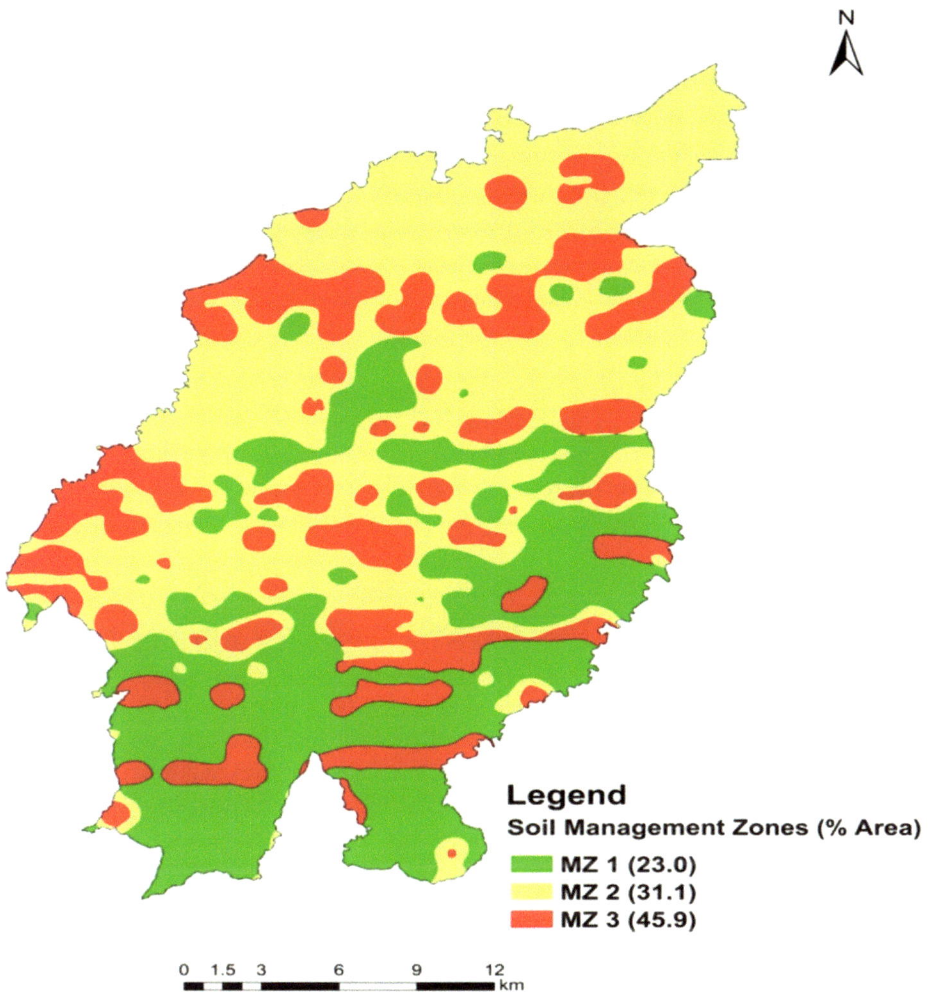

FIGURE 5.6 Management zones map for three clusters in the study area.

TABLE 5.6 Variance and Tukey multiple comparisons test for soil properties of MZs.

Management Zone	No. of points	pH	SOC	AN	AP	AK	Fe	Mn	Zn	Cu
MZ1	308	4.44^{b}	2.97^{a}	524.09^{a} (0.3)	39.91^{b} (28.5)	293.16^{b} (0.64)	63.16^{b} (0.3)	26.08^{b} (3.57)	1.12^{b} (18.8)	0.85^{b} (0.3)
MZ2	395	4.48^{b}	1.53^{c}	332.66^{c} (26.1)	40.54^{b} (22.7)	230.62^{b} (0.25)	57.11^{c} (1.77)	18.31^{c} (3.79)	0.97^{b} (31.6)	0.75^{b} (2.78)
MZ3	193	4.91^{a}	2.28^{b}	463.36^{b} (10.8)	61.22^{a} (6.2)	481.91^{a} (0)	68.71^{a} (0)	48.59^{a} (0)	2.73^{a} (1.03)	1.27^{a} (1.55)

Values in a column followed by different letters are significantly different at $P < .01$. Values in parenthesis indicate percent samples deficient for respective nutrients under different MZs.
AK, available potassium; *AN*, available nitrogen; *AP*, available phosphorus; *SOC*, soil organic carbon; and Fe, Mn, Zn, and Cu represent DTPA-extractable iron, manganese, zinc, and copper in soil, respectively.
^{a,b,c}*Represents the grouping parameters of Tukey multiple comparisons test.*

availability of SOC, AN, Fe, and Mn across MZs. There were no significant differences for these soil parameters between MZ 1 and MZ 2; however, there were significant differences between MZ 1 and MZ 3 for soil pH, AP, AK, Zn, and Cu. Agroecological conditions, soil types, and crop and management strategies may be to blame for the differences in soil qualities in the three MZs (Ortega and Santibanez, 2007; Shukla et al., 2017). In the NEHR, where elevation decreases from east to west, the study area is located on the western slope of the Purvanchal Hills. Here, the parent material, climate, and vegetation all change significantly with elevation, ranging from a warm, per humid thermic ecosystem at higher elevations to a hot, moist, subhumid hyperthermic ecosystem at lower elevations. As a result, another element that contributed to the heterogeneity in several soil properties is height. Even though the MZs showed higher average values for AP, AK, Fe, Mn, Zn, and Cu, the degree of these nutrients' deficit differed among the MZs (Table 5.6). Especially in the case of micronutrients, several authors also reported that the availability of micronutrients in the acidic region and highly weathered soils of the northeastern hilly regions are quite high (Deb Roy et al., 2020; Ray et al., 2021). With regard to AK, Fe, and Mn, the MZ 3 had no deficient zone, although the extent of the deficit was extremely low for AP and Zn. In MZ 2, nutritional deficiencies were found in the following order: Zn (31.6%) > AN (26.1%) > Mn (3.79%) > Cu (2.78%) > Fe (1.77%). MZ 1 had the greatest AP and AK deficit, followed by MZ 2. The greater levels of SOC, acidic pH, high altitude with temperature changes, and forest vegetation may be to blame for the higher mean values of these nutrients, with the exception of AN, despite the significant variability among the MZs. By using appropriate nutrient applications, this zonation idea based on the availability of vital nutrients can assist to find optimal regions for producing cereals, pulses, vegetables, and tuber crops.

The technique would also offer a practical and effective way to control nutrients scientifically. The geographical variability of soil parameters, notably macro- and micronutrients in the research region, was shown by geostatistical analysis. Therefore, farmers and other stakeholders might use the knowledge on MZs for site-specific nutrient management.

5.4 Conclusions

Monitoring regional soil fertility is essential for determining how changes in the soil resource affect maintaining agricultural production. The results of the current study showed that the soil characteristics and soil available macro- and micronutrients in India's NEHR varied widely. The soil pH was considerably inversely connected with Fe and significantly positively correlated with AP, AK, Mn, Zn, and Cu. The research area's height was positively associated with each soil characteristic. The best fit models for estimated soil properties, according to geostatistical analysis, were circular, exponential, tetraspherical, pentaspherical, and rational quadratic models. The regional variability of soil characteristics and the spatial dependency of the nutrients available were mild to moderate. In order for farmers to adopt site-specific nutrient management that fulfils the requirements of MZs to be simple, functional, understandable, and economically viable, they may use the MZ maps created using PCA and FCMs algorithm as their major guidance. For variable-rate fertilization and liming, the mean values of the soil parameters in each MZ can be utilized as a guide. As a result, fuzzy cluster analysis would decrease variability within a zone and offer an effective and efficient method for a variety of interventions for agricultural purposes, both here and elsewhere.

References

Aaron, R.S., John, F.S., Mark, A.L., James, S.S., Sven, H.J., Ariovaldo, L.J., 2004. Appropriateness of management zones for characterizing spatial variability of soil properties and irrigated corn yields across years. Agron. J. 96, 195–203.

Almasri, M.N., Kaluarchchi, J.J., 2004. Assessment of and management of long-term nitrate pollution of ground water in agriculture dominated watersheds. J. Hydrol. 295, 225–245.

Bandyopadhyay, B., Ray, P., Padua, S., Ramachandran, S., Jena, R.K., Deb Roy, P., et al., 2018. Priority zoning of available micronutrients in the soils of agroecological sub-regions of North-East India using geo-spatial techniques. Agric. Res. 7, 200–214.

Bezdek, J.C., 1981. Pattern Recognition with Fuzzy Objective Function Algorithms. Plenum, New York.

Bhattacharyya, R., Ghosh, B.N., Mishra, P.K., Mandal, B., Rao, C.S., Sarkar, D., et al., 2015. Soil degradation in India: challenges and potential solutions. Sustainability. 7, 3528–3570.

Bray, R.H., Kurtz, L.T., 1945. Determination of total, organic and available forms of phosphorus in soils. Soil. Sci. 59, 39–46.

Brown, D.G., 1998. Classification and boundary vagueness in mapping resettlements forest types. Int. J. Geogr. Inf. Sci. 12, 105–129.

Cambardella, C.A., Moorman, T.B., Novak, J.M., Parkin, T.B., Karlen, D.L., Turco, R.F., et al., 1994. Field-scale variability of soil properties in central Iowa soils. Soil. Sci. Soc. Am. J. 58, 1501–1511.

Chen, Y., He, B.H., Zhao, X.L., Zhang, L., 2010. Effect of soil erosion and water loss in Farmland on water eutrophication in Xiaojiang River Basin. J. Soil. Water Conserv. 24, 31–43.

Davatgar, N., Neishabouri, M.R., Sepaskhah, A.R., 2012. Delineation of site specific nutrient management zones for a paddy cultivated area based on soil fertility using fuzzy clustering. Geoderma 173–174, 111–118. Available from: https://doi.org/10.1016/j.geoderma.2011.12.005.

Deb Roy, P., Jena, R.K., Ray, P., Bandyopadhyay, S., Padua, S., Singh, S.K., et al., 2020. Vertical distribution of cationic micronutrients across landscape positions on Meghalayan plateau in the North-Eastern Region of India. J. Environ. Biol. 41, 1089–1098.

Dunn, J.C., 1973. A fuzzy relative of the ISODATA process and its use in detecting compact well-separated clusters. J. Cybern. 3, 32–57.

Fathi, H., Fathi, H., Moradi, H., 2014. Spatial variability of soil characteristic for evaluation of agricultural potential in Iran. Merit Res. J. Agric. Sci. Soil. Sci 2, 24–31.

Ferguson, R.B., Hergert, G.W., Schepers, J.S., Gotway, C.A., Cahoon, J.E., Peterson, T.A., 2002. Site-specific nitrogen management of irrigated maize: yield and soil residual nitrate effects. Soil. Sci. Soc. Am. J. 66, 544–553.

Flowers, M., Weisz, R., White, J.G., 2005. Yield-based management zones and grid sampling strategies: describing soil test and nutrient availability. Agron. J. 97, 968–982.

Fridgen, J.I., Kitchen, N.R., Sudduth, K.A., Drummond, S.T., Wiebold, W.J., Fraisse, C.W., 2004. Management zone analyst (MZA): software for subfield management zone delineation. Agron. J. 96, 100–108.

Ghosh, P.K., Saha, R., Gupta, J.J., Ramesh, T., Das, A., Lama, T.D., et al., 2009. Long-term effect of pastures on soil quality in acid soil of North–East India. Aust. J. Soil. Res. 47, 372–379. Available from: https://doi.org/10.1071/SR08169.

Goovaerts, P., 1998. Geostatistical tools for characterizing the spatial variability of microbiological and physico-chemical soil properties. Biol. Fertil. Soils 27, 315–334.

Hanway, J.J., Heidel, H., 1952. Soil analysis methods as used in Iowa State College Soil Testing Laboratory. Iowa Agric. 57, 1–31.

Hornung, A., Khosla, R., Reich, R., Inman, D., Westfall, D.G., 2006. Comparison of site specific management zones: soil-color-based and yield-based. Agron. J. 98, 407–415.

Jackson, M.L., 1973. Soil Chemical Analysis. Prentice Hall of India Private Limited, New Delhi.

Jaynes, D.B., Kaspar, T.C., Colvin, T.S., James, D.E., 2003. Cluster analysis of spatiotemporal corn yield patterns in an Iowa field. Agron. J. 95, 574–586.

Jena, R.K., Duraisami, V.P., Sivasamy, R., Shanmugasundaram, R., Krishnan, R., Padua, S., et al., 2015. Spatial variability of soil fertility parameters in Jirang Block of Ri-Bhoi District, Meghalaya. Clay Res. 34, 35–45.

Jena, R.K., Bandyopadhyay, S., Pradhan, U.K., Moharana, P.C., Kumar, N., Sharma, G.K., et al., 2022. Geospatial modelling for delineation of crop management zones using local terrain attributes and soil properties. Remote. Sens. 14, 2101. Available from: https://doi.org/10.3390/rs14092101.

Jolliffe, I.T., 1986. Principal Component Analysis. Springer-Verlag, New York.

Krig, D.G., 1981. Lognormal-de Wijsian Geostatistics for Ore Evaluation. Printpak (Cape) Ltd, Johannesburg.

Li, W., Wang, D., Wang, Q., Liu, S., Zhu, Y., Wu, W., 2017. Impacts from land use pattern on spatial distribution of cultivated soil heavy metal pollution in typical rural-urban fringe of northeast China. Int. J. Environ. Res. Publ. Health 14, 1–14.

Lindsay, W.L., Norvell, W.A., 1978. Development of a DTPA soil test for zinc, iron, manganese and copper. Soil. Sci. Soc. Am. J. 42, 421–448.

Liu, D.W., Wang, Z.M., Zhang, B., Song, K.S., Li, X.Y., Li, J.P., et al., 2006. Spatial distribution of soil organic carbon and analysis of related factors in croplands of the black soil region, Northeast China. Agric. Ecosyst. Environ. 113, 73–81.

Maji, A.K., Baruah, U., Dubey, P.N., Verma, T.P., Butte, P.S., Shilu, K., et al., 2004. Soil series of Nagaland (NBSSPubl. No.109), Nagpur.

Meliyo, J.L., Massawe, B.H.J., Brabers, L., Msanya, B.M., Kimaro, D.N., Mulungu, L.S., et al., 2015. Status and variability of soil micronutrients with landforms in the plague focus of western usambara mountains, Tanzania. Int. J. Soil. Sci. Plant. Nutr. 4, 389–403.

Moharana, P.C., Jena, R.K., Pradhan, U.K., Nogiya, M., Tailor, B.L., Singh, R.S., et al., 2020. Geostatistical and fuzzy clustering approach for delineation of site-specific management zones and yield-limiting factors in irrigated hot arid environment of India. Precis. Agric. 21, 426–448.

Ortega, R.A., Santibanez, O.A., 2007. Determination of management zones in corn (*Zea mays* L.) based on soil fertility. Comput. Electron. Agric. 58, 49–59.

Padua, S., Chattopadhyay, T., Bandyopadhyay, S., Ramachandran, S., Jena, R.K., Ray, P., et al., 2018. A simplified soil nutrient information system: study from the North East Region of India. Curr. Sci. 2018 (114), 1241–1249.

Page, T., Haygarth, P.M., Beven, K.J., Jones, A., Butler, T., Keeler, C., et al., 2005. The spatial variability of soil phosphorus in relation to topographic indices and important source areas: samples to assess the risks to water quality. J. Environ. Qual. 34, 2263–2277.

Pal, N.R., Pal, K., Keller, J.M., Bezdek, J.C., 2005. A possibilistic fuzzy c-means clustering algorithm. IEEE Trans. Fuzzy Syst. 13, 517–530.

Peralta, N.R., Costa, J.L., 2013. Delineation of management zones with soil apparent electrical conductivity to improve nutrient management. Comput. Electron. Agric. 99, 218–226.

Ray, P., Gogoi, S.N., Bandyopadhyay, S., Jena, R.K., Ramachandran, S., Sharma, G.K., et al., 2021. Soil-cum-nutrient management package for achieving high yield and quality of mulberry (*Morusindica* L.) leaf in acid soil of North Eastern India. J. Plant. Nutr. 44, 2501–2513.

R-Core-Team, 2013. R: A Language and Environment for Statistical Computing. R Foundation for Statistical Computing, Vienna, Austria.

Reyniers, M., Maertens, K., Vrindts, E., De Baerdemaeker, J., 2006. Yield variability related to landscape properties of a loamy soil in central Belgium. Soil. Tillage Res. 88, 262–273. Available from: https://doi.org/10.1016/j.still.2005.06.005.

Reza, S.K., Baruah, U., Sarkar, D., 2012. Spatial variability of soil properties in Brahmaputra plains of north-eastern India: a geostatistical approach. J. Indian. Soc. Soil. Sci. 60, 108–115.

Reza, S.K., Nayak, D.C., Chattopadhyay, T., Mukhopadhyay, S., Singh, S.K., Srinivasan, R., 2016. Spatial distribution of soil physical properties of alluvial soils: a geostatistical approach. Arch. Agron. Soil. Sci. 62, 972–981. Available from: https://doi.org/10.1080/03650340.2015.1107678.

Ruffo, M.L., Bollero, G.A., Hoeft, R.G., Bullock, D.G., 2005. Spatial variability of the Illinois soil nitrogen test: implications for soil sampling. Agron. J. 97, 1485–1492.

Schepers, A.R., Shanaham, J.F., Liebig, M.A., Schepers, J.S., Johnson, S.H., Luchiari, J.A., 2004. Appropriateness of management zones for characterizing spatial variability of soil properties and irrigated corn yields across years. Agron. J. 96, 195–203.

Sharma, B.D., Mukhopadhyay, S.S., Sidhu, P.S., Katyal, J.C., 2000. Pedospheric attributes in distribution of total and DTPA-extractable Zn, Cu, Mn and Fe in Indo-Gangetic plains. Geoderma 96, 131–151.

Shukla, A.K., Sinha, N.K., Tiwari, P., Prakash, C., Behera, S.K., Lenka, N.K., et al., 2017. Spatial distribution and management zones for sulphur and micronutrients in Shiwalik Himalayan region of India. Land. Degrad. Dev. 28, 959–969.

Shukla, A.K., Tiwari, P.K., Prakash, C., 2014. Micronutrients deficiencies *vis-a-vis* food and nutritional security of Indian. Indian. J. Fertilizer 10, 94–112.

Singh, R.L., 1971. India A Regional Geography. National Geographical Society of India. UBS Publishers, New Delhi, India.

Soil Survey Staff, 1999. Soil Taxonomy. A Basic System of Soil Classification for Making and Interpreting Soil Surveys, second ed. USDA, U.S. Gov. Print Office, Washington, DC.

Subbiah, B.V., Asija, C.L., 1956. A rapid procedure for the estimation of available nitrogen in soils. Curr. Sci. 25, 259–260.

Tagore, G.S., Bairagi, G.D., Sharma, R., Verma, P.K., 2014. Spatial variability of soil nutrients using geospatial techniques: a case study in soils of Sanwer tehsil of Indore district of Madhya Pradesh. ISPRS J. Photogramm. Remote. Sens. XL-8, 1353–1363.

Tian, Q., He, H., Cheng, W., Bai, Z., Wang, Y., Zhang, X., 2016. Factors controlling soil organic carbon stability along a temperate forest altitudinal gradient. Sci. Rep. 6, 18783.

Tiwari, S.C., Sashi, S., Acharya, L., 2010. Agrobiodiversity potential of Nagaland state, Northeastern India. Indian. J. Tradit. Know. 9, 350–354.

Tripathi, R., Nayak, A.K., Shahid, M., Lal, B., Gautam, P., Raja, R., et al., 2015. Delineation of soil management zones for a rice cultivated area in eastern India using fuzzy clustering. Catena 133, 128–136.

Walkley, A.J., Black, I.A., 1934. An examination of the Degtjareff method for determining soil organic matter and a proposed modification of the chromic acid titration method. Soil. Sci. 37, 29–38.

Webster, R., Oliver, M.A., 2007. Geostatistics for Environmental Scientists. John Wiley and Sons, Chichester.

Wei, X., Hao, M., Shao, M., Gale, W.J., 2006. Changes in soil properties and the availability of soil micronutrients after 18 years of cropping and fertilization. Soil. Tillage Res. 91, 120–130.

White, J.G., Zasoski, R.J., 1999. Mapping soil micronutrients. Field Crop. Res. 60, 11–26.

Xin-Zhong, W., Guo-Shun, L., Hong-Chao, H., Zhen-Hai, W., Qing-Hua, L., Xu-Feng, L., et al., 2009. Determination of management zones for a tobacco field based on soil fertility. Comput. Electron. Agric. 65, 168–175.

Yan, L., Zhou, S., Feng, L., Hong-Yi, L., 2007. Delineation site-specific management zones using fuzzy clustering analysis in a coastal saline land. Comput. Electron. Agric. 56, 174–186.

Chapter 6

Spatial mapping of soil salinity in Jubail Industrial City of the Eastern Province of Saudi Arabia using linear regression modeling

Sheik Mujabar P. and M. Khysar Pasha
Department of General Studies, Jubail Industrial College, Jubail Industrial City, Kingdom of Saudi Arabia

6.1 Introduction

Land degradation is a primary concern in several world regions undergoing rapid industrialization, economic development, and agricultural growth. The land is rapidly degrading faster than the natural soil replenishing processes. The human-induced activities on land, soil utilization, soil function, and the bio-diversity/ecosystem cause the ground to degrade faster. Lhissou et al. (2014) state that the accumulation process of salts in the top and subsurface of the earth considerably degrades the soil composition, water quality, and agricultural productivity, which negatively influences economic growth. Asfaw et al. (2018) state that salinization is one of the most common degradation forms in the arid and semiarid zones, where evaporation rates exceed the precipitation rate. The excess salinity in the soil significantly reduces the quality and productivity of land worldwide (Allbed et al., 2014; Elhag, 2016). Douaik et al. (2009) imply that the spatial variation of the salinity on the top and subsurface is exceptionally profound and controlled by several factors. These factors include soil features such as parent material, permeability, water table depth, topography, and climatic factors. Scudiero et al. (2017) state that climate change and land-use changes can intensify the saline level of lands. Abbas et al. (2013) state that yearly, around 2 Mha areas convert newly affected by salinity, resulting in severe land utilization and crop production problems. So frequent estimation and mapping of soil properties are essential to preserve the quality of lands, sustainable development, and ecofriendly exploitation of our land resources.

The Arabian Peninsula has several complex geomorphological features, primarily arid and semiarid regions in which low organic content and soil quality lead to less soil productivity. Aldakheel (2011) states that the sustainability of irrigated agriculture in arid and semiarid lands depends mainly on the soil salinity level and the irrigation water quality. Valipour (2014) says that soil salinity is a significant threat to arid and semiarid regions, severely negatively impacting several aspects of agriculture and environmental sustainability. Gorji et al. (2017) suggest that high spatial and temporal variability of soil salinity creates significant challenges in mapping and monitoring the desertification process in the Mediterranean countries. They also state that the rapid desertification process also leads to problems in mitigating severe economic and social consequences, especially in the arid and semiarid regions. So, a regular periodic soil quality assessment is essential for nations in arid and semiarid areas undergoing rapid economic growth and development.

During the last decade, several researchers (Aksoy et al., 2022; Basyoni and Aref, 2016; Biro Turk and Aljughaiman, 2020; Jabbar and Zhou, 2012; Rajendran et al., 2021; Shahid et al., 2010; Wu et al., 2018; Zewdu et al., 2017) have shown significant interest in evaluating and mapping the soil salinity in many regions around the world, especially in the arid and semiarid areas that are heavily affected by salinization. The soil salinity can be called the soil's electrical conductivity (EC). The traditional soil survey and sampling provides detailed information on the type and characteristics of the soil. Usually, the salinity of soil can be experimentally measured by collecting in situ soil samples from the field. The laboratory testing and analysis correlated with the ground testing yielded more accurate information on soil EC, compositions, and concentrations. It is also the affirmative evidence of the soil characteristics.

Remote Sensing of Soils. DOI: https://doi.org/10.1016/B978-0-443-18773-5.00039-9

However, these traditional soil survey methods require adequate spatial variability of an area and more funding and are time-consuming processes; thereby, the field surveys are confined to small areas of the land (Brunner et al., 2007).

Recent developments in remote sensing (RS) and Geographic Information System (GIS) technology have made it possible to identify and efficiently monitor the soil salinity levels precisely. Salinity mapping can be accomplished by integrating with RS and GIS at broad and small scales (Shahid et al., 2010). The wide range of multi-spectral and hyperspectral satellite data provide significant spectral information. Since poor vegetation adversely affects salinity, these spectral indices provide unique information on the soil salinity value of a land. Rajendran et al. (2021) state that the RS techniques have the potential to detect evaporite minerals, map salt crust, and saline soils in diverse landforms. The RS techniques are accomplished by observing saline regions, delivering information on the spatial distribution, and measuring their suitability for development (Bannari and Al-Ali, 2020). There are several soil studies (Al-Gaadi et al., 2021; Allbed et al., 2014; Biro Turk and Aljughaiman, 2020; Elhag, 2016; Modaihsh et al., 2007) have been performed in the Kingdom of Saudi Arabia. The Eastern Province of Saudi Arabia has significant landform features and rich bio-diverse nature.

Further, this region has undergone rapid economic growth in recent years; however, there is a limited investigation of salinity studies using the latest RS techniques. Hence, the present study aims to perform preliminary research on the salinity of soils in the Jubail Industrial City of the Eastern Province, one of the world's largest industrial hubs. The study mainly focuses on the comprehensive analysis of the effective use of Landsat 8 satellite data combined with field investigation for mapping the spatial variation of soil salinity in the semiarid region.

6.2 Materials and methods

6.2.1 Investigation area

The region designated for the present study is Jubail Industrial City (Fig. 6.1) in the Eastern Province of Saudi Arabia. It is located in the Arabian Peninsula of the Persian Gulf. It has geographic coordinates of latitude 26° 57′ 35.18″ N and longitude 49° 34′ 7.47″ E with an elevation of 7 m above the Mean Sea Level. The region subsists in a warm-humid dry climate of more than 45°C during summer with low (100 mm per year) annual rainfall (Biswas and Agarwal, 2013). The regional airflow of the Arabian Gulf influences synoptic winds. Local marine/land/air interaction affects coastal temperatures and humidifies, which causes mesoscale air mass circulations. There are several small coastal islands in the city's northern part, and it has a series of coral reefs along the coast.

The coastal and inland areas of the Eastern Province of Saudi Arabia has a series of salt pan known as "Sabkha." Sabkha is an Arabic term representing the coastal topography consisting of salt-coated sand, silt, and clay sediments. Evans et al. (1969) state that Sabkha are geologic features characterized by salt marshes and salt flats which are typically found in shallow continental shelf/marine environment with arid and semi-arid climatic conditions. Sabkha formations are supposed to be due to the sea level variations on the Arabian coast during the Quaternary periods. These Sabkha lands usually have high evaporation rates, and the soils have a rich content of insoluble salt sediments. These salt sediments can be easily identified through visual inspection of satellite imagery and field visits. Further, these Sabkha soil sediments also reflect EM radiation with different proportions, which can differ from the saline soils with vegetation cover.

The Jubail Industrial City is one of the largest industrial hubs in the world. During the last few decades, the city has been under rapid industrialization process and settlements. Several petrochemical industries and developmental projects have been operational in the city. Therefore, environmental impact assessment studies and pollution monitoring are more vital for the area. The entire process of the present study is represented in the flowchart of the methodology shown in Fig. 6.2. As the first step, the landforms in the investigation area were studied through field visits, and the soil samples were collected for the analysis described in the next section.

6.2.2 Soil sampling and analysis

The soil samples were randomly collected from different locations in the study area. The sampling locations were identified through visual inspection of the Landsat satellite image and a hand-held GPS receiver so as to cover diverse landforms for effective modeling. Soils with color/tonal variations were favored for divergent soil types. The samples were collected within the subsoil's upper layer (30 cm) with the help of a soil auger. Moreover, importance was given to sampling in the Sabkha and adjacent saline regions, as shown in Fig. 6.1. The soil samples were taken in separate plastic bags and marked with the geographical locations. The collected samples were air-dried at room temperature in the laboratory and passed through sieves of the required sizes for EC estimation. He et al. (1992) described that the EC ($EC_{1:5}$) of soil extract is the most optimum parameter for describing the salinity of soils. So, the EC ($EC_{1:5}$) of the soil samples at room temperature (25°C) is measured in the

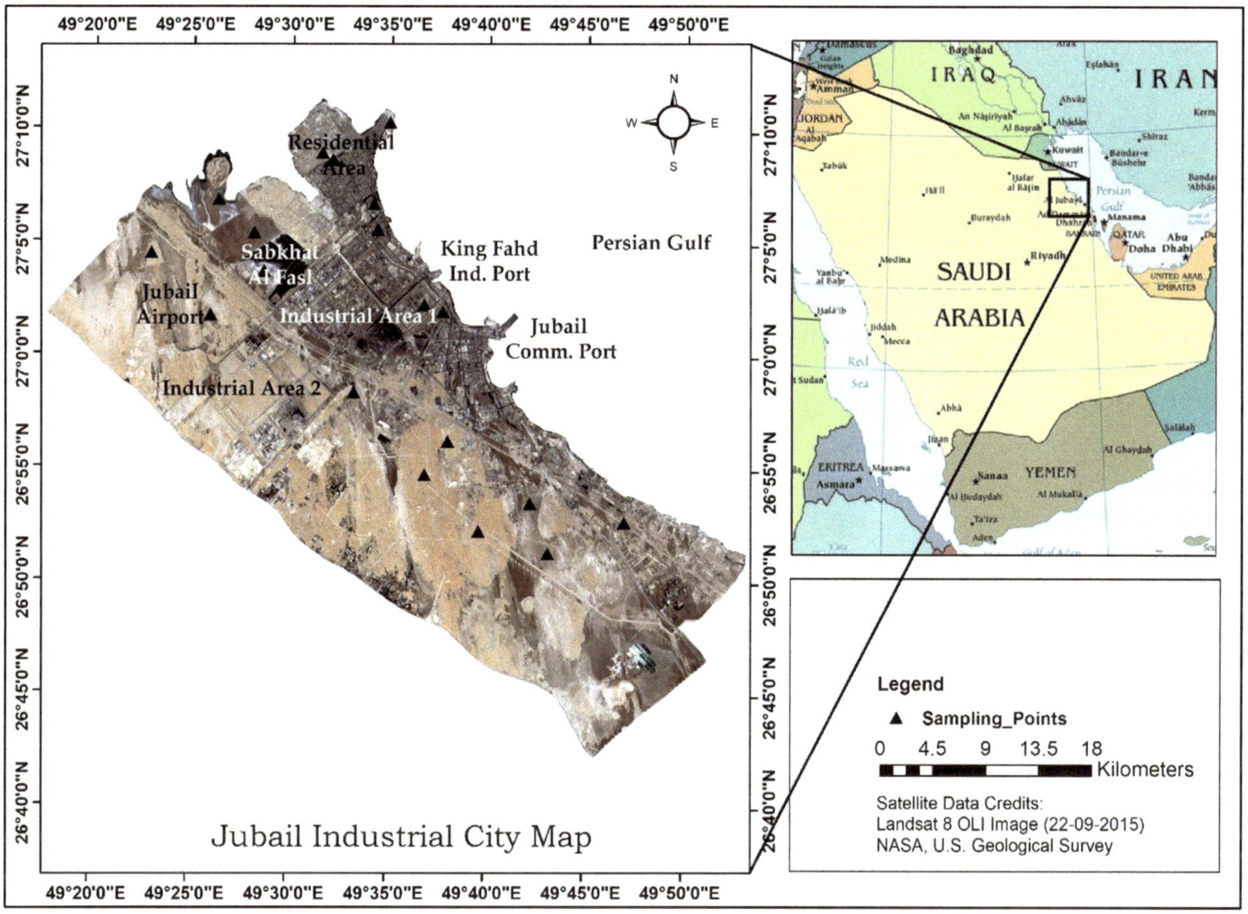

FIGURE 6.1 Study area and location of soil sampling.

laboratory as described by Hardie and Doyle (2012) and Mylavarapu et al. (2020). De-ionized water of 100 mL was added to 20 g of soil sample in a beaker and subjected to gentle mechanical stirring and shaking for 4 hours. A 0.01 M potassium chloride solution in de-ionized water was used as a standard to calibrate the conductivity meter. sympHonyTM conductivity probe was used in this study. Then, the mixture was filtered through a Whatman (No. 41, 11 cm) filter paper to collect the filtrate in a 100 mL beaker. The filtrate was then measured for its EC by a standardized conductivity meter.

6.2.3 Remote sensing data acquisition and processing

The study mainly uses the remotely sensed satellite images acquired from the Landsat 8 sensors. The Landsat satellite offers extended and reliable global records. The satellite has two onboard sensors: Operational Land Imager (OLI) and Thermal Infrared Remote Sensor. Data availability at the region, dust and clouds coverages, atmospheric conditions, etc., establishes the data acquisition. The geometrically and terrain-corrected LT1 data over the city during September 2015 has been downloaded from the Earth Explorer website (https://earthexplorer.usgs.gov/) of the United States Geological Survey. The data have already been resampled to 30 m resolution and projected with Universal Transverse Mercator (UTM) (Zone 39 N) and WGS -84 as a datum. The spectral bands 2–7 of wavelength ranging from 0.43 to 1.38 μm of the OLI have been used for the study. The region of interest was created from the satellite imagery using a shape file of Jubail Industrial City as the mask in ArcGIS 10.8 software.

6.2.3.1 Preprocessing of satellite data

The radiometric and atmospheric corrections are the significant procedures in the Preprocessing of Landsat 8 OLI data. The satellite data is subjected to radiometric and atmospheric corrections in this study using the ENVI 5.3 Software. First, the satellite data is subjected to radiometric calibration to make the Digital Numbers (DN) to "Top of

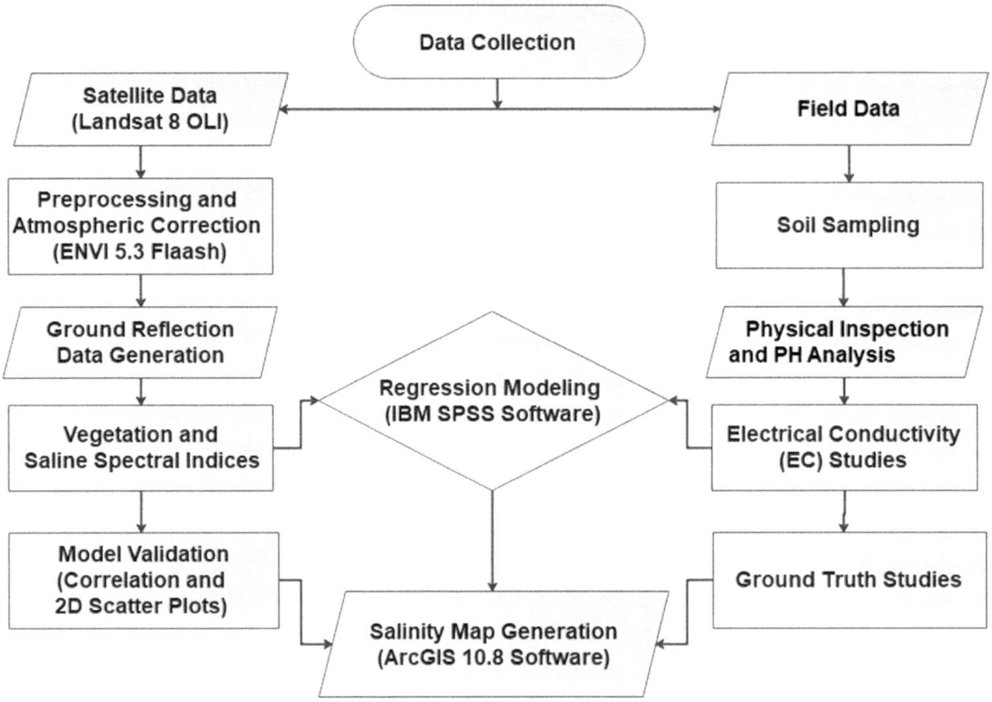

FIGURE 6.2 Flowchart of methodology.

Atmosphere" radiance data for "Fast Line-of-sight Atmospheric Analysis of Hypercubes" (FLAASH) Atmospheric Correction. The DN of the satellite data is affected by the reflectance and radiance of the surface. Various conditions of the earth's atmosphere also influence them. Different calibration algorithms are employed to correct the Landsat image and improve data quality. The radiometric calibration process involves the conversion of DN of the satellite image recorded by the onboard sensors into expressive physical parameters such as radiance or reflectance. The calibration parameter files are used for obtaining the radiometric and geometric coefficients required for the preprocessing.

Removing atmospheric interventions such as absorption and scattering of light are also significant in the preprocessing. The RS data uses the electromagnetic radiation (EM) reflected by the earth's surface and passing through the different layers of the atmosphere. So, the data includes information about the bare earth's soil and atmosphere contents. The quantitative analysis of soil salinity and its composition must explicitly remove these atmospheric influences. Eliminating the effects of water vapor and the distribution of aerosols and dust is a critical preprocessing step in the practical methodology of soil studies using the RS data. The direct measurements of these atmospheric influences on the satellite data are hardly available. Different models and techniques are available to nullify these effects. These techniques segregate the noises pixel-by-pixel basis, producing true reflectance data of the earth's surface. The FLAASH module of the ENVI 5.3 software has been utilized for this. This atmospheric correction tool corrects the wavelengths in the visible spectrum through near-infrared and shortwave infrared regions, up to 3 μm, using the MODTRAN radiation transfer codes (ENVI, 2009). The FLAASH uses a process for retrieving the amount of aerosol in the atmosphere and assessing a scene's average visibility using a dark pixel reflectance ratio method based on Kaufman et al. (1997). The Tropical Atmospheric Model and Tropospheric Aerosol Model with an inbuilt filter function file of the ENVI software have been used for the study. The Short Wave Infrared—2 band has been employed as KT Upper Channel and Red band as Lower Channel.

6.2.3.2 Generation of spectral indices

The quantity of salts in the soil is a significant factor in land vegetation cover. Asfaw et al. (2018) stated that the stressed vegetation is a secondary sign of the occurrence of excess salt in the soils. They also say that salt-affected soils are generally categorized as poorly vegetated lands. The soil salinity indices are the most preferred tool for detecting the surface soil salinity and analyzing the soil properties. The relative maximum difference value bands were selected for developing these salinity indices (Solangi et al., 2019). Though the study area is present in the semiarid region, the city's residential areas have considerable vegetation. So, the present study uses soil reflectance (salinity indices) and vegetation reflectance (vegetation indices) to map the salinity level. To generate a regression model for this region, various vegetation and saline indices were tested for evaluating and enhancing the spectral variations in the salinity of the soil. Based on the literature review (Table 6.1), we computed and processed 12 vegetation and saline spectral indices. These indices were identified and selected for this study due to their potential association with salinity detection. Table 6.1 also shows the mathematical expressions of these spectral indices. The spatial variation of these indices is shown in Fig. 6.3A and B (i–vi).

6.3 Results and discussion

6.3.1 Linear regression modeling

Linear regression modeling has been used to predict the spatial variation of soil salinity constructed (EC) from RS data and field measurements. It is a simple way to evaluate the dependence of a physical quantity on the associated functions or parameters. It can provide valuable information on making extensive modeling procedures. It can also eliminate the complications in making comprehensive modeling processes in multidimensional soil analysis. Multiple linear regression modeling is a preferred tool to evaluate salinity levels and is widely used for soil studies. In this study, multivariate regression modeling was carried out using the IBM SPSS statistical software program (Ver. 25). The soil EC prediction modeling was established based on the relation between spectral indices and estimated EC (i.e., salinity) values in the laboratory. The association between the dependent and independent variables is analyzed by the 2D-scatter plots and shown in Fig. 6.4A and B (i–vi).

The values of R^2 provide the effect and dependency level of the salinity and vegetation parameters. The 2D-scatter plots of the saline and vegetation indices show a significant correlation with the predicted salinity levels in the region. Some vegetation and saline indices negatively correlate with the land's salinity. It is also evident that including or excluding a dependent variable in soil salinity modeling is quite a difficult task, and it varies with different regions and landforms. Royston and Sauerbrei (2008) state that adding more variables in modeling may lead to over-fit and provide less accurate predictions. So, 2D-scatter plots and correlation studies are valuable in identifying the linearity of the dependent variables for the modeling process. Finally, the relationship between the independent and dependent variables has been established and given in Eq. (6.1):

$$\mathrm{EC(Salinity)} = (23.4 \times \mathrm{SAVI}) + (13 \times \mathrm{NDSI}) + (9.88 \times \mathrm{SI\_2}) + (7.38 \times \mathrm{BI}) - (3.9 \times \mathrm{SBI}) - (2.91 \times \mathrm{CRSI}) - (0.83 \times \mathrm{SI\_1}) + 4.19 \tag{6.1}$$

TABLE 6.1 Saline and vegetation spectral indices.

Indices (with acronym)	Band equation	References
Salinity index 1 (SI_1)	$(R \times G)/B$	Allbed et al. (2014)
Salinity index 2 (SI_2)	$\sqrt{(G \times R)}$	Nicolas and Walter (2006)
Salinity index 3 (SI_3)	$\sqrt{(R \times NIR)}$	Dehni and Lounis (2012)
Salinity index 4 (SI_4)	$\sqrt{(G^2 + R^2 + NIR^2)}$	Nicolas and Walter (2006)
Salinity index 5 (SI_5)	$\sqrt{(G^2 + R^2)}$	Yahiaoui et al. (2015)
Vegetation soil salinity index	$2 \times G - 5 \times (R + NIR)$	Dehni and Lounis (2012)
Brightness index	$\sqrt{(R^2 + NIR^2)}$	Khan et al. (2005)
Canopy response salinity index	$\sqrt{((NIR \times R - G \times R)/(NIR \times R + G \times R))}$	Scudiero et al. (2017)
Normalized difference salinity index	$(R - NIR)/(R + NIR)$	Khan et al. (2005)
Normalized difference vegetation index	$(NIR - R)/(NIR + R)$	Rouse (1973)
Saline adjusted vegetation index with $L = 0.5$	$(1 + L) \times (NIR - R/L) + (NIR + R)$	Martinez et al. (2001)
Sabkha index	G/R	Abuelgasim (2015)

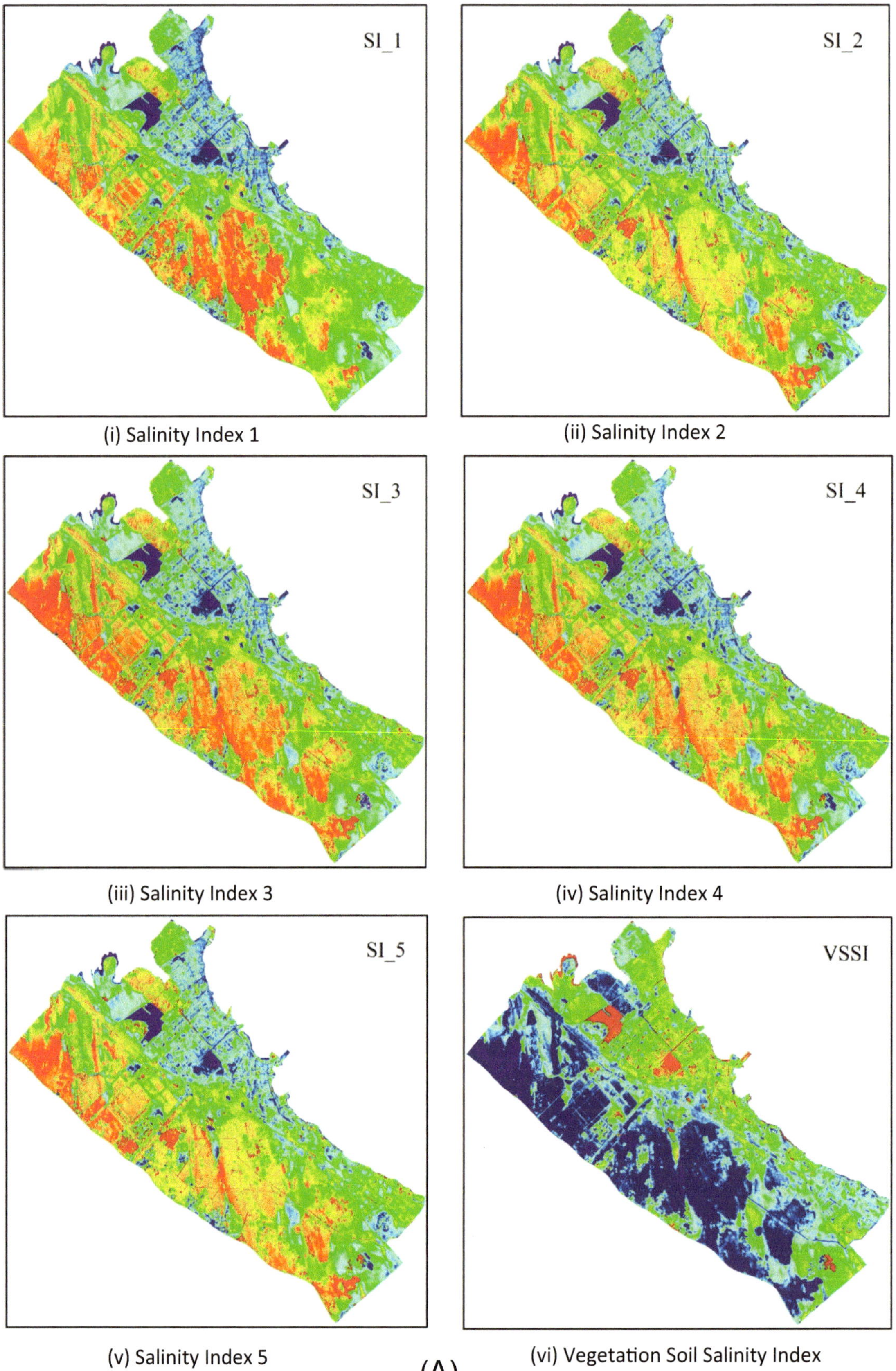

FIGURE 6.3 (A) (i–vi) Spatial variation of different salinity indices. (B). (i–vi) Spatial variation of different vegetation indices.

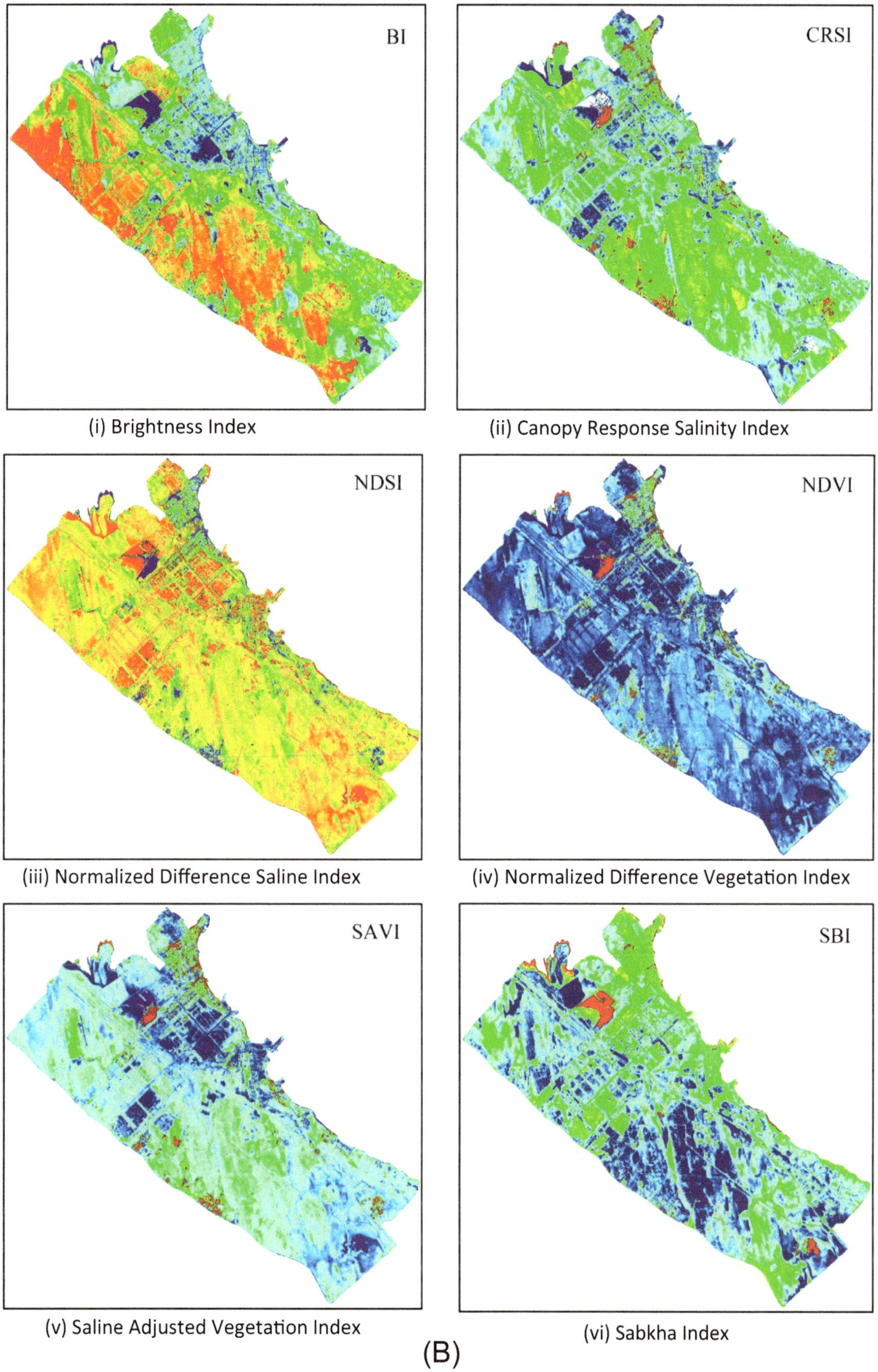

FIGURE 6.3 Continued.

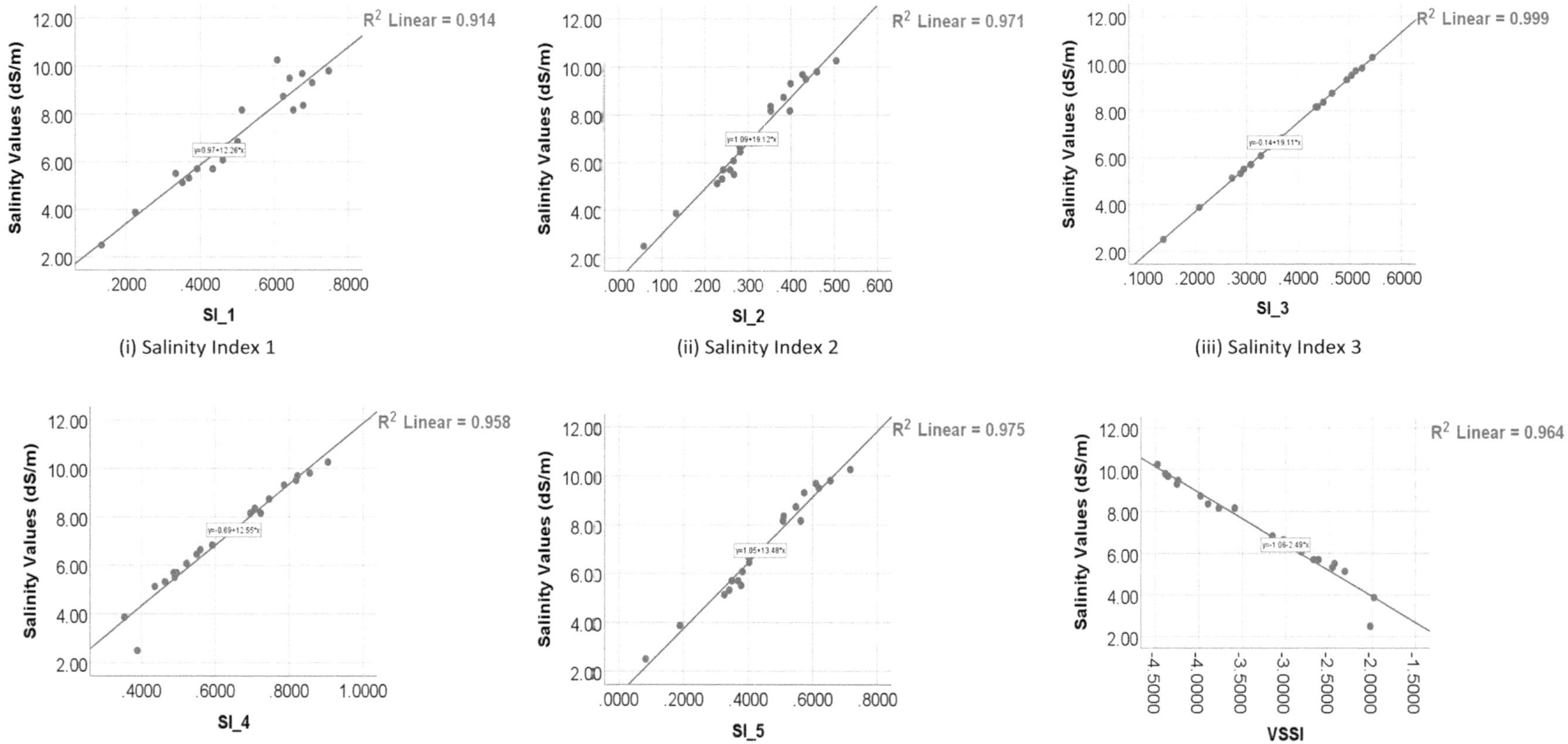

(i) Salinity Index 1

(ii) Salinity Index 2

(iii) Salinity Index 3

(iv) Salinity Index 4

(v) Salinity Index 5

(vi) Vegetation Soil Salinity Index

FIGURE 6.4 (Continued.)

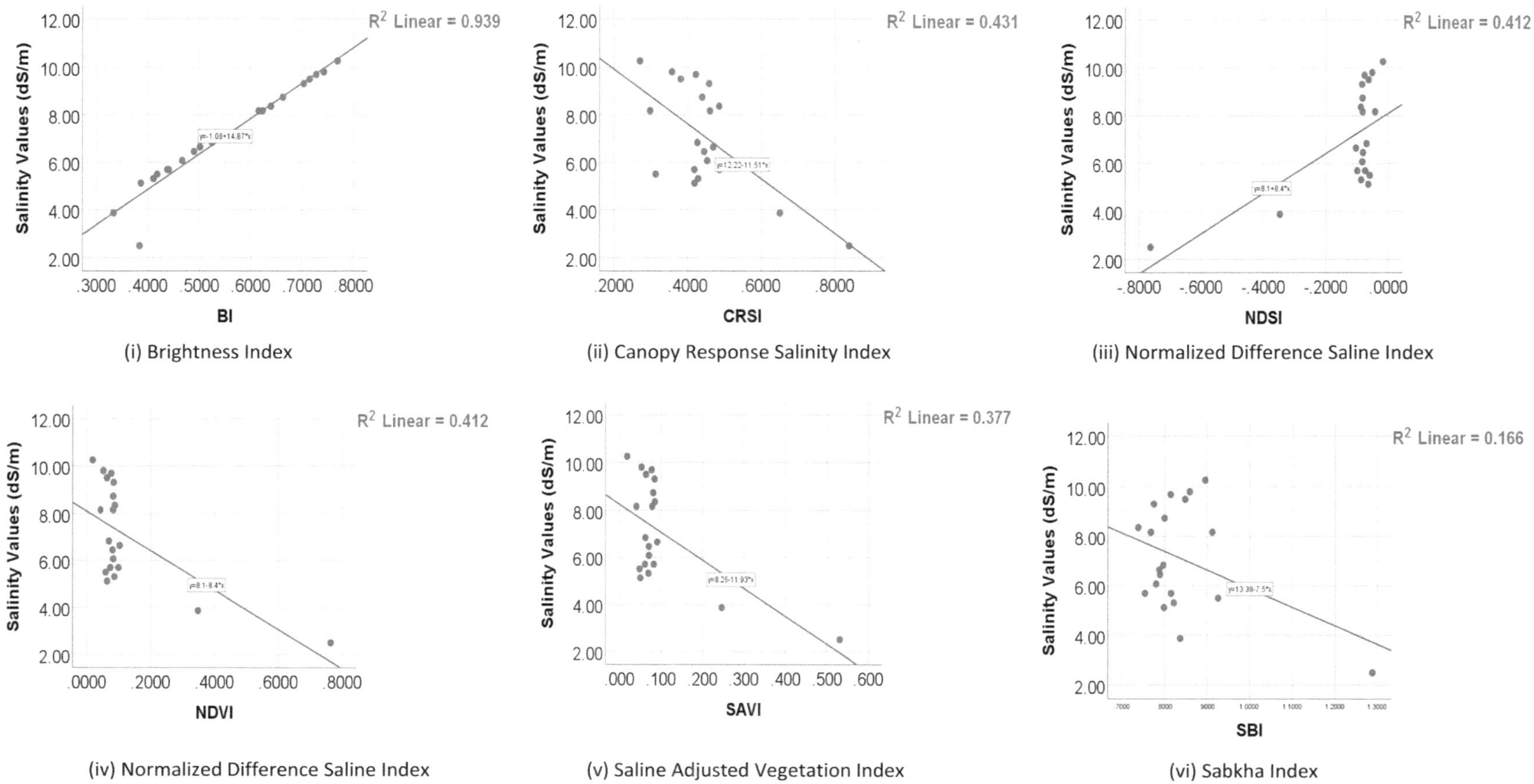

FIGURE 6.4 (A) (i–vi) 2D scatter plot of different saline indices. (B) (i–vi) 2D scatter plot of different vegetation indices.

The predicted EC values from the modeling technique are also reviewed with the ground truth studies. The relevance of the dependent variables and the generated model's reliability is examined using the validation procedure. In addition to the validation process, the final soil map is visually analyzed and interpreted per the ground investigations.

6.3.2 Model validation and analysis

The progress of a regression model for mapping soil salinity is quite complex, since the variables are dynamic and have a multidimensional relationship with the spectral indices. Different spectral indices show specific characteristic relations with the salinity of a particular area or region. Formulating a standard relationship with common variables applicable to other parts or locations is problematic. Site-specific and efficient regression models with explicit spectral indices are required to estimate the salinity of the soil precisely. The 2D-scatter plots indicate that most spectral indices significantly correlate with the independent variable (salinity). They have R^2 values from 0.16 to 0.99 and are significant with independent variables. However, including too many spectral indices may weaken the overall performance of the regression model. So, the individual relationship between the variables is insignificant in generating the model. Statistical estimation is carried out to assess their significance with independent variables.

The summary of the generated regression model for estimating the EC from the spectral indices performed by the IBM SPSS software is shown in Table 6.2. The model provides an optimum R^2 value of 0.795 with an adjusted R^2 value of 0.213. The selected parameters by the model provide an optimum significance level of 0.055. The selected dependent variables and the regression coefficients of the model are given in Table 6.3. The saline and vegetation spectral indices

TABLE 6.2 Regression model summary.

Model summary[a]

Model	*R*	*R* square	Adjusted *R* square	Std. error of the estimate	Change statistics				
					R square change	*F* change	df1	df2	Sig. *F* change
1	0.892[b]	0.795	0.213	1.68479	0.446	6.856	6	13	0.049

[a]*Dependent variable: salinity values (dS m^{-1}).*
[b]*Predictors: (constant), BI, SAVI, CRSI, SI_2, SI_3, NDSI.*

TABLE 6.3 Regression coefficients.

Coefficients[a]

Model		Unstandardized coefficients		Standardized coefficients	t	Sig.	Collinearity statistics	
		B	Std. Error	Beta			Tolerance	VIF
1	(Constant)	4.186	1.898		2.206	0.048		
	SI_1	−0.830	0.542	−0.065	−1.531	0.002	0.019	53.588
	SI_2	9.877	4.734	0.509	2.086	0.059	0.001	1785.874
	SBI	−3.880	1.215	−0.211	−3.192	0.008	0.008	131.208
	SAVI	23.427	5.883	1.206	3.982	0.002	0.000	2751.398
	NDSI	13.040	4.115	0.996	3.169	0.008	0.000	2964.908
	CRSI	−2.909	2.517	−0.166	−1.156	0.005	0.002	619.347
	BI	7.378	3.081	0.481	2.394	0.004	0.001	1209.701

[a]*Dependent variable: salinity values (dS m^{-1}).*

such as Salinity Index 1 (SI_1), Salinity Index 2 (SI_2), Sabkha Index (SBI), Saline Adjusted Vegetation Index (SAVI), Normalized Difference Saline Index (NDSI), Canopy Response Salinity Index (CRSI), and Brightness Index (BI) have been selected for making the regression model. Most of these variables have optimum significance values of around 0.05. The spectral indices such as Salinity Index 3 (SI_3), Salinity Index 4 (SI_4), Salinity Index 5 (SI_5), Vegetation Soil Salinity Index (VSSI), and Normalized Difference Vegetation Index (NDVI) have been excluded from the model and their statistical parameters are shown in Table 6.4. The normality of the regression model has been analyzed using a normal probability plot (Fig. 6.5). In the normal P—P plots, the standardized residual is compared with the normal distribution using a straight line. The probability plot implies that the variable data sets are normally distributed. It also shows that most of the data points lie around the straight line, indicating no normality issues in the predicted model. The standardized residual histogram (Fig. 6.6) also shows that the maximum scores of the data lie in the middle of the histogram and within the symmetrical curve.

6.3.3 Soil salinity map

The spatial variation of soil salinity in the Jubail Industrial City is shown in Fig. 6.7. The study reveals that the salinity level is moderate in most regions, ranging from 2 to 3 dS m^{-1}. Most of the region's inland, composed of dunes and aeolian deposits, have relatively low salinity levels. However, the inland Sabkha region has quite high salinity of around 5—10 dS m^{-1}. The study also reveals that coastal Sabkha lands (Fig. 6.8) have significantly high salinity of more than 10 dS m^{-1}. It may be endorsed that Sabkha region is experiencing the intrusion of saline water during the high tidal period, and the groundwater's salinity influences the land's salinity. Evans et al. (1969) concluded that adverse tidal effects and wind erosion with marine sedimentation during the late Quaternary period significantly flattened the coastal dunes and caused salinity in the coastal lands of the Gulf region. Al-Mhaidib (2002) states that the coastal Sabkha frequently experiences seawater intrusion, and it inundates the surrounding regions. He also emphasized that groundwater is experiencing an upward movement near the coastal Sabkha region. The adverse weather conditions and high evaporation rates prevailing in the area lead to salt deposition on the top surface of the Sabkha lands. The inland Sabkha region's drainage pattern may also contribute to the salinity.

The city's residential areas have comparatively less salinity than the rest of the city. The NDVI also indicates that dense vegetation covers the residential area (Fig. 6.9). The soil refilling and the presence of dense vegetation indicate the low salinity in this region. The refilling of sand for plantations has increased vegetation and decreased the land's salinity. The use of modern irrigation systems and management procedures in the residential area might reduce soil salinity to some extent, resulting in enhanced vegetation cover in the residential area. The soil salinity along the industrial area is also quite high. It may be attributed to the excess iron oxide contents and staining in the soil. Lilic et al. (2022) stated that the soil texture, moisture, organic matter (OM), and iron oxide may interfere with the reflected EM radiation from saline soils under normal circumstances. The difference in moisture content, OM, and the presence of iron may alter the spectral characteristics from visible to infrared spectrum. Palacios-Orueta and Ustin (1998) observed that soil texture, soil OM, and total iron content

TABLE 6.4 Excluded variables.

Excluded variables[a]

Model		Beta in	t	Sig.	Partial correlation	Collinearity statistics		
						Tolerance	VIF	Minimum tolerance
1	SI_3	1.086[b]	1.629	0.132	0.441	6.58E − 05	15179.364	6.588E − 05
	SI_4	−3.150[b]	−1.391	0.192	−0.387	6.02E − 06	165975.054	6.025E − 06
	SI_5	−5.968[b]	−2.611	0.024	−0.619	4.25E − 06	232832.522	4.295E − 06
	VSSI	−1.854[b]	−1.317	0.215	−0.369	1.58E − 05	63119.748	1.287E − 05
	NDVI	.[b]				0.000		0.000

[a]*Dependent Variable: Salinity Values (dS m^{-1}).*
[b]*Predictors in the Model: (Constant), BI, SBI, CRSI, SI_1, NDSI, SI_2, SAVI.*

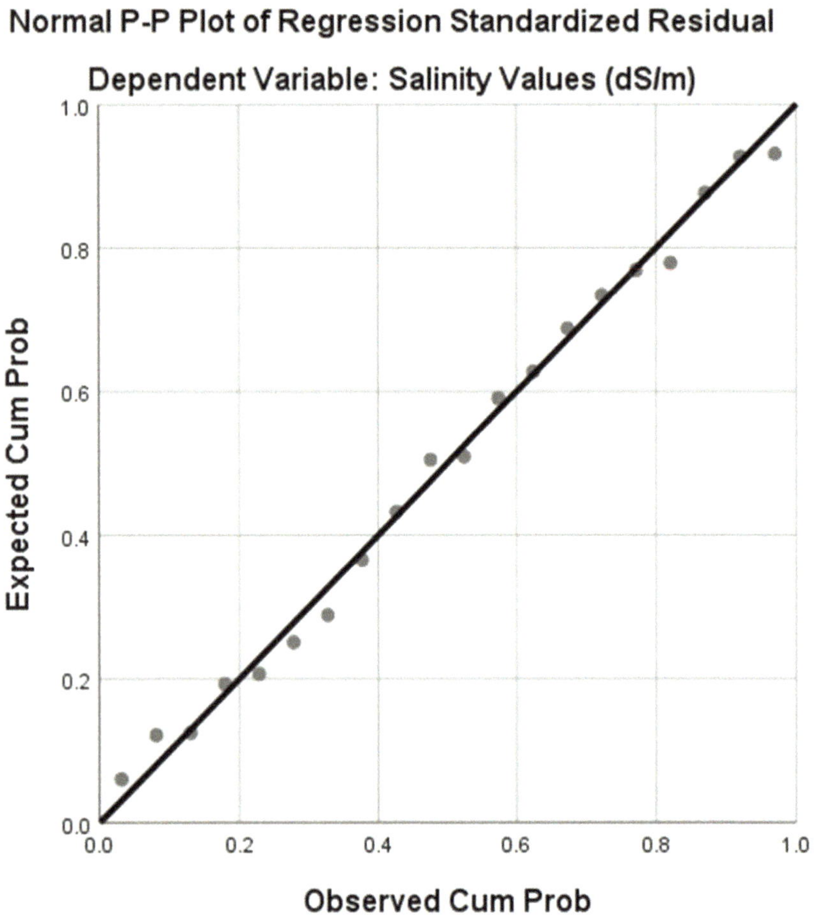

FIGURE 6.5 Normal P–P plot of standardized residual.

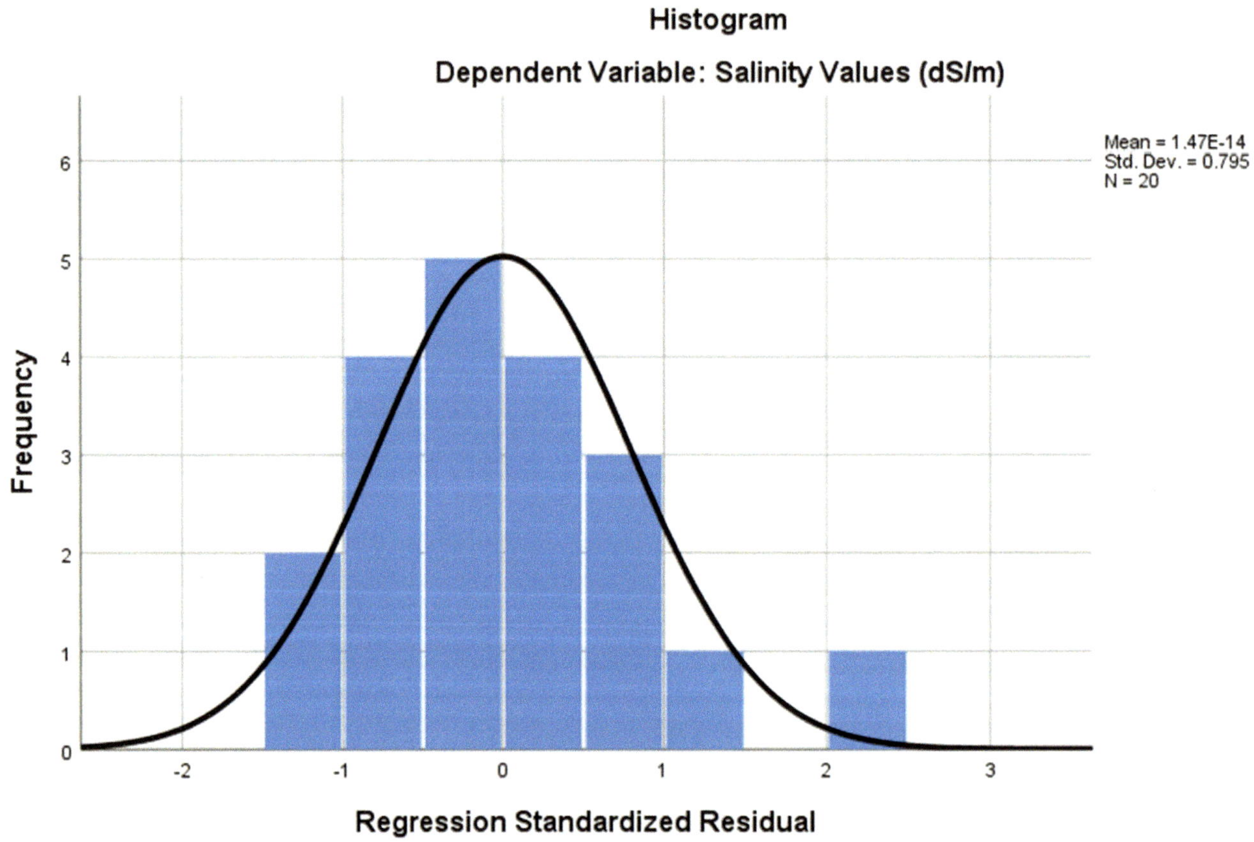

FIGURE 6.6 Frequency plot of the regression model.

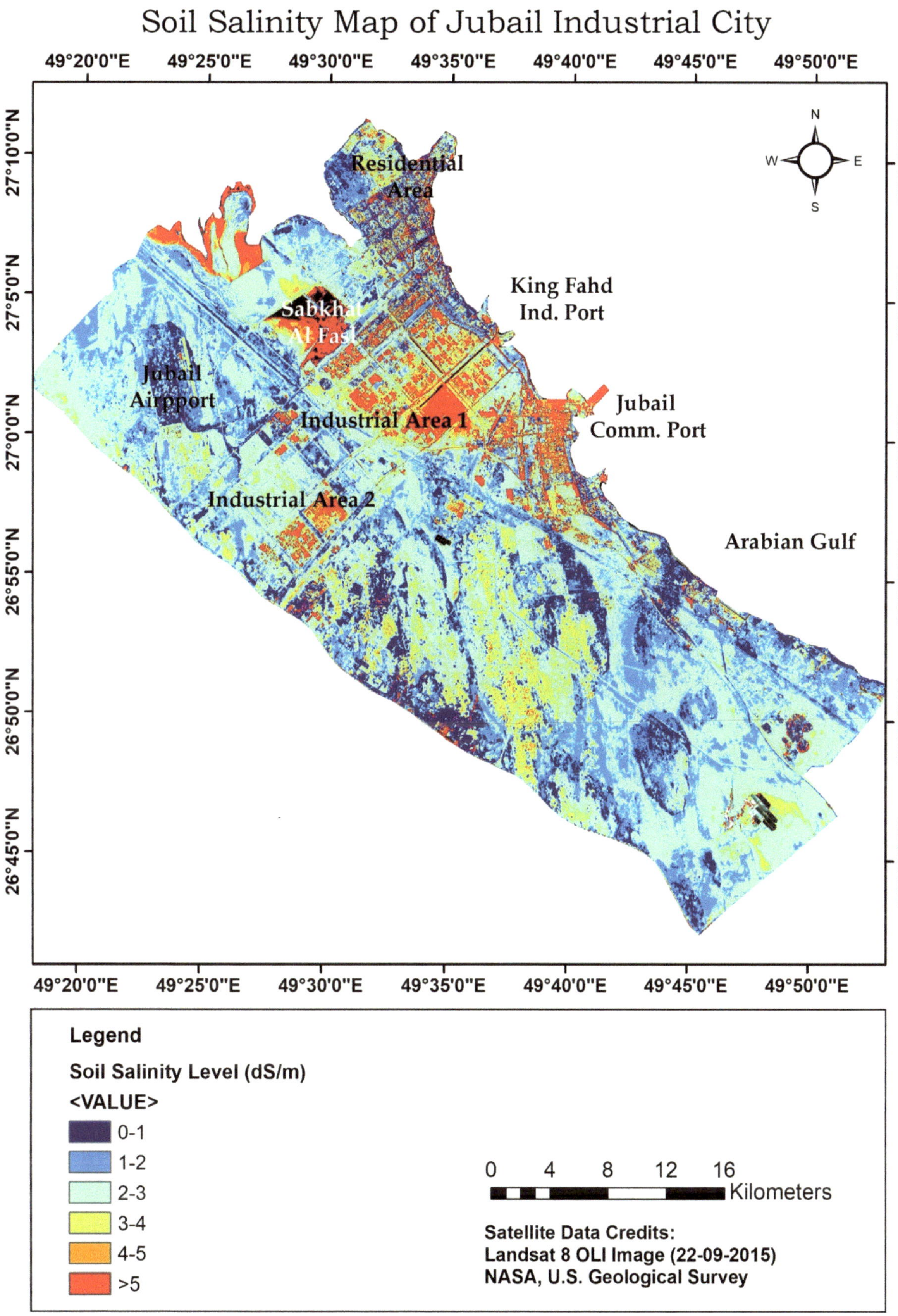

FIGURE 6.7 Soil salinity map from regression modeling.

FIGURE 6.8 Coastal "Sabkha" landform with high salinity.

FIGURE 6.9 Dense vegetation in the residential area of the city.

were the major parameters affecting the spectral response curve of the soil. These factors significantly influence the accomplishment of finding the exact surface salinity level in the city's industrial area.

The high concentration of salt ions in soil and water may also contribute to high salinity levels. The adverse weather conditions such as high temperature and evaporation rates, and the chemical composition of sand in the arid region may also enhance the soil salinity in the study area. Sheik Mujabar (2019) emphasized that there is a substantial increase in the land surface temperature along the city's industrial area. The low and infrequent rainfall increases the evaporation rates, thereby increasing the soil's salinity. The high evapotranspiration rates and rate of chemical weathering of earth compositions such as rocks, soils, and sediments in the arid regions lead to the accumulation of salts in the land. The presence of salts of sodium, nitrates, calcium, potassium, magnesium, chloride, sulfate and ammonia have significant influence on the physical properties and salinity of the soil. Thus, the present research findings are a preliminary investigation of the spatial variation of the soil salinity in the Jubail Industrial City. The study of the spatial mapping of soil salinity using the RS technology also recommends further detailed analyses on the estimation of various ions present in the soil.

6.4 Conclusion and recommendations

Mapping and monitoring the extent and severity of soil salinity is very important for agricultural and industrial developmental projects. The study demonstrates the effective utilization and potential of Landsat 8 satellite data for soil salinity assessment and mapping of the industrial city. The present study provided the salinity level and helped to understand its spatial variations in the nonagricultural arid and semiarid regions. The study's results using modern RS and GIS technology are well-mapped with the conventional EC results and benefit developmental projects and agricultural activities. The soil salinity varies significantly throughout the entire area of the city. Both natural environmental factors and human-induced activities probably trigger the spatial variation. The study also provides valuable information to the policy and decision-makers for sustainable land management strategies. The present study established salinity prediction models containing salinity and vegetation spectral indices. It can be helpful to infer the soil salinity over a large area of the city using RS data. However, the current study only represents a spatial variation of salinity during a specific period. Hence, the study also recommended further research to understand the spatial and temporal variation of the soil salinity. Multi-temporal soil mapping and investigation studies using RS data combined with field chemical analysis would enhance the better understanding of the salinity in the industrial city and avoid future adverse natural and human-induced effects on the soil.

Acknowledgments

The authors are grateful to the Managing Director of Jubail Industrial College, Jubail Industrial City, for his kind inspiration and encouragement in accomplishing the scientific research. The authors are also thankful to the Deputy Managing Director of Education and Training Affairs (ETA) and the Department of General Studies (GS) Chairman for their kind support.

Funding

This research work did not receive any grant or specific funding from agencies in the public, commercial, or not-for-profit sectors.

References

Abbas, A., Khan, S., Hussain, N., Hanjra, M.A., Akbar, S., 2013. Characterizing soil salinity in irrigated agriculture using a remote sensing approach. Phys. Chem. Earth 55–57, 43–52. Available from: https://doi.org/10.1016/j.pce.2010.12.004.

Abuelgasim, A., 2015. Mapping Sabkha habitats in the United Arab Emirates (UAE) using landsat 8 data, principal component analysis and soil salinity information. In: ACRS 2015—36th Asian Conference on Remote Sensing: Fostering Resilient Growth in Asia, Proceedings, July, 1–11. https://doi.org/10.5815/ijem.2017.04.01.

Aksoy, S., Yildirim, A., Gorji, T., Hamzehpour, N., Tanik, A., Sertel, E., 2022. Assessing the performance of machine learning algorithms for soil salinity mapping in Google Earth Engine platform using Sentinel-2A and Landsat-8 OLI data. Adv. Space Res. 69 (2), 1072–1086. Available from: https://doi.org/10.1016/j.asr.2021.10.024.

Aldakheel, Y.Y., 2011. Assessing NDVI spatial pattern as related to irrigation and soil salinity management in Al-Hassa Oasis, Saudi Arabia. J. Indian. Soc. Remote. Sens. 39 (2), 171–180. Available from: https://doi.org/10.1007/s12524-010-0057-z.

Al-Gaadi, K.A., Tola, E.K., Madugundu, R., Fulleros, R.B., 2021. Sentinel-2 images for effective mapping of soil salinity in agricultural fields. Curr. Sci. 121 (3), 384–390. Available from: https://doi.org/10.18520/cs/v121/i3/384-390.

Allbed, A., Kumar, L., Sinha, P., 2014. Mapping and modelling spatial variation in soil salinity in the Al Hassa Oasis based on remote sensing indicators and regression techniques. Remote. Sens. 6 (2), 1137–1157. Available from: https://doi.org/10.3390/rs6021137.

Al-Mhaidib, A., 2002. Sabkha soil in the Kingdom of Saudi Arabia: characteristics and treatment. J. King Abdulaziz Univ.-Eng. Sci. 14 (2), 29–80. Available from: https://doi.org/10.4197/eng.14-2.9.

Asfaw, E., Suryabhagavan, K.V., Argaw, M., 2018. Soil salinity modeling and mapping using remote sensing and GIS: the case of Wonji sugar cane irrigation farm, Ethiopia. J. Saudi Soc. Agric. Sci. 17 (3), 250–258. Available from: https://doi.org/10.1016/j.jssas.2016.05.003.

Bannari, A., Al-Ali, Z.M., 2020. Assessing climate change impact on soil salinity dynamics between 1987–2017 in arid landscape using Landsat TM, ETM + and OLI data. Remote. Sens. 12 (17), 2794. Available from: https://doi.org/10.3390/RS12172794.

Basyoni, M.H., Aref, M.A., 2016. Composition and origin of the sabkha brines, and their environmental impact on infrastructure in Jizan area, Red Sea Coast, Saudi Arabia. Environ. Earth Sci. 75 (2), 1–17. Available from: https://doi.org/10.1007/s12665-015-4913-6.

Biro Turk, K.G., Aljughaiman, A.S., 2020. Land use/land cover assessment as related to soil and irrigation water salinity over an oasis in arid environment. Open. Geosci. 12 (1), 220–231. Available from: https://doi.org/10.1515/geo-2020-0103.

Biswas, A., Agarwal, S. 2013. Environmental impact assessment for developing countries. https://books.google.com/books?hl = en&lr = &id = fPkbBQAAQBAJ&oi = fnd&pg = PP1&dq = Environmental + Impact + Assessment + for + Developing + Countries + edi ted + by + Asit + K. + Biswas&ots = MyVqFxS-BM&sig = IToUBwMuPqk02tspdiX009Qeado.

Brunner, P., Li, H.T., Kinzelbach, W., Li, W.P., 2007. Generating soil electrical conductivity maps at regional level by integrating measurements on the ground and remote sensing data. Int. J. Remote. Sens. 28 (15), 3341–3361. Available from: https://doi.org/10.1080/01431160600928641.

Dehni, A., Lounis, M., 2012. Remote sensing techniques for salt affected soil mapping: application to the Oran region of Algeria. Procedia Eng. 33, 188–198. Available from: https://doi.org/10.1016/j.proeng.2012.01.1193.

Douaik, A., Van Meirvenne, M., Tóth, T., 2009. Stochastic approaches for space-time modeling and interpolation of soil salinity. Remote. Sens. Soil. Salin.: Impact Land. Manag. 273–290. Available from: http://hdl.handle.net/1854/LU-771019.

Elhag, M., 2016. Evaluation of different soil salinity mapping using remote sensing techniques in arid ecosystems, Saudi Arabia. J. Sens. Available from: https://doi.org/10.1155/2016/7596175*2016*.

Evans, G., Schmidt, V., Bush, P., Nelson, H., 1969. Stratigraphy and geologic history of the Sabkha, Abu Dhabi, Persian Gulf. Sedimentology 12 (1–2), 145–159. Available from: https://doi.org/10.1111/j.1365-3091.1969.tb00167.x.

Gorji, T., Sertel, E., Tanik, A., 2017. Recent satellite technologies for soil salinity assessment with special focus on Mediterranean countries. Fresenius Environ. Bull. 26 (1), 196–203.

Hardie, M., Doyle, R., 2012. Measuring soil salinity. Methods Mol. Biol. 913, 415–425. Available from: https://doi.org/10.1007/978-1-61779-986-0_28.

He, H., Roach, R., Hoseney, R.C., 1992. Effect of Nonchaotropic salts on flour bread-making properties. Cereal Chem. 69 (4), 366–371. Available from: http://www.aaccnet.org/publications/cc/backissues/1992/Documents/CC1992a84.html.

Jabbar, M.T., Zhou, J., 2012. Assessment of soil salinity risk on the agricultural area in Basrah Province, Iraq: using remote sensing and GIS techniques. J. Earth Sci. 23 (6), 881–891. Available from: https://doi.org/10.1007/s12583-012-0299-5.

Kaufman, Y.J., Tanrc, D., Gordon, H.R., Nakajima, T., Lenoble, J., Frouins, R., et al., 1997. Passive remote sensing of tropospheric aerosol and atmospheric correction for the aerosol effect. J. Geophys. Res. 102 (D14), 815–831.

Khan, N.M., Rastoskuev, V.V., Sato, Y., Shiozawa, S., 2005. Assessment of hydrosaline land degradation by using a simple approach of remote sensing indicators. Agric. Water Manag. 77 (1–3), 96–109. Available from: https://doi.org/10.1016/j.agwat.2004.09.038.

Lhissou, R., El, A., Chokmani, K., 2014. Mapping soil salinity in irrigated land using optical remote sensing data. Eurasian J. Soil. Sci. 3 (3), 82–88.

Lilic, O., Budak, M., Gunal, E., Acir, N., Halbac-cotoara-zamfir, R., Alfarraj, S., et al., 2022. Soil salinity assessment of a natural pasture using remote sensing techniques in central Anatolia, Turkey. PLoS One 17 (4), 1–9.

Martinez, B., Camacho-de Coca, F., Garcia-Haro, J., Gilabert, M.A., 2001. Exploring the possibilities of a vegetation index (GESAVI) from remotely sensed data. Remote. Sens. Agri. Ecosyst. Hydrol. II 4171, 271–279. Available from: https://doi.org/10.1117/12.413935.

ENVI, 2009. ENVI Atmospheric Correction Module: QUAC and FLAASH user's guide. Module Version, 4.7. https://www.nv5geospatialsoftware.com/portals/0/pdfs/envi/flaash_module.pdf (last accessed on 26 September 2023).

Modaihsh, A.S., Ghoneim, A.M., Sallam, A.S., Mahjoub, M.O., 2007. Soil salinity, sand encroachment and erosion as indicators of land degradation in harad governorate. Saudi Arabia, September.

Mylavarapu, R., Bergeron, J., Wilkinson, N., 2020. Soil pH and electrical conductivity: a county extension soil laboratory manual. In: Edis, vol. CIR1081, Issue Series of the Department of Soil and Water Sciences. UF/IFAS Extension. https://doi.org/10.32473/edis-ss118-2020.

Nicolas, H., Walter, C., 2006. Detecting salinity hazards within a semiarid context by means of combining soil and remote-sensing data. Geoderma. https://www.sciencedirect.com/science/article/pii/S0016706105002958.

Palacios-Orueta, A., Ustin, S.L., 1998. Remote sensing of soil properties in the Santa Monica mountains I. Spectral analysis. Remote. Sens. Environ. 65 (2), 170–183. Available from: https://doi.org/10.1016/S0034-4257(98)00024-8.

Rajendran, S., Al-Kuwari, H.A.S., Sadooni, F.N., Nasir, S., Govil, H., 2021. Remote sensing of inland Sabkha and a study of the salinity and temporal stability for sustainable development: a case study from the West coast of Qatar. Sci. Total. Environ. 782, 146932. Available from: https://doi.org/10.1016/j.scitotenv.2021.146932.

Rouse, J.W., 1973. Monitoring the vernal advancement and retrogradation of natural vegetation [NASA/GSFCT Type II Report]. Greenbelt, MD: NASA/Goddard Space Flight Center. https://ntrs.nasa.gov/citations/19750020419.

Royston, P., Sauerbrei, W., 2008. Continuous-by-continuous interactions. In: Multivariable Model—Building: A Pragmatic Approach to Regression Anaylsis Based on Fractional Polynomials for Modelling Continuous Variables, pp. 172–180. https://www.wiley.com/en-gb/Multivariable + Model + Building%
3A + A + Pragmatic + Approach + to + Regression + Anaylsis + based + on + Fractional + Polynomials + for + Modelling + Cont
inuous + Variables-p-9780470028421.

Scudiero, E., Corwin, D.L., Anderson, R.G., Yemoto, K., Clary, W., Wang, Z., et al., 2017. Remote sensing is a viable tool for mapping soil salinity in agricultural lands. Calif. Agri. 71 (4), 231–238. Available from: https://doi.org/10.3733/ca.2017a0009.

Shahid, S.A., Abdelfattah, M.A., Omar, S.A.S., Harahsheh, H., Othman, Y., Mahmoudi, H., 2010. Mapping and monitoring of soil salinization remote sensing, gis, modeling, electromagnetic induction and conventional methods—case studies. In: Proceedings of the International Conference on Soils and Groundwater Salinization in Arid Countries, 97, pp. 59–97.

Sheik Mujabar, P., 2019. Spatial-temporal variation of land surface temperature of Jubail Industrial City, Saudi Arabia due to seasonal effect by using Thermal Infrared Remote Sensor (TIRS) satellite data. J. Afr. Earth Sci. 155 (March), 54–63. Available from: https://doi.org/10.1016/j.jafrearsci.2019.03.008.

Solangi, K.A., Siyal, A.A., Wu, Y., Abbasi, B., Solangi, F., Lakhiar, I.A., et al., 2019. An assessment of the spatial and temporal distribution of soil salinity in combination with field and satellite data: a case study in Sujawal district. Agronomy 9 (12). Available from: https://doi.org/10.3390/agronomy9120869.

Valipour, M., 2014. Drainage, waterlogging, and salinity. Arch. Agron. Soil. Sci. 60 (12), 1625–1640. Available from: https://doi.org/10.1080/03650340.2014.905676.

Wu, W., Zucca, C., Muhaimeed, A.S., Al-Shafie, W.M., Fadhil Al-Quraishi, A.M., Nangia, V., et al., 2018. Soil salinity prediction and mapping by machine learning regression in Central Mesopotamia, Iraq. Land. Degrad. Dev. 29 (11), 4005–4014. Available from: https://doi.org/10.1002/ldr.3148.

Yahiaoui, I., Douaoui, A., Zhang, Q., Ziane, A., 2015. Soil salinity prediction in the Lower Cheliff plain (Algeria) based on remote sensing and topographic feature analysis. J. Arid. Land. 7 (6), 794–805. Available from: https://doi.org/10.1007/s40333-015-0053-9.

Zewdu, S., Suryabhagavan, K.V., Balakrishnan, M., 2017. Geo-spatial approach for soil salinity mapping in Sego Irrigation Farm, South Ethiopia. J. Saudi Soc. Agric. Sci. 16 (1), 16–24. Available from: https://doi.org/10.1016/j.jssas.2014.12.003.

Chapter 7

Assessment of spatial variability of soil fertility parameters in Eastern Ghats of Karnataka: a case study

R. Vasundhara, S. Dharumarajan, B. Kalaiselvi, R. Srinivasan, Rajendra Hegde, M. Lalitha and V. Ramamurthy
ICAR-National Bureau of Soil Survey and Land Use Planning, Regional Centre, Bangalore, Karnataka, India

7.1 Introduction

Assessing the spatial variability of soil properties across a field is vital for efficient resource use and management. The spatial variability of soil properties is influenced by parent material, topography, land use management, and rainfall. The study area, Eastern Ghats, has a wide range of topographical variability. The interaction of extrinsic (such as soil management practices, indigenous fertility status, crop rotation, and kinds of a standing crop) and intrinsic (parent materials and climate) factors may be the cause for spatial variation in soil quality (Cambardella and Karlen, 1999). Soil fertility assessment of farms is important since it gives a sign of potential nutrient deficiencies, pH imbalance, or excess soluble salts for diversified crops (Wang et al., 2001).

It is essential to map soil fertility parameters using digital techniques and maps with high spatial resolution. It has been challenging to describe the geographical heterogeneity of soil fertility throughout a field, until the initiation of contemporary technology like geographic position system (GPS) and geographic information systems (GIS). To produce thematic soil fertility maps at the field scale, it is crucial to gather soil samples using GPS (Mishra et al., 2013; Malla et al., 2020). It aids in developing site-specific nutrient management strategies for the area. On the basis of geostatistical analysis, several studies have been conducted to describe the spatial variability of different soil properties (Liu et al., 2013; Shukla et al., 2016; Vasundhara et al., 2020). In order to examine the spatial patterns and estimate the variation at unsampled locations, a group of geostatistical techniques known as kriging interpolation methods is frequently applied. These spatial assessment techniques potentially can help better crop nutrient management.

Crops grown in rainfed farming systems are more adversely affected by nutrient deficiencies than by moisture (Srinivasa Rao et al., 2015). Few research works reports have assessed the spatial distribution of soil properties in rainfed in different regions of Karnataka at the farm level (Hegde et al., 2021). To have suitable soil management practices, farmers should know what amendments are required to optimize the productivity of the soil for specific crops (Kavitha and Sujatha, 2015). The main objective of the study is to assess the spatial variability of soil fertility parameters at the farm scale using geostatistics and spatial mapping techniques to address the deficiencies. This study area was adopted to enhance the farmer's yield by improving soil nutrient management and providing quality agricultural inputs.

7.2 Materials and methods

7.2.1 Description of the study area

The present study was carried out in Kanakapura Taluk, Ramanagara District of Karnataka, India (Fig. 7.1). The study area lies between 12° 26′ 27″ to 12° 23′ 19″ N and 77° 31′ 36″ to 77° 34′ 53″ E, covering 2123 ha, and is situated at an altitude of 584−603 m above the mean sea level. This area falls under the semiarid tract of the state and receives an average annual rainfall of 790.6 mm. The temperature typically varies from 16 to 38°C. The average potential evapotranspiration is 1496.9 mm. It belongs to the agroecological region 8.2, Deccan (Karnataka) Plateau, hot semiarid ecosystem with red loamy soils with LGP 120−150 days. Major rock formations observed in this area are granites Gneiss, and soil type is red sandy soil and red loamy soils.

Remote Sensing of Soils. DOI: https://doi.org/10.1016/B978-0-443-18773-5.00031-4

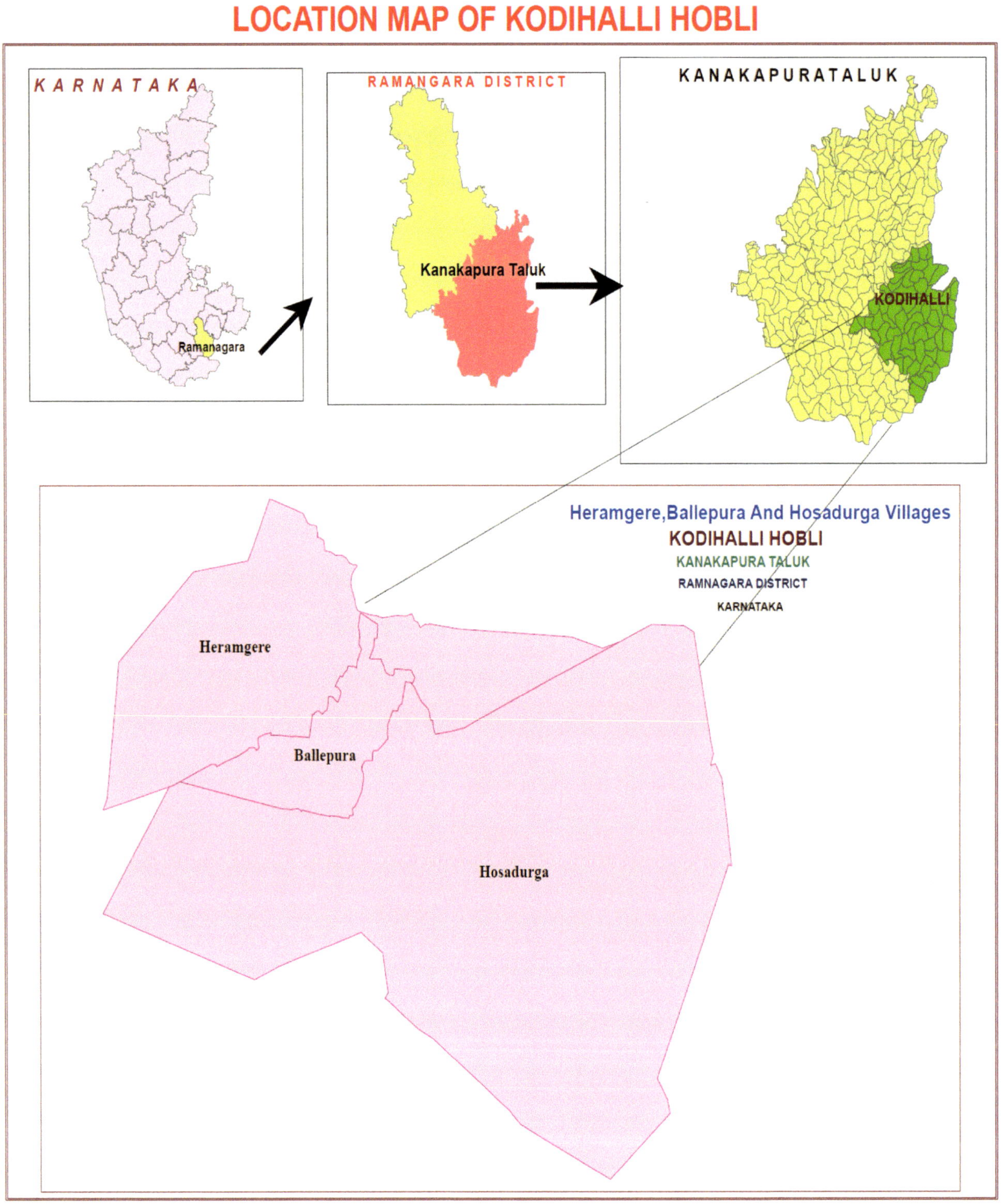

FIGURE 7.1 Location map of Kanakapura Taluk, Karnataka.

7.2.2 Soil sampling

Totally 181 surface soil samples (0–15 cm) were collected at 250 m grid intervals covering three villages excluding water bodies, settlements, and forest areas for soil fertility. Soil samples were examined for various chemical parameters, including soil pH, EC, soil organic carbon (OC), available nitrogen (N), phosphorus (P_2O_5), potassium (K_2O), sulfur, iron, copper, zinc, and boron. OC was analyzed based on Walkley-Black method (Walkley and Black, 1934). Available nitrogen was estimated by Alkaline permanganate method (Subbiah and Asija, 1956). Available phosphorus was determined Olsen method (Olsen et al., 1954), and available potassium was determined by flame photometry after extraction with a neutral normal ammonium acetate solution (pH 7.0). Available sulfur was extracted by 0.15% CaCl2 solution and the concentration of sulfur was determined by the turbidimetric method spectrophotometrically (Chesnin and Yien, 1951). Hot water-soluble Boron was estimated by azomethine-h method as outlined by Gupta (1967). Available Zn, Fe, Cu, and Mn were extracted by diethylene triamine penta acetic acid (DTPA) solution and analyzed using atomic absorption spectroscopy (Lindsay and Norvell 1978).

7.2.3 Geostatistical analysis and preparation of soil fertility maps

The spatial interpolation techniques were used to produce spatial distribution for the investigated basic soil properties using ArcGIS v.10.1 (ESRI Co, Redlands, USA). In ArcGIS, kriging can express the spatial variation and allow a variety of map outputs, and at the same time minimize the errors of predicted values (Gonzalez et al., 2014). Descriptive statistics (mean, range, standard deviation, standard error, coefficient of variation (CV), Skewness, and Kurtosis) of soil parameters were computed using SPSS Inc Software.

7.3 Results and discussion

7.3.1 Descriptive statistics

The statistics of the 12 soil variables obtained on the different soil types are summarized in Table 7.1. Most of them show a positive skewness varying between 0.05 and 7.54. The kurtosis is also highly variable. This departure of the

TABLE 7.1 Descriptive statistics of the original soil variables ($n = 181$).

Variables	Range	Mean		SD	Variance (%)	Skewness		Kurtosis	
		Statistic	SE.			Statistic	SE.	Statistic	SE.
pH	4.8–7.3	5.50	0.09	0.64	0.41	0.46	0.33	−0.18	0.64
EC (d Sm^{-1})	0.02–0.60	0.09	0.01	0.09	0.01	3.95	0.26	19.14	0.52
OC (%)	0.04–1.30	0.57	0.03	0.25	0.06	0.31	0.26	−0.21	0.52
Available N (kg ha^{-1})	107.6–399.5	179.35	6.01	52.41	2.75	1.24	0.28	2.96	0.55
Available P_2O_5 (kg ha^{-1})	1.0–52.0	19.24	1.44	13.17	173.44	0.82	0.26	−0.08	0.52
Available K_2O (kg ha^{-1})	14.6–264.0	140.75	6.01	55.10	3.04	0.05	0.26	−0.50	0.52
Available S (mg kg^{-1})	0.70–49.2	17.22	1.55	14.19	201.48	0.50	0.26	−1.10	0.52
Available B (mg kg^{-1})	0.1–0.8	0.25	0.02	0.15	0.02	1.24	0.26	1.36	0.52
Available Cu (mg kg^{-1})	0.3–2.4	1.04	0.04	0.40	0.16	0.94	0.26	0.82	0.52
Available Fe (mg kg^{-1})	5.9–48.8	18.86	0.94	8.65	74.78	1.24	0.26	2.05	0.52
Available Mn (mg kg^{-1})	8.0–49.8	25.26	0.97	8.91	79.31	0.28	0.26	−0.40	0.52
Available Zn (mg kg^{-1})	0.3–5.2	0.65	0.06	0.54	0.29	7.54	0.26	64.46	0.52

TABLE 7.2 Correlation between different soil properties.

	pH	EC	OC	N	P_2O_5	K_2O	S	B	Cu	Fe	Mn	Zn
pH	1											
EC	0.070	1										
OC	0.104	0.174	1									
N	−0.032	**0.296****	**0.626****	1								
P_2O_5	−0.190	−0.012	−0.169	−0.140	1							
K_2O	−0.112	0.048	0.243*	0.247*	−0.120	1						
S	0.067	0.109	−0.332**	−0.322**	−0.113	0.107	1					
B	−0.006	0.053	−0.233*	−0.272*	0.053	−0.043	0.211	1				
Cu	0.270*	−0.007	0.233*	0.236*	−0.398**	0.203	0.139	−0.044	1			
Fe	−0.335*	−0.185	0.024	−0.074	**0.445****	−0.045	−0.038	−0.043	−0.104	1		
Mn	−0.025	−0.042	**0.342****	**0.508****	−0.183	0.201	−0.062	−0.239*	**0.365****	0.007	1	
Zn	−0.071	**0.641****	**0.328****	**0.539****	0.088	0.135	−0.129	−0.053	−0.125	−0.036	0.227*	1

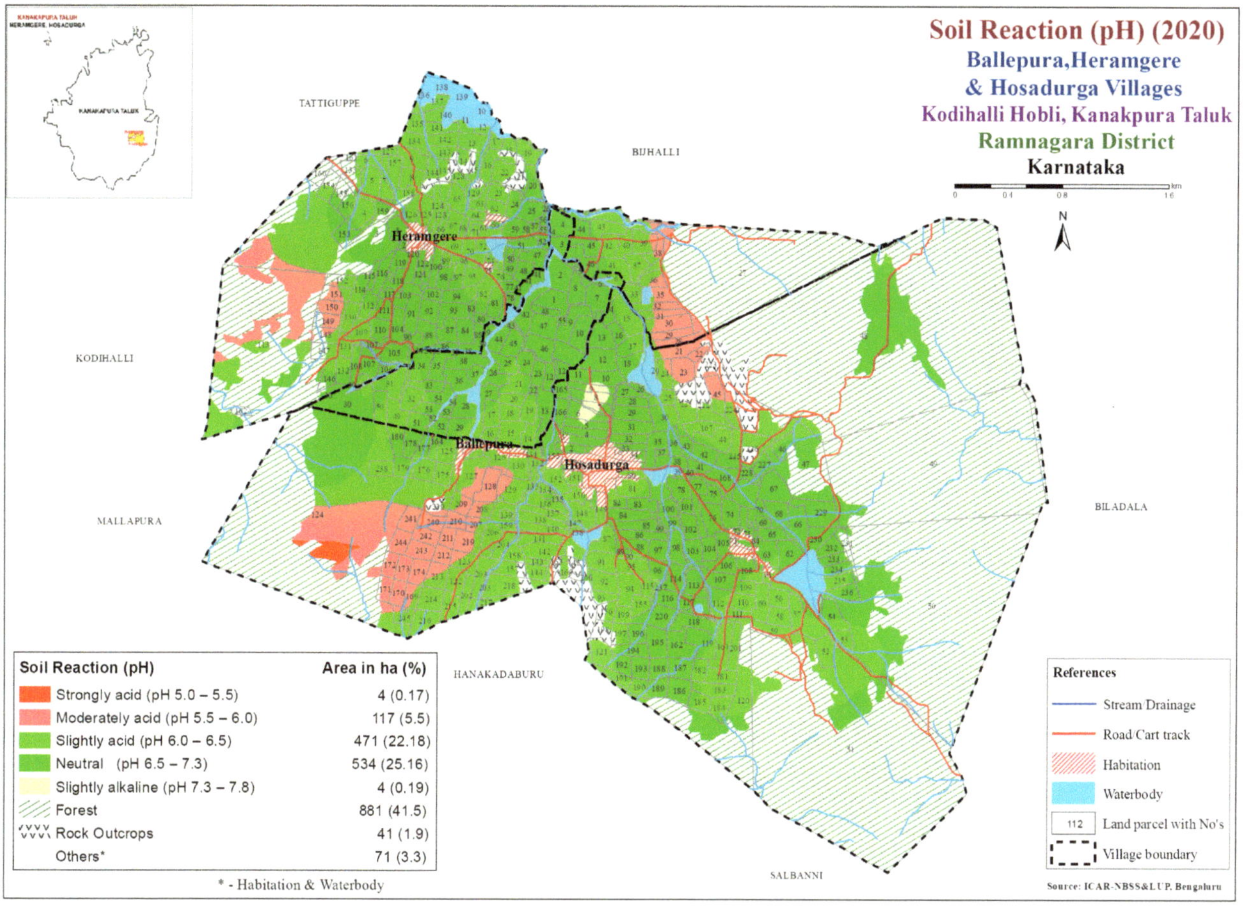

FIGURE 7.2 Soil Reaction (pH) map of the study area.

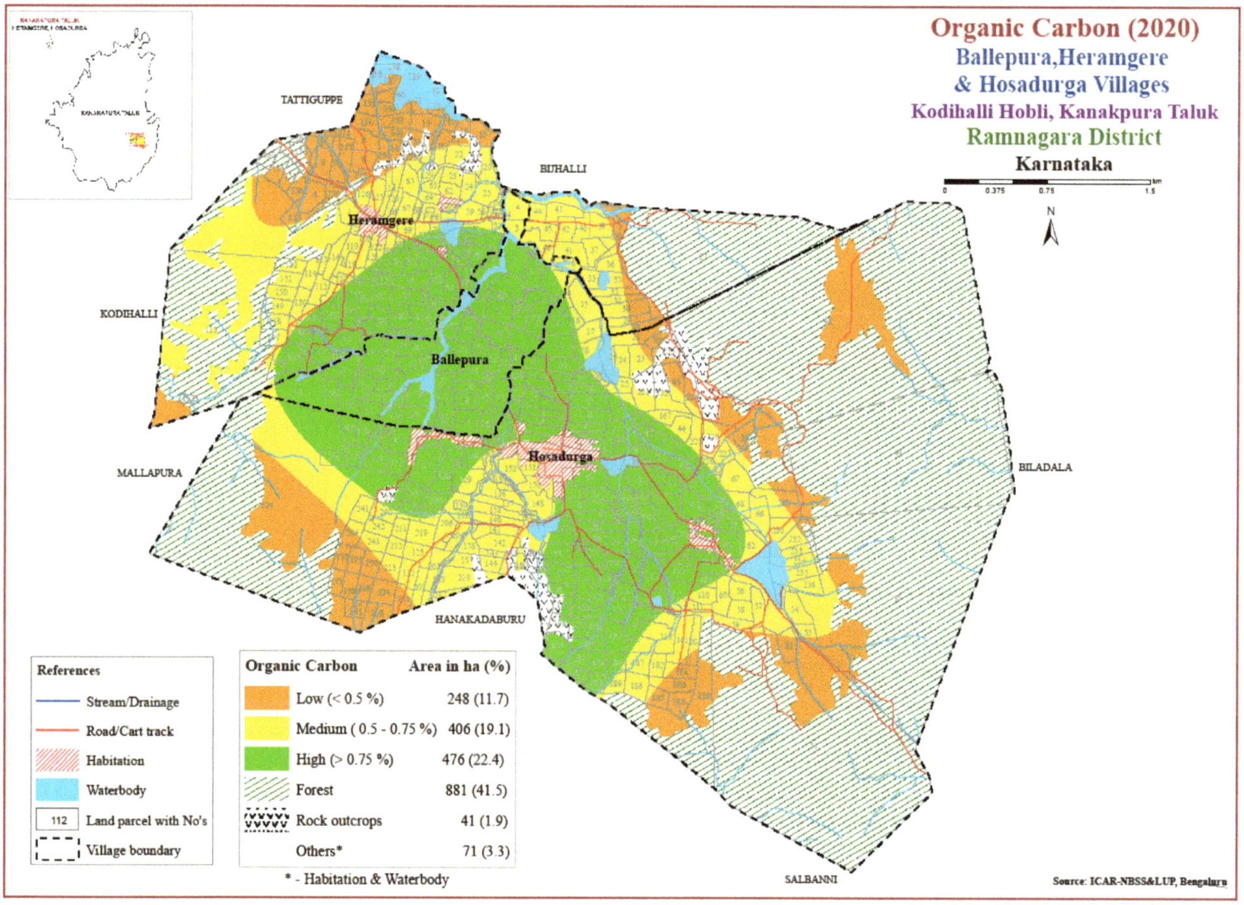

FIGURE 7.3 OC map of study area.

skewness and the kurtosis for zero means that most of these variables have slightly abnormal distribution. The CV values ranged from 0.01 to 201%. The CV <10%, between 10% and 90%, and > 90% indicates low, moderate, and high variability, respectively. The available P_2O_5 and S have high to very high CV and have high variability within soil groups of the study area (Jin et al., 2021). Whereas available Fe and Mn have a moderate CV of 74.78% and 79.31%, respectively. The variations could be attributed to various soil topography, parent material, and conventional methods of managing soil fertility management.

7.4 Correlation between soil properties

The degree of association between soil properties (pearson correlation coefficient), was computed and presented in Table 7.2. We observed a positive correlation between N with EC and OC ($r = 0.296^{**}$-$r = 0.626^{**}$), which was highly significant. There was a significant positive correlation between Fe and P_2O_5 ($r = 0.445^{**}$). Further, the analysis showed that the degree of association between Mn with OC ($r = 0.342^{**}$), N ($r = 0.508^{**}$), and Cu ($r = 0.365^{**}$) was significant ($r = 0.365^{**}$). Similarly, Zn had positive and significant association with EC ($r = 0.641^{**}$), OC ($r = 0.328^{**}$), and N ($r = 0.539^{**}$). In all other cases, the correlation was not significant.

7.4.1 Soil reaction and electrical conductivity

The pH of the soil varied from strongly acidic to neutral. About 55 ha (2.6%) area is extremely acid and 124 ha (12%) area is moderately acid (pH 5.5–6.0) and is distributed in the southwestern part of the village. About 324 ha (15.3%) area (Fig. 7.2) is under slightly acidic range (pH 6.0–6.5) and is distributed in all parts of the village. About 399 ha (18.8%) is under neutral condition (pH 6.5–7.3) and distributed in the central and eastern part of the villages. Acidic nature of soil in the study area may be due to the mineralogy of the soil, the use of acidic fertilizers, the limited input

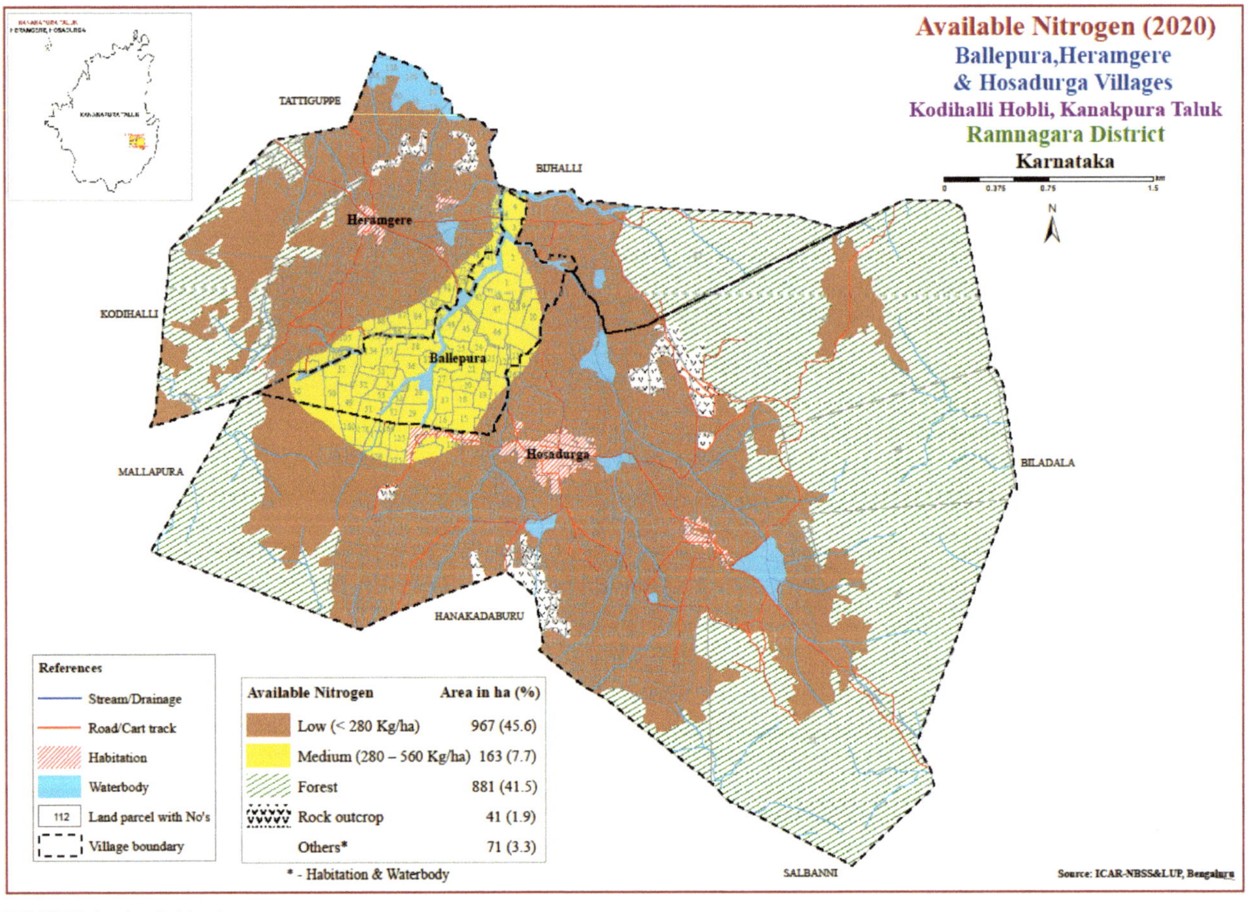

FIGURE 7.4 Available nitrogen map.

of organic materials, and the removal of basic nutrients. Normally, low pH in red soils was mainly due to iron hydroxide species, which contributed to higher hydrogen ion concentration (Dasog and Patil, 2011). High soil acidity directly affects root structure and decreases the availability of most nutrients (Havlin et al., 2010). Land usage and management also have an impact on soil pH. Over time, the pH of the soil can be lowered by adding fertilizers with nitrogen and sulfur. Introducing cover crops, minimal tillage, and diversified crop rotations can boost soil buffering capacity to limit changes in soil pH (Malla et al., 2020). The electrical conductivity of soil is very low and ranged from 0.02 to 0.6 dS m^{-1}.

7.4.2 Organic carbon

The OC of soils in the study area is varied from 0.04%–1.30% with a mean of 0.57% and classified as low (<0.5%), medium (0.5%–0.75%), and high (>0.75%). About 406 ha (19.1%) of area was classified under medium class and 476 ha area was classified as high class and is distributed in the central part of the village. While, the remaining 248 ha area marked as low in OC, which accounts 11.7% of study area (Fig. 7.3). A comparatively medium level of organic matter may be due to the regular incorporation of residues in combination application of manure and compost.

7.4.3 Available nitrogen

It has a greater impact on crop productivity, quality of crop, and in the targeted yield. The soil analysis revealed that about 45.6% of the study area was covered with a low quantity of nitrogen, and 7.7% of the area with medium nitrogen (Fig. 7.4). These variations may be due to improper soil fertility management. More than 70% of farmers used chemical fertilizers and only 7% of farmers used organic manure/farm yard manure (Leena et al., 2021). Perhaps, this would have resulted in the deficiency of the nitrogen content.

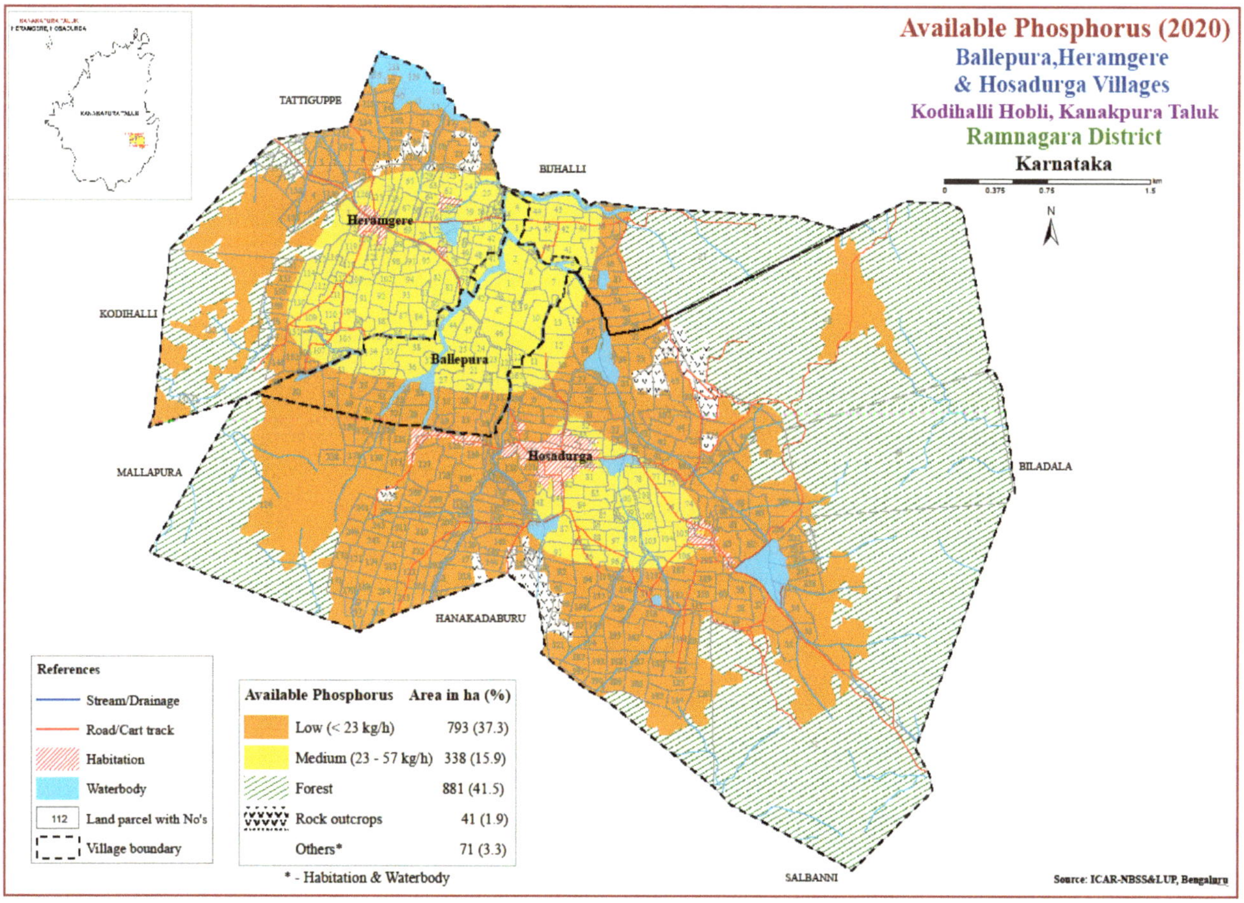

FIGURE 7.5 Available phosphorus map.

7.4.4 Available phosphorus

The soil analysis revealed that available phosphorus ranges from 1 to 52 kg ha^{-1} with mean of 19.24 kg ha^{-1}. It is recorded low in 793 ha (37.3%) area (Fig. 7.5) and is distributed in all parts of the village. There is an urgent need to increase the dose of phosphorus for all the crops by 25% over the recommended dose to realize better crop performance. About 338 ha (15.9%) area in the study area is medium and is distributed in the northwestern part of the village. It has been revealed from Fig. 7.5 that 37.3% samples falls under deficient category, and 15.09% samples in medium category of phosphorus content. It might be due to organic matter content and various soil management and agronomic practices (Bal et al., 2018).

7.4.5 Available potassium

Available potassium content of the area studied varied from 14.6 to 264.0 kg ha^{-1} with a mean of 140.75 kg ha^{-1}. It is documented low in 31.4% (668 ha), which was distributed in all parts of the area. Whereas, medium available potassium content was noticed in a small area of 463 ha (21.8%) and is distributed in the center part of the village (Fig. 7.6). Lowest available potassium was noticed in these soils; due to these, soils are poor in native soil fertility with abundant sesquioxide and low bases. Deficiency of potassium due to dominant clay mineral kaolinite, which is low in K fixation, was reported by Vasundhara (2018).

7.4.6 Available sulfur

The low availability of sulfur content was noticed in 407 ha (19.1%) of total area studied and is distributed in southeastern and western part. An area of about 417 ha (19.6%) is registered as medium and is distributed in central parts of the

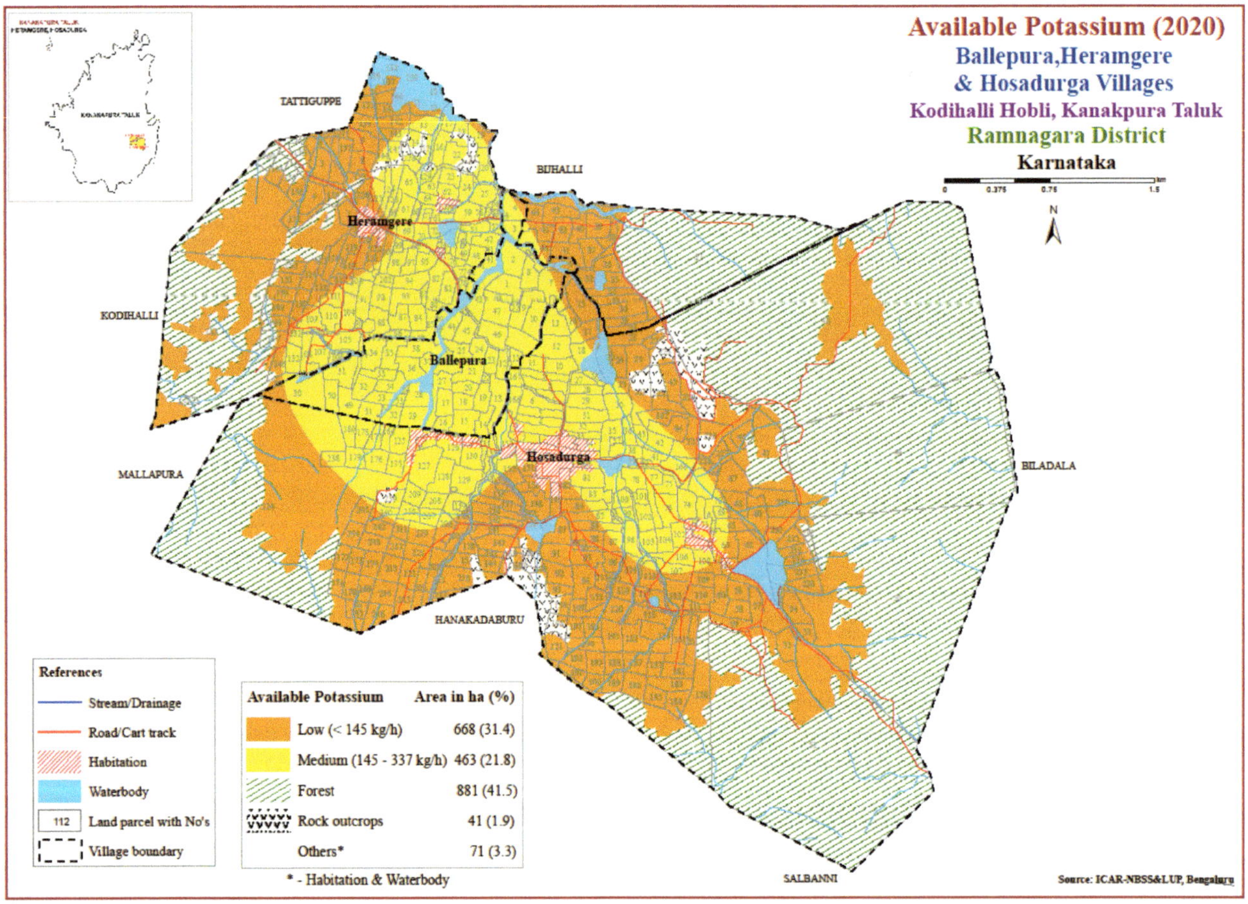

FIGURE 7.6 Available potassium map.

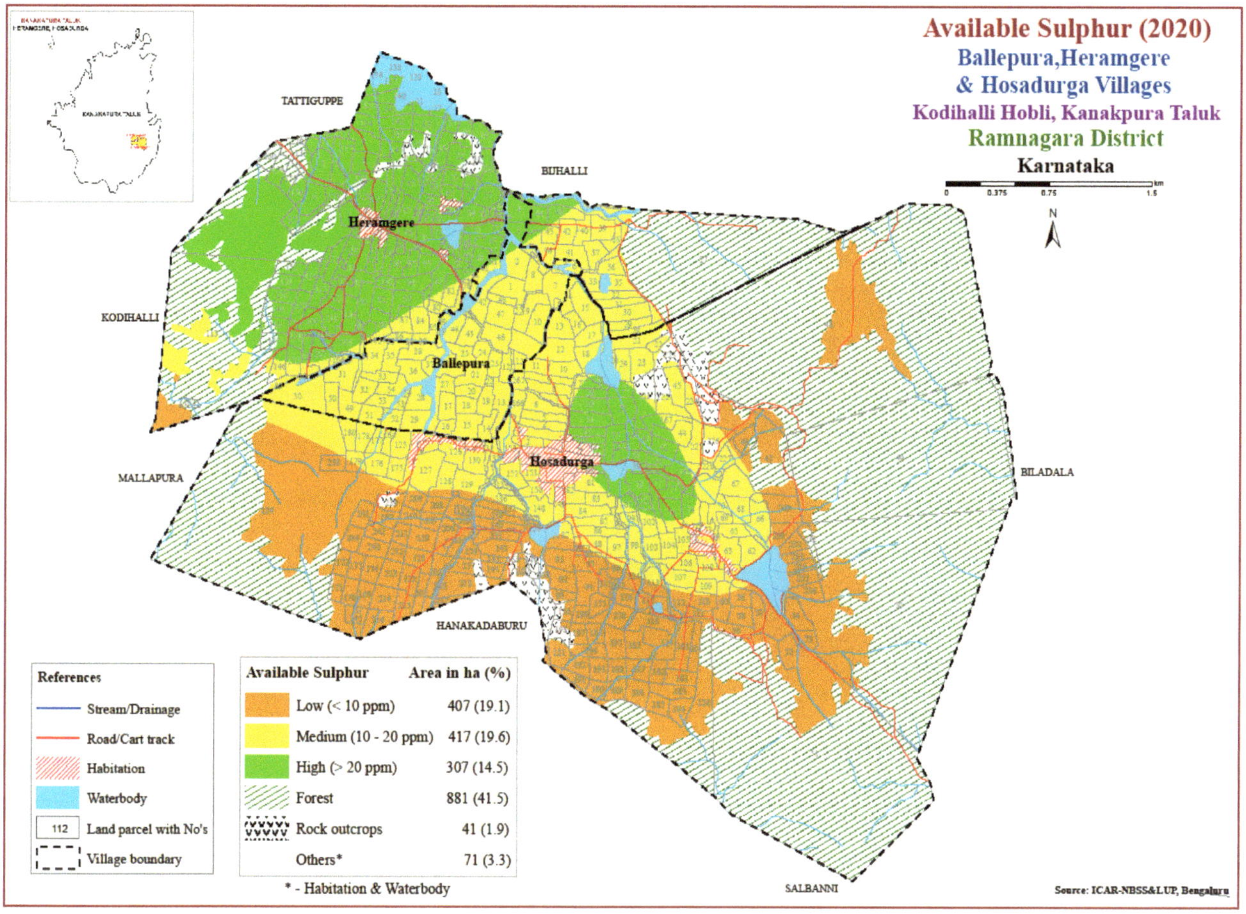

FIGURE 7.7 Available sulfur map.

village (Fig. 7.7). However, high sulfur is documented in smaller area of 307 ha (14.5%) and is distributed in the north-western part of the village. It was found that around 19.1% of the samples fall under the category of medium fertility status and 19.6% of samples were deficient in sulfur content. Deficiency of sulfur might be due to the presence of variable properties of different components of organic matter, soil and crop management practices, water use, and addition of fertilizer (Rahul et al., 2017; Singh and Mishra, 2012).

7.4.7 Available micronutrients

The status of DTPA-extractable Fe, Mn, Zn, Cu, and hot water soluble B is presented in Table 7.2 and Fig. 7.8 The available Fe, Mn, Cu, Zn, and B content varied from 5.9 to 48.8, 8.0 to 49.8, 0.3 to 2.4, 0.3 to 5.2, and 0.1 to 0.8 mg kg^{-1}, respectively. Available Fe and Mn showed moderate variance among the samples (Malla et al., 2020) except for Cu, Zn, and B; they exhibit lower variance of $<10\%$. Available iron, manganese, copper, and zinc content is sufficient in the entire area except few patches which contributes 3.6, 1.0, 1.7, and 6.0%, respectively. However, available boron content was low in the entire study area, which accounts for about 53.2%. The DTPA-extractable Fe, Mn, Zn, Cu were found to be sufficient in all most all the soils of the study area as these nutrients are well above their critical limits. The higher concentrations of available micronutrients (Fe, Mn, Cu, and Zn) might be due to higher biological activity and the chelating of organic compounds, released during the decomposition of organic matter left after harvesting of crop (Vijaya Kumar et al., 2013; Verma et al., 2005). Conversely, an intense cropping strategy without boron application could be accountable for the insufficient available boron content of the soil throughout the entire study area. The availability of boron was rated as very poor. Khadka et al. (2016) and Vasundhara et al. (2020) reported the extremely poor status of the available boron in the semiarid region of Karnataka.

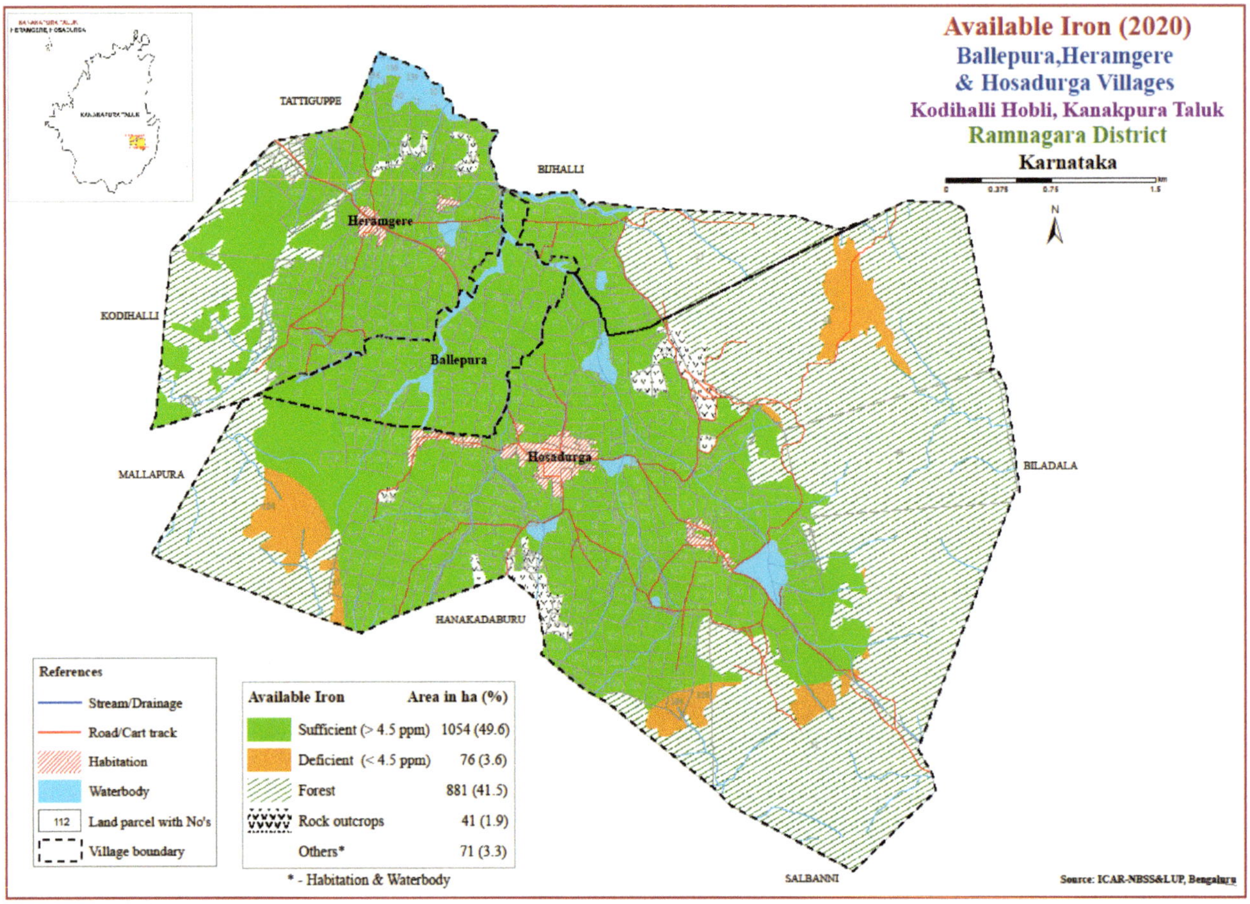

FIGURE 7.8 Available iron map.

7.5 Conclusion

The soils of study area varied from slightly acidic to neutral in reaction. The majority of soil samples have low to medium levels of OC. The availability of N, P_2O_5, K_2O, and S in these soils is low to medium. Across the parameters, the spatial variability was high dominance. Phosphorus and potassium were the deficient nutrients, which can be attributed to parent material. Available boron and sulfur were the most deficient element, indicating the need for a 25% increase in the fertilization doses for better crop performance. There is also a need to apply organic manure to the top soils to maintaining high OC. This database could help farmers and other stakeholders to address the nutrient deficiencies in the study area.

References

Bal, K., Sahu, K.K., Tedia, K., Singh, C.B., Kumar, R., 2018. Evaluation of soil fertility status of palari block under Baloda Bazar District of Chhattisgarh. J. Pharmacog. Phytochem., SP2, 243–246.

Cambardella, C.A., Karlen, D.L., 1999. Spatial analysis of soil fertility parameters. Precis. Agric. 1 (1), 5–14.

Chesnin, L., Yien, C.H., 1951. Turbidimetric determination of available sulphur. Proc. Soil. Sci. Soc. Am. 15, 149–151.

Dasog, G.S., Patil, P.L., 2011. Genesis and classification of black, red and lateritic soils of north Karnataka. In: Special Publication on Soil Science Research in North Karnataka, 76th Annual Convention of ISSS, Dharwad Chapter of ISSS, 01–10.

Gupta, U.C., 1967. A simplified method for determining hot-water soluble boron in podzol soils. Soil. Sci. 103, 424–428.

Havlin, H.L., Beaton, J.D., Tisdale, S.L., Nelson, W.L., 2010. Soil Fertility and Fertilizers an Introduction to Nutrient Management, seventh ed. PHI Learning Private Limited, New Delhi.

Hegde, R., Dharumarajan, S., Srinivasan, R., Niranjana, K.V., Savitha, H.R., Lalitha, M., et al., 2021. Spatial distribution of soil nutrients in Berambadi watershed of Chamarajanagar District, Karnataka sustainable crop production. Agropedology 31 (01), 43–51. doi.org/10.47114/j.agroped.2021.jun4.

Jin, J., Wang, L., Müller, K., Wu, J., Wang, H., Zhao, K., et al., 2021. A 10-year monitoring of soil properties dynamics and soil fertility evaluation in Chinese hickory plantation regions of southeastern China. Sci. Rep. 11, 23531. Available from: https://doi.org/10.1038/s41598-021-02947-z.

Kavitha, C., Sujatha, M.P., 2015. Evaluation of soil fertility status in various agro ecosystems of Thrissur District, Kerala, *India*. Int. J. Agric. Crop Sci 8, 328–338.

Khadka, D., Lamichhane, S., Malla, R., Joshi, S., 2016. Assessment of soil fertility status of Oilseed Research Program, Nawalpur, Sarlahi, Nepal. Int. J. Adv. Res. 4 (6), 1472–1483.

Leena, H.U., Premasudha, B.G., Panneerselvam, S., Basavaraja, P.K., 2021. Pedometric mapping for soil fertility management—a case study. J. Saudi Soc. Agric. Sci. 20, 128–135.

Lindsay, W.L., Norvell, W.A., 1978. Development of DTPA of Soil Test for Zn, Fe, Mn and Cu. J. Am. Soil. Sci. 42, 421–428.

Liu, Z.P., Shao, M.A., Wang, Y.Q., 2013. Spatial patterns of soil total nitrogen and soil total phosphorus across the entire Loess Plateau region of China. Geoderma 197–198, 67–78.

Malla, R., Shrestha, S., Khadka, D., Bam, C.R., 2020. Soil fertility mapping and assessment of the spatial distribution of Sarlahi District, Nepal. Am. J. Agric. Sci. 7 (1), 8–16.

Mishra, Antaryami, Das, D., Saren, S., 2013. Preparation of GPS and GIS based soil fertility maps for Khurda district, Odisha. Indian. Agric. 57 (1), 11–20.

Olsen, S.R., Cole, C.V., Watanabe, F.S., Dean, L.A., 1954. Estimation of available phosphorus in soils by extraction with sodium bicarbonate, Circular, 939. US Department of Agriculture, Washington, DC, p. 19.

Rahul, S.K., Amit, P.M., Aravinda, K.B.N., Hebsur, N.S., 2017. Sulphur nutrition in maize—a critical review. Int. J. Pure Appl. Biosci. 5 (6), 1582–1596.

Shukla, A.K., Behera, S.K., Lenka, N.K., Tiwari, P.K., Prakash, C., Malik, R.S., et al., 2016. Spatial variability of soil micronutrients in the intensively cultivated Trans-Gangetic Plains of India. Soil. Tillage Res. 163, 282–289.

Singh, R.P., Mishra, S.K., 2012. Available macro nutrients (N, P, K and S) in the soils of Chiraigaon block of district Varanasi (UP) in relation to soil characteristics. Indian. J. Sci. Res. 3, 97–100.

Srinivasa Rao, C., Rattan, Lal, Jasti, V.N.S.Prasad, Kodigal, A.Gopinath, Rajbir, Singh, Vijay, S.Jakkula, et al., 2015. Chapter Four—Potential and challenges of rainfed farming in India. In: Donald, L.Sparks (Ed.), Advances in Agronomy, 133. Academic Press, pp. 113–181.

Subbiah, B.V., Asija, G.L., 1956. A rapid procedure for determination of available nitrogen in soil. Curr. Sci. 25, 259–260.

Vasundhara, R., Hedge, Rajendra, Dharumarajan, S., Niranjana, K.V., 2020. Spatial assessment of soil fertility status in Shiratti sub watershed of semi-arid tropics in Southern India. Int. J. Bio-resour. Stress. Manag. 11 (2), 163–170. Available from: https://doi.org/10.23910/IJBSM/2020.11.2077.

Vasundhara, R., 2018, Assessment of soil carbon stock under arecanut and coconut plantations in Karnataka. Ph.D Thesis, University of Agricultural Sciences, GKVK.

Verma, V.K., Setia, R.K., Sharma, P.K., Charanjit, Singh, Ashok, Kumar, 2005. Micronutrient distribution in soils developed on different physiographic units of Fatehgarh Sahib district of Punjab. Agropedology 15, 70–75.

Vijaya Kumar, M., Lakshmi, G.V., Madhuvani, P., 2013. Appraisal of soil fertility in salt-affected soils of Ongole division, Prakasam district, Andhra Pradesh. J. Indian. Soc. Soil. Sci. 61, 333–340.

Walkley, A., Black, I.A., 1934. An examination of the Degtjareff method for determining soil organic matter and a proposed modification of the chromic acid titration method. Soil. Sci. 37 (1), 29–38. Available from: https://doi.org/10.1097/00010694-193401000-00003.

Wang, J., Fu, B., Qiu, Y., Chen, L., 2001. Soil nutrients in relation to land use and landscape position in the semi-arid small catchment on the loess plateau in China. J. Arid. Env. 48, 537–550.

Chapter 8

Delineation of soil management units using remote sensing and geographic information system: a case study of arid region, India

S. Dharumarajan, M. Lalitha, A. Natarajan, K.V. Niranjana, S. Srinivas and L.G.K. Naidu
ICAR-National Bureau of Soil Survey and Land Use Planning, Regional Centre, Bangalore, Karnataka, India

8.1 Introduction

The arid regions are characterized by erratic rainfall and a high rate of land degradation. Continuous overexploitation of natural resources like land, water, and forest has caused a serious threat in arid ecosystems. Issues such as limited soil moisture storage capacity, rapid soil erosion, declining groundwater levels, and a scarcity of drinking water are common in arid regions and affect millions of local population. It is critical for planners and decision-makers to develop long-term plans for land resource development and management in order to improve the quality of habitat in arid regions (Van Loon, 2015). Even though technological breakthroughs and dissemination in agriculture increased agricultural production manifold, the lack of appropriate site-specific technologies and their adoption, mismanagement, and indiscriminate use of technologies without considering the natural resources increased the severity and extent of the land degradation (Dharumarajan et al., 2018). The negative facets of the increase in food production are land degradation, declining agricultural productivity, groundwater depletion, water logging and salinization, and biodiversity losses. These factors along with the erratic climate reduced economic production, and in turn interest in farming among the farmers (Abahussain et al., 2002). Integrated natural resource management through soil, land, and water conservation practices, with special emphasis on land resource management, will ensure at least a minimum level of livelihood security even in the face of unpredictable rainfall, particularly in arid regions (Shiferaw et al., 2009). In countries such as India, efforts are being made to provide reliable and up-to-date geo-information for sustainable land resource management. Such information could only be obtained using modern research and mapping techniques and equipment (Singh et al., 2011; Woldai, 2020). It is difficult to obtain real-time information through conventional resources and these methods are time-consuming, laborious, and high-cost and workforce oriented. In modern times, satellite based remote sensing (RS) technology has been developed, which are of immense value for preparing soil management units (Rindfuss and Stern, 1998; Rogan and Chen, 2004; Voigt et al., 2007).

In the last three decades, the technologies and methods of RS have progressed significantly; this, along with an increased resolution from satellite platforms, makes these technologies appear poised to make a better impact on land resource management initiatives involved in monitoring and change detection at varying spatial ranges (Singh et al., 2010; Thakur, 2010). RS technology offers collection and analysis of data from ground-based, space, and earth-orbiting platforms, with linkages to global positioning system and geographic information system (GIS) data with promising modeling capabilities (Franklin, 2001; Thakur, 2010). As the demand for quantity and quality of information and technology continues to improve, RS is becoming more significant for the future.

The RS technique provides systematic, synoptic, rapid, and repetitive coverage of the earth's features in different windows of the electromagnetic spectrum. As a result, it offers as a unique and powerful tool for obtaining spatiotemporal information on large areas in a short time. Remotely sensed data are usually cost-effective compared to the

Remote Sensing of Soils. DOI: https://doi.org/10.1016/B978-0-443-18773-5.00010-7

conventional methods of hydrological surveys and particularly are of great significance for remote as well as data-scarce regions (Allafta et al., 2020; Casagli et al., 2023). Satellite data provide quick and useful baseline information about various factors such as geomorphology, soil types, land slope, land use/land cover, drainage patterns, lineaments, etc. (Kumar and Kumar, 2010). In addition, GIS provides an excellent framework for efficiently handling large and complex spatial data for natural resources management. RS and GIS are promising tools for assessing land degradation (Kumar et al., 2015). These tools can generate relevant maps and reliable spatial information to support decision-making. The RS and GIS datasets provide accurate information on large areas.

With this background, this resource survey was undertaken with the aim to provide site-specific database for the Muttala village, Atmakur Mandal of Anantapur district. The database was generated by using the cadastral map of the village as a base along with high resolution satellite imagery (Cartosat-1 merged IRS LISS IV). The objectives of the land resource survey are: (1) delineation of homogenous areas based on soil-site characteristics, (2) identification of problems and potentials of the area and strategies for their management, and (3) assessment of the suitability of land resources for various crops and other uses.

8.2 Material and methods

8.2.1 Study area

Muttala village is located in Atmakur mandal of Anantapur district, Andhra Pradesh. The village, with a total geographical area of 1788 hectares, lies between 77°24′7.7″ and 77°20′30.5″ East longitudes and 14°35′41.4″and 14°34′42.6″ North latitudes (Fig. 8.1). Granites-gneiss, chlorite-schist, and dykes are the major rock types observed in the village (Fig. 8.2). Dykes, popularly known as black granite, occur in many areas in the village as intrusive in the country rock.

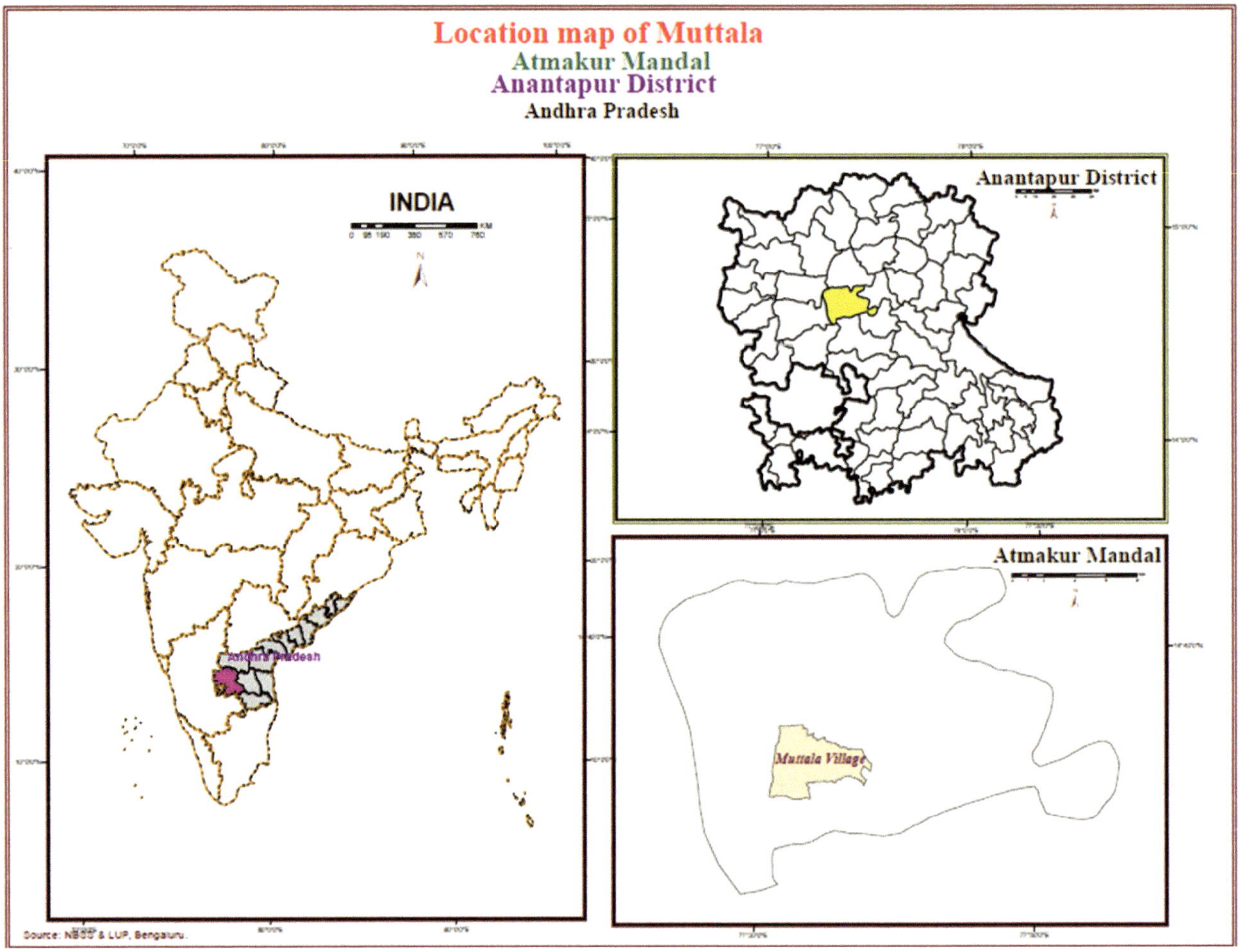

FIGURE 8.1 Location of Muttala village in Atmakur mandal of Anantapur district.

FIGURE 8.2 Geological formations in the study area.

The granite and gneiss show significant variations in their texture, color, mineral assemblage, and degree of weathering. Chlorite-schist is the second dominant rock type in the village. Due to its lighter density, the rocks are easily weathered under the prevailing climate.

8.2.2 Climate

The climate of Anantapur is warm all around the year and mainly classified as hot and dry type. The mean temperature is always above 23°C. April and May are the hottest months with mean temperatures between 32°C and 35°C. The mean maximum temperature ranges from 30°C in December and 40°C in May. Mean minimum temperatures are lowest in December and January, lying near 17°C–19°C, whereas it does not exceed 25°C in winter. The mean diurnal variation of temperature as a whole is 10°C–12°C. The rainfall is less than 550 mm and it is the driest part of the whole state. Mostly, the area gets 300–500 mm of rain during southwest monsoon (June–October). This is 55–90% of annual rainfall. September is generally the peak rainy month with about 25% of annual rain. Though the average rainfall is around 550 mm in the last 15 years, there are extremes in some years, which ranges from the lowest rainfall of 272 mm in 2006 to highest of 990 mm in 2001. There is a significant change in the amount of rainfall and also potential evapotranspiration observed in this area. These changes have a drastic effect on crop production activities in the region and have increased the risks of farming because of the change in length of growing period. The study area is a chronically drought-prone area due to its low and unpredictable nature of growing period which varies from 7 weeks in 1980–90 to 11 weeks in 2001–11.

8.2.3 Base maps used

The detailed survey of the land resources of the village was carried out by using cadastral map as a base. The cadastral map shows field boundaries with their survey numbers, location of tanks, streams, and other permanent features of the area. The RS data products from IRS LISS IV and Cartosat-1 at the scale of 1:10,000 were used in conjunction with the cadastral map to identify the landforms and other surface features. Apart from the cadastral maps and imageries, toposheets of the area (1:50,000 scale) were used for initial traversing, identification of geology and landform, drainage features, and present land use, and for the selection of transects at village level.

8.2.4 Field investigations

Preliminary traverse of the entire village was carried out with the help of cadastral map, satellite imagery, and toposheets. During the traverse, based on geology, drainage pattern, surface features, slope characteristics, and land use, landforms were identified and initial legend was prepared by studying soils in few selected places. The imageries helped in the identification and delineation of boundaries between hills, uplands and lowlands, water bodies, forest and vegetated areas, roads, habitations, and other cultural features of the area. The field boundaries and survey numbers given on the cadastral sheet were located on ground by following permanent features like roads, cart tracks, nallas, streams, tanks, etc., and wherever changes were noticed they were incorporated on the cadastral map. After this, intensive traversing of each landform unit like hills, ridges, uplands, and lowlands was carried out by involving all the soil survey party members. Based on the variability observed on the surface, transects were selected across the slope, covering all the landform units (Natarajan et al., 2016). In the selected transect, profiles were selected at closely spaced intervals to take care of any change in the land features like break in slope, erosion, gravel, stones, etc. (Fig. 8.3). In the selected sites, profiles (vertical cut showing the soil layers from the surface to parent material/hard rock) were opened up to 200 cm or to the depth limited by rock or hard substratum and studied in detail for all their morphological and physical characteristics. The soil and site characteristics were recorded for all profile sites according to a standard proforma as per the standard guidelines (Soil Survey Staff, 2008).

Apart from the transect, study profiles were studied at random, almost like in a grid pattern between transects. A total of 78 profiles were studied in the village area. Based on the soil-site characteristics, the soils were grouped into different soil series (soil series is the most homogeneous unit having similar horizons and properties and behaves similarly for a given level of management). Soil depth, texture, color, amount and nature of gravel present, nature of

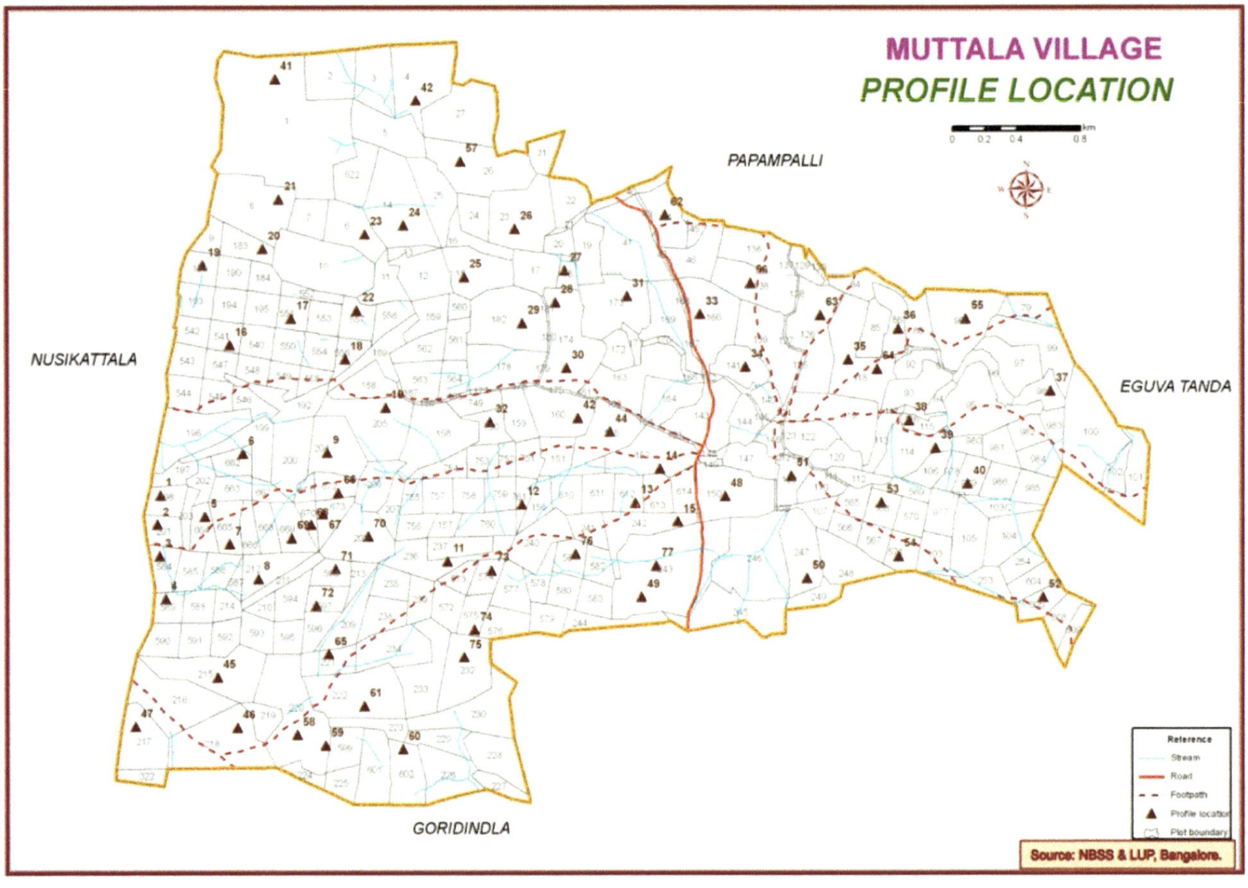

FIGURE 8.3 Profile locations over cadastral map.

substratum, and horizon sequence were the major identifying characteristics of soil series occurring in the area. Based on the above characteristics, seven soil series were identified in the area. Soil samples were collected from representative master profiles for laboratory characterization (Jackson, 1973).

8.2.5 Finalization of soil map and other thematic maps

The area under each series was further divided into phases and their boundaries were delineated on the cadastral map based on the variations observed in the texture of the surface soil, slope, erosion, presence of gravels, stoniness, etc. A soil phase is a subdivision of a soil series based mostly on surface features that affect its use and management. The delineated mapping units are shown on the map in the form of symbols. During the survey, about five transacts representing different landforms occurring in the village were identified and about 78 profiles were opened for detailed soil and site characterization. In addition, spot observations, in the form of mini pits and road cuts, terrace cuts etc., were also studied to prepare the final soil map. Out of the 41 phases mapped, 32 phases occur in the hills/upland areas and five phases occur in the lowland areas of the village. The soil-site characteristics were also interpreted for identifying the constraints and potentials like soil depth, erosion, gravelliness, salinity/sodicity, and land suitability maps for various crops generated. Fertility maps were generated using an interpolation technique.

8.3 Results and discussion

8.3.1 Soils

The major identifying characteristics are soil depth, soil color, subsurface texture, gravelliness, calcareousness, and horizon sequence (Table 8.1). A total of 7 soil series were identified in the village and mapped with 41 phases (Fig. 8.4). Out of this, five soil series occur in the hills and upland areas and two soil series occur in the lowland area. Out of the 41 phases mapped, 32 phases occur in the hills/uplands areas and 5 phases occur in the lowland areas of the village.

In the granitic uplands, out of the five soil series identified, shallow red loamy soils (M2) and moderately shallow red loamy soils (M3) occur extensively, occupying about 63.7% (1138. 9 ha) area of the upland. The other three series occupy an area of about 218.0 ha (12.2%) in the village. The two lowland series occupy about 106.9 ha (5.97%) area in the village. Rock lands and rock outcrops occur in about 150.2 ha (8.4% of the area). Similarly, gullied and severely eroded lands occur all along the footslopes which are classified as miscellaneous lands occur in about 50 ha (2.57%). As per soil Taxonomy (Soil survey staff, 2008) the soils of the village were classified into 2 orders viz. Aridisol (94.58%) and Entisol (5.42%) (Table 8.2). Profile development is very well expressed in the upland compared to the lowland soils of the area. The profile picture of major soils are presented in Fig. 8.5. Physiochemical characteristics of identified soils are presented in Table 8.3.

8.3.1.1 Series M1 (Muttala-1)

These soils are formed from granite-gneiss and occur on very gently sloping ridges and uplands with 1%–3% slopes with an elevation of 500 m above MSL. The soils are very shallow, well-drained, noncalcareous with gravelly sandy

TABLE 8.1 Identifying characteristics of soils of Muttala Village.

Series	Depth (cm)	Color	Texture	Gravel (%)	Calcareousness	Horizon sequence
M1	10–25	7.5YR4/4, 5YR3/4,5YR4/6	ls-sl	35–50	–	Ap-Cr
M2	25–50	2.5YR3/4, 2.5YR3/6	scl-sc	35–70	–	Ap-Bt-Cr
M3	50–75	2.5YR3/4, 2.5YR3/6	scl-sc	35–70	–	Ap-Bt-Cr
M4	52–50	10YR3/4, 10YR4/2,7.5YR3/4	sl-scl	35–70	–	Ap-Bw-Cr
M5	100–150	10YR3/1	c	–	Ev	Ap-Bw-Cr
M6	100–150	10YR4/3, 10YR3/4	sc	–	Ev	Ap-Bw-Cr
M7	100–150	10YR3/4, 7.5YR3/4,7.5YR4/4	sl/ls/s	–	–	Ap-AC

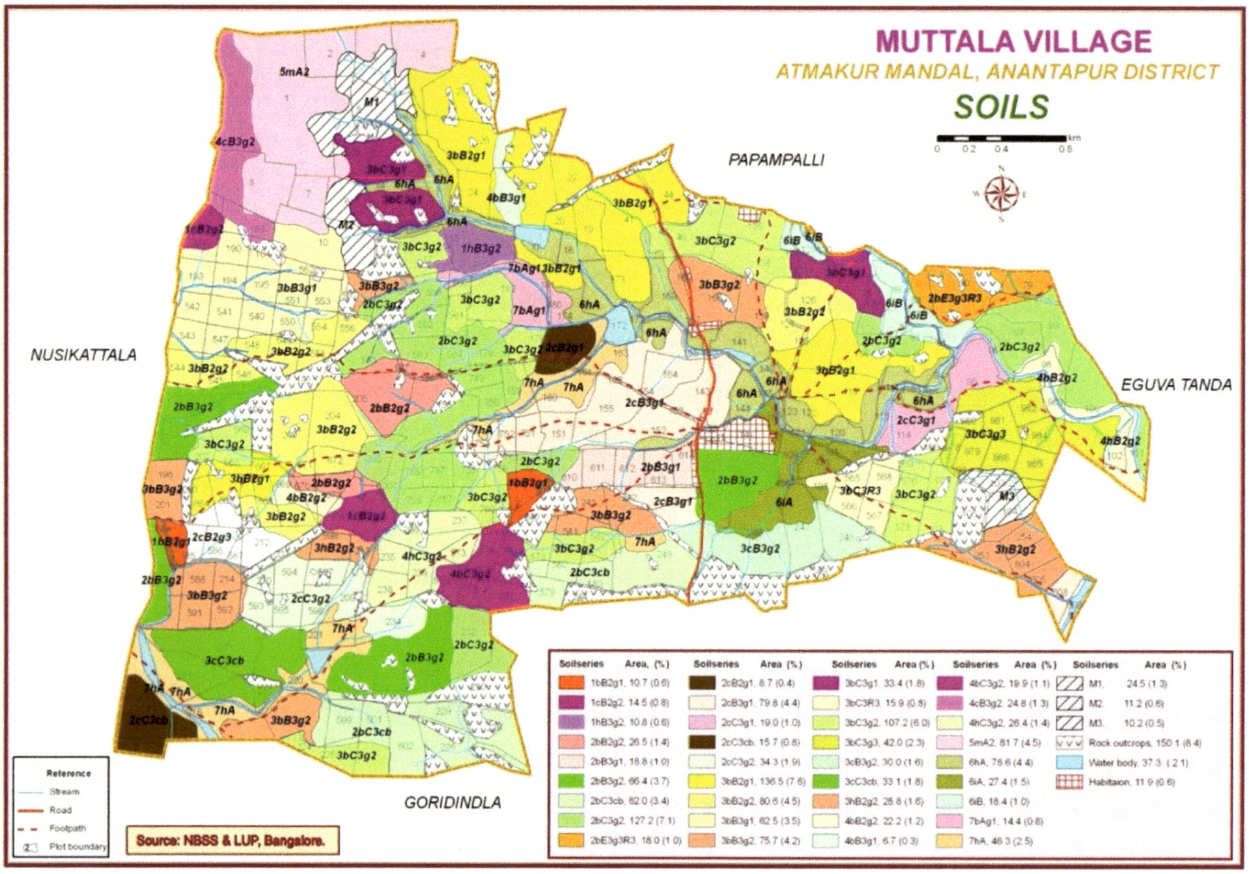

FIGURE 8.4 Distribution of soil phases or management units in Muttala village.

TABLE 8.2 Classifications of the soils.

Soil series	Family or higher taxonomic class*	Area in ha	%
Hills and uplands			
1. M1 Series	Sandy-skeletal, Lithic Ustic Torriorthents	36.12	2.02
2. M2 Series	Loamy-skeletal, Lithic Ustic Haplargids	477	27.57
3. M3 Series	Loamy-skeletal, Ustic Haplargids.	661.85	37.03
4. M4 Series	Loamy-skeletal, cal, Lithic Haplocambids	100.17	5.6
5. M5 Series	Fine, mixed, isohyperthermic, cal, Ustertic Haplocambids	81.74	4.57
Lowlands/valleys			
1. M6 Series	Fine loamy-skeletal, cal, Ustic Haplocambids	124.57	6.97
2. M7 Series	Sandy, Ustic Torripsamments	60.82	3.40

**Minorology: mixed; Temperature regime: isohyperthermic.*

loam texture. The gravel content varies from 10% to 50%. The depth of the soil ranges from 10 to 25 cm (very shallow). The color is yellowish red to dark reddish brown. The surface soils are light textured (loamy sand to sandy loam). The B horizon, if it occurs, is dark reddish brown sandy clay loam texture with weak, medium subangular blocky structure. Soil erosion is moderate to severe, which is the major constraint of the soil, since this series occurs mostly in the rocky lands/boulder areas of the village. The indiscriminate destruction of forest cover and cutting of trees anchored

FIGURE 8.5 Major soils of study area.

TABLE 8.3 Soil physiochemical properties.

Horizon	Depth (cm)	Coarse fragments (%)	Textural class	pH (1:2.5)	EC (dS m^{-1})	OC (%)
M1 series						
Ap	0–15	50	sl	5.66	0.02	0.08
M2 series						
Ap	0–17	50	sl	5.15	0.001	0.48
Bt1	17–39	40	scl	5.61	0.002	1.11
M3 series						
Ap	0–13	25	sl	7.06	0.01	1.47
Bt1	13–26	50	scl	6.79	0.01	0.71
Bt2	26–60	60	scl	6.67	0.10	0.12
M4 series						
Ap	0–12	25	sl	7.85	0.14	1.55

(Continued)

TABLE 8.3 (Continued)

Horizon	Depth (cm)	Coarse fragments (%)	Textural class	pH (1:2.5)	EC (dS m^{-1})	OC (%)
Bw1	12–25	40	scl	7.93	0.15	1.83
Bw2	25–42	40	l	8.10	0.13	0.16
M5 series						
Ap	0–14	–	cl	7.89	0.17	0.16
Bw1	14–40	–	c	8.18	0.16	0.71
Bw2	40–69	–	c	8.50	0.23	0.44
Bw3	69–94	–	c	8.58	0.28	0.56
Bw4	94–109	–	c	8.57	0.26	0.39
M6 series						
Ap	0–18	–	scl	7.68	0.20	0.03
Bw1	18–44	–	scl	8.03	0.19	0.15
Bw2	44–69	–	scl	8.20	0.22	0.95
Bw3	69–110	–	scl	8.41	0.24	0.08
M7 series						
Ap	0–10	10	sl	7.74	0.09	0.91
A2	10–26	–	scl	8.01	0.05	0.39
C1	26–53	10	s	8.16	0.05	0.15
C2	53–78	10	s	8.20	0.05	0.08
C3	78–127	–	sl	8.06	0.04	0.44

between the rocks accelerates the detachment and removal of the soil cover from these areas. This issue needs to be addressed with proper conservation measures to protect this soil from erosion. Based on location and site characteristics, the area occupied by M1 series are divided into three phases, which are briefly described below.

1bB2g1	Loamy sand surface, slope 1%–3%, moderate erosion, and 15%–35% gravels
1cB2g2	Sandy loam surface, slope 1%–3%, moderate erosion, and 35%–60% gravels
1hB3g2	Sandy clay loam surface, slope 1%–3%, severe erosion, and 35%–60% gravels

8.3.1.2 Series M2 (Muttala-2)

These soils are formed from granite-gneiss and occur on very gently sloping to gently sloping uplands with 1%–5% slopes with an elevation of 520 m above MSL. The depth of the soil ranges from 30 to 60 cm. The thickness of solum ranges from 30 to 60 cm. The thickness of A horizon varies from 9 to 17 cm. Its color is in 5 YR, 7.5 YR, and 10 YR hues with value 3–4 and chroma 3–6. Texture varies from loamy sand to sandy loam with 30%–50% gravel. The thickness of B horizon ranges from 30 to 60 cm. Color 2.5 YR and 5YR hues occur with value 3–4 and chroma 3–6. Texture varies from sandy clay loam to sandy clay with 30%–60% gravel. Based on the soil-site characteristics that influence the management needs, 12 phases were delineated in the village and brief description is given below.

2bB2g2	Loamy sand surface, slope 1%–3%, moderate erosion, and 35%–60% gravels
2bB3g1	Loamy sand surface, slope 1%–3%, severe erosion, and 15%–35% gravels
2bB3g2	Loamy sand surface, slope 1%–3%, severe erosion, and 35%–60% gravels
2bC3g2	Loamy sand surface, slope 3%–5%, severe erosion, and 35%–60% gravels
2bC3cb	Loamy sand surface, slope 3%–5%, severe erosion, and 60%–70% cobbles
2bE3R3	Loamy sand surface, slope 10%–15%, severe erosion, and 25%–30% rockiness
2cB2g1	Sandy loam surface, slope 1%–3%, moderate erosion, and 15%–35% gravels
2cB2g3	Sandy loam surface, slope 1%–3%, moderate erosion, and 60%–70% gravels
2cB3g1	Sandy loam surface, slope 1%–3%, severe erosion, and 15%–35% gravels

2cC3g1	Sandy loam surface, slope 3%–5%, severe erosion, and 15%–35% gravels
2cC3cb	Sandy loam surface, slope 3%–5%, severe erosion, and 70%–80% cobbles
2cC3g2	Sandy loam surface, slope 3%–5%, severe erosion, and 35%–60% surface gravels

8.3.1.3 Series M3 (Muttala-3)

The soils of M3 series have loamy sand to sandy loam surface layer followed by dark reddish brown to dark red gravelly sandy clay loam to gravelly sandy clay subsoil. These soils are formed from granite-gneiss and occur on very gently sloping to gently sloping uplands with 1%–5% slopes with an elevation of 510 m above MSL. The thickness of solum is from 51 to 79 cm. The thickness of A horizon varies from 10 to 15 cm. Its color is 5 YR and 7.5 YR with value 3–4 and chroma 4–6. Texture is mostly loamy sand with 10%–30% gravel. Sandy loam and sandy clay loam occurs in limited area. The thickness of Bt1 horizon ranges from 13 to 27 cm with dark reddish brown to reddish brown color gradient. Texture varies from sandy clay loam to sandy clay with 30%–60% gravel. The thickness of Bt2 horizon ranges from 24 to 42 cm with dark reddish brown to dark red color gradient. Texture varies from sandy clay loam to sandy clay with 30%–50% gravel. The Bt3 horizon, if it occurs, have reddish brown to dark red sandy clay loam to sandy clay texture having moderate, medium subangular block structure with 20% to 50% fine gravels. The eleven phases identified under this series are briefly described below

3bB2g1	Loamy sand surface, 1%–3% slope, moderate erosion, and 15%–35% gravels
3bB2g2	Loamy sand surface, 1%–3% slope, moderate erosion, and 35%–60% gravels
3bB3g1	Loamy sand surface, 1%–3% slope, severe erosion, and 15%–35% gravels
3bB3g2	Loamy sand surface, 1%–3% slope, severe erosion, and 35%–60% gravels
3bC3g1	Loamy sand surface, 3%–5% slope, severe erosion, and 15%–35% gravels
3bC3g2	Loamy sand surface, 3%–5% slope, severe erosion, and 35%–60% gravels
3bC3g3	Loamy sand surface, 3%–5% slope, severe erosion, and 60%–70% gravels
3bC3R3	Loamy sand surface, 3%–5% slope, severe erosion, and 25%–30% rockiness
3cB3g2	Sandy loam surface, 1%–3% slope, severe erosion, and 35%–60% gravels
3cC3cb	Sandy loam surface, 3%–5% slope, severe erosion, and 50%–60% cobbles
3hB2g2	Sandy clay loam surface, 1%–3% slope, moderate erosion, and 35%–60% gravels

8.3.1.4 Series M4 (Muttala-4)

The M4 series consists of shallow, well-drained calcareous soils on summits and side slopes. These soils formed from granite-gneiss and occur on very gently sloping uplands with 1%–3% slopes with an elevation of 498 m above MSL. The thickness of solum ranges from 28 to 56 cm. The thickness of A horizon varies from 10 to 16 cm. Its color is in 10 YR and 7.5 YR hues with value 3–4 and chroma 3–4. Texture varies from loamy sand to sandy clay loam with 25%–40% gravel. The thickness of B horizon ranges from 29 to 44 cm. Its color is in 10 YR, 7.5 YR, and 5 YR hues with value 3–4 and chroma 2–4. Texture varies from sandy loam to sandy clay loam with 30%–70% gravel. Brief descriptions of 5 phases of M4 series are given below

4bB2g2	Loamy sand surface, 1%–3% slope, moderate erosion, and 35%–60% gravels
4bB3g1	Loamy sand surface, 1%–3% slope, severe erosion, and 15%–35% gravels
4bC3g2	Loamy sand surface, 3%–5% slope, severe erosion, and 35%–60% gravels
4cB3g2	Sandy loam surface, 1%–3% slope, severe erosion, and 35%–60% gravels
4hC3g2	Sandy clay loam surface, 1%–3% slope, severe erosion, and 35%–60% gravels

8.3.1.5 Series M5 (Muttala-5)

The M5 series consists of deep, well-drained clayey soils on nearly level uplands. These soils are formed from chlorite-schist and occur on nearly level uplands with 0%–1% slope with an elevation of 520 m above MSL. The thickness of solum ranges from 81 to 109 cm. The thickness of A horizon varies from 14 to 18 cm. Its color is in 10 YR with value 3 and chroma 1–2. Texture is mostly clay. The thickness of B horizon ranges from 63 to 95 cm with dark grayish brown to very dark gray color gradient. Texture is clayey.

5 m A2	Clay surface, 0%–1% slope, moderate erosion

8.3.1.6 Series M6 (Muttala-6)

The M6 series consists of deep, moderately drained clayey soils on lowlands. These soils are formed from granite-gneiss parent material by the action of water and gravity and occur on nearly level lowlands with 0%–1% slope, with an elevation of 470 m above MSL. The thickness of solum ranges from 56 to 135 cm. The thickness of A horizon varies from 7 to 18 cm. Its color is in 10 YR with value 3–4 and chroma 1–2. Texture is sandy clay loam to sandy clay. The thickness of B horizon ranges from 49 to 122 cm. Its color is in 10 YR with value 3–4 and chroma 2–6. Texture is sandy clay loam to sandy clay. Brief description of different phases of M6 series are given below.

6hA	Sandy clay loam surface, 0%–1% slopes
6iA	Sandy clay surface, 0%–1% slopes
6iB	Sandy clay surface, 1%–3% slopes

8.3.1.7 Series M7 (Muttala-7)

The M7 series consists of deep, well-drained sandy soils on lowlands. These soils formed from alluvium on nearly level lowlands with 0%–1% slope with an elevation of 500 m above MSL. The thickness of solum ranges from 26 to 96 cm with 10 YR, 7.5 YR, and 5 YR with value 3–4 and chroma 3–6. Texture is loamy sand to sandy clay loam. The thickness of C horizon ranges from 43–99 cm. Its color is in 10 YR, 7.5 YR, and 5 YR with value 3–4 and chroma 2–6. Texture is sand to sandy clay loam. Brief description of two phases of M7 series are given below

7bAg1	Loamy sand surface, 0%–1% slopes and 15%–35% gravels
7hA	Sandy clay loam surface, 0%–1% slopes

8.3.2 Soil fertility status

The results showed that available nitrogen is low (< 280 kg ha^{-1}) in the entire village area which ranged from 31.4 to 203.8 kg ha^{-1}. The available phosphorus content in the soils is low (< 11 kg ha^{-1}) in 9.3 ha (0.5%) of the village area (Fig. 8.6). Nearly 163.8 ha (9.2%) of the cultivated areas in the village have medium available phosphorus content (11–22 kg ha^{-1}). Around 1614.6 ha area of the village were recorded under higher phosphorus content (> 22 kg ha^{-1}). Since nearly 90% of the area was under high phosphorus category, phosphorus application to these soils has to be

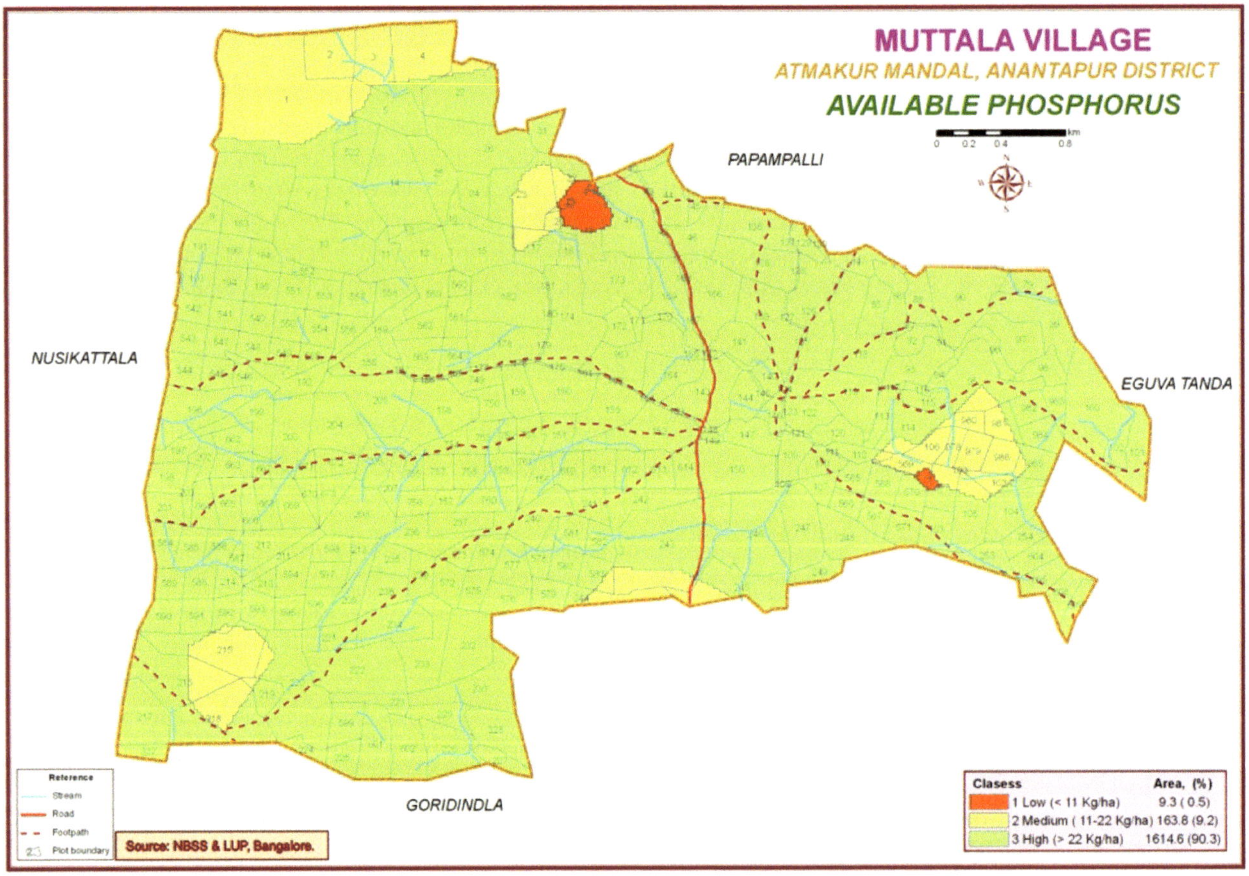

FIGURE 8.6 Available phosphorous map.

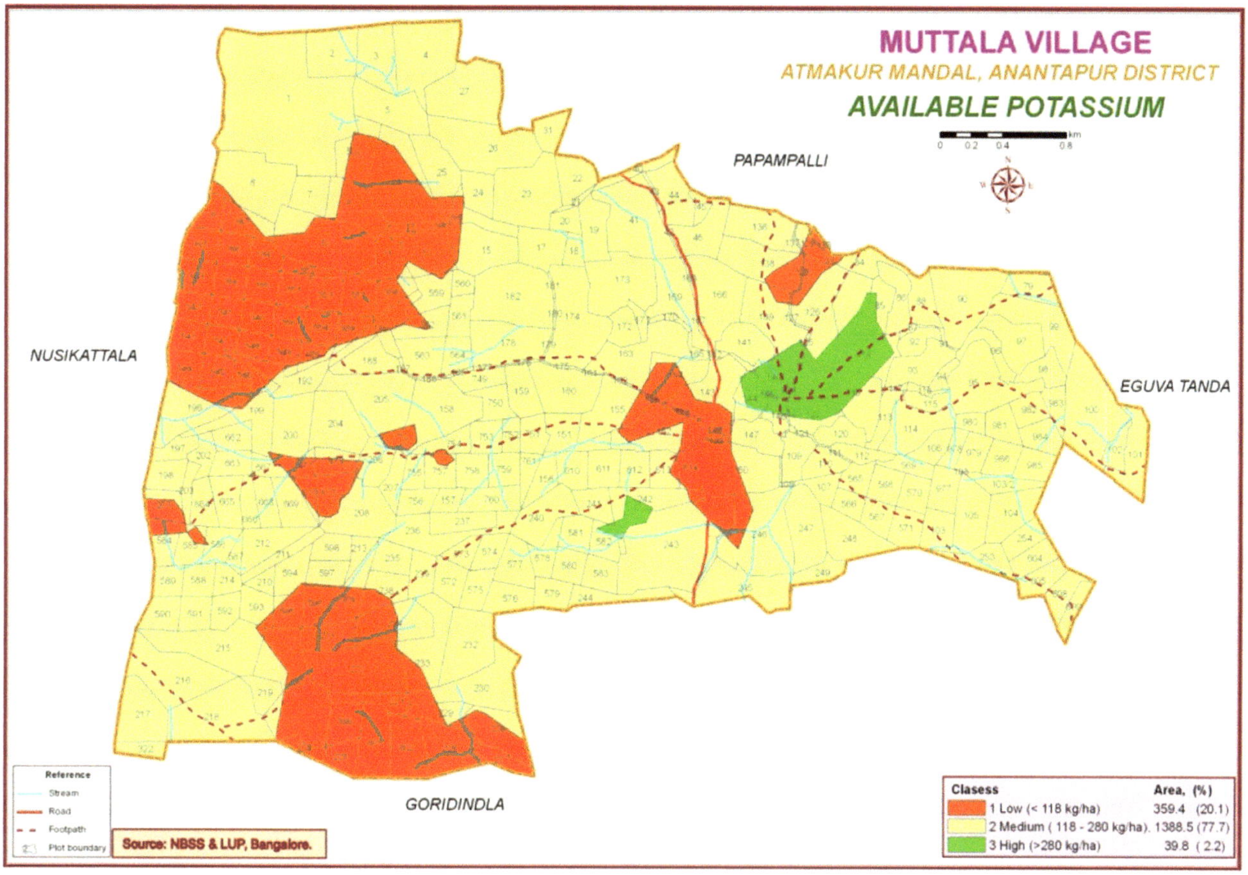

FIGURE 8.7 Available potassium map.

reduced to minimize any micro nutrient toxicity and to get better crop growth. In large parts of the area (1388.5 ha), the available potassium content in the soils is medium (118–280 kg ha^{-1}). Low potassium content is found in 354.9 ha area (20.1%) and high potassium content in 39.8 ha area (2.2%) of the village (Fig. 8.7). As oil seed crops like ground nut and castor are the dominant crops in the village, the deficiency of this nutrient needs to be taken care to realize better crop performance and pest resistance.

8.3.3 Land suitability assessment

Assessing the extent and degree of suitability of the land resources in the village for various crops is necessary to choose the right crop and variety suitable for the area. In carrying out this assessment, the specific requirements of a crop (compiled from the existing literature) are compared with the characteristics of land, and based on the extent of matching, the suitability of the area for the crop is arrived (Naidu et al., 2006). Then by comparing the relative suitability of the resources for different uses, an ideal combination of crops suitable for a particular farm within the village area can be selected.

*8.3.3.1 Land Suitability for groundnut (*Arachis hypogea*)*

It is the traditional and major oilseed crop of India as well as the world. It is predominantly a crop of tropical and subtropical climates. Groundnut is cultivated by most of the farmers in Muttala village due to lighter surface texture. Loose/friable soils facilitate good pod development. Therefore, sandy and loamy soils with fairly rich organic matter are very well suited for this crop. The suitability assessment for groundnut in the Muttala village (Fig. 8.8) showed that nearly 48% of area is moderately suitable and nearly 37% area is marginally suitable. The main constraints are gravelliness, texture, depth in uplands, and drainage in lowlands. Around 13% of the area is not suitable for groundnut cultivation due to the severe limitations of depth and erosion.

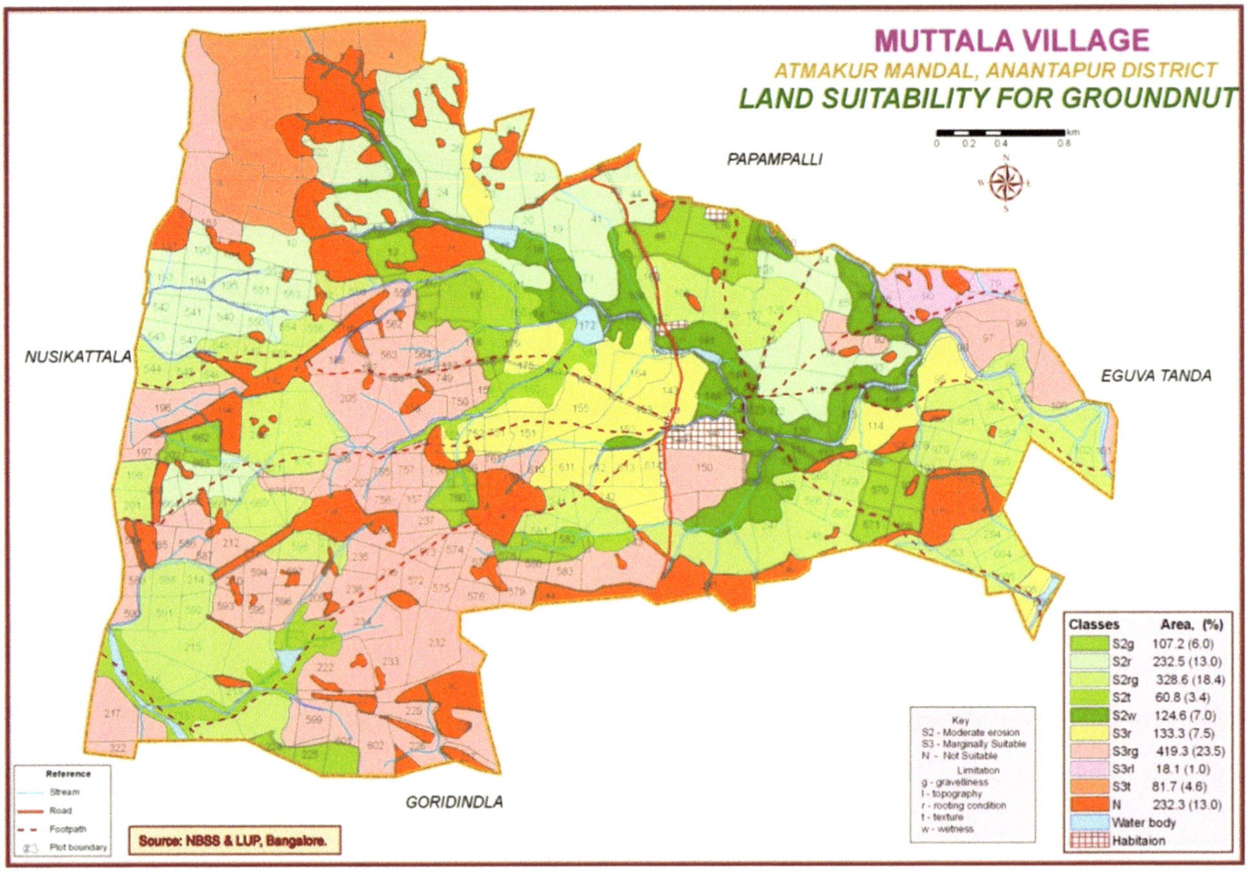

FIGURE 8.8 Land suitability map for groundnut.

8.3.3.2 Land suitability for red gram (Cajanus cajan*)*

Red gram is a highly drought resistant crop and can grow successfully in areas receiving about 65 cm annual rainfall. The crop may be grown on any type of soil but loamy soils are preferred. Very deep, well-drained, noncalcareous, and nonsaline soils are best suited for the crop. The suitability assessment for redgram in Muttala village (Fig. 8.9) showed that nearly 15% of area is moderately suitable and nearly 37% area is marginally suitable. The main constraints are texture, rooting depth, and gravelliness. Around 45% of study area is not suitable for red gram mainly due to depth.

8.3.3.3 Land suitability for mango (Mangifera indica*)*

Mango, the king of fruits, is well adapted to tropical and subtropical climate and thrives well up to 1500 m above MSL. The areas which receive bright sunny days and have low humidity during flowering are ideal for mango production (Fig. 8.10). It can be grown in all types of soils, but are sensitive to poor drainage, presence of free $CaCO_3$, high pH, extreme gravel and stoniness, sodicity, and salinity. The suitability assessment for mango in Muttala village showed that nearly 46% of area is not suitable for cultivation of mango due to rooting depth. Nearly 54% of area is marginally suitable to mango cultivation.

8.3.4 Major issues and management strategies

The continuous removal of vegetation or indiscriminate destruction of forest cover, particularly during the last few decades, has resulted in severe loss of soil and fast degradation of other natural endowments of the area (Fig. 8.11). This process was accelerated further due to the nature of the terrain, which is undulating to gently sloping in the cultivated uplands. Due to this, even during low rainfall events, the runoff rate increases significantly along the slopes of the hills and from the uplands. All this has led to severe loss of soil from the upstream areas and heavy siltation of the tanks and deposition of huge loads of sediments along the stream courses. This cascading effect of erosion is not restricted to any

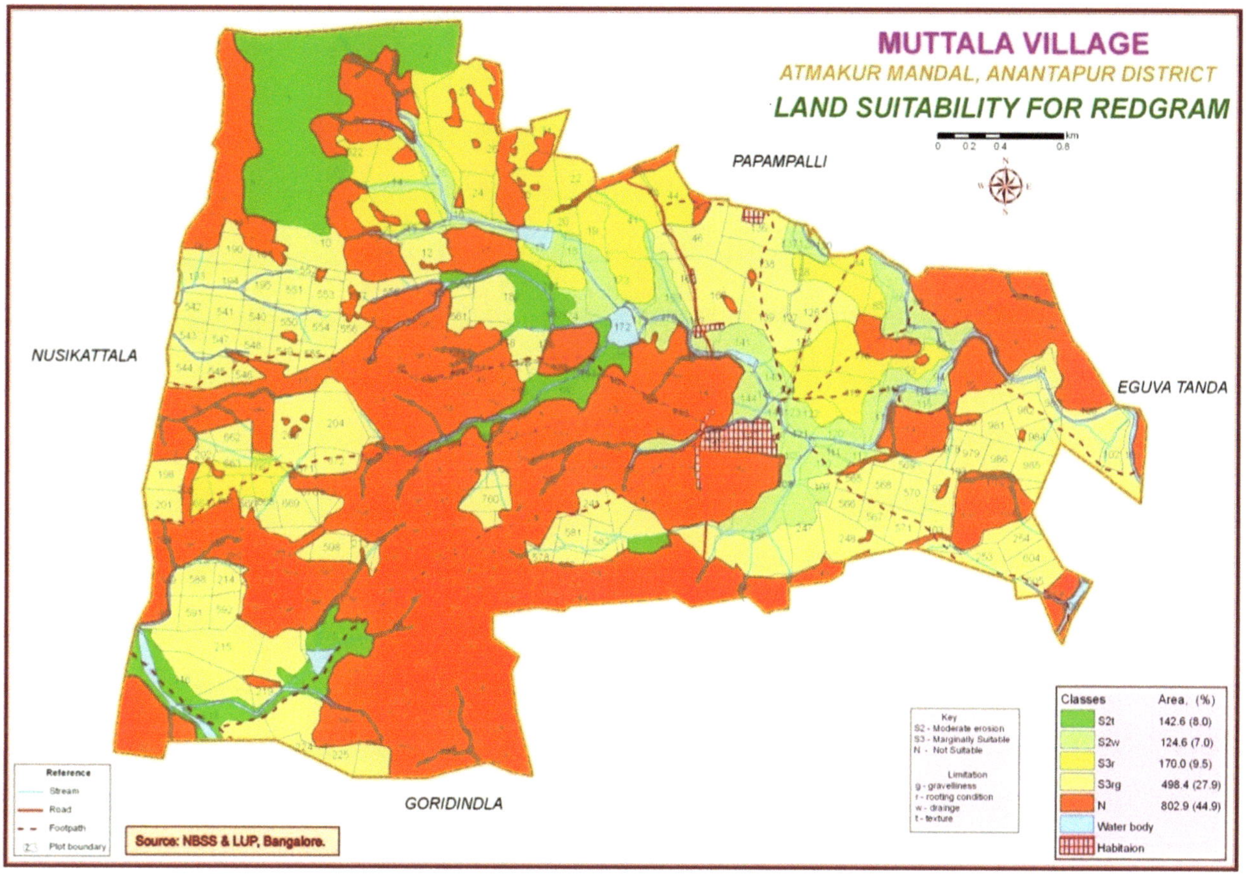

FIGURE 8.9 Land suitability map for redgram.

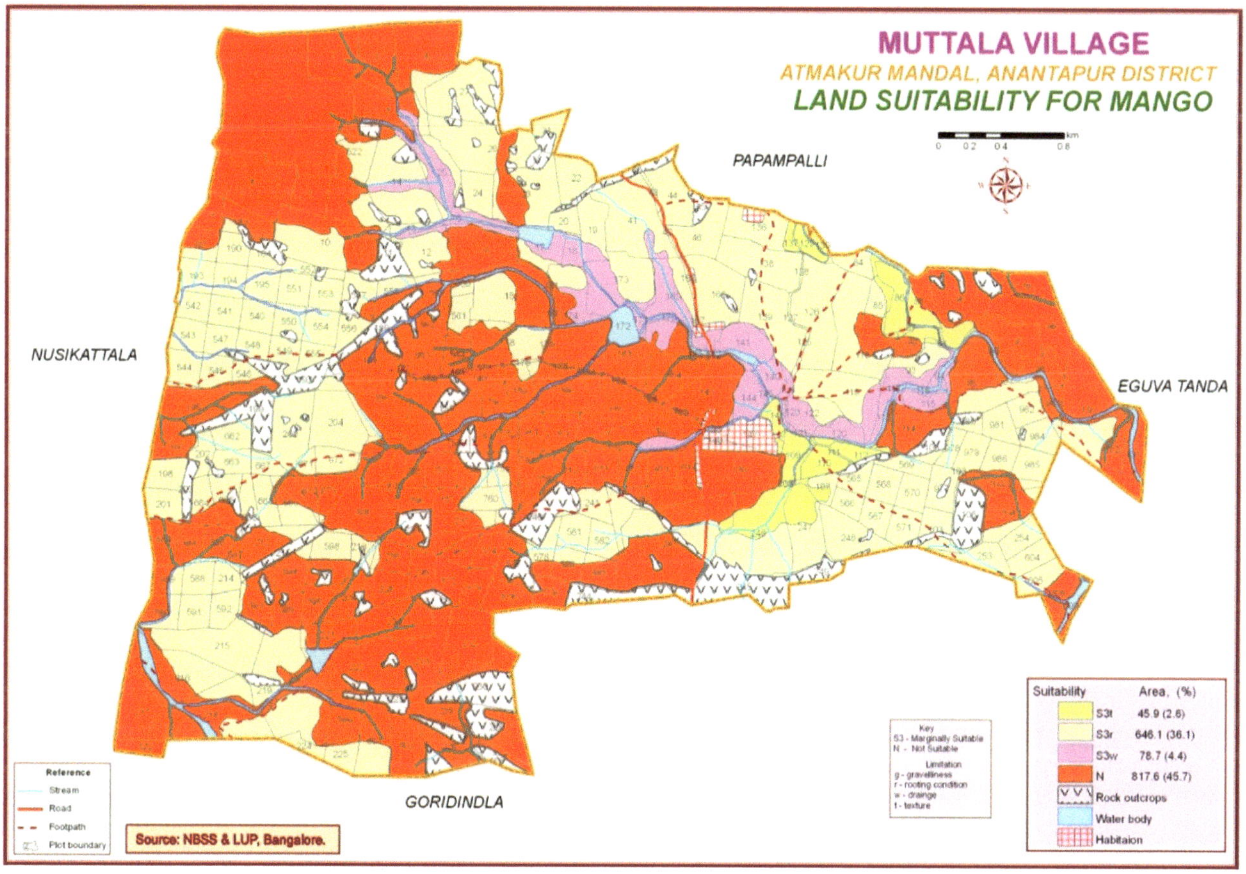

FIGURE 8.10 Land suitability map for mango.

FIGURE 8.11 Status of land resources in Muttala village (exposed gravelly faces due to complete removal of top soil by erosion, formation of initial gullies, and salinity-affected lands).

one landform in the village, but reflected in all the landforms in the village (Ursu et al., 2011). Sheet erosion is severe to very severe in all the uplands. Though the slope is not more than 5% in most of the uplands in the village, due to the absence of appropriate soil conservation measures, almost all the finer soil material is lost from the surface soil and deposited in the foot slopes and valleys. Because of this loss, the texture of the surface soil is either sandy or maximum loamy sand in nature, and to the extent, in many fields, only a concentration of stony or gravelly materials exists at the surface. The occurrence of large volume of rock fragments at the surface is a reflection of the severity of erosion, particularly of the sheet wash type in the uplands of the village (Poesen, 2018). Continued sheet erosion is the main cause of the shallow depth of the soil observed in this area. Similarly, most of the lowland soils in the village are affected by salinity. It might be due to the accumulation of salts carried by rainwater from the uplands. Because of continuous accumulation, the salinity has increased which resulted in cultivated lands became now barren. It is also one of the reasons for invasion of Prosopis in lowlands, which is the only vegetation that withstands severe salinity. Urgent soil and water conservation measures are to be initiated on a war footing for control of land degradation; the action plan is to be prepared based on the database generated through land resource inventory.

8.4 Conclusion

Lack of site-specific soil resource database is one of the important reasons for success or failure of any village development programs. Detailed site-specific soil resources database helps to identify the inherent problems and potentials of soil resources and also evaluate its suitability for various land use options. The present database in Muttala village, Atmakur mandal, was obtained by detailed soil mapping using high resolution RS data. This database strengthens the science base of watershed management by enhancing the understanding of soil-related problems and potentials, hydrological dynamics, and climatic variability; and it also helps in implementing and monitoring the watershed developmental programs.

References

Abahussain, A.A., Abdu, A.S., Al-Zubari, W.K., El-Deen, N.A., Abdul-Raheem, M., 2002. Desertification in the Arab Region: analysis of current status and trends. J. Arid. Environ. 51 (4), 521–545.

Allafta, H., Opp, C., Patra, S., 2020. Identification of groundwater potential zones using remote sensing and GIS techniques: a case study of the Shatt Al-Arab Basin. Remote. Sens. 13 (1), 112.

Casagli, N., Intrieri, E., Tofani, V., Gigli, G., Raspini, F., 2023. Landslide detection, monitoring and prediction with remote-sensing techniques. Nat. Rev. Earth Environ. 4 (1), 51–64.

Dharumarajan, S., Lalitha, M., Hegde, R., Janani, N., Rajawat, A.S., Sastry, L.N., et al., 2018. Status of desertification in South India—assessment, mapping and change detection analysis. Curr. Sci. 115 (2), 331–338.

Franklin, S.E., 2001. Remote Sensing for Sustainable Forest Management. Lewis Publishers, Boca Raton, FL.

Jackson, M.L., 1973. Soil Chemical Analysis. Prentice Hall of India Pvt. Ltd., New Delhi.

Kumar, B., Kumar, U., 2010. Integrated approach using RS and GIS techniques for mapping of ground water prospects in Lower Sanjai Watershed, Jharkhand. Int. J. Geomat. Geosci. 1 (3), 587–598.

Kumar, N., Yamaç, S.S., Velmurugan, A., 2015. Applications of remote sensing and GIS in natural resource management. J. Andaman Sci. Assoc. 20 (1), 1–6.

Naidu, L.G.K., Ramamoorthy, V., Chella, O., Hegde, R., Krishnan, P., 2006. Manual Soil Site Suitability Criteria for Major Crops, NBSS&LUP, Nagpur.

Natarajan, A., Reddy, R.S., Niranjana, K.V., Hegde, R., Srinivasan, R., Dharumarajan, S., 2016. Field Guide for Land Resources Inventory, Sujala III Project, Karnataka. NBSS Publishing; ICAR-NBSS & LUP, Bangalore.

Poesen, J., 2018. Soil erosion in the anthropocene: research needs. Earth Surf. Process. Landf. 43 (1), 64–84.

Rindfuss, R.R., Stern, P.C., 1998. Linking remote sensing and social science: the need and the challenges. People and Pixels: Linking Remote Sensing and Social Science. The National Academic Press, pp. 1–27.

Rogan, J., Chen, D., 2004. Remote sensing technology for mapping and monitoring land-cover and land-use change. Prog. Plan. 61 (4), 301–325.

Shiferaw, B.A., Okello, J., Reddy, R.V., 2009. Adoption and adaptation of natural resource management innovations in smallholder agriculture: reflections on key lessons and best practices. Environ., Dev. Sustain. 11, 601–619.

Singh, P., Thakur, J.K., Kumar, S., Singh, U.C., 2011. Assessment of land use/land cover using geospatial techniques in a semi-arid region of Madhya Pradesh, India. Geospatial Techniques for Managing Environmental Resources. Springer, pp. 152–163.

Singh, S.K., Thakur, J.K., Singh, U.K. Environmental monitoring of land cover/land use changes in Siwalik Hills, Rupnagar district of Punjab, India using remote sensing and ancillary data. In: Paper Presented at the International Conference on Global Climate Change, 2010.

Soil Survey Staff, 2008. Keys to Soil Taxonomy, eighth ed. SCS, USDA, Washigton, DC.

Thakur, J.K., 2010. Eco-hydrological wetland monitoring in a semi-arid region (A case study of Konya Closed Basin, Turkey). Faculty of International Institute for Geo Information Science and Earth Observation (ITC), Universiteit Twente, Enschede, the Netherlands.

Ursu, A., Chelaru, D.A., Mihai, F., Iordache, I., 2011. Anthropogenic landform modeling using GIS techniques case study: Vrancea region. Geogr. Tech. 13 (1), 91–100.

Van Loon, A.F., 2015. Hydrological drought explained. Wiley Interdiscip. Rev.: Water 2 (4), 359–392.

Voigt, S., Kemper, T., Riedlinger, T., Kiefl, R., Scholte, K., Mehl, H., 2007. Satellite image analysis for disaster and crisis-management support. IEEE Trans. Geosci. Remote. Sens. 45 (6), 1520–1528.

Woldai, T., 2020. The status of earth observation (EO) & geo-information sciences in Africa–trends and challenges. Geo-spatial Inf. Sci. 23 (1), 107–123.

Chapter 9

Large scale soil mapping using high-resolution remote sensing data: a case study of semi-arid area, Anantapur District, Andhra Pradesh, India

S. Dharumarajan, V. Ramamurthy, R. Vasundhara, M. Lalitha, S. Srinivas and Rajendra Hegde

ICAR-National Bureau of Soil Survey and Land Use Planning, Regional Centre, Bangalore, Karnataka, India

9.1 Introduction

Climate change and land degradation are imposing a real threat to sustainable agricultural production in India. Areas affected by salt-affected soils, sodicity, erosion, water logging, shifting cultivation, etc. are rising because of inappropriate management and overexploitation of land resources (Dharumarajan et al., 2017a). These circumstances create one of the greatest challenges for the planners, scientists, and farmers to increase yield of food grain production to 350 million tons by 2030 and 450 million tons by 2050. Soil is the foundation of all production systems, and understanding its properties, extent, and spatial distribution is critical. It is decisive to conserve soil resources in order to prolong the ecosystem. Characterization and mapping of various types of soils, as well as their interpretation, become more important in this context. According to Laake (2000), precise baseline information and methods are required to assess the quantity and quality of resources for soil mapping and characterization. Large-scale soil mapping has been done by traditional methods that are time-consuming, costly, and have low recurring value especially in complicated and inaccessible terrain. Improvement in computer and information technology has brought new set of tools, methods, instruments, and systems. Rapid developments in technologies such as remote sensing (RS), Geographic positioning system, and geographic information system (GIS) offer new scope to resource planning, and it is productively used in studying diverse aspects of soils in the spatial and temporal domains (Shrestha, 2006; Yeung and Lo, 2002).

RS and GIS are very useful in the creation of natural resource database in an effective way and also for the integration of information with the ancillary data for generating an action plan (Dharumarajan et al., 2018; Kalgapurkar Anitha et al., 2012; Lalitha et al., 2016; Sharada et al., 1993). It's been acknowledged as an important tool for inventorying land resources at the local, regional, and global levels and provides spatially precise, digital data representations of the earth's surface, which can be collective with digitized paper maps in GIS to facilitate efficient characterization and analysis of colossal amounts of data. In combination with GIS, satellite RS has established to be an exceptionally useful tool for natural resource management.

Soil maps at diverse scales are necessary to meet the requirements of preparation at various stages. Dwivedi (2001) stated that the coarse-resolution data (spatial resolution 70 m or greater) from IRS LISS-I, AWiFS, and LANDSAT-MSS sensors were beneficial for creating soil maps at a scale of 1: 250,000 or less. Medium-resolution data from LANDSAT-TM, IRS LISS-II, and SPOT-MLA are used to map soils at 1:50,000 scale. IRS-P6 (LISS-IV sensor), Cartosat-1 and Cartosat-2, and IKONOS satellite data are now being used for comprehensive soil characterization at 1:10,000 scale or larger. Srivastava and Saxena (2004) discussed the technique of large-scale soil mapping (1:12,500 scale) in a basaltic terrain with a physiographic land unit approach and differentiated soil types using toposheet and IRS-1C PAN combined data of two seasons. In a part of Basaltic terrain of Vidarbha region, Nagaraju et al. (2014) have generated a detailed soil map through the Physiography–soil relationship for preparation of action plan using

Remote Sensing of Soils. DOI: https://doi.org/10.1016/B978-0-443-18773-5.00002-8

Cartosat-1-merged IRS LISS IV data. It was found that the high-resolution satellite Cartosat-1-merged IRS LISS IV data has the potential for precise and accurate delineation of landforms by understanding the soil-forming factors for faster mapping. With the increasing availability of high-resolution satellite data in recent days, studies have been initiated to characterize soils at large scale through the physiography—land use—soil relationship. In this context, the present study was carried out to map the soils of Bukkarayasamudram mandal, Anantapur district on 1:10000 scale using high-resolution satellite data and digital elevation model.

9.2 Material and methods

9.2.1 Study area

Bukkarayasamudram mandal is situated in Anantapur district, Andhra Pradesh (Fig. 9.1). The mandal with an overall geographical area of 24,808 hectares lies between 13°37′ 51″ and 14°48′ 09″ N latitudes and 77°33′47″ E and 77°47′45″ E. It is divided into 14 revenue villages viz. Bodiganidoodi, Bukkarayasamudram, Chedulla, Chennampalle, Chennarayunupalle, Danduvaripalle, Govindapalle, Janthulur, Korrapadu, Kondakiniagraharam, Podaralla, Reddipalle, Siddaramapuram, and Siddampeta. Major rock types observed were Granites-gneiss and dykes. Most parts of mandal has 1%–5% of slope (nearly level to gently sloping). Hills and hillocks in the northern part have slope of more than 8%. The study area is mostly drained by Thadakaleru, Akuledu vanga, and Erravanga streams into Singanamalla cheruvu. The study area belongs to Agro ecological subregion 3 (AESR 3), i.e., Karnataka plateau (Rayalseema as inclusion); and is characterized by hot arid ecological subregion with deep loamy and clayey mixed red and black soils, low to medium AWC, and length of growing period (LGP) of 60–90 days.

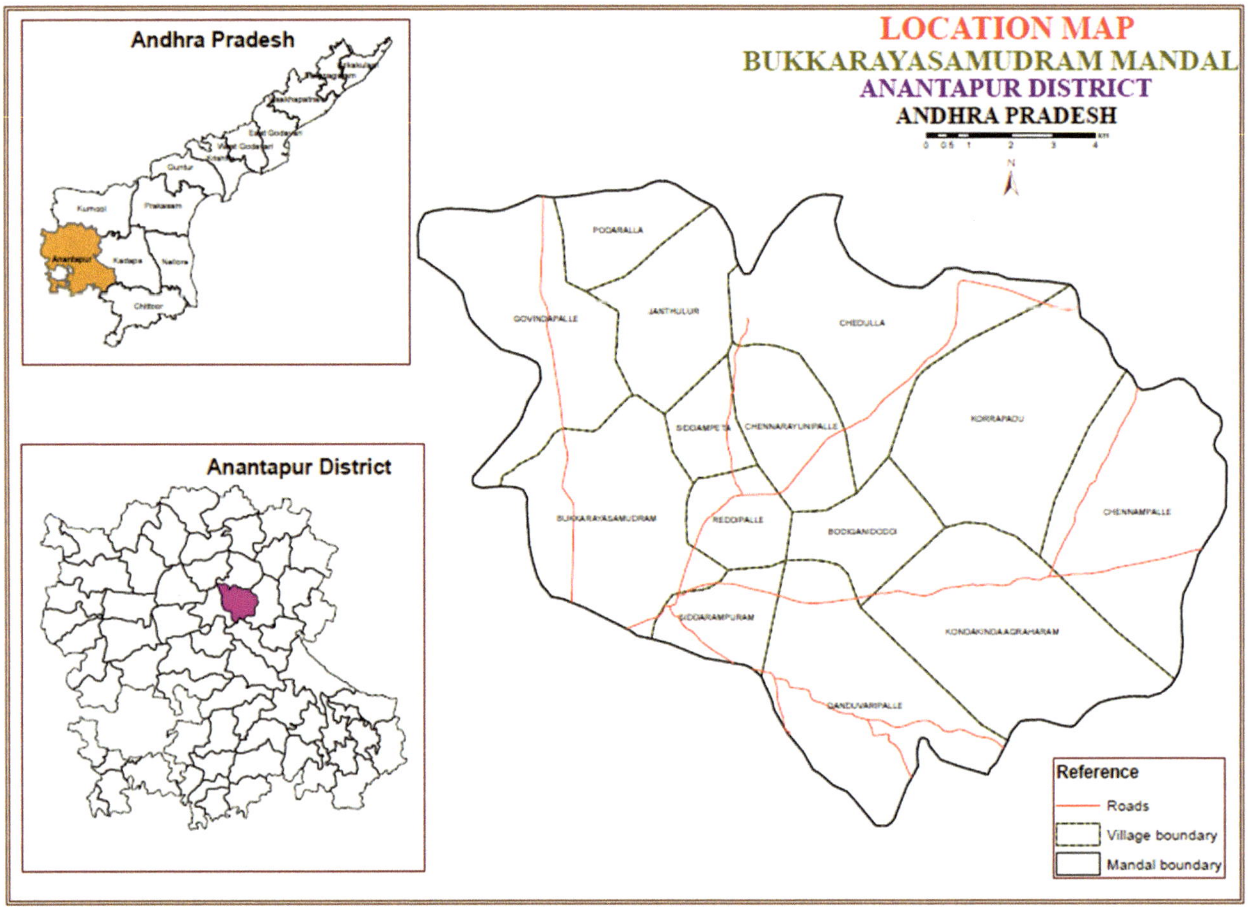

FIGURE 9.1 Location of Bukkarayasamudram mandal of Anantapur district.

9.2.2 Climate

The climate of study area is mainly classified as hot and dry type (IMD, 2013). Mean maximum temperature is ranges between 30°C and 40°C in December and May, respectively (Table 9.1). Similarly, mean minimum temperatures are lowest in December and January and lies near 17°C–19°C; whereas, it does not exceed 25°C in winter. The mean diurnal deviation of temperature as a whole is 10°C–12°C and rainfall is below 550 mm, and it is the driest part of the state. Typically, the area gets 300–500 mm of rain throughout south west monsoon (June–October) and is 55%–90% of annual rainfall. September is commonly the peak rainy month with about 25% of annual rain; despite the fact that the average rainfall is around 556 mm in the last 23 years, there are extremes in some years, which ranges from the lowest rainfall of 324 mm in 2003 to the highest of 788 mm in 1996. The area is chronically drought-prone area due to its low and impulsive nature of LGP. The LGP of study area is 11 weeks starting from August third week to October last week (Fig. 9.2).

9.2.3 Geomorphology

The classification of the geomorphic units in the study area is based on the geomorphic expression, relief, slope factor, and surface cover with soil or vegetation. The most important geomorphic units of the Anantapur district are majorly divided into five types which are Denudational hills, dissected pediments, Pediplain, Alluvium, and Channel fills (Table 9.2).

9.2.4 Survey methodology

The purpose of the resource inventory is to delineate similar areas which are expected to behave similarly to a given level of management. This was achieved in study area by comprehensive study of the village area (like slope of the land, erosion, drainage, occurrence of rock fragments, etc.) and soil characteristics (like depth, texture, color, structure, consistency, coarse fragments, porosity, soil reaction, etc.), along with grouping of similar areas based on soil-site characteristics into homogenous (management) units and showing their extent and distribution on the village cadastral map.

TABLE 9.1 Temperature and relative humidity of Bukkarayasamudram mandal, Anantapur.

Month	Mean max. temperature (°C)	Mean min. temperature (°C)	Relative humidity (%)		WS (kmh^{-1})	Sunshine (h)
			08.30	17.30		
January	29.7	16.9	77.4	36.8	6.2	7.5
February	33.5	17.4	76.1	21.7	7.9	9.9
March	36.7	22.4	64.0	25.3	8.4	9.5
April	37.1	24.3	68.4	32.9	8.4	9.1
May	38.7	25.4	65.9	32.2	11.0	7.6
June	34.9	23.8	73.0	42.1	14.9	5.4
July	35.6	24.8	72.4	37.5	18.1	6.7
August	34.4	23.9	75.4	44.5	14.1	6.4
September	33.2	23.3	81.0	49.4	8.4	6.3
October	33.1	21.6	84.4	43.5	4.2	8.7
November	29.6	20.2	87.8	55.3	5.3	4.4
December	31.3	17.4	86.9	38.0	4.8	8.5
Mean annual	34.0	21.8	76.1	38.3	9.3	7.5

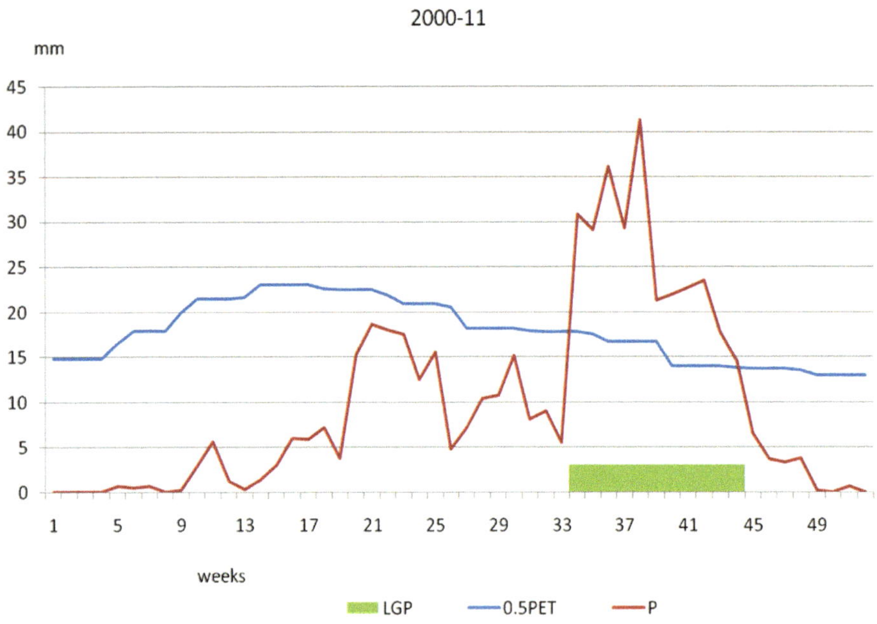

FIGURE 9.2 Length of growing period of Bukkarayasamudram mandal.

TABLE 9.2 Different geomorphic units of study area.

S. No.	Geomorphic units	Relief	Surficial expression
1	Denudational hills in granites and gneisses	Rugged surface, steep slopes along edges (45–65). Relief ranges from 100 to 600 m.	Mostly devoid of soil cover or very thin soil. This mantle of weathered zone is less than 2 m
2	Dissected pediment	Undulating land with gentle to moderate slopes (5–12 degrees) with number of bouldery outcrops amidst coarse gritty to gravelly, brown residual soils.	General rugged appearance with coarse to medium sandy soils (thickness less than 30 cm). Supports dry cultivation.
3	Pediplain	Very gently sloping to flat area (2–5 degrees).	Red to brown colored, coarse gravelly to sandy residual soils of 20–60 cm thickness. Black clayey soils of thickness from 60 cm to 1.2 m.
4	Alluvial deposits	Flat or very gentle slopes (less than 2 degrees).	Flat terrain consist of silty sands, gravel, and sands along the major rivers; thickness of alluvium ranges from 5 to 15 m with over-weathered granites and gneisses.
5	Channel fills	Flat or very gentle slopes (less than 2 degrees).	Restricted to minor streams and nalas with limited lateral spread. Composed of silt, sand, and rocks

A high-intensity survey (at 1:10,000 scale) was carried out during November–December, 2015. The survey methodology followed in the field and laboratory is as per standard procedures (AIS&LUS, 1970; Soil Survey Staff, 2008).

Base maps used

1. Cartosat-1 DEM digital data
2. IRS-P6 LISS IV digital data
3. Digitized Survey of India (SOI) toposheets (1:50,000 scale)

4. Cadastral maps in digitized format
5. *Other collateral data viz. Global spatial datasets, SRM reports, AESR map, district reports, geology maps, available LULC maps, etc.*

9.2.5 Land use map and land ecological unit generation

Land use/land cover map of the study area was created based on visual analysis of ortho-rectified IRS-P6 LISS-IV data. Landform, land use, and slope maps were created from IRS LISS IV data and Cartosat DEM and were incorporated in ArcGIS environment and Land ecological units (LEU) were prepared (Dharumarajan et al., 2018). LEU is designed using six letters and one numeral, first two letters denote physiographic regions and subregions and the third and fourth letter is dedicated to broad landforms. The fifth letter is positioned for landform unit. The only numeral in the scheme is placed for slope class, and the last letter in the scheme is designated for land use and land cover.

9.2.6 Field investigations and soil mapping

A thorough survey of the land resources in the study area was carried out by using cadastral map and LEU as a base. The cadastral map shows field boundaries with their survey numbers, location of tanks, streams, and other permanent features of the area. Based on geology, drainage pattern, surface features, and land use, transects were marked. In the selected transect, profiles were selected at closely spaced intervals to take care of any change in the land features like break in slope, erosion, gravel, stones, etc. In the selected sites, profiles were opened up to 150 cm or to the depth limited by rock or hard substratum and analyzed in detail for all their morphological characteristics. Profiles are also opened randomly to cover the entire area. Overall, 92 profiles were studied. The soil and site characteristics were recorded for all profile sites according to a standard proforma as per the standard guidelines (Soil Survey Staff, 2008). Based on the soil-site characteristics, the soils were grouped into various soil series (soil series is the most homogeneous unit having similar horizons and properties and acts similarly for a given level of management). Soil depth, texture, color, amount and nature of gravel present, nature of substratum, and horizon sequence were the major identifying characteristics of soil series occurring in the area.

9.2.7 Laboratory characterization

Soil samples were collected from representative master profiles for laboratory characterization. Surface samples were collected from farmer's fields for fertility status (macronutrients and micronutrients) analysis (Jackson, 1973).

9.2.8 Finalization of soil map

The area under each series was further divided into phases, and their boundaries were delineated on the cadastral map based on variations in surface soil texture, slope, erosion, presence of gravels, stoniness, and so on. A phase is a subdivision of a soil series based primarily on surface characteristics that influence its use and management. Soil-site characteristics such as soil depth, erosion, gravelliness, salinity/sodicity, and land suitability maps for various crops were analyzed in order to identify constraints and potentials. Fertility maps were created using GIS software and the interpolation technique (Krigging) (Figs. 9.3–9.5).

9.3 Result and discussion

9.3.1 Land use/land cover

Land use map prepared using Resourcesat-2 LISS IV data and presented in Fig. 9.8. The lands with single crop occupy 62.9% of the study area, followed by double crop with 12.2% of the total geographic area (TGA). Scrub lands occupy 6.3% of area and fallow lands which includes current fallow and permanent fallows occupy 3.5% of the area. The major crop is groundnut, followed by pigeon pea, castor, and sunflower. The major plantation crops in Bukkarayasamudram mandal are pomegranate, citrus, and guava.

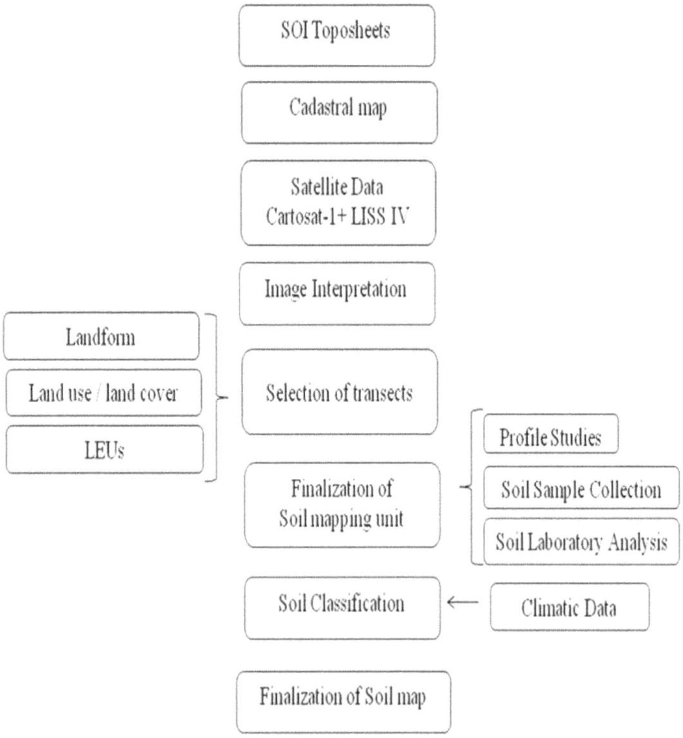

FIGURE 9.3 Overall methodology.

9.3.2 Land ecological unit

Fifteen LEUs were delineated in Bukkarayasamudram mandal based on landform, slope, and land use (Table 9.3). The LEU units are relatively homogeneous in terms of the main factors of soil formation and soil characteristics and are used as a base map for ground truth verification.

9.3.3 Soil mapping

The soils occurring in this study area are grouped under two groups: (1) Granitic-gneiss and (2) Alluvium. During the course of the survey, 8 soil series with 52 phases were identified. Out of this, five soil series were identified as Granitic-gneiss landform and three series were identified as Alluvium landform. The major series identifying characteristics are depth, soil color, texture, gravel, calcareousness, and horizon sequence (Table 9.4). Out of the eight soil series identified, Nilampalle soils (Moderately deep soils) and Chennampalle (Shallow soils) occur extensively, occupying about 35.4% of the area (9868 ha). The other dominant series in granitic landform is Dayyadakuntapalle series, which occupies an area of about 3850.0 ha (13.8%) in the mandal. The three series in alluvial landforms occupies about 7498 ha (27%) in the mandal. Rock lands and rock outcrops occur in about 255 ha (0.91% of the area) and mining lands occupy 0.2% of the study area. Dominant soil profiles of different soil series are presented in Fig. 9.6. The physiochemical characteristics of identified soil series are presented in Table 9.5. Soil map of Podralla village presented in Fig. 9.7 along with legend description in Table 9.6. Similarly soil maps of all 14 revenue villages were developed.

9.3.4 Soil characteristics

9.3.4.1 Chennampalle series

These soils are formed from granite-gneiss and occur on very gently sloping ridges and uplands with 1%–3% slopes with an elevation of 308 m above MSL. The soils are very shallow, well-drained, with loamy sand to sandy loam texture. The gravel content varies from 10 to 40%. These soils are very strongly alkaline (pH of 9.0–9.16). The soils of Chennampalle Series are fine loamy-skeletal, mixed isohyperthermic Lithic Haplargids and cover an area of 4762 ha. The productivity potential of Chennampalle series is low.

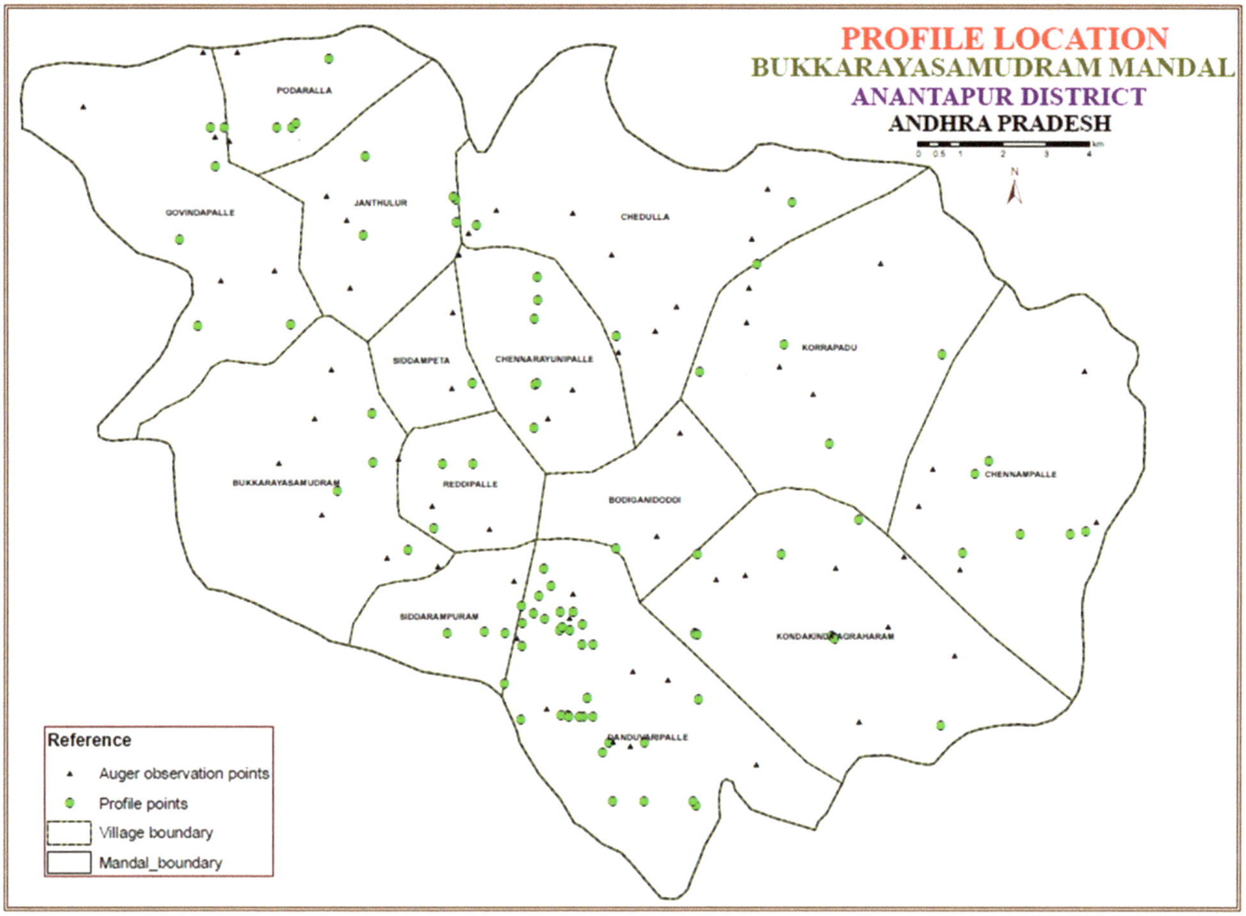

FIGURE 9.4 Profile locations over cadastral map.

9.3.4.2 *Venkatapuram series*

The Venkatapuram series are moderately shallow, well-drained soils on nearly level lands and valleys and formed from granite-gneiss. Surface soils are dark reddish brown with clay loam soil texture, and subsurface soils have sandy clay texture. The soil reaction is neutral in surface and slightly to moderately alkaline in subsurface. Organic carbon content is ranging from 0.56 to 0.74. The soils of Venkatapuram series are clayey-skeletal, mixed, isohyperthermic (cal), Typic Haplargids. The soils cover 8.7% of total geographical area and distributed mainly in Korrapadu, Chennampalle, and Danduvarpalle villages. The productivity potential is moderate.

9.3.4.3 *Dayyadakuntapalle series*

The Dayyadakuntapalle series are moderately shallow, well-drained soils formed from granite-gneiss and occurring on very gently sloping to gently sloping uplands with 1%–5% slopes, with an elevation of 326 m above MSL. These soils have dark reddish brown loamy sand surface texture with reddish brown to yellowish red sandy clay loam subsurface texture. The soils of Dayyadakuntapalle series are loamy-skeletal, mixed, isohyperthermic Ustic Haplargids. It is the third dominant soils and accounted for 13.8% of Bukkarayasamudram mandal and distributed at Bukkarayasamudrum, Kondakinda agrakaram, Chennarayanpalle, Reddipalle, Bodiganidoodi, and Siddarampuram. The productivity potential is moderate.

9.3.4.4 *Nilampalle series*

Nilampalle series are moderately deep, well-drained soils formed from granite-gneiss and occur on gently sloping uplands with 1%–5% slopes, with an elevation of 313 m above MSL. These soils are dark reddish brown to dark red in

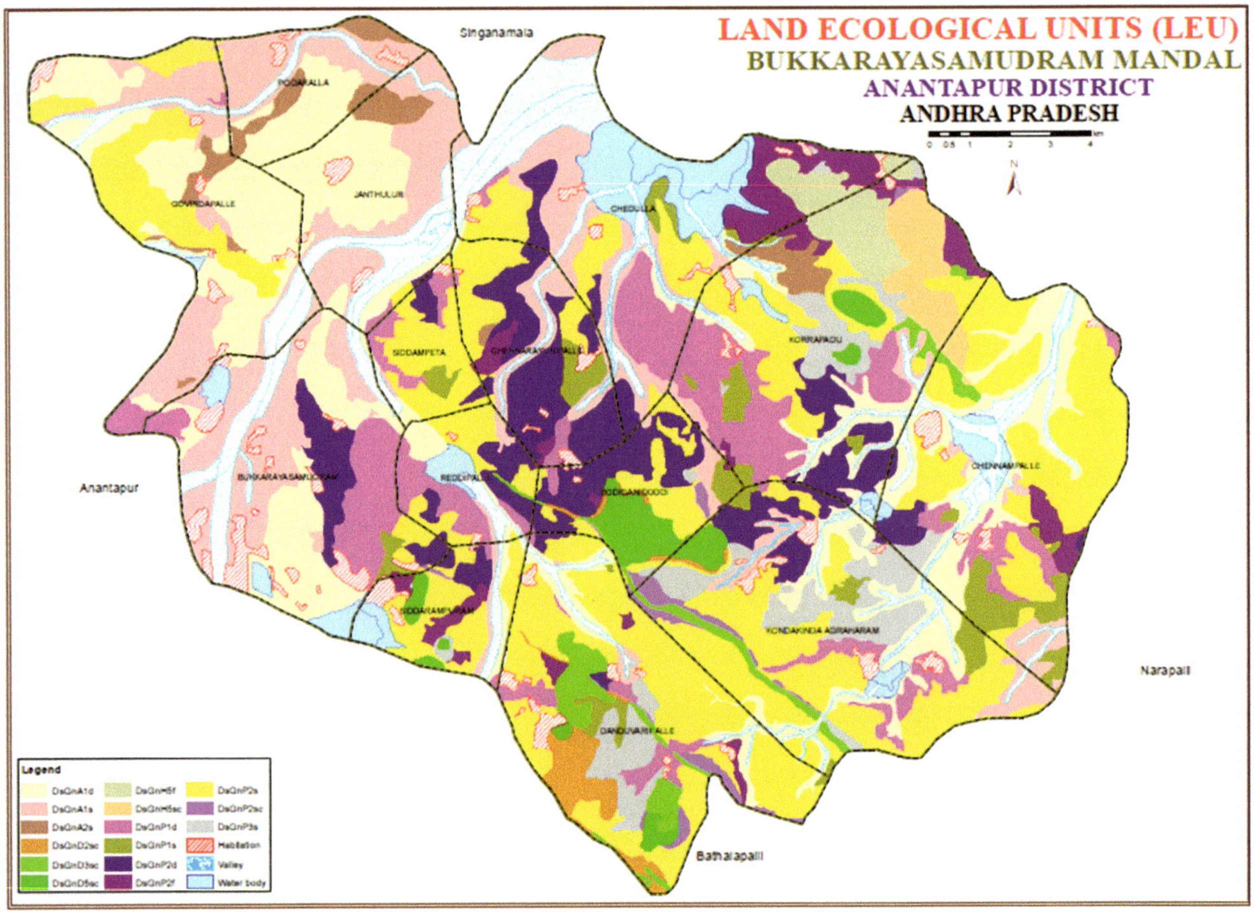

FIGURE 9.5 LEU of Bukkarayasamudram mandal, Anantapur District.

color and have sandy clay loam surface texture and sandy clay subsurface texture. Soil reaction is slightly alkaline to moderately alkaline. Organic carbon content is ranged from 0.42% to 0.84%. The soils of Nilampalle series are clayey-skeletal, mixed isohyperthermic Typic Haplargids. Nilampalle series is a dominant series and occupies about 18.3% of Bukkarayasamudram mandal and distributed throughout the mandal. The productivity potential is good.

9.3.4.5 *Rekulakuntapalle series*

Rekulakuntapalle series consists of deep, well-drained soils on formed over granite-gneiss and occur on nearly level to very gently sloping uplands with 1%–3% slope with an elevation of 314 m above MSL. These soils are yellowish red to red in color and have sandy clay loam surface texture and clay loam to sandy clay subsurface texture. Soil pH is varied from 7.26 to 7.66 and Organic carbon content is ranged from 0.39% – 0.91%. The soils of Rekulakuntapalle series are clayey-skeletal, mixed isohyperthermic Ustic Haplargids. It covers a limited area of 2% and occurs mainly in Danduvarpalle and Kondakinda Agraharam villages. The productive potential is good.

9.3.4.6 *Yadavalaparti series*

Yadavalaparti series consists of moderately shallow, well-drained clayey soils on nearly level lands. These soils are formed from alluvium parent material and occur on nearly level lowlands with 0%–1% slope, with an elevation of 336 m above MSL. These soils have sandy loam to sandy clay loam texture. The soils of Yadavalaparti series are fine loamy, mixed, isohyperthermic (cal) Typic Haplargids. Yadavalaparti series covering 9.8% of area and distributed in Bukkarayasamudrum, Chennampalle, and Siddarampuram villages of Bukkarayasamudram mandal. The productive potential is moderate.

TABLE 9.3 Description of Landscape ecological units in Bukkarayasamudram mandal.

Physiographic region	Physiographic subregion	Landforms	LEU	Description
Deccan Plateau	South Deccan plateau	Granite and gneissic complex	DsGnH5f	Strongly sloping denudational hills and ridges (forest)
			DsGnH5sc	Strongly sloping denudational hills and ridges (scrub)
			DsGnD2sc	Very gently sloping dissected pediment (scrub)
			DsGnD3sc	Gently sloping dissected pediment (scrub)
			DsGnD5sc	Steeply sloping dissected pediment (scrub)
			DsGnP1d	Nearly level pediplain (double crop)
			DsGnP1s	Nearly level pediplain (single crop)
			DsGnP2d	Very gently sloping pediplain (double crop)
			DsGnP2f	Very gently sloping pediplain (fallow)
			DsGnP2s	Very gently sloping pediplain (single crop)
			DsGnP3s	Gently sloping pediplain (single crop)
			DsGnP2sc	Very gently sloping pediplain (scrub)
			DsGnA1d	Nearly level alluvial deposits and valley fills (double crop)
			DsGnA1s	Nearly level alluvial deposits and valley fills (single crop)
			DsGnA2s	Very gently sloping alluvial deposits and valley fills (single crop)

9.3.4.7 Lolluru series

The Lolluru series consists of moderately deep, well-drained clayey soils on nearly level. These soils formed from alluvium parent material and gravity and occur on nearly level lowlands with 0%–1% slope with an elevation of 324 m above MSL. These soils are very dark grayish brown in color with clay content above 35%. The soils of Lolluru series are fine, mixed, isohyperthermic (cal) Typic Haplargids. Lolluru series covers an area of 5% and occurs mainly in Govindapalle and Podaralla villages. The productive potential is good.

9.3.4.8 Govindapalle series

The Govindapalle series consists of deep, well-drained clayey soils on nearly level lands. These soils are formed from alluvium parent material and occur on very gently sloping uplands with 1%–3% slope, with an elevation of 1096 m above MSL. These soils are dark grayish brown to very dark gray in color and have clay content above 40%. Soil reaction is moderately alkaline and organic carbon content of the soil is 0.22%–0.87%. The soils of Govindapalle series are fine, mixed isohyperthermic (cal) Ustic Haplargids. It covers an area of 11.9% of TGA and is the fourth dominant soils in Bukkarayasamudram mandal and distributed in Janthulur, Podaralla, and Govindapalle. The productive potential of Govindapalle series is medium.

9.3.5 Problems and potentials

9.3.5.1 Surface texture

Soils of Bukkaraysamudrum mandal are grouped into five surface soil textural classes. The dominant soil texture is sandy clay loam, followed by sandy loam. The surface soil texture in granite-gneissic landform is lighter (sandy loam, loamy sand) and heavier (clay and sandy clay) in alluvium. About 29% of area contains sandy clay loam texture at the

TABLE 9.4 Identifying characteristics of soils of Bukkarayasamudram mandal.

Series	Depth (cm)	Color	Texture		Gravel (%)	Calcareousness	Horizon sequence
			Surface	subsurface			
Granite-gneiss							
Chennampalle	25–50	7.5YR3/4, 5YR3/6 2.5YR2/6	ls, sl, scl	scl-sc	35–70	–	Ap-Bt-Cr
Venkatapuram	50–75	5YR3/4, 2.5YR3/4	sl, scl, sc, c	scl-c	–	es-ev	Ap-Bt-Cr
Dayyadakuntapalle	50–75	5YR3/4, 2.5YR3/4	ls, sl, scl	scl-sc	35–70	–	Ap-Bt-Cr
Nilampalle	75–100	2.5YR4/8, 2.5YR4/6, 2.5YR3/4	ls, sl, scl,	scl-sc	35–70	–	Ap-Bt-Cr
Rekulakuntapalle	100–150	2.5YR3/6, 2.5YR3/4	ls, sl	sc-scl	35–70	–	Ap-Bt-Cr
Alluvium							
Yadavalpati	50–75	10YR3/2, 10YR4/1	c, sc. scl	c-sc		e-es	AP-Bw-Ck
Lolluru	75–100	10YR3/2, 10YR4/1	c, sc	c-sc	–	es-ev	Ap-Bw-C
Govindapalle	100–150	10YR4/2,3/2	c, sc, scl	c-sc	–	es-ev	Ap-Bw-Ck

surface, and sandy loam texture observed in 20.8% of area. Clay and sandy clay texture accounts for about 23.2% of area and loamy sand occupies 11% of total study area.

9.3.5.2 Soil gravelliness

Gravels present in the surface layer are one of the important characteristic used to separate the soil series into different phases. Gravel is commonly seen in both uplands and hills/rockland areas. The lowland soils are mostly devoid of any coarse fragments. Nearly 13% of cultivated lands have slight gravelliness (15%–35%) problem in the village.

9.3.5.3 Soil erosion

Based on the intensity of erosion and also the amount of soil removed (loss of A & B horizons), three erosion classes (slight, moderate, and severe erosion) were identified in the area. Soils that are slightly eroded cover an area of about 46.5% of TGA. Moderately eroded area covers about 34.7% and severely eroded area covers about 3.7%.

9.3.5.4 Soil reaction (pH) and soil organic carbon

More than 96% of Bukkarayasamudram mandal soils are alkaline in nature and 25.2% soils are strongly to very strongly alkaline. More than 52% of area have medium organic carbon content (0.5%–0.75%), followed by low (30.7%) and high (17.2%) status (Figs. 9.8 and 9.9).

9.3.5.5 Available macronutrients

Soil Fertility status information helps the farmers to overcome the constraints affecting the fertility and productivity of soil resources. The results showed that available nitrogen is low (< 280 kg ha^{-1}) in the entire mandal and ranged from 47 to 250.9 kg ha^{-1}. The available phosphorus content in the soils ranged from 0.6 to 34.7 kg ha^{-1}. The results found

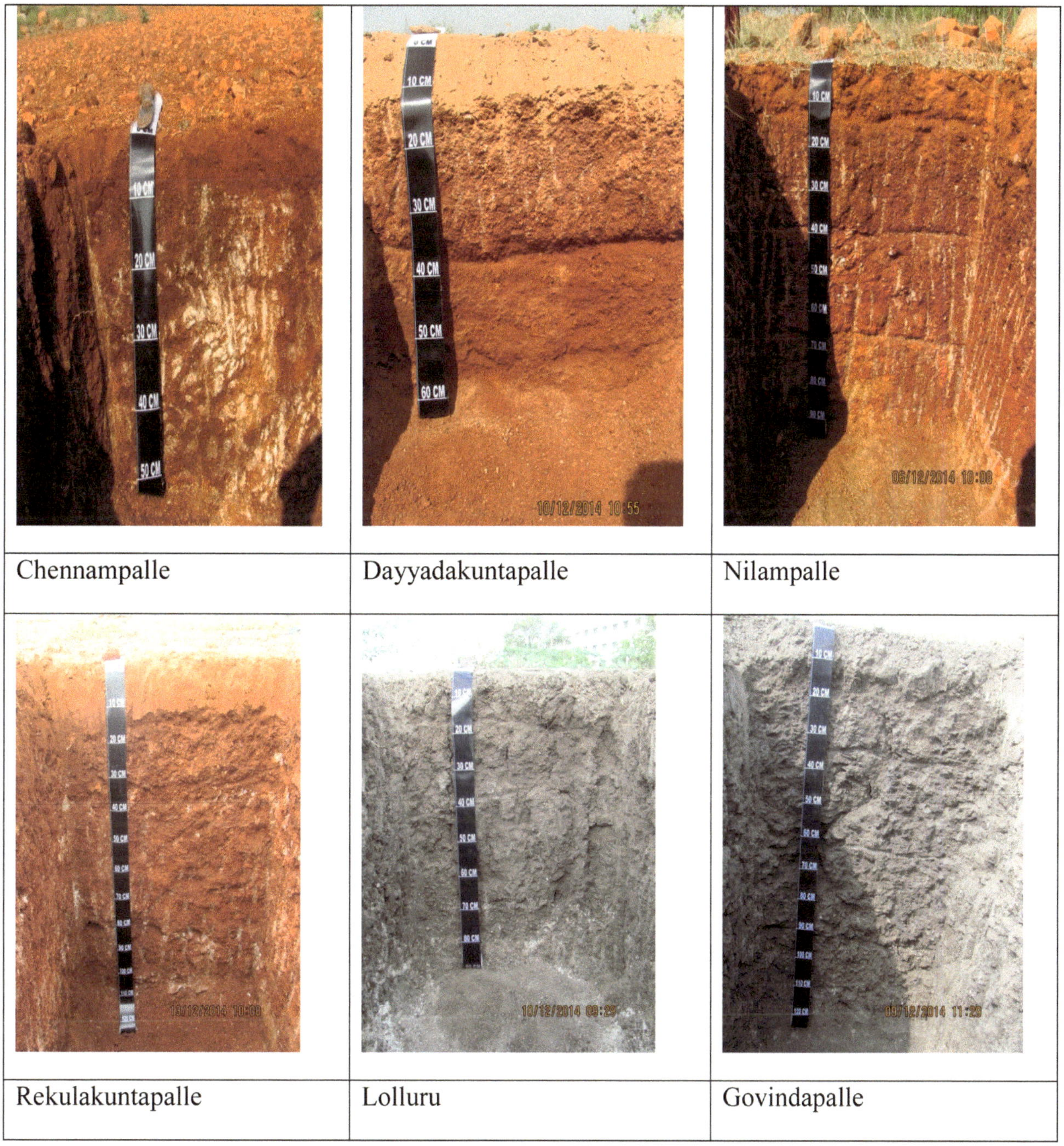

FIGURE 9.6 Major soils of Bukkaraysamudrum mandal, Anantapur district.

that the major areas are under low phosphorus category (<23 kg ha^{-1}). Available potassium content is ranged from 23.5–987 kg ha^{-1}. In large parts of the area (49.2% of area), the available potassium content is medium (140–330 kg ha^{-1}). Low potassium content is found in 23.9% of area and high potassium content in 26.8% of the mandal. Available sulfur content is ranged from 0.8–97.5 kg ha^{-1}. In about 45.5% area of mandal, the available sulfur content is high (>20 kg ha^{-1}). Low sulfur content is found in 8.4% of area (<10 kg ha^{-1})

9.3.5.6 Available micronutrients

The available iron (Fe) content of the soils ranged from 0.7 to 15.4 ppm. More than 98% of Bukkarayasamudram mandal are sufficient in the availability of Fe content and only limited area (less than 1%) are deficient. The available Mn

TABLE 9.5 Physiochemical characteristics of Bukkarayasamudram soils.

Depth (cm)	Horizon	Sand % (2.0–0.05 mm)	Silt % (0.05–0.002 mm)	Clay % (<0.002 mm)	pH (1: 2.5 water)	EC (1: 2.5) dSm^{-1}	OC%	CEC (c mol (p +) kg^{-1})
Series: Chennampalle								
0–15	Ap	74.6	9.6	15.8	9.16	0.118	0.54	19.51
15–40	Bt_1	71.1	8.9	20.0	9.01	0.358	0.84	22.92
40–50	Weathered parent material							
Series: Venkatapuram								
0–18	Ap	44.42	22.89	32.68	7.09	0.070	0.56	13.3
18–37	Bt1	45.80	16.20	38.00	8.53	0.127	0.74	21.73
37–60	Bt2	45.5	12.5	42.0	8.56	0.187	0.74	20.33
60–65	Weathered parent material							
Series: Dayyadakuntapalle								
0–11	Ap	65.99	23.21	10.80	8.31	0.85	0.42	13.3
11–36	Bt1	62.23	21.96	15.82	8.53	0.126	0.39	18.18
36–62	Bt2	33.61	34.60	31.79	8.74	0.166	0.21	16.90
62–75	Weathered parent material							
Series: Nilampalle								
0–11	Ap	46.43	39.86	13.71	8.24	1.24	0.42	19.51
11–39	Bt1	49.90	38.33	11.77	7.81	0.134	0.54	16.85
39–65	Bt2	42.30	40.34	17.36	7.82	0.126	0.49	23.06
65–83	Bt3	48.94	35.97	15.09	8.35	0.165	0.84	18.18
83–100	Weathered parent material							
Series: Rekulakuntapalle								
0–19	Ap	57.19	32.30	10.51	7.27	0.400	0.60	18.18
19–43	Bt1	34.63	33.58	31.79	7.57	0.145	0.67	27.5
43–65	Bt2	35.1	32.4	32.5	7.66	0.185	0.39	31.04
65–90	Bt3	30.63	32.58	36.79	7.26	0.100	0.49	33.7
90–128	Bt4	44.15	22.44	33.41	7.31	0.200	0.91	25.72
Series: Yadavalaparti								
0–12	Ap	76.63	12.48	10.90	8.60	0.136	1.30	13.3
12–31	Bw1	38.65	42.15	19.20	9.82	0.136	0.72	31.04
31–55	Bw2	53.21	28.12	18.67	8.56	0.187	0.57	19.51
55–75	CaCO3 layer							
Series: Lolluru								
0–14	Ap	43.45	12.45	44.2	8.50	0.126	0.32	20.84
14–35	Bw1	37.02	17.75	45.23	8.63	0.120	0.60	16.4
35–55	Bw2	51.14	13.77	35.09	8.97	0.160	0.39	12.42
55–87	Bw3	54.36	8.51	37.13	9.48	0.155	0.46	12.42

(Continued)

TABLE 9.5 (Continued)

Depth (cm)	Horizon	Sand % (2.0–0.05 mm)	Silt % (0.05–0.002 mm)	Clay % (<0.002 mm)	pH (1: 2.5 water)	EC (1: 2.5) dSm^{-1}	OC%	CEC (c mol (p +) kg^{-1})
87–95	CaCO3 layer							
Series: Govindapalle								
0–16	Ap	67.55	11.33	21.12	8.73	0.128	0.46	21.73
16–33	Bw1	33.10	26.74	40.16	8.51	0.140	0.38	35.13
33–54	Bw2	40.64	9.55	49.81	8.58	0.136	0.42	36.89
54–76	Bw3	44.18	12.34	43.48	8.59	0.140	0.87	39.52
76–110	Bw4	46.91	11.73	41.36	8.26	0.146	0.22	31.62

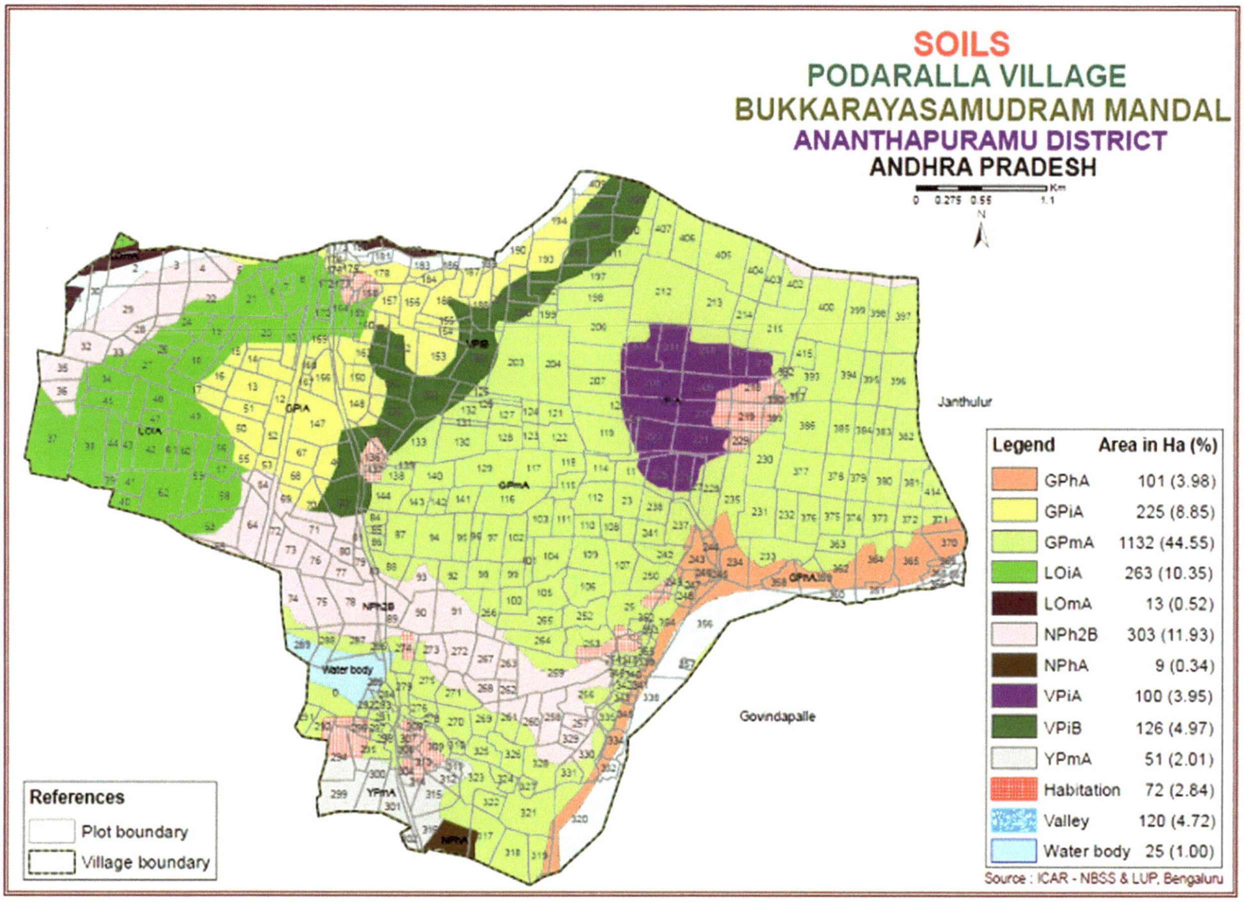

FIGURE 9.7 Soils of Podralla village.

content of the soils ranged from 1.2 to 46.5 ppm. The soils are sufficient in available Mn (> 1.0 ppm). The available copper content of the soils ranged from 0.2 to 2.2 ppm. The available zinc content of the soils ranged from 0.1 to 3.8 ppm and it is deficient in more than 79% of mandal. The available boron content of the soils ranged from 0.01 to 0.7 ppm and it is deficient in more than 79% of mandal.

TABLE 9.6 Description of soils in Podralla village, Bukkarayasamudram mandal, Anantapur district, Andhra Pradesh.

Nilampalle	NPhA	Moderately deep, well-drained, gravelly clay soils on nearly level pediplain with sandy clay loam surface and very slight erosion.
Nilampalle	NPh2B	Moderately deep, well-drained, gravelly clay soils on very gently sloping pediplain with sandy clay loam surface and moderate erosion.
Yadavalaparti	YPmA	Moderately shallow, well-drained, calcareous loamy soil on nearly level alluvial deposits with clay surface and very slight erosion
Lolluru	LOiA	Moderately deep, well-drained, calcareous clayey soils on nearly level alluvial deposits with sandy clay surface and very slight erosion
Lolluru	LomA	Moderately deep, well-drained, calcareous clayey soils on nearly level alluvial deposits with clay surface and very slight erosion
Govindapalle	GphA	Deep, moderately well-drained calcareous clayey soils on nearly level alluvial deposits and valley fills with sandy clay loam surface and very slight erosion
Govindapalle	GpiA	Deep, moderately well-drained, calcareous clayey soils on nearly level alluvial deposits and valley fills with sandy clay surface and very slight erosion
Govindapalle	GPmA	Deep, moderately well-drained, calcareous clayey soils on nearly level alluvial deposits and valley fills with sandy clay surface and very slight erosion
Venkatapuram	VpiA	Moderately shallow, well-drained, calcareous gravelly clay soils on nearly level valley fills with sandy clay surface and very slight erosion
Venkatapuram	VPiB	Moderately shallow, well-drained, calcareous gravelly clay soils on very gently valley fills with sandy clay surface and very slight erosion

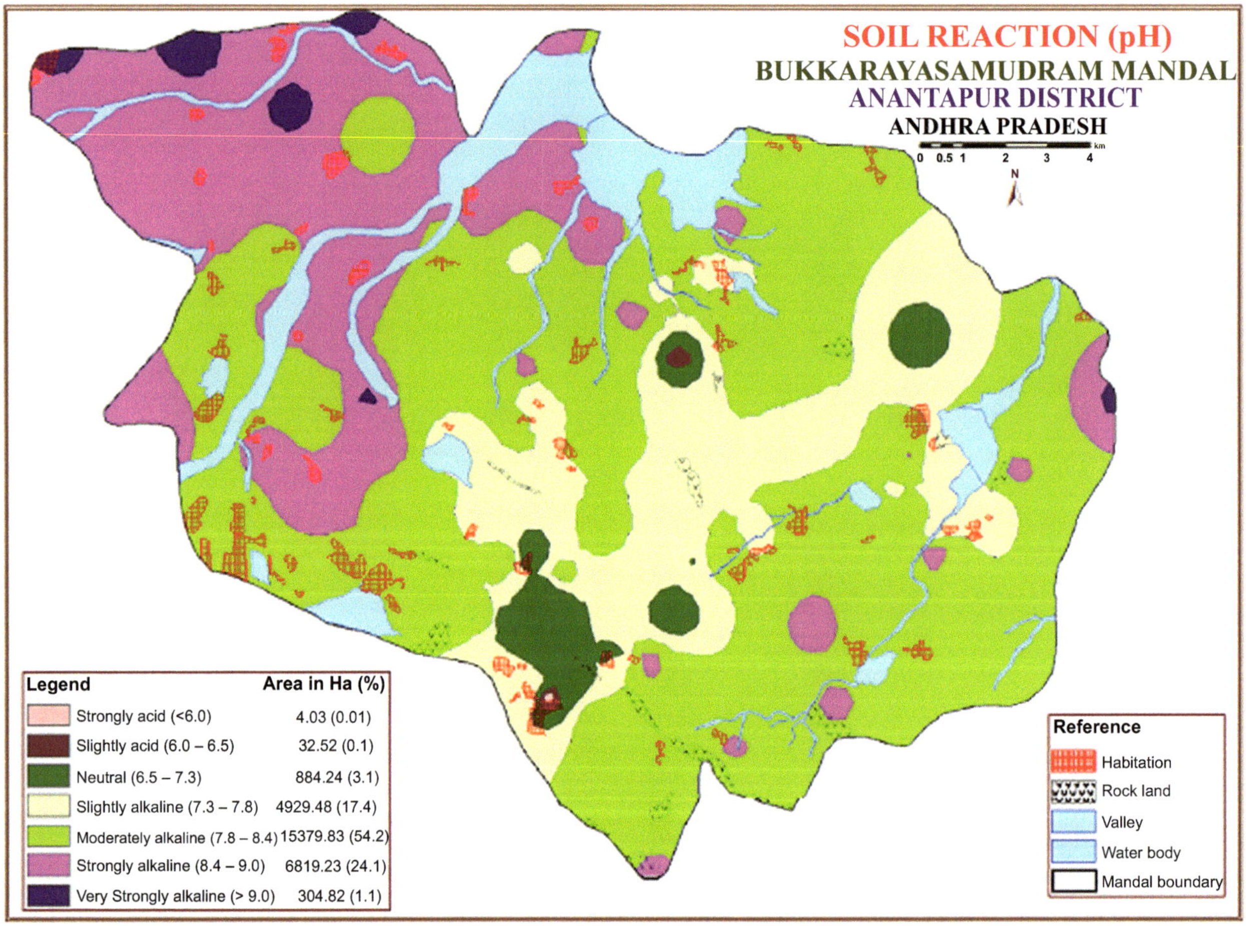

FIGURE 9.8 Soil reaction classes, Bukkarayasamudram mandal.

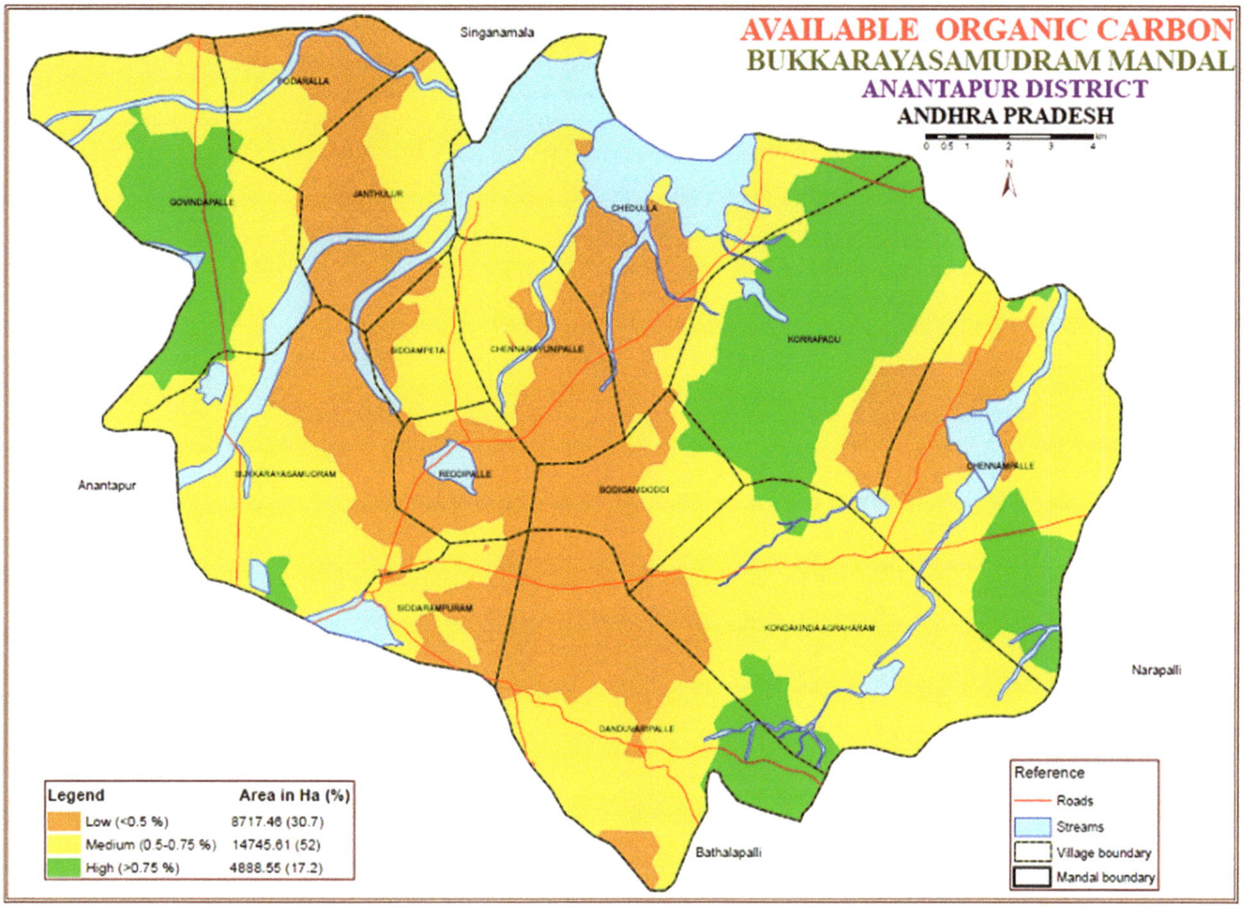

FIGURE 9.9 Soil organic carbon status of Bukkarayasamudram mandal.

9.3.6 Major issues and management strategies

The major limitations, like soil erosion, low rainfall, shallow rooting depth, and severe gravelliness, affect the productive potentials of the soil resources of Bukkarayasamudram mandal. About 40% of the study area is strongly to very strongly alkaline (>8.4 pH). Improper management practices and faulty irrigation practices are the major cause for increasing pH in this mandal. About 6% of the area is under critical situation due to very strong alkalinity, and most of these areas are barren or filled with *Prosopis*. The resource inventorisation also showed that 38% of area is occupied by shallow to moderately shallow soils, which are not suitable for deep rooted crops like mango, etc. Selection of crop based on these land resource database is viable and provides long-term benefit to the farmers. The available nitrogen content in the study area ranged from 62–141.7 kg ha^{-1}, which is very low. Adoption of integrated nutrient management practices is required to sustain the agriculture productivity in the mandal. Many farmers are practicing monocropping (groundnut) and facing crop failures regularly. Rainfed cropping system, consisting of ground nut, pigeon pea, castor, field bean, and horse gram, is very ideal for maintaining the soil health and utilizing the soil resources from different depths of soil profile in these regions.

9.4 Conclusion

Advanced RS and GIS technologies are effective in the preparation of soil mapping at phase level and also helps in evaluating the soil-site suitability. Moderately deep soils and shallow soils occur extensively, occupying about 35.4% of Bukkarayasamudram mandal. Severe soil erosion, shallow rooting depth, gravelliness, high alkalinity, and coarse soil texture are the most important factors limiting the crop selection. This present database helps planners and policymakers to devise strategies for implementation of site-specific soil and water conservation plans for conserving the natural resources.

References

AIS&LUS, 1970. Soil Survey Manual. All India Soil Land Use Survey Organisation, Indian Agricultural Research Institute, New Delhi.

Dwivedi, R.S., 2001. Soil resource mapping: a remote sensing perspective. Remote. Sens. Rev. 20, 89–122.

Dharumarajan, S., Lalitha, M., Natarajan, A., Naidu, L.G.K., Balasubramanian, R., Hegde, R., et al., 2017. Biophysical and socio-economic causes for increasing fallow land in Tamil Nadu. Soil. Use Manage 33, 487–498. Available from: https://doi.org/10.1111/sum.12361.

Dharumarajan, S., Lalitha, M., Hegde, R., Janani, N., Rajawat, A.S., Sastry, L.N., et al., 2018. Status of desertification in South India—assessment, mapping and change detection analysis. Curr. Sci. 115 (2), 331–338.

Dharumarajan, S., Lalitha, M., Ramesh Kumar, S.C., Vasundhara, R., Ramamurthy, V., Hegde, R., et al., 2018. Economic evaluation of some soils of Arid Regions of Anantapur District for groundnut cultivation. Agropedology 28 (02), 107–117.

IMD, 2013. Climate of Andhra Pradesh. India Meteorological Department, GOI, AP.

Jackson, M.L., 1973. Soil Chemical Analysis. Prentice Hall of India Pvt. Ltd., New Delhi.

Kalgapurkar Anitha, P.K., Mishra, Tripathi, K., 2012. Applicability of RS and GIS in soil and water conservation measures. Indian. J. Soil. Conserv. 40, 190–196.

Laake, P.E., 2000. Spatial tools for Laymen. Nat. Resour. Manag. experts. Int. Arch. Photogramm. Remote. Sens. 33, 1593–1600.

Lalitha, M., Dharumarajan, S., Natarajan, A., Niranjana, K.V., Srinivas, S., Naidu, L.G.K., et al., 2016. Need for site specific land resource database for integrated watershed management. Indian. J. Soil. Conserv. 44 (22), 168–176.

Nagaraju, M.S.S., Kumar, N., Srivastava, R., Das, S.N., 2014. Cadastral level soil mapping in basaltic terrain using Cartosat-1-derived products. Int. J. Remote. Sens. 35 (10), 3764–3781.

Sharada, D., RaviKumar, M.V., Venkataratnam, L., Malleswar Rao, T.Ch, 1993. Watershed Prioritisation for Soil Conservation—A GIS Approach, 8. Geocarto International, pp. 27–34.

Shrestha, H.L., 2006. Using GPS and GIS in participatory mapping of community forest in Nepal. Electron. J. Inf. Syst. Dev. Ctries. 25 (5), 1–11.

Soil Survey Staff, 2008. Keys to Soil Taxonomy, eighth ed. SCS, USDA, Washigton, DC.

Srivastava, R., Saxena, R.K., 2004. Techniques of large scale soil mapping in basaltic terrain using satellite remote sensing data. Int. J. Remote. Sens. 25 (4), 679–688.

Yeung, A.K.W., Lo, C.P., 2002. Concepts and techniques of geographic information systems. Int. J. Geogr Inf. Sci. 17 (8), 819–820.

Chapter 10

Mapping of coastal soils using remote sensing and geographic information system

S. Dharumarajan, R. Vasundhara, M. Lalitha, B. Kalaiselvi, K.N. Vidya, R. Srinivasan, N. Maddileti, S. Parvathy, Rajendra Hegde and V. Ramamurthy

ICAR-National Bureau of Soil Survey and Land Use Planning, Regional Centre, Bangalore, Karnataka, India

10.1 Introduction

The Indian coastline has a variety of coastal habitats, such as mangroves, coral reefs, sea grasses, and salt marshes, as well as geomorphological features, such as lagoons, back waters, and estuaries. The coastal terrain that has developed through time in many places offers substantial protection from natural disasters like cyclones and tsunamis. But if they are not properly managed, these natural barriers can become less effective or even disappear. Many people are concerned about the significant problems caused by changes in coastal land use and land cover brought on by human activities in recent decades (Jaiswal et al., 1999; Misra et al., 2013). These unfavorable changes are brought on, in whole or in part, by people and places being vulnerable to local climatic, economic, and sociopolitical disruptions (Nicholls and Small, 2002; Yagoub and Kolan, 2006; Kaliraj et al., 2014). The resources in coastal areas are greatly altered by waves, wind, tide, salt water, saltation, sea level rise, storm surge, cyclones, and human interference activities (Weismiller and Momin, 1977; Meyer and Turner, 1994; Li et al., 2010; Mani Murali and Dinesh Kumar, 2015). Population growth in coastal regions has a direct detrimental influence on land resources and hastens the conversion of natural land cover to built-up areas, settlements, and other recreational uses (Coppin et al., 2004; Bhatta et al., 2010).

Coastal regions, in contrast to other agricultural landscapes, have seen an increase in soil, water, and land salinization. Increased salt causes the biological health and physical qualities of the soil to be depleted, which ultimately causes the transformation of fertile soil into arid, unproductive soil. In addition, the character of soil, land, and water profiles have been altering as a result of frequent tropical cyclones, sea level rise, and human effects on land use systems. Even though technology breakthrough and dissemination in agriculture increased the agricultural production manifold, the lack of appropriate site-specific technologies, mismanagement, and indiscriminate use of technologies without considering the natural resources increased the severity and extent of the land degradation in the coastal region (Dharumarajan et al., 2018a). In addition to increasing land degradation, declining agricultural productivity, groundwater depletion, water logging, salinization, biodiversity losses, and the erratic climate in the coastal environment reduced the economic production and in turn interest in farming among the farmers.

Different organizations have taken up several efforts to enhance the agricultural productivity through introducing novel technologies for optimum utilization of natural resources. However, the efforts could not yield desired results due to unavailability of site-specific database of problems and potentials of natural resources (Dharumarajan et al., 2017). Large-scale soil survey and mapping provide site-specific database to address the field level problems. Large-scale soil mapping has mostly been done by following conventional methods. Remote sensing and geographic information system (GIS) are very useful in creation of natural resource database in an effective way and also in integration of information with the ancillary data for generating an action plan. With the increasing availability of high-resolution satellite data in recent days, studies have been initiated to characterize soils at large scale through the physiography–land use–soil relationship (Dharumarajan et al., 2018b; Lalitha et al., 2020). In this context, the present study was carried out in Mandapam block of Rameswaram, Tamil Nadu with the aim to develop site-specific soil resource database using high-resolution satellite imagery.

Remote Sensing of Soils. DOI: https://doi.org/10.1016/B978-0-443-18773-5.00029-6

10.2 Materials and methods

10.2.1 Study area

Mandapam block is located in eastern part of Ramanathapuram District, Tamil Nadu, India. Palk Strait and the Gulf of Bengal form its northern and southern borders in the east. It is located between 78.43 and 78.94 degrees east longitude and 9.16 to 9.4 degrees north latitude. An island that is a part of the Mandapam block, Rameswaram, is connected to the mainland by a road and the Pamban railway bridge. The research area's location is depicted in Fig. 10.1. The research region is largely covered by quaternary-aged fluvial, fluviomarine, aeolian, and marine sediments. Fluvial deposits with different admixtures of sand, silt, and clay are found along the Vaigai rivers' active courses. The cross-beddings in the paleochannel deposits are well-preserved and consist of fine to medium-sized, brown sands. The coastal plain deposits of sandy clay in a variety of quantities make up the majority of the marine formation. The ancient tidal flat deposits consist of black silty clay, dark clay, and mud, whereas the estuary plain is composed of medium, gray–brown sands. The fluviomarine deposits of Rameswaram Island are made up of indurated sand and dune sands. The Aeolian deposits comprise red sand, which is in nature of ancient dunes and occurs over a 2.5-km wide and 6-km long stretch and lies parallel to the sea coast. These are separated by marshy deposits of black clays. The total net area sown is 17991 ha, which is 72% of the total geographical area (TGA) (Table 10.1). Major land use in the study area is presented in Fig. 10.2.

10.2.2 Climate

A semi-arid climate prevails in the study area. Due to its proximity to the seaside, the study area experiences fewer temperature changes (Table 10.2). The region's highest average temperature varies from 30.9°C in January, when it is the

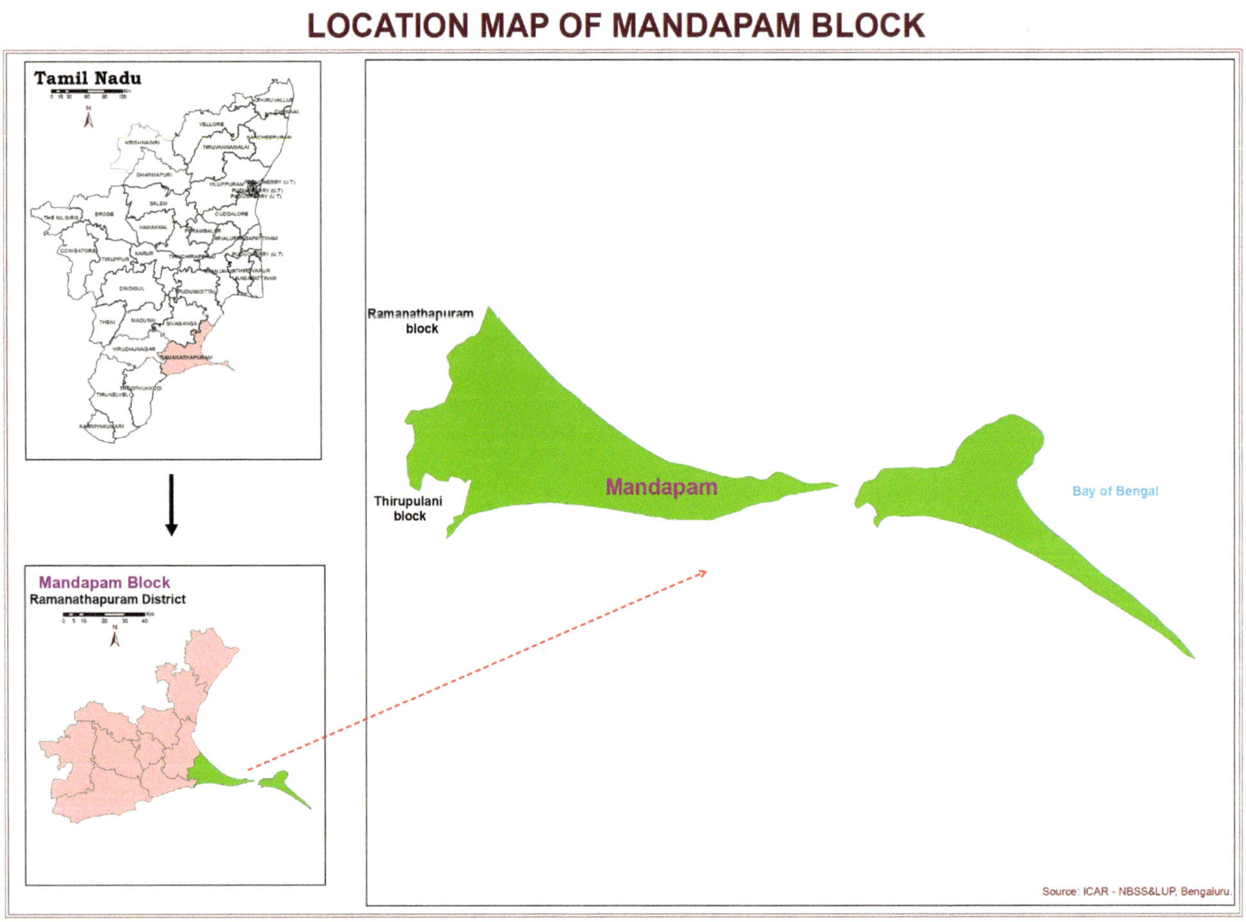

FIGURE 10.1 Location of Mandapam block, Ramanadhapuram district.

TABLE 10.1 Land use pattern of Mandapam block.

Category	Area (ha)
Forests	21
Barren and uncultivable land	1282
Land put to nonagricultural uses	1699
Cultivable waste	979
Permanent pastures and other grazing lands	174
Land under miscellaneous tree crops and groves not included in net area sown	132
Current fallows	1648
Other fallow lands	882
Net area sown	17991
Total geographical area	24808

FIGURE 10.2 Land uses of study area.

coldest, to 36°C in April, when it is the hottest. May's temperature is fairly comparable to that of April. The region receives 880 mm of rain on average yearly. This occurs during the northeast monsoon season, around 69% of the time. November is the wettest month, followed by October. During the southwest monsoon season, 11% of the total quantity of rain falls. The study area has a low relative humidity of 67% in March and a high relative humidity of 78% in September and December.

TABLE 10.2 Temperature and relative humidity of Mandapam Block, Ramanathapuram district.

Month	Temperature (°C)	Rainfall (mm)	Relative humidity (%)	Wind speed (km h^{-1})
Jan	30.9	35.8	73	17.8
Feb	32.5	25.9	70	13.3
Mar	34.8	25.8	67	9.3
Apr	36	48.3	68	7.6
May	35.9	41.2	73	11
Jun	35.6	15.5	73	14.3
Jul	35.5	20.2	72	11.7
Aug	34.8	18.2	74	11.2
Sep	35	43.8	78	9.9
Oct	34.2	221.4	76	9.8
Nov	33	237	77	13.6
Dec	31	147.4	78	18.2

10.2.3 Geomorphology

Coastal alluvial plains, coastal sandy plains, beaches, beach ridges, paleochannels, sand dunes, channel bars, tidal inlets, lagoons, tidal flats, salt pans, and marshy regions are the main geomorphological characteristics in the Mandapam block. Major features on the Mandapam mainland include beaches, paleochannels, sand dunes, coastal alluvial plains, and coastal sandy plains. Significant geomorphological characteristics may be found on the Rameswaram island. On the southwestern coastline of Rameswaram, there is a tongue-shaped spit, and it may have been formed by a littoral current that traveled from Palk Bay to the Gulf of Mannar during the northeast monsoon season (Thanickachalam and Ramachandran, 2002). Smaller coastal dunes are prevalent on the Thangachimadam coastline and the north coast of Rameswaram town. These dunes are rather huge and are located close to the island's southern coast. Two barrier ridges on the north and south sides, as well as a linear lagoon in between, make up the tail-like feature of the island that can be seen in the southeast and extending to Dhanush Kodi (Baskaran et al., 2003).

10.2.4 Land resource inventorization

The goal of the resource inventory is to identify areas that are similar to one another and respond to or are anticipated to act similarly to a certain degree of management. This was accomplished in the Mandapam block of Ramanathapuram by carefully examining the soil characteristics (such as depth, texture, color, structure, consistency, coarse fragments, porosity, soil reaction, etc.). Similar areas were then grouped based on the soil-site characteristics into homogenous (management) units, and their extent and distribution were displayed on the map. From 2019 to 2022, a high-intensity survey at 1: 10000 scale was conducted. The field and laboratory surveys were conducted using a conventional approach (AIS&LUS, 1970; Soil Survey Staff, 2008).

Base maps used

- Sentinel 2 data
- Alos DEM data
- Digitized Survey of India toposheets (1:50,000 scale or larger)
- Other collateral data viz. Global spatial datasets, SRM reports, AESR map, district reports, geology maps, available LULC maps, etc.

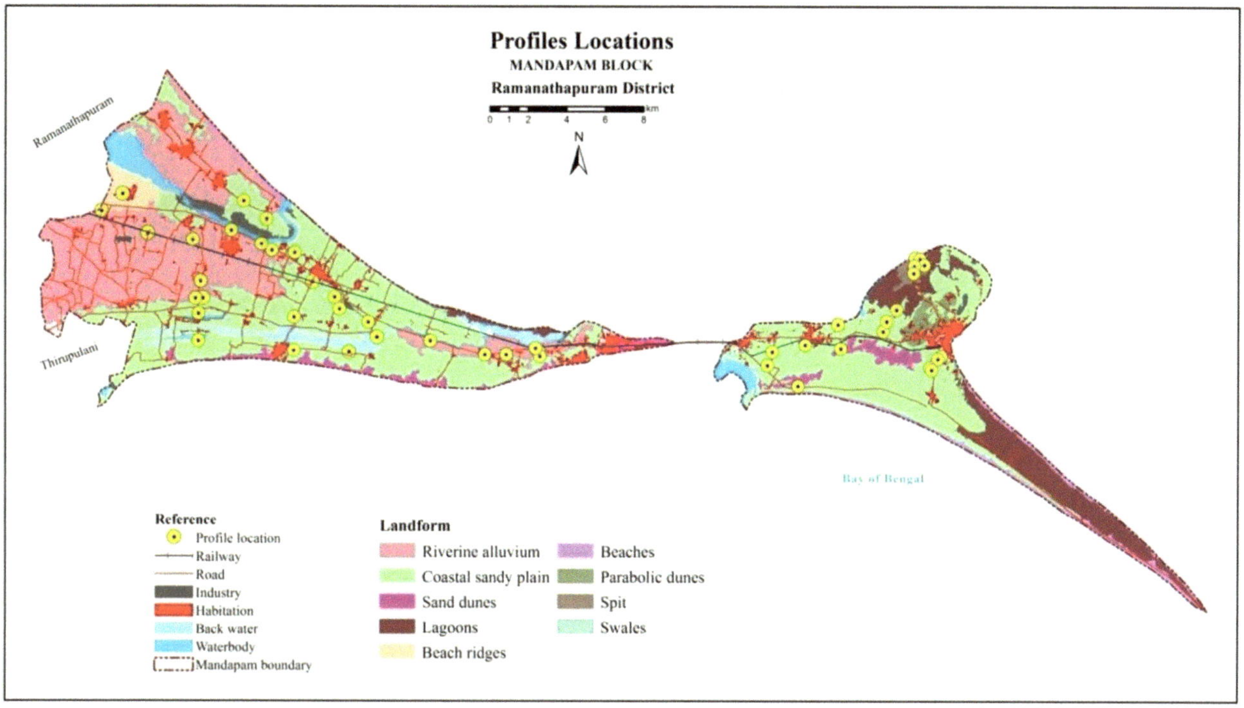

FIGURE 10.3 Soil profile locations.

10.2.5 Land use and land ecological unit map generation

Land use land cover map of the Mandapam block was generated based on visual interpretation of three season sentinel 2 data. Landform unit, land use, and slope were integrated to develop landscape ecological units (LEU) in ArcGIS environment.

10.2.6 Soil mapping

The landform map and the LEU served as the foundation for a thorough survey of the Mandapam block's land resources. In order to account for any changes in the land characteristics, such as breaks in slope, erosion, gravel, stones, etc., profiles (Fig. 10.3) were chosen in the transect at closely spaced intervals. A total of 47 profiles were examined and soil and site parameters were noted using a standard proforma in accordance with the accepted practices (Soil Survey Staff, 2008). The soils were divided into several soil series based on the soil-site features (soil series is the most homogeneous unit having similar horizons and properties and behaves similarly for a given level of management). The primary distinguishing features of the soil series found in the region were horizon sequence, substratum type, gravel quantity and type, and soil depth. Eight series were discovered in the region based on the aforementioned features. Soil samples were collected from typical master profiles for laboratory analysis (Jackson, 1973).

10.2.7 Finalization of soil map and other thematic maps

According to differences in the surface soil texture, slope, and erosion, the region under each series was further divided into phases, and their borders were established. A phase is a subdivision of a soil series that is mostly based on surface characteristics that influence how it is used and managed. Maps of soil-site parameters, such as soil depth, erosion, and soil texture, were generated in order to discover potentials and limitations. Using GIS software, pH and organic carbon maps were created using the kriging interpolation technique.

10.3 Results and discussion

10.3.1 Land use land cover and landscape ecological units

The major land use in the Mandapam block comprises of fallow lands which occupies about 41% of the study area, followed by plantations 22% TGA (Fig. 10.4). Land with single crop occupies 8.0% of the study area and double crop occupies 7.2% of the TGA. Scrublands and mangroves occupy 1.3% and 3.0% of TGA, respectively. Mandapam block is primarily cultivated for coconut, followed by rice, vegetables, and jasmine. Eleven LEUs were delineated in Mandapam block based on landform, slope, and land use. The LEU units are relatively homogeneous in terms of the main factors of soil formation and soil characteristics and are used as a base map for ground truth verification. Table 10.3 represents the description of landforms, which is used in deriving LEU. Detailed description of LEU units was presented in Table 10.4 and Fig. 10.5.

10.3.2 Soil mapping

The soils occurring in this study area are grouped under Riverine alluvium, coastal sandy plain, and sand dunes category. During the course of the survey, 8 soil series with 15 phases were identified (Tables 10.5 and 10.6 and Fig. 10.6). Out of this, four soil series were identified in riverine alluvium landform, three series were identified in coastal sandy plain, and one was identified in sand dunes. The identifying characteristics of all the eight soil series (Fig. 10.7) mapped in the block are given in Table 10.7 and series-wise physiochemical characteristics are given in Table 10.8.

FIGURE 10.4 Land use land cover map (LULC) of Mandapam block, Tamil Nadu District.

TABLE 10.3 Different landform features of Mandapam block.

S. No	Landform	Description
1	Coastal alluvial plains	A largely flat landform created by the deposition of sediment over a long period of time by one or more rivers coming from highland
2	Beaches	A beach is a narrow, gently sloping strip of land that lies along the edge of an ocean
3	Beach ridges	A wave-swept or wave-deposited ridge running parallel to a shoreline. It is commonly composed of sand as well as sediment worked from underlying beach material.
4	Paleochannel	A geological term describing a remnant of an inactive river or stream channel that has been filled or buried by younger sediment
5	Sand dunes	Any accumulation of sand grains shaped into a mound or ridge by the wind under the influence of gravity.
6	Channel bars	A bar in a river is an elevated region of sediment (such as sand or gravel) that has been deposited by the flow.
7	Tidal inlets	An opening in the shoreline through which water penetrates the land, thereby proving a connection between the ocean and bays, lagoons, and marsh and tidal creek systems.
8	Lagoons	A lake of salt water that is separated from the sea by sand or rock
9	Tidal flats	Essentially horizontal and commonly muddy or marshy land that is covered and uncovered by the rise and fall of tides
10	Salt pans	A shallow container or depression in the ground in which salt water evaporates to leave a deposit of salt.
11	Marshy lands	A wetland that is dominated by herbaceous rather than woody plant species.

TABLE 10.4 Description of LEU in Mandapam block.

Physiographic Region	Physiographic subregion	Major landform	Landform unit	LEU	Description
Coastal plain (P)	East coast (e)	Marine landform (Ma)	Coastal alluvium (Ca)	PeMaCa1a	Nearly level coastal alluvium under arable lands
				PeMaCa1w	Nearly level coastal alluvium under waste lands
			Coastal sandy plain (Cp)	PeMaCp1a	Nearly level coastal sandy plain under arable lands
				PeMaCp1w	Nearly level coastal sandy plain under waste lands
				PeMaCp1f	Nearly level coastal sandy plain under forest
				PeMaCp2w	Very gently sloping coastal sandy plain under waste lands
			PC (paleochannels)	PeMaPc1a	Nearly level Beach ridges under arable lands
			Sand dune (Sd)	PeMaSd2w	Very gently sloping sand dune under waste lands
			Coastal spits (Cs)	PeMaCs1w	Nearly level coastal sandy spits under waste lands
			Beach ridges (Br)	PeMaBr1w	Nearly level Beach ridges under waste lands
			Lagoons	PeMaBr1f	Nearly level lagoons under forest

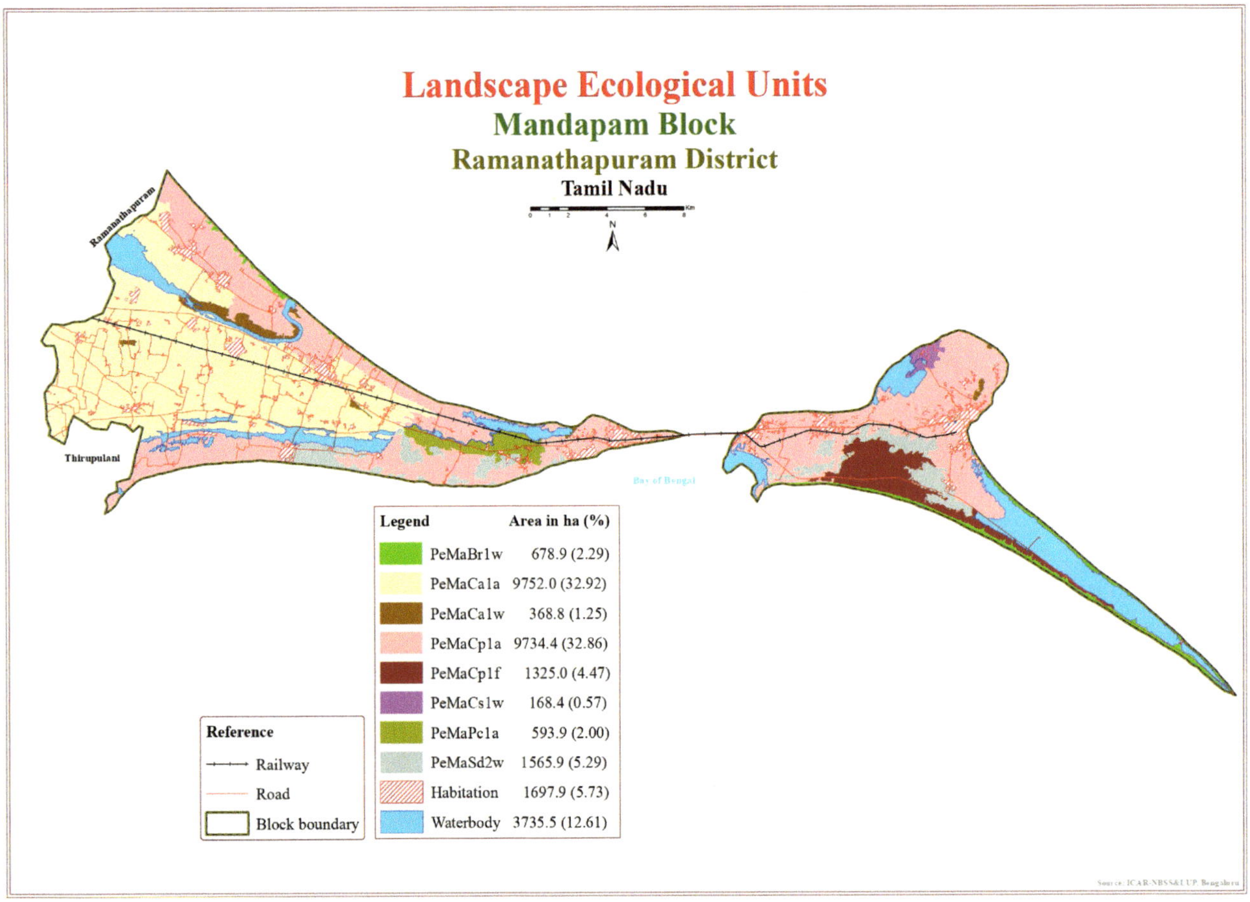

FIGURE 10.5 Landscape Ecological Units (LEU) of Mandapam block, Tamil Nadu District.

TABLE 10.5 Classification of the soils.

Sl. No	Series	Area (ha)	% TGA	Classification
1	Ragunathapuram (RP)	4528	15.3	Sandy over coarse loamy, mixed isohyperthermic Typic Ustifluvents
2	Uchipuli (UP)	3133	10.5	Sandy over coarse loamy, mixed isohyperthermicaeric endoaquepts.
3	Mandapam (MM)	593	2.0	Sandy over coarse loamy, mixed isohyperthermic (cal) aquic Calciustepts
4	Vazhuthur (VZ)	2059	7.1	Sandy over coarse loamy, mixed isohyperthermic Fluventic Haplustepts
5	Rameswaram (RM)	8712	29.4	Sandy, mixed isohyperthermic ustic Quartzipsamments
6	Vadakadu (VK)	168	0.6	Sandy, mixed isohyperthermic Typic Ustipsamments
7	Meyyampuli (MP)	1022	3.5	Sandy over coarse loamy, mixed isohyperthermic Typic Ustipsamments
8	Thankatchimadam (TM)	1565	5.3	Sandy, mixed isohyperthermic Ustic Quartzipsamments

TABLE 10.6 Description of soils in Mandapam block, Anantapur district, Andhra Pradesh.

LEU	Series	Soil phases	Description of soil characteristics
DsGnH5f	Ragunathapuram	RPAls	Very deep, well-drained, fine loamy soils on nearly level plains with loamy sand surface
DsGnH5sc	Uchipuli (UP)	UPAls	Very deep, moderately well-drained, fine loamy soils on nearly level plains with loamy sand surface
DsGnD2sc		UPAs	Very deep, moderately well-drained, coarse loamy soils on nearly level plains with sandy surface
	Mandapam (MM)	MMAsl	Very deep, well-drained, coarse loamy soils on nearly level plains with loamy sand surface
DsGnD3sc	Vazhuthur (VZ)	VZAls	Very deep, moderately well-drained, fine loamy soils on nearly level plains with loamy sand surface
DsGnD5sc		VZAs	Very deep, moderately well-drained, fine loamy soils on nearly level plains with sandy surface
DsGnP1d	Rameswaram (RM)	RMAs	Very deep, excessively well-drained, sandy soils on nearly level coastal plains with sandy surface
DsGnP1s		RMAls	Very deep, excessively drained, coarse loamy soils on nearly level coastal plains with loamy sand surface
		RMBs	Very deep, excessively drained, sandy soils on very gently sloping coastal plains with sandy surface
DsGnP2d		RMBse2	Very deep, excessively drained, sandy soils on very gently sloping coastal plains with sandy surface and moderate erosion
	Vadakadu (VK)	VKAls	Very deep, somewhat excessively drained, sandy soils on nearly level coastal plain and beaches with loamy sand surface
	Meyyampuli (MP)	MPAls	Very deep, well-drained, coarse loamy soils on nearly level coastal plain and beaches with loamy sand surface
DsGnP2f		MPAs	Very deep, well-drained, coarse loamy soils on nearly level coastal plain and beaches with sandy surface
	Thankatchimadam (TM)	TMBs	Very deep, Excessively drained, sandy soils on very gently sloping sand dunes with sandy surface
		TMBse3	Very deep, Excessively drained, sandy soils on very gently sloping sand dunes with sandy surface and severe erosion

10.3.2.1 Vadakadu series

The Vadakadu series consists of very deep, somewhat excessively drained soils formed from coastal spits, occurring on level to nearly level slope with 0%–1% gradient. The parent material is coastal alluvium. The surface soils are characterized by the brown loamy sand texture and neutral pH. The subsurface soils are dark grayish brown to very pale brown in color and loamy sand to sandy texture with neutral to strongly alkaline pH. The soils of Vadakadu Series are sandy, mixed isohyperthermic Typic Ustipsamments. The pH, CCE, and ESP values are in increasing trend with the depth of the soil; whereas, OC and CEC are in decreasing trend over the depth of the soil. EC values has followed varied trend over the depth of the soil.

10.3.2.2 Thankatchimadam series

Thankatchimadam series consists of very deep depth, excessively drained soils on very gentle sloping formed over sand dunes. The parent material is coastal sands. The surface soils are characterized by dark gray in color, sandy texture, and slightly alkaline pH. The subsurface soils are brown to strong brown in color and sandy texture with neutral pH.

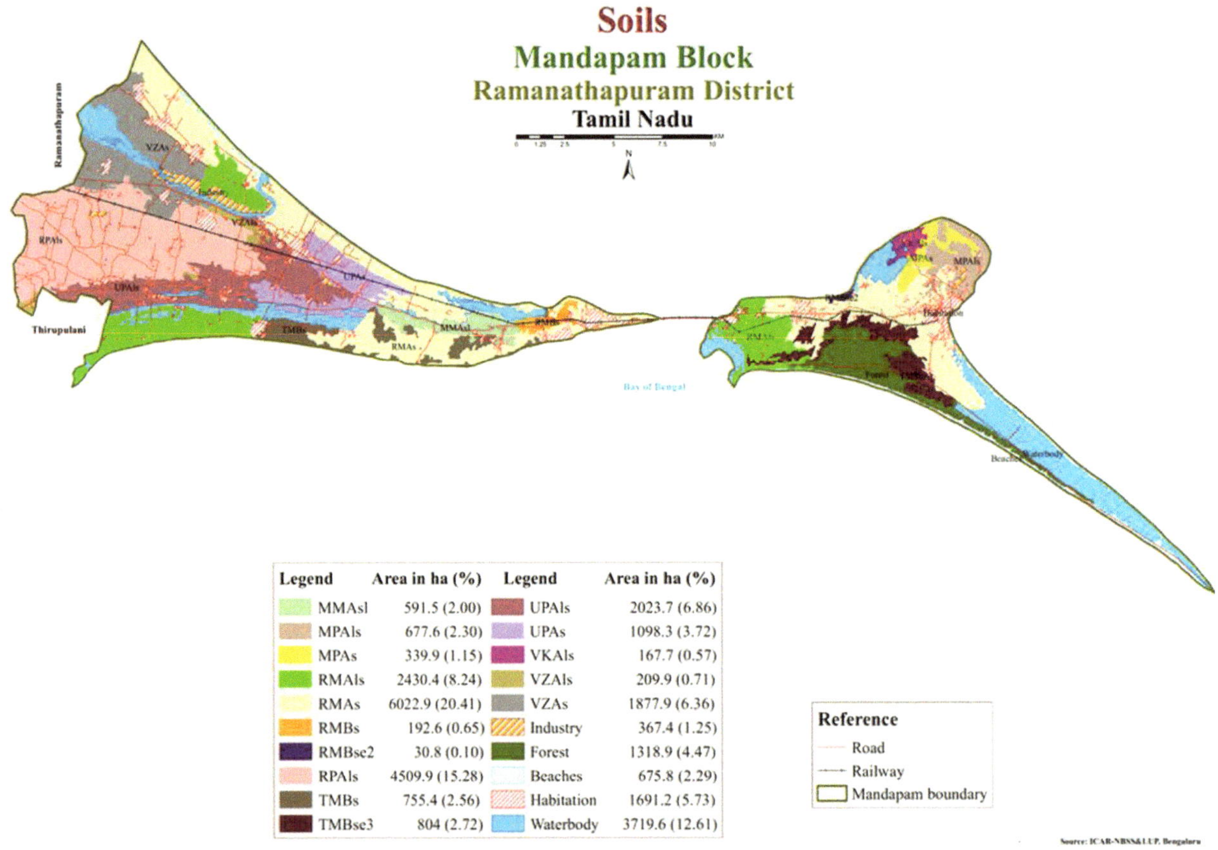

FIGURE 10.6 Distribution of soil phases or management units in Mandapam block.

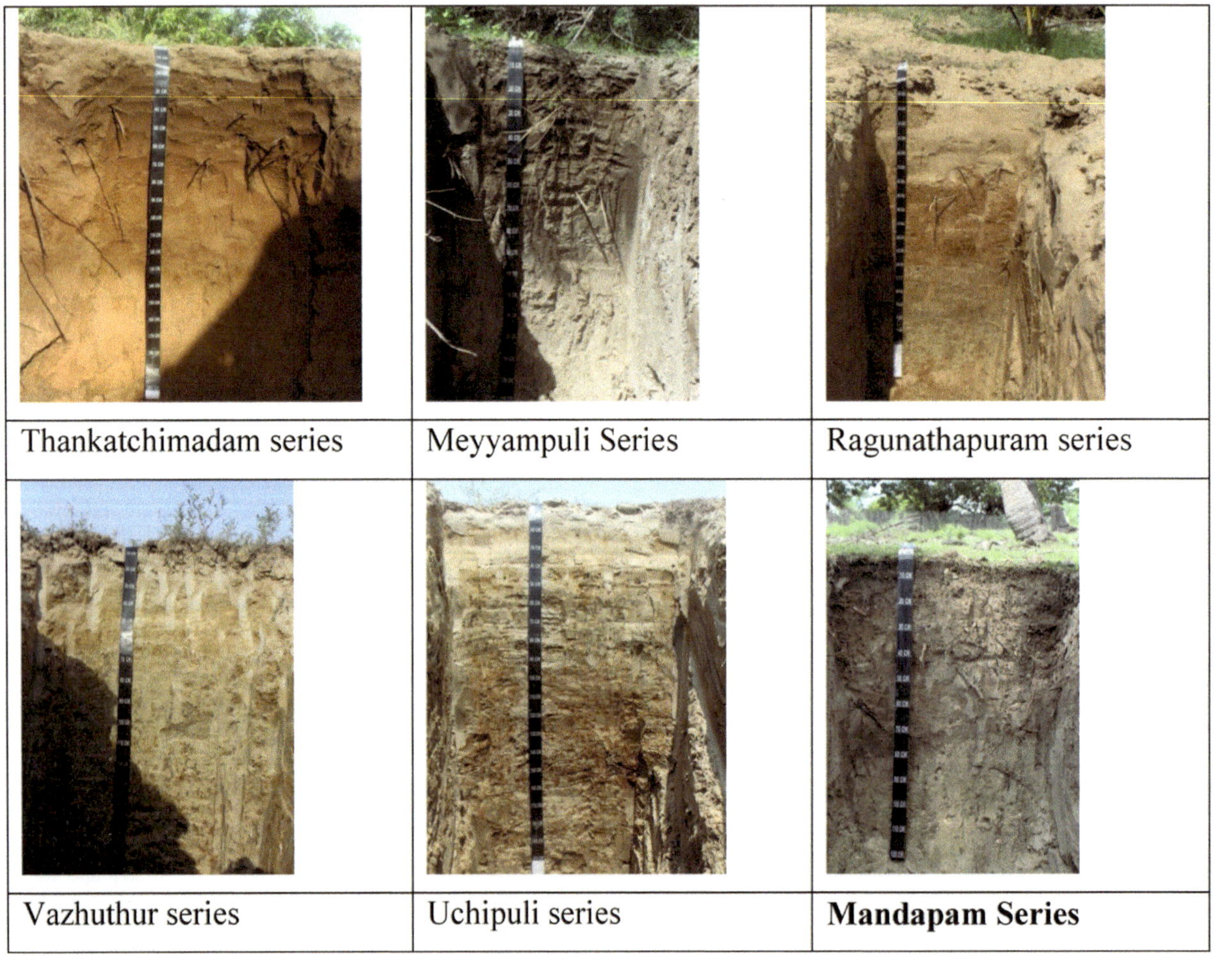

FIGURE 10.7 Major soils of Mandapam block of Ramanathapuram district.

TABLE 10.7 Identifying characteristics of soils of Mandapam block.

Series	Drainage	Color	Texture		Horizon sequence calcareousness
			Surface	subsurface	
Riverine alluvium					
Ragunathapuram (RP)	Well	10YR6/4, 7.5YR 5/8 5Y5/8	s	scl-sc	Ap-Bw
Uchipuli (UP)	Mod. well	2.5Y5/2 2.5Y4/2, 2.5Y6/2	s	Scl	Ap-A2- Bw1-Bw2- Bw3
Mandapam (MM)	Well-drained	10YR4/1, 10YR4/2,10YR6/2,	ls	sl	Ap-Bw1-Bw2-C
Vazhuthur (VZ)	Mod. well	2.5Y4/8, 5Y6/2, 2.5Y6/6	s	ls-scl	Ap-Bw1-Bw2- Bw3
Coastal sandy plain					
Rameswaram (RM)	Excessively	10YR4/6, 10YR3/6	s	s	AP-Bw-Ck
Vadakadu (VK)	Somewhat excessively	10YR3/2, 10YR4/1	ls	ls-s	AP-C1-C2-C3
Meyyampuli (MP)	Well-drained	10YR4/2, 10YR6/2	ls	ls-sl	AP-C1-C2-C3
Sand dunes					
Thankatchimadam (TM)	Excessively	10YR4/1, 7.5YR4/2,6/3	s	s	A-C1-C2-C3-C4

TABLE 10.8 Physical and chemical characteristics of soil series.

Depth (cm)	Sand (%)	Silt (%)	Clay (%)	pH (1: 2.5)	EC (1: 2.5) (dSm^{-1})	OC (%)	CCE (%)	ESP (%)	CEC (c mol (p +) kg^{-1})
Vadakadu-Sandy, mixed isohyperthermic Typic Ustipsamments									
0–20	88.82	3.00	8.18	6.99	0.082	1.39	5.25	0.59	2.94
20–80	87.44	3.89	8.67	7.36	0.046	0.38	3.63	0.93	1.86
80–109	95.28	2.41	2.31	8.19	0.051	0.27	2.50	2.22	1.67
109–138	86.19	11.51	2.30	9.23	0.066	0.27	18.88	13.75	0.98
Thankatchimadam-Sandy, mixed isohyperthermic Ustic Quartzipsamments									
0–15	95.67	0.86	3.47	8.46	0.081	0.52	1.48	2.44	3.02
15–47	94.57	3.12	2.31	6.91	0.047	0.25	0.00	3.41	2.81
47–125	97.49	0.20	2.31	6.78	0.033	0.14	0.00	3.67	1.84
125–190	97.65	0.63	1.73	6.86	0.024	0.08	0.00	3.66	1.19
Rameswaram-Sandy, mixed isohyperthermic ustic Quartzipsamments									
0–12	95.68	3.74	0.58	**6.75**	0.06	0.73	4.13	1.07	2.84
12–57	92.98	1.83	5.19	**6.16**	0.033	0.11	4.13	1.74	2.25
57–150	91.58	4.96	3.46	**6.6**	0.034	0.15	3.63	2.08	3.04
Meyyampuli-Sandy over coarse loamy, mixed isohyperthermic Typic Ustipsamments									
0–13	87.35	6.20	6.45	8.01	0.163	1.39	7.75	1.55	5.88
13–52	87.22	4.03	8.75	7.71	0.127	0.65	6.50	2.45	4.61

(Continued)

TABLE 10.8 (Continued)

Depth (cm)	Sand (%)	Silt (%)	Clay (%)	pH (1: 2.5)	EC (1: 2.5) (dSm^{-1})	OC (%)	CCE (%)	ESP (%)	CEC (c mol (p +) kg^{-1})
52–72	84.10	4.90	10.99	7.7	0.114	0.5	9.63	3.61	3.43
72–145 +	74.23	9.12	16.66	8.04	0.12	0.42	25.50	7.70	3.33
Ragunathapuram-Sandy over coarse loamy, mixed isohyperthermic Typic Ustifluvents									
0–20	94.37	2.74	2.89	6.26	**0.095**	0.55	–	1.23	1.94
20–42	93.89	2.66	3.46	6.45	0.046	0.41	–	0.34	1.30
42–72	93.08	2.88	4.04	6.61	0.035	0.32	–	0.89	1.08
72–134	82.59	1.60	15.81	6.91	0.029	0.41	–	5.86	4.97
134–180	75.95	2.77	21.27	7.94	0.221	0.33	–	13.99	10.15
180–208	72.52	2.53	24.95	9.14	0.356	0.32	–	19.93	11.66
Vazhuthur-Sandy over coarse loamy, mixed isohyperthermic Fluventic Haplustepts									
0–18	92.86	3.67	3.47	7.20	0.249	0.69	–	14.25	1.51
18–62	88.70	2.60	8.70	7.71	0.048	0.35	–	2.17	3.56
62–100	79.37	4.27	16.36	8.94	0.177	0.28	–	12.96	7.13
100–144	73.63	2.59	23.78	9.43	0.474	0.28	–	16.13	11.23
144–205	75.64	2.49	21.87	9.50	0.454	0.30	–	14.39	9.72
Uchipuli-Sandy over coarse loamy, mixed isohyperthermic aeric Endoaquepts									
0–18	95.58	1.53	2.89	5.39	0.028	0.20	–	2.01	1.73
18–40	93.57	3.54	2.89	5.10	0.017	0.08	–	1.29	1.51
40–88	76.95	2.43	20.62	6.04	**0.015**	0.06	–	0.94	7.88
88–135	74.51	2.94	22.55	7.12	0.041	0.04	–	0.83	11.02
135–185	73.86	3.56	22.57	8.19	0.052	0.06	–	1.02	11.34
185–220	72.56	4.13	23.31	8.22	0.051	0.06		1.26	13.93
Mandapam-Sandy over coarse loamy, mixed isohyperthermic (cal) aquic Calciustepts									
0–13	83.21	5.09	11.70	8.76	0.11	1.38	10.75	3.43	6.27
13–40	76.30	4.98	18.72	8.80	0.17	0.61	15.88	3.31	6.37
40–80	77.38	4.96	17.66	8.82	0.27	0.42	14.38	7.58	5.88
80–145	88.10	3.78	8.11	8.88	0.31	0.07	8.63	14.42	4.12

The soils of Thankatchimadam series are sandy, mixed isohyperthermic Ustic Quartzipsamments. In this series the pH, EC, and OC are decreasing with increasing depth

10.3.2.3 Rameswaram series

The Rameswaram series consists of deep, excessively drained soils formed from coastal sandy plain and occurring on very gently slope with 1%–3% sloping land. The parent material is coastal alluvium. The surface soils are characterized by dark yellowish brown in color, loamy sand texture, and neutral pH. The subsurface soils are dark yellowish brown in color and loamy sandy texture with neutral to slightly acidic pH. The soils of Rameswaram series are loamy sandy, mixed isohyperthermic ustic Quartzipsamments. Overall, pH, EC, OC, and CCE are in decreasing trend over the depth of the soil. But, ESP and CEC showed increasing trend over the depth of the soil.

10.3.2.4 Meyyampuli series

Meyyampuli series consists of very deep, well-drained soils formed from coastal sandy plain and occur on level to nearly level with 0%–1% slope. The parent material is coastal alluvium. The surface soils are characterized by dark grayish brown in color, loamy sand texture, and slightly alkaline pH. The subsurface soils are dark yellowish brown to light brownish gray in color and loamy sand to sandy loam texture, with neutral to slightly alkaline pH. The soils of Meyyampuli series are sandy over coarse loamy, mixed isohyperthermic Typic Ustipsamments.

10.3.2.5 Ragunathapuram series

Ragunathapuram series consists of very deep, well-drained soils formed by riverine alluvium and occur on level to nearly level (0%–1%) slope. The parent material is riverine alluvium. The surface soils are characterized by pale brown in color, sandy texture, and slightly acidic pH. The subsurface soils are light yellowish brown to strong brown in color and sandy clay loam to sandy loam in texture, with slightly acidic to very strongly alkaline pH. The soils of Ragunathapuram series are sandy over coarse loamy, mixed isohyperthermic ustic Fluvents.

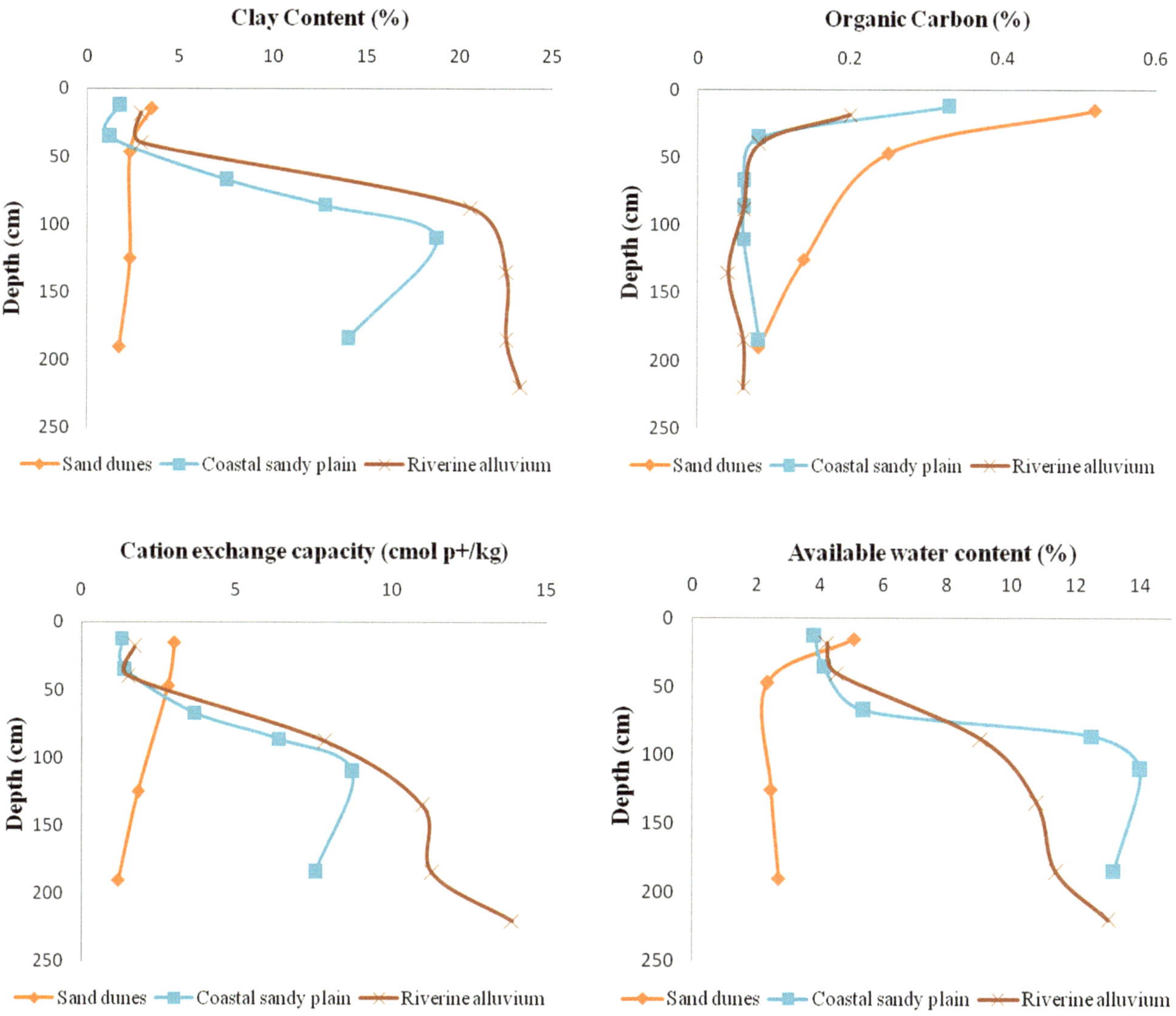

FIGURE 10.8 Soil properties under three dominant landform of Mandapam block.

10.3.2.6 Vazhuthur series

The Vazhuthur series consists of very deep, moderately well-drained fine loamy soils on level to nearly level lands. The parent material is riverine alluvium. The surface soils are characterized by the gray in color, loamy sand texture, and neutral pH. The subsurface soils are olive yellow to light olive gray in color and sandy clay loam to sandy loam in texture, with neutral to very strongly alkaline pH. The soils of Vazhuthur series are sandy over coarse loamy, mixed iso-hyperthermic Typic Ustifluvents.

10.3.2.7 Uchipuli series

Uchipuli series consists of very deep, moderately well-drained fine loamy soils formed from riverine alluvium and occurs on level to nearly level lands with 0%–1% slope. The surface soils are characterized by the light gray in color, loamy sand texture, and strongly acidic pH. The subsurface soils are very pale brown to pale red in color and sand to sandy clay loam in texture, with slightly acid to strongly alkaline pH. The soils of UP series are sandy over coarse loamy, mixed isohyperthermic aeric Endoaquepts.

10.3.2.8 Mandapam series

The Mandapam series consists of deep, well-drained, fine loamy soils on level to nearly level lands. The surface soils are characterized by the dark gray in color, loamy sand texture, and strongly alkaline pH. The subsurface soils are light gray to dark brownish gray in color and loamy sand to sandy loam in texture, with strongly alkaline pH. The soils of MM series are sandy over coarse loamy, mixed isohyperthermic (cal) aquic Calciustepts. Soil properties under three dominant landform of Mandapam block is presented in the Fig. 10.8.

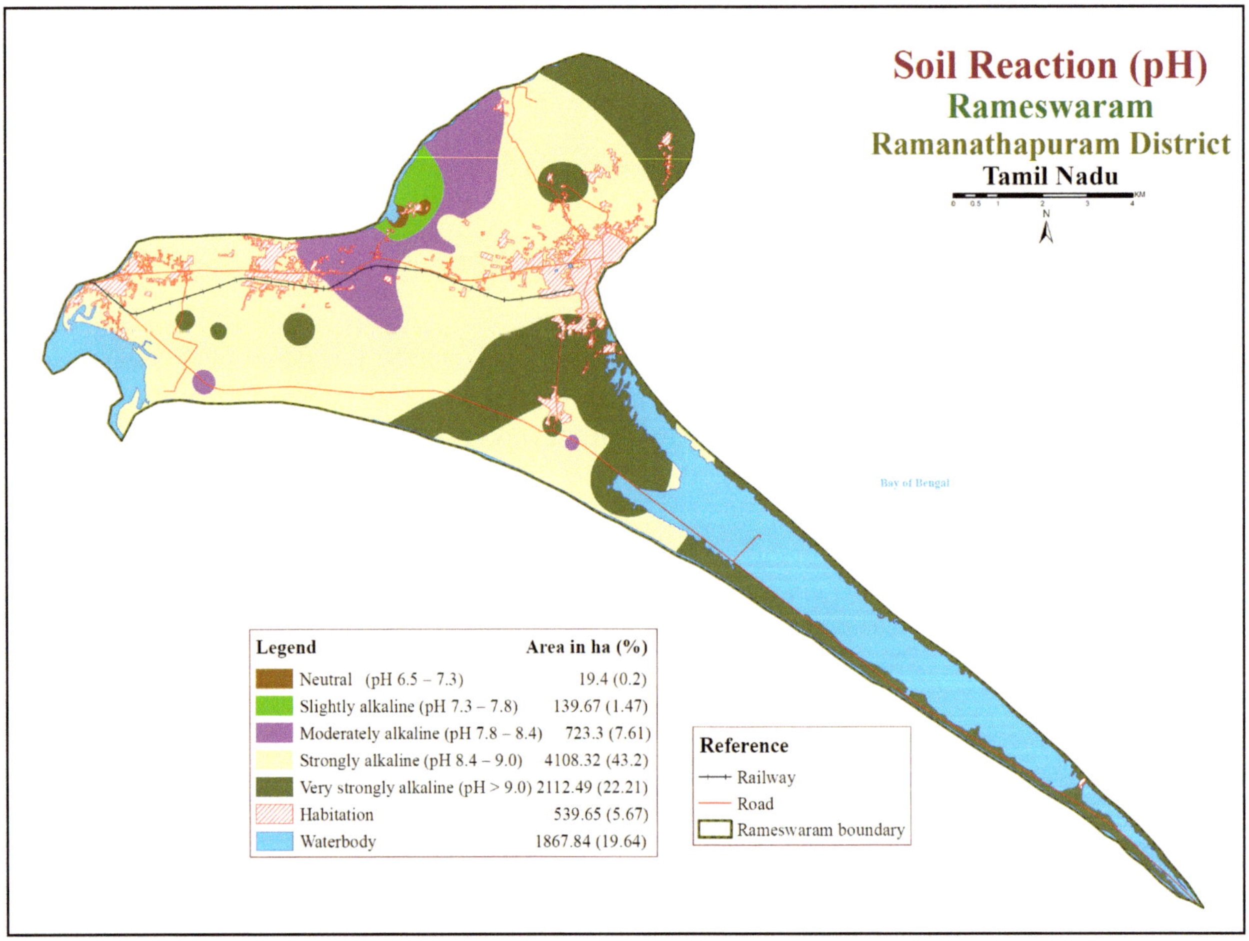

FIGURE 10.9 Soil reaction classes, Rameswaram Island.

10.3.3 Soil physicochemical properties

10.3.3.1 Soil reaction (pH)

Fig. 10.9 display soils from the Rameswaram island in various pH classes. According to the soil reaction (pH) results, 43.2% of the region is highly alkaline (pH 8.4–9.0), 22.21% is very strongly alkaline (pH >9), 7.61% is moderately alkaline (pH 7.8–8.4), roughly 1.47% is mildly alkaline (pH 7.3–7.8), and just 0.8% is neutral pH. Fig. 10.10 displays soils on the Mandapam Mainland of various pH classes. According to the soil reaction (pH), 40% of the region is neutral (pH 6.5–7.3), 20% is slightly acidic (pH 6.0–6.5), 10% is under moderately alkaline (pH 7.8–8.4), and 10% is extremely alkaline. Only the mainland contains acidic soils.

10.3.3.2 Soil organic carbon

Organic carbon content in the soils of Rameswaram island is shown in Fig. 10.11. About 29.69% of area has high organic carbon content (> 0.75%), followed by low (28.69%) and medium (16.85%) status in organic carbon. Organic carbon content in the soils of Mandapam mainland is shown in Fig. 10.12. About 64% of area has medium organic carbon content (0.5%–0.75%) and high (16%) status in organic carbon

10.3.4 Land capability maps

As the majority of the soils in the research region have a coarse texture, the soil is the main factor limiting all of the crops that may be grown there. Mandapam block soils have been divided into three land capability classes and six subclasses (Fig. 10.13). The Mandapam block has around 31% of good cultivable lands (Class II), which have modest restrictions on soil and drainage, limiting the kind of crops that may be grown there and necessitating mild conservation measures. The percentage of the block that is suitable for cultivation is about 29% (class III lands). These Class III

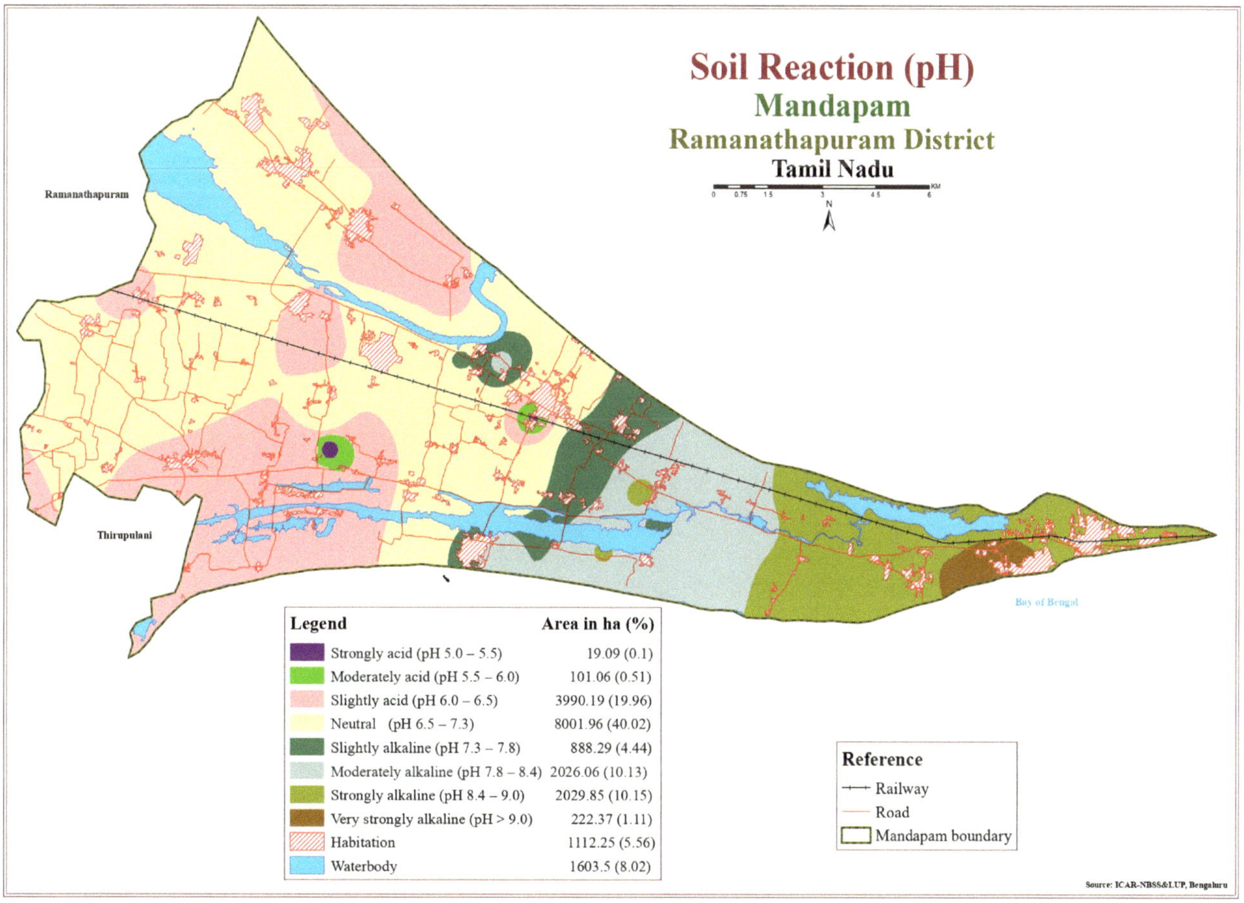

FIGURE 10.10 Soil reaction classes, Mandapam Mainland.

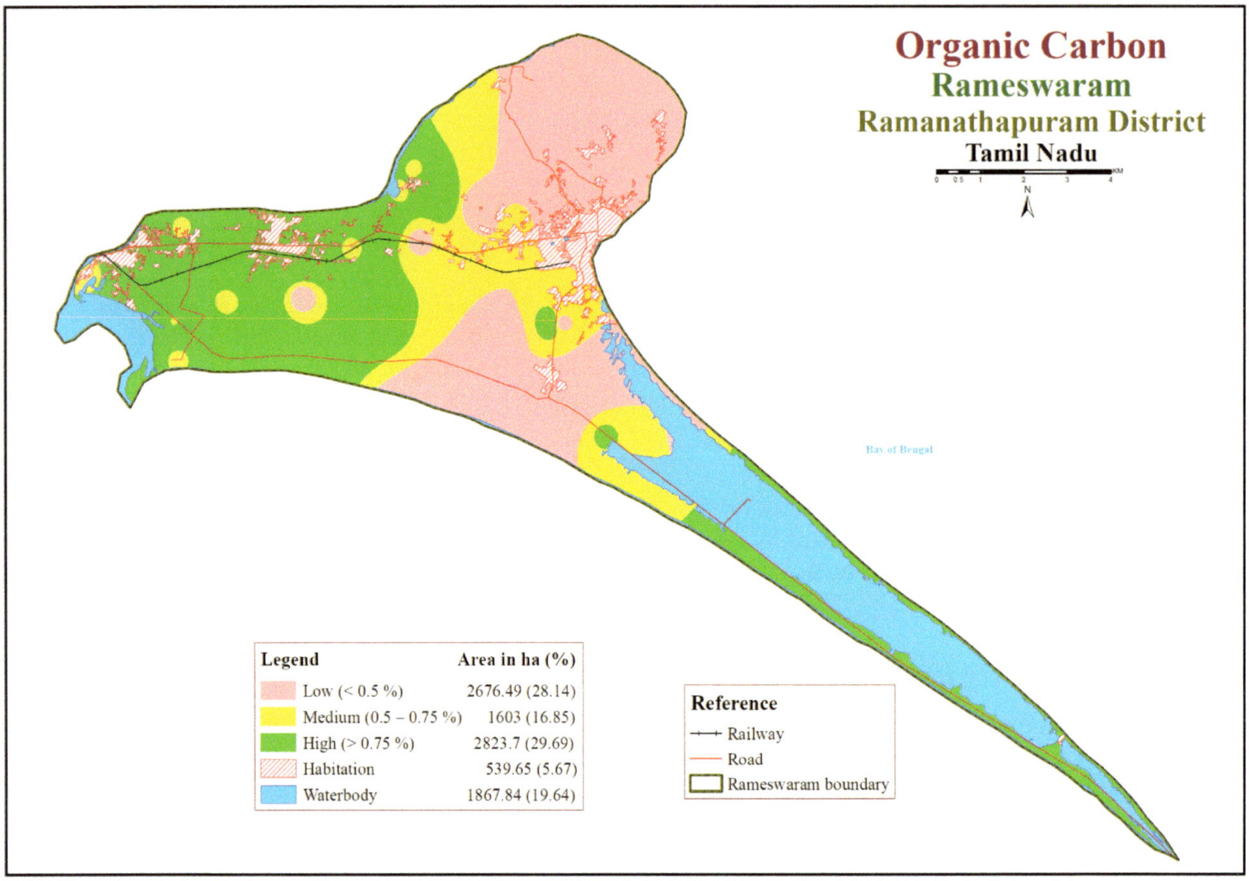

FIGURE 10.11 Soil organic carbon status of Rameswaram Island.

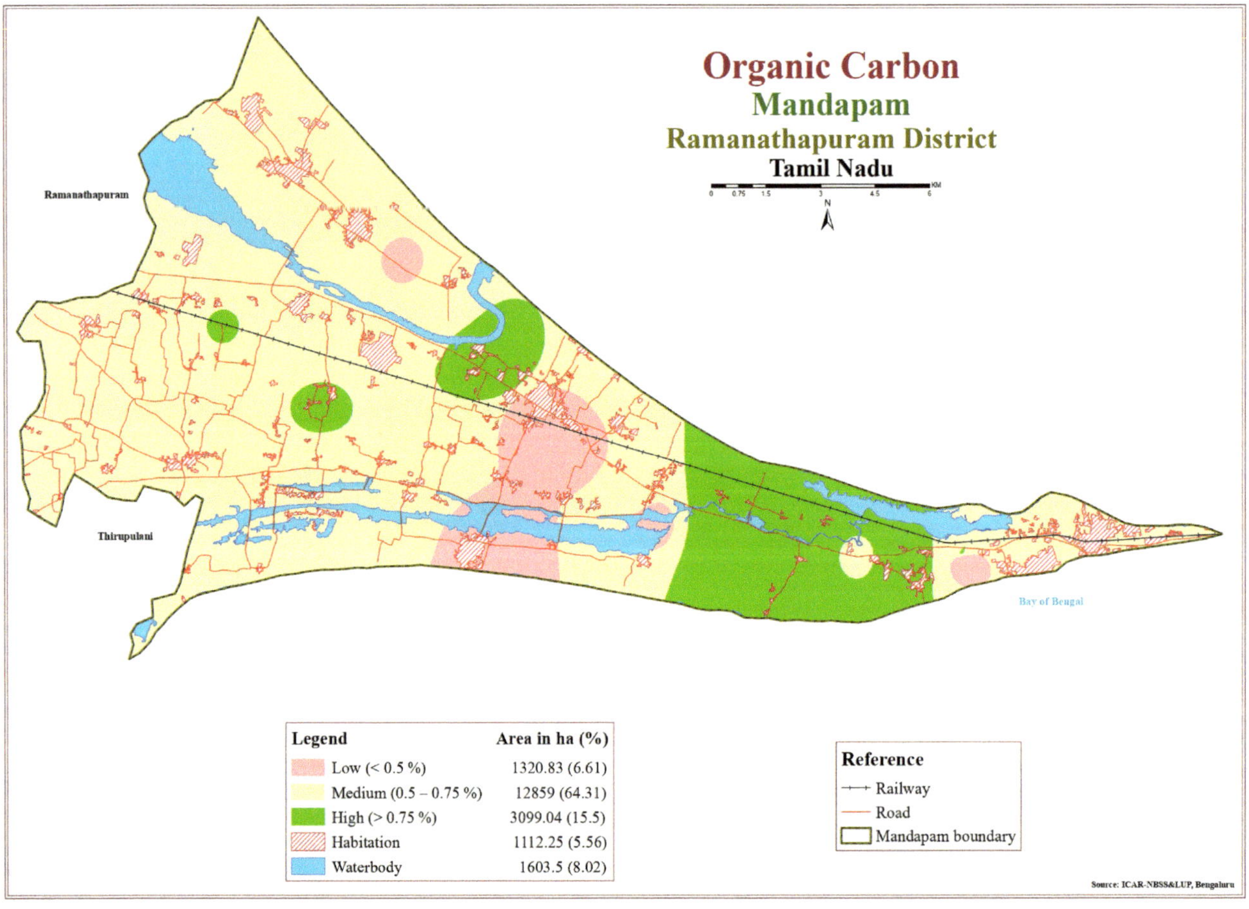

FIGURE 10.12 Organic carbon status, Mandapam Mainland.

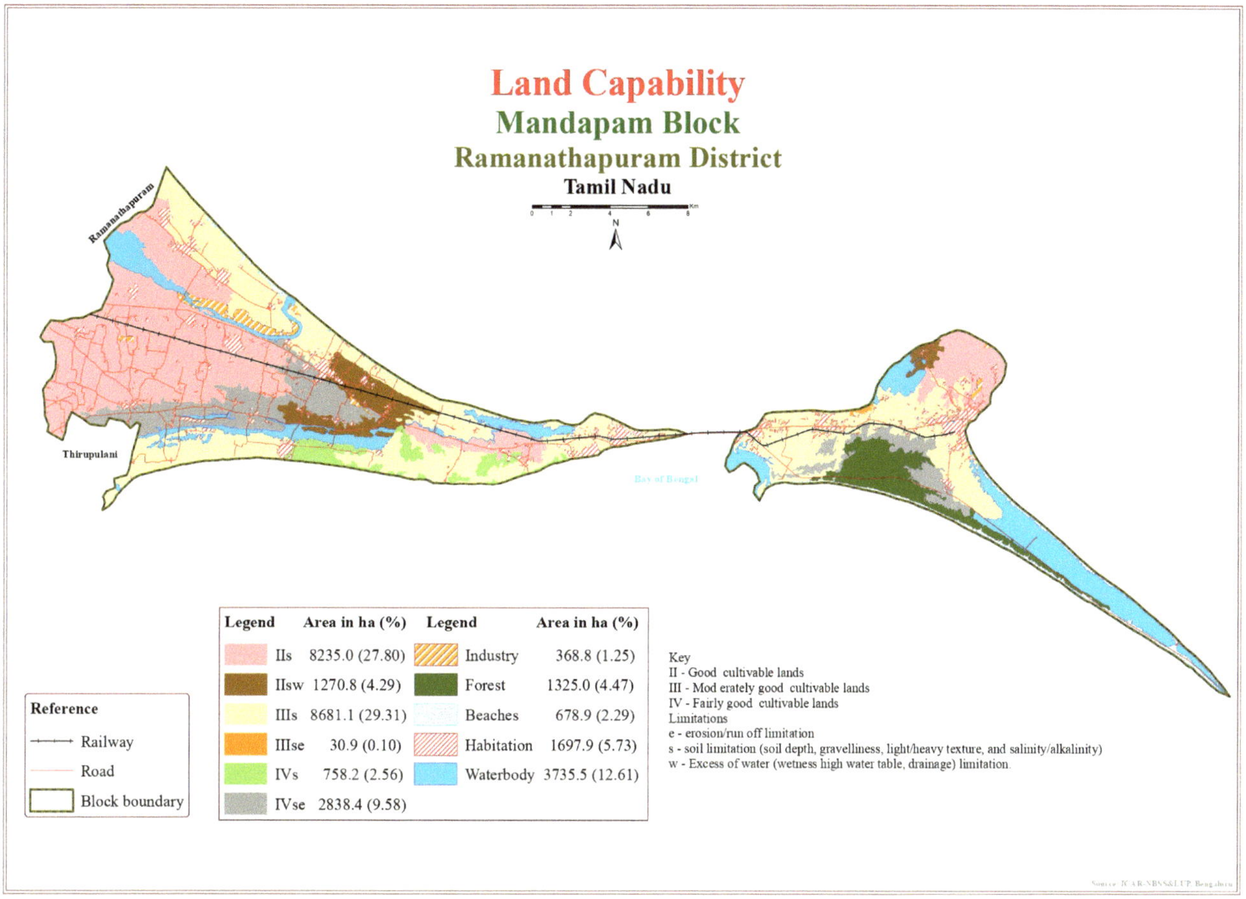

FIGURE 10.13 Land Capability Classes (LCC) of Mandapam block.

lands have a coarse texture constraint and mild erosion issues. About 12% of the land is remotely cultivable, and even those places have significant erosion issues and coarse texture limitations.

10.4 Conclusion

The present study concludes that high resolution satellite imagery sentinel 2 combined with digital elevation model helps in preparation of soil resource database at larger scale and identifies the problems and potentials of the land resources in coastal regions. This soil resource database can be used for formulating various management strategies for increasing productivity and conserving the natural resources. Further, it will form the basis for undertaking any developmental work at village/block level.

References

AIS&LUS, 1970. Soil Survey Manual. All India Soil Land Use Survey Organisation, Indian Agricultural Research Institute, New Delhi.

Baskaran, R., Anbarasu, K., Ferozekhan, M., Selvaraj, S., Rajamanickam, V., 2003. Geomorphology as a guide to mangrove afforestation in Rameshwaram Island, Tamilnadu. Indian. J. Geomorphol. 8 (1&2), 117–122.

Bhatta, B., Saraswati, S., Bandyopadhyay, D., 2010. Urban sprawl measurement from remote sensing data. Appl. Geogr. 30 (4), 731–740.

Coppin, P., Jonckheere, I., Nackaerts, K., Muys, B., Lambin, E., 2004. Review ArticleDigital change detection methods in ecosystem monitoring: a review. Int. J. Remote. Sens. 25 (9), 1565–1596.

Dharumarajan, S., Lalitha, M., Natarajan, A., Naidu, L.G.K., Balasubramanian, R., Hegde, R., et al., 2017. Biophysical and socio-economic causes for increasing fallow land in Tamil Nadu. Soil. Use Manage 33, 487–498. Available from: https://doi.org/10.1111/sum.12361.

Dharumarajan, S., Lalitha, M., Hegde, R., Janani, N., Rajawat, A.S., Sastry, L.N., et al., 2018a. Status of desertification in South India—assessment, mapping and change detection analysis. Curr. Sci. 115 (2), 331–338.

Dharumarajan, S., Lalitha, M., Ramesh Kumar, S.C., Vasundhara, R., Ramamurthy, V., Hegde, R., et al., 2018b. Economic evaluation of some soils of Arid regions of Anantapur district for groundnut cultivation. Agropedology 28 (02), 107–117.

Jackson, M.L., 1973. Soil Chemical Analysis. Prentice Hall of India Pvt. Ltd., New Delhi.

Jaiswal, R.K., Saxena, R., Mukherjee, S., 1999. Application of remote sensing technology for land use/land cover change analysis. J. Indian. Soc. Remote. 27 (2), 123–128.

Kaliraj, S., Chandrasekar, N., Simon Peter, T., Selvakumar, S., Magesh, N.S., 2014. Mapping of coastal aquifer Vulnerable Zone in the South West Coast of Kanyakumari, South India, Using GIS-Based DRASTIC Model. Monit. Assess, Environ. Available from: https://doi.org/10.1007/s10661-014-4073-2.

Lalitha, M., Kumar, Anil, Shivanand, K.S., Dharumarajan, S., Kalaiselvi, B., Arti, Koyal, et al., 2020. Land resources characterization of Kangayam Block in Tamil Nadu Uplands under Dry Semiarid Eco-region. Agropedology 30 (1), 19–30.

Li, Y., Zhu, X., Sun, X., Wang, F., 2010. Landscape effects of environmental impact on bay-area wetlands under rapid urban expansion and development policy: a case study of Lianyungang, China. Landsc. Urban. Plan. 94 (3), 218–227.

Mani Murali, R., Dinesh Kumar, P.K., 2015. Implications of sea level rise scenarios on land use/land cover classes of the coastal zones of Cochin. India. J. Environ. Mana 148, 124–133.

Meyer, W.B., Turner, B.L., 1994. Changes in Land Use and Land Cover: a Global Perspective. Cambridge University Press, Cambridge, pp. 320–342.

Misra, A., Murali, R.M., Vethamony, P., 2013. Assessment of the land use/land cover (LU/LC) and mangrove changes along the Mandovi-Zuari estuarine complex of Goa. India. Arab. J. Geosci. 8 (1), 267–279.

Nicholls, R.J., Small, C., 2002. Improved estimates of coastal population and exposure to hazards released. EOS T. AM. Geophys. Un. 83 (28), 301–305.

Soil Survey Staff, 2008. Keys to Soil Taxonomy, eighth ed. SCS, USDA, Washigton, DC.

Thanickachalam, T., Ramachandran, S., 2002. Management of coral reefs in Gulf of Mannar using remote sensing and GIS techniques, with reference to coastal geomorphology and land use. Map Asia 2002 (Bangkok, Thailand).

Weismiller, R.A., Momin, S.J., 1977. Change detection in coastal zone environments. Photogramm. Eng. Remote Sensing 43 (12), 1533–1539.

Yagoub, M.M., Kolan, G.R., 2006. Monitoring coastal zone land use and land cover changes of Abu Dhabi using remote sensing. J. Indian Soc. Remote. 34 (1), 57–68.

Chapter 11

Assessment of soil moisture adequacy for rainfed crops under semiarid agro-climatic conditions in Karnataka

U. Satish Kumar, N.L. Rajesh, B.K. Desai, B.M. Chittapur, Vidyashree and Mallikarjuna Dandu

Department of Soil and Water Conservation Engineering, College of Agricultural Engineering, UAS Campus, Raichur, Karnataka, India

11.1 Introduction

The vast semiarid region of the country succumbs to the recurring shadow of varied degrees of droughts and climate-prone degradation and imposing risks of inadequacy or lodging of excessive water in the root zone, unwarrantedly limiting the potential for agricultural growth. The evapotranspiration (ET) process that links the surface water and energy cycles with plant physiological activity is crucial, especially for water-limited ecosystems (Rodríguez-Iturbe and Porporato, 2004). Further, in such semiarid landscapes, rainfall variability causes uncertain soil moisture stress, impeding good crop development. The success of crop development and good harvest under such dwindling monsoon situations becomes risky. It needs appropriate conservation management to optimize the net available soil moisture quantity in the root zone to facilitate safer crop vegetative growth and subsequent maturity. In the case of vertisols (Usterts), which spreads to an extent of 70.3 M ha in many pockets of India (Kanwar, 1982), especially under semiarid conditions, has its conspicuous influence in providing enhanced moisture storage capacity due to its higher clay content (40%–60%) coupled with deeper profile depth (0.5–1.5 m). However, there needs to be a diagnostic analysis of soil-water interrelations to provide a site-specific plan rather than generalized solution on a watershed basis at the cadastral scale (1:8000) in implementing sustained land and water usage plan keeping because it endowed land capability and limitations simultaneously at the same time.

The study was undertaken in a representative micro-watersheds (MWS) under typical semiarid agro-climatic conditions to assess the soil moisture adequacy to support the pigeon pea crop (170–180 growth period days) under prevailing soils, net radiation, uncertain monsoonal rainfall (June–October) pattern conditions to suggest appropriate conservation and agronomical practices.

11.2 Materials and methods

11.2.1 Study area

The operational micro-watershed (4D2D8D2a and 4D2D8D2b) of 1056 ha combinedly has been located at Latitude N 16° 24′ 71″ Longitude E 77° 23′ 73″ (Fig. 11.1) and at elevation difference between 290 and 330 m above MSL. The MWS receive annual rainfall having a long-term average (period 1957 to 2020) of 666.39 mm, out of which about 87.11% of rainfall is received during dominant and dependable *Kharif* crop season (June–October months).

The process of characterization and mapping of soil resources is an interpretative method of grouping soils based on their inherent texture, profile characteristics, land features, and environmental factors that promote or limit land use or impose the risk of erosion (Verma et al., 2003; Pathak et al., 2013). The Soil Survey has been carried out at a cadastral scale (1:7920) to derive surface and profile soil characteristics leading to the assessment of the physical and fertility status of the area, land use, water sources, climate, rocks and minerals, crops through land and water resources inventorization. The ensemble of thematic maps and layers was developed using the GIS platform to depict the qualities described above. Integration of the site-specific variability and distribution of soil texture, slope, depth, erosion susceptibility, and gravelliness has paved the classification of the micro-watershed area into 12 soil mapping units according to their ability to support distinctive land use (Fig. 11.2).

Remote Sensing of Soils. DOI: https://doi.org/10.1016/B978-0-443-18773-5.00017-X

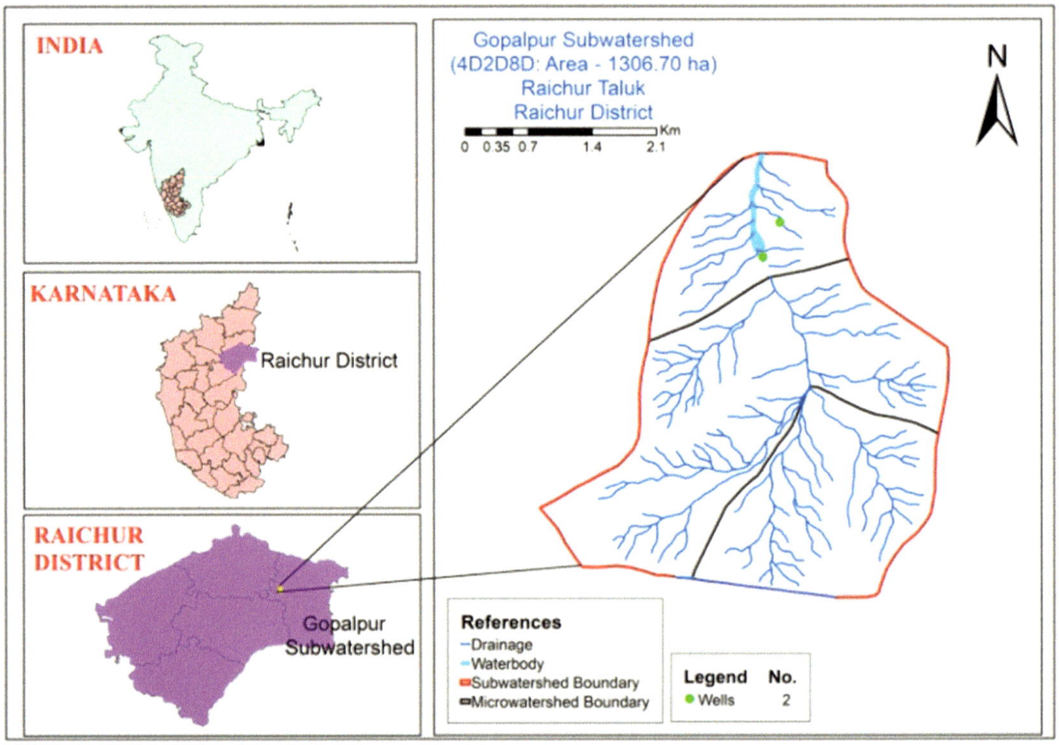

FIGURE 11.1 Location map of Jagir-Venkatapur micro-watershed.

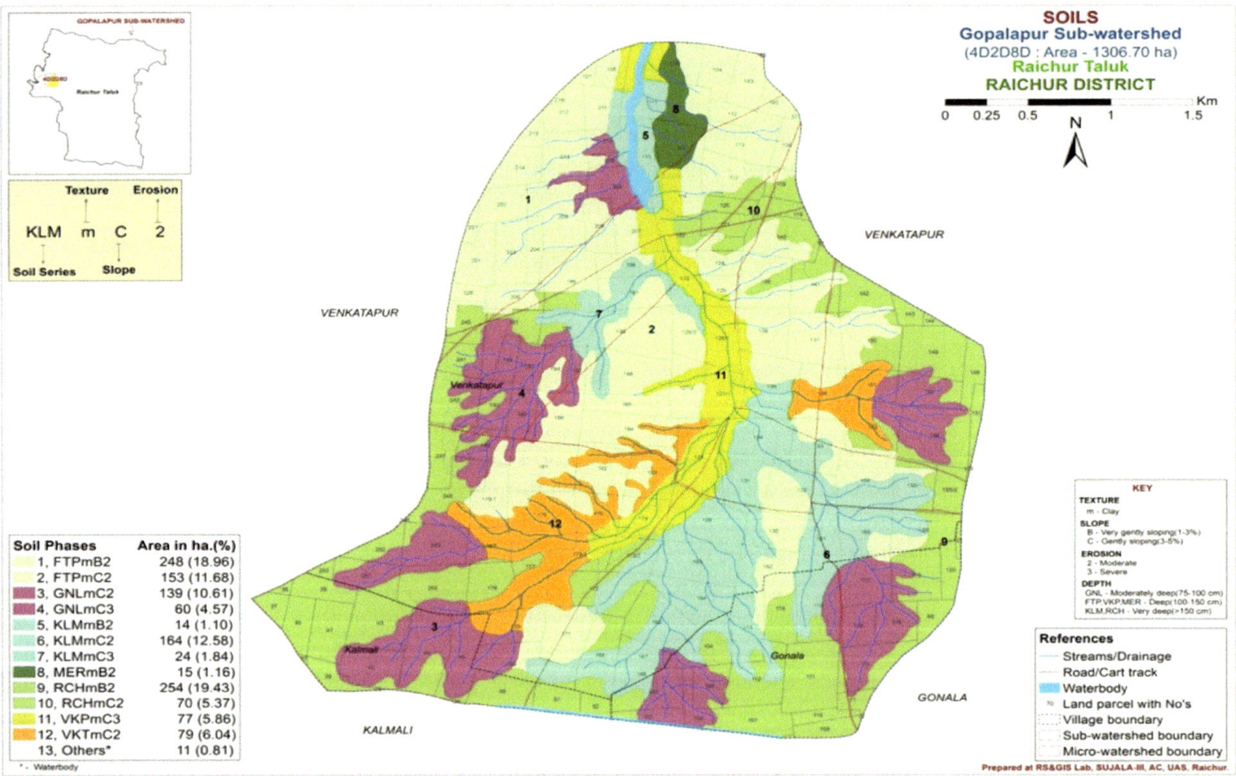

FIGURE 11.2 Delineation of the micro-watershed into 12 soil phase units through the process of land resources inventory.

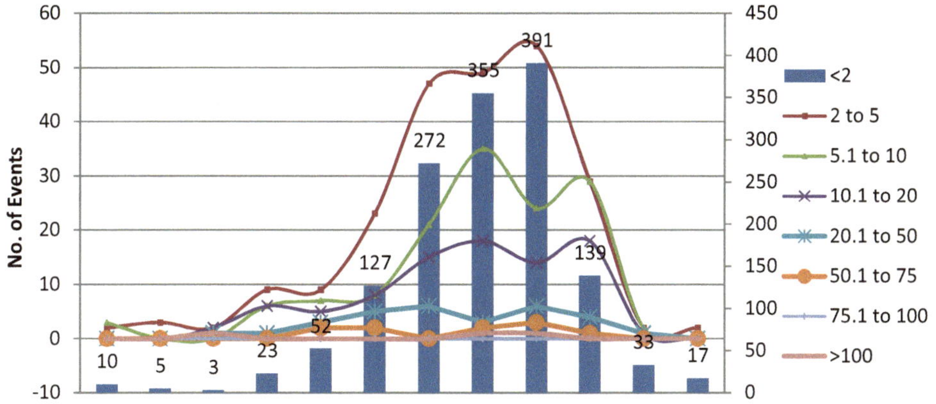

FIGURE 11.3 Distribution of rainfall event (mm) from 2009 to 2018.

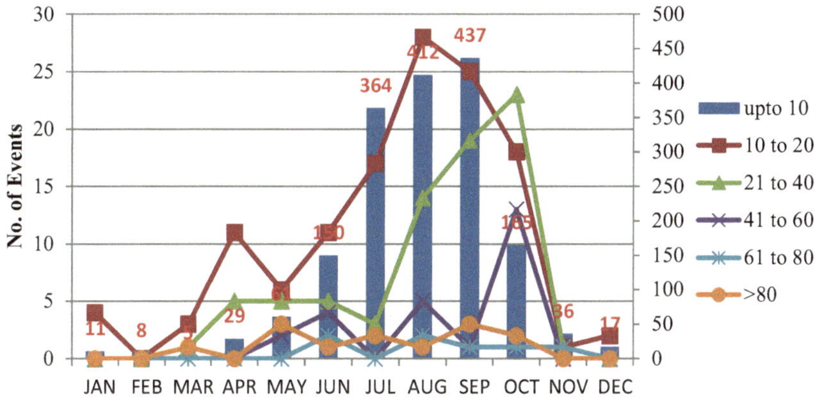

FIGURE 11.4 Variability and distribution of median rainfall event intensity (mm h^{-1}) during 2009–2019.

Further, Based on the capabilities and limitations to crop production, land has been grouped among eight capability classes. The suitability of land decreases from Class I to Class VIII, with the first four classes (I–IV) including arable lands and the remaining four classes defining the nonarable lands (Classes V to VIII). The MWS prevailed with predominantly clay soil (75 cm to > 150 cm deep) on the slope classes of 1%–3% (346 ha) and 3%–5% (708 ha). The soils are subjected to extensive erosion due to the absence of unscientific conservation measures, namely, bunding and in situ conservation measures.

The annual average rainfall of 2009–19 derived based on analysis of rainfall measured at an interval of 15 minutes shows a declined average (594.13 mm) from the long-term average (1957–2019) of 666.39 mm. Further, the occurrence of a large number of shorter interval discrete events with the modest depth of 2.5–5.0 mm (10–54 no per year) and very less frequent (<5 no) coupled with larger depth (> 30 mm) storm events demand careful management in tandem with soil and topography endowments (Fig. 11.3). The rainfall events observed to be energized normally with intensity in the range of 2–5 mm h^{-1} are more frequent whereas those above 50 mm h^{-1} are less and many times would be appearing as a part of events of larger depth (Fig. 11.4). The constant infiltration rates (mm h^{-1}) of each soil mapping units were measured through experimentation at identified locations using standard double ring infiltrometer method. The infiltration rates varied in the range of 6–15 mm h^{-1} due to the effect of different combinations of prevailing slope and crop canopy, apart from their basic soil texture both at surface and profile.

11.3 Potential evapotranspiration

The potential evapotranspiration (PET) was estimated based on the FAO-56-Modified Penman Montieth model using measured (15 minutes interval) values of net radiation (R_n), soil heat flux, ambient air temperature, humidity, atmospheric pressure, and wind speed (Batchelor, 1984; Richard et al., 1998). The usage of measured data on daily net radiation (R_n) through on-site net radiometer during the study period (2017–19) has enabled to improve the efficacy of the Modified Penman–Montieth to a larger extent in estimating PET (Figs. 11.5 and 11.6).

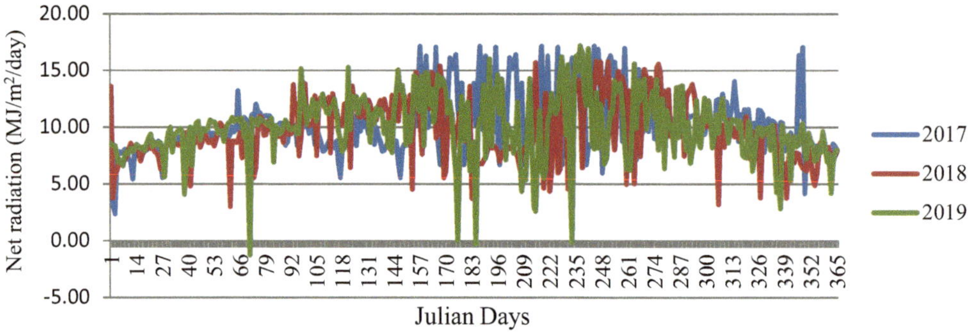

FIGURE 11.5 Variability of average daily net radiation (Rn) estimated using a net radiometer during the study period (2016–18).

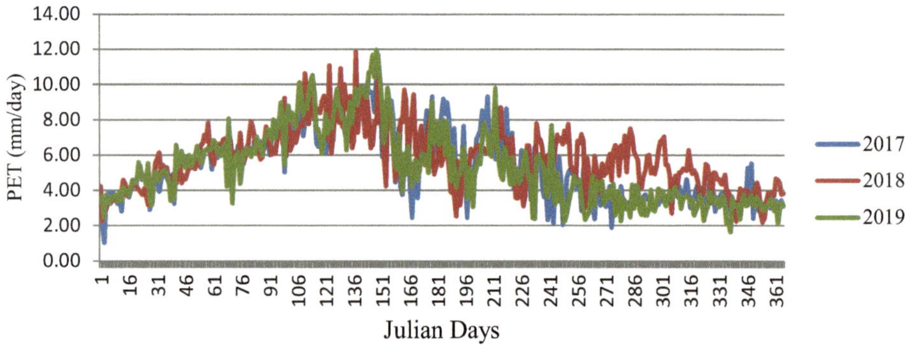

FIGURE 11.6 Variability of average daily PET estimated using net radiometer during 2016–18.

11.4 Soil moisture variability

The variability of soil moisture at spatially different but identified locations across MWS, keeping differential characters of soil phase, was measured (using Time Domain Reflectometer probe) at soil surface level followed by 30, 60, and 90 cm depths down the profile. The changes in soil moisture were observed and monitored throughout the three cropping seasons (June 17 to January 19) at the time interval of 15 days using a Time Domain Reflectometer. The profile moisture variability for two of the specific soil phase units (RCRmB2 and RCRmC2) for successive 3 years (2017–19) is shown in Figs. 11.7–11.9.

The spatial distribution of soil moisture was observed to reach its field capacity and still above during September 2017. In 2018, insufficient annual rainfall (272 mm) had prevailed to inadequate soil moisture to reach field capacity. In 2019, even with average rainfall (594.4 mm), the profile benefited from achieving the field capacity and maintaining available water above the level of 50% depletion due to the effects of storm events with moderate intensity (<30 mm h^{-1}).

11.5 Actual evapotraspiration

The time-intensive soil moisture (15 minutes interval) variation has also been measured using probes at three depths, namely, 5, 15, and 50 cm, throughout the cropping seasons continuously in all successive years of 2017, 2018, and 2019 and used to estimate the daily moisture lost to the atmosphere as ET. Fig. 11.10 depicts the daily actual ET estimated as moisture removed each day from the surface in all the days of successive three years (2017–19).

11.6 Leaf area index and crop yield

The critical leaf area index (LAI) measured intermittently in the cropping seasons of pigeon peas (*Cajanus cajan* [L.] Millsp.) is an important physiological parameter that decides crop growth and yield. It varies from crop to crop and location based on the leaf morphology, canopy structure, and climatic situations. The pigeon pea crop of medium-duration [growth across the MWS was monitored during the extended *Kharif* season (June to January)] by measuring LAI (Fig. 11.11). The LAI observations at intermittent intervals were recorded between crop sowing (170th–190th days) and harvesting (365th–30th days). The LAI increased to 40% during the flowering (280th–300th day period) stage and

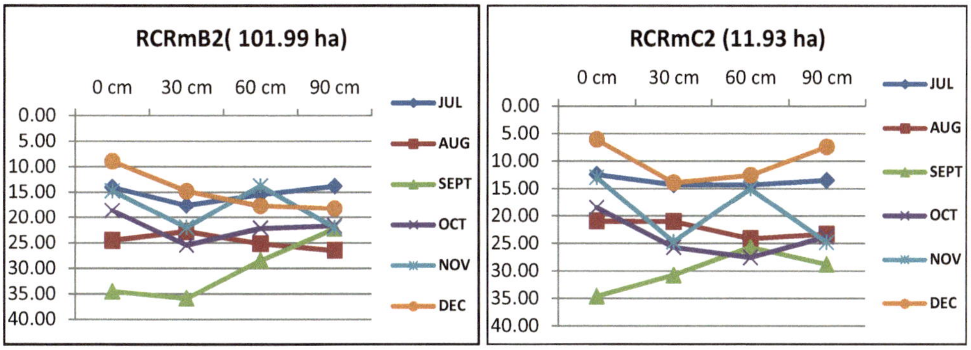

FIGURE 11.7 Variability of soil moisture across profile rooting depth (90 cm) in soil phase units RCRmB2 and RCRmC2 during 2017.

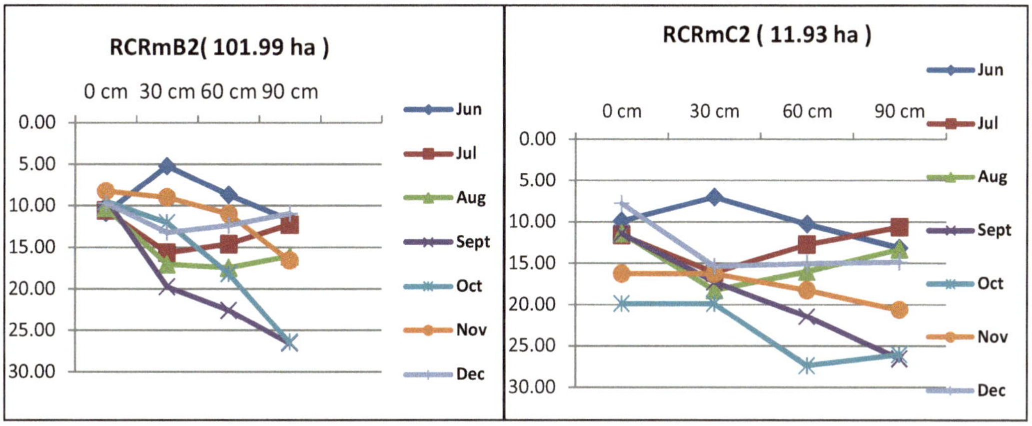

FIGURE 11.8 Variability of soil moisture across profile rooting depth (90 cm) in soil phase units RCRmB2 and RCRmC2 during 2018 SM (%).

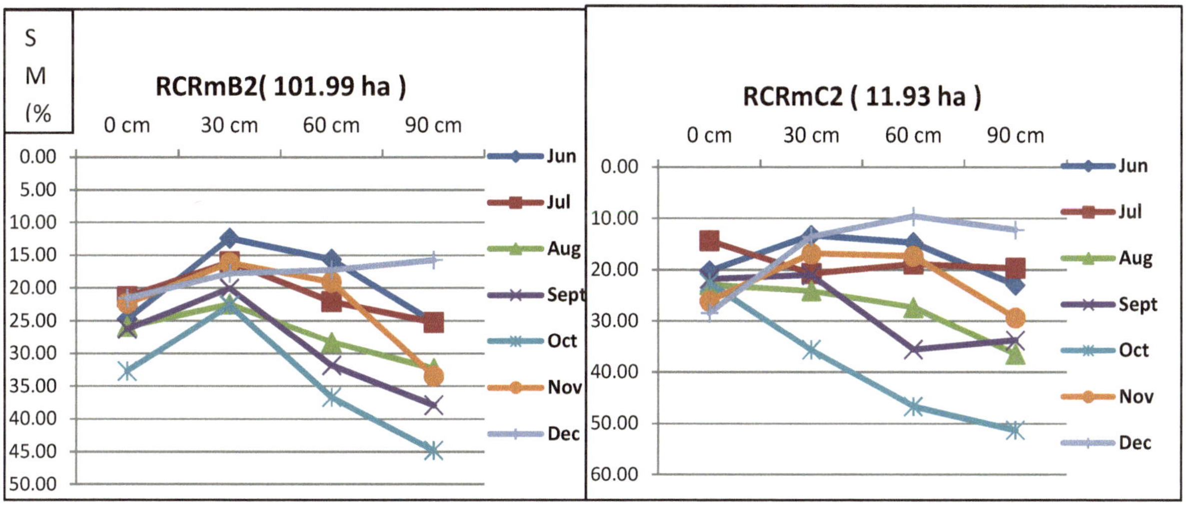

FIGURE 11.9 Variability of soil moisture across profile rooting depth (90 cm) in soil phase units RCRmB2 and RCRmC2 during 2019.

decreased afterward. The rapid growth of LAI was found to be between the 50th day and the 1st flowering phase. At the end of harvest, the red gram crop yield in each survey no across the MWS recorded during 2017 and 2019 is given in Figs. 11.12 and 11.13. It is to be noted that due to inadequate rainfall (374.4 mm) during 2018, the crop was vitiated largely across the MWS. The interpretative of the relations among the annual PET (mm), annual AET (mm), rainfall (mm), evaporative index (AET/rainfall), and dryness index (PET/rainfall) evaluated and compared (July 2017 and Jan

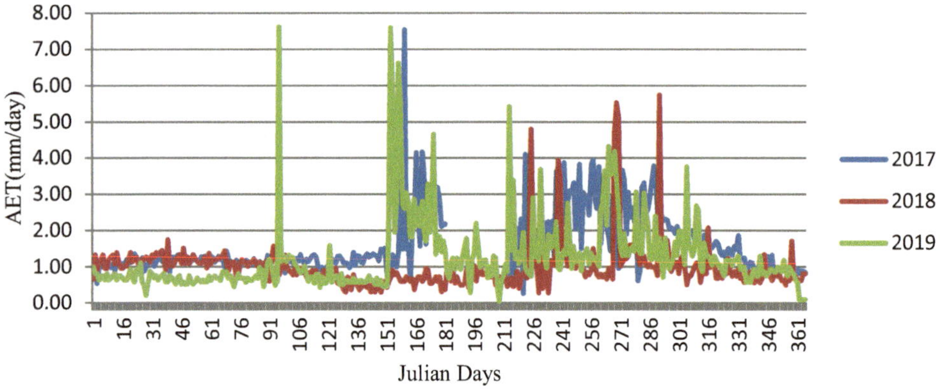

FIGURE 11.10 Derivation of actual evapotraspiration using daily surface soil moisture variability.

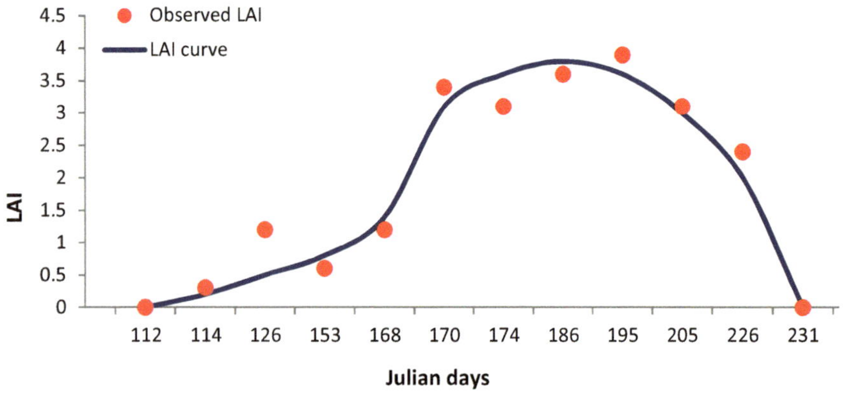

FIGURE 11.11 LAI values depicting the plant growth and corresponding yields of red gram crop in soil phase FTPmB2 during 2017.

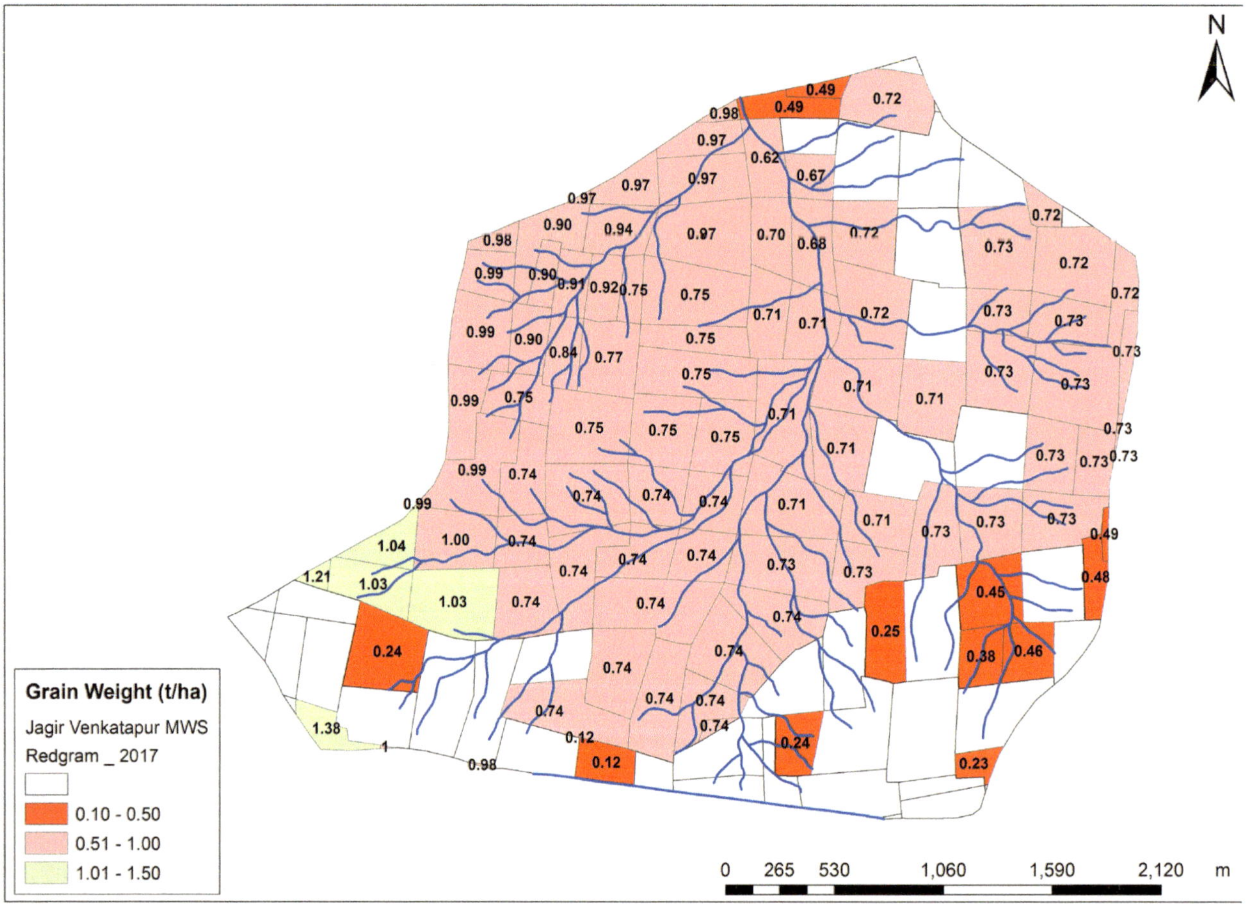

FIGURE 11.12 Distribution of Yield of pegion pea harvested during the year 2017.

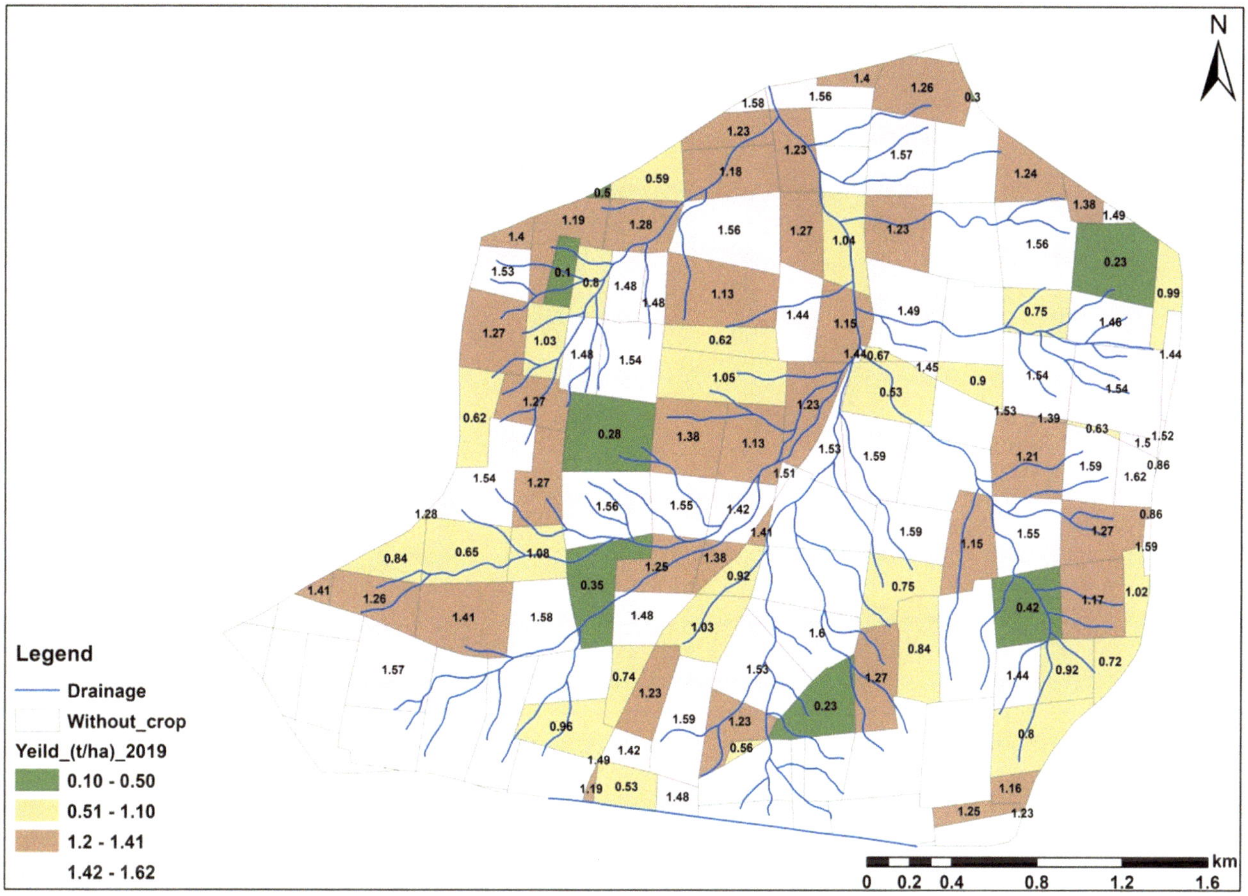

FIGURE 11.13 Distribution of Yield of pigeon pea harvested during the year 2019.

2019) to understand their effects in getting optimal pigeon pea yield cultivated across a more significant number of farmer plots (Figs. 11.12 and 11.13). The farmers could not get the minimum yield in 2018. whereas with near to long-term average annual rainfall, in 2019, the yield was appreciable with a maximum extent of 0.1 to 1.62 t ha^{-1}.

11.7 Results and discussion

The semiarid tropics (SAT) are defined in the context of vegetation zone delineation as the region within the tropics where the mean monthly rainfall exceeds the mean potential ET during 2–7 months of the year (Troll, 1965). Even though the long-term annual average rainfall of 593 mm (1957–2018), when compared to its annual distribution during the study period (2009–18), shows clear heterogeneity of monsoon arrival and withdrawal as well as in its distribution within the rainy season period of June to October months (Fig. 11.14). The *Kharif* rainfall, on average, would be 84% of the year, and the growth period of planned crops needs to be matched with this window period. The annual rainfall pattern consists of Individual storms (Figs. 11.3 and 11.4) categorized into (1) less intensive coupled with a moderate duration of 1–1.5 hours and (2) high intensive coupled with shorter duration storms which demand distinctive management options. A substantial portion of the total annual rainfall is accounted for by a few large (<5 No), but often intense, highly erosive storms demand adequate runoff management strategy in terms of energy dissipation and water harvesting. On the other hand, the storms with less intensity (<20 mm h^{-1}) and depth 2.5–5 mm h^{-1} (about 45 no) and of medium intense (20–40 mm h^{-1}) with depth in the class of 5–10 mm h^{-1} (<22 no), 10–15 mm h^{-1} (<9 no) having depth of 15–20 mm h^{-1} (<11 no) are to be targeted toward conservation to improvise soil moisture status by adopting appropriate conservation measures. For this and other related reasons, SAT regions, including the study watershed, are generally considered susceptible to high potential erosion hazards unless conservation practices are in place (El-Swaify et al., 1982). The Jagir-Venkatapur MWS typically follows similar SAT conditions in India (8% SAT of

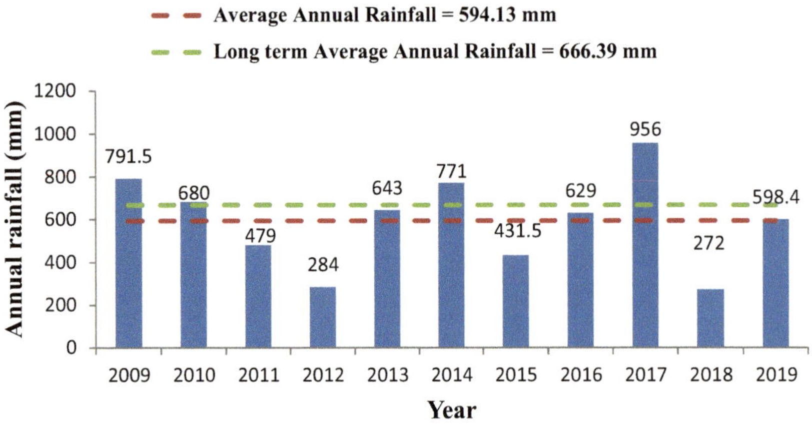

FIGURE 11.14 Distribution of annual rainfall during the period 2010–21.

Worlds distribution), wherein an average of *Kharif* rainfall (593 mm) receivable during June to October months shows that it can reasonably meet ET demand (430–470 mm) of short duration (<100 days) crops only.

The study area, consisting of clayey soil characterized as vertisols, is of deep depth (0.75–1.5 m). Their primary diagnostic features include profile inversion cracking (>1 m deep with 1–6 cm width) causing enhanced infiltration during premonsoon period (May to June months). It is observed that the abundance of cracks is responsible for high initial infiltration rates in dry *vertisols* (as much as 35 mm h^{-1}) before the monsoonal rains are received, and the rates in the saturated soils are low (<7 mm h^{-1}). Though the self-mulching observed in the surface layers is responsible for covering soil cracks, it is not the case in all prevailing *vertisols*. During the soil profile study, the soils are characterized by Gilgai micro-relief, which undergoes periodic changes with variations in the soil moisture regime. The structural development of the upper 40 cm of the profile is generally colorful, separating into subangular to angular blocky peds with shiny pressure faces. The soil phase classification (10 No of soil phase units) with clay textured soil on the slope classes of 1%–3% (346 ha) and 3%–5% (708 ha) demand careful management practices in support of absorbing "erratic rainfall spells" effectively by enhancing infiltration "opportunity time" and at the same time with provisions of smooth draining off the "runoff excess" for reducing risks of sheet erosion and also in tiding over moisture stress period between uncertain rainfall spells.

The highest temperatures in the study area are generally attained just before the onset of the rainy season (40°C–42.5°C). Prevailing daily maximum temperature variation (32°C–42.5°C) observed in the study period has given rise to the unsteady solar net radiation (R_n) values (estimated through net radiometer) vary between 17.0 MJ m^{-2} (May month) and 8.0 MJ m^{-2} day^{-1} (Cunha et al., 2014) during monsoon period (150th–310th days) which indicate that these watersheds receive a fair potential of radiation though out the year to support any crop. The PET at the start of the window period (June month) peaks at 10–12 mm day^{-1} (sunny clear sky) may dwindle to a low of 2–3 mm (cloudy and a rainy days).

Under rainfed conditions, the water requirement on a particular day has to be met only by the capacity of the soil moisture regime (90 cm deep), which in the study area varies from near to wilt (190 mm) to occasional saturation conditions on the intense rainfall (>30 mm) storm event leading to attain saturation of 510 mm (AMHC: 390 mm). The PET during September month would be about 2–6 mm, and corresponding AET demand (2–4 mm) would be generally met because profile soil moisture remains near field capacity. By the end of the monsoonal season (fag end of October or early November months), PET decreases to 6.0 to 2.0 mm. Meeting of this potential demand, however, gets restricted by water unavailability during the nonrainy season (311th to 365th days). The *vertisols*, which boost the profile storage capacity and availability, are a reliable source of moisture supply during postrainy season and act as a reserve in extending moisture supply to support the crop up to its maturity of long duration pigeon pea crop.

The soil moisture retention during rainy days and/or subsequent extent of assimilation within rooting depth (90 cm) across the MWS under all soil phases exhibits a pattern bearing on clay content and land slope. Two of the ten soil phase units, RCRmB2 (clay texture with 1%–3% slope) and RCRmC2 (clay texture with 3%–5% slope) experienced greater soil moisture variability ranging between 40% (field capacity of 43%) and 10% (close to wilt of 14%). At 90 cm depth, the range of moisture variability reduced substantially within 8%–30%, reflecting decreased unsaturated hydraulic conductivity (4×10^{-5} m day^{-1}). Interestingly, during 2018, with a low annual rainfall of 272 mm, the top soil layer (<10 cm) kept dried up (5%–12%) as against deep soil (about 90 cm), which contained quite moderate moisture level (10%–25%).

TABLE 11.1 Dryness and evaporative index of the climatic conditions prevailing in MWS.

Year	PET (mm)	AET (mm) of the year	AET Kharif season (June–December)	Rainfall (mm)	Dryness index PET/P	Evaporative index AET/P
2017	1954	926.11	586.86	956	2.04	0.61
2018	2092	442.91	384.98	272	7.69	1.42
2019	1947	456.87	341.35	598.4	3.25	0.57
Average	1997.7	608.6	437.7	608.8	4.3	0.9

The dryness index (average:4.3), a ratio of potential ET to precipitation, is indicative of the semiaridity of the study area, whereas evaporative index (average: 0.9), relates actual ET to precipitation (PET), indicative of the availability of water for growing crops under rainfed conditions (Table 11.1). The estimated daily potential ET quantity (2.0–12.0 mm) depends on the weather variables. On the other hand, actual ET (0.3–8.0 mm) measured as a change in soil moisture content depends directly on available water and the ability of clay to hold soil through the quality of adsorption. During the summer period (90–150 Julian days) with higher temperatures, find a more significant difference between PET (6–12 mm) and AET (1.5–3.0 mm), whereas during fag end of the rainy season (270–320 Julian days), the PET (2–4 mm) would be met with by the soil moisture with least stress. The stored soil moisture in a year with rainfall above the long-term annual average (666.39 mm) was found to sufficiently meet the evaporative demand till maturity (300–365 days) of the long durational crop, namely, Cotton and pigeon pea, gaining the advantage of deeper soil depth (1.0–1.5 m) and clay content (45%–60%). However, conservation measures including bunding and inter-management measures would ensure safer maturity and good crop yield (1.5–2.0 t ha^{-1}) even with an annual rainfall quantity of less than the average. The evaporative index (AET/P) during 2017 (0.9) is also indicative of fair chances of good crop yield in the case of a long-duration crop like pigeon pea or cotton (150–170 days).

The critical LAI in pigeon peas (*C. cajan* [L.] Millsp.) is an important physiological parameter that decides crop growth and yield. It varies from location to location based on the leaf morphology, canopy structure, and climatic situations. The LAI values observed for pigeon pea crop during 2017 (Max. of 4.2) and 2019 (Max. of 3.8) as against the maximum recorded of 7.8 for long-duration cultivars (Balakrishnan et al., 1997). It was observed that the moisture well supported the crop canopy to attain moderate vegetative growth. The pigeon pea yield in different pockets of micro-watershed varied in the range of 0.1 to 1.5 t ha^{-1} which improved to 0.1 to 1.9 t ha^{-1} in 2019. The soil profile wetting was found to be crucial to enhance the yield.

The results also provide insight into the prospects of the rising rainfed crop including long-duration crops such as cotton (140–160 days) and pigeon pea (150–170 days) in clay textured soils under average and above average rainfall situations. The minimum support of appropriate soil and water conservation measures will ensure the prospective crop, even under deficient rainfall, with the middle position in the study watershed, and the results can apply to other watersheds having similarities under semiarid agro-climatic conditions of Karnataka.

References

Balakrishnan, J., Natarajaratnam, K.N., Rajendran, C., 1997. Critical leaf area index in Pigeonpea. Agron. Crop Sci. 159 (16), 1–166.

Batchelor, C.H., 1984. The accuracy of evapotranspiration estimated with the FAO modified Penman equation. Irrig. Sci 5, 223–233.

Cunha, A.R. da, Schöffel, E.R., Volpe, C.A., 2014, Estimation of evapotranspiration by various net radiation estimation formulae for non-irrigated grass in Brazil. J. Water Resource Prot. 6 (15), 1425–1436.

EI-Swaify, S.A., Dangler, E.W., Armstrong, C.L., 1982. Soil erosion by water in the tropics. Research Extension Series 024, Univ. of Hawaii, Institute of Tropical Agriculture and Human Resources, Honolulu, Hawaii.

Kanwar, J.S., 1982. Problems and potential of vertisols and alfisols – the two important soils of SAT – ICRISAT experience. Tropic. Agric. Res. 15, 119–138.

Pathak, P., Sudi, R., Wani, S.P., Sahrawat, K.L., 2013. Hydrological behavior of alfisols and vertisols in the semi-arid zone: implications for soil and water management. India. J. Agric. Water Mgmt. 118 (2), 12–21.

Richard, G., Allen Luis, S., Raes, P.D., Reira, M.S. 1998. Crop evapotranspiration. FAO Irrigation and Drainage Paper 56, FAO, Rome.

Rodríguez-Iturbe, I., Porporato, A., 2004. Ecohydrology of Water-Controlled Ecosystems, Cambridge University Press, Cambridge, 442 pp.

Troll, C., 1965. Seasonal climates of the Earth. In: Rodenwaldt, E., Jusatz, H. (Eds.), World maps of climatology. Springer-Verlag, Berlin.

Verma, M.L., Sharma, R.K., Bhandari, A.R., Randev, A.K., 2003. Runoff, soil and nutrient losses and productivity under pea-tomato sequence as influenced by soil management practices. Indian J. Soil Cons 31 (2), 148–151.

Chapter 12

Morphometric analysis and prioritization of micro-watersheds of Gollalapalli subwatershed in Eastern Dry Zone of Karnataka using RS and GIS techniques

C.M. Pradeep, B.H. Prakash, K.N. Vidya, S. Brundha, S. Dharumarajan, R. Vasundhara and Rajendra Hegde
ICAR-National Bureau of Soil Survey and Land Use Planning, Regional Centre, Bangalore, Karnataka, India

12.1 Introduction

A watershed is the area of land where runoff resulting from rainfall is collected and drained through a single outlet. It is also called as catchment area or drainage basin. Generally, a watershed is constituted of natural resources, such as soil, water, and vegetation. Watershed characterization is important for proper planning and management of natural resources in sustainable development (Gajbhiye et al., 2013; Abdeta et al., 2020). Therefore, the study of watersheds plays an important role in any kind of developmental activity and sustainable management (Sangma and Guru, 2020). The watershed has been accepted as a basic unit for planning and implementation of the protective, curative, and amelioration programs. An accurate understanding of the hydrological behavior of a watershed is most important for its effective management. It has been observed that many watershed management programs that are being implemented by various government and nongovernment organizations suffer setbacks due to the inadequacy of requisite data and a lack of information on appropriate methodology for design of the hydraulic structures needed, which leads to inconsistency while arriving at appropriate technical and management decisions. Conventional watershed approaches in the past focused only on soil and water conservation measures, which did not bring in much productivity gain or contribute to improving rural livelihoods. Therefore, watershed behavior can be understood through the study of morphometric parameters, which helps to make various plans for sustainable development by using natural resources. This analysis can be achieved through the measurement of linear, aerial, and relief aspects of the watershed (Syed et al., 2017).

It is necessary to express the characteristics of the watershed in quantitative terms to understand the interrelationship between morphological characteristics and hydrological behavior. Channels in a watershed develop a tree-like structure, with the outlet stream as the main trunk and upstream having a bifurcating pattern as smaller branches. Horton (1945) established a hierarchical order stream, which provides statistical laws to describe river network composition. The relief of a watershed defines the general indication or gradient of the ground surface. Total watersheds' relief is the difference in the elevation between the most remote point of the divide line and the discharge point of a watershed. Erosion intensity increases with the value of relief.

Geomorphological analysis is the systematic description of watershed geometry and its stream system to measure the (1) linear aspects of the drainage network, (2) areal aspects of the watershed and, (3) relief aspects of the stream network. The morphological parameters directly or indirectly reflect the entire watershed area based on causative factors that are affecting runoff and sediment loss. The surface features are the fundamental unit of analysis prior to adopting any sophisticated tool to monitor the watershed responses in connection to any of the hydrologic processes acting on them. The parameters can be conveniently worked out from the SOI toposheet using the capabilities of geographical information system (GIS) tools. Chow (1964) used the drainage map of the study area to perform quantitative geomorphological analysis of both linear and aerial aspects of the drainage basin. The remote sensing (RS) and GIS techniques

Remote Sensing of Soils. DOI: https://doi.org/10.1016/B978-0-443-18773-5.00007-7

are powerful tools for assessing various terrain and morphometric parameters of the watersheds, as they provide a flexible environment for the manipulation and analysis of spatial information. Satellite images provide a synoptic view of a large area and are very useful in the analysis of drainage basin morphometry. GIS and RS techniques are proven to be proficient tools for morphometric characterization of subwatersheds (Singh, 1994; Grohmann, 2004; Sreedevi et al., 2009; Aher et al., 2014).

The Gollalapalli subwatershed in Bagepalli taluk, Chikkaballapur district of Karnataka is essential for massive development; so, its environment needs to be protected. If these crucial areas are affected by runoff and soil loss, it will lead to paralysis of various economic activities of the district. This research is critical as an effort to identify areas that are the potential areas of runoff and soil loss. Moreover, this research aims to identify the runoff-prone area in the Gollalapalli watershed based on morphometric aspects. As a result, the runoff-prone area can be managed and developed correctly through regional development policies.

12.2 Materials and methods

12.2.1 Study area

The study was carried out in the Gollalapalli subwatershed (WS-Code: 4C3G8d1A) situated in Bagepalli taluk of Karnataka, covering a total geographical area of 5610 ha, which is part of the Chitravathy (RB) subbasin and falls within the Eastern Dry Zone (KA-5) of Karnataka. The subwatershed is located between 13° 42′ 12″ to 13° 45′ 24″ N latitude and 77° 53′ 37″ to 77° 48′ 10″ E longitude, with an average elevation of 830 m above mean sea level (MSL) in the Chikkaballapur district of Karnataka. The location map of the study area is shown in Fig. 12.1.

12.2.2 Soils

Gollalapalli subwatershed is covered with shallow red loamy to red sandy soil that is often rich in soil organic carbon (average levels are between 0.56 and 1.77). The productivity can be significantly higher if the soil has a high level of organic carbon and has a good ability to retain water. In general, the soils are slightly acidic to neutral (pH range between 6.36 and 7.33). About 35% of the land is within a slope of 5%, 36% is within a slope of 10%, 25% is within a slope of 33%, and the remaining 4% is over the 33% slope. This suggests that farming with effective soil management techniques has the potential for a superior harvest of rainfall.

The Gollalapalli watershed was delineated for morphometric studies. The elevation of the subwatershed ranges from 633 to 1020 m above MSL. The minimum elevation of 633 m is found near the outlet where a gauging station exists, and the maximum elevation of 1020 m is over the hillocks on the upstream side of the watershed. The geomorphological characteristics of the micro-watershed were determined by digitizing the SoI toposheet using open source Quantum GIS (QGIS, version 3.26.1 Buenos Aires) software, and the subsequent cleaning, editing, and assigning of the topology produced a database in terms of linear, aerial, and relief aspects of the watershed.

The Gollapalli subwatershed is delineated into 11 micro-watershed boundaries using geospatial tools, viz. QGIS (Fig. 12.2). The morphological parameters either directly or indirectly represent the behavior of the entire watershed as it relates to the root causes of runoff and sediment loss. The parameters, namely size, shape, slope, relief, stream density, and stream network system, were determined from the drainage map derived from digital elevation model (DEM) by utilizing the capabilities of open-source QGIS software (Fig. 12.3). The drainage map of the study area was used for quantitative analysis of the drainage basin.

12.2.3 Geomorphological characteristics of a subwatershed

The hydrological response of a watershed is highly dependent on the rainfall characteristics and also on the drainage pattern of the watershed area. The rainfall characteristics can be measured by any nearby rain gauge stations, whereas the drainage pattern analysis, that is, the morphological characteristics, can be evaluated only through the drainage map of the area. The drainage map of the selected subwatershed area was derived from DEM.

The quantitative land-form analysis is undertaken for the developed watershed, in which flowing water and associated mass gravity movements acting over long periods of time are responsible for the development of surface geometry. The watershed is characterized by geomorphological analysis. Geomorphological analysis is the systematic description of a watershed's geometry and its stream channel system to measure the linear aspects of the drainage network, the aerial aspects of the drainage basin, and the relief aspects of the channel network and contributing ground slope, too.

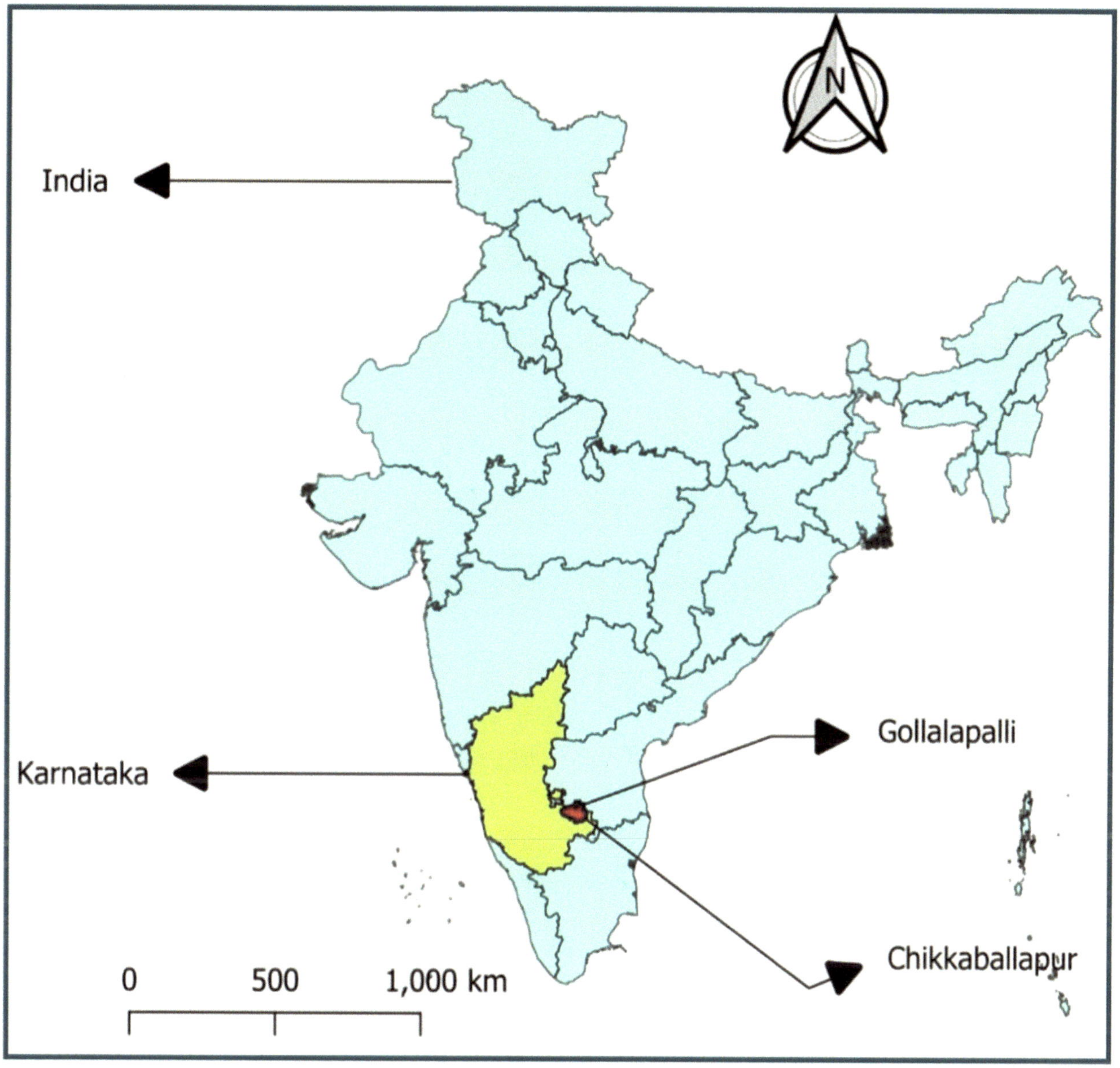

FIGURE 12.1 Location map of the study area.

The first two categories of measurement treat the projected property of a watershed on a horizontal plane, termed as "planimetric," whereas the third category of measurement counts the vertical inequalities of forms of drainage basin. All these aspects are described as follows.

12.2.3.1 Linear aspects of drainage network

It is also referred to as the linear aspect of the channel system. Morphological characteristics based on linear aspects considered in the study area were stream order, bifurcation ratio, stream length, stream length ratio, and length of overland flow.

12.2.3.1.1 Stream orders

The first step to follow in a drainage basin study is the designation of stream order, following a system introduced in the United States by Horton (1945) and slightly modified by Strahler (1957). The stream order represents the degree of stream branching within a watershed. Each length of stream is designated by its order. The smallest fingertip tributaries

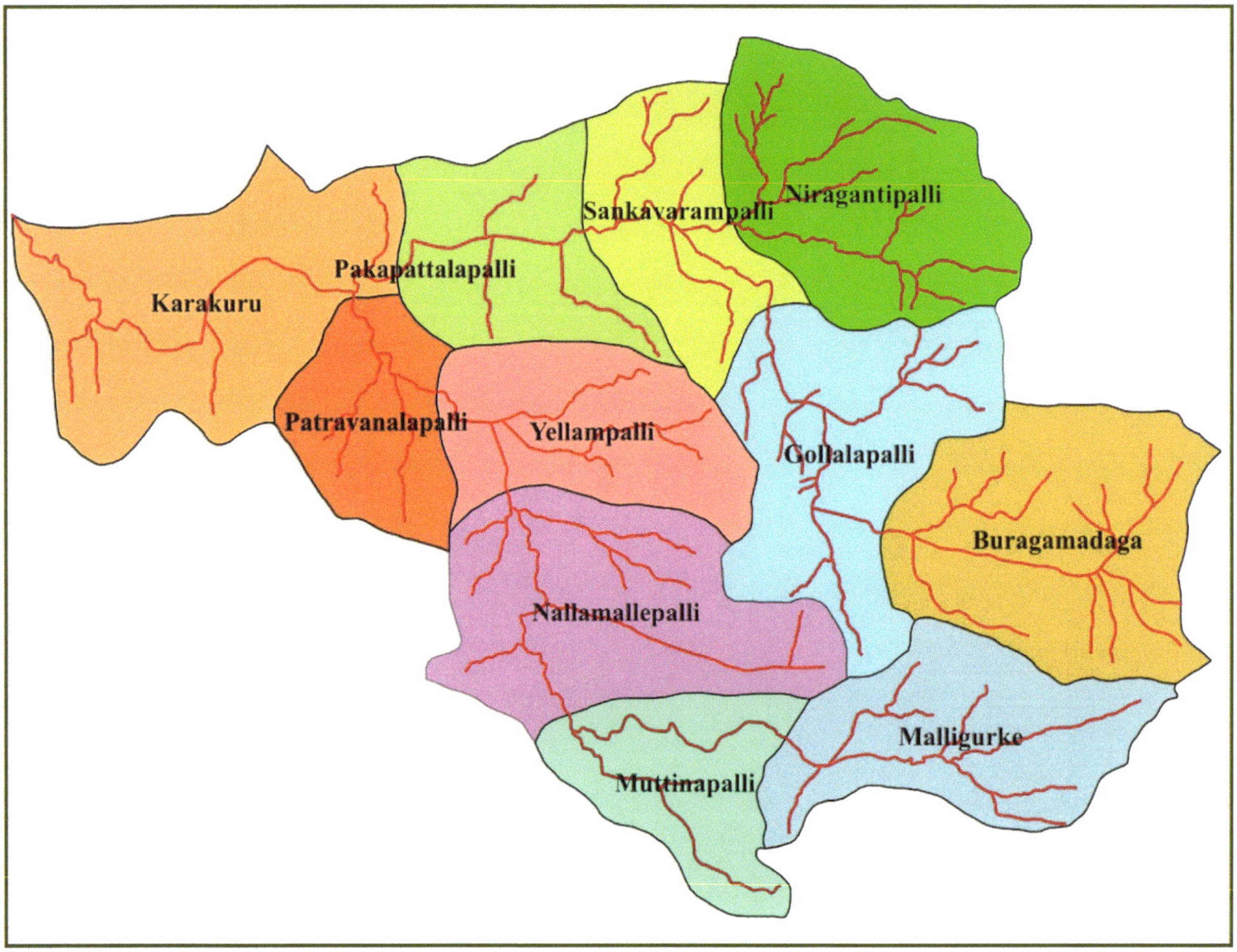

FIGURE 12.2 Location map of micro-watersheds of Gollalapalli subwatershed.

with no branches are designated as first-order streams. When two first-order streams are joined into one, then the resulting stream segment is termed as second-order stream. The third-order stream is a tributary formed by two or more second-order streams. Generally, an *n*th-order stream is formed by two or more streams of order $(n-1)$th and streams of lower order. One point should always be kept in mind is that when a low order stream segment joins to high order stream segment, the order of stream segment does not change, but the high order stream remains as it is. The highest order stream is known as the trunk or principal stream, through which all discharge from the watershed passes through to the outlet. The maximum order segment carries the sediment and flow of water at the outlet of the watershed.

12.2.3.1.2 Bifurcation ratio (R_b)

The number of segments is counted for each order, say u, and defined as N_u. The bifurcation ratio is defined as the ratio of the number of stream of any order to the number of streams of the next higher order, expressed as:

$$R_b = \frac{N_u}{N_{u+1}}$$

where, N_u, number of stream segments of order u and N_{u+1}, number of streams of next higher order $u+1$.

12.2.3.1.3 Stream lengths (L_u)

The extent of stream length in a watershed reveals the characteristics and size of different components of the drainage system and its contributing surface area. The stream length of all orders is measured using QGIS. To find out the mean

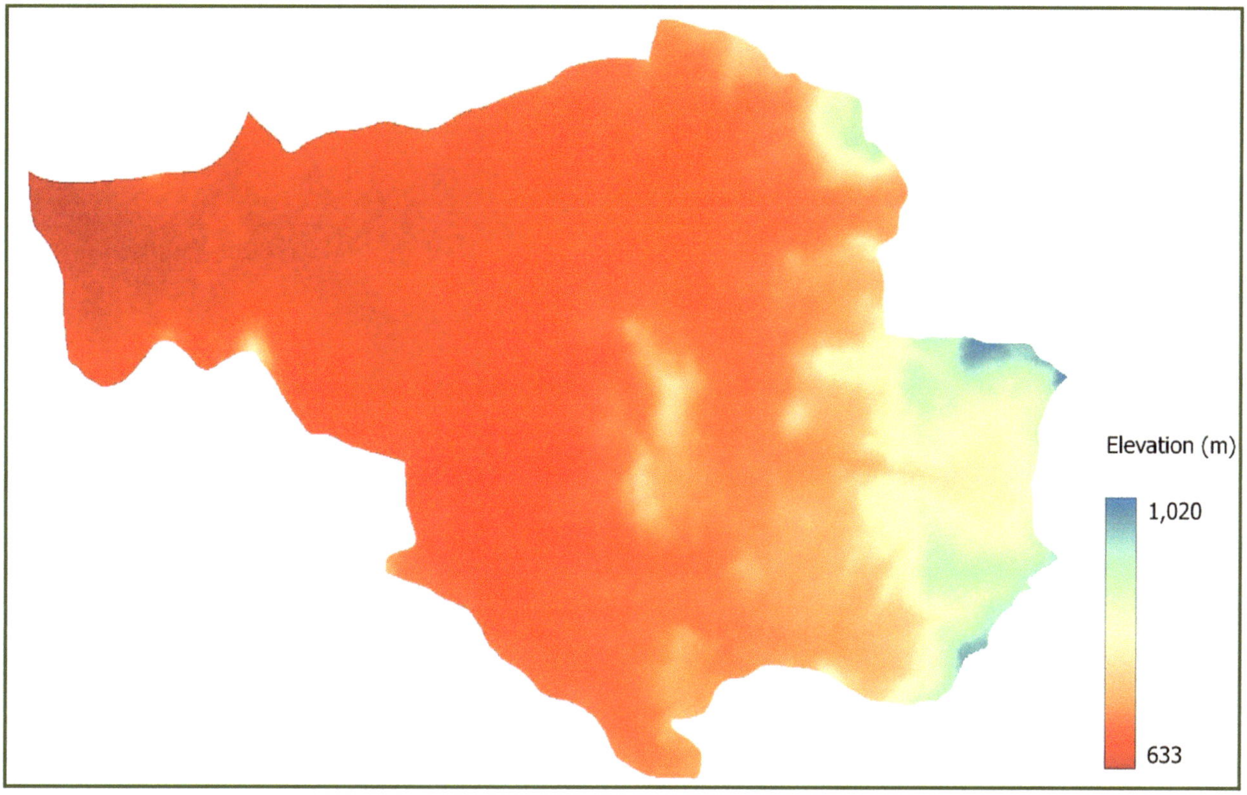

FIGURE 12.3 DEM of Gollalapalli subwatershed.

length of the channel L_u of order u, the total length is divided by the number of segments, N_u of that order, thus

$$\overline{L_u} = \frac{\sum_{i=1}^{N} L_u}{N_u}$$

where, $\overline{L_u}$, mean length of channel of order u; L_u, stream length increase with the increase of order number; and N_u, total number of stream segments of order u.

12.2.3.1.4 Stream length ratio (R_L)

Horton (1945) defined the stream length ratio, R_L, as the ratio of mean length, $\overline{L_u}$, of segments of order u to mean length of segments of the immediate lower order, $\overline{L_{u-1}}$. It is expressed as:

$$R_L = \frac{\overline{L_u}}{\overline{L_{u-1}}}$$

12.2.3.2 Areal aspects of drainage area

Many characteristics based on area measurement of the watershed are basin area, basin shape, basin shape factor, drainage density, constant of the channel maintenance, and stream frequency.

12.2.3.2.1 Area of the watershed (A)

Area of the watershed has been calculated with the help of a GIS. The areas are expressed in ha.

12.2.3.2.2 Drainage density (D_d)

Horton (1932) defined drainage density as the ratio of the total length of all streams of all orders (i.e., cumulative length of stream segments of all order) within a watershed to the total area of the watershed (A), that is,

$$D_d = \frac{\sum_{i=1}^{K}\sum_{i=1}^{N} L_u}{A_u}$$

where, D_d, drainage density; L_u, length of stream segment of order u; A, total area of watershed; K, principal order or highest stream order; and N, total number of streams.

A high value of drainage density indicates a relatively high density of streams and, thus, a rapid stream response.

12.2.3.2.3 Drainage texture (D_t)

It is the number of first-order streams per unit perimeter length of the watershed area. It is calculated by using the following formula:

$$R_t = \frac{N}{P}$$

where, N, total number of streams and P, perimeter of watershed, km.

12.2.3.2.4 Stream frequency or stream density (F)

Horton (1932) introduced this term to study basin morphology, which is the number of stream segments per unit basin area, given as:

$$F = \frac{\sum_{i=1}^{K} N_u}{A_k}$$

Where, F, stream frequency; N_u, number of stream segments of order u; A_k, basin area of principal order (k); and K, principal or highest order stream.

12.2.3.2.5 Basin shape

The basin shape is the shape of the projected surface on the horizontal plane of the basin map. The basin shape has a significant effect on stream discharge characteristics. The quantitative expression of basin shape can be characterized by: (1) form factor, (2) circulatory ratio, and (3) elongation ratio.

12.2.3.2.6 Form factor (R_f)

Horton (1932) explained the basin shape through a term called as form factor, which is defined as the ratio of basin area (A) to the square of the basin length (L_b), given as:

$$R_f = \frac{A}{L_b^2}$$

12.2.3.2.7 Elongation ratio (R_e)

Elongation ratio, R_e is defined (Schumn, 1956) as the ratio of the diameter of a circle (D_c) with the same area as the watershed to the maximum length of the watershed (L_{bm}). This parameter is used to assess whether the shape of the basin approaches a circle.

$$R_e = \frac{D_c}{L_{bm}}$$

12.2.3.2.8 Circulatory ratio (R_c)

Circulatory ratio, R_c is defined as the ratio of the circumference of a circle of the same area as the watershed to the watershed perimeter (Miller, 1953) or the ratio of basin area (A) to the area of the circle (A_c) having the same perimeter as the perimeter of the drainage basin.

$$R_c = \frac{A}{A_c}$$

12.2.3.2.9 Constant of channel maintenance (C)

Schumn (1956) defined constant of channel maintenance as the inverse of drainage density, that is,

$$C = \frac{1}{D_d} = \frac{A}{\sum_{i=1}^{K}\sum_{i=1}^{N} L_u}$$

It indicates that the magnitude of surface area of watershed needed to sustain a unit length of stream segment.

12.2.3.2.10 Length of overland flow (L_g)

Horton defined the length of overland flow as the length of flow path, projected on the horizontal plan of nonchannel flow, from a point on the drainage divide to the adjacent stream channel. The length of overland flow is calculated as one-half of the reciprocal of the drainage density, that is,

$$L_g = \frac{1}{2D_d}$$

where, L_g, length of overland flow (m) and D_d, drainage density (km^{-1}).

12.2.3.3 *Relief aspects of stream network*

Parameters involving elevation differences are important because these parameters define the potential energy or erosion potential of a watershed. These parameters are evaluated with the help of DEM and contour maps of the watersheds. The relief characteristics considered for the study are maximum watershed relief, relief ratio, relative relief, ruggedness number, and geometric number.

12.2.3.3.1 Maximum watershed relief (H)

Maximum watershed relief, H, is the elevation difference between the watershed outlet and the highest point located on the perimeter of the basin. Maximum watershed relief is obtained from the available contour maps of the watersheds. It is expressed in meters.

12.2.3.3.2 Relief ratio (R_h)

Schumn (1956) defined the relief ratio, R_r, as the maximum watershed relief divided by the maximum watershed length.

$$R_h = \frac{H}{L_b}$$

where, R_h, relief ratio; H, maximum watershed relief (m); and L_b, maximum watershed length (m).

12.2.3.3.3 Relative relief (R_r)

Melton (1957) defined relative relief, R_R, as the ratio of the maximum watershed relief to the basin perimeter length. It is computed using the following equation.

$$R_r = \frac{H}{L_p} \times 100$$

where, R_r, relative relief (%); H, maximum basin relief (m); and L_p, basin perimeter length (m).

12.2.3.3.4 Ruggedness number (N_R)

The product of relief (H) and drainage density (D_d) is called ruggedness number, that is, ruggedness number, $H.D_d$

12.2.4 Prioritization of micro-watersheds of Gollalapalli subwatershed

The morphometric characteristics like linear and relief parameters have a direct relationship with soil erodibility (Biswas et al., 1999; Nookaratnam et al., 2005; Javed et al., 2009; Rekha et al., 2011; Singh and Singh, 2014; Sujatha et al., 2015; Ahmed and Rao, 2015; Chandniha and Kansal, 2017; Abdeta et al., 2020); the highest their value shows, the most erodible the soil is in a watershed. Therefore, a micro-watershed that shows the highest value in linear and relief parameters is rated at first rank, with second higher value rated at second in rank and so on and the least value rated at last in the rank. On the contrary, areal/shape parameters have an inverse relationship with soil erodibility (Javed et al., 2009; Ahmed and Rao, 2015; Raja et al., 2017; Abdeta et al., 2020); the lowest their value, the most erodible the soil is in a watershed. Thus, a micro-watershed that showed the lowest value in areal/shape parameters is rated at first rank, with the next lower value rated at second and so on, and then the highest value is rated at the last rank. Compound method of averaging value was used in this study, as it is expected that all morphometric parameters have equal importance for final ranking (Ajay et al., 2014; Farhan, 2017). After ranking all 11 micro-watersheds based on every single parameter, the ranking values for all parameters of each micro-watershed are added and divided by the total number of parameters; in this case, it is divided by 15, and then we arrive at a compound value. Thus, the micro-watershed with the least compound value is assigned at the highest priority and denoted by number 1, with the next higher value denoted by number 2 and so on, with the micro-watershed that got the highest compound value assigned the last priority number (Ayele et al., 2017; Sheikh et al., 2017; Thapliyal et al., 2017; Chandniha and Kansal, 2017). This implies that the highest priority indicates the greatest degree of runoff, peak discharge, and soil erosion risks in that micro-watershed. Thus, it is important to plan proper land and water management practices for each of the micro-watersheds as per their sensitivity ranks.

12.3 Results and discussion

The geomorphological analysis and measurements were made from the digitized drainage pattern map of the Gollalapalli subwatershed. The micro-watershed boundary and digitized drainage pattern are shown in Fig. 12.2. The values of different geomorphological parameters were calculated as discussed in the Materials and Methods section.

12.3.1 Basic parameters of micro-watersheds

The basic parameters, such as micro-watershed area, watershed perimeter, and watershed length, of 11 micro-watersheds were calculated using QGIS (Table 12.1). The watershed area is the most important parameter, since it directly reflects the volume of water in a watershed. A larger size intercepts greater volume of rainfall, higher runoff, and peak discharge. Sometimes, the smaller size has also recorded maximum flooding and sedimentation. This is due to other watershed morphometric characteristics like stream networks, shape, length and relief parameters. In the present study, the area ranges from 306.8 to 709.9 ha (Table 12.1). Watershed perimeter is the length of the delineated watershed boundary. It also represents watershed size. The watershed perimeter ranges from 7169.4 to 16092.0 m (Table 12.1). Watershed length is the longest dimension of the basin parallel to the principal drainage channel (Schumn, 1956). The length of micro-watersheds ranges from 2610.6 to 7327.0 m, which shows the main channel of watershed in which the greatest volume of water travels (Table 12.1).

12.3.2 Linear aspects of drainage network

The linear aspects of drainage network include stream order, stream number, stream length, mean stream length, stream length ratio, and bifurcation ratio, which were determined and their results have been presented in Table 12.2.

12.3.2.1 *Stream order*

The ranking of streams has been taken according to Strahler (1964) method. The results show that the maximum stream frequency has been observed in the case of first-order streams, followed by second- and third-order streams. It is also observed that as the stream order increases, there is a decrease in stream frequency. The higher number of streams in the watershed indicates that the topography is still undergoing erosion, whereas lesser number of streams in a watershed indicates mature topography (Ahmed and Rao, 2015).

TABLE 12.1 Basic parameters of 11 micro-watersheds of Gollalapalli subwatershed.

Sl. No.	Name of micro-watershed	Micro-watershed identity code	Area (ha)	Perimeter (m)	Basin length (m)
1	Nirgantipalli	4C3G8d2B	592.5	9728.9	3775.0
2	Patravanapalli	4C3G8d1B	306.8	7190.2	2664.5
3	Karakuru	4C3G8d1A	709.9	12663.7	10298.8
4	Buragamadaga	4C3G8d3B	680.8	10846.4	5102.7
5	Nallamallepalli	4C3G8d4B	650.8	11780.6	4569.9
6	Malligurke	4C3G8d5B	510.4	11119.6	4531.1
7	Gollalapalli	4C3G8d3A	612.2	12453.1	5331.3
8	Muttinapalli	4C3G8d5A	346.3	16092.0	5779.5
9	Yellampalli	4C3G8d4A	438.4	9467.7	3840.9
10	Pakapattapalli	4C3G8d1C	394.7	9312.7	5535.4
11	Sankavarampalli	4C3G8d2A	367.4	8816.7	7327.0

TABLE 12.2 Linear aspect morphometric parameter of Gollalapalli subwatershed.

Sl.No.	Micro watershed	S_u	N_u	L_u, km	Mean L_u, km	R_b	R_{bm}	R_L	R_{Lm}
1	Nirgantipalli	1	14	6.7	0.48	3.5	2.5	–	1.40
		2	4	3.3	0.81	2.0		1.69	
		3	2	3.5	1.76	2.0		2.17	
		4	1	0.6	0.62	–		0.35	
2	Patravanapalli	1	4	3.9	0.98	2.0	2.0	–	0.97
		2	2	1.1	0.54	2.0		0.55	
		3	1	0.8	0.75	–		1.38	
3	Karakuru	1	6	4.8	0.80	3.0	2.5	–	1.64
		2	2	3.9	1.93	2.0		2.40	
		3	1	1.7	1.68	–		0.87	
4	Buragamadaga	1	12	9.3	0.78	3.0	3.5	–	4.61
		2	4	1.8	0.45	4.0		0.58	
		3	1	2.5	3.89	–		8.64	
5	Nallamallepalli	1	9	6.4	0.71	2.25	3.125	–	1.36
		2	4	4.4	1.10	4.0		1.55	
		3	1	1.3	1.29	–		1.17	
6	Malligerke	1	7	6.6	0.95	3.5	2.75	–	0.94
		2	2	2.8	1.41	2.0		1.48	
		3	1	0.6	0.55	–		0.39	
7	Gollalapalli	1	12	8.1	0.67	4.0	3.5	–	2.05
		2	3	3.3	1.10	3.0		1.62	
		3	1	2.7	2.72	–		2.48	

(Continued)

TABLE 12.2 (Continued)

Sl.No.	Micro watershed	S_u	N_u	L_u, km	Mean L_u, km	R_b	R_{bm}	R_L	R_{Lm}
8	Muttinapalli	1	3	7.8	2.59	3.0	3.0	–	1.54
		2	1	4.0	3.99	–		1.54	
9	Yellampalli	1	5	4.4	0.88	5.0	5.0	–	3.67
		2	1	3.2	3.22	–		3.67	
10	Pakapattapalli	1	4	4.0	0.99	4.0	4.0	–	3.84
		2	1	3.8	3.80	–		3.84	
11	Sankavarampalli	1	7	7.2	1.02	3.5	2.75	–	1.36
		2	2	5.3	2.64	2.0		2.58	
		3	1	0.4	0.36	–		0.14	

12.3.2.2 Stream number

Horton (1945) states that the number of stream segments in each order is known as the stream number. The number of streams ranged from 3 to 14 in first-order streams, 1 to 4 in second-order streams, 1 to 2 in third-order streams, and 1 stream in fourth-order stream (Table 12.2). Which indicates that the number of stream segments in each order forms an inverse geometric sequence with the order number, and the number of streams usually decreases as the stream order increases (Ahmed and Rao, 2015).

12.3.2.3 Bifurcation ratio

Results show that the bifurcation ratio (R_b) of the micro-watersheds ranges from 2.0 to 5.0 (Table 12.2); lower values of bifurcation ratio of micro-watersheds suggest less structural disturbance, whereas higher values indicate structurally controlled drainage pattern. The mean bifurcation ratio (R_{bm}) may be defined as the average of bifurcation ratios of all orders (Strahler, 1957). The mean R_b in micro-watersheds fluctuates from 2.0 to 5.0 and all the micro-watersheds fall under the normal basin category (Ahmed and Rao, 2015).

12.3.2.4 Stream length and mean stream length

The stream length was computed based on the law proposed by Horton (1945) for all the micro-watersheds. It can be observed from Table 12.2 that the maximum stream length was observed in first-order streams followed by second-, third-, and fourth-order streams in most of the micro-watersheds, which reveals that the stream length decreases as the stream order increases (Ahmed and Rao, 2015).

But in the case of mean stream length, more deviation was observed. Which might be due to flow of streams from high altitude, lithological variations, and moderately steep slopes (Singh and Singh, 1997; Ahmed and Rao, 2015). The length of the streams helps to understand the hydrological characteristics of the underlying rock surfaces and the degree of drainage in the watershed.

12.3.2.5 Stream length ratio

The stream length ratio of different micro-watersheds ranges from 0.14 to 8.64 (Table 12.2) and changes in stream length ratio from one order to another order indicate the late youth stage of geomorphic development of streams (Singh and Singh, 1997; Ahmed and Rao, 2015).

12.3.2.6 Relation between stream number and stream order

According to Horton's law, the plot of the logarithm of stream number (Ordinate) as a function of stream order (Abscissa) should yield a set of points lying along a straight line. For the present study, a graph was plotted for all the micro-watersheds. The graph is presented in the Fig. 12.4. The graph shows a straight line, satisfying Horton's law. It is evident that the coefficient of determination for the straight line fit for the micro-watersheds ranged from 0.9732 to 1.0,

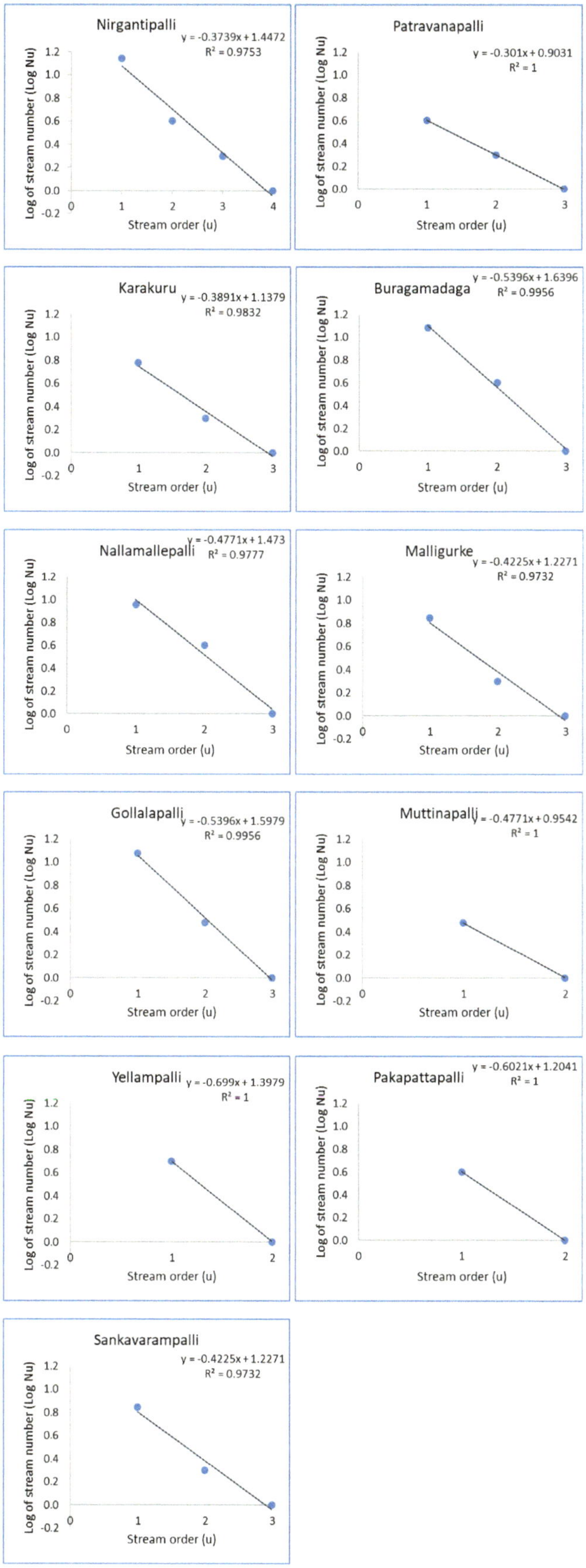

FIGURE 12.4 Relationship between stream order and stream number.

which is quite satisfactory and which exhibited a well-inversed and strong relationship between the number of streams and stream order (Abdeta et al., 2020).

12.3.2.7 Relation between cumulative stream length and stream order

An attempt has been made to establish the relationship between the cumulative stream length and stream order. The plot of the logarithm of cumulative stream length along the ordinate and stream order along the abscissa for the watershed is a straight-line fit, as shown in Fig. 12.5. The straight-line fit indicates that the ratio between cumulative stream lengths is constant throughout the successive order of a basin, and suggests that geometrical similarity is preserved in basins of increasing order. It is evident that the coefficient of determination for the straight line fit for the micro-watershed ranged from 0.5636 to 1.0, which is quite satisfactory.

12.3.3 Areal aspects of drainage area

The areal parameters, such as drainage density, stream frequency, drainage texture, form factor, circularity ratio, elongation ratio, compactness coefficient, length of overland flow, and shape factor, have been calculated and their results are summarized in Table 12.3. The study of areal aspects helped in understanding of the micro-watersheds' shape, which affects mainly stream flow hydrographs and peak flows.

12.3.3.1 Drainage density (D_d)

Drainage density depends on both the climate and the physical characteristics of the watershed. Drainage density is influenced by various factors such as resistance to erosion, infiltration capacity of the land, vegetation The significance of drainage density is to determine the time of travel by water within the watershed. In general, low values of drainage density indicate regions of highly permeable material with vegetative cover and low relief, whereas high values of drainage density indicate regions of weak and impermeable subsurface material, with spares vegetation and mountainous relief (Nautiyal, 1994). The drainage density values of 11 micro-watersheds of Gollalapalli subwatershed ranged from 1.456 to 3.486 km km^{-2} (Table 12.3), suggesting that the watershed as a whole is underlined by highly permeable material and represents low relief (Javed et al., 2009).

12.3.3.2 Drainage texture (D_t)

Drainage texture (D_t) is the total number of stream segments per perimeter of the watershed (Horton, 1945). In the present study, D_t of micro-watershed varies from 0.249 km^{-1} (Muttinapalli micro-watershed) to 2.056 km^{-1} (Nirgantipalli micro-watershed). Also, drainage texture of all the micro-watershed falls under the category of very coarse texture (D_t is less than 2), except the Nirgantipalli micro-watershed, which falls under coarse texture (D_t between 2 and 4) as classified by Smith (1950) (Table 12.3).

12.3.3.3 Stream frequency (F_s)

The values of stream frequency are directly related to stream population per unit area of the watershed (Horton, 1932). In this study, stream frequency indicates a close correlation with the drainage density values of the micro-watersheds. The higher value of stream frequency indicates high runoff (Javed et al., 2009). In this study, stream frequency in micro-watersheds varies from 1.155 (Muttinapalli watershed) to 3.376 km^{-2} (Nirgantipalli watershed). Results revealed that the Nirgantipalli micro-watershed produced more runoff compared to other micro-watersheds (Table 12.3).

12.3.3.4 Form factor (R_f)

The form factor is the ratio of watershed area to the square of watershed length. The smaller the value of the form factor, the more elongated the watershed (Strahler, 1964). A watershed with a higher form factor has a high peak discharge in a short period of time (Horton, 1945). In the present study, form factor values ranged from 0.067 (Karakuru watershed) to 0.432 (Patravanapalli watershed). However, it was found that the Karakuru watershed is more elongated compared to the Patravanapalli watershed. Therefore, the Patravanapalli watershed generates high runoff in short duration compared to the Karakuru watershed. The values of the form factors of all the micro-watershed are presented in Table 12.3.

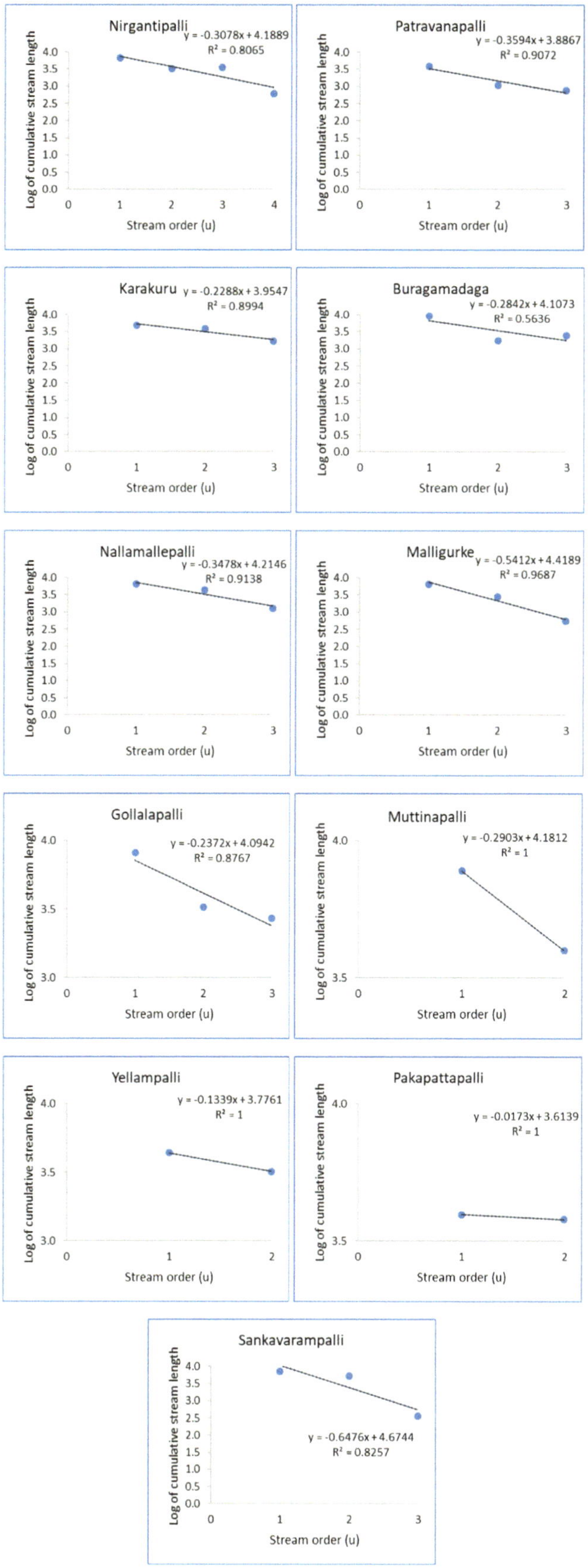

FIGURE 12.5 Relationship between stream order and stream length.

TABLE 12.3 Areal aspect morphometric parameter of Gollalapalli subwatershed.

Parameters/ micro-watershed code	Value										
	4C3G8d2B	4C3G8d1B	4C3G8d1A	4C3G8d3B	4C3G8d4B	4C3G8d5B	4C3G8d3A	4C3G8d5A	4C3G8d4A	4C3G8d1C	4C3G8d2A
D_d in km km^{-2}	2.280	1.877	1.456	2.205	1.853	1.957	2.304	3.398	1.736	1.962	3.486
D_t in km^{-1}	2.056	0.974	0.711	1.567	1.188	0.899	1.285	0.249	0.634	0.537	1.134
F_s in km^{-2}	3.376	2.282	1.268	2.497	2.151	1.959	2.614	1.155	1.369	1.267	2.722
R_f	0.416	0.432	0.067	0.261	0.312	0.249	0.215	0.104	0.297	0.129	0.068
R_e	0.728	0.742	0.292	0.577	0.630	0.563	0.524	0.363	0.615	0.405	0.295
R_c	0.787	0.746	0.556	0.727	0.589	0.519	0.496	0.168	0.615	0.572	0.594
C in km	0.439	0.533	0.687	0.454	0.540	0.511	0.434	0.294	0.576	0.510	0.287
L_g in km	0.219	0.266	0.344	0.227	0.270	0.256	0.217	0.147	0.288	0.255	0.143

Elongation ratio (R_e)

Generally, the values of the elongation ratio range between 0.6 and 1.0, which is associated with a wide variety of climate and geological properties. If the value of Re is equal to one, a watershed is equal from all side (Schumn, 1956). Re values of Nirgantipalli and Patravanapalli micro-watersheds indicated the shape to be oval, whereas the rest of the micro-watersheds represented an elongated shape (Table 12.3).

12.3.3.5 Circulatory ratio (R_c)

Circulatory ratio (R_c) is influenced by the length and frequency of the stream, geological structures, land use land cover, climatic variability, relief, and slope of the subwatersheds (Patel et al., 2013). The high R_c value indicates that the subwatersheds are more circular and are characterized by high to moderate relief and a structurally controlled drainage system, while the lower R_c values of subwatersheds indicate an elongated shape. In this study, the circulatory ratio varies from 0.168 (Muttinapalli) to 0.787 (Nirgantipalli). The high value of R_c for Nirgantipalli micro-watershed indicates that it is more or less circular in shape and is characterized by high to moderate relief, and has a structurally controlled drainage system (Ahmed and Rao, 2015); whereas all other micro-watersheds represent an elongated shape (Javed et al., 2009).

12.3.3.6 Constant of channel maintenance (C)

The constant of channel maintenance is the area necessary to maintain 1 foot of drainage channel. The constant of channel maintenance represents the drainage area required to maintain one unit of channel length; hence, it is a measure of watershed erodibility. Regions of resistant rock type, or with the surface of high permeability, or with good forest cover should have a high constant of channel maintenance and a low drainage density, and vice versa. The constant for channel maintenance varies from 0.287 (Sankavarampalli) to 0.687 (Karakuru). Lower values (less than 0.3) of the constant of channel maintenance reflect low infiltration and permeability, poor vegetal cover, and weak rock types. On the other hand, the constant of channel maintenance values is found to be higher (> 0.5) for micro-watersheds, reflecting the higher infiltration and permeability of the materials, fairly good vegetal cover, and relatively resistant rock types (Table 12.3). The rest of the micro-watersheds have moderate infiltration and permeability of the surface material, moderate vegetal cover, and moderately resistant rocks (Sakthivel et al., 2007).

12.3.3.7 Length of overland flow (L_g)

The length of overland flow is a measure of stream spacing or degree of dissection. Horton (1945) used this term to refer to the length of the runoff of the rainwater on the ground surface before it gets localized into definite channels. The length of overland flow for all micro-watersheds of Gollalapalli subwatershed was estimated and is shown in Table 12.3. In the micro-watersheds of the study area, the length of overland flow values varies from 0.143 (Sankavarampalli) to 0.344 (Karakuru). These estimations mean that the rainwater has to run over 0.143 and 0.3447 km for Sankavarampalli and Karakuru micro-watersheds, respectively, before it gets concentrated in stream channels. The smaller the values of the length of overland flow, the more the surface runoff enters the stream. In a relatively uniform terrain, less rainfall is sufficient to contribute a significant volume of surface runoff to stream discharge when the length of overland flow is small than when it is large (Sakthivel et al., 2007).

12.3.4 Relief aspects of drainage network

Relief is an important attribute of terrain in general and the drainage watershed in particular. According to Strahler (1968), relief measures are indicative of the potential energy of the drainage system because of its elevation above the MSL. The relief aspects considered for the present study include watershed relief, relief ratio, relative relief, channel gradient, and ruggedness number (Table 12.4).

12.3.4.1 Maximum basin relief (H)

Maximum basin relief (*H*) is the difference between the maximum (*Z*) and minimum (*z*) elevations in the basin and is an important factor to understand the denudation characteristics of the basin, which controls the stream surface runoff entering the gradient and therefore influences the flood pattern and the amount of sediment that can be transported (Hadley and Schumm, 1961). The values of *H* for micro-watersheds vary from 42 to 316 m (Table 12.4), which indicate low infiltration and high runoff conditions in the study area (Ahmed and Rao, 2015). All the micro-watersheds of the

TABLE 12.4 Relief aspect morphometric parameter of Gollalapalli subwatershed.

Parameters/ Micro-watershed code	Value										
	4C3G8d2B	4C3G8d1B	4C3G8d1A	4C3G8d3B	4C3G8d4B	4C3G8d5B	4C3G8d3A	4C3G8d5A	4C3G8d4A	4C3G8d1C	4C3G8d2A
Maximum elevation of the micro watershed (*Z*) in m	936	801	802	1026	782	988	847	753	795	693	743
Minimum elevation of the micro watershed (*z*) in m	676	648	632	710	663	701	679	677	656	651	665
Maximum relief of micro watershed (*H*) in m	260	153	170	316	119	287	168	76	139	42	78
Relief ratio (R_h)	0.069	0.057	0.017	0.062	0.026	0.063	0.032	0.013	0.036	0.008	0.011
Relative relief (R_r) in %	0.027	0.021	0.013	0.029	0.010	0.026	0.013	0.005	0.015	0.005	0.009
Channel gradient (C_g) in m/km	43.84	36.55	10.51	39.42	16.58	40.32	20.06	8.370	23.04	4.830	6.776
Ruggedness number (R_n)	0.593	0.287	0.247	0.697	0.221	0.562	0.387	0.28	0.241	0.082	0.272

study area are found to possess low relief (<400 m) and no micro-watersheds possess high relief (> 800 m). According to Anhurt (1970) and Gregory and Walling (1973) the mean denudation rate is directly proportional to the mean watershed relief. From these observations, it is likely that in all micro-watersheds, the mean denudation rates are relatively higher (Sakthivel et al., 2007).

12.3.4.2 Relief ratio (R_h)

When watershed relief is divided by the horizontal distance on which it is measured, it results in a dimensionless relief ratio (Schumn, 1956). It measures the overall steepness of a drainage watershed and is an indicator of the intensity of the erosion process. The relief ratio values obtained from the above relationship for the micro-watersheds of the study area are presented in Table 12.4. The maximum relief ratio value was found to be for micro-watershed Nirgantipalli (0.069), and the minimum value was found to be for micro-watershed Pakapattapalli (0.008). The areas with high relief and steep slope are characterized by high relief ratios, whereas low relief ratios are mainly due to the resistant basement rocks of the basin and low degree of slope (Vittala et al., 2004; Mahadevaswamy et al., 2011; Ahmed and Rao, 2015).

12.3.4.3 Relative relief (R_r)

Relative relief (R_r) is the ratio of the maximum watershed relief to the perimeter of the watershed (Melton, 1957). The relative relief values of micro-watersheds in the Gollalapalli subwatershed have been determined and presented in Table 12.4. The micro-watersheds of Pakapattapalli and Muttinapalli have minimum R_r (0.005), while the micro-watershed of Buragamadaga has the maximum value (0.029).

12.3.4.4 Channel gradient (C_g)

Channel gradient is an indicator of channel slope, which enables assessment of the runoff volume (Sreedevi et al., 2005; Thomas et al., 2010). The micro-watersheds of Nirgantipalli, Malligurke, Buragamadaga, Patravanapalli, Yellampalli, and Gollalapalli have a channel gradient of 43.84, 40.32, 39.42, 36.55, 23.04, and 20.06 m km^{-1}, respectively, which reflects the mountainous nature of the terrain. The channel gradient of the remaining micro-watersheds has relatively low values, which indicates that the main stream flows through the plateau (Table 12.4).

12.3.4.5 Ruggedness number (R_n)

It is the product of maximum basin relief and drainage density, where both parameters are in the same unit. An extremely high value of ruggedness occurs when both variables are large and the slope is not only steep but long as well (Strahler, 1964). In the study, the ruggedness number of each micro-watershed was computed, which varies from 0.082 in the Pakapattapalli watershed to 0.593 in the Nirgantipalli watershed (Table 12.4). The high values of the ruggedness number indicated structural complexity in association with high relief and drainage density (Ahmed and Rao, 2015).

12.3.5 Ranking and prioritization of micro-watersheds

The prioritization of the micro-watersheds on the basis of morphometric analysis and the final score of the entire eleven micro-watersheds and their ranking are shown in Table 12.5. After the ranking was done based on every single parameter, the ranking values for all the parameters of each micro-watershed were added up for each of the 11 micro-watersheds to obtain the compound value and its average. Based on the average value of compound parameters (C_p), the micro-watersheds that have the lowest C_p value have been given priority one, the next lowest have been given second priority, and so on. The compound parameter values of 11 micro-watersheds were calculated as shown in Table 12.5, and the prioritized map for the study area is shown in Fig. 12.6. The micro-watersheds were then categorized into three major priority classes, that is, high priority, medium priority, and low priority, on the basis of the range of C_p values (Table 12.6). The maximum and minimum prioritized scores of micro-watersheds are 4.07 and 7.33, respectively. High priority indicates a greater degree of erosion in a specific micro-watershed. However, it is necessary to improve soil conservation measures as per priority in particular regions.

TABLE 12.5 Prioritized rank of micro-watersheds using the morphometric parameter.

Parameters/ micro-watershed code	Value										
	4C3G8d2B	4C3G8d1B	4C3G8d1A	4C3G8d3B	4C3G8d4B	4C3G8d5B	4C3G8d3A	4C3G8d5A	4C3G8d4A	4C3G8d1C	4C3G8d2A
R_{bm}	2.50 [9]	2.00 [11]	2.50 [10]	3.50 [3]	3.13 [5]	2.75 [7]	3.50 [4]	3.00 [6]	5.00 [1]	4.00 [2]	2.75 [8]
R_{lm}	1.40 [7]	0.97 [10]	1.64 [5]	4.61 [1]	1.36 [8]	0.94 [11]	2.05 [4]	1.54 [6]	3.67 [3]	3.84 [2]	1.36 [9]
D_d	2.280 [4]	1.877 [8]	1.456 [11]	2.205 [5]	1.853[9]	1.957 [7]	2.304 [3]	3.398 [2]	1.736 [10]	1.962 [6]	3.486 [1]
D_t	2.056 [1]	0.974 [6]	0.711 [8]	1.567 [2]	1.188 [4]	0.899 [7]	1.285 [3]	0.249 [11]	0.634[9]	0.537 [10]	1.134 [5]
F_s	3.376 [1]	2.282 [5]	1.268 [9]	2.497 [4]	2.151 [6]	1.959 [7]	2.614 [3]	1.155 [11]	1.369 [8]	1.267 [10]	2.722 [2]
C	0.439 [8]	0.533 [4]	0.687 [1]	0.454 [7]	0.540 [3]	0.511 [5]	0.434 [9]	0.294 [10]	0.576 [2]	0.510 [6]	0.287 [11]
L_g	0.219 [8]	0.266 [4]	0.344 [1]	0.227 [7]	0.270 [3]	0.256 [5]	0.217 [9]	0.147 [10]	0.288 [2]	0.255 [6]	0.143 [11]
H	260 [3]	153 [6]	170 [4]	316 [1]	119 [8]	287 [2]	168 [5]	76 [10]	139 [7]	42 [11]	78 [9]
R_h	0.069 [1]	0.057 [4]	0.017 [8]	0.062 [3]	0.026 [7]	0.063 [2]	0.032 [6]	0.013 [9]	0.036 [5]	0.008 [11]	0.011 [10]
R_r	0.027 [2]	0.021 [4]	0.013 [6]	0.029 [1]	0.010 [8]	0.026 [3]	0.013 [7]	0.005 [10]	0.015 [5]	0.005 [11]	0.009 [9]
C_g	43.84 [1]	36.55 [4]	10.51 [8]	39.42 [3]	16.58 [7]	40.32 [2]	20.06 [6]	8.370 [9]	23.04 [5]	4.830 [11]	6.776 [10]
R_n	0.593 [2]	0.287 [5]	0.247 [8]	0.697 [1]	0.221 [10]	0.562 [3]	0.387 [4]	0.280 [6]	0.241 [9]	0.082 [11]	0.272 [7]
R_f	0.416 [10]	0.432 [11]	0.067 [1]	0.261 [7]	0.312 [9]	0.249 [6]	0.215 [5]	0.104 [3]	0.297 [8]	0.129 [4]	0.068 [2]
R_e	0.728 [10]	0.742 [11]	0.292 [1]	0.577 [7]	0.630 [9]	0.563 [6]	0.524 [5]	0.363 [3]	0.615 [8]	0.405 [4]	0.295 [2]
R_c	0.787 [11]	0.746 [10]	0.556 [4]	0.727 [9]	0.589 [6]	0.519 [3]	0.496 [2]	0.168 [1]	0.615 [8]	0.572 [5]	0.594 [7]
Compound parameter (C_p)	5.20	6.87	5.67	4.07	6.80	5.07	5.00	7.13	6.00	7.33	6.87
Ranking	7	3	6	10	4	8	9	2	5	1	3
Priority values	1	1	1	1	1	1	1	1	1	1	1
Priority class	High	High	High	High	High	High	High	High	High	High	High

Note: Values in square bracket indicates ranking of the micro-watersheds.

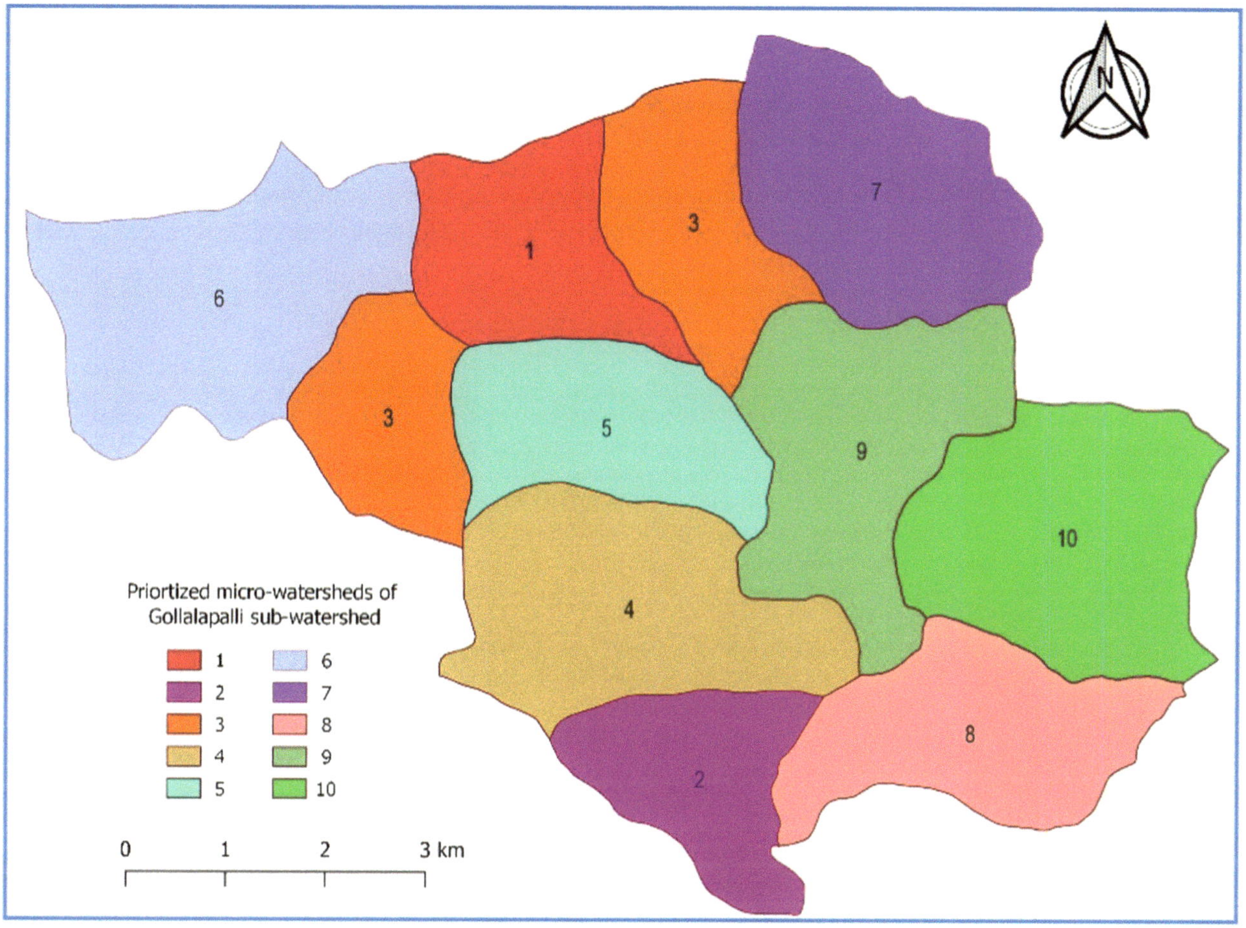

FIGURE 12.6 Prioritized rank map of the Gollalapalli subwatershed.

TABLE 12.6 Priority-wise categorized micro-watersheds.

Ranking range	Priority values	Priority class	Micro-watersheds
$C_p \leq 9.71$	1	High	Pakapattapalli, Muttinapalli, Patravanapalli, Sankavarampalli, Nallamallepalli, Yellampalli, Karakuru, Nirgantipalli, Malligerke, Gollalapalli, and Buragamadaga
$9.72 \geq C_p \leq 11.80$	2	Medium	–
$C_p \geq 11.81$	3	Low	–

12.4 Summary and conclusions

The GIS were used to analyze the drainage morphology and calculate 15 parameters in 3 aspects, such as linear, areal, and relief aspects. The morphometric parameters like drainage density, drainage texture, and stream frequency are very significant classifications of the watershed, which control and determine the pattern of various geomorphological parameters such as runoff pattern, sediment yield, etc. The drainage density of the micro-watersheds reveals that the watershed as a whole is underlined by highly permeable material and represents low relief. The drainage texture values of all the micro-watersheds represent a very coarsely structured drainage network, which results in high infiltration capacity and a low runoff rate. The shape parameters of the micro-watersheds reveal the elongated shape of the micro-watersheds and a delay in surface runoff. Based on the prioritization and ranking analysis of the Gollalapalli subwatershed, it is observed that all the micro-watersheds falls under high-priority class. The compounding parameters suggest that all the micro-watersheds fall under the high-priority zone, out of which first priority should be given to the Pakapattapalli micro-watershed, followed by Muttinapalli, Patravanapalli, Sankavarampalli, Nallamallepalli, Karakuru, Nirgantipalli, Malligerke, Gollalapalli, and Buragamadaga micro-watersheds. Based on the priority classification, the potential areas that should be considered for soil, water, and land conservation work should be given high priority to avoid any anticipated land degradation phenomena. Thus, quantification and correlation among various individual morphometric parameters gave a more logical approach to watershed prioritization. Therefore, a GIS-based approach facilitates the analysis of morphometric parameters and the exploration of the relationship between drainage morphometry and the properties of landforms, soils, and eroded lands.

References

Abdeta, G.C., Tesemma, A.B., Tura, A.L., Atlabachew, G.H., 2020. Morphometric analysis for prioritizing sub-watersheds and management planning and practices in Gidabo Basin, Southern Rift Valley of Ethiopia. Appl. Water Sci. Available from: https://doi.org/10.1007/s13201-020-01239-7.

Aher, P.D., Adinarayana, J., Gorantiwar, S.D., 2014. Quantification of morphometric characterization and prioritization for management planning in semi-arid tropics of india: a remote sensing and GIS approach. Jour. Hydrol. 511, 850–860.

Ahmed, F., Rao, K.S., 2015. Prioritization of sub-watersheds based on morphometric analysis using remote sensing and geographic information system techniques. Int. J. Remote. Sens. GIS 4 (2), 51–65.

Ajay, P., Mahmood, K., Vijay, S., Paru, P.Th, Joy, J., Nayan, P., et al., 2014. Morphometric and land use analysis for watershed prioritization in Gujarat State, India. Int. J. Sci. Eng. Res. 5 (2), 1–7.

Anhurt, F., 1970. Functional relationship between denudation, relief and uplift in large, mid-latitude drainage basin. Am. J. Sci. 268, 243–263.

Ayele, A.F., Hiroshi, Y., Katsuyuki, S., Nigussie, H., Kife, W., 2017. Quantitative analysis and implications of drainage morphometry of the Agula watershed in the semi-arid northern Ethiopia. Appl. Water Sci. 7, 3825–3840.

Biswas, S., Sudhakar, S., Desai, V.R., 1999. Prioritization of sub-watersheds based on morphometric analysis of Drainage Basin, District Midnapore, West Bengal. J. Indian. Soc. Remote. Sens. 27, 155–166.

Chandniha, S.K., Kansal, M.L., 2017. Prioritization of sub-watersheds based on morphometric analysis using geospatial technique in Piperiya watershed, India. Appl. Water Sci. 7, 329–338.

Chow, V.T., 1964. Runoff. Handbook of Applied Hydrology. McGraw-Hill, New York, pp. 14-1–14–54.

Farhan, Y., 2017. Morphometric assessment of Wadi Wala Watershed, South Jordan using. ASTER (DEM) and GIS. J. GIS 9, 158–190.

Gajbhiye, S., Mishra, S.K., Pandey, A., 2013. Prioritizing erosion-prone area through morphometric analysis: an RS and GIS perspective. Appl. Water Sci. Available from: http://doi.org/10.1007/s13201-013-0129.

Grohmann, C.H., 2004. Morphometric analysis in Geographic Information Systems: applications of free software GRASS and R star. Comput. Geosci. 30 (10), 1055–1067.

Hadley, R.F., Schumm, S.A., 1961. Sediment sources and drainage basin characteristics in upper Cheyenne River Basin. United States Geological Survey Water-Supply Paper, 1531-B, 137–196.

Horton, R.E., 1932. Drainage basin characteristics. Am. Geophys. Union Trans. 13, 348–352.

Horton, R.E., 1945. Erosional development of streams and their drainage basins: hydrophysical approach to quantitative morphology. Geol. Soc. Am. Bull. 56, 275–370.

Javed, A., Khanday, M.Y., Ahmed, R., 2009. Prioritization of sub-watersheds based on morphometric and land use analysis using remote sensing and GIS techniques. J. Indian Soc. Remote. Sens. 37, 261–274.

Mahadevaswamy, G., Nagaraju, D., Siddalingamurthy, S., Lakshmamma, Lone, M.S., Nagesh, P.C., et al., 2011. Morphometric analysis of Nanjangud Taluk, Mysore district, Karnataka, India, using GIS techniques. Int. J. Geomat. Geosci. 1 (4), 721–734.

Melton, M.A., 1957. An analysis of the relations among elements of climate, surface properties and geomorphology. Project NR 389- 042, Technical Report 11. Columbia University.

Miller, V.C., 1953. A Quantitative Geomorphic Study of Drainage Basin Characteristics on the Clinch Mountain Area, Virgina and Tennessee, Proj. NR 389–402, Tech Rep 3. Columbia University, Department of Geology, ONR, New York.

Nautiyal, M.D., 1994. Morphometric analysis of drainage basin, district Dehradun, Uttar Pradesh. J. Indian. Soc. Remote. Sens. 22 (4), 252–262.

Nookaratnam, K., Srivastava, Y.K., Venkateswarao, V., Amminedu, E., Murthy, K.S.R., 2005. Check dam positioning by prioritization of micro-watersheds using SYI model and morphometric analysis—remote sensing and GIS perspective. J. Indian. Soc. Remote. Sens. 33 (1), 25–38.

Patel, D.P., Gajjar, C.A., Srivastava, P.K., 2013. Prioritization of Malesari mini-watersheds through morphometric analysis: a remote sensing and GIS perspective. Environ. Earth Sci. 69, 2643–2656.

Raja, A., Gunasekaran, P., Ilanthirayan, A., 2017. Morphometric analysis of Kallarpatti sub-watershed, Mathur Taluk, Krishnagiri District. Int. J. Dev. Res. 7 (11), 17158–17164.

Rekha, V., George, A., Rita, M., 2011. Morphometric analysis and micro-watershed prioritization of Peruvanthanam sub-watershed, the Manimala River Basin, Kerala, South India. Environ. Res. Eng. Manag. 57.

Sakthivel, R., Manivel, M., Jawahar, N., Pugalanthi, V., Kumaran, R.D., 2007. Role of remote sensing in geomorphic mapping: a case study from Kalrayan Hills, Tamil Nadu. Indian. J. Geomorphol. 11 & 12 (1&2), 101–110.

Sangma, F., Guru, B., 2020. Watersheds characteristics and prioritization using morphometric parameters and fuzzy analytical hierarchal process (FAHP): a part of lower Subansiri sub-basin. J. Indian. Soc. Remote. Sens. Available from: https://doi.org/10.1007/s12524-019-01091-6.

Schumn, S.A., 1956. Evolution of drainage systems and slopes in badlands at Perth Amboy, New Jersey. Geol. Soc. Am. Bull. 67, 597–646.

Sheikh, U., Manzoor, S., Ayoub, M.M., 2017. Morphometric analysis and prioritization of Purinbal sub-watershed of Sindh catchment for soil and water resource management using geo-spatial tool. Int. J. Recent. Sci. Res. 8 (11), 21634–21639.

Singh, N., 1994. Remote sensing in the evaluation of morph hydrological characteristics of the drainage basin of Jojri catchment. Ann. Arid. Zone 33 (4), 273–278.

Singh, S., Singh, M.C., 1997. Morphometric analysis of Kanhar river basin. Natl Geogr. J. India 1 (43), 31–43.

Singh, N., Singh, K.K., 2014. Geomorphological analysis and prioritization of sub-watersheds using Snyder's synthetic unit hydrograph method. Appl. Water Sci. 7, 275–283.

Smith, K.G., 1950. Standards for grading textures of erosional topography. Am. J. Sci. 248 (9), 655–668.

Sreedevi, P.D., Owais, S., Khan, H.H., Ahmed, S., 2009. Morphometric analysis of a watershed of South India using SRTM data and GIS. J. Geol. Soc. India 73, 543–552.

Sreedevi, P.D., Subrahmanyam, K., Ahmed, S., 2005. The signifcance of morphometric analysis for obtaining groundwater potential zones in a structurally controlled terrain. Environ. Geol. 47 (3), 412–420.

Strahler, A.N., 1957. Quantitative analysis of watershed geomorphology. Trans. Am. Geophys. Union. 38, 913–920.

Strahler, A.N., 1964. Quantitative geomorphology of drainage basin and channel networks. In: Chow, V.T. (Ed.), Handbook of Applied Hydrology. McGraw Hill Book, New York, pp. 4–76.

Strahler, A.N., 1968. Quantitative geomorphology. In: Fairbridge, R.W. (Ed.), The Encyclopedia of Geomrophology. Reinhold Book Corporation, New York.

Sujatha, E.R., Selvakumar, R., Rajasimman, U.A.B., Victor, R.G., 2015. Morphometric analysis of sub-watershed in parts of Western Ghats, South India using ASTER DEM. Geomat. Nat. Hazards Risk 6 (4), 326–341. Available from: https://doi.org/10.1080/19475705.2013.845114.

Syed, A.A., Ikbal, J., Al, U., Saleh Aldharab, H., 2017. Measurement of watershed geometry in the central part of Udaipur sector, Rajasthan using remote sensing and GIS. Int. J. Recent. Sci. Res. 8 (11), 21786–21791.

Thapliyal, A., Panwar, A., Kimothi, S., 2017. Prioritization based on morphometric analysis in Alaknanda Basin. Glob. J. Sci. Front. Res. H. Environ. Earth Sci 17 (3), 29–34.

Thomas, J., Sabu, J., Thrivikramaji, K.P., 2010. Morphometric aspects of a small tropical mountain river system, the Southern Western Ghats, India. Int. J. Digit. Earth 3 (2), 135–156.

Vittala, S., Govindaiah, S., Honne, G.H., 2004. Morphometric analysis of sub-watersheds in the Pavagada area of Tumkur district, South India using remote sensing and GIS techniques. J. Indian Soc. Remote. Sens. 32 (4), 351–362.

Section 3

Chapter 13

Digital Soil Mapping: concepts, methods, and applications - Remote sensing and GIS perspectives

Bhabani Prasad Mondal[1,2,5], Rabi Narayan Sahoo[2], Bappa Das[3], Debashis Chakraborty[2], N. Kumar[4] and G Bhupal Raj[1]

[1]*School of Agriculture, SR University, Warangal, Telangana, India,* [2]*Division of Agricultural Physics, ICAR-Indian Agricultural Research Institute, New Delhi, India,* [3]*ICAR-Central Coastal Agricultural Research Institute, Old Goa, Goa, India,* [4]*ICAR-National Bureau of Soil Survey and Land Use Planning, Nagpur, Maharashtra, India,* [5]*Department of Agriculture, Brainware University, Kolkata, West Bengal, India*

13.1 Introduction

Soil is considered the living skin of the earth, and it acts as a medium for plant growth and helps sustain animal and human activity on earth (Wu et al., 2019). A precise understanding of the spatial distribution of soil properties and their interrelationship is necessary for the sustainable management of soil resources. For such sustainable management, accurate and detailed soil mapping is an urgent need because detailed soil mapping is helpful for land use planning, sustainable land management, environmental protection, and sustainable agriculture (Meersmans et al., 2008; Forkuor et al., 2017). Unfortunately, the traditional soil mapping techniques are laborious, time-consuming, expensive, coarser in resolution, and cannot provide the real-time or ongoing situation of land resources (McBratney et al., 2003). Thus, these polygon-based soil maps are unsuitable for site-specific management of nutrients and other resources (Castaldi et al., 2016). Therefore, a new robust method is required for detailed and accurate mapping of soil properties considering the spatial variation of these attributes for precise management of land resources. Recent progress in remote sensing (RS)-based techniques (proximal, airborne and satellite-based RS techniques) has proved their potential in predictive and quantitative mapping of various soil properties with a higher resolution and better accuracies (Ma et al., 2019). In this recent decade, digital soil mapping (DSM) has become popular in accessing and mapping world soil resources (Kidd et al., 2020). Many published scientific methodologies and case studies currently describe how and why it is employed, the techniques used, their benefits and drawbacks in operationalization, successes, and comparisons to traditional soil mapping. The last ten years have seen DSM transition from academic-based research to operational mapping of soil physical, chemical, and biological properties and soil types at various spatial scales. Nowadays, federal, state, and territorial agencies of countries like Australia and the USA employ DSM as an acknowledged and used methodology to map soil and provide information for various land resource assessment operations.

India is known as the seventh largest country in the world. It possesses 328 million hectares of geographical area, 197 million hectares of grossed cropped area, and 141 million hectares of net sown area (Government of India, 2020). India currently has the second-largest population in the world (1.37 billion people). Such immense population pressure on land resources causes severe land degradation and desertification, gradually decreasing the area of crop-growing agricultural land and increasing food demand for a projected population of 1.65 billion by 2050 (Vision, 2015). Some of India's most severe soil-related problems are desertification, low soil quality, multi-nutrient deficiencies, and stagnated crop yields. Such fatal conditions warrant a rapid assessment of soil resources for proper management (Singh and Chatterji, 2018). However, quick soil assessment and management is somewhat challenging due to significantly varied soil types nationwide, climatic conditions, geology, and topography.

Additionally, the main challenge in Indian Agriculture is small and marginal land holdings having land size of less than 2 hectares (Agriculture Census Division, 2019). Hence, the existing small-scale conventional maps covering large

Remote Sensing of Soils. DOI: https://doi.org/10.1016/B978-0-443-18773-5.00018-1

spatial extent are insufficient for managing such small land holdings. Given these facts, India would greatly benefit from adopting DSM approaches for creating fine-resolution digital soil maps for better soil management.

13.2 Evolution of digital soil mapping

Mapping of soil resources is important, providing valuable information regarding formation, spatial distribution of soil properties, and types of soil present for implementation of sustainable land use management (Pereira et al., 2017). From the historical perspective, during the middle of the 18th century, the earliest soil maps were generated to identify the homogeneous areas of similar soil attributes for suitable land management (Minasny and McBratney, 2016). Later, agro-geologists used topographic maps to portray the maps of various soil types (Miller and Schaetzl, 2014). During the 19th century, Russian concepts stressed the genetic system, while the US system stressed on the intrinsic soil properties for soil mapping. The conceptual framework or equation consisting of five factors of soil formation [i.e., climate (cl), organism (o), relief (r), parent material (p), and time (t)], mentioned by Hans Jenny (1941) made the scientific foundation of traditional soil mapping. Dokuchaiev, the Father of soil science, produced a human map of Russia based on point observation (Hartemink and Minasny, 2014). Thus, soil property mapping has been traditionally continued from earlier times. However, those conventional soil maps are qualitative, and such traditional mapping generally depends on the soil surveyors' conceptual soil-landscape model or mental model (Hudson, 1992). They have utilized their past work experience on similar types of landscapes, various landscape features, aerial photographs, digital elevation models (DEM), etc., to generate soil-type maps and databases containing descriptions of the soil profiles and the general interpretation of soil map units (Beaudette and O'Geen, 2009). The concepts of conventional soil mapping have become modernized by applying quantitative techniques in the past decade (Boettinger et al., 2010). Proximal soil sensing is one of the quantitative techniques developed in the complete phase in the early 2000s, and it needs spatial and temporal data of high resolution; thus, it may become an essential component of high-resolution soil mapping (McBratney et al., 2011). Such modern soil mapping with quantitative techniques is called DSM (McBratney et al., 2003). Various factors like the availability of digital spatial data from DEM or satellite imagery or aerial photography, high computing or supercomputing and significant processing efficiency, advancement in geographic information system (GIS) and data mining techniques (machine and deep-learning algorithms), the attraction of new spatial analysis tools by the new generation soil scientists, etc., produce a good influence for development and advancement of DSM.

In this study, we have provided a table (Table 13.1), showing the historical development of DSM (Source: Minasny and McBratney, 2016).

The evolution of soil mapping and monitoring systems in other countries like France has occurred since the 1960s. The 1960s and 1980s were defined from a soil science and pedometric perspective by so-called "conventional mapping." The first comprehensive guide, that is, "Bases et techniques d'une cartographie des sols" (meaning: Fundamentals and techniques of soil mapping), debuted at the end of the 1960s and was written by Marcel Jamagne (Jamagne, 1967). Scientists of France have developed the Commission on Soil Science and Soil Classification's collaborative effort, started traditional soil mapping in other countries predominantly African countries, and generated many databases, which were later incorporated into worldwide databases. Therefore, indirectly, the French activities during this period contributed to continental assessments of soil capability and condition, assisting in the realization of continental and worldwide DSM. (Leenaars et al., 2018). After a foundational work on DSM was published in the 2000s, that is, by McBratney et al. (2003), France hosted the First Global Workshop on DSM in Montpellier in 2004. Since then, France has assumed a more significant role in global DSM-related projects (Arrouays et al., 2020). Apart from France, the European Union also generated global soil maps earlier, like the European Soils Bureau, a Joint Research Center that created the continent's first geographical database and standardized soil maps (Arrouays et al., 2021).

13.3 Principles and concepts of DSM

The abovementioned Jenny's equation is useful for qualitative soil mapping but is unsuitable for quantitative mapping of soil classes or properties. Therefore, in the recent decade, (McBratney et al., 2003) proposed the *SCORPAN* model for expressing the quantitative relationship between soil and environmental covariates. The mathematical expression of this model is:

$$S_{c,a} = f(\mathrm{s, c, o, r, p, a, n})$$

where $S_{c,a}$ refers to soil class (S_c) or attribute (S_a) at a specific point of space in a specified time and is an empirical or quantitative function of seven environmental factors or covariates such as soil (s), climate (c), organism (o), relief (r),

TABLE 13.1 Historical development of digital soil mapping.

Period (decade)	Major developments	References
1800s	P = f (K, O, G, B)	Dokuchaev (1899)
1900s	S = f (cl, o, r, p, t)	Jenny (1941)
	S = F (addition, removal, translocation, transformation)	Simonson (1959)
1960s	Topofunctions	Walker et al. (1968)
	Integrated climate, organism, relief, parent material, time (CLORPT) model	Jenny et al. (1968)
1970s	Remote sensing application on soil	Cipra (1973)
	Soil information system (SIS)	Sadovski and Bie (1978)
	Digital soil cartography	Webster et al. (1979)
1980s	Application of geostatistics in soil	Burgess and Webster (1980)
	Application of GIS in soil	Burrough (1986)
1990s	Use of pedometrics in soil science	Odeh et al. (1992)
	Use of pedotransfer functions in soil	Wösten et al. (1995)
	Digital terrain modeling	Gallant and Wilson (1996)
2000s	Implication of data mining tools in soil	Bui and Moran (2003)
	Digital soil mapping (DSM)	McBratney et al. (2003)
	Advances in proximal soil sensing	Rossel et al. (2010)
	Digital soil morphometrics	Hartemink and Minasny (2014)

topography (t), age (a) and spatial location (n). This SCORPAN equation or model facilitates predicting any soil class or property in a spatial context by quantifying the empirical relationship between that soil class or attribute and environmental attributes. It also enables us to estimate the prediction error or uncertainties associated with the prediction model of DSM. Therefore, DSM refers to creating geographically referenced soil databases or soil maps based on quantitative relationships between spatially explicit environmental covariates and the observation and measurements made at the field level and laboratory (McBratney et al., 2003). It is also known as "predictive soil mapping," "computer-assisted soil cartography," etc. DSM consists of three main components viz., the *input*, *process* and *output*.

- The *input* of DSM generally refers to the legacy soil data or map, soil data obtained from the field laboratory-based measurements, etc.
- The *process* denotes the construction of statistical modeling for correlating the soil measurements with environmental variables, that is, spatial or nonspatial soil inference system.
- The *output* represents the spatial soil information system, illustrating the spatial distribution of soil classes or properties in the raster and prediction uncertainties.

The DSM encompasses several stages and processes, viz.

- Defining a project area
- Identifying the associated physical features of interest in the study area
- Collection of data from multiple sources and data exploration after preprocessing
- Prediction model development
- Uncertainty estimation
- Application.

A digital soil map is a graphic representation of raster data made up of pixels, and each pixel contains a specific geographic location and soil-based information. These digital maps can refine or update the existing soil survey map (Carré et al., 2007). However, digital soil maps are different from the digitized conventional soil maps because in traditional mapping, the output product is digital, but the preparation of such conventional digitized map is not related to any statistical inference system (Minasny and McBratney, 2016).

13.4 Methods of DSM

The methods of DSM include the processes of building mathematical or statistical models using several statistical algorithms, hence known as predictive mapping or modeling methods. Predictive modeling is making statistical or mathematical models that help approximate the proper relationship between soil properties or classes with environmental covariates for prediction. Predictive modeling techniques may be classified into two major categories: classification and regression. Classification methods are generally used for predicting soil classes, while the regression techniques are applied for predicting soil properties. Various statistical algorithms (predictive modeling approaches) are utilized to establish the quantitative relationship between predictor and target variables. Different machine learning (ML) based algorithms have been applied to build up predictive models for DSM using advanced programming languages like R, MATLAB®, Python, etc. A simple regression model predicts one target or dependent variable depending on the relationship between several independent or predictor variables (Zeraatpisheh et al., 2020). A linear regression model fits a linear equation in the dataset, whereas tree-based regression models are used where a nonlinear relationship exists among the variables. The multiple linear regression (MLR) model is another good option to model and predict any variable, and it has been applied to predict soil texture and other soil properties like soil organic carbon (SOC) and calcium carbonate equivalent (CCE) by several researchers (Mitran et al., 2019; Zeraatpisheh et al., 2019). Regression kriging is one of the popular methods of DSM, which refers to the hybrid model developed by combining any regression model with residuals of kriging. Kriging is also a good geostatistical approach used to predict and map various soil properties by constructing semivariogram to consider the spatial variation of those properties.

In recent years, tree-based classification models using ML algorithms have gained much more popularity than geostatistical techniques (Heung et al., 2016). The widely used tree-based model is the classification and regression tree (CART), which constructs a single tree from the given input variable to make a prediction. The advanced form of CART is the random forest (RF) model, which builds several decision trees called an ensemble of trees (forest) from the predictor variables and produces a good final prediction (Heung et al., 2016). RF is a very robust ML model capable of handling nonlinear, noisy, and large data and simultaneously providing fast computation efficiency. Another extension of the RF model is quantile random forest (QRF), which provides nonparametric estimates of the median predicted value and prediction quantiles (Dharumarajan et al., 2019). The most common hybrid approach combines RF regression and residual kriging methods to predict several soil properties through DSM (Hinge et al., 2018). Artificial neural network (ANN) is another good ML technique to handle nonlinear and complex datasets for DSM, and this neural technique consists of input layers (input variables), hidden layers, and output layers (target or output variable) (Khaledian and Miller, 2020). In recent years, the improved version of neural techniques, deep-learning (DL) techniques like convolutional neural network, has also been utilized to some extent in DSM. The best performance of predictive modeling depends on selecting the appropriate sampling techniques and adequate knowledge of the soil system (Kuhn and Johnson, 2013). The predictive model's prediction accuracy, the difference between predicted and measured value, needs to be appropriately assessed (Brus et al., 2011). Producer's accuracy, user accuracy, and overall accuracy are generally measured for estimating the prediction accuracy of soil class. Various statistical indices like the coefficient of determination (R^2), root mean square error (RMSE), concordance correlation coefficient (CCC), etc., are generally employed for measuring the accuracy of soil property prediction.

Hence, from the above discussion, it was evident that several researchers have explored models ranging from simple regression models to advanced DL-based models in DSM for quantifying the quantitative relationship between soil properties and environment covariates. Those models are MLR (Triantafilis et al., 2009), ordinary kriging (Bishop and McBratney, 2001), kriging with eternal drift (Santra et al., 2017), regression kriging (Hengl et al., 2004), CART (Breiman et al., 1984), k-nearest neighbor model (Taghizadeh-Mehrjardi et al., 2015), multinomial logistic regression (Kempen et al., 2012), SVM regression (Priori et al., 2014), and RF model (Wiesmeier et al., 2011; Dharumarajan et al., 2017; Sreenivas et al., 2016), ANN (Mosleh et al., 2016). Zeraatpisheh et al., 2019 used several machine learning-based models like cubist, RF, Regression Tree, and MLR for spatial prediction and mapping of various soil properties like SOC, CCE, and clay content from the environmental covariates derived from DEM and LANDSAT-ETM imagery. They reported that cubist and RF showed best prediction performances in predicting CCE ($R^2 = 0.30$, RMSE = 9.52) and clay content ($R^2 = 0.15$, RMSE = 7.86), while both cubist and RF performed better for accurate prediction of SOC. Digital mapping of SOC content was done by applying several statistical techniques or models like MLR (Meersmans et al., 2008), PCR and PLSR (Amare et al., 2013), kriging (Worsham et al., 2010), and regression kriging (Yigini and Panagos, 2016). The decision tree algorithm was utilized to map various soil properties in Australia (Henderson et al., 2005). In India, Santra et al. (2017) reported that the validation R^2 ranged from 0.21 to 0.30 for predicting the percent sand content using RF model in the arid parts of India, particularly for Rajasthan and parts of Gujarat states. Another study showed that the validation R^2 values varied from 0.11 to 0.44, 0.16 to 0.24, 0.35 to 0.40, and 0.66 to 0.73 for predicting SOC percent sand content, percent clay content and soil pH, respectively, for different soil layers of India using national scale digital soil maps (Reddy et al., 2021).

13.5 Uncertainty estimation in DSM

In DSM, uncertainty analysis is essential for determining whether the forecasted soil map is trustworthy enough for soil management systems or decision-making processes. Uncertainty associated with the prediction model needs to be adequately addressed. Prediction uncertainties related to DSM generally arise from some sources like positional inaccuracy, vertical anticipation of DEM, and manual error in soil property estimation. There are mainly two types of uncertainties: "aleatoric" and "epistemic". Aleatoric delays are associated with the model error, whereas epistemic uncertainties are associated with the model parameters' uncertainties (Wadoux et al., 2020). Quantile regression techniques are mostly used to quantify aleatoric uncertainty, whereas bootstrapping methods are utilized to quantify epistemic uncertainty. With techniques like the QRF, the delta or Bayesian approaches, and the mean plus variance estimate for neural network algorithms, both aleatoric and epistemic uncertainty can be quantified to offer prediction intervals (PI) (Wadoux et al., 2020). Confidence interval (CI) is usually calculated for determining the prediction uncertainty of soil classes, while PI is generally computed for measuring the prediction uncertainty of soil property. PI is generally represented by three companion maps such as lower prediction limit, mean, and upper prediction limit maps of predicted soil properties (Malone et al., 2011). Malone et al. (2011) proposed a methodological framework for quantifying uncertainties of several soil attributes like SOC, available water content, etc. Knowledge on spatial variability of soil property is required for calculating the PI. Generally, the triangulation method is applied in soil science for representing the spatial variation of soil properties based on fuzzy membership function (Odgers et al., 2015). Odgers et al. (2015) generated digital maps of soil pH, calculated the uncertainties at 90% PI, and finally showed the increment of prediction uncertainties with increasing soil depth. Previously, most of the studies ignored the analysis of uncertainties. Recent studies incorporating uncertainty assessments such as Vaysse and Lagacherie (2017) and Wadoux (2019) have both been successful in reporting PIs for RF and neural network models, respectively. CI are described in various publications and created using bootstrapped samples of the original data to train several disjoint models (Gomes et al., 2019). In Denmark, Adhikari et al. (2014) generated uncertainty maps to demonstrate the accuracy of the tree model's predictions of the study area's soil groups or categories. The confidence levels were between 0.2 and 1, with 1 being the most confident while 0.2 is the least confident. Podzols were predicted with the most confidence, with a mean value of roughly 0.72, while Podzoluvisols and Alfisols were predicted with the lowest confidence, according to pixel-by-pixel analysis of the predicted maps for the entire research region (0.48). However, we are practically lagging in assessing the uncertainty associated with DSM. Typically, businesses and commercial software do not' disclose metrics of the uncertainty related to the maps.

13.6 Conventional soil mapping versus digital soil mapping

The primary difference between conventional soil mapping (CSM) and DSM is that CSM is based on a conceptual soil-landscape model, which is qualitative. In contrast, the DSM is based on a quantitative soil-landscape model (Kempen et al., 2012). Additionally, such qualitative mental models are inflexible for quantitative studies, rarely stated straightforwardly, suffer from personal bias, and are difficult to duplicate (Jafari et al., 2012). Traditional maps have many flaws, including inadequate spatial detail and concerns with the accuracy of soil properties (Hartemink et al., 2010). The main limitation of CSM includes its irreproducibility and lack of quantification of prediction uncertainties. However, the DSM is reproducible; the digital maps can be updated at any time with new data and information availability.

Along with this, DSM quantifies uncertainties associated with the prediction model of DSM (Lark and Lapworth, 2012). With a predicted uncertainty interval, DSM can create quantitative, 2.5-D and 3-D gridded maps of soil characteristics. Maps obtained from CSM are called general-purpose maps, while the maps generated from DSM are called specific-purpose maps. Instead of producing categorical estimates of soil property values as is customary in traditional mapping methods, DSM-generated grids can depict the gradational spatial variations in soil attributes, and uncertainty ranges can give a level of confidence in the estimated attribute values. However, there are some potential drawbacks in

DSM. DSM is not a standardized technique, while the CSM is a standardized technique (NRCS US, 1993). DSM cannot represent the complex soil forming processes, while CSM can do it easily. Accurate DSM is still now challenging with the availability of limited datasets, while such limited data availability cannot hamper the accuracy of CSM.

Here, we provide a few national or regional examples that provide a good comparison between CSM and DSM and how CSM can be utilized to evaluate the predictive performance of DSM. For example, Denmark, a nation with a wealth of knowledge, is a wonderful example of how to use digital tools. The existence of a conventional soil class map for Denmark, despite its construction at a very coarse cartographic scale, offers the chance to assess the accuracy of the DSM model's predictions (Adhikari et al., 2014). They make a good comparison between the existing CSM and newly predicted DSM for different soil groups. Most of the soil mapping units' (SMUs') calculated areas for the expected soil groups were comparable to the corresponding estimated areas those same soil groups occupied on the current map. For instance, SMU 7 had an estimated 9405 km^2 area coverage, of which Luvisols were thought to occupy 75%, Cambisols 15%, and Arenosols and Gleysols each 5%. The calculated area covered by Luvisols, Cambisols, Arenosols, and Gleysols for that SMU was 70%, 12%, 5%, and 4%, respectively, demonstrating an excellent match between the two maps, that is, CSM and DSM (Adhikari et al., 2014). Although the visual or qualitative comparison was done extensively between CSM and DSM products, quantitative comparison is still not done widely. Rossiter et al. (2021) utilized a set of quantitative indicators to compare conventional and digital soil maps, but those indicators were not related to the visual comparison of the products. In France, Lemercier et al. (2022) recently evaluated the accuracy of DSM maps utilizing local scale maps. They compared local soil maps with global, national, and regional scale digital maps of various soil attributes. 'Quantitative indicators like Lin's CCC and mean squared error skill score were used to check if the soil patterns displayed by the DSM products and those shown by the existing soil maps were consistent. Since there was good agreement with conventional evaluation using spatial sets of punctual soil property measurement, evaluating DSM products with soil maps using quantitative indicators representing the closeness of projected soil patterns was relevant.

13.7 Applications of DSM

DSM has numerous applications in agriculture, from land resource management to water resource management. Various developed countries adopted many ambitious projects on DSM; the United States Department of Agriculture-Natural Resources Conservation Services (USDA-NRCS) has launched a new program called "Soils2026" to prepare legacy soil data-based digital map of 30 m resolution for the country for developing a new inventory of soil and for providing ecological sites for all parts of the United States (Thompson et al., 2020). Odgers et al. (2015) have already developed a raster-based digital map of individual soil properties at a high-resolution scale (90 m) using a polygon-based map as a legacy map from the State Soil Geographic database for the United States. In Australia, DSM has reached an operational stage, and this approach is now widely used by different agencies for real-world applications such as land suitability mapping and classification, erosion potential assessment, and ecological modeling from field scale to continental scales (Kidd et al., 2020). The most prominent use of DSM is the predictive mapping of soil properties. Across the globe, several soil scientists employed the DSM technique to predict and map several soil properties like SOM (Byrne and Yang, 2016), soil pH (Pahlavan-Rad and Akbarimoghaddam, 2018), electrical conductivity (Ranjbar and Jalali, 2016), available phosphorus (Wilson et al., 2016), soil particle size distribution (Pahlavan-Rad and Akbarimoghaddam, 2018). Mponela et al. (2020) generated high-resolution (10 m) digital maps of SOC and three major nutrients such as N, P, and K at farm scale for their effective management. The United States, Australia, China, and France also utilize their national-scale digital soil maps for various practical purposes. In France, government bodies use digital maps to assess contamination levels, agricultural policy making, private sector selling their fertilizer products, and farmers and their associated organizations for formulating fertilizer recommendations and irrigation water management (Arrouays et al., 2014). The "*Global-Soil-Map*" program and its requirements, widely used to create digital soil maps, were a significant step in improving soil data, which is now readily accessible (Poggio et al., 2021). The "*Global-Soil-Map*" format complies with the requirements of most biophysical modeling programs and fulfills the demands of many end users for ready-to-use data (Voltz et al., 2020). Nowadays, many regions of the world are covered by DSM products with similar resolution and identical targeted soil attributes due to the development of DSM applications at various scales (Minasny and McBratney, 2016; Poggio et al., 2021). For example, for the USA, three DSM items were inventoried by Rossiter et al. (2021). Similarly, in France, DSM products, that is, "*Global-Soil-Map*" products, were developed for multiple scales such as from regional, national to global scales like regional "*Global-Soil-Map*" products (Vaysse and Lagacherie 2015) to national "*Global-Soil-Map*" products (Mulder et al., 2016) and globally Soil Grids version 2.0 (Poggio et al., 2021). In central France, using around 3,200 hand-feel soil texture data, Richer-de-Forges et al. (2022) evaluated four DSM products of soil textural class prediction created at various scales (global, continental, national, and subnational scales). Adhikari et al. (2014) generated a national-level soil class map of Denmark employing information gathered from point soil observations and environmental data as predictors from the FAO-UNESCO Revised Legend in Denmark. Thy utilized a boosted decision tree-based DSM model and showed that classification error of DSM could be reduced by boosting, the tree-based model, and simultaneously the overall prediction accuracies, especially for validation dataset, could be increased from 60% to 76% for similar soil groups. The SOC, texture, pH, bulk density, and soil depth maps for the Czech Republic were developed utilizing DSM techniques in conjunction with an extensive collection of historical and present-day soil samples (Žížala et al., 2022). They showed that the combination of mapped coordinates within a buffered distance and covariates based on a mosaic of bare soils using RS data proved useful for predictive mapping, especially when training datasets maintained adequate spatial coverage.

Although DSM is a potential tool for managing soil health, natural resources, and ecosystem services, very few studies have been conducted so far on DSM of various soil properties in India (Sreenivas et al., 2014). Santra et al. (2017) have used the legacy soil map of the National Bureau of Soil Survey and Land Use Planning (NBSS&LUP) for preparing the digital map of sand content of arid western India, employing a geostatistical approach. Reddy et al. (2021) have prepared the first national-scale digital soil map of four soil properties like SOC, pH, sand, and clay content using legacy soil data and quantifying complex relationships among soil properties DEM-derived and bioclimatic variables. They have mentioned that the elevation, topographic wetness index, temperature, and rainfall were the major environmental covariates influencing the spatial distributions of the abovementioned soil properties. Such national-scale digital soil maps are useful in delineating dominant soil types, suitable crop-growing areas, affected lands, and management zones for planning purposes (Dharumarajan et al., 2017). There are several other potential applications of DSM, like land capability classification, which can be helpful for policymakers in land use planning, crop suitability mapping, and identifying major limiting factors of soil (Dharumarajan et al., 2019).

Moreover, in order to identify prospective carbon sequestration zones, one can use the spatial distribution maps (digital maps) of carbon stocks, which are important for managing global carbon emissions, carbon trading, and mitigating climate change (Dharumarajan et al., 2021). DSM could be used to make assessments of land degradation, desertification, drought risk, and soil erosion for creating appropriate region-specific action strategies (Arrouays et al., 2014). DSM can also monitor the large portions of the nation's problematic soils, including acidic soils, soils affected by salts, waterlogged soils, and degraded lands (Mitran et al., 2021). Fine-scale digital maps of soil hydraulic properties such as field capacity, permanent wilting point, and hygroscopic coefficient can assist farmers in adopting a site-specific water management strategy to optimize water use for crop plants (Santra et al., 2021). Estimation of soil moisture is another crucial aspect of scheduling irrigation of crops, and this recent digital map of surface soil moisture (SSM) of agricultural research farm of Indian Agricultural Research Institute (IARI) for different dates has been prepared by our Indian scholars using optical-thermal-microwave RS synergies. (Das et al., 2022). Digital maps can potentially be utilized for soil fertility characterization, site-specific nutrient management, and fertilizer recommendations. The major practical implication of DSM in India is its potential application in assisting the Soil Health Card mission program for the fast preparation of soil health cards for many crop growers in India.

13.8 Case study: digital soil moisture mapping

Soil moisture (SM) is a crucial parameter for water management and the hydrological cycle on the earth's surface. In agriculture, it is a critical parameter for scheduling irrigation, monitoring drought, and predicting yield. So, the estimation of SM is highly essential for agricultural production purposes. However, accurate estimation of SM is difficult due to its higher spatial and temporal variability; therefore, it is not possible through traditional techniques. The currently available RS-based global SM products are coarser in resolution (3–15 km) and unsuitable for field-level moisture mapping and monitoring application. Recently, the advancement in RS techniques has made the dynamic assessment of SM possible for various practical applications.

13.8.1 Methodology

13.8.1.1 Study area

Considering the abovementioned points, the present study was conducted at the experimental research farm of IARI, New Delhi, to produce digital maps of SSM for the IARI farm at a resolution of 30 m using the synergies of optical, thermal, and microwave RS techniques. The study area is located at 28° 37″ N to 77° 16″ E. Soils of the study site show a light loamy soil texture. The area receives mainly monsoon rainfall (June to September months). Wheat is a dominant crop in the IARI farm. The other crops are maize, pulses, vegetables, etc. The SM retrieval has been done by direct fusion of all three RS data above with the help of ML algorithms.

13.8.1.2 RS data sources

The optical, thermal, and microwave RS data products were obtained from Landsat-8 (OLI data), Landsat-8 (TRIS data), and Sentinel-1 [synthetic Aperture Radar (SAR) data], respectively. All RS-based data acquisitions were done on the same date. Landsat-8, level-1 data product, was downloaded from the United States Geological Survey website using the Earth Explorer engine (https://earthexplorer.usgs.gov/). Then, the data product was atmospherically corrected following the Dark Object Subtraction method in R software version 4.0.1. (Leutner et al., 2019). Mono window algorithm was utilized to retrieve land surface temperature (LST) using band 10 of Landsat-8 (Qin et al., 2001). The SAR data of Sentinel-1 was obtained from Copernicus Open Access Hub (https://scihub.copernicus.eu/dhus/#/home). Then, the data product of Sentinel-1 was preprocessed using SNAP (Sentinel Application Platform) for further usage. The preprocessing steps involved noise removal, radiometric correction, range doppler correction, and speckle filtering (Filipponi, 2019). These RS data were used to compute various covariates like vegetation indices, LST, backscatter coefficient, etc., which are useful for the digital mapping of SSM. For example, Landsat-8 data were utilized to derive vegetation indices like Normalized Difference Vegetation Index, Normalized Difference Water Index with the help of "SpectralIndices" function of R package.

13.8.1.3 In-situ soil moisture measurement

The ML3 Theta-Probe (Delta-T Devices, UK) was utilized to measure the SSM (5 cm top soil layer). This thermal probe is a hand-held device consisting of four rods. It needs to be inserted into the soil for direct measurement of SSM using the principle of detecting transmitted electromagnetic waves under the influence of the permittivity of soil, which influences SM. The in-situ or field level moisture measurement was done in synchronization with the date of pass (October 2020 to March 2021) of Landsat-8 and Sentinel-1 satellites.

13.8.1.4 Machine learning approaches

Four different ML-based models such as RF, Support Vector Regression (SVR), Gaussian Processes Regression (GPR), and Extreme Gradient Boosting (XGB) have been applied for digital mapping of SSM. RF is a supervised ML technique consisting of several decision trees called "forest" and trained with "bagging" method. The tuning parameters for RF are number of trees ("ntree"), number of independent variables ("mtry") at each subset, minimum node size ("min. node.size"). The study used the average of all constructed trees. The study uses the "ranger" package of R software (R 4.0.1) to develop the RF model. SVR is another ML approach that constructs the hyperplane that separates all datapoints (support vectors) into separate classes. The "kernlab" package of R studio (R 4.0.1) was utilized to develop SVR model in this study. GPR is a nonparametric, kernel-based probabilistic model, that uses the Bayesian approach to regression. It calculates the probability distribution of overall admissible functions that fit the data. "kernlab" package of R was utilized to build GPR model for soil property prediction. XGB is an ensemble ML-based model that can produce very robust and accurate results by sequentially adding weak learners in each step (Friedman, JH, 2001). It is very much flexible and customizable, allowing the optimization of several hyperparameters like "learning rate" (eta), that is, the contribution of each added learner, "number of iterations" of model (n_round or n_estimator), "depth of decision tree" (max_depth), etc., which are very much useful to reduce the bias of the model (Das et al., 2022). The "gbm" (Greenwell et al., 2019) package was used to develop this model. The statistical indices like coefficient of determination (R^2), mean biased error (MBE), RMSE, and ratio of performance to inter-quartile (RPIQ) distance were used to evaluate the predictive performance of the studied ML-based models.

13.8.2 Results and discussion

13.8.2.1 Descriptive statistics

The descriptive statistical analysis (Table 13.2) showed that SM (v/v) in both the calibration and validation dataset had nearly equal means (17.42% and 18.66%), medians (15.65% and 17.15%), standard deviations (7.37% and 7.48%), and similar ranges (3.2%–39% and 5.6%–39.2%).

TABLE 13.2 Summary of descriptive statistical analysis.

Statistical parameters	Calibration dataset	Validation dataset
Mean	17.42	18.66
Median	15.65	17.15
Standard deviation	7.37	7.48
Sample variance	59.77	56.00
Minimum	3.20	5.60
Maximum	39.00	39.20
Skewness	0.55	0.66

The datasets were normally distributed with nonzero skewness. Such commonality between calibration and validation datasets proved that both were randomly selected, well separated, and well represented the whole dataset. Such datasets are beneficial for building models using both datasets.

13.8.2.2 Feature selection

It is an important step that identifies the important variables having significant contributions to prediction and removes the redundant variables. It also reduces the complexity of the model and ensures better predictability (Hong et al., 2018). Several algorithms like Variable Importance for Projections (VIP) and Boruta algorithms can be employed to identify the variable of importance (Das et al., 2022). The present study used VIP technique to find the relative importance of covariates for different four models used in the study (Fig. 13.1A–D).

Several important variables like Sigma0_VV, normalized burn ratio index (NBRI), modified normalized difference water index (MNDWI), etc., were identified. Among all variables utilized in the study, the backscatter coefficient with vertical-vertical polarization (Sigma0_VV) is the most crucial variable for SM prediction because it is related to dielectric constant of soil, which is a function of SM (Dobson and Ulaby, 1986). The study showed that Sigma0_VV is more important than Sigma0_VH for SSM mapping, which remains consistent with earlier findings (Amazirh et al., 2018; Kumar et al., 2019).

13.8.2.3 Evaluation of model performance

The hyperparameters were optimized by following tenfold cross-validation with five times repetition of each model, and the optimization revealed that the optimized values of "ntree," "mtry," and "min-node-size" were 500, 15, and 5, respectively. For the XGB model, the final values of the tuning parameters used for modeling were "n_round," that is, number of iteration or number of trees boosted, "max_depth" (maximum tree depth), "eta" (learning rate), "gamma" (minimum loss reduction to do a split), "colsample_bytree" (fractions of the column selected for each tree), min_child_weight and "subsample" were 50, 1, 0.3, 0, 0.6, 1 and 1 respectively. For SVR model, the optimized values of the parameters like "sigma" (controls the distance of influence of a training point) and "C" (adds a penalty for misclassified point) were 0.09763221 and 1.0, respectively. The final GPR model had a sigma value of 0.1341819.

The RF exhibited the highest prediction accuracy in terms of R^2 (highest) and RMSE (least) both for calibration ($R^2 = 0.93$, RMSE = 2.76) and validation ($R^2 = 0.65$, RMSE = 5.48) datasets (Table 13.3). Such good predictability of the RF model can be attributed to the bagging technique associated with the model, which can significantly improve the predictability by considering the minute variation within the datasets (Shirzadi et al., 2018). Earlier studies also confirmed the supremacy of RF in the predictive mapping of SM (Adab et al., 2020; Liu et al., 2021; Das et al., 2022). The scatter plots of measured versus predicted showed the graphical representation of performances of all models in validation dataset (Fig. 13.2). Next to RF, SVR performed better than any other models, especially in the validation dataset ($R^2 = 0.58$, RMSE = 5.59, RPIQ = 1.84). The lowest predictability was achieved from the XGB model in terms of R2 for calibration ($R^2 = 0.61$) and validation ($R^2 = 0.44$) dataset. Interestingly, no model showed the overestimation of moisture because all the calculated MBE values for both datasets were negative. It reflects that all the models split the datasets well and trained well to predict moisture, creating multivariate models at each point (Das et al., 2022).

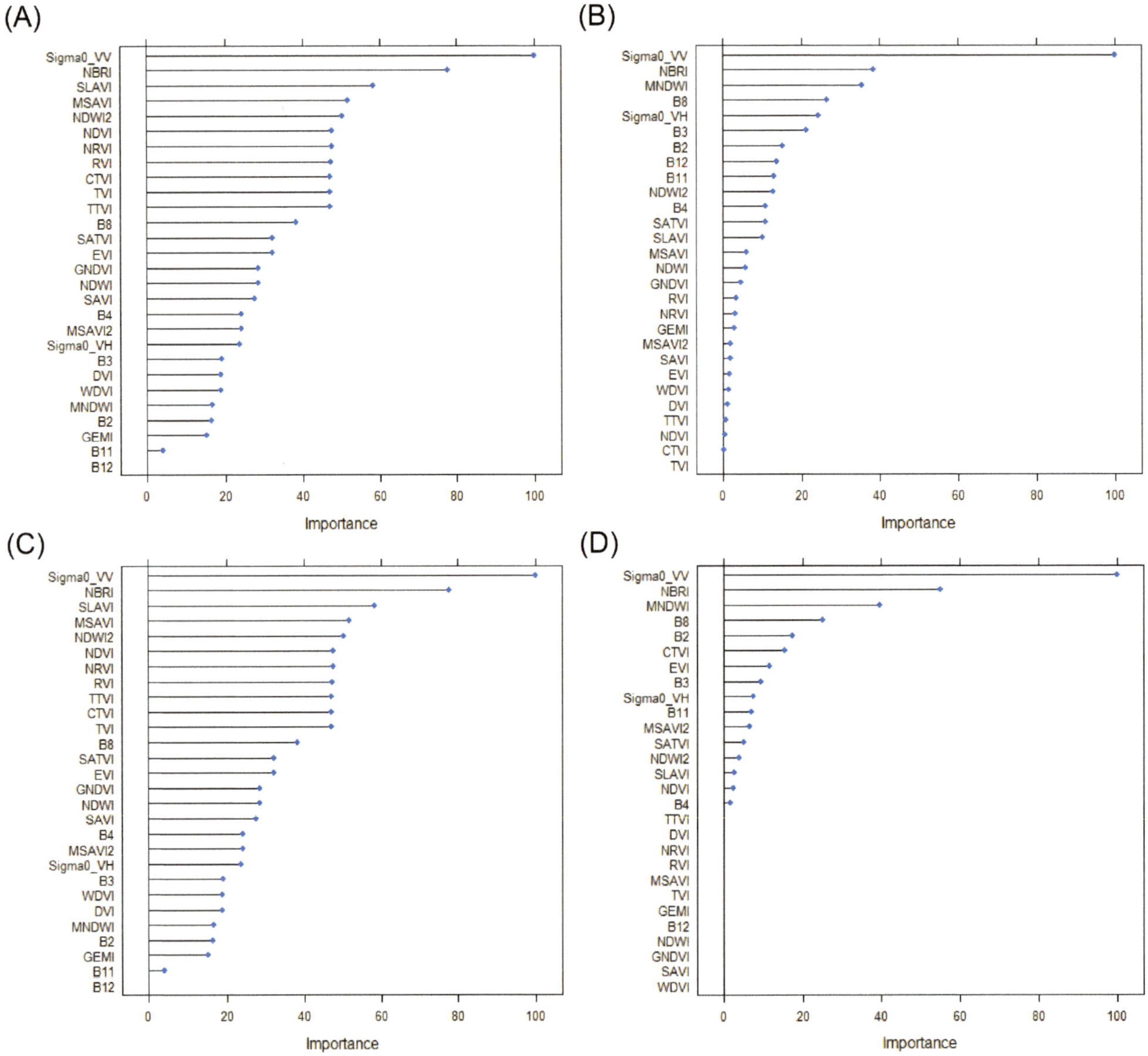

FIGURE 13.1 VIP plots (generated by RF, XGB, SVR and GPR models) showing the relative importance of variables for predictive mapping of Surface soil moisture.

TABLE 13.3 Comparing efficiency of four models for both calibration and validation datasets.

Models	Calibration				Validation			
	R^2	MBE	RMSE	RPIQ	R^2	MBE	RMSE	RPIQ
RF	0.93	−0.04	2.76	4.13	0.65	−0.54	5.48	1.88
XGB	0.61	−0.16	5.10	2.23	0.44	−0.75	6.04	1.71
SVR	0.63	−0.63	4.92	2.31	0.58	−1.28	5.59	1.84
GPR	0.69	−0.17	4.84	2.35	0.51	−1.09	6.10	1.69

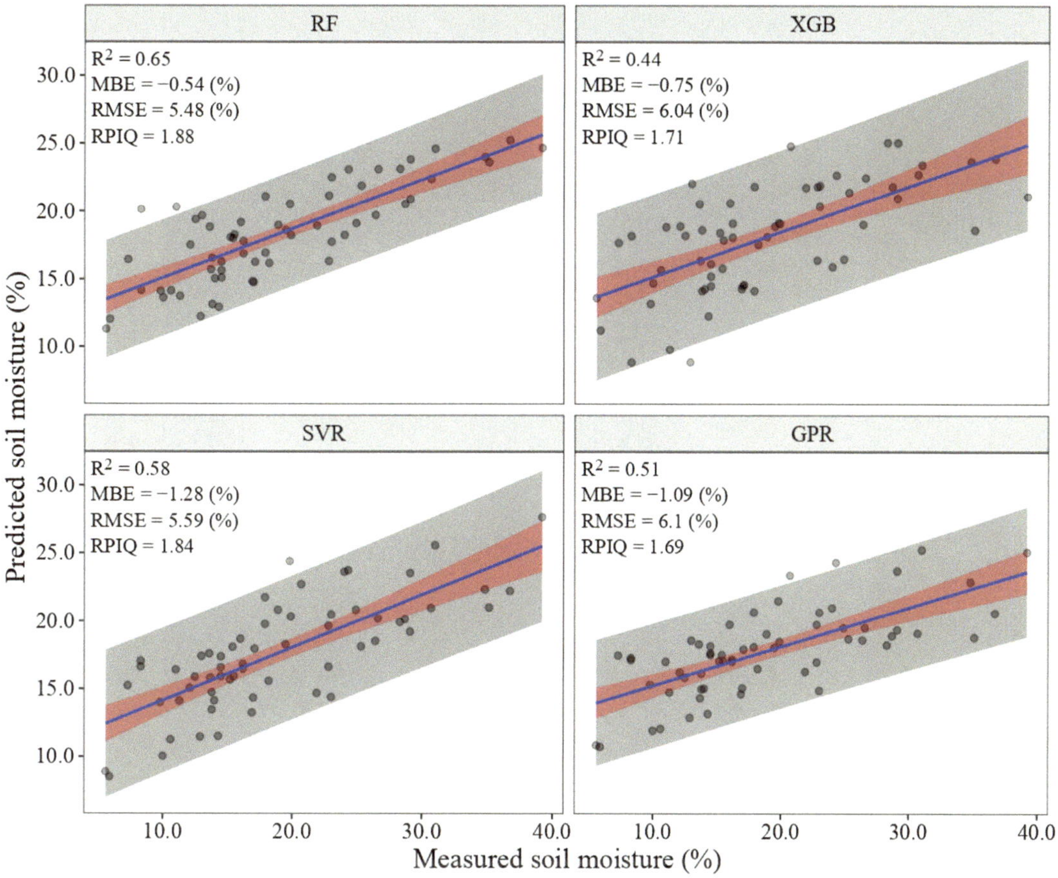

FIGURE 13.2 Scatter plots showing measured vs predicted soil moisture (%) in validation datasets for four different types of ML based models (RF, XGB, SVR and GPR models).

13.8.2.4 Spatial pattern of soil moisture

All ML models exhibited almost similar distribution of SSM in the study area (Fig. 13.3). The present study categorized the moisture classes of the study area into six categories, starting from "very dry" (less than 10% SM level) to "very wet" (more than 30% SM level). However, the earlier studies mentioned four categories of SSM such that "dry" (less than 8% SSM), "moderately wet" (8%–15% SSM), "wet" (15%–26% SSM), and "very wet" (more than 26% SSM) based on moisture content at field capacity and permanent wilting point (Santra et al., 2008; Das et al., 2022).

A major portion of the study area exhibited moderate SSM content in the range of 15%–25% (v/v), and it is mainly due to the applied irrigation to the wheat crop, the dominant crop in the study area, requiring 4–5 irrigations throughout the growth period (Das et al., 2022). As the study site is located under semiarid climatic conditions, irrigation is the main water source for crop cultivation. Such irrigation water and mild rainfall during rabi seasons, especially in January, keep the soil surface wet. As the soil has a light texture, some portions have less water-holding capacity. However, moisture uptake by high-yielding cultivars of the IARI farm is another reason for the fast depletion of SSM.

13.8.2.5 Uncertainty assessment in soil moisture mapping

The spatial distribution of uncertainty is depicted in the form of standard error (SE) maps, which are represented in Fig. 13.4.

The associated SE maps of SSM content were generated using four different ML models (GPR, RF, SVM, and XGB). The generated four SE maps of SSM content showed different trends for four different models. The SE maps of SSM measure moisture variability surrounding a coefficient of estimate. The highest variability of SSM in terms of SE was recorded for the SVM model in the range of 16%–18% of SE. The lowest SE (less than 14%), that is, less error in the prediction of SSM was observed in the case of RF model. GPR and XGB models also produced similar types of uncertainties in the prediction of SSM. It is very difficult to explain the performance of various ML models in terms of

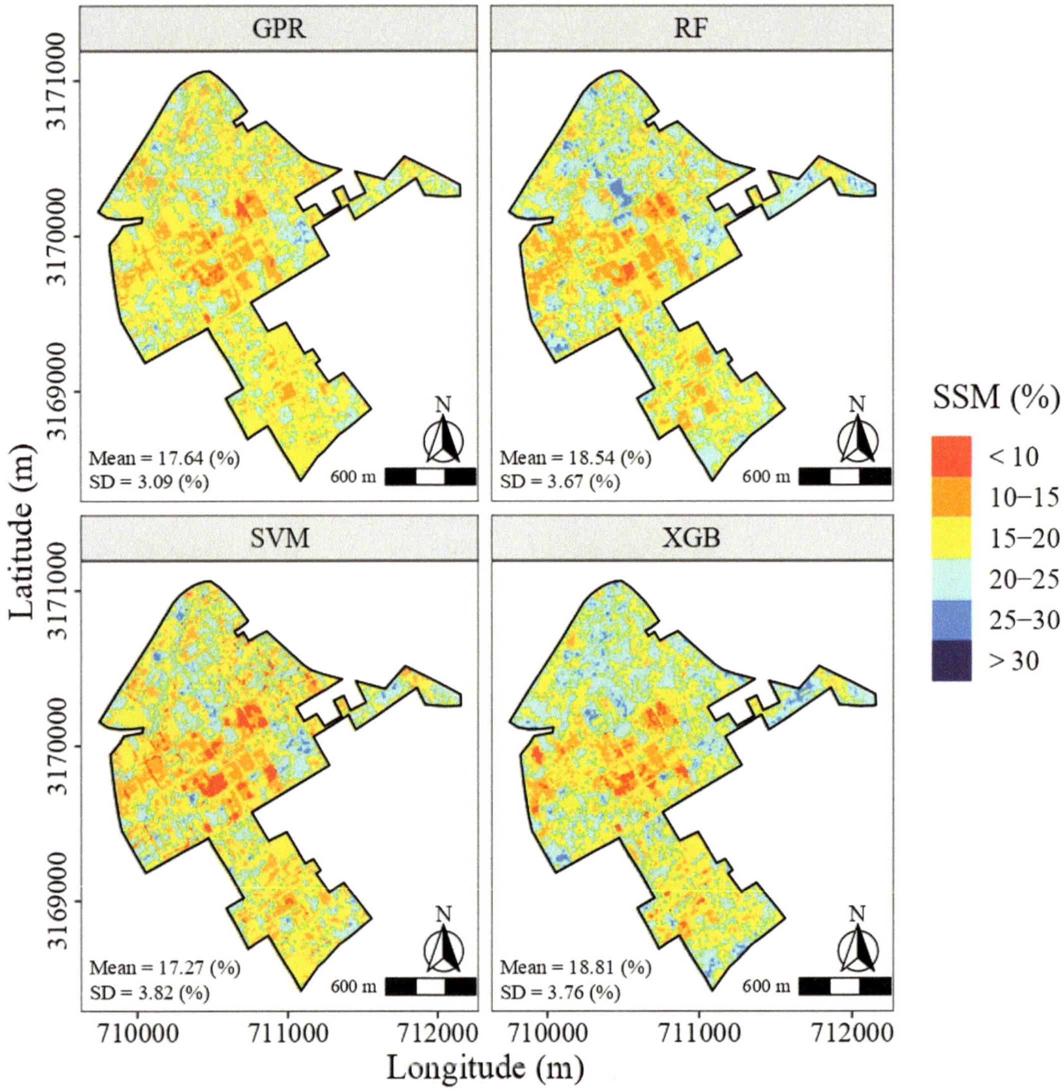

FIGURE 13.3 Spatial distribution of surface soil moisture in the study area prepared by using four ML models (GPR, RF, SVM and XGB).

accuracy and uncertainty assessment because ML-based models are empirical in nature and have region-specific applicability as compared to physical models having universal applicability requires detailed field-based observations like SM data, land surface parameters like surface roughness, vegetation indices, and other related parameters (Das et al., 2022). Another problem of ML-based models is that they cannot predict the extreme observed values of any parameter like SM. Generally, the ML-based models under estimate the very high values of observations and overestimate the very low values of observation of a parameter. Overfitting of the SM data by ML models was reported by several researchers (Bonfatti et al., 2016; Chatterjee et al., 2020). In the present study, most of the observation values were under-estimated, which was confirmed by negative bias, that is, negative MBE values. Overall, it was observed that the predicted error in the estimation of SM was comparatively higher for all ML-based models. Such error could be attributed to the high spatiotemporal variability of SM. It is well known that SM, a highly dynamic soil property, varies spatially and temporally. Moreover, there is a mismatch between the time when field observation was taken for SM in an area, and the time the satellite passes over the same area because the study area is irrigation. For example, a field-based reading of SM was taken during the morning time when the field was dry, but the satellite passed over the same area during the evening time when irrigation was given to that area; the satellite image was captured as a wet area, which is a serious mismatch for capturing the moisture data. Such mismatch in recording observation results in higher error or inaccuracies in predicting soil moisture.

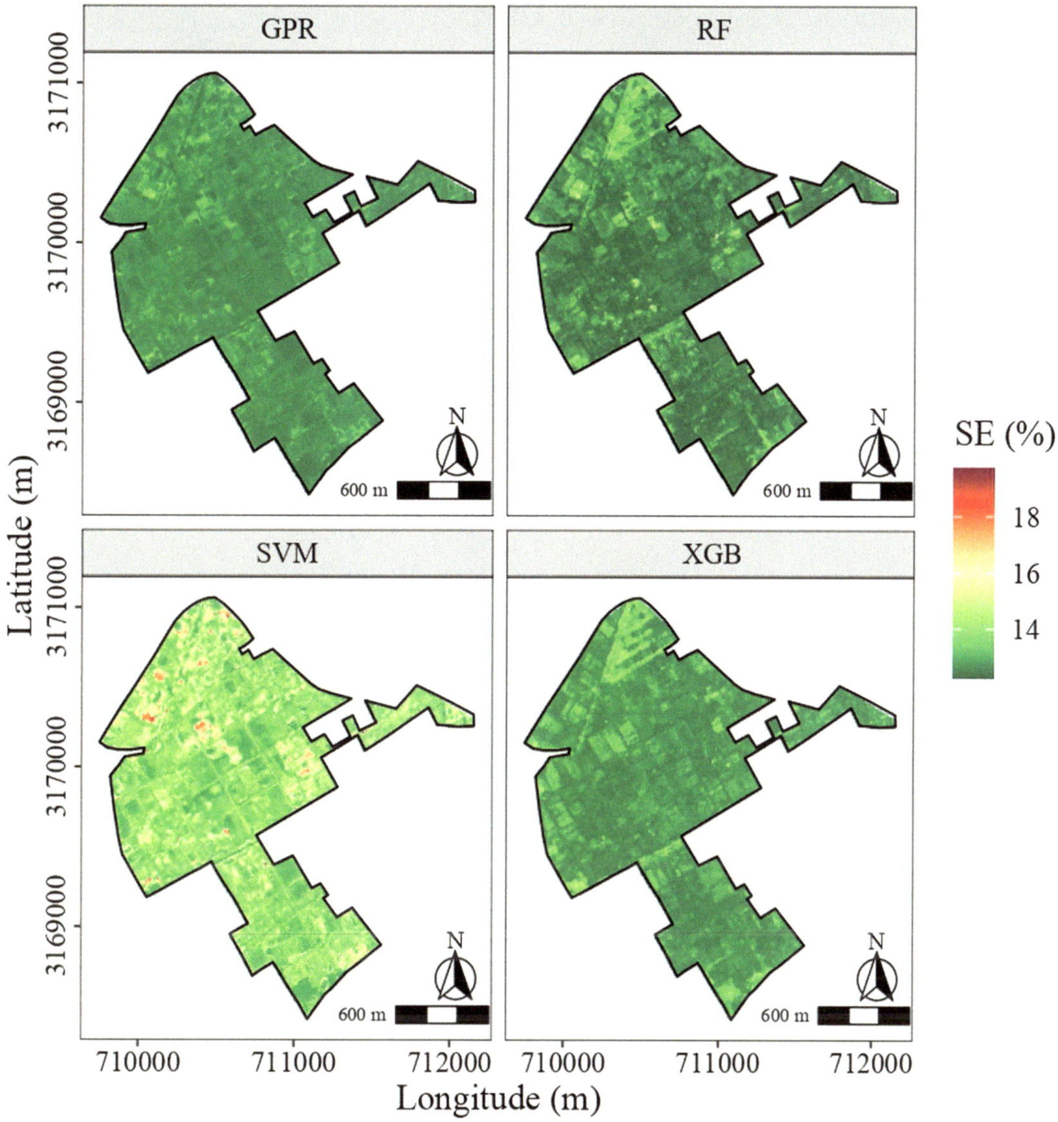

FIGURE 13.4 Uncertainty maps (standard error maps) associated with the prediction of surface soil moisture by four ML based models (GPR, RF, SVM and XGB)

13.8.3 Summary and key findings

The current study nice represented the integration of optical, thermal, and microwave RS technologies for producing digital maps of SSM of the study area using four different ML-based models. The key findings of the study are:

1. VIP technique identified the backscatter coefficient (Sigma0_VV), NBRI, and MNDWI were the key variables for SSM mapping.
2. RF model outperformed the other models (GPR, SVR, and XGB) for the predictive mapping of SSM in the study area.
3. The uncertainty analysis revealed that SVM produced more error in terms of SE as compared to the other three models. A higher error was encountered during the estimation of SSM because of the highly dynamic nature of SM in the study area. So, the spatiotemporal variability of SM must be considered before scheduling irrigation to crops.

However, the study might be limited due to the difficulty in getting all three types of RS data, especially cloud-free images and data, at the same date and time. Overall, the study demonstrated the potential of digital SSM mapping using ML techniques with better predictive accuracy. The generated high-resolution (30 m spatial resolution) SSM digital map will really be helpful for irrigation scheduling and better soil water management in the study area. Inclusion of more variables and the application of DL techniques may further improve the accuracy of digital SSM in future studies.

13.9 Limitations and future prospects of DSM

Based on the literature review, we tried to identify some of the challenges related to DSM or limitations of using ML in DSM. In the first paragraph, we pointed out the challenges associated with DSM, and then in the second paragraph, we outline some of the future aspects of DSM.

Regarding sampling designing, it is well known that the selection of a suitable sampling design plays a crucial role in obtaining a better predictive accuracy in DSM (De Gruijter et al., 2006). However, till now ML-compatible sample designs have not yet been found. One study demonstrated that when the sample size is small, it is optimal to utilize basic linear models. It also emphasized that a sufficiently high sample size is more crucial than a complicated ML method (Somarathna et al., 2017). For India, previous soil surveys have produced an enormous amount of legacy soil data, mostly available as printed books and papers. A significant barrier to DSM in India is the lack of standardized soil profile information that can be downloaded digitally. However, additional difficulties in categorizing mapping are brought on by the nature of the legacy soil data, crucial input of DSM. Balanced sets of units are necessary for the categorical mapping of soil classes or properties using ML techniques. However, the legacy soil samples create an imbalance among the classes, frequently leading to models biased toward the oversampled classes and very low predicted accuracy for the under-sampled classes (He and Garcia, 2009). Unless specifically stated, ML techniques do not consider the spatial autocorrelation present in the raw soil data. Recent studies showed that the geographically weighted regression technique can consider spatial autocorrelation in dataset for DSM (Georganos et al., 2019). The addition of pseudo-covariates can be detrimental since it prevents the study of the residuals and the development of new hypotheses from them (Hawkins, 2012). In the case of validation, because of the presence of autocorrelation in the observations and the covariates points, these procedures like random k-fold cross validation, simple random split produce noticeably overly optimistic estimates for the validation statistics (Meyer et al., 2018). The ML algorithms only utilize the information obtained from input data provided to draw digital maps of any soil properties or soil-based processes without utilizing a priori knowledge or conceptual framework of model (Koch et al., 2019). DSM techniques cannot be applied for mapping of complex soil-based processes. Hengl et al. (2014) do not provide soil maps in several under-sampled regions of the world, such as deserts and glaciers, for global mapping of several soil attributes in order to prevent extrapolation. The present case study's main flaw was the inability to get similar Sentinel-1 and cloud-free Landsat-8 launch dates, preventing SM maps with better temporal resolution. Another drawback of the present DSM of SM is the exclusion of terrain parameters, which have been reported as crucial input parameters for the digital mapping of SSM (Chatterjee et al., 2020). Suppose we again consider all the related issues in a single model to produce DSM. In that case, it will increase the complexity of a model and reduce our understanding of the basic processes occurring in soil. Therefore, an ML model that forecasts a number based on the correlation between covariates that, according to current knowledge, are uncertain should not be treated as seriously as a number that is anticipated by mechanistic processes or an accepted theory.

Based on the challenges, we tried to extract some solutions from the review of the literatures in order to draw some future perspectives on DSM. For example, most existing case studies confirmed that conditioned Latin Hypercube sampling or grid-based sampling could be used to get better prediction accuracies in DSM (Wadoux et al., 2020). To reduce the problem of overfitting or underfitting, it is suggested to calibrate a single model for predicting multiple soil properties or a single soil property at multiple soil depths (Wadoux, 2019). In the present scenario, the data-driven ML-based models provide better predictability in digital or predictive mapping of soil properties than mechanistic or geostatistical models. However, we need to improve our scientific understanding of the soil-related processes to choose a suitable ML technique to map the soil properties. Therefore, future research on DSM should include three fundamental components, that is, plausibility, interpretability, and explainability, as proposed by Roscher et al., 2020 and Lipton (2018). Plausibility denotes that models must be valid in light of scientific ideas and current knowledge in addition to being accurate. The ability to translate an abstract model or model output into human-understandable words is known as interpretability (Montavon et al., 2018). The three issues of "what is the modeled process," "how has it been modeled," and "why has this process been modeled" are the focus of explainability (Miller, 2019).

13.10 Conclusion

This chapter highlighted the basic concepts, principles, methods, and brief applications of DSM worldwide. Developed countries like Australia, USA, and France are now in leading positions for applying the DSM in solving real-world problems. Considering the alarming situation of land degradation and increasing food demand, India should adopt the DSM in its operational phase with the collaborative efforts of both government and nongovernmental bodies.

Although DSM was started in India at the research level, it is still in primitive condition to follow it in operation. So far, only two national-level digital maps have been prepared in India. However, the gradually increasing interest in DSM will show a path to shift DSM from the research to the operational stage. As the Indian agricultural land holdings are small and marginal, fine-resolution digital maps (10–30 m) are necessary to better manage soil fertility and water resources. For this, researchers in India should choose the optimum sampling strategy, sample size, and fine resolution RS product to make DSM more successful. The initiative should be taken by leading organizations like the Indian Council of Agricultural Research, NBSS&LUP, Indian Space Research Organisation, National Remote Sensing Centre, Indian Institute of Remote Sensing, Indian Institute of Technology, and other organizations in a collaborative effort to make the DSM data products easily available to the end users using online platforms for their successful application in real-world conditions.

References

Adab, H., Morbidelli, R., Saltalippi, C., Moradian, M., Ghalhari, G.A.F., 2020. Machine learning to estimate surface soil moisture from remote sensing data. Water 12, 3223.

Adhikari, K., Minasny, B., Greve, M.B., Greve, M.H., 2014. Constructing a soil class map of Denmark based on the FAO legend using digital techniques. Geoderma 214, 101–113.

Agriculture Census Division, 2019. All India Report on Number and Area of Operational Holdings, Agricultural Census (2015–16). Government of India, New Delhi.

Amare, T., Hergarten, C., Hurni, H., Wolfgramm, B., Yitaferu, B., Selassie, Y.G., 2013. Prediction of soil organic carbon for Ethiopian highlands using soil spectroscopy. Int. Sch. Res. Not. 2013.

Amazirh, A., Merlin, O., Er-Raki, S., Gao, Q., Rivalland, V., Malbeteau, Y., et al., 2018. Retrieving surface soil moisture at high spatio-temporal resolution from a synergy between Sentinel-1 radar and Landsat thermal data: a study case over bare soil. Remote. Sens. Environ. 211, 321–337.

Arrouays, D., Grundy, M.G., Hartemink, A.E., Hempel, J.W., Heuvelink, G.B.M., Hong, S.Y., et al., 2014. GlobalSoilMap. pp. 93–134. Available from: https://doi.org/10.1016/B978-0-12–800137-0.00003-0.

Arrouays, D., Poggio, L., Salazar Guerrero, O.A., Mulder, V.L., 2020. Digital soil mapping and GlobalSoilMap. Main advances and ways forward. Geoderma Reg. 21, e00265.

Arrouays, D., Mulder, V.L., Richer-de-Forges, A.C., 2021. Soil mapping, digital soil mapping and soil monitoring over large areas and the dimensions of soil security—a review. Soil. Secur. 5, 100018.

Beaudette, D.E., O'Geen, A.T., 2009. Soil-Web: an online soil survey for California, Arizona, and Nevada. Comput. Geosci. 35, 2119–2128.

Bishop, T.F.A., McBratney, A.B., 2001. A comparison of prediction methods for the creation of field-extent soil property maps. Geoderma 103, 149–160.

Boettinger, J.L., Howell, D.W., Moore, A.C., Hartemink, A.E., Kienast-Brown, S., 2010. Digital Soil Mapping: Bridging Research, Environmental Application, and Operation. Springer Science & Business Media.

Bonfatti, B.R., Hartemink, A.E., Giasson, E., Tornquist, C.G., Adhikari, K., 2016. Digital mapping of soil carbon in a viticultural region of Southern Brazil. Geoderma 261, 204–221.

Breiman, L., Friedman, J.H., Olshen, R.A., Stone, C.J., 1984. Classification and RegressionTrees. Wadsworth & Brooks Wadsworth Statistics/Probability Series.

Brus, D.J., Kempen, B., Heuvelink, G.B.M., 2011. Sampling for validation of digital soil maps. Eur. J. Soil. Sci. 62, 394–407.

Bui, E.N., Moran, C.J., 2003. A strategy to fill gaps in soil survey over large spatial extents: an example from the Murray–Darling basin of Australia. Geoderma 111, 21–44.

Burgess, T., Webster, R., 1980. Optimal interpolation and isarithmic mapping of soil properties. I. The semi-variogram and punctual kriging. J. Soil. Sci. 31, 315–331.

Burrough, P.A., 1986. Principles of Geographical Information Systems for Land Resources Assessment. Clarendon Press, Oxford.

Byrne, J.M., Yang, M., 2016. Spatial variability of soil magnetic susceptibility, organic carbon and total nitrogen from farmland in northern China. Catena 145, 92–98.

Carré, F., McBratney, A.B., Mayr, T., Montanarella, L., 2007. Digital soil assessments: beyond DSM. Geoderma 142, 69–79.

Castaldi, F., Palombo, A., Santini, F., Pascucci, S., Pignatti, S., Casa, R., 2016. Evaluation of the potential of the current and forthcoming multispectral and hyperspectral imagers to estimate soil texture and organic carbon. Remote. Sens. Environ. 179, 54–65.

Chatterjee, S., Huang, J., Hartemink, A.E., 2020. Establishing an empirical model for surface soil moisture retrieval at the U.S. climate reference network using Sentinel-1 backscatter and ancillary data. Remote. Sens. 12, 1242.

Cipra, J.E., 1973. Mapping soil associations using ERTS MSS data. LARS Tech. Rep. 117.

Das, B., Rathore, P., Roy, D., Chakraborty, D., Jatav, R.S., Sethi, D., et al., 2022. Comparison of bagging, boosting and stacking algorithms for surface soil moisture mapping using optical-thermal-microwave remote sensing synergies. Catena 217, 106485.

De Gruijter, J.J., Brus, D.J., Bierkens, M.F.P., Knotters, M., 2006. Sampling for Natural Resource Monitoring. Springer Science & Business Media, Dordrecht, NL.

Dharumarajan, S., Hegde, R., Singh, S.K., 2017. Spatial prediction of major soil properties using random forest techniques-a case study in semi-arid tropics of South India. Geoderma Reg. 10, 154–162.

Dharumarajan, S., Hegde, R., Janani, N., Singh, S.K., 2019. The need for digital soil mapping in India. Geoderma Reg. 16, e00204.

Dharumarajan, S., Hegde, R., Lalitha, M., 2021. Modelling of soil depth and hydraulic properties at regional level using environmental covariates—a case study in India. Geoderma Reg. 27, e00439. Available from: https://doi.org/10.1016/j.geodrs.2021.e00439.

Dobson, M.C., Ulaby, F.T., 1986. Active microwave soil moisture research. IEEE Trans. Geosci. Remote. Sens. 23–36.

Dokuchaev, V.V., 1899. Report to the Transcaucasian Statistical Committee on Soil Taxonomy in General and Especially for the Transcaucasia. Horizontal and Vertical Soil Zones. Office Press of the Civilian Affairs Commander-in-Chief in the Caucasus, Tifilis (in Russian).

Filipponi, F., 2019. Sentinel-1 GRD Preprocessing Workflow. Proceedings 18, 11. 10.3390/ECRS-3-06201.

Forkuor, G., Hounkpatin, O.K.L., Welp, G., Thiel, M., 2017. High resolution mapping of soil properties using remote sensing variables in south-western Burkina Faso: a comparison of machine learning and multiple linear regression models. PLoS One 12, e0170478.

Friedman, JH, 2001. Greedy function approximation: a gradient boosting machine. Ann. Stat. 29 (5), 1189–1232. Available from: https://doi.org/10.1214/aos/1013203451.

Gallant, J.C., Wilson, J.P., 1996. TAPES-G: a grid-based terrain analysis program for the environmental sciences. Comput. Geosci. 22, 713–722.

Georganos, S., Grippa, T., Gadiaga, A.N., Linard, C., Lennert, M., Vanhuysse, S., et al., 2019. Geographical random forests: a spatial extension of the random forest algorithm to address spatial heterogeneity in remote sensing and population modelling. Geocarto Int. 1, 1–12.

Gomes, L.C., Faria, R.M., de Souza, E., Veloso, G.V., Schaefer, C.E.G.R., Fernandes Filho, E.I., 2019. Modelling and mapping soil organic carbon stocks in Brazil. Geoderma 340, 337–350.

Government of India, 2020. Agricultural Statistics at a Glance 2019. Department of Agriculture Cooperation and Farmers Welfare, Directorate of Economics and Statistics, Government of India.

Greenwell, B., Boehmke, B., Cunningham, J., Developers, G.B.M., Greenwell, M.B., 2019. Package 'gbm.' R Packag. version 2.

Hartemink, A.E., Hempel, J., Lagacherie, P., McBratney, A.B., McKenzie, N.J., MacMillan, R.A., et al., 2010. GlobalSoilMap. net—a new digital soil map of the world. In: Boettinger, J.L., Howell, D.W., Moore, A.C., Hartemink, A.E., Kienast-Brown, S. (Eds.), Digital Soil Mapping: Bridging Research, Environmental Application, and Operation. Springer Science, Dordrecht, pp. 423–428.

Hartemink, A.E., Minasny, B., 2014. Towards digital soil morphometrics. Geoderma 230, 305–317.

Hawkins, B.A., 2012. Eight (and a half) deadly sins of spatial analysis. J. Biogeogr. 39, 1–9.

He, H., Garcia, E.A., 2009. Learning from imbalanced data. IEEE Trans. Knowl. Data Eng. 21, 1263–1284.

Henderson, B.L., Bui, E.N., Moran, C.J., Simon, D.A.P., 2005. Australia-wide predictions of soil properties using decision trees. Geoderma 124, 383–398.

Hengl, T., Heuvelink, G.B.M., Stein, A., 2004. A generic framework for spatial prediction of soil variables based on regression-kriging. Geoderma 120, 75–93.

Hengl, T., de Jesus, J.M., MacMillan, R.A., Batjes, N.H., Heuvelink, G.B.M., Ribeiro, E., et al., 2014. SoilGrids1km|global soil information based on automated mapping. PLOS ONE 9, e105992.

Heung, B., Ho, H.C., Zhang, J., Knudby, A., Bulmer, C.E., Schmidt, M.G., 2016. An overview and comparison of machine-learning techniques for classification purposes in digital soil mapping. Geoderma 265, 62–77.

Hinge, G., Surampalli, R.Y., Goyal, M.K., 2018. Prediction of soil organic carbon stock using digital mapping approach in humid India. Environ. Earth Sci. 77, 1–10.

Hong, H., Liu, J., Bui, D.T., Pradhan, B., Acharya, T.D., Pham, B.T., et al., 2018. Landslide susceptibility mapping using J48 Decision Tree with AdaBoost, Bagging and Rotation Forest ensembles in the Guangchang area (China). Catena 163, 399–413.

Hudson, B.D., 1992. The soil survey as paradigm-based science. Soil. Sci. Soc. Am. J. 56, 836–841.

Jafari, A., Finke, P.A., Van deWauw, J., Ayoubi, S., Khademi, H., 2012. Spatial prediction of USDA—great soil groups in the arid Zarand region, Iran: comparing logistic regression approaches to predict diagnostic horizons and soil types. Eur. J. Soil. Sci. 63 (2), 284–298.

Jamagne, M., 1967. Bases et techniques d'une cartographie des sols. Annales Agronomiques, hors-série 18, 142.

Jenny, H., 1941. Factors of soil formation. A System of Quantitative Pedology. McGraw-Hill Book Company, New York.

Jenny, H., Salem, A.E., Wallis, J.R., 1968. Interplay of soil organic matter and soil fertility with state factors and soil properties. Study Week Org. Matter Soil. Fertil. Pontif. Acad. Sci. Scr. Varia 32, 5236.

Kempen, B., Brus, D.J., Stoorvogel, J.J., Heuvelink, G.B.M., de Vries, F., 2012. Efficiency comparison of conventional and digital soil mapping for updating soil maps. Soil. Sci. Soc. Am. J. 76, 2097–2115.

Khaledian, Y., Miller, B.A., 2020. Selecting appropriate machine learning methods for digital soil mapping. Appl. Math. Model. 81, 401–418.

Kidd, D., Searle, R., Grundy, M., McBratney, A., Robinson, N., O'Brien, L., et al., 2020. Operationalising digital soil mapping–lessons from Australia. Geoderma Reg. 23, e00335.

Koch, J., Stisen, S., Refsgaard, J.C., Ernstsen, V., Jakobsen, P.R., Hojberg, A.L., 2019. Modeling depth of the redox interface at high resolution at national scale using random forest and residual gaussian simulation. Water Resour. Res. 55, 1451–1469.

Kuhn, M., Johnson, K., 2013. Applied Predictive Modeling. Springer.

Kumar, P., Prasad, R., Choudhary, A., Gupta, D.K., Mishra, V.N., Vishwakarma, A.K., et al., 2019. Comprehensive evaluation of soil moisture retrieval models under different crop cover types using C-band synthetic aperture radar data. Geocarto Int. 34, 1022–1041.

Lark, R.M., Lapworth, D.J., 2012. Quality measures for soil surveys by lognormal kriging. Geoderma 173, 231–240.

Leenaars, J.G.B., Claessens, L., Heuvelink, G.B.M., Hengl, T., Ruiperez Gonźalez, M., van Bussel, L.G.J., et al., 2018. Mapping rootable depth and root zone plant-available water holding capacity of the soil of sub-Saharan Africa. Geoderma 324, 18–36.

Lemercier, B., Lagacherie, P., Amelin, J., Sauter, J., Pichelin, P., Richer-De-Forges, A.C., et al., 2022. Multiscale evaluations of global, national and regional digital soil mapping products in France. Geoderma 425, 116052.

Leutner, B., Horning, N., Schwalb-Willmann, J. and Hijmans, R.J., 2019. RStoolbox: Tools for remote sensing data analysis (version 0.2. 4).

Lipton, Z.C., 2018. The mythos of model interpretability. Queue 16, 31–57.

Liu, Y., Qian, J., Yue, H., 2021. Combined Sentinel-1A with Sentinel-2A to estimate soil moisture in farmland. IEEE J. Sel. Top. Appl. Earth Obs. Remote. Sens. 14, 1292–1310.

Ma, Y., Minasny, B., Malone, B.P., Mcbratney, A.B., 2019. Pedology and digital soil mapping (DSM). Eur. J. Soil. Sci. 70, 216–235.

Malone, B.P., McBratney, A.B., Minasny, B., 2011. Empirical estimates of uncertainty for mapping continuous depth functions of soil attributes. Geoderma 160, 614–626.

McBratney, A., Mendonça Santos, M., Minasny, B., 2003. On digital soil mapping. Geoderma 117, 3–52. Available from: https://doi.org/10.1016/S0016-7061(03)00223-4.

McBratney, A.B., Minasny, B., Whelan, B., 2011. Defining proximal soil sensing. In: The Second Global Workshop on Proximal Soil Sensing—Montreal. pp. 15–18.

Meersmans, J., De Ridder, F., Canters, F., De Baets, S., Van Molle, M., 2008. A multiple regression approach to assess the spatial distribution of Soil Organic Carbon (SOC) at the regional scale (Flanders, Belgium). Geoderma 143, 1–13. Available from: https://doi.org/10.1016/j.geoderma.2007.08.025.

Meyer, H., Reudenbach, C., Hengl, T., Katurji, M., Nauss, T., 2018. Improving performance of spatio-temporal machine learning models using forward feature selection and target-oriented validation. Env. Model. Softw. 101, 1–9.

Miller, B.A., Schaetzl, R.J., 2014. The historical role of base maps in soil geography. Geoderma 230, 329–339.

Miller, T., 2019. Explanation in articial intelligence: insights from the social sciences. Artif. Intell. 267, 1–38.

Minasny, B., McBratney, A.B., 2016. Digital soil mapping: a brief history and some lessons. Geoderma 264, 301–311.

Mitran, T., Meena, R.S., Chakraborty, A., 2021. Geospatial technologies for crops and soils: an overview. Geospatial Technologies for Crops and Soils. Springer, Singapore, Singapore, pp. 1–48. Available from: https://doi.org/10.1007/978-981-15-6864-0_1.

Mitran, T., Solanky, V., Janakirama Suresh, G., Sujatha, G., Sreenivas, K., Ravisankar, T., 2019. Predictive mapping of surface soil texture in a semi-arid region of India through geostatistical modeling. Model. Earth Syst. Environ. 5, 645–657.

Montavon, G., Samek, W., Muller, K.-R., 2018. Methods for interpreting and understanding deep neural networks. Digit. Signal. Process. 73, 1–15.

Mosleh, Z., Salehi, M.H., Jafari, A., Borujeni, I.E., Mehnatkesh, A., 2016. The effectiveness of digital soil mapping to predict soil properties over low-relief areas. Environ. Monit. Assess. 188, 1–13.

Mponela, P., Snapp, S., Villamor, G., Tamene, L., Le, Q.B., Borgemeister, C., 2020. Digital soil mapping of nitrogen, phosphorus, potassium, organic carbon and their crop response thresholds in smallholder managed escarpments of Malawi. Appl. Geogr. 124, 102299.

Mulder, V.L., Lacoste, M., Richer-de-Forges, A.C., Arrouays, D., 2016. GlobalSoilMap France: high-resolution spatial modelling the soils of France up to two meter depth. Sci. Total. Environ. 573, 1352–1369.

NRCS US, 1993. Soil survey division staff (1993) soil survey manual. Soil conservation service. US Dep. Agric. Handb. 18, 315.

Odeh, I.O.A., Chittleborough, D.J., McBratney, A.B., 1992. Fuzzy-c-means and kriging for mapping soil as a continuous system. Soil. Sci. Soc. Am. J. 56, 1848–1854.

Odgers, N.P., McBratney, A.B., Minasny, B., 2015. Digital soil property mapping and uncertainty estimation using soil class probability rasters. Geoderma 237, 190–198.

Pahlavan-Rad, M.R., Akbarimoghaddam, A., 2018. Spatial variability of soil texture fractions and pH in a flood plain (case study from eastern Iran). Catena 160, 275–281.

Pereira, P., Brevik, E., Munoz-Rojas, M., Miller, B., 2017. Soil Mapping and Process Modeling for Sustainable Land Use Management. Elsevier.

Poggio, L., de Sousa, L.M., Batjes, N.H., Heuvelink, G.B.M., Kempen, B., Ribeiro, E., et al., 2021. SoilGrids 2.0: producing soil information for the globe with quantified spatial uncertainty. Soil 7, 217–240.

Priori, S., Bianconi, N., Costantini, E.A.C., 2014. Can γ-radiometrics predict soil textural data and stoniness in different parent materials? A comparison of two machine-learning methods. Geoderma 226, 354–364.

Qin, Z., Karnieli, A., Berliner, P., 2001. A mono-window algorithm for retrieving land surface temperature from Landsat TM data and its application to the Israel-Egypt border region. Int. J. Remote. Sens. 22, 3719–3746.

Ranjbar, F., Jalali, M., 2016. The combination of geostatistics and geochemical simulation for the site-specific management of soil salinity and sodicity. Comput. Electron. Agric. 121, 301–312.

Reddy, N.N., Chakraborty, P., Roy, S., Singh, K., Minasny, B., McBratney, A.B., et al., 2021. Legacy data-based national-scale digital mapping of key soil properties in India. Geoderma 381, 114684.

Richer-de-Forges, A.C., Arrouays, D., Poggio, L., Chen, S., Lacoste, M., Minasny, B., et al., 2022. Hand-feel soil texture observations to evaluate the accuracy of digital soil maps for local prediction of particle size distribution. A case study in central France. Pedosphere 32.

Roscher, R., Bohn, B., Duarte, M.F., Garcke, J., 2020. Explainable machine learning for scientific insights and discoveries. Ieee Access 8, 42200–42216.

Rossel, R.A.V., McBratney, A.B., Minasny, B., 2010. Proximal Soil Sensing. Springer Science & Business Media.

Rossiter, D.G., Poggio, L., Beaudette, D., Libohova, Z., 2021. How well does predictive soil mapping represent soil geography? SOIL, under discussion an investigation from the USA 10.5194/soil-2021-80.

Sadovski, A.N., Bie, S.W., 1978. Developments in soil information systems. In: Proceedings of the Second Meeting of the ISSS Working Group on Soil Information Systems, Varna/Sofia, Bulgaria, 30 May–4 June 1977. Pudoc.

Santra, P., Chopra, U.K., Chakraborty, D., 2008. Spatial variability of soil properties and its application in predicting surface map of hydraulic parameters in an agricultural farm. Curr. Sci. 95, 937–945.

Santra, P., Kumar, M., Panwar, N., 2017. Digital soil mapping of sand content in arid western India through geostatistical approaches. Geoderma Reg. 9, 56–72.

Santra, P., Kumar, M., Panwar, N.R., Yadav, R.S., 2021. Digital soil mapping: the future need of sustainable soil management. Geospatial Technologies for Crops and Soils. Springer, Singapore, Singapore, pp. 319–355. Available from: https://doi.org/10.1007/978-981-15-6864-0_9.

Shirzadi, A., Soliamani, K., Habibnejhad, M., Kavian, A., Chapi, K., Shahabi, H., et al., 2018. Novel GIS based machine learning algorithms for shallow landslide susceptibility mapping. Sensors 18, 3777.

Simonson, R.W., 1959. Outline of a generalized theory of soil genesis. Soil. Sci. Soc. Am. J. 23, 152–156.

Singh, S.K., Chatterji, S., 2018. Land resource inventory for agricultural land use planning using geospatial techniques. Geospatial Technologies in Land Resources Mapping, Monitoring and Management. Springer, pp. 163–183.

Somarathna, P.D.S.N., Minasny, B., Malone, B.P., 2017. More data or a better model? guring out what matters most for the spatial prediction of soil carbon. Soil. Sci. Soc. Am. J. 81, 1413–1426.

Sreenivas, K., Dadhwal, V.K., Kumar, S., Harsha, G.S., Mitran, T., Sujatha, G., et al., 2016. Digital mapping of soil organic and inorganic carbon status in India. Geoderma 269, 160–173.

Sreenivas, K., Sujatha, G., Sudhir, K., Kiran, D.V., Fyzee, M.A., Ravisankar, T., et al., 2014. Spatial assessment of soil organic carbon density through random forests based imputation. J. Indian Soc. Remote. Sens. 42, 577–587. Available from: https://doi.org/10.1007/s12524-013-0332-x.

Taghizadeh-Mehrjardi, R., Nabiollahi, K., Minasny, B., Triantafilis, J., 2015. Comparing data mining classifiers to predict spatial distribution of USDA-family soil groups in Baneh region, Iran. Geoderma 253–254, 67–77. Available from: https://doi.org/10.1016/j.geoderma.2015.04.008.

Thompson, J.A., Kienast-Brown, S., D'Avello, T., Philippe, J., Brungard, C., 2020. Soils2026 and digital soil mapping—A foundation for the future of soils information in the United States. Geoderma Reg. 22, e00294.

Triantafilis, J., Lesch, S.M., La Lau, K., Buchanan, S.M., 2009. Field level digital soil mapping of cation exchange capacity using electromagnetic induction and a hierarchical spatial regression model. Soil. Res. 47, 651–663.

Vaysse, K., Lagacherie, P., 2015. Evaluating digital soil mapping approaches for mapping GlobalSoilMap soil properties from legacy data in Languedoc-Roussillon (France). Geoderma Reg. 4, 20–30.

Vaysse, K., Lagacherie, P., 2017. Using quantile regression forest to estimate uncertainty of digital soil mapping products. Geoderma 291, 55–64.

Vision, C., n.d. 2050: Indian Council of Agricultural Research, New Delhi; 2015.

Voltz, M., Arrouays, D., Bispo, A., Lagacherie, P., Laroche, B., Lemercier, B., et al., 2020. Possible futures of soil-mapping in France. Geoderma Reg. 23.

Wadoux, A.M.J.C., 2019. Using deep learning for multivariate mapping of soil with quantied uncertainty. Geoderma 351, 59–70.

Wadoux, A.M.C., Minasny, B., McBratney, A.B., 2020. Machine learning for digital soil mapping: applications, challenges and suggested solutions. Earth-Sci. Rev. 210, 103359.

Walker, P.H., Hall, G.F., Protz, R., 1968. Relation between landform parameters and soil properties. Soil. Sci. Soc. Am. J. 32, 101–104.

Webster, R., Harrod, T.R., Staines, S.J. and Hogan, D.V., 1979. Grid sampling and computer mapping of the Ivybridge area, Devon (No. Tech Monograph No 12Mongograph).

Wiesmeier, M., Barthold, F., Blank, B., Kögel-Knabner, I., 2011. Digital mapping of soil organic matter stocks using Random Forest modeling in a semi-arid steppe ecosystem. Plant. Soil. 340, 7–24.

Wilson, H.F., Satchithanantham, S., Moulin, A.P., Glenn, A.J., 2016. Soil phosphorus spatial variability due to landform, tillage, and input management: a case study of small watersheds in southwestern Manitoba. Geoderma 280, 14–21.

Worsham, L., Markewitz, D., Nibbelink, N., 2010. Incorporating spatial dependence into estimates of soil carbon contents under different land covers. Soil. Sci. Soc. Am. J. 74, 635–646.

Wösten, J.H.M., Finke, P.A., Jansen, M.J.W., 1995. Comparison of class and continuous pedotransfer functions to generate soil hydraulic characteristics. Geoderma 66, 227–237.

Wu, T., Luo, J., Dong, W., Sun, Y., Xia, L., Zhang, X., 2019. Geo-object-based soil organic matter mapping using machine learning algorithms with multi-source geo-spatial data. IEEE J. Sel. Top. Appl. Earth Obs. Remote. Sens. 12, 1091–1106.

Yigini, Y., Panagos, P., 2016. Assessment of soil organic carbon stocks under future climate and land cover changes in Europe. Sci. Total. Environ. 557, 838–850.

Zeraatpisheh, M., Ayoubi, S., Jafari, A., Tajik, S., Finke, P., 2019. Digital mapping of soil properties using multiple machine learning in a semi-arid region, central Iran. Geoderma 338, 445–452.

Zeraatpisheh, M., Jafari, A., Bodaghabadi, M.B., Ayoubi, S., Taghizadeh-Mehrjardi, R., Toomanian, N., et al., 2020. Conventional and digital soil mapping in Iran: past, present, and future. Catena 188, 104424.

Žížala, D., Minařík, R., Skála, J., Beitlerová, H., Juřicová, A., Rojas, J.R., et al., 2022. High-resolution agriculture soil property maps from digital soil mapping methods, Czech Republic. Catena 212, 106024.

Chapter 14

Estimating Soil Quality Index (SQI) of arid region of south India using machine learning algorithms

S. Dharumarajan[1], G.K. Harikaran[1], M. Lalitha[1], Pravash Chandra Moharana[2], R. Vasundhara[1], B. Kalaiselvi[1], Sweta Kumari[1], Amar Suputhra[1], R. Srinivasan[1], C.M. Pradeep[1] and V. Ramamurthy[1]

[1]*ICAR-National Bureau of Soil Survey and Land Use Planning, Regional Centre, Bangalore, Karnataka, India,* [2]*ICAR-National Bureau of Soil Survey and Land Use Planning, Nagpur, Maharashtra, India*

14.1 Introduction

Deteriorating soil quality has imposed significant questions on environmental sustainability, economic growth, and agricultural productivity globally (Masto et al., 2008; Gong et al., 2015; Fathizad et al., 2020). The effects of climate change have posed a great challenge to sustaining soil quality due to the vulnerability of the soils to climate change. Its reduction is mostly caused by improper land use and soil management, inadequate precipitation, soil erosion, unsuitable topography, deforestation, and excessive grazing (Lal et al., 2003). Land deprivation and decreasing soil fertility are the main causes disturbing agricultural productivity and human welfare (Sanchez et al., 1997; Dharumarajan et al., 2017a). The main soil-environmental alarms are the reduction of nutrients, decrease in soil productivity, and loss of soil organic matter (Sanchez et al., 1997). The annual depletion of soil nutrients is about 28 million metric tons against an addition of 20 million metric tons, leaving a net gap of 8 million metric tons per annum, which is accumulating the deficiency year after year, leading to depleting soil quality (Chand, 2010). Soil quality affects plant and animal productivity, the preservation or improvement of water and air quality, and human health and habitation (Chand, 2010). The cropping pattern also has an impact on the soil quality. A fact-finding committee constituted by the Government of India in 1997 on the trend of productivity of important crops reported that there is a decline in the organic carbon content of the soils due to continuous cultivation of cereal-based cropping systems; for example, rice—wheat, rice—rice, rice—maize, etc. The continuous depletion of soil nutrients from farmlands results in adverse impacts on soil health (Chand, 2010). In arid and rain-fed regions, crop nutrition depletion was significantly greater than the advantages of the additional fertilizers (Sharma et al., 2005). The largest losers are the various soil types with poor cation exchange capacity (CEC), including alfisols, ultisols, and oxisols. The loss of soil organic carbon (SOC) in various cropping sequences in India ranged from 0.22% to 6.0%. From 1975 to 2002, inadequate fertilization with high residual sodium carbonate depleted phosphorus and potassium in dry soils in India by 7.7% and 13.4%, respectively (Singh et al., 2007). Hence, in order to reverse the trend of diminishing soil quality and maintain sustainable agriculture, assessment of soil and site characteristics relevant to their appropriateness for agricultural land use in general is required.

Continuous monitoring of the major physical, chemical, or biological features of soil is required for soil quality evaluation. The permutation of various indicators into a single index could help to understand integrated soil quality because soil quality is a multifaceted functional concept that cannot be measured directly in the laboratory or in the field (Stocking, 2003). Instead, it can be inferred from the individuality of the soil. Several approaches to soil quality evaluation have been developed, such as soil quality index (SQI) methods (Doran and Jones, 1996; Marzaioli et al., 2010; Qi et al., 2009) and geostatistical methods (Sun et al., 2003; Rahmanipour et al., 2014). Among these methods, SQI is the most commonly used methods today (Andrews et al., 2002) because they are easy to use and quantitatively flexible. SQI is the best tool to assess the soil quality of a given site and also allows comparing the situation at the plots, field, or watershed scale under varying land uses and management practices. Therefore, the expediency of the SQI as a tool to assess the management options to study within the crop experimental field or crop boundaries, or it can also be used at a

Remote Sensing of Soils. DOI: https://doi.org/10.1016/B978-0-443-18773-5.00026-0

regional scale (Sharma et al., 2005; Masto et al., 2008). Assessment of SQI using only the surface soil properties provides incomplete information as crop productivity is influenced by both surface and subsurface properties, with the latter being inherently linked to pedogenic processes (Lenka et al., 2022). Soil quality is a crucial indicator for sustainable use of land, and it can also be used to quantify the sustainability of soil use (Vasu et al., 2016; Muñoz-Rojas, 2018).

The main aim of soil quality mapping is to locate regions with low soil quality. Conventionally, mapping consists of a number of processes, such as collecting soil samples from the field survey, making field-based pedomorphological observations, followed by laboratory measurements, and finally mapping and analysis. However, this method is costly, time-consuming, and depends on the pedologist's expertise. Therefore, it is essential to develop an accurate analog of conventional soil mapping (Pahlavan-Rad et al., 2016; Zeraatpisheh et al., 2017). In the modern era of geoinformatics, digital soil mapping (DSM) has shown significant promise for overcoming some of the irregularities of conventional soil mapping (Ma et al., 2019). Geospatial environmental data (i.e., covariates) and field and laboratory methods are used in DSM techniques to create and collect soil information. A predictive model is then used to establish the relationships between covariates and soil properties in order to make predictions for unsampled locations (Lagacherie and McBratney, 2006). A DSM approach provides a number of benefits over traditional soil mapping methods. For instance, soil maps may be created automatically and with greater accuracy and precision than traditional soil maps by utilizing geospatial datasets that are increasingly available for free and with simple access. Recently, a number of general statistical and machine learning algorithms using DSM have been used to predict spatial soil quality. This includes random forest (RF) (Taghizadeh-Mehrjardi et al., 2016; Dharumarajan et al., 2020; Moharana et al., 2022a,b), fuzzy logic (Moharana et al., 2020), artificial neural networks (ANN) (Behrens et al., 2005), regression trees (RT) (Jafari et al., 2014), multiple linear regression (MLR) (Jena et al., 2022), and support vector machines (SVM) (Kovačević et al., 2010; Dharumarajan et al., 2021a). The standard research methodology involves comparing various models to see which one performs the best. However, each model has unique benefits and disadvantages for a particular area (Dharumarajan et al., 2019). Studies on Indian regions (e.g., Sreenivas et al., 2016; Dharumarajan et al., 2019; Dharumarajan et al., 2021b) reveal that out of different machine learning algorithms, the RF model was found to be most appropriate and more robust in noise and data reduction; it also has the capacity to handle both quantitative and qualitative datasets (Rodriguez-Galiano and Chica-Rivas, 2014). Considering these factors, the study was conducted in the arid region of southern India for the computation of SQI from 0 to 5, 5 to 15, 15 to 30, and 30 to 60 cm depth data. We used the quantile random forest (QRF) model, which we implemented in the current study, to predict the soil quality for unexplored locations in the arid region of South India using environmental covariates. The present study aimed to (1) integrate a range of soil quality indicators using additive-based weighted index methods in order to calculate soil quality indices (SQI); (2) evaluate the performance of the QRF model to predict soil quality parameters; and (3) integrate a range of environmental data layers using QRF the method so as to develop predicted SQI in the arid region of southern India.

14.2 Materials and methods

14.2.1 Study area

The present study was carried out in the arid region of south India. Five districts, Koppal, Bellary, Raichur (Karnataka), Anantapur, and Kurnool (Andhra Pradesh), were mapped. The study area is located between 15°14′ to 15°20′ N latitude and 76°5′ to 76°13′ longitude, with a total area of 97.8 km^2. The study area falls under the arid tract of south India and is categorized as drought-prone. The elevation in the study area ranged from 360 to 750 m. The major geological formations in the study area are granite gneiss, basalt, chlorite schist, etc.

14.2.2 Legacy data collection

Legacy soil profile datasets of the study area were collected and organized for mapping soil properties (Table 14.1). These soil profiles were studied based on the soil–physiographic relationship. A total of 137 soil legacy profiles were used for mapping soil quality parameters. Soil depth, sand, clay, pH, OC, and profile available water content (PAWC) were used for the computation of the SQI.

14.2.3 Environmental covariates

Digital terrain attributes, such as profile and plan curvature, topographic wetness index (TWI), topographic position index (TPI), aspect, slope, multiresolution index of valley bottom flatness (MrVBF), and multiresolution ridge top flatness (MrRTF) were used to represent topography factors in the *SCORPAN* model. MrVBF is useful to identify valley bottoms and depositional areas, and MrRTF is beneficial to identify high flat areas. TWI was used to represent potential hydrological

TABLE 14.1 Primary sources of soil depth data.

S. No	Project	Scale	References
1	Soil resource mapping of Karnataka (1989–98)	1: 2,50,000	Soils of Karnataka for optimizing land use. NBSS&LUP publication No.47b
2	Soils of Bellary district	1: 50,000	The Soils of Bellary District NBSS Publ.1015
3	Land resource inventory of Bukkarayasamudrum mandal of Anantapur district, Andhra Pradesh		Land resource inventory of Bukkarayasamudrum mandal of Anantapur district, Andhra Pradesh on 1:10000 scale for optimal agricultural land use planning using geospatial techniques. NBSS&LUP, Technical Bulletin No. 1105 (Dharumarajan et al., 2017b)
4	Sujala III project (2014–18)	1:10,000	Sujala III project SWS report of Gadag district

TABLE 14.2 Different covariates used for prediction of soil fertility parameters.

Predictor	Source	Resolution	*Scorpan* factor
Elevation (m)	SRTM DEM	30 m	Relief
Aspect			
Slope			
TW1			
TPI			
Profile curvature			
Plan curvature			
MrVBF			
MrRTF			
LS factor			
EVI	MOD13Q1	250 m _16 days	Organism
NDVI	MOD13Q1		
Landsat data (b1–b9)	9 bands	30 m	
Precipitation (mm)	Worldclim2	30 s	Climate
Average maximum temperature (°C)	Worldclim2	30 s	
Average minimum temperature (°C)	Worldclim2	30 s	

conditions based on slope and catchment. Normalized difference vegetation index (NDVI), enhanced vegetation index (EVI) (MOD13Q1), and Landsat 8 data were used to represent organism factors in the SCORPAN model. Annual precipitation (*prec*), average minimum (*tmin*), and average maximum temperature (*tmax*) were used to represent the climate (Table 14.2).

14.2.4 Harmonization of soil profile data

The soil profile datasets were preprocessed using equal-area spline functions (Bishop et al., 1999) to harmonize the soil depth interval. Soil properties were computed for four standard depth horizons for better comparison between their profile characteristics, for example, 0–5, 5–15, 15–30, and 30–60 cm (Malone et al., 2009; Arrouays et al., 2014). This was done using equal-area spline functions implemented in the GSIF R-package.

14.2.5 Model used

QRF model was used to map the soil parameters. QRF model is an extension of RF model which works based on the assemblage of a number of classification and regression trees by means of two levels of randomization for each tree in the forest (Breiman, 2001). The number of predictors used for fitting the tree (M_{try}), the number of tree (n_{tree}), and the minimum number of samples at the terminal node $n_{\min}$ are the three parameters that determine the fitting of RF. The RF 4.6 package in the *R* environment was used for the prediction of soil quality parameters. The accuracy of the model is set by the mean square error (MSE_{OOB}) of the aggregated OOB predictions generated from the bootstrap subset and is calculated as follows:

$$MSE_{OOB} = n^{-1}\sum_{i=1}^{n}(zi-\hat{zi}^{OOB})^2$$

Where "*n*" is the number of observations, *zi* is the average prediction of the *i*th observation, and $\hat{zi}^{OOB}$ is the average prediction for the ith observation from all trees for which the observation was OOB.

14.2.6 Model validation

To assess the effectiveness of the RF model, ten-fold cross-validation approaches with 20 repetitions n were performed. To evaluate the performance of the model, the coefficient of determination (R^2), root-mean-square error (RMSE), and mean error (ME) were determined.

$$R2 = 1 - \frac{\sum_{i=1}^{n}(pi-oi)^2}{\sum_{i=1}^{n}\left(oi-\overline{oi}\right)^2}$$

$$ME = \frac{1}{n}\sum_{i=1}^{n}(oi-pi)^2$$

$$RMSE = \sqrt{\frac{1}{n}\sum_{i=1}^{n}(oi-pi)^2}$$

Where *pi* and *oi* are predicted and observed values and $\overline{\text{pi}}$ and $\overline{\text{oi}}$ are means of predicted and observed values.

14.2.7 Calculation of SQI

The SQI is determined using an additive method based on the soil properties (Mukherjee and Lal, 2014; Dharumarajan et al., 2022). Based on the published study, threshold values for the soil characteristics in this technique were determined. Table 14.3 lists the threshold values and corresponding soil index score values. The total SQI was then determined using the following formula: add the separate index value sums.

$$\sum \text{SQI} = \sum \text{individual soil parameter index values}$$

$$\text{SQI} = \left(\sum \text{SQI} - \text{SQImin}\right)/(\text{SQImax} - \text{SQImin})$$

Where SQImin = Minimum value of SQI and SQImax = Maximum value of SQI from the total dataset.

For calculation of depth-wise SQI, pH, OC, CEC, sand, and clay were used. SQI of soils at different depths, soil depth, and PAWC were used for mapping profile SQI. Finally, SQI was mapped using Arc GIS software.

14.3 Results and discussion

14.3.1 Summary statistics of soil properties

Table 14.4 provides statistical summaries of the observed soil characteristics. The pH ranged from 4.7 to 9.2, with an average of 7.4. SOC content ranged from 0.03% to 3.0% in the study area, and the CEC varied between 2 and 68.4 C mol (+) kg^{-1}. The pH had low variability (CV $<$15%) and the PAWC and sand had medium variability, while CEC, OC, silt, and clay exhibited high variability (CV $>$35%). The soil texture in the study area was predominantly composed of sand, followed by clay and silt particles, and their average composition was 54.6%, 29.5%, and 15.9%,

TABLE 14.3 Assigned scores of SQI.

Parameter	Score	Parameter	Score
pH		OC (%)	
6.0–7.5	3	<0.5	1
5.5–6.0 & 7.5–8.5	2	0.5–1.00	2
<5.5 &> 8.5	1	>1.00	3
Clay (%)		Soil depth (cm)	
<10 & > 60	1	<50	0.33
10–20 & 40–60	2	50–100	0.66
20–40	3	>100	1.0
Sand (%)		PAWC (mm)	
<10 & > 60	1	<75	0.33
10–25 & > 50–60	2	75–100	0.66
25–50	3	>100	1.0
CEC (C mol (+) kg^{-1})			
<10	1		
10–35	2		
>35	3		

TABLE 14.4 Summary statistics of observed soil properties.

Statistics parameters	pH	OC (%)	Sand (%)	Silt (%)	Clay (%)	CEC (c mol (+) kg^{-1})	PAWC (%)
Min	4.7	0.03	13.2	2.0	3.4	2.0	8.2
Max	9.2	3.0	87.3	39.9	62.0	68.4	14.5
Mean	7.4	0.8	54.6	15.9	29.5	21.7	11.6
SD	0.9	0.6	17.3	8.4	15.4	14.5	1.8
Kurt	−0.3	1.0	−0.4	−0.6	−1.0	0.8	−0.8
Skew	0.1	1.1	−0.2	0.5	0.4	1.1	−0.5
CV (%)	12.4	78.2	31.6	53.2	52.1	66.9	15.5

respectively. The soil properties of pH, CEC, OC, clay, and sand were positively skewed, whereas sand particles were negatively skewed.

The summary statistics of soil properties at four depths were presented in Table 14.5. Average organic carbon content at 0–5, 5–15, 15–30, and 30–60 cm soil depths were 0.8, 0.8, 0.7, and 0.6%, respectively, which revealed that OC content at surface (i.e., 0–5 cm) soil depth was high compared to deeper soil depths. It is reported that cropping systems and soil management practices can significantly influence soil OC dynamics. The highest variation in surface SOC (0–5 and 5–15 cm) may be attributed to high vegetative growth, rapid root proliferation, decomposition of organic matters, and, subsequently, their trapping in clay complexes within soil aggregates, as evident from the highest amount of fine soil particles (Moharana et al., 2022a, b). Root proliferation associated with the growth could be another source of organic matter for the soils in arid region (Singh et al., 2007). The average pH ranged from 7.4 to 7.7 in four depths. The observed mean CEC values were 22.2, 22.5, 24.0, and 24.7 c mol (+) kg^{-1} at 0–5, 5–15, 15–30, and 30–60 cm depths, respectively, and the results revealed that the CEC value increased with increasing soil depth. It is

TABLE 14.5 Summary statistics of soil properties at different depths.

Properties	Depth	Min	Max	Mean	SD	Kurt	Skew	CV (%)
OC (%)	0–5	0.1	3	0.8	0.6	0.4	1	78.7
	5–15	0.01	3	0.8	0.6	0.4	1.1	80.9
	15–30	0.1	2.7	0.7	0.5	0.9	1.1	70.7
	30–60	0.05	2.3	0.6	0.4	2.1	1.5	77.5
pH	0–5	4.7	9.2	7.4	0.9	−0.4	0	12.6
	5–15	4.7	9.2	7.4	1	−0.5	0.1	13
	15–30	4.8	9.8	7.5	0.9	−0.3	0.1	12.3
	30–60	5	9.4	7.7	0.9	−0.3	−0.3	11.8
CEC (c mol (+) kg^{-1})	0–5	1.8	68.4	22.2	15.2	0.8	1.2	68.2
	5–15	1.9	68.4	22.5	15.1	0.8	1.2	67.1
	15–30	2.8	68.1	24	14.6	0.8	1.2	61
	30–60	0.2	76.6	24.7	15.7	1.1	1.2	63.4
Sand (%)	0–5	11.8	87.5	54.7	18.1	−0.5	−0.3	33
	5–15	10	87.1	54.1	17.5	−0.4	−0.3	32.4
	15–30	16.6	84	50.4	15.4	−0.2	−0.4	30.5
	30–60	9.8	90.1	46.5	14.3	0.8	−0.5	30.7
Clay (%)	0–5	3.2	61.9	30	15.9	−1	0.3	53
	5–15	3.6	62.1	30.7	15.6	−1.1	0.3	50.6
	15–30	7.9	65.1	34.2	13.6	−0.6	0.4	39.8
	30–60	4.4	78.8	37.1	13.9	0.4	0.5	37.5

well known that the CEC determines the soil's capacity to hold nutrients. All of the soil's buffering activities depend heavily on CEC. In the case of soil texture, clay and silt particles were skewed positively, whereas sand particles were skewed negatively, and the coefficient of variation was observed more in silt (47.2%–58.8%), followed by clay (37.5%–53.0%) and sand (30.5%–33%). The transfer of finer particles from the surface layers and subsequent illuviation in subsurface horizons may be the cause of the higher concentration of finer particles (silt and clay) in lower depths. Our findings are consistent with those of Verma et al. (2017) and Moharana et al. (2018).

14.3.2 Performance of QRF model

The performance of the QRF model in the prediction of various soil properties was given in Tables 14.6 and 14.7. The model could capture R^2 of 0.16–0.26 in pH, 0.28–0.41 in CEC, 0.21–0.31 in sand, and 0.20–0.44 in clay at various soil depths. Low amounts of SOC in the arid region may potentially be contributing to the poor performance. PAWC was predicted well compared to other parameters ($R^2 = 0.43$). Many studies have employed QRF and validated its accuracy in predicting and mapping different soil qualities (Liu et al., 2020; Zeraatpisheh et al., 2019; Taghizadeh-Mehrjardi et al., 2020; Dharumarajan et al., 2021a), which is consistent with the findings of this study, demonstrating that QRF is a potent instrument capable of giving performance that is among the most accurate approaches to date. The results of prediction interval coverage probability (PICP), which indicate uncertainty in estimations, indicate that 87.2%–88.8% of the observed values of the test dataset lie within the prediction interval. We would expect 90% of observations to fall inside the lower and upper prediction limits for a 90% prediction interval. All of the fractions had PICP values close to 90%, indicating that the QRF method's lower and upper prediction bounds were accurate. To put it another way, the uncertainty estimates were mostly accurate. Overall, the prediction performance of this model was

TABLE 14.6 Performance of QRF model in prediction of soil properties.

Properties	Depth (cm)	ME ± SD	RMSE ± SD	R^2 ± SD	PICP ± SD
OC (%)	0–5	0.14 ± 0.01	0.66 ± 0.01	0.14 ± 0.04	88.2 ± 1.23
	5–15	0.14 ± 0.01	0.68 ± 0.01	0.14 ± 0.03	89.6 ± 1.27
	15–30	0.07 ± 0.01	0.52 ± 0.01	0.15 ± 0.03	87.8 ± 1.04
	30–60	0.11 ± 0.01	0.47 ± 0.01	0.09 ± 0.03	87.2 ± 1.25
pH	0–5	0.03 ± 0.01	0.89 ± 0.02	0.24 ± 0.03	88.3 ± 1.07
	5–15	0.07 ± 0.01	0.83 ± 0.02	0.25 ± 0.03	87.7 ± 0.67
	15–30	0.06 ± 0.01	0.80 ± 0.02	0.26 ± 0.03	89.0 ± 1.14
	30–60	0.04 ± 0.01	0.85 ± 0.02	0.16 ± 0.04	88.1 ± 0.85
CEC (c mol (+) kg^{-1})	0–5	2.6 ± 0.2	13.14 ± 0.26	0.28 ± 0.03	88.4 ± 0.9
	5–15	2.6 ± 0.2	13.14 ± 0.26	0.28 ± 0.03	88.4 ± 0.9
	15–30	2.08 ± 0.19	12.7 ± 0.16	0.28 ± 0.02	88.2 ± 1.42
	30–60	2.68 ± 0.2	12.28 ± 0.16	0.41 ± 0.02	88.2 ± 1.29
Sand (%)	0–5	0.56 ± 0.23	15.49 ± 0.25	0.21 ± 0.02	88.6 ± 0.91
	5–15	0.85 ± 0.21	14.96 ± 0.2	0.22 ± 0.02	89.2 ± 0.95
	15–30	1.03 ± 0.19	13.29 ± 0.26	0.23 ± 0.03	89.0 ± 1.19
	30–60	0.84 ± 0.14	11.8 ± 0.22	0.31 ± 0.03	87.8 ± 0.78
Clay (%)	0–5	2.01 ± 0.20	13.64 ± 0.23	0.24 ± 0.03	89.8 ± 0.97
	5–15	2.02 ± 0.24	13.81 ± 0.22	0.20 ± 0.03	89.3 ± 1.08
	15–30	0.80 ± 0.14	12.2 ± 0.23	0.22 ± 0.03	88.5 ± 1.24
	30–60	0.51 ± 0.16	10.46 ± 0.17	0.44 ± 0.02	87.9 ± 0.96

TABLE 14.7 Performance of QRF model in prediction of total soil depth and PAWC.

Properties	ME ± SD	RMSE ± SD	R^2 ± SD	PICP ± SD
Soil depth (cm)	7.31 ± 0.76	38.7 ± 0.87	0.05 ± 0.00	89.5 ± 1.46
PAWC (mm)	−1.65 ± 0.22	10.99 ± 0.29	0.43 ± 0.03	92.1 ± 1.07

reasonably good for soil properties, which is close to several previous studies (Dharumarajan et al., 2019; Nabiollahi et al., 2021).

14.3.3 Summary statistics of predicted soil properties at four soil depths

Table 14.8 presents summary statistics of predicted soil properties for various depths. The average OC content of the study area was 0.7, 0.7, 0.6, and 0.6% at 0–5, 5–15, 15–30, and 30–60 cm soil depths, respectively. The soil pH value varied from 6 to 9.1, with a mean of 7.5. The predicted CEC varied from 5.1 to 60.7, with an average of 21.2, 22.1, 24.4, and 24.9 at 0–5, 5–15, 15–30, and 30–60 cm depths, respectively. Results show that the CEC value increases with increasing soil depth, as witnessed from the observed data. The predicted PAWC of the study area varied from 56.9 to 104.8 mm. Predicted OC, pH, CEC, and soil texture were positively skewed, except PAWC, which was negatively skewed. The CV of the predicted soil parameters ranged between 6.8% and 30.9%. Predicted maps of studied soil properties (pH, OC, CEC, sand, clay, and PAWC) using the QRF model are presented in Figs. 14.1–14.6.

TABLE 14.8 Summary statistics of predicted soil properties at different depths.

Properties	Depth (cm)	Min	Max	Mean	SD	Kurtosis	Skewness	CV (%)
OC (%)	0–5	0.3	3	0.7	0.1	109.9	6.5	16.9
	5–15	0.2	3	0.7	0.1	131.8	6.5	12.4
	15–30	0.4	2.7	0.6	0.1	141.6	7.5	16.2
	30–60	0.2	1.2	0.6	0.1	2.1	1.4	19.4
pH	0–5	6.3	9	7.5	0.5	−1	−0.1	6.8
	5–15	6.3	9	7.5	0.5	−1	−0.1	6.8
	15–30	6	9.1	7.5	0.5	−0.8	−0.1	7
	30–60	6	9.1	7.5	0.5	−0.8	−0.1	7
CEC (c mol (+) kg^{-1})	0–5	5.1	50.7	21.2	6.5	−1	0.2	30.8
	5–15	5.7	50.1	22.1	6.8	−1.2	0	30.9
	15–30	7.3	53.1	24.4	5.9	−0.2	0.2	24.1
	30–60	12.8	60.7	24.9	5	0.4	0.7	19.9
Sand (%)	0–5	27	83.4	54.5	8.3	0.2	1	15.3
	5–15	27.1	78.6	52.6	6.2	0.2	0.6	11.7
	15–30	23.5	66.1	51.9	4.7	1	−0.6	9.1
	30–60	18	57.9	46.1	3.4	5.5	−1.3	7.4
Clay (%)	0–5	10.1	55.3	28.9	7.4	−0.9	−0.1	25.4
	5–15	8.3	51	30.7	7	−1	−0.2	22.8
	15–30	21.8	57.5	34.8	3.9	0.4	0.2	11.2
	30–60	27.8	62.8	41	3.8	−0.3	0.1	9.3

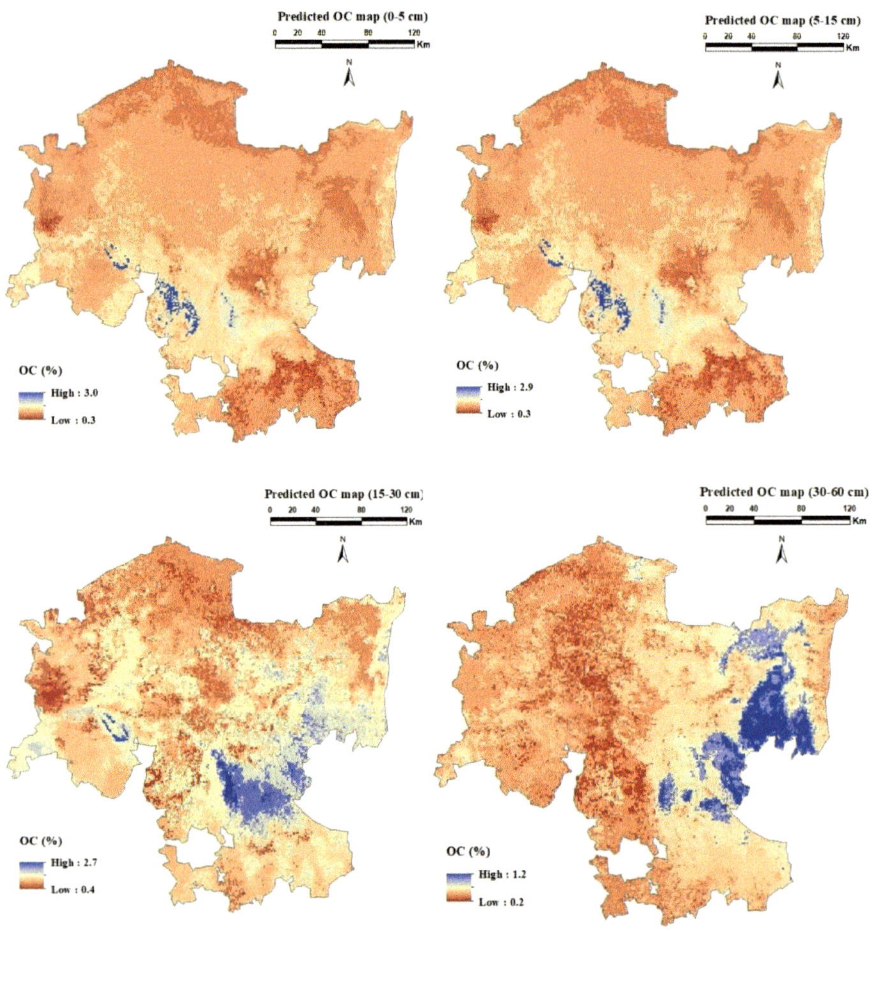

FIGURE 14.1 Predicted maps of OC in the study area of the arid region of south India.

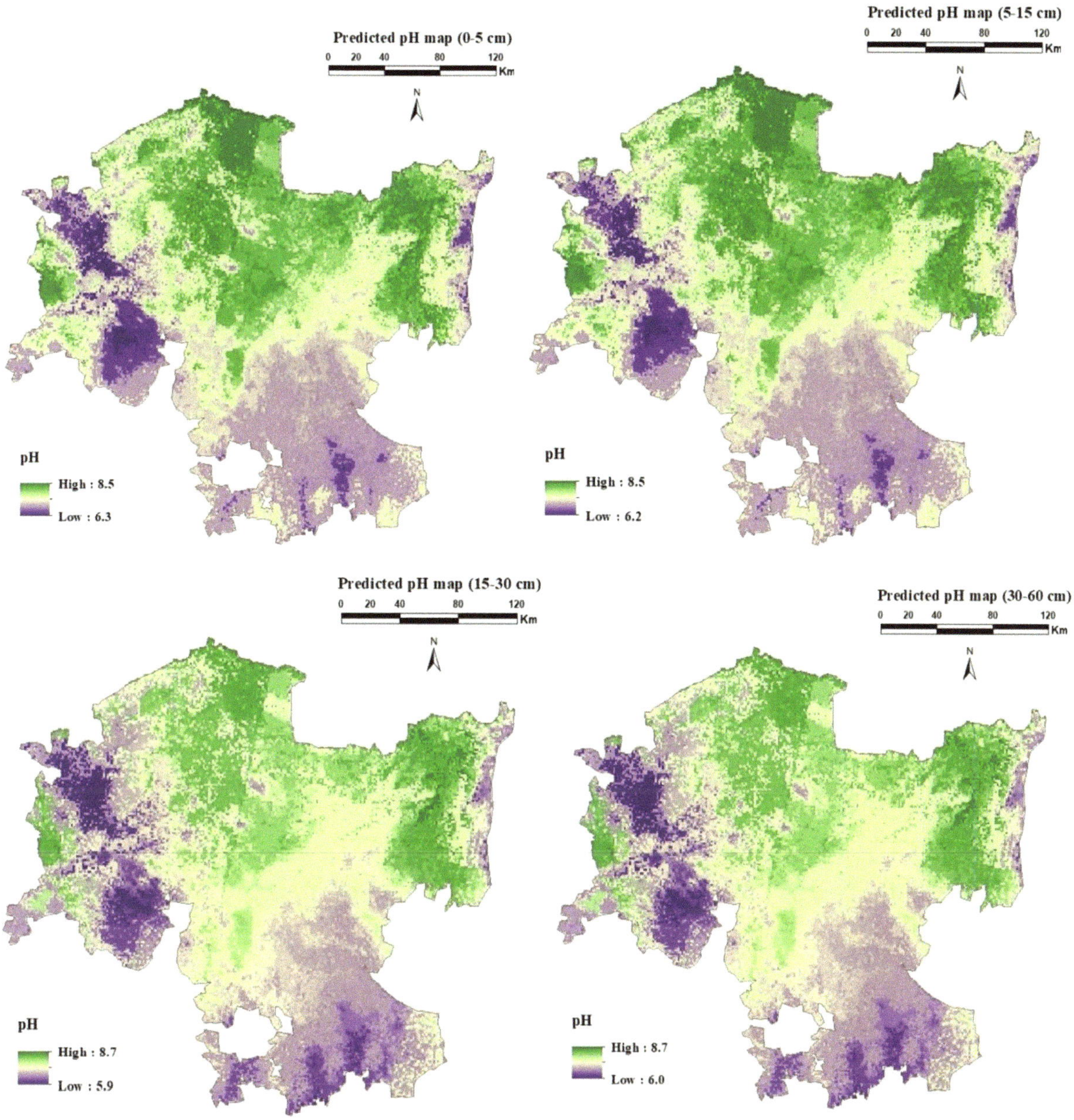

FIGURE 14.2 Predicted maps of pH in the study area of the arid region of south India.

Soil properties showed high variations in the spatially predicted maps, which were attributed to variations in environmental factors, land management, and use (Dharumarajan et al., 2021b). Such predicted maps with high spatial resolution will help to assess and monitor the soil properties, and at the same time, they will guide the preparation of a proper land use plan.

14.3.4 Soil quality index at various depths

A high SQI was observed (Fig. 14.7) in northern and southern parts of the Ananthapur and Kurnool districts of Andhra Pradesh, respectively, and low SQI was observed in Raichur, Koppal, Bellary, the southern part of Ananthapur, and the eastern part of Kurnool at both 0–5 and 5–15 cm depths. At soil depths of 15–30 cm, high SQI was observed in the eastern part of Kurnool and also in pockets of Ananthapur, Bellary, and Koppal districts. Low SQI was observed in

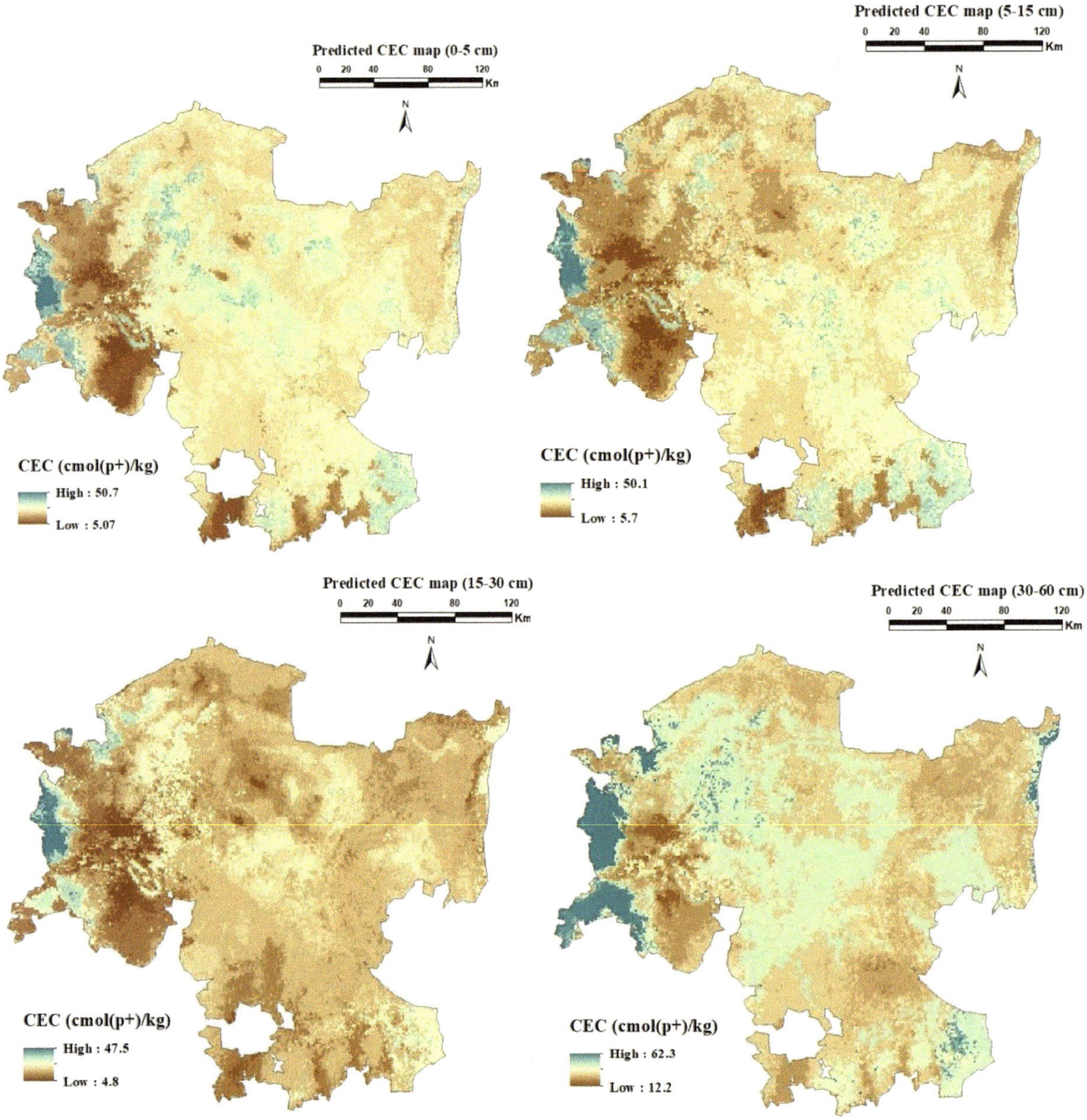

FIGURE 14.3 Predicted maps of CEC in the study area of the arid region of south India.

parts of northern Ananthapur, southern Bellary, and Koppal districts and was also distributed over the Raichur districts at 15–30 cm soil depth. Similarly, at 30–60 cm soil depth, high SQI was observed in most of the Ananthapur, Bellary, and eastern parts of Kurnool districts, whereas a low SQI was observed over the Raichur district. This situation is in accordance with the common fact that SQI varies with fertility gradient. On the other side, the lower content of soil quality concentrates in the studied regions, which correspond to lateritic soil with virtually higher elevation (Vasu et al., 2016). Low values of OM and high values of pH confirm the decline in the quality of these soils.

14.3.5 Predicted profile's SQI

Predicted profile SQI (Fig. 14.8) show high values in most of the Ananthapur and Kurnool districts of Andhra Pradesh, whereas low-profile SQI was observed in most of the Bellary, Koppal, southern parts of the Raichur, and

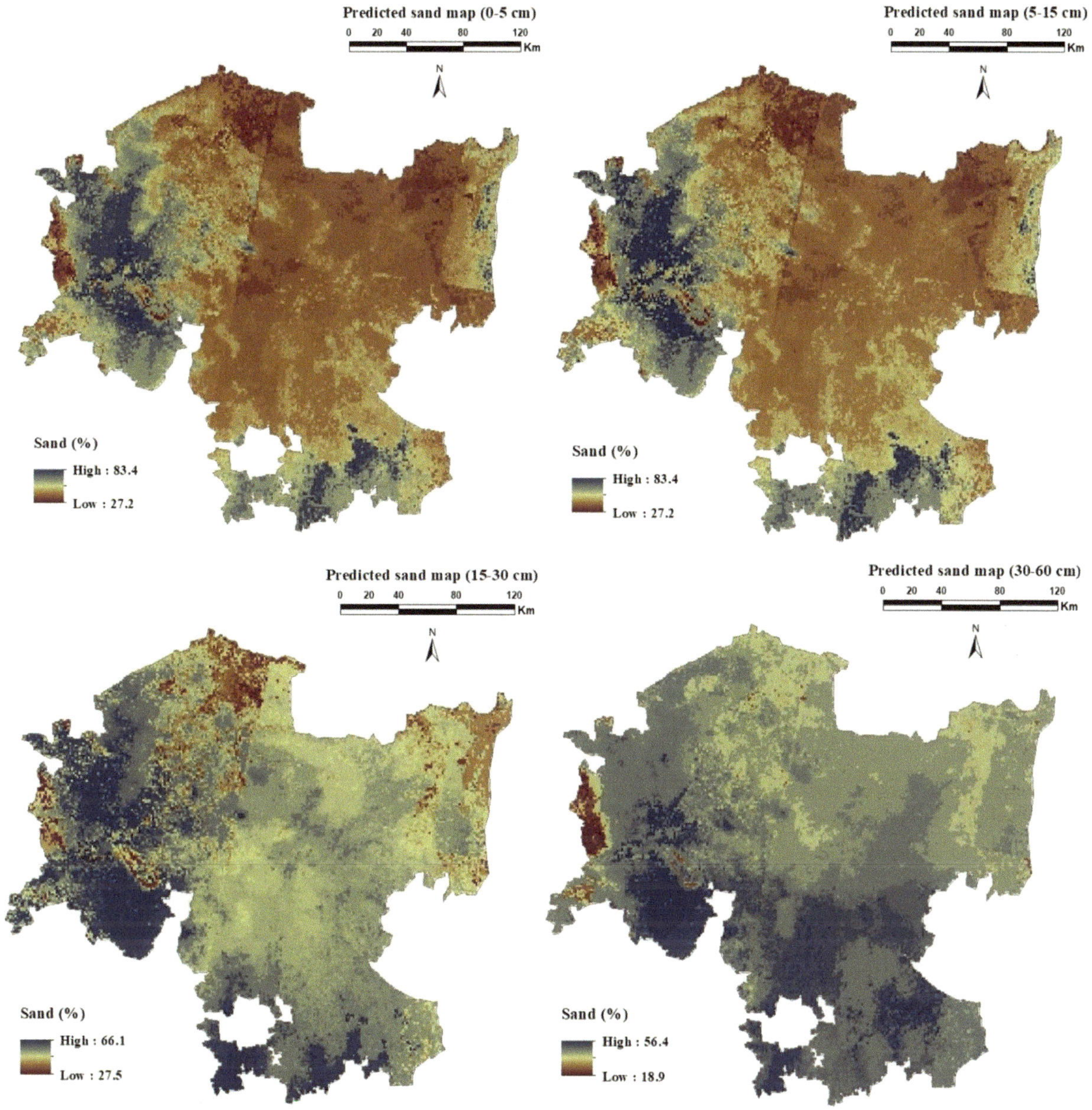

FIGURE 14.4 Predicted maps of sand content in the study area of the arid region of south India.

Ananthapur districts. This historical pattern of declining soil quality is probably due to the fast urbanization of the study region (Masto et al., 2008). For instance, Islam and Weil (2000) found soil quality deterioration brought on by changes in land use in a tropical forest environment in Bangladesh. Similar to this, studies by Zeraatpisheh et al. (2020) and Nabiollahi et al. (2018) evaluated the effects of intensive agriculture on soil quality and found that changing the land's use could lower soil quality indicators, indicating that careful consideration should be given to using the land sustainably to raise the quality index. It is suggested that this index could be used to monitor spatial changes in soil quality. The SQI approach could be used as a suitable tool to develop a quantitative procedure to evaluate the effects of land management practices on soil quality. Many researchers have successfully used these models to predict spatiotemporal chances in soil quality (Fathizad et al., 2020; Rahmanipour et al., 2014; Paul et al., 2020). Model-based SQI is reliable for predicting the spatial and temporal variation of the SQI in arid region of south India.

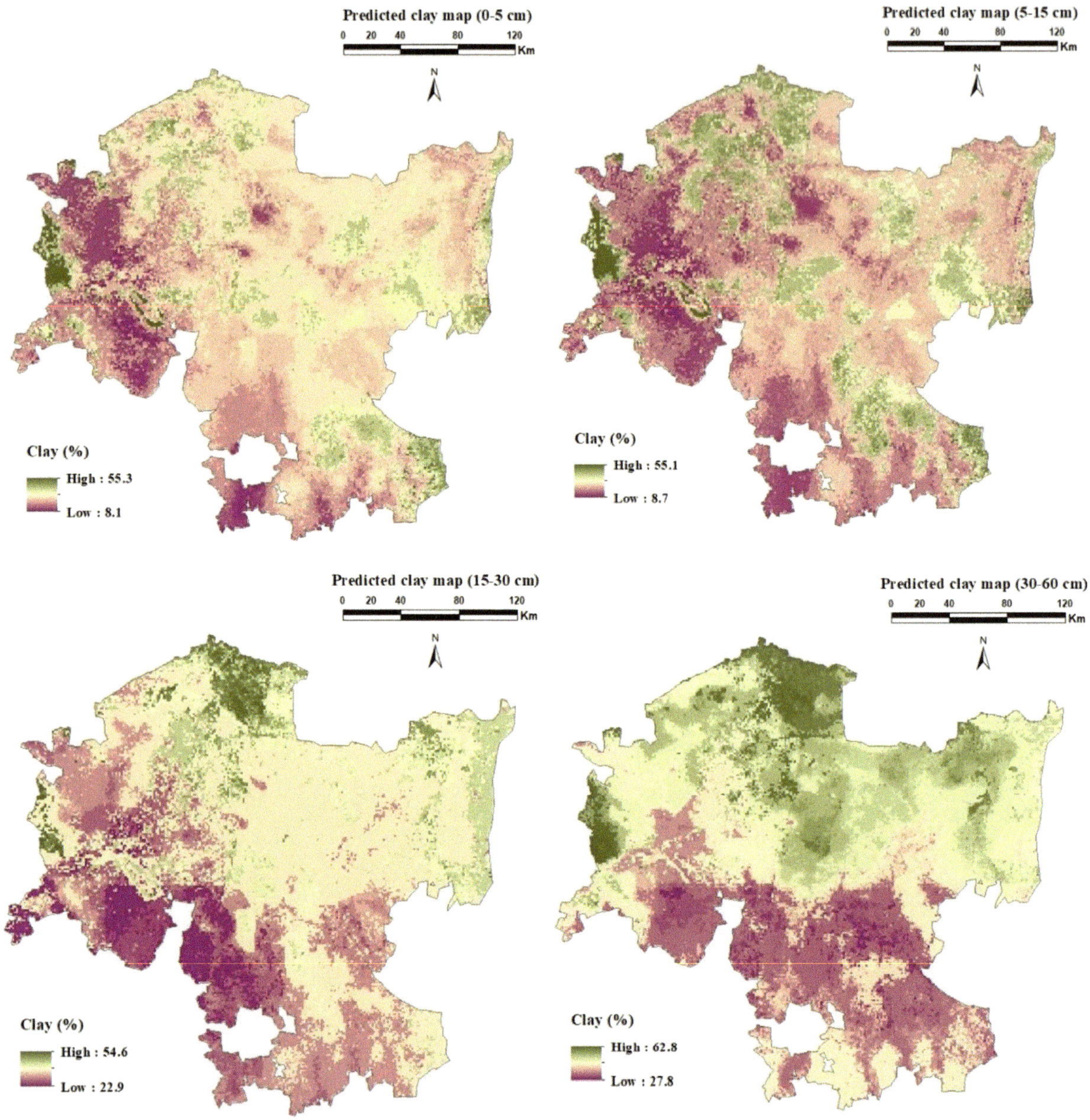

FIGURE 14.5 Predicted maps of clay content in the study area of the arid region of south India.

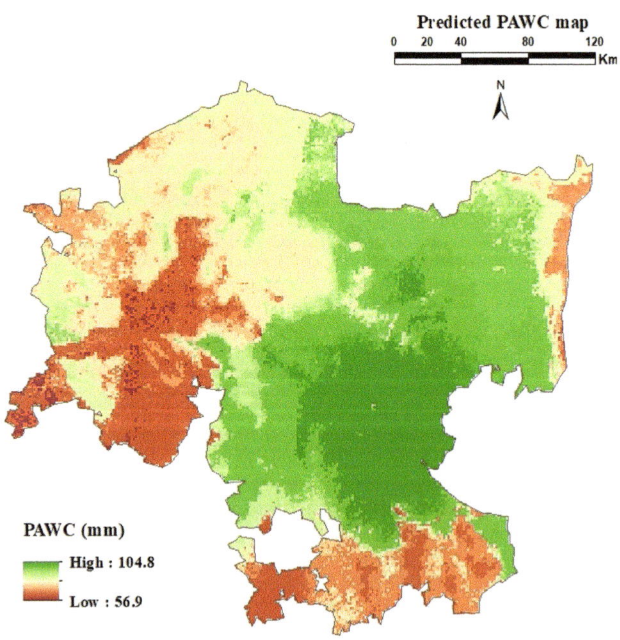

FIGURE 14.6 Predicted maps of potential available water content in the study area of the arid region of south India.

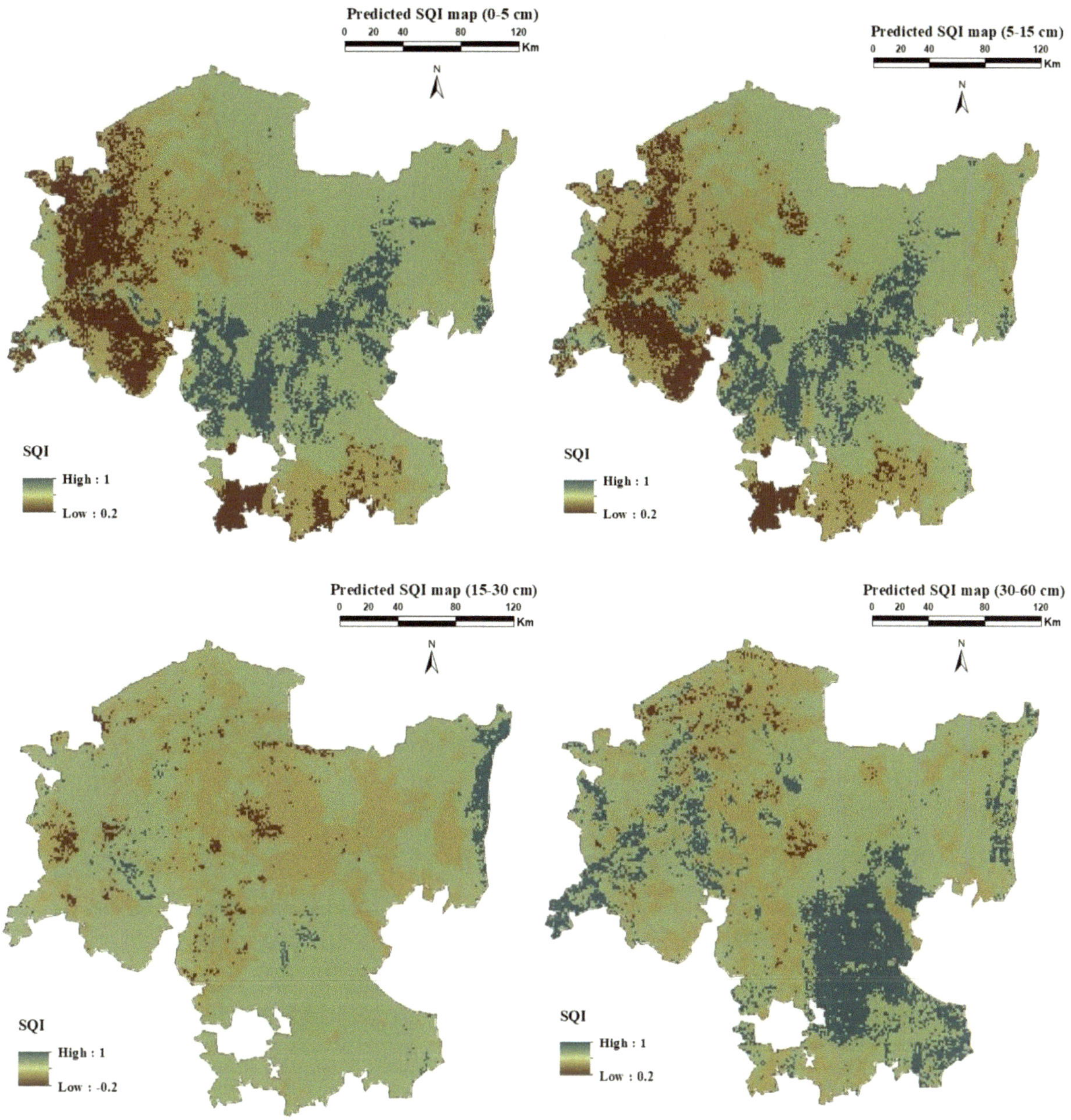

FIGURE 14.7 Predicted maps of SQI in various soil depths of the study area.

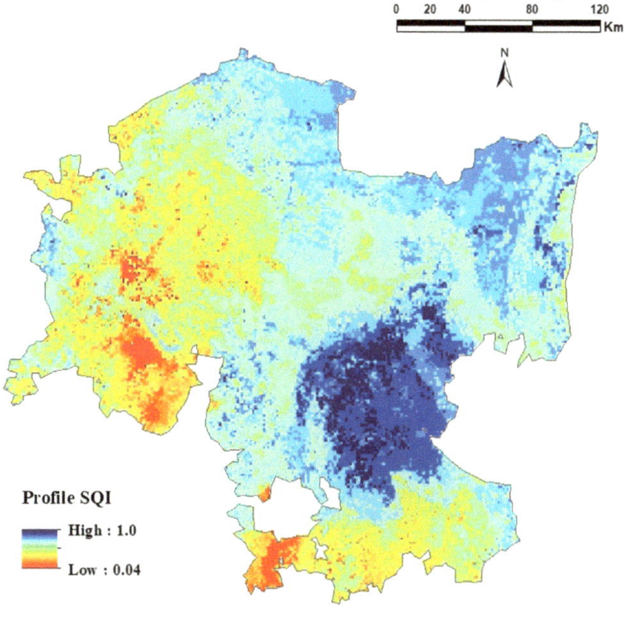

FIGURE 14.8 Predicted map of profile SQI in the study area of arid region of south India.

14.4 Conclusions

The study described here focuses on creating a SQI in an arid region in south India using machine learning algorithms and statistical assimilation of important soil quality variables. Given their functional link with environmental variables, the projected SQI based on QRF have a respectable performance. The QRF model did a good job of predicting SQI, as shown by the results of the three error criteria (R^2, ME, and RMSE). For the majority of the research region, the overall model-based soil quality indicators are considered moderate. The degree of spatial specificity in the predicted soil quality maps shows that QRF-based digital soil quality mapping could be used for identifying low-quality soil areas, which is a more beneficial and easily interpreted form of precise information for agronomists in a data-limited environment for sustainable soil management.

References

Andrews, S.S., Karlen, D.L., Mitchell, J.P., 2002. A comparison of soil quality indexing methods for vegetable production systems in Northern California. Agric. Ecosyst. Environ. 90, 25–45.

Arrouays, D., Grundy, M.G., Hartemink, A.E., Hempel, J.W., Heuvelink, G.B., Hong, S.Y., et al., 2014. GlobalSoilMap: toward a fine-resolution global grid of soil properties. Adv. Agron. 125, 93–134.

Behrens, T., Förster, H., Scholten, T., Steinrücken, U., Spies, E.-D., Goldschmitt, M., 2005. Digital soil mapping using artificial neural networks. J. Plant. Nutr. Soil Sci. 168, 21–33.

Bishop, T.F.A., McBratney, A.B., Laslett, G.M., 1999. Modelling soil attribute depth functions with equal-area quadratic smoothing splines. Geoderma 91, 27–45.

Breiman, L., 2001. Random forests. Mach. Learn. 45, 5–32.

Chand, S., 2010. Challenges of soil quality of Indian soils vis-á-vis food security. Curr. Sci. 99, 278–280.

Dharumarajan, S., Lalitha, M., Natarajan, A., Naidu, L.G.K., Balasubramanian, R., Hegde, R., et al., 2017a. Biophysical and socio-economic causes for increasing fallow land in Tamil Nadu. Soil Use Manage 33, 487–498. Available from: https://doi.org/10.1111/sum.12361.

Dharumarajan, S., Ramamurthy,V., Vasundhara, R., Lalitha, M., Srinivas, S., Hegde, R., 2017b. Land Resource Inventory of Bukkarayasamudrum mandal of Anantapur district, Andhra Pradesh on 1:10000 scale for optimal agricultural land use planning using geospatial techniques. NBSS&LUP, Technical Bulletin No. 1105.

Dharumarajan, S., Hegde, R., Janani, N., Singh, S.K., 2019. The need for digital soil mapping in India. Geoderma Reg. 16, e00204. Available from: https://doi.org/10.1016/j.geodrs.2019.e00204.

Dharumarajan, S., Vasundhara, R., Suputhra, A., Lalitha, M., Hegde, R., 2020. Prediction of soil depth in karnataka using digital soil mapping approach. J. Indian Soc. Remote Sens. 48, 1593–1600. Available from: https://doi.org/10.1007/s12524-020-01184-7.

Dharumarajan, S., Kalaiselvi, B., Suputhra, A., Lalitha, M., Vasundhara, R., Kumar, K.A., et al., 2021a. Digital soil mapping of soil organic carbon stocks in Western Ghats, South India. Geoderma Reg. 25, e00387.

Dharumarajan, S., Hegde, R., Lalitha, M., 2021b. Modelling of soil depth and hydraulic properties at regional level using environmental covariates—a case study in India. Geoderma Reg. 27, e00439. Available from: https://doi.org/10.1016/j.geodrs.2021.e00439.

Dharumarajan, S., Kalaiselvi, B., Lalitha, M., Vasundhara, R., Hegde, R., 2022. Defining fertility management units and land suitability analysis using digital soil mapping approach. Geocarto Int. 37 (20), 5914–5934. Available from: https://doi.org/10.1080/10106049.2021.1926553.

Doran, J.W., Jones, A.J., 1996. Methods for Assessing Soil Quality, vol. 49. ASA, Madison, WI, SSSA Special Publication.

Fathizad, H., Ardakani, M.A.H., Heung, B., Sodaiezadeh, H., Rahmani, A., Fathabadi, A., et al., 2020. Spatio-temporal dynamic of soil quality in the central Iranian desert modeled with machine learning and digital soil assessment techniques. Ecol. Indic. 118, 106736.

Gong, L., Ran, Q., He, G., Tiyip, T., 2015. A soil quality assessment under different land use types in Keriya river basin, Southern Xinjiang, China. Soil Tillage Res. 146, 223–229.

Islam, K.R., Weil, R.R., 2000. Land use effects on soil quality in a tropical forest ecosystem of Bangladesh. Agric. Ecosyst. Environ. 79, 9–16.

Jafari, A., Khademi, H., Finke, P.A., Van de Wauw, J., Ayoubi, S., 2014. Spatial prediction of soil great groups by boosted regression trees using a limited point dataset in an arid region, southeastern Iran. Geoderma 232, 148–163.

Jena, R.K., Bandyopadhyay, S., Pradhan, U.K., Moharana, P.C., Kumar, N., Sharma, G.K., et al., 2022. Geospatial modelling for delineation of crop management zones using local terrain attributes and soil properties. Remote Sens. 14, 2101.

Kovačević, M., Bajat, B., Gajić, B., 2010. Soil type classification and estimation of soil properties using support vector machines. Geoderma 154, 340–347.

Lagacherie, P., McBratney, A.B., 2006. Spatial soil information systems and spatial soil inference systems: perspectives for digital soil mapping. Develop. Soil Sci. 31, 3–22.

Lal, R., Iivari, T., Kimble, J.M., 2003. Soil Degradation in the United States: Extent, Severity, and Trends. CRC Press.

Lenka, N.K., Meena, B.P., Lal, R., Khandagle, A., Lenka, S., Shirale, A.O., 2022. Comparing four indexing approaches to define soil quality in an intensively cropped region of Northern India. Front. Environ. Sci. 287.

Liu, F., Zhang, G.-L., Song, X., Li, D., Zhao, Y., Yang, J., et al., 2020. High-resolution and three-dimensional mapping of soil texture of China. Geoderma 361, 114061.

Ma, Y., Minasny, B., Malone, B.P., Mcbratney, A.B., 2019. Pedology and digital soil mapping (DSM). Eur. J. Soil Sci. 70, 216–235.

Malone, B.P., McBratney, A.B., Minasny, B., Laslett, G.M., 2009. Mapping continuous depth functions of soil carbon storage and available water capacity. Geoderma 154, 138–152.
Marzaioli, R., d'Ascoli, R., De Pascale, R.A., Rutigliano, F.A., 2010. Soil quality in a Mediterranean area of Southern Italy as related to different land use types. Appl. Soil Ecol. 44, 205–212.
Masto, R.E., Chhonkar, P.K., Purakayastha, T.J., Patra, A.K., Singh, D., 2008. Soil quality indices for evaluation of long-term land use and soil management practices in semi-arid sub-tropical India. Land Degrad. Dev. 19, 516–529.
Moharana, P.C., Singh, R.S., Singh, S.K., Jena, R.K., Naitam, R.K., Verma, T.P., et al., 2018. Assessment of soil quality monitoring indicators under long term rice cultivation in hot arid Ghaggar-flood plains of India. Arch. Agron. Soil Sci. 64, 2030–2044.
Moharana, P.C., Jena, R.K., Pradhan, U.K., Nogiya, M., Tailor, B.L., Singh, R.S., et al., 2020. Geostatistical and fuzzy clustering approach for delineation of site-specific management zones and yield-limiting factors in irrigated hot arid environment of India. Precis. Agric. 21, 426–448.
Moharana, P.C., Dharumarajan, S., Kumar, N., Jena, R.K., Pradhan, U.K., Meena, R.S., et al., 2022a. Modelling and prediction of soil organic carbon using digital soil mapping in the Thar Desert Region of India. J. Indian Soc. Soil Sci. 70, 86–96.
Moharana, P.C., Meena, R.L., Nogiya, M., Jena, R.K., Sharma, G.K., Sahoo, S., et al., 2022b. Impacts of land use on pools and indices of soil organic carbon and nitrogen in the Ghaggar Flood Plains of Arid India. Land 11, 1180.
Mukherjee, A., Lal, R., 2014. Comparison of soil quality index using three methods. PLoS One 9, e105981.
Muñoz-Rojas, M., 2018. Soil quality indicators: critical tools in ecosystem restoration. Curr. Opin. Environ. Sci. Health 5, 47–52.
Nabiollahi, K., Taghizadeh-Mehrjardi, R., Eskandari, S., 2018. Assessing and monitoring the soil quality of forested and agricultural areas using soil-quality indices and digital soil-mapping in a semi-arid environment. Arch. Agron. Soil Sci. 64, 696–707.
Nabiollahi, K., Taghizadeh-Mehrjardi, R., Shahabi, A., Heung, B., Amirian-Chakan, A., Davari, M., et al., 2021. Assessing agricultural salt-affected land using digital soil mapping and hybridized random forests. Geoderma 385, 114858.
Pahlavan-Rad, M.R., Khormali, F., Toomanian, N., Brungard, C.W., Kiani, F., Komaki, C.B., et al., 2016. Legacy soil maps as a covariate in digital soil mapping: a case study from Northern Iran. Geoderma 279, 141–148.
Paul, G.C., Saha, S., Ghosh, K.G., 2020. Assessing the soil quality of Bansloi river basin, eastern India using soil-quality indices (SQIs) and Random Forest machine learning technique. Ecol. Indic. 118, 106804.
Qi, Y., Darilek, J.L., Huang, B., Zhao, Y., Sun, W., Gu, Z., 2009. Evaluating soil quality indices in an agricultural region of Jiangsu Province, China. Geoderma 149, 325–334.
Rahmanipour, F., Marzaioli, R., Bahrami, H.A., Fereidouni, Z., Bandarabadi, S.R., 2014. Assessment of soil quality indices in agricultural lands of Qazvin Province, Iran. Ecol. Indic. 40, 19–26.
Rodriguez-Galiano, V.F., Chica-Rivas, M., 2014. Evaluation of different machine learning methods for land cover mapping of a Mediterranean area using multi-seasonal Landsat images and Digital Terrain Models. Int. J. Digit. Earth 7, 492–509.
Sanchez, P.A., Shepherd, K.D., Soule, M.J., Place, F.M., Buresh, R.J., Izac, A.M.N., et al., 1997. Soil fertility replenishment in Africa: an investment in natural resource capital. In: Buresh, R.J., Sanchez, P.A., Calhoun, F. (Eds.), Replenishing Soil Fertility in Africa. SSSA and ICRAF, Madison, WI, pp. 1–46.
Sharma, K.L., Mandal, U.K., Srinivas, K., Vittal, K.P.R., Mandal, B., Grace, J.K., et al., 2005. Long-term soil management effects on crop yields and soil quality in a dryland Alfisol, Soil Tillage Res., 83. pp. 246–259.
Singh, S.K., Kumar, M., Sharma, B.K., Tarafdar, J.C., 2007. Depletion of organic carbon, phosphorus, and potassium stock under a pearl millet based cropping system in the arid region of India. Arid. Land Res. Manag. 21, 119–131.
Sreenivas, K., Dadhwal, V.K., Kumar, S., Harsha, G.S., Mitran, T., Sujatha, G., et al., 2016. Digital mapping of soil organic and inorganic carbon status in India. Geoderma 269, 160–173.
Stocking, M.A., 2003. Tropical soils and food security: the next 50 years. Science 302, 1356–1359.
Sun, B., Zhou, S., Zhao, Q., 2003. Evaluation of spatial and temporal changes of soil quality based on geostatistical analysis in the hill region of subtropical China. Geoderma 115, 85–99.
Taghizadeh-Mehrjardi, R., Nabiollahi, K., Kerry, R., 2016. Digital mapping of soil organic carbon at multiple depths using different data mining techniques in Baneh region, Iran. Geoderma 266, 98–110.
Taghizadeh-Mehrjardi, R., Schmidt, K., Amirian-Chakan, A., Rentschler, T., Zeraatpisheh, M., Sarmadian, F., et al., 2020. Improving the spatial prediction of soil organic carbon content in two contrasting climatic regions by stacking machine learning models and rescanning covariate space. Remote Sens. 12, 1095.
Vasu, D., Singh, S.K., Ray, S.K., Duraisami, V.P., Tiwary, P., Chandran, P., et al., 2016. Soil quality index (SQI) as a tool to evaluate crop productivity in semi-arid Deccan plateau, India. Geoderma 282, 70–79.
Verma, T.P., Moharana, P.C., Naitam, R.K., Meena, R.L., Kumar, S., Singh, R., et al., 2017. Impact of cropping intensity on soil properties and plant available nutrients in hot arid environment of North-Western India. J. Plant Nutr. 40, 2872–2888.
Zeraatpisheh, M., Ayoubi, S., Jafari, A., Finke, P., 2017. Comparing the efficiency of digital and conventional soil mapping to predict soil types in a semi-arid region in Iran. Geomorphology 285, 186–204.
Zeraatpisheh, M., Ayoubi, S., Jafari, A., Tajik, S., Finke, P., 2019. Digital mapping of soil properties using multiple machine learning in a semi-arid region, central Iran. Geoderma 338, 445–452.
Zeraatpisheh, M., Bakhshandeh, E., Hosseini, M., Alavi, S.M., 2020. Assessing the effects of deforestation and intensive agriculture on the soil quality through digital soil mapping. Geoderma 363, 114139.

Chapter 15

Digital soil mapping algorithm for soil quality assessment and monitoring: a case study in desert ecosystem of India

Pravash Chandra Moharana[1], Roomesh Kumar Jena[2], Brijesh Yadav[3], Ravindra Naitam[1], N. Kumar[1], Upendra Kumar Pradhan[4] and Gulshan Kumar Sharma[5]

[1]*ICAR-National Bureau of Soil Survey and Land Use Planning, Nagpur, Maharashtra, India,* [2]*ICAR-Indian Institute of Water Management, Bhubaneswar, Odisha, India,* [3]*ICAR-National Bureau of Soil Survey and Land Use Planning, Regional Centre, Udaipur, Rajasthan, India,* [4]*ICAR-Indian Agricultural Statistics Research Institute, New Delhi, India,* [5]*ICAR-Indian Institute of Soil & Water Conservation, Regional Centre, Kota, Rajasthan, India*

15.1 Introduction

As the population and food demand continuously increase, conversion of desert soils into arable lands has been largely expanded in the western plain of India over the last several decades. With a fast expanding population, climate change, and unpredictable rain patterns, many arid locations are in risk of becoming desertified, and other places are seeing the migration of desert dunes. However, because of limited water supplies and increased wind erosion, agricultural exploitation has put desert soil sustainability in risk. Due to their lower levels of plant and soil carbon storage, desert ecosystems have received significantly less attention than semiarid and wet environments (Ojima et al., 1993; Jobbagy and Jackson, 2000). The 31.7 Mha hot, arid region of India is incredibly diversified in terms of its soils, landforms, fauna, flora, water balance, and human activities (Singh et al., 2009). In the last 60 years, the irrigation network has transformed arid deserts into rich, green farmland. The sand dunes are stabilized with extensive plantations and agronomic conservation methods (Singh et al., 2007; Mathur and Sundaramoorthy, 2016). It is widely known that any activity related to a change in land use and managed for agriculture has the potential to alter the properties of the soil. Many factors contribute to these variations in properties, but the dynamics of soil organic matter have drawn a lot of attention because of their importance to SQ (Datta et al., 2015).

Maintaining SQ is crucial for both environmental preservation and the sustainability of agriculture. The state of soil resources at all scales, including those found in a field, farm, watershed, county, state, country, or the entire world, needs to be periodically assessed. Because historically, humankind has neglected its soil resources more than once; often ending in failure of the dominant society and culture (Hillel, 1991). Several ancient civilizations that relied on irrigation agriculture, such as the civilization of ancient Mesopotamia, have fallen into collapse, according to historical sources, as a result of soil deterioration caused by salinization (Hillel, 2005; Singh, 2015). Despite this well-known past, the degradation of the earth's soil resources continues to rank as one of the greatest and most pervasive threats to humanity. Due to its many functions, measuring and monitoring SQ is crucial. From crop production point of view, SQ evaluation can help farmers, farm managers, extension workers, and policymakers identify the sustainability of a given land use. One of the most pressing issues that the community of soil scientists has ever argued about is how to assess SQ. MacDonald et al. (1998) stated that SQ is a composite expression of properties and processes that interact to determine its ability to perform a number of basic functions, such as supporting crop production, buffering the environment from nutrients and other chemicals, and partitioning water and gases. Regional and national assessments of soil/land quality, which were broadly carried out, relied on personal knowledge to compensate for gaps in data and inconsistencies between geographical areas. However, current assessments should be carried out in a documented, reproducible way with a minimum of subjectivity (de Paul Obade and Lal, 2013; Fathizad et al., 2020). soil quality indexes (SQIs)

Remote Sensing of Soils. DOI: https://doi.org/10.1016/B978-0-443-18773-5.00032-6

are perhaps the most widely and often used techniques for sustainability and soil management studies among the different SQ evaluation methods developed so far (Klingebiel and Montgomery, 1961; Doran and Jones, 1996; Larson and Pierce, 1994; Andrews et al., 2002). SQIs are particularly significant to soil management practices because of their ability to use site-specific indicators of soil status that can integrate anthropogenic effects over time and over numerous types of effects (Wang and Gong, 1998; Arshad and Martin, 2002). Based on various SQ indicators and their relative importance for distinct soil functions, the estimation of the SQI, an indirect method for assessing SQ, is used (Qi et al., 2009). The development of the SQI includes the weighting of each attribute and the scoring of various indicators using a variety of scoring functions (such as Gaussian, sigmoid, *etc.*) (Mandal et al., 2010). This approach is well-liked because, in addition to its ability to recognize the complexity of soil productivity under various natural conditions as well as various farming techniques, it also has the ability to assess the crucial relationships between various soil indicators and soil productivity through the use of various mathematical models (Dobermann and Oberthur, 1997; McBratney and Odeh, 1997; Sun et al., 2003).

Modeling of the agricultural ecosystem and environment must include an accurate assessment of the regional variability of SQ. Kriging, a geostatistical approach, is generally accepted as an important spatial interpolation method in land resource inventories (Hengl et al., 2004; Bhunia et al., 2016). The majority of soil maps in India were created using traditional methodologies, and relatively little work has been done thus far to employ current spatial prediction tools (Saha et al., 2012; Pal et al., 2014; Behera and Shukla, 2015; Tripathi et al., 2015; Jena et al., 2022). Precision agriculture depends on the precise calculation of the spatial distribution of SQ, which also serves as a foundation for decision- and policy-makers when developing plans and strategies. A widely used framework for mapping the spatial distribution of soil properties is provided by the digital soil mapping (DSM) method (McBratney et al., 2003; Scull et al., 2003). The central part of DSM is that soil formation is a result of time, topography, parent material, organisms, climate, and other factors (Jenny 1941). Therefore, if information about those variables is available for the area, soil properties can theoretically be calculated. McBratney et al. (2003) introduced further variables in space and soil information derived from other investigations, as soil can be predicted from its own properties in the so-called "scorpan model." DSM techniques identify soil properties using environmental variables and quantify geographical distribution and variability based on the spatial scale of the studied area. The majority of traditional soil mapping methods rely on ground-based surveys, and they rarely give data on the spatial distribution of soil properties at the appropriate resolution over the landscape (McBratney et al., 2003; Minasny et al., 2013). Additionally, it is challenging to accurately capture the real status of the dynamic soil properties while conducting standard field surveys to map soil spatial variations, particularly at national, regional, or global scales (Minasny and Hartemink, 2011; Dharumarajan et al., 2019). Therefore, it is necessary to have robust methods and models to predict soil properties at a given location or scale. Significant advancements in machine learning algorithms and remote sensing techniques have made it possible to predict soil properties with new techniques like DSM (McBratney et al., 2003; Hengl et al., 2015). Recently, numerous machine learning techniques have been offered as alternatives to traditional soil mapping methods. These techniques include linear models, regression trees, multivariate adaptive regression splines, artificial neural networks, decision trees, and support vector machines (Lacoste et al., 2016; Taghizadeh-Mehrjardi et al., 2016). With the help of correlations between soil and environmental variables derived from terrain attributes and satellite imagery, these methods were developed to overcome the drawbacks of the traditional soil mapping approach (McBratney et al., 2003; Minasny and Hartemink, 2011). Most of the covariates used were derived from digital elevation models (DEM) and remote sensing. Some attempts were made with multispectral remote sensing (Ben-Dor, 2002; Mulder et al., 2011), hyper-spectral remote sensing (Selige et al., 2006; Lagacherie et al., 2008; Ouerghemmi et al., 2016), and radar remote sensing (Niang et al., 2014). Several studies have recently been conducted on the spatial variation of soil properties using DSM methods (Zeng et al., 2016; Ratnayake et al., 2016; Camera et al., 2017). Many studies of SQI assessment have been done in different parts of the world; however, a few studies have been done to study the effect of different environmental covariates through agricultural lands on SQI using DSM. The objectives of the present study were to: (1) provide a framework to assess SQ using different SQI scoring functions and DSM techniques for future land use planning; and (2) spatially predict SQIs maps and determine the importance of environmental covariates affecting SQ for a better understanding of the land's potential and suitability.

15.2 Materials and methods

Fig. 15.1 presents the methodology for this investigation, which is summarized here. Step 1 of the process involved identifying, measuring, and utilizing principal component analysis (PCA) to create a SQI. In Step 2, a variety of environmental variables (e.g., remote sensing, land-use and land-cover data, and digital elevation data) that were anticipated

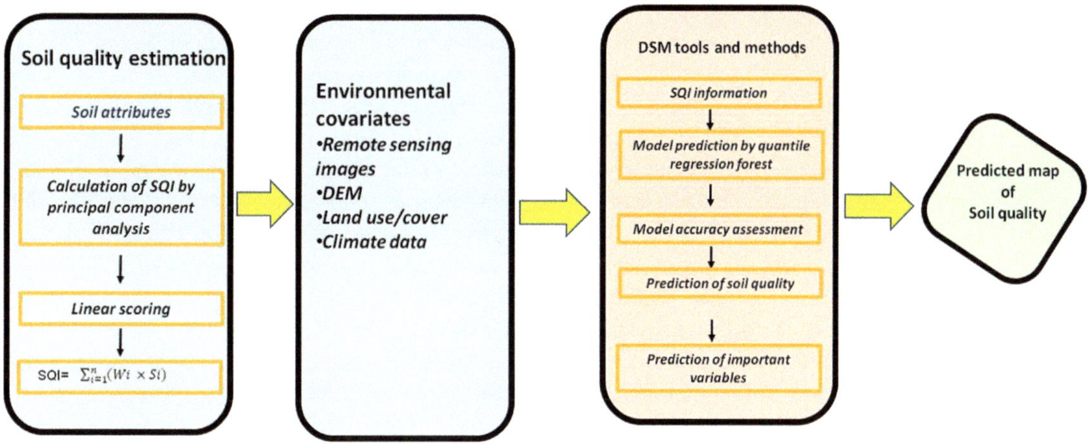

FIGURE 15.1 Flowchart of the digital mapping of SQI estimation.

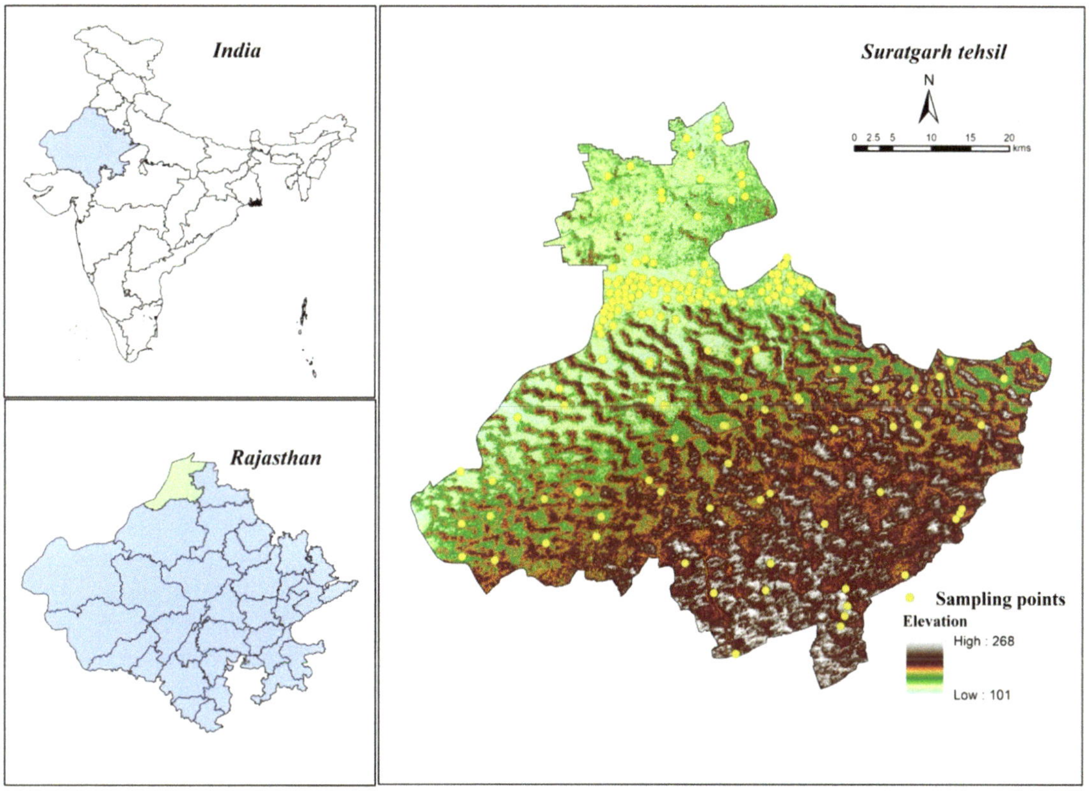

FIGURE 15.2 Spatial distribution of soil sampling points in the study area of the desert regions of India.

to affect SQ were collected. The RF algorithm was used to produce spatial predictions of SQ based on a set of environmental covariates in order to determine the correlations between SQ and the environment. Through the use of a tenfold cross-validation process, the SQ predictions were verified. Additionally, we employed quantile regression forest (QRF) to quantify the uncertainty of the predicted SQI.

15.2.1 Study area

The research area is located between 73.441°E and 74.218°E longitude and between 28.903°N and 29.686°N latitude. It is also between 160 and 200 m above mean sea level (Fig. 15.2). Dune, inter-dune, sandy arid plain, and Ghaggar flood plains are the four terrain types that comprise the research area. With extremely hot summers and chilly winters, this region experiences very meager and unpredictable rainfall. The region receives 286 mm of rain on average. The yearly rainfall in these locations ranges from 100 to 500 mm, with a coefficient of variance between 40% and 70%. During the summer, the area is characterized by strong southerly winds, which frequently result in sandstorms and sand movement. A canal system was introduced to this region in the late 1960s, and today, 56% of the arable land in the region is irrigated. Most of the land is used for agriculture, including mustard (*Brassica Juncea*), moth bean (*Vigna Aconitifolia*), cluster bean (*Cyamopsis Tetragonoloba*), chickpea (*Cicer Arietinum*), wheat (*Triticum Aestivum*), green gram (*Vigna Radiata*), and pearl millet (*Pennisetum Glaucum*), but the study region also has sand dunes, salt wetlands, and abandoned lands. Since the construction of the canal network in the late 1960s, cropping techniques based on cereals and legumes have been used for more than 60 years. Much of the natural vegetation includes *Acacia Arabica* (Babul), *Prosopis spicigera* (khejri), *Azadirachta indica* (Neem), *Prosopis juliflora*(Vilayati babul), *and Zizyus jijuba*(Ber). According to US Soil Taxonomy, common soils include Entisols and Aridisols. Deep to extremely deep, calcareous or noncalcareous, and sandy soils predominate. Sand dunes surround a shallow dip that contains gypsite-rich beds. Due to the region's scant natural vegetation, which is brought on by extremely low precipitation and rapid evaporation, as well as management practices that discourage soil organic carbon (SOC) retention in the soil, the soils of this area contain low SOC.

15.2.2 Soil Sampling and laboratory analysis

A stratified random sampling approach was used to obtain a total of 178 geo-referenced composite surface (0–30 cm depth) soil samples. A portable global positioning system was used to record the geographic coordinates. The collected soil samples were taken to the laboratory, where they were dried by air at room temperature (25°C ± 2°C). The materials were cleaned of rocks and other debris, crushed, and passed through a 2-mm sieve. Well-processed samples were stored in polythene bags until used for the analysis of soil properties. The international pipette method was used to measure the size of soil particles (Jackson, 1973). Using a digital pH and conductivity meter, soil pH and EC were measured in 1:2.5 (w/v) soil water suspensions (Jackson, 1973). Utilizing potassium dichromate oxidation and ferrous ammonium sulfate titration, SOC was measured (Walkley and Black, 1934). Following the alkaline potassium permanganate ($KMnO_4$-N) method, the amount of available nitrogen (N) was determined (Subbiah and Asija, 1956). Olsen's reagent (Olsen et al., 1954) was used to extract the available phosphorus (P), which was then measured with a UV-visible spectrophotometer. With a neutral normal ammonium acetate solution, available potassium (K) was extracted, and its concentration was measured using a flame photometer (Hanway and Heidel, 1952). Diethylene-triamine-penta-acetic acid (DTPA) solution was used to extract micronutrients like zinc (Zn), copper (Cu), iron (Fe), and manganese (Mn), and the concentrations of these micronutrients were determined using an atomic absorption spectrophotometer (Lindsay and Norvell, 1978).

15.2.3 SQ assessment

The minimum data set (MDS) framework is used as the basis for the calculation of the SQI. The PCA technique was largely used by using software to find the MDS through a series of iterative data analysis processes (Andrews et al., 2002). The variables with high factor loading and eigenvalues 1 on the principal components (PC) were considered to be the most representative of the system components. Only highly weighted criteria were taken into account for the MDS within each PC. These variables, which were designated as "key indicators," were taken into account for the SQI computation after being appropriately transformed and scored. The linear scoring technique was used to transform the values of each indicator (Andrews et al., 2002; Masto et al., 2008). Indicators were evaluated according to whether a higher value was considered "good" or "poor" in terms of impacting soil function before being assigned scores. For the "more is better" category of indicators, the highest observed value of each observation was divided by one, giving the highest observed value a score of one. The lowest observed value (in the numerator) for the "less is better" indicators was divided by each observation (in the denominator), giving the lowest observed value a score of one. The PCA findings were used to weight the MDS indicators for each observation after transformation using the linear scoring approach. A certain percentage (%) of the variation in the entire data set was explained by each PC. The weighted factors for the indicators selected under a particular PC were determined by dividing this percentage by the overall percentage of variation explained by all PCs with eigenvalues >1. After performing these steps to obtain SQI, the weighted MDS indicator scores for "n" observations (no. of indicators qualified from PCA) were summed up according to Eq. (15.5):

$$\text{SQI} = \sum_{i=1}^{n} (Wi \times Si) \tag{15.1}$$

where *Si* is the score for the subscripted variable obtained by linear scoring method and *Wi* is the weighing factor obtained by dividing the individual percent variance of a given PC with the cumulative variability of all the PCs. It was assumed that higher index scores meant better SQ or greater soil function.

15.2.4 Digital mapping of soil quality index

DSM techniques identify soil properties using environmental variables and quantify geographical distribution and variability based on the spatial scale of the studied area. The variables that characterize the soil-forming factors of climate, parent material, topography, vegetation, human activities, and time are known as environmental covariates. Numerous attempts have been made in the past ten years to select and create efficient environmental variables in accordance with the desired soil characteristics and landscapes. Terrain and climate variables have been widely implemented in DSM. The ArcGIS10 data management toolbox was used to process a DEM that was generated with the shuttle radar topographic mission (SRTM) and has a spatial resolution of 30 m. The primary and secondary derivates of DEM like elevation, slope, aspect, curvatures (plan and profile), topographic wetness index (TWI) and topographic position index (TPI), topographic ruggedness index (TRI), LS factor, Multi-resolution ridge top flatness (MrRTF) and multi-resolution index of valley bottom flatness (MrVBF), convergence index (CI), channel network distance (CND), channel network base level (CNBL), relative slope position (RSP), and total catchment area (TCA) were derived by using Saga-GIS 6.3.0 version (Table 15.1). Along with DEM attributes, other indices were also computed, which included the salinity index (SI), soil-adjusted vegetation index (SAVI), vegetation soil salinity index (VSSI), normalized difference salinity index (NDSI), normalized difference vegetation index (NDVI), salinity ratio (SR), and brightness index (BI), which were also derived from the Landsat 8. Raster data on bioclimatic variables at 30 seconds resolution were downloaded from http://worldclim.org/current for the whole world, and the respective grids for the study area were extracted from these world grids. Two bioclimatic variables were selected for use as covariates in the DSM study, and these variables are mean annual temperature (Annual_tem) and annual precipitation (Annual_ppt).

SQI of the study area and uncertainty estimations were predicted using the QRF model. The advantage of QRF over the Random Forest (RF) model is that QRF retains the value of all observations in each node and evaluates the conditional distribution based on the information, whereas RF only keeps the mean of the observations that fall into each node and ignores all other information for each node in each tree (Meinshausen, 2006). Ranger package was used to perform the QRF algorithm in the R environment for the current study (R Core Team, 2019). The Ranger package helps identify the best RF properties for running the model. Tenfold cross-validation techniques with 20 repetitions were used to evaluate the performance of the QRF model.

The prediction performance of QRF was evaluated based on three parameters: coefficient of determination (R^2), which is defined by the percentage of variation explained by the model; mean error (ME) and root-mean-square error (RMSE); and Lin's concordance correlation coefficient (CCC) (Lin, 1989), which is a measure of agreement between predicted and observed values. The good models have a R^2, the CCC is equal to or close to 1, and the RSME close to 0.

$$\text{Coefficeint of determination } \left(R^2\right) = 1 - \frac{\sum_{i=1}^{n} (\text{pi}-\text{oi})^2}{\sum_{i=1}^{n} \left(\overline{\text{pi}}-\overline{\text{oi}}\right)^2} \tag{15.2}$$

$$\text{Mean error (ME)} = \frac{1}{n}\sum_{i=1}^{n} (\text{pi}-\text{oi}) \tag{15.3}$$

$$\text{Root mean squared error (RMSE)} = \sqrt{\frac{1}{n}\sum_{(i=1)}^{n} (\text{pi}-\text{oi})^2} \tag{15.4}$$

where, pi and oi are predicted and observed values, $\overline{\text{pi}}$ and $\overline{\text{oi}}$ are means of predicted and observed values.

$$\text{Lin's concordance correlation coefficient (CCC)} = \frac{2\rho\sigma_0\sigma_p}{\sigma_0^2+\sigma_p^2+(\mu_0-\mu_p)^2} \tag{15.5}$$

μ_0 and μ_p are the means of observed and predicted values and ${\sigma_0}^2$ and ${\sigma_p}^2$ are corresponding variance. ρ is Pearson correlation coefficient between observed and predicted values.

A key component of the DSM is uncertainty representation (Arrouays et al., 2014). DSM models must be able to predict uncertainty in order to be suitable for producing maps. This capability is in addition to being

TABLE 15.1 Environmental covariates used for spatial prediction of SQ.

Auxiliary data	Land surface parameters	Definition	Reference/ source	Soil-forming factors
Terrain data	Elevation	Height above sea level (m)	SRTM DEM	R
	Aspect	The compass direction of the maximum rate of change	SRTM DEM	R
	Slope	Average gradient above flow path	SRTM DEM	R
	MrRTF index	Identifies the depositional areas	SRTM DEM	R
	MrVBF index	Identifies the depositional areas	SRTM DEM	R
	LS factor	Slope-length factor	SRTM DEM	R
	Plan curvature	Plan curvature is perpendicular to the slope	SRTM DEM	R
	Profile curvature	Profile curvature is parallel to the direction of the maximum slope	SRTM DEM	R
	Total catchment area (TCA)	Calculated the flow accumulation	SRTM DEM	R
	Topographic wetness index	DEM/Calculates slope gradient and specific catchment area based Topographic wetness index. TWI = ln (As/tan β): where As is upslope catchment area and β is the slope gradient	SRTM DEM	R, Cl
	Valley depth	relative position of the valley(Meters)	SRTM DEM	R
Remote sensing data	SI	$\mathrm{SI} = \sqrt{B4 \times B3}$	Landsat 8	PM,S
	SR	$\mathrm{SR} = \frac{B3 - B4}{B2 + B4}$	Landsat 8	PM,S
	NDSI	$\mathrm{NDSI} = \frac{B4 - B5}{B4 + B5}$	Landsat 8	PM,S
	NDVI	$\mathrm{NDSI} = \frac{B5 - B4}{B5 + B4}$	Landsat 8	O
	SAVI	$\mathrm{SAVI} = \frac{(B5 - B4)}{(B5 + B4 + 0.5)} \times 1.5$	Landsat 8	PM,S
	VSSI	$\mathrm{VSSI} = 2 \times B3 - 5(B4 + B5)$	Landsat 8	PM,S
	BI	$\mathrm{BI} = \sqrt{B3^2 + B4^2}$	Landsat 8	PM,S
Climate	Annual_tem	Annual temperature	Worldclim2	Cl
	Annual_ ppt	Precipitation	Worldclim2	Cl

CL, Climate; *PM*, parent materials; R, relief, *S*, soil, and *T*, time.
B2,3,4,5: different Landsat bands.

expected to offer correct soil predictions at a specific location (Vaysse and Lagacherie, 2017). The prediction interval coverage percentage (PICP) was used to evaluate the uncertainty of prediction. The PICP is simply the proportion of observations that are encapsulated by the corresponding prediction interval. In this study, the uncertainties of the predicted maps of SQI were estimated using QRF (Vaysse and Lagacherie, 2017). The estimation produced maps of 0.05 (lower limit) and 0.95 (upper limit) quantiles for SQI. That is to say, for every pixel location in the study area, there were values of uncertainty for the predictions of SQI. If the uncertainty estimates have been reasonably defined, the PICP should result in an estimate of 90% for a 90% prediction interval. The estimation produced by differentiation of 0.05 and 0.95 quantiles for SQI and the map produced for every pixel location of the study area.

15.2.5 Statistical analysis

The parameters of minimum, maximum, mean, kurtosis, skewness, standard deviation (SD), and coefficient of variation (CV) were among those included in the descriptive statistical analysis of the investigated soil properties. Through the use of Pearson's correlation coefficient analysis, relationships between the SQI and environmental factors were examined. R software was used to analyze the soil data.

15.3 Results and discussion

15.3.1 Characteristics of soil attribute

Table 15.2 shows the findings of a descriptive analysis of the soil's properties, including its sand, silt, and clay content, as well as its pH, EC, OC, $CaCO_3$, N, P, K, Zn, Cu, Mn, and Fe levels. The level and intensity of the data's variation are indicated by the CV, skewness, and kurtosis. The pH of the soil was found to be alkaline (8.42 ± 0.35), and its electrical conductivity was neutral (1.08 ± 5.81 dSm^{-1}). Among the determined soil properties, the EC value had the highest CV, skewness, and kurtosis (436%, 10.0, and 108, respectively), which indicated a very high degree of variation in the data. The interdunal soil had the highest EC in the soil profile, indicating an upward translocation and accumulation of soluble salts via capillary migration due to alteration of the water table caused by flooding of irrigation return flow into the wetlands. This inherent soil salinity is typical of the arid climatic conditions (Khan et al., 2003; Moharana et al., 2019). The main mechanisms for salt transport to the soil surface are convection, which results from water moving higher in response to evapotranspiration, and diffusion, which is driven by a salinity gradient with depth. Salt migration is influenced by both the texture of the soil and the overlaying of differently textured soils. The major causes of the development of secondary salinization are continuous water supply, high evaporation, the presence of possible salt sources in the profile, and the input of salts with irrigation water. Similar soil pH and EC values were found in oil palm plantation soils in the Western Ghats of coastal India (Behera et al., 2016; Mahajan et al., 2015). In the flood plain of the River Oasis in the Mongolian Altay highlands, a significant variation in EC (CV = 86%) has been observed (Jordan et al., 2018).

In the study area, OC ranged from 0.04% to 0.68% and $CaCO_3$ from 0.05% to 18.9%. It's possible that the adoption of various soil management techniques, such as variations in fertilizer application and other crop management

TABLE 15.2 Descriptive statistics of physical and chemical parameters of soil samples in the study area of desert regions of India.

Parameters	Min	Max	Mean	Median	SD	CV (%)	Skewness	Kurtosis
Sand	1.86	94.01	51.58	48.73	31.37	60.82	−0.02	−1.60
Silt	0.31	69.57	27.27	31.07	20.93	76.76	0.02	−1.62
Clay	3.10	48.65	21.10	18.84	12.18	57.71	0.49	−0.94
pH	7.70	9.70	8.42	8.41	0.35	4.13	0.50	1.27
EC	0.09	68.60	1.08	0.21	5.81	536.94	9.99	110.05
OC	0.04	0.68	0.29	0.29	0.15	52.35	0.26	−0.64
$CaCO_3$	0.05	18.90	3.47	3.09	2.44	70.35	2.50	10.24
N	75.50	240.50	137.31	135.40	26.69	19.44	0.40	1.20
P	11.10	55.33	26.34	24.70	7.80	29.59	0.81	0.68
K	125.00	380.80	255.55	260.66	52.45	20.53	0.23	−0.07
Zn	0.01	2.46	0.45	0.35	0.36	80.13	2.57	9.65
Cu	0.01	2.64	0.63	0.61	0.43	67.85	0.84	1.76
Fe	0.17	24.00	7.56	7.26	4.12	54.54	0.49	0.60
Mn	0.61	11.68	5.05	4.78	2.38	47.10	0.55	−0.17

techniques, is the cause of the variance in OC distribution in the studied areas (Behera et al., 2016; Jena et al., 2022). The maximum amount of fine soil particles in our study indicates that the highest SOC can be attributed to strong vegetative growth, rapid root proliferation, organic matter decomposition, and subsequent organic matter entrapment in clay complexes and soil aggregates. In arid areas, root proliferation coupled with growth may be an additional source of organic matter for the soils (Moharana et al., 2017). Due to poor vegetation and soil exposure to the hot sun, the lower SOC in the sand dunes area is primarily responsible for SOC oxidation (Pandey et al., 2010). Because the region's soils are classified as *Aridisols*, the study area had a higher $CaCO_3$ stock than SOC stock. *Aridisols* had more inorganic carbon stock than any other soil because of $CaCO_3$ weathering and its subsequent precipitation in the soil solum (Singh et al., 2007; Moharana et al., 2021).

In vulnerable terrestrial ecosystems, where soil fertility, plant growth, and carbon and nitrogen biogeochemical cycles are critical for environmental protection, soil nutrient distributions are intimately connected with these factors. N, P, and K availability varied from very low to very high. With an average value of 137.3 ± 26.69 kg ha^{-1}, the amount of available N ranged from 75.5 to 240.5 kg ha^{-1}. The increased anthropogenic inputs of nitrogen resulting from fertilizer application and the rapid transformation rates of inorganic nitrogen on surface soils and topography are responsible for the large variation in N. Available P varied from 11.1 to 55.3 kg ha^{-1} with an average value of 24.70 ± 7.8 kg ha^{-1}. High P levels in these soils are attributed to the arid nature of the soil. The arid soils of India's desert region had similar P findings (Verma et al., 2018). K varied widely, ranging from 125 to 380 kg ha^{-1} with a CV of 20.5%. The variance in nutrients available is attributed to a number of intrinsic (including climate, parent material, terrain, and soil type) and extrinsic (land use type, farming practices, and anthropogenic activities) environmental variables (Moharana et al., 2019). The availability of DTPA extractable Zn, Cu, Fe, and Mn was recorded with a mean value of 0.45 ± 0.36, 0.63 ± 0.43, 7.56 ± 4.12, and 5.05 ± 2.38 mg kg^{-1}, respectively. The variability indicated widespread deficiency of Zn, Cu, Mn, and Fe with a CV ranging from 47.1 to 80.1. The highest CV was recorded for Zn (80.1%) followed by Cu (67.8%), Fe (54.5%), and Mn (47.1%). The micronutrient availability and its variability are attributed to parent material, pedogenic processes, climatic conditions, and anthropogenic activities (Moharana et al., 2019).

15.3.2 Monitoring soil quality indicators

The PCA was used to analyze the whole data set to select SQ monitoring indicators for the studied arid soil (Table 15.3). For the analysis, the critical eigenvalue of 1 was used. The only five-components that had an eigenvalue greater than one were kept and interpreted. The five component model that accounted for 66.8% of the overall variation included soil characteristics as its subset. Sand, silt, clay, and OC were among the variables heavily weighted in PC1. Among these, sand was found to be the highest weighted variable in PC1 and retained for the indexing procedure by considering its active role in desert soil. In other PCs, only the single variables of Mn in PC2, EC in PC3, K in PC4, and pH in PC4 were found to be highly weighted and were retained for the MDS. Hence, the various SQ indicators retained in the final MDS consisted of sand, Mn, EC, K, and pH. The weighted factors (percent variation of each PC divided by the cumulative percent variation explained by all the PCs) for PC1, PC2, PC3, and PC4 were 0.393, 0.197, 0.154, 0.138, and 0.119, respectively. So the final PCA-based SQI equation under different years of rice cultivation was the following:

$$SQI = +0.393 \times \text{Sand (S)} + 0.197 \times \text{Mn (S)} + 0.154 \times \text{EC (S)} + 0.138 \times \text{K (S)} + 0.119 \times \text{pH(S)} \quad (15.6)$$

where S is the score derived from the homothetic transformation and coefficients are derived from weighted factors obtained from the results of PCA. After identifying the MDS indicators, every observation of each MDS indicator was transformed using a linear scoring method as suggested by Andrews et al. (2002). After transformation using a linear scoring method, the MDS indicator scores thus obtained for each observation were multiplied with the weighted factor obtained from PCA results as given in Equation (15.6). The descriptive statistics for SQIs are given in Table 15.4. The results show that the SQI ranged from 0.243to 0.747, with a mean value of0.388 in the study area. The difference between the minimum and maximum SQI is very high; this difference shows that the surface soils are more easily affected by management, environmental variables, and disturbance. The CV for the SQI was 21.2%, which shows a wide range of values across the study area and implies that the SQI has strong spatial dependence. According to CV analysis, our results indicate that SQI has a semihomogeneous spatial distribution that could be associated with the difference in land use, types of farming and management practices, and topographic position in the study area. It was observed that out of all the key indicators, pH contributed the maximum (28%), whereas Mn and K were contributed 22% and 24%, respectively, toward improving SQ (Fig. 15.3). These parameters could serve as monitoring indicators of long-term cultivation and showed significantly greater changes in management than other parameters.

TABLE 15.3 PCA of soil parameters for selection of MDS.

Parameters	PC1	PC2	PC3	PC4	PC5
Total eigenvalue	3.67	1.84	1.44	1.29	1.11
% of variance	26.23	13.14	10.28	9.24	7.92
Cumulative (%)	26.23	39.37	49.65	58.89	66.81
Eigen vectors					
Sand	**−0.94**	−0.19	0.15	0.03	−0.11
Silt	0.89	0.23	−0.16	−0.04	0.16
Clay	0.88	0.11	−0.12	0.03	0.01
pH	0.25	−0.06	0.08	−0.25	**0.71**
EC	0.07	−0.04	**0.83**	0.03	−0.15
OC	0.67	0.32	0.26	0.13	0.04
CaCO3	−0.32	−0.01	0.74	0.06	0.26
N	−0.51	0.09	−0.15	0.15	0.47
P	0.02	−0.01	0.11	0.69	−0.06
K	−0.04	0.04	−0.04	**0.79**	−0.04
Zn	−0.06	0.63	−0.02	−0.21	−0.36
Cu	0.22	0.66	0.08	0.13	0.06
Fe	0.47	0.48	−0.06	−0.09	0.34
Mn	0.26	**0.74**	−0.14	0.09	0.04

TABLE 15.4 Descriptive statistics of SQ in the study area of desert regions of India.

Parameters	SQI
Minimum	0.24
Maximum	0.75
Mean	0.39
SE	0.01
Median	0.38
SD	0.08
CV (%)	21.20
Skewness	1.33
Kurtosis	3.14

15.3.3 Relationships between the environmental covariates and SQI

Linear correlations between SQI and predictor variables are shown in Table 15.5. SQI at the soil surface is positively correlated with slope ($r = 0.170^*$), TWI ($r = 0.186^*$), and VSSI ($r = 0.181^*$). SQI is negatively correlated with CD ($r = -0.196^{**}$) and CNBL ($r = -0.243^{**}$), respectively. The correlation between climatic and topographic variables and the quantity of SQI has been documented previously (Manickam et al., 2021). Climate factors affect soil salinity, soil characteristics, vegetation cover, water retention, and soil erosion. Our results also indicated a negative association

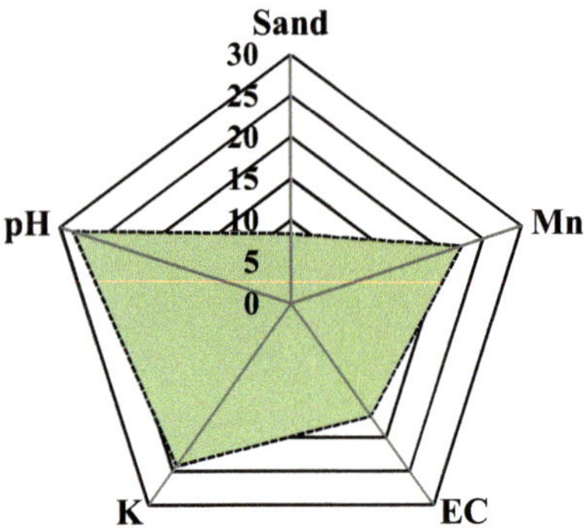

FIGURE 15.3 Contribution (%) of selected indicators to SQ in the study area of desert regions of India.

TABLE 15.5 Correlation analysis between covariates and SQI in the study area of desert regions of India.

Covariates	SQI
Annual_ppt	−0.022
Annual_tem	0.022
Aspect	−0.106
BI	−0.137
CD	−0.196[b]
CI	0.087
CNBL	−0.243[b]
CND	0.189[a]
LS_Factor	0.097
MRRTF	−0.080
MRVBF	0.070
NDSI	−0.060
NDVI	0.060
PlaC	0.000
PrC	−0.012
RSP	0.119
SAVI	0.022
SI	−0.137
Slope	0.170[a]
SR	0.121
TCA	−0.043
TPI	0.098
TRI	0.010
TWI	0.186[a]
VSSI	0.181[a]
Elevation	−0.116

[a]*Correlation is significant at the 0.05 level.*
[b]*Correlation is significant at the 0.01 level.*

between SQI and elevation and a positive relationship with valley depth, similar to the findings of Taghizadeh-Mehrjardi et al. (2014). Other terrain attributes did not show a correlation with SQI in our study area. Bioclimatic covariates have a significant negative correlation with SQI.

Annual_ppt, annual precipitation; Annual_tem, mean annual temperature; BI, brightness index; CD, closed depressions; I, convergence index; CNBL, channel network base level; CND, channel network distance; LS_Factor, length slope factor; MRRTF, multiresolution ridge top flatness; MRVBF, multiresolution index of valley bottom flatness; NDSI, normalized difference salinity index, NDVI, normalized difference vegetation index; PlaC, plan curvatures; PrC, profile curvatures; RSP, Relative Slope Position; SAVI, soil adjusted vegetation index; SI, salinity index; SR, salinity ratio, TCA, total catchment area; TPI, topographic position index; TRI, Topographic Ruggedness Index; TWI, topographic wetness index; VSSI, vegetation soil salinity index

15.3.4 Model prediction and uncertainty estimation for estimation of SQI

The performance of QRF model was evaluated by calculating statistical indicators: R^2, ME, and RMSE. Tenfold cross-validation was used to evaluate the performance of the QRF algorithms. Cross-validation provides a framework for creating several train/test splits in the dataset and guaranteeing that each data point is in the test set at least once. The procedure is simple: split the data into 10 equal-sized groups. Then, for 1–10, select group 1 to be the test set and all other (9) groups to be the training set. After that, train the model on the training set and evaluate it on the test set. Each iteration is called a fold. The advantage of this method is that it performs reliably and is unbiased on smaller data sets because the process requires much more computational effort than simple trained-and-tested (hold-out) procedures. The cross-validation results showed that the combination of different covariates explained the variability of predicted SQI (Table 15.6 and Fig. 15.4). The model could capture low variability while predicting SQI. Overall, the mean CCC

TABLE 15.6 Performance of QRF model for prediction of SQ in the study area of desert regions of India

Error statistics	Performance of QRF model
ME	0.010 ± 0.0001
RMSE	0.080 ± 0.0001
R^2	0.070 ± 0.010
PICP	90.00 ± 1.050

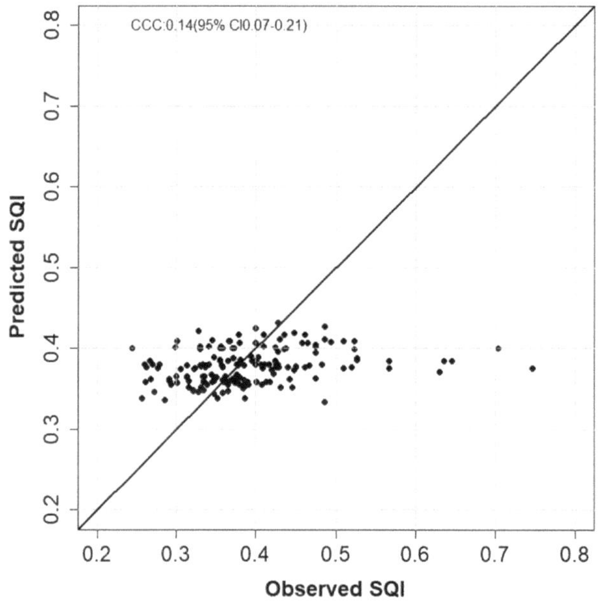

FIGURE 15.4 Predicted *vs* observed SQI by QRF model in the study area of desert regions of India.

values ranged from 0.07 to 0.21, indicating poor agreement between the predicted and observed values. The mean ME values were very close to zero, suggesting overall unbiased predictions. The performance of SQI prediction is related to its dynamic nature, and poor performance may be related to the low levels of SOC compared to soils with high organic carbon (Sharma and Singh, 2015; Dharumarajan et al., 2021). The value of the PICP of SQI was 90.0 ± 1.05. For a 90% prediction interval, we would expect 90% of observations to fall within the lower and upper prediction limits. It can be seen that all the fractions achieved PICP values very close to 90%, suggesting that these lower and upper prediction limits estimated by the quantile RF method were of an appropriate magnitude. That is to say, the uncertainty estimations were, to a large extent, reliable. This can also be indicated by the small values of the SD of the overall prediction accuracy indicators. Overall, the prediction performance of this model was low for SQI. Higher sample density is required for better findings in regions with a wide range of environmental covariates such as soil properties, climate variables, land cover, and relief factors.

15.3.5 Importance of predictor variables for predicting SQI

The parameter sensitivity analysis for the RF model is presented in Fig. 15.5. The variable's importance is measured based on "out-of-bag" samples, which means observation is not included in the model. Another detail is that they are

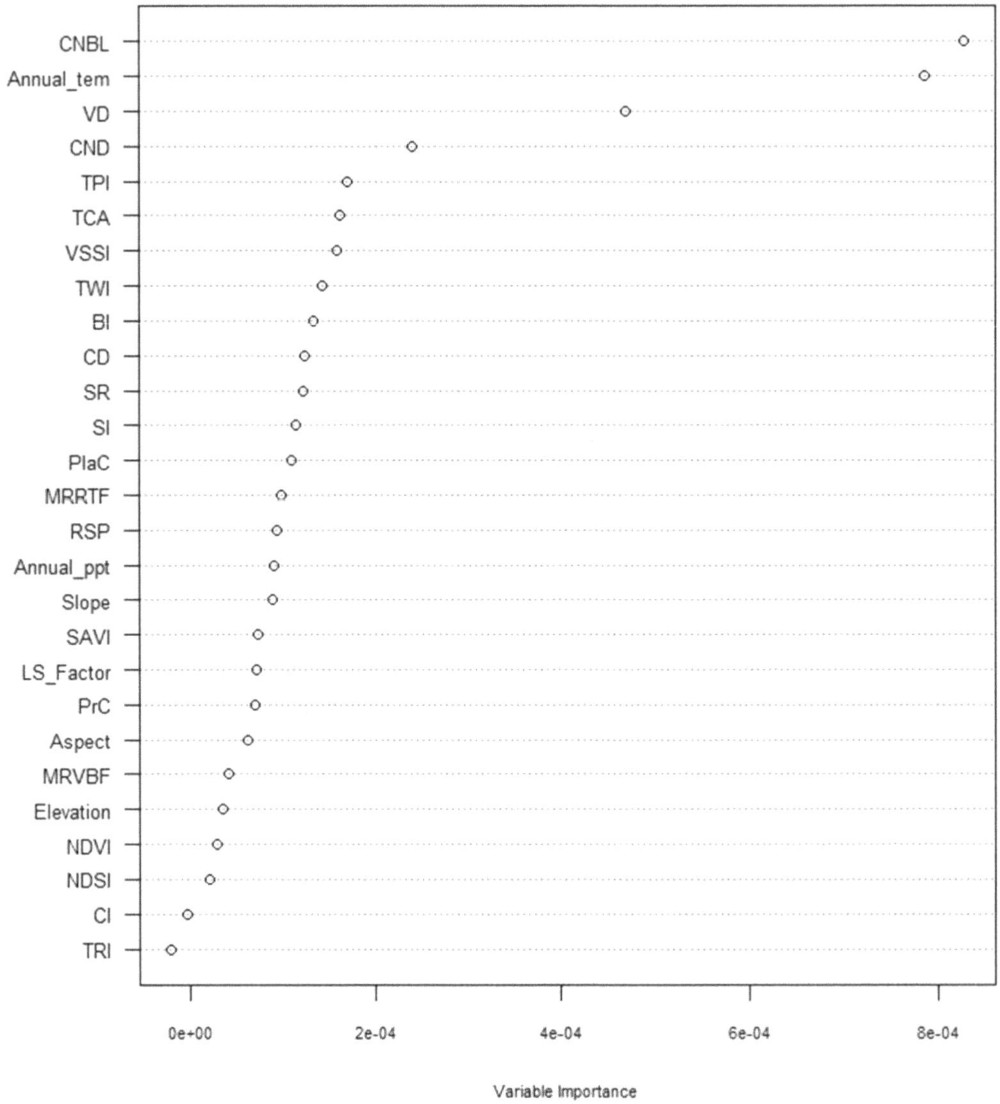

FIGURE 15.5 Importance of variables for prediction of SQI by QRF model in the study area of desert regions of India.

based on an MSE accuracy measure—in this case, the difference between when a covariate is included and excluded in a tree model. For node purity, it is the total decrease in node impurities from splitting on the variable, averaged over all trees. Channel network base level, annual temperature, valley depth, and channel network distance are the best predictors to explain the variability of SQI in our study area. In addition, the result reveals that the climatic indicators (annual temperature) contribute much more than the terrain attribute in estimating SQI in arid ecosystems. Relatively low temperature changes and low moisture conditions in the hot arid region of India lead to the relative importance of predictor variables for predicting soil properties and SQ (Dharumarajan et al., 2021). NDVI and EVI were not important variables in this study. This could be because of the sandy, arid nature of the study area or because of the spare natural vegetation and fewer agricultural practices. Overall, these results indicate that both temperature-related variables and imagery products are important for predicting SQI in sandy desert regions (Fig. 15.6).

15.3.6 Prediction of SQI

By applying the QRF model to the stack of environmental covariates, spatial predictions of SQI were computed at a 30 m spatial resolution. The predicted SQI content varied from 0.322−0.448 in the study area. Overall, SQI was predicted to be low in the south and southeast but high in the north. The lowest occurred in the deserts of the southeast, while the highest was in the Gaghar flood plain. These results revealed that the mean SQI concentration was extremely low. The main reasons for low SOC in the study area, which is part of the Thar Desert, include the high temperature, high evaporation rates, lack of precipitation, intense wind erosion, and spare natural vegetation. The lower SQI mainly found in the sand dunes area, which may be due to low vegetation and soil exposure to the scorching sun, which causes oxidation of SOC (Singh et al., 2009). The highest SQI in the Gaghar flood plain area may be due to the adoption of different soil management practices, including variation in fertilizer application and other crop management practices (Sharma and Singh, 2015; Moharana et al., 2017). Fig. 15.7 shows maps of the uncertainties in predictions of SQI. The uncertainty was expressed as lower and upper prediction limits at a 90% confidence interval. The 5% lower and 95% upper prediction limits had similar spatial patterns with the mean predictions shown in Fig. 15.6, that is, higher mean

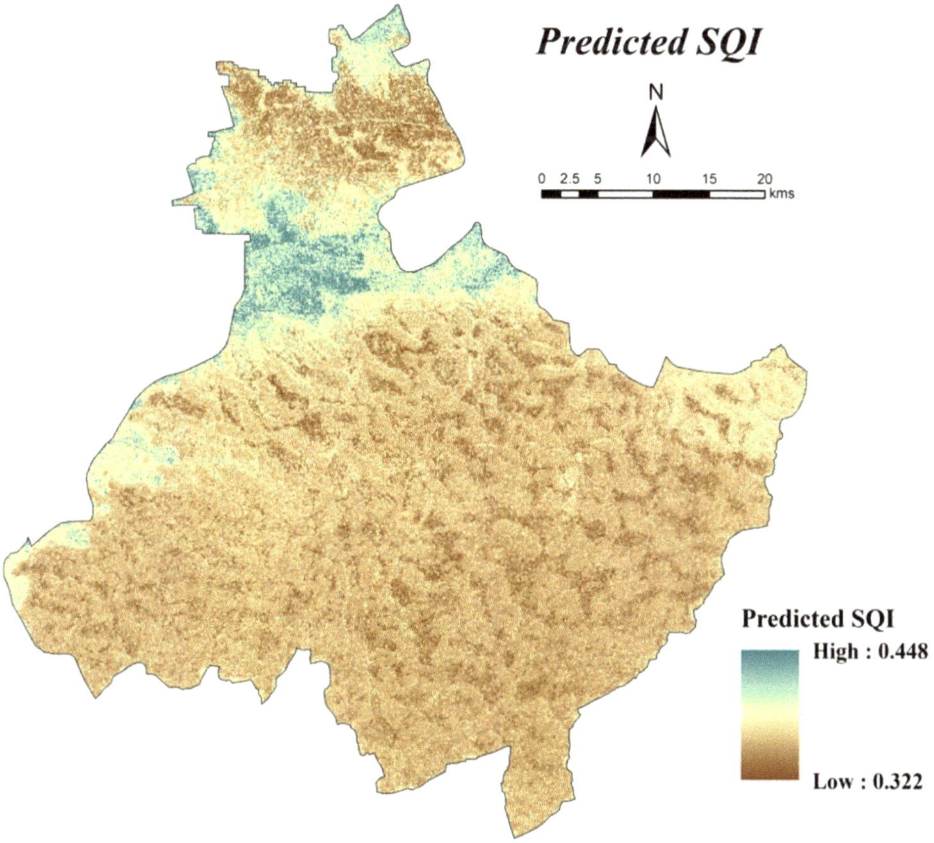

FIGURE 15.6 Map of SQI spatial prediction by QRF model in the study area of desert regions of India.

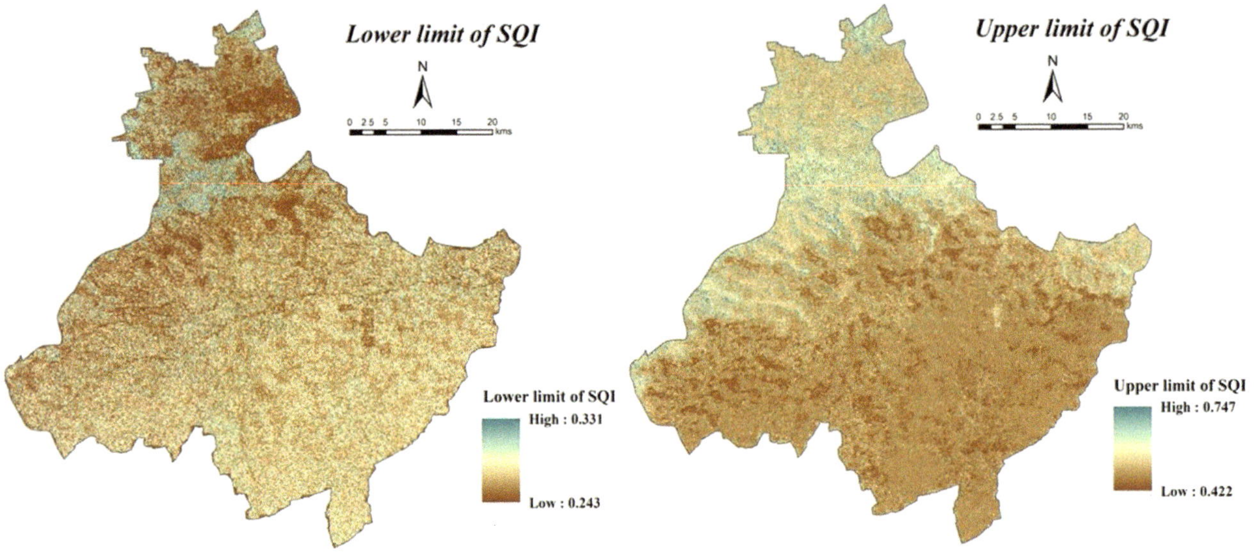

FIGURE 15.7 Maps of the lower and upper limits of 90% prediction interval of SQI in the study area of desert regions of India.

prediction values responded to higher values of the lower and upper limits. The range between 5% lower and 95% upper prediction limits appeared rather wide for SQI, suggesting that there is room to improve the current spatial predictions (Lagacherie et al., 2019).

15.4 Conclusions and future research

The objectives of this study were to (1) provide a framework for evaluating SQ using DSM techniques and the RF machine-learner and (2) evaluate the environmental covariates to determine the SQ for the hot arid regions. Our findings indicate:

- Remote sensing data combined with the RF model to estimate SQI under hot, arid climate conditions produced satisfactory results.
- According to Pearson's correlation analysis, the highest correlations with the soil variables were obtained with the covariates slope, TWI, CD, CNBL, and VSSI.
- The most important predictors obtained from the analysis were channel network base level, annual temperature, valley depth, and channel network distance to explain the variability of SQI in our study area.
- The prediction of SQI and uncertainty by QRF model was reasonable, with an R^2 of 7%. Higher sample density is required for better results in desert regions where soil and climatic patterns are complex in nature.
- The vast range of uncertainty in the desert environment shows that spatial modeling should be improved. It's possible that the current QRF model's ability to capture low variability is attributable to the covariate selection. We anticipate that these predictions can be used to assist in comprehensive planning for arid regions for monitoring SQ.

Acknowledgements

This work is part of Indian Soil Grid project supported by ICAR-National Bureau of Soil Survey and Land Use Planning, Nagpur. The authors are also thankful to laboratory staff, Regional Centre, ICAR-NBSS & LUP, Udaipur, for their assistance during the analysis of soil samples.

References

Andrews, S., Karlen, D., Mitchell, J., 2002. A comparison of soil quality indexing methods for vegetable production systems in Northern California. Agric. Ecosyst. Environ. 90, 25–45.

Arrouays, D., Grundy, M.G., Hartemink, A.E., Hempel, J.W., Heuvelink, G.B.M., Hong, S.Y., et al., 2014. Globalsoilmap: toward a fine-resolution global grid of soil properties. Adv. Agron. 125, 93–134.

Arshad, M.A., Martin, S., 2002. Identifying critical limits for soil quality indicators in agro-ecosystems. Agric. Ecosyst. Environ. 88 (2), 153–160.

Behera, S.K., Shukla, A.K., 2015. Spatial distribution of surface soil acidity, electrical conductivity, soil organic carbon content and exchangeable potassium, calcium and magnesium in some cropped acid soils of India. Land. Degrad. Dev. 26, 71–79.

Behera, S.K., Suresh, K., Rao, B.N., Mathur, R.K., Shukla, A.K., Manorama, K., et al., 2016. Spatial variability of some soil properties varies in oil palm (*Elaeis guineensis* Jacq.) plantations of west coastal area of India. Solid. Earth 7, 979–993. Available from: https://doi.org/10.5194/se-7-979-2016.

Ben-Dor, E., 2002. Quantitative remote sensing of soil properties. In: Sparks, D. (Ed.), Advances in Agronomy. 75 of Advances in Agronomy. pp. 173–243.

Bhunia, G.S., Shit, P.K., Maiti, R., 2016. Comparison of GIS-based interpolation methods for spatial distribution of soil organic carbon (SOC). J. Saudi Soc. Agric. Sci. Available from: https://doi.org/10.1016/j.jssas.2016.02.001.

Camera, C., Zomenia Zomeni, Z., Jay, S., Noller, J.S., Andreas, M., Zissimos, A.M., et al., 2017. A high resolution map of soil types and physical properties for Cyprus: a digital soil mapping optimization. Geoderma 285, 35–49.

Datta, A., Basak, N., Chaudhari, S.K., Sharma, D.K., 2015. Soil properties and organic carbon distribution under different land uses in reclaimed sodic soils of North-West India. Geoderma Reg. 4, 134–146.

de Paul Obade, V., Lal, R., 2013. Assessing land cover and soil quality by remote sensing and geographical information systems (GIS). Catena 104, 77–92.

Dharumarajan, S., Hegde, R., Janani, N., Singh, S.K., 2019. The need for digital soil mapping in India. Geoderma Reg. 16, e00204. Available from: https://doi.org/10.1016/j.geodrs.2019.e00204.

Dharumarajan, S., Kalaiselvi, B., Suputhra, A., Lalitha, M., Vasundhara, R., Anil Kumar, K.S., et al., 2021. Digital soil mapping of soil organic carbon stocks in Western Ghats, South India. Geoderma Reg. 25, e00387. Available from: https://doi.org/10.1016/j.geodrs.2021.e00387.

Dobermann, A., Oberthur, T., 1997. Fuzzy mapping of soil fertility—a case study on irrigated rice land in the Philippines. Geoderma 77 (2–4), 317–339.

Doran, J.W., Jones, A.J., 1996. Methods for assessing soil quality. Soil Science Society of America, Madison, Wisconsin, USA, p. 410, Special Publication Number 49.

Fathizad, H., Ardakani, M.A.H., Heung, B., Sodaiezadeh, H., Rahmani, A., Fathabadi, A., et al., 2020. Spatio-temporal dynamic of soil quality in the central Iranian desert modeled with machine learning and digital soil assessment techniques. Ecol. Indic. 118, 106736.

Hanway, J.J., Heidel, H., 1952. Soil analysis methods as used in Iowa State College Soil Testing Laboratory. Iowa Agric. 57, 1–31.

Hengl, T., Rossiter, D.G., Stein, A., 2004. Soil sampling strategies for spatial prediction by correlation with auxiliary maps. Aust. J. Soil. Res. 41 (8), 1403–1422.

Hengl, T., Heuvelink, G.B., Kempen, B., Leenaars, J.G., Walsh, M.G., Shepherd, K.D., et al., 2015. Mapping soil properties of Africa at 250 m resolution: random forests significantly improve current predictions. PLoS One 10, 1–26.

Hillel, D., 1991. Out of the Earth: Civilization and the life of the soil. Univ. of California Press, Los Angeles.

Hillel, D., 2005. Soil salinity: historic and contemporary perspectives. In: Proceedings of the International Salinity Forum. Riverside, California.

Jackson, M.L., 1973. Soil Chemical Analysis. Prentice Hall of India Pvt. Ltd., New Delhi.

Jena, R.K., Bandyopadhyay, S., Pradhan, U.K., Moharana, P.C., Kumar, N., Sharma, G.K., et al., 2022. Geospatial modelling for delineation of crop management zones using local terrain attributes and soil properties. Remote. Sens. 14, 2101. Available from: https://doi.org/10.3390/rs14092101.

Jenny, H., 1941. Factors of Soil Formation. McGraw Hill, New York.

Jobbagy, E.G., Jackson, R.B., 2000. The vertical distribution of soil organic carbon and its relation to climate and vegetation. Ecol. Appl. 10, 423–436.

Jordan, S.G., Jannoura, R., Jordan, G., Buerkert, A., Joergensen, R.G., 2018. Spatial variability of soil properties in the floodplain of a river oasis in the Mongolian Altay Mountains. Geoderma 330 (15), 99–106. Available from: https://doi.org/10.1016/j.geoderma.2018.05.028.

Khan, M.A., Moharana, P.C., Singh, S.K., 2003. Integrated natural resources and environmental impact assessment for sustainable development of Ganganagar district, Rajasthan. Research Report, Division of Natural Resources and Environment, Central Arid Zone Research Institute, Jodhpur, India.

Klingebiel, A.A., Montgomery, P.H., 1961. Land capability classification. U.S. Department of Agriculture, Soil Conservation Service, Agriculture Handbook No. 210, Washington DC, USA. Available at: https://www.nrcs.usda.gov/Internet/FSE_documents/nrcs142p2_052290.pdf.

Lacoste, M., Mulder, V., Richer-de-Forges, A., Martin, M., Arrouays, D., 2016. Evaluating large-extent spatial modeling approaches: a case study for soil depth for France. Geoderma Reg. 7, 137–152.

Lagacherie, P., Baret, F., Feret, J.-B., Netto, J.M., Robbez-Masson, J.M., 2008. Estimation of soil clay and calcium carbonate using laboratory, field and airborne hyperspectral measurements. Remote. Sens. Environ. 112 (3), 825–835.

Lagacherie, P., Arrouays, D., Bourennane, H., Gomez, C., Martin, M., Saby, N.P.A., 2019. How far can the uncertianty on a digital soil map be known?: A numerical experiment using pseudo values of clay content obtained from Vis-SWIR hyperspectral imagery. Geoderma 337, 1320–1328.

Larson, W.E., Pierce, F.J., 1994. The dynamics of soil quality as a measure of sustainable management. In: Doran, J.W., Coleman, D.C., Bezdicek, D. F., Stewart, B.A. (Eds.), Defining Soil Quality for a Sustainable Environment, 35. Soil Science Society of America Special Publication, Madison, WI, pp. 37–51.

Lin, L.I., 1989. A concordance correlation coefficient to evaluate reproducibility. Biometrics 45, 255–268.

Lindsay, W.L., Norvell, W.A., 1978. Development of a DTPA soil test for zinc, iron, manganese, and copper. Soil. Sci. Soc. Am. J. 42, 421–428.

MacDonald, K.B., Wang, F., Fraser, W.R., Lelyk, G.W., 1998. Broad-scale assessment of agricultural soil quality in canada using existing land resource databases and GIS. Res. Branch. Tech. Bull. 1998. -3E.

Mahajan, G.R., Manjunath, B.L., Latare, A.M., D'Souza, R., Vishwakarma, S., Singh, N.P., 2015. Fertility status of the unique coastal acid saline soils of Goa. J. Indian. Soc. Soil. Sci. 63 (2), 232–237. Available from: https://doi.org/10.5958/0974-0228.2015.00031.6.

Mandal, D., Singh, R., Dhyani, S.K., Dhyani, B.L., 2010. Landscape and land use effects on soil resources in a Himalayan watershed. Catena 81 (3), 203–208.

Manickam, L., Subramanian, D., Khandal, S., Hegde, R., 2021. Modeling and mapping of salt-affected soils through spectral indices in inland plains of semi-arid agro-ecological region. J. Indian. Soc. Remote. Sens. 49, 1475–1481.

Masto, R.E., Chhonkar, P.K., Singh, D., Patra, A.K., 2008. Alternative soil quality indices for evaluating the effect of intensive cropping, fertilisation and manuring for 31 years in the semi-arid soils of India. Environ. Monit. Assess. 136 (1–3), 419–435.

Mathur, M., Sundaramoorthy, S., 2016. Patterns of herbaceous species richness and productivity along gradients of soil moisture and nutrients in the Indian Thar Desert. J. Arid. Environ. 125, 80–87. Available from: https://doi.org/10.1016/j.jaridenv.2015.10.011.

McBratney, A.B., Odeh, I.O.A., 1997. Application of fuzzy sets in soil science: fuzzy logic, fuzzy measurements and fuzzy decisions. Geoderma 77 (2-4), 85–113.

McBratney, A.B., Mendonca-Santos, L., Minasny, B., 2003. On digital soil mapping. Geoderma 117, 3–52.

Meinshausen, N., 2006. Quantile regression forests. J. Mach. Learn. Res. 7, 983–999.

Minasny, B., Hartemink, A.E., 2011. Predicting soil properties in the tropics. Earth Sci. Rev. 106, 52–62.

Minasny, B., McBratney, A.B., Malone, B.P., Wheeler, I., 2013. Digital mapping of soil carbon. Adv. Agron. 118, 1–47.

Moharana, P.C., Naitam, R.K., Verma, T.P., Meena, R.L., Sunil Kumar, Tailor, Singh, B.L., et al., 2017. Effect of long-term cropping systems on soil organic carbon pools and soil quality in western plain of hot arid India. Arch. Agron. Soil. Sci. 63, 1661–1675.

Moharana, P.C., Singh, R.S., Singh, S.K., Tailor, B.L., Jena, R.K., Meena, M.D., 2019. Development of secondary salinity and salt migration in the irrigated landscape of hot arid India. Environ. Earth Sci. 78, 454.

Moharana, P.C., Jena, R.K., Kumar, N., Singh, R.S., Rao, S.S., 2021. Assessment of soil organic and inorganic carbon stock at different soil depths after conversion of desert into arable land in the hot arid regions of India. Carbon Manag. 12, 153–165.

Mulder, V., de Bruin, S., Schaepman, M., Mayr, T., 2011. The use of remote sensing in soil and terrain mapping—a review. Geoderma 62, 1–19.

Niang, M., Nolin, M., Jego, G., Perron, I., 2014. Digital mapping of soil texture using RADARSAT-2 polarimetric synthetic aperture radar data. Soil. Sci. Soc. Am. J. 78 (2), 673–684.

Ojima, D.S., Dirks, B.M., Glenn, E.P., Owensby, C.E., Scurlock, J.O., 1993. Assessment of C budget for grasslands and drylands of the world. Water Air Soil. Pollut. 70, 95–109.

Olsen, S.R., Cole, C.V., Watanabe, F.S., Dean, L.A., 1954. Estimation of available phosphorus in soils by extraction with sodium bicarbonate, USDA Circular, 939. U.S. Government Printing Office, Washington, DC.

Ouerghemmi, W., Gomez, C., Naceur, S., Lagacherie, P., 2016. Semi-blind source separation for the estimation of the clay content over semi-vegetated areas using VNIR/SWIR hyperspectral airborne data. Remote. Sens. Environ. 181, 251–263.

Pal, S., Manna, S., Aich, A., Chattopadhyay, B., Mukhopadhyay, S.K., 2014. Assessment of the spatio-temporal distribution of soil properties in East Kolkata wetland ecosystem (A Ramsar site: 1208). J. Earth Syst. Sci. 123 (4), 729–740.

Pandey, C.B., Singh, G.B., Singh, S.K., Singh, R.K., 2010. Soil nitrogen and microbial biomass carbon dynamics in native forests and derived agricultural land uses in a humid tropical climate of India. Plant. Soil. 333, 453–467. Available from: https://doi.org/10.1007/s11104-010-0362-x.

Qi, Y., Darilek, J.L., Huang, B., Zhao, Y., Sun, W., Gu, Z., 2009. Evaluating soil quality indices in an agricultural region of Jiangsu Province, China. Geoderma 149 (3-4), 325–334.

R Core Team, 2019. R: A language and environment for statistical computing. R Foundation for Statistical Computing, Vienna, Austria <https://www.R-project.org/>.

Ratnayake, R.R., Karunaratne, S.B., Lessels, J.S., Yogenthiran, N., Rajapaksha, R., Gnanavelrajah, N., 2016. Digital soil mapping of organic carbon concentration in paddy growing soils of Northern Sri Lanka. Geoderma Reg. 7, 167–176.

Saha, D., Kukal, S.S., Bawa, S.S., 2012. Soil organic carbon stock and fractions in relation to land use and soil depth in the degraded Shiwaliks hills of lower Himalayas. Land. Degrad. Dev. Available from: https://doi.org/10.1002/ldr.2151.

Scull, P., Franklin, J., Chadwick, O.A., McArthur, D., 2003. Predictive soil mapping: a review. Prog. Phys. Geogr. 27, 171–197.

Selige, T., Boehner, J., Schmidhalter, U., 2006. High resolution topsoil mapping using hyperspectral image and field data in multivariate regression modeling procedures. Geoderma 136 (1–2), 235–244.

Sharma, R.P., Singh, R.S., 2015. Carbon stock and its management in soils of Bhilwara district, Rajasthan. J. Indian. Soc. Soil. Sci. 63, 304–309.

Singh, A., 2015. Soil salinization and waterlogging: a threat to environment and agricultural sustainability. Ecol. Indic. 57, 128–130.

Singh, S.K., Kumar, M., Sharma, B.K., Tarfadar, J.C., 2007. Depletion of organic carbon, phosphorus and potassium stock under pearl millet based cropping sequence in arid environment of India. Arid. Land. Res. Manag. 21, 119–131.

Singh, S.K., Kumar, M., Sharma, B.K., 2009. Changes in soil properties in hot arid region of India. J. Indian. Soc. Soil. Sci. 57, 24–30.

Subbiah, B.V., Asija, G.L., 1956. A rapid method for the estimation of available nitrogen in soils. Curr. Sci. 25, 259–260.

Sun, B., Zhou, S.L., Zhao, Q.G., 2003. Evaluation of spatial and temporal changes of soil quality based on geostatistical analysis in the hill region of subtropical China. Geoderma 115 (1–2), 85–99.

Taghizadeh-Mehrjardi, R., Minasny, B., Sarmadian, F., Malone, B., 2014. Digital mapping of soil salinity in Ardakan region, central Iran. Geoderma 213, 15–28.

Taghizadeh-Mehrjardi, R., Nabiollahi, K., Kerry, R., 2016. Digital mapping of soil organic carbon at multiple depths using different data mining techniques in Baneh region, Iran. Geoderma 266, 98–110.

Tripathi, R., Nayak, A.K., Shahid, M., Raja, R., Panda, B.B., Mohanty, S., et al., 2015. Characterizing spatial variability of soil properties in salt affected coastal India using geostatistics and kriging. Arab. J. Geosci. Available from: https://doi.org/10.1007/s12517-015-2003-4.

Vaysse, K., Lagacherie, P., 2017. Using quantile regression forest to estimate uncertainty of digital soil mapping products. Geoderma 291, 55–64.

Verma, R.R., Manjunath, B.L., Singh, N.P., Kumar, A., Asolkar, T., Chavan, V., et al., 2018. Soil mapping and delineation of management zones in the Western Ghats of coastal India. Land. Degrad. Dev. 29, 4313–4322. Available from: https://doi.org/10.1002/ldr.3183.

Walkley, A., Black, I.A., 1934. An examination of the Degtjareff method for determining soil organic matter and a proposed modification of the chromic acid titration method. Soil. Sci. 37, 29–38.

Wang, X., Gong, Z., 1998. Assessment and analysis of soil quality changes after eleven years of reclamation in subtropical China. Geoderma 81 (3–4), 339–355.

Zeng, C., Yang, L., Zhu, A.X., Rossiter, D.G., Liu, J., Liu, J., et al., 2016. Mapping soil organic matter concentration at different scales using a mixed geographically weighted regression method. Geoderma 281, 69–82.

Chapter 16

Investigation on the effect of vegetation and soil spectral characteristics on land surface temperature (LST) of Al-Ahsa Oasis in the Kingdom of Saudi Arabia—a remote sensing and GIS approach

Sheik Mujabar P.
Department of General Studies, Jubail Industrial College, Jubail Industrial City, Kingdom of Saudi Arabia

16.1 Introduction

The land surface temperature (LST) is an important feature for understanding various properties such as surface energy processes, land cover assessment, heat energy flux, land evapotranspiration, climatological, and environmental processes. LST is the thermal level of the earth's surface soil, usually measured in Kelvin or Celsius scales. It is also represented by the standard air temperature measured by a shielded thermometer 1–2 m above a plane and ventilated earth surface. Several geological, geophysical, and geochemical parameters of the earth surface and atmosphere influence the LST. The surface-atmospheric interactions and soil energy flux alter the LST. The LST plays a significant role in soil moisture conditions of the land. Estimating LST and soil moisture is vital for agricultural developments and land management strategies, especially in arid and semiarid regions. The spatial and temporal variations of soil moisture and LST also relate to soil type, land use, and land cover. LST is a direct parameter that specifies the earth's surface energy balance since it is a significant feature of the physics of the earth's surface processes. It also comprises the outcomes of surface-atmosphere exchanges and energy fluxes between the atmosphere and the earth's surface. The difference between the LST and the surface air temperature controls the heat energy flux and exchange in a land surface boundary. The change in soil type and air temperature is complexly interrelated in setting the LST.

The latest advancements in satellite remote sensing technology deliver satellite-based surface temperature estimations with good precision and cost-effectiveness, becoming readily available globally (Dickinson, 1994). Satellite LST products also provide a wide spatial and temporal energy flux and kinetic temperature of the earth's surface skin (Becker and Zhao-Liang Li, 1995), that is, the aggregate surface medium viewed by the sensor to a depth of about 12 μm. Surface temperature is inferred from the thermal emission of the earth's surface and is usually the average effective radiative temperature of various canopy and soil surfaces (Hall et al., 1992; Betts et al., 1996). Numerous investigations have been conducted using remote sensing technology to measure the LST (Akhoondzadeh and Saradjian, 2008; Zareie et al., 2016). Vlassova et al. (2014) mentioned that the LST is one of the key inputs for soil-vegetation-atmosphere transfer modeling in terrestrial ecosystems. LST is associated with the carriage of heat energy flux between the land surface and the atmospheric boundary layer (Kustas and Anderson, 2009; Quattrochi and Luvall, 2004). This aspect makes LST a key factor in estimating sensible heat flux (Jia et al.) and latent heat flux, or evapotranspiration (Moran et al., 1996; Kalma et al., 2008). Truong Linh et al. (2015) state that the direct observation and estimation of LST in industrial and agricultural areas are quite complex and problematic, as it is not possible to construct a network of weather stations with sufficiently high density and continuous operation in such zones. El-Nahry and Rashash (2013) implied that climatic conditions are measured by weather stations in major locations and cities. Due to insufficient weather stations and technical constraints, the weather data does not express accurate spatial and temporal climate

Remote Sensing of Soils. DOI: https://doi.org/10.1016/B978-0-443-18773-5.00006-5

variability at microclimate conditions. In addition, the climate station may not be found in the exact location under investigation; therefore, data could be obtained from the neighboring stations. Therefore, the remotely sensed LST is significant and has greater capability to cover wide areas in any remote region. It also provides more spatial and temporal variability of thermal conditions of the earth's surface. The remotely sensed thermal infrared (TIR) data are unique sources of information to define the surface heat islands (Weng, 2009). Numerous studies have demonstrated that precise LST can be estimated from measurements of brightness temperature and emissivity in TIR sensors (Schmugge et al., 1991; Coll and Caselles, 1997; Li et al., 2013).

The present study mainly emphasizes investigating the effect of soil spectral indices on LST in the Al-Ahsa region (Fig. 16.1), one of the biggest oasis in the Kingdom of Saudi Arabia. This oasis is a major agricultural area in the Kingdom with a large amount of date plantations. The study also aims to investigate the effectiveness of the recently launched Landsat 9 satellite data for mapping the LST in the major agricultural area of the arid region. The bands of both Operational Land Imager (OLI-2) and Thermal Infrared Remote Sensor (TIRS 2) have been used for finding the LST by using the single-channel (SC) method. The TIRS-2 bands were used to find the At-sensor brightness temperature, and the OLI-2 bands were used to find surface emissivity. The influence of soil spectral indices and land use types on LST in the area are analyzed by using the 2-D scatter plots. A linear regression model using the specified spectral indices has been developed for this major agricultural area in the Kingdom. Error estimation analysis has been carried out to assess the model relationship with the soil spectral indices. The variation of LST is also analyzed with different landforms and seasonal variations.

16.2 Study area

The Al-Hassa oasis is located in the Eastern Province of Saudi Arabia between 25°20′ to 25°40′N and 49°32′ to 49°46′E with an elevation of approximately 140 m from the Mean Sea Level (MSL). The oasis appears as an "L"

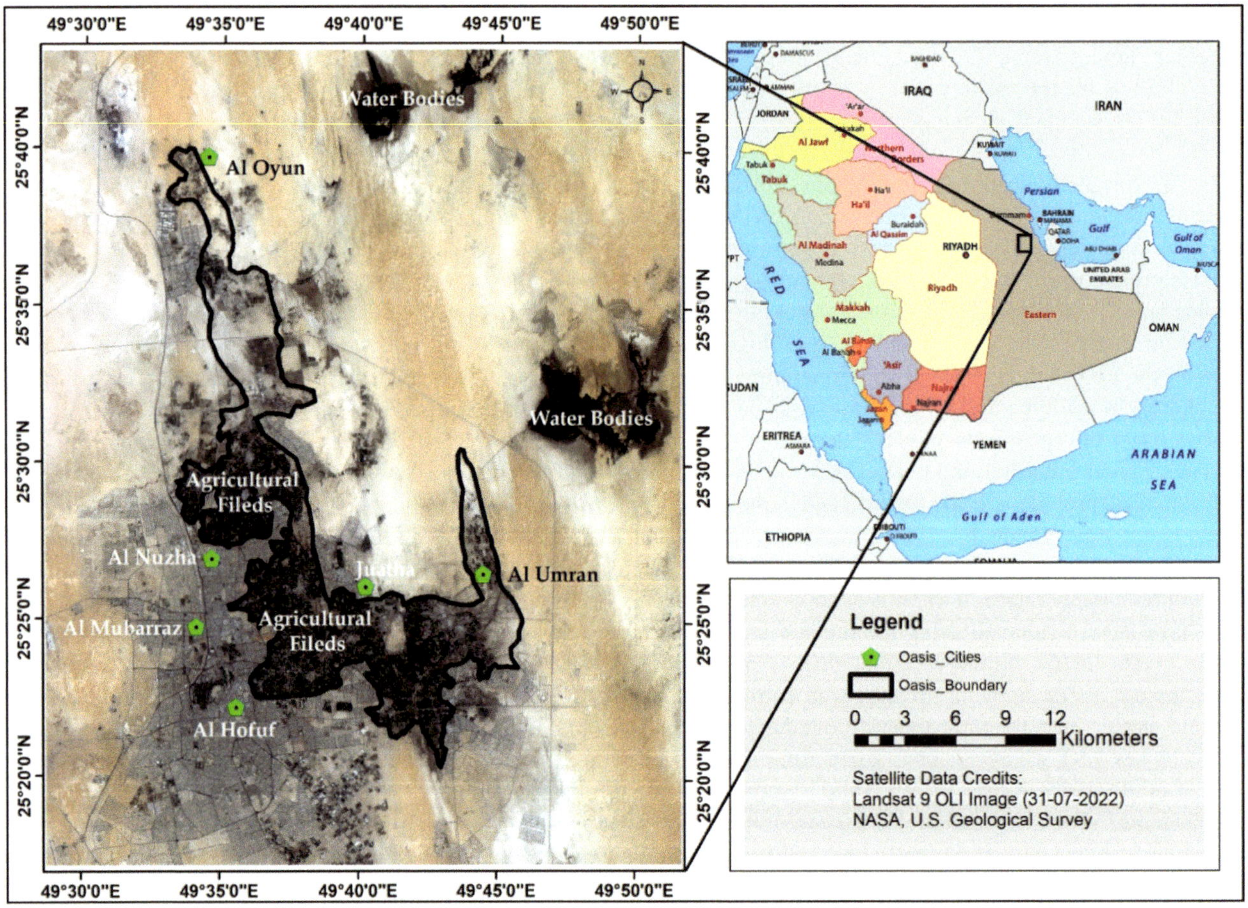

FIGURE 16.1 Location map of the Al-Ahsa oasis, KSA.

shaped region located inland around 50 km from the coast of the Persian Gulf. The oasis is about 150 km from the Dammam, the capital city of the Eastern Province. It is situated at altitudes ranging between 160 m in the west and 130 m in the east above the MSL. The oasis has nearly 10 towns and more than 60 villages. The oasis is one of the biggest irrigated in the Kingdom, with a huge number of date plantations (Al-Tahir AAS, 1999). The topography of the oasis is very gentle, with little relief and a few surrounding ridges. Vast areas of active sand dunes surround the three sides of the oasis, and the other side is covered by several settlements. The study area has vast active sand dunes (Salih, 2018). Elhadary (2014) states that these sand dunes are active dynamic, and drifting sand, at a rate of 3 $m^3 m^{-1}$, occurs from the northwest and north direction of the oasis. The western part of the oasis has dense settlements with several small cities and a large population. Agriculture is the main source of income and employment for the population.

The oasis has several underground springs and produces huge water supply, making agricultural production. Rahman (2012) states that the oasis is the major agricultural area for the Kingdom's eastern region. The agricultural area comprises about 180 km^2 of palm trees, green parks, and garden areas (Salih, 2018). The total area under cultivation within the oasis is approximately 80 km^2, of which 92% is occupied by date palms (Al-Zara, 2011). Al-Barrak and Al-Badawi (1988) state that the soil is sandy loam in texture and covers six taxonomic classes, such as Gypsiorthids, Haplaquepts, Calciorthids, Torripsamments, Salorthids, and Torriorthents. The climate of this area is typically arid, with an ambient temperature exceeding 45°C during the summer and a high evaporation rate that exceeds the average annual precipitation of approximately 488 mm (Allbed et al., 2014). The region has a high evapotranspiration rate, exceeding the yearly net rainfall ($<$120 mm $year^{-1}$). Most of the year, this leads to arid weather along the study area. Hounington-Huene (1979) and Alqarni et al. (2018) stated that high-potential evapotranspiration impacts surface temperature. The ambient air temperature along the oasis is generally from 5°C to 25°C during the winter, starting from December to February. The temperature rises steadily during summer and attains around 45°C–55°C during the peak summer.

(Al-Saeedi, 2022) The oasis has a natural aquifer known as Neogene, which is the major source of water irrigating the vast areas of date plantations. Abderrahman (1988) states that the Al-Hassa Irrigation and Drainage Project includes about 1482 km of concrete canals and drainage patterns. The terrain of the oasis has a gentle slope toward north and east directions, and most of the drainage patterns are toward these directions (Al-Barrak and Al-Badawi, 1988). Amal Allbed et al. (2014) stated that the oasis's soil is strongly saline and spread across most of the study area. They also stated that these soils' salinity varies significantly throughout the area, and several environmental and human-induced factors, including topography, drainage pattern, irrigation methodology, water quality, and agricultural practices, probably trigger the difference. Al-Saeedi (2022) states that the texture of the soil in the oasis is scattered between loamy sand, sandy, sandy loam, silty, and silty loam.

16.3 Materials and methods

16.3.1 Satellite data

The Landsat satellite provides a widespread and reliable global record on earth observation and monitoring. The Landsat 9 satellite has two onboard sensors: OLI-2 and TIRS-2. The geometrically and terrain-rectified L1TP data over the oasis during the year 2022 (Table 16.1) has been utilized for the present study. The satellite data with maximum quality have been downloaded from the Earth Explorer website (https://earthexplorer.usgs.gov/) of the United States Geological Survey. The data acquirement is based on the atmospheric conditions, data availability at the region, dust,

TABLE 16.1 Remote sensing data.

Spatial data	Sensor and bands	Date acquired on	Period (season)
Landsat 9 (OLI-2, TIRS-2)	Operational land imager bands & thermal infrared sensor band 10	04–01–2022	Winter
		09–03–2022	Pre-summer
		31–07–2022	Summer
		19–10–2022	Prewinter
SRTM DEM	C-band (5.6 cm)	11–02–2000	–

cloud coverages, etc., to provide the highest possible results for spectral studies. The data has already been resampled to 30 m resolution and projected with UTM (Zone 39 N) and WGS -84 as datum. Spectral bands of the OLI-2 and TIRS-2 have been used for the present study. The TIR radiation in the specific wavelength will excite electrons within the Quantum Well Infrared Photodetectors, creating an electrical signal to find the surface temperature (Landsat 9 Mission Brochure, 2021). The region of interest was made from satellite imagery and local maps using a shape file of Al-Ahsa oasis as the mask in ArcGIS 10.8 software.

The methodology used in the study is shown in Fig. 16.2. The SC method has been performed by using the TIRS-2 band 10 for finding the LST of the oasis. Several techniques such as the SC algorithm, multi-channel algorithm, Split–Window Algorithm, multi-angle algorithm, classification-based emissivity method, Normalized Difference Vegetation Index (NDVI)-based emissivity method, Day/night temperature independent spectral indices-based method, two-temperature method, temperature emissivity separation methods are available to retrieve the LST from remotely sensed imageries (Zhao-Liang et al., 2013). SC algorithms to estimate LST from one thermal band are advantageous over other methods when the imageries with only one thermal band are used. Yu et al. (2014) stated that the Landsat TIRS has two spectrally adjacent channels, which is suitable for the split window methods for finding the LST. Zhao-Liang et al. (2013) specified that the SC methods include a simple inversion of the RTE and can offer theoretically accurate LST retrieval. Even though the SC method can be applied to both bands of TIRS data, band 10 is preferred since the band is located in a lower atmospheric absorption region (high atmospheric transmissivity values). Yu et al. (2014) stated that the results for LST retrieval from band 11 of the TIRS sensor show larger uncertainty than band 10, which is consistent with the announced TIRS calibration issues. The OLI data bands have been used for estimating the land surface emissivity.

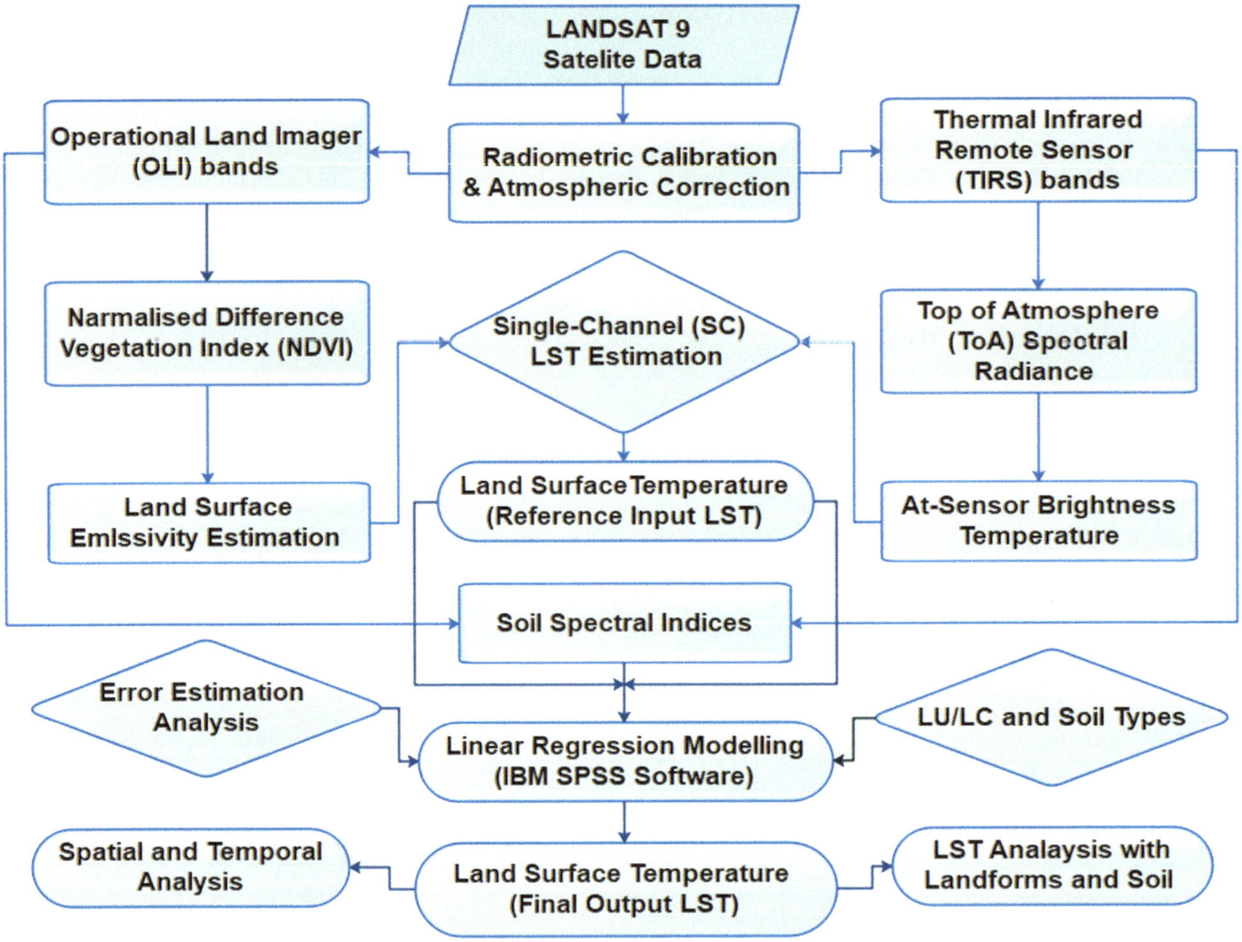

FIGURE 16.2 Flowchart of the methodology.

16.3.2 Preprocessing of satellite data

Spatial data preprocessing is essential for effectively mapping earth features and resources. The basic preprocessing procedure includes atmospheric and radiometric corrections of the satellite data. The Landsat 9 satellite data is subjected to radiometric and atmospheric corrections methods in this present investigation using the ENVI 5.3 Software. The satellite data is processed to radiometric standardization to make the Digital Numbers (DN) to "Top of Atmosphere" (ToA) radiance data with respect to "Fast Line-of-sight Atmospheric Analysis of Hypercubes (FLAASH) Atmospheric Correction." The DN of the data reflects the reflectance and radiance of the earth's surface. Various surface conditions and atmospheres may alter the actual reflectance and radiance. Numerous calibration algorithms are available to rectify these effects to improve the data quality. The radiometric calibration process modifies the DN recorded by the satellite sensors into measurable physical values in radiance or reflectance. The calibration values are obtained from the metadata of the satellite data and used in the preprocessing procedure. The earth's surface reflects electromagnetic radiation (EM) and passes through the different layers of the atmosphere. So, eliminating atmospheric interferences like absorption and scattering of light is important in the pro-processing procedure.

The estimation and analysis of LST definitely requires the exclusion of the atmospheric effects. The exclusion of the presence of water vapor and circulation of aerosol and dust particles is a critical preprocessing step in the effective methodology of soil LST studies using the RS data. The direct estimation of these atmospheric effects on spatial data is not available in a large scale, and different models are employed to remove these effects. These procedures isolate the noises by pixel by pixel and make actual reflectance or radiance of the earth's surface. The FLAASH tool of the software ENVI 5.3 is simple and user-friendly for atmospheric correction. This tool modifies the wavelengths in the visible spectrum through near infrared (NIR) and shortwave infrared (SWIR) regions, up to 3 μm, by using the MODTRAN radiation transfer codes (Flaash, 2008). The FLAASH uses a process for retrieving the amount of aerosol in the atmosphere and assessing a scene average visibility by using a dark pixel reflectance ratio method based on Kaufman et al. (1997). The tropical atmospheric model and tropospheric aerosol model with an inbuilt filter function file of the ENVI software have been used for the study. The SWIR—2 band has been employed as the KT upper channel and the red band as a lower channel.

16.3.3 Estimation of land surface temperature

16.3.3.1 Conversion to ToA spectral radiance

The Landsat sensors obtain data in the units of absolute radiance values using a 32-bit floating-point, and the level 1 product is a 16-bit raster data with DN. So, first the TIRS image bands with DN values are changed into ToA spectral radiance data by using the subsequent relation and the radiance-scaling factor given in the metadata file of the Landsat data:

$$L_\lambda = M_L \times Q_{CAL} + A_L$$

where L_λ = Spectral radiance value (measured in W $[\mathrm{m}^2 \times \mathrm{sr} \times \mu\mathrm{m}]^{-1}$), M_L = Radiance multiplicative scaling factor for the respective band. (RADIANCE_MULT_BAND_n from the metadata, $M_L = 0.0003342$ for both bands 10, 11). A_L = Radiance additive scaling factor for the band. (RADIANCE_ADD_BAND_n from the metadata, $A_L = 0.1$ for both bands 10, 11), Q_{cal} = L1 pixel value in DN.

16.3.3.2 Conversion to at-sensor brightness temperature

The brightness temperature (T_B) is the amount of radiance of the incoming EM radiation traveling skyward from the top of the earth's surface to the sensor. It is expressed in units of the temperature of an equivalent blackbody. The brightness temperature is the actual temperature the remote sensor views under an assumption of unit surface emissivity value. Then, the ToA spectral radiance data has been converted into brightness temperature (T_B) using the following relation:

$$T_B = \frac{K_2}{\ln\left(1 + \frac{K_1}{L_\lambda}\right)}$$

where T_B = ToA brightness temperature, in Kelvin. (The temperature can also be converted into Celsius by subtracting 273.) (i.e. $T_{B\ (^\circ C)} = T_{B\ (K)} - 273.15$), L_λ = Spectral radiance [Watts $(\mathrm{m}^2 \times \mathrm{sr} \times \mu\mathrm{m})^{-1}$], K_1 = Thermal conversion constant for the band (K1_CONSTANT_BAND_n from the metadata), K_2 = Thermal conversion constant for the band (K2_CONSTANT_BAND_n from the metadata). $K_1 = 774.8853$ and $K_2 = 1321.0789$.

16.3.3.3 *Land surface emissivity (ε)*

A surface's emissivity (ε) is an important parameter for finding the surface temperature. It is also valuable for calculating energy budget from the remotely sensed spatial data. It is a proportionality factor that scales blackbody radiance (Planck's law) to estimate emitted radiance, and it is the efficiency of transmitting thermal energy across the surface into the atmosphere (Sobrino et al., 2008). The OLI spectral bands of the Landsat 9 data have been processed for estimating the surface's NDVI values by using the subsequent equation:

$$\text{NDVI} = \frac{\text{NIR} - \text{RED}}{\text{NIR} + \text{RED}} = \frac{\text{Band} \quad 5 - \text{Band} \quad 4}{\text{Band} \quad 5 + \text{Band} \quad 4}$$

The NDVI values typically vary from −1 to +1, and the image is shown in Fig. 16.5. The minimum and maximum values of the NDVI are also measured. The following equation has predicted the land surface emissivity according to Yu et al. (2014):

$$\text{Emissivity } \varepsilon = 0.973 - 0.047 \text{ Pv}$$

where Pv is the proportion of vegetation or fractional vegetation cover, it can be obtained from NDVI values according to Carlson and Ripley (1997) since the NDVI values of the surface are a key factor in the assessment of proportion of vegetation (Pv). Consequently, it is measured from the below equation:

$$\text{Pv} = \left[\frac{\text{NDVI} - \text{NDVI}_{\text{min}}}{\text{NDVI}_{\text{max}} - \text{NDVI}_{\text{min}}}\right]^2$$

16.3.3.4 *Land surface temperature*

The LST of the Al-Ahsa oasis has been calculated from the At-sensor brightness temperature and surface emissivity by using the following equation:

$$\text{LST} = \frac{T_B}{\left[1 + \left(\lambda \times \frac{T_B}{\rho} \times \ln\varepsilon\right)\right]}$$

where T_B = brightness temperature, λ = Wavelength of emitted radiance, $\rho = h \times c/\sigma = 1.438 \times 10^{-2}$ mK (σ = Boltzmann constant = 1.38×10^{-23} J K^{-1}, h = Planck's constant = 6.626×10^{-34} Js, and c = velocity of light = 2.998×10^8 m s^{-1}), ε = Land surface emissivity.

16.3.4 Vegetation and soil spectral indices

The elevation of the earth's surface, vegetation covers, and land use (Figs. 16.3 and 16.4) may alter the land surface temperature of any region. Most vegetation and soil spectral indices have a significant relation with the LST. Therefore, several vegetation-based indices like NDVI and soil spectral indices like Normalized Difference Barren Index were effectively used for estimating the emissivity of land and for estimating the LST. Moisa and Gemeda (2022) stated that while these spectral indices have been used to model the LST in several studies, only a few of them have compared index results for mapping the spatial and temporal variation of LST. The vegetation and soil spectral indices are preferred features for detecting the surface soil temperature and its variations. Though the Al-Ahsa oasis is present in the semiarid region, it has significant green vegetation. So, the present study uses vegetation reflectance (vegetation indices) and soil reflectance (bare soil indices) to predict the LST. In order to generate a regression model for this region, more than 20 indices were tested by using the variation of indices with the measured LST values at different locations in different landforms. We computed and processed six major vegetation and soil spectral indices based on the variation of indices with the LST (2D scatter plots—Fig. 16.11A–F). These indices were identified and selected for this study due to their potential association with measured LST values. Table 16.2 shows the mathematical expressions of these spectral indices. The spatial variation of these selected spectral indices is shown in Figs. 16.5–16.10.

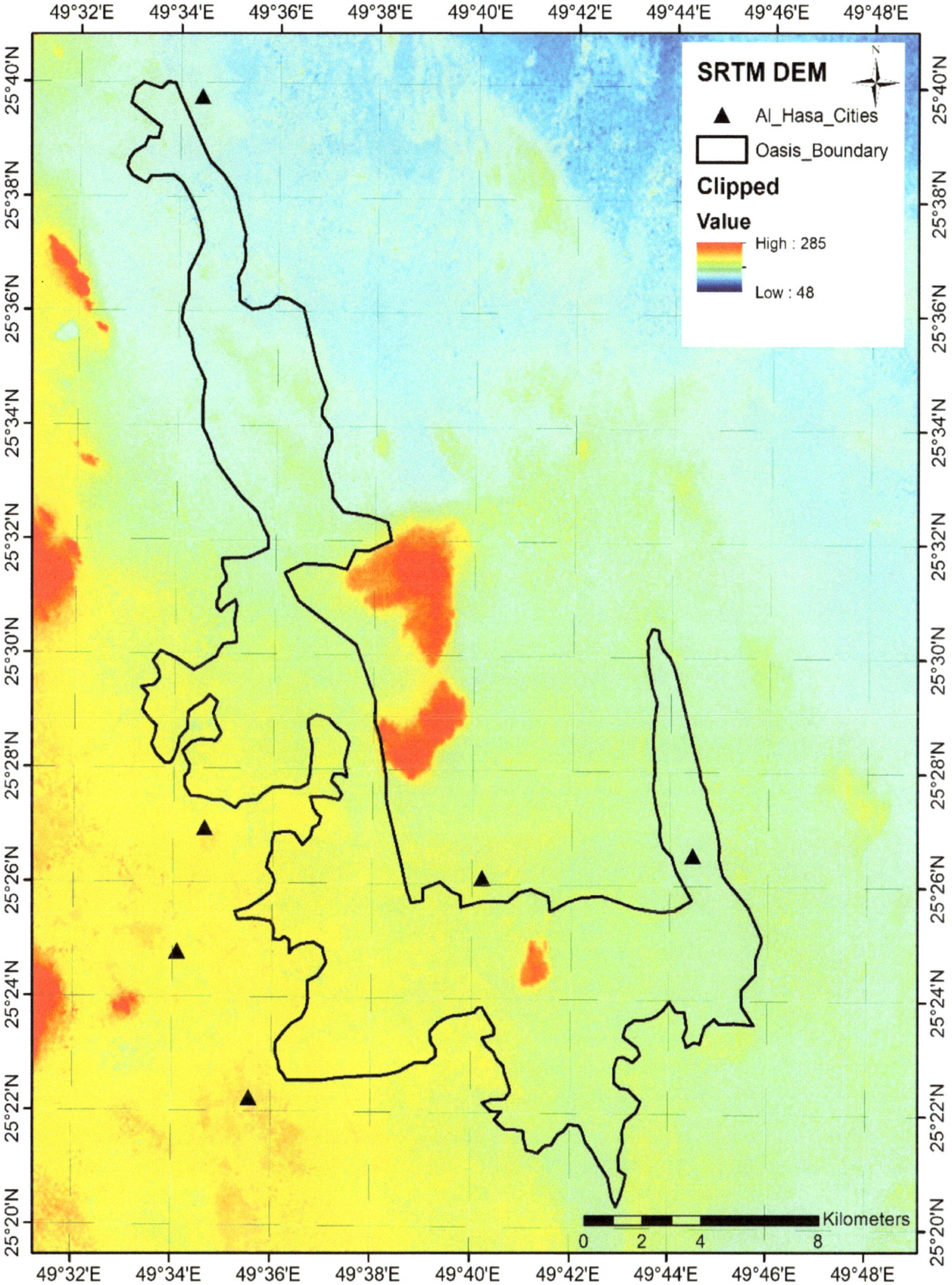

FIGURE 16.3 Digital elevation model (DEM).

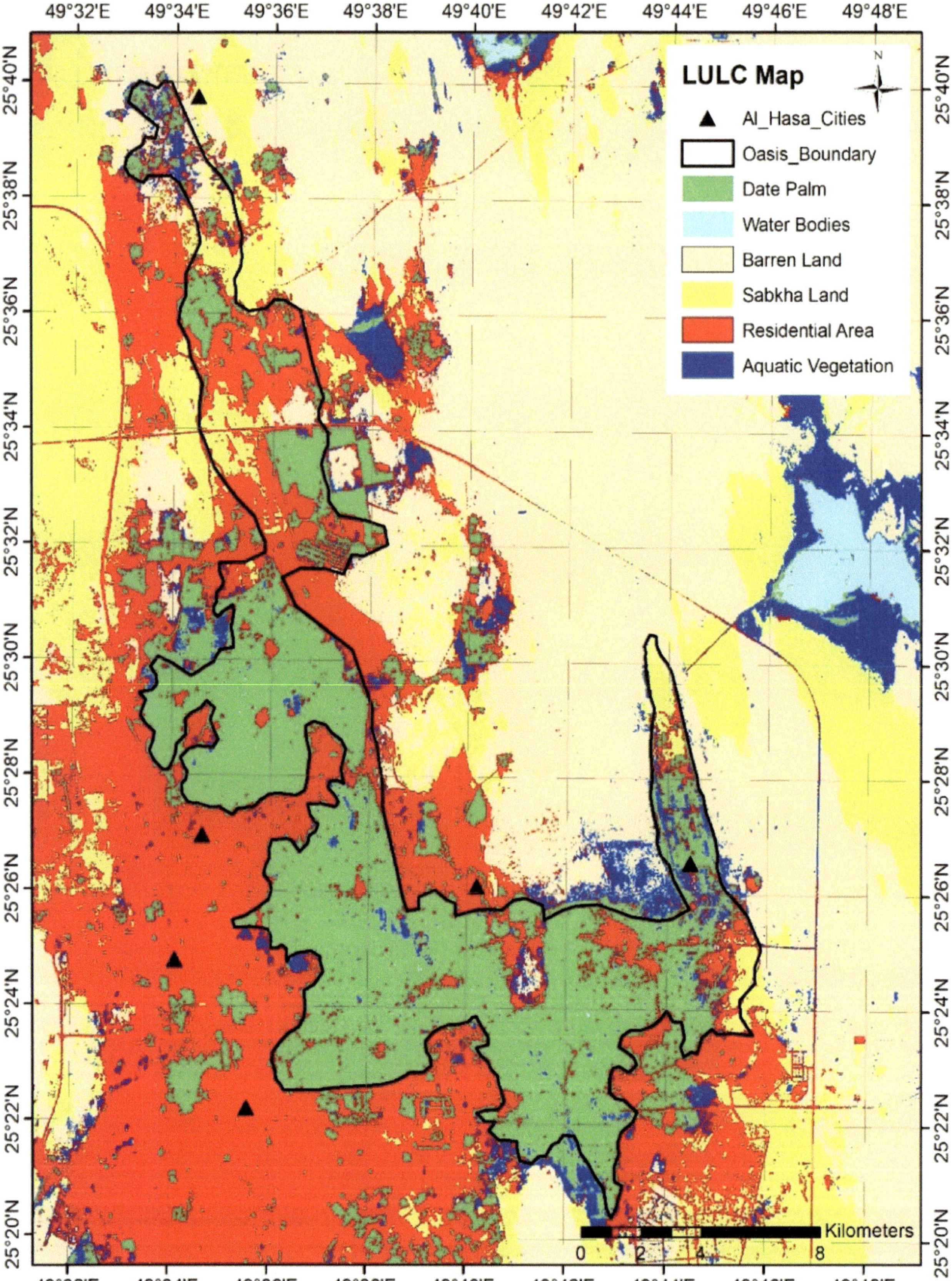

FIGURE 16.4 Land use land cover (LULC) map.

TABLE 16.2 Soil and vegetation spectral indices.

Indices (with acronym)	Band equation	References
Normalized Difference Vegetation Index (NDVI)	$(NIR - R)/(NIR + R)$	Rouse (1973)
Bare Soil Index (BSI)	$(SWIR2 + R) - (NIR + B)/(SWIR2 + R) + (NIR + B)$	Diek et al. (2017)
Normalized Difference latent heat Index (NDLI)	$(G - R)/(G + R + SWIR1)$	Liou et al. (2019)
Normalized difference Moisture Index (NDMI)	$(NIR - SWIR\ \ 1)/(NIR + SWIR1)$	Gao (1996)
Normalized difference Soil Index (NDSI)	$(SWIR2 - G)/(SWIR2 + G)$	Deng et al. (2015)
Normalized Difference Water Index (NDWI)	$(NIR - SWIR)/(NIR + SWIR)$	Gao (1996)

16.4 Results and discussion

16.4.1 Effect of spectral indices on LST

16.4.1.1 Normalized difference vegetation index

NDVI is the most used index for estimating the LST (Gutman and Ignatov, 1998; Guha et al., 2020). The correlation between LST-NDVI is helpful to understand the pattern of LST. It is calculated as the ratio between reflected radiations in the red and NIR regions. The Fig. 16.11A shows the negative correlation between the LST and NDVI in the study area, which also displays that LST values are low for regions with high NDVI and vice versa. Several studies have established a strong relationship between the NDVI, fractional vegetation cover, and the emissivities of terrestrial materials (Valor and Caselles, 1996). The NDVI image of the study area is shown in Fig. 16.5. The NDVI values range from 0.19 to 0.43. It also implies that the regions inside the oasis with date plantations have high NDVI values. The dense aquatic vegetation near the water bodies have also high index values. The sparse vegetation cover in the outer areas of the oasis also has moderate index values, whereas the residential area and barren land have comparatively lesser index values.

16.4.1.2 Bare soil index

Bare soil is the soil on the top of the earth not covered by grass, lawn, trees, or vegetation, which plays a vital role in the earth's surface process and impacts in arid and semiarid regions. The barren lands in the deserts make dust storms and impact the urban environment. Bare soil landforms are also impacted by modifying the LST of a region. Bare Soil Index (BSI) is a spectral index to identify soil variations in different landforms by using the reflection of EM radiations in the blue, red, NIR, and SWIR spectral bands. The different mineral compositions in the soil will be enhanced and highlighted red and SWIR radiations, while the blue and NIR radiations highlight the vegetation. The specific heat capacity of the bare soil is relatively lower than other landforms, and it plays a vital role in increasing the LST. Identifying bare soil is an important factor in analyzing the urban heat island (UHI) and its spatial and temporal variations. The variation of BSI of the study area is shown in Fig. 16.6. The values of BSI range from $-$ 0.22 to +0.30. BSI's lower and negative values indicate the absence of bare soil, and positive values represent the bare soil pixel in the study area.

16.4.1.3 Normalized difference latent heat index

The latent heat and surface energy fluxes emitted by the earth's surface are measured using Normalized Difference Latent Heat Index (NDLI). The latent heat is the thermal energy released or absorbed by the earth's surface without modifying the LST. This spectral index is vital for understanding the hydrological cycle. It also provides information on the role of land-air interface on the LST. The NDLI is used to study the water content and potential heat flux in different landforms. NDLI uses the spectral reflections in red, green, and SWIR region of the EM spectrum. The NDLI was first proposed by Liou et al. (2019) for estimating the latent heat flux of different surfaces using the Surface Energy Balance Algorithm for Land (SEBAL). They emphasized that the NDLI establishes an excellent correlation

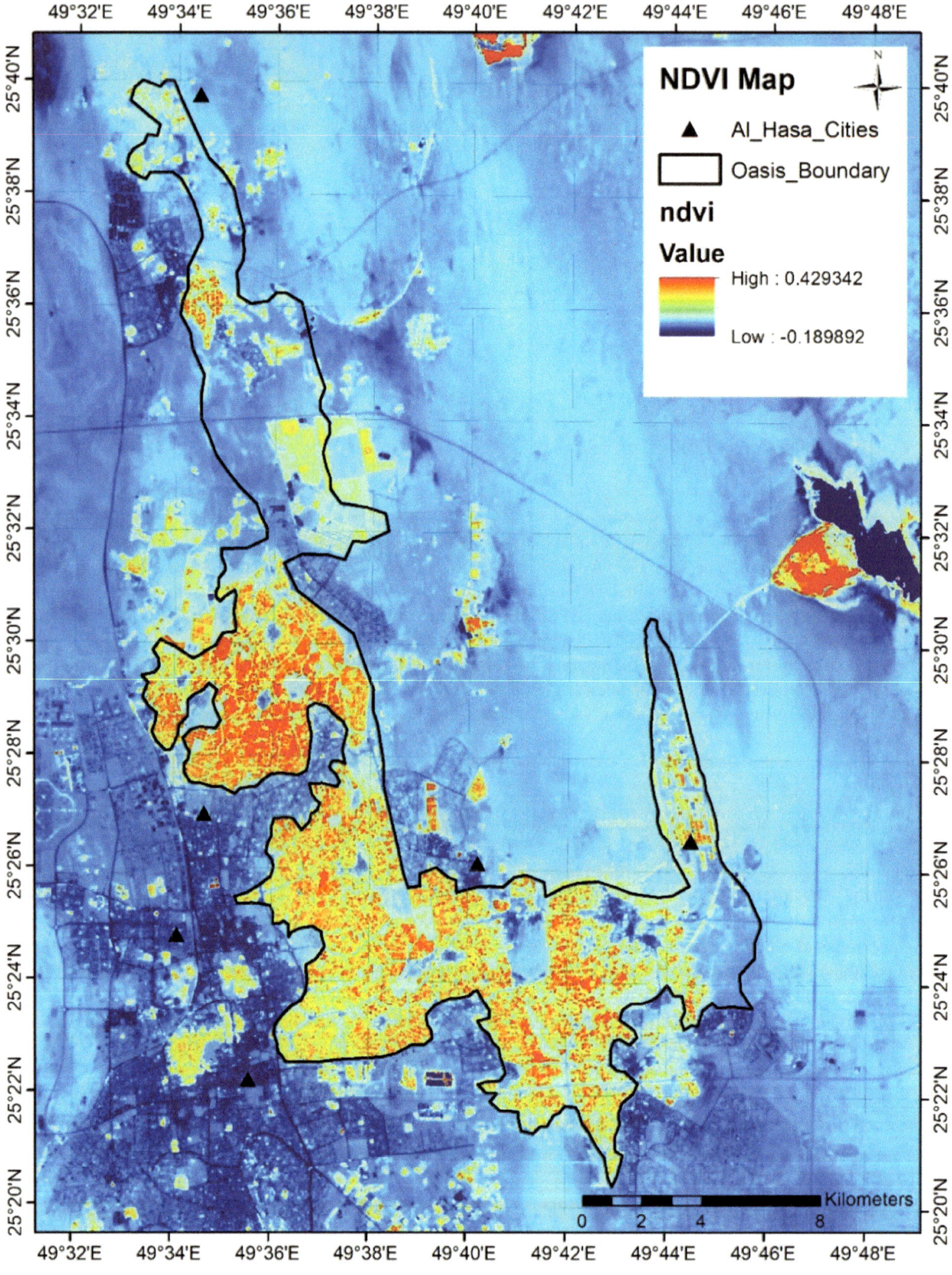

FIGURE 16.5 Normalized difference vegetation index image

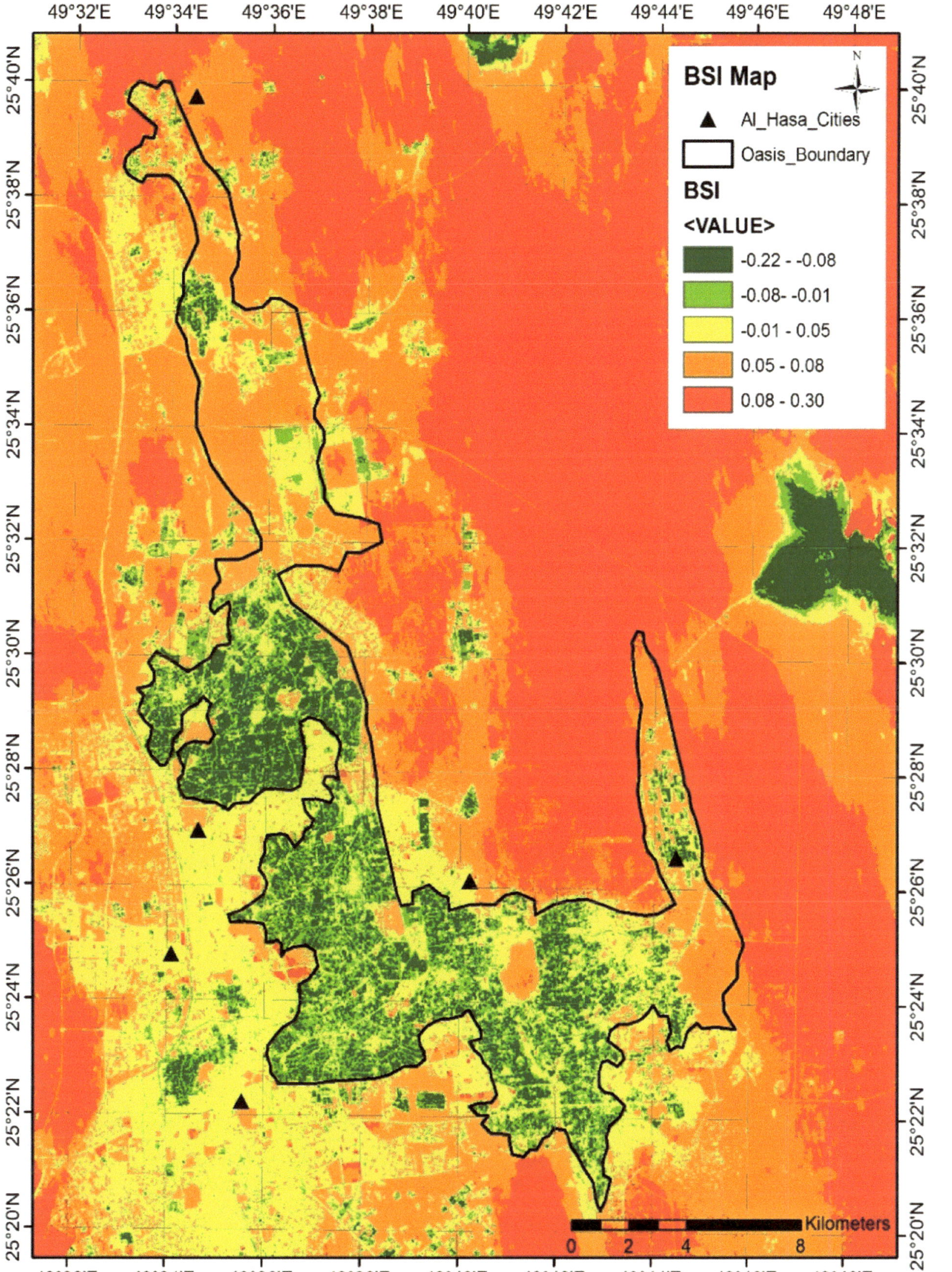

FIGURE 16.6 Bare soil index (BSI) spectral image.

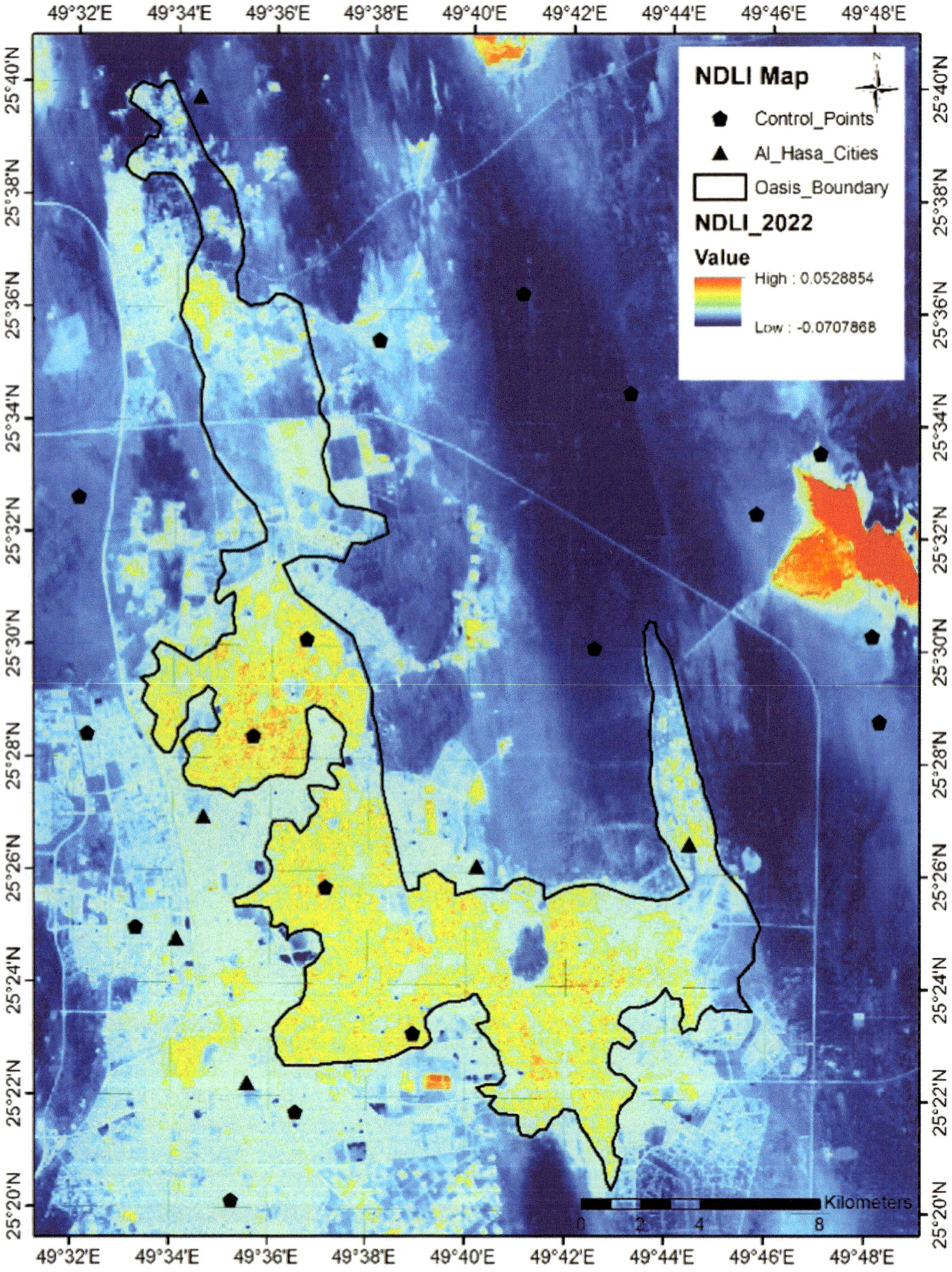

FIGURE 16.7 NDLI spectral image.

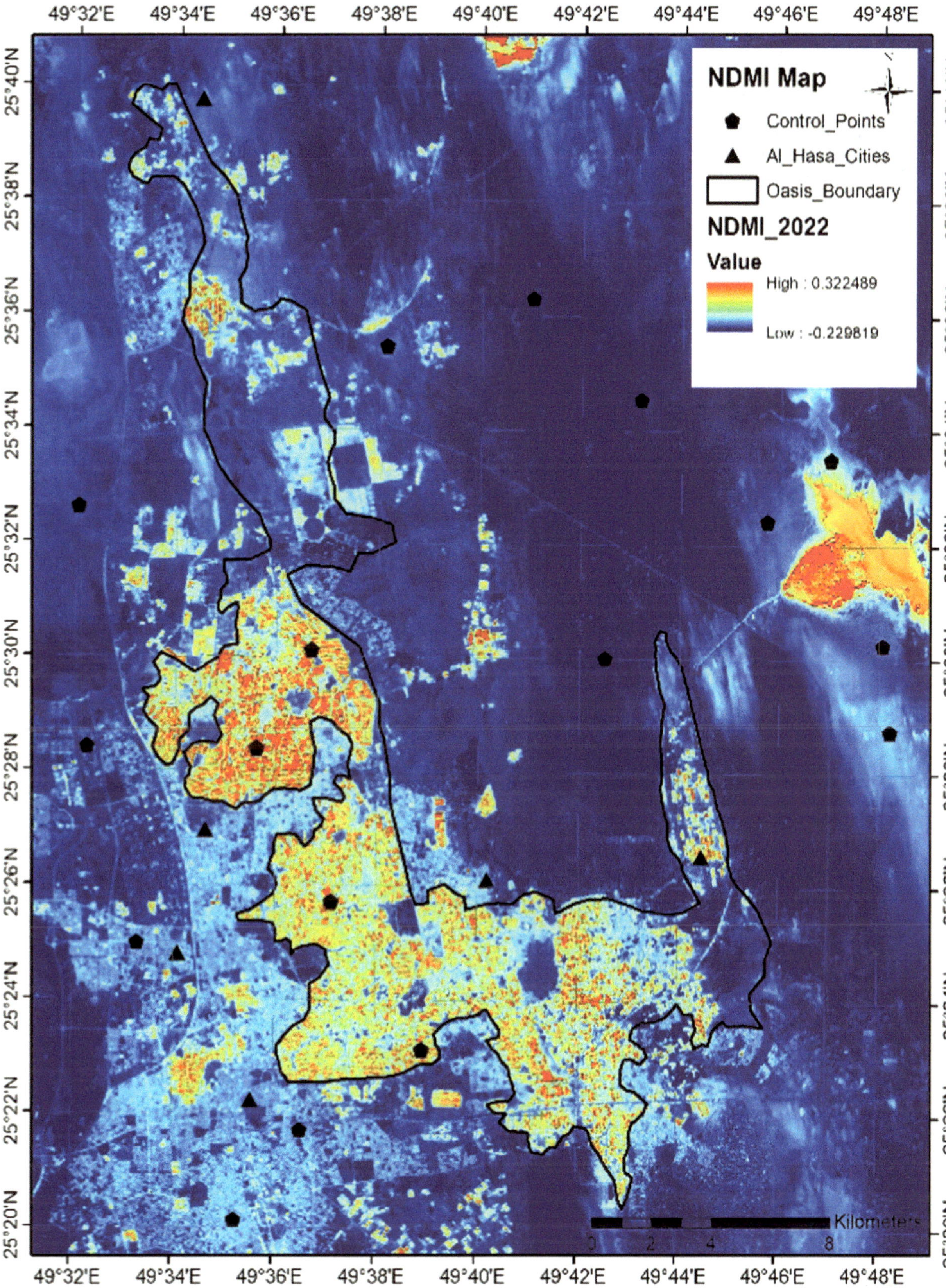

FIGURE 16.8 NDMI spectral image.

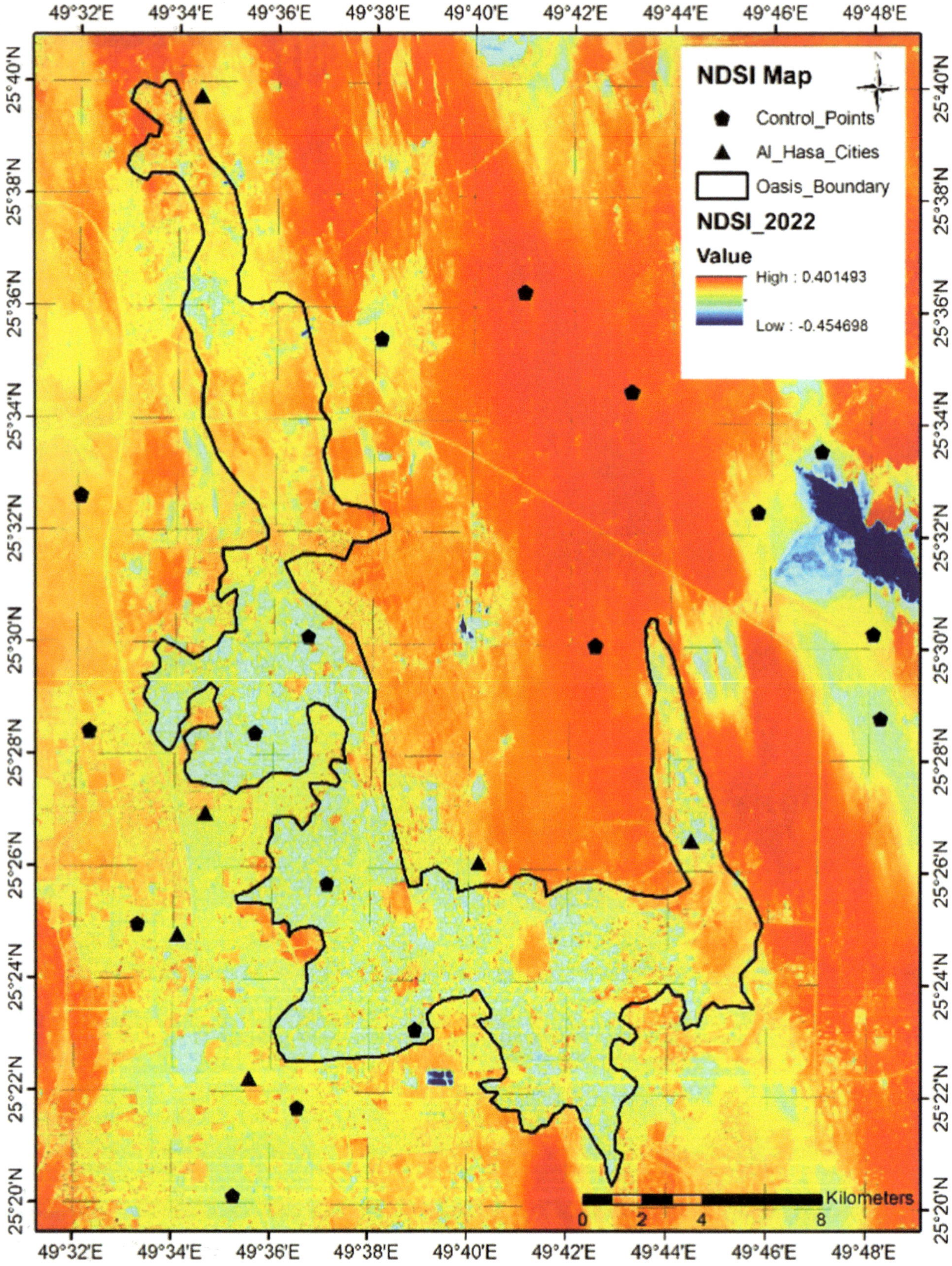

FIGURE 16.9 NDSI spectral image.

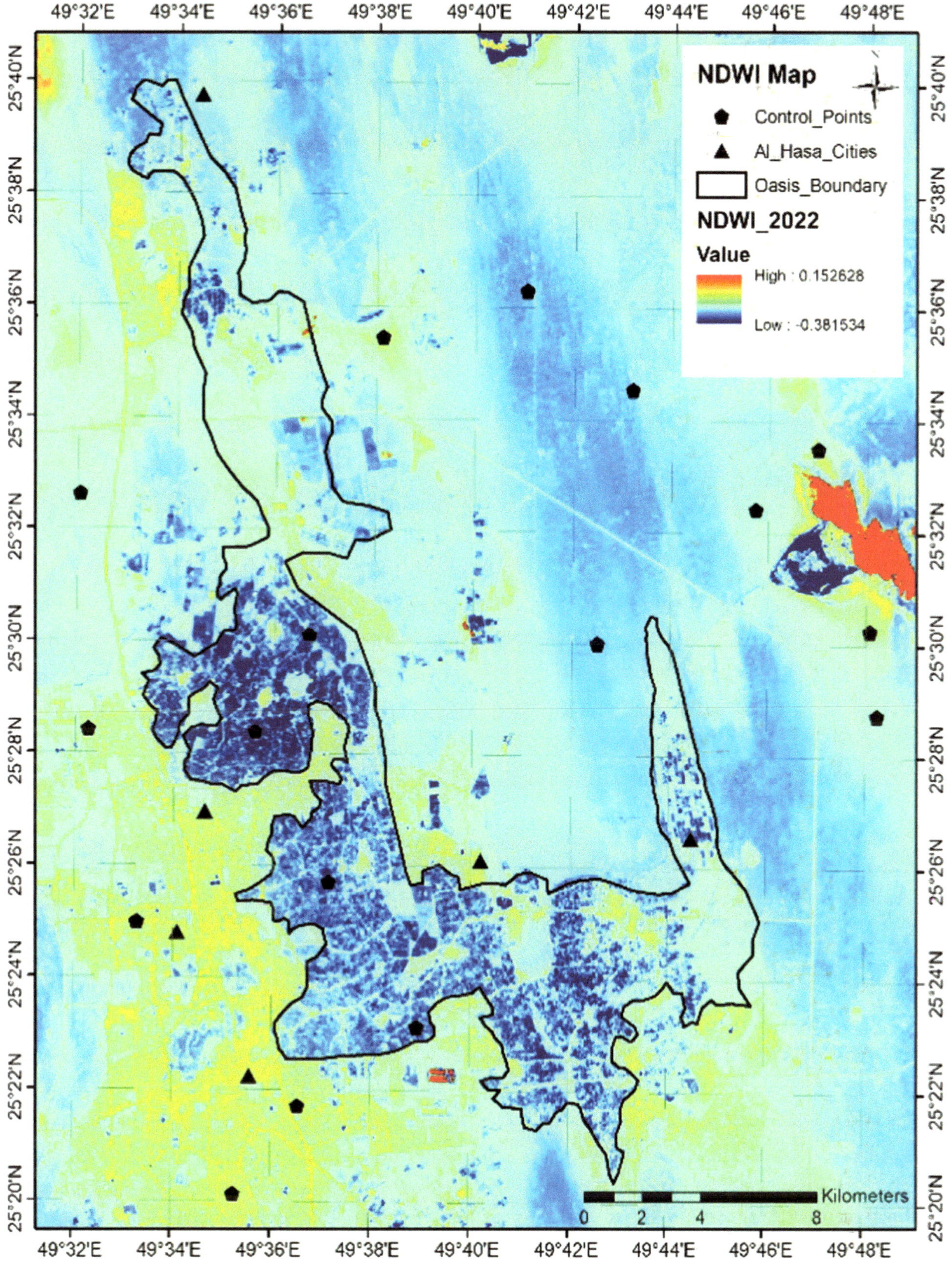

FIGURE 16.10 NDWI spectral image.

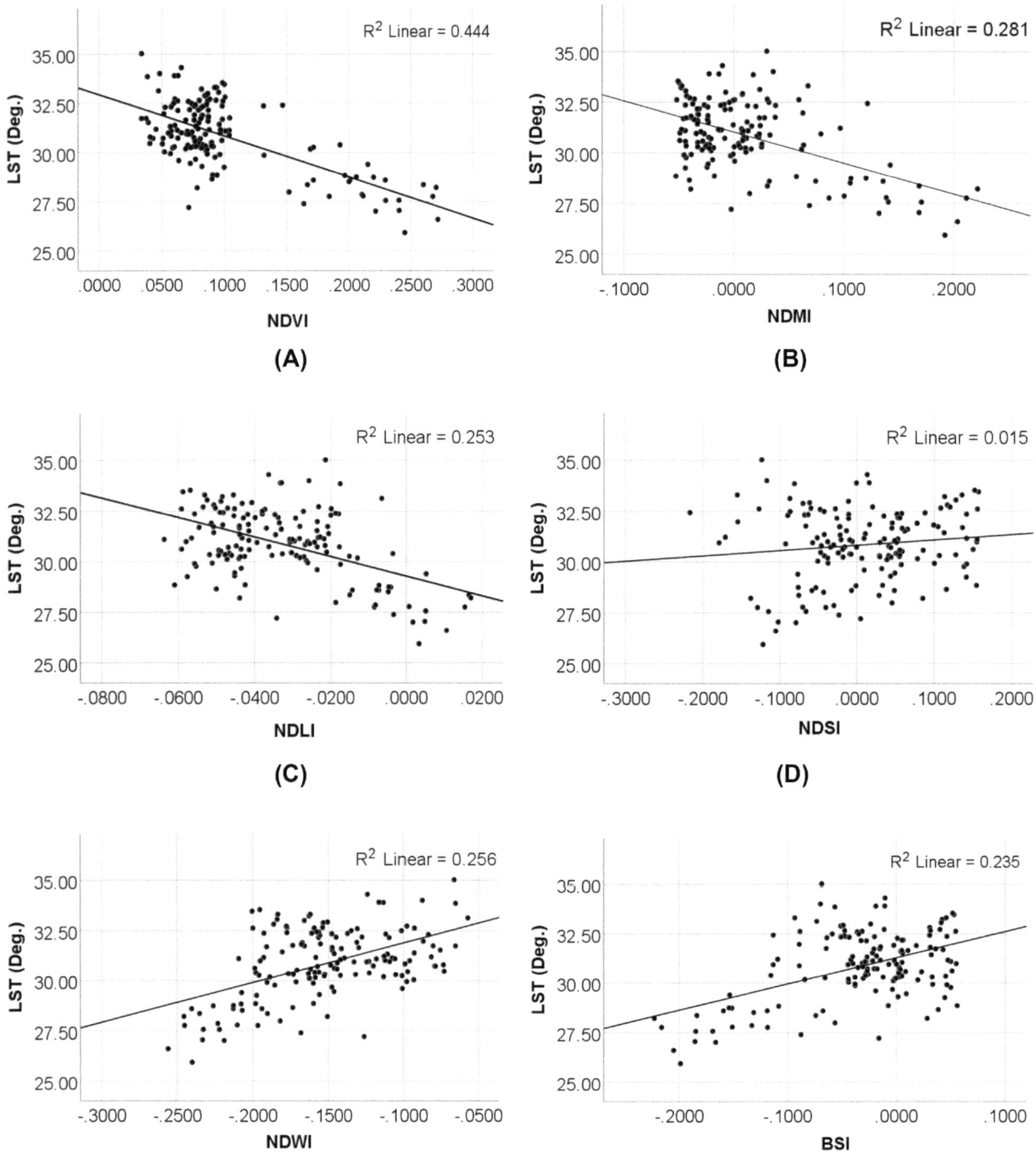

FIGURE 16.11 (A–F) Scatter plots of spectral indices with LST.

with SEBAL-derived latent heat flux with a correlation coefficient $r = 0.75$. They also determined that NDLI can be a worthy indicator to characterize the prospective latent heat flux in the earth's surface. The NDLI values are typically ranging from -1 to $+1$. The values near -1 indicate landforms with extreme drought, and $+1$ indicates extreme wet conditions of landforms. The variations of NDLI in the study area are shown in Fig. 16.7, and in the study area, it ranges from − 0.07 to + 0.05. The barren and desert lands have relatively low NDLI values, leading to high LST, whereas the oasis and dense date vegetation show relatively low NDLI values and lower surface temperatures.

16.4.1.4 Normalized difference moisture index

Normalized Difference Moisture Index (NDMI) denotes the amount of water content in the vegetated landforms. The ratio of spectral reflection of radiation between the NIR and SWIR regions measures it. The grouping of NNIR and SWIR eliminates the spectral variation caused by the internal plant structure of the green vegetation and the dry leaf matter contents, thereby increasing the accuracy of estimating the water content in the vegetation. The values are ranging from −1 to +1. The negative values of NDMI (values approaching −1) indicate the presence of barren soil. The values of nearly zero correspond to the areas of water stress. The high positive values of NDMI denote the high canopy regions without water stress. The NDMI of the study area is shown in Fig. 16.8. The NDMI in the study area changes from −0.22 to +0.32. The barren lands have less than 0 and negative values, whereas the vegetated lands in the oasis have more than +0.3, which shows the water contents of the dates vegetation in the oasis.

16.4.1.5 Normalized difference soil index

Normalized Difference Soil Index (NDSI) expressively characterizes the thermal behavior of the land surface. Bala et al. (2020) effectively used NDSI to map the surface temperature of Varanasi city of India. They emphasized that NDSI can be effectively used for mapping the LST in urban areas. They found a high correlation between LST with Normalized Difference Built-up Index and NDSI. Vibhute et al. (2018) used NDSI to understand the evaluation of soil conditions in Hulambri Taluka of Aurangabad district of Maharashtra, India, using spectral indices from hyperspectral datasets. The NDSI uses the spectral reflections of EM radiation in green and SWIR bands. The NDSI values normally range from −1 to +1, and for the study area, it ranges from −0.45 to +0.40, as shown in Fig. 16.9. The negative and lower values indicate the presence of waterbodies and the higher NDSI positive values represent the desert landforms with barren lands.

16.4.1.6 Normalized difference water index

The Normalized Difference Water Index (NDWI) is used to enhance the waterbodies in satellite data. It can also be used to study the presence of water content in green vegetation and landforms. The water bodies absorb a high amount of radiation from visible to IR of the spectrum. There are two NDWI that have been used, and one is to estimate the changes in the water content of vegetation, using NIR and SWIR wavelengths (Gao, 1996). Another is to find the changes related to the water content of water bodies using green and NIR wavelengths (McFeeters, 1996)—the present estimate of the water content in the date plantations and green vegetation in the oasis. Wu et al. (2009) used NDWI to predict the vegetation water content in wheat in China. Chen et al. (2003) used NDWI to estimate the vegetation water content for corn and soybeans. They stated that NDWI provides excellent results for estimating the water content during the corn and soybean vegetation's entire crop growing cycle. The NDWI values in the study areas ranged from −0.38 to +0.15, as shown in Fig. 16.10. The high values indicate the presence of water bodies. In contrast, the vegetation has lower values, which leads to distinguishing the presence of vegetation from waterbodies. The residential buildings and built-up areas have medium positive values between the vegetation and waterbodies.

16.4.2 Linear regression model for LST

Linear regression modeling has been used to predict the LST from the selected vegetation and soil spectral indices. It is a simple way to evaluate the dependence of a physical quantity on the associated functions or parameters. It can provide valuable information on making extensive modeling procedures. It can also eliminate the complications in making comprehensive modeling processes in multidimensional soil analysis. Multiple linear regression modeling is a preferred tool to evaluate salinity levels and is widely used for soil studies. In this study, multivariate regression modeling was carried out using the IBM SPSS statistical software program (Ver. 25). The association between the dependent and independent variables is analyzed by the 2D scatter plots and shown in Fig. 16.11A–F. Finally, the LST prediction modeling was established based on the relationship between spectral indices and measured LST values from the satellite data. The Model summary is shown in Table 16.3.

The values of R^2 provide the effect and dependency level of the LST and spectral indices values. The 2D-scatter plots of the soil and vegetation indices significantly correlate with the predicted LST values in the desert oasis region. Some vegetation and soil indices negatively correlate with the land surface's LST. It is also evident that including or excluding a dependent variable in LST modeling is relatively a tough task and varies with different regions and landforms. Royston and Sauerbrei (2008) emphasized that adding more variables in modeling may lead to over-fit and

TABLE 16.3 Regression model summary.

Model summary[b]

Model	R	R square	Adjusted R square	Std. error of the estimate	Change statistics				
					R square Change	F Change	df1	df2	Sig. F Change
1	0.725[a]	0.526	0.506	1.24	0.526	26.4	6	143	0.001

[a]*Predictors: (constant), NDVI, NDMI, NDLI, NDWI, BSI, NDSI*
[b]*Dependent variable: LST (Deg.)*

provide less accurate predictions. So, 2D scatter plots and correlation studies are valuable to identify the linearity of the dependent variables for the modeling process. Finally, the relationship between the independent and dependent variables has been established and given in Eq. (16.1):

$$\text{LST} = (486 \times \text{NDLI}) - (326 \times \text{NDVI}) - (313 \times \text{NDWI}) - (28 \times \text{NDMI}) - (16 \times \text{NDSI}) - (14 \times \text{BSI}) + 31.5 \quad (16.1)$$

The estimated LST values obtained from the modeling technique are studied with the satellite-based LST values and ground truth studies. The relevance of the dependent variables and the generated model's reliability are examined using error estimation and approximation. In addition to the validation process, the temporal and spatial variation LST values are analyzed with different soil types and landforms in the oasis. The LST map is also examined and interpreted per the existing studies.

Developing an empirical model for mapping the LST of a land surface using spectral indices is a relatively tricky method, since the different indices have a specific relationship with the LST. The spectral index has explicit characteristic associations with the LST of the particular landforms. Expressing a typical correlation with standard spectral characteristics appropriate to dissimilar landforms is also tricky. Location-specific and effective empirical models with categorical spectral indices are essential to precisely estimate the LST of the lands. The 2D-scatter plots of several spectral indices indicate substantial correlations with the LST (independent variable). However, including too many spectral indices may weaken the overall performance of the empirical modeling. The spectral indices with significant R^2 values are selected for modeling process. Further, the specific relationship among the variables is not significant for making the model. Statistical estimation is also carried out to assess their significance with the variable.

The regression model for estimating the LST is developed by using the selected spectral indices and the model summary is also shown in Table 16.3. The model provides an optimum R^2 value of 0.526 with an adjusted R^2 value of 0.506. The selected spectral indices by the model deliver an ideal significance level of 0.001. The designated dependent variables and the regression coefficients of the empirical model are given in Table 16.4. The spectral indices such as NDVI. NDSI, NDWI, NDLI, NDMI, and the BSI have optimum significance values of around 0.05. The Pearson correlation coefficients are shown in Table 16.5, which also indicates the dependency of the selected spectral indices for determining the LST. The normality of the empirical model has been analyzed using a normal probability plot (Fig. 16.12).

The standardized residual is compared with the normal distribution using a straight line in the normal P–P plots. The probability plot implies that the variable data sets are normally distributed. It also shows that most of the data points lie around the straight line, indicating no normality issues in the anticipated model. The standardized residual histogram (Fig. 16.13) also shows that the maximum data scores lie in the middle of the histogram and within the symmetrical curve. Figs. 16.14 and 16.15 show the 2D-scatter plot of measured and predicted LST in landforms, and the variation of measured and predicted LST in different landforms is also shown Figs. 16.16 and 16.17.

16.4.3 Error estimation and correlation analysis

In order to understand the effectiveness of the generated regression model, error estimation and correlation analysis between the measured and predicted values of LST has been carried out by estimating the mean bias error (MBE), root mean square error (RMSE), and the coefficient of determination (r^2) by using the following equations:

$$\text{MBE} = \frac{1}{n}\sum_{i=1}^{10}\frac{L_m - L_p}{L_p}$$

TABLE 16.4 Regression coefficients.

Coefficients[a]

Model		Unstandardized coefficients		Standardized coefficients	t	Sig.	95.0% confidence interval for B	
		B	Std. error	Beta			L. bound	U. bound
1	(Constant)	31.351	0.672		46.681	0.000	30.023	32.678
	NDVI	−325.775	120.247	−10.398	−2.709	0.008	−563.466	−88.083
	NDMI	−28.256	7.790	−0.974	−3.627	0.000	−43.654	−12.857
	NDLI	485.954	186.133	5.057	2.611	0.010	118.027	853.882
	NDWI	−312.800	114.428	−7.980	−2.734	0.007	−538.989	−86.612
	BSI	−14.266	30.053	−0.521	−0.475	0.636	−73.671	45.139
	NDSI	−15.926	14.390	−0.739	−1.107	0.270	−44.371	12.518

[a]*Dependent Variable: LST (Deg.)*

TABLE 16.5 Pearson correlation coefficients.

		LST	NDVI	NDMI	NDLI	NDWI	BSI	NDSI
Pearson correlation	LST	1.000	−0.666	−0.530	−0.503	0.505	0.485	0.122
	NDVI	−0.666	1.000	0.736	0.639	−0.840	−0.714	−0.200
	NDMI	−0.530	0.736	1.000	0.840	−0.348	−0.970	−0.773
	NDLI	−0.503	0.639	0.840	1.000	−0.119	−0.891	−0.658
	NDWI	0.505	−0.840	−0.348	−0.119	1.000	0.282	−0.223
	BSI	0.485	−0.714	−0.970	−0.891	0.282	1.000	0.818
	NDSI	0.122	−0.200	−0.773	−0.658	−0.223	0.818	1.000

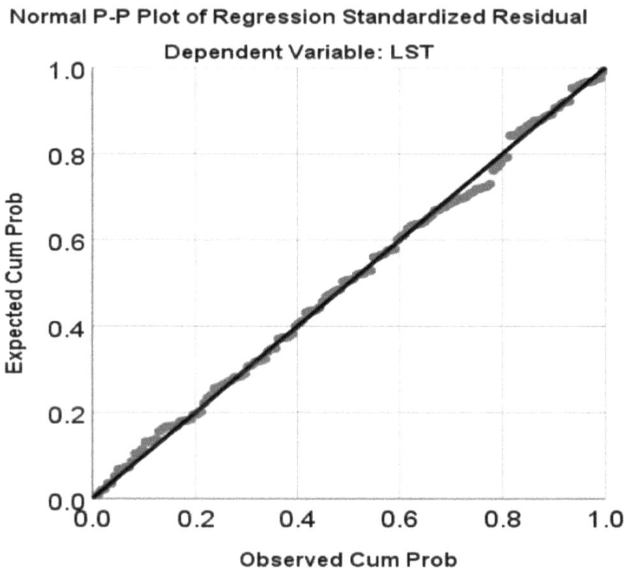

FIGURE 16.12 Normal P–P plot of standardized residual.

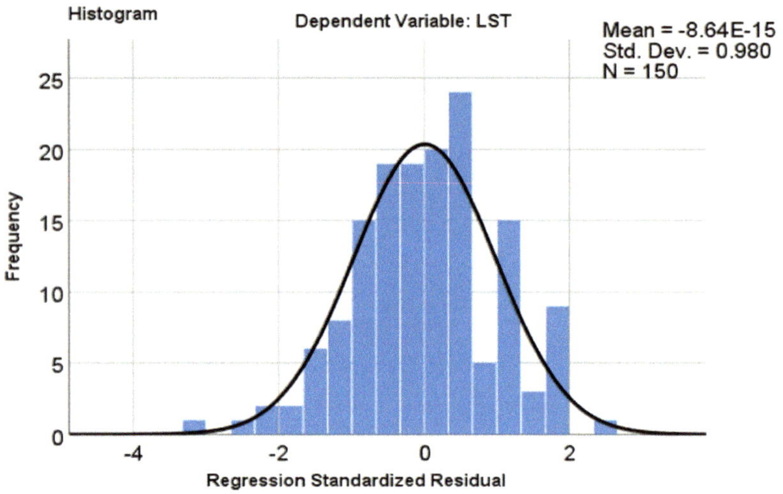

FIGURE 16.13 Frequency plot of the regression model.

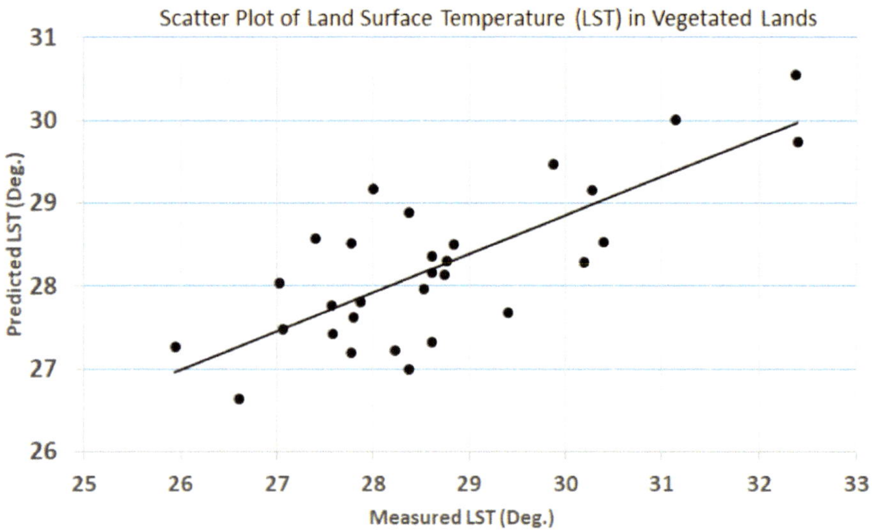

FIGURE 16.14 Scatter plot of LST in vegetated lands.

$$\text{RMSE} = \sqrt{\frac{1}{n}\sum_{i=1}^{10}\left(\frac{L_m - L_p}{L_i}\right)^2}$$

where L_m is the measured LST, and L_p is the predicted LST. The coefficient of determination (R^2) is calculated from sum of squared error and sum of squared totals by using the following equation:

$$\text{Coefficients of Determination} \quad R^2 = 1 - \frac{SSE}{SST}$$

The relationship between the measured and predicted value of the LST has also been analyzed using Pearson's correlation analysis. Correlation is a bivariate analysis that measures the strength of the association between two variables and the direction of the relationship. The correlation analysis is a technique of statistical assessment used to study the strength of a relationship between two numerically measured parameters (e.g., air temperature and LST). The most common correlation coefficient is the Pearson Correlation Coefficient used to test the linear relationships between data

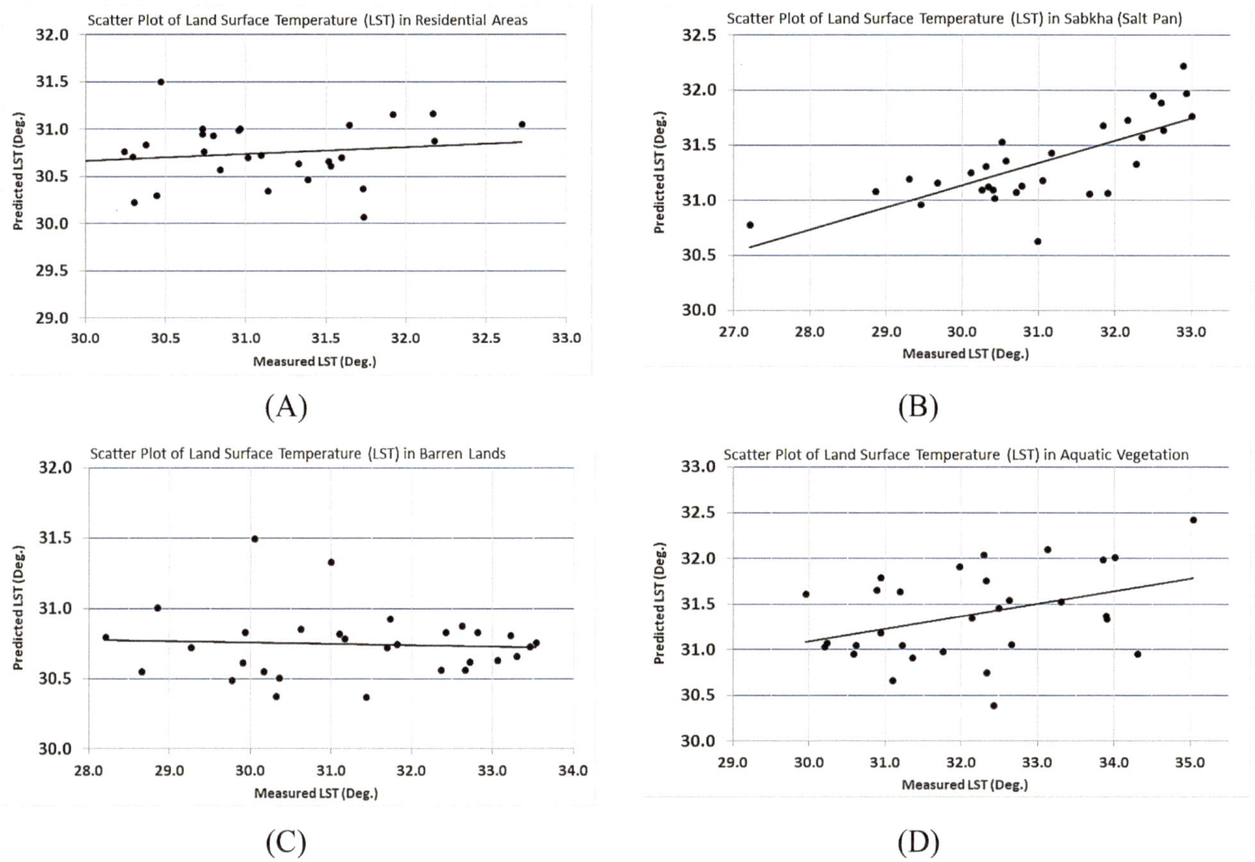

FIGURE 16.15 2D-scatter plot of measured and predicted LST in landforms.

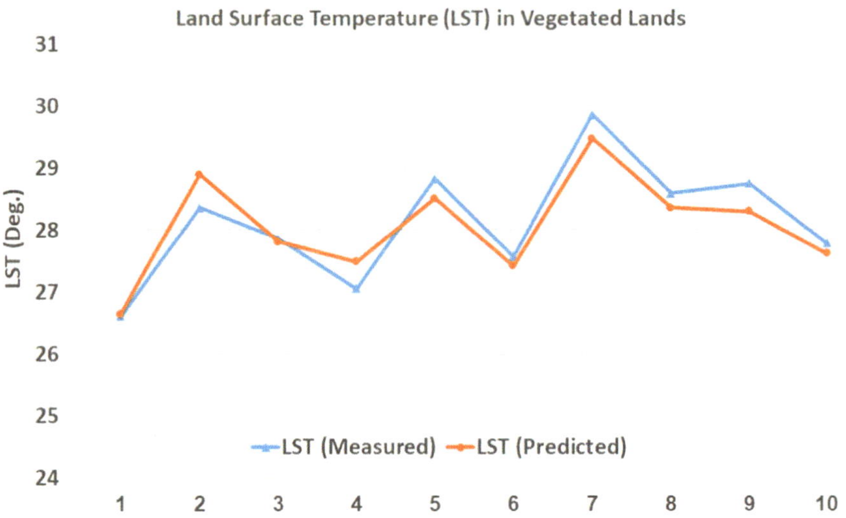

FIGURE 16.16 Measured LST vs predicted LST in vegetated lands.

and has been estimated by using the following equation to find the dependence of the two values in both space and time scale:

$$r = \frac{\sum_{i=1}^{n}(t_m - \overline{t_m})(t_p - \overline{t_p})}{\sqrt{\sum_{i=1}^{n}(t_m - \overline{t_m})^2 \sum_{i=1}^{n}(t_p - \overline{t_p})^2}}$$

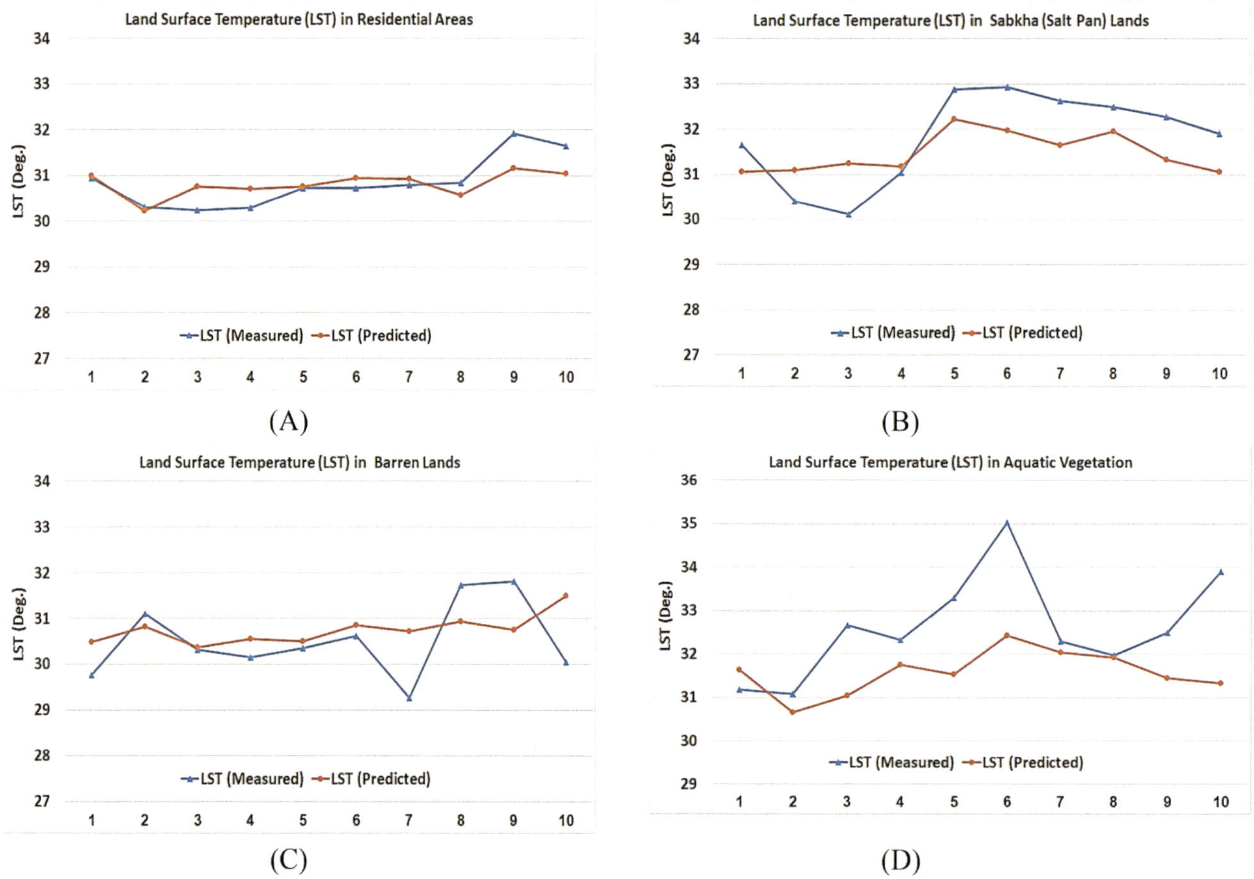

FIGURE 16.17 Variation of measured and predicted LST in different landforms.

TABLE 16.6 Error Estimation Data.

Statistical data	Mean bias error (MBE)	Root mean square error (RMSE)	Coefficients of determination (R^2)	Pearson's correlation coefficient
Value (Percentage)	0.47%	0.07%	0.09	0.89

where t_m is the measured LST, $\overline{t_m}$ is the average value, t_p is the predicted LST, and $\overline{t_p}$ is the average value. The statistical error estimation (Table 16.6) yields excellent outcomes, which vitally validated the regression modeling of LST by using the vegetation and spectral indices.

16.4.4 Spatial and temporal variations of LST

The spatial and temporal variation of the LST Al-Ahsa oasis is shown in Figs. 16.18–16.21. The study implies that during the peak winter month (January), most of the study areas have relatively very low LST ranging from 4°C to 20°C. The water bodies have only around 10°C–15°C. However, some of the land areas have very low temperatures of lower than 10°C and even less than the temperature of water bodies. The elevation of LST in the waterbodies may be attributed to the high specific heat capacity of water. The dense vegetation cover in the oasis does not show any significant reduction in the surface temperature. It may be endorsed that the lack of sunlight during the winter period limits the growth of green vegetation and causes elevated surface temperatures. The urban and dense residential areas in and

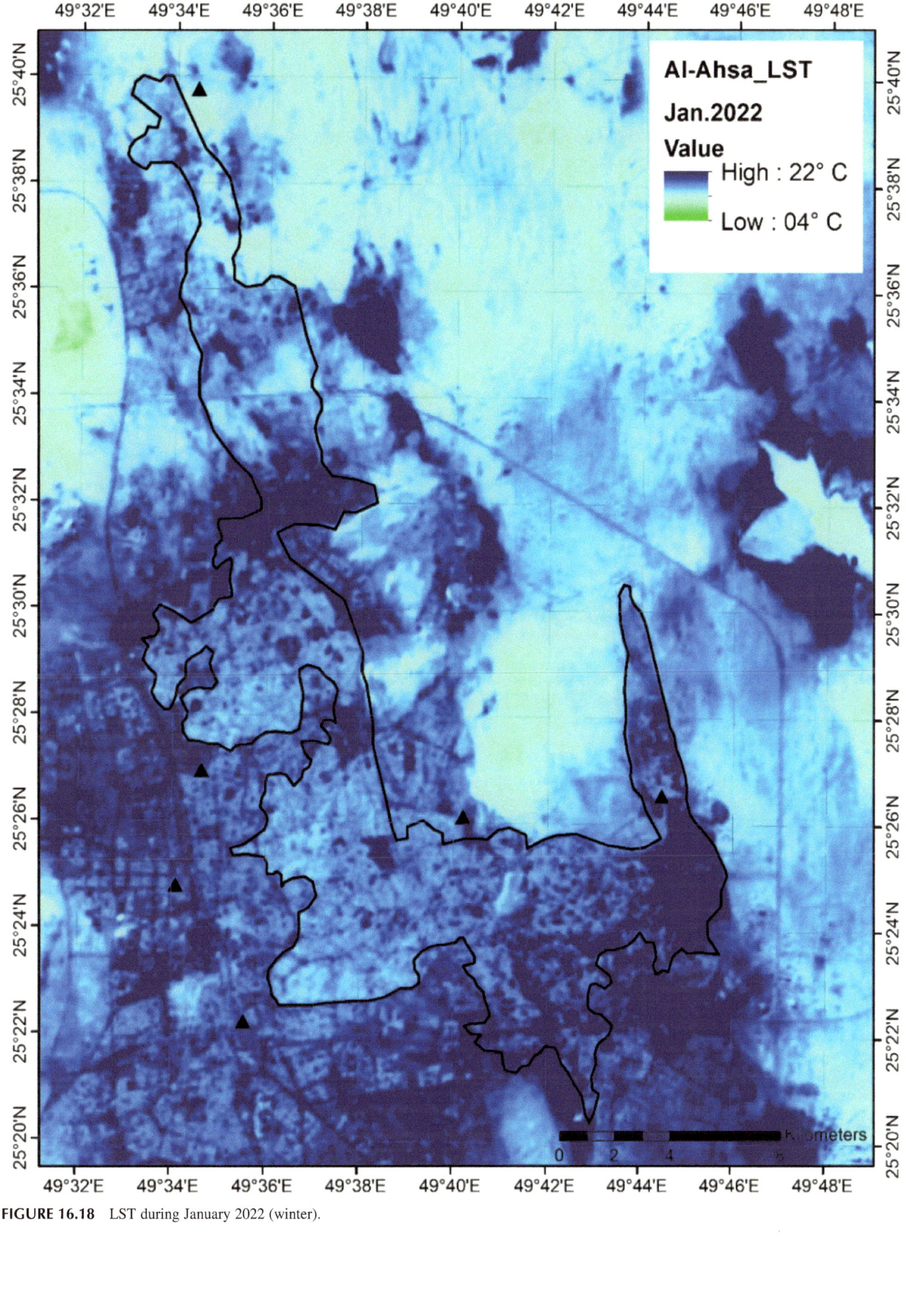

FIGURE 16.18 LST during January 2022 (winter).

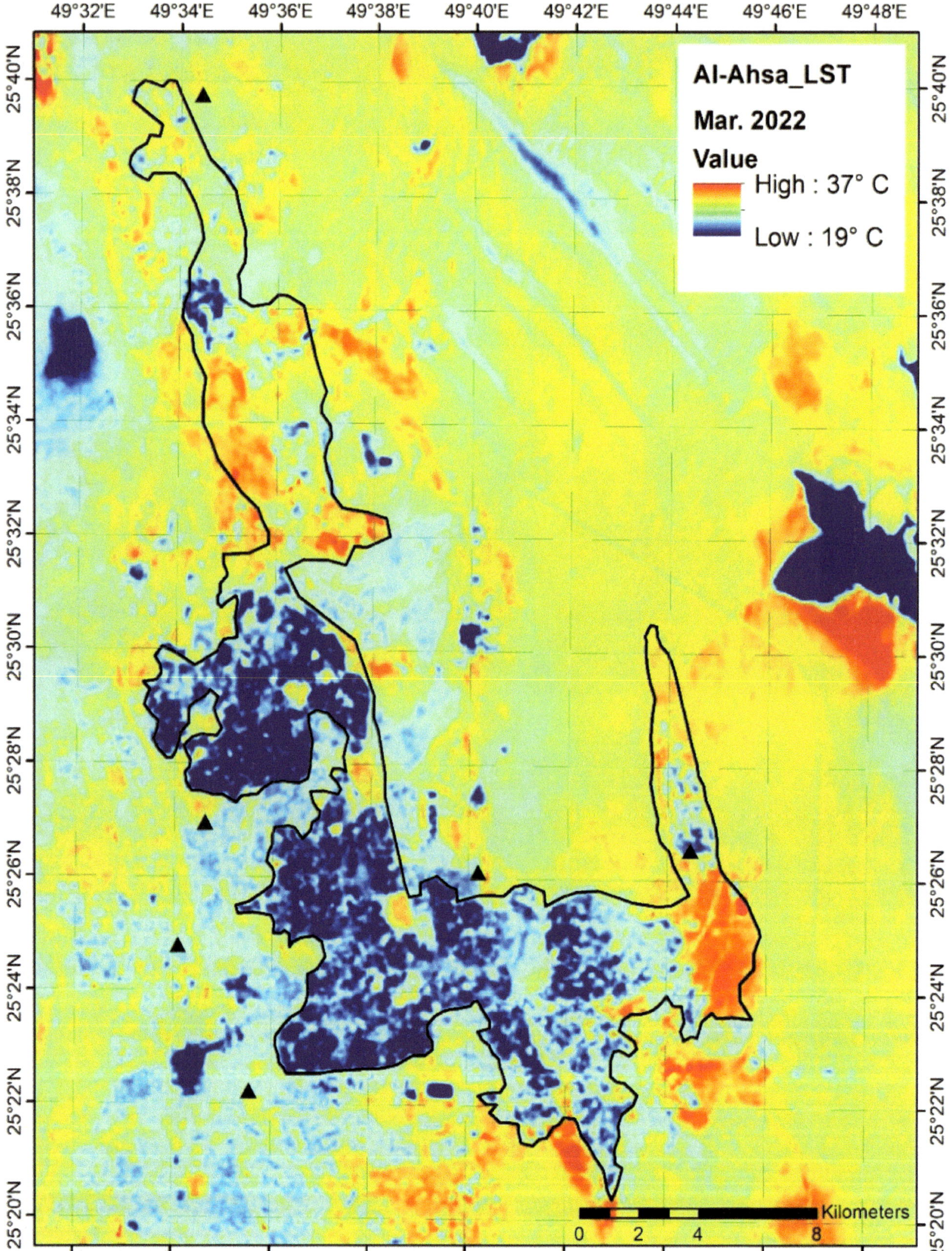

FIGURE 16.19 LST during March 2022.

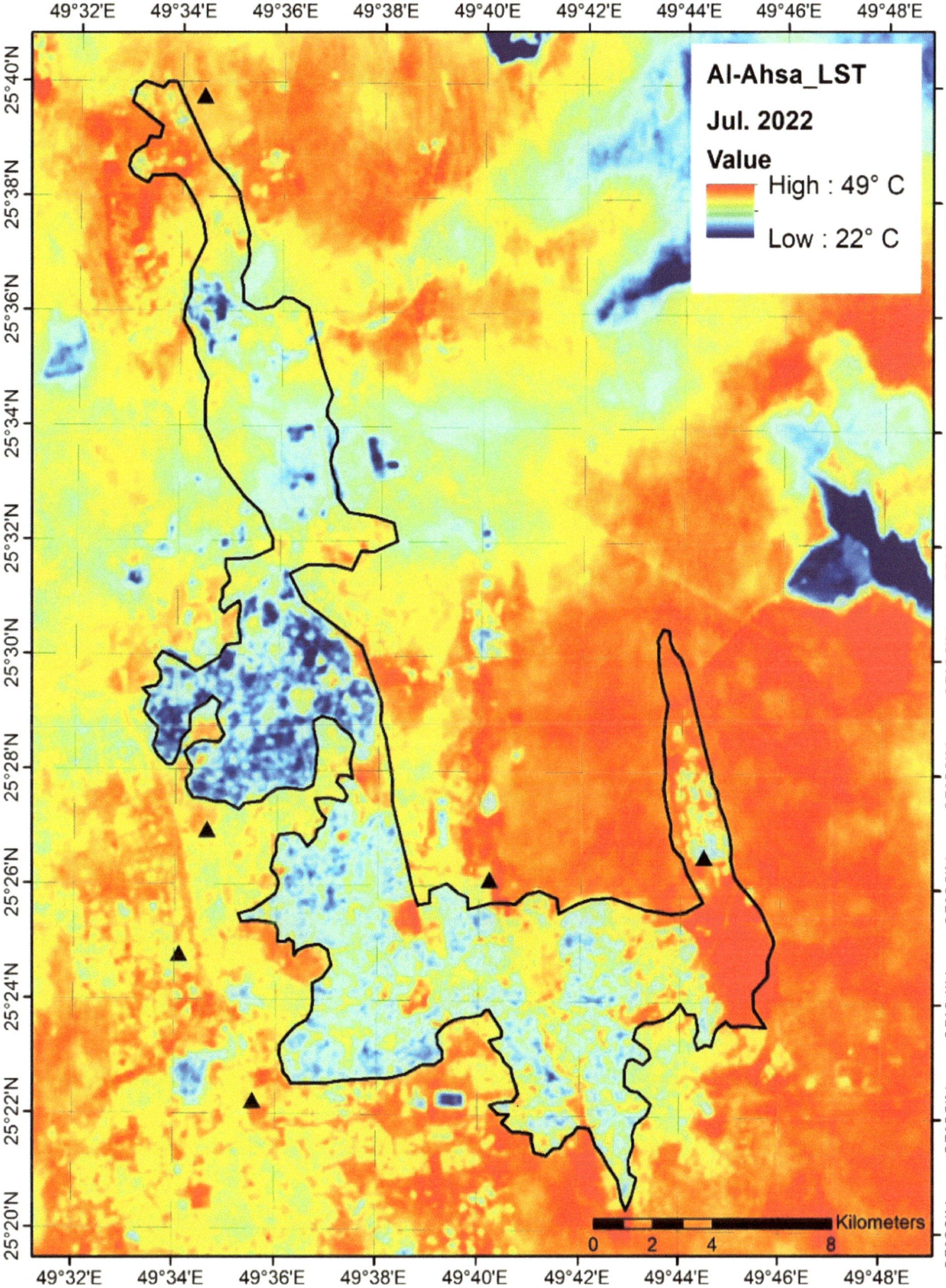

FIGURE 16.20 LST during July 2022 (summer).

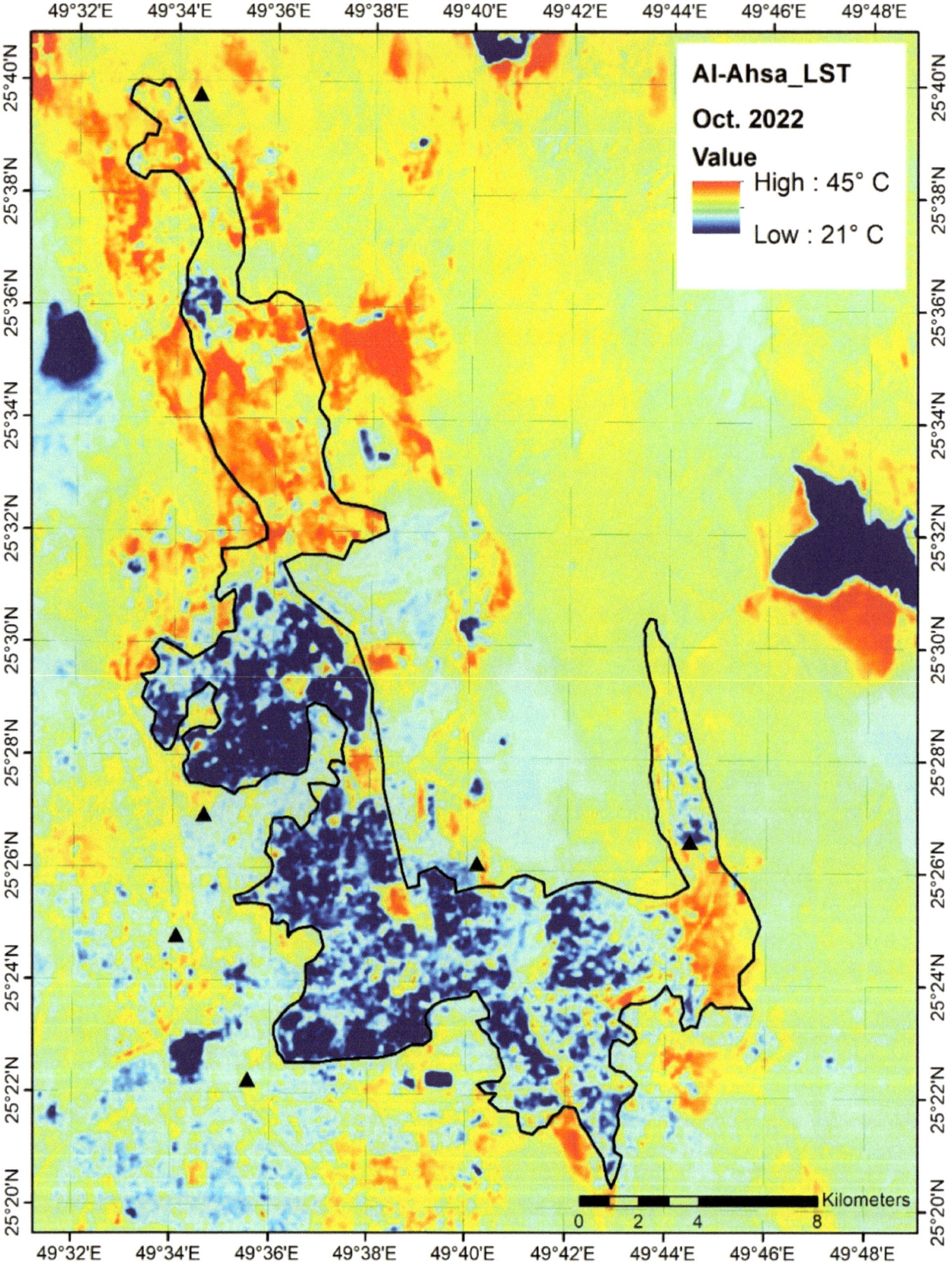

FIGURE 16.21 LST during October 2022.

around the oasis have high LST ranging from 15°C to 22°C. Dense buildings, roads, roofs, and blacktop pavements significantly influence and enhance the LST of the areas during the winter periods. The dense buildings in the residential areas block the cool winds during the peak winter periods and control the surface temperature.

During the prewinter and postsummer periods (March and October), the surface temperature in most of the study area is quite higher than the winter period and lower than that of peak summer. The dense vegetation cover has shown a significant reduction in the surface temperature. The optimum sunlight and weather enhance the growth of healthy green vegetation such as trees and grass might impact regulating the LST of the study area. Wan (1996) states that the temperature of vegetation canopies is usually lower than soil temperature, and surface temperature under shadows is lower than in sunshine areas. However, the urban areas have higher surface temperatures than the dense vegetation, ranging from 20°C to 30°C. The barren and Sabkha (salt marsh) have relatively higher temperatures of more than 30°C. During the peak summer season (July), most areas have become hot and advanced in LST. The water bodies have relatively low temperatures, which may be attributed to the high specific heat capacity. The surface temperature in the residential area of the city is high, around 35°C–40°C. Mujabar and Chandrasekar (2013) state that anthropogenic heat discharges increase the surface temperatures due to energy consumption, increased land surface coverage by artificial materials having high heat capacities and conductivities and the associated decrease in vegetation cover. The temperature in the suburban areas are moderate; however, high temperature of more than 45°C have been found in the barren lands and desert areas of the study area.

16.4.5 Effect of landforms in LST

The study implies that different landforms and soil types significantly impact the LST of the study area. The LST in the city's residential areas is quite high in almost all study periods. The oasis region with date plantation and green vegetated lands shows a substantial decrease in LST during all periods except the winter. The NDVI during this period is moderately high and negatively correlates with the LST. However, the oasis and vegetated landforms during the peak winter period show no significant reduction in LST. The LST in most of oasis is still higher than the residential areas even though they have higher NDVI values than the residential areas. The high NDVI and high LST in the vegetated landform during the winter periods lead to a positive correlation. Karnieli et al.(2019) also stated that the LST–NDVI correlation is negative in regions where water is the limiting factor for vegetation, and it becomes positive where energy is the limiting factor.

During July, the surface temperature in the barren lands and desert areas is more than 45°C. The urban areas also substantially increase the surface temperature, ranging from 35 to 40°C during summer. The urban areas have dense commercial buildings, and the land is fully covered by concrete and tar materials, which may enhance the LST of the residential areas. The coverage of dense residential buildings is backed crucially by the high LST. It is also noticed that the residential areas have LST lower than the surrounding barren lands and desert areas of the study area. Several studies show that the modification of land cover enhances the amount of absorption of solar radiation, evaporation rates, thermal storage of surfaces, and wind turbulence. Rajendran and Mani (2015) imply that the development of urbanization is related to transforming arable lands into concreted surfaces. They also state that urban growth and unsustainable land management practices greatly impact a region's local climate. However, from the present study in the study area, it is observed that the urbanization around the oasis substantially reduces the LST than the barren and desert areas. It is also found that there are no UHIs in the residential areas of the study area. It may be related to developing green vegetation and date plantations in the residential areas. Blooshi et al. (2020) have also noticed a significant reduction in the LST of residential areas of the UAE previously desert areas, due to the substantial growth of green vegetation and date plantations during the city development and urbanization process.

The elevation of the study area has been characterized by the Digital Elevation Model (DEM) shown in Fig. 16.3. The DEM shows that the vegetated landforms in the Al-Ahsa oasis and the residential areas are at higher elevation than the barren and desert lands in the northeastern part of the study area. The present study on the variation of LST shows that the LST has a strong relationship with the elevation of the lands. Khandelwal et al. (2018) also indicate a significant linear association between the elevation and LST and elevation of the Jaipur city of India. The LST shows a falling trend as the increase in elevation of the land. During winter, the LST in the vegetated and residential areas is higher than in the desert lands, as shown in Fig. 16.11. The surface temperature in the barren and desert lands in the northeastern part is much lower than that of the vegetated lands. A 10°C–15°C decrease has been observed in the desert lands.

Some regions of the desert reach low temperatures, even lower temperatures below the freezing point. Lack of sunlight and low humidity have been involved in lowering the surface temperature of the lands during the winter periods. It is also evident that the sands are good absorbers and radiators of heat energy. The soil in the desert lands acts as an effective heat distributor and quickly releases the energy back into the air. The heat from the earth easily escapes to the atmosphere due to the lack of clouds in the desert areas. This makes the desert and barren landforms much lower than the other landforms. However, during the summer season, the reverse is happening; that is, the barren and desert lands are much higher LST than the vegetated landforms, as shown in Fig. 16.13. During the summer periods, the sunlight incidents on the earth's surface at almost right angles. Large amounts of direct sunlight incident on the desert sands due to lack of clouds. This high amount of radiation heats the earth's surface and increases the LST.

The landforms' type significantly influences altering the LST of the study area. The date plantation areas and the vegetated landforms in the Al-Ahsa oasis show relatively lesser LST than the rest of the areas (Figs. 16.12 and 16.14). The land surfaces with soil, plants, and water resources have good absorption of humidity, which controls the LST considerably along the study area. Salisbury and D'Aria (1994) also state that multiple scattering and reflection of light within the surface vegetation will dramatically reduce the spectral contrast and that typical trees, greeneries, and plants have emissivities nearly equal to 1. So many incident light radiations were emitted back into the atmosphere, leading to low LST in most study periods. The urban and residential lands in the study area have experienced higher LST than the vegetated landforms. In these residential and urban lands, the surfaces are altered to tar and concrete due to construction and developmental plans, which increases heat absorption. These solid surfaces in the city will emit the gathered heat energy into the atmosphere more than other natural earth surfaces. These solid surfaces will also increase moisture evaporation more than natural surfaces, with a better property of moisture absorption.

16.5 Conclusion

The thermal properties of soil and atmospheric conditions greatly influence the surface heat energy balance and significantly modify the LST of a region. The microscopic changes on LST in many areas of the ecosphere have led to earth's climate change over the last centuries. These deviations happen due to both natural phenomena and anthropogenic activities. However, the vegetation cover, type of landforms, and soils have significantly altered the LST, which impacts a region's heat energy balance, especially in agricultural locations. So, the present studies on vegetation and soil spectral indices' effect on LST are very useful for understanding the surface heat energy balance, thermal flux, urban heat transfer models, water resource management, climate change modeling, and environmental studies. The study validates that the surface temperature is a function of landforms, vegetation covers, and soil. Even though there are no heat islands in the region's residential areas, the landforms influence the LST. Moreover, the present study recognized the effective use of Landsat 9 satellite data for spatial and temporal mapping of LST and their relations with the spectral indices. The study also supports understanding and legalese the environmental effects of agricultural landforms and soil on LST due to industrialization and urbanization.

Acknowledgments

The author is thankful to the Director and Deputy Directors of Jubail Industrial College, Jubail Industrial City, for their kind encouragement in the accomplishment of the research work. The author is also thankful to the Chairman, Deputy Chairman, and Course Director of the General Studies Department for their kind suggestions and effective support.

Funding

This research work did not receive any specific grant from funding agencies in the public, commercial, or not-for-profit sectors.

References

Abderrahman, W., 1988. Water management plan for the Al-Hassa irrigation and drainage project in Saudi Arabia. Agric. Water Manag. 13, 185–194. Available from: https://doi.org/10.1016/0378-3774(88)90153-9.

Akhoondzadeh, M., Saradjian, M.R., 2008. Comparison of land surface temperature mapping using MODIS and ASTER images in semi-arid area. Int. Arch. Photogramm. Remote. Sens. Spat. Inf. Sci.—ISPRS Arch 37, 873–876.

Al-Barrak, S., Al-Badawi, M., 1988. Properties of some salt affected soils in Al-Ahsa, Saudi Arabia. Arid. Soil. Res. Rehabil. 2, 85–95. Available from: https://doi.org/10.1080/15324988809381162.

Allbed, A., Kumar, L., Aldakheel, Y.Y., 2014. Assessing soil salinity using soil salinity and vegetation indices derived from IKONOS high-spatial resolution imageries: applications in a date palm dominated region. Geoderma 230–231, 1–8. Available from: https://doi.org/10.1016/j.geoderma.2014.03.025.

Alqarni, S., Babiker, A., Salih, A., 2018. Detection, mapping and assessment change in urban and croplands area in Al-Hassa Oasis, Eastern Region in Saudi Arabia using remote sensing and geographic information system. J. Geogr. Inf. Syst. 10, 659–685. Available from: https://doi.org/10.4236/jgis.2018.106034.

Al-Saeedi, A.H., 2022. Characterizing physical and hydraulic properties of soils in Al-Ahsa, Kingdom of Saudi Arabia. Saudi J. Biol. Sci. 29, 3390–3402. Available from: https://doi.org/10.1016/j.sjbs.2022.01.061.

Al-Zara, I., 2011. Elemental composition of groundwater and spring waters in Al-Ahsa Oasis, Eastern Region Saudi Arabia. Trends Appl. Sci. Res. 6, 1–18. Available from: https://doi.org/10.3923/tasr.2011.1.18.

Bala, R., Prasad, R., Yadav, V.P., 2020. Thermal sharpening of MODIS land surface temperature using statistical downscaling technique in urban areas. Theor. Appl. Climatol. 141, 935–946. Available from: https://doi.org/10.1007/s00704-020-03253-w.

Becker, F., Zhao-Liang Li, 1995. Surface temperature and emissivity at various scales: definition, measurement and related problems. Remote. Sens. Rev. 12, 225–253. Available from: https://doi.org/10.1080/02757259509532286.

Betts, A.K., Ball, J.H., Beljaars, A.C.M., et al., 1996. The land surface-atmosphere interaction: a review based on observational and global modeling perspectives. J. Geophys. Res. Atmos. 101, 7209–7225. Available from: https://doi.org/10.1029/95JD02135.

Blooshi, L.S., Al Abuelgasim, A., Nassar, A., Ksiksi, T., 2020. Impact of desert urbanization on urban heat islands effect. Open. J. Geol. 10, 760–770. Available from: https://doi.org/10.4236/ojg.2020.107034.

Carlson, T.N., Ripley, D.A., 1997. On the relation between NDVI, fractional vegetation cover, and leaf area index. Remote. Sens. Env. 62, 241–252. Available from: https://doi.org/10.1016/S0034-4257(97)00104-1.

Chen D., Jackson T.J., Li F., et al., 2003. Estimation of vegetation water content for corn and soybeans with a normalized difference water index (NDWI) using landsat thematic mapper data. In: International Geoscience and Remote Sensing Symposium (IGARSS). pp. 2853–2856.

Coll, C., Caselles, V., 1997. A split-window algorithm for land surface temperature from advanced very high resolution radiometer data: validation and algorithm comparison. J. Geophys. Res. Atmos. 102, 16697–16713. Available from: https://doi.org/10.1029/97jd00929.

Deng, Y., Wu, C., Li, M., Chen, R., 2015. RNDSI: a ratio normalized difference soil index for remote sensing of urban/suburban environments. Int. J. Appl. Earth Obs. Geoinf. 39, 40–48. Available from: https://doi.org/10.1016/j.jag.2015.02.010.

Diek, S., Fornallaz, F., Schaepman, M.E., de Jong, R., 2017. Barest pixel composite for agricultural areas using landsat time series. Remote. Sens. 9. Available from: https://doi.org/10.3390/rs9121245.

El-Nahry, A.H., Rashash, A., 2013. Impact of Industrial areas on surface temperature using thermal infrared remote sensing and GIS techniques. A case study of Jubail City, KSA. In: 8th Natl GISsymposium, Saudi Arabia.

Flaash, 2008. FLAASH module user's guide. Solutions .

Gao, B.C., 1996. NDWI—a normalized difference water index for remote sensing of vegetation liquid water from space. Remote. Sens. Env. 58, 257–266. Available from: https://doi.org/10.1016/S0034-4257(96)00067-3.

Guha, S., Govil, H., Diwan, P., 2020. Monitoring LST-NDVI relationship using premonsoon landsat datasets. https://doi.org/10.1155/2020/4539684.

Gutman, G., Ignatov, A., 1998. The derivation of the green vegetation fraction from NOAA/AVHRR data for use in numerical weather prediction models. Int. J. Remote. Sens. 19, 1533–1543. Available from: https://doi.org/10.1080/014311698215333.

Hall, F.G., Huemmrich, K.F., Goetz, S.J., et al., 1992. Satellite remote sensing of surface energy balance: success, failures, and unresolved issues in FIFE. J. Geophys. Res. 97. Available from: https://doi.org/10.1029/92JD02189.

Jia, L., Menenti, M., Su, Z., et al. Modeling sensible heat flux using estimates of soil and vegetation temperatures: the HEIFE and IMGRASS experiments. Springer.

Kalma, J.D., McVicar, T.R., McCabe, M.F., 2008. Estimating land surface evaporation: a review of methods using remotely sensed surface temperature data. Surv. Geophys. 29, 421–469. Available from: https://doi.org/10.1007/s10712-008-9037-z.

Karnieli, A., Ohana-Levi, N., Silver, M., et al., 2019. Spatial and seasonal patterns in vegetation growth-limiting factors over Europe. Remote. Sens. 11, 2406. Available from: https://doi.org/10.3390/rs11202406.

Kaufman, Y.J., Tanrc, D., Gordon, H.R., et al., 1997. Passive remote sensing of tropospheric aerosol and atmospheric correction for the aerosol effect. J. Geophys. Res. 102, 815–831.

Khandelwal, S., Goyal, R., Kaul, N., Mathew, A., 2018. Assessment of land surface temperature variation due to change in elevation of area surrounding Jaipur, India. Egypt. J. Remote. Sens. Sp. Sci. 21, 87–94. Available from: https://doi.org/10.1016/j.ejrs.2017.01.005.

Kustas, W., Anderson, M., 2009. Advances in thermal infrared remote sensing for land surface modeling. Agric. For. Meteorol. 149, 2071–2081. Available from: https://doi.org/10.1016/j.agrformet.2009.05.016.

Li, Z.L., Tang, B.H., Wu, H., et al., 2013. Satellite-derived land surface temperature: current status and perspectives. Remote. Sens. Environ. 131, 14–37. Available from: https://doi.org/10.1016/j.rse.2012.12.008.

Liou, Y.A., Le, M.S., Chien, H., 2019. Normalized difference latent heat index for remote sensing of land surface energy fluxes. IEEE Trans. Geosci. Remote. Sens. 57, 1423–1433. Available from: https://doi.org/10.1109/TGRS.2018.2866555.

McFeeters, S.K., 1996. The use of the Normalized Difference Water Index (NDWI) in the delineation of open water features. Int. J. Remote. Sens. 17, 1425–1432. Available from: https://doi.org/10.1080/01431169608948714.

Moisa, M.B., Gemeda, D.O., 2022. Assessment of urban thermal field variance index and thermal comfort level of Addis Ababa metropolitan city, Ethiopia. Heliyon 8, e10185. Available from: https://doi.org/10.1016/j.heliyon.2022.e10185.

Moran, M.S., Rahman, A.F., Washburne, J.C., et al., 1996. Combining the Penman-Monteith equation with measurements of surface temperature and reflectance to estimate evaporation rates of semiarid grassland. Agric. For. Meteorol. 80, 87–109. Available from: https://doi.org/10.1016/0168-1923(95)02292-9.

Mujabar, P.S., Chandrasekar, N., 2013. Beach topography and morphodynamics along the Southern Coastal Tamil of India by using beach profile analysis. Mar. Hydrophys. J. 46, 35–62.

Quattrochi, D.A., Luvall, J.C, 2004. Thermal Remote Sensing in Land Surface Processing. CRC Press, Boca Raton, FL, USA.

Rajendran, P., Mani, D.K., 2015. Estimation of spatial variability of land surface temperature using Landsat 8 Imagery. Int. J. Eng. Sci. 4, 19–23.

Rouse, J.W., 1973. Monitoring the vernal advancement and retrogradation of natural vegetation [NASA/GSFCT Type II Report]. Greenbelt, MD NASA/Goddard Sp Flight Cent.

Royston, P., Sauerbrei, W., 2008. Continuous-by-continuous interactions. Multivariable Model-Building - A Pragmatic Approach to Regression Anaylsis based on Fractional Polynomials for Modelling Continuous: A Pragmatic 172–180.

Salih, A., 2018. Classification and mapping of land cover types and attributes in Al-Ahsaa Oasis, Eastern Region, Saudi Arabia using Landsat-7 data. J. Remote. Sens. GIS 07. Available from: https://doi.org/10.4172/2469-4134.1000228.

Salisbury, J.W., D'Aria, D.M., 1994. Emissivity of terrestrial materials in the 3–5 μm atmospheric window. Remote. Sens. Env. 47, 345–361. Available from: https://doi.org/10.1016/0034-4257(94)90102-3.

Schmugge, T.J., Becker, F., Li, Z.L., 1991. Spectral emissivity variations observed in airborne surface temperature measurements. Remote. Sens. Env. 35, 95–104. Available from: https://doi.org/10.1016/0034-4257(91)90003-O.

Sobrino, J.A., Jiménez-Muñoz, J.C., Sòria, G., et al., 2008. Land surface emissivity retrieval from different VNIR and TIR sensors. IEEE Trans. Geosci. Remote. Sens. 46, 316–327. Available from: https://doi.org/10.1109/TGRS.2007.904834.

Truong Linh, N., Quang Huy, T., Jungwon, H., Dongyeob, H., 2015. Land surface temperatures of industrial complexes in Jeonnam using Landsat 7 ETM + satellite images. J. KRSA 31, 99–112.

Valor, E., Caselles, V., 1996. Mapping land surface emissivity from NDVI: application to European, African, and South American areas. Remote. Sens. Env. 57, 167–184. Available from: https://doi.org/10.1016/0034-4257(96)00039-9.

Vibhute, A.D., Dhumal, R., Nagne, A., et al., 2018. Evaluation of soil conditions using spectral indices from hyperspectral datasets, Proc—2017 2nd Int Conf Man Mach Interfacing, 2017. MAMI, pp. 1–62018-March. Available from: https://doi.org/10.1109/MAMI.2017.8308008.

Vlassova, L., Perez-Cabello, F., Nieto, H., et al., 2014. Assessment of methods for land surface temperature retrieval from landsat-5 TM images applicable to multiscale tree-grass ecosystem modeling. Remote. Sens. 6, 4345–4368. Available from: https://doi.org/10.3390/rs6054345.

Wan, Z., 1996. A generalized split-window algorithm for retrieving land-surface temperature from space. IEEE Trans. Geosci. Remote. Sens. 34, 892–905. Available from: https://doi.org/10.1109/36.508406.

Weng, Q., 2009. Thermal infrared remote sensing for urban climate and environmental studies: methods, applications, and trends. ISPRS J. Photogramm. Remote. Sens. 64, 335–344. Available from: https://doi.org/10.1016/j.isprsjprs.2009.03.007.

Wu, C., Niu, Z., Tang, Q., Huang, W., 2009. Predicting vegetation water content in wheat using normalized difference water indices derived from ground measurements. J. Plant. Res. 122, 317–326. Available from: https://doi.org/10.1007/s10265-009-0215-y.

Yu, X., Guo, X., Wu, Z., 2014. Land surface temperature retrieval from landsat 8 TIRS-comparison between radiative transfer equation-based method, split window algorithm and single channel method. Remote. Sens. 6, 9829–9852. Available from: https://doi.org/10.3390/rs6109829.

Zareie, S., Khosravi, H., Nasiri, A., Dastorani, M., 2016. Using Landsat Thematic Mapper (TM) sensor to detect change in land surface temperature in relation to land use change in Yazd, Iran. Solid. Earth 7, 1551–1564. Available from: https://doi.org/10.5194/se-7-1551-2016.

Chapter 17

Application of remote sensing in terrestrial soil organic carbon determination: a review

Kaushik Saha[1], Udayan Rudra Bhowmick[2], K.S. Anil Kumar[3], K.S. Karthika[3], Parinita Das[4] and M. Lalitha[3]

[1]*Department of Soil Science and Agriculture Chemistry, University of Agricultural Sciences, GKVK, Bangalore, Karnataka, India,* [2]*Department of Agronomy, Bidhan Chandra Krishi Viswavidyalaya, Mohanpur, Nadia, West Bengal, India,* [3]*ICAR-National Bureau of Soil Survey and Land Use Planning, Regional Centre, Bangalore, Karnataka, India,* [4]*Department of Agriculture Bioinfromatics, ICAR-Indian Agricultural Statistics Research Institute, New Delhi, India*

17.1 Introduction

Carbon sequestration may be able to lessen climate change, but carbon storage (C-storage) in soils has drawn more interest on a global scale (Crowther et al., 2016). The UN's Sustainable Development Goals (SDGs), United Nations Framework Convention on Climate Change (COP21), and FAO's Global assessment of carbon sequestration by soils (e.g., the GSOCseq programme) (GSOCMap, 2022), highlighted the "4 per 1000 Initiative" by emphasizing the importance of C-storage in soils (Lal et al., 2021). The largest sink of organic carbon on Earth's surface is soil. Soil organic carbon (SOC) pools have a significant impact on the health of terrestrial ecosystems by changing the characteristics and quality of soil (Schillaci et al., 2017). Terrestrial carbon stocks act as a living embodiment of the global carbon cycle as well as the Earth's climate since they not only "breathe" in carbon dioxide (CO_2) *via* photosynthesis but also release carbon (C) into the atmosphere through respiration (Schimel, 1995).

Soil is highly complex and diverse, with many components. Quantifying and evaluating these are difficult, and traditional approaches are highly time-consuming (Gehl and Rice, 2007). Remote sensing (RS) is comparatively more rapid, cost-effective, and nondestructive than traditional laboratory methods (Vaudour et al., 2016, Angelopoulou et al., 2019). The Earth's land surface has been observed by a number of airborne and satellite sensors over the past 50 years, and numerous methods have been developed to quantify C fluxes and stocks using remotely sensed data. The SOC content varies depending on diverse soil types, vegetation, and land use change concerns; hence, precise determination requires consistent and accurate datasets, which is both challenging and expensive (Omran, 2017). RS has been widely used in measuring ecosystem C fluxes and stocks (Xiao et al., 2014). Evaluation of SOC using RS systems and mapping could be based on data from visible near infrared-shortwave infrared (VNIR−SWIR, 400−2500 nm) based on the principles of energy-matter interaction (Schwartz et al., 2012) or on the number of informative bands in VNIR−SWIR. The multispectral and hyperspectral imaginary provide effective input data to map the spatial variability of SOC using different platforms like aerial (Castaldi et al., 2018; Denis et al., 2014), satellite (Mondal et al., 2017), and unmanned aerial systems (UAS) (Wehrhan and Sommer, 2021).

Despite the fact that hyperspectral RS is expensive and technically challenging, research involving multispectral data has depicted satisfactory precision with cost-effective acquisition (Ladoni et al., 2010; Angelopoulou et al., 2019). Consequently, the RS community is continuously approaching deep learning (DL) and machine learning (ML) techniques to optimize the use and output of RS data (Masemola et al., 2020). Using RS-derived data, FAO, IUSS, and global soil organic carbon map (GSOCMap) to employ earth observation (EO) imagery to create global maps of SOC (GSOCMap, 2022). The necessity for monitoring systems that can track and update SOC status and change estimates at scales ranging from local to global is highlighted by the growing awareness of soil carbon as a vital component in policy and action on climate change as well as for ecosystem health.

Remote Sensing of Soils. DOI: https://doi.org/10.1016/B978-0-443-18773-5.00004-1

Here, we present a thorough analysis of the accomplishments, difficulties, and opportunities of RS science and technology in estimating terrestrial SOC, including belowground and wetlands soil carbon stocks with changes in C fluxes and aboveground biomass (AGB) C sources, based on broad wavelength region (visible, infrared, and microwave) on the electromagnetic spectrum, platforms, and data-driven (or ML) approaches.

17.2 Terrestrial soil C cycle and SOC stock

The process of storing carbon (C) in various reservoirs in a liquid or solid state, such as plants and soil (terrestrial sequestration), underground pools (geological sequestration), and deep oceans (ocean sequestration or wet carbon), is known as carbon sequestration (Issa et al., 2020). According to Lal (1997), the pedosphere contributes significantly to the global carbon cycle, which is primarily made up of three types of carbon: (1) elemental (incomplete combustion products of organic matter (OM) and from geologic sources); (2) inorganic (mostly found in carbonate minerals like calcium carbonate ($CaCO_3$) and dolomite [$CaMg(CO_3)_2$]); and (3) organic (Nieder and Benbi, 2008). The term "C cycle" refers to the movement of these three types of carbon between Earth's reservoirs through biogeochemical activities such as photosynthesis, respiration, combustion, burial of organic matter, decomposition, and weathering processes (West, 2008). The first meters of mineral soils contain enormous amounts of carbon, between 1500 and 2400 Pg of organic carbon (Stockmann et al., 2013), which is about three to four times the amount of C in vegetation (450–650 PgC) and two to three times the amount in the atmosphere (~829 GtC). Additionally, peat soils and permafrost account for more than 1500 Pg of carbon, with peat soils being the densest terrestrial carbon deposit in the world, with 500 Gt of carbon in only 3% of the world's land surface (Joosten and Couwenberg, 2008).

The worldwide C pools and fluxes are greatly altered by anthropogenic activities (Bolstad and Vose, 2005), and the SOC pool may greatly impact the amount of CO_2 in the atmosphere (Crowther et al., 2016). C-storage in soils actually results from the equilibrium of CO_2 emissions (heterogenic respiration) from soils caused by the biological decomposition of plant litter and soil organic matter (SOM) with the input of dead plant material like leaf litter, root litter, and decaying wood (Janzen, 2004). Through a variety of soil organisms, plant biomass is incorporated into the soil (Kemmitt et al., 2008). C exudates from roots, litter, *etc.* start the process of mineralization and humification, which is crucial for soil C dynamics (Jimenez and Lal, 2006). These input and output variations, which may be quantified by gross primary production (GPP), net primary production (NPP), ecosystem respiration (ER), net ecosystem production (NEP), etc., define the size of the SOC pool (Liu et al., 2015). The various ecosystem activities and C fluxes that are vital to the preservation and sequestration of the terrestrial SOC stock are depicted in Fig. 17.1.

17.3 Spectral features of SOC

Spectral signatures of SOC are the reflectance or absorbance of wavelengths in the electromagnetic spectrum due to electronic transitions, vibrational stretching, and bending of structural groups of atoms, in which the reflectance spectra occur at varied energy levels and the absorbance occurs due to the overlapping of spectral bands from different atomic components (Dalal and Henry, 1986; Henderson et al., 1992). Electromagnetic radiation in the VNIR–SWIR region showed a significant impact on SOC due to characteristic vibrations of chemical bonds in molecules (Nocita et al., 2015). N–H, O–H, and C–H bond stretching and bending at NIR–SWIR (700–2500 nm) results in vibrations, electromagnetic spectrum absorbance, and a broad absorbance band at visible spectrum (400–700 nm) associated with the chromophores function of soil colour (Rossel et al., 2006). Determination of soil organic matter (SOM) through spectral signature is very complicated in nature, as the composition of SOM along with maturity and stages of decomposition of organic matter affect reflectance characteristics (Ben-Dor et al., 1997).

The presence of SOM decreases the range of soil reflectance spectra, which is typically from 1100 to 2500 nm, making it more difficult to establish the relationship between the physiochemical parameters (Ladoni et al., 2010). Spectral reflectance at visible ranges 410, 520, 540, 550, 570, and 660 nm (Rossel et al., 2006, Brown et al., 2006), at NIR 960, 1100, 1400, 1720, 1744, 1870, 1900, and 2052 nm (Shepherd and Walsh, 2002) were used to predict SOC. Henderson et al. (1992) generalised the most useful wavelength for estimation of SOC in the range of 2200–2500 nm across large geographic areas and varied parent material, as vibrations associated with the OH and OH–Al–OH bonds and also O–H and C–H bonds linked to lignin and cellulose (Ben-Dor et al., 1997). Although wavelength between 2225 and 2255 should be avoided, as it effects other soil properties. McCarty and Reeves (2006) found spectral reflectance at MIR range more informative than NIR and visible spectrum. In field studies, issues related to vegetation cover, soil moisture, and roughness need to be resolved (Nocita et al., 2015; Bartholomeus, 2009), as reflectance and absorption peaks of water (1400 and 1900) coincide with specific organic functional groups for humic acid (Chang and Laird,

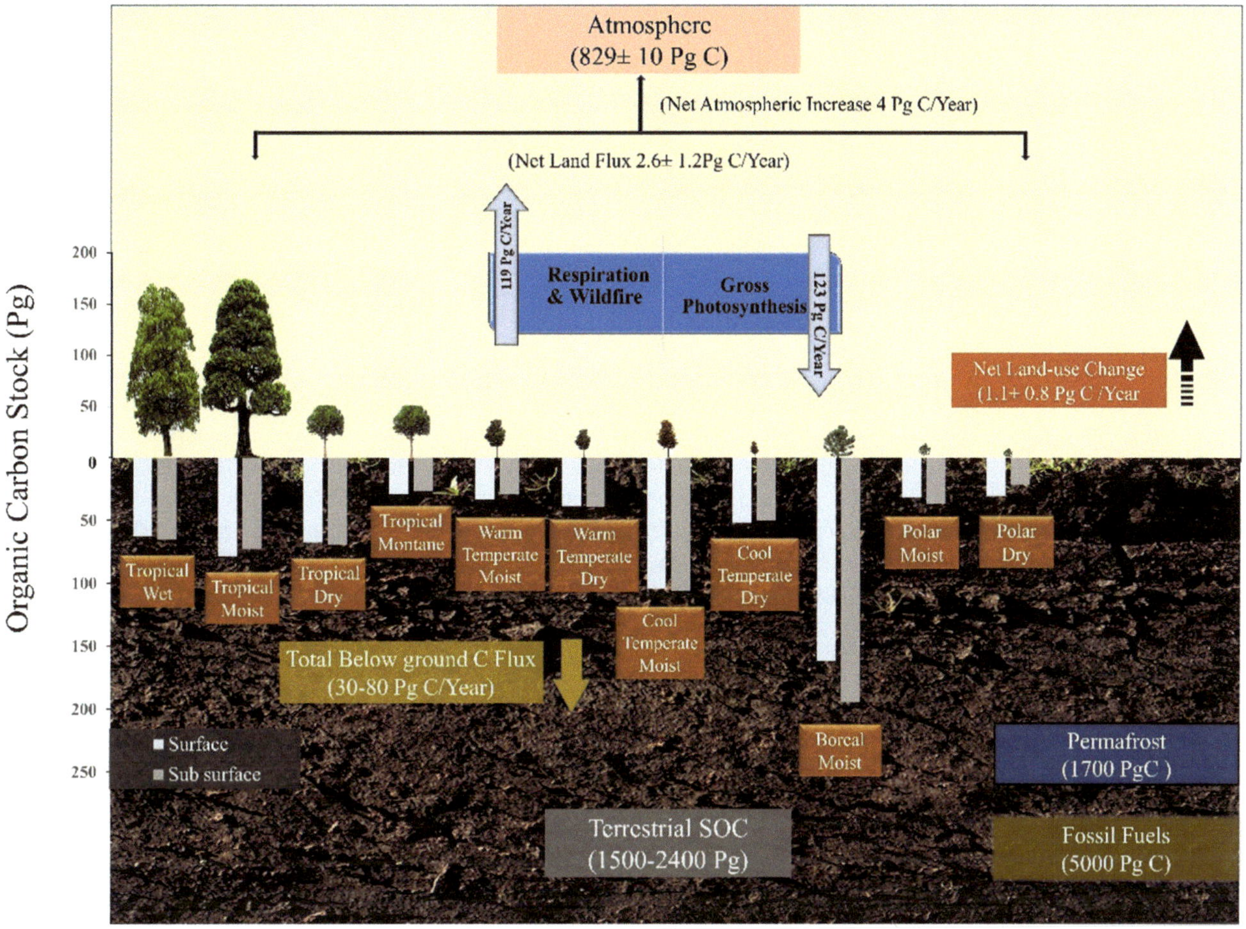

FIGURE 17.1 The global terrestrial C cycle. Total C stocks and pools are given in parentheses following Janowiak et al. (2017). Total surface, subsurface soil C stock, and AGB stock values are taken from Scharlemann et al. (2014).

2002). The relationship between the spectral signature and SOC is not necessarily linear (Peng et al., 2015), which increases the demand for multivariate statistical analysis and ML techniques (Stenberg and Rossel, 2010; Liakos et al., 2018).

17.4 Types of remote sensing sensors and platforms

17.4.1 Sensors

There are essentially two types of sensors: Active sensors, which use an energy source that emits electromagnetic energy at particular wavelengths and measures reflectance. Radar, microwave, and light detection and ranging (LiDAR) are examples of active sensors that expose the characteristics of a surface by detecting the reflected fraction of the radiated signals to incident radiation (Zhao, 2018). Passive sensors are another type of sensor, which relies on the sun's or the earth's natural radiation energy. Each part of the electromagnetic spectrum carries valuable information for a particular use in passive RS.

17.4.1.1 Resolutions

Another important aspect of any RS sensor is its spectral, radiometric, and spatial resolutions. The number of measurements made by sensors at various wavelengths within a certain spectral range is known as spectral resolution (Jensen, 2009); whereas, radiometric resolution is defined as the ability of the sensor to recognize and discriminate the smallest shift within its dynamic range, i.e., the precision at which the sensor records the data (Jafarbiglu and Pourreza, 2022).

The smallest item that a sensor can resolve, or the linear dimension on the ground that each pixel represents, is known as spatial resolution (Liang and Wang, 2019).

17.4.1.2 RS instruments for SOC determination

Some of the most common RS instruments specifically used for agriculture SOC determination are covered in this section.

RGB cameras that are reasonably inexpensive (Rossel et al., 2008) and free of atmospheric and BRDF (bidirectional reflectance distribution function) corrections (Levin et al., 2005) are often employed to estimate soil parameters, including SOC and Fe concentration (Rossel et al., 2008), by using soil color as a proxy. Multispectral RS utilizes most of the informative bands (3–10 bands) within and outside the visible region to create a false-color composite, which can effectively map surface soil SOC content (Žížala et al., 2019). The complexity of soil spectral signatures forced researchers to seek accurate and detailed estimation of SOC through hyperspectral sensors (Gomez et al., 2008; Castaldi et al., 2018; Guo et al., 2021) with a wide spectral range and the most informative bands for specific phenomena (Omidi et al., 2020). Spectral resolution and bandwidth are the key differences between hyperspectral and multispectral imaging. While hyperspectral sensors monitor hundreds or even thousands of bands (between 200 and 2500 bands with a bandwidth of 0.1– 10 nm), multispectral sensors typically measure a small number of bands (fewer than 15 bands) with bandwidths between 10 and 20 nm (Jafarbiglu and Pourreza, 2022). The thermal inferred spectral region (3–50 μm) is utilized to interpret the surface temperature based on the energy collected from both direct and indirect observations (Berni et al., 2009), although its application is quite limited (Zhai, 2019).

Active RS technologies, including LiDAR and radio detection and ranging (Radar), use the notion of broadcasting electromagnetic signals and detecting the returns. Recently developed multispectral lidars, with a small bandwidth of 1–2 nm and little divergence (less than 1 mm per meter), can be used for applications like vertical profile information of vegetation under wetland and forest SOC determination (Morsy et al., 2017). Similar to LiDAR, radar uses short radio wavelengths between 0.1 mm to 1 m, which may penetrate objects to varying degrees and provide diverse information regarding SOC in soil systems (Wang et al., 2021) and vertical distributions of SOC in soil profiles at a depth of 1.5 m (Shen et al., 2020).

17.4.2 Platforms

Platforms that carry sensors to conduct remote measurement activities. Important criteria for choosing the optimum platform include operation costs (Matese et al., 2015), sampling area (Manfreda et al., 2018), and accuracy regarding spatial and spectral resolution. This section compares and contrasts the three main RS platforms—UAS, manned aircraft, and satellites—for choosing which to use to detect particular soil properties (Fig. 17.2).

The National Aeronautics and Space Administration has been using manned aircraft as a key RS platform since the early 1970s for agricultural applications like monitoring soil moisture with numerous RS devices (Schmugge et al., 1980). According to the broad classification shown in Fig. 17.2, aircraft (manned or unmanned) can be categorized into four groups based on altitude and flight endurance: low-altitude short endurance (LASE), low-altitude long endurance (LALE), medium-altitude long endurance (MALE), and high-altitude long endurance (HALE) (Watts et al., 2012). The most popular RS systems are LALE and LASE, while HALE and MALE categories are expensive and complex (Stöcker et al., 2017). In contrast, RS satellites are in a certain range of low earth orbit, higher than any aircraft's airspace and lower than other satellites like the global navigation satellite system. An UAS is an aircraft that may fly independently or under remote control conditions (Dempsey and Rasmussen, 2010), which includes drones weighing less than 25 kg (55 lb) (FAA DroneZone, 2020). Fig. 17.2 contrasts the key traits of satellites, manned aircraft, and UAS (Manfreda et al., 2018). Since then, both satellite and airborne sensors have been extensively used for the estimation and mapping of soil variables.

17.5 Estimation of terrestrial SOC through RS

17.5.1 Source and application of remotely sensed data for estimation of terrestrial SOC stock

17.5.1.1 Spaceborne

Spaceborne remotely sensed imaging has tremendous potential as an enabling tool in SOC characterization of surface soils, from the launch of the first multispectral imagery satellite in the 1980s (Frazier and Cheng, 1989) to spaceborne Hyperion systems (Wang et al., 2022) to date. Auxiliary variables like brightness, wetness, and vegetation are highly

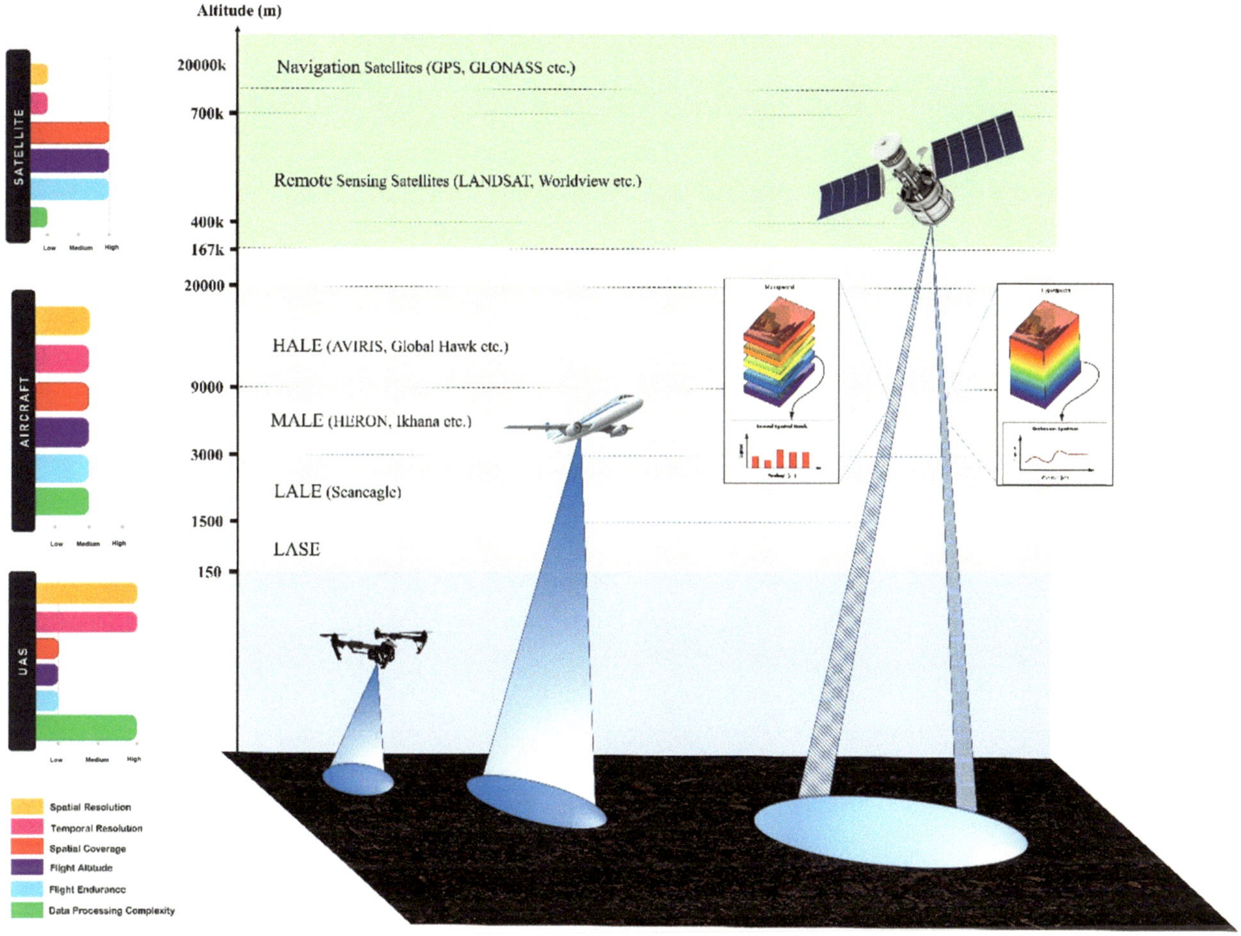

FIGURE 17.2 RS platforms based on their altitude. HALE, MALE, LALE, and LASE are the main classification of aircraft platforms, where UAS are a subclass of LASE. Comparison of satellites, manned aircraft, and UAS based on its strength and weakness following the classification of Manfreda et al. (2018).

correlated with SOC distribution (Mondal et al., 2017) and combined geostatistical methods with predicter variables help to improve SOC estimation through the development of environmental models and generate cost-effective, high-quality digital SOC maps (Wang et al., 2018; Guo et al., 2021; Zhou et al., 2020). The prediction performance of varied spaceborne multispectral and hyperspectral sensors is described in Table 17.1.

SOC could be estimated using Hyperion spectral signatures in soils with high OC (> 1%) (Gomez et al., 2008); multispectral and hyperspectral satellite imagers including EnMAP, PRISMA, HyspIRI, EO-1 ALI, Hyperion, Landsat 8 OLI and Sentinel-2 MSI (Castaldi et al., 2016; Vaudour et al., 2019); and multispectral remotely sensed data (MODIS) for topsoil (Peón et al., 2017b) (Table 17.1). Žížala et al. (2019) discovered that all spaceborne sensors (Sentinel-2, Landsat-8, and PlanetScope) had relatively similar prediction accuracy. Multisensor data mining technology proved to be an excellent alternative to single sensor-based technology for accurate prediction of SOC (Le et al., 2021), and recent studies integrating Sentinel 1 and Sentinel 2 and Sentinel 2 and Sentinel 3 (Zhou et al., 2020, 2022) performed better in terms of prediction capacity. Gholizadeh et al. (2018b) showed the advantages of Sentinel-2 high-quality data on fluctuations in SOC over airborne sensors, although sampling strategy plays a pivotal role (Steinberg et al., 2016).

There is currently a demand for precise and reliable SOC prediction utilizing low-cost ML models for agricultural soils using active multisensor data on a global and national level. DL-based hyperspectral data (Margenot et al., 2020; Yang et al., 2019; Wijewardane et al., 2018); neural networking (NN) using multispectral band combinations (Ayoubi et al., 2011; Were et al., 2015); RBF framework using MODIS data (Dai et al., 2014; Li et al., 2016) were used for SOC retrieval. XGBoost ML technique using Sentinel-1 (S1) SAR data and Sentinel-2 (S2) multispectral data (Nguyen

TABLE 17.1 Prediction performance of space borne remotely sensed SOC in agriculture soils.

Sensor	Algorithm/multivariate method	Spectral range (nm)	R^2	RMSE	References
Hyperion	PLSR	400–2500	0.51	0.73	Gomez et al. (2008)
Hyperion	PLSR	400–2500	0.62	0.22	Castaldi et al. (2016)
Landsat ETM +	ANNSK	450–2350	0.63	0.27	Mirzaee et al. (2016)
EnMAP	PLSR	420–2500	0.25–0.67	0.20–0.48	Castaldi et al. (2016)
PRISMA	PLSR	400–2500	0.26–0.65	0.21–0.48	Castaldi et al. (2016)
HyspIRI	PLSR	380–2510	0.23–0.60	0.22–0.48	Castaldi et al. (2016)
EnMAP	autoPLSR	420–2500	0.67	2.8	Steinberg et al. (2016)
HYP	PLSR	427–2355	0.87	3.69	Peón et al. (2017a)
L5TM	PLSR	485–2223	0.75	5.10	Peón et al. (2017a)
MOD	PLSR	413–2130	0.88	3.50	Peón et al. (2017a)
Sentinel-2	PLSR	–	0.56	0.23	Castaldi et al. (2016)
Sentinel-2	SVM	440–2200	–	0.08–0.24	Gholizadeh et al. (2018b)
Sentinel-2	PLSR	440–2200	0.56	1.23	Vaudour et al. (2019)
Sentinel-2 MSI	CB	9 VINR-3SWIR	0.68	–	Žížala et al. (2019)
Landsat-8 OLI	SVM	5 VNIR-3SWIR	0.65	–	Žížala et al. (2019)
PlanetScope	CB	4VNIR	0.66	–	Žížala et al. (2019)
Gaofen 1	ELM	–	0.84		Guo et al. (2021)
S1 + S2 + DEM	BRT	–	0.44	0.50	Zhou et al. (2020)

ANNSK, artificial neural network-simple kriging; *BRT*, boosted regression tree; *CB*, cubist; *ELM*, extreme learning machine; *PLSR*, partial least squares regression; *SVM*, support vector machine.

et al., 2022), boosted regression tree ML technique (Zhou et al., 2020), and long short-term memory ML technique (Wang et al., 2022) were used for SOC prediction.

17.5.1.2 Airborne

Airborne hyperspectral imaging technology has caught the interest of numerous researchers working on the forecasting of soil properties (Hbirkou et al., 2012) specifically SOC (Paz-Kagan et al., 2015; Vaudour et al., 2016), due to its high sensitivity to soil variables than NIR (Gholizadeh et al., 2018a; Guo et al., 2019; Nouri et al., 2017; Vaudour et al., 2016; Zhang et al., 2019a). Airborne hyperspectral data's high spectral and spatial resolution provides abundant spectral information while minimizing prediction error brought on by atmospheric and spatial interpolation (Franceschini et al., 2015), making it superior to satellite multispectral data for topsoil observation over a detailed study region (Ben-Dor et al., 2009; Gholizadeh et al., 2018b; Zhang et al., 2019a). Research conducted on a global basis with the prediction capacity of sensors along with ML algorithms is summarized in Table 17.2.

Prediction of SOC was carried out with the use of Airborne Hyperspectral Sensor 160 (AHS 160) and three ML algorithms: penalized spline regression (PSR), support vector machine (SVM), and partial least square regression (PLSR) (Stevens et al., 2008, 2010). Atmospheric correction and improvement of sensors' stability and sensitivity could improve the accuracy of the prediction (Brook and Dor, 2011). Soil roughness and vegetation cover affect SOC prediction (Hbirkou et al., 2012), and filtering soil moisture impacts can lead to better accuracy. Studies on bare soils under controlled conditions provide more accurate data (Denis et al., 2014), although it is hard to find non-vegetated areas on

TABLE 17.2 Prediction performance of airborne remotely sensed SOC in agriculture soils.

Sensor	Algorithm/multivariate method	Spectral range (nm)	R^2	RMSE	References
AHS-160	PLSR, SVMR	430–2540	0.53–0.89	3.13–6.22	Stevens et al. (2010)
HyMap	PLSR	450–2500	0.34–0.83	0.76–1.10	Hbirkou et al. (2012)
ProSpec TIR V-S	PLSR	400–2500	0.33	3.82	Franceschini et al. (2015)
AISA-Eagle	PLSR	400–1000	0.44	4.05	Vaudour et al. (2016)
AHS-160	PLSR, SLR, SMLR	430–2540	0.27–0.60	6.44–8.70	Peón et al. (2017a)
AISA Dual system	PLSR	400–2450	0.73	8.4	Homolová et al. (2014)
APEX	PLSR	400–2500	–	4.3	Castaldi et al. (2018)
HyMap	PLSR	450–2500	0.73–0.85	0.19–0.25	Vohland et al. (2016)
SASI, CASI	SVMR		–	0.06–0.12	Gholizadeh et al. (2018b)
SpecTIR LLC	RF	380–1000, 900–1700	0.66	–	Hong et al. (2020a)
SpecTIR LLC	PSLR	380–1000, 900–1700	0.54	0.33	Guo et al. (2021)
SpecTIR LLC	RF	380–1000, 900–1700	0.76	0.206	Hong et al. (2020b)
HySpex	PLSR	400–1000 1000–2500	0.78	–	Ward et al. (2020)

RF, Random forest; *SMLR*, sparse multinomial logistic regression; *SVMR*, support vector machine regression.

a large experimental site. The residual spectral unmixing approach was useful in partially covered fields (Guerschman et al., 2009; Bartholomeus et al., 2011) and airborne prism experiment masked the vegetation impact (Diek et al., 2016) while predicting SOC.

It is expensive and time-consuming to accurately estimate SOC using airborne sensors because model calibration and alignment between distant and laboratory spectra require field soil data. To address this drawback, researchers used alternative soil spectral libraries, such as LUCAS (Castaldi et al., 2018) and AISA dual system (Homolová et al., 2014). Selection of the most informative spectral signatures needs to be discovered in order to improve model's prediction accuracy and robustness (Lu et al., 2019; Sarathjith et al., 2016; Vohland et al., 2016), as a large volume of airborne hyperspectral data has a tendency to increase complexity within the models. Moreover, the selection of ML algorithm is similarly important, and a number of studies have demonstrated the comparison of varied models such as MLP, NN (Žížala et al., 2017), and BPNN (Khanal et al., 2018) along with considerable prediction efficiency in the prediction of SOC in agriculture field.

17.5.1.3 UAS

UAS have the ability to meet the needs of particular problems associated with space and airborne sensors, particularly those linked to atmospheric and surface conditions, and give the chance to avoid or reduce some of the drawbacks. The use of these platforms for soil ecosystem monitoring and SOC prediction is still very limited, and research conducted on a global basis is summarized in Table 17.3.

UAS-based RS was utilized in SOC forecasting for smaller regions (Aldana-Jague et al., 2016), and to map the digital SOC, back propagation neural networks, extreme learning machine (ELM), and SVM were used as three ML techniques (Guo et al., 2020; Biney et al., 2021). Topographic position index was used as a covariate to predict surface SOC

TABLE 17.3 Prediction performance of Unmanned aerial system remotely sensed SOC in agriculture soils.

Sensor	Algorithm/multivariate method	Spectral range (nm)	R^2	RMSE	References
Mini-MCA6	SVM	450–1050 nm	0.95	0.21	Aldana-Jague et al. (2016)
MicaSense RedEdgeTM 3	PLSR, SVM, ANN, ELM	475–840 nm	0.41–0.80	1.16–1.96	Guo et al. (2020)
MicaSense Altum dual sensor	RF, SVMR	475 nm to 11 μm	0.11–0.27	0.29–0.34	Biney et al. (2021)
Mini-MCA 6	MLR, LOOCV	550–900 nm	0.79–0.91	0.06–0.11	Wehrhan and Sommer (2021)

in a heterogeneous soil landscape (Wehrhan and Sommer, 2021) (Table 17.3). Though there have been very few studies, UAS-based RS has a great deal of potential for SOC prediction, monitoring, and mapping in the near future.

17.5.2 Estimation of wetland surface and subsurface SOC through RS method

Wetlands are excellent carbon sinks due to anaerobic conditions and less efficient electron acceptors (Mn and Fe), which cause the accumulation of organic matter for longer periods (Köchy et al., 2015). According to Yu et al. (2012), 4%–6% of the world's wetland regions retain nearly 20%–25% of the SOC stock, making them more susceptible to ongoing monitoring for maintenance of the atmospheric C balance. Covariables representing terrain, climate, soil type, and land cover are used at the regional scale to predict SOC for depths under 50 cm (Yang et al., 2016; Wang et al., 2018), while spectral variables are used at the field scale (Žížala et al., 2019).

17.5.2.1 Climatic and topographic covariables

Mean annual temperature (MAT) and mean annual precipitation (MAP) are mostly used for climatic covariates (Yang et al., 2019; Han et al., 2020), as both of these factors have a significant impact on SOC distribution (Albaladejo et al., 2013). Significant variables for SOC prediction were vegetation (Yang 2019); climate (Holmquist et al., 2018; Han et al., 2020; Ren et al., 2020); and terrain gradient (Ren et al., 2020). Latitude was found to be a regulating factor in SOC stability (Han et al., 2020), as higher temperature increases SOC decomposition and overall atmospheric CO_2 flux, whereas lower NPP results in low SOC input in temperate areas (Davidson et al., 2000).

The impact of MAT was shown to be lessened in subsurface soils (Yang, 2019 and Ren et al., 2020). For subsurface SOC determination, slope, elevation, compound topographic index, and wetness index capture significant patterns of soil erosiveness and retention of SOC (Taghizadeh-Mehrjardi et al., 2016) whereas several studies found a significant relationship between elevation, slope, topographic wetness index (Yang et al., 2016), landscape positions (Ren et al., 2020), and altitude (Han et al., 2020) with SOC content.

17.5.2.2 Soil covariates

SOC in wetland landscapes is highly dependent on soil salinity (Zhang et al., 2019b, Holmquist et al., 2018, Qu et al., 2019), soil types (Holmquist et al., 2018), soil texture (Grinand et al., 2017), pH (Han et al., 2020), bulk density (BD) and soil grain size (Yang 2019, Macreadie et al., 2017). In marine wetlands, tides control the spatial distribution of soil salinity. The salinity affects CO_2 and CH_4 emission fluxes (Luo et al., 2019; Zhang et al., 2019b), as a sharp increase in salinity from 6 to 9 ppt could be responsible for 50%–80% more CO_2 emissions (Wang et al., 2019). Soil pH and iron and aluminium chemistry are associated with anaerobic wetland soils, thus affecting SOC stability (Rasmussen et al., 2018, Han et al., 2020). Some recent studies have monitored BD through remotely sensed data with considerable accuracy ($R^2 = 0.50–0.80$) (Pittman and Hu, 2020; Aitkenhead and Coull, 2020; Salehi Hikouei et al., 2021), which can correlate with SOC prediction by using different ML algorithms.

17.5.2.3 Vegetative and spectral covariates

The lower magnitudes of the reflectance in wetlands present one difficulty in using spectral covariables (Sharma et al., 2022), which force the researcher to find the ideal spectral signature of SOC in wetland ecosystems. In wetland ecosystems, spectral reflectance represents vegetation biomass, which eventually contributes to the SOC stock.

Spectral reflectance from the VIS-NIR-SWIR bands along with vegetation indices like NDVI and biophysical parameters like NDMI, leaf area index, and LST (land surface temperature) have been extensively used as proxies for the prediction of SOC content in wetlands (Ding et al., 2018; Yang et al., 2016, Zhang et al., 2019a, Ren et al., 2020, Peng et al., 2015). A negative correlation between SOC and spectral reflectance in tidal wetlands is observed by Zhang et al. (2019a) and Ding et al. (2018) (Table 17.4).

17.5.3 Application of RS in terrestrial C stock management

The terrestrial C cycle circumscribes mainly the exchange of C among the geosphere, terrestrial biosphere, pedosphere, and atmosphere of the earth system, which constitute the largest C flux between the terrestrial biosphere and atmosphere. Data from Lidar, microwave, and passive optical RS have all been used to determine the amount of aboveground biomass (AGB) and C exchange through heterotopic soil respiration.

17.5.3.1 Estimation of AGB

AGB is estimated by passive optical RS techniques. Spatial resolutions range from low to high resolution in its estimation. Medium-resolution data (30 m) from Landsat TM/ETM + /OLI, Sentinel-2 Multispectral Imager, and Terra/Aqua ASTER are the widely used ones. Coarse-resolution data (250–8000 m) from AVHRR, SPOT, and MODIS are utilized

TABLE 17.4 Estimation of wetland surface and subsurface SOC and prediction performance of varied RS approaches.

Covariates	Sensor/ source of data	Algorithm/ multivariate method	R^2	RMSE (g kg^{-1})	Findings	References
Vegetation	Sentinel-1	SEM	–	1.63	Improved prediction accuracy of SOM up to 100 cm of soils with vegetation dynamics through path analysis	Yang and Guo (2019)
Climatic and terrain factor	Landsat 8	RF, SVM, IDW, OK	0.52–0.67	5.26–6.64	A total of 4 climatic and 14 terrain factors used for SOC determination of SOC up to 100 cm, and RF-performed best accurate model	Ren et al. (2020)
Terrain and wetness index	Landsat 8	ANN, SVM, kNN, RF	0.38–0.46	0.07–0.1	ANN-performed best model and wetness index were the most useful auxiliary data	Taghizadeh-Mehrjardi et al. (2016)
Salinity	World View and Quick Bird	ANN, SVM, RF, and k-NN	0.29–0.37	6.86–7.18	First attempt to examine the potential of object-based modelling and produced a higher accuracy than the pixel-based method where SVM performed best model	Zhang et al. (2019b)
Climatic and topographic factors	Landsat 8	NN	0.62	8.1	Integrates legacy survey data for peatland restoration surveys in Scotland	Pittman and Hu (2020)
Spectral	EO-1 Hyperion	ACO-iPLS, RF-SVM, RF	0.83–0.92	0.27–0.40	Integrates wavelength transformation strategies and SVM produces the best overall results	Ding et al. (2018)

ACO-iPLS, Ant colony optimization-interval partial least squares; *IDW*, inverse distance weighted; *kNN*, K-nearest neighbor; *OK*, ordinary kriging; *RF–SVM*, recursive feature elimination-support vector machine.

for the estimation of woods (Sibanda et al., 2015), grasslands (Friedl et al., 1994), and Arctic tundra (Shoshany and Karnibad, 2011). However, high spatial resolution data (5 m), which includes high-resolution satellite imagery from QuickBird, IKONOS, and WorldView-2/3, is frequently used for calculating local tree biomass (Leboeuf et al., 2007; Palace et al., 2008; Fuchs et al., 2009) and grass biomass (Sibanda et al., 2017).

Microwave radiometers and scatterometers including synthetic aperture radar (SAR) are used to overcome the limitations of passive optical RS in the estimation of AGB. The other options include Radargrammetry (Vastaranta et al., 2012), space-borne SAR with multiband and multi-polarization data (Ni et al., 2016), Lidar (Simard et al., 2011), and terrestrial laser scanning (Liang et al., 2016). It is not only useful for allometric equation development (Stovall et al., 2017), but also creates the opportunity to estimate 3D structural models (Disney et al., 2018). Because of its global coverage, variety of spatial resolutions, repeat visits, and affordability, passive optical RS is possibly the best tool for estimating biomass at regional to global scales, although it is challenging to accurately estimate vegetation canopy attributes due to saturation of the spectral signal in high biomass density (Xiao et al., 2019).

17.5.3.2 *Estimation of carbon fluxes*

Estimation of carbon fluxes can be done by measuring GPP, NPP, and NDVI as primary covariates. Spectral VIs were capable of quantifying GPP accurately for different ecosystems ranging from North American ecosystems (Rahman et al., 2005) to African tropical savannas (Sjostrom et al., 2011), dry to humid tropical forests in Southeast Asia, and the Amazon (Huete et al., 2008).

RS data can be used by both prognostic and diagnostic models to enhance the quantification of C fluxes. ER is a significant source of carbon dioxide (CO_2) flux from the land to the atmosphere, and as a result, it plays a significant role in the rise of atmospheric CO_2 and thus the resulting global climate change. So, C fluxes could be estimated by estimating ER (Xiao et al., 2019), LST (Rahman et al., 2005), and biochemical mechanisms in autotrophic and heterotrophic respiration (Wehr et al., 2016). LST produced from the advanced microwave scanning radiometer for EOS sensor is an effective substitute for soil respiration, including heterotrophic and root respiration, in a wide range of boreal forest, grassland, and tundra sites in the boreal and arctic biomes (Kimball et al., 2009). The correct selection of wavelengths in the SWIR region plumes to the planetary boundary layer, near-surface flow, sunlight, and cloud-free regions, thereby alleviating the limitations of the transmission model's accuracy. It also allows better identification of CO_2 signals in the measured spectrum compared to signals from other atmospheric or surface components, reducing system retrieval errors (Xiao et al., 2019).

17.6 Terrestrial SOC cycle in India: an estimation through RS

Bhattacharyya et al. (2000) produced the first in depth report on SOC stock in India and estimated 9.5 Pg SOC at a depth of 0–0.3 m. The SOC and soil inorganic carbon (SIC) were mapped at 250 m resolution in India using climate, land cover, rock type, soil type, NDVI, and irrigation status as covariates (Sreenivas et al., 2016). Several studies on predictions and modelling were carried out to estimate soil carbon. The SOC, SIC, and total carbon (TC) pool sizes of India were projected to be 22.720.93, 12.831.35, and 35.551.87 Pg, respectively, using random forest (RF)-based spatial prediction. Vegetation indices (NDVI, RDVI, MSAVI, and MNLI) through Landsat 8 OLI (Bhunia et al., 2019), regression kriging model (Kumar et al., 2016), hybrid interpolation model (Mondal et al., 2017), land-surface temperature, and land use and land cover pattern as a covariate (Momo and Devi, 2022) were used to predict SOC stocks, and environmental covariates through geographically weighted regression kriging with linear regression kriging were used to predict TC stocks in the 0–30 cm soil depth of two southern states, i.e., Karnataka and Telangana, in India using field observations (Mitran et al., 2018).

Forest soils make up a bigger portion of SOC, and OC pool of Indian forest soils was 6.8 Pg C up to 1 m by RS (Chhabra et al., 2003). AGB, belowground biomass, litter biomass, dead wood, and the SOC pool make up the majority of the forest's carbon stocks. Researchers have used the passive sensor such as Moderate Resolution Imaging Spectroradiometer (MODIS) (Pandey et al., 2019), Landsat TM (Tripathi et al., 2015), Landsat OLI (Kashung et al., 2018; Bordoloi et al., 2019). Using a combined technique of field inventory and vegetation indices generated from Landsat OLI, Bordoloi et al. (2022) evaluate and predict the aboveground biomass, carbon stock, and carbon sequestration potential of varied land uses in Northeast India. The projected AGB had an accuracy of $R^2 = 0.79$ and an RMSE of 51.04 Mg ha^{-1}, ranging from 14.32 to 185.95 Mg ha^{-1} on average. The results of an assessment of AGB by Rizvi et al. (2020) using high-resolution multispectral remote sensing data (LISS-IV, spatial resolution- 5.8 m) were 7.854 and 9.946 million tons of AGB and total biomass carbon, respectively. Hari et al. (2021) used stepwise multiple linear regression (SMLR) and local geographically weighted regression (GWR) models to map and quantify above-ground carbon (AGC) stocks over Coimbatore and the Nilgiris Biosphere of Tamil Nadu.

Blue carbon is another part of C-storage in the form of biomass and sediments from mangrove forests, tidal marshes, and seagrasses. Indian mangrove has an estimated cover of 4662.56 km^2 along the nine coastal states and three union territories, and studies regarding the estimation of carbon stock combining ground-based and RS approaches were done for the western ghats (Bindu et al., 2020) and parts of the eastern ghats of mangrove forest (Anand et al., 2020).

17.7 Limitations of RS for SOC monitoring

Regardless of any approach, there are some limitations of RS techniques such as low signal-to-noise ratio (Minu et al., 2016), low spectral resolution (Gomez et al., 2008), geometric and atmospheric interventions (Jakob et al., 2017), estimations for only few first centimeters of topsoil *etc*., no consideration of soil management practices such as crop rotation, conservation tillage practices, fertilization level, ploughing depth, and organic amendments though these affect SOC content (Noirot-Cosson et al., 2016; Chenu et al., 2019; Crystal-Ornelas et al., 2021), use of one or more than one algorithm in datasets to perform any comparisons of their performances through a systematic way in the prediction of SOC (Castaldi et al., 2019). Performance of algorithms according to regions needs to be checked properly.

Satellite platforms provide a prefixed temporal resolution that cannot be redesigned, leading to malformed images affected by weather conditions, while airborne applications require a higher operational cost that is somehow complex and not affordable to farmers (Angelopoulou et al., 2019). Although UASs can be operated at a more affordable cost, their payload capacity and flight duration are limited. Moreover, several external factors such as soil moisture, structure, roughness, vegetation cover, and changes in atmospheric conditions may lead to overestimation and inaccurate predictions of SOC content (Guerschman et al., 2009). Apart from all these, infrastructural capabilities are also important to proliferate the full potential of optical technologies and AI techniques.

17.8 Policymaking and future research direction

Policymakers around the globe are now focusing on soil ecosystems in the current context of climate change by incorporating SOC monitoring, as it is closely associated with soil quality and function. Although there is no sole segment for "soil" in the SDGs of the United Nations till now, the soil ecosystem has been monitored by land-related SDGs either directly or indirectly (Angelopoulou et al., 2019). Modern technologies such as RS can be implemented in plenty of soil-related policies to take a big step in the field of improved soil management practices. In addition to this, these data and services will provide well-designed management practices that will also improve the status of agricultural soil and restrict land degradation through the application of variable rate fertilization (Schwartz et al., 2012; Vaudour et al., 2016).

In the future, with higher accessibility of free satellite sensor images, strategies may be provided to enhance agriculture management and incorporate them into their decision-making processes. Being a reliable source, RS technologies have been put forward by the global soil partnership and its intergovernmental technical panel on soils through the publication of the GSOCMap (FAO, 2017). Andries et al., 2021 proposed a brief framework for the monitoring, reporting, and verification or MRV of SOC data generated from satellites, airbornes, and UASs. Incorporation of SOC monitoring through RS into policies may be very important as well as impactful for pursuing future research activities due to its cost-effective nature (Aldana-Jague et al., 2016; Andries et al., 2021). To create systems that are sustainable, scalable, and economically feasible, governments and NGOs should rely on freely available, fine spatial and temporal resolution satellite sensor imagery, such as that from Landsat and Sentinel. A portion of these difficulties and ideas could be valid in different regions of the planet; however, because of safety issues, the absence of information and information quality issues, the delicate climate, and the absence of assets in these areas, embracing RS ways to deal with agriculture productivity is critically required.

17.9 Conclusion

The goal of this review was to demonstrate the advancements made in the last decade in the use of RS approaches in SOC calculations. The stability of various SOC stocks, including AGB, carbon fluxes from heterogenic respiration, and wetland SOC stocks, was emphasized. In this, we have reviewed the distinctive characteristics of several sensing platforms as well as a variety of sensing techniques for tackling the significant issues involved in the SOC's routine monitoring and reporting of wide areas. The precise assessment of SOC stock using data-driven methodologies was found to be a significant limiting factor in most of the studies. We have also included current developments in ML and DL techniques that are extensively utilized to improve the overall accuracy and robustness of the models. We have highlighted the use of various sensing platforms as a crucial facilitator for SOC determination due to their affordability, rapidness, and spatial coverage. The developments in UAS, notably drone technologies, are limited to SOC estimation, despite the possibility that they could prove useful in the current situation after expanding the spatial coverage. Policymakings at the national as well as international levels should incorporate the significance of SOC monitoring, as SOC plays a direct or indirect role in the SDGs as well as global climate change. In light of the aforementioned, it is proposed that the incorporation of RS technologies should be seen as essential to produce monitoring solutions that are accurate and efficient with high spatial resolution to make decisions regarding land use.

References

Aitkenhead, M., Coull, M., 2020. Mapping soil profile depth, bulk density and carbon stock in Scotland using remote sensing and spatial covariates. Eur. J. Soil. Sci. 71 (4), 553–567.

Albaladejo, J., Ortiz, R., Garcia-Franco, N., Navarro, A.R., Almagro, M., Pintado, J.G., et al., 2013. Land use and climate change impacts on soil organic carbon stocks in semi-arid Spain. J. Soils Sediment. 13 (2), 265–277.

Aldana-Jague, E., Heckrath, G., Macdonald, A., van Wesemael, B., Van Oost, K., 2016. UAS-based soil carbon mapping using VIS-NIR (480–1000 nm) multi-spectral imaging: potential and limitations. Geoderma 275, 55–66.

Anand, A., Pandey, P.C., Petropoulos, G.P., Pavlides, A., Srivastava, P.K., Sharma, J.K., et al., 2020. Use of hyperion for mangrove forest carbon stock assessment in Bhitarkanika forest reserve: a contribution towards blue carbon initiative. Remote. Sens. 12 (4), 597.

Andries, A., Morse, S., Murphy, R.J., Lynch, J., Mota, B., Woolliams, E.R., 2021. Can current earth observation technologies provide useful information on soil organic carbon stocks for environmental land management policy? Sustainability 13 (21), 12074.

Angelopoulou, T., Tziolas, N., Balafoutis, A., Zalidis, G., Bochtis, D., 2019. Remote sensing techniques for soil organic carbon estimation: a review. Remote. Sens. 11 (6), 676.

Ayoubi, S., Shahri, A.P., Karchegani, P.M., Sahrawat, K.L., 2011. Application of artificial neural network (ANN) to predict soil organic matter using remote sensing data in two ecosystems. Biomass Remote. Sens. Biomass 181–196.

Bartholomeus, H., 2009. The influence of vegetation cover on the spectroscopic estimation of soil properties. Wageningen University and Research.

Bartholomeus, H., Kooistra, L., Stevens, A., van Leeuwen, M., van Wesemael, B., Ben-Dor, E., et al., 2011. Soil organic carbon mapping of partially vegetated agricultural fields with imaging spectroscopy. Int. J. Appl. Earth Obs. Geoinf. 13 (1), 81–88.

Ben-Dor, E., Inbar, Y., Chen, Y., 1997. The reflectance spectra of organic matter in the visible near-infrared and short-wave infrared region (400–2500 nm) during a controlled decomposition process. Remote. Sens. Environ. 61 (1), 1–15.

Ben-Dor, E., Chabrillat, S., Demattê, J.A.M., Taylor, G.R., Hill, J., Whiting, M.L., et al., 2009. Using imaging spectroscopy to study soil properties. Remote. Sens. Environ. 113, S38–S55.

Berni, J.A., Zarco-Tejada, P.J., Suárez, L., Fereres, E., 2009. Thermal and narrowband multispectral remote sensing for vegetation monitoring from an unmanned aerial vehicle. IEEE Trans. Geosci. Remote. Sens. 47 (3), 722–738.

Bhattacharyya, T., Pal, D.K., Velayutham, M., Chandran, P., Mandal, C., 2000. Total carbon stock in Indian soils: issues, priorities and management. Land Resource Management for Food and Environmental Security. Soil Conservation Society of India 1–46.

Bhunia, G.S., Shit, P.K., Pourghasemi, H.R., Edalat, M., 2019. Prediction of soil organic carbon and its mapping using regression analyses and remote sensing data in GIS and R. Spatial Modeling in GIS and R for Earth and Environmental Sciences. Elsevier, pp. 429–450.

Bindu, G., Rajan, P., Jishnu, E.S., Joseph, K.A., 2020. Carbon stock assessment of mangroves using remote sensing and geographic information system. Egypt. J. Remote. Sens. Space Sci. 23 (1), 1–9.

Biney, J.K.M., Saberioon, M., Borůvka, L., Houška, J., Vašát, R., Chapman Agyeman, P., et al., 2021. Exploring the suitability of UAS-based multispectral images for estimating soil organic carbon: comparison with proximal soil sensing and spaceborne imagery. Remote. Sens. 13 (2), 308.

Bolstad, P.V., Vose, J.M., 2005. Forest and pasture carbon pools and soil respiration in the southern Appalachian Mountains. For. Sci. 51 (4), 372–383.

Bordoloi, R., Das, B., Yam, G., Deka, S., Tripathi, O.P., 2019. Carbon stock assessment i n different land use sectors of Ziro valley, Arunachal Pradesh using geospatia l approach. J. Geomat. 13 (2), 262–270.

Bordoloi, R., Das, B., Tripathi, O.P., Sahoo, U.K., Nath, A.J., Deb, S., et al., 2022. Satellite based integrated approaches to modelling spatial carbon stock and carbon sequestration potential of different land uses of Northeast India. Environ. Sustain. Indic. 13, 100166.

Brook, A., Dor, E.B., 2011. Supervised vicarious calibration (SVC) of hyperspectral remote-sensing data. Remote. Sens. Environ. 115 (6), 1543–1555.

Brown, D.J., Shepherd, K.D., Walsh, M.G., Dewayne Mays, M., Reinsch, T.G., 2006. Global soil characterization with VNIR diffuse reflectance spectroscopy. Geoderma 132, 273–290.

Castaldi, F., Palombo, A., Santini, F., Pascucci, S., Pignatti, S., Casa, R., 2016. Evaluation of the potential of the current and forthcoming multispectral and hyperspectral imagers to estimate soil texture and organic carbon. Remote. Sens. Environ. 179, 54–65.

Castaldi, F., Chabrillat, S., Jones, A., Vreys, K., Bomans, B., Van Wesemael, B., 2018. Soil organic carbon estimation in croplands by hyperspectral remote APEX data using the LUCAS topsoil database. Remote. Sens. 10 (2), 153.

Castaldi, F., Hueni, A., Chabrillat, S., Ward, K., Buttafuoco, G., Bomans, B., et al., 2019. Evaluating the capability of the Sentinel 2 data for soil organic carbon prediction in croplands. ISPRS J. Photogramm. Remote. Sens. 147, 267–282.

Chang, C.W., Laird, D.A., 2002. Near-infrared reflectance spectroscopic analysis of soil C and N. Soil. Sci. 167 (2), 110–116.

Chenu, C., Angers, D.A., Barré, P., Derrien, D., Arrouays, D., Balesdent, J., 2019. Increasing organic stocks in agricultural soils: knowledge gaps and potential innovations. Soil. Tillage Res. 188, 41–52.

Chhabra, A., Palria, S., Dadhwal, V.K., 2003. Soil organic carbon pool in Indian forests. For. Ecol. Manag. 173 (1–3), 187–199.

Crowther, T.W., Todd-Brown, K.E., Rowe, C.W., Wieder, W.R., Carey, J.C., Machmuller, M.B., et al., 2016. Quantifying global soil carbon losses in response to warming. Nature 540 (7631), 104–108.

Crystal-Ornelas, R., Thapa, R., Tully, K.L., 2021. Soil organic carbon is affected by organic amendments, conservation tillage, and cover cropping in organic farming systems: a meta-analysis. Agric., Ecosyst. Environ. 312, 107356.

Dai, F., Zhou, Q., Lv, Z., Wang, X., Liu, G., 2014. Spatial prediction of soil organic matter content integrating artificial neural network and ordinary kriging in Tibetan Plateau. Ecol. Indic. 45, 184–194.

Dalal, R.C., Henry, R.J., 1986. Simultaneous determination of moisture, organic carbon, and total nitrogen by near infrared reflectance spectrophotometry. Soil. Sci. Soc. Am. J. 50 (1), 120–123.

Davidson, E.A., Trumbore, S.E., Amundson, R., 2000. Soil warming and organic carbon content. Nature 408 (6814), 789–790.

Dempsey, M.E., Rasmussen, S. 2010. Eyes of the army–US Army roadmap for unmanned aircraft systems 2010–2035. US Army UAS Center of Excellence, Ft. Rucker, Alabma, vol. 9.

Denis, A., Stevens, A., Van Wesemael, B., Udelhoven, T., Tychon, B., 2014. Soil organic carbon assessment by field and airborne spectrometry in bare croplands: accounting for soil surface roughness. Geoderma 226, 94–102.

Diek, S., Schaepman, M.E., De Jong, R., 2016. Creating multi-temporal composites of airborne imaging spectroscopy data in support of digital soil mapping. Remote. Sens. 8 (11), 906.

Ding, J., Yang, A., Wang, J., Sagan, V., Yu, D., 2018. Machine-learning-based quantitative estimation of soil organic carbon content by VIS/NIR spectroscopy. PeerJ 6, e5714.

Disney, M.I., Boni Vicari, M., Burt, A., Calders, K., Lewis, S.L., Raumonen, P., et al., 2018. Weighing trees with lasers: advances, challenges and opportunities. Interface Focus. 8 (2), 20170048.

FAA DroneZone, 2020. FAA Drone Zone <https://faadronezone.faa.gov/#/> (accessed Feb. 20, 2021).

FAO, 2017. Soil Organic Carbon the Hidden Potential. ISBN 9789251096819. http://www.fao.org/3/a-i6937e.pdf%20 (accessed 17.07.22).

Franceschini, M.H.D., Demattê, J.A.M., da Silva Terra, F., Vicente, L.E., Bartholomeus, H., de Souza Filho, C.R., 2015. Prediction of soil properties using imaging spectroscopy: considering fractional vegetation cover to improve accuracy. Int. J. Appl. Earth Obs. Geoinf. 38, 358–370.

Frazier, B.E., Cheng, Y., 1989. Remote sensing of soils in the eastern Palouse region with Landsat Thematic Mapper. Remote. Sens. Environ. 28, 317–325.

Friedl, M.A., Schimel, D.S., Michaelsen, J., Davis, F.W., Walker, H., 1994. Estimating grassland biomass and leaf area index using ground and satellite data. Int. J. Remote. Sens. 15 (7), 1401–1420.

Fuchs, H., Magdon, P., Kleinn, C., Flessa, H., 2009. Estimating aboveground carbon in a catchment of the Siberian forest tundra: combining satellite imagery and field inventory. Remote. Sens. Environ. 113 (3), 518–531.

Gehl, R.J., Rice, C.W., 2007. Emerging technologies for in situ measurement of soil carbon. Clim. Change 80 (1), 43–54.

Gholizadeh, A., Saberioon, M., Ben-Dor, E., Borůvka, L., 2018a. Monitoring of selected soil contaminants using proximal and remote sensing techniques: background, state-of-the-art and future perspectives. Crit. Rev. Environ. Sci. Technol. 48 (3), 243–278.

Gholizadeh, A., Žižala, D., Saberioon, M., Borůvka, L., 2018b. Soil organic carbon and texture retrieving and mapping using proximal, airborne and Sentinel-2 spectral imaging. Remote. Sens. Environ. 218, 89–103.

Gomez, C., Rossel, R.A.V., McBratney, A.B., 2008. Soil organic carbon prediction by hyperspectral remote sensing and field vis-NIR spectroscopy: an Australian case study. Geoderma 146 (3–4), 403–411.

Grinand, C., Le Maire, G., Vieilledent, G., Razakamanarivo, H., Razafimbelo, T., Bernoux, M., 2017. Estimating temporal changes in soil carbon stocks at ecoregional scale in Madagascar using remote-sensing. Int. J. Appl. Earth Obs. Geoinf. 54, 1–14.

GSOCMap, <http://54.229.242.119/GSOCmap/> (accessed 22.07.22).

Guerschman, J.P., Hill, M.J., Renzullo, L.J., Barrett, D.J., Marks, A.S., Botha, E.J., 2009. Estimating fractional cover of photosynthetic vegetation, non-photosynthetic vegetation and bare soil in the Australian tropical savanna region upscaling the EO-1 Hyperion and MODIS sensors. Remote. Sens. Environ. 113 (5), 928–945.

Guo, L., Zhang, H., Shi, T., Chen, Y., Jiang, Q., Linderman, M., 2019. Prediction of soil organic carbon stock by laboratory spectral data and airborne hyperspectral images. Geoderma 337, 32–41.

Guo, L., Fu, P., Shi, T., Chen, Y., Zhang, H., Meng, R., et al., 2020. Mapping field-scale soil organic carbon with unmanned aircraft system-acquired time series multispectral images. Soil. Tillage Res. 196, 104477.

Guo, L., Sun, X., Fu, P., Shi, T., Dang, L., Chen, Y., et al., 2021. Mapping soil organic carbon stock by hyperspectral and time-series multispectral remote sensing images in low-relief agricultural areas. Geoderma 398, 115118.

Han, L., Wan, Z., Guo, Y., Song, C., Jin, S., Zuo, Y., 2020. Estimation of Soil Organic Carbon Storage in Palustrine Wetlands, China. Int. J. Environ. Res. Public. Health 17 (13), 4646.

Hari, M., Srinivasan, S., Rajasekaran, A., Tyagi, B., 2021. Above ground carbon stock mapping over Coimbatore and Nilgiris Biosphere: a key source to the C sink. Carbon Manag. 12 (4), 411–428.

Hbirkou, C., Pätzold, S., Mahlein, A.K., Welp, G., 2012. Airborne hyperspectral imaging of spatial soil organic carbon heterogeneity at the field-scale. Geoderma 175, 21–28.

Henderson, T.L., Baumgardner, M.F., Franzmeier, D.P., Stott, D.E., Coster, D.C., 1992. High dimensional reflectance analysis of soil organic matter. Soil. Sci. Soc. Am. J. 56, 865–872.

Holmquist, J.R., Windham-Myers, L., Bliss, N., Crooks, S., Morris, J.T., Megonigal, J.P., et al., 2018. Accuracy and precision of tidal wetland soil carbon mapping in the conterminous United States. Sci. Rep. 8 (1), 1–16.

Homolová, L., Schaepman, M.E., Lamarque, P., Clevers, J.G., de Bello, F., Thuiller, W., et al., 2014. Comparison of remote sensing and plant trait-based modelling to predict ecosystem services in subalpine grasslands. Ecosphere 5 (8), 1–29.

Hong, Y., Guo, L., Chen, S., Linderman, M., Mouazen, A.M., Yu, L., et al., 2020a. Exploring the potential of airborne hyperspectral image for estimating topsoil organic carbon: effects of fractional-order derivative and optimal band combination algorithm. Geoderma 365, 114228.

Hong, Y., Chen, S., Chen, Y., Linderman, M., Mouazen, A.M., Liu, Y., et al., 2020b. Comparing laboratory and airborne hyperspectral data for the estimation and mapping of topsoil organic carbon: feature selection coupled with random forest. Soil. Tillage Res. 199, 104589.

Huete, A.R., Restrepo-Coupe, N., Ratana, P., Didan, K., Saleska, S.R., Ichii, K., et al., 2008. Multiple site tower flux and remote sensing comparisons of tropical forest dynamics in Monsoon Asia. Agric. For. Meteorol. 148 (5), 748–760.

Issa, S., Dahy, B., Ksiksi, T., Saleous, N., 2020. A review of terrestrial carbon assessment methods using geo-spatial technologies with emphasis on arid lands. Remote. Sens. 12 (12), 2008.

Jafarbiglu, H., Pourreza, A., 2022. A comprehensive review of remote sensing platforms, sensors, and applications in nut crops. Comput. Electron. Agric. 197, 106844.

Jakob, S., Zimmermann, R., Gloaguen, R., 2017. The need for accurate geometric and radiometric corrections of drone-borne hyperspectral data for mineral exploration: Mephysto—a toolbox for pre-processing drone-borne hyperspectral data. Remote. Sens. 9 (1), 88.

Janowiak, M., Connelly, W.J., Dante-Wood, K., Domke, G.M., Giardina, C., Kayler, Z., et al., 2017. Considering forest and grassland carbon in land management. Gen. Tech. Rep., Wash. Off. 95.

Janzen, H.H., 2004. Carbon cycling in earth systems—a soil science perspective. Agric., Ecosyst. Environ. 104 (3), 399–417.

Jensen, J.R., 2009. Remote Sensing of the Environment: An Earth Resource Perspective 2/e. Pearson Education, India.

Jimenez, J.J., Lal, R., 2006. Mechanisms of C sequestration in soils of Latin America. Crit. Rev. Plant. Sci. 25 (4), 337–365.

Joosten, H., Couwenberg, J., 2008. Peatlands and carbon. Assessment on Peatlands, Biodiversity and Climate Change 99–117.

Kashung, Y., Das, B., Deka, S., Bordoloi, R., Paul, A., Tripathi, O.P., 2018. Geospatial technology based diversity and above ground biomass assessment of woody species of West Kameng district of Arunachal Pradesh. For. Sci. Technol. 14 (2), 84–90.

Kemmitt, S.J., Lanyon, C.V., Waite, I.S., Wen, Q., Addiscott, T.M., Bird, N.R., et al., 2008. Mineralization of native soil organic matter is not regulated by the size, activity or composition of the soil microbial biomass—a new perspective. Soil. Biol. Biochem. 40 (1), 61–73.

Khanal, S., Fulton, J., Klopfenstein, A., Douridas, N., Shearer, S., 2018. Integration of high resolution remotely sensed data and machine learning techniques for spatial prediction of soil properties and corn yield. Comput. Electron. Agri. 153, 213–225.

Kimball, J.S., Jones, L.A., Zhang, K., Heinsch, F.A., McDonald, K.C., Oechel, W.C., 2009. A satellite approach to estimate land–atmosphere CO_2 exchange for Boreal and Arctic biomes using MODIS and AMSR-E. IEEE Trans. Geosci. Remote. Sens. 47 (2), 569–587.

Köchy, M., Hiederer, R., Freibauer, A., 2015. Global distribution of soil organic carbon–Part 1: masses and frequency distributions of SOC stocks for the tropics, permafrost regions, wetlands, world. Soil 1 (1), 351–365.

Kumar, P., Pandey, P.C., Singh, B.K., Katiyar, S., Mandal, V.P., Rani, M., et al., 2016. Estimation of accumulated soil organic carbon stock in tropical forest using geospatial strategy. Egypt. J. Remote. Sens. Space Sci. 19 (1), 109–123.

Ladoni, M., Bahrami, H.A., Alavipanah, S.K., Norouzi, A.A., 2010. Estimating soil organic carbon from soil reflectance: a review. Precis. Agric. 11 (1), 82–99.

Lal, R., 1997. Residue management, conservation tillage and soil restoration for mitigating greenhouse effect by CO2-enrichment. Soil. Tillage Res. 43 (1–2), 81–107.

Lal, R., Bouma, J., Brevik, E., Dawson, L., Field, D.J., Glaser, B., et al., 2021. Soils and sustainable development goals of the United Nations: an International Union of Soil Sciences perspective. Geoderma Reg. 25, e00398.

Le, N.N., Pham, T.D., Yokoya, N., Ha, N.T., Nguyen, T.T.T., Tran, T.D.T., et al., 2021. Learning from multimodal and multisensor earth observation dataset for improving estimates of mangrove soil organic carbon in Vietnam. Int. J. Remote. Sens. 42 (18), 6866–6890.

Leboeuf, A., Beaudoin, A., Fournier, R.A., Guindon, L., Luther, J.E., Lambert, M.C., 2007. A shadow fraction method for mapping biomass of northern boreal black spruce forests using QuickBird imagery. Remote. Sens. Environ. 110 (4), 488–500.

Levin, N., Ben-Dor, E., Singer, A., 2005. A digital camera as a tool to measure colour indices and related properties of sandy soils in semi-arid environments. Int. J. Remote. Sens. 26 (24), 5475–5492.

Li, Q.Q., Zhang, X., Wang, C.Q., Li, B., Gao, X.S., Yuan, D.G., et al., 2016. Spatial prediction of soil nutrient in a hilly area using artificial neural network model combined with kriging. Arch. Agron. Soil. Sci. 62 (11), 1541–1553.

Liang, S., Wang, J. (Eds.), 2019. Advanced Remote Sensing: Terrestrial Information Extraction and Applications. Academic Press.

Liakos, K.G., Busato, P., Moshou, D., Pearson, S., Bochtis, D., 2018. Machine learning in agriculture: a review. Sensors 18 (8), 2674.

Liang, X., Kankare, V., Hyyppä, J., Wang, Y., Kukko, A., Haggrén, H., et al., 2016. Terrestrial laser scanning in forest inventories. ISPRS J. Photogramm. Remote. Sens. 115, 63–77.

Liu, Y.Y., Van Dijk, A.I., De Jeu, R.A., Canadell, J.G., McCabe, M.F., Evans, J.P., et al., 2015. Recent reversal in loss of global terrestrial biomass. Nature Climate Change 5 (5), 470–474.

Lu, B., Liu, N., Li, H., Yang, K., Hu, C., Wang, X., et al., 2019. Quantitative determination and characteristic wavelength selection of available nitrogen in coco-peat by NIR spectroscopy. Soil. Tillage Res. 191, 266–274.

Luo, M., Huang, J.F., Zhu, W.F., Tong, C., 2019. Impacts of increasing salinity and inundation on rates and pathways of organic carbon mineralization in tidal wetlands: a review. Hydrobiologia 827 (1), 31–49.

Macreadie, P.I., Ollivier, Q.R., Kelleway, J.J., Serrano, O., Carnell, P.E., Ewers Lewis, C.J., et al., 2017. Carbon sequestration by Australian tidal marshes. Sci. Rep. 7 (1), 1–10.

Manfreda, S., McCabe, M.F., Miller, P.E., Lucas, R., Pajuelo Madrigal, V., Mallinis, G., et al., 2018. On the use of unmanned aerial systems for environmental monitoring. Remote. Sens. 10 (4), 641.

Margenot, A., O'Neill, T., Sommer, R., Akella, V., 2020. Predicting soil permanganate oxidizable carbon (POXC) by coupling DRIFT spectroscopy and artificial neural networks (ANN). Comput. Electron. Agri. 168, 105098.

Masemola, C., Cho, M.A., Ramoelo, A., 2020. Towards a semi-automated mapping of Australia native invasive alien Acacia trees using Sentinel-2 and radiative transfer models in South Africa. ISPRS J. Photogramm. Remote. Sens. 166, 153–168.

Matese, A., Toscano, P., Di Gennaro, S.F., Genesio, L., Vaccari, F.P., Primicerio, J., et al., 2015. Intercomparison of UAV, aircraft and satellite remote sensing platforms for precision viticulture. Remote. Sens. 7 (3), 2971–2990.

McCarty, G.W., Reeves, J.B., 2006. Comparisons of near infrared and mid infrared diffuse reflectance spectroscopy for field-scale measurement of soil fertility parameters. Soil. Sci. 171, 94–102.

Minu, S., Shetty, A., Gopal, B., 2016. Review of preprocessing techniques used in soil property prediction from hyperspectral data. Cogent. Geosci. 2 (1), 1145878.

Mirzaee, S., Ghorbani-Dashtaki, S., Mohammadi, J., Asadi, H., Asadzadeh, F., 2016. Spatial variability of soil organic matter using remote sensing data. Catena 145, 118–127.

Mitran, T., Mishra, U., Lal, R., Ravisankar, T., Sreenivas, K., 2018. Spatial distribution of soil carbon stocks in a semi-arid region of India. Geoderma Reg. 15, e00192.

Momo, N., Devi, T.T., 2022. Assessment of land surface temperature and carbon sequestration using remotely sensed satellite data in the Imphal-West district, Manipur, India. J. Earth Syst. Sci. 131 (4), 1–25.

Mondal, A., Khare, D., Kundu, S., Mondal, S., Mukherjee, S., Mukhopadhyay, A., 2017. Spatial soil organic carbon (SOC) prediction by regression kriging using remote sensing data. Egypt. J. Remote. Sens. Space Sci. 20 (1), 61–70.

Morsy, S., Shaker, A., El-Rabbany, A., 2017. Multispectral LiDAR data for land cover classification of urban areas. Sensors 17 (5), 958.

Nguyen, T.T., Pham, T.D., Nguyen, C.T., Delfos, J., Archibald, R., Dang, K.B., et al., 2022. A novel intelligence approach based active and ensemble learning for agricultural soil organic carbon prediction using multispectral and SAR data fusion. Sci. Total. Environ. 804, 150187.

Ni, W., Sun, G., Zhang, Z., Yu, H., 2016. Retrieval of forest aboveground biomass through the synergy of X-band (TeraSAR-X/TandDEM-X) and L-band (PALSAR-2) In SAR data. 2016 IEEE International Geoscience and Remote Sensing Symposium (IGARSS). IEEE, pp. 5304–5306.

Nieder, R., Benbi, D.K., 2008. Carbon and Nitrogen in the Terrestrial Environment. Springer Science & Business Media.

Nocita, M., Stevens, A., van Wesemael, B., Aitkenhead, M., Bachmann, M., Barthès, B., et al., 2015. Soil spectroscopy: an alternative to wet chemistry for soil monitoring. Adv. Agron. 132, 139–159.

Noirot-Cosson, P.E., Vaudour, E., Gilliot, J.M., Gabrielle, B., Houot, S., 2016. Modelling the long-term effect of urban waste compost applications on carbon and nitrogen dynamics in temperate cropland. Soil. Biol. Biochem. 94, 138–153.

Nouri, M., Gomez, C., Gorretta, N., Roger, J.M., 2017. Clay content mapping from airborne hyperspectral Vis-NIR data by transferring a laboratory regression model. Geoderma 298, 54–66.

Omidi, R., Moghimi, A., Pourreza, A., El-Hadedy, M., Eddin, A.S., 2020. Ensemble Hyperspectral Band Selection for Detecting Nitrogen Status in Grape Leaves. 2020 19th IEEE International Conference on Machine Learning and Applications (ICMLA). IEEE, pp. 286–293.

Omran, E.S.E., 2017. Rapid prediction of soil mineralogy using imaging spectroscopy. Eurasian Soil. Sci. 50 (5), 597–612.

Palace, M., Keller, M., Asner, G.P., Hagen, S., Braswell, B., 2008. Amazon forest structure from IKONOS satellite data and the automated characterization of forest canopy properties. Biotropica 40 (2), 141–150.

Pandey, P.C., Koutsias, N., Petropoulos, G.P., Srivastava, P.K., Ben Dor, E., 2021. Land use/land cover in view of earth observation: data sources, input dimensions, and classifiers—a review of the state of the art. Geocarto Int. 36 (9), 957–988.

Paz-Kagan, T., Zaady, E., Salbach, C., Schmidt, A., Lausch, A., Zacharias, S., et al., 2015. Mapping the spectral soil quality index (SSQI) using airborne imaging spectroscopy. Remote. Sens. 7 (11), 15748–15781.

Peng, Y., Xiong, X., Adhikari, K., Knadel, M., Grunwald, S., Greve, M.H., 2015. Modeling soil organic carbon at regional scale by combining multispectral images with laboratory spectra. PLoS One 10 (11), e0142295.

Peón, J., Fernández, S., Recondo, C., Calleja, J.F., 2017a. Evaluation of the spectral characteristics of five hyperspectral and multispectral sensors for soil organic carbon estimation in burned areas. Int. J. Wildland Fire 26 (3), 230–239.

Peón, J., Recondo, C., Fernández, S., Calleja, J.F., De Miguel, E., Carretero, et al., 2017b. Prediction of topsoil organic carbon using airborne and satellite hyperspectral imagery. Remote. Sens. 9 (12), 1211.

Pittman, R., Hu, B., 2020. Estimation of soil bulk density and carbon using multi-source remotely sensed data. ISPRS Ann. Photogramm., Remote. Sens. Spat. Inf. Sci. 3, 541–548.

Qu, W., Li, J., Han, G., Wu, H., Song, W., Zhang, X., 2019. Effect of salinity on the decomposition of soil organic carbon in a tidal wetland. J. Soils Sediment. 19 (2), 609–617.

Rahman, A.F., Sims, D.A., Cordova, V.D., El-Masri, B.Z., 2005. Potential of MODIS EVI and surface temperature for directly estimating per-pixel ecosystem C fluxes. Geophys. Res. Lett. 32 (19).
Rasmussen, C., Heckman, K., Wieder, W.R., Keiluweit, M., Lawrence, C.R., Berhe, A.A., et al., 2018. Beyond clay: towards an improved set of variables for predicting soil organic matter content. Biogeochemistry 137 (3), 297–306.
Ren, Y., Li, X., Mao, D., Wang, Z., Jia, M., Chen, L., 2020. Investigating spatial and vertical patterns of wetland soil organic carbon concentrations in China's Western Songnen plain by comparing different algorithms. Sustainability 12 (3), 932.
Rizvi, R.H., Handa, A.K., Sridhar, K.B., Singh, R.K., Dhyani, S.K., Rizvi, J., et al., 2020. Spatial analysis of area and carbon stocks under Populus deltoides based agroforestry systems in Punjab and Haryana states of Indo-Gangetic Plains. Agrofor. Syst. 94 (6), 2185–2197.
Rossel, R.A.V., Walvoort, D.J.J., McBratney, A.B., Janik, L.J., Skjemsta, J.O., 2006. Visible, near infrared, mid infrared or combined diffuse reflectance spectroscopy for simultaneous assessment of various soil properties. Geoderma 131, 59–75.
Rossel, R.V., Fouad, Y., Walter, C., 2008. Using a digital camera to measure soil organic carbon and iron contents. Biosyst. Eng. 100 (2), 149–159.
Salehi Hikouei, I., Kim, S.S., Mishra, D.R., 2021. Machine-learning classification of soil bulk density in salt marsh environments. Sensors 21 (13), 4408.
Sarathjith, M.C., Das, B.S., Wani, S.P., Sahrawat, K.L., 2016. Variable indicators for optimum wavelength selection in diffuse reflectance spectroscopy of soils. Geoderma 267, 1–9.
Scharlemann, J.P., Tanner, E.V., Hiederer, R., Kapos, V., 2014. Global soil carbon: understanding and managing the largest terrestrial carbon pool. Carbon Manag. 5 (1), 81–91.
Schillaci, C., Acutis, M., Lombardo, L., Lipani, A., Fantappie, M., Märker, M., et al., 2017. Spatio-temporal topsoil organic carbon mapping of a semi-arid Mediterranean region: the role of land use, soil texture, topographic indices and the influence of remote sensing data to modelling. Sci. Total. Environ. 601, 821–832.
Schimel, D.S., 1995. Terrestrial ecosystems and the carbon cycle. Glob. Change Biol. 1 (1), 77–91.
Schmugge, T.J., Jackson, T.J., McKim, H.L., 1980. Survey of methods for soil moisture determination. Water Resour. Res. 16 (6), 961–979.
Schwartz, G., Ben-Dor, E., Eshel, G., 2012. Quantitative analysis of total petroleum hydrocarbons in soils: comparison between reflectance spectroscopy and solvent extraction by 3 certified laboratories. Appl. Environ. Soil. Sci. 2012.
Sharma, R., Mishra, D.R., Levi, M.R., Sutter, L.A., 2022. Remote sensing of surface and subsurface soil organic carbon in Tidal Wetlands: a review and ideas for future research. Remote. Sens. 14 (12), 2940.
Shen, L., Gao, M., Yan, J., Li, Z.L., Leng, P., Yang, Q., et al., 2020. Hyperspectral estimation of soil organic matter content using different spectral preprocessing techniques and PLSR method. Remote. Sens. 12 (7), 1206.
Shepherd, K.D., Walsh, M.G., 2002. Development of reflectance spectral libraries for characterization of soil properties. Soil. Sci. Soc. Am. J. 66, 988–998.
Shoshany, M., Karnibad, L., 2011. Mapping shrubland biomass along Mediterranean climatic gradients: the synergy of rainfall-based and NDVI-based models. Int. J. Remote. Sens. 32 (24), 9497–9508.
Sibanda, M., Mutanga, O., Rouget, M., 2015. Examining the potential of Sentinel-2 MSI spectral resolution in quantifying above ground biomass across different fertilizer treatments. ISPRS J. Photogramm. Remote. Sens. 110, 55–65.
Sibanda, M., Mutanga, O., Rouget, M., Kumar, L., 2017. Estimating biomass of native grass grown under complex management treatments using worldview-3 spectral derivatives. Remote. Sens. 9 (1), 55.
Simard, M., Pinto, N., Fisher, J.B., Baccini, A., 2011. Mapping forest canopy height globally with spaceborne lidar. J. Geophys. Res.: Biogeosci. 116 (G4).
Sjostrom, M., Ardo, J., Arneth, A., Boulain, N., Cappelaere, B., Eklundh, L., et al., 2011. Exploring the potential of MODIS EVI for modeling gross primary production across African ecosystems. Remote. Sens. Environ. 115 (4), 1081–1089.
Sreenivas, K., Dadhwal, V.K., Kumar, S., Harsha, G.S., Mitran, T., Sujatha, G., et al., 2016. Digital mapping of soil organic and inorganic carbon status in India. Geoderma 269, 160–173.
Steinberg, A., Chabrillat, S., Stevens, A., Segl, K., Foerster, S., 2016. Prediction of common surface soil properties based on Vis-NIR airborne and simulated EnMAP imaging spectroscopy data: prediction accuracy and influence of spatial resolution. Remote. Sens. 8 (7), 613.
Stenberg, B., Rossel, R.A., 2010. Diffuse reflectance spectroscopy for high-resolution soil sensing. Proximal Soil Sensing. Springer, Dordrecht, pp. 29–47.
Stevens, A., van Wesemael, B., Bartholomeus, H., Rosillon, D., Tychon, B., Ben-Dor, E., 2008. Laboratory, field and airborne spectroscopy for monitoring organic carbon content in agricultural soils. Geoderma 144 (1–2), 395–404.
Stevens, A., Udelhoven, T., Denis, A., Tychon, B., Lioy, R., Hoffmann, L., et al., 2010. Measuring soil organic carbon in croplands at regional scale using airborne imaging spectroscopy. Geoderma 158 (1–2), 32–45.
Stöcker, C., Bennett, R., Nex, F., Gerke, M., Zevenbergen, J., 2017. Review of the current state of UAV regulations. Remote. Sens. 9 (5), 459.
Stockmann, U., Adams, M.A., Crawford, J.W., Field, D.J., Henakaarchchi, N., Jenkins, M., et al., 2013. The knowns, known unknowns and unknowns of sequestration of soil organic carbon. Agric., Ecosyst. Environ. 164, 80–99.
Stovall, A.E., Vorster, A.G., Anderson, R.S., Evangelista, P.H., Shugart, H.H., 2017. Non-destructive aboveground biomass estimation of coniferous trees using terrestrial LiDAR. Remote. Sens. Environ. 200, 31–42.
Taghizadeh-Mehrjardi, R., Nabiollahi, K., Kerry, R., 2016. Digital mapping of soil organic carbon at multiple depths using different data mining techniques in Baneh region, Iran. Geoderma 266, 98–110.
Tripathi, O.P., Deka, J., Yumnam, J.Y., Mahanta, P., 2015. Improvement in estimation of aboveground biomass of albizia lebbeck using fraction reflectance of landsat TM data. Int. J. Plant. Environ. 1 (01), 99–105.
Vastaranta, M., Holopainen, M., Karjalainen, M., Kankare, V., Hyyppä, J., Kaasalainen, S., et al., 2012. SAR radargrammetry and scanning LiDAR in predicting forest canopy height. 2012 IEEE International Geoscience and Remote Sensing Symposium. IEEE, pp. 6515–6518.

Vaudour, E., Gilliot, J.M., Bel, L., Lefevre, J., Chehdi, K., 2016. Regional prediction of soil organic carbon content over temperate croplands using visible near-infrared airborne hyperspectral imagery and synchronous field spectra. Int. J. Appl. Earth Obs. Geoinf. 49, 24–38.

Vaudour, E., Gomez, C., Loiseau, T., Baghdadi, N., Loubet, B., Arrouays, D., et al., 2019. The impact of acquisition date on the prediction performance of topsoil organic carbon from Sentinel-2 for croplands. Remote. Sens. 11 (18), 2143.

Vohland, M., Harbich, M., Ludwig, M., Emmerling, C., Thiele-Bruhn, S., 2016. Quantification of soil variables in a heterogeneous soil region with VIS–NIR–SWIR data using different statistical sampling and modeling strategies. IEEE J. Sel. Top. Appl. Earth Obs. Remote. Sens. 9 (9), 4011–4021.

Wang, B., Waters, C., Orgill, S., Gray, J., Cowie, A., Clark, A., et al., 2018. High resolution mapping of soil organic carbon stocks using remote sensing variables in the semi-arid rangelands of eastern Australia. Sci. Total. Environ. 630, 367–378.

Wang, F., Kroeger, K.D., Gonneea, M.E., Pohlman, J.W., Tang, J., 2019. Water salinity and inundation control soil carbon decomposition during salt marsh restoration: an incubation experiment. Ecol. Evol. 9 (4), 1911–1921.

Wang, H., Zhang, X., Wu, W., Liu, H., 2021. Prediction of soil organic carbon under different land use types using sentinel-1/-2 data in a small watershed. Remote. Sens. 13 (7), 1229.

Wang, S., Guan, K., Zhang, C., Lee, D., Margenot, A.J., Ge, Y., et al., 2022. Using soil library hyperspectral reflectance and machine learning to predict soil organic carbon: assessing potential of airborne and spaceborne optical soil sensing. Remote. Sens. Environ. 271, 112914.

Ward, K.J., Chabrillat, S., Brell, M., Castaldi, F., Spengler, D., Foerster, S., 2020. Mapping soil organic carbon for airborne and simulated enmap imagery using the LUCAS soil database and a Local PLSR. Remote. Sens. 12 (20), 3451.

Watts, A.C., Ambrosia, V.G., Hinkley, E.A., 2012. Unmanned aircraft systems in remote sensing and scientific research: classification and considerations of use. Remote. Sens. 4 (6), 1671–1692.

Wehr, R., Munger, J.W., McManus, J.B., Nelson, D.D., Zahniser, M.S., Davidson, E.A., et al., 2016. Seasonality of temperate forest photosynthesis and daytime respiration. Nature 534 (7609), 680–683.

Wehrhan, M., Sommer, M., 2021. A parsimonious approach to estimate soil organic carbon applying unmanned aerial system (UAS) multispectral imagery and the topographic position index in a heterogeneous soil landscape. Remote. Sens. 13 (18), 3557.

Were, K., Bui, D.T., Dįçk, Ø.B., Singh, B.R., 2015. A comparative assessment of support vector regression, artificial neural networks, and random forests for predicting and mapping soil organic carbon stocks across an Afromontane landscape. Ecol. Indic. 52, 394–403.

West, K., 2008. Essential Chemistry: Carbon Chemistry. Chelsea House An imprint of Infobase Publishing, p. 117.

Wijewardane, N.K., Ge, Y., Wills, S., Libohova, Z., 2018. Predicting physical and chemical properties of US soils with a mid-infrared reflectance spectral library. Soil. Sci. Soc. Am. J. 82 (3), 722–731.

Xiao, J., Ollinger, S.V., Frolking, S., Hurtt, G.C., Hollinger, D.Y., Davis, K.J., et al., 2014. Data-driven diagnostics of terrestrial carbon dynamics over North America. Agric. For. Meteorol. 197, 142–157.

Xiao, J., Chevallier, F., Gomez, C., Guanter, L., Hicke, J.A., Huete, A.R., et al., 2019. Remote sensing of the terrestrial carbon cycle: a review of advances over 50 years. Remote. Sens. Environ. 233, 111383.

Yang, R.M., 2019. Interacting effects of plant invasion, climate, and soils on soil organic carbon storage in coastal wetlands. J. Geophys. Res.: Biogeosci. 124 (8), 2554–2564.

Yang, R.M., Guo, W.W., 2019. Modelling of soil organic carbon and bulk density in invaded coastal wetlands using Sentinel-1 imagery. Int. J. Appl. Earth Obs. Geoinf. 82, 101906.

Yang, M., Xu, D., Chen, S., Li, H., Shi, Z., 2019. Evaluation of machine learning approaches to predict soil organic matter and pH using Vis-NIR spectra. Sensors 19 (2), 263.

Yang, R.M., Zhang, G.L., Liu, F., Lu, Y.Y., Yang, F., Yang, F., et al., 2016. Comparison of boosted regression tree and random forest models for mapping topsoil organic carbon concentration in an alpine ecosystem. Ecol. Indic. 60, 870–878.

Yu, J., Wang, Y., Li, Y., Dong, H., Zhou, D., Han, G., et al., 2012. Soil organic carbon storage changes in coastal wetlands of the modern Yellow River Delta from 2000 to 2009. Biogeosciences 9 (6), 2325–2331.

Zhai, M., 2019. Inversion of organic matter content in wetland soil based on Landsat 8 remote sensing image. J. Vis. Commun. Image Rep. 64, 102645.

Zhang, Y., Guo, L., Chen, Y., Shi, T., Luo, M., Ju, Q., et al., 2019a. Prediction of soil organic carbon based on Landsat 8 monthly NDVI data for the Jianghan Plain in Hubei Province, China. Remote. Sens. 11 (14), 1683.

Zhang, C., Mishra, D.R., Pennings, S.C., 2019b. Mapping salt marsh soil properties using imaging spectroscopy. ISPRS J. Photogramm. Remote. Sens. 148, 221–234.

Zhao, T., 2018. Remote Sensing of Water Stress in Almond Trees Using Unmanned Aerial Vehicles. University of California, Merced.

Zhou, T., Geng, Y., Chen, J., Liu, M., Haase, D., Lausch, A., 2020. Mapping soil organic carbon content using multi-source remote sensing variables in the Heihe River Basin in China. Ecol. Indic. 114, 106288.

Zhou, Y., Zhao, X., Guo, X., Li, Y., 2022. Mapping of soil organic carbon using machine learning models: combination of optical and radar remote sensing data. Soil. Sci. Soc. Am. J. 86 (2), 293–310.

Žížala, D., Zádorová, T., Kapička, J., 2017. Assessment of soil degradation by erosion based on analysis of soil properties using aerial hyperspectral images and ancillary data, Czech Republic. Remote. Sens. 9 (1), 28.

Žížala, D., Minařík, R., Zádorová, T., 2019. Soil organic carbon mapping using multispectral remote sensing data: prediction ability of data with different spatial and spectral resolutions. Remote. Sens. 11 (24), 2947.

Chapter 18

Geospatial analysis of heavy metal contamination in soil and groundwater: a case study

Prarthana P. Hazarika[1], B.K. Medhi[1], R.K. Thakuria[2], A.N. Kondareddy[1] and Saurav Das[3]
[1]*Department of Soil Science, Assam Agricultural University, Jorhat, Assam, India,* [2]*Horticultural Research Station, Assam Agricultural University, Jorhat, Assam, India,* [3]*Department of Agronomy and Horticulture, University of Nebraska, Lincoln, NE, United States*

18.1 Introduction

Heavy metals pose a signficant threat to human health. Contaminants with concentrations above the standard permissible limit can cause chronic health disorders. The acute toxicity of heavy metals, coupled with their widespread distribution in soil and groundwater has become challenging to devise an effective remediation approach (Zaynab et al., 2022). A notable concern in contemporary environmental studies is the biomagnification of heavy metals in the food chain, exacerbated by increased anthropogenic deposition. As the global population grows, ensuring the provision of safe food and water with minimal environmental repercussions has become paramount (Raju, 2022). Chronic exposure to heavy metals and metalloids can cause severe damage to vital organs like the kidneys, liver, lungs, brain, and bones (Balali-Mood et al., 2021). Assessing heavy metals in soil and groundwater is critical to identify risk-prone zones and better predict the fate and transport of heavy metals in the environment and ecosystem. In most developing countries, communities primarily depends on groundwater and surface water for daily needs. Evaluation of heavy metals in soil and groundwater and their horizontal distribution with the class categorization from minimum to maximum toxicity over a large area can help planners, policymakers, and other government stakeholders to make informed decisions.

The Geographic Information System (GIS) has advanced remarkably since the pioneering efforts of individuals like Dr. John Snow. In 1854, Snow meticulously plotted cholera cases in London's Soho district on a map, leading him to identify a contaminated water pump as the outbreak's epicenter. This groundbreaking work laid the foundation for modern epidemiological studies. Today, GIS offers pivotal insights into prevailing pollution levels and facilitates the demarcation of high-risk zones using intricate mapping techniques (Mohammadi et al., 2020). Mapping the spatial distribution of contaminants in groundwater is the basis for the identification of heavy metal loads, this offers researchers a reference to assess the potential contamination threats and devise appropriate mitigation measures. Geostatistical approaches for surface interpolations are emerging as the most sought-after methods for researchers. Models like Inverse Distance Weighted (IDW) Interpolation, Global Polynomial Index, Local Polynomial Index, Kriging, Kernel Smoothing, and Diffusion Kernel, help researchers interpret and validate the results based on key parameters. This book chapter will introduce basic concepts in geospatial and geostatistical analysis for mapping heavy metal contamination and will explain the process with an example case study.

18.2 Case study

The case study was performed in Nagaon district, Assam, India (Fig. 18.1), to evaluate the geospatial methodology in mapping the distribution of heavy metals, including cadmium (Cd), chromium (Cr), copper (Cu), zinc (Zn), and manganese (Mn).

Remote Sensing of Soils. DOI: https://doi.org/10.1016/B978-0-443-18773-5.00035-1

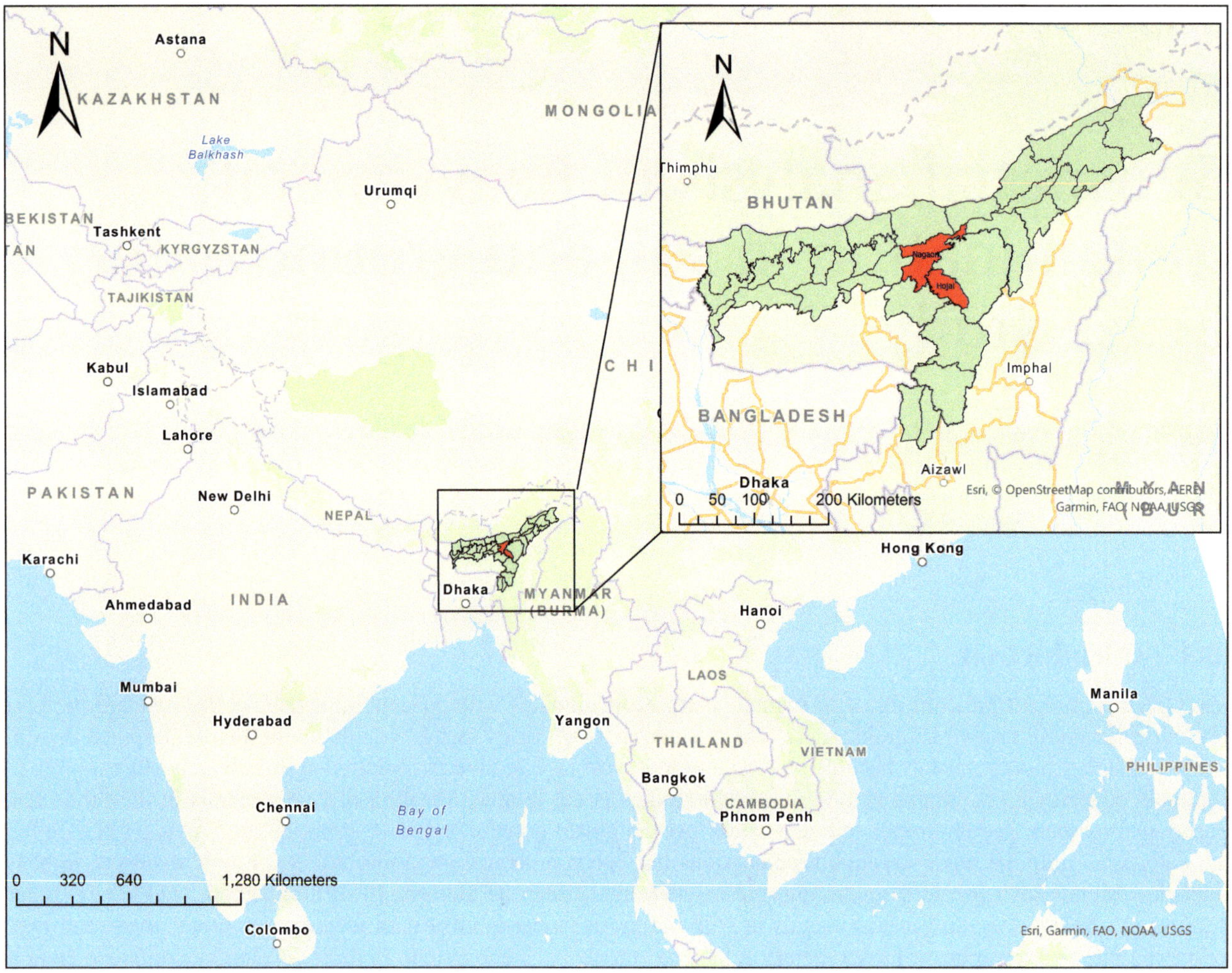

FIGURE 18.1 (A) A map of India highlights Nagaon district in red. (B) A closer view emphasizes Nagaon district, also in red. Note: Historically, the Nagaon and Hojai districts were unified as "Nagaon District". On 15 August 2015, Hojai was separated and established as an independent district. However, on 31 December 2022, Hojai was reintegrated into Nagaon District. For the purposes of this study, we considered the entire Nagaon district, inclusive of both Nagaon and Hojai.

18.3 Study site

Nagaon is an administrative district in the Indian state of Assam. The district covers an area of 3993 km^2 and lies between 25°45′ and 26°45′ North latitudes and 91°50′ and 93°20′ East longitudes. To the north, the district is bounded by the river Brahmaputra, beyond which lies the Sonitpur district. The south borders the Karbi Anglong and Dima Hasao districts, while to the east, it's flanked by the Golaghat and Karbi-Anglong districts. The Morigaon district, once a part of Nagaon, lies to its west. The district's topography is predominantly a flat plain with minor undulation. Average annual rainfall is 1726.90 mm with 78% humidity. The districts consist of four Urban Local Bodies i.e.,the Nagaon Municipal Board with 26 wards (or blocks/divisions), Kampur Town Committee with 5 wards, Raha Town Committee with 7 wards, and Dhing Municipal Board with 10 wards. The district headquarters lies at Nagaon Sadar. As per the 2011 census, the total population of the entire district was 1,57,084.

18.4 Sampling methodology

For this study, the district systematicallydivided into 160 grids, each measuring 5×5 km^2, ensuring comprehensive representaiton of the entire area (Fig. 18.2). From every grid, one soil (depth 0–15 cm) and one groundwater sample were collected following the standard sampling protocols. The geographic location of each sampling site was recorded using a handheld global positioning system unit (Garmin eTrex 32x) (Fig. 18.2).

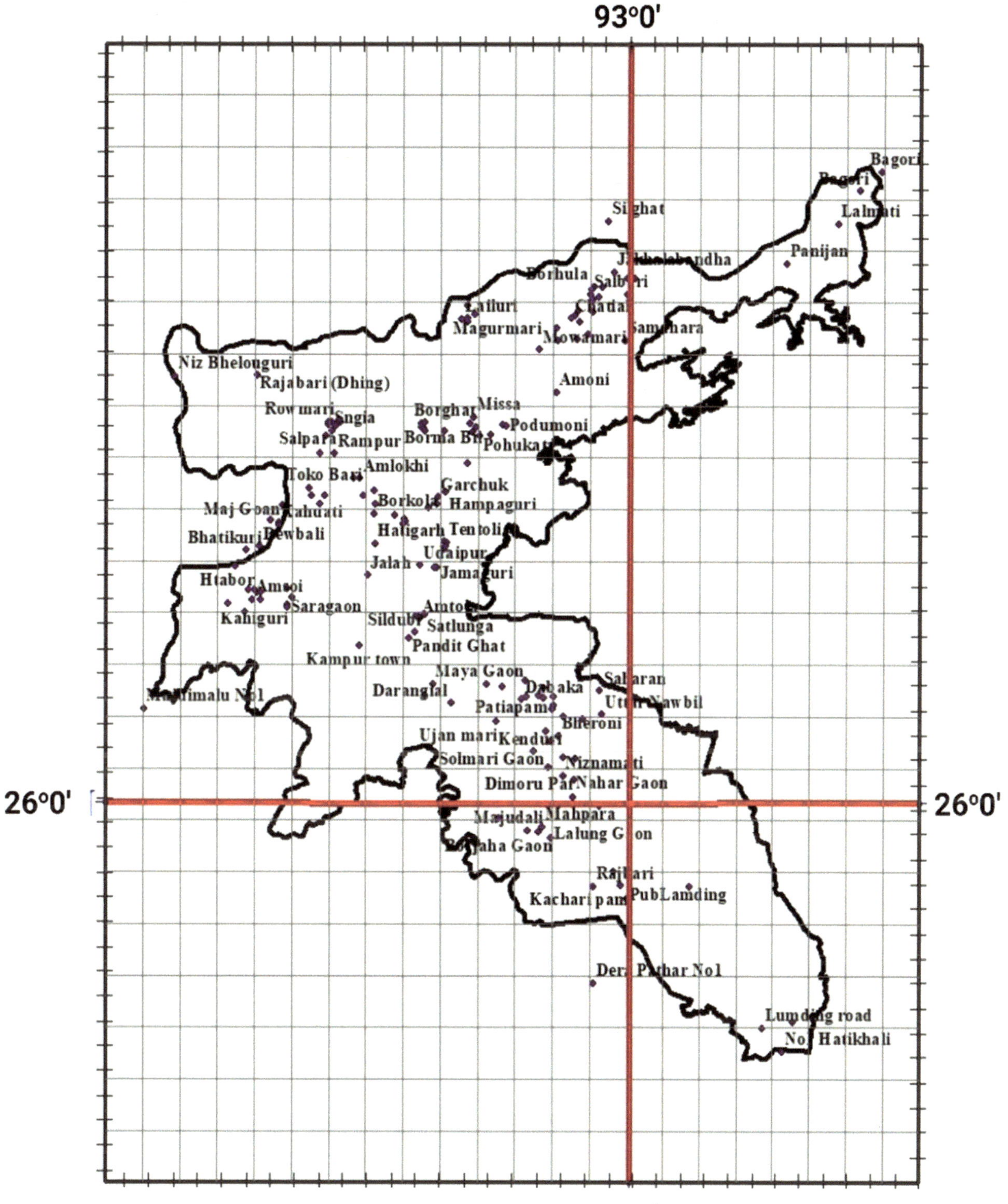

FIGURE 18.2 Sampling grids. Each sampling site is represented as a purple dot in the map. Each sampling grid is of size 5×5 km^2.

18.5 Analysis of soil and water samples

Heavy metals, including Cadmium (Cd), Chromium (Cr), Copper (Cu), Zinc (Zn), and Manganese (Mn) in soil samples were determined by HNO_3-$HClO_4$ acid digestion method and then analyzed in atomic absorption spectrophotometer (Model iCE 3500, Thermofisher) using acetylene flame as per the procedure described by Watson and Isaac (1990). For groundwater samples, the analysis of these metals was carried out in line with the methodologies detailed by Islam et al. (2015).

18.6 Geostatistical approaches

In assessing the distribution pattern of heavy metals, the Inverse Distance–Weighted (IDW) interpolation and trend analyses are widely used. IDW is an interpolation technique that estimates the unknown values based on their distance from known data points. Essentially, it works on the assumption that the points close to one another are more alike or similar compared to the points that are farther apart. As such, IDW allocates more weight to the nearest known data points, and this weight decreases as the distance increases (Fig. 18.3A). Several studies used the IDW for measuring and mapping soil and groundwater pollution (Gong et al., 2014; Ouabo et al., 2020; Qi et al., 2020; Qiao et al., 2018; Xie et al., 2011). Trend analysis, in contrast, provides a three-dimensional perspective of spatial data. Here, sampling points are plotted on the x and y coordinates, and their associated values are plotted on the Z coordinates, which gives a height or magnitude . Polynomial regression models are then fitted through a scatter plot on the projected planes to understand the trend and spatial distribution of the data (Fig. 18.3B).

18.7 Results & discussions

18.7.1 Heavy metal concentrations and interpolation

In our study, concentration of heavy metals, including Cr, Cu, Zn, and Mn, were found to be within the accepted safety limits for soil. However, Cd was exception and had higher concentration than the standard permissible limits. Cadmium (Cd) is a nonessential trace element widely distributed in the environment and can originate from both geogenic (natural) and anthropoegenic (human-induced) sources. Higher level of Cd is carcinogenic to humans. Cd concentration in the soil of the Nagaon district ranged from 0.13 mg kg^{-1} to 1.3 mg kg^{-1}, with a mean of 0.415 mg kg^{-1} (Fig. 18.4a). This surpasses the World Health Organization's (WHO) recommended maximum permissible limit for soil, which is 0.31 mg kg^{-1}. Disturbingly, about 70% of the areas of the Nagaon district had a Cd concentration considerably exceeding the permissible limits, indicating significant environmental and human health threats. The higher values of Cd in the soil can be correlated to the intensive application of phosphorus fertilizers in the soil (Wang et al., 2020; Zhao et al., 2021). Furthermore, the natural geology of the region, characterized by granite, gneiss, sandstone and shales also contributes to the higher concentration of Cd in soil (Central Ground Water Board, 2013), (Srinivasarao et al., 2014). In groundwater, Cd concentration ranged from 0.004 mg L^{-1} to 0.029 mg L^{-1} (Fig. 18.4b), with a mean concentration of 0.009. Worringly, a total of 15% of the studied area was above the permissible limits of 0.01 mg L^{-1}. The bulk of the groundwater and soils contaminated with Cd are concentrated in the central region of the district, specifically in Nagaon town. Additional areas of notable contamination include Rupahi town to the northwest and Hojai and Doboka town to the south (Fig. 18.4.).

Chromium (Cr) concentration in the soils of the Nagaon district ranged from 0.60 to 4.41 mg kg^{-1}, with a mean of 1.85 mg kg^{-1} (Fig. 18.5A). These level are well within the safe limits (<100 mg kg^{-1}). The absence of Cr-containing rocks in the soil, such as ultramafic and serpentine (Geological Survey of India, 2009), may be the primary reason for a considerably low amount of Cr in the soil (Oliveira, 2012). Moreover, the limited industrial activity in and around the district also be a reason for less Cr in the soil, as Poznanović Spahić et al. (2019) identified industrial sources as primary contributors to Cr contamination. The small amount of Cr present may be from vehicle exhaust smoke (Yang et al., 2020). Cr's concentration (mg L^{-1}) in the groundwater ranged from 0.015 mg L^{-1} to 0.067 mg L^{-1} (Fig. 18.5B), with a mean of 0.042 mg L^{-1}. It is worth highlighting that, approximately 10% of the groundwater samples were above the permissible limit of Cr (0.05 mg L^{-1}). The majority of high chromium concentrations in both soil and groundwater were observed in Nagaon and Sauguri towns, located centrally. Elevated levels were also noted to the west of Kaliabor town in the north, as well as in Hojai and Doboka towns to the south (Fig. 18.5).

Copper (Cu), an essential nutrient for plants, its concentration in natural soils highly dictated by the characteristics of the parent material. In the soils of the Nagaon district, the Cu concentration ranged from 0.66 to 11.46 mg kg^{-1}, with a mean concentration of 3.97 mg kg^{-1} (Fig. 18.6A). The presence of Cu in soil is typically bound within clay minerals as a cation and is closely associated with organic matter. The concentration of Cu was below the maximum

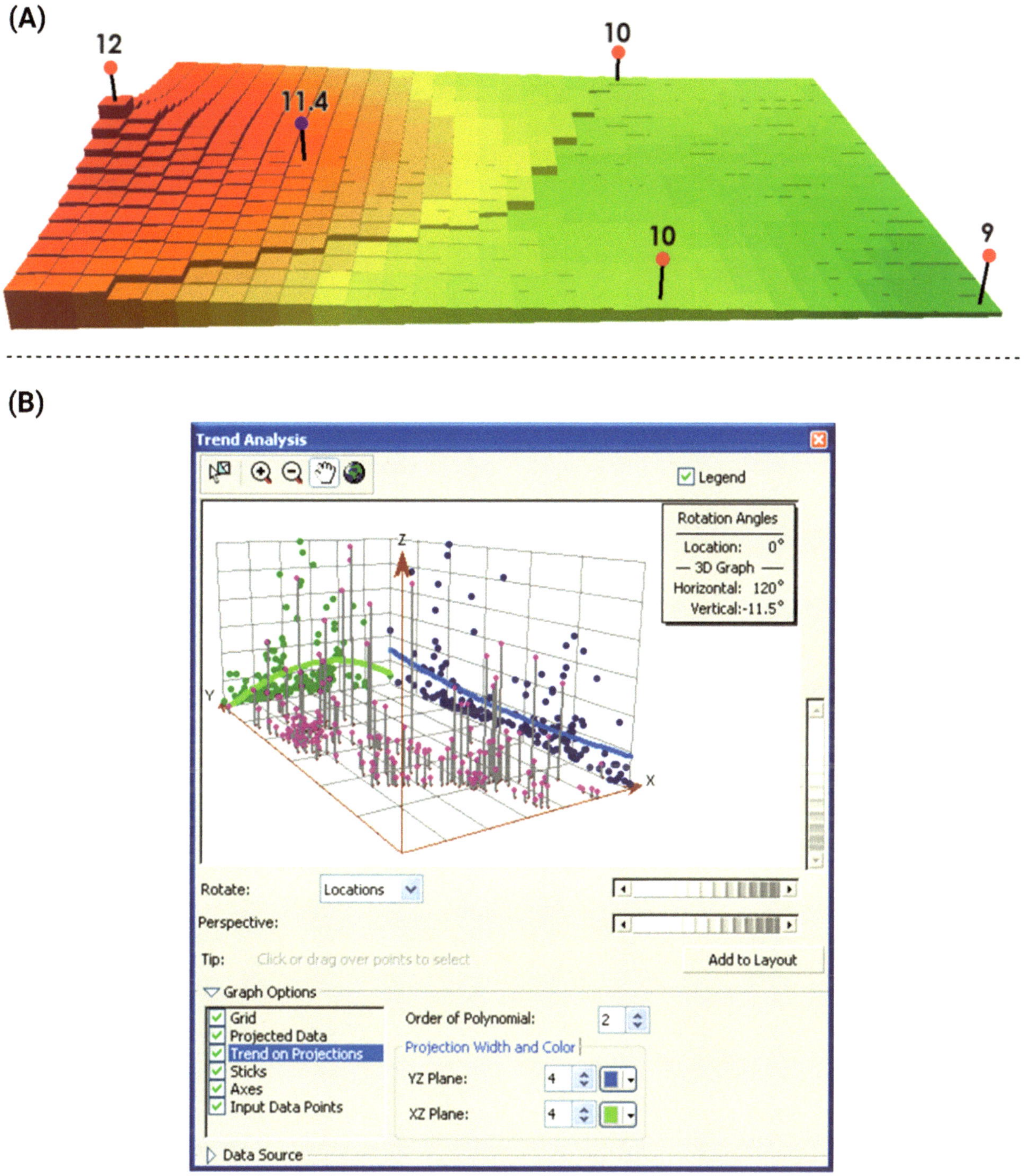

FIGURE 18.3 (A) Inverse distance weighted interpolation, where the purple dot/sample point (unknown value) was predicted based on the red dot/points (known values) as a function of distance; (B) Trend analysis. *(A) Image source: https://gisgeography.com/inverse-distance-weighting-idw-interpolation/; (B) Image source: https://desktop.arcgis.com/en/arcmap/latest/extensions/geostatistical-analyst/trend-analysis.htm.*

permissible limit of 36 mg kg^{-1}, which may be due to the formation of the soil from parent material poor in Cu. However, there was a relatively higher concentration of Cu in the southern part of the district, probably due to the presence of chalcopyrite minerals in the soil (Majumdar, 2010). The Cu concentration in the groundwater ranged from 0.10 mg L^{-1} to 1.84 mg L^{-1} (Fig. 18.6B), with a mean of 0.62 mg L^{-1}. The Cu concentration in groundwater depends on anthropogenic sources like mining, farming, manufacturing operations, and the discharge of municipal or industrial

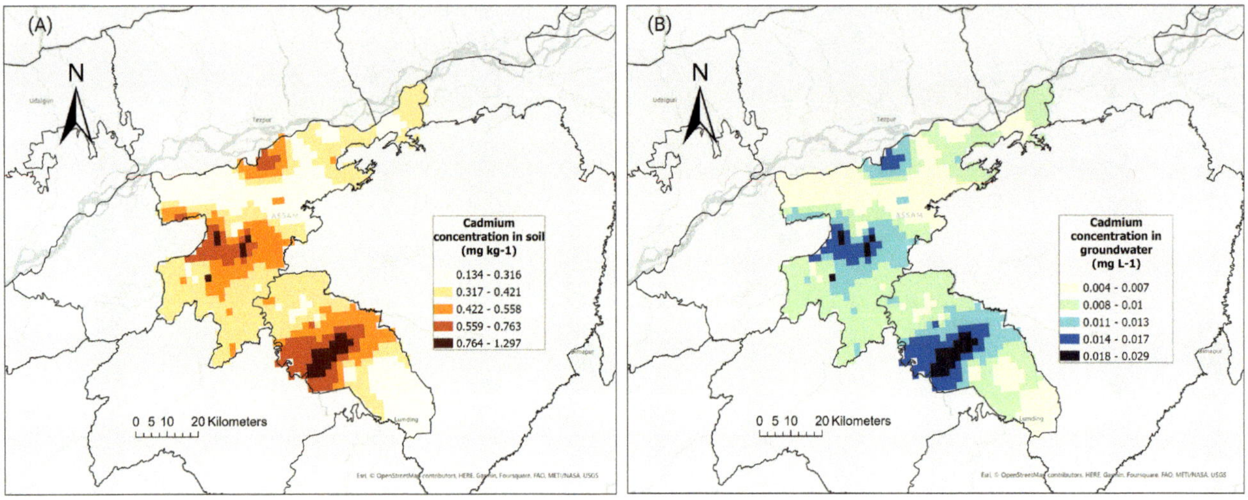

FIGURE 18.4 Inverse distance weighted (IDW) interpolated map for the cadmium concentration in (A) soil and (B) groundwater. Each color represents a range of concentrations, as mentioned in the legend. The concentration for the soil is represented as mg $kg^{-1}$and for groundwater as mg L^{-1}.

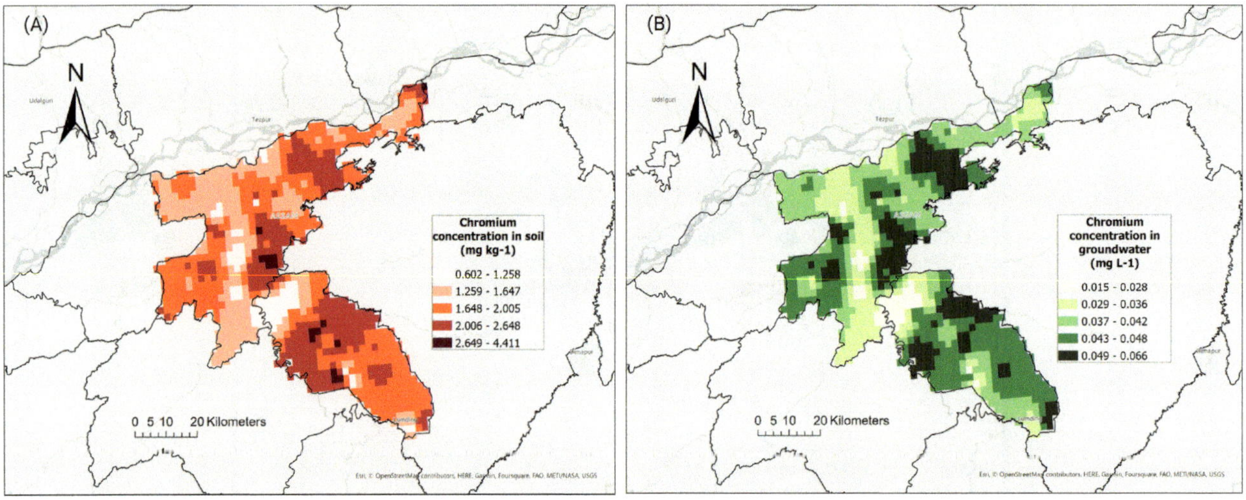

FIGURE 18.5 Inverse distance–weighted (IDW) interpolated map for the chromium concentration in (A) soil and (B) groundwater. Each color represents a range of concentrations, as mentioned in the legend. The concentration for the soil is represented as mg kg^{-1} and for groundwater as mg L^{-1}.

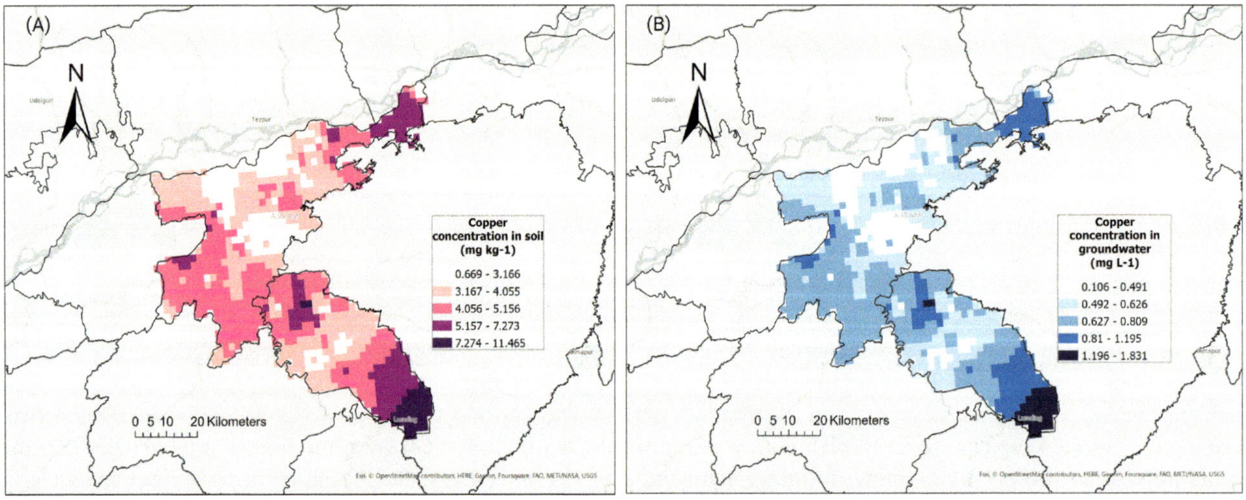

FIGURE 18.6 Inverse distance–weighted interpolated map for the copper concentration in (A) soil and (B) groundwater. Each color represents a range of concentrations, as mentioned in the legend. The concentration for the soil is represented as mg kg^{-1} and for groundwater as mg L^{-1}.

wastewater into rivers and lakes (Punia et al., 2021). Nagaon district is not a mining district, and the prevalence of industries is also lower, so the Cu content in the groundwater was well below the permissible limits. About 3% of the entire district was above the permissible limit set by the EPA (1.3 mg L^{-1}), which may be due to the corrosion of the copper pipes used in the water supplies and other activities (Manne et al., 2022).

Zinc (Zn) is an essential micronutrient required by plants in trace amounts. However, various human activities have altered the safe Zn concentration in the environment. The district's Zn concentration ranged from 0.42 to 1.0 mg kg^{-1} with a mean concentration of 0.710 mg kg^{-1} (Fig. 18.7A). The amount of Zn present in the soil mainly depends on the parent materials of the soil. The recommended permissible limit of Zn for soil by the WHO is 50 mg kg^{-1}, and 100% of the area was below the standard permissible limit. The low Zn content of the soil was mainly due to the parent materials, viz., granite and gneiss, which are low in Zn content (García-Navarro et al., 2021). The Zn concentration in the groundwater ranged from 0.04 to 0.10 mg L^{-1}, with a mean of 0.069 mg L^{-1} (Fig. 18.7B); 100% of the area was within the safe maximum permissible limit in water (0.2 mg L^{-1}). The Zn content was relatively low, as Zn levels in alluvial soils are generally low (Okoli et al., 2016). Most of high concentrated zinc soil are recorded from the Nagaon town in central part of district, while the majority of the higher conctrated groundwater were scatteer acorss the district.

Manganese (Mn) is the fifth most abundant metal in the Earth's crust, and an acidic environment favors its availability. The concentration of Mn in the soil ranged from 12.38 to 38.09 mg kg^{-1} (Fig. 18.8A), with a mean of

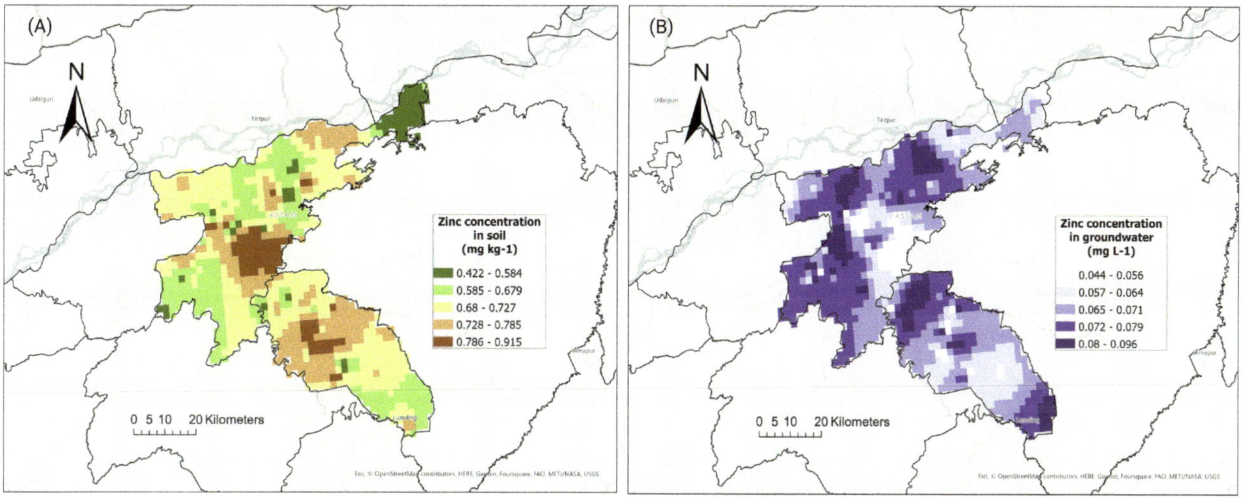

FIGURE 18.7 Inverse distance–weighted (IDW) interpolated map for the zinc concentration in (A) soil and (B) groundwater. Each color represents a range of concentrations, as mentioned in the legend. The concentration for the soil is represented as mg kg^{-1} and for groundwater as mg L^{-1}.

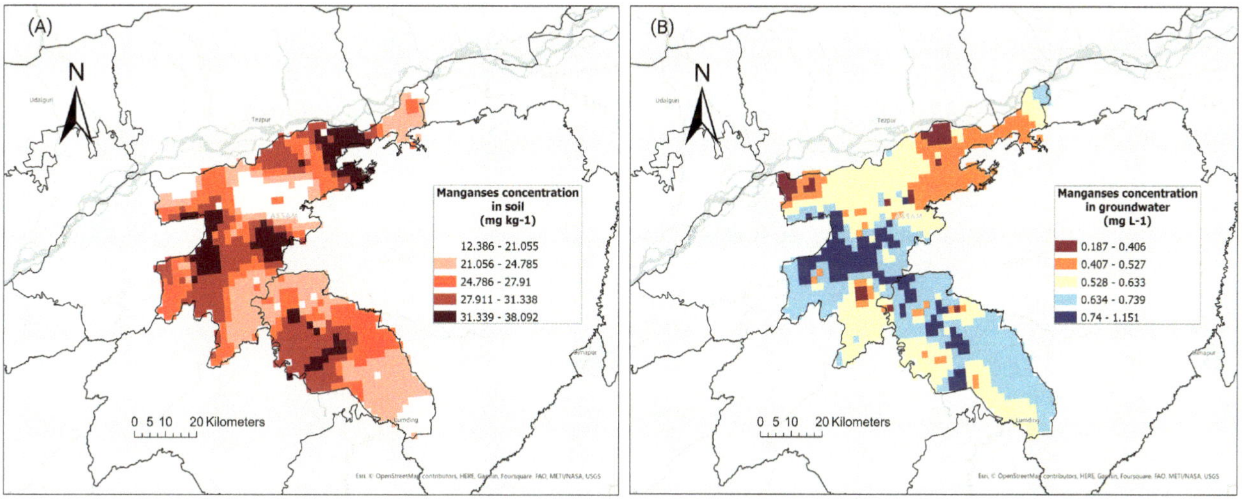

FIGURE 18.8 Inverse distance–weighted (IDW) interpolated map for the manganese concentration in (A) soil and (B) groundwater. Each color represents a range of concentrations, as mentioned in the legend. The concentration for the soil is represented as mg kg^{-1} and for groundwater as mg L^{-1}.

26.87 mg kg^{-1}. The Mn concentration in the soil was within the permissible limit suggested by WHO i.e. 2000 mg kg^{-1}. However, given the district's predominantly alluvial soil type (Central Ground Water Board, 2013; Rusydi et al., 2021) and its acidic pH (Sintorini et al., 2021), there is an inherent potential for high Mn availability and mobility within the soil matrix. This suggests that future applications of inorganic fertilizer, mining operations, and the discharge of industrial effluents should be approacched with caution, considering the environmental implications. The Mn concentration in the groundwater ranged from 0.18 to 1.15 mg L^{-1}, with a mean of 0.627 mg L^{-1}. Groundwater had a higher Mn concentration than the permissible limits of 0.1 mg L^{-1}, indicating a higher risk to the health and environment. The elevated level of Mn can be attributed to the alluvial soils of the region, which typically exhibit higher Mn content (Fernando and Lynch, 2015; Rusydi et al., 2021). Higher infiltration rate due to a higher percentage of sand and silt than clay (Central Ground Water Board, 2013), coupled with a low pH (acidic) of the soil (Sintorini et al., 2021), further favored the leaching of Mn to the groundwater. All the groundwater in the district exhibited contamination with Mn. The central part of the district, in particular, recorded exceptionally high levels of contamination, with concentrations ranging from >0.7 mg L^{-1} to 1.15 mg L^{-1}.

18.8 Trend analysis

A three-dimensional trend analysis of Cd in the soil showed a distinct spatial distribution across the Nagaon district. Specifically, the central region, encompassing the major municipal area and the Nagaon Sadar (district headquarters), exhibited the highest Cd levels, while the district's peripheries recorded comparatively lower concentrations on both the XZ and YZ axes (Fig. 18.9A). The central region, characterized by increased disposal of municipal waste and more pronounced industrial footprints, emerges as a potential hotspot for Cd accumulation. Factos such as municipal wastes, construction activities, and increased vehicular emissions, known to be significant sources of Cd, and likely contributed to elevated level in the central part of the district (Silva et al., 2016). Additionally, the rapidly changing land use pattern and higher use of agricultural chemicals toward the central part of the district, mainly in the Rupohi and Samaguri areas (Saikia and Sahariah, 2020), may also be contributing reasons. In terms of groundwater, Cd distribution followed a second-order polynomial pattern on both the XZ and YZ axes, mirroring the soil trends with elevated concentrations in the district's central part compared to its boundaries (Fig. 18.9B). Significantly higher soil Cd levels in the central region and potential leaching into the groundwater is the primary reason of higher groundwater Cd concentration. Factors such as the extensive application of inorganic fertilizers and anthropogenic activities predominantly in the district's central region also acted as catalysts, driving up the groundwater Cd concentrations.

In analyzing the trend of Cr distribution in soil, the XZ direction exhibited a linear progression, consistent with a first-order polynomial. This suggests a uniform Cr concentration when traversing from east to west. In contrast, the YZ axis revealed a curvilinear pattern, aligned with a second-order polynomial, demonstrating the lowest Cr concentrations in the district's center and higher concentrations towards its peripheries (Fig. 18.9C). This consistent Cr concentration across the east-west trajectory can likely be attributed to the absence of Cr-rich rocks, such as ultramafic and serpentine formations (Geological Survey of India, 2009; Oliveira, 2012). However, the presence of sandstone in higher amounts toward the southern part of the district (Central Ground Water Board, 2013) may be a reason for the higher availability of Cr than that of the central part, as Cr is often found related to sandstone (Bertolo et al., 2011). The trend of Cr in the groundwater showed a straight line following the first-order polynomial in both the YZ and XZ axes (Fig. 18.9D). The limited presence of Cr-bearing minerals in the region, coupled with the scarcity of Cr-intensive industries like paint manufacturing, largely account for the groundwater's benign Cr concentrations (Mani Tripathi and Chaurasia, 2020).

Copper (Cu) concentrations in the soil exhibited a distinct pattern, following a second-order polynomial trend across both the XZ and YZ axes (Fig. 18.9E). This pattern indicates a higher concentration of Cu toward the border of the district than in the center, both while moving from east to west and moving from north to south. Curved lines following the second-order polynomial were seen in both axes while observing the Cu distribution trend in the groundwater (Fig. 18.9F). Concentrations peaked at the district's edges in both the east-west and north-south directions, with the central region recording the lowest values. The presence of chalcopyrite minerals, likely influences these patterns, contributing to the elevated Cu levels in both soil and groundwater (Majumdar, 2010).

Zinc (Zn) concentration in the soil showed second-order polynomial curved lines in both the XZ and YZ directions (Fig. 18.10A), indicating an elevated concentration of Zn in the district's central region compared to its peripheries. The higher application of agricultural chemicals, especially fertilizers, may explain the higher Zn concentration in the central part of the district (Manzeke et al., 2019). This is further supported by the evolving land use patterns with commercial agriculture on the rise in the central part of the district, including places like Rupohi andSamaguri (Saikia and Sahariah, 2020). In terms of groundwater, the Zn distribution showcased a linear decline from west to east on the YZ

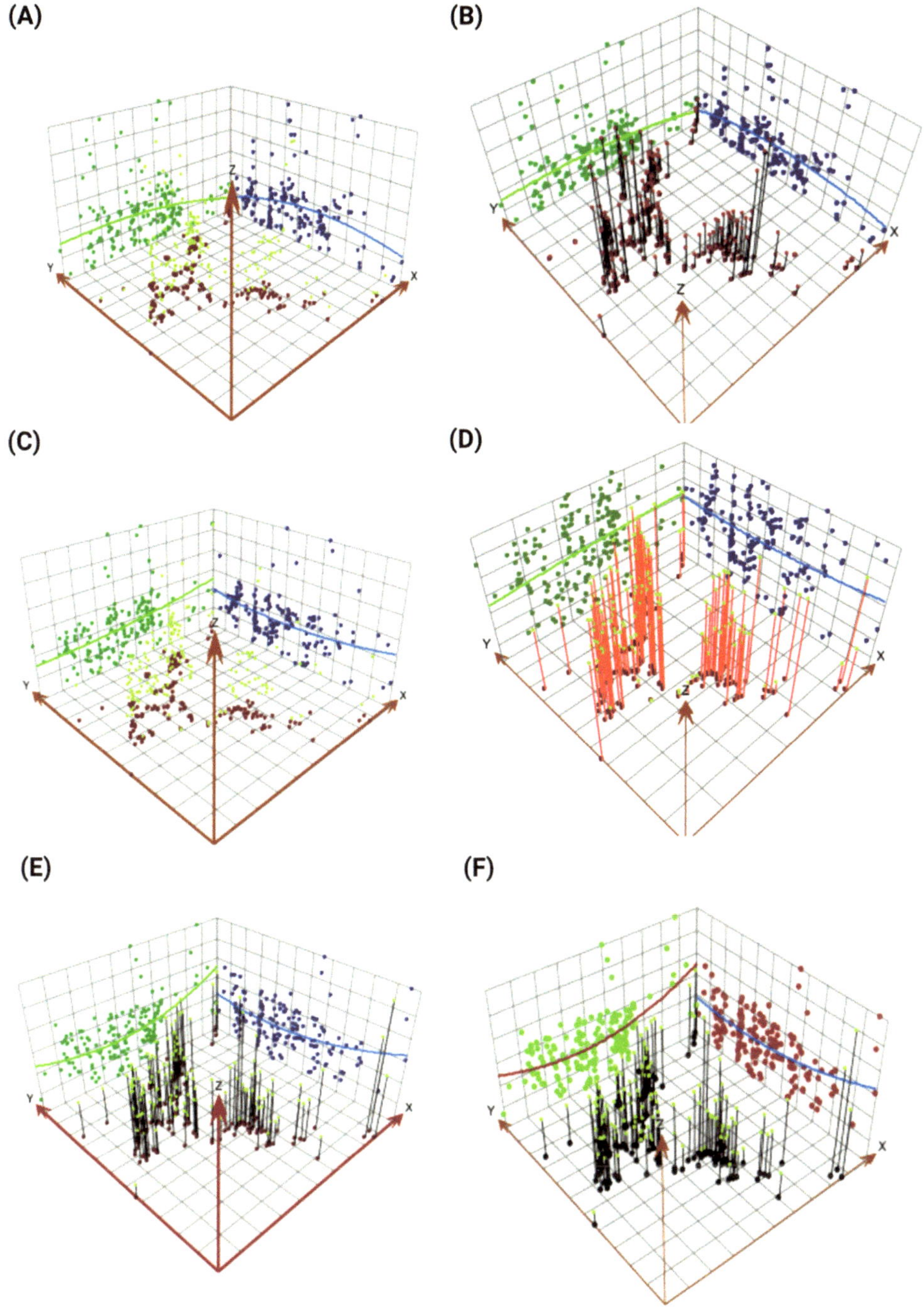

FIGURE 18.9 Three-dimensional trend analysis for (A) cadmium in soil, (B) cadmium in groundwater, (C) chromium in soil, (D) chromium in groundwater, (E) copper in soil, and (F) copper in groundwater in the Nagaon district, Assam, India.

axis. Conversely, the XZ axis reflected a curved trend with the district's center exhibiting a more pronounced Zn concentration than its boundaries (Fig. 18.10B). The western part of the district, including Dhing and its surrounding areas, is intensively cultivated with high applications of fertilizers and pesticides (Saikia and Sahariah, 2020), which may be the probable reason for a higher concentration of Zn in the area (Onwe et al., 2020). Additionally, Zn is deficient in most of the soils in the state, which encourages farmers to use Zn fertilizers extensively (Reza et al., 2021), which likely contributes to potential increase in the Zn concentration in groundwater.

For Mn concentrations in the soil, a second-order polynomial trend was evident across both XZ and YZ axes, suggesting that the district's central region has a more pronounced Mn presence relative to its peripheries (Fig. 18.10C). The central region of the district is characterized by recent alluvial deposits, which might have inherited higher Mn concentrations from parent materials. Additionally, certain soil conditions, such as pH or periodic waterlogging

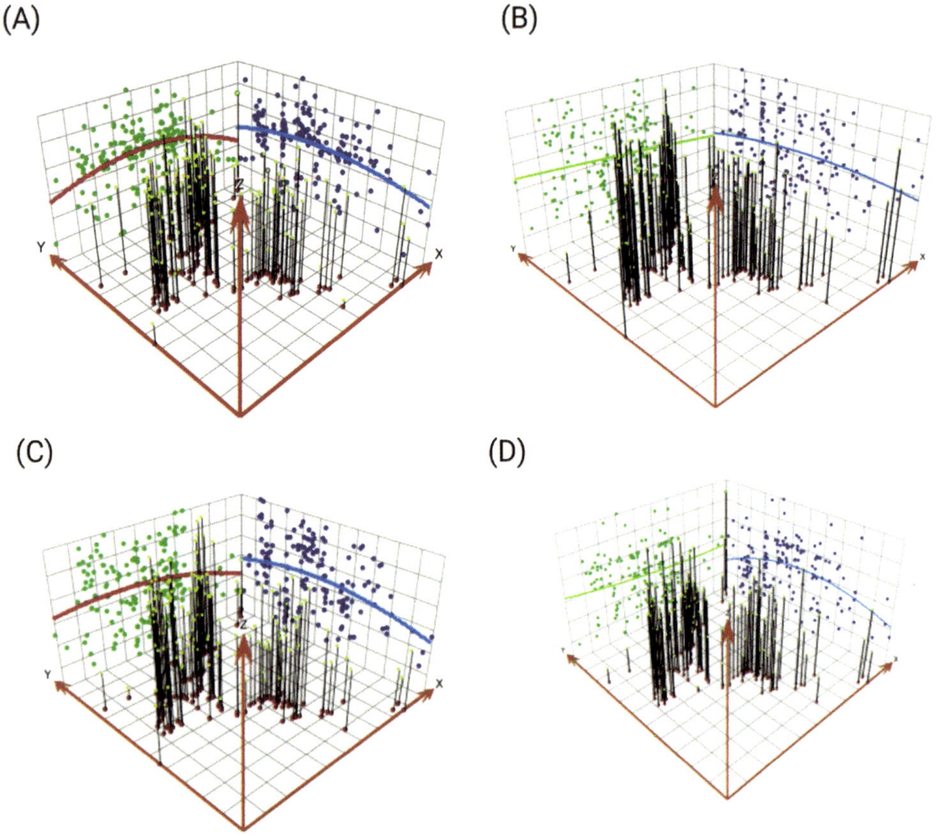

FIGURE 18.10 Trend analysis for (A) zinc in soil, (B) zinc in groundwater, (C) manganese in soil, and (D) manganese in groundwater in the Nagaon district, Assam, India.

(consider the high precipitation of the region), could enhance Mn availability and concentration in this region (Central Ground Water Board, 2013; Koranteng-Addo et al., 2011). Regarding groundwater Mn distribution, the data showed a linear, first-order polynomial trend on the YZ axis, while a curvilinear second-order polynomial trend was observed on the XZ axis. The latter emphasizes a more pronounced Mn influence in the district's center compared to its boundaries (Fig. 18.10D).

18.9 Summary

Spatial variability maps, facilitated by Geographic Information System (GIS) techniques, establish a foundational database for the sustainable management of groundwater resources with respect to heavy metal concentrations. The Nagaon district presents a region of concern due to elevated levels of Cd in the soil and heightened concentrations of Cd, Cr, and Mn in groundwater. This situation necessitates immediate interventions.Through geospatial analyses, the distribution of these heavy metals has been delineated, and areas at greater risk have been identified. The central region of the district, being a significant agricultural and industrial hub, exhibits higher heavy metal concentrations compared to its surrounding areas. To ensure future environmental stability, it is imperative to regulate the application of agrochemicals and fertilizers and to manage industrial waste discharges in this central zone. Given the reliance on groundwater as the primary source for irrigation, especially for summer paddy cultivation, there is an inherent risk of these metals entering the food chain. This poses potential health threats if timely and effective remediation measures are not implemented. Employing digital mapping techniques for soil and groundwater, alongside varied interpolation methodologies and toxicity indices, can provide insights into contamination sources and their distribution patterns. Such comprehensive assessments are essential for informed decision-making in soil and water resource management.

References

Balali-Mood, M., Naseri, K., Tahergorabi, Z., Khazdair, M.R., Sadeghi, M., 2021. Toxic mechanisms of five heavy metals: mercury, lead, chromium, cadmium, and arsenic. Front. Pharmacol. 12.

Bertolo, R., Bourotte, C., Marcolan, L., Oliveira, S., Hirata, R., 2011. Anomalous content of chromium in a Cretaceous sandstone aquifer of the Bauru Basin, state of São Paulo, Brazil. J. S. Am. Earth Sci. 31, 69–80. Available from: https://doi.org/10.1016/j.jsames.2010.10.002.

Central Ground Water Board, 2013. Ground water information booklet: Nagaon District, Assam.

Fernando, D.R., Lynch, J.P., 2015. Manganese phytotoxicity: new light on an old problem. Ann. Bot. 116, 313–319. Available from: https://doi.org/10.1093/aob/mcv111.

García-Navarro, F.J., Jiménez-Ballesta, R., Garcia-Pradas, J., Amoros, J.A., Perez de los Reyes, C., Bravo, S., 2021. Zinc concentration and distribution in vineyard soils and grapevine leaves from Valdepeñas designation of origin (Central Spain). Sustainability 13, 7390. Available from: https://doi.org/10.3390/su13137390.

Geological Survey of India, 2009. Geology and mineral resources of Assam.

Gong, G., Mattevada, S., O'Bryant, S.E., 2014. Comparison of the accuracy of kriging and IDW interpolations in estimating groundwater arsenic concentrations in Texas. Environ Res 130, 59–69. Available from: https://doi.org/10.1016/j.envres.2013.12.005.

Islam, M.S., Ahmed, M.K., Raknuzzaman, M., Habibullah-Al-Mamun, M., Masunaga, S., 2015. Metal speciation in sediment and their bioaccumulation in fish species of three urban rivers in Bangladesh. Arch Environ Contam Toxicol 68, 92–106. Available from: https://doi.org/10.1007/s00244-014-0079-6.

Koranteng-Addo, E.J., Owusu-Ansah, E., Boamponsem, L.K., Bentum, J.K., Arthur, S., 2011. Levels of zinc, copper, iron and manganese in soils of abandoned mine pits around the Tarkwa gold mining area of Ghana. Adv. Appl. Sci. Res. 2, 280–288.

Majumdar, D., 2010. Need to intensify base metal exploration activities in Mikir Hills, northeastern India. Curr. Sci. 99, 656–662.

Mani Tripathi, S., Chaurasia, S., 2020. Detection of chromium in surface and groundwater and its bio-absorption using bio-wastes and vermiculite. Eng. Sci. Technol. Int. J. 23, 1153–1161. Available from: https://doi.org/10.1016/j.jestch.2019.12.002.

Manne, R., Kumaradoss, M.M.R.M., Iska, R.S.R., Devarajan, A., Mekala, N., 2022. Water quality and risk assessment of copper content in drinking water stored in copper container. Appl Water Sci 12, 27. Available from: https://doi.org/10.1007/s13201-021-01542-x.

Manzeke, M.G., Mtambanengwe, F., Watts, M.J., et al., 2019. Fertilizer management and soil type influence grain zinc and iron concentration under contrasting smallholder cropping systems in Zimbabwe. Sci Rep 9, 6445. Available from: https://doi.org/10.1038/s41598-019-42828-0.

Mohammadi, A., Faraji, M., Nemati Mansour, S., Abdolahnejad, A., Miri, M., 2020. Spatial analysis of heavy metals in surface soil, NW Iran. Int. J. Environ. Anal. Chem. 0, 1–10. Available from: https://doi.org/10.1080/03067319.2020.1858069.

Okoli, N., Uzoho, B., Onweremadu, E., Ugochukwu, N., Irokwe, I., 2016. Zinc fractionation of soils of different parent materials and their relationships with some soil properties. Malaysian J. Soil Sci. 20, 49–66.

Oliveira, H., 2012. Chromium as an environmental pollutant: insights on induced plant toxicity. J. Bot. 2012, e375843. Available from: https://doi.org/10.1155/2012/375843.

Onwe, O.H., Adesiji, A.R., Agbese, E.O., 2020. Effect of chemical fertilizers on groundwater quality in an unconfined aquifer. 2nd International Civil Engineering Conference (ICEC 2020). Department of Civil Engineering, Federal University of Technology, Minna, Nigeria.

Ouabo, R.E., Sangodoyin, A.Y., Ogundiran, M.B., 2020. Assessment of ordinary kriging and inverse distance weighting methods for modeling chromium and cadmium soil pollution in e-waste sites in Douala, Cameroon. J. Health Pollut. 10, 200605. Available from: https://doi.org/10.5696/2156-9614-10.26.200605.

Poznanović Spahić, M.M., Sakan, S.M., Glavaš-Trbić, B.M., Tančić, P.I., Škrivanj, S.B., Kovačević, J.R., et al., 2019. Natural and anthropogenic sources of chromium, nickel and cobalt in soils impacted by agricultural and industrial activity (Vojvodina, Serbia). J Environ Sci Health A Tox Hazard Subst Environ Eng 54, 219–230. Available from: https://doi.org/10.1080/10934529.2018.1544802.

Punia, A., Bharti, R., Kumar, P., 2021. Impact of mine pit lake on metal mobility in groundwater. Environ Earth Sci 80, 245. Available from: https://doi.org/10.1007/s12665-021-09559-w.

Qi, Z., Gao, X., Qi, Y., Li, J., 2020. Spatial distribution of heavy metal contamination in mollisol dairy farm. Environ. Pollut. 263, 114621. Available from: https://doi.org/10.1016/j.envpol.2020.114621.

Qiao, P., Lei, M., Yang, S., Yang, J., Guo, G., Zhou, X., 2018. Comparing ordinary kriging and inverse distance weighting for soil as pollution in Beijing. Environ Sci Pollut Res 25, 15597–15608. Available from: https://doi.org/10.1007/s11356-018-1552-y.

Raju, N.J., 2022. Arsenic in the geo-environment: a review of sources, geochemical processes, toxicity and removal technologies. Environ. Res. 203, 111782. Available from: https://doi.org/10.1016/j.envres.2021.111782.

Reza, S.K., Baruah, U., Mukhopadhyay, S., Dutta, D., Bandyopadhyay, S., Sahoo, A., 2021. Quantifying spatial variability of available zinc in alluvial soils of Brahmaputra Plains, India. J. Indian Soc. Soil Sci. 69, 28–36. Available from: https://doi.org/10.5958/0974-0228.2021.00017.7.

Rusydi, A.F., Onodera, S.-I., Saito, M., Ioka, S., Maria, R., Ridwansyah, I., et al., 2021. Vulnerability of groundwater to iron and manganese contamination in the coastal alluvial plain of a developing Indonesian city. SN Appl. Sci. 3, 399. Available from: https://doi.org/10.1007/s42452-021-04385-y.

Saikia, B., Sahariah, D., 2020. Agricultural landuse pattern of nagaon district, Assam: present status and changing scenario. PalArch's Journal of Archaeology of Egypt/Egyptology 17, 7661–7668.

Silva, E., Fonseca, F., Alleoni, L., Nascimento, S., Grazziotti, P., Nardis, B., 2016. Availability and toxicity of cadmium to forage grasses grown in contaminated soil. Int. J. Phytoremediation 18, . Available from: https://doi.org/10.1080/15226514.2016.114622500–00.

Sintorini, M.M., Widyatmoko, H., Sinaga, E., Aliyah, N., 2021. Effect of pH on metal mobility in the soil. IOP Conf. Ser.: Earth Environ. Sci. 737, 012071. Available from: https://doi.org/10.1088/1755-1315/737/1/012071.

Srinivasarao, Ch, Rama Gayatri, S., Venkateswarlu, B., Jakkula, V.S., Wani, S.P., Kundu, S., et al., 2014. Heavy metals concentration in soils under rainfed agro-ecosystems and their relationship with soil properties and management practices. Int. J. Environ. Sci. Technol. 11, 1959–1972. Available from: https://doi.org/10.1007/s13762-013-0350-9.

Wang, X., Liu, W., Li, Z., Teng, Y., Christie, P., Luo, Y., 2020. Effects of long-term fertilizer applications on peanut yield and quality and plant and soil heavy metal accumulation. Pedosphere 30, 555–562. Available from: https://doi.org/10.1016/S1002-0160(17)60457-0.

Watson, M.E., Isaac, R.A., 1990. Analytical instruments for soil and plant analysis. Soil Testing and Plant Analysis. John Wiley & Sons, Ltd, pp. 691–740. Available from: https://doi.org/10.2136/sssabookser3.3ed.c26.

Xie, Y., Chen, T., Lei, M., Yang, J., Guo, Q., Song, B., et al., 2011. Spatial distribution of soil heavy metal pollution estimated by different interpolation methods: accuracy and uncertainty analysis. Chemosphere 82, 468–476. Available from: https://doi.org/10.1016/j.chemosphere.2010.09.053.

Yang, Y., Khan, Z., Ahmad, K., Arshad, N., Rehman, S., Fahad Ullah, M., et al., 2020. Does the chromium element in forages and fodders grown in contaminated pasture lands cause toxicity in livestock: assessing the potential risk. Rev. Chim. 71, 397–405. Available from: https://doi.org/10.37358/RC.20.7.8257.

Zaynab, M., Al-Yahyai, R., Ameen, A., Sharif, Y., Ali, L., Fatima, M., et al., 2022. Health and environmental effects of heavy metals. J. King Saud Univ.—Sci. 34, 101653. Available from: https://doi.org/10.1016/j.jksus.2021.101653.

Zhao, H., Guan, J., Liang, Q., Zhang, X., Hu, H., Zhang, J., 2021. Effects of cadmium stress on growth and physiological characteristics of sassafras seedlings. Sci Rep 11, 9913. Available from: https://doi.org/10.1038/s41598-021-89322-0.

Chapter 19

Soil biodiversity and climate change impacts—a geospatial approach

Pankaj Dahiya[1], Satender Kumar[2], M.L. Khichar[1], Sachin Kumar[3], Harwinder Singh Buttar[4] and Arvind Dhaloiya[5]

[1]*Department of Agricultural Meteorology, CCS Haryana Agricultural University, Hisar, Haryana, India,* [2]*Department of Soil Science, CCS Haryana Agricultural University, Hisar, Haryana, India,* [3]*Department of Agronomy, CCS Haryana Agricultural University, Hisar, Haryana, India,* [4]*Department of Plant Pathology, Punjab Agricultural University, Ludhiana, Punjab, India,* [5]*Department of Soil and Water Engineering, Punjab Agricultural University, Ludhiana, Punjab, India*

19.1 Introduction

Climate change influences specific species, their interactions with other living things, and their habitats and environments, which results in structural and operational changes in ecosystems and services. The increasingly prevalent global climate change puts natural ecosystems and biodiversity at risk (Singh and Dhaloiya, 2022; Díaz et al., 2019). Although the consequences of climate change are vast and variable, the accumulating evidence suggests that different people will respond to it in different ways depending on their relative vulnerability owing to modifications in exposure, sensitivity, and adaption capability (Beever et al., 2016; Foden and Young, 2016; Glick et al., 2011; Kovach et al., 2019).

According to a number of studies, changes in interactions among species brought about by climate change affect biodiversity and the effectiveness of terrestrial ecosystems (Walther et al., 2002; Gottfried et al., 2012; Langley and Hungate 2014). Less research, nevertheless, has been done on soil communities (Schimel et al., 2007; de Vries et al., 2012). To maintain and shape the characteristics of ecosystems, plants and soil organisms interact in various ways. In actuality, plants and soil microbial organisms' interactions can impact the diversity, quantity, and makeup of plants and animals (Berg et al., 2010; Van der Putten et al., 2013). In contrast, entire soil fauna, including pathogens, symbiotic mutualists, and decomposers, negatively affect plant performance, while plant–microbial interactions are beneficial when the benefits offered by the soil community boost plant performance in terms of biomass output and survival. A research goal is to understand the response of soil microbe–microbe and soil microbe–plant interactions to climate change. This would provide insight into essential ecosystem processes like soil carbon storage and net primary productivity. These interactions are important in defining ecosystem characteristics (Ostle et al., 2009; Berg et al., 2010; Fischer et al., 2014).

Natural and human-caused climate change are both contributing causes significantly affecting biodiversity, agricultural production, and food security. Most indigenous and specialized species are in danger of extinction. Because it supplies primary healthcare for more than 60%–80% of people on the earth and nutrition for all living things, there are grounds for concern about species extinction. Although climate change has been acknowledged as impacting biodiversity and food security, diminutive research has been done in comparison to the global scale of the problem. Therefore, this chapter aims to discover, evaluate, and synthesize the relationship between climate change, biodiversity, and food security.

19.2 Soil biodiversity: its reality

Biodiversity is described as "variability among living organisms from all sources, including, for example, terrestrial, marine, and other aquatic ecosystems and the ecological complexes to which they belong; this encompasses diversity within species, between species, and within ecosystems." Functional diversity defines the biological significance of

Remote Sensing of Soils. DOI: https://doi.org/10.1016/B978-0-443-18773-5.00028-4

individual or groups of species within an ecosystem. All of the organisms that reside in the soil collectively are referred to as "biodiversity." Bacteria, fungi, protozoa, nematodes, meso- and macrofauna, including acari, springtails, and enchytraeids, are examples of microorganisms viz., insects and earthworms (Wallwork, 1970; Swift et al., 1979). In an added study, collembola were categorized as fungivores, while fungi, bacteria, and microbial biomass were categorized as detritivores (Hunt et al., 1987; Chen et al., 1996; Briones et al., 1999). Bacteriovores were defined as bacteria and protozoa (Ingham et al., 1986; Hunt et al., 1987). Specialists who forage above ground (such as in the litter or in the layer of fine woody debris) and below ground can be distinguished based on their preferred habitats.

Various research reports suggest that soil biodiversity is key to the global ecosystem. Disturbance regimes vary drastically due to the mutual effects of climate change and biological intrusions. Large-scale aspects influencing microbial activity are abiotic limitations brought on by the climate. The seasonal subtleties of microbial populaces can show the inclusive impact of climate on soil microorganisms. These changes result from the two primary climatic variables, viz., temperature and moisture, which have a bearing on microbial communities' growth, activity, and composition. Microbial communities can have a range of individual characteristics, including growth and activity rates.

As a result, soil biodiversity is recognized as a key factor in the global system, and the current chapter is based on this viewpoint.

19.3 Soil organisms and ecological function

For all ecosystems to continue functioning sustainably, soil organisms have a dramatic role in various crucial functions. Along with changes in the water regimes and soil's physical characteristics, the primary forces behind carbon sequestration, greenhouse gas (GHG) emission, regulation of the dynamic behavior of soil organic matter, and biogeochemical cycles are soil organisms. This improves the portion and effectiveness of plant nutrient uptake, enhancing their health and growth habits. Soil organisms interact with one another in a soil food chain that supplies food for the tropic level below. Usually, roots and other organic components that have decayed make up the basis of a soil food web (detritus). The integrity of the soil food web is necessary to operate an ecological system properly. Soil microorganisms work with plant roots, fight infections, weathering, and solubilization of minerals, and contribute to the soil aggregation and structure in addition to the decomposition of organic matter, release of nutrients in plant-available forms, and eradication of harmful materials. The chronology of microbial metabolism is crucial to comprehend how human impact affects the environment. The average soil microbial biomass turnover rates are between 0.2 and 6 years, as opposed to over 40 years for the widely held organic matter.

19.3.1 Nutrient cycling

19.3.1.1 Carbon cycle

The most extensively researched nutrient cycle in recent years is the carbon cycle because of its link to climate change. When carbon dioxide (CO_2) is transformed into an organic form via photosynthesis, it enters the carbon cycle. Although many microbiological organisms, including bacteria, cyanobacteria, and algae, can perform photosynthesis, plants are usually thought of as performing it. This fixed carbon can travel up the trophic chain in the carbon cycle as photosynthetic species, or "primary producers," are fed upon by "primary consumers" like herbivores, who are, in turn, consumed by "secondary consumers," and so on.

19.3.1.2 Nitrogen cycle

The soil biota plays a vital role in the nitrogen cycle. Most creatures on the earth, including plants, cannot utilize gaseous nitrogen. It entails free-living "nitrogen fixing" microbes like cyanobacteria, bacteria, and actinomycetes, and symbiotic microbes like Rhizobium (which forms nodules root in legumes). They convert atmospheric nitrogen into ammonia, utilized by plants, or convert a significant percentage of this ammonia into other forms accessible to plants, such as nitrite (NO_2^-) or nitrate (NO_3^-). The denitrification process involves turning nitrogen compounds like nitrates and nitrites back into nitrogen gas. The nitrogen cycle has imperative agricultural and environmental connotations since nitrogen is frequently the restrained nutrient for crop development and can be a source of the GHG viz., N_2O.

19.3.1.3 Sulfur cycle

The amino acids cysteine and methionine include sulfur, vitamins, and other metabolites. Although uncommon in cells, it is essential to all biological processes. Sulfur may be converted by microbes from its most oxidized form (sulfate or

SO_4) to its most reduced form (sulfate or SO_2), much like nitrogen and carbon (sulfide or H_2S). The functions of various distinct procaryote kinds make up the sulfur cycle. Two independent groups of procaryotes convert H_2S to S and S to SO_4, respectively. The first group consists of purple and green sulfur bacteria that are anoxygenic photosynthetic and employ H_2S as an electron source in cyclic photophosphorylation. The second group, "colorless sulfur bacteria," derives energy from S and H_2S. Both SO_4 and S possibly will act as electron acceptors for respiration; H_2S is produced during respiration by sulfate-reducing bacteria, parallel to the denitrification process. This process leads to H2S production emitting typical odor in anaerobic marshy areas. Plants and bacteria absorb sulfur in the form of SO_4 for usage and then reduce it to form sulfide.

19.3.2 Formation and weathering of soil

Soil is necessary for our continued existence, and weathering is crucial to the soil-building process. Or, to put it another way, we must safeguard the soil since weathering gives us life. Scientists describe soil as a substance rich in organic matter accumulating in the topmost few inches of the earth's surface. While generally, soil is referred to as loose material on the earth's surface, that is not the scientific terminology for soil.

In soil, air, and water are present to varying degrees, along with soil minerals, organic matter, and empty space at 45%, 5%, and 50%, respectively, in fluctuating proportions. Clay minerals and quartz predominate in soil, with feldspar and small rock fragments in smaller proportions. Soil mineral concentration varies. Since saprotrophic and mutualistic fungi have been shown to accelerate the weathering of minerals across ecological and evolutionary timescales (Hoffland et al., 2004).

19.3.2.1 Waste recycling

Saprotrophic organisms/decomposers carry out the decomposition process by feeding on dead or decaying organic matter (shredded leaves and twigs). Saprotrophs acquire their nutrients and energy from organic substrates in this heterotrophic process. Bacteria and fungi are considered the principal decomposers, but soil invertebrates like earthworms also play a role in decomposition. Sometimes millipedes and collembolan, are misidentified as decomposers. These are called detritivores due to the inability to digest the large spectrum of chemicals that fungi and bacteria can. The most common decomposers are bacteria for dead organisms, while for plant litter, fungi.

19.4 Causes of soil biodiversity

Both vertically and horizontally, soils exhibit signs of significant geographical variation. Soil characteristics differ both within and between fields. When we look at soil at a microscopic level, we can see much more spatial variation—the pores in soil range in size from 0.2 μ to 2 mm. Pores with a diameter of 2–100 μ protect most of the bacteria from predators. The physical and chemical properties of soil, viz., the status of pH and oxygen, can change dramatically over a short distance range. It is not difficult to conceive that there is a place for all forms of organisms when the geographical variety of prospective living places is linked to the diversity of food and energy sources available, which are also subject to temporal variation in moisture and temperature.

19.5 What is climate change?

Climate change occurs due to long-term variations in weather and temperature. Since 1800s, burning fossil fuels, such as coal, oil, and gas, has been the primary contributor to climate change, producing gases able to trap heat. Nevertheless, some of the changes may be due to natural processes.

19.5.1 Causes and effect of climate change?

The greenhouse effect, a typical element of the climate system, ensures that the earth will continue to be a habitable planet. When the atmosphere's GHGs, which retain the sun's heat, are generally open to incoming solar radiation, this phenomenon takes place. The atmosphere allows the sun's light to travel to the earth's surface. Solar radiation is absorbed by the surface or reflected back to space as infrared radiation. While a portion of the radiation is emitted, GHG molecules absorb another part before being reemitted in all directions. The troposphere and the crust of the earth are warmed as a result. H_2O (water vapor) and CO_2 are the main contributors to greenhouse effect (CO_2). Despite their low concentrations, GHGs like methane (CH_4), nitrous oxide (N_2O), chlorofluorocarbons (CFCs), and others can nonetheless significantly increase global temperatures because of the duration and ability to trap heat in the atmosphere. The average earth temperature, without greenhouse effect, would be −0.4 °F (−18°C), as opposed to the present 59 °F (15°C).

19.5.2 Effect of climate change on soil environment

Abiotic limitations brought on by the climate are important factors in microbial activity. The seasonal dynamics of microbial populations can show how the climate has an overall impact on soil microbes. Microbial communities can have varying growth and activity rates, for example. In general, increased microbial activity is correlated with rising air temperatures. It is anticipated that variations in soil temperature will impact the dynamics of nitrification and denitrification in soil, which are driven by microbial activity. Soil environment changes can affect the structure metabolic activity of the microorganisms to cycle nutrients, along with the existence and virulence of soil-borne diseases like *Salmonella typhimurium*.

In the winter, microbes' activity and growth typically decline due to the decreased temperature. Normally, many microorganisms are killed at unreasonably high temperatures. Specific chemical engineers might be able to endure such hardship by changing into inactive forms that can sustain temperatures higher than those experienced by active individuals. It is critical to remember that on various microbial groups, the effects of temperature are mainly unknown. Changes in the microbe community composition are commonly related to seasonal fluctuations in the microbial activity in soil.

19.5.3 Soil biodiversity and climate change

In general, increased microbial activity is correlated with rising air temperatures. Microbial activity and growth often decline in the winter because of the decreased temperature. Individual soil ecosystems may have different seasonal patterns, though. In polar soils, for instance, when temperatures are low, microbial biomass rises in the late winter (Schadt et al., 2003). Even temperature responses may differ depending on the chemical engineers in the microbial community and the temperature range in question if there is a general positive correlation between temperature and microbial growth and activity. Overall, high temperatures render a large number of microorganisms unviable. It has been established that long-term temperature upsurges, mainly brought about by climate change, affect soil microbial respiration.

Soil microbial respiration plays a crucial part in regulating organic matter decomposition, due to carbon storage. During efficient respiration, more organic matter decomposes with the release of CO. At any rate, the ideal conditions for the enzymatic activity of microbe differ from place to place based on the species composition in the region under consideration (De Santo et al., 1993). The availability of organic matter is influenced by factors other than temperature, such as soil moisture, frequency of wet/dry and freeze/thaw cycles, and soil aggregation. This, in turn, can have an impression on the composition and activity of microbial communities. The function of biological regulator communities is shaped by temperature and moisture. Nematodes and microarthropods play a key role in assessing how average temperature increases brought on by climate change or other regional effects, such as fires, may affect ecosystems. On these nematodes, the main effects are known. Whether nematodes are sensitive to soil temperature and moisture depends on their metabolic condition (Ruess et al., 1999). Since they may form cysts or become dormant, this group of organisms has a unique survival technique for harsh climatic circumstances that enables them to resist severe temperature and moisture fluxes in the soil (Wall and Virginia, 1999).

For chemical engineers, soil moisture has both short-term and long-term effects. Bacteria's physiological state is directly affected by soil moisture (Harris, 1980), and this could make it more difficult for them to breakdown different organic compounds. Climate variables including air carbon dioxide concentration (~300 ppm), temperature (~ 3°C), and precipitation (wet and dry) interrelate in an old-field context to change the number and community structure of soil bacteria and fungi (Castro et al., 2010). In warmed treatments, fungal abundance increased, while bacteria in warmed plots increased with enhanced atmospheric (CO_2) and decreased under ambient atmospheric conditions (CO_2). Acidobacteria and Proteobacteria differed in relative abundance depending on the amount of precipitation, with Acidobacteria falling and Proteobacteria increasing in wet versus dry conditions.

Climate changes can significantly impact the physiology of earthworms through variations in soil temperature and moisture. According to research, temperature and soil moisture affect earthworm growth and activity seasonally. Earthworms tend to either become less abundant, dig deeper into the soil, or go into a state of dormancy or diapause when the soil becomes excessively dry (Holmstrup, 2001; Booth et al., 2000). On the other hand, high temperatures and high moisture content in the soil are favorable for growth. The ideal temperature and soil water latent for food ingestion in *Lumbricus terrestris* are approximately 22°C and 7 kPa, respectively. These findings imply that in wet soils, there is less digging and more widespread eating because more soil and organic materials are consumed, whereas somewhat drier, undisturbed conditions favor tunneling and soil profile investigation (Bolton and Phillipson,1976; Scheu,1987; and Daniel,1991). As the climate has warmed, the abundance of the majority of collembola species, with *Hypogastrura tullbergi*, *Lepidocyrtus lignorum*, and *Isotoma anglicana*, has decreased (Dollery et al., 2006). The soil fauna underwent relatively little change at elevated temperatures even after 6 years of rising temperature treatment (Haimi et al., 2005). Soil temperature and moisture are the crucial abiotic variables influencing the biology of microarthropods (springtails and mites) and population abundance (Cassagne et al., 2003; Roy and Roy, 2006). The maximum temperature at which one can survive is roughly 50°C, whereas 20°C is the ideal average temperature (Vannier, 1994). At large, organisms feeding on litter's surface can endure temperatures higher than their counterparts residing deeper in the soil. Temperatures between 35°C and 40°C have been shown to cause the majority of springtails and mites to perish (Choudhuri, 1963).

19.5.3.1 *Impact of soil biodiversity on CO_2*

Carbon dioxide, being chief GHG, contributes to climate change (CO_2). Over and above CO_2, soil biota is regulated by other GHGs, such as methane (CH_4) and nitrous oxide (N_2O), which are created throughout the carbon and nitrogen cycles, respectively. Although these gases have far lower fluxes than CO2, they are much more effective than GHGs (21 times and 310 times, respectively). With its quick and affordable availability, a healthy soil's ability to store carbon could be one of the strategies used to mitigate climate change. Furthermore, the soil carbon pool is vulnerable to warming, resulting in increased carbon loss to the atmosphere and detrimental feedback on the carbon cycle (Huntingford et al., 2000). Hence, how soil carbon and atmospheric CO_2 interact is still a mystery.

Nonetheless, it is considered that the soil biota makes the most significant contribution to this interface. Consequently, soil biological activity can significantly impact the global carbon cycle. The distribution of nutrients above and below the soil crust can be considerably altered by climate change, brought about by increased atmospheric CO_2. For example, the rise in atmospheric CO2 can lead to more significant plant growth because CO2 is a crucial component of photosynthesis. This might quicken the rate of litter production and change the chemical composition of the litter, which might change how easily it can be digested. These alterations will affect the types of organic compounds that soil bacteria can utilize or access (Zak et al., 2000). The overall carbon supply and the nitrogen flow between plants and microorganisms may be affected by changed litter formation as a result (Berntson and Bazzaz, 1997).

Furthermore, increased root growth might be brought on by rising CO2, which would be highly detrimental to soil structure and biota. Therefore, it is evident that biological activity in soil can have a significant impact on global carbon sequence. Owing to hundreds of megatons of carbon fluxes per year between soil and atmosphere, soils now hold around twice as much carbon found in the atmosphere. To better comprehend the carbon feedback between soil and atmosphere and whether or how to manage it, we must have a thorough grasp of the carbon cycle.

19.5.3.2 *Biodiversity in the soil at its most extreme*

Nearly everywhere has been found to have microbes. Extremes in temperature, humidity, and solar radiation are only a few of the situations covered by this. The soil environment can be equally hostile for microorganisms, although they have been observed to thrive in demanding conditions like blazing volcanic springs. Across the world, soil communities have been found in various challenging environments, from the freezing Antarctic to the hot desert, including areas with a lot of UV exposure. In such circumstances, communities with low levels of biodiversity are frequently found. The limited biodiversity often permits a better understanding of the interactions and connections between species and their physiochemical environments, which facilitates research. Consequences and perceptions from these systems can then be used to understand further intricate designs. Communities often have little to no functional redundancy due to severe conditions, often having low levels of biodiversity. Due to this, they are more susceptible to disturbance events than other organisms in the community. The loss of a vibrant function from society can profoundly affect other organisms, perhaps leading to significant adjustments in the population's structure and the rate of environmental processes.

19.5.3.2.1 Temperature

Hot On Earth, Death Valley is among the hottest region. The summertime daytime highs are 45°C and 93°C-plus ground temperatures (Douglas 2006). Even though the biodiversity of the soil in Death Valley is far lower than that of temperate soils, extremophilic bacteria may nonetheless thrive there. Regarding the geochemical cycle, microorganisms have been discovered to function in hot deserts. According to Walvoord et al. (2003), a significant bioavailable nitrogen reservoir has built up in the subsurface zones of hot deserts (up to 104 Kg N ha^{-1} hectare, in nitrate form). Usual sources of nitrate under desert conditions include fixation of atmospheric N_2 by N-fixing organisms, nitrate in precipitation, and aeolian deposition of nitrate salts. This shows that ecosystem activity, controlled by bacteria, may persist even in the very hot desert setting.

Cold The soils of Antarctica are the iciest on the planet (Campbell and Claridge 1981). In some parts of Antarctica, mean monthly air temperatures barely climb above 0°C in December and January and never exceed −20°C in July and August. These soils are the lone ones that harbor nematodes at the top of the food chain, so their food webs are oddly primary. Because nematodes are sea organisms, soil moisture impacts their survival more than cold. Nematodes, however, can experience "anhydrobiosis," a kind of cryptobiosis that occurs when an organism loses all access to water and is characterized by the total stoppage of all metabolic activity within the organism.

Organisms are resilient in this stage, enabling them to persist for a long time until circumstances improve. Nematodes do not function in this state regarding nutrient cycling, along with microorganisms' relatively slow metabolic activity during low temperatures and the absence of metabolic activity during the cooler months. This contributes to the enormously slow frequency of biogeochemical cycles in the Antarctic environment. Microarthropods use a variety of strategies to endure the bitter cold in habitats that experience extreme cold, including those in Antarctica. To keep their body fluids liquefied, they use carbohydrates as cryoprotectants and, in rare instances, antifreeze proteins. This can be facilitated by removing or hiding ice-nucleating chemicals from their bodily fluids (Sinclair and Stevens 2006). Such techniques avert the freezing of body fluids and the cellular damage that would follow.

Dry The world's driest soils, as well as the world's coldest soils, are found in the McMurdo Dry Valleys of Antarctica. Outside Antarctica, Chile's Atacama Desert is the driest place on earth. There may be decades without rain in some areas when annual rainfall is less than 5 mm (Warren-Rhodes et al., 2006). There is, however, just a small quantity of biodiversity in this area. Cyanobacteria capable of photosynthesis can still be found in regions where the annual water availability is less than 75 hours. This seems to be the upper limit of photosynthesis regarding water availability. It is determined that the organic carbon within the soil communities in such ecosystems has a turnover duration of 320 decades because severe water restrictions have a negative impact on photosynthesis and metabolism (Warren-Rhodes et al., 2006). In contrast, arable soil turnover is frequently below 100 years (Yamashita et al., 2006).

Solar radiation The soil crust receives extreme solar energy. Just 0.3% of solar radiation can pass through even sandy soils with reasonably large pores of 2 mm depth, and in silty clay soil, this transmittance drops to just 0.2% at that depth (Benvenuti 1995). Although UV is harmful to life, restricting photosynthesis is nevertheless necessary for photosynthetic organisms. Inhibition of fungal hyphal development by UV light increased with decreasing wavelength, according to research by Hughes et al. (2003). Pigments are created by organisms when they are exposed to sunlight for protection against the harmful effects of UV. Being closer to the soil surface causes collembola to exhibit higher levels of melanization, which is a direct effect of their increased UV exposure (Hopkin 1997). It may come as a surprise Arrange et al. (1993) discovered several of the underground bacteria exhibiting comparable UV tolerance compared to surface microbes. This may be because the soil system is so dynamic that soil bacteria cannot rely on UV protection and may become exposed after bouts of erosion. To survive in such circumstances, subsurface creatures must have the necessary defense mechanisms.

19.5.4 Mechanisms for coping with extremes

19.5.4.1 Cryptobiosis

The cryptobiotic state of the organism is characterized by an undisclosed metabolism and the presence of morphological and physiological changes. Certain organisms, including several species of bacteria and nematodes, can pass in a cryptobiotic state at any stage of their lifecycle when environmental conditions become severe to maintain their active existence. Under such conditions, organisms become resistant to environmental factors. In such form, Nematodes have demonstrated resistance to cryogenic temperatures, desiccation (under 0% relative humidity), vacuum, and nematicides. Nonetheless, they can rapidly regain activity on reoccurrence of appropriate environmental settings (Freckman and Womersley, 1983).

19.5.4.2 Use of proteins

Proteins have demonstrated the ability to protect organisms against either heat or cold extremes. Heat shock proteins are proteins that assist in protecting organisms against the impacts of heat. Such proteins act as "chaperones" for other proteins, which can assist in folding and preventing denaturation, otherwise produced by heat. On the other hand, cold shock proteins help in protecting organisms against the cold. Cold shock proteins have a less defined activity mode than heat shock proteins, though many can bind with DNA for chromosome assembly or act as RNA 'chaperones. Some creatures that can survive in subzero temperatures can also produce antifreeze proteins. They act by attaching to ice crystals with the ability to develop and prevent expansion.

19.5.4.3 Cyst formation

Many microorganisms, such as bacteria, protozoa, and fungi, can encyst and create "cysts." Cysts are latent and hardy creatures with slight or no metabolism and thick cell walls. Cysts are more resistant to temperature, pH, and desiccation extremes than other types of bacteria. The organisms can emerge from their cyst form and resume metabolizing, growth, and reproduction if the conditions are favorable again.

19.6 Mitigation strategies

Soil management strategies have a direct impact on soil biodiversity along with the ecosystem that ensues. Organic matter input is the primary mechanism underlying changes in soil biodiversity as management methods are intensified. Organic matter fuels the soil food web dominated by bacteria (low C:N) or fungus (high C:N). In grasslands, higher litter inputs support microbial communities dominated by fungi (Bardgett et al., 1997) in addition to larger nematode diversity (Bardgett et al., 1997; Wasilewska, 1994) and microarthropods (Siepel, 1996). Increased microbial activity may improve biological regulators, reducing the prevalence of nematodes and soil pathogens (Freckman, 1988; Griffiths et al., 1994). The microbial populations are repressed in intensively managed (fertilized) grasslands or croplands, and a shift to communities dominated by opportunistic microorganisms occurs (Lovell et al., 1995; Bardgett et al., 1993; Bardgett and Leemans, 1996). As a result, it is more probable that fauna that feeds on opportunistic bacteria would prosper.

Soil tillage activities disrupt fungal hyphae and larger earthworm species, such as anecic earthworms that frequent the soil surface in search of plant material to eat (Emmerling et al., 2002). In contrast to soils under organic management, biomass and the number of anecic earthworms are decreased by a ratio of 1.3–3 (Pfiffner and Mader, 1997; Hoschitz and Kaufmann, 2004). Conventional management also has a negative impact on soil drainage and aeration.

19.6.1 Some techniques for preserving soil biodiversity include

19.6.1.1 Light soil sealing/mulching

The process of mulching involves adding a layer of mulch to the soil's surface in order to reduce erosion and boost fertility. Crops are normally covered with mulch during early growing season and can spread as necessary. It first aids the soil's ability to hold heat and moisture, warming it. Mulching can be done from various substances, such as manure, sewage sludge, compost, rubber, plastic films, and organic leftovers (such as crop residue, hay, or bark).

19.6.1.2 Use of organic wastes (compost, manure, and sludge)

The organic matter content of the soil can be improved by applying animal dung and sludge. Organic leftovers should decompose before being used on the land for agricultural purposes. Adding carbon-rich chemicals to the soil temporarily immobilizes available nitrogen, which microorganisms require for growth and development.

19.6.1.3 Fertilizers

High concentrations of various inorganic nitrogenous fertilizers provide rapidly available nitrogen, hence boosting the activity of microbes. The amount of organic matter continues to decline, and eventually, the soil loses its structure and water-holding capacity with accelerated breakdown of ordinary organic inputs and organic matter.

19.6.2 Crop management

19.6.2.1 Choice of the crop's species

High concentrations of various inorganic nitrogen fertilizers provide easy-to-use nitrogen, increasing microbes' activity. The amount of soil organic matter continues to decline because of the accelerated breakdown of poor-quality organic inputs and soil organic matter. Eventually, the soil loses its structure and water-holding capacity.

19.6.2.2 Crop rotations

Crop rotations can also assist in preventing disease and pest development by changing the communities of biological regulators linked with each crop.

19.6.3 Landscape management

19.6.3.1 Hedgerows and grassy field margins

Hedges or grassy strips along the border of arable farms provide a steady environment, food, and a safe microclimate for soil animals thriving near highly maintained fields. Due to their modest movement, hedgerows, like biological regulators, are better for soil organisms. The 10% of all soil-inhabitant organisms found in farms along field margins.

19.7 Remote sensing of biodiversity

Ecosystem processes and functions may be negatively impacted by anthropogenic leads to the loss of biodiversity (Hooper et al., 2012). Yet, assessing biodiversity is challenging since the world is so big and so many variables are at play that change over time and place. There are various methods to monitor biodiversity (Gotelli and Colwell, 2001), but many are time-consuming, expensive, and arbitrary (Bonar et al., 2011). Additionally, there are diverse national programs to assess biodiversity; hence, it becomes tedious to monitor biodiversity globally (Skidmore et al., 2015). The global GEO BON (Group on Earth Observations Biodiversity Observation Network) launched an endeavor to integrate and harmonize prevailing biodiversity monitoring activities by supporting the development of new observing systems to coordinate biodiversity monitoring efforts and prioritization of needs (Scholes et al., 2012). GEO BON determined a list of "Essential Biodiversity Variables" to aid the foundation for biodiversity monitoring projects (Pereira et al., 2013).

By measuring the disparity in spectral data produced by plants' biochemical, structural, or phenological properties, RS can predict species richness. Turner (2014). Also, future space-based hyperspectral missions include the Italian Hyperspectral Imager, the German Environmental Mapping and Analysis Program, and NASA's Hyperspectral Infrared Imager (Lee et al., 2015; Dhaloiya et al., 2022). German DLR Earth Sensing Imaging Spectrometer (Eckardt et al., 2015), the French HYPXIM (Michel et al., 2011), and the Japanese Hyperspectral Imager Suite (Iwasaki et al., 2011) aid in better assessment and tracking of global biodiversity the. A wide variety of sensors and approaches are used in RS. The passive RS detects reflected sunlight as multispectral or hyperspectral images across several spectral bands. For more than 20 years, to monitor biodiversity, researchers have been investigating the role of passive RS (Kuenzer et al., 2014; Schäfer et al., 2016). These strategies may be divided into three groups: (1) Plant or species mapping, (2) habitat mapping, and (3) creation of association among biodiversity and RS (Nagendra, 2001). The third group is the most significant and is termed as the spectrum variation hypothesis (Palmer et al., 2002), claiming that variation in spectral reflectance is associated with biodiversity (Cavender-Bares et al., 2017). Taxonomic variety (species' quantity and relative abundance), functional diversity (species' ecological responsibilities), phylogenetic diversity (species' evolutionary changes), and genetic diversity (species' genetic variation) are all aspects of biodiversity (Cisneros et al., 2015; Lyashevska and Farnsworth, 2012). By measuring the spectral data variation brought on by the biochemical, structural, or phenological traits of plants, RS has the potential to forecast species richness.

19.8 Remote sensing in assessing the soil biodiversity

RS is a valuable tool to assess and map soil biodiversity. There are several ways in RS that can be used for this purpose. Vegetation indices viz., Normalized Difference Vegetation Index and Enhanced Vegetation Index can be employed to indirectly evaluate soil biodiversity by measuring the amount of vegetation cover and the photosynthetic activity of plants (Kumar et al., 2021; Hooda et al., 2020; Kumar et al., 2019). A higher level of vegetation cover and photosynthetic activity generally indicates better soil quality and greater biodiversity. Hyperspectral imaging can be used to identify the spectral signatures of different types of soil and vegetation. It can recognize distinct soil types and create maps of their distribution. It can also be used to identify changes in soil composition and quality. Thermal imaging can measure the temperature of the soil surface and identify areas with a higher level of microbial activity or organic matter content. This information is critical in assessing soil health and biodiversity. LiDAR technology can create highly detailed 3D maps of soil and vegetation, which can be used to identify microhabitats and areas with high or low biodiversity.

Overall, RS is a vital tool to assess soil biodiversity and can provide information on the distribution and health of different soil types. This information is crucial for making informed decisions about land management and conservation efforts.

19.9 Conclusion

It is difficult to nail down the feedback reactions that soil microorganisms impact global warming. Hence, microbes significantly impact climate and ecosystems' response to climate change, be it a positive response with the microbial community contributing to climate change or a negative feedback response with decelerated effects of climate change. Due to its substantial CO_2 and CH_4 emissions, dependency of soil carbon stocks on respiration rates, and early sensitivity to high air temperatures, soil respiration is crucial to these impacts. Studying the long-standing response of soil respiration to climate change and the potential of terrestrial forests to absorb more CO_2 from the atmosphere may help us better comprehend the overall consequences of climate change. RS and GIS are crucial methods in monitoring the effect of climate change on soil biodiversity on a wide scale. In attempts to reduce GHG emissions and adapt to anticipated consequences of climate change, the focus must be driven toward microscopic life below the surface to better understand the forthcoming macroscopic and global changes. Quantifying these consequences is crucial since climate change is projected to significantly affect soils by altering soil biodiversity. In any event, any adaptation measures to slow down climate change are anticipated to increase soil biodiversity, soil functionality, and related services.

19.10 Future research strategies

I. How temperature and precipitation variations, in addition to biodiversity, soil ecosystems, and related processes, are impacted by climate change. Risk assessment has an impact on how the soil works and how services are provided.
II. Studying how changes in land usage affect the environment.
III. Studies on how humidity affects the pace of decomposition.

References

Arrange, A.A., Phelps, T.J., Benoit, R.E., White, D.C., 1993. Survival of subsurface microorganisms exposed to UV radiation and hydrogen peroxide. Appl. Environ. Microbiol. 59, 3545–3550.

Bardgett, R., Leemans, D., 1996. Soil microbial activity on exposed mountain ridges in Snowdonia (Eryri), North Wales. Soil. Biol. Biochem. 28 (10), 1533–1536.

Bardgett, R.D., Frankland, J.C., Whittaker, J.B., 1993. The effects of agricultural management on the soil biota of some upland grasslands. Agric., Ecosyst. Environ. 45 (1–2), 25–45.

Bardgett, R.D., Leemans, D.K., Cook, R., et al., 1997. Seasonality of soil biota of grazed and ungrazed hill grasslands. Soil. Biol. Biochem. 29 (8), 1285–1294.

Beever, E.A., O'Leary, J., Mengelt, C., West, J.M., Julius, S., Green, N., et al., 2016. Improving conservation outcomes with a new paradigm for understanding species' fundamental and realized adaptive capacity. Conserv. Lett. 9, 131–137. Available from: https://doi.org/10.1111/conl.12190.

Benvenuti, S., 1995. Soil light penetration and dormancy of Jimsonweed. Weed Sci. 43, 389–393.

Berg, M.P., Kiers, E.T., Driessen, G., Van Der Heijden, M., Kooi, B.W., Kuenen, F., et al., 2010. Adapt or disperse: understanding species persistence in a changing world. Glob. Change Biol. 16, 587–598.

Berntson, G.M., Bazzaz, F.A., 1997. Regenerating temperate forest mesocosms in elevated CO2: belowground growth and nitrogen cycling. Oecologia 113 (1), 115–125.

Bolton, P.J., Phillipson, J., 1976. Burrowing, feeding, egestion and energy budgets of Allolobophora-Rosea (Savigny) (Lumbricidae). Oecologia. 23, 225–245.

Bonar, S.A., Fehmi, J.S., Mercado-Silva, N., 2011. An overview of sampling issues in species diversity and abundance surveys. In: Magurran, A.E., McGill, B.J. (Eds.), Biological Diversity: Frontiers in Measurement and Assessment. Oxford University Press, New York, pp. 11–24.

Booth, L.H., Heppelthwaite, V., McGlinchy, A., 2000. The effect of environmental parameters on growth, cholinesterase activity and glutathione S-transferase activity in the earthworm (Apporectodea caliginosa). Biomarkers. 5, 46–55.

Briones, M.J.I., Ineson, P., Sleep, D., 1999. Use of □13C to determine food selection in collembolan species. Soil. Biol. Biochem. 31, 937–940.

Campbell, I.B., Claridge, G.G.C., 1981. Soil research in the Ross Sea region of Antarctica. J. R. Soc. N. Zealand 11, 401–410.

Cassagne, N., Gers, C., Gauquelin, T., 2003. Relationships between Collembola, soil chemistry and humus types in forest stands (France). Biol. Fert. Soils. 37, 355–361.

Castro, H.F., Classen, A.T., Austin, E.E., Norby, R.J., Schadt, C.W., 2010. Soil microbial community response to multiple experimental climate change drivers. Appl. Environ. Microb. 76, 999–1007.

Cavender-Bares, J., Gamon, J.A., Hobbie, S.E., Madritch, M.D., Meireles, J.E., Schweiger, A.K., et al., 2017. Harnessing plant spectra to integrate the biodiversity sciences across biological and spatial scales. Am. J. Bot. 104, 966–969.

Chen, B., Snider, R.J., Snider, R.M., 1996. Food consumption by collembola from northern Michigan deciduous forest. Pedobiologia. 40, 149–161.

Choudhuri, D.K., 1963. Temperature and its effect on the three species of the genus Onychiurus Collembola. Proc. Zool. Soc. 16, 97–117.

Cisneros, L.M., Fagan, M.E., Willig, M.R., 2015. Effects of human-modified landscapes on taxonomic, functional and phylogenetic dimensions of bat biodiversity. Divers. Distrib. 21, 523–533.

Daniel, O., 1991. Leaf-litter consumption and assimilation by juveniles of *Lumbricus-terrestris* L. (Oligochaeta, Lumbricidae) under different environmental-conditions. Biol. Fertil. Soils. 12, 202–208.

De Santo, A.V., Berg, B., Rutigliano, F.A., et al., 1993. Factors regulating early–stage decomposition of needle litters in five different coniferous forests. Soil. Biol. Biochem. 25 (10), 1423–1433.

de Vries, F.T., Liiri, M.E., Bjornlund, L., Bowker, M.A., Christensen, S., Setala, H.M., et al., 2012. Land use alters the resistance and resilience of soil food webs to drought. Nat. Clim. Change 2, 276–280.

Dhaloiya, A., Hooda, R.S., Kumar, D., Malik, A., Kumar, A., 2022. Geoinformatics-based assessment of gross irrigation requirement of different crops grown in the south-western region of Haryana, India, Current Directions in Water Scarcity Research, 7. Elsevier, pp. 299–316. Available from: https://doi.org/10.1016/B978-0-323-91910-4.00018-2.

Díaz, S., Settele, J., Brondízio, E., Ngo, H.T., Guèze, M., Agard Trinidad, J., et al., 2019. Summary for Policymakers of the Global Assessment Report on Biodiversity and Ecosystem Services of the Intergovernmental Science-policy Platform on Biodiversity and Ecosystem Services. Secretariat of the Intergovernmental Science-Policy Platform on Biodiversity and Ecosystem Services, Bonn, Germany.

Dollery, R., Hodkinson, I.D., Jonsdottir, I.S., 2006. Impact of warming and timing of snow melt on soil microarthropod assemblages associated with Dryas-dominated plant communities on Svalbard. Ecography 29, 111–119.

Douglas, S., 2006. Nature: "Life in Death Valley." http://www.pbs.org/wnet/nature/deathvalley.

Eckardt, A., Horack, J., Lehmann, F., et al., 2015. DESIS (DLR Earth Sensing Imaging Spectrometer for the ISS-MUSES platform). In: Geoscience and Remote Sensing Symposium (IGARSS), 2015 IEEE International. IEEE, pp. 1457–1459.

Emmerling, C., Schloter, M., Hartmann, A., et al., 2002. Functional diversity of soil organisms—a review of recent research activities in Germany. J. Plant. Nutr. Soil. Sci. 165 (4), 408–420.

Fischer, D.G., Chapman, S.K., Classen, A.T., Gehring, C.A., Grady, K.C., Schweitzer, J.A., et al., 2014. Marschner review: plant genetic effects on soils under climate change. Plant. Soil. 37, 91 –19.

Foden, W.B., Young, B.E., 2016. IUCN SSC Guidelines for Assessing Species' Vulnerability to Climate Change. Cambridge, England and Gland, Switzerland. Available from: https://doi.org/10.1038/scientificamerican0315-S1.

Freckman, D.W., Womersley, C., 1983. Physiological adaptations of nematodes in Chihuahuan desert Soil. In: Lebrun, P.H. (Ed.), New Trends in Soil Biology. Dieu-Brichard Publishers, Louvain, pp. 396–404.

Freckman, D.W., 1988. Bacterivorous nematodes and organic–matter decomposition. Agri., Ecosyst. Environ. 24 (1–3), 195–217.

Glick, P., Stein, B.A., Edelson, N.A., 2011. Scanning the Conservation Horizon: A Guide to Climate Change Vulnerability Assessment.

Gotelli, N.J., Colwell, R.K., 2001. Quantifying biodiversity: procedures and pitfalls in the measurement and comparison of species richness. Ecol. Lett. 4, 379–391.

Gottfried, M., Pauli, H., Futschik, A., Akhalkatsi, M., Barancok, P., Alonso, J.L.B., et al., 2012. Continent-wide response of mountain vegetation to climate change. Nat. Clim. Change 2, 111–115.

Griffiths, B.S., Ritz, K., Wheatley, R.E., 1994. Nematodes as indicators of enhanced microbiological activity in a Scottish organic farming system. Soil. Use Manage 10 (1), 20–24.

Haimi, J., Laamanen, J., Penttinen, R., Räty, M., Koponen, S., 2005. Impacts of elevated CO2 and temperature on the soil fauna of boreal forests. Appl. Soil. Ecol. 30, 104–112.

Harris, R.F., 1980. Effect of water potential on microbial growth and activity. Water potential relations in soil microbiology Special publication no 9. Soil Society of America, Madison, Wisconsin, 23–95.

Hoffland, E., Kuyper, T.W., Wallander, H., Plassard, C., Gorbushina, A.A., Haselwandter, K., et al., 2004. The role of fungi in weathering. Front. Ecol. Environ. 2, 258–264.

Holmstrup, M., 2001. Sensitivity of life history parameters in the earthworm Aporrectodea caliginosa to small changes in soil water potential. Soil. Biol. Biochem. 33, 1217–1223.

Hooda Arvind, R.S., Sheoran, H.S., Kumar, D., Bhardwaj, S., 2020. RS-based regional crop identification and mapping: a case study of Barwala sub-branch of Western Yamuna Canal in Haryana (India. Indian. J. Tradit. Knowl. 19 (1), 182–186. Available from: https://doi.org/10.56042/ijtk.v19i1.30865.

Hooper, D.U., Adair, E.C., Cardinale, B.J., Byrnes, J.E., Hungate, B.A., Matulich, K.L., et al., 2012. A global synthesis reveals biodiversity loss as a major driver of ecosystem change. Nature 486, 105–108.

Hopkin, S.P., 1997. Biology of the Springtails (Insecta: Collembola). Oxford University Press.

Hoschitz, M., Kaufmann, R., 2004. Soil nematode communities of Alpine summits–site differentiation and microclimatic influences. Pedobiologi 48, 313–320.

Hughes, K.A., Lawley, B., Newsham, K.K., 2003. Solar UV B radiation inhibits the growth of Antarctic terrestrial fungi. Appl. Environ. Microbiol. 69, 1488–1491.

Hunt, H.W., Coleman, D.C., Ingham, E.R., Ingham, R.E., Elliott, E.T., Moore, J.C., et al., 1987. The detrital food web in a shortgrass prairie. Biol. Fertil. Soils. 3, 57–68.

Huntingford, C., Cox, P.M., Lenton, T.M., 2000. Contrasting responses of a simple terrestrial ecosystem model to global change. Ecol. Model. 134, 41–58.

Ingham, E.R., Trofymow, J.A., Ames, R.N., Hunt, H.W., Morley, C.R., Moore, J.C., et al., 1986. Trophic interactions and nitrogen cycling in a semi-arid grassland soil. I. Seasonal dynamics of the natural populations, their interactions and effects on nitrogen cycling. J. Appl. Ecol. 23, 597–614.

Kovach, R.P., Dunham, J.B., Al-Chokhachy, R., Snyder, C.D., Letcher, B.H., Young, J.A., et al., 2019. An integrated framework for ecological drought across riverscapes of North America. Bioscience 69, 418–431. Available from: https://doi.org/10.1093/biosci/biz040.

Kuenzer, C., Ottinger, M., Wegmann, M., Guo, H., Wang, C., Zhang, J., et al., 2014. Earth observation satellite sensors for biodiversity monitoring: potentials and bottlenecks. Int. J. Remote. Sens. 35, 6599–6647.

Kumar, D., Arvind, Nain, A.S., Darshana, Arya, S., Bhardwaj, S., et al., 2019. Soil loss estimation using geo-spatial technology in north western trai region of India. J. Agrometeorol. 21, 182–188.

Kumar, D., Arvind, Nain, A.S., Singh, A., Mor, A., Bhardwaj, S., 2021. Geo-spatial technology application for prioritization of land resources in Udham Singh Nagar District of Uttarakhand, India. Indian. J. Tradit. Knowl. (IJTK) 20 (2), 595–603. Available from: https://doi.org/10.56042/ijtk.v20i2.30770.

Langley, J.A., Hungate, B.A., 2014. Plant community feedbacks and long-term ecosystem responses to multi-factored global change. AoB Plants 6, 12.

Lee, C.M., Cable, M.L., Hook, S.J., Green, R.O., Ustin, S.L., Mandl, D.J., et al., 2015. An introduction to the NASA Hyperspectral InfraRed Imager (HyspIRI) mission and preparatory activities. Remote. Sens. Environ. 167, 6–19.

Lovell, R.D., Jarvis, S.C., Bardgett, R.D., 1995. Soil microbial biomass and activity in long–term grassland—effects of management changes. Soil. Biol. Biochem. 27 (7), 969–975.

Lyashevska, O., Farnsworth, K.D., 2012. How many dimensions of biodiversity do we need? Ecol. Indic. 18, 485–492.

Michel, S., Gamet, P., Lefevre-Fonollosa, M.-J., 2011. HYPXIM—A hyperspectral satellite defined for science, security and defence users. In: Hyperspectral Image and Signal Processing: Evolution in Remote Sensing (WHISPERS), 2011 3rd Workshop on. IEEE, pp. 1–4.

Nagendra, H., 2001. Using remote sensing to assess biodiversity. Int. J. Remote. Sens. 22, 2377–2400.

Ostle, N.J., et al., 2009. Integrating plant–soil interactions into global carbon cycle models.

Palmer, M.W., Earls, P.G., Hoagland, B.W., White, P.S., Wohlgemuth, T., 2002. Quantitative tools for perfecting species lists. Environmetrics 13, 121–137.

Pereira, H.M., Ferrier, S., Walters, M., Geller, G.N., Jongman, R., Scholes, R.J., et al., 2013. Essential biodiversity variables. Science 339, 277–278.

Pfiffner, L., Mader, P., 1997. Effects of biodynamic, organic and conventional production systems on earthworm Populations. Biol. Agric. Hortic. 111555, 3–10.

Roy, S., Roy, M.M., 2006. Spatial distribution and seasonal abundance of soil mites and collembola in grassland and Leucaena plantation in a semi-arid region. Trop. Ecol. 47, 57–62.

Ruess, L., Michelsen, A., Schmidt, I.K., et al., 1999. Simulated climate change affecting microorganisms, nematode density and biodiversity in subarctic soils. Plant. Soil. 212 (1), 63–73.

Schadt, C.W., Martin, A.P., Lipson, D.A., et al., 2003. Seasonal dynamics of previously unknown fungal lineages in tundra soils. Science. 301 (5638), 1359–1361.

Schäfer, E., Heiskanen, J., Heikinheimo, V., Pellikka, P., 2016. Mapping tree species diversity of a tropical montane forest by unsupervised clustering of airborne imaging spectroscopy data. Ecol. Indic. 64, 49–58.

Scheu, S., 1987. Microbial activity and nutrient dynamics in earthworm casts (Lumbricidae). Biol. Fert. Soils. 5, 230–234.

Schimel, J., Balser, T.C., Wallenstein, M., 2007. Microbial stress-response physiology and its implications for ecosystem function. Ecology 88, 1386–1394.

Scholes, R.J., Walters, M., Turak, E., Saarenmaa, H., Heip, C.H., Tuama, É.Ó., et al., 2012. Building a global observing system for biodiversity. Curr. Opin. Environ. Sustain. 4, 139–146.

Siepel, H., 1996. Biodiversity of soil micro arthropods: the filtering of species. Biodivers. Conserv. 5 (2), 251–260.

Sinclair, B.J., Stevens, M.I., 2006. Terrestrial microarthropods of Victoria Land and Queen Maud Mountains, Antarctica: Implications of climate change, Soil Biol. Biochem., 38. pp. 3158–3170.

Singh, J.P., Dhaloiya, A., 2022. Technology for smart agriculture under climate change scenario. In: Existing Climate Change Scenario and its Arising Risks. Mahima Publications, BHU, Varanasi, UP, India, pp. 101–108. 978-81-953029-0-1.

Skidmore, A.K., Pettorelli, N., Coops, N.C., Geller, G.N., Hansen, M., Lucas, R., et al., 2015. Agree on biodiversity metrics to track from space: ecologists and space agencies must forge a global monitoring strategy. Nature. 523, 403–405.

Swift, M.J., Heal, O.W., Anderson, J.M., 1979. Decomposition in Terrestrial Ecosystems–Studies Ecology, 5. Blackwell Scientific, Oxford.

Turner, W., 2014. Sensing biodiversity. Science 346, 301–302.

Van der Putten, W.H., et al., 2013. Plant–soil feedbacks: the past, the present and future challenges. J. Ecol. 101, 265–276.

Vannier, G., 1994. The thermo biological limits of some freezing intolerant insects –The super cooling and thermo stupor points. Acta Oecol. 15, 31–42.

Wall, D.H., Virginia, R.A., 1999. Controls on soil biodiversity: insights from extreme environments. Appl. Soil. Ecol. 13, 137–150.

Wallwork, J.A., 1970. Ecology of Soil Animals. McGraw Hill, London, p. 283.

Walther, G.R., Post, E., Convey, P., Menzel, A., Parmesan, C., Beebee, T.J.C., et al., 2002. Ecological responses to recent climate change. Nature 416, 389–395.

Walvoord, M.A., Phillips, F.M., Stonestrom, D.A., Evans, R.D., Hartsough, P.C., Newman, B.D., et al., 2003. A reservoir of nitrate beneath desert soils. Science 302, 1021–1024.

Warren-Rhodes, K., Rhodes, K., Pointing, S., Ewing, S., Lacap, D., Mez-Silva, B., et al., 2006. Hypolithic cyanobacteria, dry limit of photosynthesis, and microbial ecology in the hyperarid Ata.

Wasilewska, L., 1994. The effect of age of meadows on succession and diversity in soil nematode communities. Pedobiologia 38 (1), 1–11. world. Global Change Biology 16:587–598.

Yamashita, T., Flessa, H., John, B., Helfrich, M., Ludwig, B., 2006. Organic matter in density fractions of waterstable aggregates in silty soils: effect of land use. Soil. Biol. Biochem. 38, 3222–3234.

Zak, D.R., Pregitzer, K.S., Curtis, P.S., Holmes, W.E., 2000. Atmospheric CO2 and the composition and function of soil microbial communities. Ecol. Appl. 10, 47–59.

Chapter 20

Mapping of soil physical and chemical properties in Vemagal Hobli of Kolar District, Karnataka, India

G.K. Harikaran[1], S. Dharumarajan[2], R. Vasundhara[2], M. Lalitha[2], S. Parvathy[2] and G. Ushakiran[2]

[1]*Department Soil Science and Agricultural Chemistry, University of Agricultural Sciences, Bangalore, India,* [2]*ICAR-National Bureau of Soil Survey and Land Use Planning, Regional Centre, Bangalore, Karnataka, India*

20.1 Introduction

In the modern world, complex problems are confronted by a wide range of concerns. Due to the growing population, it is essential to ensure a safe environment for all living things to thrive. Degradation of land is a widespread problem that has adverse effects on the environment, biodiversity, and human livelihoods. Consequently, ecosystem services and functions, such as soil fertility, food security, and other essential elements, cannot be maintained. Sustainable land management requires accurate assessment and forecasting of the geographical distribution of soil qualities. In order to maintain a sustainable present and a safe future, scientists, environmentalists, and policymakers have always considered soil a highly important component. Modern land appraisal, land suitability analysis, and land resource management require increasingly precise and accurate soil maps that contain more details on soil qualities and their spatial distribution (Gobin et al., 2001). There is a great need for quantitative, geographical soil data, which is crucial for understanding the state of the soil resources and carrying out restoration efforts successfully. However, the information generated through conventional methods may not be enough to meet expectations due to a few constraints such as time-consuming, laborious, lack of spatial continuity, lack of a multidimensional approach, lack of relationship between soil variables, etc.

The limitations of traditional methods can be overcome by integrating them with modern technologies. One such cutting-edge technique is digital soil mapping (DSM), which creates geographically referenced soil databases using correlations between measurements taken in the field and the laboratory and spatially explicit environmental data with the help of machine learning (ML) models (McBratney et al., 2003). The DSM is based on the SCORPAN approach (McBratney et al., 2003), a descendant of the CLORPT (Climate, Organism, Relief, Parent Material, Time) model (Jenny, 1941). According to this approach, soil property (S) at a certain place can be obtained from the empirical quantitative function of environmental variables: soil (s), climate (c), organisms (o), terrain (r), parent material (p), age (a), and spatial position (n), that is, S = f. (s, c, o, r, p, a, n). To determine the soil characteristics and soil classes for unidentified locations, a suitable model is fitted to the association between environmental variables and soil properties or soil classes. This ML-assisted prediction provides accurate quantitative spatial soil information in a short period of time and at a lower cost than conventional methods.

Various works on ML—aided digital mapping of soil properties were carried out worldwide. Using the random forest (RF) model, Dharumarajan et al. (2017) mapped the regional variation of pH, electrical conductivity (EC), and organic carbon (OC). The RF model was developed using surface sample data from field surveys. According to the RF model's variable relevance ranking, the most significant predictors of OC were enhanced vegetation index (EVI) and Normalized difference vegetation index (NDVI), whereas EC and pH were best predicted by drainage and NDVI respectively. Pouladi et al. (2019) compared the performance of kriging, Cubist (Cu), RF, and regression-kriging for the prediction of SOM content utilizing proximate soil sensors, topography data, and remote sensing data. The findings indicated that kriging, followed by regression-kriging, had the best performance.

Remote Sensing of Soils. DOI: https://doi.org/10.1016/B978-0-443-18773-5.00022-3

Zeraatpisheh et al. (2019) studied the spatial prediction of soil OC and clay content of topsoil using Cu, RF, multiple linear regression (MLR), and Regression Tree (RT). The most accurate prediction for clay content was from the RF and Cu models, whereas both models performed best for predicting Soil organic carbon (SOC) content. Dharumarajan et al. (2021) used 5352 soil profile datasets to map the total profile permanent wilting point (TPWP) and total profile field capacity (TFC) in Karnataka using regression kriging and an RF model. Based on soil texture classes, a class pedotransfer function was created, and for a soil depth of 1 m, total FC and PWP were calculated for the profile. The predicted TFC and TPWP varied from 58 to 521 mm and 35 to 342 mm respectively.

In this regard, the present work was conducted with an aim to develop predicted and uncertainty maps of various soil physicochemical properties at 4 different depths (0–25, 25–50, 50–75, and 75–100 cm) in Vemagal Hobli, Kolar district, Karnataka. The specific objectives of the study are as follows:

1. To compare different ML models [RF, Support Vector Machine (SVM), Cu, and Artificial Neural Network (ANN)] for prediction of soil physicochemical properties.
2. To map soil physicochemical properties along with uncertainty using the best ML model.

20.2 Material and methods

20.2.1 Study site description

The study site constitutes Vemagal Hobli (block), which is located in Kolar district, Karnataka. Fig. 20.1 represents the location map of the study site. It covers an area of about 13,948 ha and is geographically located at 13°9′27.47″ N latitude to 13°18′17.13″ N latitude and 77°56′5.32″ E longitude to 78°5′10.43″ E longitude. The study area belongs to the Eastern Dry Zone of Karnataka, as per the National Agriculture Research Project. According to the ICAR agro ecological subregion, it is located in Central Karnataka plateau, hot, moist, and semiarid eco-subregion.

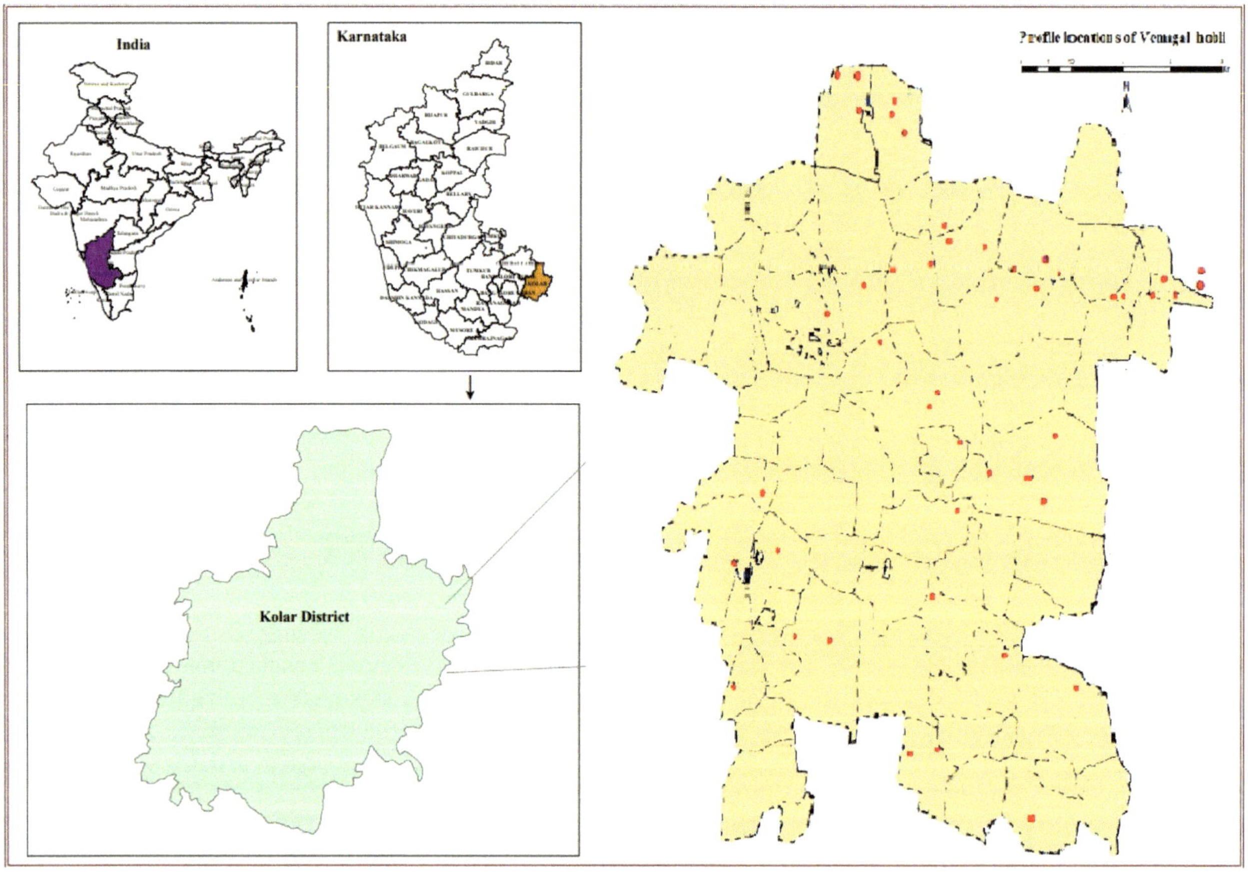

FIGURE 20.1 Location map of Vemagal Hobli, Kolar district, Karnataka, India.

The study site is occupied by granite and granite-gneiss. The elevation ranges from 820 to 1100 m above the mean sea level. According to Indian Meteorological Department data, the study site is distributed in a tropical climate with a maximum temperature ranging from 30°C to 36°C and a minimum temperature ranging from 15°C to 22°C. The annual rainfall is about 600–700 mm. The major soil orders are *Alfisols* and *Inceptisols*. The site falls under the ustic soil moisture regime and the isohyperthermic soil temperature regime. The main cropping season is kharif and crops like mango (*Mangifera indica*), finger millet (*Eleusine coracana*), red gram (*Cajanus cajan*), tomato (*Solanum lycopersicum*), and mulberry (*Morus albus*) are commonly cultivated in this region. The major vegetation in this area is Eucalyptus (*Eucalyptus globulus*), Tamarind (*Tamarindus indica*), Pongamia (*Pongamia pinnata*), etc.

20.2.2 Field study

The soil profile locations were identified using high-resolution satellite imagery (Sentinel-2). Field examination was carried out from 28.12.2021 to 19.02.2022 and about 46 profile locations were considered for study. Profile locations were selected in transects based on variations in the physiography and cropping pattern of the study site. These soil profiles were studied to a depth of 200 cm or to the depth limited by bedrock.

20.2.3 Soil sampling and analysis

The soil samples were collected horizon-wise from each profile, covering the variability of the study area. A total of 183 samples were collected from 46 identified profiles. The samples for bulk density (BD) estimation were collected separately in a metallic core of known volume, which was inserted in the soil horizon. The collected samples were shade-dried in the laboratory and then ground using a wooden pestle and mortar. After grinding, the samples were passed through a 2 mm sieve to separate coarse fragments (gravel, pebbles, roots, etc.). For OC estimation, the soil samples were finely grinded and then sieved using a 0.2-mm sieve.

The processed soil samples were analyzed in a laboratory for physicochemical properties using standard analytical methods. BD was determined gravimetrically by using the core method. The initial weight of the moist core samples was recorded and then placed in a hot air oven overnight for drying. After drying, the oven dry weight of the samples was recorded, and BD was then calculated using the recorded weights and volume of the core. Particle size analysis was carried out by the international pipette method (Jackson, 1973). Air-dry soil particles of <2 mm size fraction were treated with H_2O_2 for dissolution of organic matter and completely dispersed using sodium hexa-meta phosphate with ultrasound sonicator. The sand particles were separated by passing the dispersed solution through a 53-micron sieve, and the obtained solution after sieving was analyzed for silt and clay fractionation.

Soil pH was determined using the potentiometric method (Jackson, 1973) by taking freshly stirred 1:2.5 soil: water suspension. EC was determined using the conductivity bridge method (Jackson, 1973) by taking a 1:2.5 soil: water supernatant solution. SOC was determined by the wet digestion method (Walkley and Black, 1934), where a known weight of powdered soil sample was treated with a known volume of standard potassium dichromate and concentrated H_2SO_4. The unreacted $K_2Cr_2O_7$ was analyzed with standard ferrous ammonium sulfate using the ferroin indicator. Cation exchange capacity (CEC) was determined by the ammonium acetate leaching method described by Jackson (1973) using a neutral, normal ammonium acetate solution. A known quantity of soil was saturated with the ammonium ions, and repeated leaching was done with alcohol and NH_4Cl through filter paper. The NH_4^+ ions adsorbed in the exchangeable soil surface were determined through the micro-kjeldahl distillation technique.

Available water capacity (AWC) was estimated by using the pedo-transfer functions given by Dharumarajan et al. (2019).

$$\text{Field capacity (FC)} = 39.18 - 0.041(\text{clay}) - 0.371(\text{sand}) + 0.257(\text{CEC})$$

$$\text{Permanent wilting point (PWP)} = 8.227 + 0.168(\text{clay}) - 0.101(\text{sand}) + 0.217(\text{CEC})$$

$$\text{AWC } (\%) = \text{FC} - \text{PWP}$$

AWC values in % were converted into mm by using the formula mentioned below.

$$\text{AWC (mm)} = (\text{AWC } (\%) \times \text{BD} \times \text{depth})/10$$

20.2.4 Preparation of basic datasets

In addition to the 46 soil profiles studied during the survey, legacy datasets from 53 profiles of the study area were also collected from the National Bureau of Soil Survey and Land Use Planning, Regional Centre, Hebbal, Bangalore. The data collected during the present study was compiled with the legacy data. A total of about 99 profiles were considered for the prediction of soil physicochemical properties. Depth wise (0–25, 25–50, 50–75, and 75–100 cm) distribution of all soil physicochemical properties was assessed using a weighted average method.

20.2.5 Environmental attributes considered

A different set of environmental variables (Table 20.1) was used for prediction of soil physicochemical properties. The digital elevation model (DEM) was obtained from shuttle radar topography mission (SRTM) with a 30-m resolution. The derivatives of DEM or the digital terrain attributes like multi-resolution index of valley bottom flatness (MrVBF), multi-resolution index of ridge top flatness (MrRTF), topographic position index (TPI), topographic wetness index (TWI), aspect, channel network base level, channel network distance, analytical hill shading, closed depressions, convergence index (CI), plan curvature, profile curvature, relative slope position (RSP), slope, total catchment, and valley depth were derived by using system of automated geoscientific analyses, that is, Saga-GIS 2.3.1 version. NDVI and

TABLE 20.1 Different variables used for the prediction of soil physicochemical properties.

Variables	Source	Resolution
Sentinel—2 bands (b2-b12)	11 bands	10–60 m
Elevation (m)	SRTM DEM	30 m
Aspect	SRTM DEM	30 m
Channel network base level	SRTM DEM	30 m
Channel network distance	SRTM DEM	30 m
Analytical hill shading	SRTM DEM	30 m
Closed depressions	SRTM DEM	30 m
Convergence index	SRTM DEM	30 m
MrRTF	SRTM DEM	30 m
MrVBF	SRTM DEM	30 m
Plan curvature	SRTM DEM	30 m
Profile curvature	SRTM DEM	30 m
Relative slope position	SRTM DEM	30 m
Slope	SRTM DEM	30 m
TPI	SRTM DEM	30 m
TWI	SRTM DEM	30 m
Total catchment	SRTM DEM	30 m
Valley depth	SRTM DEM	30 m
NDVI	MOD13Q1	250 m_16 days
EVI	MOD13Q1	250 m_16 days
Land surface temperature	MOD11A1	1 km
Precipitation (mm)	WorldClim 2 data	30 seconds
Average temperature (°C)	WorldClim 2 data	30 seconds

EVI, Enhanced vegetation index; ***MrRTF***, multi-resolution index of ridge top flatness; ***MrVBF***, Multi-resolution index of valley bottom flatness; ***NDVI***, normalized difference vegetation index; ***SRTM DEM***, shuttle radar topography mission digital elevation model; ***TPI***, topographic position index; ***TWI***, topographic wetness index.

EVI were obtained from moderate resolution imaging spectroradiometer (MODIS) data (MOD13Q1). Land surface temperature (LST) imagery was obtained from MODIS data (MOD11A1). Sentinel 2 bands, namely band 2–12, were also used as variable input. Precipitation and average temperature datasets were obtained from World Climate Version 2 data. All the datasets and environmental variables were intersected at the sampling points for the prediction of soil physicochemical properties.

20.2.5.1 Modeling and mapping

In the present study, four ML models—RF, SVM, Cu, and ANN—were compared for the prediction of soil physicochemical properties.

RF model is an extension of the RT model, which works based on the assemblage of a number of classifications and RTs by means of two levels of randomization for each tree in the forest (Breiman, 2001). It is the most used model at present since it improves prediction accuracy and reduces model over-fitting, and it is nonsensitive to missing data and has the capacity to handle large numbers of both quantitative and categorical data (Dharumarajan et al., 2017). The number of trees (n_{tree}), minimum number of samples at the terminal node (n_{min}), and number of predictors used for fitting the tree (M_{try}) will determine the fitting of the model. For the present study, the RF package (version 4.6–14) was used for running the RF model in an R environment. This package helps in performing classification and regression operations.

SVM maintains all covariates to define a maximal margin, which is the margin of tolerance, using the support vectors (observations) and to separate or fit data linearly. The margin is the distance from the decision surface, which ensures the high generalization ability of the algorithm, making the results more applicable to the unseen data (Pradhan, 2013). In addition, this approach applies kernel functions to map nonlinear vectors to a very high-dimensional space for solving nonlinear problems. For performing statistical functions, the e1071 package (version 1.7–9) was used in the R environment.

Cu is a rule-based algorithm that has recently increased in popularity among digital soil mappers (Miller et al., 2015). It initially creates a tree structure from a pool of provided covariates, and then it collapses paths through the tree to create rules using boosting training. Each rule contains a MLR model for predicting the target variable under the conditions of the respective rule. The final model located at the terminal nodes shows a collection of MLR models for calculating predicted values. In addition, this ensemble model adds boosting to improve prediction accuracy. Ensemble learning combines models produced by multiple repetitions of the same algorithm, which usually obtains stronger predictive performance than results produced from any of the models individually. In the present study, the Cu package (version 0.3.0) was used for running the Cu model in R environment. This package performs rule and instance-based regression modeling.

The ANN algorithm was inspired by the functioning of human brain through a network of neurons; this algorithm is composed of connections of components operating in parallel. It can be trained to perform a specific function and obtain a certain output by adjusting the values of the connection weights between the components. The network is adjusted based on a comparison of the secured output and targeted outcome until the sum of square differences between them becomes minimum. The attractiveness of ANN comes from the remarkable information processing characteristics of the biological system, such as nonlinearity, high parallelism, robustness, fault and failure tolerance, learning, the ability to handle imprecise and fuzzy information, and their capability to generalize. An ANN can be regarded as a class of universal approximators that implements a nonlinear input–output mapping of a general nature (Prieto et al., 2013). For training the ANN model in the R environment, the neuralnet package (version 1.44.2) was used.

The models were validated using 10-fold cross-validation techniques with 20 repetitions. The predictive performance of models was evaluated using uncertainty indicators such as coefficient of determination (R^2) and root-mean-squared error (RMSE). For good models, R^2 should be equal to or close to 1, whereas RMSE should be low or near 0.

1. Coefficient of determination $(R2) = 1 - \frac{\sum_{i=1}^{n} (pi - oi)^2}{\sum_{i=1}^{n} \left(oi - \overline{oi}\right)^2}$
2. Root-mean-squared error (RMSE) = $\sqrt{\frac{1}{n}\sum_{i=1}^{n} (oi - pi)^2}$, where *pi* and *oi* are predicted and observed values.

Uncertainty analysis was performed at a 90% confidence interval using the quantile random forest (QRF) model. The prediction uncertainty was evaluated using the prediction interval (PI) coverage percentage (PICP). A 30-m grid was created, and all the environmental covariates were extracted for that grid. Among the different models used for the study, the best model was used to predict soil physicochemical properties at a 30-m grid. Predicted properties were then mapped using the point-to-raster tool of ArcGIS version 10.7.1.

20.3 Results and discussion

20.3.1 Summary statistics: observed soil physicochemical properties

Table 20.2 shows the summary statistics of observed soil physical properties. The mean observed BD was found to increase with depth, with a coefficient of variation ranging from 10.13% to 12.5%. Higher variability in BD was observed at 50–75 cm depth. The mean of the observed sand and silt content was found to decrease and increase with depth, respectively. The mean clay content exhibited an irregular trend with depth. The kurtosis was found to be positive in all depths for sand and clay content. The sand content was negatively skewed with depth, whereas silt and clay content were found to be positively skewed with depth. The variability in silt content was found to be decreasing with depth. The mean observed AWC was found to exhibit an irregular trend with depth. The kurtosis was positive at different depths. The observed AWC was found to be negatively skewed at all depths.

The summary statistics of the observed soil chemical properties is represented in Table 20.3. The mean observed soil pH increased with depth, ranging from 6.11 to 6.64. The mean observed EC was found to be lower, with a range of 0.09 to 0.12 dS m^{-1}. The observed EC exhibited positive kurtosis and skewness at all depths. The mean observed OC content exhibited a decreasing trend with depth, and the coefficient of variation was found to be ranging from 53.59% to 57.62%. The mean observed CEC range was found to be lower (9.91 to 12.40 cmol (p +) kg^{-1}) with positive kurtosis and skewness at all depths.

TABLE 20.2 Summary statistics of observed soil physical properties

Property	Depth (cm)	Min	Max	Mean	SD	Kurtosis	Skewness	CV (%)
BD (g cm^{-3})	0–25	1.08	1.59	1.58	0.16	1.07	−0.12	10.13
	25–50	1.24	1.65	1.59	0.17	−0.79	−0.32	10.69
	50–75	1.28	1.70	1.60	0.20	2.07	0.74	12.5
	75–100	1.40	1.75	1.61	0.18	−0.71	0.12	11.18
Sand (%)	0–25	21.17	85.73	59.66	12.02	0.58	−0.74	20.15
	25–50	10.73	73.64	49.51	10.88	2.06	−0.63	21.98
	50–75	10.85	70.40	46.55	10.16	2.41	−0.84	21.83
	75–100	11.35	64.57	45.70	10.53	1.56	−0.74	23.04
Silt (%)	0–25	2.50	36.79	13.51	7.25	0.18	0.62	53.66
	25–50	2.20	35.47	14.37	6.91	−0.12	0.48	48.09
	50–75	3.75	36.97	15.03	6.96	0.13	0.44	46.31
	75–100	3.48	36.97	15.56	7.07	0.08	0.35	45.44
Clay (%)	0–25	5.17	61.69	26.79	11.34	0.43	0.44	42.33
	25–50	7.09	65.24	36.14	11.27	0.70	0.32	31.18
	50–75	18.34	68.72	38.20	10.98	0.51	0.70	28.74
	75–100	10.18	67.80	37.98	11.65	1.11	0.29	30.67
AWC (mm)	0–25	10.02	71.35	38.61	10.44	2.10	−0.19	27.04
	25–50	10.14	61.81	44.13	10.83	1.16	−0.85	24.54
	50–75	13.30	65.30	45.90	11.02	1.20	−0.59	24.01
	75–100	10.84	64.80	45.82	11.40	1.50	−0.74	24.88

TABLE 20.3 Summary statistics of observed soil chemical properties.

Property	Depth (cm)	Min	Max	Mean	SD	Kurtosis	Skewness	CV (%)
pH	0–25	4.21	8.70	6.11	0.85	0.37	0.29	13.91
	25–50	4.31	9.12	6.32	0.92	0.36	0.43	14.56
	50–75	4.26	9.12	6.53	0.96	0.07	0.14	14.7
	75–100	4.06	8.78	6.64	1.00	− 0.19	− 0.04	15.06
EC (dS m^{-1})	0–25	0.01	0.61	0.09	0.08	21.85	4.13	88.89
	25–50	0.01	0.65	0.09	0.08	30.65	4.68	88.89
	50–75	0.01	0.72	0.10	0.11	17.84	3.74	110
	75–100	0.02	1.12	0.12	0.17	20.52	4.39	141.67
OC (%)	0–25	0.08	1.29	0.49	0.27	0.48	0.98	55.1
	25–50	0.12	1.03	0.38	0.20	2.16	1.31	52.63
	50–75	0.12	0.86	0.32	0.17	1.76	1.24	53.13
	75–100	0.09	0.85	0.30	0.16	2.74	1.52	53.33
CEC (meq/100 g)	0–25	3.16	30.45	9.91	5.71	3.18	1.80	57.62
	25–50	2.95	32.61	11.98	6.42	1.35	1.37	53.59
	50–75	1.94	33.18	12.40	6.71	1.53	1.40	54.11
	75–100	1.94	33.10	12.18	6.73	2.34	1.63	55.26

20.3.2 Correlation analysis

Pearson correlation analysis was performed between environmental covariates used for prediction and surface physicochemical properties. Table 20.4 represents the Pearson correlation coefficients between surface soil physical properties and environmental covariates. At the 1% level, a significant correlation was found to exist between BD and band 9, sand content and MrRTF, silt content and band 2, silt content and band 3, silt content and LST, silt content and valley depth, and clay content and band 12. At the 5% level, a significant correlation was found to exist between BD and CI, BD and plan curvature, silt content and channel network base level, silt content and MrRTF, silt content and precipitation, clay content and band 11, clay content and valley depth, and AWC and MrRTF. Among various properties and covariates, a strong positive correlation (0.45**) was observed between silt content and valley depth at the 1% level, whereas a strong negative correlation (−0.42**) was observed between silt content and LST at the 1% level.

The Pearson correlation analysis between environmental covariates and surface chemical properties is shown in Table 20.5. At 1% level, significant correlation was found to exist between pH and band 6, pH and band 8, pH and channel network distance, pH and RSP, and EC and total catchment area. At 5% level, significant correlation was found to exist between pH and band 7, EC and TWI, OC and band 4, OC and band 5, OC and band 12, OC and MrRTF, and OC and TPI. Strong positive correlation (0.46**) was observed between EC and total catchment area, whereas a strong negative correlation (−0.42**) was observed between pH and RSP.

20.3.3 Comparison of models performance

Physical properties like BD, sand content, silt content, clay content, and AWC were predicted using different ML models. Table 20.6 shows the performance of four models in predicting soil physical properties. The results showed that in the case of BD, the order of performance of algorithms is RF > SVM > Cubist > ANN in terms of R^2 and RMSE. The performance of RF is better in the prediction of BD, with R^2 ranging from 23% to 27% and RMSE ranging from 0.15 to 0.16 Mg m^{-3}. However, Katuwal et al. (2020) recorded R^2 of 62% and RMSE of 0.10 Mg m^{-3} in the prediction of BD with RF. The order of performance of algorithms in the prediction of sand content is RF > ANN > Cubist > SVM in terms of R^2. In the case of silt and clay content, the order is noticed as RF > Cubist > ANN > SVM with

TABLE 20.4 Correlation coefficients between surface soil physical properties and environmental covariates.

Properties/covariates	BD	Sand	Silt	Clay	AWC
B2	0.05	0.05	0.42[b]	− 0.27	0.26
B3	0.04	0.04	0.39[b]	− 0.24	0.24
B4	0.05	0.15	0.23	− 0.28	0.12
B5	0.02	0.18	0.19	− 0.29	0.07
B6	− 0.17	0.14	0.08	− 0.20	− 0.11
B7	− 0.14	0.18	0.05	− 0.22	− 0.13
B8	0.03	0.07	0.22	− 0.19	0.09
B9	0.44[b]	− 0.08	− 0.06	0.12	0.20
B10	0.10	0.04	0.03	− 0.06	0.05
B11	0.10	0.23	0.23	− 0.37[a]	0.11
B12	0.16	0.26	0.23	− 0.41[b]	0.13
NDVI	0.10	0.05	0.07	0.02	0.10
EVI	− 0.06	− 0.20	0.23	0.10	0.15
LST	0.07	0.07	− 0.42[b]	0.14	− 0.24
DEM	0.15	0.04	− 0.28	0.10	− 0.09
Aspect	0.17	− 0.13	− 0.01	0.15	0.14
CNBL	0.18	− 0.08	− 0.30[a]	0.24	− 0.05
CND	− 0.01	0.27	− 0.15	− 0.22	− 0.17
AHS	− 0.01	0.18	0.15	− 0.28	0.02
CD	− 0.16	− 0.02	0.17	− 0.06	0.02
CI	0.30[a]	− 0.14	0.09	0.11	0.29
MrRTF	− 0.01	− 0.38[b]	0.32[a]	0.25	0.29[a]
MrVBF	− 0.25	− 0.11	0.02	0.11	− 0.09
Plan curvature	0.32[a]	− 0.10	0.02	0.11	0.23
Profile curvature	0.25	− 0.07	0.08	0.03	0.19
RSP	− 0.04	0.24	− 0.21	− 0.15	− 0.22
Slope	0.17	0.11	− 0.19	− 0.02	− 0.04
TPI	0.17	− 0.12	− 0.08	0.17	0.08
TWI	− 0.08	− 0.15	0.13	0.10	0.06
Total catchment	− 0.11	− 0.19	0.17	0.12	0.06
Valley depth	− 0.02	0.07	0.45[b]	− 0.31[a]	0.22
Precipitation	0.03	0.05	0.38[a]	− 0.25	0.21
Average temperature	− 0.07	0.01	− 0.19	0.08	− 0.14

AHS, Analytical hill shading; *B2–B12*, Bands of sentinel-2 satellite imagery; *CD*, closed depressions, *CI*, convergence index; *CNBL*, channel network base level; *CND*, channel network distance; *DEM*, digital elevation model; *EVI*, enhanced vegetation index; *LST*, land surface temperature; *MrRTF*, multi-resolution index of ridge top flatness; *MrVBF*, multi-resolution index of valley bottom flatness; *NDVI*, normalized difference vegetation index; *RSP*, relative slope position; *TPI*, topographic position index, *TWI*, topographic wetness index.
[a]*Significant correlation at 5% level.*
[b]*Significant correlation at 1% level.*

TABLE 20.5 Correlation coefficients between surface soil chemical properties and environmental covariates.

Properties/covariates	pH	EC	OC	CEC	OC stock	CSP
B2	0.02	0.16	−0.18	0.14	−0.01	−0.01
B3	−0.06	0.03	−0.22	0.15	−0.01	−0.01
B4	−0.10	−0.03	−0.31[a]	0.03	−0.04	−0.04
B5	−0.19	−0.09	−0.35[a]	−0.02	−0.05	−0.05
B6	−0.40[b]	−0.23	−0.22	0.02	−0.02	−0.02
B7	−0.37[b]	−0.22	−0.14	0.01	−0.04	−0.04
B8	−0.39[b]	−0.27	−0.19	0.14	0.10	0.11
B9	−0.09	−0.07	0.09	−0.09	0.19	0.19
B10	−0.11	0.01	−0.03	−0.11	−0.06	−0.06
B11	−0.24	0.07	−0.28	−0.08	−0.09	−0.09
B12	−0.19	0.06	−0.32[a]	−0.09	−0.07	−0.07
NDVI	−0.28	−0.02	−0.06	−0.10	−0.01	−0.01
EVI	−0.03	0.02	−0.05	0.10	0.07	0.07
LST	0.02	0.19	−0.12	−0.22	−0.01	−0.01
DEM	−0.28	−0.21	0.16	−0.17	−0.05	−0.05
Aspect	0.03	0.02	0.20	−0.11	−0.05	−0.05
CNBL	−0.16	−0.17	0.11	−0.09	0.09	0.09
CND	−0.40[b]	−0.17	0.15	−0.26	−0.31[a]	−0.31[a]
AHS	0.21	0.10	−0.14	0.001	−0.08	−0.08
CD	−0.08	0	−0.20	0.08	−0.16	−0.02
CI	0.18	0.22	0.04	0.20	0.30[a]	0.36[a]
MrRTF	−0.14	−0.18	0.33[a]	0.28	−0.01	0.10
MrVBF	−0.13	0.09	−0.04	0.01	−0.25	−0.08
Plan curvature	0.19	0.07	0.17	0.17	0.32[a]	0.33[a]
Profile curvature	−0.12	0.001	0.19	−0.12	0.25	0.16
RSP	−0.42[b]	−0.19	0.22	−0.31[a]	−0.04	−0.33[a]
Slope	0.09	−0.09	0.23	−0.22	0.17	0.01
TPI	−0.01	0.02	0.30[a]	−0.04	0.17	0.09
TWI	0.16	0.31[a]	−0.15	0.26	−0.08	0.05
Total catchment	0.15	0.46[b]	−0.09	0.23	−0.11	0.1
Valley depth	0.17	0.13	−0.21	0.26	−0.02	0.04
Precipitation	0.07	0.02	−0.07	0.26	0.03	−0.07
Average temperature	0.08	0.02	0.13	−0.002	−0.07	−0.10

AHS, Analytical hill shading; *B2–B12*, Bands of sentinel-2 satellite imagery; *CD*, closed depressions; *CI*, convergence index, *CNBL*, channel network base level; *CND*, channel network distance; *DEM*, digital elevation model; *EVI*, enhanced vegetation index; *LST*, land surface temperature; *MrRTF*, multi-resolution index of ridge top flatness; *MrVBF*, multi-resolution index of valley bottom flatness, *NDVI*, normalized difference vegetation index; *RSP*, relative slope position; *TPI*, topographic position index, *TWI*, topographic wetness index.
[a]*Significant correlation at 5% level.*
[b]*Significant correlation at 1% level.*

TABLE 20.6 Performance of different models for prediction of soil physical properties.

Properties	Depth (cm)	SVM		RF		Cubist		ANN	
		R^2	RMSE	R^2	RMSE	R^2	RMSE	R^2	RMSE
BD ($g\ cm^{-3}$)	0–25	0.21	0.17	0.23	0.15	0.22	0.16	0.18	0.19
	25–50	0.22	0.16	0.23	0.15	0.22	0.18	0.17	0.21
	50–75	0.26	0.18	0.27	0.16	0.21	0.21	0.22	0.23
	75–100	0.23	0.17	0.27	0.15	0.25	0.19	0.21	0.23
Sand (%)	0–25	0.08	12.37	0.15	11.90	0.13	12.33	0.12	12.29
	25–50	0.11	10.80	0.17	10.73	0.14	10.78	0.16	10.81
	50–75	0.15	9.74	0.16	9.63	0.13	10.19	0.14	10.33
	75–100	0.13	10.53	0.17	10.40	0.14	10.89	0.16	10.52
Silt (%)	0–25	0.37	5.97	0.45	5.50	0.40	6.17	0.39	5.86
	25–50	0.42	5.48	0.51	4.99	0.50	5.07	0.43	5.45
	50–75	0.37	5.63	0.50	5.12	0.49	5.15	0.43	5.53
	75–100	0.33	5.94	0.43	5.50	0.36	5.90	0.35	5.86
Clay (%)	0–25	0.24	10.22	0.26	10.15	0.21	11.10	0.22	10.67
	25–50	0.24	10.08	0.31	9.77	0.28	10.05	0.29	9.94
	50–75	0.33	9.23	0.37	8.97	0.35	9.22	0.32	9.63
	75–100	0.27	10.31	0.39	10.14	0.35	10.17	0.28	10.60
AWC (mm)	0–25	0.32	9.22	0.45	8.41	0.40	9.21	0.36	9.79
	25–50	0.29	9.62	0.35	9.52	0.27	11.18	0.31	12.29
	50–75	0.23	10.31	0.37	10.06	0.28	12.02	0.32	13.39
	75–100	0.23	10.58	0.38	10.19	0.35	12.29	0.35	14.74

respect to R^2. It was found that RF showed satisfactory prediction performance in the case of silt content with R^2 ranging from 43% to 51% than clay content ($R^2 = 26\%–39\%$) and sand content ($R^2 = 15\%–17\%$). The prediction of silt content was better than the previous study by Dharumarajan et al. (2020) with R^2 ranging from 29%–49% using Quantile Regression Forest (QRF) model. The performance of RF in predicting clay content was found to be slightly increasing with depth. For prediction of AWC, the order of performance of algorithms was found to be RF > Cubist > ANN > SVM. The performance of RF in the prediction of AWC is better with $R^2 = 35\%–45\%$ and RMSE = 8.41–10.19 mm. RF predicted variability of AWC, particularly good in 0–25 cm depth than subsequent depths.

Chemical properties like pH, EC, SOC, and CEC were predicted using different ML algorithms. The depth-wise performance of four different algorithms in the prediction of chemical properties is shown in Table 20.7. The results showed that in the case of pH and EC prediction, the order of performance of algorithms is RF > SVM > Cubist > ANN in terms of R^2. The performance of RF in the prediction of pH and EC was found to be better with R^2 and RMSE values of 23%–25% and 0.74–0.91, 30%–44% and 0.05–0.09 dS m^{-1}, respectively. However, Dharumarajan et al. (2017) recorded variability of 30% and 62% in the prediction of pH and EC, respectively. It has been stated that 20% more accurate pH mapping was achieved with the RF model than other linear regression models (Hengl et al., 2015). For prediction of SOC, the order of performance of algorithms is RF > Cubist > SVM > ANN with respect to R^2. The RF model's performance for prediction of OC is low (i.e., $R^2 = 16\%–19\%$), which may be due to the dynamic nature of the study area because of variability in cropping patterns. In the case of CEC, the order of performance of algorithms is RF > ANN > Cubist > SVM with respect to R^2. The RF model explained 23%–32% of the variability of CEC in different depth intervals. A similar result was recorded by Dharumarajan et al. (2020) with an R^2 of 31%–56% using the RF model in the prediction of CEC.

TABLE 20.7 Performance of different models for prediction of soil chemical properties.

Properties	Depth (cm)	SVM		RF		Cubist		ANN	
		R^2	RMSE	R^2	RMSE	R^2	RMSE	R^2	RMSE
pH	0–25	0.27	0.76	0.29	0.74	0.21	0.84	0.16	0.86
	25–50	0.21	0.84	0.23	0.83	0.16	0.94	0.14	0.94
	50–75	0.17	0.91	0.23	0.86	0.16	0.97	0.16	1.00
	75–100	0.20	0.94	0.25	0.91	0.19	1.01	0.13	1.04
EC (dS m^{-1})	0–25	0.26	0.06	0.30	0.05	0.26	0.07	0.18	0.08
	25–50	0.36	0.06	0.44	0.05	0.28	0.07	0.26	0.08
	50–75	0.29	0.08	0.31	0.07	0.16	0.10	0.20	0.11
	75–100	0.33	0.12	0.34	0.09	0.26	0.15	0.20	0.17
OC (%)	0–25	0.16	0.26	0.17	0.25	0.13	0.28	0.13	0.27
	25–50	0.14	0.20	0.16	0.18	0.11	0.21	0.12	0.21
	50–75	0.14	0.17	0.16	0.15	0.11	0.17	0.15	0.19
	75–100	0.14	0.16	0.19	0.15	0.18	0.16	0.15	0.17
CEC (meq/100 g)	0–25	0.09	5.53	0.23	5.24	0.22	6.10	0.19	5.62
	25–50	0.17	6.04	0.30	5.51	0.25	6.72	0.22	6.02
	50–75	0.21	6.16	0.32	5.77	0.21	7.13	0.25	6.24
	75–100	0.18	6.10	0.28	5.92	0.21	7.27	0.23	6.42

Based on uncertainty indicators (R^2 and RMSE), it was found that the RF model exhibited better predictive performance than other three models in the prediction of soil physicochemical properties. Apart from this aspect, it also improved the accuracy of the prediction compared to other models (Breiman, 2001; Dharumarajan and Hegde, 2022).

20.3.4 Uncertainty analysis

Uncertainty analysis in the field of ML refers to 90% PI. Uncertainty mainly arises due to noise in observations (variability), training data, and model used (Kläs & Vollmer, 2018). Uncertainty can be indicated by an indicator called PICP, which indicates the percentage of all observed values fitting within their prediction limits. The commonly used confidence interval for PICP is 90%, which means that the true value can be found 9 out of 10 times within the range of values. For the current study, uncertainty analysis was performed for soil physicochemical properties like BD, sand content, silt content, clay content, AWC, pH, OC, and CEC. For physical properties, the PICP ranged from 80.6 to 90.5 (Table 20.8), whereas in the case of chemical properties, the range was from 86.7 to 91.4 (Table 20.9).

20.3.5 Variable importance ranking

Environmental variables play an important role in improving the prediction accuracy of soil properties. Based on how accurate or inaccurate the prediction would be if one or more variables were removed from the model, the RF model predicts the relative importance of variables considered (Prasad et al., 2006). The importance rankings of variables in the RF model in the prediction of soil physicochemical properties are depicted in Figs. 20.2 and 20.3. In the case of prediction of BD, band 3 of sentinel-2 imagery and LST were found to be the most important covariates. Channel network base level and channel network distance were important covariates in the prediction of sand content. For prediction of silt and clay content, total catchment area and slope and total catchment area and TWI were found to be important

TABLE 20.8 Uncertainty analysis of soil physical properties.

Property	Depth (cm)	PICP value
BD ($Mg\ m^{-3}$)	0–25	83.8
	25–50	87.8
	50–75	86.4
	75–100	85.0
Sand (%)	0–25	87.6
	25–50	86.8
	50–75	90.5
	75–100	88.3
Silt (%)	0–25	89.3
	25–50	88.5
	50–75	89.8
	75–100	89.4
Clay (%)	0–25	87.0
	25–50	86.6
	50–75	89.0
	75–100	88.5
AWC (mm)	0–25	80.6
	25–50	85.4
	50–75	86.4
	75–100	86.9

TABLE 20.9 Uncertainty analysis of soil chemical properties.

Property	Depth (cm)	PICP value
pH	0–25	88.8
	25–50	89.3
	50–75	87.8
	75–100	87.4
OC (%)	0–25	87.6
	25–50	87.3
	50–75	86.7
	75–100	87.2
CEC ($cmol(p+)\ kg^{-1}$)	0–25	87.7
	25–50	91.4
	50–75	89.5
	75–100	88.2

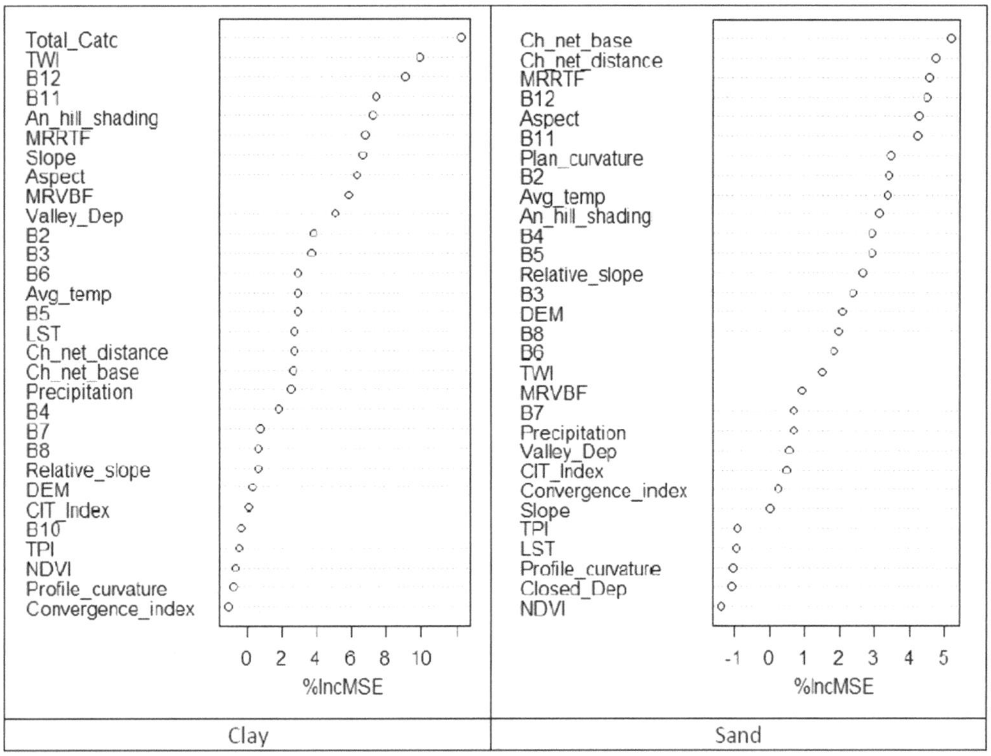

FIGURE 20.2 Variable importance ranking—Clay and sand content.

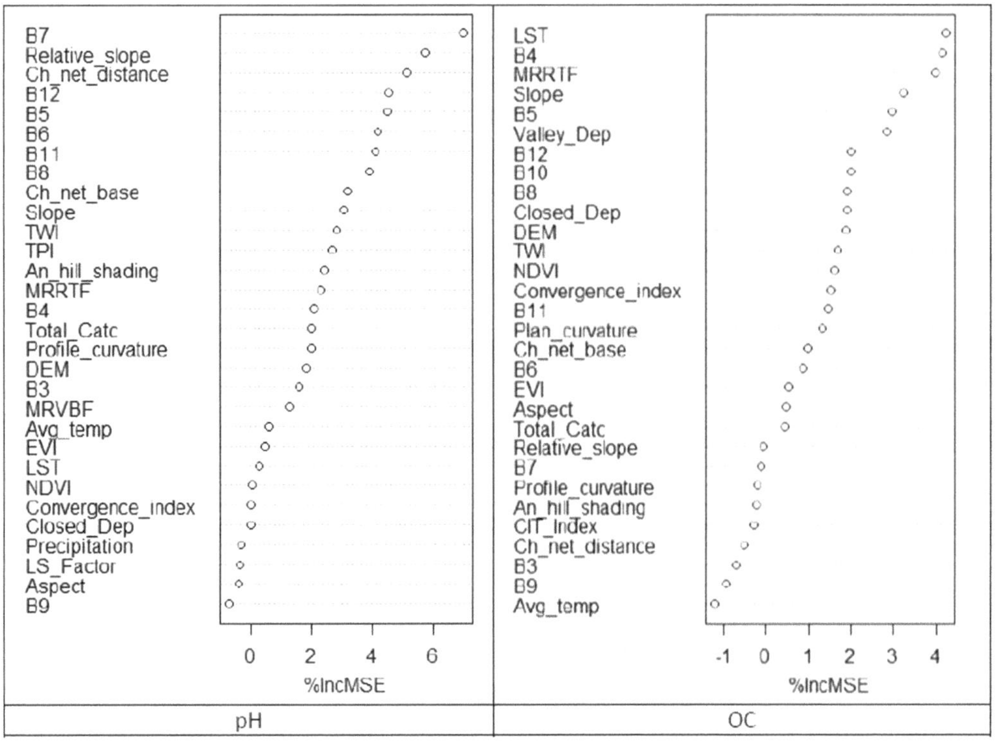

FIGURE 20.3 Variable importance ranking—pH and OC.

covariates, respectively. Soil moisture influences clay content, as it plays an important role in clay formation through the process of hydrolysis. Dharumarajan et al. (2020) found elevation to be the most important covariate in the prediction of clay content. The MrRTF and MrVBF were found to be the most important covariates in the prediction of AWC. AWC is low in high flat areas due to less clay content because of erosion processes and transportation, whereas it is high in low flat areas like valley due to more accumulation of clay content because of deposition process.

In the case of pH and EC prediction, band 7 of sentinel-2 imagery and RSP and aspect and MrRTF were found to be important covariates, respectively. Areas with a higher slope (i.e., hills and mountains) may have a low pH due to the removal of basic cations along the surface due to erosion and runoff. LST was found to be the top-most covariate in the prediction of SOC, as high temperatures affect the decomposition of organic matter. Aspect and analytical hill shading were found to be the most important covariates in the prediction of CEC. Soils in hilly areas show high CEC due to their high organic matter content.

20.3.6 Prediction of soil physicochemical properties

Table 20.10 shows the summary statistics of predicted soil physical properties. The mean BD value was found to be ranging from 1.55 to 1.59 $Mg\ m^{-3}$. The low coefficient of variation (1.91%) in surface horizon indicates less variability in BD when compared with subsequent layers. The kurtosis was found to be negative (−1.07) in

TABLE 20.10 Summary statistics of predicted soil physical properties.

Property	Depth (cm)	Min	Max	Mean	SD	Kurtosis	Skewness	CV (%)
BD ($Mg\ m^{-3}$)	0–25	1.42	1.67	1.57	0.03	5.97	−1.65	1.91
	25–50	1.41	1.70	1.58	0.07	−1.07	0.11	4.43
	50–75	1.40	1.72	1.59	0.07	0.10	0.07	4.40
	75–100	1.40	1.77	1.55	0.04	2.03	−1.26	2.58
Sand (%)	0–25	50.62	69.84	61.60	2.27	0.23	0.19	3.69
	25–50	39.94	55.86	48.52	1.31	9.11	2.80	2.70
	50–75	41.40	50.66	47.07	0.87	0.70	−0.39	1.85
	75–100	25.50	49.50	46.17	1.99	6.69	−1.97	4.31
Silt (%)	0–25	5.70	17.14	7.27	1.13	11.89	2.42	15.54
	25–50	6.26	18.41	8.97	1.03	19.96	3.03	11.48
	50–75	5.28	18.72	8.93	1.43	3.96	0.64	16.01
	75–100	5.50	18.72	11.83	3.08	−1.09	−0.21	26.04
Clay (%)	0–25	19.87	38.54	30.91	2.49	−0.31	−0.59	8.06
	25–50	31.23	48.44	41.61	2.68	0.36	−0.86	6.44
	50–75	33.39	48.44	43.92	2.16	1.95	−1.16	4.92
	75–100	36.55	53.60	45.85	1.56	3.66	−0.77	3.40
AWC (mm)	0–25	19.99	36.84	23.66	4.45	−1.12	0.83	18.81
	25–50	23.76	49.11	42.97	2.11	1.77	0.87	4.91
	50–75	36.11	51.59	45.94	2.38	−0.94	0.30	5.18
	75–100	31.30	51.96	41.70	4.84	−0.03	−0.89	11.61

TABLE 20.11 Summary statistics of predicted soil chemical properties.

Property	Depth (cm)	Min	Max	Mean	SD	Kurtosis	Skewness	CV (%)
pH	0–25	5.58	6.88	6.33	0.29	−0.71	−0.07	4.58
	25–50	5.54	7.02	6.45	0.28	0.01	−0.93	4.34
	50–75	5.60	7.40	6.50	0.34	−0.70	0.02	5.23
	75–100	5.50	7.56	6.56	0.37	−0.61	0.10	5.64
EC (dS m^{-1})	0–25	0.06	0.17	0.10	0.003	231.68	4.87	3.00
	25–50	0.07	0.12	0.10	0.002	60.38	−7.22	2.00
	50–75	0.06	0.15	0.10	0.002	189.43	−6.88	2.00
	75–100	0.06	0.17	0.10	0.005	54.38	6.11	5.00
OC (%)	0–25	0.29	0.81	0.46	0.08	0.69	0.39	17.39
	25–50	0.25	0.48	0.41	0.03	0.58	−0.38	7.32
	50–75	0.18	0.41	0.31	0.03	0.59	−0.66	9.68
	75–100	0.15	0.53	0.28	0.03	3.70	1.36	10.71
CEC (cmol(p+) kg^{-1})	0–25	6.29	11.92	8.96	0.86	−0.42	−0.45	9.60
	25–50	7.80	19.61	10.35	1.19	8.05	1.98	11.50
	50–75	7.20	15.42	10.89	1.59	1.59	1.47	14.60
	75–100	7.45	16.05	10.26	1.69	2.80	1.93	16.47

25–50 cm depth and positive at other depths. The BD was found to be positively skewed at 25–50 and 50–75-cm depths and negatively skewed in other depths. The average values of predicted sand, silt, and clay content were found to be in the range of 46.17%–61.60%, 7.27%–11.83%, and 30.91%–45.85%, respectively. The mean sand content followed a decreasing trend with depth, whereas the mean clay content followed an increasing trend with depth due to the illuviation process. The mean silt content exhibited an irregular trend with depth. The sand content exhibited positive kurtosis at different depths, whereas the silt and clay content exhibited negative kurtosis at 75–100 and 0–25-cm depths, respectively. The skewness was found to be negative at different depths in the case of clay content, whereas the sand and silt content exhibited negative skewness at 50–75, 75–100, and 75–100-cm depths, respectively. The coefficient of variation was found to decrease in the case of clay content (8.06 –3.40%), whereas the sand and silt content exhibited irregular trend. The mean value of the predicted AWC ranged from 23.66 to 45.94 mm. The mean AWC value was found to be increasing, which may be due to increased clay content since clay holds more water. The coefficient of variation (18.81%) is higher in the surface layer (i.e., 0–25 cm) than in subsequent layers. It was negatively skewed at 75–100-cm depth.

The summary statistics of predicted chemical properties are shown in (Table 20.11). The predicted pH and EC ranged from 5.50 to 7.56 and 0.06 to 0.17 dS m^{-1} respectively. The mean value of predicted pH was found to increase with depth (6.33–6.56), whereas the mean EC value was found to be same (0.1 dS m^{-1}) at different depths. The predicted pH indicates that the soils of the study area were moderately acidic to neutral. The soils were found to be nonsaline and exhibited less variability in EC throughout the study area. The predicted OC content ranged from 0.15% to 0.81%, and the mean value was found to be decreasing with an increase in depth (0.46%–0.28%). The surface layer was found to exhibit a higher coefficient of variation (17.39%) compared to other subsequent layers. This may be due to the variability in cropping pattern in different locations of the study area. The mean value of predicted CEC ranged from 8.96 to 10.89 cmol (p^+) kg^{-1}. The coefficient of variation exhibited an increasing trend in CEC among different depths (9.60%–16.47%). Figs. 20.4–20.7 represent the predicted and uncertainty maps of surface soil physicochemical properties.

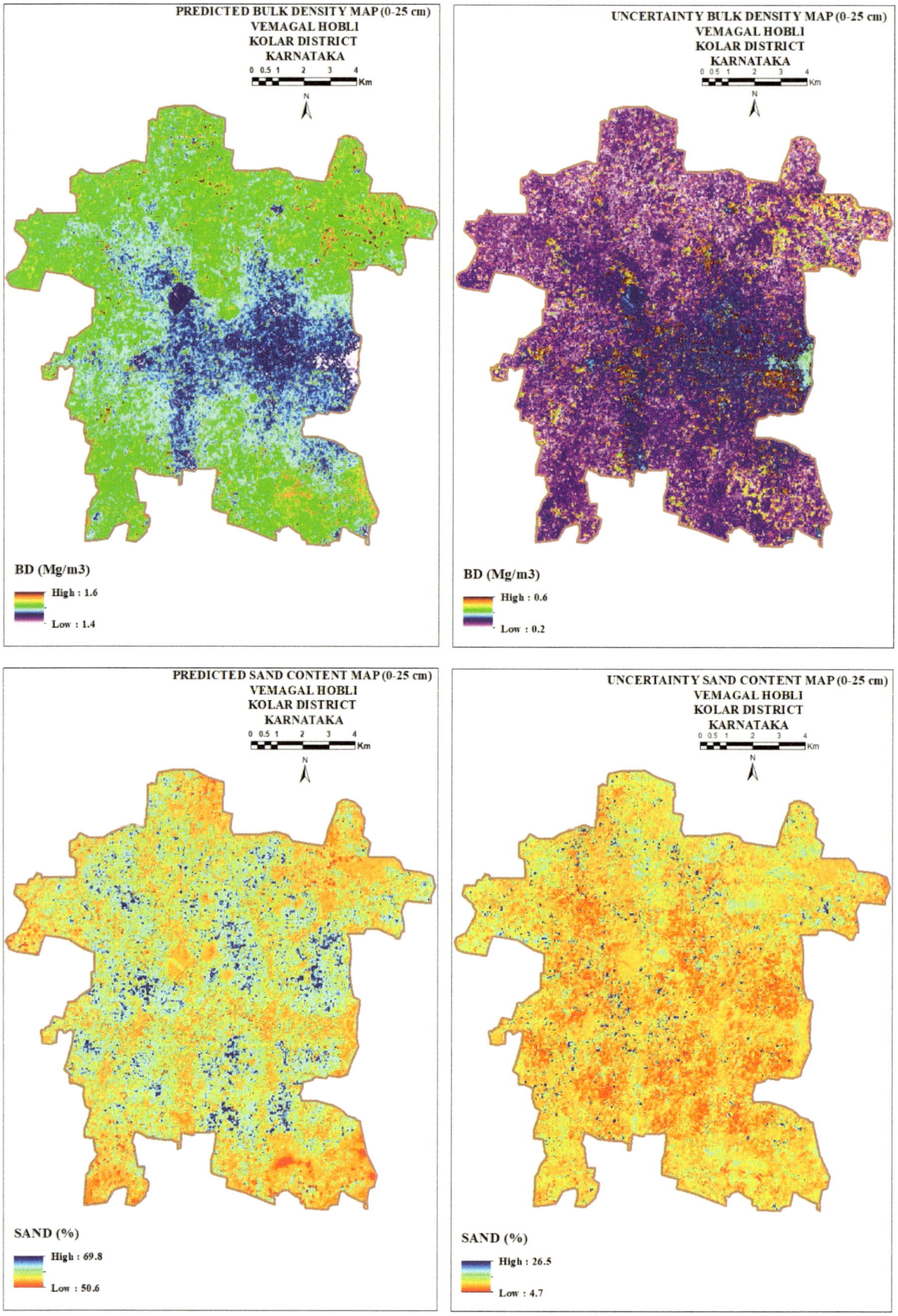

FIGURE 20.4 Predicted and uncertainty maps of bulk density and sand content (0–25 cm).

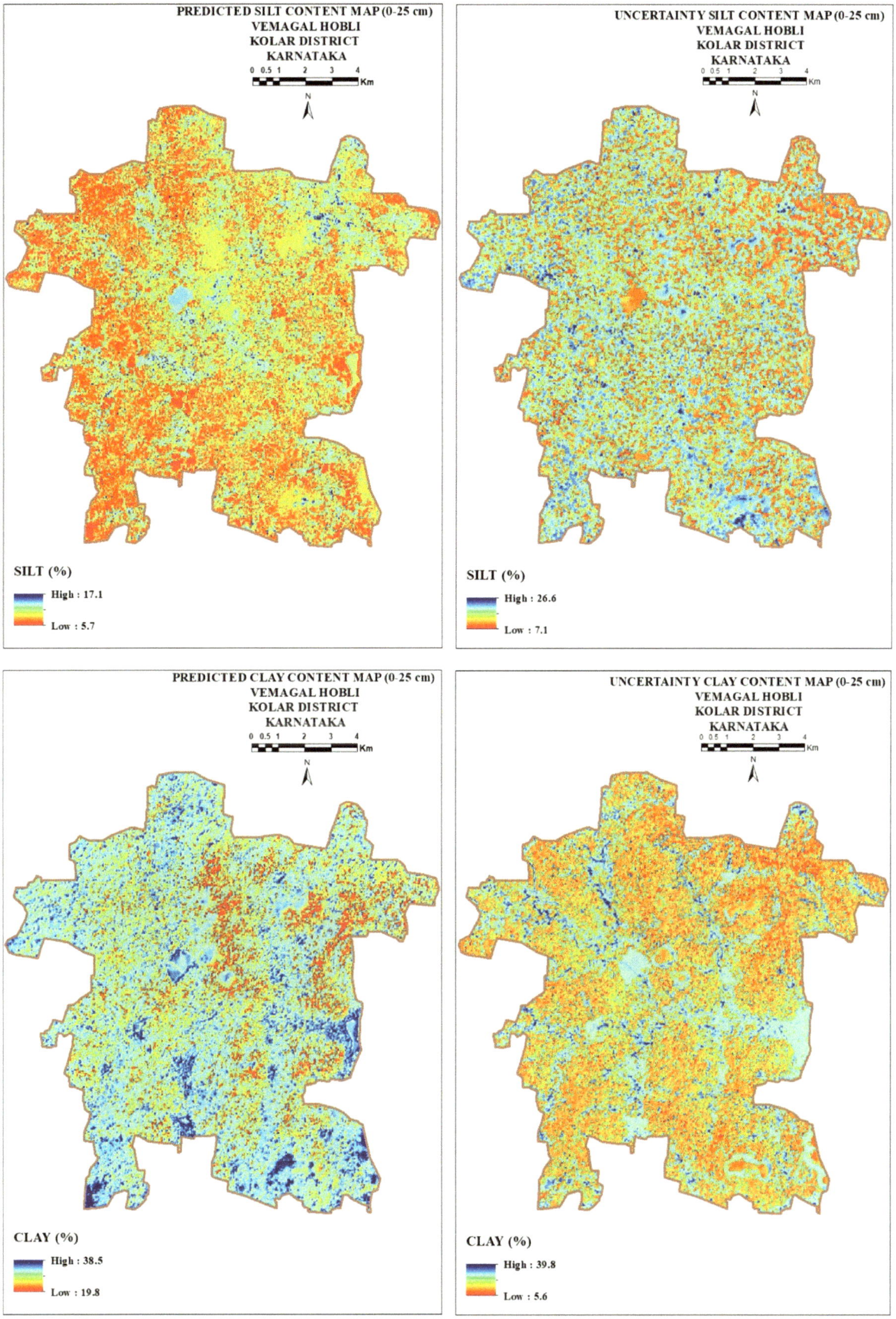

FIGURE 20.5 Predicted and uncertainty maps of silt and clay content (0–25 cm).

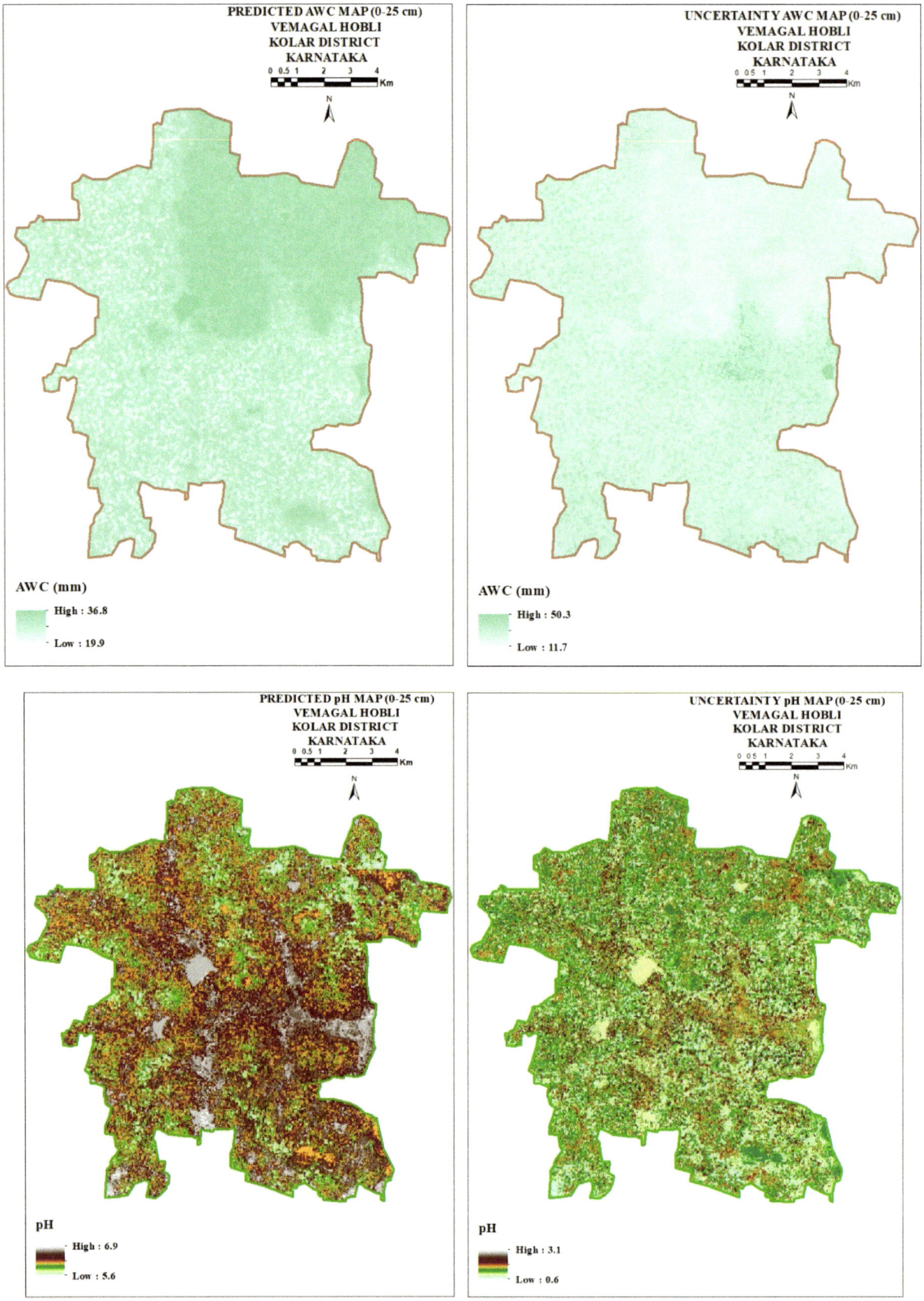

FIGURE 20.6 Predicted and uncertainty maps of AWC and pH (0–25 cm).

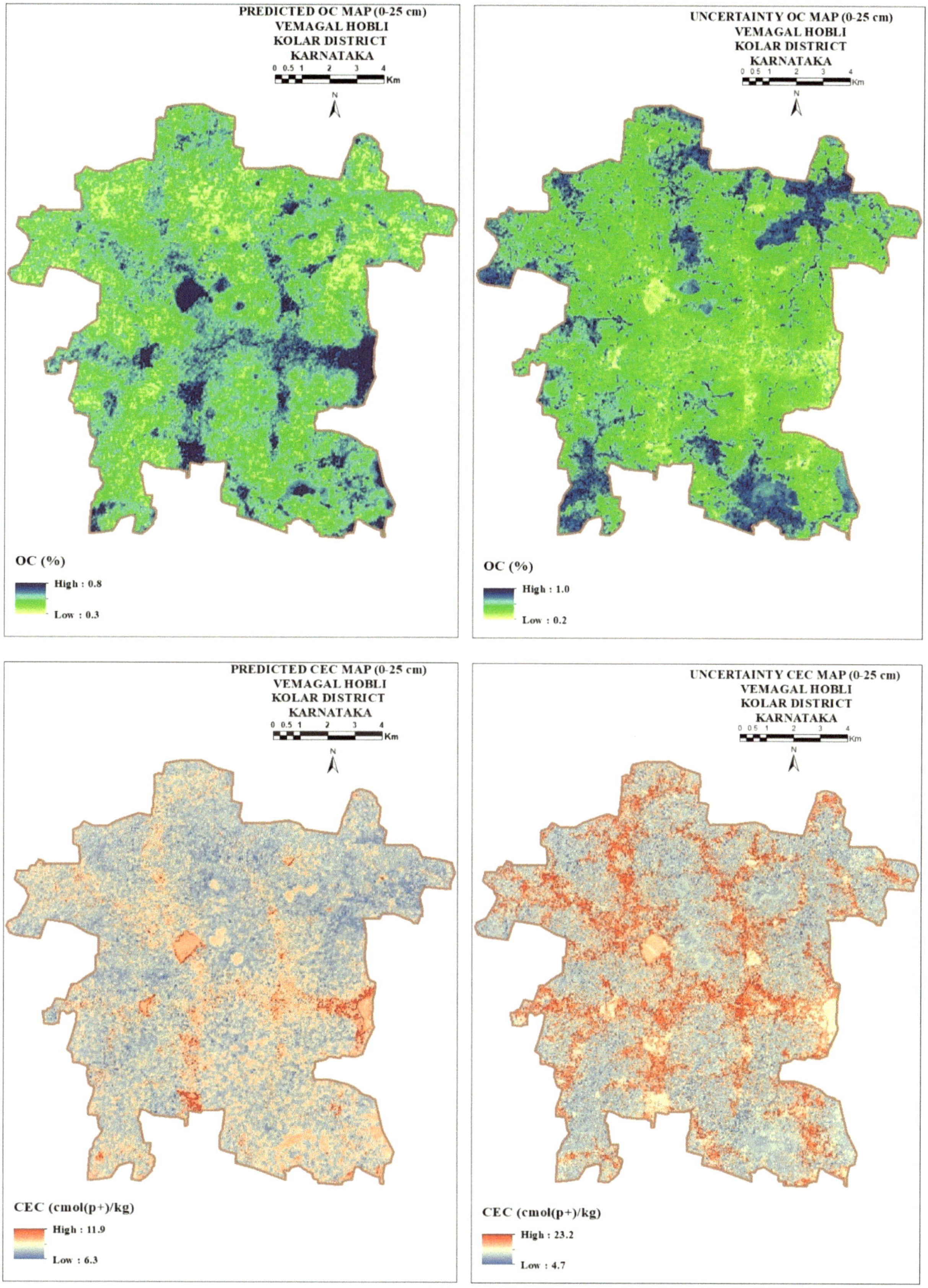

FIGURE 20.7 Predicted and uncertainty maps of OC and CEC (0–25 cm).

20.4 Conclusions

From the study, it can be seen that the soils of upland were found to exhibit a higher BD, which may be due to the presence of coarser particles because of erosion of fine fractions. Soils with high clay content were mostly distributed in the lowland regions of the study area. Low pH soils were mostly distributed in the upland areas which are usually characterized by high rainfall, causing the leaching of basic cations and hence resulting in a lesser CEC. Soils in lowland areas exhibited neutral pH and moderate OC content. These findings indicate that the soil's physicochemical properties were greatly influenced by the topography and other environmental factors such as precipitation, temperature, etc. With respect to uncertainty estimation, very low uncertainty was observed in the prediction of BD at the surface. The prediction of sand and clay content exhibited low to moderate uncertainty, whereas high uncertainty was evident in the case of silt content. The reason for high uncertainty might be due to the existence of limited datasets and due to the different landforms prevailing in the study area. This implies the necessity of collecting more samples, especially in areas of high uncertainty, in order to improve prediction accuracy by reducing the errors.

References

Breiman, L., 2001. Random forests. Mach. Learn. 45 (1), 5–32. Available from: https://doi.org/10.1023/A:1010933404324.

Dharumarajan, S., Hegde, R., 2022. Digital mapping of soil texture classes using Random Forest classification algorithm. Soil. Use Manage 38 (1), 135–149. Available from: https://doi.org/10.1111/sum.12668.

Dharumarajan, S., Hegde, R., Singh, S.K., 2017. Spatial prediction of major soil properties using random forest techniques—a case study in semi-arid tropics of South India. Geoderma Reg. 10, 154–162. Available from: https://doi.org/10.1016/j.geodrs.2017.07.005.

Dharumarajan, S., Hegde, R., Lalitha, M., Kalaiselvi, B., Singh, S.K., 2019. Pedo-transfer functions for predicting soil hydraulic properties in semi-arid regions of Karnataka Plateau, India. Curr. Sci. 116 (7), 1237–1246.

Dharumarajan, S., Kalaiselvi, B., Suputhra, A., Lalitha, M., Hegde, R., Singh, S.K., et al., 2020. Digital soil mapping of key GlobalSoilMap properties in Northern Karnataka Plateau. Geoderma Reg. 20, 1–8.

Dharumarajan, S., Hegde, R., Lalitha, M., Vasundhara, R., 2021. Predicting and mapping of soil hydraulic properties in Karnataka. J. Indian. Soc. Remote. Sens. 49, 1623–1631.

Gobin, A., Campling, P., Feyen, J., 2001. Soil-landscape modelling to quantify spatial variability of soil texture. Phys. Chem. Earth Part. B 26 (1), 41–45.

Hengl, T., Heuvelink, G.B., Kempen, B., Leenaars, J.G., Walsh, M.G., Shepherd, K.D., et al., 2015. Mapping soil properties of Africa at 250 m resolution: random forests significantly improve current predictions. PLoS One 10 (6), 1–26.

Jackson, M.L., 1973. Soil Chemical Analysis. Prentice Hall of India (pvt) Ltd., New Delhi.

Jenny, H., 1941. Factors of Soil Formation: A System of Quantitative Pedology. Hans Jenny McGraw Hill, p. 281.

Kläs, M., Vollmer, A.M., 2018. Uncertainty in machine learning applications: a practice-driven classification of uncertainty. International Conference on Computer Safety, Reliability and Security. Springer, Cham, pp. 431–438.

McBratney, A.B., Mendonca Santos, M.L., Minasny, B., 2003. On digital soil mapping. Geoderma 117 (1–2), 3–52.

Miller, B.A., Koszinski, S., Wehrhan, M., Sommer, M., 2015. Impact of multi-scale predictor selection for modeling soil properties. Geoderma 239, 97–106.

Pouladi, N., Moller, A.B., Tabatabai, S., Greve, M.H., 2019. Mapping soil organic matter contents at field level with cubist, random forest and kriging. Geoderma 342, 85–92.

Pradhan, B., 2013. A comparative study on the predictive ability of the decision tree, support vector machine and neuro-fuzzy models in landslide susceptibility mapping using GIS. Comput. Geosci. 51, 350–365.

Prasad, A.M., Iverson, L.R., Liaw, A., 2006. Newer classification and regression tree techniques: bagging and random forests for ecological prediction. Ecosystems 9, 181–199.

Prieto, A., Atencia, M., Sandoval, F., 2013. Advances in artificial neural networks and machine learning. Neurocomputing 121, 1–4.

Walkley, A., Black, I.A., 1934. An estimation of the method for determining soil organic matter and a proposed modification of the chromic acid titration method. Soil. Sci. 37, 29–38.

Zeraatpisheh, M., Ayoubi, S., Jafari, A., Tajik, S., Finke, P., 2019. Digital mapping of soil properties using multiple machine learning in a semi-arid region, central Iran. Geoderma 338, 445–452.

Chapter 21

Digital library software tool for land use planning under Karnataka Watershed Management System, India

S. Srinivas, Rajendra Hegde, S. Dharumarajan, K. Karunya Lakshmi, K.V. Seema, D.H. Venkatesh, K.V. Archana, A.G. Devendra Prasad, R. Vasundhara, B. Kalaiselvi and R. Srinivasan
ICAR-National Bureau of Soil Survey and Land Use Planning, Regional Centre, Bangalore, Karnataka, India

21.1 Introduction

In India, the soil resource information has a large diversity, showing a large variability in factors of soil formation that have wider applicability in an agriculturally progressive country like India (Bhattacharyya et al., 2013; Velayutham and Pal, 2004). To develop an information system together with soils and landscapes, land resource inventories (LRI) denote a soil survey with any spatially explicit information about the distribution of soil types or properties. Information about soil properties and behavior is widely recognized as vital for making decisions on proper land use and management, especially soil survey interpretations and land evaluations for specific land uses. Advancement in space and information technology, especially in the field of remote sensing, has become an important tool in soil resource inventory. The LRI database helps in preparing optimum land use plans for the micro-watersheds that help in restoring ecological balance and improving production on a sustainable basis (Hegde et al., 2018). The LRI gives information about land use/land cover to other uses that need to obtain reliable data about land resources for planning, management, conservation, and optimum use for further development in a sustainable manner (Shaw et al., 2015; Akike and Samanta, 2016). Kumar et al. (2015) reported that remote sensing, GIS, and GPS offer an abundant opportunity to monitor and manage natural resources at multitemporal, multispectral, and multispatial resolution. Soil information is generated for an assortment of land capability classifications, land irrigability assessments, crop suitability studies, watershed management, prioritization of watersheds, etc. (Ali and Kotb, 2010). The soil survey organizations have been concerned to publicize their work and make sure it is useful to potential users.

In India, there is an absolute lack of crop advisories and crop loss due to seasonal droughts and floods every year. For the necessity of farmers, LRI work as a knowledge based expert system mainly useful for better utilization of farm resources for their economic goals. The cost and time reduction and increase in the detailed information for soil survey are possible through the integration of remote sensing within a GIS database (Bandyopadhyay et al., 2009). GIS is a tool that allows for the processing of spatial data into information (Samarakoon, 2005) and has the ability to manipulate spatial data and corresponding attribute information in different types at high speed, which is unmatched with manual methods (Rashed et al., 2006). It provides a digital representation of landforms, which could be used in hydrological modeling, and the database available in a GIS environment facilitates the assimilation of different thematic datasets to understand interrelationships. Many GIS-based watershed applications have been developed since the early 1990s due to advances in desktop GIS capabilities, programming languages, and data availability (Strager et al., 2010). GIS-aided LRI software named as digital library (DL) was developed as a public library for agricultural land-use and planning under specific micro-watersheds of Karnataka state and is constructed to evaluate suitability of the watershed under multiple objectives of each category and weighted factors such as soil series image, soil characteristics, and query (soil and fertility, village-based query, etc.). It makes use of new techniques such as custom-built individualization services for all uses with a specially designed computer system. Availability via the geospatial techniques/DL with liberal use restrictions means that anyone can use the data, perhaps in a novel and unpredicted way.

Remote Sensing of Soils. DOI: https://doi.org/10.1016/B978-0-443-18773-5.00009-0

Recently, the digital mapping used DSM, which mapped "soil map units" defined by the NZSC taxa (Hewitt, 2010) in general to subgroup level, along with further observations regarding soil texture, depth, drainage, and stoniness. LiDAR is one of the advanced remote sensing information system that makes nationwide natural resource inventories in the form of high-resolution natural resource maps (Blanco et al., 2016). Climate modelers have abstracted the FAO Soil Map of the World (FAO, 1974) to a 0.5 degrees $\times$ 0.5 degrees grid called the "Zobler" data set, which is relevant to soil properties including texture, depth, and slope (Post, 2000). Several years of development have resulted in improving the performance of agriculture through digital libraries. One of the great illustrations of regional soil databases developed together by UNEP, IUSS, ISRIC, and FAO is the SOTER methodology (Engelen and Wen, 1995), which is available online from ISRIC and as CD-ROMs under FAO's Land and Water Digital Media Series for Latin America and the Caribbean (SOTERLAC) (FAO, 1998b) and Central and Eastern Europe (SOVEUR) (FAO, 1998c). A soil and physiographical database has been produced for northeast Africa (FAO, 1998d) and for North and Central Eurasia (FAO, 1999) and is available on CD-ROM. Hellweger and Maidment (1999) developed an automated procedure in ARC/INFO and ARC/View to produce the connection of hydrological elements using the geographic data in the Tenkiller reservoir watershed in Oklahoma, USA. In the context of rural development planning, Gramya Vikas was developed as a web-based distributed collaboration model to assist the rural extension community in their own decision-making processes in a more interactive, integrated, and coordinated manner (Ponnusamy, 2007; Adinarayana et al., 2008). Similarly, DL requires the integration and application of a wide range of methods and disciplines such as data management, information retrieval, library and information sciences, human-computer interaction, and digital curation to provide services confined to providing information for users (Innocenti et al., 2010; Baoji et al., 2007).

A geospatial desktop application of LRI called DL is a user-friendly inventory for land use planning research relevant to Karnataka watershed management in India, with special emphasis on agricultural development and planning. Digital libraries offer the potential to improve and provide adequate information to a specific user community, which may be widely dispersed over large geographic locations. Sharifabadi (2006) also noted that digital libraries are not a mere equivalence of digitized collections with information management tools. According to Arms (2000), a digital library is a collection of managed information with assorted services, stored in digital formats and accessible over a network. DLs are meant for the acquisition, storage, preservation, retrieval, and dissemination of resources through digital technologies (Gani and Magoi, 2014). It represents a computerized information system for acquiring, storing, organizing, searching, and disseminating information resources in digital forms to user communities. Warr and Hangsing (2009) conceptualized a DL as an assemblage of digital computing, storage, and communications machinery together with the content and software needed to reproduce, emulate, and extend the services provided by conventional libraries. The objective of study was to develop DL to integrate both scientific and heuristic knowledge of land evaluation exercises and to suggest suitable crops for a given land unit with standard procedures for easy modifications and updates at the micro-watershed level.

21.2 Material and Methods

21.2.1 Study area

The DL was demonstrated over a micro-watershed located at Handaraki micro-watershed (Motanahalli subwatershed) in the northeastern part of Karnataka in Yadgir Taluk, Yadgir District, Karnataka state, India (Fig. 21.1). The study area lies the geographical extend between 16°53′–16°54′ N latitude and 77°25′–77°26′ E longitude and covers the catchment area of 5.39 km^2. The climate is hot, moist, and semiarid with a mean average annual rainfall of 684 mm. Out of the total rainfall, 520 mm is received during the southwest monsoon (June to September), 140 mm from the northeast monsoon (October to early December), and the remaining 24 mm during the rest of the year. The winter season of the area prevails during the period of December - January, with the lowest optimum temperature of 17°C. During peak summer, temperature rises up to 40°C. The average potential evapotranspiration (PET) is 887 mm and varies from a low of 4 mm in December to 39 mm in May. The PET is always higher than precipitation. Generally, the length of the crop growing period is 150 days, starting from the first week of June to the third week of October.

21.2.2 Survey methodology

The detailed survey at 1:7920 scale was carried out in a 539 ha area. The methodology for LRI was as per the guidelines given in the Soil Survey Manual (IARI, 1971). The detailed survey of the land resources (Fig. 21.2) in the subwatersheds was carried out by using digitized cadastral maps.

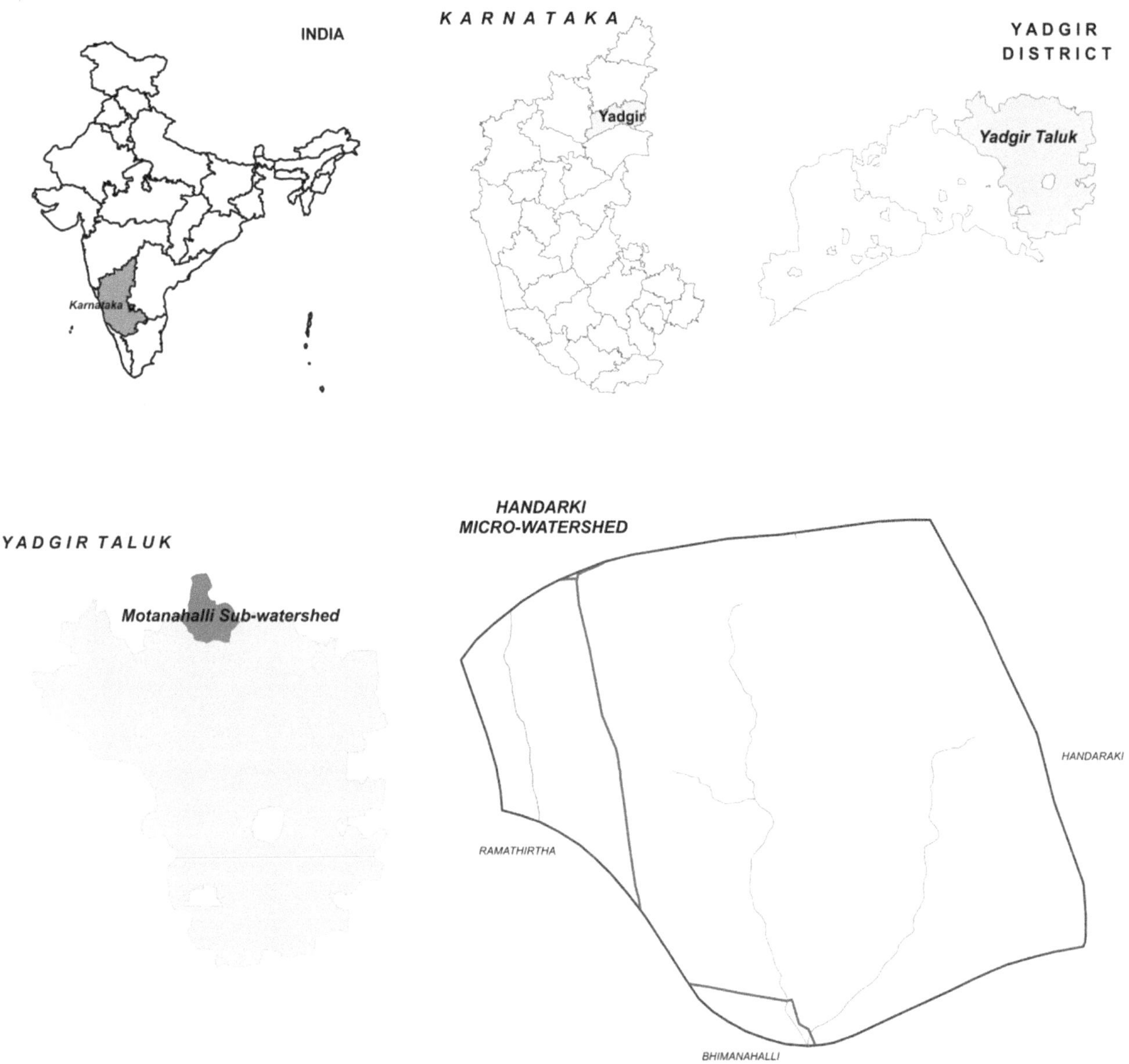

FIGURE 21.1 Location map of the Handaraki micro-watershed.

Apart from the cadastral map, the fused remote sensing data of Cartosat-1 (2.5 m) and LISS-IV (5.8 m) at the scale of 1:7920 were used in conjunction with the cadastral map to identify the landscapes, landforms, and other surface features. The cadastral map was overlaid on the satellite imagery to identify the parcel boundaries and other permanent features. The fused imagery helped in the identification and delineation of boundaries between hills, uplands and lowlands, water bodies, forest/vegetated areas, roads, habitations, and other cultural features of the area (Fig. 21.3).

21.3 Conceptual design of digital library

DL integrates and generalizes data from different university partners and uses various dynamic knowledge rules, which contain scientific and heuristic information that can be used for better planning and management of land resources. The system is a conglomeration of multiple technologies like geographic information system (GIS), remote sensing (RS), global positioning system (GPS), hydrological modeling, soft-computing tools, etc.

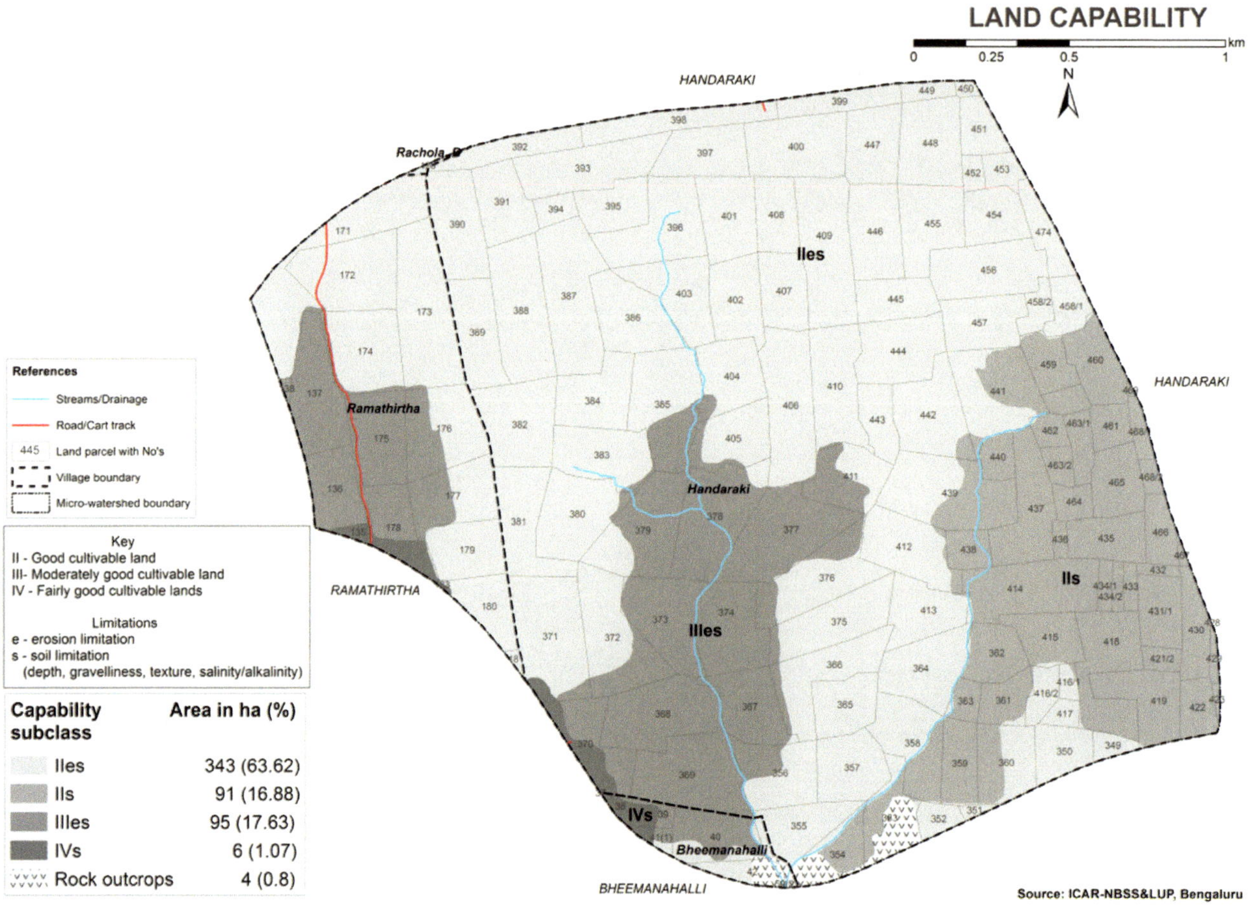

FIGURE 21.2 Land capability—the Handaraki micro-watershed.

21.4 Database used

In the present study, a land suitability evaluation for watersheds has been carried out. The parameters taken into consideration were slope, soil depth, soil texture, land use/land cover, organic carbon content, phosphorus, soil reaction, potassium, magnesium, calcium, boron, sulfur, zinc, iron, manganese, nitrogen, overlay of plot' and soil information, and wells. Satellite images for the study area have been classified for land use/land cover map preparation. Toposheet and ancillary data have been used for slope maps and soil property determination.

An integrated land suitability potential (LSP) index was computed considering the various parameters of land suitability. These categories were arrived at by integrating the various layers with corresponding weights in a GIS. The flow chart demonstrates the step-wise activity in delineating agriculture potential zones in a watershed-based land resource data sets and agro-geographic analysis. Detailed information pertaining to the nature, extent, and distribution of different kinds of soils occurring in the Handaraki micro-watershed was used as case study.

21.5 Digital library working mechanism

A survey number–based query performs the necessary database operation to retrieve the plot-based information from the maintained database "Sujala" and generalizes the data into reports (Figs. 21.4 and 21.5). DL begins with the login page to validate the authorized users to use the functionality of the application (Fig. 21.6). DL allows an authorized user to select the area of interest (Fig. 21.7). The interactive page based on the user's selection displays the PMF (Published Map Format) file for the corresponding area, along with options for plot-based query and various other map viewing controls. DL generates different types of informative reports for the users' analysis, such as soil suitability crops, proposed crops, climatic graphs with annual rainfall data, fertility recommendations, etc. User

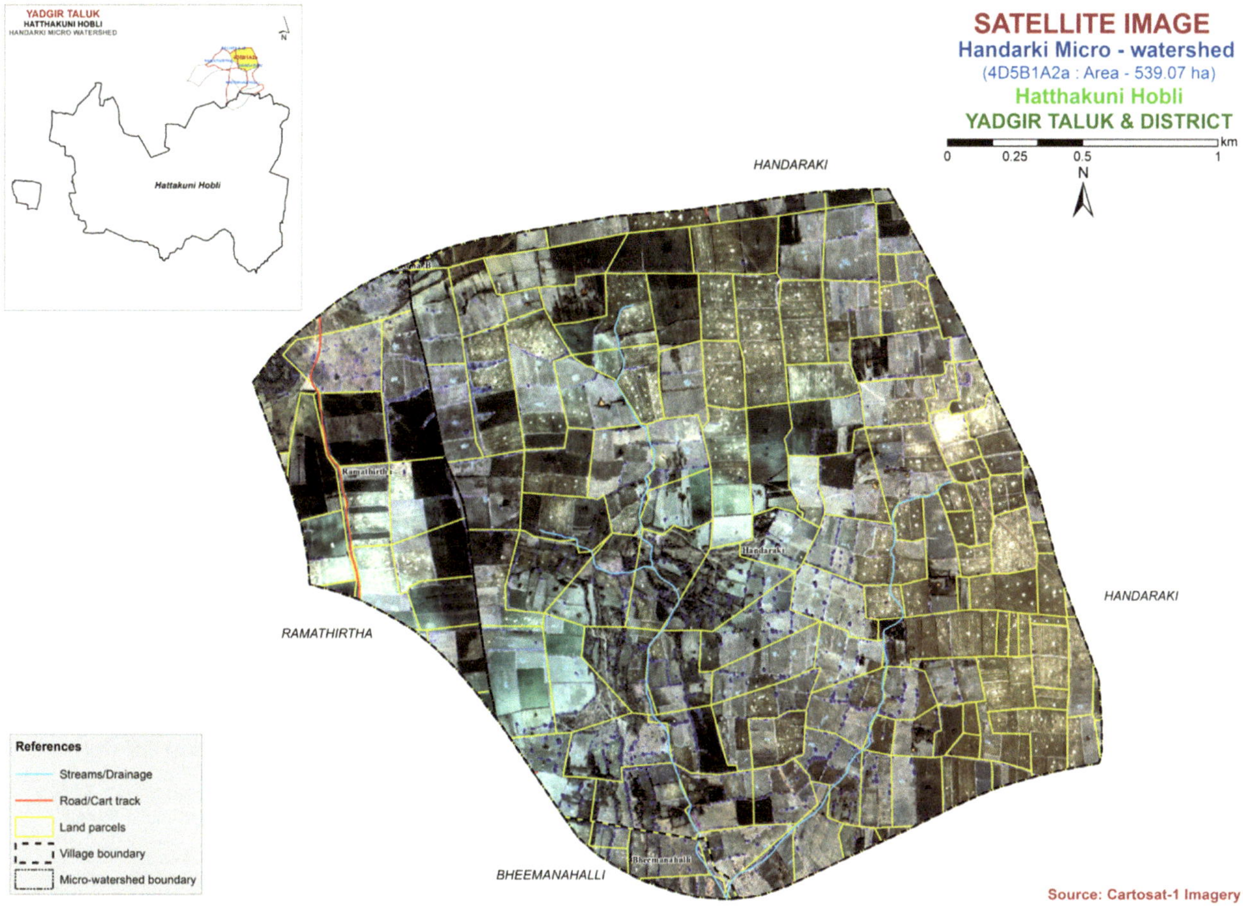

FIGURE 21.3 Cadastral map overlaid on IRS PAN + LISS IV merged imagery of the Handaraki micro-watershed.

informative page (query—soil and fertility) displays criteria-based information (fertility, suitability, soil properties, etc.,) for the selected micro-watershed. The soil information page for the chosen micro-watershed area displays data as images for selected soil type and generates handy soil lab data reports (Fig. 21.8). DL generates village-wise informative supplements for the users. Rossiter and Hengl (2001) developed a low-cost desktop GIS for new surveys made on stereo vertical air photos, which have the potential to convert image interpretation directly into geometrically correct and geo-referenced map.

21.5.1 Technical requirement

Software requirement specifications were Visual Studio 2015 Professional edition, Arc Reader 10.3, MS Access 2007, and SAP Crystal Reports for Visual Studio 2015. Operating system used Windows 7 SP1 (X86 and X64) or Windows 8 (X86 and X64) and DOT NET Framework 4.5 (64 bit). Hardware requirements are Core (TM) i7 2.20 GHz or fast processor, with 4GB RAM (1.5 GB, if running on a virtual machine).

21.6 Digital library features

21.6.1 Prerequisites

Crystal Report Viewer (Runtime) was necessary to display the soil site information (plot-based and village-based). Place the "Report" Folder with soil site information (.rpt files) at the data location (Sujala.mdb). Select the Drive for Data location (Sujala.mdb, Reports, etc.) that has all user permissions.

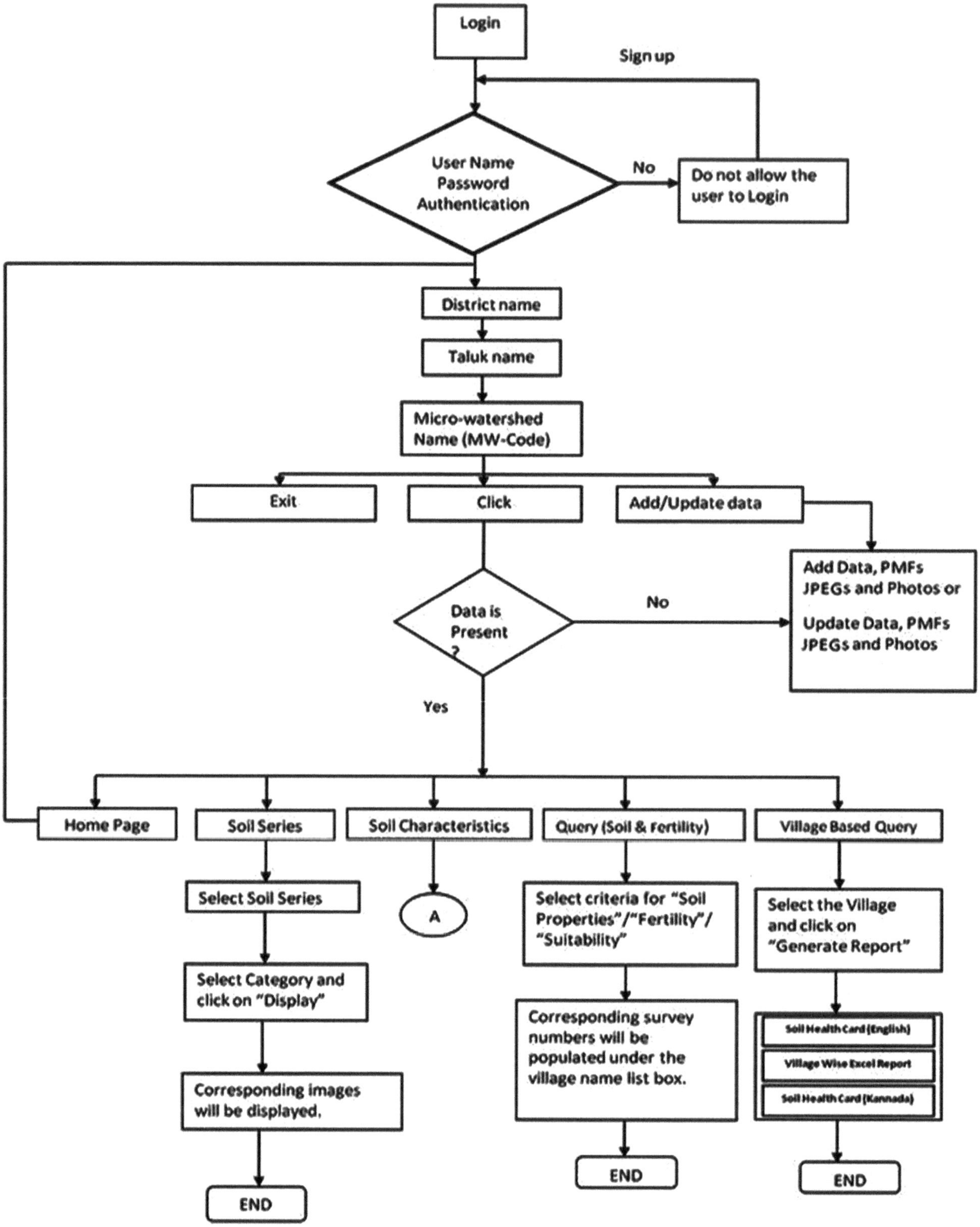

FIGURE 21.4 Schematic flow chart of DL.

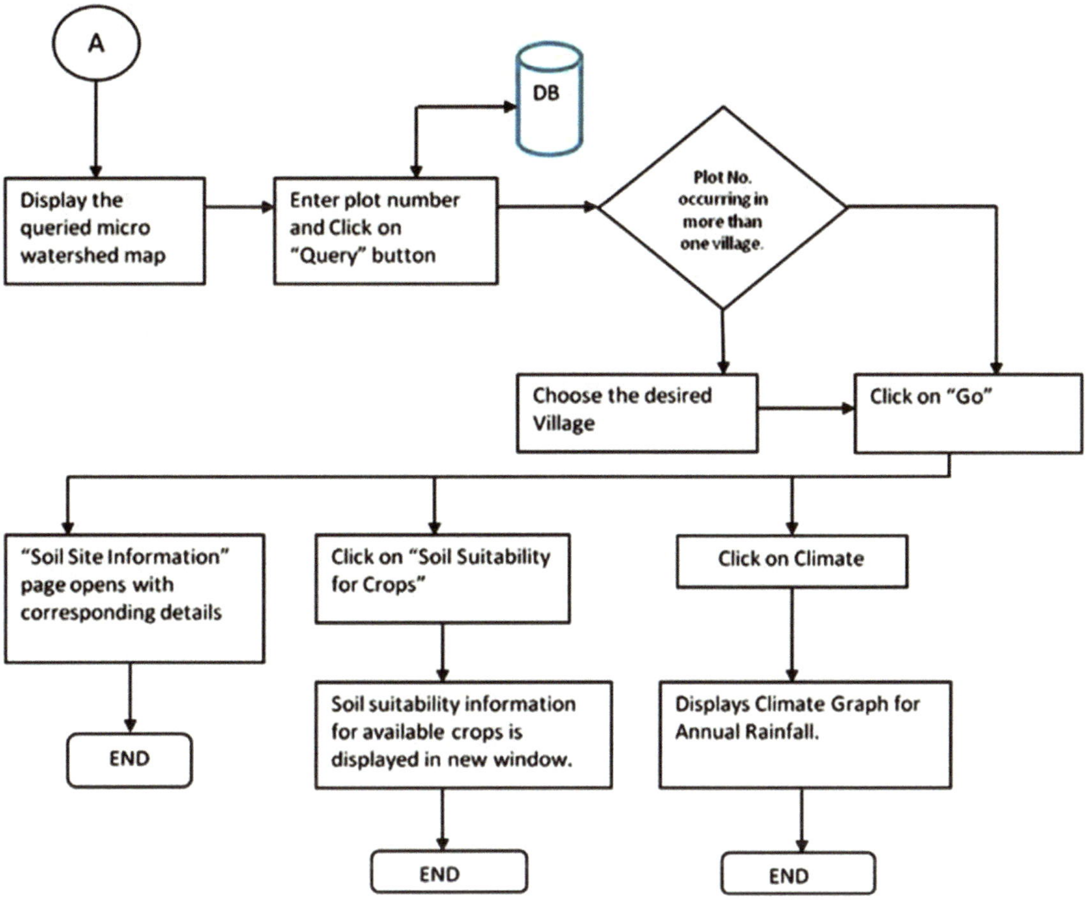

FIGURE 21.5 Working mechanism of plot-based query.

FIGURE 21.6 Login page.

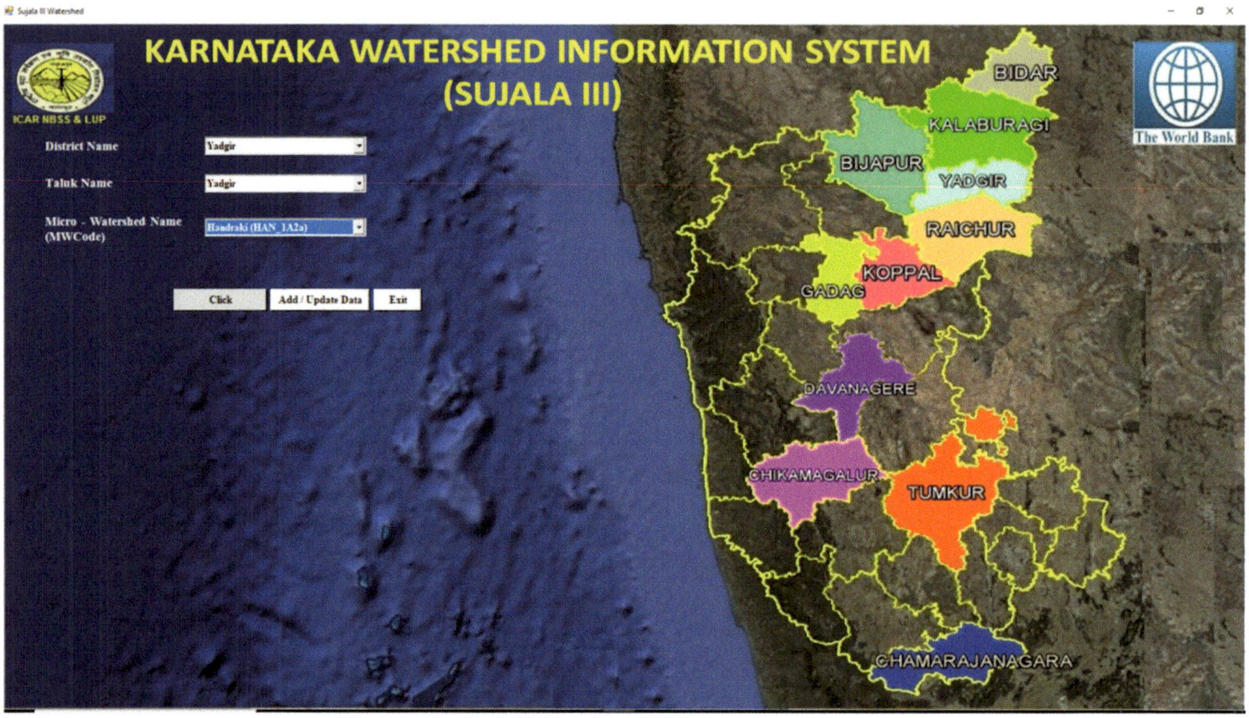

FIGURE 21.7 Home pages.

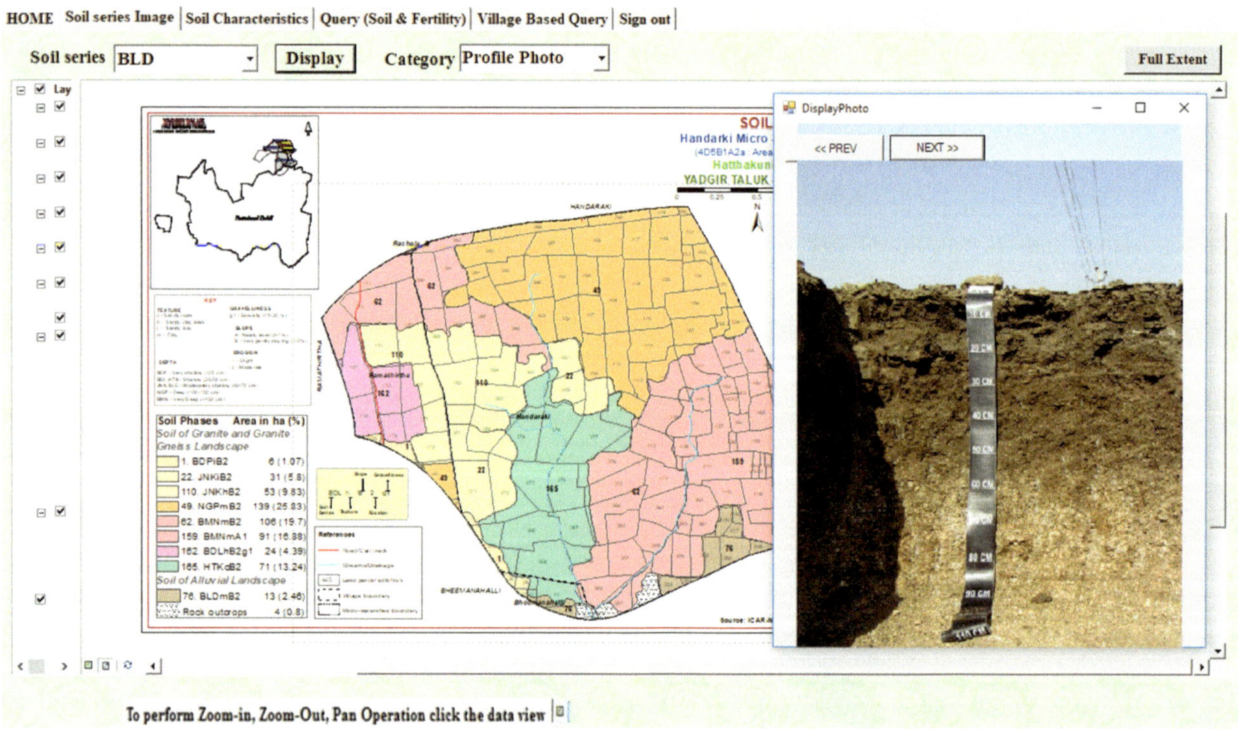

FIGURE 21.8 Soil series with profile photo of Handaraki MWS.

21.6.2 Facts

It does not require users to have GIS software for its use, Arc Reader is a freely downloadable software used to access. pmf (Published Map Format) files packed with File Geo-database, and this is used to query the map. Presently, MS Access database is used for other data sets (suitability, fertility, and wells). Images and thematic maps (.jpg) are stored in folders in structured format. Presently, it is being developed for use as a web application.

21.6.3 Organization of database

DL uses MS Access to create, store, and organize the database and data. For example, the database "Sujala.mdb," which consists of 20 tables, contained parameters (such as boron, calcium, copper, electrical conductivity, iron, magnesium, manganese, nitrogen, organic carbon, phosphorus, potassium, soil reaction, sulfur, zinc, union_plot_soil, Well_info, Traditional_Soil_Name, climate, PCP_Details, and village) as per the requirements of the user. The data should be organized in the following structure as shown for the micro-watershed "Handaraki (HAN_2A1a)" under "Yadgir district and Taluk."

21.6.4 Software installation

The steps to be followed to install the application DL are: (1) double click on "DL-LRI" to install the application, (2) browse to the file location, and (3) after successful completion of the installation procedure, locate the configuration file named "Attribute Query.exe." The user must change the source of the database in the configuration file to the directory where the database "Sujala" exists.

21.6.5 Steps to execute built-in data

Copy and paste the Sujala.mdb to a user-specific drive, for example, F: Drive or available Drive, in which it executes the data using DL. Copy the District Folder from G: Drive (for example, Yadgir) and paste it to another system's drive, where Sujala.mdb is already pasted. After copying and pasting the district folder and Sujala.mdb, change the App. Config file (which will be found after the installation of the executable file). To change the data source for Sujala.mdb, click on the App.Config file from the installed location.

21.7 Result and discussion

DL is designed and developed by using modern scientific methodologies to achieve holistic, sustainable watershed management aspects, as shown in schematic flow diagram shown in Fig. 21.4. DL maintains various types of databases, such as soils, climate, nutritional and physical status of soil, hydrological models, and spatial/nonspatial user database model interfaces, to produce significant reports. Reports can be utilized for the land appraisal. The client organizations need to specify only the Published Map Formats (PMF) File and data of interest as inputs to the application. Users can continuously update their databases when required.

21.7.1 Add/update data

The Add/Update Data page of the application allows the user to add and update the mentioned data as (1) data, (2) PMFs, (3) JPEGs, (4) photographs, (5) climate graphs, (6) map unit description, and (7) proposed crop plan. Add/Update page, depicting multiple data options by adding new data and updating existing data in Excel file. It allows the user to add corresponding data for selected tables. After successful insertion of data, it is required to add/update PMFs. New PMFs allow the user to add new PMF package for the selected micro-watershed. For adding new JPEGs or updating the existing ones, creation of "Final JPEGs" folder is required for the first time. The Add/Update JPEG page allows the user to select JPEGs to add to the specified data location. While adding the JPEGs to the corresponding micro-watershed, the following standard names should be followed: (1) soil macro- and micro-nutrients (boron, copper, iron, organic carbon, soil reaction, phosphorus, potassium, electrical conductivity (salinity), sulfur, zinc, manganese, magnesium, nitrogen, and calcium); (2) soil characteristics (land capability, depth, erosion, gravel, slope, texture, land management unit (LMU), soil water characteristic (SWC), and soil phase); and (3) crops [amla (*Phyllanthus emblica*), cashew (*Anacardium occidentale*), custard apple (*Annona reticulata*), guava (*Mangifera indica*), jackfruit (*Artocarpus heterophyllus*), jamun (*Syzygium cumini*), lime (*Citrus* sp.), mango (*M. indica*), musambi (*Citrus limetta*), sapota

(*Manilkara zapota*), tamarind (*Tamarindus indica*), chilly (*Capsicum annuam*), groundnut (*Arachis hypogaea*), redgram (*Cajanus cajan*), sorghum (*Sorghum bicolor*), maize (*Zea mays*), cotton (*Saccharum officinarum*), coconut (*Cocos nucifera*), sunflower (*Helianthus annus*), and *Bengal gram* (*Cicer arietinum*)].

The photograph button allows users to add and update existing photographs and photo types (Pedon Form) for the selected micro-watershed based on soil series. The climatic button allows the user to add and update the climate graph, along with the climate data. The map unit description button allows the users to add/update map unit descriptions. The selection of the new map with its scale unit and description showed in image format to the user for better understanding. At the end of the data entry, the added or updated new proposed crop plan and the updated plan of the already existing, proposed crop plan should be saved successfully in JPEG format. After successfully updating all the data for a selected MWCode, the soil characteristics page will query for a selected plot number.

21.7.2 Soil characteristics

The soil characteristics screen provides the following functions (Fig. 21.9). The user can zoom in and zoom out the of the pmf. Along with scanning, switch between data view and layout view is provided to view the map in more detail. The Query button allows the user to query the file geo-database plot-wise and the Go button generates the soil site information page for the queried plot number (Fig. 21.10). The Full Extent button displays the micro-watershed map in full extent. The Map Unit Description button displays all the soil phase information. Soil and land use/land cover maps and AESR maps, in conjunction with information on socioeconomic and production systems, are used in various steps of land evaluation, thereby facilitating land allocation based on agricultural land use plans (Bhattacharyya et al., 2009, 2013).

21.7.3 Soil site information report

The soil site information report contains soil health information for the entered plot number. The English/Kannada Language button displays the formal soil site information card (plot-wise) in both English and Kannada. The Soil

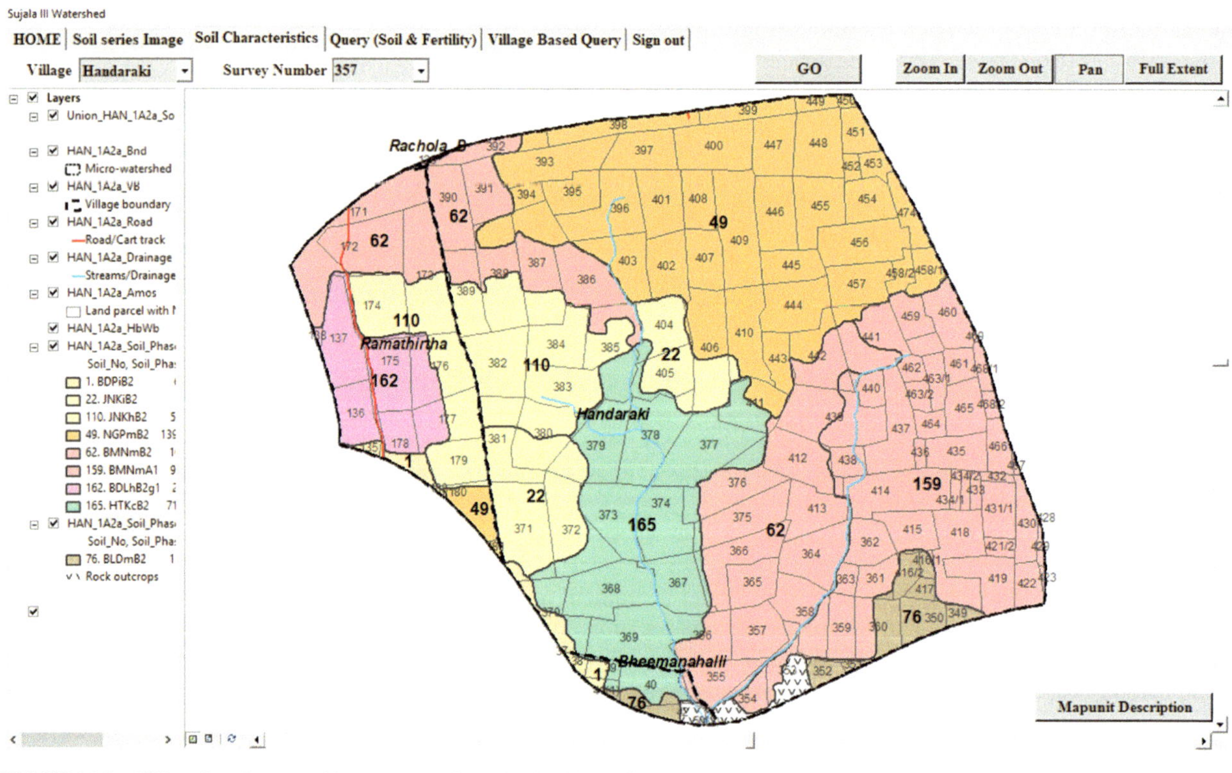

FIGURE 21.9 Village-based query with survey no. of the Handaraki MWS.

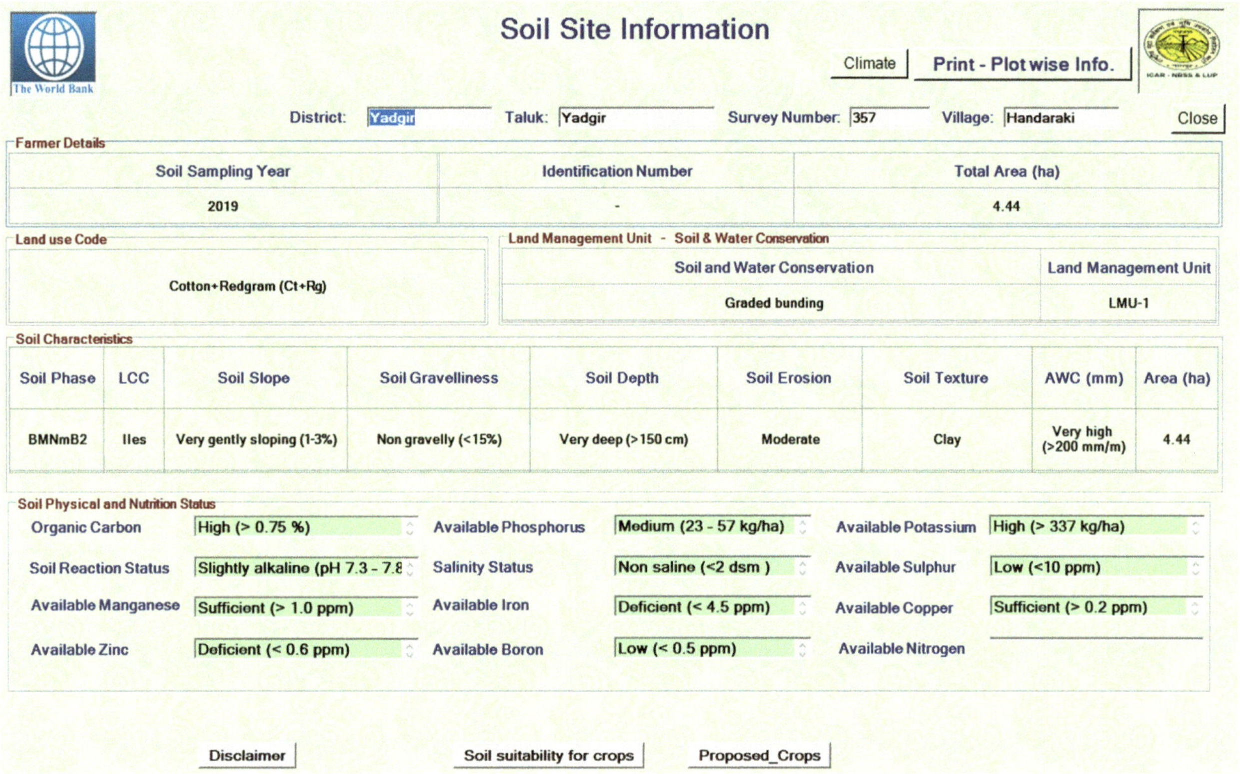

FIGURE 21.10 Soil site information of the Handaraki MWS.

Suitability for Crops button displays crop suitability information for the selected plot number (Figs. 21.11 and 21.12). Using the GIS platform and the soil site suitability criteria (Naidu et al., 2006), crop suitability maps were generated for field and horticulture crops and categorized as highly suitable (S1), moderately suitable (S2), marginally suitable (S3), and unsuitable (N) for production of these crops. The Climate button displays the climate graphs with the annual rainfall data (Fig. 21.13). The Print button displays soil site and fertility information (plot-wise). Available climatic information with regard to temperature, rainfall, and length of growing period (Naidu et al., 1998) was also used for crop suitability assessment.

Based on soil site characteristics and crop requirements, the soil suitability classes were developed as per the criteria suggested by Sys et al. (1993) and NBSS and LUP (1994). Similarly, Bhagia et al. (2016) developed a national inventory for the rabi pulse crop in different parts of India using multidate satellite images with pixel accuracy ranging from 50% to 100%, which helped to create a crop map with its spatial extents.

21.7.4 Fertilizer suggestion for a specific crop

To give survey number–wise fertilizer recommendations, the available nitrogen, phosphorus, and potassium status will be segregated as low, medium, and high status using the nutrient ratings. The recommended dose of fertilizer (RDF) for a particular crop can be modified according to the fertility status level. The modification in RDF will be done as per the package of practices (GKVK Packages of Practices, 2013) of 125% RDF for low status soil nutrient, 100% RDF for medium status soil nutrient, and 75% RDF for high status soil nutrient. Based on the modified RDF for particular crops, such as ragi (*Eleusine coracana*), maize (*Z. mays*), horse gram (*Macrotyloma uniflorum*), coconut (*C. nucifera*), and arecanut (*Areca catechu*) etc. to the specific survey number, the fertilizer doses for different fertilizer combinations like fertilizer formulation I—DAP: Urea: MOP and fertilizer formulation II—17:17:17 Complex: Urea: MOP are calculated, and with these, the database has been updated. If we select any survey number in DL, with respect to the fertility status of the survey number, it will display the fertilizer recommendation for different crops (Fig. 21.14).

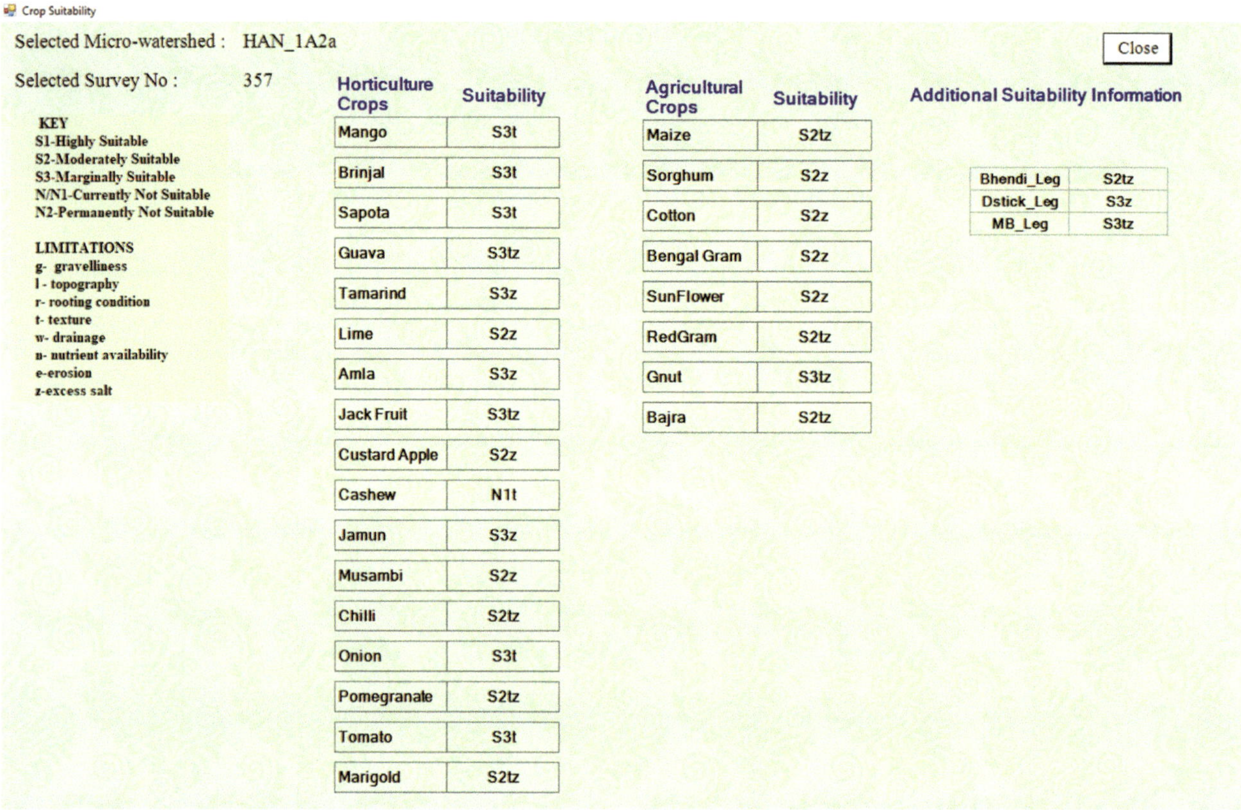

Crop Suitability

Selected Micro-watershed : HAN_1A2a

Selected Survey No : 357

Close

KEY
S1-Highly Suitable
S2-Moderately Suitable
S3-Marginally Suitable
N/N1-Currently Not Suitable
N2-Permanently Not Suitable

LIMITATIONS
g- gravelliness
l - topography
r- rooting condition
t- texture
w- drainage
n- nutrient availability
e-erosion
z-excess salt

Horticulture Crops	Suitability
Mango	S3t
Brinjal	S3t
Sapota	S3t
Guava	S3tz
Tamarind	S3z
Lime	S2z
Amla	S3z
Jack Fruit	S3tz
Custard Apple	S2z
Cashew	N1t
Jamun	S3z
Musambi	S2z
Chilli	S2tz
Onion	S3t
Pomegranate	S2tz
Tomato	S3t
Marigold	S2tz

Agricultural Crops	Suitability
Maize	S2tz
Sorghum	S2z
Cotton	S2z
Bengal Gram	S2z
SunFlower	S2z
RedGram	S2tz
Gnut	S3tz
Bajra	S2tz

Additional Suitability Information

Bhendi_Leg	S2tz
Dstick_Leg	S3z
MB_Leg	S3tz

FIGURE 21.11 Crop suitability of Handaraki MWS.

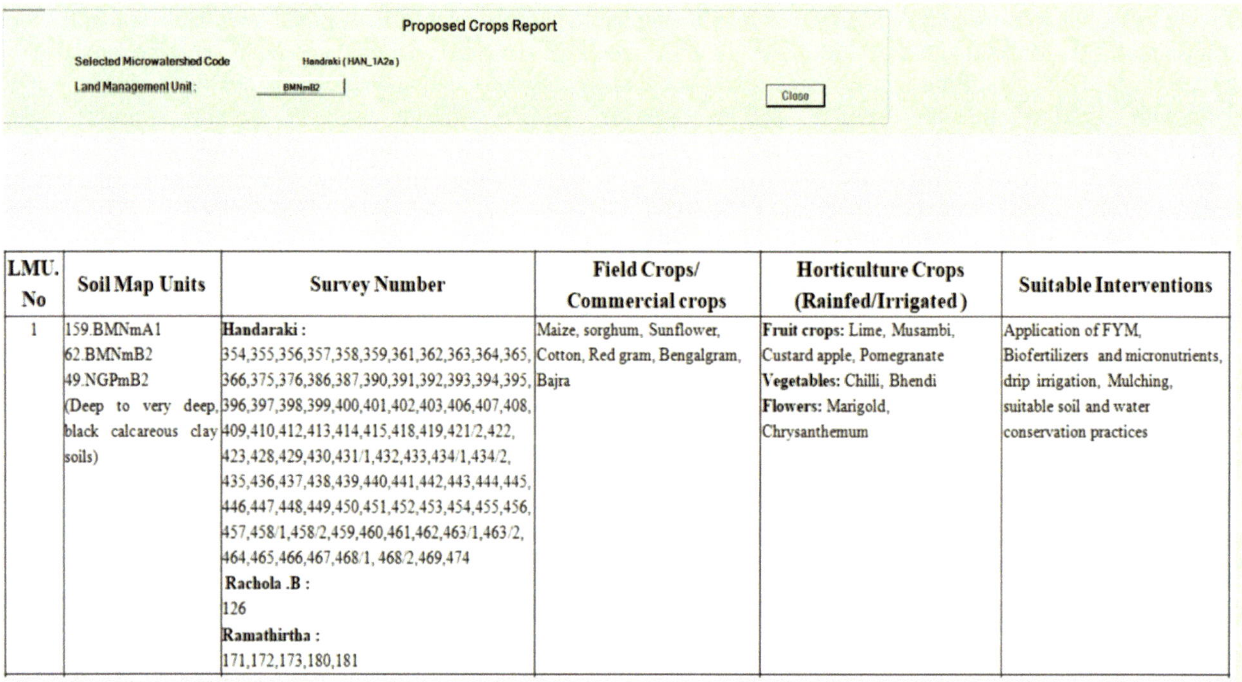

Proposed Crops Report

Selected Microwatershed Code Handraki (HAN_1A2a)

Land Management Unit : BMNmB2

Close

LMU. No	Soil Map Units	Survey Number	Field Crops/ Commercial crops	Horticulture Crops (Rainfed/Irrigated)	Suitable Interventions
1	159.BMNmA1 62.BMNmB2 49.NGPmB2 (Deep to very deep, black calcareous clay soils)	**Handaraki :** 354,355,356,357,358,359,361,362,363,364,365, 366,375,376,386,387,390,391,392,393,394,395, 396,397,398,399,400,401,402,403,406,407,408, 409,410,412,413,414,415,418,419,421/2,422, 423,428,429,430,431/1,432,433,434/1,434/2, 435,436,437,438,439,440,441,442,443,444,445, 446,447,448,449,450,451,452,453,454,455,456, 457,458/1,458/2,459,460,461,462,463/1,463/2, 464,465,466,467,468/1, 468/2,469,474 **Rachola .B :** 126 **Ramathirtha :** 171,172,173,180,181	Maize, sorghum, Sunflower, Cotton, Red gram, Bengalgram, Bajra	**Fruit crops:** Lime, Musambi, Custard apple, Pomegranate **Vegetables:** Chilli, Bhendi **Flowers:** Marigold, Chrysanthemum	Application of FYM, Biofertilizers and micronutrients, drip irrigation, Mulching, suitable soil and water conservation practices

FIGURE 21.12 Proposed crop reports of the Handaraki MWS.

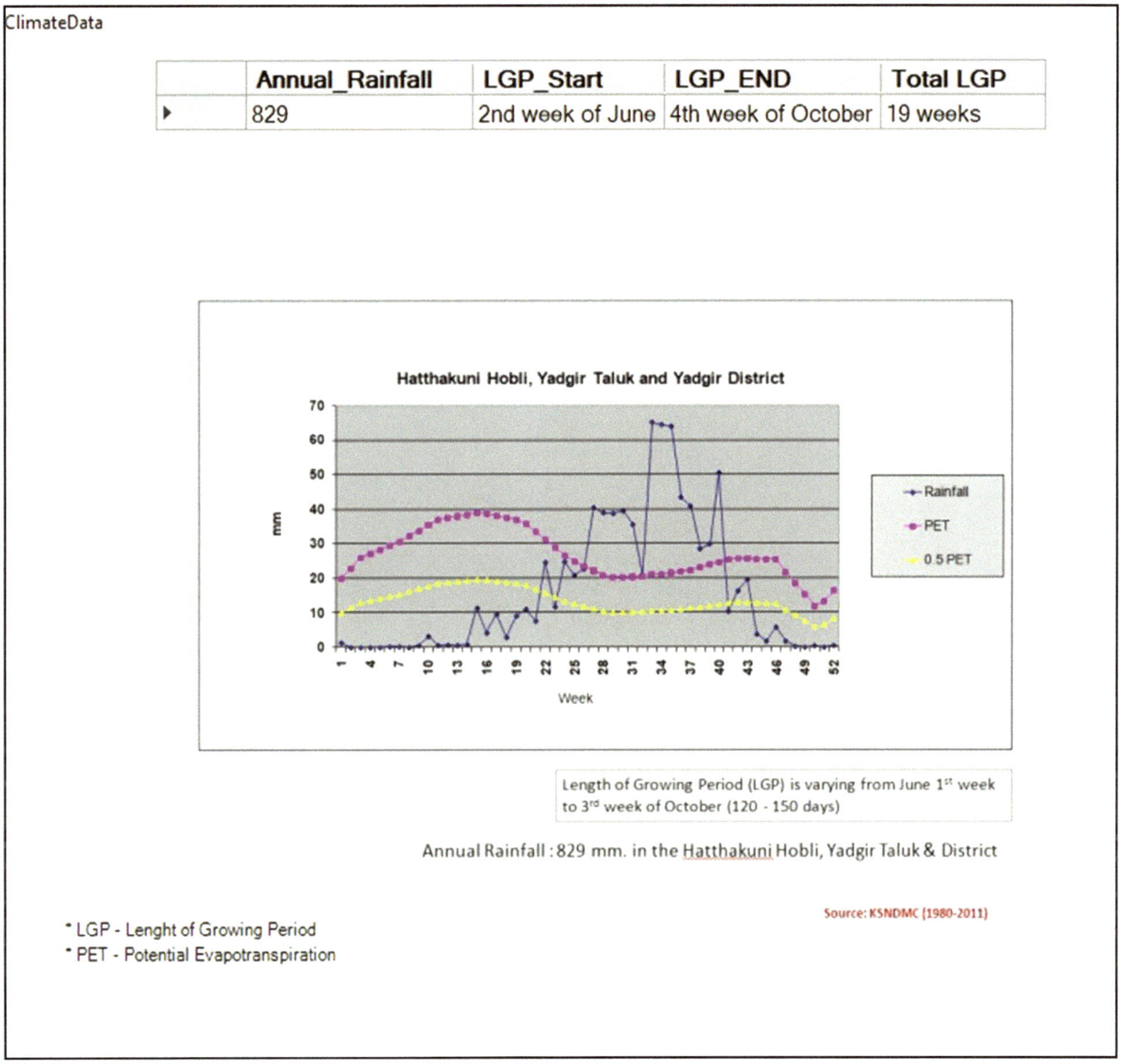

FIGURE 21.13 Climate data of the Handaraki MWS.

21.7.5 Query (soil and fertility): micro watershed–wise

The Query (Soil and Fertility) page allows the user to query the map based on fertility, soil properties, and suitability (Fig. 21.15). Users can take a printout of that map and open the village-wise result of the query in a Word document.

21.7.6 Village-based query

The Village-Based Query page allows the user to generate detailed village-wise report and village-wise soil health card report for the selected micro-watershed in English/Kannada (Fig. 21.16).

21.7.7 Soil health card generation

The village-level query page provides scientific information on their land parcels to farmers about their land resources at the farmers contact centers (Figs. 21.17 and 21.18).

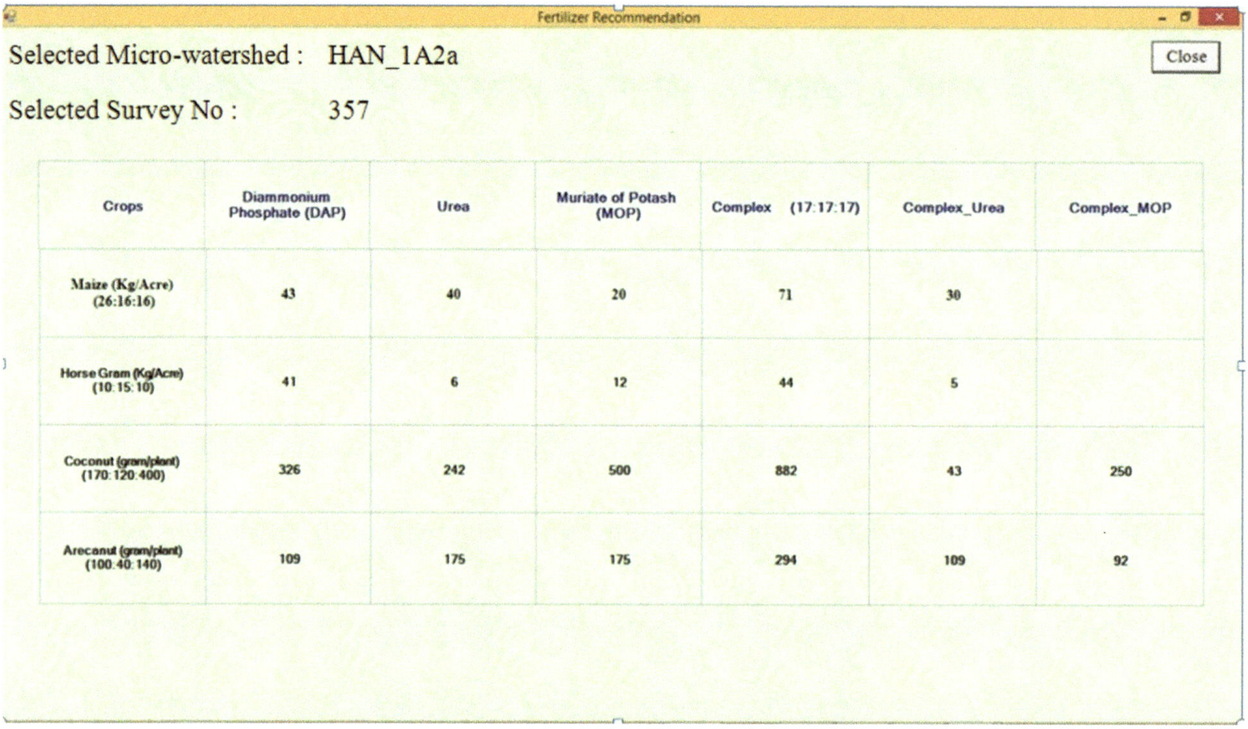

Fertilizer Recommendation

Selected Micro-watershed : HAN_1A2a

Selected Survey No : 357

Close

Crops	Diammonium Phosphate (DAP)	Urea	Muriate of Potash (MOP)	Complex (17:17:17)	Complex_Urea	Complex_MOP
Maize (Kg/Acre) (26:16:16)	43	40	20	71	30	
Horse Gram (Kg/Acre) (10:15:10)	41	6	12	44	5	
Coconut (gram/plant) (170:120:400)	326	242	500	882	43	250
Arecanut (gram/plant) (100:40:140)	109	175	175	294	109	92

FIGURE 21.14 Fertilizer recommendation for the crop of Handaraki MWS.

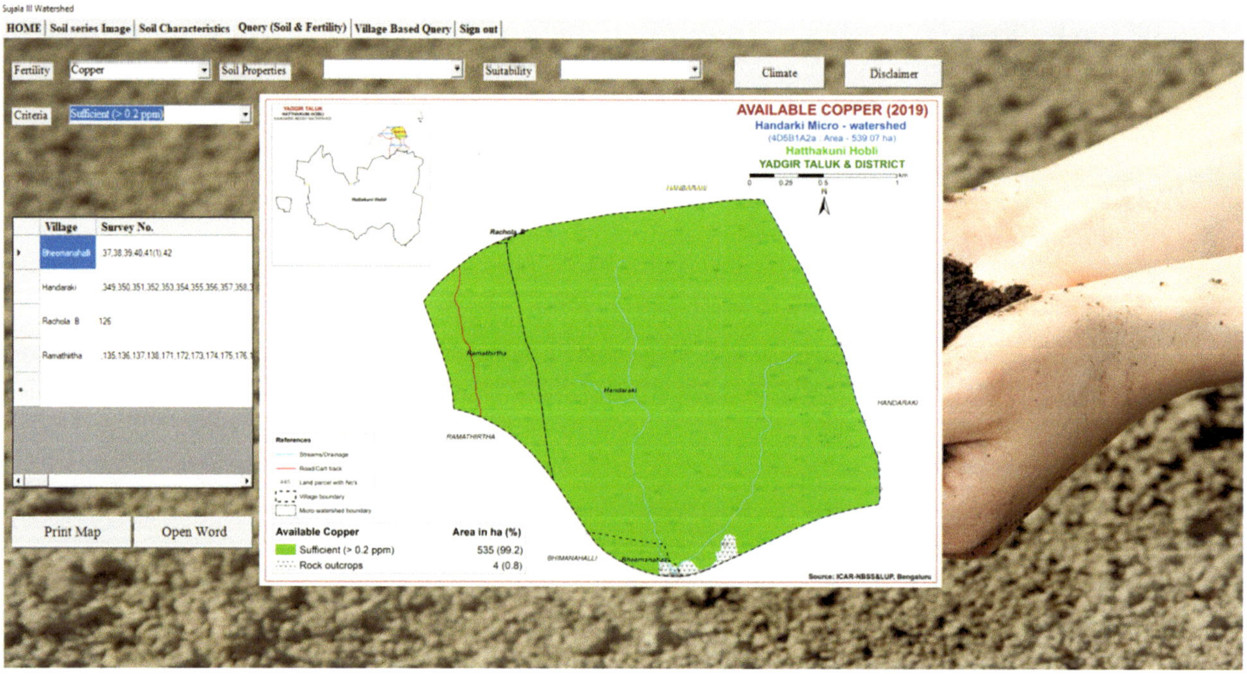

FIGURE 21.15 Soil suitability of Handaraki MWS.

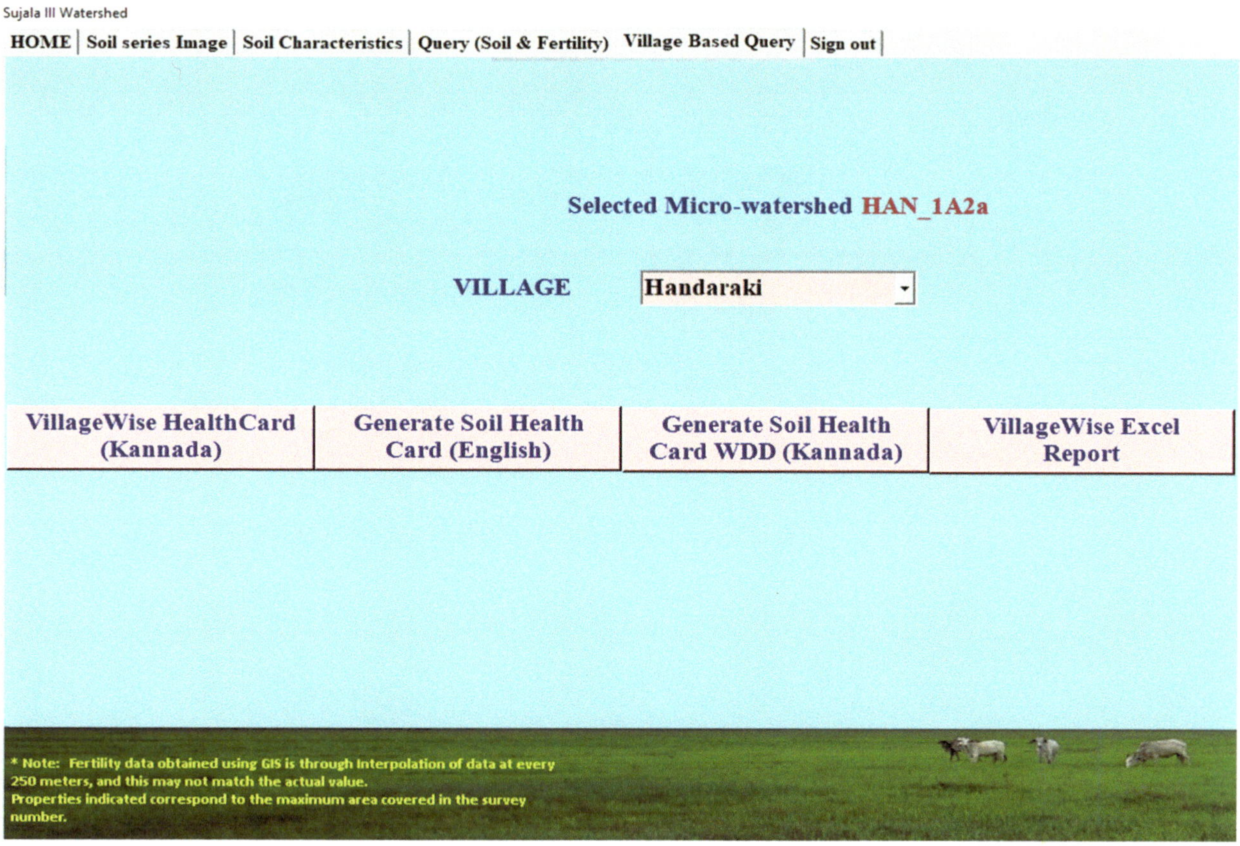

FIGURE 21.16 Village-wise report of photo Handaraki MWS.

Govt. of Karnataka
Sujala – III
Watershed Development Department
Govt.of Karnataka Kaveri Bhavana,
Bengalore-560 009
And
ICAR - National Bureau of Soil Survey and Land Use Planning,Regional centre,Hebbal,
Bangalore -560 024
Contact :080-23412242 E-Mail:nbssrcb@gmail.com
ICAR - NBSS & LUP

Farmer's Name	
Gender: Male/Female	
Microwatershed Name	Handaraki (4D5B1A2a)
Address	Handaraki Yadgir Taluk, yadgir District
Soil sampling year	2019
Survey/Hissa No	350
Area in (Acre/Gunta)	7 Acre
Annual Rainfall (mm)	829
*Note: Survey Number total area	

Details Of Land Surface And Soil Properties

Depth	Moderately shallow (50-75 cm)
Texture	Clay
Gravelliness (%)	Non gravelly(<15 %)
Slope (%)	Very gently sloping (1-3%)
Soil Erosion	Moderate Erosion
Land capability classes	good cultivable land with moderate limitations of soils
Available Water Capacity	Low (51-100 mm/m)
Soil & Water Conservation Plan	Graded bunding
Traditional Soil Name	Moderately shallow Black Loamy soil

Land Resource Cards

Soil Nutrient status

Soil reaction (pH)	Electrical Conductivity (EC)	Organic Carbon (OC)	Available Nitrogen (N)	Available phosphorus (P_2O5)	Available Potassium (K2O)	Available Sulphur (S)	Available Zinc (Zn)	Available Boron (B)	Available Iron (Fe)	Available Manganese (Mn)	Available Copper (Cu)
Neutral	Non saline	High	High	Medium	High	Low	Deficient	Low	Deficient	Sufficient	Sufficient

Disclaimer: Fertility data obtained from 250 meters,and may not match the actual value.Properties indicated correspond to the maximum area covered in the survey number. For complete details please refer the LRI reports /atlases of the watershed area.

Colour Code: Green : High Rating Yellow : Medium Rating Red : Low Rating

Suggested Crop Plan Based on Land Resorce Information

Suitability	Suitable Crops	Limitations	Suggested Interventions
Highly suitable	-	-	Application of FYM, Organic manures and micronutrients, drip irrigation, Mulching, suitable soil and water conservation practises
Moderately suitable	Amla, Chilly,Custard apple,Chrysanthemum, Marigold,Onion,Tomato, Brinjal,Bhendi,Bajra,Maize	Rooting conditions	
	Cotton,Groundnut,Sorghum	Rooting conditions, Texture	
Marginally suitable	Cashew,Guava,Jamun,Jackfruit, Lime,Musambi,Pomegranate, Sapota,Mulberry,Redgram, Sunflower,Drumstick	Rooting conditions	
	Bengalgram	Texture	
Currently not suitable	Mango, Tamarind	Rooting conditions	-

Farmers Helpline Centeres: Agricultural Problems-1800-425-3553,Varuna Mitra-9243345433, Horticulture Helpline-1800-4257910 and Krishi marata vahini-1800-425-1552

Issued Month & Year: January 2020

FIGURE 21.17 English soil health card of survey number 350, Handaraki (K) Village, Handaraki MWS.

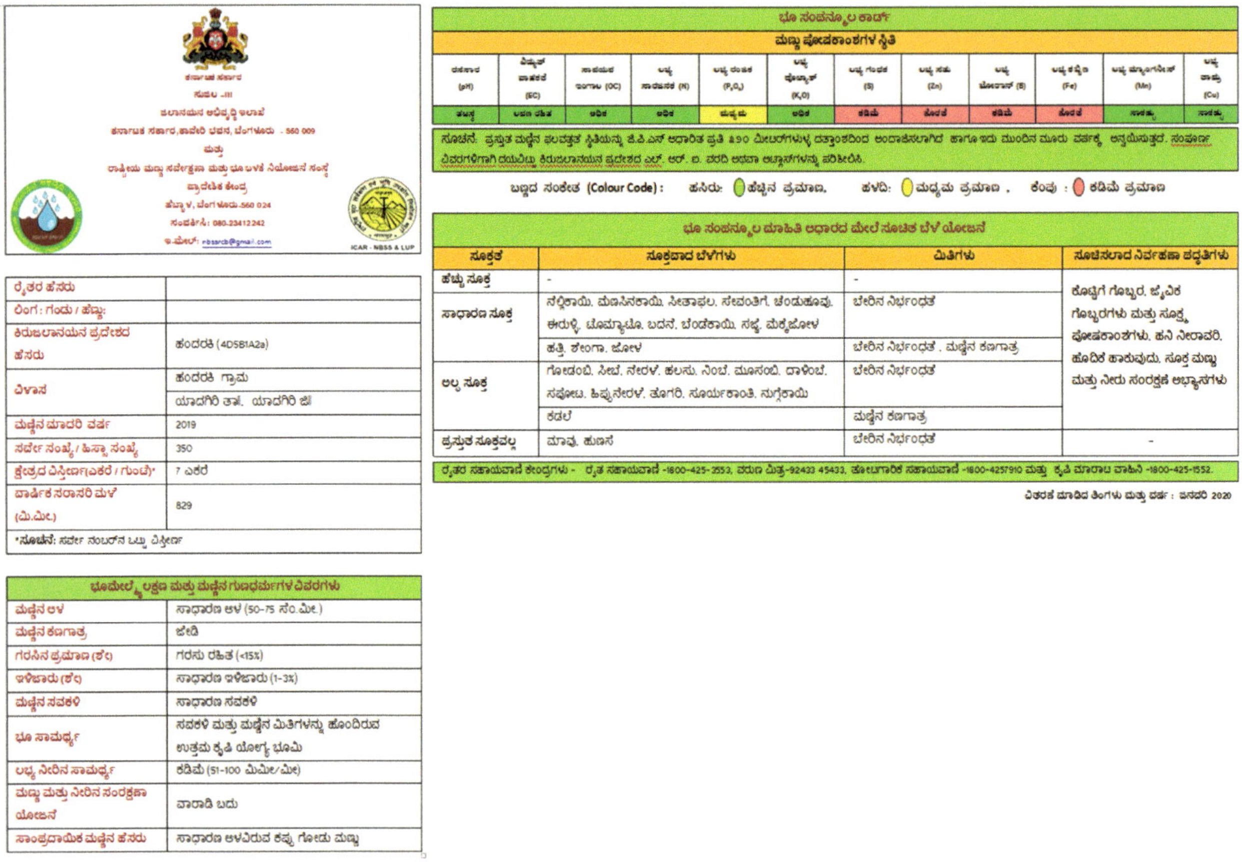

FIGURE 21.18 Kannada soil health card of survey number 350, Handaraki (K) Village, Handaraki MWS.

Furthermore, a village-wise Excel report generation service was also provided to the client, which facilitates comparative analysis among different possible planning strategies to accomplish the optimal solution.

21.8 Conclusion

The research on DL for farm-level agricultural planning combines knowledge-based crop systems with GIS tools. DL provides crop recommendations based on soil nutrition, physical status, and climatic information, which in turn improves the crop production rate. Hence, it is useful for farmers to have a knowledge-based expert system like the "Digital Library" for better utilization of farm resources. The prototype of the DL system successfully demonstrates the "Handaraki" micro-watershed. GIS-based DL has been developed under Microsoft Visual Studio platform using DOT NET framework. It uses published map formats for user interaction, generated using ARC MAP, and maintains a database created using MS Access. The DL provides the users with the advantages of interacting, analyzing, uploading, and downloading the data.

References

Adinarayana, J., Tewari, G., Azmi, S., Sudharsan, D., 2008. Gramya Vikas—a distributed collaboration model for rural development planning. J. Comput. Electron. 62, 128–140.

Akike, S., Samanta, S., 2016. Land use/land cover and forest canopy density monitoring of wafi-Golpu project area, Papua New Guinea. J. Geosci. Environ. Prot. 4, 1–14.

Ali, R.R., Kotb, M.M., 2010. Use of satellite data and GIS for soil mapping and capability assessment. Nat. Sci. 8 (8), 104–115.

Arms, W.Y., 2000. Digital libraries. Key Concepts in the Architecture of the Digital Library. MIT Press, Cambridge.

Bandyopadhyay, S., Jaiswal, R.K., Hegde, V.S., Jayaraman, V., 2009. Assessment of land suitability potentials for agriculture using a remote sensing and GIS based approach. Int. J. Remote. Sens. 30 (4), 879–895.

Baoji, W., Jinnuo, Z., Ruiqing, X., Qingsheng, L., 2007. Innovation in agricultural digital library services based on Web 2.0. N. Zealand J. Agric. Res. 50, 1255–1260.

Bhagia, N., Bairagi, G.D., Sandeep, P., Gajendra, P., Vyas, S.P., Naveen, K.G.N., et al., 2016. National inventory of rabi pulses in India using remotely sensed data. J Indian Soc Rem Sens. Available from: http://link.springer.com/article/10.1007/s12524-016-0603-4.

Bhattacharyya, T., Pal, D.K., Mandal, C., Chandran, P., Ray, S.K., Sarkar, D., et al., 2013. Soils of India: historical perspective, classification and recent advances. Curr. Sci. 104 (10), 1308–1323. ISSN 0011–3891.

Bhattacharyya, T., Sarkar, D., Sehgal, J.L., Velayutham, M., Gajbhiye, K.S., Nagar, A.P., et al., 2009. Soil taxonomic database of India and the States (1: 250,000 scale), NBSSLUP Publ. 143. National Bureau of Soil Survey and Land Use Planning, Nagpur 266.

Blanco, A.C., Tamondong, A., Perez, A.M., Ang, M.R. C., Paringit, E., Alberto, R., et al. (2016). Nationwide natural resource inventory of the Philippines using LiDAR: strategies, progress, and challenges. International Archives of the Photogrammetry, Remote Sensing and Spatial Information. XLI-B6: 105–109.

van Engelen, V.W.P., Wen, T.-T. (Eds.) 1995. Global and National Soils and Terrain Digital Databases (SOTER): Procedures Manual. ISRIC, Wageningen, Netherlands.

FAO, 1998c. The Soil and Terrain Database for Northeastern Africa (CD-room). FAO, Rome.

FAO, 1998d. Soil and Terrain Database, Land Degradation Status and Soil Vulnerability Assessment for Central and Eastern Europe (CD-ROM). FAO, Rome.

FAO, 1999. Soil and Physiographic Database for North and Central Eurasia (CD-ROM). FAO, Rome.

Gani, E., Magai, J.S., 2014. The emergence of Digital libraries services in Northwest Nigerian universities; challenges and prospects. Libr. Philos. Pract. (e-journal) 1–21.

Hegde, Rajendra, Niranjana, K.V., Srinivas, S., Danorkar, B.A., Singh, S.K., 2018. Site-specific land resource inventory for scientific planning of Sujala watersheds in Karnataka. Curr. Sci. 115, 4.

Hellweger, F.L., Maidment, D.R., 1999. Definition and connection of hydrologic elements using geographic data. J. Hydrol. Eng. 41, 10–18.

Hewitt, A.E., 2010. New Zealand soil classification, Landcare Research Science Series. No. 1, third edn. Manaaki Whenua Press, Lincoln.

IARI, 1971. Soil Survey Manual. All India Soil and Land Use Survey. IARI, New Delhi, p. 121.

Innocenti, P., Vullo, G., Ross, S., 2010. Towards a digital library policy and quality interoperability framework: the DL.org project. N. Rev. Inf. Netw. 15, 29–53. ISSN 1361–4576.

Kumar, N., Yamac, S.S., Velmurugan, A., 2015. Applications of remote sensing and GIS in natural resource management. J. Andaman Sci. Assoc. 20 (1), 1–6.

Naidu L.G.K., Ramamutrhy, V., Challa, O., Rajendra, H., Krishnan, P. 2006. Manual soil site criteria for major crops. NBSS Publication No. 129, NBSS & LUP, Nagpur, 118.

Naidu, L.O.K., Reddy, R.S., Sah, K.D., Bhaskar, B.P., Datta, D., Niranjana, K.V., et al., 1998. Mapping of agroecological zones of AP through soil resource data. Indian. J. Agric. Sci. 68, 661–665.

NBSS & LUP. 1994. Proceedings National meet on soil site suitability criteria for different crops, NBSS and LUP, Nagpur, 30.

Packages of Practices. 2013. Published by UAS, Bangalore and Govt. of Karnataka, Bangalore.

Ponnusamy, T., 2007. Knowledge based expert system for agricultural planning. MA thesis. University of Adelaide, Australia, pp. 1–136.

Post, W.M., 2000. Global soil types by 0.5-degree grid (modified Zobler). Oak Ridge National Laboratory, Oak Ridge, TN, http://www.daac.ornl.gov/.

Rashed, M., Idris, Y., Shaban, M., (2006). Integrative approach of GIS and remote sensing to represent the hydrogeological and hydrochemical conditions of Wadi Qena-Egypt. 2nd International Conferences on Water Resources and Arid Environments., Saudi Arabia, 26–29.

Rossiter D.G., Hengl, T. (2001). Creating geometrically-correct photointerpretations, photomosaics, and base maps for a project GIS. International Institute for Geo-Information Science and Earth Observation (ITC), Netherlands.

Samarakoon, L., (2005). Basic geographic information system (GIS) vector/raster model. ISPRS Workshop on Remote Sensing and GIS for Watershed Managment, Laos, 2–5 December.

Sharifabadi, S.R., 2006. How digital libraries can support e learning. Electron. Libr. 24 (3), 389–401. Available from: https://doi.org/10.1108/02640470610671231.

Shaw, S.K., Panda, S., Londhe, S., Vinod, V.H., 2015. Remote sensing and GIS for natural resource inventory of hilly terrain-a case study of Kakrorasa subwatershed, Ranchi district, Jharkhand. Int. J. Sci. Eng. Appl. Sci. 1 (3), 122–131.

Strager, M.P., Fletcher, J.J., Strager, J.M., Yuill, C.B., Eli, R.N., Todd, P.J., et al., 2010. Watershed analysis with GIS: the watershed characterization and modeling system software application. Comput. Geosci. 36, 970–976.

Sys, C., Van Ranst, E., Debaveye Ir, J., Beernaert, F. 1993. Land evaluation, part III. Crop requirements. Agriculture Publication, No.7, General Administration for Development Cooperation. Brussels, Belgium, 166.

Velayutham, M., Pal, D.K., 2004. Classification, Indian. In: Lal, R. (Ed.), Encyclopedia of Soil Science. Marcel Dekker, New York, pp. 1–3.

Warr, H., Hangsing, P., 2009. Open source digital library software: a literature review. Proceedings of the National Seminar on Preservation and Conservation of Information Resources in Knowledge Society: Issues, Challenges and Trends 238–258. Available from: https://doi.org/10.2139/ssrn.1639383.

Chapter 22

Remote sensing as a potential tool for advancing digital soil mapping

Prava Kiran Dash[1,2]

[1]Department of Soil Science, College of Agriculture, Odisha University of Agriculture and Technology, Bhubaneswar, Odisha, India, [2]Regional Research and Technology Transfer Station, Mahisapat, Dhenkanal, Odisha, India

22.1 Introduction

The growing demand for food and non-food commodities is likely to increase by 75−100 % globally by the year 2050 (Choudhury et al., 2019). Global crises of food and rising food prices in international markets have already been noticed indicating the threat of food insecurity in future. On the other hand, other issues like "declining soil fertility" and "stagnant yield" are rising because of intensive agriculture, introduction of high-yielding varieties, imbalanced application of fertilizers, insufficient application of organic manures, less recycling of crop residues etc. (Sahrawat et al., 2010). No doubt, the increased occurrence of climate change events also imposes additional instability on the crop production systems.

Raising crop yields is even more challenging in the scenario of rapidly declining arable lands owing to use of agricultural lands for non-agricultural purposes such as, urbanization, industrialization, etc. Moreover, irrational use of agri-inputs, continuous overexploitation of inherent soil fertility, indiscriminate use of the finite soil resource, and erratic land use without following suitable management practices have led to severe land degradation and development of a wide variety of problematic (saline, alkaline, acidic, eroded) soils all over the world (Singh et al., 2019). Thus, soaring food demand, population pressure, land degradation, and allied soil issues warrent rapid soil assessments so that soils can be well restored and best used in a sustainable manner (Dash et al., 2022; Sarmah et al., 2019; Vedadri and Naidu, 2018).

Keeping in view the widespread challenges related to soils, there is a worldwide demand for developing systematic soil resource databases to address "soil security" (McBratney et al., 2014). A wide range of disciplines and subdisciplines are also requiring precise soil information for dealing with different global issues related to food, water, energy, environment, biodiversity, and human health (Grunwald et al., 2011; Hartemink, 2015). Site-specific soil studies are essential for restoring soil quality, enhancing or at least maintaining the present level of crop productivity, and making the crop production system more sustainable over the years to come. In these circumstances, making use of advanced geospatial technologies including global positioning system (GPS), geographic information system (GIS), remote sensing (RS), and proximal sensing (PS) will help continuous assessment, monitoring, and management of natural resources in a systematic manner (Mitran et al., 2021). Assuredly, geospatial technologies are great tools to assist decision support systems in agriculture.

Soil mapping is a vital aspect of soil management. For example, soil maps can guide where to apply more or less of any certain nutrient from external source, where to take necessary innervations for remedying problematic soils, and to decide which crop is most suitable for a certain area (Dharumarajan et al., 2017). In light of this, digital soil mapping (DSM) is an advanced tool to prepare continuous spatial soil maps for soil properties, classes, and functions (Dharumarajan et al., 2019). Over the past few years, DSM (otherwise known as predictive soil mapping) has been widely adopted as an advanced tool for prediction of soil properties at locations that are not sampled based on the observed values at the sampled sites and other predictive factors (Dash et al., 2022). Thus, DSM markedly reduces the tedious, costly, and time-consuming processe of field surveys and lab analysis, besides facilitating easy accessibility by the end users (Lagacherie and McBratney, 2006).

Remote Sensing of Soils. DOI: https://doi.org/10.1016/B978-0-443-18773-5.00021-1

Unlike conventional soil maps, digital soil maps provide us with estimates of accuracy and uncertainty (McBratney et al., 2003; Minasny and McBratney, 2016). DSM is indeed a strong supporting tool for present-day precision agriculture (Minasny and Hartemink, 2011; Zhang et al., 2017). Keeping the essentiality of DSM in view, not only the national organizations but also the international organizations such as the International Union of Soil Sciences (IUSS), International Soil Reference and Information Centre (ISRIC), and Food and Agriculture Organization (FAO) are developing soil resource databases through global programs such as GlobalSoilMap (https://www.isric.org/projects/globalsoilmapnet), SoilGrids 2.0 (https://soilgrids.org/), and Global Soil Partnership (http://www.fao.org/globalsoilpartnership/). Recently, initial versions of global extent digital soil maps notably SoilGrids 2.0 have also been prepared at a 250 m spatial resolution using the World Soil Information Service (WoSIS) database developed by ISRIC (Poggio et al., 2021). An elevated number of DSM attempts has also been observed at local and regional levels in the past two decades (Wadoux et al., 2020).

Remote sensing products are key inputs for DSM (Zeraatpisheh et al., 2020). Relivant ground information for DSM can be well extracted from remotely sensed satellite imageries. Many soil properties can be measured through information acquired by sensors either directly or indirectly using different pedo-transfer functions (Santra et al., 2017). Band values and soil or vegetation indices calculated from reflectance values are potential resources to be used as covariates in DSM (Reddy et al., 2021). Over the past few years, increased availability of broad-spectrum RS products including digital elevation models (DEMs) has facilitated spatial soil mapping. Moreover, fine-resolution of RS products enable preparing soil maps at fine resolutions making them more suitable for local or plot-level applications (Zhang et al., 2017). Details regarding the concept of DSM, RS resources, and DSM applications will be discussed in detail in the following sections.

22.2 Digital soil mapping

22.2.1 Spatial variability of soil properties

Soils show high degrees of spatial variability for soil physicochemical properties as well as soil nutrients owing to both natural and anthropogenic processes (Iticha and Takele, 2019). Spatial variations for soil nutrients can also arise because of prevalent pedogenic processes and dynamic interactions among natural, environmental, and other human-induced factors (Liu et al., 2016). Spatial variability causes uneven patterns in soil fertility levels, which is the reason behind the decreased efficiency of fertilizers being applied uniformly across the field irrespective of the within-field variations (Behera et al., 2018). Uniform management of fields often results in overapplication of inputs in areas with high concentrations and underapplication in areas with low concentrations.

Over the past few decades, intensive cultivation of high-yielding varieties and the imbalanced application of synthetic fertilizers have brought about multinutrient deficiencies, soil and environmental pollution, besides adversely impacting human health (Sahrawat et al., 2010). On this account, assessing spatial variabilities of physicochemical properties and evaluating fertility status is crucial for establishing variable-rate fertilizer applications. Site-specific soil management is not only helpful for correcting nutrient deficiencies but also helps cutting down cost of cultivation by avoiding overapplication of fertilizers (Franzen and Peck, 1995). Precise application of synthetic fertilizers is also critical for reducing their environmental loads (Shukla et al., 2016).

Farmers, especially those of small land holding settings are not always capable of getting their field soils tested. To deal with this problem, DSM is a time-sensitive tool that can provide advisories to large numbers of farmers for spatial management of soil properties. Technically, DSM makes the best possible use of limited soil observations from laboratory analysis and relevant information from available covariates (Minasny and McBratney, 2016). Digital soil maps thus prepared at larger mapping extents can serve as important resources for policy planners and land managers. Overall, DSM holds promise to serve several stakeholders including farmers, land managers, policymakers, and resource planners.

22.2.2 Digital soil mapping approaches

The process of mapping soil properties using several computation methods and modeling approaches to generate spatially continuous soil maps is known DSM (Santra et al., 2021). Broadly, there are two types of DSM approaches viz., (1) geostatistical approach and (2) SCORPAN approach.

22.2.2.1 Geostatistical approach (spatial interpolation)

Geostatistics-based spatial interpolation is a popular conventional technique for preparing digital soil maps (Oliver, 1987; Oliver and Webster, 2014). In this approach, estimates of soil attributes at the unsampled locations are calculated based on the spatial autocorrelation observed among the sampled points. Types of spatial interpolation techniques are deterministic (inverse distance weighting, spline), stochastic (ordinary kriging, universal kriging), and hybrid approaches (cokriging, regression kriging). Kriging (Krige, 1966) is a commonly practiced geostatistical method for predicting soil properties with minimized error variances using the methodology of semivariogram modeling (Goovaerts, 1997; Oliver and Webster, 2015). A diagrammatic representation of DSM using the geostatistical approach has been presented in Fig. 22.1.

Among different forms of kriging, ordinary kriging is a popularly used technique for the prediction of soil properties without using covariate information (Rosemary et al., 2017; Saby et al., 2006; Umali et al., 2012). However, geostatistical methods are incapable of dealing with nonlinear relationships among soil and auxiliary variables (Wadoux et al., 2020). Need of larger sample size, assumptions for data distribution, requirement of data transformation, and incompetency to handle complex datasets are the other underlying limitations of the geostatistical spatial interpolation approach necessitating adoption of advanced machine learning approaches for DSM (Lamichhane et al., 2019; Wadoux et al., 2019).

22.2.2.2 SCORPAN approach

The SCORPAN concept originates from Jenny's classical CLORPT theory (Jenny, 1941) and it describes the quantitative relationship between soil property or class with the auxilliary variables (covariates) (McBratney et al., 2003). Three major components of SCORPAN-based DSM are input, process, and output (Lagacherie and McBratney, 2006). The "input" refers to data generated from field and laboratory observational methods, which are "processed" through spatial soil inference systems (predictive models) to finally produce spatially continuous soil maps as "outputs". Here, "SCORPAN" refers to the set of environmental covariates (also known as predictive factors or auxiliary variables), which are used for establishing quantitative relationships with the target soil property or class to develop regression equations. The SCORPAN approach as proposed by McBratney et al. (2003) is equated as Eq. (22.1).

$$S_a(\text{or})S_c = f(S, C, O, R, P, A, N) + e \tag{22.1}$$

Where, S_a (soil property) or S_c (soil class) is represented as a function (f) of 7 different environmental covariates viz., S (soil), C (climate), O (organism), R (relief), P (parent material), A (age of weathering), and N (spatial position). Here, "e" stands for the error component.

The quantitative relationship between soil and covariates can be established by machine learning (ML) algorithms. ML refers to the process of learning from the existing dataset and making new predictions without any human intervention. ML algorithms overcome the limitations of geostatistical methods as they can handle complex non-linear datasets (Khaledian and Miller, 2020). ML involves the process of regression and classification of input data using the principle of spatial association, establishing mathematical relationships, building predictive models, and predicting attribute

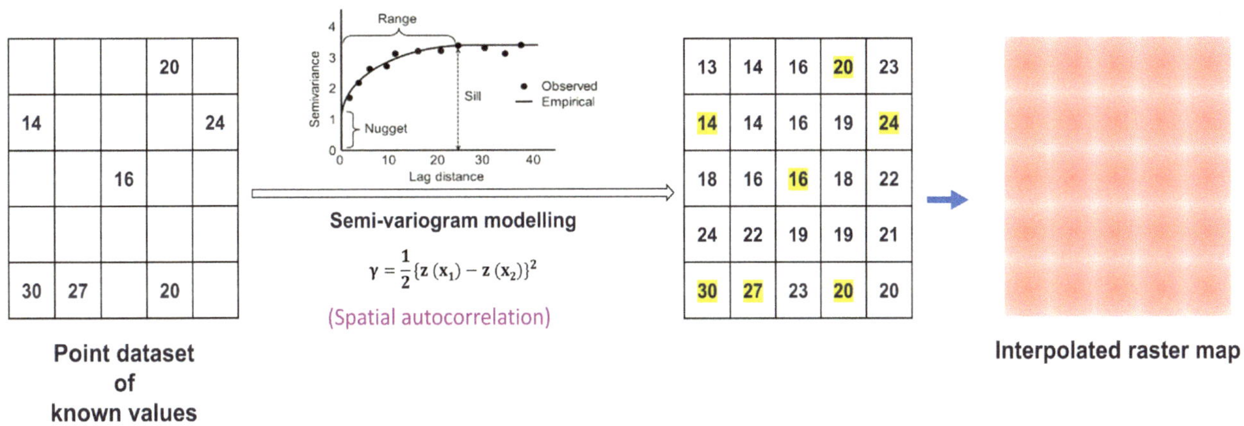

FIGURE 22.1 Diagrammatic representation of geostatistical approach of the DSM.

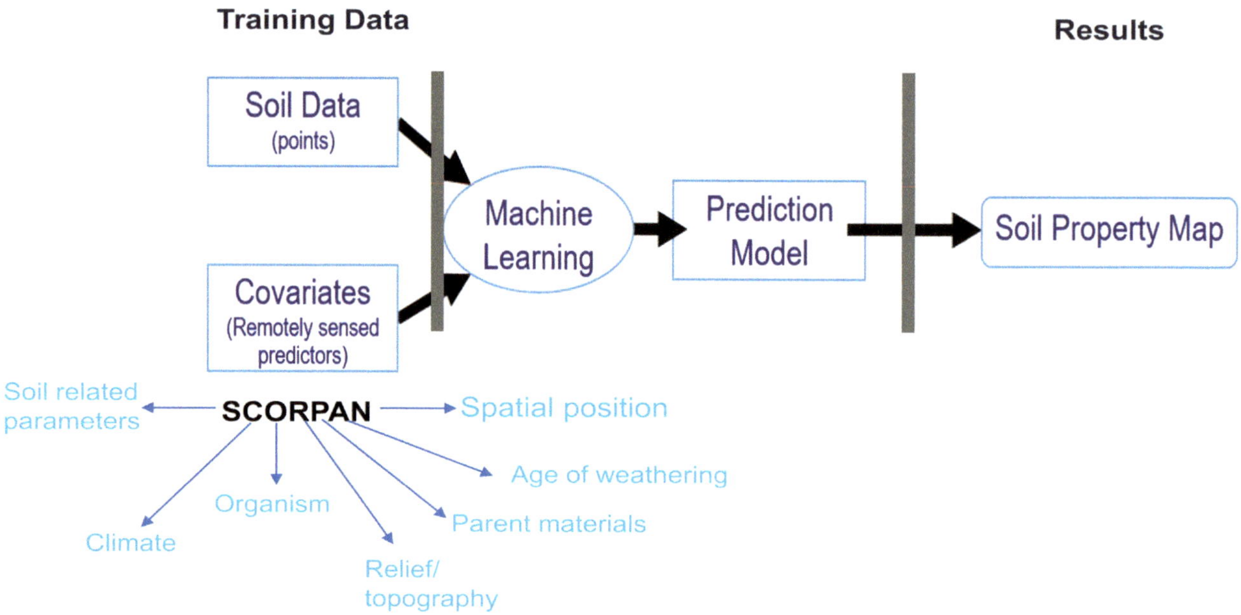

FIGURE 22.2 Diagrammatic representation of the SCORPAN approach of DSM.

values for the unsampled locations (Santra et al., 2017). A diagrammatic representation of DSM using SCORPAN approach has been presented in Fig. 22.2.

Examples of advanced ML algorithms are random forest (RF), cubist, artificial neural network (ANN), support vector machine (SVM), and other techniques. ML algorithms being capable of modeling non-linear datasets have gained more attention over the geostatistical methods over the past few years (Chen et al., 2022; Wadoux et al., 2020; Zhang et al., 2017). On top of this, ML algorithms also allow users to extrapolate the trained algorithms beyond the sampling boundaries using information from covariates (Wang et al., 2021; Saurette et al., 2022).

22.3 Remote sensing

Depending on physical features and chemical properties, different objects on earth's surface reflect, reradiate, or emit different amounts of electromagnetic energy at various wavelengths. The measurement of reflected, radiated, or emitted electromagnetic radiation forms the basis for studying the earth's surface features. RS is a quick and effective approach that may gather and analyze information on the spectral characteristics of earth surfaces without actually coming into direct contact with them. RS can be performed from a variety of distant locations from the target such as from a satellite, a manned aircraft, an unmanned aerial vehicle (UAV), or a drone. Sensing, which is performed in high resolution by sensors placed in a small aircraft moving at a relatively smaller height from the earth (3–5 km) than the satellites is termed as "aerial photography" (Mulla, 2013). Relatively higher immunity to cloudiness and the provision of scheduling the time of image acquisition are the advantages of aerial photography (Mitran et al., 2021). However, when sensing is performed by the sensors in a close proximity to the target such as by a handheld spectroradiometer, it is known as proximal sensing. In case of proximal sensing, platforms are generally handheld, mounted in tractors, or may be managed by robotics for rapid collection of high-resolution/precise data about the target object.

RS is a useful technique for studying inaccessible or difficultly accessible areas. Remotely sensed images are captured by specialized cameras, typically called "sensors" and scientists/technicians analyze the data to interpret them. When sensors gather electromagnetic radiation from natural sources that is reflected or transmitted, it is known as passive RS. But when sensors extract reflected/emitted responses from objects that are exposed to radiation from artificial energy sources such as RADAR (Radio Detection and Ranging), it is known as active RS. Different imaging approaches use varieties of sensors including visible, infrared, thermal, multispectral, and hyperspectral. RS products (usually satellite imageries) after proper preprocessing with necessary corrections (such as geometric correction, cloud correction, null value corrections, etc.) are used to extract necessary information from the pixel values. After proper retrieval, remotely sensed products are stored, managed, and analyzed by computer based technologies like GIS. GIS is an efficient tool for handling spatial information for example spatial soil information, climatic attributes, etc. Together, RS and GIS are the emerging technologies of the present era providing vast scope and opportunities for natural resource management.

Different types of sensing persist based on the wavelength band(s) of the electromagnetic spectrum that the onboard sensors can sense. Sensing performed in the visible (0.4–0.7 μm), near-infrared (0.7–1.1 μm) and short-wave infrared (1.1–3.0 μm) wavelength bands of the electromagnetic spectrum is known as optical RS. Optical RS measures reflectance energy and is vital for studying soil and crops. However, the requirement of cloud and atmospheric corrections is one of the major limitations associated with optical RS. Another form of RS is thermal RS, which works in the wavelength ranges of 3.0–5.0 μm (mid-infrared) and 8.0–15.0 μm (thermal infrared) and senses emitted energy from the objects. Thermal infrared RS can gather information about land and crop canopy temperatures, although it also gets affected by atmospheric temperature and evapotranspiration (Gerhards et al., 2019). RS performed at the wavelength range of 0.1 cm to 1 m is known as microwave RS. Microwave RS outperforms optical and thermal RS as it is independent of day or night conditions, rain, haze, snow, fog, sunny, smoky, or cloudy conditions. Microwaves are supposed to penetrate the vegetation and dry soil conditions as well (Mitran et al., 2021). A microwave radiometer is a passive device that records the natural microwave emission from the earth, whereas RADAR is an active microwave RS system that transmits electromagnetic radiation in the microwave range of wavelengths as a source of illumination to detect remote targets and receives the echoes coming back from the target as means of terrain information. In synthetic aperture radar (SAR) imaging, microwave pulses are transmitted by an antenna towards the earth's surface. The microwave energy scattered back to the spacecraft is measured. Some of the satellites carrying microwave sensors are Sentinel-1, Radarsat-2, RISAT-1, and ALOS-PALSAR.

When sensors sense in multiple (typically 3–15) numbers of spectral bands of the electromagnetic spectrum, for example, visible, near-infrared, short-wave infrared, thermal infrared, etc., it is known as multispectral scanning. Some of the multispectral satellite sensors include Landsat, Sentinel-2, Resourcesat, Quickbird, and SPOT. When repetitively larger numbers (usually 100–200) of narrow wavelength spectral bands are used for sensing the target object, it is known as hyperspectral RS. Hyperspectral RS is gaining more popularity these days because of its ability to detect in detail the land, water, and crop features. Reflectance data over a wide spectral range are generally collected at small spectral increments (typically 10 nm) in hyperspectral RS (Mulla, 2013). EO 1 Hyperion is a hyperspectral imager. The resolution of the RS products is a key factor that determines their useability and application for different purposes. For instance, a coarse resolution imagery might be useful for country-level planning and monitoring, but for managing resources at a farm level, fine resolution RS products are essential.

The benefits of RS include the ability to gather data in a timely and cost-effective manner over vast spatial extents, characterize spatial variability of ground objects regularly, track changes over time (temporal variability), and the potential to support decision-making in many fields. RS has been a proven technology for a variety of sectors including agriculture, climate, telecommunication, transport, and military. Particularly in the agriculture sector, RS technology help generating valuable information on soil physicochemical properties, nutrients, soil moisture, terrains, crop canopy, climate, parent materials, and mineralogy (Pareeth et al., 2019). To a considerable extent, RS technology can be utilised for measuring a few soil properties directly such as soil carbon, moisture, temperature, mineralogy, etc. RS is also a vital supporting tool for crop monitoring and management viz., crop stress monitoring, precise input application, mapping crop phenology, cropped area mapping, crop growth monitoring, disease and pest management, and yield forecasting. Vast data on soil and terrain gets generated by RS almost every now and then using which, DSM can be advanced by many folds.

22.4 Remote sensing as a potential tool for DSM

Satellites including ASTER, MODIS, Landsat 8, 9, ResourceSat, Sentinel-1, Sentinel-2, WorldView-2, SPOT-6,7, Cartosat, KOMPSAT-3, RADARSAT-2, RADARSAT Constellation, RISAT, and ALOS-2 have been effectively used in various soil and crop studies. DEMs derived from Cartosat, ALOS, and SRTM are also useful resources for DSM. Some of the leading space organizations across the world are NASA/USGS (USA), ESA, METI of Japan, ISRO (India), Digital Globe (USA), KARI (Korea), CSA (Canada), and JAXA (Japan). A brief account on different satellites, sensors and their providers has been presented in Table 22.1. Many of these satellite imageries can be acquired online from various platforms and service providers. One such platform is USGS EarthExplorer (https://earthexplorer.usgs.gov/).

There are different ways in which the information gathered by RS can be used for DSM. One way is to measure soil properties directly i.e., using the remotely sensed data as a primary data source for DSM. This approach is especially suitable if legacy soil data is scarce or unavailable. Alternatively, if legacy soil data is sufficient, soil and terrain

TABLE 22.1 List of major satellite data available for soil and crop studies.

Satellite/sensor	Spatial resolution	Provider/developer
Landsat 7 ETM + [b]	15, 30, 60 m	NASA, USA
ASTER	15–90 m	NASA, USA and Govt. of Japan
ASTER GDEM Version 3	30 m	METI of Japan and the NASA, USA
MODIS[a,b]	250, 500, 1000 m	NASA, USA
Landsat 8 (OLI: B1-B9; TIRS: B10–11)	OLI: 15 m (panchromatic), 30 m (multispectral); TIRS: 100 m	NASA/USGS, USA
ResourceSat	(LISS-IV sensor: 5.8 m; LISS-III: 23.5 M, AWiFS: 56 m)	ISRO, India
Sentinel-1	5 m (Strip Map Mode)	ESA
Sentinel-2	10, 20, 60 m	ESA
WorldView-2	0.46, 1.84 m	Digital Globe, USA
Rapid Eye[b]	6.5 m	Rapid Eye AG, Germany
Ikonos[b]	80 cm (panchromatic), 3.2 m (multispectral)	Digital Globe, USA
SPOT-6,7	1.5 m (panchromatic), 6 m (multispectral)	CNES (Centre national d'études spatiales—the French space agency), SSTC (Belgian scientific, technical and cultural services) and the Swedish National Space Board (SNSB)
Cartosat[c]	2.5 m (Cartosat-1), 0.8 m (Cartosat-2), 0.28 m PAN and 1.12 m multispectral (Cartosat-3)	ISRO, India
CartoDEM[d]	2.5, 10 m	ISRO, India
Landsat 9 (OLI-2: B1-B9; TIRS-2: B10–11)	OLI-2: 15 m (panchromatic), 30 m (multispectral); TIRS-2: 100 m	NASA/USGS, USA
KOMPSAT-3	0.7 m (panchromatic), 2.8 m (multispectral)	Korea Aerospace Research Institute (KARI), Korea
RADARSAT-2 [MW], RADARSAT Constellation [MW]	3–100 m	Canadian Space Agency (CSA), Canada
RISAT[MW]	50 m (RISAT-1), 1 m (RISAT-2), 0.5 × 0.3 (RISAT-2B)	ISRO, India
ALOS-2[bMW] (PALSAR-2 sensor)	3, 6, 10 m (strip map mode), 60, 100 m (scanSAR mode)	JAXA, Japan
EO 1 Hyperion Hyperspectral Imager[b]	30 m	Goddard Space Flight Centre (GSFC), NASA, USA
SRTM[bMW]	30, 90 m	NASA, USA

ALOS, Advanced land observing satellite (ALOS-1); *ASTER,* Advanced spaceborne thermal emission and reflection radiometer; *ASTER GDEM,* ASTER Global Digital elevation model; *AWiFS,* advanced wide field sensor; *DEM,* digital elevation model; *ESA,* European space agency, *ETM +*, enhanced thematic mapper plus scanner; *JAXA,* Japan Aerospace Exploration Agency; *KOMPSAT,* Korea multiple-purpose satellite; *Landsat,* land satellite; *LISS,* linear imaging self-scanning; *METI,* ministry of economy, trade, and industry; *MODIS,* moderate resolution imaging spectroradiometer; [MW], performs microwave remote sensing; *NASA,* National Aeronautics and Space Administration, *OLI,* operational land imager; *PALSAR*: phased array type l-band synthetic aperture radar; *RADARSAT,* RADAR satellite; *RISAT,* RADAR imaging satellite; *SPOT,* Satellite Pour l'Observation de la Terre (in French) meaning "Satellite for observation of Earth"; *SRTM,* Shuttle Radar Topography Mission; *TIRS,* thermal infrared sensor; USGS, United States Geological Survey.
[a]*MODIS itself is not a satellite, MODIS is an instrument on board both the Terra and Aqua satellites.*
[b]*Retired/not operational at present.*
[c]*The name Cartosat is a combination of cartography and satellite.*
[d]*CartoDEM is an Indian national DEM generated from Cartosat-1 stereo data.*

attributes derived from RS can be used as secondary variables (covariates) to complement the existing soil data generated by field and laboratory observational methods.

RS as a primary data source: Remotely sensed information of soil properties collected with good spatial coverage (generally in sparsely vegetated areas) can be directly used as primary data in DSM (Mulder et al., 2011). Some soil properties including soil texture, iron content, soil moisture, soil organic carbon, soil salinity, mineralogy, and carbonate content can be measured by statistical and chemometric analysis of spectroscopic data making use of the extensive spectral libraries (Santra et al., 2021). Nevertheless, RS is a great tool to identify homogeneous and heterogeneous soil—landscape units before conventional soil surveys and soil sampling. Moreover, continuous and large areal coverage of satellite-based RS can help reduce the efforts needed for soil sampling, especially in the inaccessible or difficultly assessable areas.

RS as a secondary data source (covariates): Sensor information from spaceborne, airborne, and in situ measurements using optical, thermal, and active microwave instruments provides necessary information for soil predictions. Taking field, laboratory measurements, or proximally sensed information as primary data, the RS products (satellite imageries or DEM) can be used as secondary variables for DSM to improve the prediction accuracy (Ferhatoglu and Miller, 2022). Band values (reflectance or emittance values), band ratios, soil and vegetation indices, etc. can be used as potential environmental covariates in DSM (Dharumarajan et al., 2019). DEMs are also useful resources for obtaining digital terrain attributes (DTA), which can easily be integrated as covariates in DSM. SRTM is one of the most popularly used DEM for terrain attributes (Dash et al., 2022; Dharumarajan et al., 2019). Integrating both RS and DEM derivatives is helpful for improving prediction (Wadoux et al., 2020).

For predicting soil properties at unvisited (unsampled) sites, a mathematical relationship is established among soil information from the sampled sites and the corresponding covariates. Based on the established mathematical expressions or regression equations, predictions are made for the locations not being sampled, and thereby spatially continuous digital soil maps are prepared (Dash et al., 2022). Over the years, several robust mathematical algorithms like neural networks, support vector regression, decision tree-based models, etc., have been found promising for their use in DSM (Heung et al., 2016; Khaledian and Miller, 2020). A list of remotely sensed covariates used in DSM studies in India has been presented in Table 22.2.

Shukla et al. (2018) used synthetic aperture radar data from RISAT-1 to derive soil and parent material-related information while predicting soil class. Temperature is considered as an important predictor for prediction of soil organic carbon (Sreenivas et al., 2016). Mitran et al. (2018) used MODIS-derived land surface temperature (LST) data as an auxiliary covariate for DSM. Vegetation and land use particulars are important predictors of soil chemical properties, especially soil carbon (Minasny et al., 2013).

Vegetation indices are potential predictors for various soil properties as they are linked to agroecological, climatic, and terrain characteristics. Few of the commonly used vegetation indices are the Normalized Difference Vegetation Index (NDVI), Enhanced Vegetation Index, Renormalized Difference Vegetation Index, Soil Adjusted Vegetation Index (SAVI), Transformed SAVI, Modified SAVI (MSAVI), Modified Nonlinear Vegetation Index, Leaf Area Index, Bare Soil Index, Photochemical Reflectance Index, Crop Water Stress Index, and Vegetation Temperature Condition Index. Relief or topography also greatly affects soil properties (Dash et al., 2019a,b,c). Different DEMs including SRTM DEM, Cartosat-1 DEM, ASTER DEM, and ALOS DEM can be used for extracting DTAs such as elevation, slope, aspect, curvature, flow accumulation, flow direction, topographic position index, terrain wetness index, etc.

22.5 Regional extent soil organic carbon mapping using remote sensing and digital terrain attributes: a case study

Study area: The study area is Keonjhar district of Odisha, India with a geographical area of 8303 km^2. The study area extends between 21.01° N to 22.16° N latitude and 85.18° E to 86.37° E longitude. A location map of the study area has been presented in Fig. 22.3.

Materials and methods: A total of 1024 surface (0—30 cm) soil samples were randomly collected. SOC was determined by the procedures of Walkley and Black (1934). Out of the 1024 surface soil samples, 800 were used for model calibration and 224 were used for model validation. Sampling locations have been presented in Fig. 22.4.

A total of 574 environmental covariates consisting of 184 RS variables and 390 DEM derivatives (DTAs: digital terrain attributes) were used as environmental covariates. The RS variables were the spectral bands of Sentinel-2 satellite imagery and their associated vegetation indices including CLG (chlorophyll index-green), CM (clay minerals index), GDVI (green difference vegetation Index), MSAVI (modified soil-adjusted vegetation index), NDI5 (normalized

TABLE 22.2 Remotely sensed environmental covariates used in previous DSM studies in India.

Predicted factor	Environmental covariates used	Sources of environmental covariates	Reference
Soil depth	Land use map	ASTER image	Kuriakose et al. (2009)
	Slope, aspect, curvature, distance from the stream, TWI	20 m horizontal-resolution DEM	
Soil depth	Elevation, slope, aspect, general curvature, TWI, distance from the stream	Cartosat-1 DEM	Sarkar et al. (2013)
	LULC map	Cartosat-1 and Resourcesat – 1 LISS-IV	
SOC	Elevation	Altimeter corrected DEM	Sreenivas et al. (2014)
	NDVI	MODIS	
TC, TN, P_{av}	LULC map	Resourcesat-1 LISS IV	Kumar and Singh (2016)
	Elevation, slope, TWI, WPI, UCA, STI	Cartosat-1 stereo digital data	
SOC, SIC	NDVI map	MODIS	Sreenivas et al. (2016)
pH, EC, SOC	NDVI, EVI and LULC	MODIS	Dharumarajan et al. (2017)
	Band data (Band 1 (Green), Band 2 (Red) and Band 3 (NIR))	Resourcesat-2 LISS IV	
	Elevation, slope, aspect, TWI, TPI, plan & profile curvature, MrVBF, MrRTF	SRTM DEM	
Sand content	Elevation, slope, altitude above channel network, hill shade, plan curvature, profile curvature, TRI, TWI	SRTM DEM (90 m)	Santra et al. (2017)
TC	LST	MODIS	Mitran et al. (2018)
	Elevation, slope, slope angle	SRTM DEM (90 m)	
	NDVI, NPP	MODIS	
TN	BSI, spectral reflectance, band ratio	WorldView-2, Pleiades-1A, Landsat-8	Xu et al. (2017)
	Vegetation indices	WorldView-2, Pleiades-1A, Landsat-8	
	Elevation, aspect, flow accumulation, flow direction, slope	ASTER DEM	
pH, EC	Elevation, slope, flow direction	CartoDEM (30 m)	Okonkwo et al. (2018)
	NDVI	MODIS	
	LULC	GlobeLand30	
	Elevation, slope gradient, slope aspect, profile curvature, plan curvature, slope length, TWI, MrVBF, MrRTF	SRTM DEM	
	NDVI	MODIS	
SOC, N_{av}, P_{av}	CI, SI, HI, RI, BI	Landsat-8	Kalambukattu et al. (2018)
	Slope, aspect, SPI, TWI	CartoDEM (30 m)	
Soil class	Slope, elevation, aspect	ALOS Global digital surface model	Shukla et al. (2018)
	Surface reflectance of bands 2–7	Landsat 8 OLI	
	NDVI	MODIS	
	Reflectance, CTI, IMI	Landsat 8, SAR data (RISAT-1, Sentinel-1A), ALOS DEM	
	CTI, slope, curvature	ALOS DEM	

(Continued)

TABLE 22.2 (Continued)

Predicted factor	Environmental covariates used	Sources of environmental covariates	Reference
pH, EC, SOC	Elevation	CartoDEM	Chhipa et al. (2019)
Sand, silt, clay	Elevation, Slope	CartoDEM (30 m)	Mitran et al. (2019)
	NDVI, GSI and Band 2, 3, 4, 5 6 and 7 of Landsat 8 OLI	Landsat 8 OLI	
pH, EC, SOC	Band data	Sentinel 2 imagery (13 bands)	Dharumarajan et al. (2019)
	NDVI and EVI	MODIS data	
	Elevation, slope, aspect, TWI, TPI, plan & profile curvature, MrVBF, MrRTF	Cartosat-1 DEM (10 m)	
SOC, TN	NDVI	Landsat 8 (OLI and TIRS)	Bangroo et al. (2020)
	Elevation, slope, CTI, SPI, STI	Cartosat DEM	
Sand, clay, pH, SOC	Slope, aspect, elevation, flow accumulation, hill shading, channel network base level, convergence index, cross-Sectional curvature, longitudinal curvature, LS Factor, relative slope position, TRI, TWI, Valley depth, vertical distance to channel network	SRTM DEM (90 m)	Reddy et al. (2021)

BI, Brightness index; *BSI*, bare soil index; *CI*, coloration index; *CTI*, compound topographic index; *EC*, electrical conductivity; *EVI*, enhanced vegetation index; *HI*, hue index; *IMI*, integral moisture index; *LULC*, land use land cover; *LST*, land surface temperature; *MrRTF*, multiresolution ridge top flatness; *MrVBF*, multiresolution index of valley bottom flatness; N_{av}, available nitrogen; *NDVI*, normalized difference vegetation index; *NPP*, net primary productivity; P_{av}, available phosphorus; *RI*, redness index; *SI*, saturation index; *SIC*, soil inorganic carbon; *SOC*, soil organic carbon; *SPI*, stream power index; *STI*, sediment transport index; *TC*, total carbon; *TPI*, topographic position index; *TRI*, terrain ruggedness index; *TN*, total nitrogen; *TWI*, terrain wetness index; *UCA*, upslope catchment area; *WPI*, water power index.

difference residue index-5), NDRI7 (normalized difference residue index-7), NDRI (normalized difference residue index), NDSVI (normalized difference senescent vegetation index), NDTI (normalized difference tillage index), NDVI (normalized difference vegetation index), NSI (soil salinity index), SATVI (soil-adjusted total vegetation index), and SAVI (soil adjusted vegetation index).

DTAs were derived from SRTM DEM of 30 m spatial resolution. DEM derivatives like aspect, cross-sectional curvature, longitudinal curvature, plan curvature, profile curvature, relative elevation, slope, eastness, and northness at different analysis scales (Miller et al., 2015) were also used as covariates. All covariates were resampled to 10 m before modeling. A few of the covariates used in this study have been presented in Fig. 22.5.

Random forest–embedded feature selection method was employed to screen the relevant covariates. Random forest algorithm was used as the predictive model. Lin's concordance correlation coefficient (CCC) and root-mean-square error (RMSE) of independent validation were determined as evaluation matrices.

Salient outcomes:

Descriptive statistics of SOC revealed that SOC content in the study area varied from 0.09 % (low) to 1.38 % (high) with mean and standard deviation values of 0.73 (medium) and 0.29 %, respectively. The coefficient of variability, skewness, and kurtosis were observed to be 39.3 %, −0.09, and −0.71, respectively. A prediction map for SOC was prepared at a spatial resolution of 10 m (Fig. 22.6). Independent validation with the independent validation dataset consisting of 224 points revealed that Lin's CCC value of prediction was 0.68 and the corresponding RMSE value of prediction was 0.195. SOC content was relatively higher in the north-west part than in the south-east part of the study area. The higher concentration of SOC in the north-west part could be because of the higher vegetation cover in the same, as also represented by the NDVI map of the study area (Fig. 22.5). Also, since the study area was topographically very diversified (Fig. 22.5), elevation was observed as one of the major predictors of SOC.

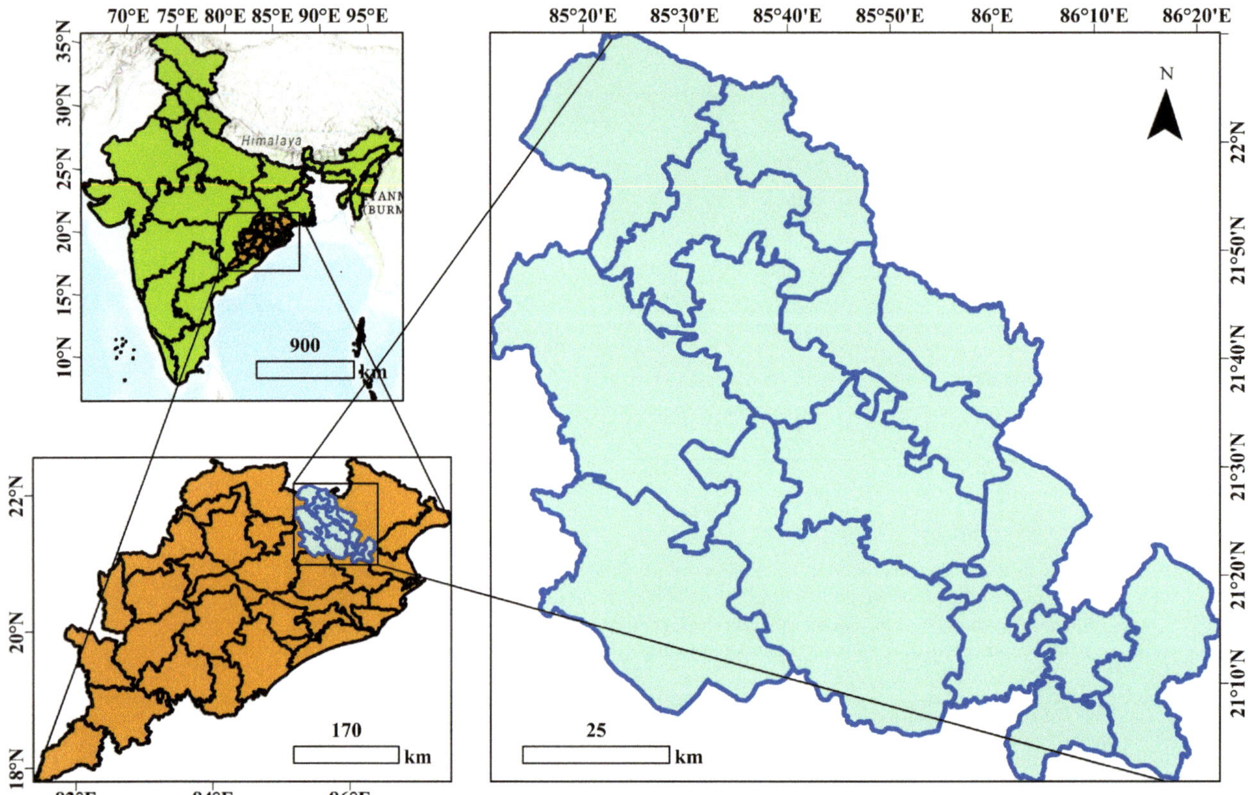

FIGURE 22.3 Location map of the study area (Keonjhar district). Outlines within the district indicate the administrative block boundaries of the district.

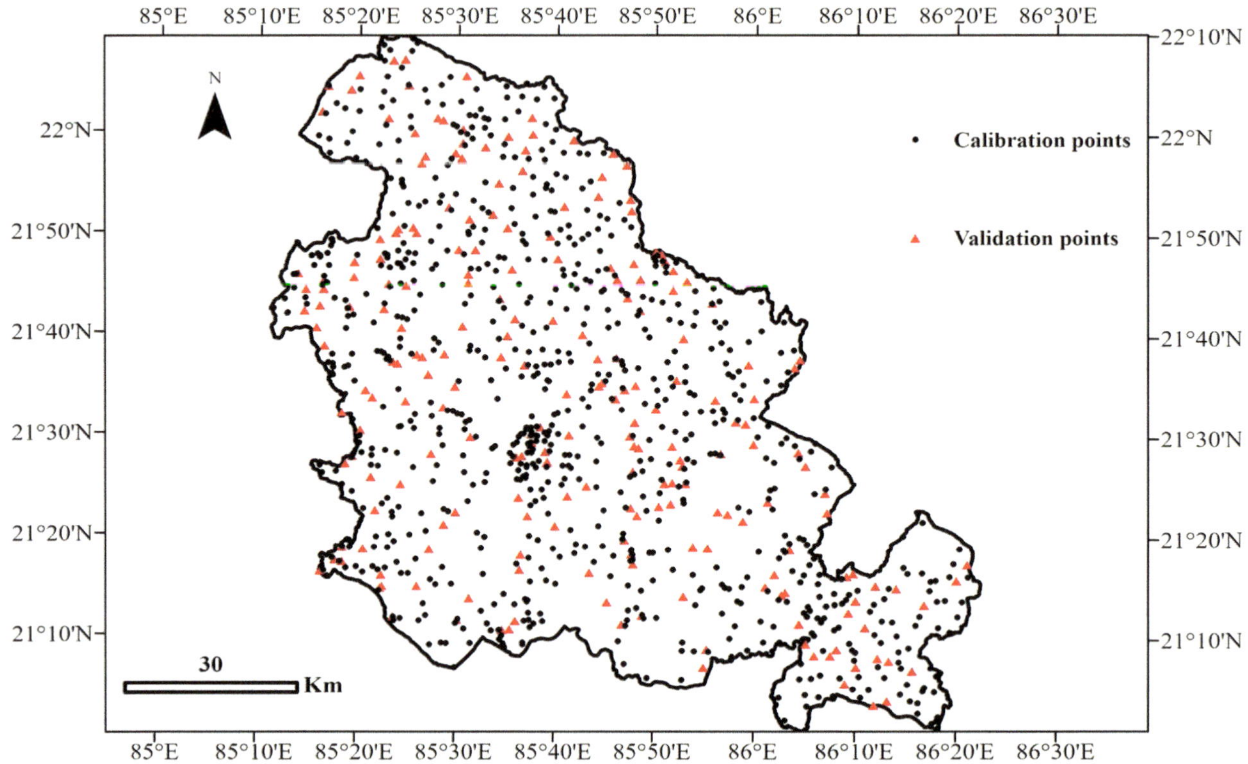

FIGURE 22.4 Sampling locations of the study area.

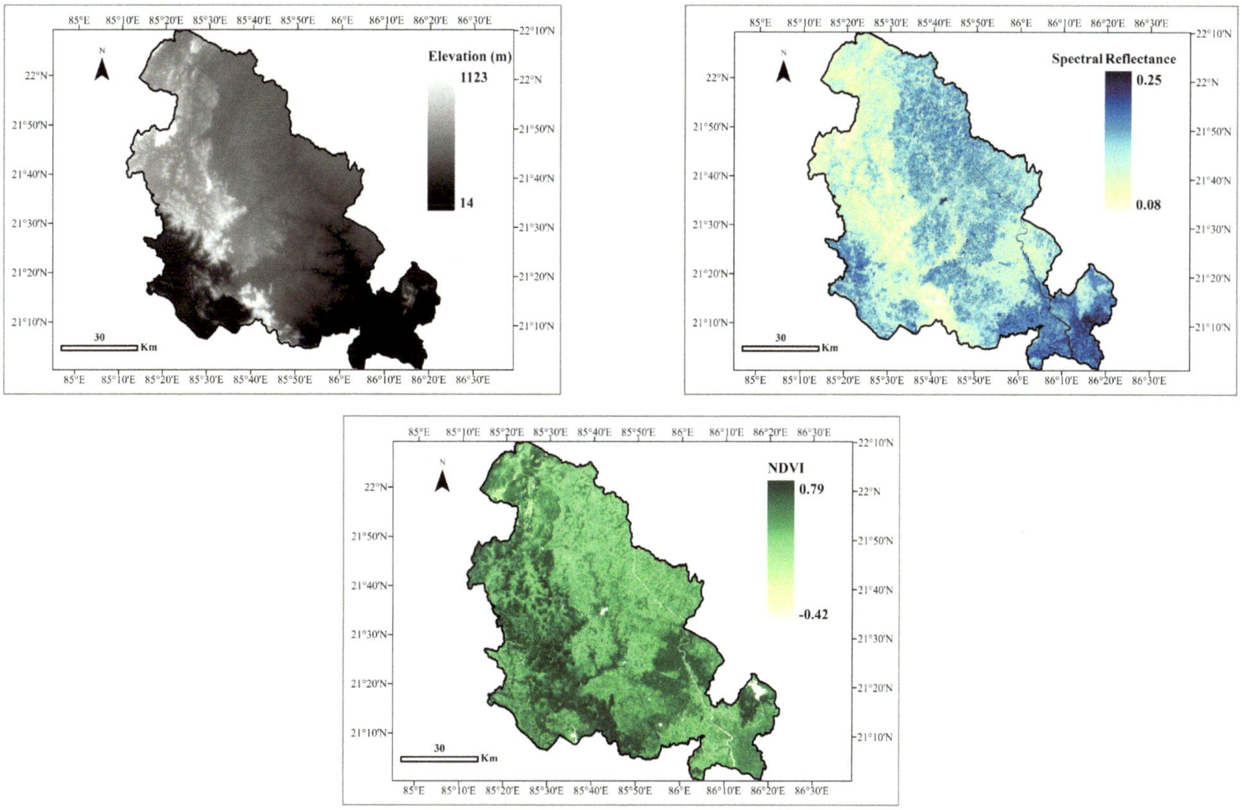

FIGURE 22.5 Selected covariates of the study area.

FIGURE 22.6 SOC prediction map for Keonjhar district, Odisha.

22.6 Challenges and opportunities of RS for DSM

Vegetation is a barrier for sensing soil that limits the useability of proximal sensing techniques to be implemented at a spaceborne level. Spectral signatures of urban areas, roads, and water surfaces are also difficult to interpret from spectral values. Hence, techniques such as spectral unmixing need to be improved for removing the spectral influence of vegetation and other objects from the signal while measuring soil attributes. Many times, RS applications are still limited by the resolution of imageries since coarse resolution products are not sufficient for monitoring spatial and temporal variabilities of soils at farm level. Therefore, obtaining more and more RS products with high spatial, spectral, radiometric, and temporal resolutions will be beneficial for soil and crop studies.

Despite higher spatial coverage, RS also involves large uncertainties because of the indirect nature of measurement. Also, most RS applications are still limited to surface soils only. Till now, proximal sensing dominates RS in terms of measuring soil properties because of complex terrains, vegetation covers, poor resolution of sensors, sensor noise, and atmospheric distortions (Mulder et al., 2011). Cloud coverage and other atmospheric phenomena are the major limitations of optical RS, which are usually overcome by microwave RS. Nevertheless, with the availability of more multispectral and hyperspectral sensors, RS can potentially measure various soil properties and help generate reliable primary soil datasets for DSM.

In this era of big data, where there is an increase in data availability, there also arises the issue of data selection and management. For example, very often choosing too many covariates limits the prediction performance of DSM (Chandrashekar and Sahin, 2014). Therefore, RS data users must be aware of choosing right RS products for specific applications.

22.7 Potential applications of RS-assisted DSM

Fine-resolution RS products generate fine-resolution digital soil and terrain maps with high accuracy. Accurate digital soil maps hold great relevance in terms of natural resource management. Digital soil maps can potentially be used for soil moisture monitoring, irrigation management, nutrient management, soil quality assessment, soil health monitoring, heavy metal assessment, assessment of soil biodiversity etc. DSM can also support carbon stock assessment, mapping of carbon sequestration potential (holds relevance in terms of mitigating climate change, global carbon management, and carbon trading), and other soil functions. DSM services can further be extended to land resource management such as wasteland management, land use planning, assessment of land degradation, desertification, problematic soils, etc. Besides this, digital soil maps can directly or indirectly assist watershed management, drought management, and protecting ecosystem services. Future DSM attempts should focus on preparing digital soil maps at fine spatial resolutions. DSM efforts should also focus on higher sampling density, optimized sampling strategies, robust algorithms, uncertainty analysis, and data availability to end users.

22.8 Conclusion

Site-specific soil studies are essential for making the crop production system more sustainable in a long run. DSM is a potential tool for site-specific assessment, monitoring, and management of the precious soil resource. Digital soil maps are great decision-supporting assets for farmers, land managers, policy makers, and resource planners. In recent years, remotely sensed products have been extensively used for generating high-quality, fine-resolution digital soil maps. RS can directly measure a few of the soil and terrain attributes. Also, many remotely sensed products are being exclusively used as environmental covariates or predictors in DSM. ASTER, MODIS, Landsat 8, 9, ResourceSat, Sentinel-1, Sentinel-2, WorldView-2, SPOT-6,7, Cartosat, KOMPSAT-3, RADARSAT-2, RADARSAT Constellation, RISAT, and ALOS-2 are a few of the satellites, whose products have been effectively used in various soil and crop studies. DEMs prepared from CartoSat, ALOS, and SRTM have been engaged for extracting terrain/topography related information for DSM. Despite tremendous advancements, applicability of optical and thermal RS for soil studies is often restricted by cloud coverage and atmospheric distortions. To overcome these limitations, multispectral and hyperspectral sensors have gained more attention in recent times. Moreover, in this era of RS and big data, data acquisition is no longer an issue. But data choice and management are the current challenges for the RS data users. DSM can be well supported by the RS technology to further assist various aspects of the agricultural sector including but not limited to variable rate fertilizer applications, soil health monitoring, managing problematic soils, and land use planning. Further improvement in the existing DSM techniques is also required to produce high-quality soil maps for better soil and crop monitoring and management.

References

Bangroo, S., Najar, G., Achin, E., Truong, P.N., 2020. Application of predictor variables in spatial quantification of soil organic carbon and total nitrogen using regression kriging in the North Kashmir forest Himalayas. Catena 193, 104632.

Behera, S.K., Mathur, R.K., Shukla, A.K., Suresh, K., Prakash, C., 2018. Spatial variability of soil properties and delineation of soil management zones of oil palm plantations grown in a hot and humid tropical region of southern India. Catena 165, 251–259.

Chandrashekar, G., Sahin, F., 2014. A survey on feature selection methods. Computers & Electrical Engineering, 40th-Year Commemorative Issue 40, 16–28. Available from: https://doi.org/10.1016/j.compeleceng.2013.11.024

Chen, S., Arrouays, D., Mulder, V.L., Poggio, L., Minasny, B., Roudier, P., et al., 2022. Digital mapping of GlobalSoilMap soil properties at a broad scale: a review. Geoderma 409, 115567.

Chhipa, V., Stein, A., Shankar, H., Alidoost, F., 2019. Assessing and transferring soil health information in a hilly terrain. Geoderma 343, 130–138.

Choudhury, S.G., Banerjee, T., Das, K., Sahoo, A.K., Nayak, D.C., Singh, S.K., 2019. Soil resource characterization and classification under different toposequences in eastern extension of Chhotanagpur Plateau. J. Indian. Soc. Soil. Sci. 67, 1–11.

Dash, P.K., Mishra, A., Saren, S., 2019a. Characterization and taxonomic classification of soils under a toposequence located in Eastern India. Environ. Ecol. 37, 1240–1249.

Dash, P.K., Mishra, A., Saren, S., 2019b. Vertical distribution of available nutrients in an Eastern Indian Catena. Ann. Plant. Soil. Res. 21, 320–325.

Dash, P.K., Nayak, B., Mishra, A., Saren, S., 2019c. Geomorphological characterization of soils under a toposequence located in Eastern India. Environ. Ecol. 37, 1349–1359.

Dash, P.K., Panigrahi, N., Mishra, A., 2022. Identifying opportunities to improve digital soil mapping in India: a systematic review. Geoderma Reg. e00478.

Dharumarajan, S., Hegde, R., Singh, S., 2017. Spatial prediction of major soil properties using random forest techniques-a case study in semi-arid tropics of South India. Geoderma Reg. 10, 154–162.

Dharumarajan, S., Hegde, R., Janani, N., Singh, S., 2019. The need for digital soil mapping in India. Geoderma Reg. 16, e00204.

Ferhatoglu, C., Miller, B.A., 2022. Choosing feature selection methods for spatial modeling of soil fertility properties at the field scale. Agronomy 12, 1786.

Franzen, D.W., Peck, T.R., 1995. Field soil sampling density for variable rate fertilization. J. Prod. Agric. 8, 568–574.

Gerhards, M., Schlerf, M., Mallick, K., Udelhoven, T., 2019. Challenges and future perspectives of multi-/Hyperspectral thermal infrared remote sensing for crop water-stress detection: a review. Remote. Sens. 11, 1240.

Goovaerts, P., 1997. Geostatistics for Natural Resources Evaluation. Oxford University Press on Demand.

Grunwald, S., Thompson, J., Boettinger, J., 2011. Digital soil mapping and modeling at continental scales: finding solutions for global issues. Soil. Sci. Soc. Am. J. 75, 1201–1213.

Hartemink, A.E., 2015. On global soil science and regional solutions. Geoderma Reg. 5, 1–3.

Heung, B., Ho, H.C., Zhang, J., Knudby, A., Bulmer, C.E., Schmidt, M.G., 2016. An overview and comparison of machine-learning techniques for classification purposes in digital soil mapping. Geoderma 265, 62–77.

Iticha, B., Takele, C., 2019. Digital soil mapping for site-specific management of soils. Geoderma 351, 85–91.

Jenny, H., 1941. Factors of Soil Formation. McGraw-Hill Book Co, New York.

Kalambukattu, J.G., Kumar, S., Arya Raj, R., 2018. Digital soil mapping in a Himalayan watershed using remote sensing and terrain parameters employing artificial neural network model. Environ. Earth Sci. 77, 1–14.

Khaledian, Y., Miller, B.A., 2020. Selecting appropriate machine learning methods for digital soil mapping. Appl. Math. Model. 81, 401–418.

Krige, D., 1966. Two-dimensional weighted moving average trend surfaces for ore valuation. J. South. Afr. Inst. Min. Metall. 13–38.

Kumar, S., Singh, R.P., 2016. Spatial distribution of soil nutrients in a watershed of Himalayan landscape using terrain attributes and geostatistical methods. Environ. Earth Sci. 75, 1–11.

Kuriakose, S.L., Devkota, S., Rossiter, D., Jetten, V., 2009. Prediction of soil depth using environmental variables in an anthropogenic landscape, a case study in the Western Ghats of Kerala, India. Catena 79, 27–38.

Lagacherie, P., McBratney, A., 2006. Spatial soil information systems and spatial soil inference systems: perspectives for digital soil mapping. Dev. Soil. Sci. 31, 3–22.

Lamichhane, S., Kumar, L., Wilson, B., 2019. Digital soil mapping algorithms and covariates for soil organic carbon mapping and their implications: a review. Geoderma 352, 395–413.

Liu, W., Du, P., Zhao, Z., Zhang, L., 2016. An adaptive weighting algorithm for interpolating the soil potassium content. Sci. Rep. 6, 1–11.

McBratney, A.B., Santos, M.M., Minasny, B., 2003. On digital soil mapping. Geoderma 117, 3–52.

McBratney, A., Field, D.J., Koch, A., 2014. The dimensions of soil security. Geoderma 213, 203–213.

Miller, B.A., Koszinski, S., Wehrhan, M., Sommer, M., 2015. Impact of multi-scale predictor selection for modeling soil properties. Geoderma 239–240, 97–106. Available from: https://doi.org/10.1016/j.geoderma.2014.09.018.

Minasny, B., Hartemink, A.E., 2011. Predicting soil properties in the tropics. Earth-Sci. Rev. 106, 52–62.

Minasny, B., McBratney, A., Malone, B., Wheeler, I., 2013. Digital soil mapping of carbon. Adv. Agron. 118, 1–47.

Minasny, B., McBratney, A.B., 2016. Digital soil mapping: a brief history and some lessons. Geoderma 264, 301–311.

Mitran, T., Mishra, U., Lal, R., Ravisankar, T., Sreenivas, K., 2018. Spatial distribution of soil carbon stocks in a semi-arid region of India. Geoderma Reg. 15, e00192.

Mitran, T., Solanky, V., Janakirama Suresh, G., Sujatha, G., Sreenivas, K., Ravisankar, T., 2019. Predictive mapping of surface soil texture in a semi-arid region of India through geostatistical modeling. Model. Earth Syst. Environ. 5, 645–657.

Mitran, T., Meena, R.S., Chakraborty, A., 2021. Geospatial technologies for crops and soils: an overview. Geospatial Technologies for Crops and Soils 1–48.

Mulder, V.L., de Bruin, S., Schaepman, M.E., Mayr, T.R., 2011. The use of remote sensing in soil and terrain mapping—a review. Geoderma 162, 1–19. Available from: https://doi.org/10.1016/j.geoderma.2010.12.018.

Mulla, D.J., 2013. Twenty five years of remote sensing in precision agriculture: key advances and remaining knowledge gaps. Biosyst. Eng. 114, 358–371.

Okonkwo, E.I., Corstanje, R., Kirk, G.J., 2018. Digital soil assessment for quantifying soil constraints to crop production: a case study for rice in Punjab, India. Soil. Use Manag. 34, 533–541.

Oliver, M., 1987. Geostatistics and its application to soil science. Soil. Use Manag. 3, 8–20.

Oliver, M., Webster, R., 2014. A tutorial guide to geostatistics: computing and modelling variograms and kriging. Catena 113, 56–69.

Oliver, M.A., Webster, R., 2015. Basic Steps in Geostatistics: The Variogram and Kriging. Springer.

Pareeth, S., Karimi, P., Shafiei, M., De Fraiture, C., 2019. Mapping agricultural landuse patterns from time series of Landsat 8 using random forest based hierarchial approach. Remote. Sens. 11, 601.

Poggio, L., Sousa, De, Batjes, L.M., Heuvelink, N.H., Kempen, G., Ribeiro, B., et al., 2021. SoilGrids 2.0: producing soil information for the globe with quantified spatial uncertainty. Soil 7, 217–240.

Reddy, N.N., Chakraborty, P., Roy, S., Singh, K., Minasny, B., McBratney, A.B., et al., 2021. Legacy data-based national-scale digital mapping of key soil properties in India. Geoderma 381, 114684.

Rosemary, F., Indraratne, S., Weerasooriya, R., Mishra, U., 2017. Exploring the spatial variability of soil properties in an Alfisol soil catena. Catena 150, 53–61.

Saby, N., Arrouays, D., Boulonne, L., Jolivet, C., Pochot, A., 2006. Geostatistical assessment of Pb in soil around Paris, France. Sci. Total. Environ. 367, 212–221.

Sahrawat, K., Wani, S., Pardhasaradhi, G., Murthy, K., 2010. Diagnosis of secondary and micronutrient deficiencies and their management in rainfed agroecosystems: case study from Indian semi-arid tropics. Commun. Soil. Sci. Plant. Anal. 41, 346–360.

Santra, P., Kumar, M., Panwar, N., 2017. Digital soil mapping of sand content in arid western India through geostatistical approaches. Geoderma Reg. 9, 56–72.

Santra, P., Kumar, M., Panwar, N., Yadav, R., 2021. Digital soil mapping: the future need of sustainable soil management. Geospatial Technologies for Crops and Soils, Springer Nature Singapore Pte Limited. 319–355.

Sarkar, S., Roy, A.K., Martha, T.R., 2013. Soil depth estimation through soil-landscape modelling using regression kriging in a Himalayan terrain. Int. J. Geogr. Inf. Sci. 27, 2436–2454.

Sarmah, T., Dutta, S., Karmakar, R., Tamuly, D., Dutta, N., 2019. Characterization and classification of some alluvium-derived rice and associated non-rice soils of Jorhat district of Assam. J. Indian. Soc. Soil. Sci. 67, 379–388.

Saurette, D.D., Berg, A.A., Laamrani, A., Heck, R.J., Gillespie, A.W., Voroney, P., et al., 2022. Effects of sample size and covariate resolution on field-scale predictive digital mapping of soil carbon. Geoderma 425, 116054. Available from: https://doi.org/10.1016/j.geoderma.2022.116054.

Shukla, A.K., Behera, S.K., Lenka, N.K., Tiwari, P.K., Prakash, C., Malik, R., et al., 2016. Spatial variability of soil micronutrients in the intensively cultivated Trans-Gangetic Plains of India. Soil. Tillage Res. 163, 282–289.

Shukla, G., Garg, R.D., Srivastava, H.S., Garg, P.K., 2018. An effective implementation and assessment of a random forest classifier as a soil spatial predictive model. Int. J. Remote. Sens. 39, 2637–2669.

Singh, A., Rahman, B., Thakur, R., 2019. Assessment of land suitability and irrigation potential for improving cropping intensity and ensuring food security in North East India. J. Indian Soc. Soil. Sci. 67, 281–290.

Sreenivas, K., Sujatha, G., Sudhir, K., Kiran, D.V., Fyzee, M., Ravisankar, T., et al., 2014. Spatial assessment of soil organic carbon density through random forests based imputation. J. Indian Soc. Remote. Sens. 42, 577–587.

Sreenivas, K., Dadhwal, V., Kumar, S., Harsha, G.S., Mitran, T., Sujatha, G., et al., 2016. Digital mapping of soil organic and inorganic carbon status in India. Geoderma 269, 160–173.

Umali, B.P., Oliver, D.P., Forrester, S., Chittleborough, D.J., Hutson, J.L., Kookana, R.S., et al., 2012. The effect of terrain and management on the spatial variability of soil properties in an apple orchard. Catena 93, 38–48.

Vedadri, U., Naidu, M., 2018. Characterization, classification and evaluation of soils in semi-arid ecosystem of Chillakur mandal in SPSR Nellore district of Andhra Pradesh. J. Indian. Soc. Soil. Sci. 66, 1–19.

Wadoux, A.M.-C., Brus, D.J., Heuvelink, G.B., 2019. Sampling design optimization for soil mapping with random forest. Geoderma 355, 113913.

Wadoux, A.M.-C., Minasny, B., McBratney, A.B., 2020. Machine learning for digital soil mapping: applications, challenges and suggested solutions. Earth-Sci. Rev. 210, 103359.

Walkley, A., Black, I.A., 1934. An examination of the Degtjareff method for determining soil organic matter, and a proposed modification of the chromic acid titration method. Soil. Sci. 37, 29–38.

Wang, J., Zhao, X., Zhao, D., Triantafilis, J., 2021. Selecting optimal calibration samples using proximal sensing EM induction and γ-ray spectrometry data: an application to managing lime and magnesium in sugarcane growing soil. J. Environ. Manage 296, 113357. Available from: https://doi.org/10.1016/j.jenvman.2021.113357.

Xu, Y., Smith, S.E., Grunwald, S., Abd-Elrahman, A., Wani, S.P., 2017. Evaluating the effect of remote sensing image spatial resolution on soil exchangeable potassium prediction models in smallholder farm settings. J. Environ. Manag. 200, 423–433.

Zeraatpisheh, M., Jafari, A., Bodaghabadi, M.B., Ayoubi, S., Taghizadeh-Mehrjardi, R., Toomanian, N., et al., 2020. Conventional and digital soil mapping in Iran: past, present, and future. Catena 188, 104424.

Zhang, G., Feng, L., Song, X., 2017. Recent progress and future prospect of digital soil mapping: a review. J. Integr. Agric. 16, 2871–2885.

Section 4

Chapter 23

Morphometric analysis of Jajmau watershed in Kanpur, India using remote sensing and geographical information system

Nisha Sahu, Jayant Kumar Saha, N.K. Sinha, Narayan Lal, Asha Sahu and Ashok Kumar Patra
ICAR-Indian Institute of Soil Science, Bhopal, Madhya Pradesh, India

23.1 Introduction

The water demand is increasing with the increase of the growing population worldwide. This lead affects both surface and subsurface supplies, resulting in the exploitation of good quality water. Furthermore, the distribution of water is not even over the region. It creates a severe water scarcity in many areas and water surplus in others. It indicates that the reservoirs need to be carefully managed. In order to do so, decision-makers should know about the basin's hydrological characteristics through morphometric properties. Morphometric analysis is the science of measurement and mathematical presentation of the earth's shape, size, surface, and dimensions of its landform (Reddy et al., 2002). The morphometric analysis is performed by measuring all three aspects: linear, areal, and relief (gradient) of the watershed area (Ali et al., 2016). It affects both geological dynamics and geomorphological mechanisms. The information obtained from morphometric analysis of watersheds is thus a crucial tool for water resource management. Each watershed defined with a single outlet is distinct regarding rainfall, slope, geology, soil type, and land use pattern. The quantitative analysis of morphometric properties of a watershed by using various formulae may help evaluate tectonic activities in any area (Pavano et al., 2016). It provides information about the formation of various processes occurring on the land's surface, which can be better reflected by measuring linear, aerial, and relief aspects (Banerjee et al., 2015). Hence, there is involvement of a large set of parameters, and it becomes difficult sometimes to identify the parameters which play an important role. According to Strahler's classification system, the stream with no tributaries is designated as first order, the stream formed by joining two first-orders is designated as second-order, and so on. For instance, drainage density positively correlates with flood behavior and erosion rate (Gregory and Walling, 1968).

However, most earlier studies were manually traced and used measurements and calculations based on topographic sheets, which were tedious and time-consuming. Recently, a new technique based on the integration of remote sensing (RS) and geographic information systems (GIS) has been developed (Kant et al., 2015). With the advancement of RS technologies in the GIS environment, the analysis has become much easier and quicker for large areas. This advanced approach can improve the accuracy of the analysis and save time. Satellite and aerial photographs can help prepare thematic layers in morphometric analysis.

On the other hand, extraction of drainage from the digital elevation model (DEM) is quite easy. The IS model builder using geoprocessing tools (ArcGIS 10.4) with thematic layers such as aspect, slope, hill shade, relief, and drainage density generates a precise drainage network. The results extracted from the model can be used for morphometric analysis. Many researchers have conducted morphometric analysis to evaluate basin morphometry using RS and GIS techniques (Chopra et al., 2005 and Magesh and Chandrasekar, 2012).

Recently, many researchers have identified that RS and GIS are important tools for analyzing drainage networks using different data sets (Pankaj and Kumar, 2009). These tools can manage and process spatial information in huge amounts more efficiently and accurately. Hydrologists and geomorphologists must address some natural hazards like soil erosion, flood, landslide,

Remote Sensing of Soils. DOI: https://doi.org/10.1016/B978-0-443-18773-5.00016-8

extreme surface runoff, etc. This new approach can improve the results' accuracy and save time. The main advantage of remote sensed images is the synoptic view of a large area with distinct topography. The present study employed the same concept to understand the geometry of the Jajmau basin for the conservation and management of water resources in a sustainable manner.

23.2 Materials and methods

23.2.1 Study area

The study area, Jajmau (26.38 to 26.41 N latitudes and 80.42 to 80.47 E longitudes), is located in one of the industrial areas of Kanpur, Uttar Pradesh, India (Fig. 23.1). Kanpur is a tanneries hub in India as well as in World comprises of small and large scale industries. It is situated on the left bank of the Ganga River and the right bank of its tributary Panku River. Jajmau is a severely chromic-polluted area. About 400 tanneries are located at Jajmau. For over three decades, treated and untreated wastewater has been used for irrigation in agricultural lands comprising nearly 10 villages. The elevation of this region is almost leveled with a slope percentage between 1 and 3. The average annual optimum temperature ranges from 22°C to 25°C, and two types of monsoons such as southwest (June to September) and northeast (October to December) prevail in this region. Jajmau experiences extremes of temperature, with a minimum of 0°C in the winters between December and February, while it goes up to 48°C in summers between May and June. The climate is subtropical and humid, with an average annual rainfall of 800 mm. The monsoon rainfall is the main source for storage sectors and occurs maximum during southwest monsoon periods.

23.2.2 Analysis

The data employed is the 30 m resolution Cartosat-1 DEM obtained from the Bhuvan website. These data are freely available and used by my researchers for morphometric analysis. The Cartosat-DEM was imported in ArcMap 10.4, and the study area was clipped using Arctool Box. The hydrology tools were used to fill DEM and generate flow

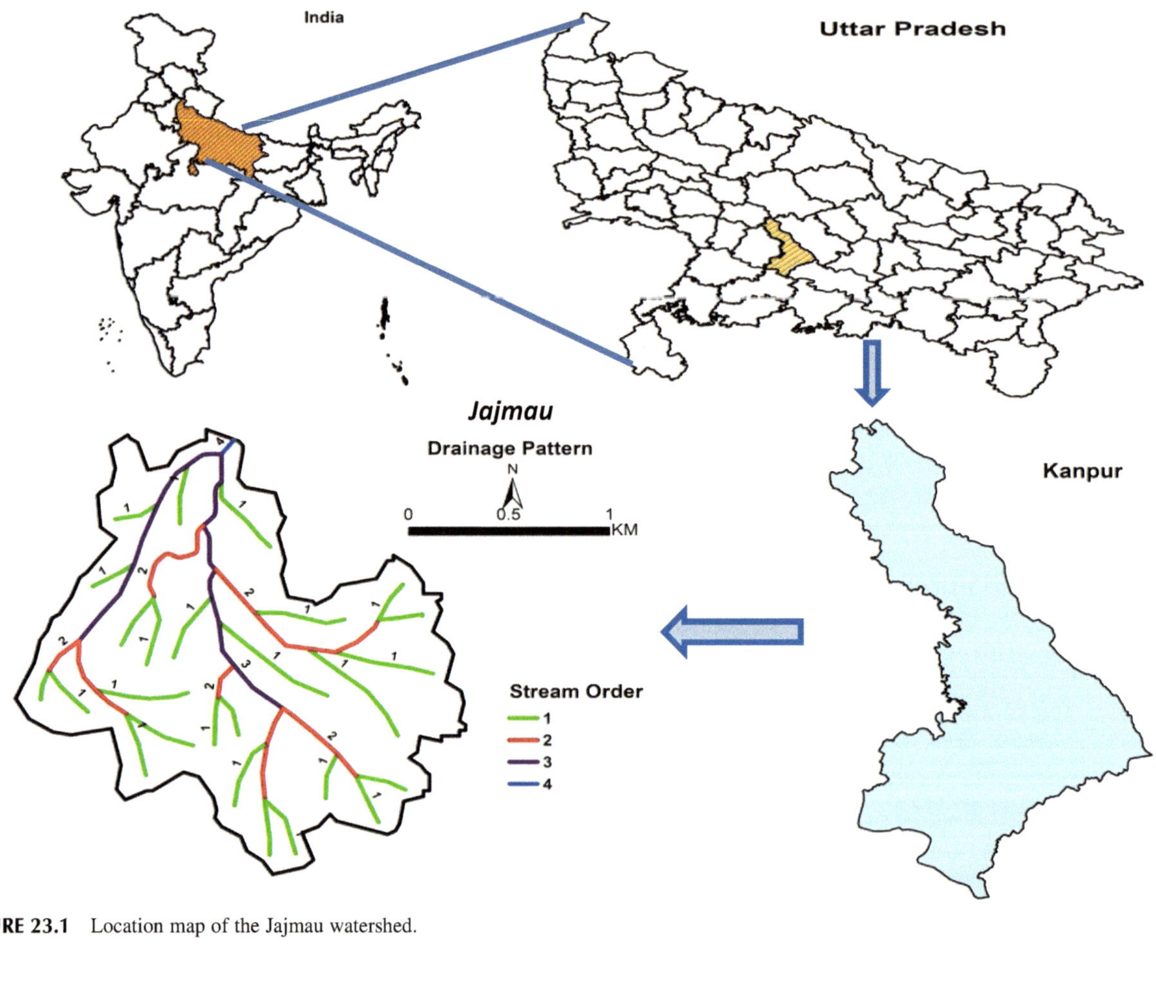

FIGURE 23.1 Location map of the Jajmau watershed.

direction, flow accumulation, conditional, stream order, stream to feature for converting raster streams to vector streams, basin, raster to polygon for converting raster basins to vector basins. The stream orders were extracted from Cartosat-1 DEM and updated following a classification by Strahler (1952). Similarly, the stream length is also calculated in the GIS platform. Apart from these, Carto-DEM was processed to delineate the watershed based on the pour point. The results obtained with DEM data were projected onto the WGS 84 UTM system. The morphometric parameters analyzed were grouped into three categories related to their orientation in space. The linear morphometric aspects involve several one-dimensional (X-axis) stream properties, and aerial aspects affect two-dimensional (X and Y axis), whereas relief aspects consider all three-dimensional parameters. They are linear, aerial, and relief aspects based on the formulae shown in Table 23.1, and the methodology followed for the study is shown in Fig. 23.2. The aspect and flow accumulation of the Jajmau basin is shown in Fig. 23.3. Soil pH and Soil EC maps are shown in Fig. 23.4.

TABLE 23.1 Formulae adopted for computation of morphometric parameters.

S. no.	Morphometric parameters	Formula/relationship	Reference
Linear aspects			
1.	Stream order (U)	Hierachical rank	Strahler (1964)
2.	Stream length (Lu)	Length of stream	Horton (1945)
3.	Mean stream length (Lsm)	Lsm = Lu/Nu, where, Lu = total stream length of order "u," Nu = total no. of stream segments of order "u"	Strahler (1964)
4.	Stream length ratio (RL)	RL = Lu/Lu − 1, where, Lu = total stream length of order "u," Lu − 1 = the total stream length of its next lower order	Horton (1945)
5.	Bifurcation ratio (Rb)	Rb = Nu/Nu + 1, where Nu = total no. of stream segments of order "u," Nu + 1 = No. of stream segments of the next higher order	Schumm (1956)
6.	Mean bifurcation ratio (Rbm)	Rbm = average of the bifurcation ratio of all order	Strahler (1957)
Aerial aspects			
7.	Drainage texture (Dt)	Dt = Nu/P, where Nu is the total no. of streams of all order, P is the perimeter of the basin in km	Horton (1945)
8.	Texture ratio (Rt)	Rt = N1/P, where N1is the total no. of first order streams	Horton (1932)
9.	Drainage density (D)	D = Lu/A, where Lu is the total stream length of all orders, A is the total area of the basin in km^2	Horton (1932)
10.	Stream frequency (Fs)	Fs = Nu/A, where Nu is the total number of streams of all order, A is basin area in km^2	Horton (1932)
11.	Form factor (Ff)	$Ff = A/Lb^2$ where A is basin area in km^2, Lb^2 is the square of the basin length in km	Horton (1932)
12.	Circulatory ratio (Rc)	$Rc = 4\pi A/P^2$ where A is basin area in km^2, P is the perimeter of the basin in km	Miller (1953)
13.	Elongation ratio (Re)	$Re = 2sqrt(A/\pi)/Lb$, A is basin area in km^2, Lb is basin length	Schumm (1956)
14.	Length of overland flow (Lg)	$Lg = 1/(D \times 2)$, where D is the drainage density	Horton (1945)
15.	Constant channel maintenance (C)	Inverse of drainage density	Schumm (1956)
16.	Shape index (S)	Reciprocal of form factor	Horton (1932)
Relief aspects			
17.	Relief (H)	(Elevation of basin mouth) − (Elevation of highest point on the basin)	Schumm (1956)
18.	Relief ratio (Rr)	Rr = H/Lb where H is relief, Lb is basin length	Schumm (1956)
19.	Relative Relief (Rh)	Rh = H/P where H is relief, P is perimeter	Melton (1957)
20.	Ruggedness number (Rn)	$Rn = H \times D$, where H is relief, D is drainage density	Strahler (1964)

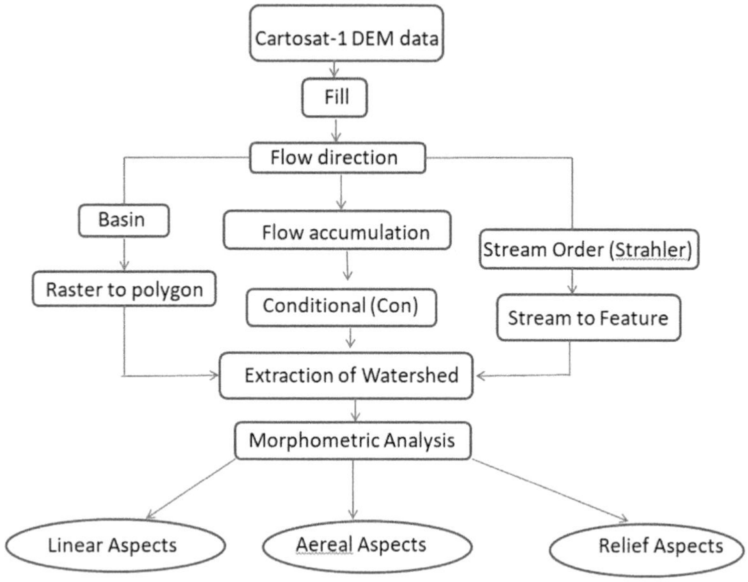

FIGURE 23.2 Methodology for morphometric analysis.

23.3 Results and discussion

23.3.1 Linear aspects

The linear aspects of a watershed are mainly related to the characteristics of its drainage network, which in turn are influenced by the local topography. The determination of stream order is the first step in the morphometric analysis. The researchers consider several ordering methods of drainage (Gravelius, 1914; Horton, 1945; Strahler, 1957; Scheidegger, 1965; Shreve, 1967). The classification based on Strahler for stream ordering is adopted in this study. The Jajmau watershed is of fourth order. The total stream number is 51, while the number of streams of each order is shown in Tables 23.2 and 23.3. The stream number is obtained using the method followed by Horton (1945) "law of stream numbers," which shows an inverse relationship between the stream order and stream number. This indicates that the total number of streams decreases with the increase in stream order. Finally, the water volume capacity of the basin is controlled by the basin order. They provide a good indicator of permeability and infiltration of the study area.

Stream length is one of the most important parameters in morphometric analysis and indicates the surface runoff characteristics. The stream length decreases with the increase in stream order, following Horton's (1945) second law of stream length (Fig. 23.5). Mean stream length reveals the property related to the size of drainage network components and its contributing drainage basin surfaces (Khare et al., 2014). In general, the mean stream length of a channel is greater than that of the lower order. The value varied from 349.78 to 96.05 meters. The stream length ratio varied from 0.03 to 0.79. This indicates the presence of resistant rocks, nearly leveled topography, and low slope of the basin. These results are significant in investigating basin characteristics such as the drainage pattern. These parameters are closely related to the drainage network and its associated surfaces (Strahler, 1957). The bifurcation ratio is the ratio of a number of stream segments of a given order to the number of segments of the next higher order. It is also an indicator of the shape of the basin. The value ranges between 1.18 and 11, with a mean bifurcation ratio of 4.72. An elongated basin has a higher value compared to a circular basin. Many researchers indicate that the mean bifurcation ratio of watersheds without differential geological control ranges between three and five (Kant et al., 2015). In the case of the Jajmau basin catchment area, the mean stream length is 349.78 and 96.05 meters for the first and fourth orders, respectively. The cumulative stream length for the study basin is 987.75 meters.

23.3.2 Aerial aspects

The aerial morphometric parameters are relative to geological structures, climatic conditions, and catchment denudation history. The values of the aerial morphometric parameters for the Jajmau watershed are shown in Table 23.4. The total area of the drainage basin is 300.43 ha. The length of the basin is 2126.52 m, and the basin perimeter, which is the outer boundary of the drainage basin, is 9459 m using ArcGIS software. The value of area, length, and perimeter indicates that the slope of the basin is low with level to nearly level topography. Drainage texture and drainage density are

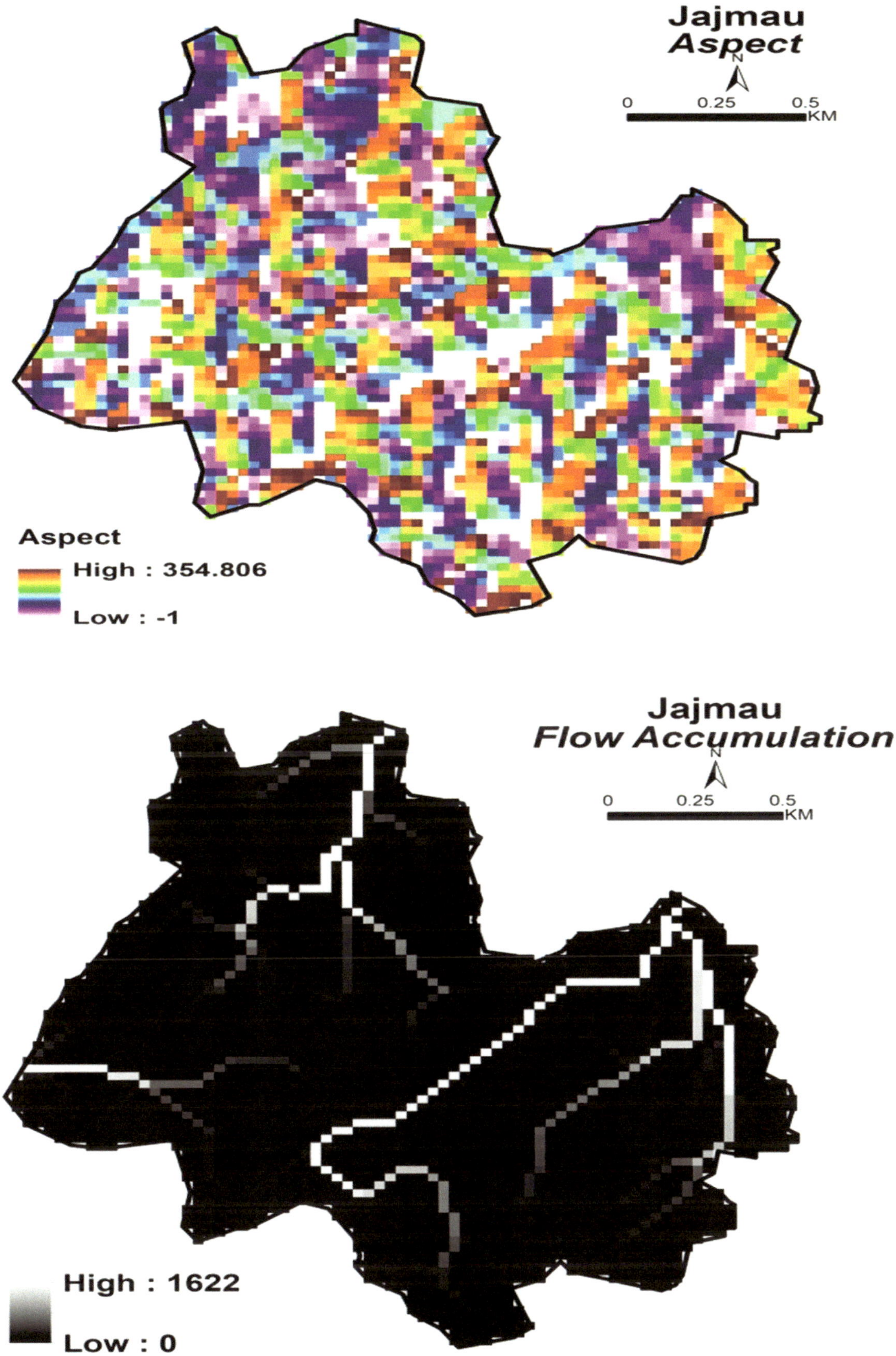

FIGURE 23.3 Aspect and flow accumulation of Jajmau catchment area.

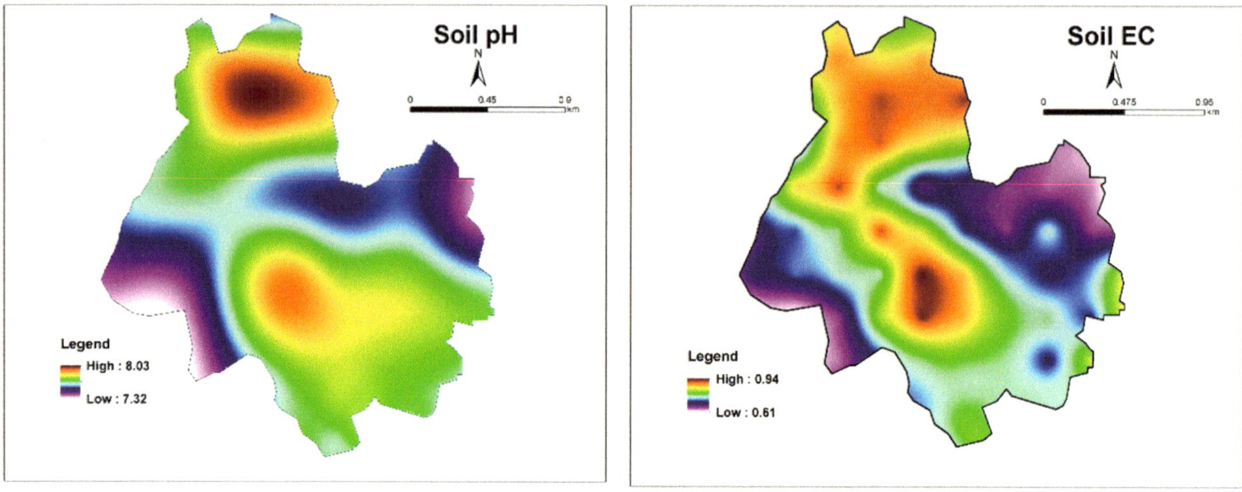

FIGURE 23.4 Soil pH and Soil EC of the study area.

TABLE 23.2 Stream order, stream number, and bifurcation ratio of the study area.

Stream order	Stream number	Bifurcation ratio
1	26	2
2	13	1.18
3	11	11
4	1	–
Total	51	Avg. 4.72

TABLE 23.3 Linear aspects of the drainage network of the study area.

Stream order	Stream length (m)	Mean stream length (m)	Cumulative stream length (m)	Stream length ratio
1	9094.34	349.78	349.78	-
2	3635	279.61	629.39	0.39
3	2885.42	262.31	891.70	0.79
4	96.05	96.05	987.75	0.03
Total	15710.81	987.75		1.21

important parameters influenced by terrain and soil of the basin. According to Smith (1954), drainage texture was classified into five classes, that is, very course (<2), course (2–4), moderate (4–6), fine (6–8), and very fine (>8). This watershed has a value of 5.39, which falls under the moderate category. The values also indicate that the basin has good permeability and infiltration capacity, low runoff with maximum groundwater recharge, and old mature alluvium topography. Basin has high drainage density, that is 5.23, indicating high permeability and low relief. Generally, drainage density increases with decreasing infiltration capacity of the underlying rocks or decreasing soil transmissivity (Montgomery and Dietrich, 1992). It also depends on the type of rocks. Another important parameter is the texture ratio, which is 2.75, and the value of stream frequency is 16.97. Stream frequency depends on the lithology of the basin.

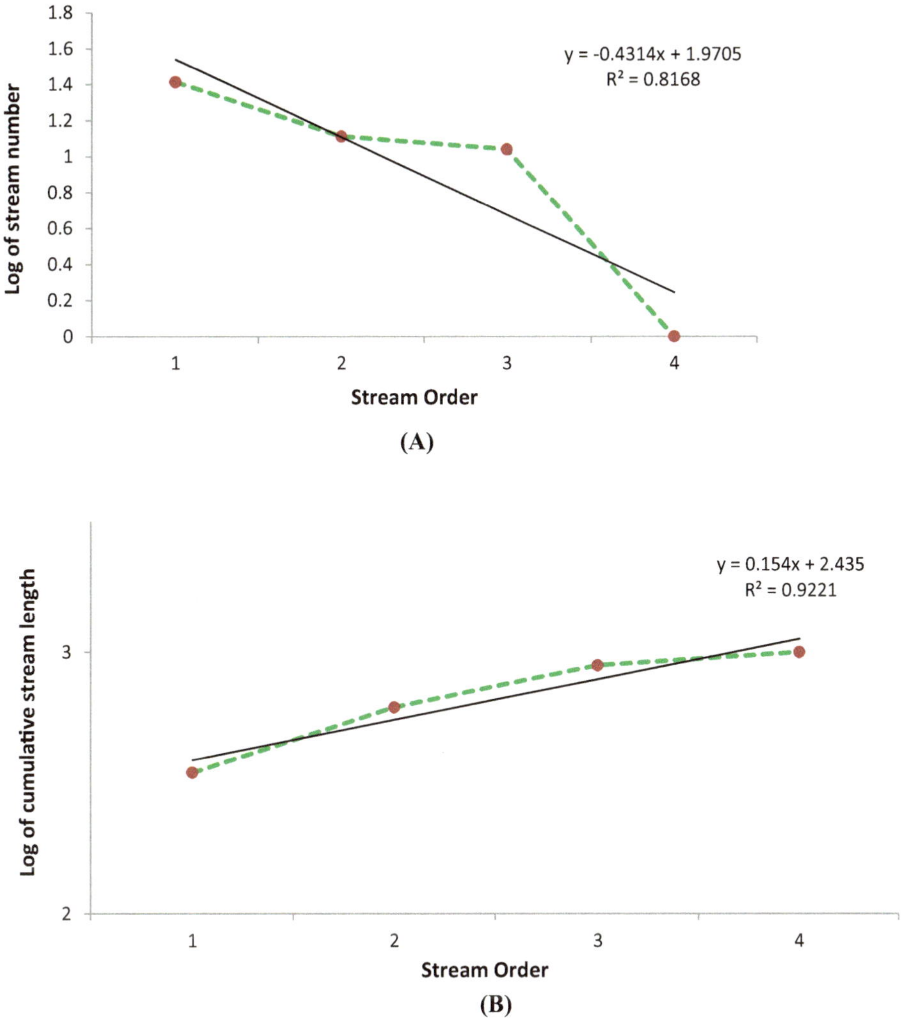

FIGURE 23.5 (A) Regression of logarithm of stream number and stream order (B) Regression of logarithm of cumulative stream length and stream order.

High stream frequency indicates low relief, slow runoff, and flooding is less likely to occur in such areas. Form factor was calculated by the ratio of the basin area to the square of the basin length, and it also indicates the shape of the watershed (Schumm, 1956). The form factor value of the basin is 0.66, suggesting the shape is elongated with a flatter peak for a long duration. Miller (1953) defined a dimensionless circularity ratio as the ratio of the basin area to the area of a circle having the same perimeter as the basin. It depends on land use/land cover, climate, relief, and slope of the basin. The value is 0.414 of the basin, which indicates that the basin is strongly elongated with low discharge of runoff and high permeability of the soil. The elongation ratio is the ratio of the diameter of the circle, which represents the area of the basin to its length (Schumm, 1956). The value is 1.33, which indicates low relief and a nearly leveled slope. Mustafa and Yusuf (1999) concluded that the flow of water in elongated shape basins is distributed over a longer period than in circular ones. Therefore, this basin shape is less prone to flood and lower erosion.

Length of overland flow is one of the most important variables affecting basin development. The water flows to the ground before it gets concentrated into a specific stream channel and is significantly affected by infiltration of the soil. The value of length of overland flow is 0.09. The constant of channel maintenance of the basin is 0.19, which indicates low runoff. The shape index value of the basin is 1.51.

23.3.3 Relief aspects

The relief aspects of the basin are related to three-dimensional features, that is, area, volume, and altitude of landforms. The important relief parameters have been analyzed using standard formulae, and the results are presented in Table 23.4. Basin relief is an important parameter in understanding the denudational characteristics of the basin and is

TABLE 23.4 Aerial and Relief aspects of the study area.

Morphometric characteristics	Estimated values
Aerial	
Area	300.43 ha
Perimeter	9459 m
Length of basin	2126.52 m
Drainage texture	5.39
Texture ratio	2.75
Drainage density	5.23
Stream frequency	16.97
Form factor	0.66
Circulatory ratio	0.414
Elongation ratio	1.33
Length of overland flow	0.09
Constant of channel maintenance	0.19
Shape index	1.51
Relief	
Relief	0.025 km
Relief ratio	0.012
Relative relief	0.003
Ruggedness number	0.13

defined as the difference between the highest and lowest elevation of the watershed. The value of basin relief is 0.025 km. The relief ratio increases with a decrease in drainage area. It indicates the intensity of erosion and overall steepness of a drainage basin. The relief ratio of the basin is 0.012, which is quite low and indicates the presence of a gentle slope.

23.4 Conclusions

Morphometric analysis helps policy and decision makers to understand various characteristics of a catchment area such as basin shape, size, runoff, slope, and infiltration capacity. The application of RS and GIS based on DEM has been very useful in precisely estimating the catchment area's linear, areal, and relief aspects. Based on the current analysis, the highest stream order present in the catchment is of the fourth order.

Conflict of interests/competing interests

There was no conflict of interest.

References

Ali, S.A., Alhamed, M., Ali, U., 2016. Morphometric analysis of Abdan Basin, Almahfid Basement Rock, Yemen: using remote sensing and GIS. Int. J. Adv. Remote Sens. GIS. pp. 1605–1617.

Banerjee, A., Singh, P., Pratap, K., 2015. Morphometric evaluation of Swarnrekha watershed, Madhya Pradesh, India: an integrated GIS based approach. Appl. Water Sci. Available from: https://doi.org/10.1007/s13201-015-0354-3.

Chopra, R., Dhiman, R.D., Sharma, P.K., 2005. Morphometric analysis of sub-watersheds in Gurdaspur district, Punjab using Remote sensing and GIS techniques. J. Indian Soc. Remote. Sens. 33 (4), 531–539.
Gravelius, H., 1914. Flusskunde, Goschen'sche Verlagshandlung. GJ göschen, Berlin.
Gregory, K., Walling, D., 1968. The variation of drainage density within a catchment. Hydrol. Sci. J. 13, 61–68.
Horton, R.E., 1945. Erosional development of streams and their drainage basins; hydrophysical approach to quantitative morphology. Geol. Soc. Am. Bull. 56, 275–370.
Kant, S., Meshram, S., Dohare, R., Singh, S., 2015. Morphometric analysis of Sonar sub-basin using SRTM data and geographical information system (GIS). Afr. J. Agric. Res. 10, 1401–1406.
Khare, D., Mondal, A., Mishra, P.K., Kundu, S., Meena, P.K., 2014. Morphometric analysis for prioritization using remote sensing and GIS techniques in a hilly catchment in the State of Uttrakhand, India. Indian. J. Sci. Tech. 7 (10), 1650–1662.
Magesh, N.S., Chandrasekar, N., 2012. GIS model-based morphometric evaluation of Tamiraparani subbasin, Tirunelveli district, Tamil Nadu, India. Arab. J. Geosci. Available from: https://doi.org/10.1007/s12517-012-0742-z.
Melton, M.A., 1957. An Analysis of the Relations Among Elements of Climate, Surface Properties, and Geomorphology. Columbia University, New York.
Miller, V.C. 1953. A quantitative geomorphic study of drainage basin characteristics in the Clinch Mountain area, Virginia and Tennessee. Project NR 389042, Tech. Rept. 3. Columbia.
Montgomery, R., Dietrich, W.E., 1992. Channel initiation and the problem of landscape scale. Science 255, 826–830.
Mustafa, S., Yusuf, M.I., 1999. A Textbook of Hydrology and Water Resources, first ed. Jenas Prints and Publishing Company, Abuja, pp. 60–75.
Pankaj, A., Kumar, P., 2009. GIS-based morphometric analysis of five major sub-watersheds of Song River, Dehradun district, Uttarakhand with special reference to landslide incidences. J. Indian. Soc. Remote. 37 (1), 157–166.
Pavano, F., Pazzaglia, F.J., Catalano, S., 2016. Knickpoints as geomorphic markers of active tectonics: a case study from northeastern Sicily (southern Italy). Lithosphere-US 8 (6), 633–648. Available from: https://doi.org/10.1130/L577.
Reddy, O.G.E., Maji, A.K., Gajbhiye, K.S., 2002. GIS for morphometric analysis of drainage basins. GIS India 4, 9–14.
Scheidegger, A.E., 1965. The algebra of stream-order numbers. US Geol. Surv. Prof. Pap. 525, 187–189.
Schumm, S.A., 1956. Evolution of drainage systems and slopes in badlands at Perth Amboy, New Jersey. Geol. Soc. Am. Bull. 67, 597–646.
Shreve, R.L., 1967. Infinite topologically random channel networks. J. Geol. 75, 178–186.
Smith, K.G., 1954. Standards for grading texture of erosional topography. Am. J. Sci. 248, 655–668.
Strahler, A.N., 1952. Dynamic basis of geomorphology. Geol. Soc. Am. Bull. 63, 923–938. University, Department of Geology, ONR, Geography Branch, N Y.
Strahler, A.N., 1957. Quantitative analysis of watershed geomorphology. Eos Trans. Am. Geophys. Union. 38, 913–920.

Chapter 24

Prediction of topsoil organic carbon stocks in forest ecosystem of the Northeastern China using remote sensing data

Shuai Wang[1,2,3], K. Adhikari[4], Xinxin Jin[1], Zhenxing Bian[1], Fengkui Qian[1], Chunlan Han[1], Yan Wang[5], Chongwei Gan[1], Zicheng Wang[1] and Xingyu Zhang[1]

[1]*College of Land and Environment, Shenyang Agricultural University, Shenyang, Liaoning Province, China,* [2]*Institute of Bio- and Geosciences, Agrosphere (IBG-3), Forschungszentrum Jülich GmbH, Juelich, Germany,* [3]*Key Laboratory of Ecosystem Network Observation and Modeling, Institute of Geographic Sciences and Natural Resources Research, Chinese Academy of Sciences, Beijing, China,* [4]*USDA-ARS, Grassland, Soil and Water Research Laboratory, Temple, TX, United States,* [5]*College of Tourism and Geography, Jiujiang University, Jiujiang, China*

24.1 Introduction

The forest ecosystem is the largest ecosystem on land and the largest pool of soil organic carbon (SOC), accounting for 56% of the total terrestrial carbon pool (Don et al., 2011). It maintains 86% of the global vegetation carbon pool and 73% of the global soil carbon pool (He et al., 2017). Therefore, understanding the role of forest ecosystems in the carbon cycle is a foundation for studying the carbon cycle of the land-air system and even the global carbon cycle (Martin et al., 2011). Forest ecosystems are more productive than other terrestrial ecosystems and are the largest contributors to the primary productivity of the biosphere (Litton et al., 2007).

Accurately simulating the spatial distribution pattern of SOC stocks and clarifying the main environmental variables in the region is of great significance for the sustainable utilization and effective management of forest resources. The spatial variation of SOC stocks is the result of the comprehensive effect of many factors, such as nature and anthropogenic factors (Adhikari et al., 2020). It is known that obtaining regional SOC information by intensive sampling is rather unrealistic due to the high cost and time required (Gomes et al., 2019; Wang et al., 2020a). Therefore, there is always a challenge to accurately predict and map SOC stock variations in large areas. Recent development, especially in the field of geographic information system (GIS), Global Positioning System (GPS), and remote sensing (RS)—a "3 S" technology—has made it possible to rapidly collect and process high-resolution environmental data for multiple agri-environmental applications (Croft et al., 2012). The "3 S" technology has been widely applied in predicting global and regional SOC stocks' spatial distribution and identifying potential changes and their dynamics under future land use and climate change scenarios (Gomez et al., 2008; Bhunia et al., 2019; Reyes Rojas et al., 2018). RS provides real-time high-resolution environmental data that can be combined with limited soil measurements to establish a mathematical relationship between measured properties (e.g., SOC) and its controlling factors for spatial prediction and mapping (Croft et al., 2012; Minasny et al., 2013; Zhang et al., 2016).

RS data is widely used in soil investigation, soil sampling, and soil property characterization because of its lower cost, faster speed, and higher accuracy (Rasel et al., 2017). Gomez et al. (2008) compared the performance of visible and near-infrared reflectance (vis-NIR) hyperspectral proximal signals and RS data in predicting SOC. Chen et al. (2000) analyzed the potential relationship of the R, G, and B bands with SOC to estimate topsoil SOC distribution. Bhunia et al. (2019) estimated the SOC stock using RS data and showed that SOC and NDVI values could explain nearly 71% of the SOC distribution variance. Although a number of prediction techniques have been used to estimate SOC stocks (Lamichhane et al., 2019), only a few studies have used tree-based algorithms such as boosted regression

Remote Sensing of Soils. DOI: https://doi.org/10.1016/B978-0-443-18773-5.00011-9

trees (BRT) (Wang et al., 2020a). Tree-based machine learning algorithms are widely used in spatial predictions (Elith et al., 2008; Colin et al., 2017) because of their advantages of high prediction accuracy, easy interpretation, and the ability to express the nonlinear relationship (Carslaw and Taylor, 2009; Elith and Leathwick, 2017).

The BRT model can improve prediction accuracy and stability through continuous random selection and self-learning. (Zhang et al., 2016). Different from the linear model, the BRT model does not need to consider the interaction between independent variables in the process of building the model. The contribution and response curves of output independent variables are relatively intuitive and easy to explain. Based on these advantages, the BRT model was widely used in ecology (Elith and Leathwick, 2017), marine science (Froeschke and Froeschke, 2011), environmental science (Carslaw and Taylor, 2009), RS (Colin et al., 2017), medicine (Zhang et al., 2019), and soil science (Wang et al., 2020a). However, there are only a few studies on using the BRT model combined with RS technology to predict SOC stocks. The specific research objectives were to: (1) establish the BRT model with three different combinations of environmental variables to predict SOC stocks; (2) explore the potentials of RS−related variables in the prediction of SOC stocks; and (3) evaluate the model's performance and applications.

24.2 Materials and methods

24.2.1 Study area

The northeast forest region is the largest natural forest region in China, which is mostly distributed across the Greater Khingan Range, Lesser Khingan Mountains, and Changbai Mountain in Northeast China (Liaoning Province, Jilin Province, and Heilongjiang Province) (119°01′ − 135°06′ E, 38°48′ − 53°55′ N) (Fig. 24.1). The elevation of the study

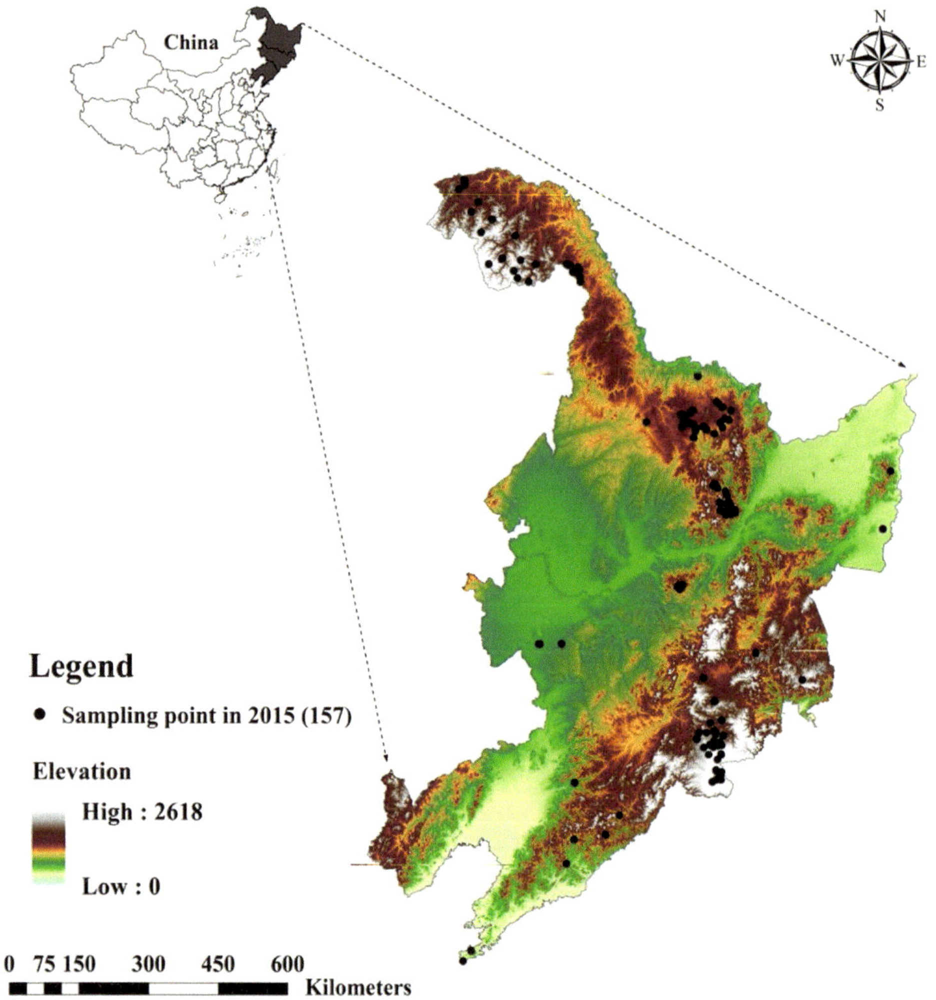

FIGURE 24.1 Location map of sampling points superimposed on a digital elevation model of the study area.

area ranges from 0 m in the southwest to 2665 m in the northeast. It is dominated by coniferous deciduous broad-leaved mixed forest in the middle temperate zone. According to the China Statistical Yearbook (China National Bureau of Statistics, 2015), there are about 305,000 km^2 of forests in the region, accounting for 37% of the total forest area in China. The timber volume reached 3.2 billion cubic meters, accounting for one third of the total timber volume in China. Among them, Heilongjiang Province has 2.4 billion cubic meters, accounting for a quarter of the national timber volume; 700 million cubic meters are in Jilin Province; and 100 million cubic meters are in Liaoning Province.

The northeast forest region crosses the middle temperate zone and the cold temperate zone from south to north. It belongs to a temperate monsoon climate with four distinct seasons (spring, summer, autumn, and winter), and is warm and rainy in summer and cold and dry in winter. The mean annual temperature (MAT) is about—20°C in winter, and more than 25°C in summer. From the southeast to the northwest, the mean annual precipitation (MAP) drops from 1000 mm to less than 300 mm and transitions from humid and semihumid areas to semiarid areas. The main soil types were Cambosols (82% of the study area), Gleyosols (7% of the study area), and Isohumosols (6% of the study area), and the rest were Argosols, Primosols, Anthosols, Histosols, Andosols, and Halosols (Chinese Soil Taxonomy Research Group, 2001),

24.2.2 Field sampling and laboratory analysis

Because of the large area it covers, intensive soil sampling in the study area was not feasible due to cost and time constraints. Therefore, we applied a purposeful sampling method (Zhu et al., 2008) to design the sampling scheme. Firstly, the environmental factors closely related to SOC in the topsoil, namely, normalized difference vegetation index (NDVI), MAP, MAT, topographic wetness index (TWI), and elevation, were selected and pretreated to unify the coordinate system and spatial resolution.

Secondly, the fuzzy c-means clustering method was used to cluster the selected environmental variables (NDVI, MAP, MAT, TWI, and elevation), and a total of 33 clusters or landscape units were finally obtained. Thirdly, local soil experts were invited to identify 3–5 typical soil sampling sites in each landscape unit, considering other factors such as road accessibility. At each sampling location (157 locations), 1 kg mixed soil samples and 100 cm^3 undisturbed soil cores were collected from 0–20 cm depth. The landforms and other environmental information were also recorded together with the longitude and latitude information using a handheld GPS. The analysis and determination of all soil samples were completed in the Key Laboratory of Agricultural Resources and Environment of Liaoning Province. The SOC content was determined by a CN element analyzer (2400 II CHNS/O Elemental Analyzer, Perkin-Elmer, USA). The undisturbed soil cores were placed in an oven at 105°C for 48 hours to measure the bulk density (BD) required to calculate SOC stocks.

24.2.3 Environmental variables

In this study, a total of 12 environmental variables from different platforms and institutions were collected, representing RS-related variables, topographic variables, and climatic variables, which were resampled to raster data with 30 m spatial resolution for subsequent modeling and analysis.

24.2.3.1 RS-related variables

The RS image data of growing seasons (July to September 2016) with cloud cover less than 10% covering the whole research area were downloaded from the Resources and Environment Science Data Center, Chinese Academy of Science (https://www.resdc.cn/). The images were preprocessed for solar radiation correction, atmospheric correction, geometric correction, terrain correction, and image registration (Wang et al., 2020a). In this study, a total of five RS-related variables were selected, namely soil-adjusted vegetation index (SAVI), normalized NDVI, near-infrared band (B_{NIR}), red band (B_{RED}), and green band (B_{GREEN}). NDVI could reflect the background influence of plant canopy, such as soil, wet ground, snow, dead leaves, roughness, etc. and is related to vegetation coverage. Huete (1988) proposed SAVI based on NDVI to reduce the impact of soil background. B_{NIR}, B_{RED}, and B_{GREEN} could represent the growth status and biomass of vegetation in the region and were widely used in the inversion of soil properties (Friedman et al., 2000). NDVI and SAVI were calculated as Eqs. (24.1) and (24.2):

$$NDVI = (B_{NIR} - B_{RED})/(B_{NIR} + B_{RED}) \tag{24.1}$$

$$SAVI = (B_{NIR} - B_{RED})(1 + L)/(B_{NIR} + B_{RED} + L) \; L = 0.5 \tag{24.2}$$

24.2.3.2 Topographic variables

Topographic variables were considered the most widely used environmental predictors in the spatial simulation of soil properties. In this study, five topographic attributes, namely elevation, slope aspect (SA), slope gradient (SG), TWI, and plan curvature (PC), were derived from the digital elevation model (DEM). Elevation, SA, SG, and PC were generated in ArcGIS 10.2 (Environmental Systems Research Institute, Redlands, CA, 2012), while TWI was derived in SAGA GIS software (Olaya, 2004). The DEM data was downloaded from the Resources and Environment Science Data Center, Chinese Academy of Science (https://www.resdc.cn/).

24.2.3.3 Climatic variables

Climate variables included 30-year MAP and MAT and were downloaded from the Resources and Environment Science Data Center, Chinese Academy of Science (https://www.resdc.cn/). The MAT and MAP raster data were generated from point measurements recorded at 673 meteorological stations in China for 30 years (calculated by monthly average value) using Kriging interpolation at 1000 m grid resolution. Then, in ArcGIS 10.2 software, the nearest neighbor principle was used to resample the climate data to 30 m resolution for subsequent modeling.

24.2.4 Prediction model

We used the BRT model to predict SOC stocks in the study area, the model training and subsequent analysis were done in the R environment (R Development Core Team, 2013). BRT is a self-learning method derived from the classification regression tree algorithm with boosting (Colin et al., 2017; Elith and Leathwick, 2017; Wang et al., 2020a). BRT can capture the marginal effect among variables (Carslaw and Taylor, 2009) to minimize potential impacts on the model due to changes in its size. For different data formats and complex data, the BRT model has strong flexibility and learning ability (Colin et al., 2017). In the process of model construction, the BRT model does not need to consider the interaction and correlation between independent variables, and has significant advantages in studying the interaction between environmental variables and soil attributes under complex environmental conditions (Zhang et al., 2019). Based on the above advantages, the BRT model has been successfully used in many scientific research fields, such as ecology science, environmental science, epidemiology, and soil science, among others (Wang et al., 2020a).

In the BRT model, four model parameters need to be defined: learning rate (LR), tree complexity (TC), bag fraction (BF), and number of trees (NT). LR represents the contribution of each tree to the final fitting model (Carslaw and Taylor, 2009). TC is the complexity of the tree, that is, the maximum interaction level between predictors (Elith et al., 2008). BF represents the proportion of data used in the data set—the more data the model uses, the less randomness it has (Colin et al., 2017). Although the BRT model can avoid overfitting by expanding the operation of the model, it is still necessary to set NT. The number of NT can be finally determined according to the combination of LR and TC (Elith and Leathwick, 2017). In order to obtain the best prediction performance of the model, a 10-fold cross-validation was used to test the best combination of parameter settings. Through multiple models, LR $(0.02-0.15)$, TC $(10-15)$, BF $(0.65-0.75)$, and NT $(8000-3000)$ were covered. Finally, LR, TC, BF, and NT were set to 0.03, 12, 0.75, and 1800, respectively. We constructed three different BRT models (Model A, Model B, and Model C) for prediction. Model A was a full-variable model that used all the variables as predictors of SOC stock; Model B used only topographic and climatic variables; and Model C only used RS-related variables.

24.2.5 Model evaluation

The predictive performance of the BRT model was evaluated using a 10-fold cross-validation method. Four statistical indices were calculated, and they were mean error (ME), root-mean-square error (RMSE), coefficient of determination (R^2), and Lin's concordance correlation coefficient (LCCC) (Lin, 1989) [Eqs. (24.3)–(24.6)]:

$$\mathrm{ME}=\frac{1}{n}\sum_{i=1}^{n}(x_i-y_i) \tag{24.3}$$

$$\mathrm{RMSE}=\sqrt{\frac{1}{n}\sum_{i=1}^{n}(x_i-y_i)^2} \tag{24.4}$$

$$R^2 = \frac{\sum_{i=1}^{n} (x_i - \overline{y}_i)^2}{\sum_{i=1}^{n} (y_i - \overline{y}_i)^2} \tag{24.5}$$

$$\text{LCCC} = \frac{2r\partial_x\partial_y}{\partial_x^2 + \partial_y^2 + (\overline{x} + \overline{y})^2} \tag{24.6}$$

where x_i, y_i, $\overline{x}$, $\overline{y}$,∂_x, and ∂_y represent the predicted value, observed value, the average of the predicted values, the average of the observed values, the variance of the predicted values, and the variance of the observed values at point i; r is the correlation coefficient; n is the number of data.

24.3 Results

24.3.1 Exploratory data analysis

Table 24.1 showed the statistical results between the SOC stocks and environmental variables at each sampling point. SOC stocks ranged from 2.99 kg m^{-2} to 17.44 kg m^{-2}, with an average of 8.85 kg m^{-2}. Since SOC stocks were unevenly distributed on both sides of the average values, we log-transformed SOC stocks for the analysis. The standard deviation (SD) of the SOC stock measurement was 2.95 kg m^{-2}, which was lower than the average value and indicated that it presents a stable SOC inventory variance.

In addition, the Pearson's correlation coefficient between SOC stocks and various environmental variables was calculated to further reveal their previous relationship. The results showed that there was a significant correlation between SOC stocks and RS-related variables. The correlation coefficients between SOC stocks, SAVI, NDVI B_{GREEN}, B_{RED}, and B_{NIR} were 0.49, 0.62, −0.53, −0.32, and −0.24, respectively. In addition, SOC stocks were significantly positively correlated with altitude, SG, and MAP, while TWI and MAT were significantly negatively correlated (Table 24.2).

24.3.2 Model performance and predicted maps

In this study, three BRT models with different combinations of variables were constructed to evaluate their performance in predicting topsoil SOC stocks in forest ecosystems. Among the three models, Model A (the full variable model), Model B (the topographic and climatic variables model), and Model C (the only RS-related variables model) could

TABLE 24.1 Summary statistics of soil samples at 0–20 cm and associated environmental variables.

Category	Property	Unit	Min.	Median	Mean	Max.	SD	Skewness	Kurtosis
Soil	SOC stocks	kg m^{-2}	2.99	8.28	8.85	17.44	2.95	0.34	−0.87
RS-related variables	B_{GREEN}	Digital number	0.00	43.87	73.85	226.17	65.31	0.66	−0.91
	B_{RED}	Digital number	2.18	163.78	167.43	267.14	69.34	−0.62	−0.87
	B_{NIR}	Digital number	1.37	42.18	68.49	224.36	65.18	0.89	−0.48
	NDVI		0.1	0.42	0.38	0.72	0.22	−0.42	1.75
	SAVI		0.14	0.46	0.43	0.8	0.26	−0.43	2.04
Topographic variables	Elevation	m	4.8	53.64	64.98	278.95	48.31	1.57	3.67
	SG	Degree	0	6.96	9.88	33.35	8.4	0.99	0.39
	SA	Degree	0	215.3	193.67	362.17	101.1	−0.46	−0.96
	PC		−1.72	0	0.03	1.56	0.47	−0.43	2.76
	TWI		2.46	4.33	4.82	10.79	1.53	1.04	1.27
Climatic variables	MAT	Degree celsius	9.21	10.6	10.56	10.63	0.22	−1.37	2.76
	MAP	mm	600.64	656.42	654.97	605.33	3.98	1.57	1.81

TABLE 24.2 Pearson's correlation coefficients between ln-transformed soil organic carbon (SOC) stocks with all environmental variables.

Property	SOC stocks	B_{GREEN}	B_{RED}	B_{NIR}	NDVI	SAVI	Elevation	SG	SA	PC	TWI	MAT
B_{GREEN}	−0.53**											
B_{RED}	−0.32**	0.36**										
B_{NIR}	−0.24**	0.66*	0.31**									
NDVI	0.62**	−0.64**	0.52**	−0.39**								
SAVI	0.49**	−0.55**	0.44**	−0.32**	0.61**							
Elevation	0.47**	−0.31**	−0.23*	−0.41**	0.19	0.18						
SG	0.37**	−0.44**	−0.13	−0.42**	0.09	0.10	0.57**					
SA	0.09	−0.04	0.07	−0.08	−0.07	−0.12	0.16	0.13				
PC	0.06	−0.05	−0.09	−0.06	0.11	0.14	0.18	0.06	−0.05			
TWI	−0.38**	0.43**	0.27*	0.34**	−0.04	−0.09	−0.42**	−0.63**	−0.26*	−0.18		
MAT	−0.28**	0.27**	−0.03	0.26*	−0.17*	−0.22**	−0.34**	−0.31**	0.12	0.08	0.23*	
MAP	0.22**	−0.14	0.26*	−0.12	0.16*	0.27**	0.33**	0.39**	0.13	−0.06	−0.29*	−0.16

Note: *$P < 0.05$; **$P < 0.01$.

explain 64%, 49%, and 55% variations in SOC stocks, respectively. The results showed Model A (ME = 0.01, RMSE = 0.08, R^2 = 0.64, and LCCC = 0.82) had the best prediction performance, followed by Model C (ME = 0.05, RMSE = 0.13, R^2 = 0.55, and LCCC = 0.80) and Model B (ME = 0.07, RMSE = 0.19, R^2 = 0.49, and LCCC = 0.67). Surprisingly, the prediction performance of Model A and Model C was significantly better than that of Model B without RS-related variables. To further explore the application potential of RS-related variables, the SD of 100 iterations of the models was calculated, and Table 24.3 listed the validation indices. We found that Model A performed the best, and the predicted map of SOC stocks with its corresponding prediction uncertainty as SD of the 100 iterations in shown in Fig. 24.2. We observed that Model A produced the lowest mean SD (0.28 kg m^{-2}) (Fig. 24.2A) and showed an average predicted SOC stocks of 6.89 kg m^{-2} in the study area. The blank part in Fig. 24.2 referred to the rest of the ecosystems, except for forest ecosystems that were not included in the study. In addition, the scatter diagram of predicted and observed values was also calculated. Further, the BRT model had good prediction performance in predicting topsoil SOC stocks because its fitting line and 1:1 line were nearly coincident (Fig. 24.3).

24.3.3 Relative importance of environment variables

Model A (full environmental variables model) was iterated 100 times, and the mean relative importance of each environmental variable was calculated. In order to facilitate comparison, this study digitized the total number of environmental variables' relative importance to 100%. It can be found from Fig. 24.4 that B_{GREEN}, NDVI, elevation, MAP, and SAVI were the main environmental variables controlling the spatial variation of SOC stocks in the region, accounting for 71% of the total relative importance. Among the three categories of environmental variables, RS-related variables showed the most important relative importance (54%), followed by topographic variables (31%) and climate variables (15%).

24.4 Discussion

24.4.1 Controllers of SOC stocks

We found that the RS data were effective in controlling SOC stock distribution in the forest ecosystem, and results were consistent with many previous studies (e.g., Wang et al., 2018; Dvornikov et al., 2021). In the cold-temperate mountainous forest areas of Japan, Li et al. (2010) found that the SOC stock was closely related to aboveground vegetation properties. Ground vegetation residues and roots were the main sources of forest SOC stocks and it affected the spatial variation pattern in natural ecosystems, which was further confirmed in this study. NDVI and B_{GREEN} were considered the RS variables affecting the spatial variation of SOC stocks because they could characterize the density and

TABLE 24.3 Summary statistics of the predictive quality of the three boosted regression trees (BRT) models in predicting soil organic carbon (SOC) stocks with 100 runs in the forest ecosystem of Northeastern China.

Model	Index	Min.	1st quartile	Median	Mean	3rd quartile	Max.
Model A	ME	−0.06	−0.01	0.02	0.01	0.03	0.04
	RMSE	0.08	0.08	0.07	0.08	0.09	0.09
	R^2	0.58	0.60	0.65	0.64	0.63	0.63
	LCCC	0.79	0.8	0.81	0.82	0.86	0.82
Model B	ME	−0.13	−0.03	0.04	0.07	0.09	0.12
	RMSE	0.2	0.2	0.2	0.19	0.2	0.21
	R^2	0.47	0.48	0.48	0.49	0.53	0.54
	LCCC	0.69	0.68	0.7	0.67	0.68	0.66
Model C	ME	−0.09	−0.06	0.03	0.05	0.06	0.07
	RMSE	0.13	0.14	0.14	0.13	0.14	0.13
	R^2	0.57	0.56	0.59	0.55	0.57	0.59
	LCCC	0.72	0.81	0.78	0.80	0.76	0.74

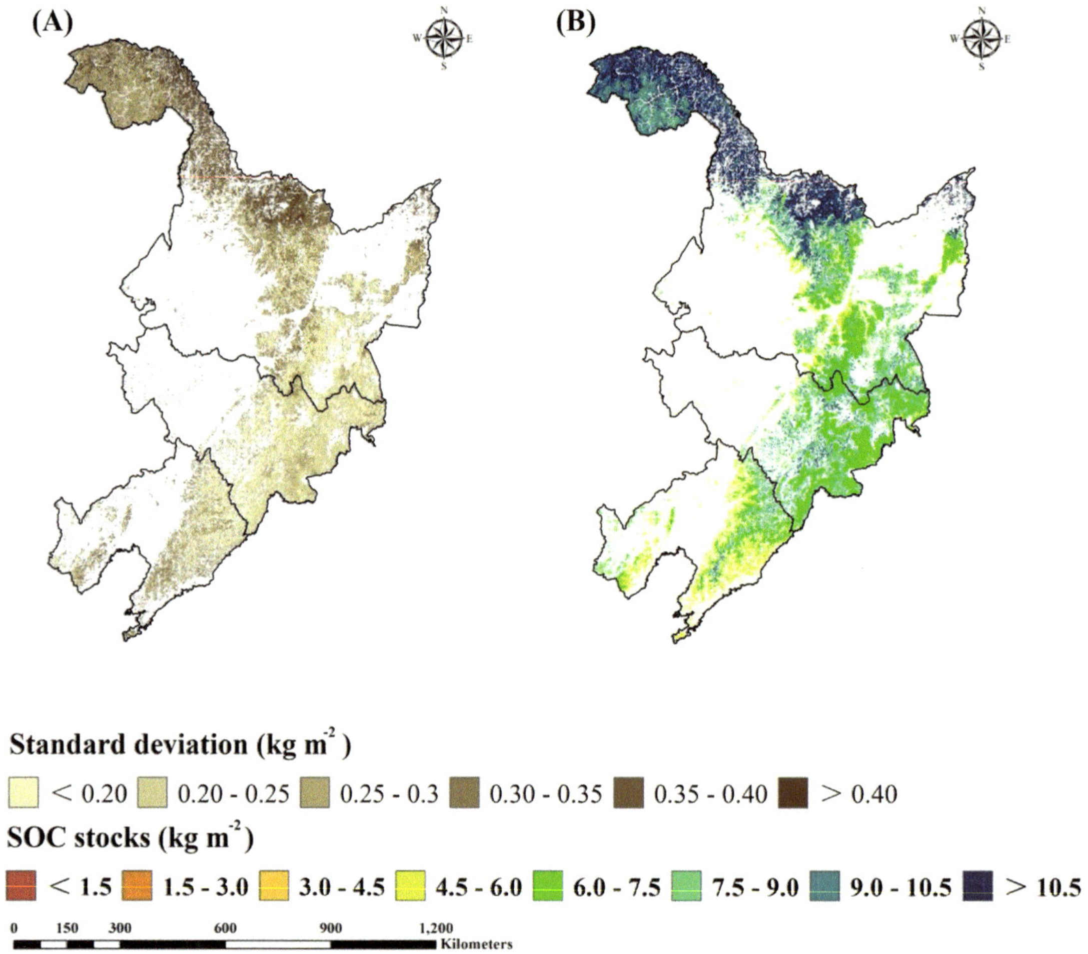

FIGURE 24.2 Average standard deviation map and average spatial distribution map of SOC stocks obtained by Model A (full environmental variables model) iterated 100 times in the forest ecosystem of northeastern China.

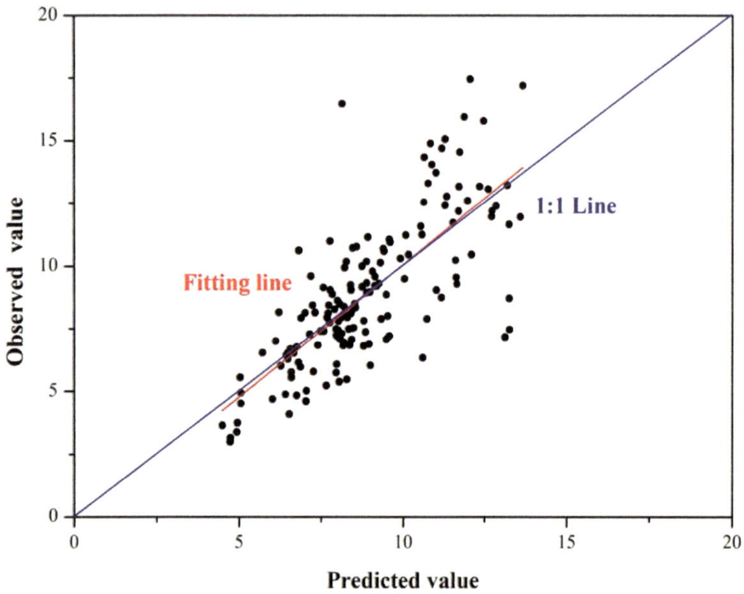

FIGURE 24.3 Scatter diagram between predicted and observed SOC stocks obtained by Model A (full environmental variables model) iterated 100 times in the forest ecosystem of northeastern China.

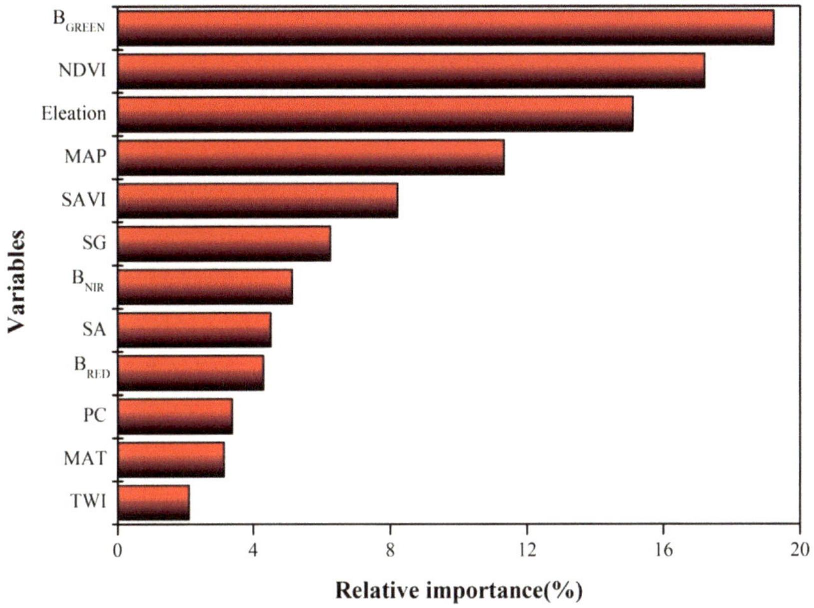

FIGURE 24.4 Relative importance of prediction variables in the full variable model (Model A).

biomass of vegetation to a certain extent in previous studies (Huete, 1988;Wang et al., 2020b; Odebiri et al., 2022). SAVI was very sensitive to changes in the soil background. These studies indicated that RS data can be used to map SOC stocks in forest ecosystems with good natural vegetation coverage. The RI of B_{GREEN} was significantly higher than that of NDVI and SAVI in the all-variable model, although NDVI was considered by many scholars to be a more effective and powerful predictor of SOC.

In the northeast forest area of China, Wang et al. (2020b) applied the BRT model and 12 environment variables to simulate the topsoil forest SOC stocks and considered that B_{GREEN}, NDVI, and SAVI were the key environmental variables affecting the spatial variation of SOC stock. Qi et al. (2019) obtained a similar conclusion in Lushun, the northeast coast of China. They concluded that SAVI was an efficient and low-cost environmental variable in forest-dominated areas. Odebiri et al. (2022) used RS-related variables and machine learning algorithms to simulate the topsoil SOC stocks of KwaZulu Natal province, South Africa. At the Ranthambhore Tiger Reserve Forest area, Kumar et al. (2016) believed that the spatial distribution pattern of SOC stock was closely related to vegetation type and coverage.

Topographic variables were considered to be a better environmental predictor for predicting SOC stocks (Adhikari and Hartemink, 2015). In this study, elevation, SG, SA, and PC were selected to represent the main topographic conditions in the region. However, it was observed that the relative importance of topographic variables was lower than that of RS-related variables (Fig. 24.4). Similar conclusions were obtained in previous studies (Fernández-Romero et al., 2014; Wang et al., 2020a; Qi et al., 2019). Wang et al. (2018) concluded that in areas with high vegetation coverage, the role of terrain variables and SOC stock may be integrated by RS and other environmental factors to some extent. The full variable model (Model A) had elevation as the most important topographic factor for the spatial variation of SOC stocks in the topsoil of the forest ecosystem. MAT and MAP, as important climatic variables, were widely used in the prediction of soil properties, especially in the spatial simulation of SOC stocks. (Adhikari et al., 2014; Gomes et al., 2019). MAP played a greater role in predicting the SOC stocks of forest ecosystems than MAT in the all-variable model. In mountain ecosystems, temperature can accelerate or reduce the release of organic carbon by regulating the decomposition and accumulation of organic matter, resulting in the spatial differentiation of SOC stocks. (Wang et al., 2022).

This study revealed that RS-related variables were efficient and low-cost indicators to predict topsoil SOC stocks in high-density vegetation-covered areas. It could be attributed to the significant relationship between the vegetation coverage of forest ecosystems and SOC stocks. This conclusion had been verified in many previous research conclusions (Kumar et al., 2016; Wang et al., 2020a,b; Dvornikov et al., 2021; Odebiri et al., 2022). In the Medinipur Block, Paschim Medinipur District, and West Bengal in India, Bhunia et al. (2019) used the bare soil index (BSI) and NDVI combined with a multivariable expression model to predict the spatial distribution of SOC. The result showed nearly 71% of SOC variance could be explained by BSI and NDVI values. Zhou et al. (2020) investigated the ability of multi-temporal Sentinel-1A data in SOC prediction by comparing the predictive performance of random forest (RF) and BRT models in the Heihe River Basin in northwestern China. The results showed that adding multitemporal RS images can significantly improve the prediction performance of the model.

24.4.2 Research limitations

Although the BRT model constructed in this study was not compared with other statistical or machine learning models, the lower SD value showed that the BRT model had a relatively stable performance in predicting the topsoil SOC stocks of the northeast forest ecosystem. There were still some uncertain elements to be explained in this study. Firstly, due to the large study area, different people were assigned to accomplish soil sampling, which might have introduced human bias as sampling errors or measurement errors. Moreover, the number of samples collected might not be representative of the study area. Secondly, environmental variables were obtained from different platforms and sources, and during the GIS analysis, they might have lost some spatial signature. Thirdly, the forest SOC stocks estimated were only limited to the topsoil depth (0–20 cm), but there were usually more SOC stocks at a deeper level, and there might be estimation errors. Lastly, due to the altitudinal influence, RS data usually had shadows in high-altitude areas, resulting in large reflectivity errors.

24.5 Conclusions

By constructing BRT models with different combinations of environmental variables, the potential of RS-related variables to predict SOC stocks in the topsoil of forest ecosystems in Northeast China was explored. The results showed that the only RS-related variable model had no worse performance than the full variable model, which further revealed that the constructed method was an effective and powerful method of topsoil SOC stock prediction. The selection of RS-related environmental variables had good practical significance and economic value in predicting the topsoil SOC stocks in the forest-dominated areas. Therefore, this study suggests that appropriate RS-related variables should be selected in future SOC stock prediction and modeling research, especially in forest ecosystems with high vegetation cover. Moreover, results from this study can be of great significance to forest ecosystem carbon cycles, soil fertility, ecological modeling, and environmental protection on a regional scale.

Acknowledgment

This work was funded by the China Postdoctoral Science Foundation (Grant No. 2019M660782), National Science and Technology Basic Resources Survey Program of China (Grant No. 2019FY101300), Doctoral research start-up fund project of Liaoning Provincial Department of Science and Technology (Grant No. 2021-BS-136), China Scholarship Council (201908210132), and Young Scientific and Technological Talents Project of Liaoning Province (Grant No. LSNQN201910 and LSNQN201914).

References

Adhikari, K., Hartemink, A.E., 2015. Digital mapping of topsoil carbon content and changes in the Driftless Area of Wisconsin, USA. Soil. Sci. Soc. Am. J. 79 (1), 155–164.

Adhikari, K., Hartemink, A.E., Minasny, B., Bou Kheir, R., Greve, M.B., Greve, M.H., 2014. Digital mapping of soil organic carbon contents and stocks in Denmark. PLoS One 9 (8), e105519.

Adhikari, K., Mishra, U., Owens, P.R., Libohova, Z., Wills, S.A., Riley, W.J., et al., 2020. Importance and strength of environmental controllers of soil organic carbon changes with scale. Geoderma 375, 114472.

Bhunia, G.S., Kumar Shit, P., Pourghasemi, H.R., 2019. Soil organic carbon mapping using remote sensing techniques and multivariate regression model. Geocarto Int. 34 (2), 215–226.

Carslaw, D.C., Taylor, P.J., 2009. Analysis of air pollution data at a mixed source location using boosted regression trees. Atmos. Environ. 43 (22–23), 3563–3570.

Chen, F., Kissel, D.E., West, L.T., Adkins, W., 2000. Field-scale mapping of surface soil organic carbon using remotely sensed imagery. Soil. Sci. Soc. Am. J. 64 (2), 746–753.

Colin, B., Clifford, S., Wu, P., Rathmanner, S., Mengersen, K., 2017. Using boosted regression trees and remotely sensed data to drive decision-making. Open. J. Stat. 7 (5), Article-number.

Croft, H., Kuhn, N.J., Anderson, K., 2012. On the use of remote sensing techniques for monitoring spatio-temporal soil organic carbon dynamics in agricultural systems. Catena 94, 64–74.

Don, A., Schumacher, J., Freibauer, A., 2011. Impact of tropical land-use change on soil organic carbon stocks–a meta-analysis. Glob. Change Biol. 17 (4), 1658–1670.

Dvornikov, Y.A., Vasenev, V.I., Romzaykina, O.N., Grigorieva, V.E., Litvinov, Y.A., Gorbov, S.N., et al., 2021. Projecting the urbanization effect on soil organic carbon stocks in polar and steppe areas of European Russia by remote sensing. Geoderma 399, 115039.

Elith, J., Leathwick, J. 2017. Boosted regression trees for ecological modeling. R Documentation. <https://cran.r-project.org/web/packages/dismo/vignettes/brt.pdf> (accessed 12.06.11.).

Elith, J., Leathwick, J.R., Hastie, T., 2008. A working guide to boosted regression trees. J. Anim. Ecol. 77 (4), 802–813.

Fernández-Romero, M.L., Lozano-García, B., Parras-Alcántara, L., 2014. Topography and land use change effects on the soil organic carbon stock of forest soils in Mediterranean natural areas. Agric. Ecosyst. Environ. 195, 1–9.

Friedman, J., Hastie, T., Tibshirani, R., 2000. Additive logistic regression: a statistical view of boosting (with discussion and a rejoinder by the authors). Ann. Stat. 28 (2), 337–407.

Froeschke, J.T., Froeschke, B.F., 2011. Spatio-temporal predictive model based on environmental factors for juvenile spotted seatrout in Texas estuaries using boosted regression trees. Fish. Res. 111 (3), 131–138.

Gomes, L.C., Faria, R.M., de Souza, E., Veloso, G.V., Schaefer, C.E.G., Fernandes Filho, E.I., 2019. Modelling and mapping soil organic carbon stocks in Brazil. Geoderma 340, 337–350.

Gomez, C., Rossel, R.A.V., McBratney, A.B., 2008. Soil organic carbon prediction by hyperspectral remote sensing and field vis-NIR spectroscopy: an Australian case study. Geoderma 146 (3–4), 403–411.

He, N., Wen, D., Zhu, J., Tang, X., Xu, L., Zhang, L., et al., 2017. Vegetation carbon sequestration in Chinese forests from 2010 to 2050. Glob. Change Biol. 23 (4), 1575–1584.

Huete, A.R., 1988. A soil-adjusted vegetation index (SAVI). Remote. Sens. Environ. 25 (3), 295–309.

Kumar, P., Pandey, P.C., Singh, B.K., Katiyar, S., Mandal, V.P., Rani, M., et al., 2016. Estimation of accumulated soil organic carbon stock in tropical forest using geospatial strategy. Egypt. J. Remote. Sens. Space Sci. 19 (1), 109–123.

Lamichhane, S., Kumar, L., Wilson, B., 2019. Digital soil mapping algorithms and covariates for soil organic carbon mapping and their implications: a review. Geoderma 352, 395–413.

Lin, L., 1989. A concordance correlation coefficient to evaluate reproducibility. Biometrics 45, 255–268.

Litton, C.M., Raich, J.W., Ryan, M.G., 2007. Carbon allocation in forest ecosystems. Glob. Change Biol. 13 (10), 2089–2109.

Li, P., Wang, Q., Endo, T., Zhao, X., Kakubari, Y., 2010. Soil organic carbon stock is closely related to aboveground vegetation properties in cold-temperate mountainous forests. Geoderma 154 (3–4), 407–415.

Martin, M.P., Wattenbach, M., Smith, P., Meersmans, J., Jolivet, C., Boulonne, L., et al., 2011. Spatial distribution of soil organic carbon stocks in France. Biogeosciences 8 (5), 1053–1065.

Minasny, B., McBratney, A.B., Malone, B.P., Wheeler, I., 2013. Digital mapping of soil carbon. In: Sparks, D.L. (Ed.), Advances in Agronomy. Academic Press, pp. 1–47.

Odebiri, O., Mutanga, O., Odindi, J., Naicker, R., 2022. Modelling soil organic carbon stock distribution across different land-uses in South Africa: a remote sensing and deep learning approach. ISPRS J. Photogramm. Remote. Sens. 188, 351–362.

Olaya, V.F., 2004 A gentle introduction to Saga GIS. The SAGA User Group eV: Göttingen, Germany.

Qi, L., Wang, S., Zhuang, Q., Yang, Z., Bai, S., Jin, X., et al., 2019. Spatial-temporal changes in soil organic carbon and pH in the Liaoning Province of China: a modeling analysis based on observational data. Sustainability 11 (13), 3569.

R Development Core Team, 2013. R: a language and environment for statistical computing. R Foundation for Statistical Computing, Vienna, Austria. <http://www.Rproject.org/>.

Rasel, S.M., Groen, T.A., Hussin, Y.A., Diti, I.J., 2017. Proxies for soil organic carbon derived from remote sensing. Int. J. Appl. Earth Obs. Geoinf. 59, 157–166.

Reyes Rojas, L.A., Adhikari, K., Ventura, S.J., 2018. Projecting soil organic carbon distribution in central Chile under future climate scenarios. J. Environ. Qual. 47 (4), 735–745.

Wang, S., Gao, J., Zhuang, Q., Lu, Y., Gu, H., Jin, X., 2020b. Multispectral remote sensing data are effective and robust in mapping regional forest soil organic carbon stocks in a northeast forest region in China. Remote. Sens. 12 (3), 393.

Wang, B., Gray, J.M., Waters, C.M., Anwar, M.R., Orgill, S.E., Cowie, A.L., et al., 2022. Modelling and mapping soil organic carbon stocks under future climate change in south-eastern Australia. Geoderma 405, 115442.

Wang, B., Waters, C., Orgill, S., Gray, J., Cowie, A., Clark, A., et al., 2018. High resolution mapping of soil organic carbon stocks using remote sensing variables in the semi-arid rangelands of eastern Australia. Sci. Total. Environ. 630, 367–378.

Wang, S., Zhuang, Q., Jin, X., Yang, Z., Liu, H., 2020a. Predicting soil organic carbon and soil nitrogen stocks in topsoil of forest ecosystems in northeastern china using remote sensing data. Remote. Sens. 12 (7), 1115.

Zhang, L., Zhang, L., Du, B., 2016. Deep learning for remote sensing data: a technical tutorial on the state of the art. IEEE Geosci. Remote. Sens. Mag. 4 (2), 22–40.

Zhang, Z., Zhao, Y., Canes, A., Steinberg, D., Lyashevska, O., 2019. Predictive analytics with gradient boosting in clinical medicine. Ann. Transl. Med. 7, 7.

Zhou, T., Geng, Y., Chen, J., Liu, M., Haase, D., Lausch, A., 2020. Mapping soil organic carbon content using multi-source remote sensing variables in the Heihe River Basin in China. Ecol. Indic. 114, 106288.

Zhu, A.X., Yang, L., Li, B., Qin, C., English, E., Burt, J.E., et al., 2008. Purposive sampling for digital soil mapping for areas with limited data. In: Hartemink, A.E. (Ed.), Digital Soil Mapping with Limited Data. Springer, pp. 33–245.

Chapter 25

Soil organic carbon: measurement and monitoring using remote sensing data

Saurav Das and Deepak Ghimire
Department of Agronomy and Horticulture, University of Nebraska, Lincoln, NE, United States

25.1 Background

Soil organic carbon (SOC) is a critical indicator of soil quality and health. Carbon (C) sequestration has been regarded as one of the potential climate mitigation approaches (Amelung et al., 2020). Soil C sequestration is the process by which atmospheric CO_2 is captured by plants and stored in the soil. SOC can be directly correlated to soil quality as it determines many soil properties and ecosystem services, including crop productivity (Maharjan et al., 2021), soil structure (Blanco-Canqui and Lal, 2007), water holding capacity and infiltration (Olness and Archer, 2005), and net removal of CO_2 from the atmosphere for climate mitigation goals. The SOC concentration can be used as a proxy indicator to determine the impact of major agricultural practices and the sustainability of an agricultural production system. It can be used to inform policymakers, develop strategies for agricultural sustainability, and monitor the environmental impact of agricultural development.

Soil is the largest terrestrial carbon reservoir, containing two times more C than the atmosphere ($\sim$830 Pg C), holding 1500 Pg C (1 Pg = 10^{15} gm) to a depth of 1 m and 2400 Pg C to a depth of 2 m (Lal, 2008; Paustian et al., 2016). Thus, increasing net soil C storage by a tenth of a percent represents substantial C sink potential. Most of the soil C is stored as SOC. A small change in soil SOC stock can significantly impact the global C cycle. Thus, the magnitude and direction of SOC change are of prime importance in meeting global climate goals. The release and sequestration of C from the soil are substantially altered depending on the land use, parent material, soil drainage, texture, nutrient availability, soil management practices, and climate. Since the development of agricultural practices, 116 Pg C has been lost to the atmosphere (Sanderman et al., 2017). Agronomic management practices such as tillage (Haddaway et al., 2017), intensive cropping systems (Varvel and Wilhelm, 2010), and the application of large quantities of organic and inorganic fertilizer, all have a significant impact on soil C dynamics (Kaur et al., 2008; Maharjan et al., 2021).

Soil organic C plays a pivotal role in both the productivity and sustainability of the terrestrial ecosystem and global C cycling. Agricultural practices and land-use change account for 20% and 14% of greenhouse gas emissions, respectively. However, conservation practices, including cover crops, reduced tillage, residue management, and manure management, in suitable agroclimatic conditions can help in C sequestration and improve productivity, fertility, and biophysicochemical properties of soil (Paustian et al., 2016). The improvement in SOC content with management changes is very slow, and it takes several years to see notable changes. It requires continuous monitoring of changes in agricultural management practices to predict and estimate SOC dynamics. Heterogenic characteristics of soil and multifaceted complexation of organic and inorganic constituents in the soil create a large spatiotemporal variation from site to site and even within the same field, making qualitative and quantitative measurement a laborious process. Chronological changes in land use, vegetation, and climatic parameters create uncertainties in SOC estimation. The remote sensing (RS) technique is significantly advantageous for monitoring soil C status because of its ability to capture large spatial areas along with historical vegetation and land cover data records. RS is cost-effective, and it reduces laborious in situ measurement. There are various kinds of satellites, including Landsat, Terra, and Aqua, which many researchers have used to estimate and inform SOC (Wu, 2012; Gupta et al., 2014). Temporal changes in soil C stock can be monitored over a large spatial area at regular intervals using satellite data, which are nowadays freely accessible in most cases. Applying RS with process-based simulation models makes it an effective approach to monitoring C fluxes across a range of ecosystems, guiding policymakers and researchers to formulate site-specific strategies to improve SOC

Remote Sensing of Soils. DOI: https://doi.org/10.1016/B978-0-443-18773-5.00024-7

sequestration and create a sustainable agricultural production system. This book chapter summarizes the basic principles of RS in measuring soil C change and discusses the methods' advantages and challenges. The chapter will also introduce the concepts of indirect modeling approaches for SOC measurement, C credits, and climate-smart commodities.

25.2 Remote sensing

Remote sensing of soils has become a tool of importance for assessing and mapping the soil environment from afar ever since the launch of the first commercial satellite (ERTS-1, known as LANDSAT-1) in 1972. Since then, tremendous progress has been made in data acquisition and processing techniques. RS tools can provide quantitative information based on the electromagnetic spectral reflectance of every ground picture element or pixel. In general, RS is defined as "the science and art of obtaining information about an object, area, or phenomenon through the analysis of data acquired by a device that is not in contact with the object, area, and phenomenon in the investigation" (Lillesand et al., 2015). The difference in reflectance characteristics of minerals, materials, or physical surfaces, such as soils, vegetation, and water, is leveraged in RS to identify specific features or properties. The unique reflectance characteristics of materials and soil surfaces enable the interpretation of their properties directly or indirectly (through inference). The wavelength of the electromagnetic spectrum used in soil analysis varies from visible, near-infrared, and far infrared (e.g., thermal infrared) to microwave (Fig. 25.1).

Based on the number of spectral bands, RS is broadly classified into panchromatic, multispectral, and hyperspectral. Panchromatic sensors acquire data in a single band sensitive over a broad range of wavelengths, such as the visible spectrum. It records only black-and-white images, essentially capturing varying shades of gray based on the intensity of radiation reflected or emitted by objects. Multispectral sensors acquire data in a relatively small number of spectral bands (typically ranging from three to ten) where the spectral reflectance across a range of electromagnetic fields is summarized. Data from multispectral sensors is typically used in agriculture and natural resources, where most of the spectral regions are selected based on understanding the relationship between spectral information and earth surface features. Multispectral sensors are useful in highlighting the variations in land cover and soil properties. Hyperspectral sensors, on the other hand, acquire data in numerous spectral bands (ranging from ten to hundreds) with narrow bandwidths. Data from hyperspectral sensors is typically used to capture and identify features that are not detected in the limited number of spectral bands in multispectral data (Goetz et al., 1985; Thenkabail et al., 2000).

Over the years, soil science has significantly improved to consider both permanent and dynamic soil properties in preparing the map. A systemic soil survey combines remotely sensed images and field data to create digital soil maps (DSM). RS has emerged as a powerful method for mapping soil dynamic properties such as changes in soil temperature, moisture, vegetation, and structural change, which are critical in estimation of SOC. Existing SOC estimation methods are largely based on the grid soil sampling method, which, while thorough, is not spatially continuous and can be resource intensive. Studies showed high spatial heterogeneity in SOC, requiring a sampling interval of <50 m to capture the spatial variability. Thus, a spatially extended method is needed to improve the inventories of SOC. It is important to monitor land use and management changes to determine and model the spatiotemporal changes in SOC. The main advantages of RS applications in soil science studies include: (1) a non-invasive means of obtaining data on soil properties, (2) data covering large geographical areas, (3) access to information from regions otherwise hard to reach, (4) data that encompasses multiple attributes, and (5) a reduction in traditional resource-intensive soil samplings. It's noteworthy that, the capacity of an RS platform varies depending on its spatial, spectral, temporal, and radiometric resolution (Jensen, 2007).

25.2.1 Spatial resolution

Spatial resolution is the dimension of an individual pixel, or picture element, within a digital image and the corresponding area it represents on the Earth's surface. A single value usually represents it. For example, a spatial resolution of 30 cm means that one pixel in the image represents a ground area measuring 30 cm by 30 cm. The finer resolution provides better discrimination of features on the ground (Fig. 25.2). Higher resolution satellite can more accurately predict land cover classifications and changes, which are vital for soil mapping and SOC estimation. Guo et al. (2019) demonstrated that the prediction accuracy of SOC is inversely correlated with spatial resolution. In the study, they evaluated the model accuracy across varying resolutions ranging from 10 to 60 m, noting a reduction in accuracy as resolution decreased. Similarly, Garosi et al. (2022) used variable resolution from 2 to 30 m for testing model accuracy and showed that 10 m pixel-size sentinel-2 images in combination with topographic attributes of couture-DEM produced the best SOC prediction. Both studies identified pixel size or spatial resolution as an important variable affecting the model's accuracy or SOC estimation.

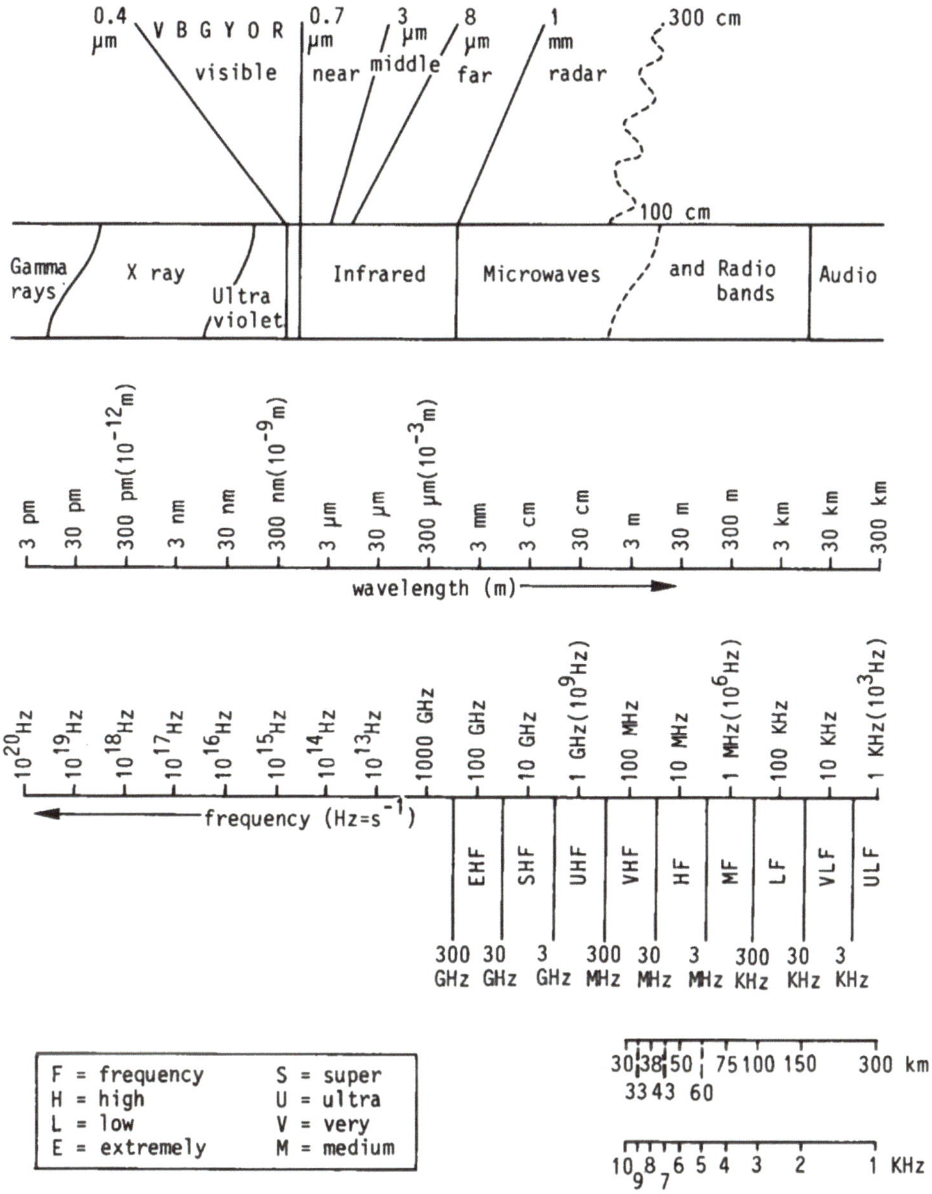

FIGURE 25.1 The electromagnetic spectrum. *V, B, G, Y, O, R,* violet, blue, green, yellow, orange, and red, respectively (Mulders, 1987). *From: Mulders, M.A., 1987. Remote Sensing in Soil Science. Elsevier.*

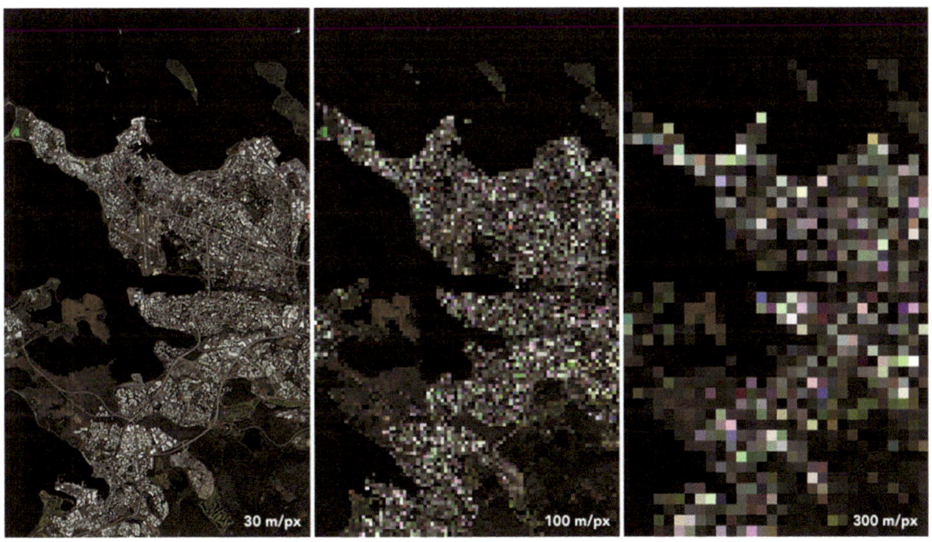

FIGURE 25.2 Landsat 8 image of Reykjavik, Iceland, acquired on July 7, 2019, illustrating pixel resolution differences. *From: NASA Earth Observatory.*

TABLE 25.1 List of a few hyperspectral sensor systems.

Sensor	Wavelength range (nm)	Bandwidth (nm)	Number of bands
AVRIS	400–2500	10	224
TRWIS III	367–2328	5.9	335
HYDICE	400–2400	10	210
CASI	400–900	1.8	288
OKSI	400–1000	10	61
MAIS	440–1180	20/600	71
GERIS	400–2500	25/120/1	63
MIVIS	430–1270	20/50/8/4	102
OMIS-I	460–1250	10	68
OMIS-II	460–1250	64/16/32/8	128
HYPERION	400–2500	10	220
HySI	421–964	<20	64

25.2.2 Spectral resolution

Spectral resolution describes a sensor's capacity to differentiate between fine distinctions in wavelengths. The narrower the bands, the higher the spectral resolution. However, having an ultra-narrow band isn't always advantageous. For instance, certain spectral regions (say red, blue, and green) can be optimal for the differentiation of specific characteristics of objects or cover types, and having multiple narrow bands within that specific spectral region will only add cost to data acquisition and processing. The key is to choose a spectral resolution that amplifies the contrast between the target object and its surroundings and can effectively capture spectral shifts. Multispectral sensors have 3–10 bands, while hyperspectral sensors have hundreds of bands (Table 25.1). For example, the Airborne Visible/Infrared Imaging Spectrometer captures information in 224 spectral bands (Table 25.1). The higher spectral resolution allows the exploitation of well-defined narrow absorption spectral features associated with functional groups in the very near-infrared (VNIR) and shortwave infrared (SWIR) ranges, such as clay minerals. Castaldi et al. (2016) showed a decrease in SOC estimation accuracy when bandwidth was changed from 10 to 40 to 160 nm. The study found that the benefit of higher spectral resolution can be partially offset by the amplification of noise when increasing the number of spectral bands (Castaldi et al., 2016). Their findings indicated that data acquired between 400 and 2400 nm with a bandwidth of 40 nm could provide similar results to higher spectral resolution because of the lower signal-to-noise ratio of the 40 nm bandwidth (Fig. 25.3).

25.2.3 Temporal resolution

Temporal resolution refers to the frequency at which satellite revisits the same observational point on the Earth's surface (Fig. 25.4). This resolution depends on the orbit, sensor characteristics, and swath width (the width of the ground area observed in a single pass). Geostationary satellites are positioned to rotate synchronously with the Earth, resulting in a particularly fine temporal resolution as they constantly observe the same location. The temporal resolution of polar satellites can vary from 1 to 16 days (Fig. 25.4). For example, the Moderate Resolution Imaging Spectroradiometer (MODIS) sensor onboard NASA's Terra and Aqua satellites has a temporal resolution of 1–2 days, allowing the sensor to track the daily changes of the earth. The Operational Land Imager (OLI) on the NASA/USGS Landsat 8 satellite has a narrower swath width and a temporal resolution of 16 days. In an agricultural application, the temporal domain is as important as the spatial scales, particularly when analyzing the variability of SOC. Local transfer processes (erosion) and past and present management practices significantly affect SOC dynamics (Goidts and van Wesemael, 2007). A complex combination of topography, soil type, climate, and biological processes leads to high SOC variability in time and space, and thus capturing both spatio-temporally continuous images is important.

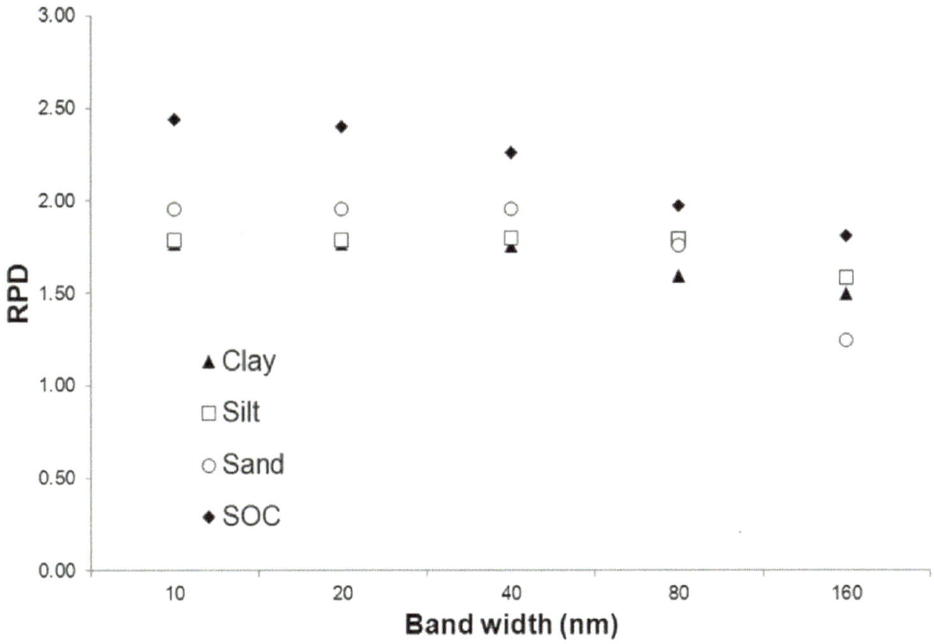

FIGURE 25.3 Ratio of performance to deviation (RPD) of clay, sand, silt, and SOC estimation obtained from the resampled spectra from 10, 20, 40, 80, and 160 nm bandwidth. *From: Castaldi, F., Palombo, A., Santini, F., Pascucci, S., Pignatti, S., et al., 2016. Evaluation of the potential of the current and forthcoming multispectral and hyperspectral imagers to estimate soil texture and organic carbon. Remote Sens. Environ. 179, 54–65. doi: 10.1016/j.rse.2016.03.025.*

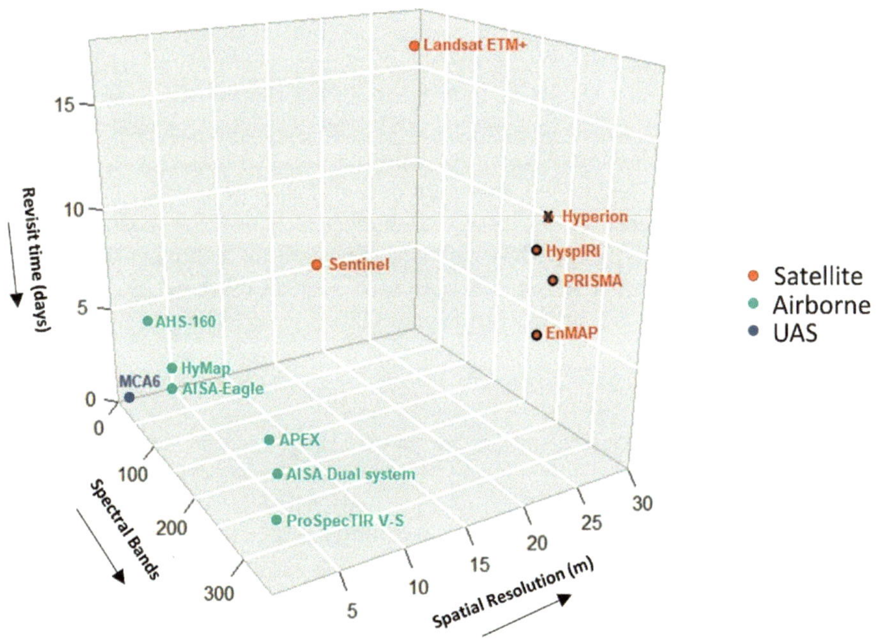

FIGURE 25.4 An overview of spatial, spectral, and temporal resolution of different RS systems applied in SOC estimation. The black-outlined symbols indicate the forthcoming remote sensing systems, while the x mark indicates that Hyperion is not operational (Angelopoulou et al., 2019). *From: Angelopoulou, T., Tziolas, N., Balafoutis, A., Zalidis, G., Bochtis, D., 2019. Remote sensing techniques for soil organic carbon estimation: a review. Remote Sens. 11 (6), 676. doi: 10.3390/rs11060676.*

25.2.4 Radiometric resolution

Radiometric resolution is the amount of information or the number of energy bits present in a pixel. Each bit records in an exponent of power 2, and an 8-bit resolution is 2^8, meaning the sensor has 256 potential digital values (0–255) to store information. The higher the radiometric resolution, the more information can be stored, allowing better discrimination between the slightest change/difference in energy. A comparative study using 12-bit LISS-III and 7-bit LISS-III sensors showed 3% higher overall accuracy in land cover estimation with higher radiometric resolution. The higher radiometric resolution of 12-bit LISS-III also provided 4%–6% higher classification accuracy for specific crops like cotton, black gram, chilies, and sugar cane (Rama Rao et al., 2007).

25.3 Sources of remote sensing SOC data

25.3.1 Spaceborne

From the beginning of public satellite RS technology in 1972 until the mid-2010s, most of the satellite data acquired was from multispectral sensors. Most of these data were obtained from the Landsat satellites equipped with the thematic mapper (TM), enhanced thematic mapper (ETM +), and the Operational Land Imager (OLI) with 30 m resolution. Another noteworthy satellite from this period is the "Satellite Pour l'Observation de la Terre" (SPOT). Launched in 1986, equipped with the sensor Haute Résolution Visible (HRV), had a spatial resolution of 20 m. The SPOT HRV bands have been instrumental in several SOC estimation studies (Agbu et al., 1990; Vaudour et al., 2013) and similarly Landsat bands (Nanni and Demattê, 2006; Huang et al., 2007).

Since the early 2000s, hyperspectral images have been made available from the Hyperion sensor with 30 m resolution onboard the satellite Earth Observing 1 (2000–2017), and from the Compact High-Resolution Imaging Spectrometer (CHRIS) with 17 m resolution under the Project for On-Board Autonomy (PROBA-1) microsatellite (2001–ongoing). Since 2019, the *PRecursore IperSpettrale della Missione Applicativa* (PRISMA) with 239 spectral bands between 400 and 2505 nm has delivered images with 30 m resolution (Loizzo et al., 2018). Simulation modeling for SOC has been carried out using the hyperspectral satellite-based images from Hyperion (400–2500 nm) (Gomez et al., 2008b; Jaber et al., 2011), CHRIS–PROBA (400–1050 nm) (Casa et al., 2013a), and PRISMA (400–2500 nm) (Castaldi et al., 2016; Mzid et al., 2022). Recently, on April 1, 2022, the Environmental Mapping and Analysis Program (EnMap) was launched. This German hyperspectral satellite mission, slated to operate until 2027, is currently in the commission phase, and only a few studies have tested the potential of EnMap sensors for the assessment of SOC content (Steinberg et al., 2016; Ward et al., 2020). There are a few upcoming satellite missions, including the Copernicus Hyperspectral Imaging Mission for the Environment and the Spaceborne Hyperspectral Applicative Land and Ocean Mission. However, as of now, no published studies have utilized data from these upcoming platforms.

Field-scale SOC modeling using high spatial resolution data from commercial satellites remains relatively limited. Notable mentions include the use of IKONOS multispectral data, which provides a 3.2 m resolution in visible bands and also covers the near-infrared (NIR) spectrum (Sullivan et al., 2005), multispectral PlanetScope with 3 m resolution (Žížala et al., 2019) and Worldview-2 with 2.5 m resolution (Samsonova et al., 2018). The launch of Sentinel 2-A & -B in 2015 and 2017 expanded the SOC simulation study with chronological multispectral data (13 bands) with 10–20 m resolution. The sentinel 2 has a swath width of 290 km and a temporal resolution or revisit time of 5 days. Several SOC simulation studies have used the Sentinel 2 data (Gholizadeh et al., 2018; Vaudour et al., 2019b; a; Castaldi et al., 2019; Dvorakova et al., 2020; Castaldi, 2021; Wang et al., 2021a). Only a few studies used the Sentinel—1 synthetic aperture radar (SAR) in their approach (Wang et al., 2021b). A few studies also used MODIS (Sothe et al., 2022) and Sentinel—3, which have a coarse resolution of around 200–500 m (Odebiri et al., 2022).

25.3.2 Airborne

In airborne remote sensing (ARS), downward- or sideward-looking sensors are mounted on an aircraft or drone to capture an image of the earth's surface. A high spatial resolution of <20 cm can be obtained in an airborne RS, which is a significant advantage over satellite RS. However, low coverage and a high cost per unit area of ground coverage can be disadvantages of ARS. Airborne radiometric data has been extensively used in digital soil mapping (Lundén et al., 2001) and estimating SOC (Selige et al., 2006; Rawlins et al., 2009). Improvements in laboratory spectroscopy or reference spectral databases have increased the number of studies using VNIR (400–1200 nm)/shortwave infrared (SWIR) (1200–2500 nm) hyperspectral airborne imaging for soil properties (Gomez et al., 2008a; Stevens et al., 2010). Most of the studies used hyperspectral airborne sensors such as DAIS—7915 (400–12.6 mm; 79 bands) (Ben-Dor et al., 2002), AHS—160 (430–2540 nm) (Stevens et al., 2010), airborne MIVIS (430–1270 nm, spatial resolution 4.8 m) (Casa et al., 2013b), CASI and SASI (Žížala et al., 2017), HYMPA (400–2500 nm, 126 bands) (Gomez et al., 2008a), and SpecTIR LLC—A series (380–1000 nm) & X-series (900–1700nm) (Hong et al., 2020).

25.4 Measurement and monitoring of SOC

Direct measurement of SOC is highly dependent on sampling protocols and spatial variability. Often, a large number of samples is required to reduce spatial variability and errors. Sampling depth is also crucial for properly evaluating soil C content changes (the IPCC recommends a minimum depth of 30 cm). Several C sequestration methods require deep sampling to confirm and determine the positive effect. However, the machinery required for deep soil samples is costly.

The cost associated with collecting, processing, and storing soil samples and soil C measurements can make large-scale direct analysis an expensive approach. It was estimated that detecting meaningful changes in soil C stocks across forest ecosystems in Finland (i.e., 3000 plots at the national scale) might cost 4 million euros for one sampling campaign (e.g., baseline measurement spanning one year) (Mäkipää et al., 2008). Thus, there is a need to evaluate these costs against the value of soil C sequestered and search for tradeoffs between the costs involved and alternative SOC estimation methods, including different modeling approaches. RS provides an efficient, spatially distributed, scale-appropriate, and resource-efficient measurement of SOC content.

25.4.1 Measurement

The remote sensing (RS) method for SOC measurement predominantly relies on soil reflectance. This reflectance is influenced by several factors including, soil chemical structure, mineralogy, organic matter content, moisture, and physical structure, such as surface roughness and particle size (Stoner and Baumgardner, 1981; Dalal and Henry, 1986; Cozzolino and Morón, 2003). Hyperspectral bands, with their detailed spectral readings, can capture the distinct absorption patterns of SOC and minerals. For limited details, wider bands and characteristics of multispectral sensors can be employed. The visible spectrum often displays broad absorption patterns, typically associated with iron oxides and soil organic matter. In the near-infrared (NIR) and short-wave infrared (SWIR) regions, specific absorption features associated with cellulose and lignin are evident around 2100 and 1120 nm, respectively (Mcmorrow et al., 2004). The SWIR spectrum (spanning 1200–2500 nm) further reveals narrow vibrational absorption characteristics, linked to carbonates and clay minerals at approximately 2300 and 2200 nm (Ben-Dor et al., 2003). Strong absorption features due to water can be seen at around 1400 and 1900 nm, with weaker features at around 950 and 1200 nm (Rossel and Behrens, 2010). It is critical to note that the absorption features observed in soil are often the result of biochemical groups associated with hydroxyl, carboxyl, and amine functional groups (Rossel and Behrens, 2010).

Spectroscopic models derive the mathematical relationship between the soil spectra and constituents and develop a spectral library. For estimation, the standardized measurement from the spectral library is used as a reference to predict or determine the spectral profile of unknown soil samples and determine the soil constituents (Smith et al., 2020). Statistical techniques like partial least square regression (Waes et al., 2005) and principal components regression (Brown et al., 2006), along with artificial neural network analysis (Ingleby and Crowe, 2001), are applied to develop a distinct relationship between soil reflectance and soil properties.

Remote sensing has proven invaluable in mapping the changes in SOC over extensive geographical scales. Such mapping can be accomplished by correlating data from drones, aircraft, or satellite with well-documented soil attributes or by analyzing vegetation dynamics and their corresponding impact on SOC. Changes in land use and vegetation pattern, as visible in remote imagery, can be used in calculating the net primary productivity and SOC sequestration potential (Miller and Michalak, 2020). Satellite imagery can also be used as covariates in digital soil mapping, where the soil properties and satellite information are used to predict the SOC at various depths correlating with ground point observation (Maynard and Levi, 2017). A combination of remote and in situ point data will be critical in predicting or creating high-resolution and accurate SOC maps. The model derived from point-source measurement of SOC and correlated with spectral reflectance on remotely sensed imagery can provide complete datasets for extrapolating or estimating SOC in unsampled locations for spatial continuity. The high temporal and spatial resolution of remote imagery offers an unprecedented possibility to study and monitor SOC change over space-time dynamics reflecting vegetation change, climate change, and land-use.

Remote sensing is increasingly recognized as a pivotal technique for estimating Soil Organic Carbon (SOC). However, as with many advanced methodologies, it comes with its own set of challenges. One of the predominant challenges in harnessing RS techniques for SOC assessment lies in the intricacies associated with data acquisition, preprocessing necessities, and the requisite technical expertise. In addition, satellite-borne sensors can only detect surface reflectance, which is correlated to surface SOC. Subsurface SOC can be computed indirectly from the surface and subsurface SOC models, but accuracy can sometimes be a tradeoff. Another complicating factor is the variability in soil moisture across different regions, which can potentially impede the sensors' capability to accurately register soil reflectance. Spectral data must be preprocessed to reduce noise and errors before SOC estimation. The integrity of sensor data can be compromised by various factors, including atmospheric disturbances, Rayleigh and Mie scattering, nonselective scattering, cloud interference, variations in the bidirectional reflectance distribution function, sensor drift over time and space, and alignment inaccuracies (Stevens et al., 2010).

Integration of RS data with different geostatistical methods has shown potential in predicting SOC variability and developing high-quality maps (Mirzaee et al., 2016). In a study by Schillaci et al. (2017), used the stochastic gradient treeboost (SGT) with the climatic data from WORLDCLIM, land cover from CORINE, soil texture, and topography to estimate the topsoil (0–30 cm) SOC stock of Mediterranean semiarid area. The study found that the SGT model showed better predictability and accuracy than published global estimates and data from the International Soil Reference and Information Center. The study described annual rainfall, soil texture, land use, mean annual temperature, and Landsat 7 ETM + panchromatic Band 8 as the most critical predictors of SOC stock. The panchromatic band of Land 7 ETM + showed better predictability compared to NDVI indices.

25.4.2 Monitoring

The extent to which SOC stocks respond to immediate climate fluctuations and long-term global warming driven by climate change remains an area of uncertainty (Bradford et al., 2016). Warmer temperatures increase microbial activity and thus SOC loss (Hicks Pries et al., 2017), while warmer temperatures with higher CO_2 levels may also increase plant productivity and carbon inputs to soil (Ziegler et al., 2017; Dietzen et al., 2019), potentially offsetting any increase in SOC loss. Low soil moisture conditions reduce microbial activity and soil C inputs from plants; the balance of these two determines the net SOC gain. Accurate monitoring of SOC change is critical to determining the global C balance. Land use, crop variety, soil moisture, and potential yield are critical factors determining the net gain in SOC. Over the years, repeated soil sampling programs have been conducted to monitor the change in many countries, including England and Wales, Denmark, Belgium, New Zealand, and the USA. The traditional method mostly relies on resampling previously sampled locations after a certain period. However, the land use and management changes between the sampling periods create uncertainties in the prediction of true drivers of SOC change. Studies have shown that agricultural practices like tillage and continuous cropping with crop residue removal can significantly change the SOC content (Bricklemyer et al., 2007; Álvaro-Fuentes et al., 2009). Thus, it is important to monitor and verify the changes in agronomic practices to better predict and estimate the SOC. Bricklemyer et al. (2002) used Landsat ETM + imagery in classifying fallow fields with > 95% accuracy, and Viña et al. (2003) differentiated between conventional and conservation tillage with a 77% overall accuracy in Nebraska using IKONOS imagery. Bricklemyer et al. (2006) estimated soil disturbance using boosted classification tree analysis of Landsat ETM + images. Moran et al. (2002) used a combination of structural (ERS-2 SAR) and spectral (Landsat TM) data to monitor agricultural crop and soil conditions. Terrestrial data informed by Soil Moisture Active Passive L-band microwave observation, land cover and vegetation inputs from the MODIS, visible infrared imaging radiometer suite, and the Goddard Earth Observing System Model, land assimilation model, can be used to calculate the net ecosystem carbon exchange and SOC estimation (Jones et al., 2017). RS is successfully used in long-term monitoring and reporting of changes in land management, including bare ground, fallow period, cover crop, and residue retention (Rogge et al., 2018; Gallo et al., 2018).

25.5 Spectral library

Remote sensing and reflectance spectroscopy have emerged as an important, rapid, and low-cost alternative to the traditional approach. However, more widespread adoption of the RS method will require a large spectral reference database to train and test the models and accurately estimate the soil properties and SOC. Spectral measurement of soil at mid-IR (2500–25,000 nm) and vis-NIR (400–2500 nm) can be used to estimate multiple soil properties, including minerals and organics. For example, the features related to various components of SOM can inferred in the mid-to thermal-infrared range (2500–25,000 nm) (Shepherd and Walsh, 2002).

A soil spectral library represents a meticulously organized repository containing soil samples, their associated spectral readings, reference analytical measurements, and comprehensive metadata detailing the samples as well as the methodologies employed for sampling and measurement. Though it's not a new concept and has been used for geological science (Clark and Roush, 1984), there are only a few spectral libraries for soil science. Shepherd and Walsh (2002) generated a spectral library of >1000 archived topsoil samples from eastern and southern Africa using a laboratory measurement approach. Brown et al. (2006) constructed a global spectral library with >4000 samples from all 50 U.S. states and five continents by scanning the archived samples and using the boosted regression tree statistical method to predict SOC content. More recently, soil spectral libraries spanning large geographical extents are being developed (Viscarra Rossel et al., 2016; Sila et al., 2016; England and Viscarra Rossel, 2018; Dangal et al., 2019).

Some of the most used soil spectral libraries are:

a) **KSSL-SSL**—MIR and NIR spectral library includes >75,000 MIR spectra collected on soil samples, primarily from the United States. Freely available on request (URL: https://www.nrcs.usda.gov/wps/portal/nrcs/main/soils/research/).
b) **ICRAF-ISRIC**—A globally distributed soil spectral library for Mid-infrared diffuse reflectance spectra. The library contains >4000 profiles from the soil collection of the ISRIC World Soil Reference Collection. Data are available under the CC-BY 4.0 license (URL: https://www.worldagroforestry.org/sd/landhealth/soil-plant-spectral-diagnostics-laboratory/soil-spectra-library).
c) **LUCAS-SSL**—Vis-NIR scans of >20,000 samples. (URL: https://esdac.jrc.ec.europa.eu/resource-type/soil-point-data)
d) **AFSIS1-SSL**—African Soil Information Service (AfSIS), MIR, and VisNIR scans of 18,500 georeferenced soil samples from 19 countries in Sub-Saharan Africa. Data are available under the ODbL license. (URL: http://africa-soils.net/services/data/soil-databases/).

25.6 Modeling SOC change

Regular quantification of SOC presents challenges due to its variability across the soil profile, influenced by factors such as land use change, climatic conditions, and shifts in agronomic management. Directly mapping SOC through RS might be challenging in locations where the soil surface is partially or wholly covered by vegetation, cloud, construction, or other land use or in regions with significant surface moisture variability. A synergistic approach that integrates both simulation models and RS can enhance SOC estimation. Specifically, simulation models can predict SOC adjustments following changes in land use or management practices, while optical RS offers the capability to map SOC across diverse spatial and temporal scales. Depending on the complexity of the system and data availability, one can opt for either empirical or process-based models for soil C quantification. Among the most commonly used carbon models are the CENTURY, its daily version DayCent, the Rothamsted Carbon Model (RothC), the Erosion Depositional Carbon Model (EDCM), Biome-BGC, CO2FIX, DNDC, FORCARB, ORCHIDEE, LPJ, CASA, and Agro-IBIS. These models vary in complexity, spatial and temporal resolution, and specific applications, and are vital tools in understanding the dynamics and sequestration potential of soil carbon.

Process-based models can integrate the data from diverse sources including, management practices, land cover, climate, and soil characteristics. Model input variables range from maximum and minimum temperature, monthly precipitation, soil texture, drainage, initial SOC level, water holding capacity, cropping system, fertilization, harvesting, grazing, land-cover data, and natural disturbances. Output variables of the model include net primary productivity, grain yield, C decomposition, C exchange rates between the atmosphere and soil system, biomass removal by harvest, and the C pool in the vegetation & soils.

Recently, there has been a significant increase in the adoption of machine learning (ML) models for estimating SOC, especially using RS data. Unlike process-based physical models, ML models can enhance learning procedures because of the complex nonlinear relationship between the variables. Several studies have successfully incorporated the use of deep-learning (neural networks: convolutional neural network, recurrent neural network, artificial neural network) RS-based models for predicting SOC (Kuang et al., 2015; Yang et al., 2019; Odebiri et al., 2021).

25.7 Challenges with remote sensing SOC estimation

a) **Challenges in Spatial and Temporal Consistency of SOC Estimation via Remote Sensing**
While remote sensing offers promising tools for SOC estimation, several challenges persist. Predominantly, many SOC models are empirical in nature, capturing a specific relationship between soil constituents and reflectance that may only be applicable at a particular spatial and temporal scale, and often for a specific instrument (Bartholomeus et al., 2008; Lobsey et al., 2017). The reflectance properties of soil are inherently dynamic, fluctuating both temporally—due to variations in soil type and mineralogy—and spatially, influenced by factors such as moisture content, tillage practices, vegetation cover, and the presence of crop residues (Ladoni et al., 2010).
b) **Complexities Introduced by Sensor Variability in SOC Predictions**
Sensor characteristics, calibration, and image processing vary from one sensor to another, including spectral resolution, sampling wavelength, and the numbers of bands. This introduces complexities in prediction and pose challenges for consistent statistical quality measurement.
c) **Influence of External Factors on Recorded SOC Variations**
Furthermore, even if SOC contents remain unchanged, differences can be recorded due to variations in illumination, terrain changes, and atmospheric attenuation. The angle and intensity of sunlight, influenced by different times of day, seasons, or cloud cover, can markedly affect how satellite or aerial sensors interpret the soil's reflectance. In

addition to light variations, alterations in the landscape, whether they are natural processes like erosion or human-induced changes like construction, can modify surface reflectance, thereby affecting perceived SOC content. The spectral signature that a sensor detects can be modified by atmospheric factors such as humidity, aerosol content, and other pollutants.

d) **Challenges in Hyperspectral Data Analysis for SOC Estimation**

The analysis of hyperspectral data introduces unique challenges, particularly when applied to soil organic carbon (SOC) estimation. One of the primary issues is the sheer volume of the data. With an increased number of spectral bands, hyperspectral datasets are considerably larger, demanding advanced computational resources and specialized algorithms for efficient processing. In addition to size considerations, the data are also influenced by atmospheric absorption. This phenomenon can distort the true spectral signature of the soil, necessitating rigorous preprocessing steps to correct and calibrate the data. Moreover, the high spectral dimensionality of hyperspectral data can lead to redundancies and require techniques to extract relevant information effectively. Navigating these intricacies is essential for ensuring that hyperspectral data provides accurate insights into SOC content.

e) **Methodological Inconsistencies and Their Impact on Study Comparability**

A significant challenge in SOC estimation arises from the lack of comparability between studies, primarily stemming from variations in method selection and evaluation criteria. The choice of measurement methods plays a pivotal role, as predictions and interpretations can vary widely based on the techniques employed (Angelopoulou et al., 2017; Romero et al., 2018).

25.8 Carbon market and climate-smart commodities

Carbon markets are trading systems for selling and buying carbon credits. A carbon credit is a tradable security that grants the owner permission to release one ton of CO_2 or greenhouse gas equivalent (CO_2e). Currently, there are two types of carbon markets: compliance and voluntary. Compliance markets are created because of national, regional, and/or international policy or regulatory requirements. Voluntary carbon markets cater to the voluntary issuance, purchase, and trade of carbon credits. The backbone of these markets is often private enterprises embarking on carbon-centric projects or governmental schemes aligning with specific carbon standards. The most commonly referred to and known compliance market is the Emissions Trading Systems (ETS) or, commonly referred to as, "Cap and Trade" (CAT) system. Under this framework, governments delineate a cap on total permissible emissions, distributing permits or allowances among regulated businesses. Entities that overshoot their emission limits are obligated to acquire additional permits, whereas those under the cap can monetize their surplus by selling. Implementation of CAT is still challenging, and only 11 states out of 50 in the United States have adopted such a market (https://www.c2es.org/content/multi-state-initiatives/). The essence of carbon markets, particularly compliance markets, is to establish clear environmental benchmarks, incentivize emission mitigation, and ensure accountability through systematic tracking and reporting. The recent US Inflation Reduction Act, promulgated on August 16, 2022, accentuates this global trend, introducing provisions to reward corporations that effectively store or recycle their greenhouse gases.

The Clean Development Mechanism (CDM), adopted under the Kyoto Protocol in 1997, is a well-known example of an international compliance market for C trading. Under the CDM, countries are divided into industrialized and developed economies. Industrialized countries, or Annex 1, operate their emission trading markets. If a country emits less than its target amount, it can sell its surplus allowance to countries that did not achieve its Kyoto goals through an emission reduction purchase agreement. The first commitment of the Kyoto Protocol ended in 2012. During the 2012 United Nations Climate Change Conference held in Doha, Qatar, the Doha Amendment to the Kyoto Protocol was adopted. This amendment established the second commitment period, which began in 2013 and lasted until 2020. However, the Doha Amendment only entered into force on December 31, 2020, after a sufficient number of parties had ratified it (UNFCC, 2022).

Soil carbon can significantly contribute to climate mitigation and resilience efforts. Recent studies suggest that improved agricultural soil management can mitigate between 4% and 6% of annual US emissions (EDF, 2022a). In response to these findings, there is a rapid expansion in the voluntary carbon market and protocols to pay farmers for sequestering carbon in agricultural soil as a climate mitigation strategy. These protocols take different approaches to quantifying SOC and net GHG removals. Three dominant approaches to SOC quantification have emerged: Remote Sensing (RS), periodic soil sampling, and process-based modeling. While RS offers scalability for monitoring, reporting, and verifying SOC, its broad implementation necessitates further research and specialized expertise. Advanced RS tools can monitor key agricultural metrics, such as crop yields and the adoption of conservation techniques like no-till farming and cover cropping. These metrics can then inform process-based models to ensure precise SOC change estimations at the farm level. Consequently, this data-driven approach can tailor incentives or rewards for farmers based on verifiable practices. Innovative tools like the Operational Tillage Information System (OpTIS) and Descartes Lab Products for Carbon Monitoring exemplify this trend. Leveraging data from multiple earth-observing satellites, these tools annually map farmland conditions to monitor vital indicators like cover crop growth, crop residues, and tillage practices. Such data are invaluable for model parameterization and for addressing potential challenges, such as determining the additionality and leakage associated with SOC sequestration projects under different protocols (EDF, 2022b). Both the US government and large supply chain entities, including giants like Walmart, are striving to cultivate an environmentally responsible, economically viable, and socially sustainable supply chain. This approach prioritizes produce cultivated using soil management practices that improve carbon sequestration, offering these products enhanced market opportunities and potential financial incentives. A testament to this commitment is the US Department of Agriculture's recent investment of $2.8 billion in the partnership for climate-smart commodities. This initiative seeks to establish a sustainable market that capitalizes on the benefits of GHG reduction, ensuring that producers receive direct and tangible benefits.

25.9 Outlook

Soil organic carbon as a commodity for future climate mitigation goals will require local, regional, and global-scale data to rationalize agricultural productivity in a balanced way to reduce future environmental threats. Remote sensing will be critical in developing spatiotemporal data inventories to support and guide different policies and protocols aimed at achieving future climate objectives. However, a truly comprehensive understanding will emerge only when in-situ data acquired by soil sampling is seamlessly integrated with remotely sensed imagery to develop a holistic and standard measurement approach. The scientific community is already rallying behind efforts to curate a global spectral library, a move geared towards introducing uniformity in measurements, analyses, and reporting protocols. Advancements in sensor technology are poised to further amplify the capabilities of remote sensing, particularly in the domain of precise SOC stock estimations, which are crucial for carbon accounting and trading markets. As carbon solidifies its status as a prized commodity in the coming years, the relevance and application of remote sensing in tracking and monitoring SOC are set to witness a significant upswing.

Conflict of interests

The authors declare no conflict of interest.

References

Agbu, P.A., Fehrenbacher, D.J., Jansen, I.J., 1990. Soil property relationships with SPOT satellite digital data in East Central Illinois. Soil Sci. Soc. Am. J. 54 (3), 807–812. Available from: https://doi.org/10.2136/sssaj1990.03615995005400030031x.

Álvaro-Fuentes, J., López, M.V., Arrúe, J.L., Moret, D., Paustian, K., 2009. Tillage and cropping effects on soil organic carbon in Mediterranean semiarid agroecosystems: testing the century model. Agric., Ecosyst. Environ. 134 (3), 211–217. Available from: https://doi.org/10.1016/j.agee.2009.07.001.

Amelung, W., Bossio, D., de Vries, W., Kögel-Knabner, I., Lehmann, J., et al., 2020. Towards a global-scale soil climate mitigation strategy. Nat Commun 11 (1), 5427. Available from: https://doi.org/10.1038/s41467-020-18887-7.

Angelopoulou, T., Dimitrakos, A., Terzopoulou, E., Zalidis, G., Theocharis, J., et al., 2017. Reflectance spectroscopy (Vis-NIR) for assessing soil heavy metals concentrations determined by two different analytical protocols, based on ISO 11466 and ISO 14869-1. Water Air Soil Pollut. 228 (11), 436. Available from: https://doi.org/10.1007/s11270-017-3609-9.

Angelopoulou, T., Tziolas, N., Balafoutis, A., Zalidis, G., Bochtis, D., 2019. Remote sensing techniques for soil organic carbon estimation: a review. Remote Sens. 11 (6), 676. Available from: https://doi.org/10.3390/rs11060676.

Bartholomeus, H.M., Schaepman, M.E., Kooistra, L., Stevens, A., Hoogmoed, W.B., et al., 2008. Spectral reflectance based indices for soil organic carbon quantification. Geoderma 145 (1–2), 28–36.

Ben-Dor, E., Goldlshleger, N., Benyamini, Y., Agassi, M., Blumberg, D.G., 2003. The spectral reflectance properties of soil structural crusts in the 1.2- to 2.5-μm spectral region. Soil Sci. Soc. Am. J. 67 (1), 289–299. Available from: https://doi.org/10.2136/sssaj2003.2890.

Ben-Dor, E., Patkin, K., Banin, A., Karnieli, A., 2002. Mapping of several soil properties using DAIS-7915 hyperspectral scanner data - a case study over clayey soils in Israel. Int. J. Remote Sens. 23 (6), 1043–1062. Available from: https://doi.org/10.1080/01431160010006962.

Blanco-Canqui, H., Lal, R., 2007. Soil structure and organic carbon relationships following 10 years of wheat straw management in no-till. Soil Tillage Res. 95 (1), 240–254. Available from: https://doi.org/10.1016/j.still.2007.01.004.

Bradford, M.A., Wieder, W.R., Bonan, G.B., Fierer, N., Raymond, P.A., et al., 2016. Managing uncertainty in soil carbon feedbacks to climate change. Nat. Clim. Change 6 (8), 751–758. Available from: https://doi.org/10.1038/nclimate3071.

Bricklemyer, R.S., Lawrence, R.L., Miller, P.R., 2002. Documenting no-till and conventional till practices using Landsat ETM + imagery and logistic regression. J. Soil Water Conserv. 57 (5), 267–271.

Bricklemyer, R.S., Lawrence, R.L., Miller, P.R., Battogtokh, N., 2006. Predicting tillage practices and agricultural soil disturbance in north central Montana with Landsat imagery. Agri., Ecosyst. Environ. 114 (2), 210–216. Available from: https://doi.org/10.1016/j.agee.2005.10.005.

Bricklemyer, R.S., Lawrence, R.L., Miller, P.R., Battogtokh, N., 2007. Monitoring and verifying agricultural practices related to soil carbon sequestration with satellite imagery. Agric., Ecosyst. Environ. 118 (1), 201–210. Available from: https://doi.org/10.1016/j.agee.2006.05.017.

Brown, D.J., Shepherd, K.D., Walsh, M.G., Dewayne Mays, M., Reinsch, T.G., 2006. Global soil characterization with VNIR diffuse reflectance spectroscopy. Geoderma 132 (3), 273–290. Available from: https://doi.org/10.1016/j.geoderma.2005.04.025.

Casa, R., Castaldi, F., Pascucci, S., Basso, B., Pignatti, S., 2013a. Geophysical and hyperspectral data fusion techniques for in-field estimation of soil properties. Vadose Zone J. 12 (4), . Available from: https://doi.org/10.2136/vzj2012.0201vzj2012.0201.

Casa, R., Castaldi, F., Pascucci, S., Palombo, A., Pignatti, S., 2013b. A comparison of sensor resolution and calibration strategies for soil texture estimation from hyperspectral remote sensing. Geoderma 197–198, 17–26. Available from: https://doi.org/10.1016/j.geoderma.2012.12.016.

Castaldi, F., 2021. Sentinel-2 and Landsat-8 multi-temporal series to estimate topsoil properties on croplands. Remote Sens. 13 (17), 3345. Available from: https://doi.org/10.3390/rs13173345.

Castaldi, F., Palombo, A., Santini, F., Pascucci, S., Pignatti, S., et al., 2016. Evaluation of the potential of the current and forthcoming multispectral and hyperspectral imagers to estimate soil texture and organic carbon. Remote Sens. Environ. 179, 54–65. Available from: https://doi.org/10.1016/j.rse.2016.03.025.

Castaldi, F., Hueni, A., Chabrillat, S., Ward, K., Buttafuoco, G., et al., 2019. Evaluating the capability of the Sentinel 2 data for soil organic carbon prediction in croplands. ISPRS J. Photogramm. Remote Sens. 147, 267–282. Available from: https://doi.org/10.1016/j.isprsjprs.2018.11.026.

Clark, R.N., Roush, T.L., 1984. Reflectance spectroscopy: quantitative analysis techniques for remote sensing applications. J. Geophys. Res.: Solid Earth 89 (B7), 6329–6340. Available from: https://doi.org/10.1029/JB089iB07p06329.

Coleman, K., Jenkinson, D.S., Crocker, G.J., Grace, P.R., Klír, J., et al., 1997. Simulating trends in soil organic carbon in long-term experiments using RothC-26.3. Geoderma 81 (1), 29–44. Available from: https://doi.org/10.1016/S0016-7061(97)00079-7.

Cozzolino, D., Morón, A., 2003. The potential of near-infrared reflectance spectroscopy to analyse soil chemical and physical characteristics. J. Agric. Sci. 140 (1), 65–71. Available from: https://doi.org/10.1017/S0021859602002836.

Dalal, R.C., Henry, R.J., 1986. Simultaneous determination of moisture, organic carbon, and total nitrogen by near infrared reflectance spectrophotometry. Soil Sci. Soc. Am. J. 50 (1), 120–123. Available from: https://doi.org/10.2136/sssaj1986.03615995005000010023x.

Dangal, S.R.S., Sanderman, J., Wills, S., Ramirez-Lopez, L., 2019. Accurate and precise prediction of soil properties from a large mid-infrared spectral library. Soil Syst. 3 (1), 11. Available from: https://doi.org/10.3390/soilsystems3010011.

Del Grosso, S., Parton, W., Mosier, A., Hartman, M., Brenner, J., et al., 2001. Simulated interaction of carbon dynamics and nitrogen trace gas fluxes using the DAYCENT model. Modeling Carbon and Nitrogen Dynamics for Soil Management. CRC Press, Boca Raton, FL, pp. 303–332.

Dietzen, C.A., Larsen, K.S., Ambus, P.L., Michelsen, A., Arndal, M.F., et al., 2019. Accumulation of soil carbon under elevated CO2 unaffected by warming and drought. Global Change Biol. 25 (9), 2970–2977. Available from: https://doi.org/10.1111/gcb.14699.

Dvorakova, K., Shi, P., Limbourg, Q., van Wesemael, B., 2020. Soil organic carbon mapping from remote sensing: the effect of crop residues. Remote Sens. 12 (12), 1913. Available from: https://doi.org/10.3390/rs12121913.

EDF. 2022a. Soil carbon credits have potential but need clear standards. Environmental Defense Fund. <https://www.edf.org/soilcarbon> (accessed 15.10.22).

EDF. 2022b. Agricultural soil carbon credits: making sense of protocols for carbon sequestration and net greenhouse gas removals.

England, J.R., Viscarra Rossel, R.A., 2018. Proximal sensing for soil carbon accounting. Soil 4 (2), 101–122. Available from: https://doi.org/10.5194/soil-4-101-2018.

Gallo, B.C., Demattê, J.A.M., Rizzo, R., Safanelli, J.L., Mendes, W.D.S., et al., 2018. Multi-temporal satellite images on topsoil attribute quantification and the relationship with soil classes and geology. Remote Sens. 10 (10), 1571. Available from: https://doi.org/10.3390/rs10101571.

Garosi, Y., Ayoubi, S., Nussbaum, M., Sheklabadi, M., 2022. Effects of different sources and spatial resolutions of environmental covariates on predicting soil organic carbon using machine learning in a semi-arid region of Iran. Geoderma Reg. 29, e00513. Available from: https://doi.org/10.1016/j.geodrs.2022.e00513.

Gholizadeh, A., Žižala, D., Saberioon, M., Borůvka, L., 2018. Soil organic carbon and texture retrieving and mapping using proximal, airborne and Sentinel-2 spectral imaging. Remote Sens. Environ. 218, 89–103. Available from: https://doi.org/10.1016/j.rse.2018.09.015.

Goetz, H., Vane, G., Solomon, J. E., Rock, B. N. 1985. Imaging Spectrometry for Earth Remote Sensing. *Science*. https://doi.org/10.1126/science.228.4704.1147.

Goidts, E., van Wesemael, B., 2007. Regional assessment of soil organic carbon changes under agriculture in Southern Belgium (1955–2005). Geoderma 141 (3), 341–354. Available from: https://doi.org/10.1016/j.geoderma.2007.06.013.

Gomez, C., Lagacherie, P., Coulouma, G., 2008a. Continuum removal versus PLSR method for clay and calcium carbonate content estimation from laboratory and airborne hyperspectral measurements. Geoderma 148 (2), 141–148. Available from: https://doi.org/10.1016/j.geoderma.2008.09.016.

Gomez, C., Viscarra Rossel, R.A., McBratney, A.B., 2008b. Soil organic carbon prediction by hyperspectral remote sensing and field vis-NIR spectroscopy: an Australian case study. Geoderma 146 (3), 403–411. Available from: https://doi.org/10.1016/j.geoderma.2008.06.011.

Guo, L., Shi, T., Linderman, M., Chen, Y., Zhang, H., et al., 2019. Exploring the influence of spatial resolution on the digital mapping of soil organic carbon by airborne hyperspectral VNIR Imaging. Remote Sensing 11 (9), 1032. Available from: https://doi.org/10.3390/rs11091032.

Gupta, G., Singh, J., Pandey, P.C., Tomar, V., Rani, M., et al., 2014. Geospatial strategy for estimation of soil organic carbon in tropical wildlife reserve. In: Srivastava, P.K., Mukherjee, S., Gupta, M., Islam, T. (Eds.), Remote Sensing Applications in Environmental Research. Springer International Publishing, Cham, pp. 69–83.

Haddaway, N.R., Hedlund, K., Jackson, L.E., Kätterer, T., Lugato, E., et al., 2017. How does tillage intensity affect soil organic carbon? A systematic review. Environ. Evid. 6 (1), 30. Available from: https://doi.org/10.1186/s13750-017-0108-9.

Hicks Pries, C.E., Castanha, C., Porras, R.C., Torn, M.S., 2017. The whole-soil carbon flux in response to warming. Science 355 (6332), 1420–1423. Available from: https://doi.org/10.1126/science.aal1319.

Hong, Y., Chen, S., Chen, Y., Linderman, M., Mouazen, A.M., et al., 2020. Comparing laboratory and airborne hyperspectral data for the estimation and mapping of topsoil organic carbon: feature selection coupled with random forest. Soil Tillage Res. 199, 104589. Available from: https://doi.org/10.1016/j.still.2020.104589.

Huang, X., Senthilkumar, S., Kravchenko, A., Thelen, K., Qi, J., 2007. Total carbon mapping in glacial till soils using near-infrared spectroscopy, landsat imagery and topographical information. Geoderma 141 (1), 34–42. Available from: https://doi.org/10.1016/j.geoderma.2007.04.023.

Ingleby, H., Crowe, T., 2001. Neural network models for predicting organic…. Google Scholar. Outlook 43. <https://scholar.google.com/scholar_-lookup?title = Neural%20network%20models%20for%20predicting%20organic%20matter%20content%20in%20Saskatchewan%20soils&publication_year = 2001&author = H.%20Ingleby&author = T.%20Crowe> (accessed 11.09.22).

Jaber, S.M., Lant, C.L., Al-Qinna, M.I., 2011. Estimating spatial variations in soil organic carbon using satellite hyperspectral data and map algebra. Int. J. Remote Sens. 32 (18), 5077–5103. Available from: https://doi.org/10.1080/01431161.2010.494637.

Jensen, J.R., 2007. Remote Sensing of the Environment: An Earth Resource Perspective, 2nd Edition Pearson Prentice Hall, Upper Saddle River.

Jones, L.A., Kimball, J.S., Reichle, R.H., Madani, N., Glassy, J., et al., 2017. The SMAP Level 4 carbon product for monitoring ecosystem land–atmosphere CO2 exchange. IEEE Trans. Geosci. Remote Sens. 55 (11), 6517–6532. Available from: https://doi.org/10.1109/TGRS.2017.2729343.

Kaur, T., Brar, B.S., Dhillon, N.S., 2008. Soil organic matter dynamics as affected by long-term use of organic and inorganic fertilizers under maize–wheat cropping system. Nutr. Cycl. Agroecosyst. 81 (1), 59–69. Available from: https://doi.org/10.1007/s10705-007-9152-0.

Kuang, B., Tekin, Y., Mouazen, A.M., 2015. Comparison between artificial neural network and partial least squares for on-line visible and near infrared spectroscopy measurement of soil organic carbon, pH and clay content. Soil Tillage Res. 146, 243–252. Available from: https://doi.org/10.1016/j.still.2014.11.002.

Ladoni, M., Bahrami, H.A., Alavipanah, S.K., Norouzi, A.A., 2010. Estimating soil organic carbon from soil reflectance: a review. Precis. Agric. 11 (1), 82–99.

Lal, R., 2008. The urgency of conserving soil and water to address 21st century issues including global warming. J. Soil Water Conserv. 63 (5), 140A–141A. Available from: https://doi.org/10.2489/jswc.63.5.140A.

Lillesand, T., Kiefer, R.W., Chipman, J., 2015. Remote Sensing and Image Interpretation. John Wiley & Sons.

Liu, S., Bliss, N., Sundquist, E., Huntington, T.G., 2003. Modeling carbon dynamics in vegetation and soil under the impact of soil erosion and deposition. Global Biogeochem. Cycles 17 (2). Available from: https://doi.org/10.1029/2002GB002010.

Liu, S., Tan, Z., Chen, M., Liu, J., Wein A., et al. 2012. The General Ensemble Biogeochemical Modeling System (GEMS) and its applications to agricultural systems in the United States: Chapter 18: 309–323. doi: 10.1016/B978-0-12-386897-8.00018-8.

Lobsey, C.R., Viscarra Rossel, R.A., Roudier, P., Hedley, C.B., 2017. RS-local data-mines information from spectral libraries to improve local calibrations. Eur. J. Soil Sci. 68 (6), 840–852. Available from: https://doi.org/10.1111/ejss.12490.

Loizzo, R., Guarini, R., Longo, F., Scopa, T., Formaro, R., et al. 2018. Prisma: the Italian hyperspectral mission. In: IGARSS 2018—2018 IEEE International Geoscience and Remote Sensing Symposium, pp. 175–178.

Lundén, B., Wang, G., Wester, K., 2001. A GIS based analysis of data from Landsat TM, airborne geophysical measurements, and digital maps for geological remote sensing in the Stockholm region, Sweden. Int. J. Remote Sens. 22 (4), 517–532. Available from: https://doi.org/10.1080/01431160050505838.

Maharjan, B., Das, S., Nielsen, R., Hergert, G.W., 2021. Maize yields from manure and mineral fertilizers in the 100-year-old Knorr–Holden Plot. Agronomy J. . Available from: https://doi.org/10.1002/agj2.20713.

Mäkipää, R., Häkkinen, M., Muukkonen, P., Peltoniemi, M. 2008. The costs of monitoring changes in forest soil carbon stocks. https://jukuri.luke.fi/handle/10024/514760 (accessed 10.09.22).

Maynard, J.J., Levi, M.R., 2017. Hyper-temporal remote sensing for digital soil mapping: Characterizing soil-vegetation response to climatic variability. Geoderma 285, 94–109. Available from: https://doi.org/10.1016/j.geoderma.2016.09.024.

Mcmorrow, J.M., Cutler, M.E.J., Evans, M.G., Al-Roichdi., A., 2004. Hyperspectral indices for characterizing upland peat composition. Int. J. Remote Sens. 25 (2), 313–325. Available from: https://doi.org/10.1080/0143116031000117065.

Miller, S.M., Michalak, A.M., 2020. The impact of improved satellite retrievals on estimates of biospheric carbon balance. Atmos. Chem. Phys. 20, 323–331. Available from: https://doi.org/10.5194/acp-20-323-2020.

Mirzaee, S., Ghorbani-Dashtaki, S., Mohammadi, J., Asadi, H., Asadzadeh, F., 2016. Spatial variability of soil organic matter using remote sensing data. CATENA 145, 118–127. Available from: https://doi.org/10.1016/j.catena.2016.05.023.

Moran, M.S., Hymer, D.C., Qi, J., Kerr, Y., 2002. Comparison of ERS-2 SAR and Landsat TM imagery for monitoring agricultural crop and soil conditions. Remote Sens. Environ. 79 (2), 243–252. Available from: https://doi.org/10.1016/S0034-4257(01)00276-0.

Mulders, M.A., 1987. Remote Sensing in Soil Science. Elsevier.

Mzid, N., Castaldi, F., Tolomio, M., Pascucci, S., Casa, R., et al., 2022. Evaluation of agricultural bare soil properties retrieval from Landsat 8, Sentinel-2 and PRISMA satellite data. Remote Sens. 14 (3), 714. Available from: https://doi.org/10.3390/rs14030714.

Nanni, M.R., Demattê, J.A.M., 2006. Spectral reflectance methodology in comparison to traditional soil analysis. Soil Sci. Soc. Am. J. 70 (2), 393–407. Available from: https://doi.org/10.2136/sssaj2003.0285.

Odebiri, O., Mutanga, O., Odindi, J., 2022. Deep learning-based national scale soil organic carbon mapping with Sentinel-3 data. Geoderma 411, 115695. Available from: https://doi.org/10.1016/j.geoderma.2022.115695.

Odebiri, O., Odindi, J., Mutanga, O., 2021. Basic and deep learning models in remote sensing of soil organic carbon estimation: a brief review. Int. J. Appl. Earth Obs. Geoinform. 102, 102389. Available from: https://doi.org/10.1016/j.jag.2021.102389.

Olness, A., Archer, D., 2005. Effect of organic carbon on available water in soil. Soil Science 170 (2), 90–101.

Paustian, K., Lehmann, J., Ogle, S., Reay, D., Robertson, G.P., et al., 2016. Climate-smart soils. Nature 532 (7597), 49–57. Available from: https://doi.org/10.1038/nature17174.

Rama Rao, N., Garg, P.K., Ghosh, S.K., 2007. Evaluation of radiometric resolution on land use/land cover mapping in an agricultural area. Int. J. Remote Sens. 28 (2), 443–450. Available from: https://doi.org/10.1080/01431160600733181.

Rawlins, B.G., Marchant, B.P., Smyth, D., Scheib, C., Lark, R.M., et al., 2009. Airborne radiometric survey data and a DTM as covariates for regional scale mapping of soil organic carbon across Northern Ireland. Eur. J. Soil Sci. 60 (1), 44–54. Available from: https://doi.org/10.1111/j.1365-2389.2008.01092.x.

Rogge, D., Bauer, A., Zeidler, J., Mueller, A., Esch, T., et al., 2018. Building an exposed soil composite processor (SCMaP) for mapping spatial and temporal characteristics of soils with Landsat imagery (1984–2014. Remote Sens. Environ. 205, 1–17. Available from: https://doi.org/10.1016/j.rse.2017.11.004.

Romero, D.J., Ben-Dor, E., Demattê, J.A.M., Souza, A.B.e, Vicente, L.E., et al., 2018. Internal soil standard method for the Brazilian soil spectral library: performance and proximate analysis. Geoderma 312, 95–103. Available from: https://doi.org/10.1016/j.geoderma.2017.09.014.

Rossel, R.V., Behrens, T., 2010. Using data mining to model and interpret soil diffuse reflectance spectra. Geoderma 158 (1–2), 46–54. Available from: https://doi.org/10.1016/j.geoderma.2009.12.025.

Samsonova, V.P., Meshalkina, J.L., Blagoveschensky, Y.N., Yaroslavtsev, A.M., Stoorvogel, J.J., 2018. The role of positional errors while interpolating soil organic carbon contents using satellite imagery. Precision Agric 19 (6), 1085–1099. Available from: https://doi.org/10.1007/s11119-018-9575-4.

Sanderman, J., Hengl, T., Fiske, G.J. 2017. Soil carbon debt of 12,000 years of human land use. Proc. Natl. Acad. Sci. 114 (36), 9575–9580. doi: 10.1073/pnas.1706103114.

Schillaci, C., Lombardo, L., Saia, S., Fantappiè, M., Märker, M., et al., 2017. Modelling the topsoil carbon stock of agricultural lands with the stochastic gradient treeboost in a semi-arid Mediterranean region. Geoderma 286, 35–45. Available from: https://doi.org/10.1016/j.geoderma.2016.10.019.

Selige, T., Böhner, J., Schmidhalter, U., 2006. High resolution topsoil mapping using hyperspectral image and field data in multivariate regression modeling procedures. Geoderma 136 (1), 235–244. Available from: https://doi.org/10.1016/j.geoderma.2006.03.050.

Shepherd, K.D., Walsh, M.G., 2002. Development of reflectance spectral libraries for characterization of soil properties. Soil Sci. Soc. Am. J. 66 (3), 988–998. Available from: https://doi.org/10.2136/sssaj2002.9880.

Sila, A.M., Shepherd, K.D., Pokhariyal, G.P., 2016. Evaluating the utility of mid-infrared spectral subspaces for predicting soil properties. Chemom. Intell. Lab. Syst. 153, 92–105. Available from: https://doi.org/10.1016/j.chemolab.2016.02.013.

Smith, P., Soussana, J., Angers, D., Schipper, L., Chenu, C., et al., 2020. How to measure, report and verify soil carbon change to realize the potential of soil carbon sequestration for atmospheric greenhouse gas removal. Glob. Chang. Biol. 26 (1), 219–241. Available from: https://doi.org/10.1111/gcb.14815.

Sothe, C., Gonsamo, A., Arabian, J., Snider, J., 2022. Large scale mapping of soil organic carbon concentration with 3D machine learning and satellite observations. Geoderma 405, 115402. Available from: https://doi.org/10.1016/j.geoderma.2021.115402.

Steinberg, A., Chabrillat, S., Stevens, A., Segl, K., Foerster, S., 2016. Prediction of common surface soil properties based on vis-NIR airborne and simulated EnMAP imaging spectroscopy data: prediction accuracy and influence of spatial resolution. Remote Sens. 8 (7), 613. Available from: https://doi.org/10.3390/rs8070613.

Stevens, A., Udelhoven, T., Denis, A., Tychon, B., Lioy, R., et al., 2010. Measuring soil organic carbon in croplands at regional scale using airborne imaging spectroscopy. Geoderma 158 (1), 32–45. Available from: https://doi.org/10.1016/j.geoderma.2009.11.032.

Stoner, E.R., Baumgardner, M.F., 1981. Characteristic variations in reflectance of surface soils. Soil Sci. Soc. Am. J. 45 (6), 1161–1165. Available from: https://doi.org/10.2136/sssaj1981.03615995004500060031x.

Sullivan, D.G., Shaw, J.N., Rickman, D., 2005. IKONOS imagery to estimate surface soil property variability in two alabama physiographies. Soil Science Society of America Journal 69 (6), 1789–1798. Available from: https://doi.org/10.2136/sssaj2005.0071.

Thenkabail, P.S., Smith, R.B., De Pauw, E., 2000. Hyperspectral Vegetation Indices and Their Relationships with Agricultural Crop Characteristics. Remote Sensing of Environment 71 (2), 158–182. Available from: https://doi.org/10.1016/S0034-4257(99)00067-X.

UNFCC. 2022. The Doha Amendment. <https://unfccc.int/process/the-kyoto-protocol/the-doha-amendment> (accessed 15.10.22).

Varvel, G.E., Wilhelm, W.W., 2010. Long-term soil organic carbon as affected by tillage and cropping systems. Soil Sci. Soc. Am. J. 74 (3), 915–921. Available from: https://doi.org/10.2136/sssaj2009.0362.

Vaudour, E., Bel, L., Gilliot, J.M., Coquet, Y., Hadjar, D., et al., 2013. Potential of SPOT multispectral satellite images for mapping topsoil organic carbon content over peri-urban croplands. Soil Sci. Soc. Am. J. 77 (6), 2122–2139. Available from: https://doi.org/10.2136/sssaj2013.02.0062.

Vaudour, E., Gomez, C., Fouad, Y., Lagacherie, P., 2019a. Sentinel-2 image capacities to predict common topsoil properties of temperate and Mediterranean agroecosystems. Remote Sens. Environ. 223, 21–33. Available from: https://doi.org/10.1016/j.rse.2019.01.006.

Vaudour, E., Gomez, C., Loiseau, T., Baghdadi, N., Loubet, B., et al., 2019b. The impact of acquisition date on the prediction performance of topsoil organic carbon from Sentinel-2 for croplands. Remote Sens. 11 (18), 2143. Available from: https://doi.org/10.3390/rs11182143.

Viña, A., Peters, A.J., Ji, L., 2003. Use of multispectral ikonos imagery for discriminating between conventional and conservation agricultural tillage practices. Photogramm. Eng. Remote Sens. 69 (5), 537–544. Available from: https://doi.org/10.14358/PERS.69.5.537.

Viscarra Rossel, R.A., Behrens, T., Ben-Dor, E., Brown, D.J., Demattê, J.A.M., et al., 2016. A global spectral library to characterize the world's soil. Earth-Sci. Rev. 155, 198–230. Available from: https://doi.org/10.1016/j.earscirev.2016.01.012.

Waes, C.V., Mestdagh, I., Lootens, P., Carlier, L., 2005. Possibilities of near infrared reflectance spectroscopy for the prediction of organic carbon concentrations in grassland soils. J. Agri. Sci. 143 (6), 487–492. Available from: https://doi.org/10.1017/S0021859605005630.

Wang, K., Qi, Y., Guo, W., Zhang, J., Chang, Q., 2021a. Retrieval and mapping of soil organic carbon using Sentinel-2A spectral images from bare cropland in autumn. Remote Sens. 13 (6), 1072. Available from: https://doi.org/10.3390/rs13061072.

Wang, H., Zhang, X., Wu, W., Liu, H., 2021b. Prediction of soil organic carbon under different land use types using Sentinel-1/-2 data in a small watershed. Remote Sens. 13 (7), 1229. Available from: https://doi.org/10.3390/rs13071229.

Ward, K.J., Chabrillat, S., Brell, M., Castaldi, F., Spengler, D., et al., 2020. Mapping soil organic carbon for airborne and simulated EnMAP imagery using the LUCAS soil database and a local PLSR. Remote Sens. 12 (20), 3451. Available from: https://doi.org/10.3390/rs12203451.

Wu, C., 2012. Use of a vegetation index model to estimate gross primary production in open grassland. JARS 6 (1), 063532. Available from: https://doi.org/10.1117/1.JRS.6.063532.

Yang, M., Xu, D., Chen, S., Li, H., Shi, Z., 2019. Evaluation of machine learning approaches to predict soil organic matter and pH using vis-NIR spectra. Sensors 19 (2), 263. Available from: https://doi.org/10.3390/s19020263.

Ziegler, S.E., Benner, R., Billings, S.A., Edwards, K.A., Philben, M., et al. 2017. Climate warming can accelerate carbon fluxes without changing soil carbon stocks. Front. Earth Sci. 5. <https://www.frontiersin.org/articles/10.3389/feart.2017.00002> (accessed 18.09.22).

Žížala, D., Minařík, R., Zádorová, T., 2019. Soil organic carbon mapping using multispectral remote sensing data: prediction ability of data with different spatial and spectral resolutions. Remote Sens. 11 (24), 2947. Available from: https://doi.org/10.3390/rs11242947.

Žížala, D., Zádorová, T., Kapička, J., 2017. Assessment of soil degradation by erosion based on analysis of soil properties using aerial hyperspectral images and Ancillary Data, Czech Republic. Remote Sens. 9 (1), 28. Available from: https://doi.org/10.3390/rs9010028.

Chapter 26

Spatial prediction of soil organic carbon and its stocks using digital soil mapping approach

B. Kalaiselvi, Ranabir Chakraborty, S. Dharumarajan, K.S. Anil Kumar and Rajendra Hegde
ICAR-National Bureau of Soil Survey and Land Use Planning, Regional Centre, Bangalore, Karnataka, India

26.1 Introduction

A reliable estimate of the amount of organic carbon stock in soils is helpful in a variety of issues, such as climate change, the water retention capacity of soil, soil fertility, and soil buffering capacity. The soil organic carbon (SOC) is a crucial indicator of the quality of the soil (de Moraes Sa, Lal, 2009; Stevens et al., 2010). Some writers have even asserted that variations in SOC are primarily responsible for observed variances in soil quality (Tan et al., 2007). It is a significant indicator of the physical, chemical, and biological qualities of soil, all of which contribute to soil productivity (Lal, 2001; Rossi et al., 2009). Therefore, a higher amount of SOC is equivalent to better soil quality, and vice versa. The amount of carbon stock on the planet is around 1.5 gigatons (Scharlemann et al., 2014). The SOC is one of the five major carbon pools on the planet; the other four are the oceans, lithosphere, atmosphere, and biomass (Batjes and Sombroek, 1997). The SOC pool is the most active carbon storage system on earth and its total carbon content is equal to that of the biosphere and the atmosphere put together (Wang et al., 2009). Soil has an indispensable role in the carbon sequestration processes that are responsible for approximately 75% of the carbon storage in the terrestrial biosphere (Scharlemann et al., 2014). Soil carbon sequestration is a process that involves removing carbon dioxide from the air and then safely storing it in the ground for a considerable period of time. This is done in order to prevent the gas from being released back into the atmosphere (Lal, 2004a,b). Carbon sequestration dynamics are affected by a wide variety of thermodynamic parameters, biome characteristics, different land uses and land management systems (Batjes, 1998). Depending on the proportional rates of carbon absorption and breakdown by soil organisms, the soil may operate as both a source and a sink of carbon. This is because the soil contains both living and dead organisms. The estimation of SOC stock and an understanding of different biophysical processes that may impact the distribution of SOC in soils require data on the spatial variability of the environment. Among the greenhouse gases, carbon dioxide (CO_2) plays the greatest role in global warming (UNFCCC, 1997). Parent material, climate, land use and soil management, drainage, soil texture, availability of nutrients, etc. have a substantial impact on the release and sequestration of carbon in the soil (FAO, 2007; Rutunga et al., 2007). According to Perez et al. (2007), high temperatures in sub-Saharan Africa and the tropics expedite the breakdown of SOM, which in turn lowers the amount of carbon that is sequestered by the soil. It is anticipated that anaerobic soils, which are often cool and wet, would be able to retain a greater quantity of carbon than upland aerobic soils due to slower decomposition (Sinsabaugh, 2010). Variations in temperature, moisture and activity of soil biota influence the amount of carbon that is sequestered. The Kyoto Protocol recognized the SOC as a key variable and mandated that signatory governments monitor SOC changes and take actions to maintain or build up the SOC (IPCC, 2000; Lal, 2003; Steffen et al., 1998). However, estimating the amounts of SOC and its changes faces several hurdles, such as mismatching of analytical data on a geographical and temporal scale, soil heterogeneity, a lack of bulk density data, etc. (Dungait et al., 2012; Wielopolski et al., 2011). Scattered methodologies and unequal precision make it more difficult to identify and measure the geographical distribution of SOC on a local, regional or global scale (Mulder et al., 2011; Reeves, 2010). The estimation of the spatial and temporal dynamics of carbon sources and sinks at the local, regional, and global scales requires the development of novel methodologies (Liu et al., 2004). The SOC may

Remote Sensing of Soils. DOI: https://doi.org/10.1016/B978-0-443-18773-5.00003-X

be examined in a laboratory setting as well as out in the field. In recent years, an interest in technologies such as remote sensing, geographical information systems (GIS), and digital soil mapping (DSM) has been spurred because of their cost-effectiveness, precision, and monitoring ability across large stretches of area. In the GIS platform, geographically referenced datasets are combined for modeling and decision-making applications, whereas remote sensing provides synoptic views, vast area repetition coverage, and continuous spatial information. Remote sensing and GIS together may prove indispensable in the construction of national C inventory databases and the tracking of changes inf C pool composition. However, SOC evaluation that is based on surface reflectance has its limits. Recently, DSM has been recognized as an empirical method for identifying soil properties, including SOC as a function of the SCORPAN factors. This will be detailed in the next sections of this chapter.

26.2 The development of methodologies for calculating SOC

26.2.1 Laboratory methods

The acidic dichromate oxidation of SOC developed by Walkley and Black (1934) is simple, quick and requires minimum equipment. But, the accuracy of estimation varies due to land use, soil depth, and texture (De Vos et al., 2007). Dry combustion devices, such as the CHNS analyzer, burn the C at a high temperature (around 900°C−1100°C) in the presence of O_2 and detect the CO_2 by gas chromatography (Nelson and Sommers, 1996). On the other hand, the weight loss on ignition approach is a dry combustion methodology that measures the weight loss of soil samples ignited at 430°C for 24 hours. The SOC measured in laboratories is expressed on a weight basis, which requires conversion to a volume basis to delineate the geographical distribution of SOC. The bulk density data of soil is needed to perform this transformation. Estimation of bulk density needs fresh core samples acquired in the field. The volume of the soil core is measured from its radius (r), height (h), and the mass of the soil sample per unit volume of core gives the bulk density of the soil (g cm^{-3} or Mg m^{-3}). The estimation of SOC in the laboratory frequently requires destructive sampling, which is then followed by preliminary processing to minimize the heterogeneity of the soil sample. In addition, soil BD may be promptly estimated in situ using near-infrared or mid-infrared spectroscopy, or with a gamma ray probe (Bellon-Maurel and McBratney, 2011; Croft et al., 2012). Despite the need for extensive sample preparation (collecting, grinding, sifting, and drying), spectroscopic techniques offer a fast and nondestructive method for measuring SOC with a greater precision, saving money and processing time. Croft et al. (2012) and Rossel et al. (2006) both found that spectroscopic techniques were superior to other methods. The spectroscopic methods may prove convenient for calibrating satellite and aerial data (Stevens et al., 2010). The testing of SOC and the evaluation of BD both take time and require a significant amount of labor (Chatterjee et al., 2009; Lal, 2006). In addition, generating an inventory of soil carbon across a large region using any of the laboratory methods is time-consuming and extremely expensive.

26.2.2 Field-based methods

Assessment of SOC that is conducted outside of a laboratory may be both lucrative and relevant in situations when a laboratory is not accessible (Reeves, 2010). In situ computation of the SOC pool is possible with the inelastic neutron scattering (INS) approach (Wielopolski et al., 2011). In this method, thermal neutrons are generated that interact with soil components, allowing for in situ computation of the SOC pool. The INS method is nondestructive, has the ability to assess enormous soil volumes, does not require a considerable number of BD calibrations, and can be utilized in either the scanning mode or the static mode (Wielopolski et al., 2011). Other field-based methods include laser-induced breakdown spectroscopy, dry combustion, and diffuse reflectance Fourier transform infrared spectroscopy. However, these are destructive methods that only offer point data (McCarty et al., 2002). Calculations of SOC and data collection could also be done on the go, thanks to sensors mounted on tractors (Croft et al., 2012). Even if these field methods cost less than lab-based methods, they do not significantly lower the cost of generating a soil carbon map across a large region.

26.2.3 Aerial and satellite measurements

As more historical data becomes readily available, as the price of data continues to drop, and as the resolution of satellite platforms continues to improve, remote sensing is becoming a more practical tool for use in applications in the public and private sectors of the economy (GOFC-GOLD, 2012; Rowland et al., 2007). It is possible to assess SOC remotely by first determining the spectrum reflectance of the soil and then developing algorithms that relate spectral features to the SOC that is detected in a particular location. On the other hand, distant sensors that are carried by aircraft or satellites pick up just the surface reflectance that is related to SOC. It is possible that it will be necessary to compute subsurface SOC in a roundabout way by utilizing models that connect surface and subsurface SOC. Other aspects to take into account when calculating SOC using airborne and satellite-borne sensors are the fact that soil moisture levels can vary in different places, as can the presence of vegetation, which can impede the capacity of remote sensors to capture soil reflectance. When crops are not present in the field or have just started to grow, agricultural systems may make use of airborne or spaceborne sensors to monitor the reflectance of the soil (de Paul Obade and Lal, 2013). A significant match was found between the data collected from ground-based sensors and the data collected from optical sensors that were installed on aerial platforms to simulate the SOC (Stevens et al., 2010). On the other hand, Gomez et al. (2008) discovered a weaker relationship between the SOC that was predicted by the Hyperion satellite and the SOC that was determined on the ground or in the laboratory. They attributed this finding to a low signal-to-noise ratio. The analysis of spectral reflectance data is necessary in order to get rid of noise and mistakes (Coppin et al., 2004). The reflected light that reaches the satellite includes many altered signals that could be caused by several factors, including Rayleigh, Mie and nonselective scattering, absorption by clouds, sensor drift, and bidirectional reflectance distribution functions. To minimize the errors, atmospheric, geometric and radiometric corrections are performed (Coppin et al., 2004; de Paul Obade and Lal, 2013; Moran et al., 1997; Stevens et al., 2010). In order to reduce the likelihood of interference of this kind, surface and subsurface soil samples can be transported to a laboratory and analyzed using proximal remote sensing. The proximal sensing allows the sensor's front optic to be in contact with the object or within a limit of 2 m and minimizes noise. For the purpose of data collection, either a portable spectro-radiometer or a spectro-radiometer attached to a tractor can be used. The adoption of remote sensing technology was intended to create an accurate and cost-effective carbon inventory for a large region. However, collecting soils from different depths and transporting them into the laboratory significantly increases the capital requirements. Eventually, this defeats the purpose of considering remote sensing technology. Moreover, it is essential to develop robust statistical methods, regardless of whether proximal or distant remote sensing techniques are utilized. After methods have been developed, models are constructed that establish a connection between the concentration of SOC at different depths and reflectance. Statistical analysis, such as regression analysis or analysis of variance, may be applied in order to validate remotely sensed objects by making use of data collected from actual fields (ANOVA). If there is a significant disparity between the modeled and real values, the model is adjusted and reapplied to obtain a new SOC estimate. This new estimate is then examined, and the process of developing the model is repeated till the outcomes are satisfactory (de Paul Obade and Lal, 2013). Comparisons or discussions are done over how accurate the final SOC values are.

26.2.4 Estimation of SOC using standard modeling

Generic models are developed by applying a collection of model parameters with the expectation that these models will provide reliable SOC estimations throughout extensive geographic regions. GIS-based models for monitoring SOC at various scales include the CENTURY, RothC, GEMS, DayCent, and EDCM (de Paul Obade and Lal, 2013) (Fig. 26.1). The "CENTURY" model, presented by Parton et al. (1987), is an ecosystem model that focuses on plant productivity. In order to accurately depict biogeochemistry, the CENTURY model requires an in-depth description of the crop species and the management practices that are used. Carbon, nitrogen, and phosphorus cycle dynamics across many ecosystems can be reproduced by both GEMS and EDCM, which are both more than a century old and process-based modeling systems and both models were developed with the purpose of exploring how soil erosion and deposition might affect ecosystems (Liu et al., 2011; Zhao et al., 2010).

The factors influencing crop productivity are used as inputs in the Rothamsted Carbon Model (RothC), which belongs to the class of soil decomposition models (Coleman and Jenkinson, 1996). EPIC and CQUESTER are two other SOC forecast models (Lal, 2009). Different models need a different set of inputs; some of the most common ones are climatic factors, for example, monthly rainfall, monthly minimum and maximum temperatures, and evaporation; soil factors, for example, texture, bulk density, permeability, water holding capacity, soil and carbon level; management factors, for example, cropping system, cultivation practices, atmospheric N deposition, fertilization, and land cover; and natural disturbances, such as soil erosion or forest fire (de Paul Obade and Lal, 2013; Dieye et al., 2012; Liu et al., 2008; Parton et al., 1987). The most significant output variables are the net primary production, grain yield, rate of

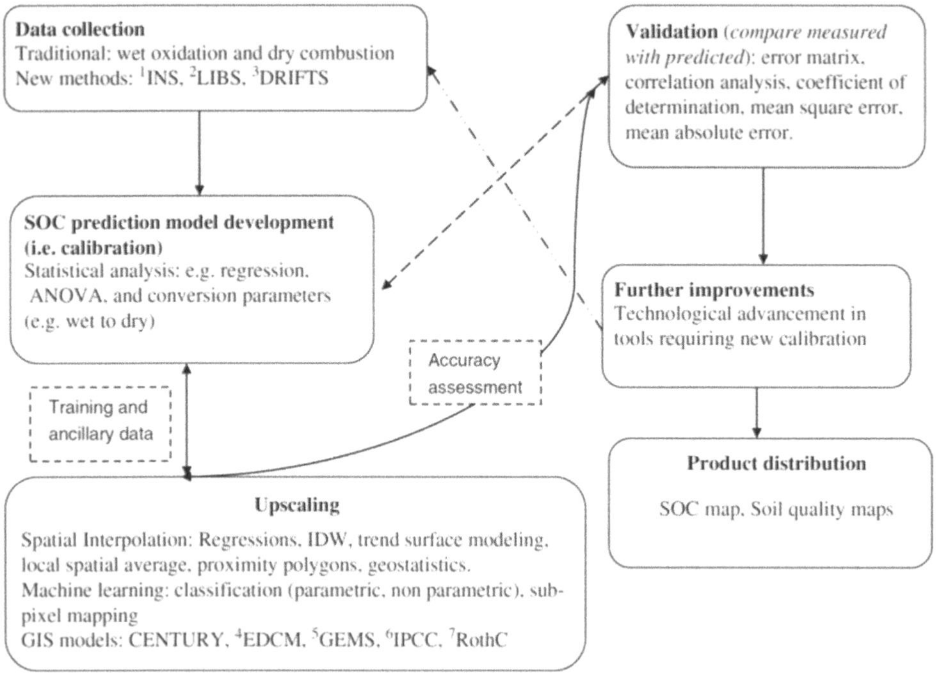

FIGURE 26.1 Steps involved in mapping soil carbon sequestration at regional scales. *DRIFTS*, Diffuse reflectance Fourier transform infrared spectroscopy; *GEMS*, general ensemble biogeochemical modeling system; *IDW*, inverse distance weighting; *INS*, inelastic neutron scattering; *RothC*, Rothamsted carbon model. *Adapted from de Paul Obade, V., Lal, R., 2013. Assessing land cover and soil quality by remote sensing and geographical information systems (GIS). Catena 104, 77–92.*

carbon breakdown, rates of carbon exchange between ecosystems and the atmosphere, amount of biomass removed by harvesting, amount of carbon stored in plants and soils, etc. (de Paul Obade and Lal, 2013). Predictability in terms of geography and time may also be achieved with generic models (Liu et al., 2004, 2008). The major drawback of using generic models is that they performed well only in the soil systems in which they were developed, and therefore entail inputs from limited historical data. This is a significant limitation of these types of models (Dungait et al., 2012; Liu et al., 2011). Ogle et al. (2007) discovered that when the CENTURY model was employed at a regional level rather than a local one, it gave more accurate results for agricultural management of SOC. In addition, the utilization of secondary data in generic models has the potential to amplify the degree of uncertainty in SOC estimations (Jones et al., 2007; Stevens et al., 2010).

26.2.5 The drawbacks of conventional SOC estimate methods

Establishing precise, rapid, and methodical methods for measuring SOC at wide regional scales is a challenging endeavor. Interpolation accuracy can be affected by factors such as the requirement for a high sample density when working with point data as well as the significant geographical variance in bulk density (Moran et al., 1997; van Wesemael et al., 2011). It is necessary to test SOC and evaluate BD, but doing so takes a lot of time and requires a lot of work (Chatterjee et al., 2009; Lal, 2006). Furthermore, using a laboratory approach to create an inventory of soil carbon across a large region is ridiculously expensive. In spite of the fact that aerial and satellite-derived reflectance data have the potential to create geographically continuous SOC maps with high temporal resolutions, the development of these maps has been slowed down by obstacles such as the high costs of data collection, the requirements for preprocessing, and the complexity of the technology (Croft et al., 2012; Roy et al., 2006). Evaluation of SOC changes based on remote sensing is made even more challenging by the inconsistent geographical, spectral, and temporal resolutions of the data collected by remote sensing instruments (Lobell, 2010). When comparing maps, one ought to utilize a method that is resolution-independent to reduce the error brought on by discrepancies in the spatial resolution of the various sensors. Due to changes in chemical or physical factors such as soil mineralogy, soil moisture, soil texture, and particle size, it is challenging to capture pure spectrum reflectance signals of SOC. As a result, monitoring SOC levels is difficult (Croft et al., 2012; Wielopolski et al., 2011). On the other hand, the downside of generic models is that they can only be applied to the soil systems for which they were built, and they only accept a limited amount of historical data as inputs. This is the case even though these models were supposed to be universally applicable.

26.3 The scope of digital soil mapping

Through the compilation of soil maps at various resolutions, soil science has the potential to establish a spatial data infrastructure with the advances of computer and information technologies. On the other hand, the existing soil databases do not include nearly enough information, nor are they precise enough for the accurate and widespread use of soil data. This lack of soil spatial data is primarily attributable to the time-consuming and expensive traditional soil survey processes (Lagacherie and McBratney, 2006; Dharumarajan et al., 2019), in addition to a worldwide decline in funding for soil scientific research since the 1980s (Lagacherie and McBratney, 2006). Throughout the globe, people are using DSM to create detailed maps of soil type and features (Malone et al., 2017). Both FAO and ITPS have accepted the DSM methodology as a method for reliably producing digital soil maps, which is essential for sustainable and resilient land management. DSM is an interdisciplinary field that requires knowledge of soil science, GIS, data science, geostatistics, and cartography. It frequently incorporates technological and computational inputs such as the global positioning system, remote sensing, field scanners, and other similar tools, as well as the GIS, digital elevation models (DEMs), geostatistical interpolations, inference techniques, and data mining (Lenka and Divya, 2020). The International Working Group on Digital Soil Mapping describes DSM as the "creation and population of geographically referenced soil databases generated at a given resolution using field and laboratory observation methods coupled with environmental data via quantitative relationships" (Lenka and Divya, 2020). DSM uses semi-automated algorithms to analyze visible soil information as well as other data in order to build a digital soil map that is both more accurate and more cost-effective. The final product of the DSM is a "raster" program-based map, which would replace earlier soil maps that were based on polygons (Mendonça-Santos et al., 2008). The spatial distribution of soil classes or soil attributes can be shown using digital soil maps, which can also take into account the uncertainty inherent in soil survey predictions (Lenka and Divya, 2020). These maps are kept and altered inside a framework known as GIS, which enables the examination and interpretation of vast amounts of data (Malone et al., 2017). High-resolution digital mapping of SOC is a challenging endeavor, but it is essential for SOC evaluation and monitoring (Moharana et al., 2022).

26.3.1 The SCORPAN approach

The scientific basis for soil mapping is provided by Jenny's conceptual model of soil formation (pedogenesis), which states that the process of soil formation (S) is controlled by five factors: climate (cl), organisms (o), terrain (r), parent material type (p), and time (t):

$$S = f(cl, o, r, p, t)$$

Despite its use in traditional soil mapping systems, this paradigm is not well suited to DSM because of quantitative data representation and spatial data interpolation. The SCORPAN soil formation model was developed by McBratney et al. (2002) in order to describe soil and associated environmental elements using spatial-based interpolation with quantitative expressions. This was done with the intention of making DSM more useful. The following equation represents the "SCORPAN" soil formation model:

$$Sc, a = f(s, c, o, r, p, a, n) + \varepsilon$$

where

Sc, a = Soil at a certain location and time (expressed either as soil classes, Sc, or as soil characteristics, Sa), is a numerical empirical function of the seven environmental factors.
s = Soil data of the location as found by Jenny's model
c = Climate of the location
o = Organisms residing in that location
r = Relief
p = Nature of the parent material
a = Age of the soil
n = Spatial location defined in (x, y) coordinates
ε = Standard error due to spatial and geographical variability

In nature, the soil formation process is impacted by a variety of environmental factors, which could influence it either directly or indirectly. Over the course of the past decade, several sources of data relating to climate, flora, terrain, parent materials, anthropogenic activities, and time have been easily accessible at appropriate spatial scales, and digital mapping techniques have been used to maximize the prediction of specific soil attributes. This has allowed for a greater understanding of the relationships between climate, vegetation, terrain, parent materials, and human activities (Zhang et al., 2017). Wiesmeier et al. (2019) carried out an in-depth investigation on the factors that have an effect on the SOC storage capacity and found clay content, bulk density, depth, and mineralogy are some of the qualities that influence a soil's ability to retain organic carbon. Actual SOC levels may be negatively impacted by pervasive environmental variables such as precipitation, temperature, and sun radiation. These factors can all contribute to this effect. However, the actual levels of OC in soil are also controlled by agronomic and soil management activities such as plant productivity, rotation strategy, erosion prevention, tillage, residue clearance, and drainage (Ingram and Fernandes, 2001; Rabbinge and Van Ittersum, 1994). To improve the selection of important environmental factors for evaluating SOC stock, it is essential to have a detailed understanding of the function that modifies the carbon sequestration process of the soil (Lamichhane et al., 2019).

The SCORPAN factors, also known as environmental covariates, are made up of digitally available and spatially populated data (Fig. 26.2). Some examples of these data include DEM, photographs taken by remote sensing, a radiometric data, geological map of the study area, legacy soil maps, and so on (Malone et al., 2017). It is essential that the DSM accurately portrays the uncertainty of the prediction (Arrouays et al., 2014). It is generally assumed that the residuals component of the formulation possesses some spatial structure. This is for a few reasons, such as an inaccurate definition of f(), inadequate features employed in the deterministic component, and a lack of interactions between attributes (Malone et al., 2017). This generic formulation is known as the "SCORPAN Kriging Approach," where the kriging component is the process of determining the spatial trend of residuals (using variograms) and utilizing kriging to estimate residuals at non-visited sites. The application of SCORPAN kriging is only possible in areas where soil point data are available (Malone et al., 2017).

26.3.1.1 Soil

The (cl, o, r, p, t) model that Jenny developed is put to use in order to generate soil information for a particular location. The uppermost layer of the earth's surface is composed of organic components, minerals, gases, and liquids, all of which work in tandem to maintain and sustain life. The majority of its components are composed of unconsolidated rocks and humus. It is possible to characterize soil using covariates that were generated from geo-referenced point data, spectral data provided by remote sensing satellites or aerial drones or aircraft, legacy soil maps, and other sources. The

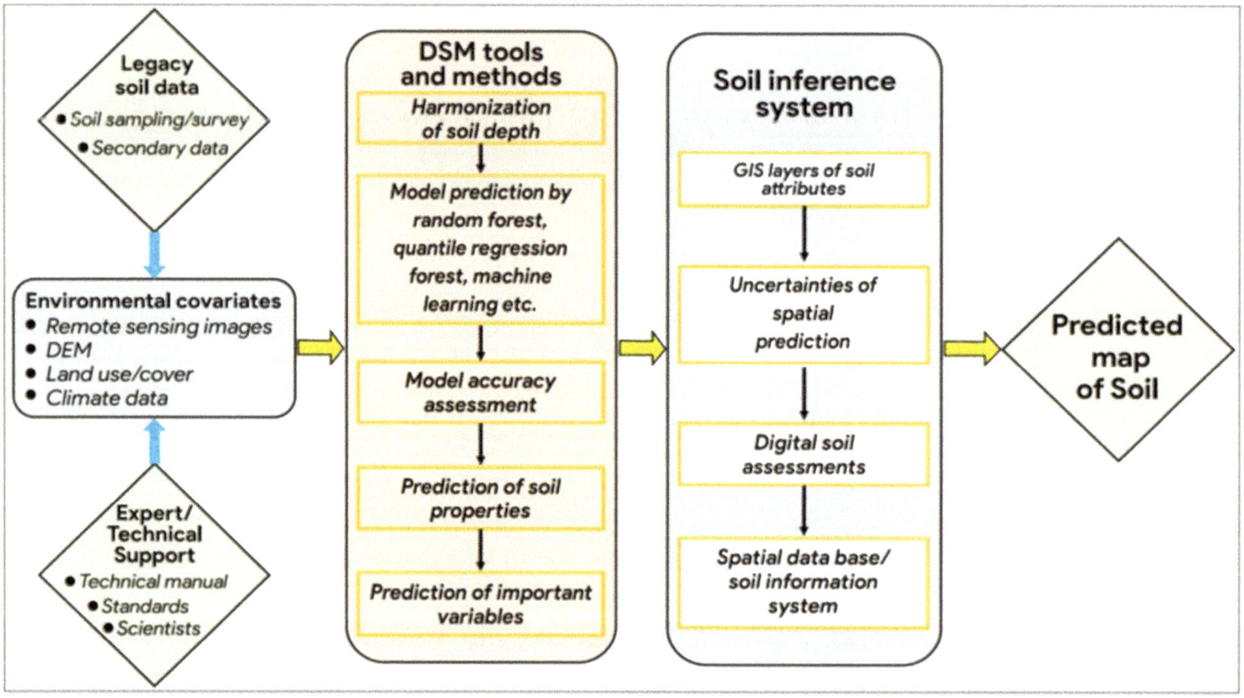

FIGURE 26.2 A schematic representation of the DSM method, which is used to make soil property predictions (Moharana and Dharumarajan, 2022). *Adapted from Moharana, P. C., Dharumarajan, S., 2022. Digital soil mapping for precise land.*

use of soil sensing technologies allows for the collection of soil data for DSM as well as the acquisition of historical soil data through the use of soil maps and information data on soil profiles.

26.3.1.2 Climate

The long-term average of an area's predominant weather components, such as temperature, atmospheric pressure, relative humidity, precipitation, wind speed, etc., is known as climate. Soil formation is greatly determined by the prevailing climate of that area because climate primarily influences processes of chemical and physical weathering. The outputs of climatic models are used to collect climate data for DSM in a specific location. The improvement of technologies used in remote sensing allows one to predict the climatic conditions of an area by measuring some climatic components, such as surface temperature, precipitation, soil moisture, and evapo-transpiration, remotely through satellites (Petropoulos et al., 2012). Multiple studies have concluded that precipitation, in addition to soil and atmospheric moisture, is the most important indicator for predicting SOC. For instance, in New South Wales, Australia, Hobley et al. (2015) found that precipitation had a positive correlation with SOC and vapor pressure deficit, and mean annual relative humidity was useful for mapping SOC in the upper 10 cm of soil. Similarly, precipitation was the most important element in explaining SOC stock in the top 30 cm of soils in semiarid rangelands in eastern Australia (Wang et al., 2018). In the same region, Gray et al. (2015) reported that temperature, the composition of the parent material, and the presence or absence of vegetative cover were the most relevant factors that determined SOC stocks. Adhikari et al. (2014) also observed that precipitation was the most important factor for the forecasting of SOC in Denmark. On the other hand, soil temperature determined the rate of organic carbon oxidation. High temperatures stimulate bacterial activity and hasten organic matter's decomposition. Hobley et al. (2015) came to the conclusion that temperature had a detrimental effect on the SOC levels. At the continental scale, the aridity index and annual mean temperature were shown to be the most critical variables of SOC deposition in European topsoil (Rial et al., 2017a).

26.3.1.3 Organisms

The physiochemical transformations of minerals contained in rock, which play an essential role in the processes of soil formation, are influenced by the specific species composition of the soil biota. There are several different ways in which vegetation cover may enhance SOC, such as the above-ground biomass added to soil, root exudation and root biomass, prevention of erosion by protecting the soil from falling raindrops, anchorage to soil particles, soil aggregation, and the preservation of soil moisture (Paul, 2016; Weil and Brady, 2016). The normalized difference vegetation index (NDVI), the enhanced vegetation index (EVI), and landsat-8 were applied as predictors of SOC for DSM (Moharana et al., 2022). Compared to individual bands, certain vegetation traits and indices, such as NDVI, give more accurate predictions of soil quality parameters, including SOC (Mahmoudabadi et al., 2017). Land use and land cover data are widely used as indicators of organism activity for estimating SOC concentrations. For instance, Rial et al. (2017a) observed that land cover is a significant local-scale predictor of SOC. Similarly, land use was the most influential predictor of SOC stocks not only in a semiarid steppe ecosystem in China (Wiesmeier et al., 2011) but also in the humid region of northeast India (Hinge et al., 2018). According to Hinge et al. (2018), there is a strong correlation between the amount of plant cover and surface organic carbon content (Bui et al., 2009; Wang et al., 2018). In the Shivalik range of northwest India, Kalambukattu et al. (2018) used Landsat 8–based spectral indices, for example, brightness index, coloration index, hue index, saturation index, and redness index, along with some DEM-derived terrain features, for predicting SOC and other soil nutrients. Wilson and Lonergan (2013) recommended using long-term vegetation cover rather than a single snapshot from a single year for a better prediction of SOC from vegetation cover. The consistency of greenness on the ground surface may be seen via time-series photography that spans several years or even decades, and this constancy may be a better indicator of SOC (Lamichhane et al., 2019).

26.3.1.4 Relief

The topographical representation of a geographical region's highest and lowest heights is referred to as the area's relief. This term is also sometimes used interchangeably with the term "terrain." Because of gravity and slopes, the topography of a location has a considerable impact on the quality of the soil in that location. The lowlands have darker soils that are richer in organic matter and minerals than the higher altitudes. This is because of the washing out of topsoil at higher altitudes and its accumulation at lower elevations. As a consequence of this, the soil at higher altitudes is lighter in color and lower in organic components and nutrients. DSM receives information about the topography from the outputs of digital terrain models. In DSM, the use of topographical data that was obtained using DEM as a covariate is commonplace. The digital elevation model along with its different components and derivatives were used as environmental covariates for preparing the digital soil map (Moharana et al., 2022). This included slope, elevation, aspect, curvatures, topographic wetness index, topographic position index, LS factor, multiresolution ridge top flatness, and multiresolution index of valley bottom flatness (MrVBF). In a study that was carried out in eastern China, it was discovered that DEM was a major predictor of SOC (Ma et al., 2017). In order to accurately forecast SOC concentrations in the Mediterranean region, topographic considerations proved to be more important than climatological indicators (Schillaci et al., 2017a). In the humid parts of northeast India, it was discovered that elevation was largely responsible for the variation of soil carbon (Hinge et al., 2018). DEM-derived indicators such as the MrVBF and the SAGA moisture index were found to be the second-most significant determinants in forecasting SOC in Denmark; precipitation, however, had the greatest influence (Adhikari et al., 2014).

26.3.1.5 Parent materials

The underlying geological material from which the soil horizons are formed is referred to as the parent material. It has a sizeable influence on the characteristics of the soil, including its color, texture, pH level, and mineral concentration. There is a substantial relationship between the composition of parent materials and soil qualities, such as soil texture and clay mineralogy. These relationships, in turn, influence other soil metrics that describe a soil's potential to store organic carbon (Weil and Brady, 2016). Parent materials seem to have a greater influence on the SOC content with increasing depth in the soil profile (Gray et al., 2015). In New South Wales, Australia, where the climate ranges from semiarid to subtropical, Hobley et al. (2015) found that geology was the primary factor in determining soil depth, whereas land use was the primary factor in determining SOC content and vertical distribution. In a similar vein, geology was shown to be the most accurate predictor of regional SOC (Guo et al., 2015). Soils developed from mafic rocks with low silica content had a high SOC level, while soils derived from siliceous felsic rocks had a low SOC level (Gray et al., 2015). This was elucidated from the silica index, which was used by the authors to determine the silica concentration of parent materials.

26.3.1.6 The age of the soil

The pedogenesis process is influenced by the age of the soil, which may be thought of as an alternate version of the time factor. It is the component of soil development that is passive but the most important. The formation of soil is a process that occurs continuously. Recent soil depositions have a lighter hue, are mostly composed of primary minerals, and show little indication of active soil development. The older, darker soils have a greater concentration of secondary and tertiary minerals, as well as a greater concentration of organic matter. According to some authors, the growth of 1 cm of soil requires two hundred to four hundred years (Delgado-Baquerizo et al., 2020), but it varies depending on the latitude of the site. When compared to temperate zones, the pace of soil development in wet tropical conditions is substantially faster. The age of the soil in DSM is calculated with the use of statistical techniques and spatial inference models.

26.3.1.7 Spatial location

The (x, y) coordinates are used to define the position of any geographic region in three-dimensional space. Statistical techniques and spatial inference models are utilized in order to ascertain the precise geographical location of soil for digital mapping.

26.4 Tools and algorithms for digital mapping of the soil

Digital mapping methodologies that range from basic linear statistical models (Moore et al., 1993) to highly developed, and intricate machine learning ones have been utilized to map the geographical distribution of SOC. This has been accomplished through the use of digital mapping (Minasny et al., 2013; Zhang et al., 2017).

26.4.1 Geostatistical tools

A subfield of statistics known as geostatistics, which employs geostatistical methods to examine geographical or spatio-temporal data, is known as geostatistics. A number of methodologies, including kriging, splines, simulation, semivariance functions, and variograms, are utilized in the process of geostatistical modeling. Within the context of this type of modeling, a number of statistical processes are utilized in order to estimate and reduce the estimation error of data that is either ambiguous or absent. Because of this, geostatistical modeling is taking on an increasingly significant role in DSM.

26.4.1.1 The theory of regional variability

In kriging, a geostatistical method known as kriging is frequently utilized for the purpose of the spatial interpolation of datasets. The fundamental premise of this line of thought is that the process of spatial interpolation needs to be founded on stochastic model parameters that take into account the varying patterns exhibited by the points that initially made up the datasets.

26.4.1.2 Semivariance

The semivariance is a kriging characteristic that reflects the degree of connection between points on a surface. Although it only accounts for half of the variance of all possible point-to-point distance discrepancies, its value increases when points are compared to progressively distant points (planet.uwc.ac.za).

26.4.1.3 Variogram

In geostatistical modeling, the variogram measures the degree to which two soil samples obtained from the same location differ in different soil attributes based on the distance between them and their distances from a reference point. In DSM, the variogram measures the degree to which two soil samples obtained from the same location differ in different soil attributes based on the distance between them and their distances from a reference point (Lenka and Divya, 2020).

26.4.1.4 Splines

Spline is a mathematical function that is defined by piecewise polynomials and satisfies some continuity condition by going through the given datapoints; spline interpolation is frequently preferred over polynomial interpolation due to its ability to define a greater number of parameters, including "smoothing" of the original data; it yields comparable results even when using polynomials of low degree while avoiding Runge's phenomenon for higher degrees; and spline interpolation is frequently preferred over polynomial interpolation because of its ability to (Yang et al., 2022).

26.4.1.5 Kriging

Georges Matheron (1960) developed the kriging technique based on Danie G. Krige's master's thesis. Kriging is a sort of regression that gives data estimates with the fewest squares (Remy et al., 2011). It derives an approximated surface model from the spatial description of a dispersed set of data points using z-scores. One advantage of this sort of interpolation is that it produces not only an interpolated spatial model, but also an estimate of the uncertainty of each point in that model. Kriging interpolation is mostly dependent on actual sample data points, rather than a previously assumed model, and provides more weight to sample points close to a place than to those farther away. To reduce sampling bias, clusters are given less weight than single points.

26.4.2 Machine learning

26.4.2.1 Regression kriging

The method of spatial prediction known as regression kriging (RK) is a hybrid technique that combines regression of the target soil attribute on environmental parameters with regression residual interpolation (Glen; Hengl et al., 2004, Yigini and Panagos, 2016). The RK is an effective method for forecasting the geographical distributions of soil characteristics (Dorji et al., 2014; Ma et al., 2017). However, some literature indicates that combining RK with other models gives more precision. For example, Guo et al. (2015) found that the combination of the random forest (RF) algorithm with RK performed better than only RF. Similarly, the neural network (NN) model performed better when residuals were added through kriging (Mirzaee et al., 2016). Thus, the RK approach is gaining popularity in DSM since it integrates traditional geostatistical techniques with linear models or even advanced machine learning algorithms for mapping soil attributes (Lamichhane et al., 2019).

26.4.2.2 Random forest

RF is an ensemble learning approach that creates a large number of decision trees during training. It is a multiclass classifier that is robust to noise, has strong discriminative performance, and can train and classify quickly (Breiman, 2001; Zhou, 2012). It is an ensemble learning method that works by building a lot of decision trees at training time and, for classification tasks, selecting the output that was chosen by the majority of trees (Ho, 1995). In the review of Lamichhane et al. (2019), RF was the most commonly used machine learning tool among the DSM studies, and it was an effective predictive model for estimating SOC. In a comparative study, RF outperformed support vector machines (SVM), NN, and multiple linear regression (MLR) in a subarctic peatland in Sweden (Siewert, 2018). In another study, RF and Boosted Regression Tree (BRT) were at par in predicting SOC (Yang et al., 2016) in northeast China. In general, RF performs better in larger study areas that include a variety of landscape elements, and its performance may be enhanced by including kriging of the residuals (Lamichhane et al., 2019).

26.4.2.3 Boosted random forest/boosted regression tree

Because RF has fewer decision trees, it may not provide a valid result if the training sample size is small; as a result, boosting is added to RF (Mishina et al., 2014). Most boosting strategies involve learning many relatively weak classifiers for a given distribution and then combining them to produce a single, more robust classifier. Weak learners' performance is taken into account when BRT decision trees are weighted. A set of weak learners is added one by one, and the misclassified data of the previous weak learner is given more weight in the bootstrap training sample of the newly added weak learner. Thus, future weak learners will focus more on the examples that previous weak learners misclassified. After adding each weak learner, "re-weighting" is done, which refers to re-adjusting the weights of the weak learners as the weak learner's added letter is more reliable than the previous one (Zhou, 2012; Kearns, 1988). BRTs are immune to outliers and missing data (De'Ath, 2007; Elith et al., 2008) and can therefore estimate the SOC stock over a large area with great accuracy (Ottoy et al., 2017). However, in contrast to RF models, BRTs need more parameter adjustment to obtain the necessary prediction accuracy, which may lead to model overfitting to the data. In most studies, BRT models were shown to be more successful in predicting SOC than other models such as SVM, ANN, and MLR. (Lamichhane et al., 2019).

26.4.2.4 Cubist

Cubist is a sophisticated regression tree-based model that predicts values using linear regression models rather than discrete values. The linear regression models are found on the tree's terminal leaves. These models are constructed with predictors taken from earlier splits. In addition to this, each level of the tree also contains intermediate linear models. A prediction is created at the tree's terminal node using the linear regression model, but it is "smoothed" by utilizing the prediction from the node that came before it, which is also something that happens recursively up the tree (Kuhn et al., 2012). Quite bit of research has demonstrated that the Cubist model is accurate when it comes to forecasting SOC (Gray and Bishop, 2016; Rudiyanto et al., 2018). This approach is useful for analyzing complex nonlinear interactions and does not involve any data reduction on the part of the user. In addition, the proportionate contribution of different elements to SOC prediction may be evaluated based on the number of times those factors are used in the rulesets (Adhikari et al., 2013). A cubist prediction model was employed in order to estimate the geographical distribution of 0−30 cm of SOC supply in South Korea. This was accomplished with the use of a geographical prediction model and validation by Park et al. (2021). Its capacity to "mine" nonlinear data relationships is one of the reasons for its considerable allure (Malone et al., 2017). The overall cubist prognosis at both depths was optimistic, with mean R^2 values of 0.78 and 0.62, respectively (Park et al., 2021).

26.4.2.5 Support vector machine

A SVM is a supervised learning strategy that is based on statistical learning and may be used for data classification and regression (Cortes and Vapnik, 1995). To do this, the necessary kernel function must be specified (Li et al., 2020). When it came to estimating SOC, SVR modeling performed far better than both RF and BRT (Rossel and Behrens, 2010). According to Were et al. (2015), SVR was the most reliable predictor of SOC, with the least propensity to overestimate its value. According to Somarathna et al.'s (2016) findings, SVR performed better than MLR and Cubist. However, some authors reported that RF and boosted regression trees give more accurate estimations of SOC stock than SVM (Wang et al., 2018; Rudiyanto et al., 2018).

26.4.2.6 The neural network

An artificial intelligence technique, known as a NN, attempts to replicate the way in which the human brain functions. It is a technique of "deep learning" that uses interconnected nodes or neurons to create a layered structure that is analogous to the human brain. It offers a structure that enables computers to constantly learn from their errors and improve their performance (Aws.amazon.com). NN methods make use of a number of hidden nodes or layers in order to facilitate the formation of new associations between the input and output data and to strengthen existing ones (Campbell and Wynne, 2011). In a Himalayan watershed in India, Kalambukattu et al. (2018) found that NN has the potential to be a useful approach. Similarly, Mirzaee et al. (2016) discovered that NN outperformed MLR in projecting SOC in Iran's East-Azerbaijan region.

26.4.2.7 Quantile forest regression

The quantile regression forest (QRF) is a more sophisticated variation of the RF model. RF just preserves the mean of the observations and discards redundant data, but QRF keeps the value of all observations in this node and assesses the conditional distribution depending on the data (Meinshausen and Ridgeway, 2006; Moharana et al., 2022) applied the QRF model to forecast SOC and for uncertainty estimations. These variations in research results show that the DSM does not include a magic bullet that may answer all of the problems regarding the predictive accuracy of its tests (Guevara et al., 2018). It is possible that factors other than model selection have a greater impact on SOC forecast accuracy, such as the size of the sample used for training or the reliability of the underlying predictor variables (Somarathna et al., 2017; Li, 2010). This is despite the fact that various predictive algorithms have varying capacities. The evolution of digital mapping algorithms from simple linear statistical models to geostatistical hybrid techniques to complex and intricate machine learning and deep learning algorithms resulted in more and more accurate estimation and mapping of the geographical distribution of SOC (Minasny et al., 2013; Zhang et al., 2017).

26.5 Estimation of soil organic carbon stock through DSM

The biggest terrestrial reservoir of sequestered carbon is SOC, and it plays a significant role in the C dynamics that occur on a global scale (Batjes, 1996; Chhabra et al., 2003). According to Jobbagy and Jackson (2000), the first 3 m of soil contain roughly 2344 Pg of carbon, of which 1502 Pg in the first 1-m, and 491 and 351 Pg in the second and third meters, respectively. According to Dexter et al. (2008), SOC influences several physical, chemical, and biological characteristics, and these have a significant effect on agricultural sustainability and resilience. For the purpose of defining carbon dynamics and carbon trade, baseline data as well as information on SOC storage are crucial (Stockmann et al., 2013). This is because the sequestration of atmospheric CO_2 in soils has been identified as a realistic strategy for reducing the effects of global warming (IPCC, 2007). It is necessary to map and quantify the spatial distribution of SOC stocks in order to find the potential of soil as a sink for SOC and to achieve higher levels of carbon sequestration (Mishra et al., 2009).

Quantifying SOC stocks involves knowledge of the geographic variability of SOC stocks in a landscape, which necessitates the identification of its regulating elements, including land use and land cover types (Smith, 2008; Saha et al., 2011). The geographical distribution of soil quality characteristics, the majority of which are evaluated in the lab, may be determined by field sampling. This sort of data is used to make predictions about the condition of the soil in areas that have not been inspected. Digital soil maps can give information on the dynamic soil properties as well as the uncertainty associated with such forecasts. The DSM of different parameters, including SOC stocks, is a four-step process that was summarized by Minansy et al. (2010) and depicted in Fig. 26.3.

In the initial phase of the process, known as "data input," base maps are generated by collecting and calibrating all variables relevant to the area of interest. Topographical features, gamma radiometric imaging, multi- and hyperspectral photos, land use, geology, and historical soil mapping are all examples of covariates that reveal soil formation state parameters.

The estimation of the soil attributes is the second step in the spatial soil inference method. These soil attribute estimations are expressed as estimates, and the associated uncertainty is also stated. Establishing quantitative correlations between point soil measurements and regionally spread components is used to generate them. The following format is used for digital mapping of soil:

$$f(s, c, o, r, p, a, n) + e = S$$

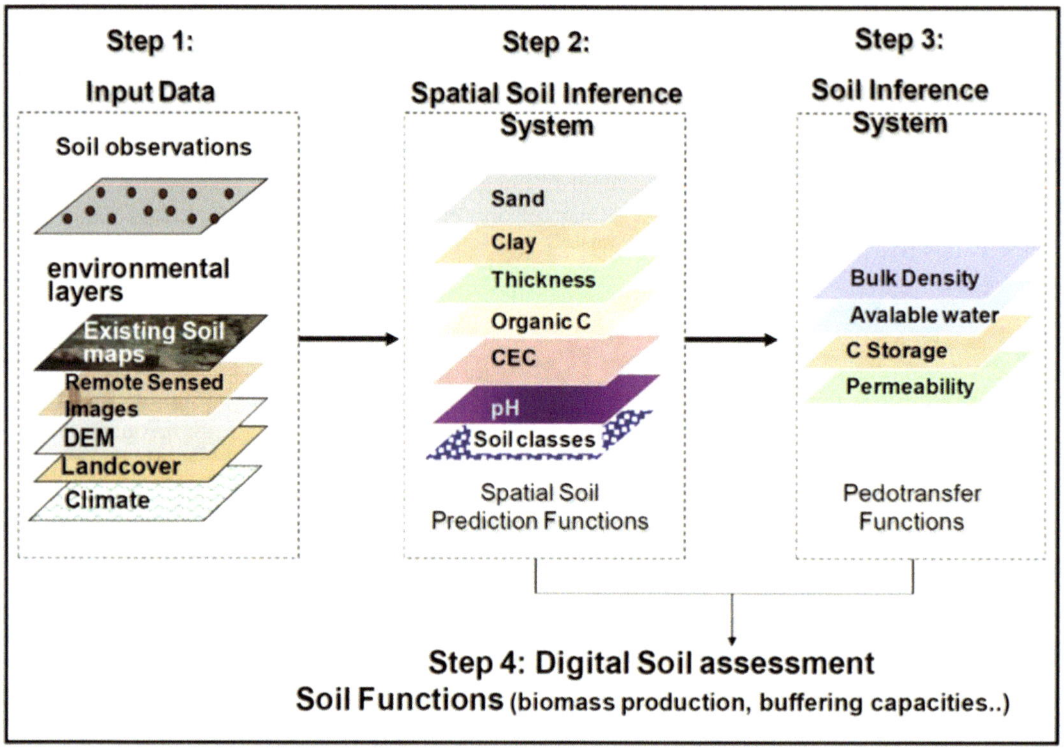

FIGURE 26.3 Three essential steps are involved in digital soil mapping (Minansy et al., 2010).

In the third step, soil parameters that are determined globally are used to make predictions about soil functions that are more challenging to define. These functions include available soil water storage, carbon density, and phosphate fixation. This is achieved by the use of integrated pedotransfer functions inside a soil inference system (McBratney et al., 2002). These soil processes have a considerable impact on the ability of soils to carry out a variety of ecosystem services, including those that are both provisioning and regulating in nature. The entire uncertainty of the forecast is determined by merging the uncertainties associated with the input data, spatial inference, and soil functions into a single calculation. There is a fourth step that comes before the evaluation, which was recently developed by Carré et al. (2007). This acknowledges the importance of using the information to guide policymakers and land managers on the appropriate actions to take.

26.6 Estimation of the bulk density of soil

The majority of estimation procedures for soil carbon give data for organic carbon per unit mass of soil. But for the computation of soil carbon stock, three more pieces of information are required: area, soil depth, and bulk density (BD) for each soil layer.

$$\text{SOC stock(Mg)} = \text{Area(ha)} \times \text{soil depth(m)} \times \text{bulk density}\left(\text{Mg m}^{-3}\right) \times \text{organic carbon(\%)}$$

The area of estimation can be calculated from DSM, but the layer-wise BD data has to be measured separately. The direct way is to measure the bulk density of a fresh soil sample using a core sampler. But, for a large study area, this method is highly expensive. If a readily available database is available, the BD can be estimated from this database. For instance, Yang et al. (2022) extracted the BD and coarse fraction (CF) values from the available Global Soil Information database, with a precision of 1 km (Hengl et al., 2017) and thereafter modeled the BD and CF as three-dimensional variables in this database with gradient tree boosting with a tenfold cross-validation and found a prediction r^2 value of 0.76 for BD and 0.56 for CF. The authors used an equal-area smoothing spline function (Bishop et al., 1999) to compute the SOC content at three different predetermined depth intervals.

$$\text{SOC stocks}_i = \sum_{i=1}^{n} SOC_i \times BD_i \times D_i \times (100 - CF_i)$$

where SOC stocks are given in kilograms per square meter, SOC content is given in grams per kilogram, BD is given in kilograms per cubic meter, D is the depth of the soil layer in meters, CF is given as a percentage (%), and i is the ith layer.

Alternatively, bulk density could be estimated from the pedotransfer function of mechanical composition and organic matter. Several authors proposed various pedotransfer functions for predicting soil BD. For instance, Park et al. (2021) estimated the mineral bulk density (BD_{min}) from the sand content and depth using restricted database and pedotransfer functions (PTFs) based on a prior study by Hong et al. (2013).

$$BD_{min} = 1.017 + 0.0032 \times s \quad \text{and} + 0.054 \times \log(\text{mean} \quad \text{depth})$$

As the organic matter differs significantly in terms of bulk density from the mineral particles, an adjustment for organic matter is needed, for which the authors used the model developed by Adams (1973).

$$\text{BD}\left(\text{g cm}^{-3}\right) = \frac{100}{\frac{OM}{BD_{\text{OM}}} + \frac{100 - OM}{BD_{\text{min}}}}$$

where OM stands for the mass proportion of organic matter (in grams per hundred grams), BD_{OM} stands for the average bulk density of organic matter (in grams per cubic centimeter), and BD_{min} stands for the bulk density of minerals (g cm^{-3}).

From this BD data, authors calculated the SOC density and carbon stocks, which are often used interchangeably in literature (Minasny et al., 2006), but differ in scale and unit. SOC density is the mass of mass per unit area for a given depth, which was calculated by Park et al. (2021) using the following formula:

$$\text{SOC density}\left(\text{ton C ha}^{-1}\right) = \sum_{i=1}^{k} \frac{B_i \times C_i \times D_i}{10}(1 - CF_i)$$

where B_i is the bulk density (g cm^{-3}), C_i is the carbon concentration (g kg^{-1}), D_i is the thickness (cm), and CF_i the gravel and stone content (%).

Similarly, the SOC stock was calculated by Park et al. (2021) by summing up the product of the mean SOC density per grid and the number of grids.

$$\text{SOC stock(MT)} = \sum_{i=1}^{n} (SOC_i \times A_i) \times 10^{-3}$$

where, SOC_{di} is SOC density (ton C ha^{-1}), A_i is the area of the grid cell and 10^{-3} is the conversion factor.

26.7 Few case studies of estimating SOC concentration and SOC stock using DSM approaches in India

SOC has the ability to improve the soil in terms of its physical, chemical, and biological characteristics, which would ultimately lead to an increase in soil security. Quantification of SOC is necessary in order to calculate the soil's ability to absorb carbon and its rate of SOC transformation (Yang et al., 2009). In India, in the last decade, researchers started concentrating on DSM after the inception of the GlobalSoilMap concept. Few attempts were made to spatially map SOC and its stock using geostatistical and machine learning models (Mishra and Gorai, 2020; Mitran et al., 2018; Dharumarajan et al., 2017, 2020, 2021). Dharumarajan et al. (2017) mapped major soil properties (pH, electrical conductivity and organic carbon) of Bukkarayasamudrum mandal of Anantapur district, India, using a RF model using 116 surface samples (0–30 cm). Owing to the proven outperformance of the RF model on the handling of large datasets and its inertia towards overfitting and noise features, it was selected for modeling and spatial mapping. IRS LISS IV data—three bands, terrain attributes, vegetation indices, a land use land cover map, and 1:50000 scale legacy soil data—were deployed as environmental covariates. The predicted organic carbon ranged between 0.24% and 1.03% with the coefficient of determination (R^2) and Lin's Concordance coefficient (CCC) 0.23 and 0.38, respectively. The vegetation indices NDVI and EVI were found to be potential covariates in the prediction of SOC, followed by spectral reflectance and terrain attributes. Out of this experience, Dharumarajan et al. (2020) experimented with mapping the SOC at regional level for different depth levels (0–5, 5–15, 15–30, 30–60, 60–100, and 100–200 cm) as per GlobalSoilMap specifications for Gadag and Koppal districts of the Northern Karnataka Plateau, Karnataka. In order to predict depth-wise SOC, the soil profile information was harmonized using an equal-area quadratic spline function for six standard depths. High-resolution sentinel-2 imageries, DEM derivatives such as elevation, slope, aspect, topographic wetness index, topographic position index, plan and profile curvature, multiresolution index of valley bottom flatness, multiresolution ridge top flatness, and vegetation factors like NDVI and EVI were used as environmental covariates. The model's performance on OC prediction was low ($R^2 = 0\%-27\%$), and elevation was identified as the dominant factor controlling SOC. The low level of organic carbon content in the study region is attributed to the poor performance.

Digital soil carbon mapping for the Western Ghats was accomplished using Landsat satellite data, terrain attributes, and bioclimatic variables as environmental covariates. Equal-area quadratic splines were used as a depth distribution function for harmonizing the legacy data into standard depth intervals (0–5, 5–15, 15–30, 30–60, 60–100, and 100–200 cm). The QRF algorithm was used to predict the SOC stocks and the uncertainty associated with them. The study demonstrated that the QRF model could capture maximum variability for the prediction of SOC, while the prediction was more accurate for the surface layer ($R^2 = 31\%-43\%$) than the subsurface soil layers ($R^2 = 7\%-21\%$). This was attributed to the low organic carbon content of the subsurface layer and the sparse dataset for the deepest layers (Dharumarajan et al., 2021). However, the DSM approach faces a significant challenge, which is the difficulty of applying these models to temporal domains of SOC change that are distinct from the geographical regions in which the models were calibrated.

References

Adams, W.A., 1973. The effect of organic matter on the bulk and true densities of some uncultivated podzolic soils. J. Soil Sci. 24 (1), 10–17.

Adhikari, K., Hartemink, A.E., Minasny, B., Bou Kheir, R., Greve, M.B., Greve, M.H., 2014. Digital mapping of soil organic carbon contents and stocks in Denmark. PLoS one 9 (8), e105519.

Adhikari, K., Kheir, R.B., Greve, M.B., Bøcher, P.K., Malone, B.P., Minasny, B., et al., 2013. High-resolution 3-D mapping of soil texture in Denmark. Soil. Sci. Soc. Am. J. 77 (3), 860–876.

Arrouays, D., Grundy, M.G., Hartemink, A.E., Hempel, J.W., Heuvelink, G.B.M., Hong, S.Y., et al., 2014. GlobalSoilMap: toward a fine-resolution global grid of soil properties. Adv. Agron. 125, 93–134.

Aws.amazon.com, 2022. Available from: <https://aws.amazon.com/what-is/neural-network>.

Batjes, N.H., 1996. Total carbon and nitrogen in the soils of the world. Eur. J. Soil Sci. 47 (2), 151–163.

Batjes, N.H., 1998. Mitigation of atmospheric CO2 concentrations by increased carbon sequestration in the soil. Biol. Fertil. Soils 27 (3), 230–235.

Batjes, N.H., Sombroek, W.G., 1997. Possibilities for carbon sequestration in tropical and subtropical soils. Glob. Change Biol. 3 (2), 161–173.

Bellon-Maurel, V., McBratney, A., 2011. Near-infrared (NIR) and mid-infrared (MIR) spectroscopic techniques for assessing the amount of carbon stock in soils–Critical review and research perspectives. Soil. Biol. Biochem. 43 (7), 1398–1410.

Bishop, T.F.A., McBratney, A.B., Laslett, G.M., 1999. Modelling soil attribute depth functions with equal-area quadratic smoothing splines. Geoderma 91 (1–2), 27–45.

Breiman, L., 2001. Random forests. Mach. Learn. 45 (1), 5–32.

Bui, E., Henderson, B., Viergever, K., 2009. Using knowledge discovery with data mining from the Australian Soil Resource Information System database to inform soil carbon mapping in Australia. Glob. Biogeochem. Cycles 23 (4).

Campbell, J. B., Wynne, R. H., 2011. Introduction to Remote Sensing. Guilford Press, New York, USA, 5th edition. Google-Books-ID: NkLmDjSS8TsC.

Carré, F., McBratney, A.B., Mayr, T., Montanarella, L., 2007. Digital soil assessments: beyond DSM. Geoderma 142 (1–2), 69–79.

Chatterjee, A., Lal, R., Wielopolski, L., Martin, M.Z., Ebinger, M.H., 2009. Evaluation of different soil carbon determination methods. Crit. Rev. Plant. Sci. 28 (3), 164–178.

Chhabra, A., Palria, S., Dadhwal, V.K., 2003. Soil organic carbon pool in Indian forests. For. Ecol. Manag. 173 (1–3), 187–199.

Coleman, K., Jenkinson, D.S., 1996. RothC-26.3—a model for the turnover of carbon in soil. Evaluation of Soil Organic Matter Models. Springer, Berlin, Heidelberg, pp. 237–246.

Coppin, P., Jonckheere, I., Nackaerts, K., Muys, B., Lambin, E., 2004. Review ArticleDigital change detection methods in ecosystem monitoring: a review. Int. J. Remote Sens. 25 (9), 1565–1596.

Cortes, C., Vapnik, V., 1995. Support-vector networks. Mach. Learn. 20, 273–297.

Croft, H., Kuhn, N.J., Anderson, K., 2012. On the use of remote sensing techniques for monitoring spatio-temporal soil organic carbon dynamics in agricultural systems. Catena 94, 64–74.

de Moraes Sa, J.C., Lal, R., 2009. Stratification ratio of soil organic matter pools as an indicator of carbon sequestration in a tillage chronosequence on a Brazilian Oxisol. Soil. Tillage Res. 103 (1), 46–56.

de Paul Obade, V., Lal, R., 2013. Assessing land cover and soil quality by remote sensing and geographical information systems (GIS). Catena 104, 77–92.

De Vos, B., Lettens, S., Muys, B., Deckers, J.A., 2007. Walkley–Black analysis of forest soil organic carbon: recovery, limitations and uncertainty. Soil. Use Manag. 23 (3), 221–229.

Delgado-Baquerizo, M., Reich, P.B., Bardgett, R.D., et al., 2020. The influence of soil age on ecosystem structure and function across biomes. Nat. Commun. 11, 4721. Available from: https://doi.org/10.1038/s41467-020-18451-3.

De'Ath, G., 2007. Boosted trees for ecological modeling and prediction. Ecology 88 (1), 243–251.

Dharumarajan, S., Hegde, R., Singh, S.K., 2017. Spatial prediction of major soil properties using random Forest techniques – a case study in semi-arid tropics of South India. Geoderma Reg. 10C, 154–162. Available from: https://doi.org/10.1016/j.geodrs.2017.07.005.

Dexter, A.R., Richard, G., Arrouays, D., Czyż, E.A., Jolivet, C., Duval, O., 2008. Complexed organic matter controls soil physical properties. Geoderma 144, 620–627. Available from: https://doi.org/10.1016/j.geoderma.2008.01.022.

Dharumarajan, S., Hegde, R., Janani, N., Singh, S.K., 2019. The need for digital soil mapping in India. Geoderma Reg. 16, e00204.

Dharumarajan, S., Kalaiselvi, B., Suputhra, A., Lalitha, M., Hegde, R., Singh, S.K., et al., 2020. Digital soil mapping of key GlobalSoilMap properties in northern Karnataka plateau. Geoderma Reg. 20. Available from: https://doi.org/10.1016/j.geodrs.2019.e00250.

Dharumarajan, S., Kalaiselvi, B., Suputhra, A., Lalitha, M., Vasundhara, R., Kumar, A., et al., 2021. Digital soil mapping of soil organic carbon stocks in Western Ghats, South India. Geoderma Reg. 25, e00387.

Dieye, A.M., Roy, D.P., Hanan, N.P., Liu, S., Hansen, M., Toure, A., 2012. Sensitivity analysis of the GEMS soil organic carbon model to land cover land use classification uncertainties under different climate scenarios in senegal. Biogeosciences 9 (2), 631–648.

Dorji, T., Odeh, I.O., Field, D.J., Baillie, I.C., 2014. Digital soil mapping of soil organic carbon stocks under different land use and land cover types in montane ecosystems, Eastern Himalayas. For. Ecol. Manag. 318, 91–102.

Dungait, J.A., Hopkins, D.W., Gregory, A.S., Whitmore, A.P., 2012. Soil organic matter turnover is governed by accessibility not recalcitrance. Glob. Change Biol. 18 (6), 1781–1796.

Elith, J., Leathwick, J.R., Hastie, T., 2008. A working guide to boosted regression trees. J. Anim. Ecol. 77 (4), 802–813.

FAO, 2007. Climate change and food security: a framework document. Summary. FAO Rome.

Glen, S., 2022. Kriging: Definition, Limitations. From StatisticsHowTo.com: Elementary Statistics for the Rest of Us! Available from: <https://www.statisticshowto.com/kriging/>.

GOFC-GOLD, 2012. Global Observations of Forest Cover–Global Observations of Land Cover Dynamics, Regional Networks.

Gomez, C., Rossel, R.A.V., McBratney, A.B., 2008. Soil organic carbon prediction by hyperspectral remote sensing and field vis-NIR spectroscopy: an Australian case study. Geoderma 146 (3–4), 403–411.

Gray, J.M., Bishop, T.F.A., 2016. Change in soil organic carbon stocks under 12 climate change projections over New South Wales, Australia. Soil. Sci. Soc. Am. J. 80 (5), 1296–1307.

Gray, J.M., Bishop, T.F., Wilson, B.R., 2015. Factors controlling soil organic carbon stocks with depth in eastern Australia. Soil. Sci. Soc. Am. J. 79 (6), 1741–1751.

Guevara, M., Olmedo, G.F., Stell, E., Yigini, Y., Aguilar Duarte, Y., Arellano Hernandez, C., et al., 2018. No silver bullet for digital soil mapping: country-specific soil organic carbon estimates across Latin America. Soil 4 (3), 173–193.

Guo, P.T., Li, M.F., Luo, W., Tang, Q.F., Liu, Z.W., Lin, Z.M., 2015. Digital mapping of soil organic matter for rubber plantation at regional scale: an application of random forest plus residuals kriging approach. Geoderma 237, 49–59.

Hengl, T., Heuvelink, G.B., Stein, A., 2004. A generic framework for spatial prediction of soil variables based on regression-kriging. Geoderma 120 (1–2), 75–93.

Hengl, T., Mendes de Jesus, J., Heuvelink, G.B.M., Ruiperez Gonzalez, M., Kilibarda, M., Blagotíc, A., et al., 2017. SoilGrids250m: global gridded soil information based on machine learning. PLoS One 12 (2), e0169748.

Hinge, G., Surampalli, R.Y., Goyal, M.K., 2018. Prediction of soil organic carbon stock using digital mapping approach in humid India. Environ. Earth Sci. 77 (5), 1–10.

Ho, T.K., 1995. Random decision forests (PDF). In: Proceedings of the 3rd International Conference on Document Analysis and Recognition. Montreal, QC, 14–16 August 1995, pp. 278–282.

Hobley, E., Wilson, B., Wilkie, A., Gray, J., Koen, T., 2015. Drivers of soil organic carbon storage and vertical distribution in Eastern Australia. Plant. Soil. 390 (1), 111–127.

Hong, S.Y., Minasny, B., Han, K.H., Kim, Y., Lee, K., 2013. Predicting and mapping soil available water capacity in Korea. PeerJ 1, e71.

Ingram, J.S.I., Fernandes, E.C.M., 2001. Managing carbon sequestration in soils: concepts and terminology. Agri. Ecosyst. Environ. 87 (1), 111–117.

IPCC, 2000. Land Use, Land-use Change, and Forestry. Cambridge University Press, Cambridge.

IPCC, 2007. Climate Change 2007: Impacts, Adaptation and Vulnerability. Working Group II Contribution to the 4th Assessment Report. Cambridge University Press, Cambridge.

Jobbagy, E.G., Jackson, R.B., 2000. The vertical distribution of soil organic carbon and its relation to climate and vegetation. Ecol. Appl. 10, 423–436.

Jones, J.W., et al., 2007. Integrating stochasticmodels and in situ sampling for monitoring soil carbon sequestration. Agric. Syst. 94, 52–62.

Kalambukattu, J.G., Kumar, S., Arya Raj, R., 2018. Digital soil mapping in a Himalayan watershed using remote sensing and terrain parameters employing artificial neural network model. Environ. Earth Sci. 77 (5), 1–14.

Kearns, M., 1988. Thoughts on Hypothesis Boosting, Unpublished manuscript (Machine Learning Class Project, December 1988).

Kuhn, M., Weston, S., Keefer, C., Coulter, N., 2012. <https://cran.microsoft.com/snapshot/2014-10-10/web/packages/Cubist/vignettes/cubist.pdf>.

Lagacherie, P., McBratney, A.B., 2006. Spatial soil information systems and spatial soil inference systems: perspectives for digital soil mapping. Dev. Soil. Sci. 31, 3–22.

Lal, R., 2001.World Cropland Soils as a Source or Sink for Atmospheric Carbon. In: Sparks, D.L. (Ed.), Advances in Agronomy, vol. 71. Elsevier, pp. 145–191.

Lal, R., 2003. Offsetting global CO2 emissions by restoration of degraded soils and intensification of world agriculture and forestry. Land. Degrad. Dev. 14, 309–322.

Lal, R., 2004a. Carbon emission from farm operations. Environ. Int. 30, 981–990.

Lal, R., 2004b. Soil carbon sequestration to mitigate climate change. Geoderma 123, 1–22.

Lal, R., 2006. Bulk density measurements for assessment of soil carbon pools. In: Lal, R. (Ed.), Carbon Sequestration in Soils of Latin America. Food Products Press, Binghamton NY, pp. 491–516.

Lal, R., 2009. Challenges and opportunities in soil organic matter research. Eur. J. Soil Sci. 60, 158–169.

Lamichhane, S., Kumar, L., Wilson, B., 2019. Digital soil mapping algorithms and covariates for soil organic carbon mapping and their implications: a review. Geoderma 352, 395–413.

Lenka, B., Divya, R.K., 2020. An introduction to digital soil mapping. In: Naresh, R.K. (Ed.), Advances in Agriculture Sciences, vol. 25. AkiNik Publications, New Delhi, India.

Li, Y., 2010. Can the spatial prediction of soil organic matter contents at various sampling scales be improved by using regression kriging with auxiliary information? Geoderma 159 (1–2), 63–75.

Li, X., Luan, F., Wu, Y., 2020. A comparative assessment of six machine learning models for prediction of bending force in hot strip rolling process. Metals 10 (5), 685.

Liu, S., Kaire, M., Wood, E., Diallo, O., Tieszen, L.L., 2004. Impacts of land use and climate change on carbon dynamics in south-central Senegal. J. Arid. Environ. 59 (3), 583–604.

Liu, J., Liu, S., Loveland, T.R., Tieszen, L.L., 2008. Integrating remotely sensed land cover observations and a biogeochemical model for estimating forest ecosystem carbon dynamics. Ecol. Model. 219 (3–4), 361–372.

Liu, S., Tan, Z., Li, Z., Zhao, S., Yuan, W., 2011. Are soils of Iowa USA currently a carbon sink or source? Simulated changes in SOC stock from 1972 to 2007. Agric., Ecosyst. Environ. 140 (1–2), 106–112.

Lobell, D.B., 2010. Remote sensing of soil degradation: introduction. J. Environ. Qual. 39 (1), 1–4.

Ma, Y., Minasny, B., Wu, C., 2017. Mapping key soil properties to support agricultural production in Eastern China. Geoderma Reg. 10, 144–153.

Mahmoudabadi, E., Karimi, A., Haghnia, G.H., Sepehr, A., 2017. Digital soil mapping using remote sensing indices, terrain attributes, and vegetation features in the rangelands of northeastern Iran. Environ. Monit. Assess. 189 (10), 1–20.

Malone, B.P., Styc, Q., Minasny, B., McBratney, A.B., 2017. Digital soil mapping of soil carbon at the farm scale: a spatial downscaling approach in consideration of measured and uncertain data. Geoderma 290, 91–99.

Matheron, G., 1960. Krigeage d'un panneaurectangulaire par sapériphérie. Note géostatistique 28.

McBratney, A.B., Minasny, B., Cattle, S.R., Vervoort, R.W., 2002. From pedotransfer functions to soil inference systems. Geoderma 109 (1–2), 41–73.

McCarty, G.W., Reeves, J.B., Reeves, V.B., Follett, R.F., Kimble, J.M., 2002. Mid-infrared and near-infrared diffuse reflectance spectroscopy for soil carbon measurement. Soil. Sci. Soc. Am. J. 66 (2), 640–646.

Meinshausen, N., Ridgeway, G., 2006. Quantile regression forests. J. Mach. Learn. Res. 7 (6).

Mendonça-Santos, M., Santos, H., Dart, R., Pares, J., 2008. Digital mapping of soil classes in Rio de Janeiro State, Brazil: data, modelling and prediction. In: Hartemink, A.E., McBratney, A., Mendonça-Santos, M.D. (Eds.), Digital Soil Mapping with Limited Data. Springer, Dordrecht. Available from: https://doi.org/10.1007/978-1-4020-8592-5_34.

Minasny, B., McBratney, A.B., Mendonça-Santos, M., Odeh, I.O.A., Guyon, B., 2006. Prediction and digital mapping of soil carbon storage in the Lower Namoi Valley. Soil. Res. 44 (3), 233–244.

Minansy, B., Alex, B., McBratney, Malone, B., Sulaeman, Y., 2010. Digital mapping of soil carbon. In: 19th World Congress of Soil Science, Soil Solutions for a Changing World. 1–6 August 2010, Brisbane, Australia.

Minasny, B., McBratney, A.B., Malone, B.P., Wheeler, I., 2013. Digital mapping of soil carbon. Adv. Agron. 118, 1–47.

Mirzaee, S., Ghorbani-Dashtaki, S., Mohammadi, J., Asadi, H., Asadzadeh, F., 2016. Spatial variability of soil organic matter using remote sensing data. Catena 145, 118–127.

Mishina, Y., Tsuchiya, M., Fujiyoshi, H., 2014. Boosted random forest. In: Proceedings of the 9th International Conference on Computer Vision Theory and Applications (VISAPP-2014), pp. 594–598.

Mishra, G., Gorai, A.K., 2020. Horizontal and vertical profiling of soil organic carbon stock in Nagaland, North East India. Curr. Sci. 119 (4), 632–640.

Mishra, U., Lal, R., Slater, B., Calhoun, F., Liu, D., Van Meirvenne, M., 2009. Predicting soil organic carbon stock using profile depth distribution functions and ordinary kriging. Soil. Sci. Soc. Am. J. 73 (2), 614–621.

Moharana, P.C., Dharumarajan, S., 2022. Digital Soil Mapping for Precise Land Management.

Moharana, P.C., Dharumarajan, S., Kumar, N., Jena, R.K., Pradhan, U.K., Meena, R.S., et al. 2022. Modelling and Prediction of Soil Organic Carbon Using Digital Soil Mapping in the Thar Desert Region of India.

Mitran, T., Mishra, U., Lal, R., Ravisankar, T., Sreenivas, K., 2018. Spatial distribution of soil carbon stocks in a semi-arid region of India. Geoderma Reg 00192. Available from: https://doi.org/10.1016/j.geodrs.2018.e00192.

Moore, I.D., Gessler, P.E., Nielsen, G.A.E., Peterson, G.A., 1993. Soil attribute prediction using terrain analysis. Soil Sci. Soc. Am. J. 57 (2), 443–452.

Moran, M.S., Inoue, Y., Barnes, E.M., 1997. Opportunities and limitations for image-based remote sensing in precision crop management. Remote. Sens. Environ. 61 (3), 319–346.

Mulder, V.L., De Bruin, S., Schaepman, M.E., Mayr, T.R., 2011. The use of remote sensing in soil and terrain mapping—a review. Geoderma 162 (1–2), 1–19.

Nelson, D.W., Sommers, L.E., 1996. Total carbon, organic carbon, and organic matter. Methods Soil Anal.: Part 3 Chem. Methods 5, 961–1010.
Ogle, S.M., Breidt, F.J., Easter, M., Williams, S., Paustian, K., 2007. An empirically based approach for estimating uncertainty associated with modelling carbon sequestration in soils. Ecol. Model. 205 (3–4), 453–463.
Ottoy, S., De Vos, B., Sindayihebura, A., Hermy, M., Van Orshoven, J., 2017. Assessing soil organic carbon stocks under current and potential forest cover using digital soil mapping and spatial generalisation. Ecol. Indic. 77, 139–150.
Park, S.J., Kwon, S.I., Kim, S.H., Shim, J., Lee, Y.H., Oh, T.K., 2021. Estimation of Soil Organic Carbon (SOC) Stock in South Korea Using Digital Soil Mapping Technique.
Parton, W.J., Schimel, D.S., Cole, C.V., Ojima, D.S., 1987. Analysis of factors controlling soil organic matter levels in Great Plains grasslands. Soil. Sci. Soc. Am. J. 51 (5), 1173–1179.
Paul, E.A., 2016. The nature and dynamics of soil organic matter: plant inputs, microbial transformations, and organic matter stabilization. Soil. Biol. Biochem. 98, 109–126.
Perez, C., Roncoli, C., Neely, C., Steiner, J.L., 2007. Can carbon sequestration markets benefit low-income producers in semi-arid Africa? Potentials and challenges. Agric. Syst. 94 (1), 2–12.
Petropoulos, G.P., Arvanitis, K., Sigrimis, N., 2012. Hyperion hyperspectral imagery analysis combined with machine learning classifiers for land use/ cover mapping. Expert. Syst. Appl. 39 (3), 3800–3809.
Planet.uwc.ac.za, 2022. Introduction to Spatial analysis—understanding semivariance. Available from: <http://planet.uwc.ac.za/nisl/gis/spatial/ chap_1_37.htm>.
Rabbinge, R., Van Ittersum, M.K., 1994. Tension between aggregation levels. In: Fresco, L.O., et al., (Eds.), The Future of the Land. John Wiley & Sons, Chichester, pp. 31–40.
Reeves II, J.B., 2010. Near-versus mid-infrared diffuse reflectance spectroscopy for soil analysis emphasizing carbon and laboratory versus on-site analysis: where are we and what needs to be done? Geoderma 158 (1–2), 3–14.
Remy, et al., 2011. Applied Geostatistics with SGeMS: A User's Guide. Cambridge University Press.
Rial, M., Cortizas, A.M., Rodríguez-Lado, L., 2017a. Understanding the spatial distribution of factors controlling topsoil organic carbon content in European soils. Sci. Total. Environ. 609, 1411–1422.
Rossel, R.V., Behrens, T., 2010. Using data mining to model and interpret soil diffuse reflectance spectra. Geoderma 158 (1–2), 46–54.
Rossel, R.V., Walvoort, D.J.J., McBratney, A.B., Janik, L.J., Skjemstad, J.O., 2006. Visible, near infrared, mid infrared or combined diffuse reflectance spectroscopy for simultaneous assessment of various soil properties. Geoderma 131 (1–2), 59–75.
Rossi, J., Govaerts, A., De Vos, B., Verbist, B., Vervoort, A., Poesen, J., et al., 2009. Spatial structures of soil organic carbon in tropical forests—a case study of Southeastern Tanzania. Catena 77 (1), 19–27.
Rowland, J., Wood, E., Tieszen, L.L., 2007. Review of Remote Sensing Needs and Applications in Africa. USGS Center for Earth Resources Observation and Science (EROS).
Roy, D.P., Lewis, P., Schaaf, C.B., Devadiga, S., Boschetti, L., 2006. The global impact of clouds on the production ofMODIS bidirectional reflectance model-based composites for terrestrial monitoring. IEEE Geosci. Remote. Sens. Lett. 3, 452–456.
Rutunga, V., Janssen, B.H., Mantel, S., Janssens, M., 2007. Soil use and management strategy for raising food and cash output in Rwanda. J. Food Agri. Environ. 5 (3/4), 434.
Rudiyanto, Minasny, B., Setiawan, B.I., Saptomo, S.K., McBratney, A.B., 2018. Open digital mapping as a cost-effective method for mapping peat thickness and assessing the carbon stock of tropical peat lands. Geoderma 313, 25–40. https://doi.org/10.1016/j.geoderma.2017.10.018.
Saha, D., Kukal, S.S., Sharma, S., 2011. Landuse impacts on SOC fractions and aggregate stability in typic ustochrepts of Northwest India. Plant. Soil. 339 (1), 457–470.
Scharlemann, J.P.W., Tanner, E.V.J., Hiederer, R., Kapos, V., 2014. Global soil carbon: Understanding and managing the largest terrestrial carbon pool. Carbon Manag 81–91.
Schillaci, C., Acutis, M., Lombardo, L., Lipani, A., Fantappie, M., Märker, M., et al., 2017a. Spatio-temporal topsoil organic carbon mapping of a semi-arid Mediterranean region: the role of land use, soil texture, topographic indices and the influence of remote sensing data to modelling. Sci. Total. Environ. 601, 821–832.
Siewert, M.B., 2018. High-resolution digital mapping of soil organic carbon in permafrost terrain using machine learning: a case study in a sub-Arctic peatland environment. Biogeosciences 15 (6), 1663–1682.
Sinsabaugh, R.L., 2010. Phenol oxidase, peroxidase and organic matter dynamics of soil. Soil. Biol. Biochem. 42 (3), 391–404.
Smith, P., 2008. Land use change and soil organic carbon dynamics. Nutr. Cycl. Agroecosyst. 81, 169–178. Soc. Am. J. 73, 614–621.
Somarathna, P.D.S.N., Malone, B.P., Minasny, B., 2016. Mapping soil organic carbon content over New South Wales, Australia using local regression kriging. Geoderma Reg. 7 (1), 38–48.
Somarathna, P.D.S.N., Minasny, B., Malone, B.P., 2017. More data or a better model? Figuring out what matters most for the spatial prediction of soil carbon. Soil. Sci. Soc. Am. J. 81 (6), 1413–1426.
Steffen, W., et al., 1998. The terrestrial carbon cycle: implications for the Kyoto Protocol. Science 280, 1393–1394.
Stevens, A., Udelhoven, T., Denis, A., Tychon, B., Lioy, R., Hoffmann, L., et al., 2010. Measuring soil organic carbon in croplands at regional scale using airborne imaging spectroscopy. Geoderma 158 (1–2), 32–45.
Stockmann, U., Adams, M.A., Crawford, J.W., Field, D.J., Henakaarchchi, N., Jenkins, M., et al., 2013. The knowns, known unknowns and unknowns of sequestration of soil organic carbon. Agric. Ecosyst. Environ. 164, 80–99.

Tan, Z., Liu, S., Li, Z., Loveland, T.R., 2007. Simulated responses of soil organic carbon stock to tillage management scenarios in the Northwest Great Plains. Carbon Balance Manag. 2 (1), 1–10.

UNFCCC, 1997. Kyoto Protocol to the United Nation Framework Convention on Climate Change (UNFCCC).

van Wesemael, B., Paustian, K., Andrén, O., Cerri, C.E., Dodd, M., Etchevers, J., et al., 2011. How can soil monitoring networks be used to improve predictions of organic carbon pool dynamics and CO2 fluxes in agricultural soils? Plant. Soil. 338 (1), 247–259.

Walkley, A., Black, I.A., 1934. An examination of the Degtjareff method for determining soil organic matter, and a proposed modification of the chromic acid titration method. Soil. Sci. 37 (1), 29–38.

Wang, L., Okin, G.S., Caylor, K.K., Macko, S.A., 2009. Spatial heterogeneity and sources of soil carbon in southern African savannas. Geoderma 149 (3–4), 402–408.

Wang, B., Waters, C., Orgill, S., Gray, J., Cowie, A., Clark, A., et al., 2018. High resolution mapping of soil organic carbon stocks using remote sensing variables in the semi-arid rangelands of eastern Australia. Sci. Total. Environ. 630, 367–378.

Weil, R., Brady, N.C., 2016. The Nature and Properties of Soils, 15th edition. Pearson Education Were, K., Bui, D.T., Dick, O.B., Singh, B.R., 2015. A comparative assessment of support vector regression, artificial neural networks, and random forests for predicting and mapping soil organic carbon stocks across an Afromontane landscape. Ecol. Indic. 52, 394–403.

Were, K., Bui, D.T., Dick, Ø.B., Singh, B.R., 2015. A comparative assessment of support vector regression, artificial neural networks, and random forests for predicting and mapping soil organic carbon stocks across an Afromontane landscape. Ecol. Indic. 52, 394–403.

Wielopolski, L., Chatterjee, A., Mitra, S., Lal, R., 2011. In situ determination of soil carbon pool by inelastic neutron scattering: comparison with dry combustion. Geoderma 160 (3–4), 394–399.

Wiesmeier, M., Barthold, F., Blank, B., Kögel-Knabner, I., 2011. Digital mapping of soil organic matter stocks using Random Forest modeling in a semi-arid steppe ecosystem. Plant. Soil. 340 (1), 7–24.

Wiesmeier, M., Urbanski, L., Hobley, E., Lang, B., von Luetzow, M., Marin-Spiotta, E., et al., 2019. Soil organic carbon storage as a key function of soils-a review of drivers and indicators at various scales. Geoderma 333, 149.

Wilson, B.R., Lonergan, V.E., 2013. Land-use and historical management effects on soil organic carbon in grazing systems on the Northern Tablelands of New South Wales. Soil. Res. 51 (8), 668–679.

Yang, Y., Fang, J., Smith, P., Tang, Y., Chen, A., Ji, C., et al., 2009. Changes in topsoil carbon stock in the Tibetan grasslands between the 1980s and 2004. Glob. Change Biol. 15 (11), 2723–2729.

Yang, R.M., Zhang, G.L., Liu, F., Lu, Y.Y., Yang, F., Yang, F., et al., 2016. Comparison of boosted regression tree and random forest models for mapping topsoil organic carbon concentration in an alpine ecosystem. Ecol. Indic. 60, 870–878.

Yang, R.M., Liu, L.A., Zhang, X., He, R.X., Zhu, C.M., Zhang, Z.Q., et al., 2022. The effectiveness of digital soil mapping with temporal variables in modeling soil organic carbon changes. Geoderma 405, 115407.

Yigini, Y., Panagos, P., 2016. Assessment of soil organic carbon stocks under future climate and land cover changes in Europe. Sci. Total. Environ. 557, 838–850.

Zhang, G.L., Feng, L.I.U., Song, X.D., 2017. Recent progress and future prospect of digital soil mapping: a review. J. Integrat. Agric. 16 (12), 2871–2885.

Zhao, S., Liu, S., Li, Z., Sohl, T.L., 2010. Federal land management, carbon sequestration, and climate change in the southeastern US: a case study with Fort Benning. Environ. Sci. Technol. 44 (3), 992–997.

Zhou Z.-H., 2012. Ensemble Methods: Foundations and Algorithms. Chapman and Hall/CRC, p. 23. The term boosting refers to a family of algorithms that are able to convert weak learners to strong learners.

Chapter 27

Soil salinity assessment and mapping using spectral indices and geostatistical techniques - concepts and reviews

Mukesh Kumar Mehla[1,2], Ashish Kumar[2], Sanjay Kumar[2] and R.K. Jhorar[2]

[1]*Department of Soil and Water Engineering, College of Technology and Engineering, Maharana Pratap University of Agriculture and Technology, Udaipur, Rajasthan, India,* [2]*Department of Soil and Water Engineering, College of Agricultural Engineering and Technology, Chaudhary Charan Singh Haryana Agricultural University, Hisar, Haryana, India*

27.1 Introduction

According to the FAO, soil salinization and alkalization affect nearly 953 Mha (8%) of the world's land area (Szabolcs, 1989). The active process of soil salinity is influenced by the type of soil as well as by hydrological, climatic, geochemical, agricultural, social, and economic factors (Abrol et al., 1988). Salinization has also resulted in a shortage of good-quality water resources and is becoming a prominent issue in arid and semiarid zones (Kumar et al., 2020). Soil properties, like moisture, texture, mineral composition, and surface roughness, and type of vegetation cover, such as halophyte and non-halophyte and salt-tolerant and non-salt tolerant plants, impact the degree, type, and distribution of soil salinity (Abrol and Bhumbla, 1971). Conventional techniques, such as laboratory analysis of soil samples to measure solute concentrations and electrical conductivity (EC), are accurate but time-consuming and costly as many samples will be required to account for spatial variability (Farifteh et al., 2008). While geostatistical methods and modeling are more effective, affordable, and quick tools for measuring soil salinity when combined with ground data.

Black-and-white and color aerial images have been used to identify soils affected by soil salinity since the 1960s. Since then, remote sensing data and techniques have revolutionized the way we monitor and map soil salinity. Satellite data has a great potential to provide a cost-effective alternative to traditional methods for monitoring salinization. An efficient technique to convey spatial distribution data on soil salinity is through soil salinization mapping. Remote sensing and geographic information system (GIS) technology have recently ushered in a new age in the area of dynamic soil salinity monitoring. Readily available images can be used to study spatial and temporal changes in salinity over a large area (Dwivedi, 1996). This data can be analyzed using a GIS system to obtain meaningful data, make predictions, and monitor and assess current conditions. GIS software provides us with tools to harness, manipulate, and evaluate large chunks of remotely sensed data using computer processing power. Different high-resolution sensors on satellites provide us with higher-resolution data for analysis to recognize more details for better results and precision. Soil salinity can be mapped directly by reflectance from bare soil or the salt crust and indirectly by vegetative coverage and health using remote sensing (Abbas et al., 2013). Chaturvedi et al. (1983) used remotely sensed brightness and thermal layers to delineate salty soils. Menenti et al. (1986) used a Landsat image and, from three bands of TM (1, 5, and 7), detected minerals that can be the origin of salts in the soil. Naseri (1998) used unsupervised and supervised classification algorithms and delineated soil salinity at a regional scale from three bands of MSS and four bands of TM images. Spectral signatures obtained from Landsat images along with ground truth data were used to differentiate the wasteland categories (salt-affected and waterlogged) using statistical methods in India (Saha et al., 1990).

Geostatistics is the branch of statistics that deals with the statistics of the earth, and it is used extensively in many fields. At first, geostatistics was synonymous with kriging, a geostatistical version of interpolation. It has widened into many statistical methods and techniques for studying spatial variations, including kriging, ordinary kriging, co-kriging, inverse distance-weighted (IDW), etc. The geostatistical analysis involves generating a continuous surface representing

Remote Sensing of Soils. DOI: https://doi.org/10.1016/B978-0-443-18773-5.00019-3

a particular attribute in a particular area. Geostatistical analysis tools have become essential to GIS application software like Arc GIS, ERIDAS QGIS, GRASS GIS, etc. This software uses data layers and datasets to run geostatistical analyses. Geostatistics has been used for the characterization of soil salinity in irrigated areas in different regions of the world (Abbas et al., 2013; Mamattursun et al., 2016; Al-Khaier, 2003). Remotely sensed data layers, like the digital elevation model of terrain, land use and land cover, and soil, water, and vegetation indices (NDVI, NDSI, NSWI, NDSI, etc.), are used in soil salinity assessment. These datasets are readily available at various large scales for different parts of the world.

27.1.1 Vegetation Indices

Vegetation in salt-affected areas has distinct characteristics, that is, sparse vegetation present in an area and halophytic plants that naturally flourish in salty soil and have adapted to it. Crop spectral signatures can be used to identify soil salinity and the effect of salinity on yield (Leone et al., 2007; Naumann et al., 2009; Poss et al., 2006). For predicting and mapping soil salinity, vegetation has been utilized as an indirect indicator using various spectral indices, such as the vegetation index (VI), NDVI, soil-adjusted vegetation index (SAVI), ratio vegetation index, green vegetation index, wetness index, and soil brightness index, among others. NDVI is a widely used VI for mapping soil salinity by monitoring halophytic plants. Vegetation exhibits higher reflectance in the NIR range and better absorption in the visible range of the EM spectrum, making NDVI a suitable VI for soil salinity measurement. Reflectance between the visible and NIR bands is divided by the sum of the reflectance in the NIR range based on reflectance in the NIR area, and normalized values between -1 and 1 are derived (Deering, 1975). NDVI value for bare soils is 0.1 and 0.9 for healthy vegetation; there is an important link between NDVI, EC, and SAR values. Bannari et al. (2008) stated that the NDVI can be used to anticipate soil salinity and hence salt stress, which results in a significant drop in plant growth and development. Nevertheless, some researchers have argued that the presence of vegetation can cause spectral confusion with the reflectance properties of salt (Metternicht and Zinck, 2003; Zhang et al., 1997). Since NDVI is linked with other factors including chlorophyll concentration, biomass, and leaf area, some people believe that is reliable. Liu and Huete (1995) developed the Enhanced VI, a modified version of the NDVI in which the atmospheric and canopy background noise were reduced.

In locations with different soil backgrounds or little canopy cover, the SAVI can be used to eliminate the effect of soil-induced variation (Huete, 1988). SAVI values can be either positive, negative, or zero, and the image can be classified accordingly. An increase in vegetation is indicated by a positive SAVI value, whereas a decline is indicated by a negative SAVI value, with zero indicating no change. Zhang et al. (1997) proposed four soil-adjusted salinity indices (SASIs), which help characterize halophyte plants in an area. When compared to other chosen vegetation indices like the NDVI and SAVI, SASIs produced better results. Many different vegetation indices have been proposed for different applications and under different conditions, some of them are listed in Table 27.1.

27.1.2 Soil indices

Salinity indices have given promising results for soil salinity prediction (Elhag, 2016). Employing vegetation indices in dense vegetation areas yields good results, but they cannot work on bare soils. The comparatively straightforward calculation and dependability of spectral soil salinity indicators have enabled significant scientific potential in the field of

TABLE 27.1 Commonly used vegetation indices and their equations.

S.No.	Indices	Equation	Reference
1.	Normalized differential vegetation index	$\mathrm{NDVI} = \frac{(\mathrm{NIR} - R)}{(\mathrm{NIR} + R)}$	Deering (1975)
2.	Enhanced vegetation index	$\mathrm{EVI} = \frac{2.5 \times (\mathrm{NIR} - R)}{(\mathrm{NIR} + 6R - 7.5B + 1)}$	Liu and Huete (1995)
3.	Soil-adjusted vegetation index	$\mathrm{SAVI} = \frac{(\mathrm{NIR} - R)}{(\mathrm{NIR} + R + L) \times (1 + L)}$	Huete (1988)
4.	Ratio vegetation index	$\mathrm{RVI} = \frac{\mathrm{NIR}}{R}$	Major et al. (1990)
5.	Differential vegetation index	$\mathrm{DVI} = \mathrm{NIR} - R$	Clevers (1988)
6.	Vegetation soil salinity index	$\mathrm{VSSI} = (2 \times G) - 5 \times (R + NIR)$	Dehni & Lounis (2012)

B, Blue band; *G*, green band; *L*, green vegetation cover; *NIR*, near-infrared band, and *R*, red band.

TABLE 27.2 Frequently used soil indices and their equations.

S.No.	Indices	Equation	Reference
1.	Normalized differential salinity index	$\text{NDSI} = \frac{(R-\text{NIR})}{(R+\text{NIR})}$	Khan et al. (2005)
2.	Brightness index	$\text{BI} = \sqrt{(R^2+\text{NIR})}$	Khan et al. (2005)
3.	Salinity index-1	$\text{SI} = \sqrt{(B \times R)}$	Khan et al. (2005)
4.	Salinity index-2	$\text{SI} = \sqrt{(G \times R)}$	Khan et al. (2005)
5.	Salinity index-3	$\text{SI} = \sqrt{(G^2+R^2+\text{NIR}^2)}$	Nicolas and Walter (2006)
6.	Salinity index-4	$\text{SI} = \sqrt{(G^2+R^2)}$	Nicolas and Walter (2006)
7.	Salinity index-5	$\text{SI} = \frac{B}{R}$	Nicolas and Walter (2006)
8.	Salinity index-6	$\text{SI} = \frac{(B-R)}{(B+R)}$	Nicolas and Walter (2006)
9.	Salinity index-7	$\text{SI} = \frac{(G \times R)}{B}$	Bannari et al. (2008)
10.	Salinity index-8	$\text{SI} = \frac{(B \times R)}{G}$	Abbas et al. (2013)
11.	Salinity index-9	$\text{SI} = \frac{(\text{NIR} \times R)}{G}$	Abbas et al. (2013)

B, Blue band; *G*, green band; *NIR*, near-infrared band and *R*, red band.

soil salinity research. Bouaziz et al. (2011) found that vegetation indices like SAVI, NDVI, EVI, etc., show a lower correlation with the EC because of low vegetation cover. They observed higher correlations with EC for soil salinity indices. Spectral signatures and reflectance in various spectral regions can be used to develop different spectral salinity indices that can be used to map salinity and discover salt minerals. Nicolas and Walter (2006) proposed three soil salinity indices to detect and map soil salinity problems in a semiarid environment, which were very promising and showed a strong correlation with measured values. However, they substantially underestimated the salinity of areas with high surface salt levels. Utilizing information from the LISS-II sensor and IRS-1B satellite, three salinity indices, brightness index, normalized difference salinity index (NDSI), and salinity indices, were developed for the evaluation of hydrosalinized soil degradation in Pakistan (Khan et al., 2005). From these indices, NDSI was found to yield superior results. Bannari et al. (2008) developed three indices, SI-1, SI-2, and SI-3, using EO-1 ALI spectral bands that can be utilized to distinguish even very mild and moderate soil salinity. SI-3 had the highest correlation (46.9%), although the results of this index are insufficient to draw firm conclusions. Likewise, the researcher proposes several soil indices for soil salinity assessment, given in Table 27.2.

27.1.3 Geostatistics

Satellite images and geostatistical techniques can be used in tandem for soil salinity studies and surveys. Different geostatistical methods are applied when large-scale sampling is not feasible and uneconomical. Hosseini et al. (1994) created a salinity map using statistical techniques to analyze soil salinity using techniques like ordinary kriging, Co-kriging, moving average of inverse distance, and thin plate smoothing splines (TPSS). TPSS and kriging methods have been found to have higher accuracy. Mapping soil salinity using classical methods and statistics does not consider the influence of neighboring data points. The geostatistical spatial model considers the neighboring effect, and some commonly used Geostatistical methods are kriging, co-kriging, spline, and IDW (Fig. 27.1).

Uts et al. (1998) calibrated the four-electrode probe (FEP) for inexpensive and indirect salinity determinations from EC in a plot with 50-m square grid sampling. Cross-validation showed significant similarity in EC semi-variograms obtained from FEP measurements and laboratory-measured soil EC results. Pozdnyakova and Zhang (1999) used geostatistical methods (kriging and co-kriging) to estimate SAR in a 3375-ha area. Compared to kriging, co-kriging uses fewer data to improve the estimations significantly, reducing mean squared error and variance up to 70%, with a correlation of about 60% and reducing costs up to 80%. Nawar et al. (2011) used a geostatistical approach for mapping soil salinity in the El-Tina plains of Egypt and proposed that ordinary kriging with spherical semi-variogram model (MPE and MSPE close to 0 and RSSPE close to 1) was the best model to generate a soil salinity map. Nezami and Alipour, 2012 mapped the salinity of the south Qazvin plain by sampling topsoil samples from 0 to 30 cm and validating the accuracy of several interpolation methods (kriging, co-kriging, spline, and IDW). Results showed that the co-kriging method had the lowest mean absolute error (0.5) and root-mean-square error (108) compared to the other methods. Mamattursun et al. (2016) used geostatistics to characterize the spatial and temporal variability of soil salinity in an

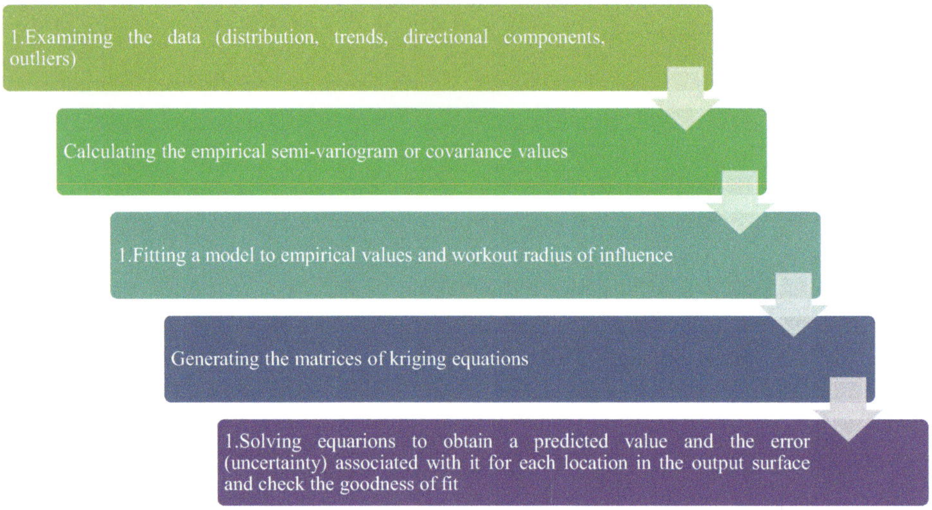

FIGURE 27.1 Main steps involved in forming a geostatistical model.

irrigated area. The spherical model was the best variogram model for describing the salinity in the depths of 0–3 cm and 3–20 cm. The order of salinity in the root zone for various land use types was as follows: grassland > cropland > forestland. Zewdu et al. (2017) used multitemporal LAND SAT images from 1984, 1995, and 2010. Using geospatial tools, they developed models from EC_e vs. NDSI and thematic layers to map soil salinity. Regression analysis revealed a correlation coefficient of 66% for the given model. Shi et al. (2005) used geostatistical techniques to provide salinity distribution maps in a coastal saline field in China. Mohammad et al. (2010) investigated the distribution of soil pH, salinity, and plant cover using geospatial techniques. They suggested that kriging and co-kriging methods were better for the IDW spatial distribution of soil properties.

27.1.4 Soil salinity mapping

Soil salinity is a serious threat we are facing across the globe. There are various soil salinity estimates at various scales (Allbed and Kumar, 2013; IP and Bhumbla, 1971; Keshavarzi and Sarmadian, 2012; Marcos et al., 2018; Metternicht and Zinck, 2003; Morgan et al., 2018). Soil salinity currently affects more than 1 billion hectares of area all over the world in more than 100 countries, and these numbers are continuously going upward (Abbas et al., 2013; Szabolcs, 1989). Most studies give only a rough approximation of the salt-affected area.

Harmonized World Soil Database is the only database that currently offers data on soil salinity that covers the entire world (FAO, 2012). Most of it is based on the dated FAO/UNESCO Soil Map of the World, which was produced between 1970 and 1981. The lack of spatial resolution and the outdated data show the need for newer and higher spatial–resolution soil salinity maps. Proper soil salinity maps will be crucial for its management and planning the required steps to lessen or even prevent financial losses due to damage to agriculture. Ivushkin et al. (2019) used the Google Earth engine to produce global soil salinity maps using a random forest classifier for the years 1986, 2000, 2002, 2005, 2009, and 2016, by combining soil properties maps with thermal infrared imagery and a large set of field observations (Figs. 27.2 and 27.3). The resulting maps had a validation accuracy of 67%–70% and revealed a clear trend of increasing salinity, with roughly 1 billion hectares of land affected by soil salinity.

Traditional laboratory techniques for measuring salinity, such as EC_e from saturated paste, are labor-intensive, expensive, and only provide point data (Triantafilis et al., 2012). The apparent soil electrical conductivity (EC_a), which depends on elements like clay, mineralogy, and moisture, can be easily measured by electromagnetic induction (EMI)-based instruments. To quickly map soil characteristics related to salinity, EMI instruments can be used. EM38 is a shallow measuring instrument that has been used to map salinity across the farm after first enabling site selection for calibration (Buchanan and Triantafilis, 2009; Li et al., 2013). This includes the use of mechanized and mobilized sensing systems, which enable rapid field-scale collection of sizable amounts of EC_a data. At the field scale, the EMI method is a cost-effective and rapid method for assessing soil salinity on a large scale (Liu et al., 2016; Metternicht and Zinck, 2003). EMI instruments are experimental apparatuses for soil salinity assessment at the field level. Narjary et al. (2019) evaluated the spatial variability of EC_a and mapped soil salinity (vertically and horizontally) while considering two trend surface parameters (easting and northing) in an 11-ha area located in Panipat district in Haryana, India. For the

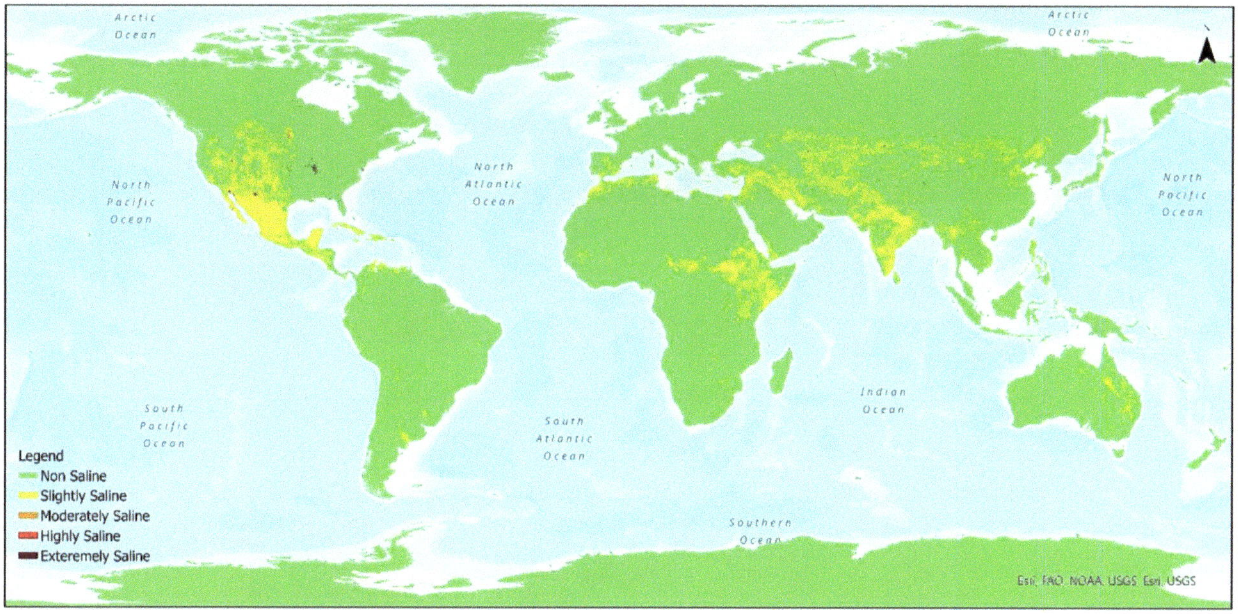

FIGURE 27.2 Global soil salinity map for the year 2016. *Adopted from Ivushkin, K., Bartholomeus, H., Bregt, A. K., Pulatov, A., Kempen, B., and De Sousa, L., 2019. Global mapping of soil salinity change. Remote. Sens. Environ. 231, 111260.*

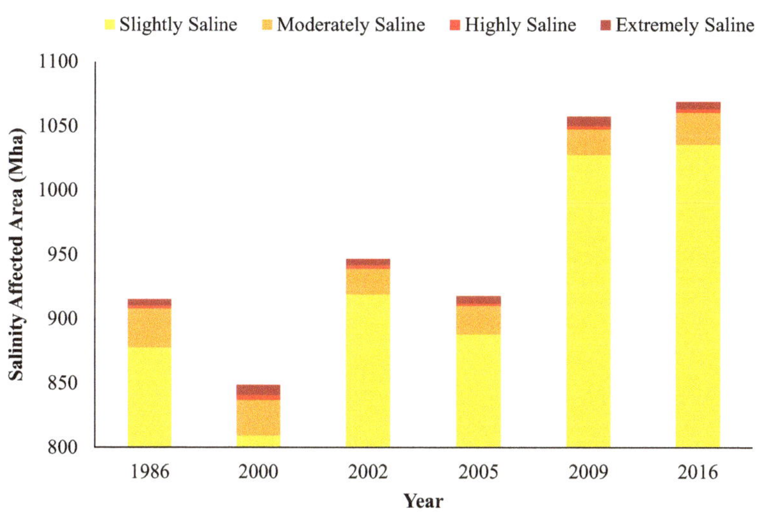

FIGURE 27.3 Salt-affected area over the years globally. *Adopted from Ivushkin, K., Bartholomeus, H., Bregt, A. K., Pulatov, A., Kempen, B., and De Sousa, L., 2019. Global mapping of soil salinity change. Remote. Sens. Environ. 231, 111260.*

survey, an EM-38 instrument was used, which is based on the EMI method. It was found that this instrument is a reliable and quick way to characterize soil salinity in general (Fig. 27.4). Salinity variation with depth has been shown to be more easily and efficiently determined using a single calibration equation in two dimensions. The apparent conductivity (EM_V and EM_H) and the sodium (Na^+) and chloride (Cl^-) ions had a strong correlation, so the soil salinity measured using the EM-38 can be considered reliable (Narjary et al., 2017).

27.1.5 Case studies

Remotely sensed data can be used for soil salinity assessment using spectral indices and field measurements. RS and GIS techniques show great promise for modeling and mapping soil salinity. Alqasemi et al. (2021) created a model employing field measurements and RS-based spectral indices to estimate soil salinity in Abu Dhabi (UAE). Significant relationships between NDVI and soil salinity ($R^2 = 0.43$) and between LST/NDVI and LST/BSI are also found, with correlations between the two variables of $R^2 = 0.14$ and 0.05 for LST/BSI and LST/NDVI, respectively. The regression

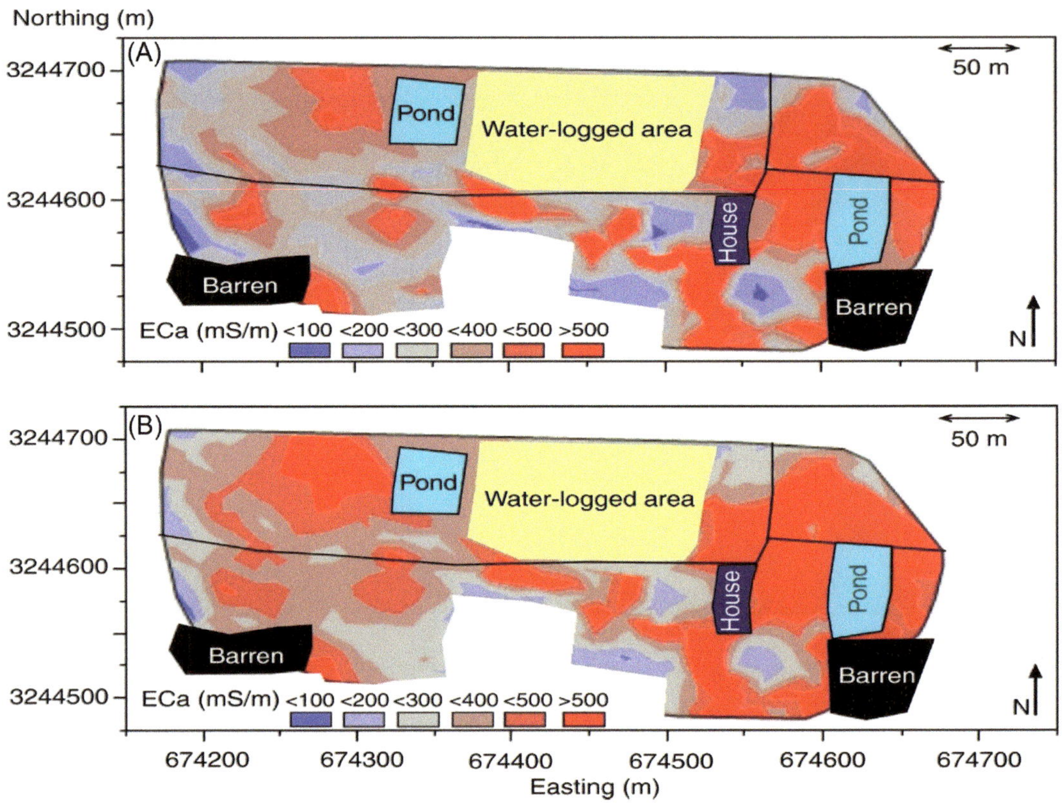

FIGURE 27.4 Measured EC_a (mS/m) using an EM38 in (A) horizontal (EC_{ah}) and (B) vertical (EC_{av}) modes of operation. *Adopted from Narjary, B., Meena, M.D., Kumar, S., Kamra, S.K., Sharma, D.K., Triantafilis, J., 2019. Digital mapping of soil salinity at various depths using an EM38. Soil. Use Manag. 35 (2), 232–244.*

model's median predictive power for NDVI BSI is $R^2 = 0.79$. According to research, the growth of vegetation will be negatively impacted by any rise in soil salinity. Additionally, this study revealed a weak link between soil salinity and LST at night, suggesting that salinity will rise as the temperature rises.

Combining these remotely sensed data with field-based EC_e variables was a promising option for monitoring soil salinity at farm scale in Ethiopia (Asfaw et al., 2018). The SI showed the highest correlation with EC_e, resulting in the best fit with $R^2 = 0.78$. While spectral indices and remote sensing approach is good for large-scale soil salinity assessment, there is a need for a more practical approach at the field level.

In order to map the soil salinity in the Neretva River valley, Zovko et al. (2018) conducted a study. Their goal was to propose a practical method based on Vis-NIR spectroscopy and geostatistics. The research was conducted on 5068 ha of agricultural land in the valley of the Neretva River (Croatia). Topsoil (0–30 cm) samples were taken at 246 locations using a grid sampling scheme (500 m) and their chemical and physical characteristics were examined. Additionally, measurements of the visible and NIR spectrums, with wavelengths between 350 and 2500 nm, were made on soil samples. The spectral data was preprocessed, and partial least squares regression (PLSR) was used to predict EC_e. To measure and map the spatial variation of soil salinity, the first significant latent variable, which explained 85% of the total variance, was chosen. The results and performances of the univariate and multivariate geostatistical approaches were remarkably similar. Regarding the two multivariate approaches, the one that only included the spectral index as a covariate produced more accurate and impartial results.

Triki Fourati et al. (2017) used remote sensing and geostatistical techniques to map the spatial distribution of this salinity over the southern-east region of Tunisia and study the impact of environmental factors on soil salinity. To illustrate geomorphological factors like elevation, slope, plan curvature, profile curvature, and aspect, data from the Advanced spaceborne thermal emission and reflection radiometer was used. The slope and elevation primarily influence the salt concentration in the soil, according to a Pearson correlation between these variables and soil electrical conductivity (EC_{soil}). The high potential of short-wave infrared bands to detect salty soils was also demonstrated by spectral analysis. Simple regression, minimum distance classification, and OK were used to map the soil salinity in southern

Tunisia. The results demonstrated that the common kriging technique, with a root-mean-square error of 1.83 and a mean error of 0.018, offers the most accurate performance to identify and classify saline soils over the study area, in order to investigate the spatial variability of soil salinity.

Masoud et al. (2019) conducted research in the Dakhla Oasis in Egypt's central Western Desert with the goal of defining an efficient monitoring method using remote sensing techniques. A single band, band ratio and combination, spectral index, linear spectral unmixing (LSU), and mixture-tuned matched filtering (MTMF) were used to estimate soil salinity using spectral data from a Landsat 8 OLI image of one scene that was acquired close to the ground sampling time. The southern region's lowest VES resistivity coincided with the most salinized zones. The main contributors to the salinity difference between the southern and northern regions were identified as the thickness of the surface layer and the amount of clay. In 2001, increasing salinization was detected in the LST retrieved from the thermal band data of the OLI image and another Landsat ETM $^{+}$ image. With 76% accuracy, LSU outperformed the MTMF model (67%) and the band combination and spectral indices (55% at most) in predicting salinity.

Yusuf et al. (2020) studied soil salinity assessment at Kano River Irrigation Project Phase-I. The grid sampling technique was used to collect 50 subsurface soil samples at a depth of 0–15 cm from plots with three different management practices. The following parameters, including EC, pH, and exchangeable bases (Ca^{2+}, Mg^{2+}, Na^{+}, and K^{+}), were determined using laboratory techniques for soil salinity. For the purpose of analyzing the spatial variation of soil properties, geostatistics and semivariogram analysis were used. The chemical properties of the soil were moderately to slightly acidic, according to the results. According to USDA standards, the mean values of EC and CEC were rated as low, whereas ESP was rated as high, indicating a sodicity issue. Additionally, the results showed that the model's dependence on pH, ESP, CEC, Ca^{2+}, Na^{+}, and Mg^{2+} was stronger, while that of K^{+} and EC was weaker, by combining the benefits of remote sensing and GIS technology.

Wang et al. (2020) studied how to develop a temporal and spatial soil salinity risk assessment method based on an integrated scoring method. The findings indicate that: (1) The plant senescence reflectance index, deep soil water content (D_{wat}), and impact of shallow soil salinity (SH_{sal}) are the main factors influencing the risk of soil salinity during the wet season. SH_{sal}, LUCC, and D_{wat} are the main factors that affect the risk of soil salinity during the dry season. (2) The risk of salinization during the wet season was found to be very high, with the majority of risks falling into the moderate, high, and very high categories. However, the risk of salinity is primarily characterized as low and moderate during the dry season. (3) In the ELWNNR, areas with moderate risk during the wet season easily change to low-risk and risk-free areas during the dry season (May to August), and areas with high risk during the wet season easily change to areas with moderate risk. The ELWNNR's overall change in salinity risk generally exhibited a significant relationship with variations in lake water volume, indicating that these variations have a significant impact on the likelihood that soil salinity will occur.

Hateffard et al. (2022) used a combination of ensemble machine learning and multivariate geostatistics on three salt-affected soil indicators to map and evaluate salt-affectedness on arable land in Hungary with high spatial resolution (i.e., alkalinity, EC, and sodium adsorption ratio). The results showed that ensemble modeling outperformed the base learners for alkalinity and sodium adsorption ratio with R^2 values of 0.43 and 0.96, respectively, while only the random forest prediction was suitable for EC. Ensemble modeling was done with five base learners (i.e., random forest, tremendous gradient boosting, support vector machine, neural network, and generalized linear model). For each chosen salt-affected soil indicator, tenfold cross-validation was used to assess how well the spatial predictions and uncertainty quantifications performed (like pH value, EC, and sodium adsorption ratio, for which, the the root-mean-square values were 0.11, 0.86, and 0.22, respectively).

Conclusions

Salinity development is a complex and dynamic process with enormous impact and consequences. Detection and assessment of soil salinity are essential to have better prospects about its extent and severity and to control it early. The kriging method for detecting problematic areas is a good tool for managing problematic soils. A quick and affordable alternative for tracking and mapping changes in soil salinity is to use remote sensing and GIS. To locate and map saline areas, a lot of data from sensors and satellites is used. The multispectral satellite sensor is more frequently used because of its low acquisition cost and capability to accurately map the surface salinity. Due to its spatial and spectral resolution, it has, however, a limited set of capabilities. Surface reflectance is influenced by surface roughness, color, moisture, and salt content. High salt concentrations result in specific vegetation types and growth patterns, as well as easily recognizable salt efflorescence and crust on bare soils. We can detect salinity using different vegetation indices and salinity indices from remotely sensed images. Several statistical methods are available, such as regression models, local interpolation techniques such as the IDW or thin-plate splines, or geostatistical techniques such as kriging. We can use kriging, co-kriging, spline, IDW, etc. Geostatistical methods can be used to generate continuous salinity maps from limited data. Kriging and co-kriging methods were better than the IDW method in general. The use of field-based survey methods is difficult and expensive. Combining limited field sampling with geostatistical analysis can yield relatively inexpensive, accurate, large-scale results. Applying geostatistics to salinity assessment and mapping can produce better insight into soil salinization. EMI measurements and estimates made from them are appropriate for areas where there are dominant salts in the soil and provide a cost-effective alternative to traditional laboratory analysis techniques.

References

Abbas, A., Khan, S., Hussain, N., Hanjra, M.A., Akbar, S., 2013. Characterizing soil salinity in irrigated agriculture using a remote sensing approach. Phys. Chem. Earth, Parts A/B/C 55, 43–52.

Abrol, I.P., Yadav, J.S.P., Massoud, F.I. 1988. Salt-affected soils and their management (Issue 39). Food and Agriculture Org.

Al-Khaier, F., 2003. Soil salinity detection using satellite remote sensing. Geo-Information Science and Earth Observation. http://www.itc.nl/library/papers_2003/msc/wrem/khaier.pdf

Abrol, I.P., Bhumbla, D.R., 1971. Saline and alkali soils in India-their occurrence and management, World Soil Resources Report, 41. FAO, Rome, Italy, pp. 42–51.

Allbed, A., Kumar, L., 2013. Soil salinity mapping and monitoring in arid and semi-arid regions using remote sensing technology: a review. Adv. Remote. Sens. 2 (December), 373–385. Available from: https://doi.org/10.4236/ars.2013.24040.

Alqasemi, A.S., Ibrahim, M., Al-Quraishi, A.M.F., Saibi, H., Kaplan, G., 2021. Detection and modeling of soil salinity variations in arid lands using remote sensing data. Open. Geosci. 13 (1), 443–453.

Asfaw, E., Suryabhagavan, K.V., Argaw, M., 2018. Soil salinity modeling and mapping using remote sensing and GIS: The case of Wonji sugar cane irrigation farm. Ethiopia. J. Saudi Soc. Agric. Sci. 17 (3), 250–258.

Bannari, A., Guedon, A.M., El-Harti, A., Cherkaoui, F.Z., El-Ghmari, A., 2008. Characterization of slightly and moderately saline and sodic soils in irrigated agricultural land using simulated data of advanced land imaging (EO-1) sensor. Commun. Soil. Sci. Plant. Anal. 39 (19–20), 2795–2811.

Bouaziz, M., Matschullat, J., Gloaguen, R., 2011. Improved remote sensing detection of soil salinity from a semi-arid climate in Northeast Brazil. C. R. Geosci. 343 (11–12), 795–803.

Buchanan, S., Triantafilis, J., 2009. Mapping water table depth using geophysical and environmental variables. Groundwater 47 (1), 80–96.

Chaturvedi, L., Carver, K.R., Harlan, J.C., Hancock, G.D., Small, F.V., Dalstead, K.J., 1983. Multispectral remote sensing of saline seeps. IEEE Trans. Geosci. Remote. Sens. 3, 239–251.

Clevers, J.G.P.W, 1988. The derivation of a simplified reflectance model for the estimation of leaf area index. Remote Sens. Environ. 25 (1), 53–69.

Deering, D.W., 1975. Measuring "forage production" of grazing units from Landsat MSS data. In: Proceedings of the Tenth International Symposium of Remote Sensing of the Envrionment, 1169–1198.

Dehni, A., Lounis, M., 2012. Remote sensing techniques for salt affected soil mapping: Application to the Oran region of Algeria. Procedia Eng. 33 (September 2015), 188–198. Available from: https://doi.org/10.1016/j.proeng.2012.01.1193.

Dwivedi, R.S., 1996. Monitoring of salt-affected soils of the Indo-Gangetic alluvial plains using principal component analysis. Int. J. Remote. Sens. 17 (10), 1907–1914.

Elhag, M., 2016. Evaluation of different soil salinity mapping using remote sensing techniques in arid ecosystems, Saudi Arabia. J. Sens. 2016. Available from: https://doi.org/10.1155/2016/7596175

Fao/Iiasa/Isric/Isscas/Jrc, 2012. Harmonized World Soil Database (Version 1.2). *FAO, Rome, Italy and IIASA, Laxenburg, Austria.*

Farifteh, J., Van der Meer, F., Van der Meijde, M., Atzberger, C., 2008. Spectral characteristics of salt-affected soils: a laboratory experiment. Geoderma 145 (3–4), 196–206.

Hateffard, F., Balog, K., Tóth, T., Mészáros, J., Árvai, M., Kovács, Z.A., et al., 2022. High-resolution mapping and assessment of salt-affectedness on arable lands by the combination of ensemble learning and multivariate geostatistics. Agronomy 12 (8), 1858.

Huete, A.R., 1988. A soil-adjusted vegetation index (SAVI). Remote. Sens. Environ. 25 (3), 295–309.

IP, A., Bhumbla, D.R., 1971. Saline and alkali soils in India: their occurrence and management.

Ivushkin, K., Bartholomeus, H., Bregt, A.K., Pulatov, A., Kempen, B., De Sousa, L., 2019. Global mapping of soil salinity change. Remote. Sens. Environ. 231, 111260.

Keshavarzi, A.L. I., Sarmadian, F., 2012. Mapping of spatial distribution of soil salinity and alkalinity in a semi-arid region. Ann. Wars. Univ. Life Sci.-SGGW Land Reclam. 44 (1).

Khan, N.M., Rastoskuev, V.V., Sato, Y., Shiozawa, S., 2005. Assessment of hydrosaline land degradation by using a simple approach of remote sensing indicators. Agric. Water Manag. 77 (1–3), 96–109.

Kumar, A., Mehla, M.K., Kumar, S., 2020. A review: use of saline water in irrigation for agriculture. Ind. J. Pure App. Biosci. 8 (1), 262–271.

Leone, A.P., Menenti, M., Buondonno, A., Letizia, A., Maffei, C., Sorrentino, G., 2007. A field experiment on spectrometry of crop response to soil salinity. Agric. Water Manag. 89 (1–2), 39–48.

Li, H.Y., Shi, Z., Webster, R., Triantafilis, J., 2013. Mapping the three-dimensional variation of soil salinity in a rice-paddy soil. Geoderma 195, 31–41.

Liu, H.Q., Huete, A., 1995. A feedback based modification of the NDVI to minimize canopy background and atmospheric noise. IEEE Trans. Geosci. Remote. Sens. 33 (2), 457–465.

Liu, G., Li, J., Zhang, X., Wang, X., Lv, Z., Yang, J., et al., 2016. GIS-mapping spatial distribution of soil salinity for eco-restoring the Yellow River Delta in combination with Electromagnetic Induction. Ecol. Eng. 94, 306–314. Available from: https://doi.org/10.1016/j.ecoleng.2016.05.037.

Major, D.J., Baret, F., Guyot, G., 1990. A ratio vegetation index adjusted for soil brightness. Int. J. Remote. Sens. 11 (5), 727–740.

Mamattursun, E., Mihrigul, A., Li, X., 2016. Geostatistical analysis of variations in soil salinity in a typical irrigation area in Xinjiang, northwest China. Sci. Cold Arid Reg. 8 (2), 147–155. Available from: https://doi.org/10.3724/SP.J.1226.2016.00147.Geostatistical.

Marcos, M., Sharifi, H., Grattan, S.R., Linquist, B.A., 2018. Spatio-temporal salinity dynamics and yield response of rice in water-seeded rice fields. Agric. Water Manag. 195, 37–46. Available from: https://doi.org/10.1016/j.agwat.2017.09.016.

Masoud, A.A., Koike, K., Atwia, M.G., El-Horiny, M.M., Gemail, K.S., 2019. Mapping soil salinity using spectral mixture analysis of landsat 8 OLI images to identify factors influencing salinization in an arid region. Int. J. Appl. Earth Obs.Geoinf. 83, 101944.

Menenti, M., Lorkeers, A., Vissers, M., 1986. An application of thematic mapper data in Tunisia. ITC J. 1, 35–42.

Metternicht, G.I., Zinck, J.A., 2003. Remote sensing of soil salinity: potentials and constraints. Remote. Sens. Environ. 85 (1), 1–20. Available from: https://doi.org/10.1016/S0034-4257(02)00188-8.

Mohammad, Z.-M., Taghizadeh-Mehrjardi, R., Akbarzadeh, A., 2010. Evaluation of geostatistical techniques for mapping spatial distribution of soil pH, salinity and plant cover affected by environmental factors in Southern Iran. Not. Sci. Biol. 2 (4), 92–103.

Morgan, R.S., El-Hady, M.A., Rahim, I.S., 2018. Soil salinity mapping utilizing sentinel-2 and neural networks. Indian. J. Agric. Res. 52 (5).

Narjary, B., Jangra, P., Abhishek, R., Kumar, N., Raju, R., Thimappa, K., et al., 2017. Quantitative assessment of soil salinity using electromagnetic induction technique and geostatistical approach. J. Soil. Salinity Water Qual. 9 (2), 156–166.

Narjary, B., Meena, M.D., Kumar, S., Kamra, S.K., Sharma, D.K., Triantafilis, J., 2019. Digital mapping of soil salinity at various depths using an EM38. Soil. Use Manag. 35 (2), 232–244.

Naseri, M.Y., 1998. Characterization of salt-affected soils for modelling sustainable land management in semi-arid environment: a case study in Gorgon Region, Northeast Iran. Ghent University.

Naumann, J.C., Young, D.R., Anderson, J.E., 2009. Spatial variations in salinity stress across a coastal landscape using vegetation indices derived from hyperspectral imagery. Plant. Ecol. 202 (2), 285–297.

Nawar, S., Reda, M., Farag, F., El-Nahry, A., 2011. Mapping soil salinity in El-Tina plain in Egypt using geostatistical approach. In *GI_Forum 2011 : GI community in focus* (pp. 81–90). Wichmann. http://gispoint.de/fileadmin/user_upload/paper_gis_open/537509011.pdf.

Nezami, M.T., Alipour, Z.T., 2012. Preparing of the soil salinity map using geostatistics method in the Qazvin Plain. J. Soil Sci. Environ. Manage. 3 (2), 36–41. Available from: https://doi.org/10.5897/JSSEM11.031.

Nicolas, H., Walter, C., 2006. Detecting salinity hazards within a semiarid context by means of combining soil and remote-sensing data. Geoderma 134 (1–2), 217–230.

Poss, J.A., Russell, W.B., Grieve, C.M., 2006. Estimating yields of salt-and water-stressed forages with remote sensing in the visible and near infrared. J. Environ. Qual. 35 (4), 1060–1071.

Pozdnyakova, L., Zhang, R., 1999. Geostatistical analyses of soil salinity in a large field. Precis. Agric. 1, 153–165.

Saha, S.K., Kudrat, M., Bhan, S.K., 1990. Digital processing of Landsat TM data for wasteland mapping in parts of Aligarh District (Uttar Pradesh), India. Remote. Sens. 11 (3), 485–492.

Shi, Z., Li, Y., Wang, R.C., Makeschine, F., 2005. Assessment of temporal and spatial variability of soil salinity in a coastal saline field. Environ. Geol. 48 (2), 171–178.

Szabolcs, I., 1989. Salt-Affected Soils. CRC Press, Inc.

Triantafilis, J., Wong, V., Monteiro Santos, F.A., Page, D., Wege, R., 2012. Modeling the electrical conductivity of hydrogeological strata using joint-inversion of loop-loop electromagnetic data. Geophysics 77 (4), WB99–WB107.

Triki Fourati, H., Bouaziz, M., Benzina, M., Bouaziz, S., 2017. Detection of terrain indices related to soil salinity and mapping salt-affected soils using remote sensing and geostatistical techniques. Environ. Monit. Assess. 189 (4), 1–11.

Utset, A., Ruiz, M.E., Herrera, J., De Leon, D.P., 1998. A geostatistical method for soil salinity sample site spacing. Geoderma 86 (1–2), 143–151. Available from: https://doi.org/10.1016/S0016-7061(98)00037-8.

Wang, Z., Zhang, F., Zhang, X., Chan, N.W., Kung, H., Zhou, X., et al., 2020. Quantitative evaluation of spatial and temporal variation of soil salinization risk using GIS-based geostatistical method. Remote. Sens. 12 (15), 2405.

Yusuf, B.L., Mustapha, A., Yusuf, M.A., Ahmed, M., 2020. Soil salinity assessment using geostatistical models in some parts of Kano River Irrigation Project Phase I (KRPI). Model. Earth Syst. Environ. 6 (4), 2225–2234.

Zewdu, S., Suryabhagavan, K.V., Balakrishnan, M., 2017. Geo-spatial approach for soil salinity mapping in Sego Irrigation Farm, South Ethiopia. J. Saudi Soc. Agric. Sci. 16 (1), 16–24.

Zhang, M., Ustin, S.L., Rejmankova, E., Sanderson, E.W., 1997. Monitoring Pacific coast salt marshes using remote sensing. Ecol. Appl. 7 (3), 1039–1053.

Zovko, M., Romić, D., Colombo, C., Di Iorio, E., Romić, M., Buttafuoco, G., et al., 2018. A geostatistical Vis-NIR spectroscopy index to assess the incipient soil salinization in the Neretva River valley, Croatia. Geoderma 332, 60–72.

Chapter 28

Soil salinity mapping using multisensor data employing machine-learning technique: a case study from indo-gangetic plain

Ellur Rajath, Justin George Kalambukattu, Dipanwita Haldar and Suresh Kumar
Agriculture and Soil Department, Indian Institute of Remote Sensing (IIRS), Indian Space Research Organization (ISRO), Dehradun, Uttarakhand, India

28.1 Introduction

Accumulation of salts in the low layer of soils is one of the serious land degradation processes leading to a decline in productivity of land and desertification, which can be natural or through anthropogenic activities (Nachshon, 2018). Around 953 M ha of salt-affected soils are distributed in 120 countries (Wicke et al., 2011) and productivity of these soils has reduced by 7%–8% at a global scale. Sodicity is affecting >50% of these soils, with the highest area in Australia. While 20% of the global area containing dry and semiarid regions of Asia and Pacific is affected by salinity, and about 39.9 M ha of the salt-affected area is contributed by waterlogging along with secondary salinization in irrigated regions (Yadav, 2003). Based on state-wise area statistics, 6.7 M ha of land was affected by salts in India, out of which 2.9 M ha is due to salinity and the rest due to sodicity (Mandal et al., 2011).

Inappropriate use of irrigation water and old techniques of irrigation, use of salt-rich irrigation water, and fertilizers that add a countable amount of salts to soils are the main human-induced causes for soil salinity development, together with some natural causes such as salt-rich parent material, shallow groundwater tables with high salts, high temperature, and low rainfall. As a result, the area under salinity is increasing every year, which makes it crucial to map and manage affected areas. Conventional methods of soil sample collection and analysis at the point scale are not very good approaches for large spatial extents. Many studies have shown that remote sensing can be a promising technology to map soil salinity in spatial extent (Lesch et al., 1992; Chanzy, 1993; Al-khaier, 2003; Douaoui et al., 2006; Mandal et al., 2009; Dehni and Lounis, 2012; Lhissoui et al., 2014; Gorji et al., 2015; Hoa et al., 2019; Periasamy and Ravi, 2020).

Several mature approaches involving optical remote sensing data have been formulated and successfully used to map the severity and distribution of soil salinity (Al-khaier, 2003; Khan et al., 2005; Douaoui et al., 2006; Elnaggar and Noller, 2010; Dehni and Lounis, 2012; Lhissoui et al., 2014; Azabdaftari and Sunar, 2016; Hu et al., 2019; Taghadosi et al., 2019a). These techniques heavily rely on the purity of the salt, salt coating, soil moisture, and optical contrast with other surface features. Optical remote sensing can easily detect moderate to highly saline areas, while initial stages of salinization and low salinity cause detection failure (McGowen and Mallyon, 1996). This kind of failure always leads to over- or underestimation of saline areas.

The potential of microwave radar data to detect soil salinity was also explored by some authors, and this radar data is independent of weather conditions (Taylor et al., 1996; Shao et al., 2003; Aly et al., 2004; Srineevas et al., 2007; Lasne et al., 2008; Gong et al., 2014). The principle behind the use of microwave data is the influence of the dielectric constant on backscattering (Srineevas et al., 2007). The intended variable dielectric constant is highly sensitive to the physicochemical properties of the soil, especially moisture and electrical conductivity. So, at a particular polarization, the radar backscattering coefficient is principally determined by the target's scattering characteristics and electrical conductivity (Periasamy and Ravi, 2020). A complex SAR product expresses the dielectric permittivity (ε^*) as real (ε') and

Remote Sensing of Soils. DOI: https://doi.org/10.1016/B978-0-443-18773-5.00013-2

imaginary part (ε″). The real part is related to polarization characteristics and is significantly attenuated by soil moisture. The imaginary part is related to dielectric loss (Li et al., 2014) and influenced by the chemical properties of soil, like salinity.

The disadvantages of optical data can be overcome by incorporating radar data, which has the advantages of weather independence and subsoil penetration and can improve the accuracy of salinity mapping. Researchers have reported significant improvements in the accuracy of salinity assessment (Wu et al., 2020), land cover mapping (McGowen and Mallyon, 1996), irrigation, and moisture assessment (Fieuzal et al., 2011; He et al., 2014).

The detrimental effect of soil salinity on soil physicochemical properties hinders plant growth and production (Brady and Weil, 2008), which are dependent on tolerance exhibited by each crop. Salinity-induced alterations in plant properties, such as leaf area index, biomass, water content, and chlorophyll concentrations, are blamed for the reduction in growth and production. These changes lead to variations in spectral properties (Salisbury, 1991; Encarni et al., 2014) as well as the backscattering properties of crops (Attema and Ulaby, 1978; Bindlish and Barros, 2001; El-Hendawy et al., 2019).

So the integration of optical and microwave data to assess the salt-affected soils in alluvial plain of Mathura district of Uttar Pradesh is the main objective of the present study. Along with it, efforts were also made to determine the effect of salinity levels on wheat spectral and backscattering properties.

28.2 Study area and data collection

28.2.1 Study area

The study was carried out in the Mathura district of Uttar Pradesh. Mathura is one of the western districts of Uttar Pradesh and forms part of the Yamuna basin. It is surrounded by Aligarh and Etah district in the east, Bharatpur district of Rajasthan in the west, Agra district in the south, and Aligarh and Gurgaon districts in Haryana. The area is located between 26°76′ to 27°62′ north latitudes and 77°31′ to 77°59′ east longitudes (Fig. 28.1), with an average elevation of 174 m. The study area is dominated by wheat and mustard in the rabi season and paddy in the kharif season.

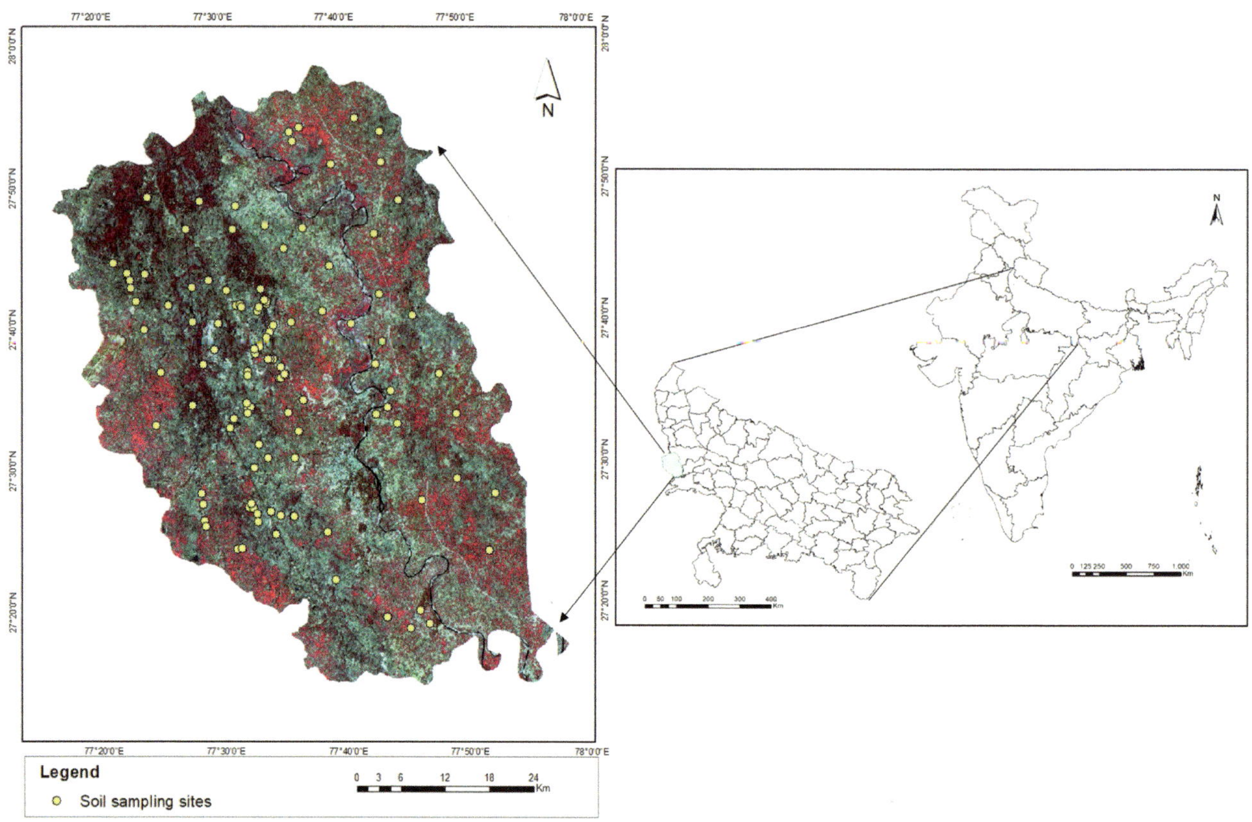

FIGURE 28.1 Location map of the study area with soil sampling locations overlaid.

28.2.1.1 Climate and soils

The study area experiences severe summers, cold winter, and dryness throughout the year except during the southwest monsoon period from the first week of July to September, making the area a semiarid climatic region. The mean annual temperature was observed to be 24.4°C. In summers, the maximum temperature reaches up to 45°C, while the minimum temperature in winter will be 14°C. The average annual rainfall is 600 mm, the maximum of which is received from July to September.

The area forms a part of the Gangetic plain, and the river Yamuna passes through the district. Most of the soils have developed over alluvium of different ages. The soils of the area are characterized as recent alluvial soils belonging to various categories such as typic ustifluents, typic endoaquepts, aeric fluvaquents, and udic ustifluents, which occur along the Yamuna in a recent flood plain. The old alluvial belongs to typic haplustepts, typic haplustalfs, typic natrustalfs, typic halaquepts, and calciustepts (Sharma and Bhargava, 1988).

28.3 Satellite data used and preprocessing

28.3.1 Sentinel 1

To study the effect of soil salinity on the backscattering properties of wheat, microwave images of five days were acquired between December 2018 and April 2019 (Table 28.1). The images acquired on April 8, 2019, corresponding to field survey, were used for salinity mapping. The downloaded images were subjected to a sequence of preprocessing steps (Fig. 28.2) using SNAP software (SNAP: ESA Sentinel Application Platform v2.0.2, http://step.esa.int). Preprocessing is intended to improve the image data by reducing undesirable noise and distortions and enhancing some image qualities that are crucial for subsequent applications. The precise location of a satellite will wander over time regardless of how well-tuned its orbit is, so it is necessary to go for orbit correction. This is done using the orbit correction file in level-1 data recorded using GNSS. It is essential to transform SAR digital numbers stored inside SAR picture products into radar backscatter. Based on the product's metadata, SNAP will automatically determine the type of input product provided and what conversion needs to be performed. The presence of speckles in the data poses an obstacle to further utilization of the data. To remove these speckles, Lee Sigma adaptive filter with a 5 × 5 window size was used (Misra and Ajmera, 2017). Geographic coordinates are not available for the Sentinel-1 GRDH picture. Future applications will require the conversion of pictures (either slant or ground range geometry) into a map coordinate system. This was achieved with terrain correction using a 1-Sec SRTM DEM. Thus, processed data was used for further analysis in the study.

28.3.2 Sentinel 2

Cloud-free optical data (level 1 C) acquired by Sentinel 2 constellation was downloaded corresponding to five days (Table 28.2) from Copernicus datahub to study spectral changes in wheat due to salinity. Images acquired on April 5, 2019, corresponding to survey dates, were used for salinity prediction and mapping. The images were atmospherically

TABLE 28.1 Characteristics of acquired sentinel 1A IW Level-1 GRD product of study area.

Date/parameters	21 Dec 2018	14 Jan 2019	7 Feb 2019	3 Mar 2019	8 Apr 2019
Incidence near (°)	30.81	30.81	30.81	30.81	30.81
Incidence far (°)	46.20	46.20	46.20	46.2	46.20
First line time	12:54:53	12:54:52	12:54:51	12:54:51	12:54:52
Last line time	12:55:22	12:55:21	12:55:20	12:55:20	12:55:21
Polarization	VV &VH	VV &VH	VV &VH	VV &VH	VV &VH
Pass	Ascending	Ascending	Ascending	Ascending	Ascending
Azimuth look	1	1	1	1	1
Range look	5	5	5	5	5
Range spacing (m)	10	10	10	10	10
Azimuth spacing (m)	10	10	10	10	10

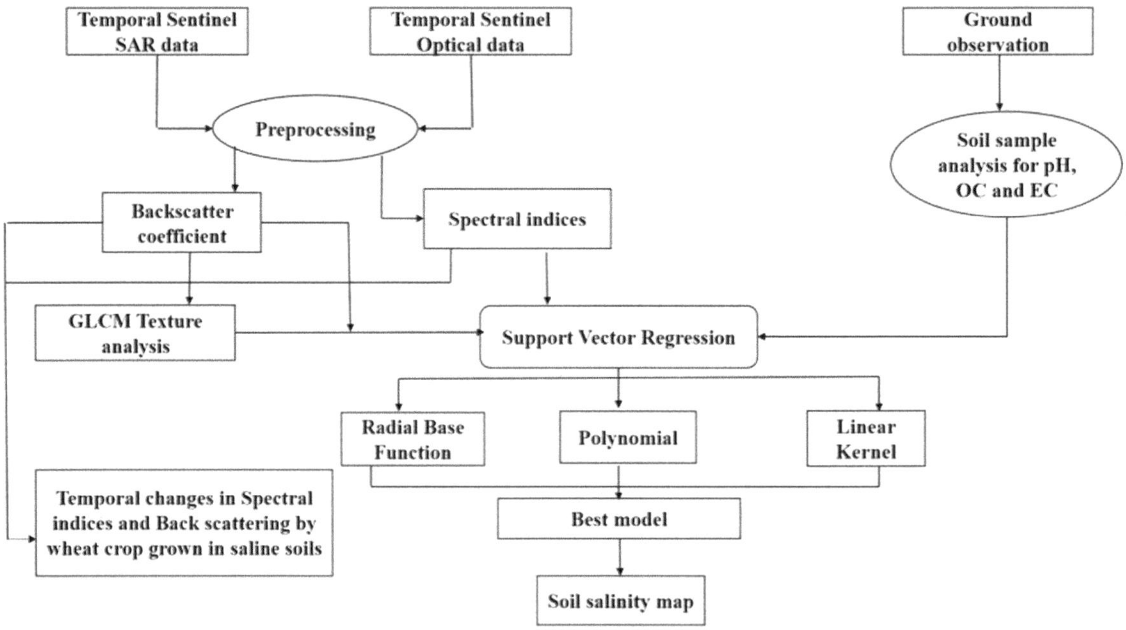

FIGURE 28.2 Flow chart of methodology adopted in the study.

TABLE 28.2 Characteristics of acquired sentinel 2 product of study area.

Parameter/dater	Sensor	Sensing time	Processing level	Product type
21 Dec 2018	2A	5:32:31	Level 1C	S2MSI1C
10 Jan 2019	2A	5:32:11	Level 1C	S2MSI1C
9 Feb 2019	2A	5:29:51	Level 1C	S2MSI1C
6 Mar 2019	2A	5:27:09	Level 1C	S2MSI1C
5 Apr 2019	2B	5:26:49	Level 1C	S2MSI1C

corrected using the DOS 1 tool of the SCP plugin in QGIS (Ver. 2.18.13) software (Congedo, 2021). Reflected energy from earth's surface reaching the satellite mounted sensors is attenuated due to the various components present in earth's atmosphere. So, atmospheric correction is needed to remove these attenuations.

28.4 Methodology

28.4.1 Survey and soil sample collection

A random soil sample collection strategy was followed, and 115 soil samples at 0–15 cm depth were collected from the study area in a five-day field survey program (April 25–30, 2019). The locations of soil samples collected are presented in Fig. 28.1. Since the study involved salinity mapping and analyzing the effect of salinity on wheat properties in the optical microwave region of the spectrum, ground truth points of wheat along with soil samples were collected.

28.4.2 Preprocessing and chemical analysis of soil samples

The determination of the physicochemical properties of saline and alkali soils serves as the basis for their characterization. The electrical conductivity of saturation paste extract of soil (ECe) samples was determined according to the protocol of Rhoades et al. (1989). Further samples were classified into slight, medium, and strong salinity classes (Table 28.3) based on pH and ECe values (Mandal et al., 2009).

28.4.3 Generation of spectral indices

28.4.3.1 Vegetation indices

The atmospherically corrected Sentinel 2 reflectance images were used for generating vegetation indices for the area. The indices were explored for studying the effect of varying levels of soil salinity on the temporal spectral properties of wheat crops grown in the area. normalized difference vegetation index (NDVI), canopy response salinity index (CRSI), ratio vegetation index (RVI), and soil-adjusted vegetation index (SAVI) were generated as per the details given in Table 28.4.

28.4.3.2 Salinity indices

Several salinity indices were calculated from the atmospherically and radiometrically corrected Sentinel 2 image for April. The study area was dominated by a rice–wheat cropping system, and most of the rabi wheat was harvested. The soil was exposed to a maximum extent during April, and cloud-free images were available. So, these images were used to calculate various salinity indices. These indices were used with several textural features for salinity mapping using SVR. Table 28.4 shows the summary of salinity indices used in the study.

28.4.4 GLCM textural features extracted from SAR image

Using image-based texture features, which define the spatial distribution of radar brightness values in the image, it is possible to distinguish between salt-affected areas. Textural features were discovered by computing histogram-based features from first-order statistics and second-order statistics retrieved from the gray-level co-occurrence matrix (GLCM), which takes into account the relative position of the pixels with respect to one another. Textural analysis was carried out on sigma-nought images (VV and VH polarizations) for the 5th date image. This was performed using

TABLE 28.3 Criteria for salinity classification (Mandal et al. 2009).

Severity	ECe (dSm^{-1})
Slight	0-8
Medium	8-30
Strong	>30

TABLE 28.4 Various bands/indices and GLCM textural parameters used in study.

Optical		GLCM textural Parameters	
NDVI[a]	Salinity index I[b]	Contrast[c]	Variance[d]
CRSI[e]	Salinity index II[c]	Dissimilarity[c]	Correlation[c]
RVI	Salinity index III[a]	Homogeneity[c]	
SAVI	Salinity index IV[c]	Angular Second moment[c]	
	Salinity index V[a]	Energy[c]	
	Salinity index VI[c]	Entropy[c]	
	Brightness index[a]	Mean[c]	
	NDSI[a]	Maximum[c]	

[a]*Scudiero, Skaggs, and Corwin (2015)*
[b]*Khan et al. (2001)*
[c]*Douaoui, Nicolas, and Walter (2006)*
[d]*Alhammadi and Glenn (2008)*
[e]*Strzelecki (1998)*

SNAP software with a 9×9 window size and displacement in all 4 directions and various textural parameters were extracted (Table 28.4).

28.4.5 Temporal analysis of spectral indices and SAR backscattering of wheat crop

Initially, based on the electrical conductivity of the soil samples analyzed, 115 observations were categorized into slight, medium, and strong salinity classes (Table 28.3) as described by Mandal et al. (2009). Values of spectral indices such as NDVI, SAVI, CRSI, and RVI were calculated and averaged for all three categories of soil samples, as shown in Fig. 28.4. Similarly, SAR backscatter values were obtained and averaged in VV and VH polarization for all categories of soil samples, as shown in Fig. 28.5.

28.4.6 Support vector regression for salinity map generation

Support vector machine is one of the most effective machine learning methods used to predict a variety of biophysical characteristics from remote sensing data, including soil organic carbon, soil salinity, and biomass (Hoa et al., 2019). It is based on the computation of a linear regression function in a high-dimensional feature space, first proposed by Vapnik et al. (1997) as the following equation:

$$f(x) = \sum_{n=1}^{n} \left(\lambda_i - \lambda_i^*\right) \ k \ (x_i, x) + b$$

Where λ_i, λ_i^* denote Lagrange multipliers and $k(x_i, x)$ is a kernel function.

Most researchers use the SVM technique for classification and the SVR for regression. Estimation and selection of suitable parameters for SVR (C, γ) make this technique complex. Linear separation is done by transforming the data into a new, higher feature space, which can be achieved by various kernel functions, viz., linear (LN), polynomial (PL), and (Gaussian) radial base function (RBF). A good-fit model and high accuracy can be obtained by estimating and selecting various parameters for each kernel.

28.4.6.1 Multicollinearity analysis

When there are high correlations among independent variables, these independent variables may compete to explain much of the same variance, making it challenging to distinguish their contributions to the dependent variable (Zainodin and Yap, 2013).

For identifying the nature of relationships, multicollinearity, and selection of predictor variables for modeling, correlation analysis was done between EC values, various bands/indices, and textural parameters with each other (Fig. 28.3). Initially, 34 predictor variables, that is, spectral indices and textural features generated using the 5th date images of Sentinel 1 and Sentinel 2, respectively, including Blue, Green, Red, NIR, VV, and VH backscatter images, were considered for salinity mapping. A certain degree of multicollinearity was observed between variables, so variables exhibiting correlation coefficient values greater than 0.75 with each other (Zainodin and Yap, 2013) and lesser correlation with ECe were eliminated from further analysis.

Point values of uncorrelated predictor variables were used for SVM regression analysis in R software, employing the SVM function of the e1071 library (Meyer et al., 2001). The three types of kernels, namely LN, PL, and Gaussian RBF, were used for model development, and the best model from each kernel was selected and used for spatial mapping of soil salinity in the form of EC_e maps. NDVI thresholds for vegetation, urbanity, and water body were determined, and these features were masked out from the final ECe maps predicted.

28.5 Results and discussion

28.5.1 Soil pH and ECe

All 115 soil samples were analyzed for pH and EC_e, and they were classified into slight, moderate, and strong salinity classes based on their EC_e values. Out of these, 2 samples belonged to strong salinity, 32 to moderate salinity, and the remaining were categorized under slight salinity. The highest and lowest EC_e found were 33.6 and 1.87 dSm^{-1}, with an average EC_e 6.701 dSm^{-1} and a standard deviation of 6.63. The main reasons for salinity in the study area were due to poor quality groundwater that was used for irrigation, the nature of parent material, and semiarid climatic conditions (Kumar et al., 2015). Extreme salinity was observed in some areas where salt encrustation was visible, and the fields were devoid of any vegetation.

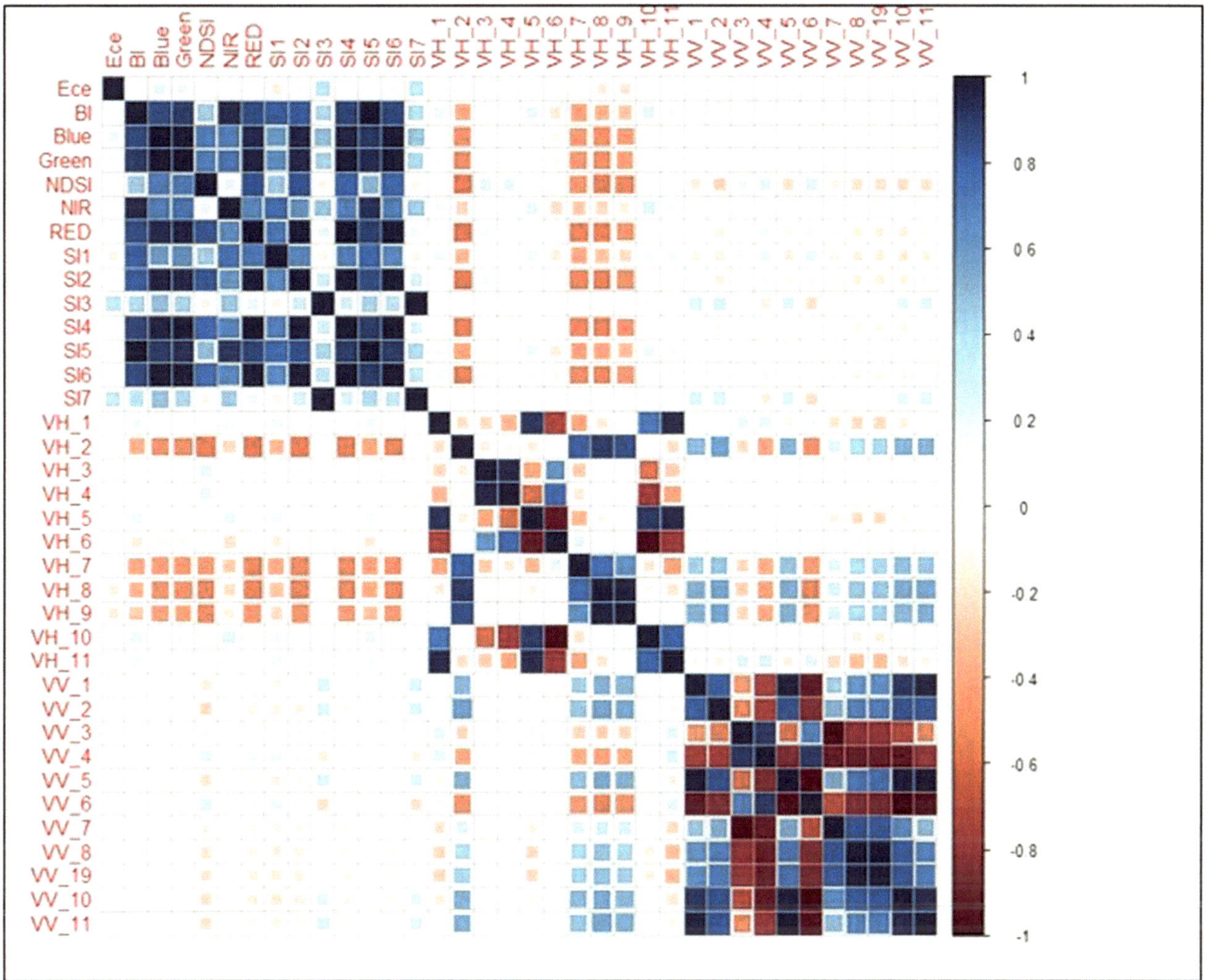

ECe	Electrical Conductivity	SI2	Salinity Index II	VV & VH 4	Dissimilarity
BI	Brightness Index	SI3	Salinity Index III	VV & VH 5	Energy
Blue	Blue band of Sentinel 2	SI4	Salinity Index IV	VV & VH 6	Entropy
Green	Green band of Sentinel 2	SI5	Salinity Index V	VV & VH 7	Correlation
NDSI	Normalized Difference Salinity Index	SI6	Salinity Index VI	VV & VH 8	Mean
NIR	Near Infrared Band	VV & VH 1	Angular Second Moment	VV & VH 9	Variance
RED	Red band of sentinel 2	VV & VH 2	Backscatter	VV & VH 10	Homogeneity
SI1	Salinity Index I	VV & VH 3	Contrast	VV & VH 11	Maximum

FIGURE 28.3 Correlation matrix showing the relationship between various variables.

28.5.2 Temporal study of spectral and backscattering properties of wheat in saline soils

28.5.2.1 Effect of salinity on spectral properties of wheat

Temporal variation in the spectral properties of wheat was studied using various vegetation indices (Table 28.4) generated for all the 5-day images of Sentinel 2. In most of the areas of Mathura district, crop sowing was done in December 2018. Since then, the NDVI values had increased up until February 2019, and thereafter, values decreased due to the maturity and harvest of the wheat crop (Fig. 28.4).

All the optical indices, viz., NDVI, SAVI, RVI, and CRSI, showed a similar pattern for temporal changes (Fig. 28.4). Wheat crops grown in slightly saline areas showed high values, followed by crops grown in medium salinity and strong salinity classes, thus revealing that increased salinity levels have significantly altered the spectral properties of the crop canopy. Reduction in the concentration of leaf chlorophyll and other photosynthetic pigments, reduced photosynthetic efficiency, and damage to leaf structure are the major logical considerations under salinity stress, which is known to cause changes in the pattern of various vegetation indices under different salinity levels (Scudiero et al., 2017). The Blue and Green bands are also sensitive to chlorophyll-a and b, α and β-carotenes, and lutein. Similarly, the Red and NIR bands used in the generation of different vegetation indices in this study are sensitive to leaf structure as well as leaf water content, the main plant characteristics adversely affected by soil salinity (Poss et al., 2010).

Munir Ahmad et al. (2013) observed that salinity stress and other abiotic stresses caused the early maturity of wheat and preparation for various growth stages. In our study, it was observed that the jointing stage happened to occur way earlier in wheat grown in high salinity than in medium or slight salinity (Chu et al., 2016; Rameeh and Gerami, 2015). As a result, peak vegetation values were observed earlier in February (Fig. 28.4). Accelerated development of the shoot apex for a short time decreased spikelet primordia numbers and led to early terminal spikelet stages and anthesis (Ashraf and Ashraf, 2016). This caused a sharp decrease in vegetation indices after the month of February.

28.5.2.2 Effect of salinity on backscattering properties of wheat

Wheat crops grown under different levels of salinity had shown significant variation in backscattering under both VH and VV polarizations (Fig. 28.5). The first date recorded the highest backscattering in VH polarization due to volume

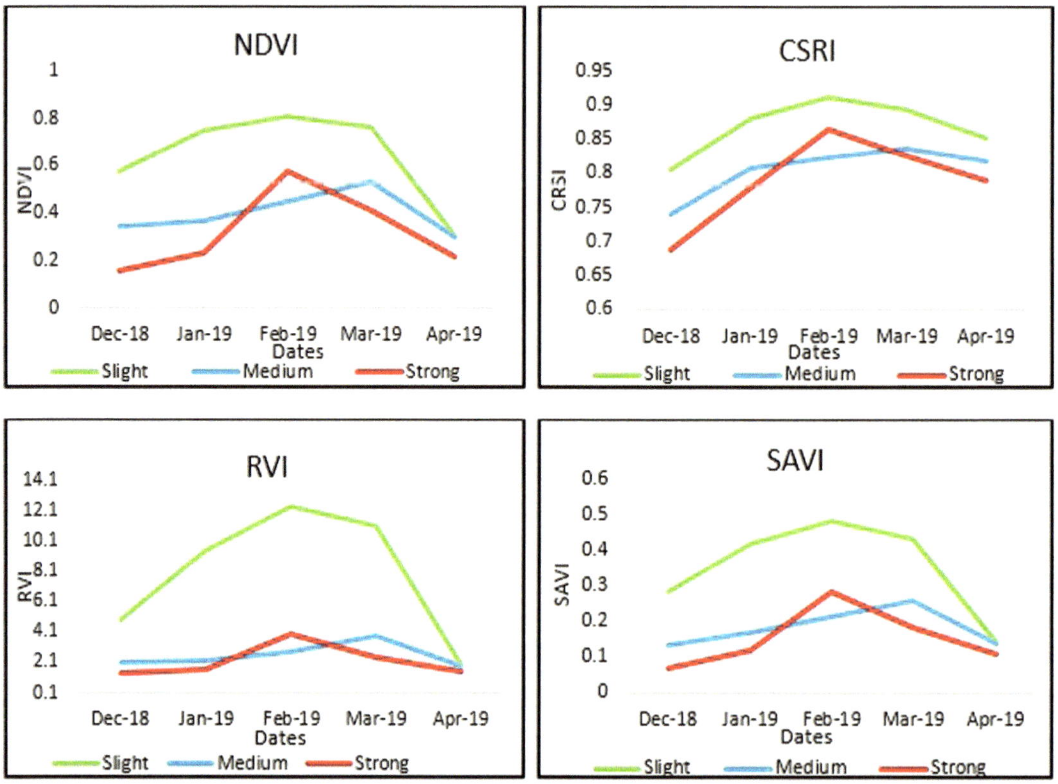

FIGURE 28.4 Temporal changes in optical indices of the wheat crop.

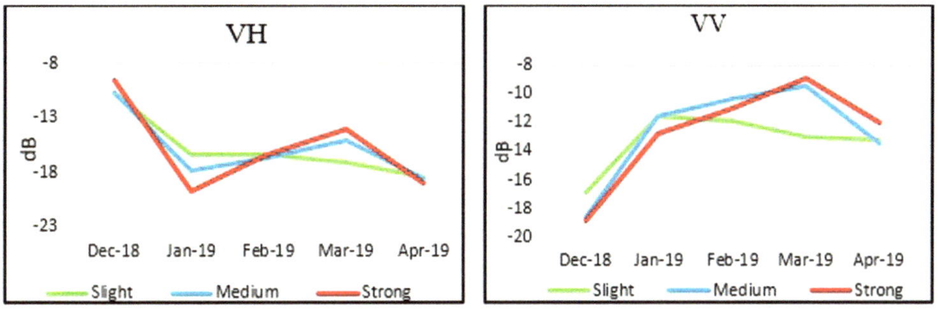

FIGURE 28.5 Backscatter in VV and VH polarization by wheat crop grown.

scattering by plowed land, which was quite rougher (Vreugdenhil et al., 2018), whereas the same condition led to depolarization of incident microwave and thus lower backscattering in VV polarization. As the crop started growing, the effect of salinity became more pronounced, and backscattering in VH polarization has decreased gradually in all the salinity classes. Among the different salinity classes, a comparatively higher backscatter was seen with the wheat crops grown in slightly saline soils, followed by medium and strong salinity during January 2019. But this trend was reversed after February, when wheat under strong and medium salinity continued to show an increase in backscatter up to March, and then it decreased. But the backscatter corresponding to wheat crop grown under slight salinity started decreasing after January and this decreasing trend continued till harvest.

In the case of VV polarization, the initial low backscattering gradually increased in crops under all the salinity classes until the second date. The crops under strong saline soils had lower backscatter values on the second date in comparison to the other two classes. Thereafter, the backscatter of crops under medium and strong classes continued to increase, whereas the crops under slight class showed a declining trend (Fig. 28.5).

In wheat crops, as biomass increases, the sensitivity of the *C*-band decreases. Backscattering has been reported to decrease with an increase in biomass for the wheat crop (Haldar et al., 2014). Backscattering by wheat at different dates, that is, different growth stages, has followed the same trend as studied by Vreugdenhil et al. (2018). There is no study found regarding the effect of salinity on the backscattering of any crop or vegetation to validate the obtained results. But the biomass accumulation of wheat grown under strong salinity will be generally lower than that of crops grown under medium or slight salinity. The temporal profile of backscattering showed a decreasing trend in the case of crops grown under slight salinity conditions after the 2nd date (Fig. 28.5). Whereas, the crops under medium and strong salinity classes showed an increasing trend after the 2nd date. This variability in backscattering in VV and VH polarization by wheat might be attributed to variation in the vigor of the crop with salinity levels.

28.5.3 Salinity mapping

To characterize the severity level of salinity in salt-affected soils, ECe has been considered the best parameter. Support vector regression was employed for salinity mapping using various salinity indices generated for the 5th date optical image and backscatter images of VV and VH polarization and GLCM textural parameters, as discussed in earlier sections.

28.5.3.1 Model building and validation

28.5.3.1.1 Removal of multicollinearity

A total of 9 salinity indices, Blue, Green, Red, and NIR bands of the optical image, and 20 textural parameters of VV and VH backscatter images were initially used as predictor variables. Based on the analysis of the correlation matrix (Fig. 28.3) generated, variables exhibiting multicollinearity and leading to data redundancy were identified and eliminated from further processes. After the removal of 12 correlated variables, brightness index, NDSI, NIR, salinity index 1, 3, and 7, VH and VV backscatter, VH ASM, VH contrast, VH dissimilarity, VH energy, VH entropy, VH GLCM correlation, VH GLCM mean, VH GLCM variance, VH homogeneity, VH max, VV ASM, VV contrast, VV dissimilarity, and VV GLCM were selected for the development of different models.

28.5.3.1.2 Model parameterization, accuracy assessment, and salinity map generation

The data corresponding to 115 field observations was divided into training and testing data (70:30). These training and testing data were used for model training and testing, respectively. For each kernel (linear, polynomial, and radial basis

TABLE 28.5 Results of regression analysis and all explanatory variables.

SVR kernel type	Train/test	R^2	RMSE	Selected parameters
RBF	Train	0.99	0.53	C = 1000, γ = 0.02
	Test	0.87	1.47	
Polynomial degree 1	Train	0.69	3.2	C = 0.9, γ = 1, coefficient = 0
	Test	0.68	2.3	
Polynomial degree 2	Train	0.94	1.38	C = 1, γ = 0.1, coefficient = 8
	Test	0.80	1.94	
Polynomial degree 3	Train	0.66	5.22	C = 0.001, γ = 0.01, coefficient = 10
	Test	0.68	3.56	
Linear	Train	0.93	1.80	C = 2, γ = 0.1
	Test	0.87	1.60	

function) of ε-SVR, various parameters were optimized, and the selected parameters of each kernel and their estimated best values are given in Table 28.5. The results of the regression analysis and accuracy assessment are presented in Table 28.5. Accuracy was assessed based on R^2 and root-mean-square error (RMSE) values obtained from different kernels of ε-SVR method. Using the best parameters of each kernel identified during optimization and validation, salinity maps were generated for the entire area (Fig. 28.6).

GLCM parameters improve prediction accuracy because they radar brightness data that can be used to deduce surface structural properties and their relationships with the surrounding environment. Effects of soil salinity are expressed in some regular tone, texture, and patterns (Mandal, 2016); as a result, there exists a strong linear correlation between textural parameters and various salts of soils (Hoa et al., 2019). Ren et al. (2016) found a linear relation between textural parameters and Na, Ec, and salinity ($R^2 > 0.87$) and also a good relation with Cl ($R^2 > 0.73$).

The best performance of the RBF kernel was observed with parameters $C = 1000$ and $\gamma = 0.02$, RMSE of 0.53 and 1.47, and R^2 of 0.99 and 0.87, respectively, for training and test data. Taghadosi et al. (2019b) also reported similar results of best performance by the RBF kernel (RMSE 1.99 and 0.35 and R^2 of 0.73 and 0.97 for train and test data, respectively). Kumar et al. (2022) also found that the RBF kernel of SVR outperformed other kernels with a coefficient of determination of 0.98 for testing dataset, and the highest area of salt-affected soils was estimated with the RBF kernel. For the polynomial kernel, the 2nd degree polynomial performed best with parameters $C = 1$, $\gamma = 0.1$ and Coef0 = 8 yielding R^2 of 0.94 and 0.80, RMSE of 1.38 and 1.94 for training and test data, respectively. In the case of the linear kernel, the values of parameters that performed well, producing R^2 of 0.93 and 0.87 as well as RMSE of 1.80 and 1.60 for training and test data, were $C = 1$ and $\gamma = 0.1$ (Table 28.5). The optimized and tested models were finally used for the spatial mapping of soil salinity (Fig. 28.6). Results are in agreement with the findings of (Wu et al., 2018; Hoa et al., 2019; Taghadosi et al., 2019b; Taghadosi et al., 2019a; Wu et al., 2020).

Concerning the prediction capacity of kernels, RBF and polynomial kernels were capable of predicting salinity up to 60.42 and 51.59 dS m^{-1} respectively (Fig. 28.7). Polynomial 1, 2, and linear kernels were capable of predicting salinity up to 30.99, 25.46, and 29.16 dS m^{-1}, respectively, which were far less than RBF predictions. Fig. 28.7 clearly shows that for a very highly saline patch in the study area, the RBF, as well as polynomial 2 kernels, predicted much higher salinity compared to polynomial 1 and 3 degree and linear kernels, which were underpredicting the high values. Accumulations of white salts and encrustations (Fig. 28.7), which have higher reflectivity (Taghadosi et al., 2019a), and in medium and slight saline soils, which are normal in appearance, might have caused wide variations in the predicted values (Fig. 28.6).

The perusal of the results of this study investigating spatial prediction and mapping of soil salinity using parameters derived from optical and microwave data of Sentinel 1 and 2 employing support vector regression reveals that

1. GLCM textural parameters and backscatter images of Sentinel-1 SAR data have strong potential to estimate soil salinity.

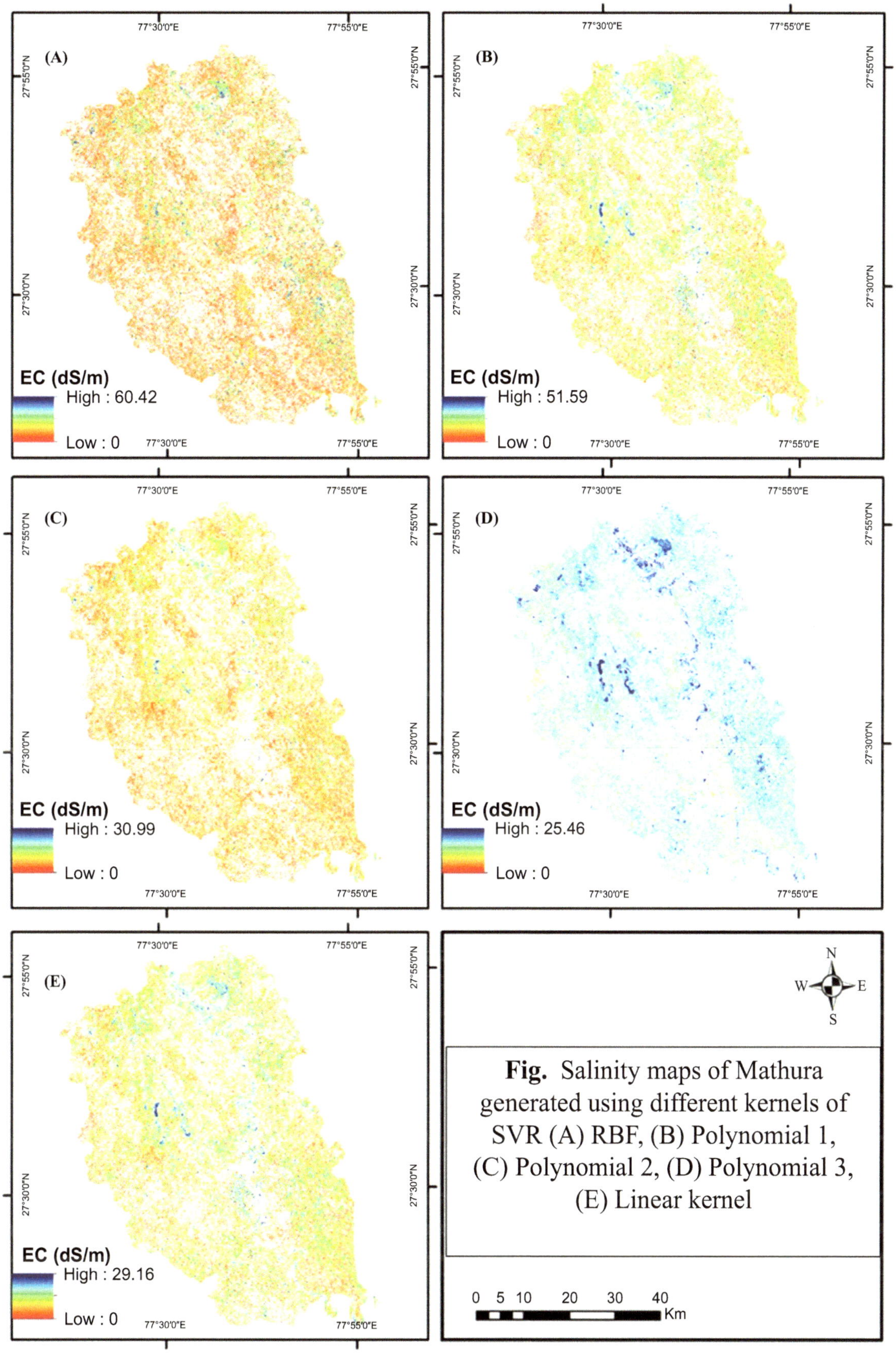

FIGURE 28.6 Salinity maps of Mathura generated using different kernels.

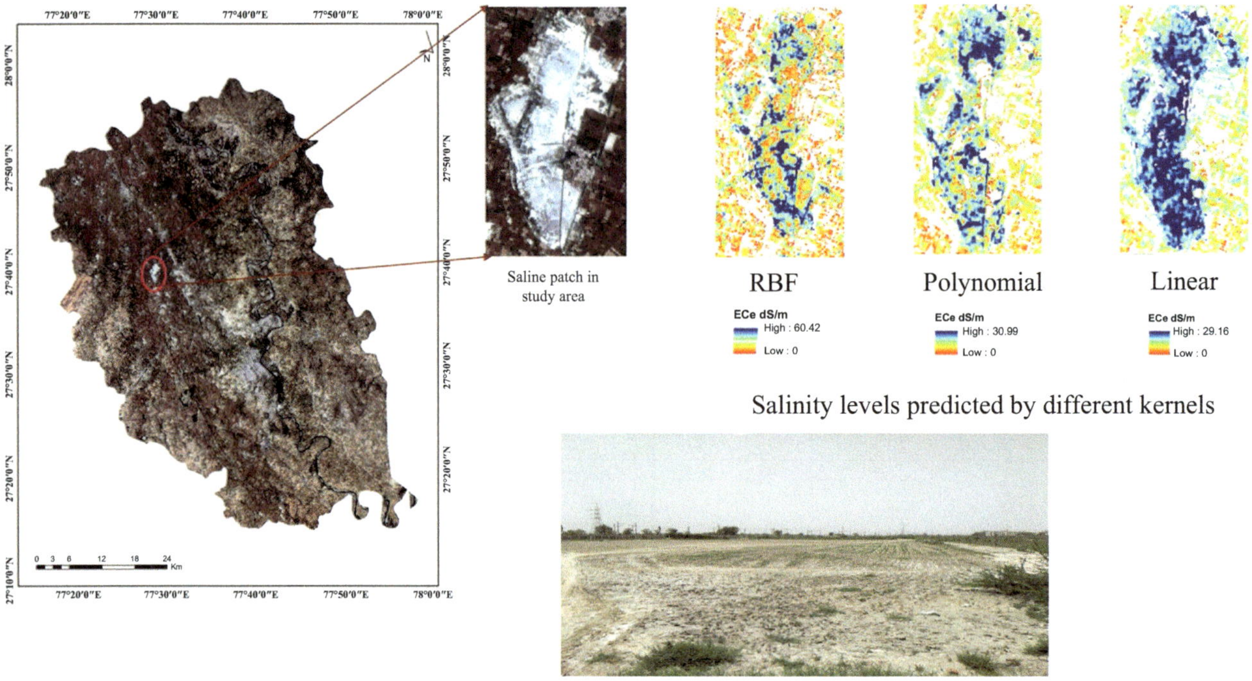

FIGURE 28.7 Comparison of salinity levels predicted by different kernel.

2. This study finds that the combined use of various salinity indices and textural parameters of Sentinel-1 and 2, respectively, has an added advantage in predicting soil salinity using ε-SVR.
3. This investigation portrays that ε-SVR is a strong regression tool for salinity model building, especially when a nonlinear sort of relationship exists between salinity and various remote sensing–derived parameters.
4. It was found that most of the salt-affected soils were concentrated at the sides of the rivers. And all the kernels showed a similar trend, but the severity of the salt infestation varied.

So, the findings of this study show that soils near the river channel have higher chances of soil salinity. This might be due to river water quality and the type of cropping system followed in the region. Strategies like introducing crops that are tolerant to salinity and require little water can help improve the socioeconomic conditions of the farmer. And strategies for managing salt-affected soil should be implemented on a priority basis in lands near the river course.

28.6 Conclusion

This study was aimed at knowing the impact of soil salinity on the spectral and backscattering properties of wheat as well as the feasibility of the combined use of optical and microwave data for soil salinity mapping in Mathura district, located in the Indian state of Uttar Pradesh. Wheat crops grown under different salinity levels showed a significant difference in spectral and backscattering properties. Spectral properties were studied using various vegetation indices and backscattering properties with σ naught images. Vegetation indices had shown a decreasing trend with an increase in salinity levels. Backscattering in VH polarization decreased from the 1st to the 2nd date; this decreasing trend was continued till harvest with wheat grown in slight salinity, but backscattering of wheat in medium and slight salinity increased up to the 4th date; the rate of increase was more in strong salinity than medium, and thereafter decreased with the maturity of the crop. The same trend was observed with VV polarization, except for a sharp rise in backscattering on the 2nd date.

Since the salts in soils affect the dielectric constant and, in turn, manipulate microwave data, soil salinity mapping with the combined use of optical and microwave data may be a very feasible option. The RBF kernel of the support vector regression model with RMSE 0.53 and 1.47 and R^2 0.99 and 0.87 for train and test data, respectively, was found to be superior in comparison to linear and polynomial kernels for predictions as well as spatial mapping of soil salinity. This concludes that the use of sentinel 1 and 2 data with comparatively high spatial as well as spectral resolution may provide a very feasible method to map soil salinity using the SVR method.

Further attempts may be initiated to study the interaction of soil salts with microwave energy in detail, as well as the underlying mechanisms. Similarly, the interaction of microwave energy with various ion species predominant on salt soils may also be explored, and the performance of various other machine learning techniques needs to be tested for applicability in soil salinity mapping and related studies.

Acknowledgements

Authors are thankful to the Director, Indian Institute of Remote Sensing (IIRS), for providing necessary facilities and other supports to carry out the research work.

References

Ahmad, A., Shahzad, A., Iqbal, M., Asif, M., Hirani, A.H., 2013. Morphological and molecular genetic variation in wheat for salinity tolerance at germination and early seedling stage. Aust. J. Crop. Sci. 7 (1), 66–74.

Al-khaier, F., 2003. Soil salinity detection using satellite remote sensing. Afr. J. Environ. Sci. Technol. 2 (4), 7–20.

Aly, Z., Bonn, F., Magagi, R., 2004. Modelling the backscattering coefficient of salt-affected soils: application to wadi el-natrun bottom. Egypt 372.

Ashraf, M.A., Ashraf, M., 2016. Growth stage-based modulation in physiological and biochemical attributes of two genetically diverse wheat (*Triticum aestivum* L.) cultivars grown in salinized hydroponic culture. Environ. Sci. Pollut. Res. 23 (7), 6227–6243. Available from: https://doi.org/10.1007/s11356-015-5840-5.

Attema, E.P.W., Ulaby, F.T., 1978. Vegetation modeled as a water cloud. Radio. Sci. 13 (2), 357–364. Available from: https://doi.org/10.1029/RS013I002P00357.

Azabdaftari, A., Sunar, F., 2016. Soil salinity mapping using multitemporal landsat data. Int. Arch. Photogram., Remote. Sens. Spat. Inf. Sci.—ISPRS Arch. 41 (July), 3–9. Available from: https://doi.org/10.5194/isprsarchives-XLI-B7-3-2016.

Bindlish, R., Barros, A.P., 2001. Parameterization of vegetation backscatter in radar-based, soil moisture estimation. Remote. Sens. Environ. 76 (1), 130–137.

Brady, N.C., Weil, R.R., 2008. The Nature and Properties of Soils, 14th Pearson Prentice Hall, Upper Saddle River, N.J.

Chanzy, A., 1993. Basic soil surface characteristics derived from active microwave remote sensing. Remote. Sens. Rev. 7 (3–4), 303–319. Available from: https://doi.org/10.1080/02757259309532183.

Chu, L., Huang, C., Liu, Q., Liu, G., 2016. Estimation of winter wheat phenology under the influence of cumulative temperature and soil salinity in the Yellow River Delta, China, using MODIS time-series data. Int. J. Remote. Sens. 37 (10), 2211–2232. Available from: https://doi.org/10.1080/01431161.2015.1131871.

Congedo, L., 2021. Semi-automatic classification plugin: a python tool for the download and processing of remote sensing images in QGIS. J. Open. Source Softw. 6 (64), 3172. Available from: https://doi.org/10.21105/joss.03172.

Dehni, A., Lounis, M., 2012. Remote sensing techniques for salt affected soil mapping: application to the Oran region of Algeria. Procedia Eng. 33 (December), 188–198. Available from: https://doi.org/10.1016/j.proeng.2012.01.1193.

Douaoui, A.E.K., Nicolas, H., Walter, C., 2006. Detecting salinity hazards within a semiarid context by means of combining soil and remote-sensing data. Geoderma 134 (1–2), 217–230. Available from: https://doi.org/10.1016/j.geoderma.2005.10.009.

El-Hendawy, S., Al-Suhaibani, N., Dewir, Y.H., Elsayed, S., Alotaibi, M., Hassan, W., et al., 2019. Ability of modified spectral reflectance indices for estimating growth and photosynthetic efficiency of wheat under saline field conditions. Agronomy 9 (1), 35. Available from: https://doi.org/10.3390/agronomy9010035.

Elnaggar, A.A., Noller, J.S., 2010. Application of remote-sensing data and decision-tree analysis to mapping salt-affected soils over large areas. Remote. Sens. 2 (1), 151–165. Available from: https://doi.org/10.3390/rs2010151.

Encarni, I.H., Melendez-Pastor, I., Navarro-Pedreño, J., Gómez, I., 2014. Spectral indices for the detection of salinity effects in melon plants. Sci. Agricola 71 (4), 324–330. Available from: https://doi.org/10.1590/0103-9016-2013-0338.

Fieuzal, R., Duchemin, B., Jarlan, L., Zribi, M., Baup, F., Merlin, O., et al., 2011. Combined use of optical and radar satellite data for the monitoring of irrigation and soil moisture of wheat crops. Hydrol. Earth Syst. Sci. 15 (4), 1117–1129. Available from: https://doi.org/10.5194/HESS-15-1117-2011.

Gong, H., Shao, Y., Brisco, B., Hu, Q., Tian, W., 2014. Modeling the dielectric behavior of saline soil at microwave frequencies. Can. J. Remote. Sens. 39 (1), 17–26. Available from: https://doi.org/10.5589/M13-004.

Gorji, T., Tanik, A., Sertel, E., 2015. Soil salinity prediction, monitoring and mapping using modern technologies. Procedia Earth Planet. Sci. 15, 507–512. Available from: https://doi.org/10.1016/J.PROEPS.2015.08.062.

Haldar, D., Chakraborty, M., Manjunath, K.R., Parihar, J.S., 2014. Role of polarimetric SAR data for discrimination/biophysical parameters of crops based on canopy architecture. Int. Arch. Photogramm., Remote. Sens. Spat. Inf. Sci.—ISPRS Arch. 40 (8), 737–744. Available from: https://doi.org/10.5194/isprsarchives-XL-8-737-2014.

He, B., Xing, M., Bai, X., 2014. A synergistic methodology for soil moisture estimation in an Alpine Prairie using radar and optical satellite data. Remote. Sens. 6 (11), 10966–10985. Available from: https://doi.org/10.3390/RS61110966.

Hoa, P.V., Giang, N.V., Binh, N.A., Hai, L.V.H., Pham, T.D., Hasanlou, M., et al., 2019. Soil salinity mapping using SAR Sentinel-1 data and advanced machine learning algorithms: a case study at Ben Tre Province of the Mekong River Delta (Vietnam). Remote. Sens. 11 (2), 128. Available from: https://doi.org/10.3390/rs11020128.

Hu, J., Peng, J., Zhou, Y., Xu, D., Zhao, R., Jiang, Q., et al., 2019. Quantitative estimation of soil salinity using UAV-borne hyperspectral and satellite multispectral images. Remote. Sens. 11 (7). Available from: https://doi.org/10.3390/rs11070736.

Khan, N.M., Rastoskuev, V.V., Sato, Y., Shiozawa, S., 2005. Assessment of hydrosaline land degradation by using a simple approach of remote sensing indicators. Agric. Water Manag. 77 (1–3), 96–109. Available from: https://doi.org/10.1016/j.agwat.2004.09.038.

Kumar, A., Sharma, A.K., Rani, A., 2015. Transport of solutes under transient flow conditions—a case study—Yamuna river sub basin (Kosi Kalan to Agra. Int. Soil. Water Conserv. Res. 3 (3), 209–223. Available from: https://doi.org/10.1016/j.iswcr.2015.06.004.

Kumar, N., Reddy, G.P.O., Nagaraju, M.S.S., Naitam, R.K., 2022. Remote Sensing and Machine Learning for Identification of Salt-affected Soils. Data Sci. Agric. Nat. Resour. Manag. 267–287. Available from: https://doi.org/10.1007/978-981-16-5847-1_13.

Lasne, Y., Paillou, P., Freeman, A., Farr, T., McDonald, K.C., Ruffié, G., et al., 2008. Effect of salinity on the dielectric properties of geological materials: implication for soil moisture detection by means of radar remote sensing. IEEE Trans. Geosci. Remote. Sens. 46 (6), 1674–1688. Available from: https://doi.org/10.1109/TGRS.2008.916220.

Lesch, S.M., Rhoades, J.D., Lund, L.J., Corwin, D.L., 1992. Mapping Soil Salinity Using Calibrated Electromagnetic Measurements. Soil. Sci. Soc. Am. J. 56 (2), 540–548. Available from: https://doi.org/10.2136/SSSAJ1992.03615995005600020031X.

Lhissoui, R., Harti, A.El, Chokmani, K., 2014. Mapping soil salinity in irrigated land using optical remote sensing data. Eurasian J. Soil. Sci. (Ejss) 3 (2), 82. Available from: https://doi.org/10.18393/ejss.84540.

Li, Y.Y., Zhao, K., Ren, J.H., Ding, Y.L., Wu, L.L., 2014. Analysis of the dielectric constant of saline-alkali soils and the effect on radar backscattering coefficient: a case study of soda alkaline saline soils in western Jilin province using RADARSAT-2 data. Sci. World J. 2014. Available from: https://doi.org/10.1155/2014/563015.

Mandal, A.K., 2016. Mapping and characterization of salt affected and waterlogged soils in the Gangetic plain of central Haryana (India) for reclamation and management. Cogent. Geosci. Available from: https://doi.org/10.1080/23312041.2016.1213689.

Mandal, A.K., Sharma, R.C., Singh, G., 2009. Assessment of salt affected soils in India using GIS. Geocarto Int. 24 (6), 437–456. Available from: https://doi.org/10.1080/10106040902781002.

Mandal, A.K., Reddy, G.P.O., Ravisankar, T., 2011. Digital database of salt affected soils in India using Geographic Information Digital database of salt affected soils in India using Geographic Information System. J. Soil. Salin. Water Qual. 3 (1), 16–29. January 2018.

McGowen, I., Mallyon, S., 1996. Detection of dryland salinity using single and multi-temporal Landsat imagery. In: Proceedings of the 8th Australasian Remote Sensing Conference, Canberra, pp. 26–34.

Meyer, D., Meyer, D., Wien, T.U., 2001. Support vector machines. The interface to LIBSVM in package e1071. Online-Documentation of the package e1071 for 'R.

Misra, A., Ajmera, D., 2017. Analysis of adaptive and advanced speckle filters on SAR data. IOSR J. Comput. Eng. 19 (1), 48–54. Available from: https://doi.org/10.9790/0661-1901024854.

Nachshon, U., 2018. Cropland soil salinization and associated hydrology: trends, processes and examples. Water 10 (8), 1030. Available from: https://doi.org/10.3390/W10081030.

Periasamy, S., Ravi, K.P., 2020. A novel approach to quantify soil salinity by simulating the dielectric loss of SAR in three-dimensional density space. Remote. Sens. Environ. 251, 112059. Available from: https://doi.org/10.1016/J.RSE.2020.112059.

Poss, J.A., Russell, W.B., Bonos, S.A., Grieve, C.M., 2010. Salt tolerance and canopy reflectance of kentucky bluegrass cultivars. HortScience 45 (6), 952–960. Available from: https://doi.org/10.21273/hortsci.45.6.952.

Rameeh, V., Gerami, M., 2015. Soil salinity effects on phenological traits, plant height and seed yield in rapeseed genotypes. Soil Sci. Annu. 66 (1), 17–25. Available from: https://doi.org/10.1515/ssa-2015-0014.

Ren, J., Li, X., Zhao, K., Fu, B., Jiang, T., 2016. Study of an on-line measurement method for the salt parameters of soda-saline soils based on the texture features of cracks. Geoderma 263, 60–69. Available from: https://doi.org/10.1016/J.GEODERMA.2015.08.039.

Rhoades, J.D., Manteghi, N.A., Shouse, P.J., Alves, W.J., 1989. Estimating soil salinity from saturated soil-paste electrical conductivity. Soil. Sci. Soc. Am. J. 53 (2), 428–433. Available from: https://doi.org/10.2136/sssaj1989.03615995005300020019x.

Salisbury, F.B., 1991. Système Internationale: the use of SI units in plant physiology. J. Plant. Physiol. 139 (1), 7–7.

Scudiero, E., Corwin, D.L., Anderson, R.G., Yemoto, K., Clary, W., Wang, Z., et al., 2017. Remote sensing is a viable tool for mapping soil salinity in agricultural lands. Calif. Agric. 71 (4), 231–238. Available from: https://doi.org/10.3733/ca.2017a0009.

Shao, Y., Hu, Q., Guo, H., Lu, Y., Dong, Q., Han, C., 2003. Effect of dielectric properties of moist salinized soils on backscattering coefficients extracted from RADARSAT image. IEEE Trans. Geosci. Remote. Sens. 41 (8), 1879–1888. Available from: https://doi.org/10.1109/TGRS.2003.813499.

Sharma, R.C., Bhargava, G.P., 1988. Landsat imagery for mapping saline soils and wet lands in North-West India. Int. J. Remote. Sens. 9 (1), 39–44. Available from: https://doi.org/10.1080/01431168808954835.

Srineevas, K., Venkataratnam, L., Rao, P.V.N., 2007. Dielectric properties of salt-affected soils. Int. J. Remote. Sens. 16 (4), 641–649. Available from: https://doi.org/10.1080/01431169508954431.

Taghadosi, M.M., Hasanlou, M., Eftekhari, K., 2019a. Retrieval of soil salinity from Sentinel-2 multispectral imagery. Eur. J. Remote. Sens. 52 (1), 138–154. Available from: https://doi.org/10.1080/22797254.2019.1571870.

Taghadosi, M.M., Hasanlou, M., Eftekhari, K., 2019b. Soil salinity mapping using dual-polarized SAR Sentinel-1 imagery. Int. J. Remote. Sens. 40 (1), 237–252. Available from: https://doi.org/10.1080/01431161.2018.1512767.

Taylor, G.R., Mah, A.H., Kruse, F.A., Kierein-Young, K.S., Hewson, R.D., Bennett, B.A., 1996. Characterization of saline soils using airborne radar imagery. Remote. Sens. Environ. 57 (3), 127–142. Available from: https://doi.org/10.1016/0034-4257(95)00239-1.

Vapnik, V., Golowich, S.E., Smola, A., 1997. Support vector method for function approximation, regression estimation, and signal processing. Adv. Neural Inf. Process. Syst. 281–287.

Vreugdenhil, M., Wagner, W., Bauer-Marschallinger, B., Pfeil, I., Teubner, I., Rüdiger, C., et al., 2018. Sensitivity of Sentinel-1 backscatter to vegetation dynamics: an Austrian case study. Remote. Sens. 10 (9), 1–19. Available from: https://doi.org/10.3390/rs10091396.

Wicke, B., Smeets, E., Dornburg, V., Vashev, B., Gaiser, T., Turkenburg, W., et al., 2011. The global technical and economic potential of bioenergy from salt-affected soils. Energy Environ. Sci. 4 (8), 2669–2681. Available from: https://doi.org/10.1039/C1EE01029H.

Wu, W., Zucca, C., Muhaimeed, A.S., Al-Shafie, W.M., Fadhil Al-Quraishi, A.M., Nangia, V., et al., 2018. Soil salinity prediction and mapping by machine learning regression in Central Mesopotamia, Iraq. Land. Degrad. Dev. 29 (11), 4005–4014. Available from: https://doi.org/10.1002/ldr.3148.

Wu, W., Muhaimeed, A.S., Al-Shafie, W.M., Al-Quraishi, A.M.F., 2020. Using radar and optical data for soil salinity modeling and mapping in Central Iraq. Environ. Remote. Sens. GIS Iraq 19–40. Available from: https://doi.org/10.1007/978-3-030-21344-2_2.

Yadav, J.S.P., 2003. Managing soil health for sustained productivity. Indian. Soc. Soil. Sci. 51 (4), 448–465.

Zainodin, H.J., Yap, S.J., 2013. Overcoming multicollinearity in multiple regression using correlation coefficient. AIP Conf. Proc. 1557 (1999), 416–419. Available from: https://doi.org/10.1063/1.4823947.

Chapter 29

Soil health assessment and spatial characterization using remote sensing

Saurav Das[1] and Dinesh Panday[2]
[1]Department of Agronomy and Horticulture, University of Nebraska, Lincoln, NE, United States, [2]Rodale Institute—Pocono Organic Center, Long Pond, PA, United States

Abbreviations

C	carbon
RS	remote sensing
SOC	soil organic carbon
SOM	soil organic matter

29.1 Introduction

Coordinated efforts toward defining and quantifying soil health and ecosystem services have recently increased due to the potential benefit of soil health in agriculture, environmental quality, and economics. Soil health is broadly defined as the continued capacity of the soil to function as a vital ecosystem to support and sustain crop productivity, maintain environmental quality, support ecosystem services, and promote human and plant health. Soil is a complex system at the intersection of the atmosphere, lithosphere, hydrosphere, and biosphere and is critical for developing a sustainable agricultural production system. The concept of soil health emerged in the early 2000s. Now, it has been linked to the "One Health" concept, which connects the health of humans, animals, and the environment (Lehmann et al., 2020). The multidimensionality of the soil health concept allows for soil management to be aligned with sustainability and climate mitigation goals. Soil health provides a foundation for evaluating the ecosystem functions both at spatial and temporal scales. A most important achievement of the soil health framework is the addition of the most needed biological perspective to soil management and processes to address the long-term sustainability of crop production (Lehmann et al., 2020).

Soil provides goods and services (ecosystem services), including food production by controlling nutrient cycling, water flow, water quality by buffering contamination, air quality by regulating greenhouse gas emissions, and climate change mitigation by sequestering carbon (C). Soil ecosystem services are a functional response to the complex interaction of bio-physicochemical properties, land use, and agronomic management. It is important to manage soil health status for improved ecosystem services, including crop production, C sequestration (Das et al., 2022; Lal, 2016), soil organic matter (SOM), water and nutrient availability (Ankenbauer and Loheide II, 2017; Hudson, 1994), soil resilience to climate change (drought and heat waves) (Ankenbauer and Loheide II, 2017; Lal, 2016), disease suppression (Janvier et al., 2007; Raaijmakers and Mazzola, 2016), and soil erosion reduction (Das et al., 2022).

Quantifying soil health status is important for proper management and determining functional correlations with ecosystem services. Soil health is measured using multiple indicators representing nutrient cycling, biogeochemistry of carbon and nitrogen, microbial diversity, water and air quality, and soil structure, which can be grouped as physical, chemical, and biological indicators. The interrelationship of inherent bio-physicochemical soil health properties with land-use change, agronomic management, and interannual climatic variations makes it challenging at spatial and temporal measurement scales. Studies have shown that soil management practices like tillage, cover crops, and residue

Remote Sensing of Soils. DOI: https://doi.org/10.1016/B978-0-443-18773-5.00034-X

management can significantly affect soil health. Hence, a large continuum scale of measurement is needed in predicting soil health status and goods and services.

There is a collective effort toward developing indices to assess soil health holistically (Andrews et al., 2004; Bünemann et al., 2018; de Paul Obade and Lal, 2016). Developing a universal index will require an accounting of soil heterogeneity, land-use change, and climatic variation in space and time. Current in-situ measurement approaches are resource-intensive and do not provide a global understanding of soil health status. Restoration of the current cropland with an aim toward sustainable global food production and climate mitigation will require an evaluation of the success or failure of management practices. A universal index for measuring soil health status will be required to meet the global challenges. Thus, a soil health assessment must bridge the variability among soil types, land use, and climatic variability (Rinot et al., 2019). High-resolution digital soil maps based on in situ and remotely sensed data will be critical in measuring/predicting soil health status and developing global indices. The current chapter will summarize recent developments in remote sensing (RS) approaches for measuring soil health and ecosystem services, their principles, and their future perspectives. The chapter will also highlight the emerging concept of ecosystem services as commodities for future markets and discuss the possible application of RS techniques in developing markets.

29.2 Current approaches and development in soil health assessment

Soil health is a functional response to multidimensional interactions between the environment, plants, soil, and agronomic management. The changes in management practices reflect changes in the measured soil health parameters and the related ecosystem services. Soil health status is assessed by measuring different bio-physiochemical indicators, which (1) represent physical, chemical, and biological properties and processes of the soil system; (2) are sensitive to changes in soil function caused by management, land-use change, and climate change; (3) are manageable, effective, and cost-effective measurements that can be conducted at relevant space and timescale for decision-making; (4) represent the environmental vulnerability of the soil such as soil erosion risk and water and air contamination risk; and (5) represent ecosystem services such as productivity, air and water quality, soil nutrient availability, and biodiversity (Rinot et al., 2019). Some of the most used indicators in soil health assessments are SOM, soil organic carbon (SOC), aggregate stability, bulk density, soil texture, soil respiration, autoclaved citrate extractable protein, permanganate oxidizable carbon or active carbon, water extractable organic carbon and -nitrogen, water infiltration, water holding capacity, and different soil enzyme related to carbon, nitrogen, phosphorus, and sulfur nutrient cycle. However, the availability of multiple indicators also makes soil health assessment a complicated process and hard to track over extended scales of spatial and temporal variation and is also a limiting factor in the wide adoption of soil health management. Many studies are putting efforts into identifying a few critical indicators that can be used to define the soil health status in space and time (Bagnall et al., 2023; Soil Health Institute, 2022a; Soil Health Institute, 2022b)

A robust soil health assessment will depend on (1) the selection and measurement of critical indicators relevant to soil attributes of interest, which can either reflect or forecast alterations in soil due to management changes; (2) direct and indirect quantification of these indicators should able to provide a comprehensive status of the soil health and the management impact on soil ecosystem services; and (3) taking a weighted average of these scores either regionally or globally, should able to determine the soil health status on a large spatial scale and should help in designing informed management practices.

29.3 Remote sensing for soil health assessment and ecosystem services

Measuring the changes in the soil environment with respect to agronomic management can help determine and predict soil health status. In the classical approach, soil samples are collected from the site of interest under different treatments or natural conditions using a grid or zone sampling method to measure the bio-physicochemical parameters and determine the potential soil health status. However, such measurement is tedious and resource-intensive and often requires a large number of samples to capture the soil's heterogeneity; and this is where the different spectrophotometric methods are critical. Visible near-infrared reflectance spectroscopy has been successfully used to determine several soil properties (Ben-Dor et al., 2002; Gomez et al., 2008; Viscarra Rossel et al., 2009). Unique spectral signatures of the soil components and plants can be leveraged in remote sensing methods to create spatiotemporally extended digital soil maps related to land use change and specific soil properties like SOM (Ertlen et al., 2015; Heller Pearlshtien and Ben-Dor, 2020; Wang et al., 2010), SOC (Angelopoulou et al., 2019; Ladoni et al., 2010), and moisture (Mohanty et al., 2017; Wang and Qu, 2009).

Remote sensing plays an important role in studying complex environmental interactions and has been widely used in quantifying and mapping the soil's inherent properties and ecosystem services. A major benefit of the remote sensing method is the ability to capture a large continuum dataset with varying spatial and temporal resolution through frequent observation. Advancements in sensor technology and earth observation will continue to contribute extensively to research on the modeling, mapping, and valuation of ecosystem goods and services using remote sensing. Integrating remote sensing with in situ measurement has enabled measuring how different agronomic management and land use changes affect soil health properties over time. Hereafter, the chapter will discuss the measurement of a few selected soil health indicators using remote sensing.

29.4 Soil health indicators

29.4.1 Soil organic matter and carbon

SOM is a central and critical indicator in soil health measurement. SOM defines several ecosystem services, including crop production, water retention and availability (Lal, 2020), higher cation exchange capacity (Parfitt et al., 1995), high nutrient retention in the root zone, greater soil buffering capacity, improved structure and stable aggregates, and sustaining biological activity and diversity. SOM can influence environmental processes on a global scale. Soil is the largest reservoir of the C, which modifies carbon dioxide concentration in the atmosphere and can influence global warming. SOM may be lost under an intensive cropping system because of the increased C mineralization rate, especially following intensive tillage, excessive crop removal, and erosion (Ding et al., 2002; McLauchlan, 2006). Estimating SOM in chronological order with broad spatial coverage is important to define and develop a sustainable agroecosystem and carbon offset. RS, especially using spectrophotometric data, is becoming one of the most sought-after methods for monitoring and measuring SOM change on a regional and global scale. The principle behind the use of RS for SOM measurement is the distinct reflectance characteristics of soil with varying organic matter content. Studies have shown a negative correlation for SOM with soil reflectances, and have illustrated SOM is sensitive to the visible near-infrared (VNIR) (400 ~ 1100 nm) and shortwave infrared (SWIR) (1100 - 2500 nm) spectral regions (Zhang et al., 2021). Satellites equipped with multispectral sensors such as Landsat, Sentinel-2, and MODIS can be used for SOM estimation. Several studies have used the RS data to estimate and model the SOM change (Mirzaee et al., 2016; Wang et al., 2018; Yuzugullu et al., 2020; Zhang et al., 2021). However, the prediction of SOM is challenging and susceptible to being influenced by environmental covariates including soil types (texture), seasonal conditions (soil moisture), and remotely sensed image features (pixel data) (Zhang et al., 2021). A concurrent SOM measurement from the field samples and correlating them with spectral data can be useful in training the model for better accuracy. Advanced machine-learning methods such as random forests, support vector machines, and neural networks can handle complex interaction spectral bands and soil properties and thus can be used to improve estimation accuracy (Zhang et al., 2021). Mirzaee et al. (2016) found that using Landsat 7 ETM + spectral images with an artificial neural network (ANN) for ordinary kriging can significantly improve the model's prediction. Mapping high-precision SOM maps is challenging due to the difficulty of selecting appropriate satellite data sources and prediction algorithms. Combining data from multiple sensors can provide a more comprehensive spectral range and better differentiation between SOM and other soil properties. Zhang et al. (2021) compared complete- and standard-band datasets from Sentinel-2 A and MODIS images using Google Earth Engine and created a predictive model with random forest, an artificial neural network, and support vector regression for SOM mapping. The results showed that the model based on the full-band Sentinel-2A data was best performing in SOM mapping, and the random forest was the best predictive model with the lowest root-mean-squared error (RMSE). Further, time series analysis of satellite images can help in predicting periods with minimum soil moisture, which can reduce its confounding effects on SOM prediction. Microwave sensors in SMOS and SMAP satellites can provide data on soil moisture, which can be used in correcting SOM predictions (Entekhabi, 2010). The incorporation of data on soil types, geology, and topography can further improve the model's accuracy. Geospatial databases such as SoilGrids and national soil databases can be helpful in improving the prediction (Hengl et al., 2017). The establishment of libraries that catalog spectral signatures for various soil textures can help in distinguishing between the effects of textures and organic matter on reflectance and can help in further improving the accuracy of the models (Brown et al., 2006). In the future, advances in sensor technology and data processing algorithms, coupled with ground-truthing, will enhance the precision of remote sensing-based SOM assessment.

Soil organic carbon serves as a cornerstone in the global carbon cycle and provides an index of soil health and broader ecosystem functionality. Its role underscores the significance of accurate and expansive monitoring, particularly in addressing overarching challenges tied to climate change and agricultural dynamics. The remote sensing method and

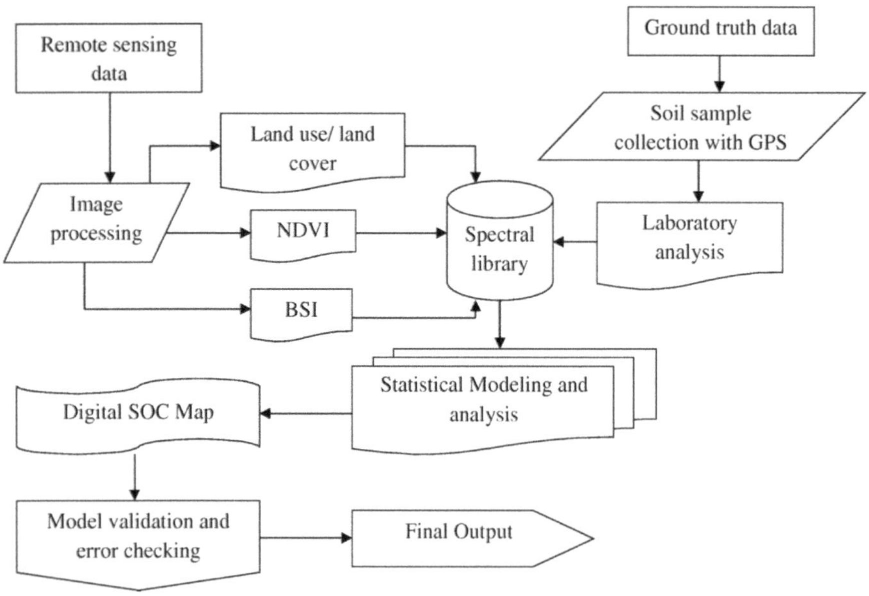

FIGURE 29.1 SOC mapping using remote sensing techniques and a multivariate regression model. *From Bhunia, G.S., Kumar Shit, P., Pourghasemi, H.R., 2019. Soil organic carbon mapping using remote sensing techniques and multivariate regression model. Geocarto Int. 34, 215–226. https://doi.org/10.1080/10106049.2017.1381179.*

in situ measurement have become integrated tools for SOC prediction and quantification (Fig. 29.1). Satellites like Landsat, Sentinel-2, and WorldView have multispectral sensors that capture data useful for SOC estimation. Several studies have used multispectral and hyperspectral remotely sensed images in SOC quantification and prediction (Angelopoulou et al., 2019; Bhunia et al., 2019; Chen et al., 2005). Development in machine learning and different geostatistical models has further improved the application of RS in developing more accurate algorithms for SOC predictions (Angelopoulou et al., 2019; Mondal et al., 2017; Odebiri et al., 2021). One such advancement is Digital Soil Mapping (DSM), which makes use of data sources, including remote sensing, terrain attributes, land use data, and other environmental covariates to derive spatial and temporal prediction of SOC.

29.4.2 Carbon flux and C sequestration

Carbon sequestration and storage efficiency can be estimated through the quantification of net ecosystem exchange (NEE) of CO_2 flux and can be used to determine the amount of atmospheric C stored in an ecosystem. Calculation of vegetative indices such as normalized difference vegetation index (NDVI), emissivity difference vegetation index (EDVI), water band index (WBI), and leaf area index (LAI) from the remotely sensed images can help in predicting the net CO_2 flux of the ecosystem (Dagg and Lafleur, 2010; Hao et al., 2012; Kross et al., 2013). Carbon sequestration provides an indirect measurement of emission reduction. Key carbon flux parameters include gross primary productivity (GPP), net primary productivity (NPP), and net ecosystem productivity (NEP). In definition, GPP refers to the total amount of C fixed by plants through photosynthesis per unit time and per unit area, and NPP is the difference between the GPP and the loss of C through plant respiration (GPP - plant respiration). Net ecosystem production is the NPP minus the respiration of the soil heterotrophic organisms, which is an indicator of the rate of net carbon flux and carbon storage (Zhao et al., 2022).

Landowners are increasingly asked about the impact of their management practices with respect to C sequestration to protect areas with high carbon densities and increase stewardship to improve soil C sequestration and storage. Measuring carbon fluxes from the agricultural field and estimating the seasonal carbon budget is critical in evaluating the sustainability of management practices with respect to crop production, carbon uptake, and greenhouse gas emissions. However, C fluxes and ecosystem services are complex, and are an integrated system of multiple variables. Process-based ecosystem models can be useful in calculating mechanistic processes and tradeoffs. However, vegetation dynamics and environmental changes are highly variable and determining factors in C fluxes. Remote sensing provides a unique opportunity to monitor changes in vegetation, soil moisture, soil temperature, and agronomic management and integrate these variables in a process-based model to improve prediction accuracy. MODIS and Landsat data can be used to estimate photosynthetic activity by calculating NDVI and Enhanced Vegetation Index (EVI), which directly relates to carbon uptake by terrestrial ecosystems (Huete et al., 2002). Soil Moisture Active Passive (SMAP) satellite can detect the changes in soil moisture, which impacts soil respiration - a key component of terrestrial carbon flux. LIDAR instruments in ICESat-2 and GEDI can be used to determine vertical forest structure information, which can be used to estimate above-ground biomass and carbon storage (Dubayah et al., 2020). In addition, the networks of flux towers, like FLUXNET, can be used to measure CO_2 exchange and to validate satellite-based estimation (Baldocchi et al., 2001). For instance, Turner et al. (2005) combined MODIS-derived vegetation indices with in situ flux tower measurements to enhance the estimation of Gross Primary Production (GPP) and Net Primary Production (NPP) across diverse sites (Turner et al., 2005). This synergy between satellite and ground data has proven invaluable in addressing the intricacies of carbon dynamics. In a comprehensive review, Schimel et al. (2015) underscored the potential of amalgamating data from optical, radar, and lidar remote sensing modalities to provide a nuanced understanding of the terrestrial carbon cycle (Schimel et al., 2015). Such a multisensor perspective compensates for the limitations inherent to individual data sources. This sentiment is echoed by Jung et al. (2011), who harmonized eddy covariance measurements from the FLUXNET network with satellite insights, painting a detailed spatial-temporal canvas of global carbon dioxide fluxes (Jung et al., 2011). Similarly, Xiao et al. (2004) adeptly harnessed both optical (MODIS) and microwave (SRTM) satellite data, and supplemented with ground observations, to model GPP in tropical forests (Xiao et al., 2005). In the challenging terrains of the Alaskan Arctic, Fisher et al. (2014) demonstrated the merits of fusing datasets from diverse remote sensing platforms, including optical, radar, and lidar, with on-ground metrics to delineate carbon flux dynamics (Fisher et al., 2014). Collectively, these studies affirm the indispensability of a comprehensive, multipronged approach, interweaving various remote sensing datasets with in situ measurements, in the quest for accurate carbon flux predictions. Niu et al. (2021) developed a "carbon and exchange between vegetation, soil, and atmosphere—ecosystem service (CEVSA—ES)" model integrating the remotely sensed leaf area index (LAI) with estimated productivity provision (primary productivity), carbon sequestration, water retention, and soil erosion. The simulated results from this study showed 95%, 92%, 76%, and 65% goodness of fit, explaining the variabilities of GPP, ecosystem respiration, net ecosystem productivity, and evapotranspiration. The CEVSA-ES model could explain 96% and 81% of the spatiotemporal variation of the observed annual productivity provision and carbon sequestration (Fig. 29.2). Several studies integrated the remotely sensed data with the process-based model to improve accuracy in predicting C flux and C sequestration (Su et al., 2022; Turner et al., 2004). Some studies have also used remotely sensed data in calculating the vegetation index and used the same in the estimation of GPP and C fluxes (Rahman et al., 2005; Xiao and Moody, 2004).

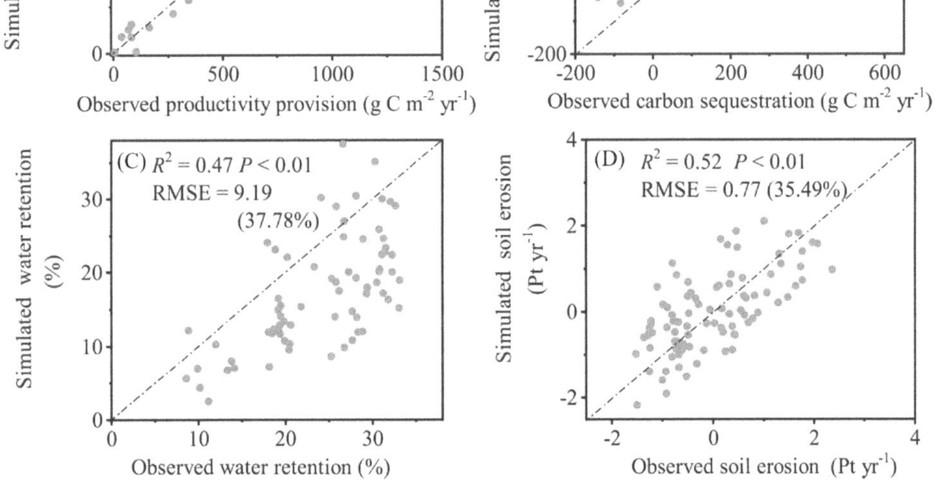

FIGURE 29.2 Performance of the CEVSA-ES model in estimating (A) productivity provision (g C m^{-2} year^{-1}), (B) carbon sequestration (g C m^{-2} year^{-1}), (C) water retention (%), and (D) soil erosion (Pt year^{-1}). CEVSA-ES, carbon and exchange between vegetation-, soil-, and atmosphere-ecosystem service. *From Niu, Z., He, H., Peng, S., Ren, X., Zhang, L., Gu, F., et al., 2021. A process-based model integrating remote sensing data for evaluating ecosystem services. J. Adv. Model. Earth Syst. 13, e2020MS002451. https://doi.org/10.1029/2020MS002451.*

29.4.3 Soil nutrient availability

Measuring soil nutrient availability and fertility is pivotal in establishing a sustainable agricultural production system and environmental quality. The success of precision agriculture depends on an efficient and accurate in-field determination of soil nutrients. The traditional grid soil sampling method for determining in-field nutrient variability is time consuming and resource expensive. A soil map at a scale of 1:31680 (a map unit size of 4.0 ha or 10 ac) has an average cost of 4.2 USD ha^{-1} or 1.7 USD/ac, and a map at a scale of 1:3000 could be 47 USD ha^{-1} (19 USD/ac) (Ge et al., 2011). Measurement of non-photosynthetic vegetation, green vegetation, NDVI, and soil spectral mixture analysis using remotely sensed data can help predict decline and improvement in soil nutrient availability (Numata et al., 2003). Remote sensing data, especially hyperspectral and multispectral images, can be used as auxiliary variables in predicting spatial variability in soil nutrients. Different narrowband and broadband vegetation indices such as NDVI, EVI, fPAR, and LAI can be used as indicators of productivity in a crop's growing season, as these can be used to determine the variation in phenology and photosynthetic potential of a crop. These variables can also indicate in-season nutrient deficiency in crops. The NDVI has been used in several studies to predict nutrient deficiency and subsequent management (Burns et al., 2022; Crain et al., 2012; Edalat et al., 2019). Remote sensing data can also provide information on soil nutrient availability. Song et al. (2018) used hyperspectral images (115 bands) from the Chinese environmental 1 A satellite as auxiliary variables in predicting the total soil nitrogen, available phosphorus, and soil-available potassium. The study used different machine learning models with the auxiliary variables for better prediction and found that the back-propagation neural network, combined with ordinary kriging, demonstrated the best predictive accuracy in mapping the spatial variation in soil nutrients. Zhang et al. (2010) used relative elevation, surface gradient, surface roughness, river dynamic index, and NDVI in describing soil total nitrogen. The study showed that soil nutrient distribution was affected by terrain and vegetation. The use of regression kriging improves the accuracy of mapping the spatial distribution. Li et al. (2021) used Gaofen-1 satellite 8m resolution multispectral images with four bands, blue, green, red, and near-infrared, to calculate the variables like NDVI, difference vegetation index, ratio vegetation index (RVI), and renormalized difference vegetation index to determine the vegetation growth, cover, yield, and crop health status, and used the variables in predicting the forest soil nutrient status (Fig. 29.3). Li et al. (2021) used the RS data as auxiliary variables with the artificial neural network model and found remote sensing variables performed better in predicting the available soil nitrogen, phosphorus, and organic matter in the upper soil layers (0–20 cm). The predictive efficiency decreases with an increase in depth, but the study found it could be improved using the terrain hydrology variables.

29.4.4 Soil moisture

Soil moisture is the most critical parameter in land-surface hydrology, which determines agricultural productivity and the well-being of animal, plant, and human health. Microwave remote sensing has been widely used in soil moisture measurement. The large difference between the dielectric constant of water and dry soil is a major advantage of using the passive microwave technique for soil moisture estimation. Low-frequency bands like X (frequency range: 8.0 to 12.0 GHz), C (frequency range: 4.0 to 8.0 GHz), and L (frequency range: 1.0 to 2.0 GHz) are widely used to detect soil surface moisture and vegetative water content. Low-frequency L bands (SMAP satellite) are mainly sensitive to surface soil moisture (Calvet et al., 2011). Several satellite-based L-band radiometers and radars, including SMOS (launched by the European Space Agency in 2009, 1.4 GHz), AQUARIUS Ocean Salinity (launched by NASA in 2011, 1.413 GHz [passive], 1.26 GHz [active]), and SMAP (launched by NASA in 2015, 1.41 GHz [passive] and 1.26 GHz [active]), were used for global monitoring of near-surface (0–5 cm) soil moisture content. The SMOS and SMAP passive radiometers currently provide 35- to 60-km-resolution soil moisture data globally at 2- to 3-day intervals (Mohanty et al., 2017). Several studies effectively used the L-band in soil moisture estimation (Escorihuela et al., 2010; Saleh et al., 2007; Schwank et al., 2010). However, there are several challenges also, including, dense vegetation can obscure soil, making it difficult for microwave signals to penetrate and reach the soil surface; roughness of the soil surface can affect the backscatter in active microwave remote sensing leading to inaccurate estimation; soil texture can also affect the dielectric properties and soil moisture affecting remote sensing, and the remote sensing are primarily captured soil moisture in the top few centimeters of the soil (Naeimi et al., 2009).

29.4.5 Soil salinity

Soil salinity is a sign of soil degradation and overall health. It can significantly reduce crop yield and negatively affect human nutrition, livelihoods, and ecosystem health. Soil salinity can destroy the soil structure, reduce water infiltration

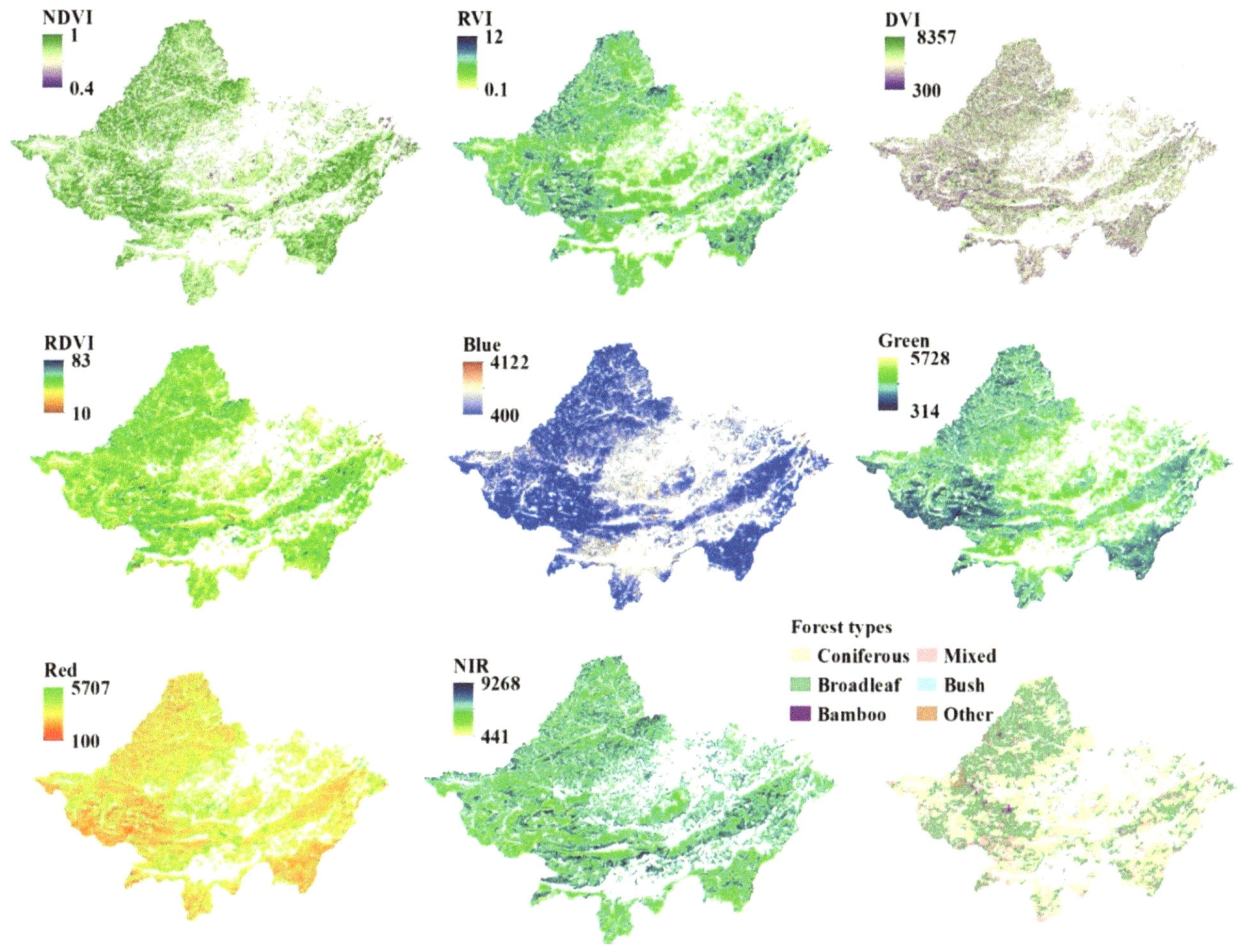

FIGURE 29.3 Remote sensing variables derived from GF- 1. *From Li, Y., Zhao, Z., Wei, S., Sun, D., Yang, Q., Ding, X., 2021. Prediction of regional forest soil nutrients based on gaofen-1 remote sensing data. Forests 12, 1430. https://doi.org/10.3390/f12111430.*

and conductance, increase erosion potential, influence soil pH and nutrient availability, and contaminate the potable water source (Sahab et al., 2021). Soil salinity is a major problem in semiarid and arid areas with low precipitation, high evapotranspiration, a high water table, and a high water-soluble salt content (Avdan et al., 2022; Wang et al., 2019). Approximately 950 million ha of land on a global scale are affected by soil salinity (Avdan et al., 2022). As an economic loss, it is estimated that secondary salinization costs $750 million per year for the Colorado River Basin in the USA, $300 million per year for the Punjab and Northwest Frontier Provinces in Pakistan, and $208 million per year for the Murray–Darling Basin in Australia for management (Metternicht and Zinck, 2003). Measuring, monitoring, and managing soil salinity for land restoration, reclamation, and sustainable agroecosystems is important.

In general, soil salinity is measured as electrical conductivity (EC), a laboratory-based method. However, collecting soil samples and performing laboratory analysis is time-consuming, not cost-effective, and cannot capture large-scale spatial variability. Remote sensing provides cost-effective, dynamic monitoring and modeling of soil salinity at a spatiotemporal scale, which can predict regions highly susceptible to soil salinity and develop subsequent management. Remote sensing also allows correlating anthropogenic covariates, such as land-use change, to better predict future scenarios. The Landsat-8 satellite (spatial resolution of 30 m) (Avdan et al., 2022) and Sentinel-2 MSI (spatial resolution of 10 m) (Farahmand and Sadeghi, 2020; Taghadosi and Hasanlou, 2021) have been effectively used in salinity studies.

A study by Avdan et al. (2022) used data from two middle (Landsat-8 OLI and Sentinel- 2 MSI) and one high spatial resolution (PlanetScope) sensors to calculate salinity indices and develop a salinity prediction model (Fig. 29.4). The study showed that the higher spatial resolution of PlanetScape showed the highest model correlation with observed and predicted values. A study by Gorji et al. (2017) used Landsat-8 TM/ETM + images to calculate the salinity index (SI), which was correlated with the EC values using regression analysis to create the soil salinity maps. The study found

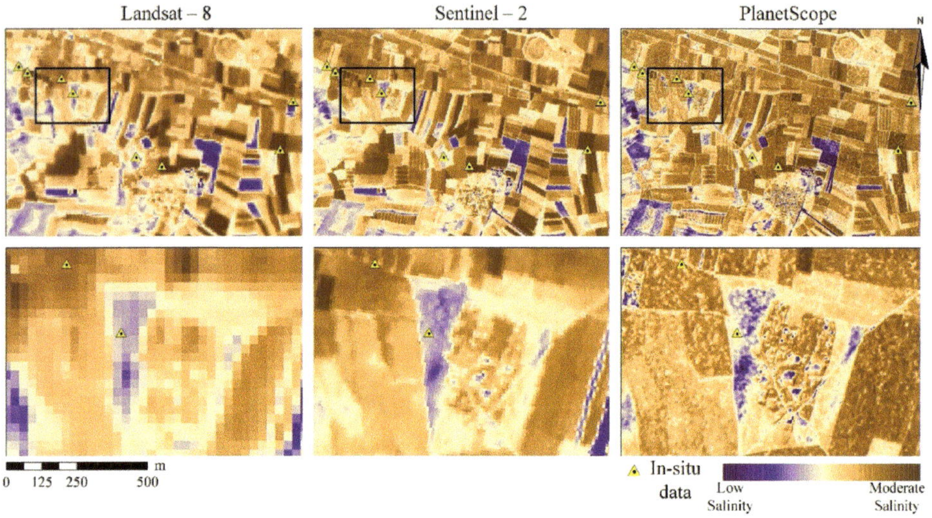

FIGURE 29.4 Predicted soil salinity map, developed with the data obtained from Landsat-8, Sentinel-2, and PlanetScoep (left to right). *From Avdan, U., Kaplan, G., Küçük Matcı, D., Yiğit Avdan, Z., Erdem, F., Tuğba Mızık, E., et al., 2022. Soil salinity prediction models constructed by different remote sensors. Phys. Chem. Earth, Parts A/B/C 128, 103230. https://doi.org/10.1016/j.pce.2022.103230.*

that the SI, calculated as $\sqrt{\text{Band } 1 \times \text{Band } 3}$ showed the best correlation with the observed data. The study found most of the soil salinity is caused by nature when correlated with CORINE land cover data.

29.5 Soil health management

29.5.1 No-till and crop residue management

Reduced tillage and no-till (NT) are two of the most tested methods for soil erosion control and fall under the major principle of soil health, "reduce the soil disturbances." Continuous intensive tillage can significantly disrupt soil aggregate stability, increase topsoil erosion by wind and water, and reduce soil fertility. Conventional tillage can also increase SOM mineralization and reduce soil C storage (Six et al., 2000). Several studies have found that NT can improve water infiltration, and soil water content improves soil macroaggregate structure and reduces soil erosion (Blanco-Canqui et al., 2009; Sainju, 2021; Stone and Schlegel, 2010). It is important to continuously monitor changes in agronomic management to determine the success of such practices. Hyperspectral images with reflectance from visible and near-infrared regions of the electromagnetic spectrum can provide cost-effective measurement and monitoring for field operations such as fertilization, pesticide spraying, tillage, and residue management. A study by Yang et al. (2003) used hyperspectral images with 71 wave bands in the spectral range 400–950 nm and used classification and regression trees to determine conventional tillage, reduced tillage, and NT. The study used 18 replicated tillage treatment plots. They found that classification and regression trees can be used with the hyperspectral image to distinguish the tillage practices with an accuracy of 0.89 (or 89%).

Crop residue on the soil surface plays a significant role in surface energy balance, NPP, nutrient cycling, and C sequestration (Blanco-Canqui and Lal, 2009; Turmel et al., 2015). Tillage accelerates soil erosion by increasing soil exposure to wind and rain and increasing C oxidation by soil aeration and temperature. Thus, most agricultural land has less soil C than native land. Crop residue management is an integral part of soil health management, which is used along with conservation tillage or NT to improve soil health. Long-term use of NT and proper crop residue management can increase soil organic C and improve soil structure and aggregation compared to conventional tillage. The standard method of measuring the crop residue cover in a particular field is visual estimation along the transect line (Laflen et al., 1981). This technique is laborious and prone to operator bias. The current method also fails to capture the spatial variability of residue over a large farm or region. A reliable, rapid, and accurate quantification method is needed for precision farming systems and management decisions. Remote sensing can be critical in estimating/quantifying crop residue, as it can leverage the spectral signature of the crop, soil, and different plant litter components with the indices developed using visible, near-infrared, and shortwave infrared bands, like the NDVI, NDRE, normalized difference tillage index (NDTI), normalized difference index (NDI), and normalized differential senescent vegetation index (NDSVI). These indices are based on the relative difference in broadband reflectance for soil and crop residues. Few RS methods use the specific spectral signature of chemical compounds like nitrogen, cellulose, and lignin, which are highly associated with plant litter (at 1730, 2100, and 2300 nm) (Daughtry et al., 2005). The typical reflectance pattern

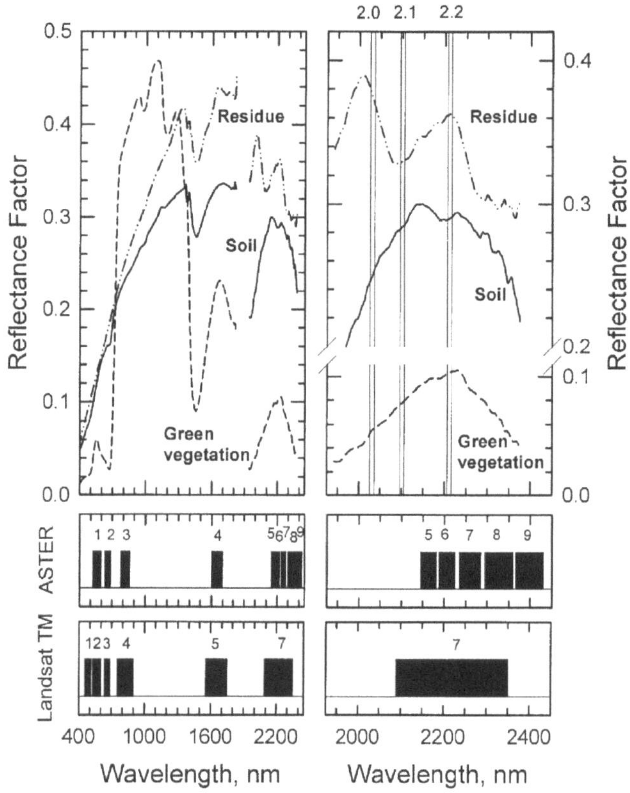

FIGURE 29.5 Typical reflectance pattern for green vegetation, soil, and corn residue. Landsat 7-Tm, ASTER, and CAI bands. *From Daughtry, C.S.T., Hunt Jr. E.R., Doraiswamy III P.C., McMurtrey, J.E., 2005. Remote sensing the spatial distribution of crop residues. Agron. J. 97, 864–871. https://doi.org/10.2134/agronj2003.0291.*

of the soil, green vegetation, and crop residue is shown in Fig. 29.5. A study by Daughtry et al. (2005) leveraged the greenness index like NDVI (related to green vegetation) and lignin and cellulose absorption indices like the cellulose absorption index (CAI) and lignin cellulose absorption index (LCAI) to distinguish between the green cover and crop residue in the agricultural field. The study found that CAI and LCAI are linearly related to crop residue. The study also showed that the spectra of tilled soil could be separated from NT soil based on the residue cover percentage. Serbin et al. (2009) developed a shortwave infrared normalized difference residue index (SINDRI), using Advanced Spaceborne Thermal Emission and Reflectance (ASTER) bands 6 and 7 for the detection of crop residue. The study found that SINDRI can be used as an alternative to CAI and LCAI in measuring crop residue cover, but the accuracy is lower compared to CAI.

29.5.2 Cover crops

Cover crops are considered one of the potential strategies to improve soil health and soil carbon. Studies have shown that cover can improve soil biological community, soil aggregate stability, SOC, increase water infiltration rate, and reduce soil erosion (Blanco-Canqui et al., 2015; Stewart et al., 2018). Soil health improvement policy, climate mitigation goals, and C sequestration demand have incentivized cover crop adoption. It is important to monitor the adoption of cover crops and estimate biomass for their potential benefits. Remote sensing can be used to identify cover crop adoption and estimate biomass (Hively et al., 2015; Hunt et al., 2011; Prabhakara et al., 2015). A study by Prabhakara et al. (2015) used ten different indices, including NDVI, green normalized difference vegetation index (GNDVI), triangular vegetation index (TVI), and visible atmospherically resistant index (VARI), calculated from a 16-band CROPSCAN sensor, to determine the cover crop growth on six fields that were planted to barley, rye, ryegrass, and triticale. The study found NDVI has a strong relationship with ground cover, and TVI was most accurate in estimating the cover crop biomass. Several studies used RS data to determine the spatiotemporal distribution and adoption of cover crops (Kc et al., 2021), the identification of cover crop species (Cruz-Ramírez et al., 2012), and termination dates (Gao et al., 2020)

29.6 Future perspective and Ecosystem Service Monetization

Conservation of the soil system and improving ecosystem services are priorities for the 21st century. The provision of ecosystem services as a commodity to economically reward resource managers is one of the most acclaimed policies to improve awareness and adoption of soil conservation practices and adoption. Awareness of climate conservation is gradually improving and subsequently expanding ecosystem markets. The last decade has seen the rapid development of research on the topic of ecosystem services and the economic value of ecosystem goods and services among decision-makers and the public (Bagstad et al., 2013; Holzman, 2012). Following the Millennium Ecosystem Assessment release in 2005, a significant increase in the number of scientific publications on the subject has been observed (Costanza, 2020; Mooney et al., 2004; Wang et al., 2021). However, ecosystem services are a complex interactive output of land use, climate, soil management, vegetation, and topo-geography along with socioecological aspects of human dimensions, which makes them hard to estimate and predict. Over the years, many scientific efforts have been made to measure soil heterogeneity and ecosystem services in situ. However, the variability in soil and the scale sought to determine the potential capacity of agroecology are very hard to accomplish using resource-intensive conventional sampling and analysis methods. Remote sensing is a potential alternative with wide applicability for assessing complex soil–plant–atmospheric interactions. The spectral signature of different chemical compounds and minerals in nature can be successfully used in estimating and predicting ecosystem services. The RS data can be used as auxiliary variables in developing comprehensive process-based ecosystem models. However, the field is still developing and needs the integration of in-situ measurement, modeling efforts, and statistical and machine learning approaches to improve ecosystem prediction and estimation implementation.

References

Andrews, S.S., Karlen, D.L., Cambardella, C.A., 2004. The soil management assessment framework. Soil. Sci. Soc. Am. J. 68, 1945–1962. Available from: https://doi.org/10.2136/sssaj2004.1945.

Angelopoulou, T., Tziolas, N., Balafoutis, A., Zalidis, G., Bochtis, D., 2019. Remote sensing techniques for soil organic carbon estimation: a review. Remote. Sens. 11, 676. Available from: https://doi.org/10.3390/rs11060676.

Ankenbauer, K.J., Loheide II, S.P., 2017. The effects of soil organic matter on soil water retention and plant water use in a meadow of the Sierra Nevada, CA. Hydrol. Process. 31, 891–901. Available from: https://doi.org/10.1002/hyp.11070.

Avdan, U., Kaplan, G., Küçük Matcı, D., Yiğit Avdan, Z., Erdem, F., Tuğba Mızık, E., et al., 2022. Soil salinity prediction models constructed by different remote sensors. Phys. Chem. Earth, Parts A/B/C 128, 103230. Available from: https://doi.org/10.1016/j.pce.2022.103230.

Bagnall, D.K., Rieke, E.L., Morgan, C.L.S., Liptzin, D.L., Cappellazzi, S.B., Honeycutt, W.C., 2023. A minimum suite of soil health indicators for North American agriculture. Soil Security . Available from: https://doi.org/10.1016/j.soisec.2023.100084.

Bagstad, K.J., Semmens, D.J., Waage, S., Winthrop, R., 2013. A comparative assessment of decision-support tools for ecosystem services quantification and valuation. Ecosyst. Serv. 5, 27–39. Available from: https://doi.org/10.1016/j.ecoser.2013.07.004.

Baldocchi, D., Falge, E., Gu, L., Olson, R., Hollinger, D., Running, S., et al., 2001. FLUXNET: a new tool to study the temporal and spatial variability of ecosystem-scale carbon dioxide, water vapor, and energy flux densities. Bull. Am. Meteorol. Soc. 82 (11), 2415–2434. Available from https://doi.org/10.1175/1520-0477(2001)082 < 2415:FANTTS > 2.3.CO;2

Ben-Dor, E., Patkin, K., Banin, A., Karnieli, A., 2002. Mapping of several soil properties using DAIS-7915 hyperspectral scanner data—a case study over clayey soils in Israel. Int. J. Remote. Sens. 23, 1043–1062. Available from: https://doi.org/10.1080/01431160010006962.

Bhunia, G.S., Kumar Shit, P., Pourghasemi, H.R., 2019. Soil organic carbon mapping using remote sensing techniques and multivariate regression model. Geocarto Int. 34, 215–226. Available from: https://doi.org/10.1080/10106049.2017.1381179.

Blanco-Canqui, H., Lal, R., 2009. Crop residue removal impacts on soil productivity and environmental quality. Crit. Rev. Plant. Sci. 28, 139–163. Available from: https://doi.org/10.1080/07352680902776507.

Blanco-Canqui, H., Mikha, M.M., Benjamin, J.G., Stone, L.R., Schlegel, A.J., Lyon, D.J., et al., 2009. Regional study of no-till impacts on near-surface aggregate properties that influence soil erodibility. Soil. Sci. Soc. Am. J. 73, 1361–1368. Available from: https://doi.org/10.2136/sssaj2008.0401.

Blanco-Canqui, H., Shaver, T.M., Lindquist, J.L., Shapiro, C.A., Elmore, R.W., Francis, C.A., et al., 2015. Cover crops and ecosystem services: insights from studies in temperate soils. Agron. J. 107, 2449–2474. Available from: https://doi.org/10.2134/agronj15.0086.

Brown, D.J., Shepherd, K.D., Walsh, M.G., Dewayne Mays, M., Reinsch, T.G., 2006. Global soil characterization with VNIR diffuse reflectance spectroscopy. Geoderma 132 (3–4), 273–290. Available from: https://doi.org/10.1016/j.geoderma.2005.04.025.

Bünemann, E.K., Bongiorno, G., Bai, Z., Creamer, R.E., De Deyn, G., de Goede, R., et al., 2018. Soil quality—a critical review. Soil. Biol. Biochem. 120, 105–125. Available from: https://doi.org/10.1016/j.soilbio.2018.01.030.

Burns, B.W., Green, V.S., Hashem, A.A., Massey, J.H., Shew, A.M., Adviento-Borbe, M.A.A., et al., 2022. Determining nitrogen deficiencies for maize using various remote sensing indices. Precis. Agric. 23, 791–811. Available from: https://doi.org/10.1007/s11119-021-09861-4.

Calvet, J.-C., Wigneron, J.-P., Walker, J., Karbou, F., Chanzy, A., Albergel, C., 2011. Sensitivity of passive microwave observations to soil moisture and vegetation water content: L-band to W-band. IEEE Trans. Geosci. Remote. Sens. 49, 1190–1199. Available from: https://doi.org/10.1109/TGRS.2010.2050488.

Chen, F., Kissel, D.E., West, L.T., Rickman, D., Luvall, J.C., Adkins, W., 2005. Mapping surface soil organic carbon for crop fields with remote sensing. J. Soil. Water Conserv. 60, 51–57.

Costanza, R., 2020. Valuing natural capital and ecosystem services toward the goals of efficiency, fairness, and sustainability. Ecosyst. Serv. 43, 101096. Available from: https://doi.org/10.1016/j.ecoser.2020.101096.

Crain, J., Ortiz-Monasterio, I., Raun, B., 2012. Evaluation of a reduced cost active NDVI sensor for crop nutrient management. J. Sens. 2012, e582028. Available from: https://doi.org/10.1155/2012/582028.

Cruz-Ramírez, M., Hervás-Martínez, C., Jurado-Expósito, M., López-Granados, F., 2012. A multi-objective neural network based method for cover crop identification from remote sensed data. Expert. Syst. Appl. 39, 10038–10048. Available from: https://doi.org/10.1016/j.eswa.2012.02.046.

Dagg, J., Lafleur, P., 2010. An application of plot-scale NDVI in predicting carbon dioxide exchange and leaf area index in heterogeneous subarctic tundra. Can. J. Remote. Sens. 36, S111–S123. Available from: https://doi.org/10.5589/m10-019.

Das, S., Berns, K., McDonald, M., Ghimire, D., Maharjan, B., 2022. Soil health, cover crop, and fertility management: Nebraska producers' perspectives on challenges and adoption. J. Soil. Water Conserv. . Available from: https://doi.org/10.2489/jswc.2022.00058.

Daughtry, C.S.T., Hunt Jr., E.R., Doraiswamy III, P.C., McMurtrey, J.E., 2005. Remote sensing the spatial distribution of crop residues. Agron. J. 97, 864–871. Available from: https://doi.org/10.2134/agronj2003.0291.

de Paul Obade, V., Lal, R., 2016. Towards a standard technique for soil quality assessment. Geoderma 265, 96–102. Available from: https://doi.org/10.1016/j.geoderma.2015.11.023.

Ding, G., Novak, J.M., Amarasiriwardena, D., Hunt, P.G., Xing, B., 2002. Soil organic matter characteristics as affected by tillage management. Soil. Sci. Soc. Am. J. 66, 421–429. Available from: https://doi.org/10.2136/sssaj2002.4210.

Dubayah, R., Blair, J.B., Goetz, S., Fatoyinbo, L., Hansen, M., Healey, S., et al., 2020. The global ecosystem dynamics investigation: high-resolution laser ranging of the Earth's forests and topography. Sci. Remote Sens. 1, 100002. Available from: https://doi.org/10.1016/j.srs.2020.100002.

Edalat, M., Naderi, R., Egan, T.P., 2019. Corn nitrogen management using NDVI and SPAD sensor-based data under conventional vs. reduced tillage systems. J. Plant. Nutr. 42, 2310–2322. Available from: https://doi.org/10.1080/01904167.2019.1648686.

Entekhabi, D., et al., 2010. The soil moisture active passive (SMAP) mission. Proc. IEEE 98 (5), 704–716.

Ertlen, D., Schwartz, D., Brunet, D., Trendel, J.-M., Adam, P., Schaeffer, P., 2015. Qualitative near infrared spectroscopy, a new tool to recognize past vegetation signature in soil organic matter. Soil. Biol. Biochem. 82, 127–134. Available from: https://doi.org/10.1016/j.soilbio.2014.12.019.

Escorihuela, M.J., Chanzy, A., Wigneron, J.P., Kerr, Y.H., 2010. Effective soil moisture sampling depth of l-band radiometry: a case study. Remote. Sens. Environ. 114, 995–1001. Available from: https://doi.org/10.1016/j.rse.2009.12.011.

Farahmand, N., Sadeghi, V., 2020. Estimating soil salinity in the dried lake bed of Urmia lake using optical Sentinel-2 images and nonlinear regression models. J. Indian Soc. Remote. Sens. 48, 675–687. Available from: https://doi.org/10.1007/s12524-019-01100-8.

Fisher, J.B., Sikka, M., Oechel, W.C., Huntzinger, D.N., Melton, J.R., Koven, C.D., et al., 2014. Carbon cycle uncertainty in the Alaskan Arctic. Biogeosciences 11, 4271–4288. Available from: https://doi.org/10.5194/bg-11-4271-2014.

Gao, F., Anderson, M.C., Hively, W.D., 2020. Detecting cover crop end-of-season using VENμS and Sentinel-2 satellite imagery. Remote. Sens. 12, 3524. Available from: https://doi.org/10.3390/rs12213524.

Ge, Y., Thomasson, J.A., Sui, R., 2011. Remote sensing of soil properties in precision agriculture: a review. Front. Earth Sci. 5, 229–238. Available from: https://doi.org/10.1007/s11707-011-0175-0.

Gomez, C., Viscarra Rossel, R.A., McBratney, A.B., 2008. Soil organic carbon prediction by hyperspectral remote sensing and field vis-NIR spectroscopy: an Australian case study. Geoderma 146, 403–411. Available from: https://doi.org/10.1016/j.geoderma.2008.06.011.

Gorji, T., Sertel, E., Tanik, A., 2017. Monitoring soil salinity via remote sensing technology under data scarce conditions: a case study from Turkey. Ecol. Indic. 74, 384–391. Available from: https://doi.org/10.1016/j.ecolind.2016.11.043.

Hao, F., Zhang, X., Ouyang, W., Skidmore, A.K., Toxopeus, A.G., 2012. Vegetation NDVI linked to temperature and precipitation in the upper catchments of yellow river. Env. Model. Assess. 17, 389–398. Available from: https://doi.org/10.1007/s10666-011-9297-8.

Heller Pearlshtien, D., Ben-Dor, E., 2020. Effect of organic matter content on the spectral signature of iron oxides across the VIS–NIR spectral region in artificial mixtures: an example from a red soil from Israel. Remote. Sens. 12, 1960. Available from: https://doi.org/10.3390/rs12121960.

Hengl, T., Heuvelink, M., Gonzalez, Kilibarda, M., Blagotić, A., Shangguan, W., et al., 2017. SoilGrids250m: Global gridded soil information based on machine learning. PLOS ONE 12 (2)e0169748. Available from: https://doi.org/10.1371/journal.pone.0169748.

Hively, W.D., Duiker, S., McCarty, G., Prabhakara, K., 2015. Remote sensing to monitor cover crop adoption in southeastern Pennsylvania. J. Soil. Water Conserv. 70, 340–352. Available from: https://doi.org/10.2489/jswc.70.6.340.

Holzman, D.C., 2012. Accounting for nature's benefits: the dollar value of ecosystem services. Env. Health Perspect. 120, a152–a157. Available from: https://doi.org/10.1289/ehp.120-a152.

Hudson, B.D., 1994. Soil organic matter and available water capacity. J. Soil. Water Conserv. 49, 189–194.

Huete, A., Didan, K., Miura, T., Rodriguez, E., Gao, X., Ferreira, L., 2002. Overview of the radiometric and biophysical performance of the MODIS vegetation indices. Remote Sens. Environ 83 (1–2), 195–213. Available from: https://doi.org/10.1016/S0034-4257(02)00096-2.

Hunt, E.R., Hively, W.D., McCarty, G.W., Daughtry, C.S.T., Forrestal, P.J., Kratochvil, R.J., et al., 2011. NIR-green-blue high-resolution digital images for assessment of winter cover crop biomass. GISci. Remote. Sens. 48, 86–98. Available from: https://doi.org/10.2747/1548-1603.48.1.86.

Janvier, C., Villeneuve, F., Alabouvette, C., Edel-Hermann, V., Mateille, T., Steinberg, C., 2007. Soil health through soil disease suppression: which strategy from descriptors to indicators. Soil. Biol. Biochem. 39, 1–23. Available from: https://doi.org/10.1016/j.soilbio.2006.07.001.

Jung, M., Reichstein, M., Margolis, H. A., Cescatti, A., Richardson, A. D., Arain, M. A., et al., 2011. Global patterns of land-atmosphere fluxes of carbon dioxide, latent heat, and sensible heat derived from eddy covariance, satellite, and meteorological observations. J. Geophys. Res. Biogeosci. 116(G3). Available from: https://doi.org/10.1029/2010JG001566.

Kc, K., Zhao, K., Romanko, M., Khanal, S., 2021. Assessment of the spatial and temporal patterns of cover crops using remote sensing. Remote. Sens. 13, 2689. Available from: https://doi.org/10.3390/rs13142689.

Kross, A., Seaquist, J.W., Roulet, N.T., Fernandes, R., Sonnentag, O., 2013. Estimating carbon dioxide exchange rates at contrasting northern peatlands using MODIS satellite data. Remote. Sens. Environ. 137, 234–243. Available from: https://doi.org/10.1016/j.rse.2013.06.014.

Ladoni, M., Bahrami, H.A., Alavipanah, S.K., Norouzi, A.A., 2010. Estimating soil organic carbon from soil reflectance: a review. Precis. Agric. 11, 82–99.

Laflen, J.M., Amemiya, M., Hintz, E.A., 1981. Measuring crop residue cover. J. Soil. Water Conserv. 36, 341–343.

Lal, R., 2016. Soil health and carbon management. Food Energy Secur. 5, 212–222. Available from: https://doi.org/10.1002/fes3.96.

Lal, R., 2020. Soil organic matter and water retention. Agron. J. 112, 3265–3277. Available from: https://doi.org/10.1002/agj2.20282.

Lehmann, J., Bossio, D.A., Kögel-Knabner, I., Rillig, M.C., 2020. The concept and future prospects of soil health. Nat. Rev. Earth Env. 1, 544–553. Available from: https://doi.org/10.1038/s43017-020-0080-8.

Li, Y., Zhao, Z., Wei, S., Sun, D., Yang, Q., Ding, X., 2021. Prediction of regional forest soil nutrients based on Gaofen-1 remote sensing data. Forests 12, 1430. Available from: https://doi.org/10.3390/f12111430.

McLauchlan, K., 2006. The nature and longevity of agricultural impacts on soil carbon and nutrients: a review. Ecosystems 9, 1364–1382. Available from: https://doi.org/10.1007/s10021-005-0135-1.

Metternicht, G.I., Zinck, J.A., 2003. Remote sensing of soil salinity: potentials and constraints. Remote. Sens. Environ. 85, 1–20. Available from: https://doi.org/10.1016/S0034-4257(02)00188-8.

Mirzaee, S., Ghorbani-Dashtaki, S., Mohammadi, J., Asadi, H., Asadzadeh, F., 2016. Spatial variability of soil organic matter using remote sensing data. CATENA 145, 118–127. Available from: https://doi.org/10.1016/j.catena.2016.05.023.

Mohanty, B.P., Cosh, M.H., Lakshmi, V., Montzka, C., 2017. Soil moisture remote sensing: state-of-the-science. Vadose Zone J. 16. Available from: https://doi.org/10.2136/vzj2016.10.0105.

Mondal, A., Khare, D., Kundu, S., Mondal, S., Mukherjee, S., Mukhopadhyay, A., 2017. Spatial soil organic carbon (SOC) prediction by regression kriging using remote sensing data. Egypt. J. Remote. Sens. Space Sci. 20, 61–70. Available from: https://doi.org/10.1016/j.ejrs.2016.06.004.

Mooney, H.A., Cropper, A., Reid, W., 2004. The millennium ecosystem assessment: what is it all about. Trends Ecol. Evol. 19, 221–224. Available from: https://doi.org/10.1016/j.tree.2004.03.005.

Naeimi, V., Scipal, K., Bartalis, Z., Hasenauer, S., Wagner, W., 2009. An improved soil moisture retrieval algorithm for ERS and METOP scatterometer observations. IEEE Trans. Geosci. Remote Sens. 47 (7), 1999–2013. Available from: https://doi.org/10.1109/TGRS.2008.2011617.

Niu, Z., He, H., Peng, S., Ren, X., Zhang, L., Gu, F., et al., 2021. A process-based model integrating remote sensing data for evaluating ecosystem services. J. Adv. Model. Earth Syst. 13. Available from: https://doi.org/10.1029/2020MS002451.

Numata, I., Soares, J.V., Roberts, D.A., Leonidas, F.C., Chadwick, O.A., Batista, G.T., 2003. Relationships among soil fertility dynamics and remotely sensed measures across pasture chronosequences in Rondônia, Brazil. Remote. Sens. 87, 446–455. Available from: https://doi.org/10.1016/j.rse.2002.07.001.

Odebiri, O., Odindi, J., Mutanga, O., 2021. Basic and deep learning models in remote sensing of soil organic carbon estimation: a brief review. Int. J. Appl. Earth Obs. Geoinf. 102, 102389. Available from: https://doi.org/10.1016/j.jag.2021.102389.

Parfitt, R.L., Giltrap, D.J., Whitton, J.S., 1995. Contribution of organic matter and clay minerals to the cation exchange capacity of soils. Commun. Soil. Sci. Plant. Anal. 26, 1343–1355. Available from: https://doi.org/10.1080/00103629509369376.

Prabhakara, K., Hively, W.D., McCarty, G.W., 2015. Evaluating the relationship between biomass, percent groundcover and remote sensing indices across six winter cover crop fields in Maryland, United States. Int. J. Appl. Earth Obs. Geoinf. 39, 88–102. Available from: https://doi.org/10.1016/j.jag.2015.03.002.

Raaijmakers, J.M., Mazzola, M., 2016. Soil immune responses. Science 352, 1392–1393. Available from: https://doi.org/10.1126/science.aaf3252.

Rahman, A.F., Sims, D.A., Cordova, V.D., El-Masri, B.Z., 2005. Potential of MODIS EVI and surface temperature for directly estimating per-pixel ecosystem C fluxes. Geophys. Res. Lett. 32. Available from: https://doi.org/10.1029/2005GL024127.

Rinot, O., Levy, G.J., Steinberger, Y., Svoray, T., Eshel, G., 2019. Soil health assessment: a critical review of current methodologies and a proposed new approach. Sci. Total. Environ. 648, 1484–1491. Available from: https://doi.org/10.1016/j.scitotenv.2018.08.259.

Sahab, S., Suhani, I., Srivastava, V., Chauhan, P.S., Singh, R.P., Prasad, V., 2021. Potential risk assessment of soil salinity to agroecosystem sustainability: current status and management strategies. Sci. Total. Environ. 764, 144164. Available from: https://doi.org/10.1016/j.scitotenv.2020.144164.

Sainju, U.M., 2021. The benefits of the no-till system on soil health and crop yields in dryland cropping systems. Soil. Res. 60, 399–411. Available from: https://doi.org/10.1071/SR21188.

Saleh, K., Wigneron, J.-P., Waldteufel, P., de Rosnay, P., Schwank, M., Calvet, J.-C., et al., 2007. Estimates of surface soil moisture under grass covers using L-band radiometry. Remote. Sens. Environ. 109, 42–53. Available from: https://doi.org/10.1016/j.rse.2006.12.002.

Schimel, D., Pavlick, R., Fisher, J.B., Asner, G.P., Saatchi, S., Townsend, P., et al., 2015. Observing terrestrial ecosystems and the carbon cycle from space. Glob. Change Biol. 21 (5), 1762–1776. Available from: https://doi.org/10.1111/gcb.12822.

Schwank, M., Wiesmann, A., Werner, C., Mätzler, C., Weber, D., Murk, A., et al., 2010. ELBARA II, an L-band radiometer system for soil moisture research. Sensors 10, 584–612. Available from: https://doi.org/10.3390/s100100584.

Serbin, G., Hunt, E.R., Daughtry, C.S.T., McCarty, G.W., Doraiswamy, P.C., 2009. An improved ASTER index for remote sensing of crop residue. Remote. Sens. 1, 971–991. Available from: https://doi.org/10.3390/rs1040971.

Six, J., Paustian, K., Elliott, E.T., Combrink, C., 2000. Soil structure and organic matter I. Distribution of aggregate-size classes and aggregate-associated carbon. Soil. Sci. Soc. Am. J. 64, 681–689. Available from: https://doi.org/10.2136/sssaj2000.642681x.

Soil Health Institute, 2022a. Soil Health Institute Announces Recommended Measurements for Evaluating Soil Health, Soil Health Institute. <https://soilhealthinstitute.org/news-events/soil-health-institute-announces-recommended-measurements-for-evaluating-soil-health/> (accessed 26.10.22).

Soil Health Institute, 2022b. Recommended Measurements for Scaling Soil Health Assessments, Soil Health Institute. <https://soilhealthinstitute.org/our-work/initiatives/measurements/> (accessed 26.10.22).

Song, Y.-Q., Zhao, X., Su, H.-Y., Li, B., Hu, Y.-M., Cui, X.-S., 2018. Predicting spatial variations in soil nutrients with hyperspectral remote sensing at regional scale. Sensors 18, 3086. Available from: https://doi.org/10.3390/s18093086.

Stewart, R.D., Jian, J., Gyawali, A.J., Thomason, W.E., Badgley, B.D., Reiter, M.S., et al., 2018. What we talk about when we talk about soil health. Agric. Environ. Lett. 3, 180033. Available from: https://doi.org/10.2134/ael2018.06.0033.

Stone, L.R., Schlegel, A.J., 2010. Tillage and crop rotation phase effects on soil physical properties in the west-central great plains. Agron. J. 102, 483–491. Available from: https://doi.org/10.2134/agronj2009.0123.

Su, Y., Zhang, W., Liu, B., Tian, X., Chen, S., Wang, H., et al., 2022. Forest carbon flux simulation using multi-source data and incorporation of remotely sensed model with process-based model. Remote. Sens. 14, 4766. Available from: https://doi.org/10.3390/rs14194766.

Taghadosi, M.M., Hasanlou, M., 2021. Developing geographic weighted regression (GWR) technique for monitoring soil salinity using sentinel-2 multispectral imagery. Env. Earth Sci. 80, 75. Available from: https://doi.org/10.1007/s12665-020-09345-0.

Turmel, M.-S., Speratti, A., Baudron, F., Verhulst, N., Govaerts, B., 2015. Crop residue management and soil health: a systems analysis. Agric. Syst. 134, 6–16. Available from: https://doi.org/10.1016/j.agsy.2014.05.009.

Turner, D.P., Ollinger, S.V., Kimball, J.S., 2004. Integrating remote sensing and ecosystem process models for landscape- to regional-scale analysis of the carbon cycle. BioScience 54, 573–584. Available from: https://doi.org/10.1641/0006-3568(2004)054[0573:IRSAEP]2.0.CO;2.

Turner, D.P., Ritts, W.D., Cohen, W.B., Maeirsperger, T.K., Gower, S.T., Kirschbaum, A.A., et al., 2005. Site-level evaluation of satellite-based global terrestrial gross primary production and net primary production monitoring. Glob. Change Biol. 11 (4), 666–684. Available from: https://doi.org/10.1111/j.1365-2486.2005.00936.x.

Viscarra Rossel, R.A., Cattle, S.R., Ortega, A., Fouad, Y., 2009. In situ measurements of soil colour, mineral composition and clay content by vis–NIR spectroscopy. Geoderma 150, 253–266. Available from: https://doi.org/10.1016/j.geoderma.2009.01.025.

Wang, L., Qu, J.J., 2009. Satellite remote sensing applications for surface soil moisture monitoring: a review. Front. Earth Sci. China 3, 237–247. Available from: https://doi.org/10.1007/s11707-009-0023-7.

Wang, J., He, T., Lv, C., Chen, Y., Jian, W., 2010. Mapping soil organic matter based on land degradation spectral response units using Hyperion images. Int. J. Appl. Earth Obs. Geoinf. 12, S171–S180. Available from: https://doi.org/10.1016/j.jag.2010.01.002. Suppl. Issue "Spatial analysis-modeling, methodol applications".

Wang, X., Zhang, F., Kung, H., Johnson, V.C., 2018. New methods for improving the remote sensing estimation of soil organic matter content (SOMC) in the Ebinur Lake Wetland National Nature Reserve (ELWNNR) in northwest China. Remote. Sens. Environ. 218, 104–118. Available from: https://doi.org/10.1016/j.rse.2018.09.020.

Wang, J., Ding, J., Yu, D., Ma, X., Zhang, Z., Ge, X., et al., 2019. Capability of Sentinel-2 MSI data for monitoring and mapping of soil salinity in dry and wet seasons in the Ebinur Lake region, Xinjiang, China. Geoderma 353, 172–187. Available from: https://doi.org/10.1016/j.geoderma.2019.06.040.

Wang, B., Zhang, Q., Cui, F., 2021. Scientific research on ecosystem services and human well-being: a bibliometric analysis. Ecol. Indic. 125, 107449. Available from: https://doi.org/10.1016/j.ecolind.2021.107449.

Xiao, J., Moody, A., 2004. Photosynthetic activity of US biomes: responses to the spatial variability and seasonality of precipitation and temperature. Glob. Change Biol. 10, 437–451. Available from: https://doi.org/10.1111/j.1365-2486.2004.00745.x.

Xiao, X., Zhang, Q., Saleska, S., Hutyra, L., De Camargo, P., Wofsy, S., et al., 2005. Satellite-based modeling of gross primary production in a seasonally moist tropical evergreen forest. Remote Sens. Environ. 94 (1), 105–122. Available from: https://doi.org/10.1016/j.rse.2004.08.015.

Yang, C.-C., Prasher, S.O., Enright, P., Madramootoo, C., Burgess, M., Goel, P.K., et al., 2003. Application of decision tree technology for image classification using remote sensing data. Agric. Syst. 76, 1101–1117. Available from: https://doi.org/10.1016/S0308-521X(02)00051-3.

Yuzugullu, O., Lorenz, F., Fröhlich, P., Liebisch, F., 2020. Understanding fields by remote sensing: soil zoning and property mapping. Remote. Sens. 12, 1116. Available from: https://doi.org/10.3390/rs12071116.

Zhang, M., Zhang, M., Yang, H., Jin, Y., Zhang, X., Liu, H., 2021. Mapping regional soil organic matter based on Sentinel-2A and MODIS imagery using machine learning algorithms and google earth engine. Remote. Sens. 13, 2934. Available from: https://doi.org/10.3390/rs13152934.

Zhang, M., Liu, H., Zhang, M., Yang, H., Jin, Y., Han, Y., et al., 2021. Mapping soil organic matter and analyzing the prediction accuracy of typical cropland soil types on the northern songnen plain. Remote Sens. 13 (24), 5162. Available from: https://doi.org/10.3390/rs13245162.

Zhang, S., Wang, Z., Zhang, B., Song, K., Liu, D., Li, F., et al., 2010. Prediction of spatial distribution of soil nutrients using terrain attributes and remote sensing data. Trans. Chin. Soc. Agric. Eng. 26, 188–194.

Zhao, J., Liu, D., Cao, Y., Zhang, L., Peng, H., Wang, K., et al., 2022. An integrated remote sensing and model approach for assessing forest carbon fluxes in China. Sci. Total. Environ. 811, 152480. Available from: https://doi.org/10.1016/j.scitotenv.2021.152480.

Chapter 30

Mapping of Mesquite (*Prosopis juliflora*) invasion in salt-affected soils of semiarid tropics: a case study

M. Lalitha[1], S. Dharumarajan[1], K.S. Anil Kumar[1], R. Srinivasan[1], S. Kaliraj[2], P. Senthilvalavan[3], Rajendra Hegde[1] and S.K. Singh[1]

[1]*ICAR-National Bureau of Soil Survey and Land Use Planning, Regional Centre, Bangalore, Karnataka, India,* [2]*National Centre for Earth Science Studies (NCESS), Ministry of Earth Sciences, Govt. of India, Thiruvananthapuram, Kerala, India,* [3]*Annamalai University, Annamalai Nagar, Chidambaram, Tamil Nadu, India*

30.1 Introduction

Invasive species occasionally threaten biodiversity, hydrology, and ecosystem function in the areas where they have been introduced (Chandrasekaran and Swamy, 2011, MacDougall and Turkington, 2005). *Prosopis juliflora* (family Mimosoideae), which is a plant species native to Jamaica, was introduced to South India in 1877 to meet the fuelwood requirements of the rural poor (Reddy, 1978) and for the restoration of the degraded lands (Mwangi and Swallow, 2005). However, the current coverage of the species in India exceeds approximately 5.55 million hectares ($\sim$1.8% of TGA), spread over arid and semiarid parts of Andhra Pradesh, Gujarat, Haryana, Karnataka, Madhya Pradesh, Maharashtra, Punjab, Rajasthan, Tamil Nadu, and Uttar Pradesh states (Kathiresan, 2006). There are different species and several hybrids of *Prosopis*, which are all collectively known as mesquite. It is a weed of national significance and is regarded as one of Australia's worst weeds.

The mesquite was introduced to Tamil Nadu state when aerial seeding was carried out in wastelands, which receive an annual average rainfall range of 150–750 mm. However, the mesquite trees escaped from their introduced range, and invasions have been recorded even in the rice-growing tracts of the Cauvery River delta region, where tanks and reservoirs are the significant sources of irrigation. Any patch of neglected, eroded, overgrazed, or drought-affected land is vulnerable to invasion by mesquite plants (Keblawy, 2012). The invasive behavior of naturally regenerated mesquite stands threatens fertile agricultural lands and wastelands in India (Pasiecznik et al., 2001, 2004, 2006). Mesquite has the most extensive root system of any plant in the world. The roots can grow to depths of 80 m and extend laterally to more than 30 m to extract soil moisture (Thorp and Lynch, 2001). Mesquite requires about 1000 kg of water to produce one kg of dry wood. The vast demand for soil moisture by mesquite plants reduces the water level in their surroundings and alters hydrology (Natarajan et al., 2007).

Mesquite can survive and grow in alkaline and saline soils with a salinity level equal to seawater (Felker et al., 1981). With this background, the present study aimed to delineate, characterize and compare the mesquite-invaded soils with those in adjacent uninvaded fallow lands (with similar environmental conditions to invaded areas).

30.2 Materials and methods

The study lies between 78° 29′ 12″ and 78° 44′ 43.3″ E longitude and 9° 40′ 41″ and 10° 2′ 7″ N latitude and belongs to the Kalayarkoil block of Sivagangai District, Tamil Nadu. Large areas of tank-irrigated lowlands in the district have gone out of cultivation due to the heavy invasion of mesquite (Natarajan et al., 2007). The climate is semiarid tropical, with a mean annual rainfall of 786 mm. The mean maximum and minimum temperatures of the study area are 31.6°C and 25.5°C. The soils developed from alluvial deposits. Multidate (three-season) RESOURCESAT-2 LISS IV data for

Remote Sensing of Soils. DOI: https://doi.org/10.1016/B978-0-443-18773-5.00012-0

2011–2013 were used to delineate and map mesquite-invaded areas in the study area. Toposheets, Forest Survey of India maps, and Tamil Nadu state soil maps (Natarajan et al., 1997) were used as references. A tentative map showing areas invaded by mesquite was prepared by online visual interpretation using ArcGIS 10.2.2 software and finalized after a ground truth check.

Four random sites were selected—two from mesquite-invaded areas and two from uninvaded fallow land (similar situation without mesquite)—to compare the soil properties. Two soil profiles were opened at each site at a depth of 150 cm, and the morphological characteristics of the soils were recorded in the standard format (Soil Survey Staff, 2010). Horizon-wise samples were collected from soil profiles for laboratory analysis. The air-dried and sieved (2 mm) samples were analyzed for soil reaction (1:2.5 soil–water suspension), electrical conductivity, organic carbon, particle size classes, CEC, and the sum of exchangeable cations, following standard procedures as outlined in Jackson (1973).

Descriptive statistics of the analyzed soil data, viz., minimum, maximum, mean, and standard deviation, were analyzed in a Microsoft Excel spreadsheet. For comparison, the weighted average concentrations were computed using Microsoft Excel for analyzed samples at different soil depths (i.e., 0–15, 15–30, 30–50, 50–100, and 100–150 cm). Analysis of variance (ANOVA) was computed using the split-plot design in the SAS computer program. Location was treated as a whole plot treatment and soil depth as a split-plot factor. The Duncan multiple range test compared soil properties at similar soil depths across locations.

30.3 Results and discussion

30.3.1 Delineation of lands invaded by mesquite

Mesquite-invaded areas had a dark red to brownish red color and a dark pink tone on false-color RESOURCESAT-2 LISS IV composite images because the invasive mesquite remains evergreen throughout the year. The vegetation has lower spectral reflectance in the red region and is high in the near-infrared (NIR) due to its total biomass compared to stressed plants (Felix Rembold et al., 2015). This color difference helped delineate mesquite-invaded areas and separate them from other vegetation. Mesquite invasion was associated with high soil moisture conditions available along the banks of water bodies (Fig. 30.1). Mesquite invasion extent was found in an area of 4475 ha (8% of TGA), and the invasion ratio of mesquite was 16%. Most of the tank beds have been invaded by mesquite plants. The invasion intensity was observed to be high in the lower portion of the study area compared to the upstream region.

30.3.2 Morphological and physicochemical properties of soil

The soils were deep to very deep, ranging from 109 to more than 150 cm, and the color was dark grayish brown (10 YR 4/2) to light brownish gray (2.5 Y 6/2). The dark color might be due to organic matter characteristics, which could be dissolved under high pH and ESP conditions and add dark color to the soil (Reyes-Reyes et al., 2003). The surface and subsurface soil textures varied from sandy clay loam to sandy clay. The clay content was increasing with depth due to clay illuviation. The morphological studies showed that soils are highly calcareous, with the presence of amorphous lime in the substratum. The soil structure in the A- and B-horizons of pedons was massive, implying a considerable reduction in soil permeability. The alteration process has led to a well-defined cambic horizon in these soils because of the high exchangeable sodium percentage. Overall, no difference was observed between the profile morphology of soils under mesquite and adjacent open sites, except for calcareousness.

The soils invaded by mesquite had significant changes in their physicochemical properties concerning their depth and location. The soil pH in the study area ranged from 8.03 to 10.3. Significantly, high soil pH (9.78 and 9.39) was found under mesquite as compared to the open site (8.52) (Table 30.1). Since mesquite is relatively tolerant of soil pH, the profuse growth observed under high pH is due to its adaptive nature (Felker et al., 1981). Soil electrical conductivity ranged from 0.1 to 4 d Sm^{-1}. The significantly high soil salinity under mesquite was observed compared to the open site. The electrical conductivity increased with depth, irrespective of location (Fig. 30.2).

The surface organic carbon content of the soils at different locations varied from 0.09% to 0.94%, decreasing with depth. The organic carbon content in the upper 50 cm of soil under mesquite and the open site was nearly the same (Reyes-Reyes et al., 2002). The distribution pattern of organic carbon reported by Barth and Klemmedson (1978) and Herrera-Arreola et al. (2007) indicated no improvement in soil organic carbon under mesquite and adjacent perennial grasslands. The high decomposition rate of mesquite leaf residue would confer no comparative advantage (Lalitha and Kumar, 2016).

The cation exchange capacity (CEC), exchangeable Ca, Na, and K, was high in the soils under the mesquite. The high production of biomass and subsequent decomposition in a low leaching environment have affected the

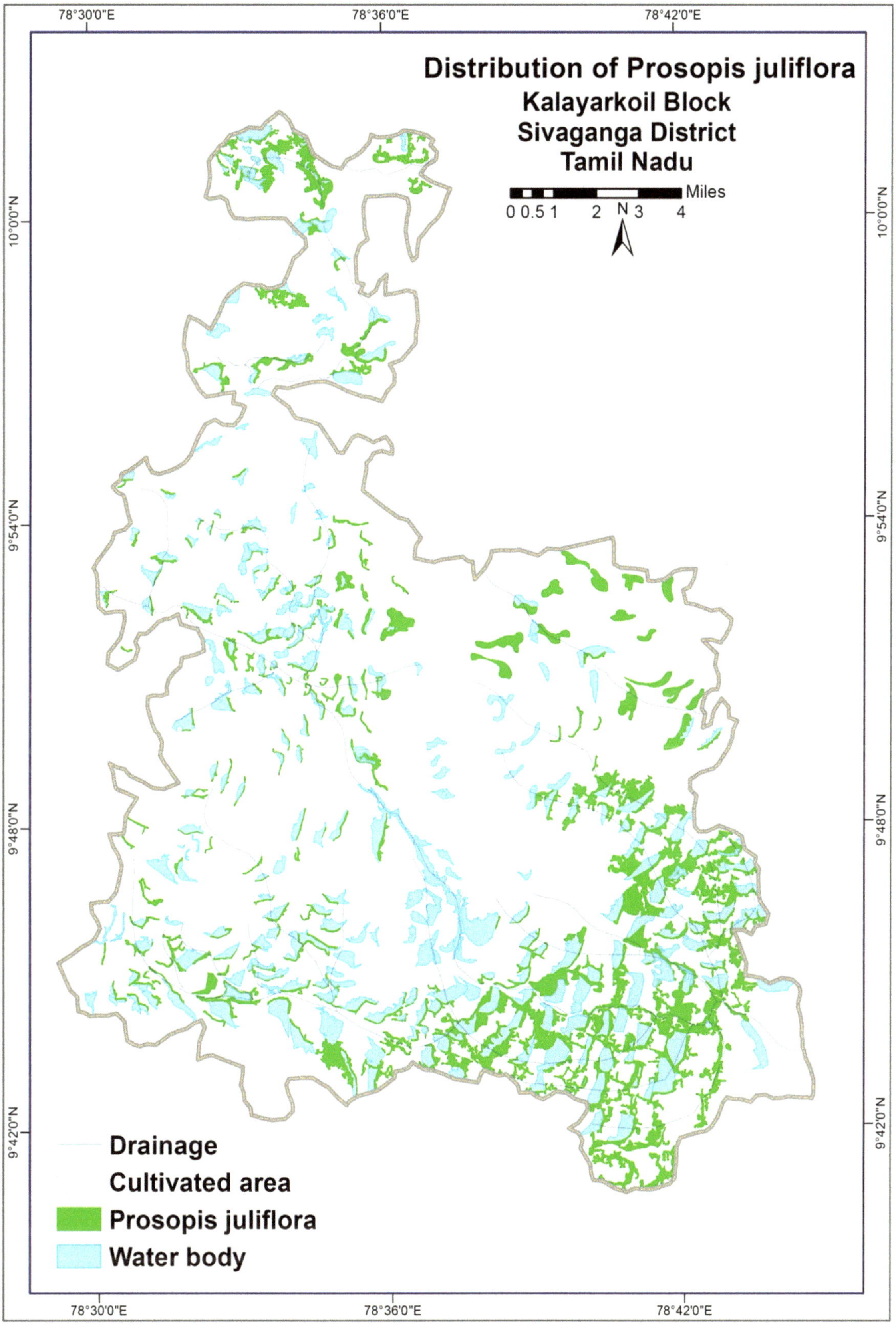

FIGURE 30.1 The distribution of *Prosopis juliflora* in the study area.

TABLE 30.1 Soil properties at four different depths under the *P. juliflora tree* and adjacent open areas.

	Depth	pH	EC	OC	Clay	CEC	Ca	Mg	Na	K	ESP	Free CaCO3
	(cm)		(dS m^{-1})	(%)	(%)	------------(c mol (+) kg^{-1})------------					(%)	(%)
Mesquite-invaded area **(Site-I)**	0–15	9.51	1.89	0.55	33.85	14.24	8.85	1.80	2.94	0.41	20.53	2.62
	15–30	9.70	3.20	0.41	40.20	21.80	13.75	2.50	4.90	0.60	22.40	5.30
	30–50	9.70	3.59	0.28	40.59	23.51	14.71	2.64	5.45	0.66	23.06	7.94
	50–100	9.70	3.90	0.18	40.90	24.90	15.50	2.75	5.90	0.70	23.60	10.10
	100–150	10.30	4.00	0.17	41.80	26.90	15.90	3.40	6.00	0.70	24.10	11.80
Mean		**9.78a**	**3.32a**	**0.32cb**	**39.47a**	**22.27ba**	**13.74b**	**2.62b**	**5.04a**	**0.61a**	**22.74a**	**7.55a**
Mesquite-invaded area **(Site-II)**	0–15	9.10	1.34	0.52	35.20	24.50	16.50	3.00	4.50	0.40	18.30	0.81
	15–30	9.40	1.51	0.44	43.40	26.80	17.50	2.50	6.10	0.50	22.70	1.54
	30–50	9.55	1.60	0.39	43.61	27.18	17.54	2.31	6.67	0.51	24.50	1.58
	50–100	9.40	1.98	0.24	43.46	28.52	18.35	2.50	6.98	0.62	24.40	2.45
	100–150	9.49	2.15	0.09	44.62	28.88	19.16	2.28	6.75	0.50	23.29	3.48
Mean		**9.39a**	**1.72b**	**0.34cb**	**42.06a**	**27.18a**	**17.81a**	**2.52b**	**6.20a**	**0.51ba**	**22.64a**	**1.97c**
Open site (without mesquite invasion) **(Site-I)**	0–15	8.10	0.14	0.56	34.20	18.10	13.00	2.75	1.00	0.26	5.50	0.85
	15–30	8.30	0.70	0.46	35.93	19.10	13.67	3.08	1.17	0.29	6.03	1.58
	30–50	8.58	1.00	0.39	38.36	19.78	13.25	2.80	2.60	0.36	13.02	2.23
	50–100	8.70	0.97	0.34	40.00	22.39	14.26	2.88	3.88	0.48	17.27	3.30
	100–150	8.90	1.17	0.21	44.80	23.70	15.00	2.75	4.50	0.60	18.90	5.61
Mean		**8.52b**	**0.80c**	**0.39b**	**38.7a**	**20.61b**	**13.84b**	**2.85b**	**2.63b**	**0.40b**	**12.15b**	**2.71cb**
Open site (without mesquite invasion) **(Site-II)**	0–15	8.03	0.18	0.94	35.63	14.33	6.93	4.67	1.08	0.40	7.50	0.80
	15–30	8.32	0.12	0.61	35.70	15.40	8.66	3.94	1.40	0.38	9.02	1.18
	30–50	8.44	0.10	0.55	35.74	19.52	8.92	7.72	2.10	0.30	10.74	1.55
	50–100	8.75	0.14	0.42	38.33	22.20	10.60	7.65	3.07	0.60	13.79	1.94
	100–150	9.05	0.25	0.32	42.30	29.76	14.76	9.52	5.04	0.30	16.82	4.81
Mean		**8.52b**	**0.16d**	**0.57a**	**37.54a**	**20.24b**	**9.97c**	**6.70a**	**2.54b**	**0.40b**	**11.58b**	**2.06c**
LSD	0.46	0.52	0.08	6.90	5.33	2.90	1.47	1.76	0.13	5.63	2.05	

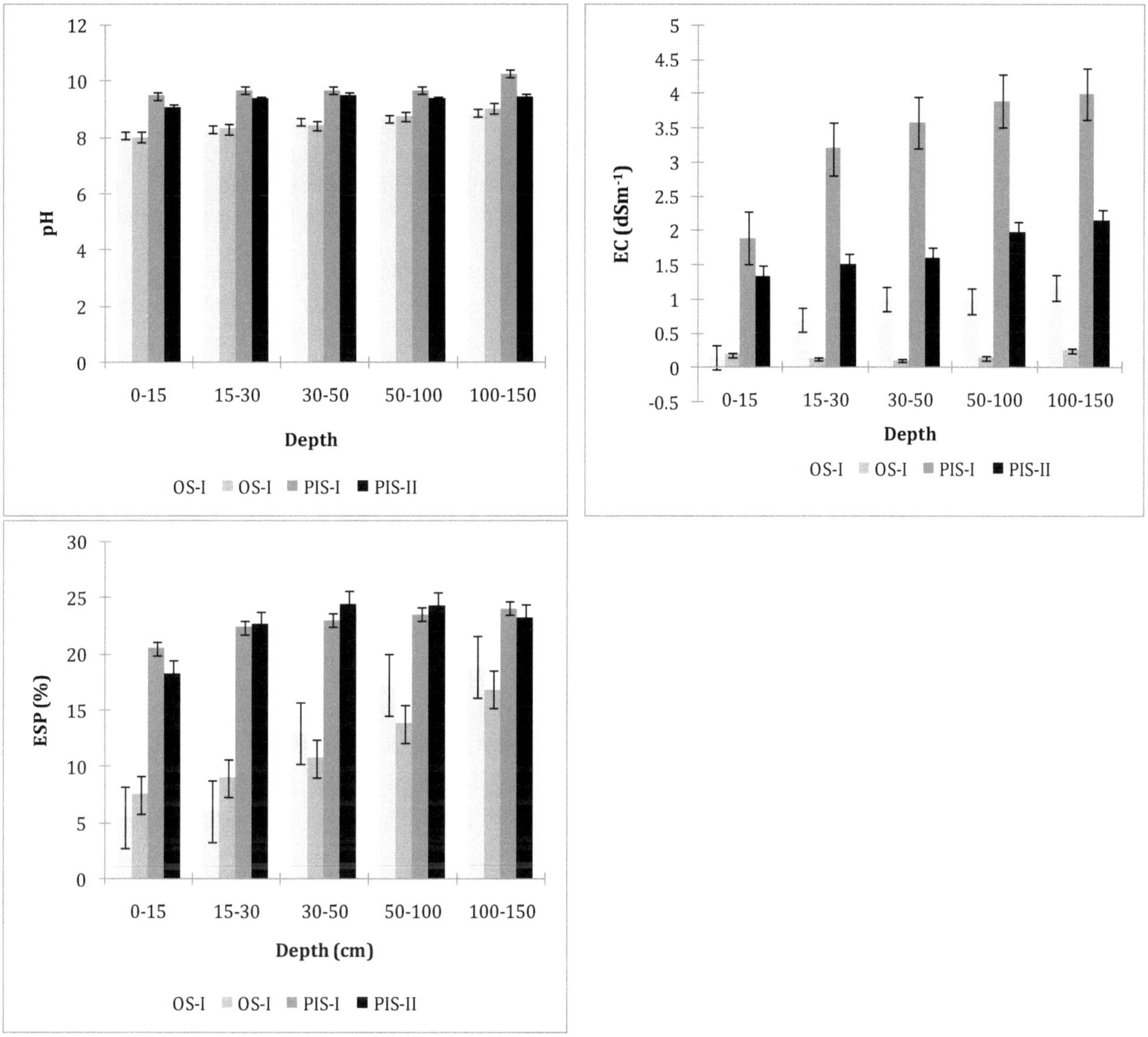

FIGURE 30.2 The soil pH, EC, and ESP under mesquite and open sites at different soil depths and locations.

accumulation of Ca, Na, and K in the soil invaded by mesquite canopies (Table 30.1). The ESP of the soil under mesquite (ESP = 22.74 and 22.64%) was significantly higher than the soil of the open site (ESP = 12.15 and 11.58%). Frias-Hernandez (1999) found that mesquite exhibits necrosis for sodium only at a concentration of about 18,000 mg/L, and mesquite was a more salt-tolerant species because it can exclude Na uptake by roots, keeping foliage concentration of Na low (Cline et al., 1986).

30.3.3 Soil sedimentation and mesquite invasion

There are about 40,000 tanks in Tamil Nadu, many of which were built between the 18th and 19th centuries, and in southern Tamil Nadu, tank irrigation and the interlinking of tanks through new channels were made before British rule (Breckenridge, 1985). Irrigated paddy production could only have been possible through the past construction of these tanks. The networks of tanks were constructed in such a way that the overflow of one tank would recharge the next tank, and subsequently, it would discharge into the river. Minerals and salt weathered from parent material are eroded from catchment areas in the upstream portion and are also carried away with water and deposited in the lower stream portion of the catchments or the peripheral portion of the water spread in the tank. Water and sediments enriched with salts are deposited incrementally in tanks in the lower stream portion. Removing silt from the tank bed and applying it to the field was common practice, yielding good crop production. However, later, due to many reasons (encroachment on the tank foreshore, malfunction of the supply channel, reduced dependency on the tank for irrigation water, and shift to nonagricultural income), the practice was discontinued, which resulted in the accumulation of salts. The soil properties deteriorated under high salt concentrations. However, the salt in the silt deposition kept increasing to the extent that it supports very minimal cultivation, with paddy as only one option. Since paddy is a high-water requirement crop, its cultivation depends mainly on uncertain rainfall. Besides, the increased cost of inputs and a decline in crop productivity in such soils resulted in a reduced margin in profits. Despite the availability of management strategies, resource-poor farmers have left the land fallow for a long time. Besides economic development and urbanization in rural areas, modern underground irrigation systems, the lack of interest of rural youth in agriculture, and reduced dependence on tanks for irrigation resulted in the discontinuation of paddy cultivation. These uncultivated lands were gradually invaded by mesquite. Mesquite grows luxuriously even in such poor soil conditions and provides a dense cover because it is a sodium excluder, which increases the adaptability of the species to strongly alkaline soils (Villagra et al., 2010). After the invasion, tank management practices like desilting were utterly stopped in the study area.

30.4 Conclusion

The data presented herein demonstrates that around 90% of the tank area has been invaded by mesquite. The soils were found to be very strongly alkaline. Because of high pH and exchangeable sodium, the cultivated lands are kept fallow, and mesquite is invading due to its adaptive behavior in strongly alkaline soil. The invasion is high in the low stream portion compared to the upstream portion because of the sequential deposition of eroded silt and favorable moisture conditions. Although there are many amelioration strategies available for sodic soil reclamation and significant progress has been achieved in improving the methods, a great deal of attention has been paid to the removal of mesquite because of its economics. Therefore, the government has to take a decision on a war footing basis to remove all the mesquite along with their roots, which occupied the tank beds, using heavy machinery and to support the farmers through various developmental programs to maintain the tank beds as well as supply channels. Hence, a comprehensive, integrated management program has to be implemented based on the soil–plant–water–environment relationship to address the issues of sustaining the irrigated ecosystem.

References

Barth, R.C., Klemmedson, J.O., 1978. Shrub induced spatial patterns of dry matter, nitrogen, and organic carbon. Soil Sci. Soci. Am. J. 42, 804–809.

Breckenridge, C.A., 1985. Social storage and the extension of agriculture in South India 1350–1750. In: Dalapicccola, A.L., et al., (Eds.), Vijayanagara City and Empire: New Currents of Research. Wiesbaden GMBH.

Chandrasekaran, S., Swamy, P.S., 2011. Proceedings of the Sixth Biennial Conference on Nature, Economy and Society: Understanding the Linkages. Centre for Economic and Social Studies, Hyderabad.

Cline, G., Rhodes, D., Felker, P., 1986. Micronutrient, phosphorus and pH influences on growth and leaf tissue nutrient levels of *Prosopis alba* and *Prosopis glandulosa*. For. Ecol. Manag. 6, 81–93.

Felix, R., Ugo, L., Ng, W.-T., Gadain, H., Michele, M., Clement, A., 2015. Mapping areas invaded by *Prosopis juliflora* in Somaliland on Landsat 8 imagery, Proc. SPIE 9637, Remote Sens. Agric., Ecosyst. Hydrol., 17. Available from: 10.1117/12.2193133, http://doi.org/10.1117/12.2193133.

Felker, P., Clark, P.R., Laag, A.E., Pratt, P.F., 1981. Salinity tolerance of the tree legumes Mesquite (*Prosopis glandulosa*, var Torreyana, *P. velutina* and *P. articulata*) Algarrobo (*P. chilensis*) Kiawe (*P. pallida*) and Tamarugo (*P. tamarugo*) grown in sand culture on nitrogen-free media. Plant. Soil. 61, 311 317.

Frias-Hernandez, J.T., Aguilar-Ledezma, A.L., Olalde-Portugal, V., Balderas-Lopez, G., Gutierrez-Juarez, J., Alvarado-Gil, J.J., et al., 1999. Soil characteristics in the Semiarid Highlands of central Mexico as affected by mesquite trees (*Prosopis laevigata*). Arid. Soil. Res. Rehab. 13, 305–312.

Herrera-Arreola, G., Herrera, Y., Reyes-Reyes, B.G., Dendooven, L., 2007. Mesquite (*Prosopis juliflora* (Sw.) DC.), huisache (*Acacia farnesiana* (L.) Willd.) and catclaw (*Mimosa biuncifera* Benth.) and their effect on dynamics of carbon and nitrogen in soils of the semi-arid highlands of Durango México. J. Arid. Environ. 69, 583–598.

Jackson, M.L., 1973. Soil Chemical Analysis. Constable and Co. Ltd. Prentice Hall of India Pvt. Ltd, New Delhi, pp. 10–114.

Kathiresan, R.M., 2006. Invasion of *Prosopis juliflora* in India. Problems Posed by the Introduction of *Prosopis* sp. in Selected Countries. Food and Agricultural Organization of the United Nations (FAO) Plant Production and Protection Division, Rome, Italy.

Keblawy, A., 2012. Impacts of native and exotic prosopis species on native plants in arid lands of the UAE. In: International Conference on Ecology, Agriculture and Chemical Engineering, Phuket, Thailand.

Lalitha, M., Kumar, P., 2016. Soil carbon fractions influenced by temperature sensitivity and land use management. Agrofor. Syst. 90 (6), 961–964.

MacDougall, A.S., Turkington, R., 2005. Are invasive species the drivers or passengers of change in degraded ecosystems? Ecology 86, 42–55.

Mwangi, E., Swallow, B., 2005. Invasion of *Prosopis juliflora* and local livelihoods: a case study from the Lake Baringo area of Kenya. ICRAF Working Paper No.3. World Agroforestry Centre, Nairobi.

Natarajan, A., Janakiraman, M., Manoharan, S., Balasubramaniyan, V., Murugappan, K., Udayakumar, J., et al., 2007. Land resources of Sivagangai Block, Sivagangai District, Tamil Nadu. Report Submitted by Soil Survey and Land Use Organization, Coimbatore and Tirunelveli, National Bureau of Soil Survey and Land Use Planning. Bangalore, Tamil Nadu Agricultural University, Coimbatore Agricultural Engineering Department, Chennai.

Natarajan, A., Reddy, P.S.A., Sehgal, J., Velayutham, M., 1997. Soil resources of Tamil Nadu for land use planning. NBSS Publ. 46b (Soils of India Series). National Bureau of Soil Survey and Land Use Planning, Nagpur, India.

Pasiecznik, N.M., Felker, P., Harris, P.J.C., Harsh, L.N., Cruz, G., Tewari, J.C., et al., 2001. The *Prosopis juliflora-Prosopis pallida* Complex: A Monograph. HDRA, Coventry.

Pasiecznik, N.M., Harris, P.J.C., Smith, S.J., 2004. Identifying Tropical Prosopis Species: A Field Guide. HDRA, Coventry, UK.

Pasiecznik, N.M., Harris, P.J.C., Trenchard, E.J., Smith, S.J., 2006. The dilemma of Prosopis: is there a rose between the thorns? Biocontrol N. Inf. 27 (1), 1–26.

Reddy, C.V.K., 1978. *Prosopis juliflora*, the precocious child of the plant world. Indian For. 104 (14), 18.

Reyes-Reyes, G., Baron-Ocampo, L., Cuali-Alvarez, I., Frias-Hernandez, J.T., Olalde- Portugal, V., Varela, L., et al., 2002. C and N dynamics in soil from the central highlands of Mexico as affected by mesquite (*Prosopis* sp.) and huisache (*Acacia tortuosa*): a laboratory investigation. Appl. Soil. Ecol. 19, 27–34.

Reyes-Reyes, G.B., Zamora-Villafranco, E., Reyes-Reyes, M.L., Frías-Hernandez, J.T., Olalde-Portugal, V., Dendooven, L., 2003. Decomposition of leaves of Huizache (*Acacia tortuosa*) and mesquite (*Prosopis* sp.) in soil of the highlands of Mexico. Plant. Soil. 256, 359–370.

Soil Survey Staff, 2010. Keys to Soil Taxonomy, eleventh ed. SCS, USDA, Washington, DC.

Thorp, J.R., Lynch, R.A. 2001. The Determination of Weeds of National Significance. National Weeds Strategy Executive Committee, Launceston.

Villagra, P., Vilela, A., Giordano, C., Alvarez, J., 2010. Ecophysiology of prosopis species from the Arid Lands of Argentina: what do we know about adaptation to stressful environments? In: Ramawat, K.G. (Ed.), Desert Plants. Springer, Berlin/Heidelberg, pp. 321–340.

Chapter 31

Spatial crop planning for sustainable resource use and conservation of ecological resources—a case study from Vattavada, Munnar

S. Dharumarajan[1], R. Vasundhara[1], B. Kalaiselvi[1], M. Lalitha[1], M. Rameshan[2], R. Srinivasan[1], K.S. Karthika[1], N. Maddileti[1], Sweta Kumari[1], Amar Suputhra[1], C.M. Pradeep[1] and Rajendra Hegde[1]
[1]*ICAR-National Bureau of Soil Survey and Land Use Planning, Regional Centre, Bangalore, Karnataka, India,* [2]*IHRML Project, Landscape Level Project Management Unit, Munnar, Kerala, India*

31.1 Introduction

The High Range Mountain Landscape (Anjunad and adjoining areas) in the Southern Western Ghats is valued for its vastness of biodiversity, ecosystem services, and ecological functions, which ultimately sustain thousands of lives. The demographically driven demands for natural resources and the diversion of forest areas for faster economic growth have intensified the stress on ecologically balanced structures. Extensive planting of exotic and invasive tree species in the grasslands and Shola forests of the high ranges has caused irrecoverable loss of biodiversity and ecosystems in the landscape. Conversion to monoculture, mainly in the form of plantations, altered the hydrological processes, declined the groundwater resources, increased the risk of pests and diseases and loss of soil nutrients and soil biodiversity as well (Baltodano, 2000; Evans, 2001; Bowyer, 2006; Liu et al., 2018). The unscientific and unsustainable crop cultivation practices for grabbing high profits in the short term are leading to the loss of germplasm for local resistant varieties and traditional cultivars. Moreover, disappearing habitats have also accelerated the rate of human—wildlife conflicts.

The survival of biodiversity in the landscape intricately depends on land utilization outside the protected areas. It is quite evident that drastic changes in cropping patterns have occurred in the landscape with nonjudicious use of pesticides and fertilizers, unsustainable utilization of water resources, etc., without considering the terrain, edaphic, weather variability, soil nutrient status, and ecological aspects of the landscape (Murugesan et al., 2018; Rahman, 2020). Current agricultural practices have given the least importance to spatial crop planning, which has resulted in the loss of traditional seed varieties and heterogeneity in cultivation, high runoff of productive soil, scarcity of water, low productivity, and increased human-wildlife conflict. The recent increased occurrence of landslides in the agricultural lands at Mankulam, Adimali, Devikulam, Munnar, Athirappally, and Kuttampuzha is evidence of anomalous cropping systems and improper land utilization. The fundamental reasons behind the present anomalies in agriculture date back to the British era. The thick forests of the region were undisturbed until the establishment of the British Empire and colonization in our country. They utilized the local human strength to fell the huge trees and plunder the natural forest resources, along with killing wild animals. The present anomalies in agriculture are a result of the land reforms, tenancy reforms, and peasant migration that had happened in the state during the early years.

31.1.1 Changes in land use from forest to nonforest

The forests in the Western Ghats have been destroyed by clear-felling for several plantations like tea, coffee, and teak during the period from 1860 to 1950. The states of Karnataka, Kerala, and Tamil Nadu in the southern region of the Western Ghats lost around 40% of the original vegetation cover as a result of land use conversion during 1920—1990 (Menon and Bawa, 1998). Even at present, less than 7% of the area is under primary vegetation cover in the Western Ghats, and nearly 15% of the Ghats is under the Protected Area System.

Remote Sensing of Soils. DOI: https://doi.org/10.1016/B978-0-443-18773-5.00023-5

31.1.2 Conversion to tea plantations

The history of tea plantation dates its records to the beginning of 17th century, and around 50 acres of dense forest in Idukki were cleared for cultivation of tea in the 18th century. Tea has been the major crop there since 1877, and there has not been any other significant land use change since then.

31.1.3 Conversion to cardamom hill reserve

Cardamom is widely cultivated in Kumily, Annavilasam, Vandanmedu, Nedumkandam, Udumbanchola, Shanthanpara, and Poopara countys of Idukki district. The Cardamom Hill Reserve (CHR) is situated in the cardamom hills of Idukki district, which came into existence as a result of the royal proclamation of the Kingdom of Travancore in April 1822 to promote cardamom cultivation and extends about 334 sq. miles. The total extent of the Cardamom Hill area was 1071.9746 km^2 (264,885 acres), of which 568.6 km^2 (140,500 acres) are assigned under various land assignment rules. The balance of 503.38 km^2 (124,386 acres) was leased to farmers for cardamom cultivation. The CHR still holds dense forest; however, several threats are being faced by the CHR, causing deterioration in its evergreen forest cover. The cardamom plantations were worst affected by the landslides and floods of 2018, indicating a need to relook into the areas of its cultivation and its suitability. Non-cardamom cultivation, illegal land conversion, and large-scale destruction of trees in the CHR have been blamed for flash floods and landslips in Idukki and siltation problems in the Idukki dam (Martha et al., 2019; Jones et al., 2021).

31.1.4 Conversion to rubber plantations

Forests were converted to rubber plantations, mainly during the period of the British invasion (Ross, 2017). Rubber flourished under the British, mainly for export. Traditional rubber plantation areas are in Kerala, Karnataka, and Tamil Nadu, where the rubber plantations were on side slopes of lateritic midland hills, plateaus, mounds, or uplands of Western Ghats at elevations ranging from 30 to 600 m, temperature ranging from 23°C to 28°C and rainfall 1500 to 4000 mm. The rubber plantations are mainly restricted to the lower portions of hill ranges.

31.1.5 Changes in cultivation from food crops to cash crop/commercial crops on steep slopes

The traditional land use in the Ghats after the major reforms has been paddy cultivation in the valleys, supplemented by the cultivation of millets and legumes on the hill slopes. Hill slope agriculture used to be largely of the shifting slash-and-burn type, but this has gradually been changed to the cultivation of terraces. The large-scale shift to cultivation of soil-eroding monoculture plantations of tea, coffee, cardamom, rubber, banana, pineapple, and timber on steep slopes led to increased surface run-off and low seepage deep into the soil (Sidle et al., 2006; Wishnie et al., 2007). The cultivation of crops on steep slopes increased soil erosion and runoff leading to soil degradation (Lalitha et al., 2021; Sprenger et al., 2013; Smethurst et al., 2015). Cultivation of tuber crops such as tapioca, ginger, and yams in paddy lands began in recent years.

Spatial crop planning, with the most appropriate crop choices for the area with sectoral demand and taking into consideration all these factors, has to evolve, which will improve the ecological integrity of the landscape.

31.2 Material and methods

31.2.1 Study area

Vattavada located at 10°14′42″ N latitude and 77°8′39″ E longitude (Fig. 31.1). It is a rain-shadow village and is situated in the Devikulam taluk of Idukki district of Kerala, along the eastern side of the Western Ghats, at a height of 6500 feet above mean sea level. The region in and around Vattavada varies in height from 1450 to 2695 m above mean sea level. Vattavada enjoys a salubrious climate. The temperature ranges between 5°C and 20°C in winter and 12°C and 18°C in summer. Temperatures as low as −4°C have been recorded in the border region of Vattavada. The mean maximum daily temperature is at its lowest during the monsoon months, with the highest temperature being 19°C (Fig. 31.2).

31.2.2 Base maps and image interpretation

The detailed survey of the land resources inventory was carried out using Sentinel-2 satellite imagery. Satellite data was visually interpreted using image interpretation elements and all the available collateral data, with local knowledge. The delineated physiographic units were used as a base for mapping soils.

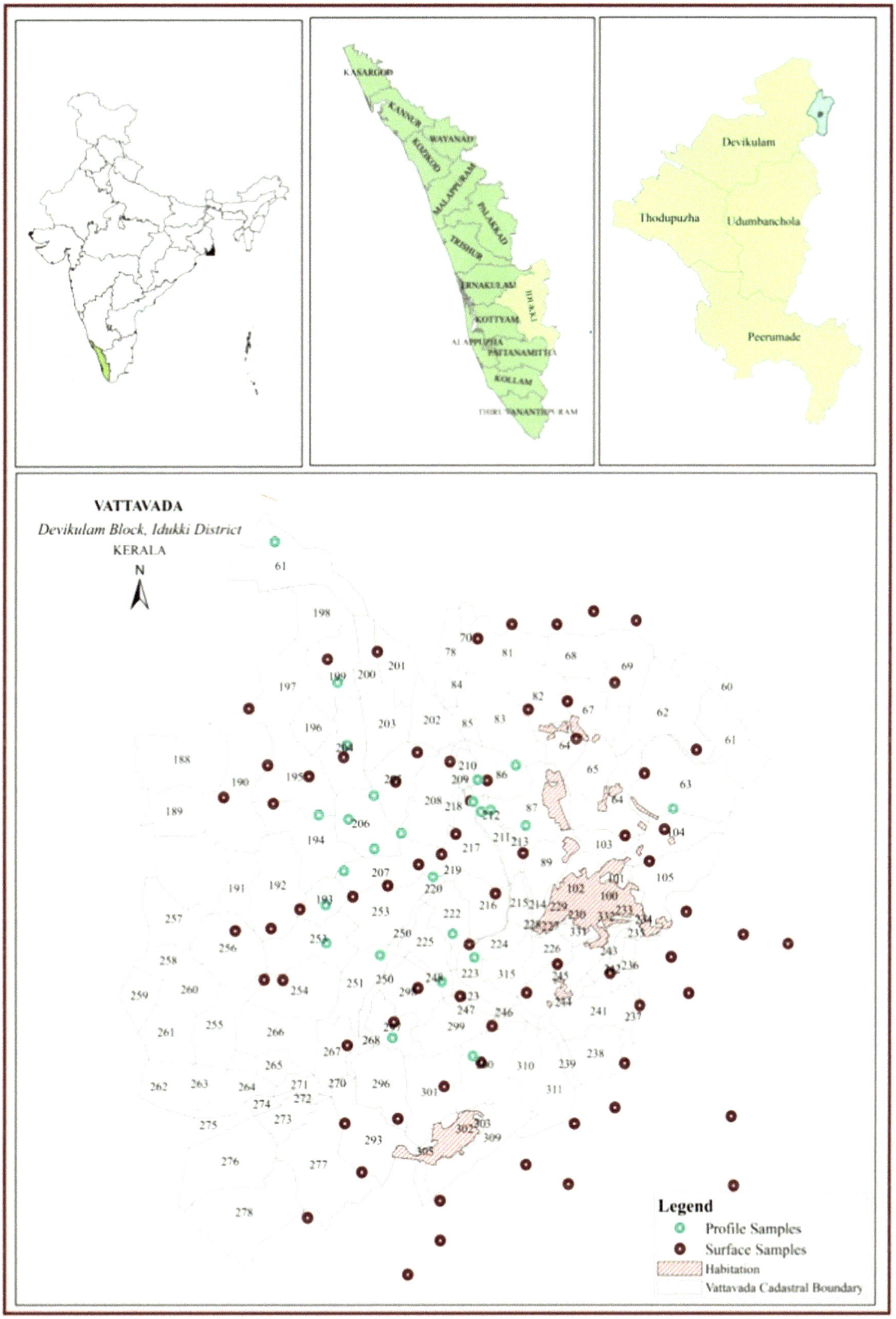

FIGURE 31.1 Location of sampling points in Vattavada.

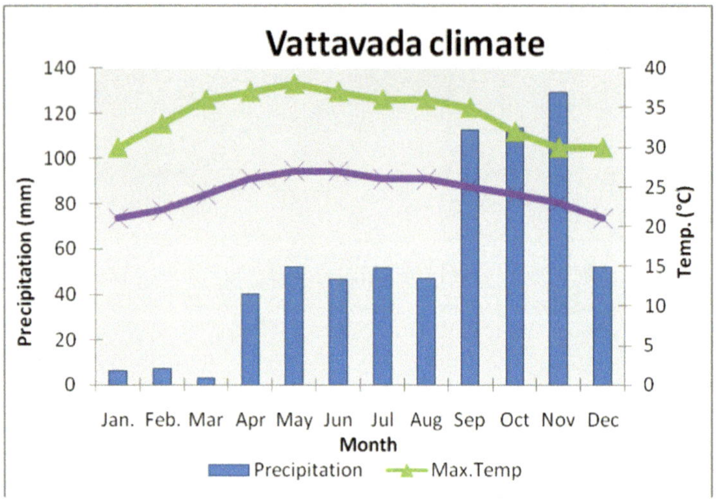

FIGURE 31.2 Average monthly rainfall and temperature in Vattavada.

31.2.3 Field investigation and soil mapping

About 33 soil profiles were studied in transects. The soil and site characteristics were recorded for all profile sites on a standard pro forma as per the guidelines given in the USDA Soil Survey Manual. Based on the soil characteristics, the soils were grouped into different soil series. Soil depth, texture, color, kind of horizon and horizon sequence, and amount and nature of gravel present are used as the major differentiating characteristics for identifying soil series occurring in the area. The area under each soil series was further separated into soil phases, and their boundaries were delineated on the cadastral map based on the variations observed in the texture of the surface soil, slope, erosion, presence of gravel, stoniness, etc. A soil phase is a subdivision of soil series based mostly on surface features that affect its use and management. In addition, about 101 surface samples were also collected and analyzed for major and micronutrients.

31.2.3.1 Laboratory characterization

Soil samples for each series were collected from representative master profiles for laboratory characterization by following the methods outlined in the Laboratory Manual. Surface soil samples collected for fertility status (major and micronutrients) were analyzed in the laboratory. By linking the soil fertility data to the survey numbers through GIS, soil fertility maps were generated using geospatial techniques.

31.2.3.2 Preparation of spatial crop plan

The soil and land resource units (soil phases) of the study area were assessed for their suitability for growing food, fodder, fiber, and other horticulture crops by following the procedure as outlined in FAO (1983) and Naidu et al. (2006). Crop requirements were developed for each crop from the available research data. The soil and land characteristics were matched with the crop requirements to arrive at the spatial crop plan.

31.3 Results and discussion

31.3.1 Soils of Vattavada

Vattavada has vegetable farming on around 1600 ha and intensive *Eucalyptus grandis* plantations. The major land uses in the Vattavada gram panchayat are annual vegetable crops like carrots, potatoes, butter beans, green peas, garlic, beetroot, cabbage, and fruit crops like strawberries, bananas, and passion fruit. The agricultural area is surrounded by Munnar Reserve Forest and tea and eucalyptus plantations. Agricultural fields are bench terraced for convenient field operation as the slope varies from very gently sloping (1%–3%) to very steeply sloping (33%) and has a short range of slope length (50–300 m). Soil erosion varied from moderate to severe based on slope variability, surface compactness, and land use. The soils of the side slope are prone to severe erosion and rapid to very rapid runoff, whereas the soils of the lower part are slightly to moderately erosion-prone. Regarding drainage, the upland soils are well to very well drained, but the lower land soils are somewhat poor in drainage due to the deposition of finer particles in low-lying areas as a result of erosion. Soil color varied from 5YR to 10YR with varying values and chroma. As a whole, the upland soils are red, varying from 5YR to 7.5YR and the low-lying area soils are observed in a dark grayish color (10YR) which may be because of frequent deposition of finer clay particles and submergence as they are vulnerable to seepage water from the nearby streams. While the continuous and huge litter fell from plantations (eucalyptus/tea) and forests, the surface layer of the uplands soils was dark (10YR). A wide range of textural classes were noticed in Vattavada, viz., sandy clay loam, clay loam, silty loam, sand, clay, loamy sand, sandy clay, etc. Though the soils did not show any specific pattern of textural classes, clayey soils were observed in the low-lying and lower parts of side slopes, ascribed to frequent deposition of finer particles. In general, the soils were nongravelly (0%–15%) with a fine- to medium-sized subangular blocky structure, whereas the forest and organic soil layers have a granular or crumby soil structure. Due to high rainfall and humid weather, the soils were devoid of calcareousness (effervescences). Some upland soils were found with an illuvial clay layer (argillic endopedons), and the majority of the low-lying soils were observed with the presence of mottles in the control section. Even in some soils, pressure faces were also noticed owing to the burden of surface soil layers.

Based on soil characteristics, the soils of Vattavada are grouped into six soil series and mapped into 36 soil mapping units (Table 31.1 and Figs. 31.3–31.5).

TABLE 31.1 Soil physicochemical properties of the Vattavada soil series

Soil series	Depth(cm)	Sand(%)	Silt(%)	Clay(%)	pH	OC (%)	CEC
V1	0–21	54.28	17.24	28.48	6.47	1.95	14.15
	21–37	38.69	25.93	35.38	6.94	1.69	17.04
V2	0–12	54.94	10.03	35.03	6.38	1.49	25.8
	12–36	42.92	13.89	43.19	6.4	1.23	25.32
	36–63	43.21	18.61	38.18	6.36	1.07	22.32
V3	0–17	61.68	19.49	18.83	6.78	2.4	14.96
	17–42	59.79	17.68	22.53	7.01	1.61	15.5
	42–64	56.66	23.9	19.44	7.24	1.11	11.0
	64–87	53.93	19.34	26.73	7.31	1.65	12.53
V4	0–20	32.7	21.2	46.1	4.7	1.08	14.16
	20–40	39.39	16.89	43.72	5.47	0.59	12.36
	40–63	52.81	13.12	34.07	5.23	1.78	9.48
	63–89	44.2	19.75	36.05	5.76	1.9	12.0
	89–150	37.7	23.39	38.91	5.85	2.36	15.0
V5	0–16	48.37	15.28	36.35	5.83	0.93	16.94
	16–29	55.1	15.55	29.35	5.72	0.43	13.08
	29–49	43.7	15.45	40.85	5.47	1.39	23.55
	49–77	52.99	18.65	28.36	5.49	1.43	17.4
	77–89	57.88	16.47	25.65	5.34	1.0	11.95
	89–152	54.41	17.81	27.78	5.11	1.82	15.2
V6	0–15	69.95	12.95	17.1	5.04	1.39	8.47
	15–40	59.98	14.71	25.31	5.61	1.43	10.92
	40–61	70.11	3.99	25.9	5.99	0.58	12.0
	61–79	54.98	16.06	28.96	5.89	0.66	12.36
	79–95	64.18	12.35	23.47	5.83	0.58	12.12
	95–112	56.91	15.82	27.27	5.05	0.89	15.2
	112–150	40.71	23.69	35.6	4.87	1.74	19.14

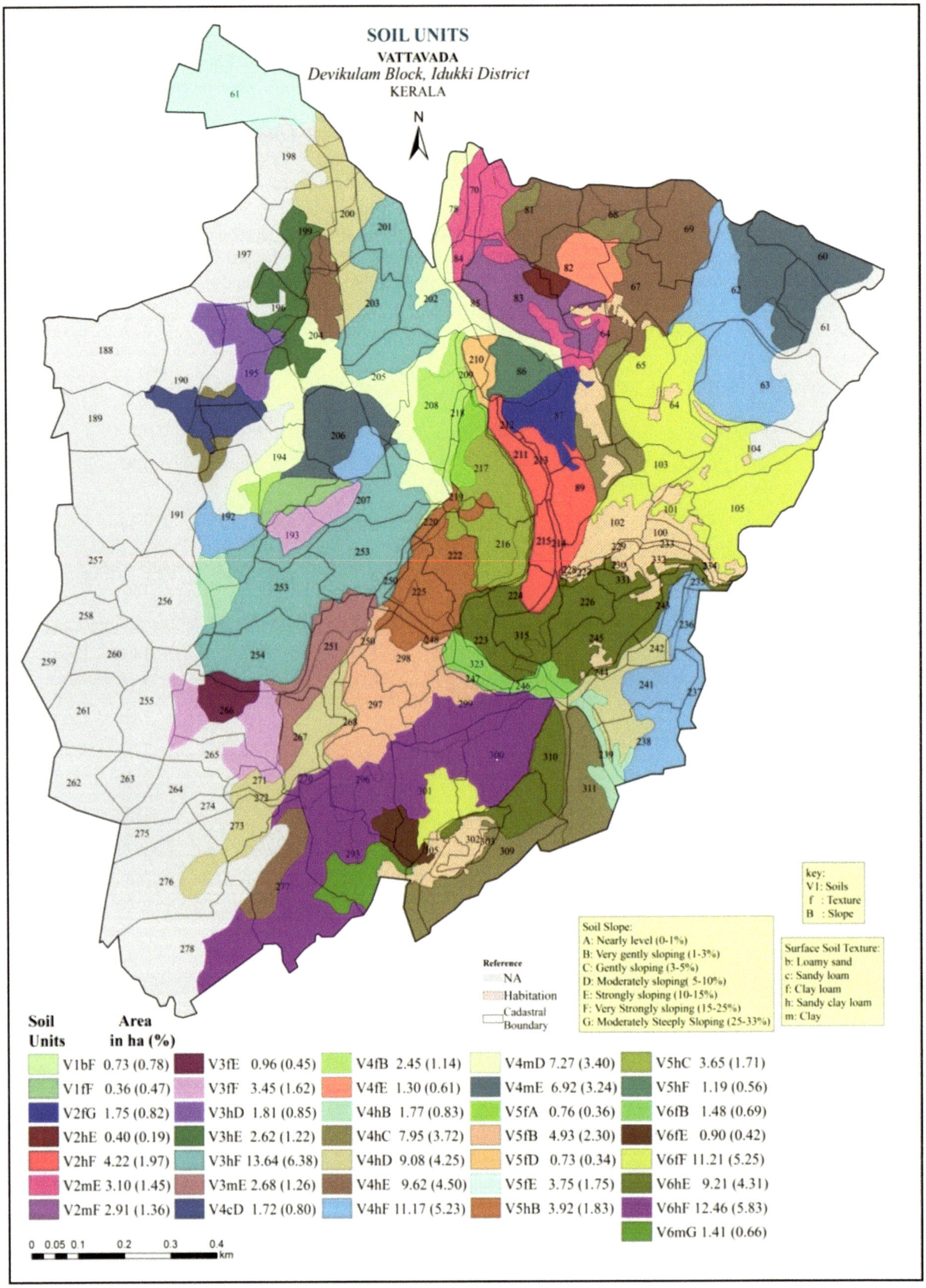

FIGURE 31.3 Soil map of Vattavada village.

FIGURE 31.4 Dominant soils and landscape of the study area.

31.3.1.1 V1: Shallow (25–50 cm) loamy brown soils

V1 soils are shallow, excessively drained soils on sloping lands. The surface soils are characterized by a dark brown color, sandy clay loam texture, and slightly acidic pH. The subsurface soils are very dark brown in color, and clay loam in texture, with a slightly acidic pH. Organic carbon content ranges from 1.69% to 1.95%, and CEC ranges from 14.15 to 17.04 C mol (+) kg^{-1}. These are classified as fine-loamy, mixed, isothermic, Typic Dystrudepts.

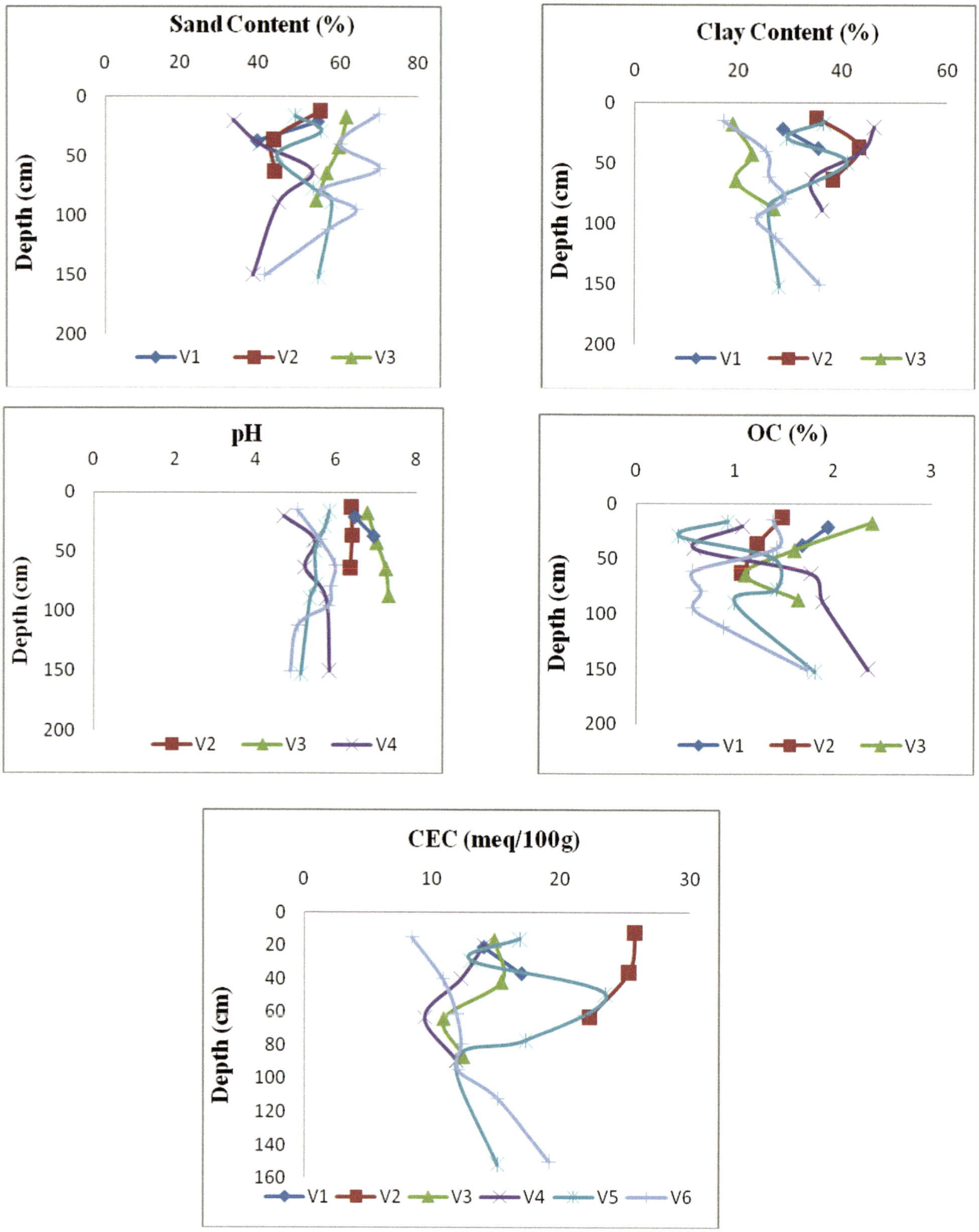

FIGURE 31.5 Depth-wise distribution of the soil properties in soil series identified in Vattavada Gram panchayat.

31.3.1.2 V2: Moderately shallow (50–75 cm) clayey soils

Soils belonging to the V2 series are found in uplands with very steep slopes and excessive drainage. The surface soils are characterized by their brown color, sandy clay texture, and slightly acidic pH. The subsurface soils are very dark brown to reddish brown, clay loam to clay in texture, and have a slightly acidic pH. Organic carbon and CEC range from 1.07% to1.49% and 22.32 to 25.8 C mol (+) kg^{-1} respectively. These are classified as fine-loamy, mixed, isothermic, Typic Dystrudepts.

31.3.1.3 V3: Moderately deep (75–100 cm) loamy soils

V3 soils are very deep, excessively drained soils found in areas with moderately steep to steep slopes. The surface soils are characterized by a dark brown color, a clayey texture, and a very strongly acidic pH. The subsurface soils are dark brown to very reddish brown in color, sandy clay loam to sandy loam in texture, and have a neutral pH. Soil pH is increasing with increasing depth (6.38–7.31). Organic carbon and CEC range from 1.1% to 2.4% and 11.0 to 15.5 C mol (+) kg^{-1}, respectively. These are classified as fine-loamy, mixed, isothermic, Fluventic Dystrudepts.

31.3.1.4 V4: Deep (100–150 cm) loamy soils

V4 soils are very deep soils that are excessively drained and are found in hills and uplands with very steep slopes. The surface soils are characterized by a dark brown color, sandy loam texture, and a strongly acidic pH. The subsurface soils are dark brown to very dark reddish brown in color, sandy clay loam to clay loam in texture, and have a strongly acidic to moderately acidic pH. Similar to V3 soils, the soil pH in V4 soils increases with depth. Organic carbon content ranges from 0.59% to 2.36%. These are classified as clayey, kaolintic, acid, isothermic, subactive, Ustic Kandihumults.

31.3.1.5 V5: Very deep (150+ cm), well-drained fine loamy soils

These soils are found on gently sloping land with good drainage and moderate permeability. The surface soils are characterized by a dark brown color, sandy clay texture, and moderately acidic pH. The subsurface soils are dark brown to very dark reddish brown in color, sandy clay loam to clayey in texture, and have a strongly acidic to moderately acidic pH. These are classified as fine, loamy, mixed, isothermic, Fluventic Dystrudepts.

31.3.1.6 V6: Very deep (150+ cm), imperfectly drained loamy soils

These soils are found in lowlands, having somewhat poor drainage and moderately rapid permeability. The surface soils are characterized by a dark grayish-brown color, sandy loam texture, and a strongly acidic pH. The subsurface soils are very dark grayish brown to black in color and sandy clay loam to clay loam in texture. These are classified as fine-loamy, mixed, isothermic, subactive, Ustic Kandihumults.

31.3.2 Surface physicochemical properties

The pH ranged from 4.45 to 6.94 with a mean value of 5.55, which indicates that the majority of the soils (84.32 ha) are strongly acidic (Table 31.2 and Fig. 31.6). About 33.09% of the soils in Vattavada are moderately acidic. A total

TABLE 31.2 Summary statistics of soil physicochemical properties of Vattavada.

	Soil pH	EC (dS/m)	OC (%)
Minimum	4.45	0.03	0.93
Maximum	6.94	0.49	6.19
Mean	5.55	0.10	2.18
Skewness	0.34	3.74	1.65
Kurtosis	−0.13	15.85	4.03
SD	0.55	0.07	0.89
CV	0.09	0.73	0.41

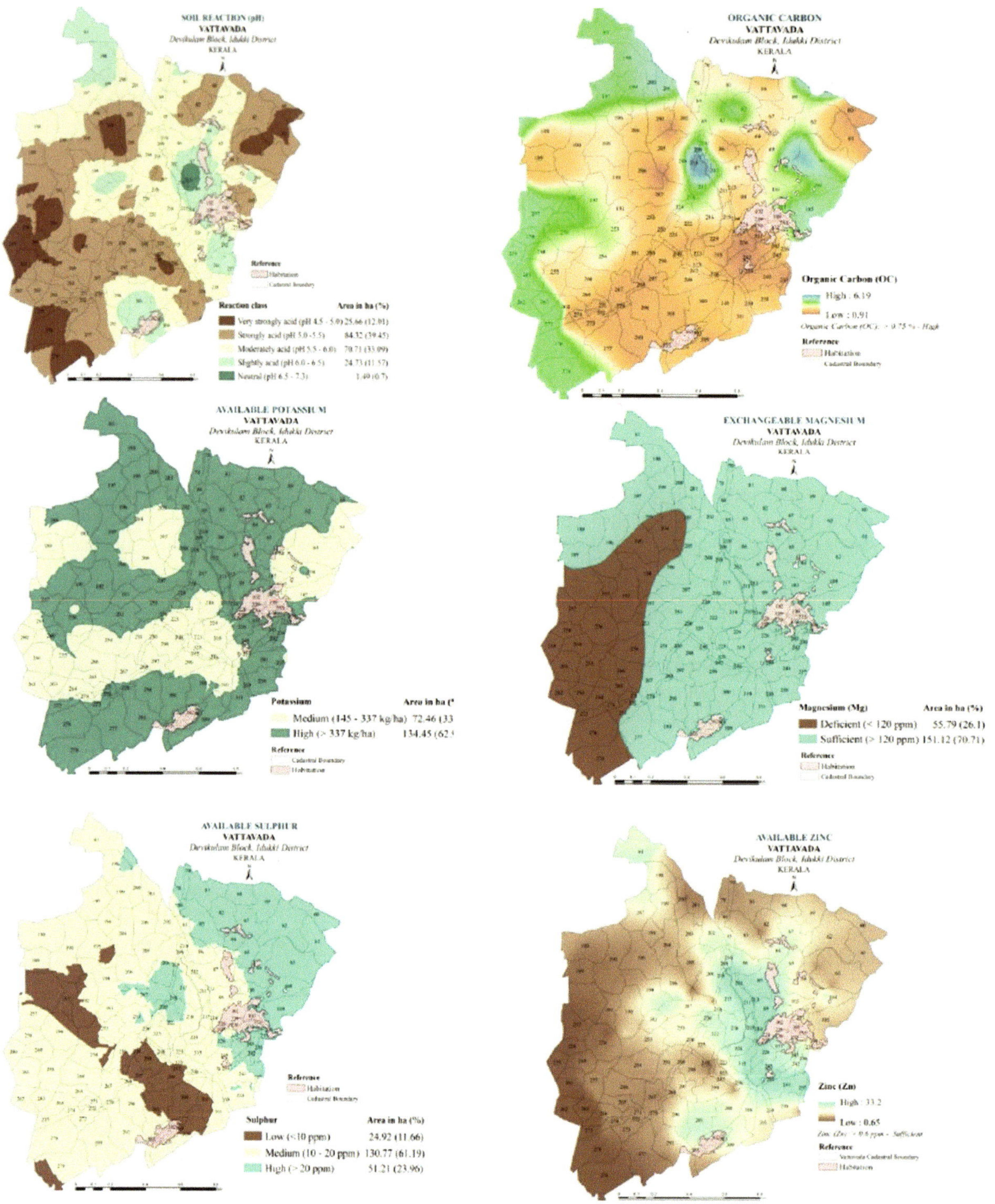

FIGURE 31.6 Thematic maps of the soil nutrient status of Vattavada.

area of 25.66 ha is covered by very strongly acidic soils. The electrical conductivity varied between 0.02 and 0.23 dS/m with a mean value of 0.07 dS/m. The soils on the extreme northeastern stretch of the Vattavada panchayat are estimated to have higher electrical conductivity than the rest of the area. Organic carbon content ranged from 0.93 to 6.19% with a mean value of 2.18%, which signifies that the majority of the soils of Vattavada is rich in organic carbon content (Dharumarajan et al., 2021).

31.3.3 Macronutrients (P, K, S, Ca, and Mg)

The available phosphorous content ranged from 12.05 to 454.70 kg/ha. The mean value is estimated to be 187.87 kg/ha. Available phosphorous content is found to be high in 198.28 ha, medium in 4.36 ha, and low in 4.27 ha. The potassium content varied between 41 and 1035 kg/ha, with an average value of 394.25 kg/ha. 62.91% (134.45 ha) of the total area of Vattavada panchayat has a high potassium content. The sulfur value ranged from 1.66 to 109.16 ppm, with a mean of 18.78 ppm. The available sulfur varied as high, medium, and low at 51.21, 130.77, and 24.91 ha, respectively (Table 31.3). The mean calcium and magnesium content in the soil is estimated to be 1328.69 and 184.15 ppm, and the majority of the soils are sufficient in these nutrients except for the southwestern region of the gram panchayat.

31.3.4 Micronutrients (Cu, Fe, Mn, Zn, and B)

Copper content in the soil ranged from 0.30 to 6.46 ppm, with a mean value of 2.40 ppm. Iron content varies between 10.48 and 109.40 ppm, with an average value of 42.89 ppm. The mean manganese content in the soil is 24.68 ppm, with minimum and maximum values of 6.04 and 79.90 ppm, respectively. The zinc content ranged from 0.65 to 33.20 ppm, with a mean value of 3.74 ppm (Table 31.4). The boron content in the soils varied from 0.02 to 0.51 ppm, with a mean value of 0.27 ppm. The entire Vattavada panchayat soil is found to be sufficient in manganese, iron, copper, and boron (Fig. 31.6).

31.3.5 Land suitability analysis

The Vattavada gram panchayat was found to be suitable for the cultivation of a variety of vegetables and certain fruit crops (mainly banana). The land suitability for a number of the horticulture crops can be attributed to favorable climatic and edaphic factors (Fig. 31.7). An area of about 103.57 ha (47.97%) has soils that are highly suitable (class S1) for the

TABLE 31.3 Summary statistics of soil macronutrients of Vattavada.

	P (kg/ha)	K (kg/ha)	S (ppm)	Ca (ppm)	Mg (ppm)
Minimum	12.05	41.00	1.66	66.00	26.55
Maximum	454.70	1035.00	109.16	2498.46	528.83
Mean	187.87	394.25	18.78	1328.69	184.15
Skewness	0.27	0.86	4.28	0.18	1.34
Kurtosis	−0.79	0.07	29.32	−0.48	1.67
SD	107.18	236.15	12.82	537.91	101.53
CV	0.57	0.59	0.68	0.40	0.55

TABLE 31.4 Summary statistics of soil micronutrients of Vattavada.

	Cu (ppm)	Fe (ppm)	Mn (ppm)	Zn (ppm)	B (ppm)
Minimum	0.30	10.48	6.04	0.65	0.02
Maximum	6.46	109.40	79.90	33.20	0.51
Mean	2.40	42.89	24.68	3.74	0.27
Skewness	0.73	0.92	1.51	4.95	−0.14
Kurtosis	0.28	−0.043	3.01	32.81	0.02
SD	1.29	24.15	14.77	4.09	0.11
CV	0.54	0.56	0.59	1.09	0.40

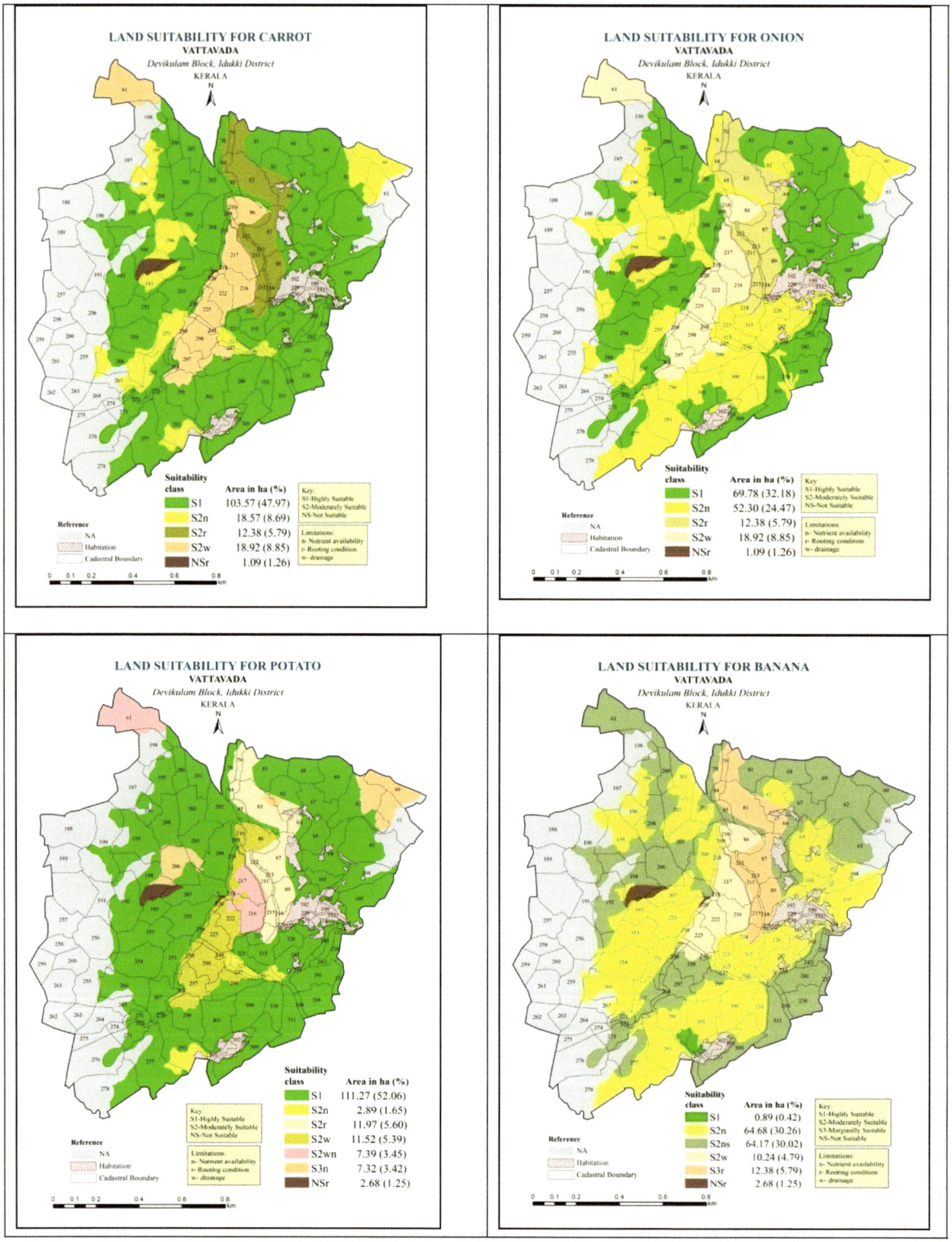

FIGURE 31.7 Soil suitability maps.

cultivation of beetroot and carrot. Under moderately suitable lands (class S2), 18.57 ha (8.64%) have soils with limitations in nutrient availability, 12.38 ha (5.79%) have rooting limitations, and 18.92 ha (8.85%) suffer from drainage limitations that are mainly confined to the center and northern tip of the area. A small area of 1.09 ha (1.26%) with rooting limitations is not suitable for growing beetroot and carrot. For growing crucifers like cabbage and cauliflower and Alliaceae crops like onion and garlic in the Vattavada panchayat, 69.78 ha (32.18%) were found to be highly suitable with minor or no limitations. Under moderately suitable lands (class S2), 52.30 ha (24.47%) have limitations in nutrient availability, 12.38 ha (5.79%) have rooting limitations, and 18.92 ha (8.85%) suffer from drainage limitations. A small area of 1.09 ha (1.26%) with rooting limitations is not suitable for growing cabbage and garlic.

An area of about 69.78 ha (32.18%) in the panchayat has soils that are highly suitable (class S1) for growing green peas. They are distributed mainly in the northeastern, central, and southeastern parts. They have minor or no limitations for growing green peas. About 36.62 ha (17.13%) have soils that are moderately suitable (class S2) with limitations of nutrient availability, 12.38 ha (5.79%) have rooting limitations, and 18.92 ha (8.85%) have drainage limitations. The marginally suitable (class S3) lands cover about 15.68 ha (7.34%) of area, with nutrient availability limitations. A small area of about 1.09 ha (1.26%) is not suitable for growing green peas and occurs in the central part of the panchayat. They have very severe limitations in rooting conditions. A large area of about 112.27 ha (52.06%) has soils that are highly suitable (class S1) for growing potatoes. They are distributed mainly in the northeastern, central, southern, and southeastern parts. They have minor or no limitations for growing potatoes. About 2.89 ha (1.65%) have soils that are moderately suitable (class S2), with limitations in nutrient availability; 11.97 ha (5.60%) have rooting limitations; 11.52 ha (5.39%) have drainage limitations; and 7.39 ha possess both drainage and nutrient limitations. An area of 7.32 ha is found to be marginally suitable with limitation in nutrient availability. A small area of about 2.68 ha (1.25%) has very severe limitations in rooting conditions, is not suitable for potato cultivation, and occurs in the central part of the panchayat. Among the fruit crops, bananas were found to be suitable for cultivation. About 0.89 ha (0.42%) have soils that are highly suitable (class S1) with no limitations for growing bananas. Under moderately suitable (class S2) land, 64.68 ha (30.26%) have soils that have limitations in nutrient availability, 64.13 ha (30.02%) have nutrient and texture limitations, and 10.24 ha (4.79%) have drainage limitations. An area of 12.38 ha is marginally suitable with rooting limitations. A small area of about 2.68 ha (1.25%) has rooting limitations, is not suitable for growing bananas, and occurs in the central part of the panchayat.

31.3.6 Alternate Land Use and Suggestions

Vattavada has seven slope classes, and about 29.45% of the area (69.24 ha) is hilly under the moderately steep sloping class (15%–25%), followed by the rolling or strongly sloping (10%–15%) class in an area of 41.46 ha (19.40%). Moderately sloping lands (5%–10%) covered an area of 20.62 ha (9.65%), and very gently sloping lands (1%–3%) covered an area of 14.54 ha (6.80%). Soils on side slopes are prone to severe erosion and rapid to very rapid runoff if not terraced. Hence, only with suitable soil and water conservation measures should cultivation practices be carried out in this area. This could be done through structural or vegetative soil and water conservation measures. Thus, cultivation of crops can be carried out even on lands that are hilly with 33% slope (steep sloping lands with 25%–33% slope cover an area of 3.16 ha in Vattavada) with suitable interventions such as bench terracing to overcome the limitations of slope. In the area, presently cultivated crops include carrots, potatoes, butter beans, green peas, garlic, beetroot, cabbages, and fruit crops like strawberries, bananas, and passion fruit. The cultivation of these crops can be continued as long as it is carried out in terraced fields. In the case of forests, staggered contour trenching has been advised as the major intervention. Care should be taken that the depth of the trenches should not go beyond the solum depth to the loose saprolite material with free iron and aluminum oxides and kaolin, which can create landslides with the entry of water with great velocity due to gravitation and mass.

Eucalyptus citriodora/grandis plantations do exist at steeper slopes greater than 25%, which in turn will not support any undergrowth or biological activity due to their antagonistic effect on other native flora and fauna. These are very useful for making the water stand for longer before percolation or seepage. Felling the eucalyptus trees and line planting with native nonbranching tree species trailed with pepper or avocado trees in the border area and field crops inside could provide more area under farming and thus could be used for cultivation. However, the tree stumps need to be removed so as to prevent the regrowth of trees. It can be understood that the plantation can only be phased out eventually. Agroforestry with annual crops is advised to be the best option in these areas. Here are certain suggestions to be included for alternate land use options in the study area (Table 31.5).

TABLE 31.5 Alternate land use options for different slope gradient.

Sl. No.	Elevation and slope classes	Alternate land use options	Remarks
1	>33% (very steep slopes)	Bamboo with vetiver Spices: Cardamom in lower terraces	Bamboo can be an ideal crop for areas with steep slopes of >33%. It is ideal for soil erosion control. The dense and widespread root system of bamboos helps filter the water and holds the earth in place. Thus, this plant can serve as a slope stabilizer, preventing soil erosion. Tree-based farming, along with protective cover crops, is an ideal solution. Vetiver can be planted for better ground cover during the initial days of canopy establishment. An added advantage of bamboo is that it is grown as a border crop. It helps in restricting the elephant menace to the borders and not to actual field-cropped areas or agricultural fields.
2	Hilly with steep sloping (25%–33%)	Coffee (Robusta variety in high elevations) along with silver oak, trailed with pepper; spice: pepper and cardamom in lower terraces; agroforestry: bamboo with vetiver or lemongrass; annual crops: potato, carrot, beans, peas, cabbage, beetroot, cauliflower, strawberry, and passion fruit	Coffee, along with silver oak as shade trees, is an example of an agroforestry system. These crops can be cultivated up to a slope of 30%, ensuring proper drainage. Vetiver (*Cenchrus* sp.) and lemongrass (*Citronella* sp.) provide proper ground coverage, ensuring protection of soil from erosion Bench terracing with regular contour trenches
3	Hilly with moderately steep sloping (15%–25%)	Coffee (Robusta and Arabica), along with silver oak and cocoa; spice: pepper and also cinnamon trees; horticultural crops: avocado and citrus; medicinal plants: lemongrass; annual crops: potato, carrot, beans, peas, cabbage, beetroot, cauliflower, strawberry, and passion fruit	Coffee-based agroforestry systems (with shade trees of silver oak trailed with pepper) can be ideal on this slope too. Lemongrass can be grown for ground protection Inaccessible areas with bamboo forestry and lemongrass Bench terracing with regular contour trenches
4	Strongly sloping /Rolling (10%–15%)	Fruit crops: banana, plum, avocado, passion fruit; spice crops: turmeric; medicinal plant: lemongrass; lower terraces: pineapple; annual crops: potato, carrot, beans, peas, cabbage, beetroot, cauliflower, strawberry, and passion fruit	Fruit crops like avocado and passion fruit are highly promising crops, with pineapple in lower tier Lemongrass can be grown for ground protection Spices: ginger and turmeric only within protected borders with tree species Bench terracing with regular contour trenches
5	Moderately sloping (5%–10%)	Fruit crop: mango, apple, citrus crops, like oranges, and lemons, pomegranate, jackfruit, passion fruit, grapes, and pineapple; spice: ginger and turmeric; medicinal plant: lemongrass; annual crops: potato, carrot, beans, peas, cabbage, beetroot, cauliflower, strawberry, and passion fruit	These fruit crops can be grown up to 10% slope. Ginger can be cultivated to an 8% slope only inside protected borders with tree species. Ginger and lemongrass can be better ground cover factors in controlling soil erosion Pineapple can also be cultivated on this slope in contour trenches, as its cultivation includes minimum practices and thereby less disturbance to the soil. However, in the initial days, there needs to be proper ground cover. Bench terracing with regular contour trenches

(*Continued*)

TABLE 31.5 (Continued)

Sl. No.	Elevation and slope classes	Alternate land use options	Remarks
6	Gently sloping (3%–5%)	Fruit crops: mango, jackfruit, banana, guava, sapota, and pineapple; annual crops: potato, carrot, beans, peas, cabbage, beetroot, cauliflower, strawberry, and passion fruit	Along the terraces, cultivation of these crops can be continued. Pineapple can also be cultivated on this slope in contour trenches, as its cultivation includes minimum practices and thereby less disturbance to the soil. However, in the initial days, there needs to be proper ground cover.
7	Very gently sloping (1%–3%) and nearly level (0%–1%)	Potato, carrot, beans, peas, cabbage, cauliflower, beetroot, and strawberry.	Along the small ridges and furrows made, cultivation of these crops can be continued. Field crops with borders planted with plantation crops: coconut, areca nut, or nutmeg, if there is no water shortage and proper drainage is possible. If not, put ginger, turmeric, pineapple, and bamboos on borders to protect from wild animals, and lemongrass for binding soil against runoff

31.4 Recommendations for further research or studies

1. Explore and promote a multidisciplinary approach to landslide risk reduction and practice.
2. Map and document all disasters in the area, assess trends and vulnerabilities, and recommend mitigation measures. Hazard mapping should be done to locate areas commonly prone to landslides.
3. The relationship between rainfall and landslides in the area needs to be further analyzed and studied.
4. A well-validated rainfall threshold model is yet to be developed for all critical regions. For building a rainfall threshold–based model, intensity duration-based thresholds, dates of past landslide events, and corresponding representative daily/hourly rainfall data are required. In India, except for some cases, these datasets are not easily available for most of the mountainous regions.
5. The threshold model itself does not provide information on the spatial occurrence of potential landslides; it has to be combined with landslide susceptibility to forecast the spatiotemporal initiation of landslides.
6. Information on the precise time of a landslide based on instrumentation and real time monitoring is mostly lacking.

Funding

The authors thank the GOI-GEF-UNDP for funding the activity "Spatial Crop Planning for Sustainable Resource Use and Conservation of Ecological Resource Use"

References

Baltodano, J., 2000. Monoculture forestry: a critique from an ecological perspective. Tree Trouble: a Compilation of Testimonies on the Negative Impact of Large-scale Monoculture Tree Plantations Prepared for the 6th COP of the FCCC. Friends of the Earth International, pp. 2–10.

Bowyer, J.L., 2006. Forest plantations threatening or saving natural forests? Arborvitae (IUCN/WWF For. Conserv. Newsl.) 31, 8–9.

Dharumarajan, S., Kalaiselvi, B., Suputhra, A., Lalitha, M., Vasundhara, R., Kumar, K.A., et al., 2021. Digital soil mapping of soil organic carbon stocks in Western Ghats, South. India. Geoderma Reg. 25, e00387.

Evans, J., 2001. Biological Sustainability of Productivity in Successive Rotations (Working Paper No. FP/2). Mead, D.J. (Ed.). FAO, Rome, Italy.

Jones, S., Kasthurba, A.K., Bhagyanathan, A., Binoy, B.V., 2021. Impact of anthropogenic activities on landslide occurrences in southwest India: an investigation using spatial models. J. Earth Syst. Sci. 130, 1–18.

Liu, C.L.C., Kuchma, O., Krutovsky, K., 2018. Mixed-species versus monocultures in plantation forestry: development, benefits, ecosystem services and perspectives for the future. Glob. Ecol. Conserv. 15, 1–13.

Lalitha, M., Kumar, K.S.A., Nair, K.M., Dharumarajan, S., Koyal, A., Khandal, S., et al., 2021. Evaluating pedogenesis and soil Atterberg limits for inducing landslides in the Western Ghats, Idukki District of Kerala, South India. Nat. Hazards 106, 487–507. Available from: https://doi.org/10.1007/s11069-020-04472-0.

Martha, T.R., Roy, P., Khanna, K., Mrinalni, K., Vinod Kumar, K., 2019. Landslides mapped using satellite data in the Western Ghats of India after excess rainfall during August 2018. Curr. Sci. 117, 804. Available from: https://doi.org/10.18520/cs/v117/i5/804-812.

Menon, S., Bawa, K.S., 1998. Deforestation in the tropics: reconciling disparities in estimates for India. Ambio, 27, 567–577.

Naidu, L.G.K., Ramamoorthy, V., Chella, O., Hegde, R., Krishnan, P., 2006. Manual Soil Site Suitability Criteria for Major Crops, NBSS&LUP, Nagpur.

Murugesan, J., Gangai, P., Selvam, K., 2018. Patterns of crop concentration, crop diversification and crop combination in Tiruchirappalli District, Tamil Nadu. IJIRST 4. Available from: https://bit.ly/3cNTFSZ.

Rahman, M.N., 2020. Crops pattern change and agricultural diversification: a case study of Domar Upazila, Nilphamari. Int. J. Agric. Sc. Food Technol. 6 (1), 022–029.

Ross, C., 2017. Colonialism, rubber, and the rainforest. Ecology and Power in the Age of Empire: Europe and the Transformation of the Tropical World (Oxford) 99–135. Available from: https://doi.org/10.1093/acprof:oso/9780199590414.003.0004.

Sidle, R.C., Ziegler, A.D., Negishi, J.,N., Nik, A.R., Siew, R., Turkelboom, F., 2006. Erosion processes in steep terrain—truths, myths, and uncertainties related to forest management in Southeast Asia. For. Ecol. Manag. 224, 199–225.

Smethurst, P.J., Almeida, A.C., Loos, R.A., 2015. Stream flow unaffected by Eucalyptus plantation harvesting implicates water use by the native forest streamside reserve. J. Hydrol.: Reg. Stud. 3–187–19.

Sprenger, M., Oelmann, Y., Weihermuller, L., Wolf, S., Wicke, W., Potvin, C., 2013. Tree species and diversity effects on soil water seepage in a tropical plantation. For. Ecol. Manag. 309, 76–86.

Wishnie, M.H., Dent, D., Mariscal, E., Deago, J., Cedeno, N., Ibarra, D., et al., 2007. Initial performance and reforestation potential of 24 tropical tree species planted across a precipitation gradient in the Republic of Panama. For. Ecol. Manage. 243, 39–49.

Chapter 32

Prediction of soil nutrients using visible-near-infrared reflectance laboratory spectroscopy

C.G. Kusuma[1], S.A. Bhoomika[1] and S. Dharumarajan[2]

[1]*Department of Soil Science and Agricultural Chemistry, University of Agricultural Sciences, Bangalore, Karnataka, India,* [2]*ICAR-National Bureau of Soil Survey and Land Use Planning, Regional Centre, Bangalore, Karnataka, India*

32.1 Introduction

Soil is a dynamic natural body that is composed of organic and inorganic constituents that control agricultural productivity and ecosystem functioning with their physicochemical and nutrient properties, which show a large variability from site to site or even within the same field. Hence, there is a need for quantitative and qualitative estimation of soil components, but measuring the soil's physical and chemical properties using conventional methods is an expensive, time-consuming, and energy-consuming process that involves the use of chemicals that are harmful to human health and the environment (Conforti et al., 2018; Adeline et al., 2017). Thus, various studies have stressed the requirement for a fast, accurate method that would make the analysis of samples feasible.

Visible-near-infrared reflectance (Vis-NIR) spectroscopy is a physical, nondestructive, rapid, and reproducible method that provides inexpensive prediction of soil physical, chemical and biological properties according to their reflectance in the wavelength range from 400 to 2500 nm (Reeves et al., 2001; Dunn et al., 2002). Reflectance signals are produced by the vibration of bonds between atoms of C, H, O, N, P and S, as well as by weak overtones and combinations of fundamental vibrations caused by stretching and bending of functional groups OH and NH, which dominate the Vis-NIR region of the electromagnetic spectrum. Such NIR spectroscopy absorption reflects the molecular structure and composition, which gives a theoretical basis for quantitative and qualitative analysis of the physical and chemical composition of soil samples (Bo et al., 2010).

Several preprocessing techniques are employed to improve the efficiency of soil property prediction with Vis-NIR spectral data and these techniques are used to remove noise, transform soil spectra, emphasize spectral features and extract useful information for quantitative predictive models. The preprocessing of spectra includes smoothing, normalization, scatter correction, and continuum removal (Dotto et al., 2018). Preprocessing methods like Savitzky–Golay (SG) smoothing, first derivative (FD), multiplicative scatter correction (MSC), standard normal variate (SNV) transformation, etc. are used.

Many multivariate methods for soil property prediction have been used in the development of higher-quality models. Among multivariate methods, PLSR models are rapid, easy to derive and interpret, insensitive to collinearity (Ghosh et al., 2020), and relatively easy to understand the principles behind the relationship between soil properties and the spectral contribution to the regression models.

32.1.1 Importance and need of spectroscopic studies

Knowledge of soil properties and mapping is one of the keys to decision-making in precision agriculture and also to maintaining a sustainable environment. Soil variability due to climatic changes, crop management practices, etc. is one of the reasons for soil erosion and land degradation. These need real-time assessment for its management and maintain soil health. Chemical and physical characteristics are important in evaluating soil fertility and understanding soil dynamics (Cozzolino et al., 2003). Many individual studies have shown the ability of Vis-NIR spectra to provide reliable information on different physical, chemical and biological soil properties. (Nawar et al., 2019; Shi et al., 2013).

Remote Sensing of Soils. DOI: https://doi.org/10.1016/B978-0-443-18773-5.00001-6

Soil assessment using visible, near infrared, and shortwave infrared spectroscopy, although it cannot replace laboratory chemical analysis, is a fast, cost-effective, environmentally friendly, nondestructive, reproducible, pollution-free, and repeatable analytical technique (Nocita et al., 2015). In order to use time and resources judiciously and effectively, spectroscopy offers a viable alternative. Also, it plays a catalytic role in soil studies by accelerating the rate of soil analysis and increasing the accuracy of assessment. The availability of commercial spectroscopic equipment and software packages for multivariate calibration has made Vis-NIR spectroscopy widely used for soil characterization and soil parameter prediction purposes. The rapid development of portable and handheld spectroradiometer allows the analyses to be conducted in situ (Cozzolino et al., 2013). Prediction methods based on diffuse reflectance spectroscopy involve minimal or no sample preparation, and avoid the use of environmentally harmful extractants in the laboratory (Reeves et al., 1999); they are also easily adaptable for proximal sensing with Vis-NIR.

Therefore, in this chapter, we have made an attempt to review the potential of the prediction of soil nutrients using Vis-NIR laboratory spectroscopy with the following objectives:

1. To list the need to use spectroscopic techniques in soil analysis
2. To highlight the potential of Vis-NIR spectral data in predicting soil physiochemical, macronutrient, and micronutrient availability.
3. To convey the major limitations in adapting spectroscopic studies
4. To suggest the way forward for progress in laboratory spectroscopic studies

32.1.2 Practical implications of visible-near-infrared reflectance spectroscopy

Implementing the Vis-NIR model will help conduct a soil nutrient content survey that can be used to develop spot fertilization in cropland (Al-Mashhadani and Chandrasekaran, 2020). It can also be applied to conduct classification, localization, and mapping on land to determine the suitability of the land and its plants (Chebrolu et al., 2017). It can be used in determining the cropland's ground-bearing capacity for selecting the appropriate type of tractor, tillage, and energy resources in the cropland (Fernandez et al., 2015). Accurate models can be embedded in agricultural machines to carry out precise agricultural processes and minimize the wasteful use of resources to increase food production (Bulan and Sitorus, 2022). Vis-NIR spectroscopy has also been used for general assessment of soil organic matter composition, soil quality, fertility, mapping, and classification (Viscarra- Rossel et al., 2010).

32.2 Methodology

In this chapter, we have reviewed around 35 research publications on spectroscopic studies of soils from different regions in laboratories using the Vis-NIR region of electromagnetic spectra. Unique keywords were defined, and the most common scientific search engines were used: Science Direct, Google Scholar, and Web of Science. In the review of publication, emphasis was given to:

- different soil properties
- spectral range (Vis-NIR)
- instrument: spectroradiometer and its model
- preprocessing technique employed
- models used for prediction
- statistical parameters used in accuracy assessment.

32.2.1 Sample preparation

To optimize the reproducibility and accuracy of spectroscopic predictions, sample preparation is needed. The collected soil samples are sieved to <2 mm particle diameter. A substantial effect was found in spectral absorbance on grinding the soil aggregates. The accuracy of infrared predictions for the same soil properties varies with the size of particles (Nduwamungu et al., 2009).

32.2.2 Instrumentation

Under standard conditions, the prediction of the same soil properties can vary according to the instrument used due to its specification differences like instrument stability, resolution, etc. (Reeves, 2010). Most of the researchers have used low-cost portable instruments for Vis-NIR, with noise and wavelength range limitations. The diffuse reflectance infrared Fourier transform is the most commonly used technique in soil analysis (Kuang et al., 2012). When the radiation is directed onto the soil sample, radiation of a given wavelength gets absorbed, reflected, or scattered to different degrees. The dependence of the extent of absorbance or reflectance on the soil composition forms the basis of the prediction of soil property status by employing diffuse reflectance spectroscopy.

32.2.3 Spectral data preprocessing

To reduce the noise-to-signal ratio, normalization and stabilization of soil spectral data during preprocessing become important. Some commonly followed preprocessing techniques that we found while reviewing are as follows:

Reflectance (uncorrected), SNV, and MSC were used to transform the data. SG smoothing with FD and second derivative helps in reducing baseline variation, and this technique also increases the resolutions of the spectral peak (Ghosh et al., 2020). The SNV method is useful in removing the multiplicative interference, and the MSC technique helps in eliminating additive or multiplicative signal effects (Dotto et al., 2018).

32.2.4 Spectroscopic modeling

To model the relationship between spectral signatures and soil properties, multivariate regression methods were used. Some of the multivariate techniques like partial least squares regression (PLSR), multivariate adaptive regression splines (MARS), random forest (RF) (Lalitha et al., 2022; Dharumarajan et al., 2023), support vector machines (SVM) (Dharumarajan et al., 2021; Mahajan et al., 2021), boosted trees, principal component regression, artificial neutral network (ANN) (George et al., 2020), and multiple linear regression are popularly used for soil quantitative analysis. Among these, the PLSR model is the most popular method in quantitative soil analysis (Reeves et al., 1999; Viscarra Rossel et al., 2006) as it is rapid, easy to derive and interpret, provides easy access to software, facilitates an easy understanding of the relationship between soil properties and spectral data, and is insensitive to collinearity (Haaland and Thomas, 1988). Modified PLSR (Dinakaran et al., 2016) is found to be a more stable and accurate model than the other standard PLSR algorithms for agriculture applications.

32.2.5 Model evaluation

Model accuracy and precision are assessed using statistical parameters like the coefficient of determination (R^2), bias-mean error, root-mean-square error (RMSE), the ratio of SD of the reference soil property values to RMSE, the ratio of performance to deviation (RPD), and the concordance correlation coefficient (CCC). According to ratio of performance to interquartile distance (RPIQ), the higher the RPIQ value, the better the model's ability to predict. It considers both prediction errors and observed value fluctuations. Among all, R^2 is commonly reported in many studies, and it is used as a crude measure in the prediction of precision. It is classified as excellent ($R^2 > 0.95$), successful ($0.95 \leq R^2 \leq 0.90$), moderately successful ($0.90 \leq R^2 < 0.80$), and moderately useful ($0.80 \leq R^2 < 0.70$) (Malley et al., 2004). Likewise, in the review process, we found the classification range based on RPD denotes >2 as good, 1.4–2 as acceptable, and <1.4 as poor and RPIQ should fall between 1.5 and 3 for an acceptable model.

The following equations were used in the evaluation of the model.

$$R^2_{\mathrm{val}} = 1 - \frac{\sum_{(i=1)}^{n} (pi - oi)^2}{\sum_{(i=1)}^{n} (oi - \overline{oi})^2}$$

$$ME_{\mathrm{val}} = \frac{1}{n}\sum_{i=1}^{n} (oi - pi)$$

$$RMSE_{\mathrm{val}} = \sqrt{\frac{1}{n}\sum_{i=1}^{n} (oi - pi)^2}$$

where $pi \wedge oi$ are the predicted and observed values and $\overline{pi} \wedge \overline{oi}$ are the means of the predicted and observed values.

$$\mathrm{CCC} = \frac{2\rho\sigma_0\sigma_p}{\sigma_{0+}^2 \sigma_p^2 + \left(\mu_0 - \mu_p\right)^2}$$

μ_0 and μ_p are the means of observed and predicted values and $\sigma_0^2 and \sigma_p^2$ are corresponding variances; ρ is Pearson correlation coefficient between observed and predicted values.

$$\text{Ratio of performance to deviation (RPD)} = \frac{\text{SD}}{\text{RMSE}}$$

where SD is the standard deviation of observed values.

$$\text{Ratio of Performance to Interquartile Distance } (\text{RPIQ}_{\text{val}}) = \frac{1\text{Q}}{\text{RMSE}}$$

where IQ is the difference between the third quartile Q3 and the first quartile Q1. A larger RPIQ value indicates improved model performance.

32.3 Results and discussion

We have tried to review some of the studies conducted on the prediction of soil nutrients using Vis-NIR laboratory spectroscopy, and presented a few highlights for soil properties like:

- Physiochemical properties include soil pH, EC, CEC, and SOC, which are soil main attributes (Table 32.1).
- Soil macronutrients like N, P, K, Ca, Mg, and S (Table 32.2).
- Soil micronutrients like Fe, Mn, Cu, Zn, B, Co, Mo, Cl, and Ni (Table 32.3).

TABLE 32.1 **Selected articles present the application of different methods in the estimation of soil physiochemical properties using Vis-NIR spectroscopy.**

Author	Soil parameter	Method of statistic	Statistical parameter
Reeves et al. (2001)	SOC	PLSR	$R^2 = 0.46–0.98$
Dunn et al. (2002)	(Top soil) pH EC ESP CEC SOC (Subsoil) pH EC ESP CEC	PLSR	$R^2 = 0.8$, RPD = 2.3 $R^2 = 0.37$, RPD = 1.2 $R^2 = 0.68$, RPD = 1.8 $R^2 = 0.9$, RPD = 3.3 $R^2 = 0.66$, RPD = 1.7 $R^2 = 0.71$, RPD = 2.0 $R^2 = 0.63$, RPD = 1.6 $R^2 = 0.8$, RPD = 2.2 $R^2 = 0.79$, RPD = 2.1
Terhoeven-Urselmans et al. (2008)	pH SOC C:N	Modified PLSR	$R^2 = 0.93$ $R^2 = 0.97$ $R^2 = 0.95$
Debaene et al. (2014)	SOC	PLSR	$R^2 = 0.76$, RMSEP = 0.04, RPD = 1.80
Gupta et al. (2018)	SOC	PLSR	$R^2 = 0.59–0.7$
Vibhute et al. (2018)	SOCpH	PLSR	$R^2 = 0.89$ $R^2 = 0.82$, RMSEP = 2.12
Gomez et al. (2008)	SOC	PLSR	RPD = 1.43
Askari et al., (2015)	SOC C:N	PLSR	$R^2 = 0.94$ $R^2 \geq 0.7$
Amin et al. (2020)	SOC	PLSR	$R^2 = 0.85$
George et al. (2020)	SOC	ANN	$R^2 = 0.92$, MSE = 0.24
Moura-Bueno et al. (2021)	SOC	Model 1–6	$R^2 = 0.37–0.88$
Mozaffari et al. (2022)	pH EC	PLSR and SMLR	$0.65 \leq R^2 \leq 0.71$ $0.55 \leq R^2 \leq 0.63$

TABLE 32.2 Selected articles present the application of different methods in the estimation of macronutrient concentrations in soil using Vis-NIR spectroscopy.

Authors	Method of statistic	Nutrients	Statistical parameters
Confalonieri et al. (2001)	MPLS-R	P K	$R^2 = 0.23–0.37$ $R^2 = 0.48–0.82$
Chang and Laird (2002)	PLS	N	$R^2 = 0.85$
Dunn et al. (2002)	PLSR	Ca Mg Na	$R^2 = 0.86$ $R^2 = 0.85$ $R^2 = 0.64$
He et al. (2007)	PLS	N P K	$R^2 = 0.93$ $R^2 = 0.47$ $R^2 = 0.68$
Terhoeven-Urselmans et al. (2008)	Modified PLSR	N P K Mg	$R^2 = 0.97 R^2 = 0.81 R^2 = 0.92 R^2 = 0.88$
Reeves et al. (2009)	PLS	P K Ca Mg S	$R^2 = 0.099$ $R^2 = 0.624$ $R^2 = 0.807$ $R^2 = 0.937$ $R^2 = 0.646$
Kadupitiya et al. (2010)	SMLR	N P K	$R^2 = 0.66–0.78$ RPD = 1.15 to 2.45
Shao et al. (2011)	LS-SVM	P K	$R^2 = 0.69$ $R^2 = 0.69$
Luce et al. (2012)	PLSR	N	$R^2 = 0.62$
Shaddad et al. (2016)	PLSR	P K	$R^2 = 0.77$ $R^2 = 0.55$
Ray et al. (2017)	SMLR	K	$R^2 = 0.81$
Vibhute et al. (2018)	PLSR	N P K	$R^2 = 0.68$ $R^2 = 0.98$ $R^2 = 0.82$
Kulig et al. (2020)	PLSR	P K Mg	$R^2 = 0.48$, RRMSE = 19% $R^2 = 0.54$, RRMSE = 7.25 $R^2 = 0.51$
Mahajan et al. (2021)	PLSR, PCR, MARS, SVR and RF	N P K Ca Mg S	RPD = 1.48–2.06
Mondal et al. (2021)	RF	S	$R^2 = 0.71$ RMSE = 8.86 RPD = 1.18 RPIQ = 1.69
Mozaffari et al. (2022)	PLSR, SMLR	K Mg Ca	$0.65 \leq R^2 \leq 0.71$ $0.65 \leq R^2 \leq 0.71$ $0.55 \leq R^2 \leq 0.63$

TABLE 32.3 Selected articles present the application of different methods in the estimation of micronutrient concentrations in soil using Vis-NIR spectroscopy.

Authors	Nutrients	Method of statistic	Statistical parameter
Reeves et al. (2009)	Al Co Cu Fe Mn Mo Ni Zn	PLS	$R^2 = 0.515$, RPD = 1.4 $R^2 = 0.910$, RPD = 3.3 $R^2 = 0.358$, RPD = 1.2 $R^2 = 0.594$, RPD = 1.6 $R^2 = 0.370$, RPD = 1.3 $R^2 = 0.199$, RPD = 1.1 $R^2 = 0.994$, RPD = 13.4 $R^2 = 0.155$, RPD = 1.1
Minasny et al. (2009)	Cl	PLS	$R^2 = 0.76$, RPD = 2.0
Kadupitiya et al. (2010)	Fe Mn Zn Cu	SMLR	$R^2 = 0.66–0.78$ RPD = 1.15 to 2.45
Sarathjith et al. (2016)	Zn Fe B Al	DWT-SVR	RPD = 1.49–1.41 RPD = 2.43–2.91 RPD = 1.63
Gupta et al. (2018)	Fe Zn	PLSR	$R^2 = 0.62–0.74$ $R^2 = 0.38–0.47$
Vibhute et al. (2018)	Fe	PLSR	0.94
Mahajan et al. (2021)	Zn B	PLSR, PCR, MARS, SVR and RF	RPD = 1.48–2.06
Mozaffari et al. (2022)	Fe Mn Cu Zn	PLSR and SMLR	$0.65 \leq R^2 \leq 0.71$

32.3.1 Physiochemical properties

The soil organic carbon pool often affects other soil properties such as soil pH, clay flocculation, available water, etc.; hence, its study, mapping, and monitoring have always been given priority. In the works we have reviewed, it shows that the PLSR method was considered a common algorithm for predicting SOC with different soil types, different data preprocessing, and different model evaluation methodologies. Many studies have been conducted on the application of Vis-NIR spectral techniques to soil organic carbon, as depicted in Table 32.1. The results showed that SOC was predicted with high accuracy ($R^2 = 0.98–0.89$ and RPD = 1.43–1.8) (Reeves et al., 2001; Gomez et al., 2008; Terhoeven-Urselmans et al., 2008; Askari et al., 2015; Moura-Bueno et al., 2021) and medium accuracy ($R^2 = 0.59–0.76$) because of low availability, and the effect of environment covariate prediction was found to be comparatively lower.

Cation exchange capacity is an important property that affects nutrient availability, soil buffering capacity, and soil fertility. In our review, the inconsistent results were found with $R^2 = 0.34–0.9$ (Dunn et al., 2002), which may be attributed to chemical methods that are used in the laboratory to determine CEC. Low R^2 values are the result of an insufficient number of samples representing specific locations (Viscarra Rossel et al., 2006)

Soil pH can be predicted using Vis-NIR spectroscopy by correlating it with other soil properties like SOM and clay. In many research works, the soil pH was predicted with high accuracy ($R^2 = 0.8–0.93$) (Terhoeven-Urselmans et al., 2008) and acceptable accuracy ($0.65 \leq R^2 \leq 0.71$) (Mozaffari et al., 2022). Thus, soil pH can be predicted through indirect correlation methodology, although such correlations can change from one set of soil to another, requiring the development of local regression models (Soriano-Disla et al., 2014).

Electrical conductivity is a major parameter that indicates soil salinity. It is known that the concentration of electrically charged water-soluble salts influences plant growth and soil structure (Rayment and Lyons, 2011). It may be influenced by soil and climatic properties. The results show poor prediction accuracy ($R^2 = 0.37$) (Dunn et al., 2002), which is similar to the results obtained by Viscarra Rossel et al. (2006), in which it is said that for soil properties like EC and SAR, as they are based on the soil solution rather than the soil matrix, the accurate prediction will be a difficult task.

32.3.2 Soil macronutrients

There are limited studies on the characterization of soil fertility properties by diffuse infrared spectroscopy.

Soil nitrogen is an important parameter determining agricultural production. In the study conducted by Luce et al. (2012), it was found that the N supply was predicted with marginal accuracy with $R^2 = 0.62$. And high accuracy was found in studies of Terhoeven-Urselmans et al. (2008) and He et al. (2007) with $R^2 = 0.86-0.97$. And it was found that the accuracy of prediction was nonreliable to acceptable based on RPD values (RPD = 1.48−2.06). In 2021, Mahajan and coworkers made a comparison where, relative to total nitrogen (N) determinations, there were limited data available for nitrogen fractions, including organic and mineral N fractions. However, mineral N prediction using MIR was effectively demonstrated by Soriano-Disla et al. (2014).

As reported by Maleki et al., (2006), Phosphate ions are not spectrally active in Vis-NIR region of the spectrum. Thus, the model for total P depends on the correlations with IR-active soil compounds. The availability of P in soil depends on its sorption and also soil pH; from the results of the reviews, we have found that, because of these reasons, the prediction of available soil P was poor in accuracy with $R^2 = 0.23-0.47$ in most of studies (Confalonieri et al., 2001; He et al., 2007; Reeves et al., 2009). Also, sufficiently high accuracy was reported in studies of Terhoeven et al. (2008), Vibhute et al. (2018), Mahajan et al. (2021).

In the study, poor results were observed in the prediction of exchangeable K with $R^2 = 0.55$ (Kulig et al., 2020; Shaddad et al., 2016). However, better predictions were found with $R^2 = 0.68$ (He et al., 2007; Reeves et al., 2009; Shao et al., 2011) and the best results were observed with $R^2 = 0.82-0.92$ (Terhoeven-Urselmans et al., 2008; Vibhute et al., 2018). The better results, including in the NIR region, may be attributed to the absorbance of illite clay minerals at NIR frequencies.

Ca in soil is available in different forms from different sources, and these sources are characterized using strongly infrared-active materials; hence, we can see a good model prediction (Soriano-Disla et al., 2014). A good prediction of Ca content in soil was observed with R^2 in the range of 0.86−0.9 (Dunn et al., 2002).

The spectral signatures of Mg are mostly due to Ca−Mg−carbonates (dolomite), which is present as a hydrated cation in heavy clays and substituted for Al in the octahedral structures in mica (Soriano-Disla et al., 2014). Mg was predicted with high accuracy, with $R^2 = 0.937-0.85$ (Reeves et al., 2009; Dunn et al., 2002), and low accuracy was reported by Kulig et al., 2020; Mozaffari et al., 2022 with $R^2 = 0.51-0.71$.

Very few studies are reported on the prediction of S. Mondal et al., 2021 and Reeves et al., 2009 in their studies found medium predictability with $R^2 = 0.646-0.7$, RMSE = 8.86, and RPD = 1.18, and (Mahajan et al., 2021) reported comparatively higher predictability of S using Vis-NIR with RPD = 1.48−2.06.

32.3.3 Soil micronutrients

Only a few articles are reported on soil micronutrient prediction by spectroscopic studies. As these micronutrient concentrations in soil are low, the spectral peaks of these small concentrations may go unnoticed. Among all the total Cu, Fe, and Zn, extractable Fe and Mn are found in most of the research works. As the concentrations of elements like Co, Zn, Mn, Cu, and Ni are in relation to carbonate and iron minerals, they can be predicted by correlating the values (Wu et al., 2005; Wu et al., 2007).

Due to the formation of inactive ionic salts like NaCl through various factors like rainwater, fertilizers, and saline, or sodic conditions, the total Na prediction by NIR spectroscopy becomes difficult. The study conducted by Soriano-Disla et al. (2014) showed moderately useful predictions of Na content with $R^2 = 0.75$, RPD = 2.

The Vis-NIR resulted in the best predictions for Co and Ni with $R^2 = 0.910$ and RPD = 3.3 and $R^2 = 0.994$ and RPD = 13.4, respectively (Reeves et al., 2009), and Cu, Fe, Zn, and Mn, with $0.65 \le R^2 \le 0.71$ (Mozaffari et al., 2022). Predictions were medium for extractable boron, with RPD = 1.48−1.63 (Sarathjith et al., 2016; Mahajan et al., 2021). Moderately useful for Cl by using MIR with $R^2 = 0.76$, RPD = 2.0 (Minasny et al., 2009) and Al using NIRS with $R^2 = 0.515$, RPD = 1.4.

By this, we can conclude that predicted Vis-NIR reflectance data based on soil physiochemical, macronutrient, and micronutrient properties can be a promising technique with advantages like cost and time efficiencies; also, the environment's health can be maintained with less use of chemicals, and it also aids in soil mapping for larger areas.

32.4 Limitations of visible-near-infrared reflectance spectroscopy

Soil property prediction through Vis-NIR spectral data is an indirect method, and there is a need to develop a calibration model with respect to the reference method. This requires a large number of samples to fit all the variability in the spectra, and it is a time-consuming and cumbersome process.

This technique, due to its low absorption coefficients, has low sensitivity, which causes the detection limit to be higher (Menezes et al., 2009).

Spectroscopic technology demands a high initial investment for instrumentation, and there is limited knowledge on the methodology followed and the relationship between soil parameters and spectral signatures. Many efforts are made in the prediction of soil chemical properties, but there are limited studies on soil biological properties.

Some soil properties pose limitations for prediction. For example, Nocita et al. (2012) found that when the calibration values of dry samples were applied to moist samples, the model error increased linearly with the increase in moisture content. On this basis, it was concluded that soil moisture is a limiting factor for the use of Vis-NIR spectroscopy in the prediction of soil properties.

In many studies, it has been found that environmental covariates are not taken into consideration, which may be the reason for the variation in prediction accuracy.

32.5 Way forward

Future work should focus on developing more reliable models for different soil types and zones with a wide range of soil characteristics.

Also, there is great potential to exploit this technology in digital soil mapping and landscape modeling. Soil spectroscopic techniques can be used for classifying (Viscarra-Rossel and Webster, 2012) and mapping the soil types along different topo-sequences, and they can be further used for land use planning (Dematte et al. (2004).

There is a great opportunity to develop a soil spectral library along with a soil bank that can store the soil sample for future analysis (Nocita et al., 2015) so that we can cut off the expense of collecting soil samples too often. There is a need to focus on building the portable Vis-NIR spectroradiometer (Christy, 2008) so that they can reduce the transportation cost and sample handling in situ. A way has been found to monitor and control the quality of soil organic additions (Tremblay et al. 2009), which can be exploited in precision and sustainable agriculture.

32.6 Conclusion

By reviewing the studies conducted on soil property prediction using Vis-NIR laboratory spectroscopy, we conclude that with the use of spectral technology developed using various models and machine learning algorithms, the soil properties can be predicted using a rapid, cost-effective, and nondestructive technique that is way easier, simpler, and more applicable. It has the potential to be used as a rapid assessment and predictive technique for soil properties: physical, chemical, biological and nutrient status. Ultimately, a well-developed soil spectral database and spectral libraries would be a potent solution to address soil management issues and maintain a sustainable environment.

References

Adeline, K.R.M., Gomez, C., Gorretta, N., Roger, J.M., 2017. Predictive ability of soil properties to spectral degradation from laboratory Vis-NIR spectroscopy data. Geoderma 288, 143–153.

Al-Mashhadani, Z., Chandrasekaran, B. 2020. Survey of agricultural robot applications and implementation. In: 2020 11th IEEE Annual Information Technology, Electronics and Mobile Communication Conference (IEMCON) 0076–0081.

Amin, I., Fikrat, F., Mammadov, E., Babayev, M., 2020. Soil organic carbon prediction by Vis-NIR spectroscopy: case study the Kur-Aras Plain, Azerbaijan. Commun. Soil Sci. Plant Anal. 51 (6), 726–734.

Askari, M.S., Cui, J., O' Rourke, S.M., Holden, N.M., 2015. Evaluation of soil structural quality using VIS-NIR spectra. Soil Till Res. 146, 108–117. Available from: https://doi.org/10.1016/j.still.2014.03.006.

Bo, S., Rossel, R.A.V., Mouazen, A.M., Wetterlind, J., 2010. Chapter five—Visible and near infrared spectroscopy in soil science. Adv. Agron. 107 (107), 163–215.

Bulan, R., Sitorus, A., 2022. Vis-NIR spectra combined with machine learning for predicting soil nutrients in cropland from Aceh Province, Indonesia. Case Stud. J. Environ. Chem. Eng. 6, 100268.

Christy, C.D., 2008. Real-time measurement of soil attributes using on-the-go near infrared reflectance spectroscopy. Comput. Electron. Agric. 61, 1019.

Confalonieri, M., Fornasier, F., Ursino, A., Boccardi, F., Pintus, B., Odoardi, M., 2001. The potential of near infrared reflectance spectroscopy as a tool for the chemical characterisation of agricultural soils. J. Near Infrared Spectrosc. 9 (2), 123–131.

Cozzolino, D., Moron, A., 2003. The potential of near-infrared reflectance spectroscopy to analyse soil chemical and physical characteristics. J. Agric. Sci. 140, 65–71.

Conforti, M, Matteucci, G, Buttafuoco, G, 2018. Using laboratory Vis-NIR spectroscopy for monitoring some forest soil properties. J. Soils Sediments. 18 (3), 1009–1019. Available from: https://doi.org/10.1007/s11368-017-1766-5.

Cozzolino, D., Cynkar, W.U., Dambergs, R.G., Shah, N., Smith, P., 2013. In situ measurement of soil chemical composition by near-infrared spectroscopy: a tool toward sustainable vineyard management. Commun. Soil Sci. Plant Anal. 44, 1610–1619.

Chang W, C, Laird A, D, 2002. Near-infrared reflectance spectroscopic analysis of soil C and N. Soil Sci. 167 (2), 110–116. Available from: http://dx.doi.org/10.1097/00010694-200202000-00003.

Chebrolu, N., Lottes, P., Schaefer, A., Winterhalter, W., Burgard, W., Stachniss, C., 2017. Agricultural robot dataset for plant classification, localization and mapping on sugar beet fields. Int. J. Robot Res. 36 (10), 1045–1052.

Debaene, G., Pikula, D., Niedzwiecki, J., 2014. Use of VIS-NIRS for land management classification with a support vector machine and prediction of soil organic carbon and other soil properties. Cien Investi Agr. 41 (1), 21–32.

Dematte, J.A.M., Campos, R.C., Alves, M.C., Fiorio, P.R., Nanni, M.R., 2004. VisibleNIR reflectance: a new approach on soil evaluation. Geoderma 121, 95–112.

Dharumarajan, S., Lalitha, M., Gomez, C., Vasundhara, R., Kalaiselvi, B., Hegde, R., 2021. Prediction of soil hydraulic properties using VIS-NIR spectral data in semi-arid region of Northern Karnataka Plateau. Geoderma Reg e00475.

Dharumarajan, S., Gomez, C., Lalitha, M., Kalaiselvi, B., Vasundhara, R., Hegde, R., 2023. Soil order knowledge as a driver in soil properties estimation from Vis-NIR spectral data—Case study from northern Karnataka (India). Geoderma Reg. 32, e00596.

Dinakaran, J., Bidalia, A., Kumar, A., Hanief, M., Meena, A., Rao, K.S., 2016. Near-infrared-spectroscopy for determination of carbon and nitrogen in Indian soils. Commun. Soil Sci. Plant Anal. 47 (12), 1503–1516.

Dotto, A.C., Dalmolin, R.S.D., TEN Caten, A., Grunwald, S., 2018. A systematic study on the application of scatter-corrective and spectral-derivative preprocessing for multivariate prediction of soil organic carbon by Vis-NIR spectra. Geoderma 314, 262. 27.

Dunn, B.W., Batten, G.D., Beecher, H.G., Ciavarella, S., 2002. The potential of near-infrared reflectance spectroscopy for soil analysis—a case study from the Riverine Plain of south-eastern Australia. Aust. J. Exp. Agric. 42 (5), 607–614.

Fernandez, R., Montes, H., Salinas, C., 2015. VIS-NIR, SWIR and LWIR imagery for estimation of ground bearing capacity. Sensors 15 (6), 13994–14015.

George, K.J., Kumar, S., Raj, R.A., 2020. Soil organic carbon prediction using visible–near infrared reflectance spectroscopy employing artificial neural network modelling. Curr. Sci 119 (2), 377–381.

Ghosh, A.K., Das, B.S., Reddy, N., 2020. Application of VIS-NIR spectroscopy for estimation of soil organic carbon using different spectral preprocessing techniques and multivariate methods in the middle Indo-Gangetic plains of India. Geoderma Reg 23, e00349.

Gomez, C., Rossel, R.A.V., Mcbratney, A.B., 2008. Soil organic carbon prediction by hyperspectral remote sensing and field vis-NIR spectroscopy: an Australian case study. Geoderma 146 (3–4), 403–411.

Gupta, A., Vasava, H.B., Das, B.S., Choubey, A.K., 2018. Local modeling approaches for estimating soil properties in selected Indian soils using diffuse reflectance data over visible to near-infrared region. Geoderma 325, 59–71.

Haaland, D.M., Thomas, E.V., 1988. Partial least-squares methods for spectral analyses. 1. Relation to other quantitative calibration methods and the extraction of qualitative information. Anal. Chem. 60, 1193–1202.

He, Y., Huang, M., Garcia, A., Hernandez, A.A.N.D., Song, H., 2007. Prediction of soil macronutrients content using near-infrared spectroscopy. Comput Electron Agric. 58 (2), 144–153.

Kadupitiya, H.K., Sahoo, R.N., Ray, S.S., Chakraborty, D., Ahmed, N., 2010. Quantitative assessment of soil chemical properties using visible (VIS) and near-infrared (NIR) proximal hyperspectral data. Trop. Agri. 158, 41–60.

Kuang, B., Mahmood, H.S., Quraishi, M.Z., Hoogmoed, W.B., Mouazen, A.M., VAN Henten, E.J., 2012. Sensing soil properties in the laboratory, in situ, and on-line. A review. Adv. Agron. 114, 155–223.

Kulig, B., Zagorda, M., Lepiarczyk, A., Oleksy, A., Kolodziejczyk, M., Slizowska, A., 2020. Application of NIR spectroscopy to evaluate the soil fertility at the field scale. J. Biotechnol. Biores. 2–4.

Lalitha, M., Dharumarajan, S., Gomez, C., Hegde, R., Koyal, A., Khandal, S., et al., 2022. Spectral assessment of soil properties in semi-arid tropical regions of southern Karnataka Plateau. Arch. Agron. Soil Sci. 1–16.

Mahajan, G.R., Das, B., Gaikwad, B., Murgaonkar, D., Desai, A., Morajkar, S., et al., 2021. Monitoring properties of the salt-affected soils by multivariate analysis of the visible and near-infrared hyperspectral data. Catena 198, 105041.

Maleki, M.R., Van Holm, L., Ramon, H., Merckx, R., De Baerdemaeker, J., Mouazen, A.M., 2006. Phosphorus sensing for fresh soils using visible and near infrared spectroscopy. Biosyst. Eng. 95, 425–436.

Malley, D.F., Martin, P.D., Ben-Dor, E., 2004, Application in analysis of soils. In Near-Infrared Spectroscopy in Agriculture, Agronomy 44. Roberts, C. A., Workman, J., Jr., Reeves, J. B., III, Eds. American Society of Agronomy, Crop Science Society of America, Soil Science Society of America: Madison, WI: 729–784.

Menezes, J.C., Ferreira, A.P., Rodrigues, L.O., Bras, L.P., Alves, T.P., 2009. Chemometrics role within the PAT context: examples from primary pharmaceutical manufacturing, Comprehensive Chemometrics, Volume 4. Elsevier, pp. 313–355.

Minasny, B., Tranter, G., Mcbratney, A.B., Brough, D.M., Murphy, B.W., 2009. Regional transferability of mid-infrared diffuse reflectance spectroscopic prediction for soil chemical properties. Geoderma 153, 155–162.

Mondal, B.P., Sahoo, R.N., Ahmed, N., Singh, R.K., Das, B., Mridha, N., et al., 2021. Rapid prediction of soil available sulphur using visible near-infrared reflectance spectroscopy. Indian J. Agri. Sci. 91 (9), 1328–1332.

Moura-Bueno, J.M., Dalmolin, R.S.D., Horst-Heinen, T.Z., Grunwald, S., Ten Caten, A., 2021. Environmental covariates improve the spectral predictions of organic carbon in subtropical soils in southern Brazil. Geoderma 393, 114981.

Mozaffari, H., Moosavi, A.A., Ostovari, Y., Nematollahi, M.A., Rezaei, M., 2022. Developing spectrotransfer functions (STFs) to predict basic physical and chemical properties of calcareous soils. Geoderma 428, 116174.

Nawar, S., Mouazen, A., 2019. On-line vis-NIR spectroscopy prediction of soil organic carbon using machine learning. Soil Tillage Res. 190, 120–127.

Nduwamungu, C., Ziadi, N., Tremblay, G.F., Parent, L.E., Tremblay, G.F., Thuri Es, B., 2009. Near-infrared reflectance spectroscopy prediction of soil properties: Effects of sample cups and preparation. Soil Sci. Soc. Am. J. 73, 1896–1903.

Nocita, M., Stevens, A., Noon, C., Wesemael, B., V.A.N., 2012. Prediction of soil organic carbon for different levels of soil moisture using Vis-NIR spectroscopy. Geoderma 199, 37–42.

Nocita, M., Stevens, A., VAN Wesemael, B., Aitkenhead, M., Bachmann, M., Barthès, B., et al., 2015. Chapter Four—Soil spectroscopy: an alternative to wet chemistry for soil monitoring. Adv. Agron. 132, 139–159.

Ray, S.S., Singh, J.P., Parihar, J.S., 2017. Soil discrimination and parameter estimation using hyperspectral data and multivariate analysis techniques. J. Agril. Phys. 17 (1), 58–72.

Rayment, G.E., Lyons, D.J., 2011. Soil Chemical Methods–Australasia. CSIRO Publishing, Collingwood, Australia.

Reeves III, J.B., 2010. Near- versus mid-infrared diffuse reflectance spectroscopy for soil analysis emphasizing carbon and laboratory versus on-site analysis: where are we and what needs to be done? Geoderma 158, 3–14.

Reeves III, J.B., Smith, D.B., 2009. The potential of mid-and near-infrared diffuse reflectance spectroscopy for determining major-and trace-element concentrations in soils from a geochemical survey of North America. Appl. Geochem. 24 (8), 1472–1481.

Reeves III, J.B., Mccarty, G.W., Meisinger, J.J., 1999. Near infrared reflectance spectroscopy for the analysis of agricultural soils. J. Near Infrared Spectrosc. 7, 179–193.

Reeves III, J.B., Mccarty, G.W., Reeves, V.B., 2001. Mid-infrared diffuse reflectance spectroscopy for the quantitative analysis of agricultural soils. J. Agric. Food Chem. 49, 766–772.

Sarathjith, M.C., Das, B.S., Wani, S.P., Sahrawat, K.L., Gupta, A., 2016. Comparison of data mining approaches for estimating soil nutrient contents using diffuse reflectance spectroscopy. Curr. Sci. 1031–1037.

Shaddad, S.M., Madrau, S., Castrignano, A., Mouazen, A.M., 2016. Data fusion techniques for delineation of site-specific management zones in a field in UK. Precis. Agric. 17, 200–217.

Shao, Y., He, Y., 2011. Nitrogen, phosphorus, and potassium prediction in soils, using infrared spectroscopy. Soil Res 49 (2), 166–172.

Shi, T., Cui, L., Wang, J., Fei, T., Chen, Y., Wu, G., 2013. Comparison of multivariate methods for estimating soil total nitrogen with visible/near-infrared spectroscopy. Plant Soil. 366, 363–375.

Soriano-Disla, J.M., Janik, L.J., Viscarra Rossel, R.A., Macdonald, L.M., Mclaughlin, M.J., 2014. The performance of visible, near-, and mid-infrared reflectance spectroscopy for prediction of soil physical, chemical, and biological properties. Appl. Spectrosc. Rev. 49 (2), 139–186.

ST. Luce, M., Ziadi, N., Nyiraneza, J., Tremblay, G.F., Zebarth, B.J., Whalen, J.K., et al., 2012. Near infrared reflectance spectroscopy prediction of soil nitrogen supply in humid temperate regions of Canada. Soil Sci. Soc. Am. J. 76, 1454–1461.

Terhoeven-Urselmans, T., Schmidt, H., Joergensen, R.G., Ludwig, B., 2008. Usefulness of near-infrared spectroscopy to determine biological and chemical soil properties: Importance of sample pretreatment. Soil Biol. Biochem. 40 (5), 1178–1188.

Tremblay, M.E., Nduwamungu, C., Parent, L.E., Bolinder, M.A., 2009. Biological stability of C and N in organic products and crop residues using FT-NIR reflectance spectroscopy. Commun. Soil Sci. Pl. Anal. (in press).

Vibhute, A.D., Kale, K.V., Mehrotra, S.C., Dhumal, R.K., Nagne, A.D., 2018. Determination of soil physicochemical attributes in farming sites through visible, near-infrared diffuse reflectance spectroscopy and PLSR modeling. Ecol. Process. 7 (1), 1–12.

Viscarra Rossel, R.A., Rizzo, R., Dematte, J.A.M., Behrens, T., 2010. Spatial modeling of a soil fertility index using visible–near-infrared spectra and terrain attributes. Soil Sci. Soc. Am. J. 74, 1293–1300.

Viscarra Rossel, R.A., Walvoort, D.J.J., Mcbratney, A.B., Janik, L.J., Skjemstad, J.O., 2006. Visible, near infrared, mid infrared or combined diffuse reflectance spectroscopy for simultaneous assessment of various soil properties. Geoderma 131, 59–75.

Viscarra Rossel, R.A., Webster, R., 2012. Predicting soil properties from the Australian soil visible near infrared spectroscopic database. Eur. J. Soil Sci 63, 848–860.

Wu, Y., Chen, J., Ji, J., Gong, P., Liao, Q., Tian, Q., et al., 2007. A mechanism study of reflectance spectroscopy for investigating heavy metals in soils. Soil Sci. Soc. Am. J. 71, 918–926.

Wu, Y., Chen, J., Wu, X., Tian, Q., Ji, J., Qin, Z., 2005. Possibilities of reflectance spectroscopy for the assessment of contaminant elements in suburban soils. Appl. Geochem. 20, 1051–1059.

Section 5

Chapter 33

Assessment of soil erosion in a tropical mountainous river of the Western Ghats, Southern India, using GIS-based RUSLE, SDR and AHP techniques

Asna Nizar[1], B. Upendra[1,2], M. Ciba[1], G. Sreenivasulu[3] and K. Anoop Krishnan[1]

[1]*National Centre for Earth Science Studies (NCESS), Thiruvananthapuram, Kerala, India,* [2]*Department of Applied Chemistry, Cochin University of Science and Technology (CUSAT), Kochi, Kerala, India,* [3]*Department of Geology, Yogi Vemana University, Kadapa, Andhra Pradesh, India*

33.1 Introduction

Soil erosion process is the carrying away of the topmost layer of terrain owing to various components like climate, topography, vegetation, soil structure, and composition (Panagos et al., 2015; GUERRA et al., 2017; Thomas et al., 2018a,b). Erosion is primarily a two-stage process that involves separating individual soil particles from the soil mass and moving them from one location to another by one of the four agents of erosion, like water, wind, ice, or gravity acting on it (Morgan, 2005; Mitasova, 2013; Fernández-Raga et al., 2017). A third stage called deposition happens when adequate energy is not accessible to move a particle. In this process, soil particles of finer size erode more easily than coarse particles, i.e., silt erodes easier than sand (Ampontuah et al., 2006; Mitasova, 2013). Therefore, soil erosion can be defined as a process of soil particle (sediment) separation, transit, and deposition, as shown below.

Soil ⟹ Detachment ⟹ Transportation ⟹ Deposition

Though soil erosion is a natural phenomenon occurring on most land areas on earth, its level and strength have been significantly amplified by anthropogenic activities over many centuries (Brown and Wolf, 1984; Garcia-Ruiz et al., 2015; Boakye et al., 2020a,b; Behera et al., 2020). It has a dual significance when land degradation is considered: not only does it affect the productivity of the land in areas suffering erosion, but it can also have negative off-site impacts in areas that receive eroded soil, like contributing to reservoir sedimentation and the occurrence of dust storms that are capable of traveling hundreds of kilometers from source areas. Negative off-site impacts can also threaten the productivity of agricultural systems. Positive off-site impacts, like when erosion leads to the movement of soil from steep slopes that are difficult to cultivate to flatter, more easily accessible valley bottoms, can accrue too. Due to rapid soil erosion brought on by the aforementioned conditions, large areas of land have been permanently turned into infertile surfaces. These damaged land areas are now a source of water contamination in the open ocean. In the downstream portions of rivers, soil that has been eroded from highland areas has deposited, aggravating the situation. This has increased the area of the rivers' flood plains, decreased the clearance beneath bridges and culverts, and sedimented reservoirs.

Out of India's entire 329 million hectares of geographical area, it is estimated that 68 million hectares are severely degraded, and 167 million hectares are subject to significant water and wind erosion. This includes 40 million hectares of land degraded by gullies and ravines, shifting cultivation, waterlogging, salinity, and alkalinity, shifting river courses, and desertification, in addition to the 127 million hectares damaged by soil erosion (Roy et al., 2022). Narayana and Babu (1983) have estimated that about 5334 Mt (16.4 t ha^{-1}) of soil is detached annually in India, out of which 29% is transported by rivers into the sea and 10% is accumulated as depositions, leading to a significant loss of storage capacity in reservoirs. Numerous erosion models, including physical, empirical, and conceptual ones, were utilized in

Remote Sensing of Soils. DOI: https://doi.org/10.1016/B978-0-443-18773-5.00027-2

regional-scale assessments to estimate and build the best erosion control plans (De Vente and Poesen, 2005; Mitasova, 2013; Prasannakumar et al., 2012; Bekele and Gemi, 2021; Chuma et al., 2022). Each model has unique qualities and application ranges (Boggs et al., 2001; Lu et al., 2004; Dabral et al., 2008; Tian et al., 2009).

The RUSLE model estimates the annual average erosion of soil for ground slopes by using Geographic Information System (GIS) (Renard, 1997; Yitayew et al., 1999; Lu et al., 2004; Prasannakumar et al., 2012; Atoma, 2018; Boakye et al., 2020a,b; Tsegaye and Bharti, 2021). The RUSLE equation evaluates the annual average soil loss from a unit of land by combining geophysical and land cover elements (Ciesiolka et al., 1995; Boggs et al., 2001; Tsegaye and Bharti, 2021; Chuma et al., 2022). The development of GIS and remote sensing technology allowed a more precise estimation of the input variables (TOLIIriSm, 1995; Park et al., 2011; Ganasri and Ramesh, 2016; Atoma, 2018). The RUSLE algorithm was developed practically, not mathematically, and does not accurately depict the erosion process. Due to the possibility that some of the eroded soil may be deposited on the way to the watershed outlet or another downstream place of interest, it is therefore impossible to directly estimate the amount of sediment reaching these areas. The RUSLE solely calculates soil loss in highland locations; it does not calculate sediment deposition (Ampontuah et al., 2006; Thomas et al., 2018a,b). Sediment deposition typically takes place at the bottom of a slope or at a change in grade, where the slope becomes gentler (Ming et al., 2007). Sediment transport is the transport of solid particles (sediment), which often results from a combination of the gravitational force acting on the sediment and/or the motion of the liquid in which the soil is entrained. Unusual rainfall seasons, particularly those with higher than average rainfall and normally strong storms, may generate greater sediment than estimated. To account for these processes, the sediment delivery ratio (SDR) and sediment yield (SY) for a given catchment are used to estimate the total transported soil particles to the catchment outlet. The analytical hierarchy process (AHP) integrated with GIS is found to be the most significant tool for assessing the environmental vulnerabilities (Saaty, 1980). This technique has drawn a lot of interest from diverse decision-makers and has proven useful in numerous research on natural hazard assessment, such as erosion hazard mapping (De Vente and Poesen, 2005; Wu and Wang, 2007; Alexakis et al., 2013; Thomas et al., 2018a,b; Panagos et al., 2015; Halefom and Teshome, 2019; Boakye et al., 2020a,b; Behera et al., 2020; Tsegaye and Bharti, 2021; Chuma et al., 2022) and landslide vulnerability mapping (Dai and Lee, 2002; Huabin et al., 2005; Akgün and Bulut, 2007; Mancini et al., 2010; Rozos et al., 2011; Kanwal et al., 2017; Ullah et al., 2022).

The development of soil erosion models has facilitated the ability to learn about soil erosion accurately and effectively, particularly for conservation planning and associated land management strategies, which require site-specific, accurate, and detailed environmental (e.g., trends and status of soil erosion) and socioeconomic information supplemented by the practical experiences of land users and experts. As a result, the purpose of this research is to evaluate the present state of soil erosion and create a map of soil erosion severity based on remote sensing data and local perceptions for optimal soil conservation planning and sustainable land management techniques (Chuma et al., 2022; Ming et al., 2007; Prasannakumar et al., 2012; Roy et al., 2022; Yesuph and Dagnew, 2019).

Despite the fact that the study estimates the erosional and depositional characteristics of the basin, it has some limitations. Due to the unavailability of data, this study evaluates virgin flow erosion rates only and does not take into account dam sedimentation rates or sedimentation from reservoirs. In spite of ensuring maximum possibilities, it is possible that some variations will occur due to the estimation method chosen, the extraction method used, or human error. Moreover, validation of estimated rates is not taken into consideration due to the unavailability of data.

33.2 Study area

The present study has been conducted in a small, tropical river basin that spreads in one of the densely populated areas on the western flank of the Southern Western Ghats. The Vamanapuram river basin (VRB) is 88 km long with a basin area of 680 km^2 and lies between 8°35′ and 8°50′N latitude and 76°40′ and 77°15′E longitude in the districts of Thiruvananthapuram and Kollam (Fig. 33.1). This seventh-order river system with a dendritic drainage pattern drains into the Arabian Sea after flowing through varied geologic and physiographic terrains (Aju et al., 2021). In recent years, the VRB has experienced rapid population growth and economic development, which has made it a significant environmental region both quantitatively and qualitatively. *Climate*: The VRB is a west-flowing river system in Kerala, and it is experiencing a humid and tropical climate. Generally, the area receives a mean rainfall of 34.3 mm during the winter, 406.2 mm during the summer, 970.8 mm during the southwest monsoon, and 56.11 mm during the northeast monsoon. *Geology*: Geologically, a major part of the basin is occupied by khondalite suite of rocks. Apart from this, the occurrence of charnockite is also noticed in the area. The terrain is often intruded by basic and ultrabasic rocks at certain places. *Soils*: The extreme upstream region of the river basin has clayey soil. As it comes down to the middle region, there is a mixture of clayey soil, gravelly loam soil, and loamy soil. Towards the downstream, the basin has gravelly,

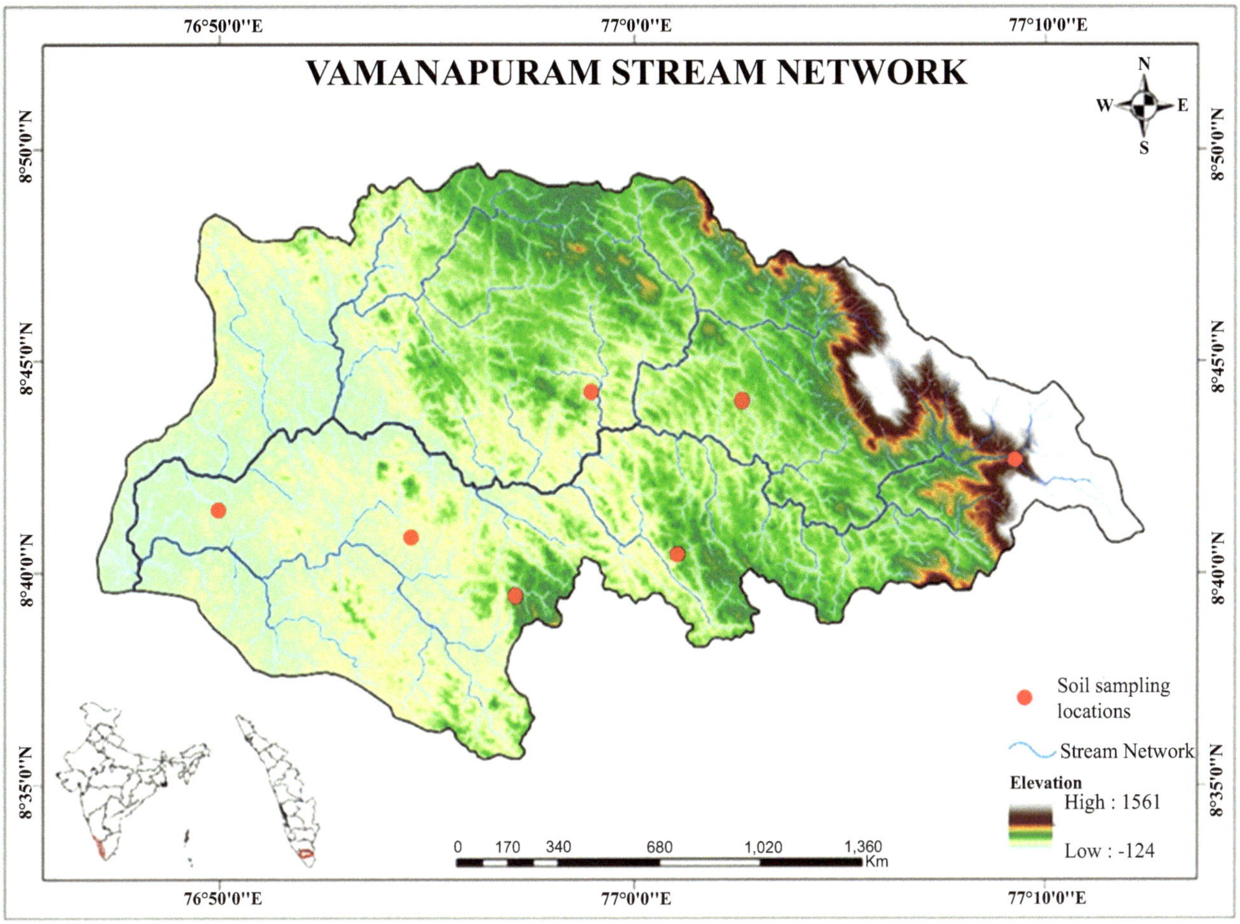

FIGURE 33.1 Study area map with soil sampling locations.

clayey soil. *Geomorphology*: While the lowland region is made up of gently undulating surfaces, the highland region is composed of mountainous terrain and undulating surfaces. The lower lateritic plateau, denudation-related hills, valleys, and water bodies are the dominant geomorphic features. *Land use/land cover (LULC)*: There are many different types of land use in the VRB, including agricultural land, forests, plantations, built-up areas, barren ground, open bushes, vegetation, and water bodies. The vegetation class outperforms the other classes. The predominant LULC types in the highlands include open bushes, bare terrain, plantations, and evergreen woods. The midlands are mostly covered in vegetation, with a few isolated areas of arid desert. The lowland is home to built-up areas, greenery, and plantations, as well as small plots of arable land. The region is seeing substantial economic and population growth. The government was forced to build a few check dams in the upstream and downstream parts of the river basin due to the increased demand for irrigation and groundwater to meet household needs in urban centers.

33.3 Methodology

33.3.1 RUSLE

The RUSLE is a widely used empirical model for the assessment of gross soil erosion. It is assessed by evaluating mainly five factors: rainfall erosivity, soil erodibility, topography, cover management, and support practices. It evaluates the long-time average soil losses in runoff from specific field areas in specified cropping and management systems using these factors, where possible values at a specific place can be numerically expressed by the equation (Renard, 1997; Wischmeier and Smith, 1978):

$$A = R \times K \times LS \times C \times P \tag{33.1}$$

where, A, average annual soil erosion per unit area (t ha^{-1} $year^{-1}$); R, rainfall-runoff erosivity factor (MJ mm ha^{-1} h^{-1} $year^{-1}$); K, soil erodibility factor (t ha h ha^{-1} MJ^{-1} mm^{-1}); LS, slope-length and slope-steepness/topographic factor (dimensionless); C, coverage and management factor (dimensionless, ranging between 0 and 1.5); P, support practice factor (dimensionless, ranging between 0 and 1).

In the present work, Eq. (33.1) is carried out in the ArcGIS 10.3 environment for the computation of particular factors, and the average annual soil loss of VRB is estimated. Various data are obtained from different sources and combined to evaluate the inputs and estimate the gross soil erosion factor (A). They are explained in the subsequent sections.

33.3.2 Generation of factors of RUSLE and their data processing

33.3.2.1 R: Rainfall-runoff erosivity factor

The R factor assesses the influence of raindrops and offers comparative data on the volume and rate of runoff that are most probably related to the precipitation. Soil loss is highly proportional to the R factor. Increased rain erosivity determines the overland water flow's greater erosive potential. Where the magnitude and length of the precipitation exceed the capability of the soil to absorb the precipitation, soil erosion by flowing water occurs subsequently. The R factor depends on characteristics such as the physical properties of the rainfall, such as the raindrop size and its distribution, terminal velocity, kinetic energy, etc. Spatial coverage of pluviographic data at 15-minutes time intervals or less is required for the sensible estimation of the annual or monthly rainfall erosivity EI_{30} index, which is the multiplication of total storm energy and maximum 30-min intensity divided by 100 (Wischmeier and Smith, 1978). However, it is very difficult to obtain such data in developing countries where the specific infrastructures are not present or are not well maintained (Yu and Rosewell, 2001; Cohen et al., 2005; Shamshad et al., 2008). Therefore, the R-factor is estimated by using monthly and/or annual rainfall data, since it is easily available for longer periods. Babu et al. (2004) developed an equation for Indian river basins using data from several rain-gauge stations spread across various zones to determine linear relationships among average annual rainfall as well as estimated EI30 values for various zones across India. The iso-erodent maps are also sketched for both annual and seasonal EI30 values:

$$\mathbf{R} = 81.5 + 0.38\,(\mathrm{MAP})\,(340\ \mathrm{mm} \le \mathrm{MAP} \le 3500\ \mathrm{mm}) \tag{33.2}$$

where, MAP, mean annual precipitation (mm); R, rainfall-runoff erosivity factor (MJ mm ha^{-1} h^{-1} y^{-1}).

For the present study, 15 years (2001–16) of monthly rainfall data are collected from the Indian Meteorological Department (IMD) for 4 rain-gauge stations that are spatially distributed across the VRB. The R-factor is computed using Eq. (33.2) by using the rainfall data for each rain-gauge station, namely: Thiruvananthapuram (76.9366°E, 8.5241°N), Thiruvananthapuram airport (76.9198°E, 8.4834°N), Nedumangad (77.0046°E, 8.608°N), and Varkala (76.7163°E, 8.7379°N) with respect to each year. It is then interpolated spatially from these stations across the VRB by using the inverse distance weighting (IDW) technique in ArcGIS (Fig. 33.2).

33.3.2.2 K: Soil erodibility factor

Soil erodibility is the estimate of the ability of different types of soil to resist erosion or transport based on their physical characteristics designated by erodibility factor K. Soils can be compared quantitatively for a given rainfall condition on the basis of their variation in shear strength, texture, infiltration capacity, aggregate stability, organic content, and chemical content. Generally, sand, sandy loam, and loam textured soils tend to be less erodible than silt, very fine sand, and certain clay textured soils, as they have a greater resistance to erosion due to their high content of organic matter and improved soil structure. Many soil erosion assessments of South Indian river basins implement the soil erodibility nomograph to compute the K-factor, using an algebraic approximation of the nomograph Eq. (33.3) given by Wischmeier and Smith (1978), where the silt fraction does not exceed 70%.

$$\mathbf{K} = \frac{2.8 \times 10^{-5} \times (12 - \mathrm{OM}) \times \mathrm{M}^{1.14} + 4.3 \times 10^{-1} \times (\mathrm{s} - 2) + 3.3 \times 10^{-1} \times (\mathrm{p} - 3)}{100} \tag{33.3}$$

where, K, soil erodibility factor (t ha h ha^{-1} MJ^{-1} mm^{-1}); OM, percent organic matter; M, product of the primary particle size fractions [i.e., (% modified silt or the 0.002–0.1 mm size fraction) × (% silt + % modified sand or the 0.1–2.0 mm)]; s, the structure; and p, permeability classes of the soil.

The K values for VRB are calculated with Eq. (33.3) using the soil texture analytical data at different locations across the basin as shown in Fig. 33.1. Different soil types and their respective textures of VRB are recognized with

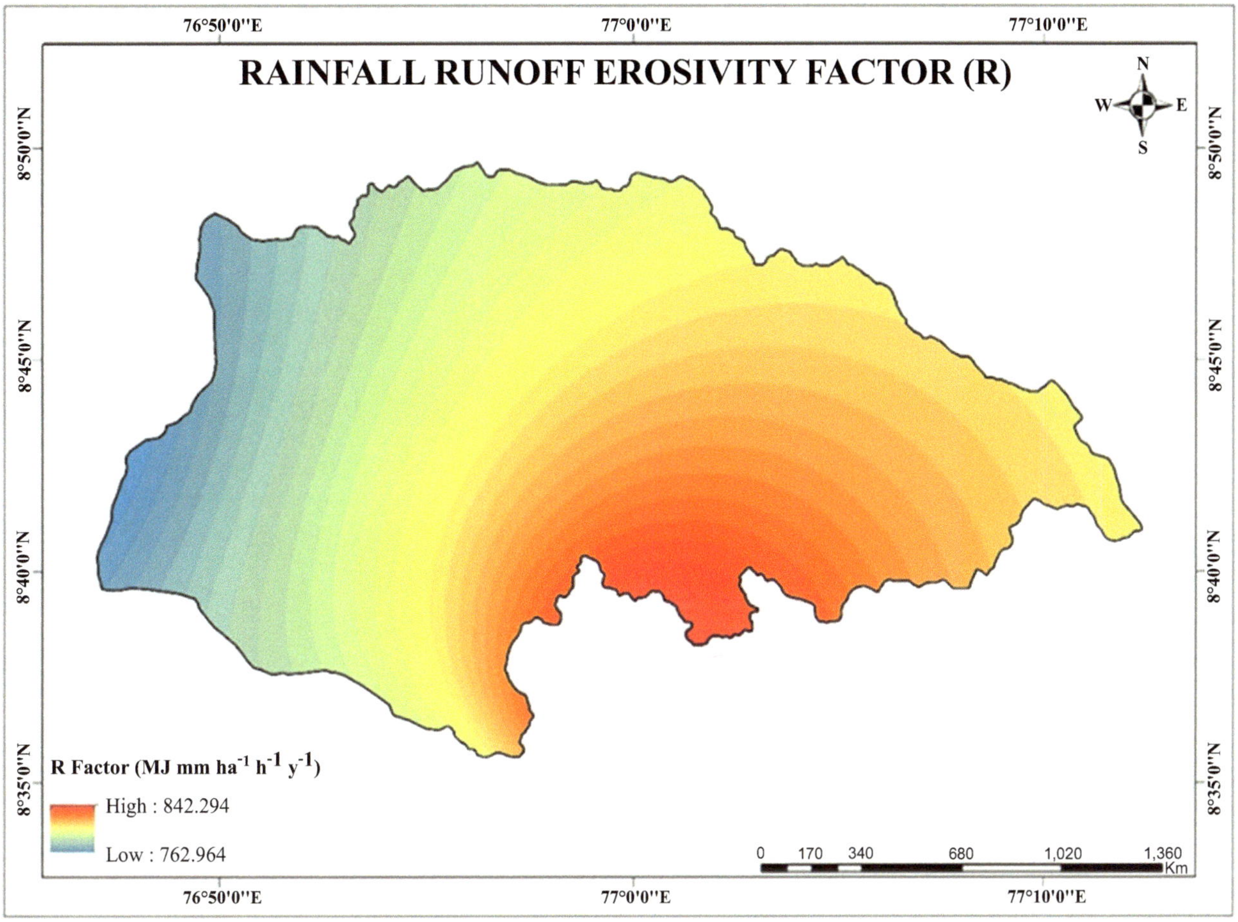

FIGURE 33.2 Rainfall runoff erosivity factor.

respect to the sand, slit, and clay proportions, and the K values are estimated for each location using Eq. (33.3), assigned to the locations where the soil samples have been collected for analysis and interpolated by the IDW technique in ArcGIS10.3 (Fig. 33.3).

33.3.2.3 LS: Topographic factor

The LS, or topographic factor, is the combination of slope length and slope steepness. The slope length (L) is calculated as the ratio of soil loss from the field slope length under evaluation to that from the 22.13 m length plots, and the slope steepness factor is calculated as the ratio of soil loss from the field slope gradient to that from a 9% slope under the same conditions (S). Soil erosion increases as the runoff velocity increases with steepness, permitting further soil to be transported and detached with runoff. Wischmeier and Smith (1978) used measurements from harvested land on slopes ranging from 3 to 18% and lengths between 10 and 100 m to create the following equation for the LS factor:

$$\mathbf{LS} = \left(\frac{\lambda}{22.13}\right)^m 65.41\sin^2\theta + 4.56\sin\theta + 0.065 \tag{33.4}$$

where, *LS*, slope-length and slope-steepness factor; λ, field slope length (meters); θ, angle of slope (radians); and m, exponent varying from 0.2 to 0.5 [$m = 0.5$ for $>$ 5% slope; 0.4 for 3.5%–4.9% slope; 0.3 for 1%–3.4% slope; 0.2 for $<$ 1% slope.]

L and *S* values in Eq. (33.4) are separately calculated in the raster calculator by taking the λ value as flow accumulation multiplied by cell size and divided by 22.13 and θ calculated in radians for VRB. The flow accumulation and slope values are developed from the Advanced Spaceborne Thermal Emission and Reflection Radiometer (ASTER) Global

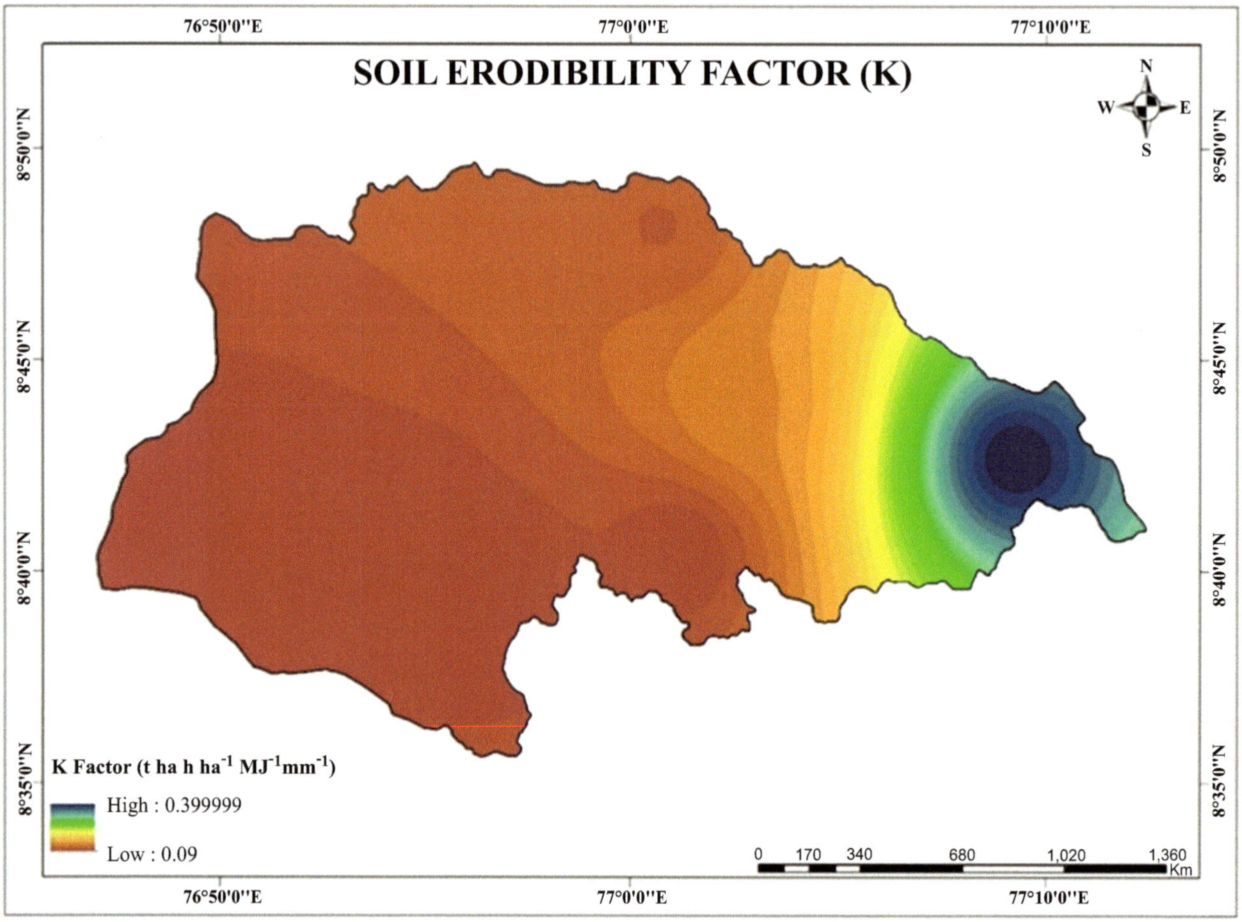

FIGURE 33.3 Soil erodibility factor.

Digital Elevation Model (GDEM) of 2013 with 30 m resolution, downloaded from the USGS Earth Data and the exponent (m) is assigned according to the slope percent values (Fig. 33.4).

33.3.2.4 C: Cover and management factor

The cover and management factor compares the amount of soil lost from a piece of land that is cropped under certain circumstances to the same amount lost from a clean-tilled, continuous fallow (Wischmeier and Smith, 1978). The C factor is used to determine the relative effectiveness of crop/vegetation management systems in terms of preventing soil loss. Satellite remote sensing data sets are used for the assessment of the C factor in the present study of VRB due to the diversity of land cover patterns with spatial and temporal variations (Van der Knijff et al., 2000).

$$\mathbf{C} = \exp\left[-\alpha \times \frac{\text{NDVI}}{\beta - \text{NDVI}}\right] \tag{33.5}$$

where, *C*, cover and management factor; α and β, unitless parameters; and *NDVI*, normalized difference vegetation index.

The shape of the curve relating to normalized difference vegetation index (NDVI) and the C-factor is determined by the unitless parameters α and β, respectively. According to a study by Van der Knijff et al. (2000), choosing values of 2 and 1 for the parameters α and β, respectively, produces better results than assuming a linear connection. This formula is successfully used to calculate the C-factor for regions with comparable topography and weather (Upendra et al., 2022; Thomas et al., 2018a; Prasannakumar et al., 2012) along South Indian rivers. The NDVI, which is an indicator of vegetation vigor and health, is scaled to approximate the C factor. The amount of vegetation in a given area affects the

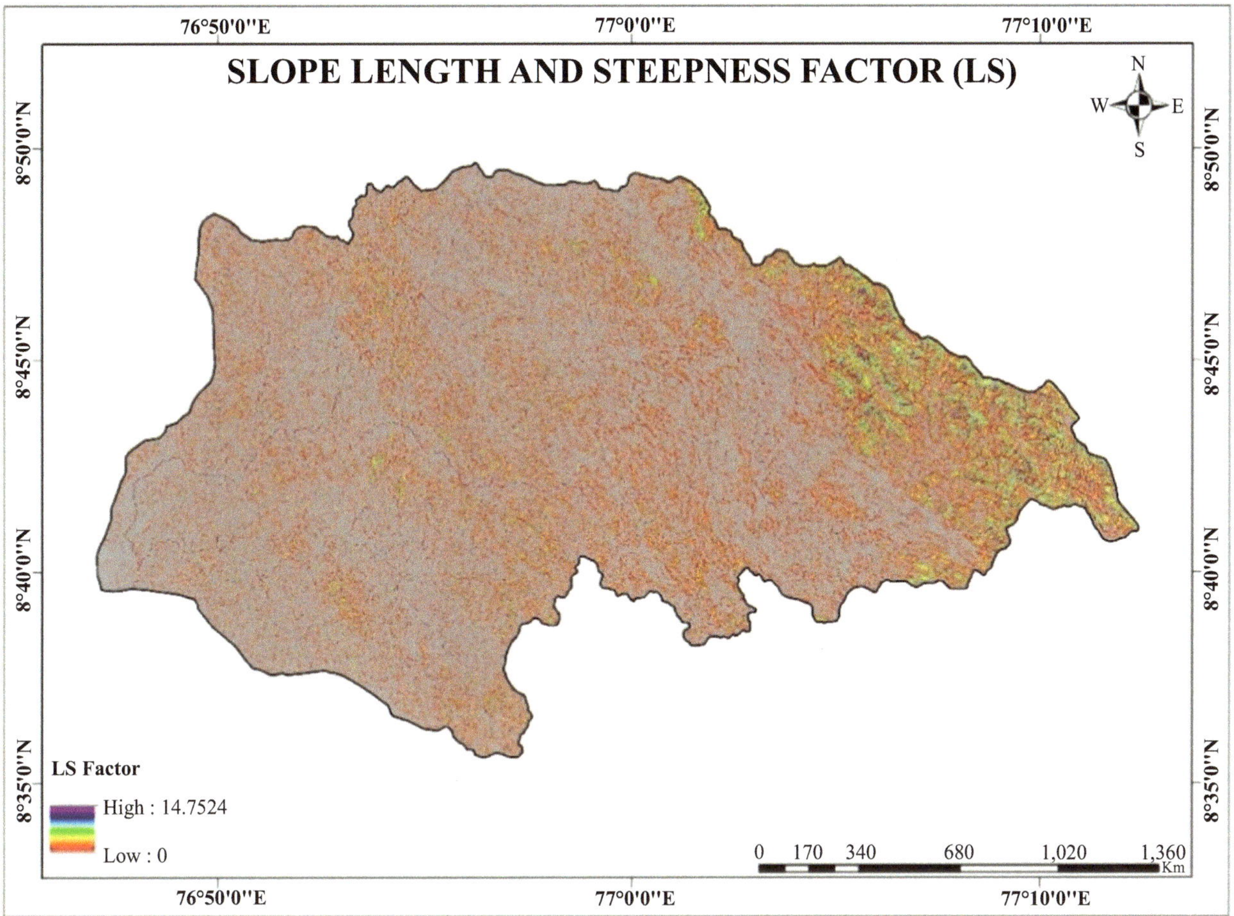

FIGURE 33.4 Topographic factor.

NDVI of that location. It is the image transformation that is evaluated using the normalized difference between the visible red (VR) and near-infrared (NIR) bands (Rouse et al., 1973; Sabins, 1986) and is expressed by Eq. (33.6)

$$\text{NDVI} = \frac{\text{NIR} - \text{Red}}{\text{NIR} + \text{Red}} = \frac{\text{Band 5} - \text{Band 4}}{\text{Band 5} + \text{Band 4}} \tag{33.6}$$

The possible range of values for NDVI lies between -1 and 1. In the present study, the C factor layer is prepared using the Landsat 8–9 Operational Land Imager (OLI) and Thermal Infrared Sensor (TIRS) satellite data corresponding to the recent years 2020–22 downloaded from the USGS EarthExplorer site of VRB. The image is geometrically corrected and analyzed using ArcGIS10.3 (Fig. 33.5).

33.3.2.5 P: Support practice factor

The P is defined as the ratio of erosion with a particular installment to the subsequent erosion with up- and downslope tillage (Renard, 1997). The values of the P-factor range from 0 to 1, in which the higher value is allocated to areas with no preservation practices and the lowest value corresponds to an area with a more effective conservation practice, such as strip or contour cropping. In the present study of VRB, the P-factor map is derived from the land use/land cover demarcated with associated support practices. LULC of VRB is developed by downloading Landsat 8–9 Operational Land Imager (OLI) and Thermal Infrared Sensor (TIRS) images for 2020–22 from USGS EarthExplorer and performing supervised classification using ArcGIS10.3 using the maximum likelihood method (Fig. 33.6). The P values are assigned to each class of the LULC (Table 33.1), and the spatial map is generated by converting polygons to rasters.

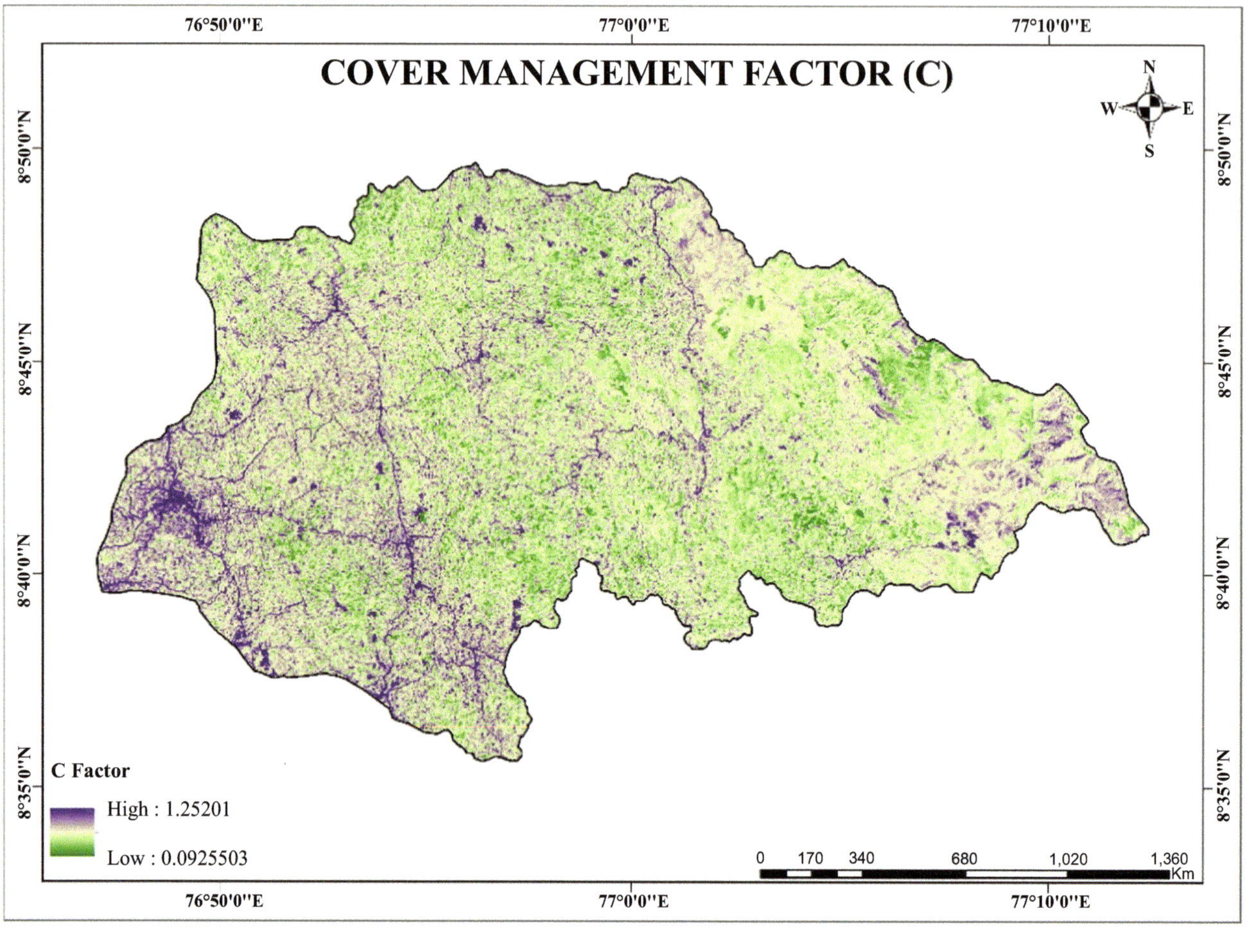

FIGURE 33.5 Cover management factor.

All the input factors of RUSLE are projected from the Geographic Coordinate System (WGS1984 datum) to the Universal Transverse Mercator (zone 43 N) in order to co-register to a common reference with a grid cell size of 30 m by taking care of the processing extend in order to minimize geolocation errors and to improve the horizontal accuracy. The quantitative output gross soil erosion factor (A) is computed on a pixel-by-pixel basis (using the raster calculator function) in ArcGIS10.3 after generation of all the input factors of RUSLE (Fig. 33.7A).

33.3.3 Sediment delivery ratio (SDR) and sediment yield (SY)

By definition, sediment delivery describes the interactions between the many processes involved: erosion in the upstream regions and sediment yield (SY) in the downstream areas. The SDR was defined by Walling (1983) as the ratio of gross erosion inside the basin to sediment delivered at the catchment outlet, which is expressed as follows:

$$\mathbf{SDR} = \mathrm{SY}/\mathrm{A} \tag{33.7}$$

Where SY represents sediment yield (t ha^{-1} $year^{-1}$) as well as A represents long-term average yearly soil loss (t ha^{-1} $year^{-1}$). The catchment's SDR is affected by several factors like geomorphology, environment (Walling, 1983), and diverse interrelation among SDR and watershed attributes, like basin length (Lb), basin area (Aw), slope, and basin relief (Rw) are suggested by numerous scholars (e.g., Williams and Berndt, 1972, 1977; Boyce, 1975; Vanoni, 1975). Nevertheless, the SDR curves, which represent power function correlations between Aw and SDR, are the most frequently utilized technique for estimating the SDR:

$$\mathrm{SDR} = \alpha \mathrm{A}_w{}^{\beta} \tag{33.8}$$

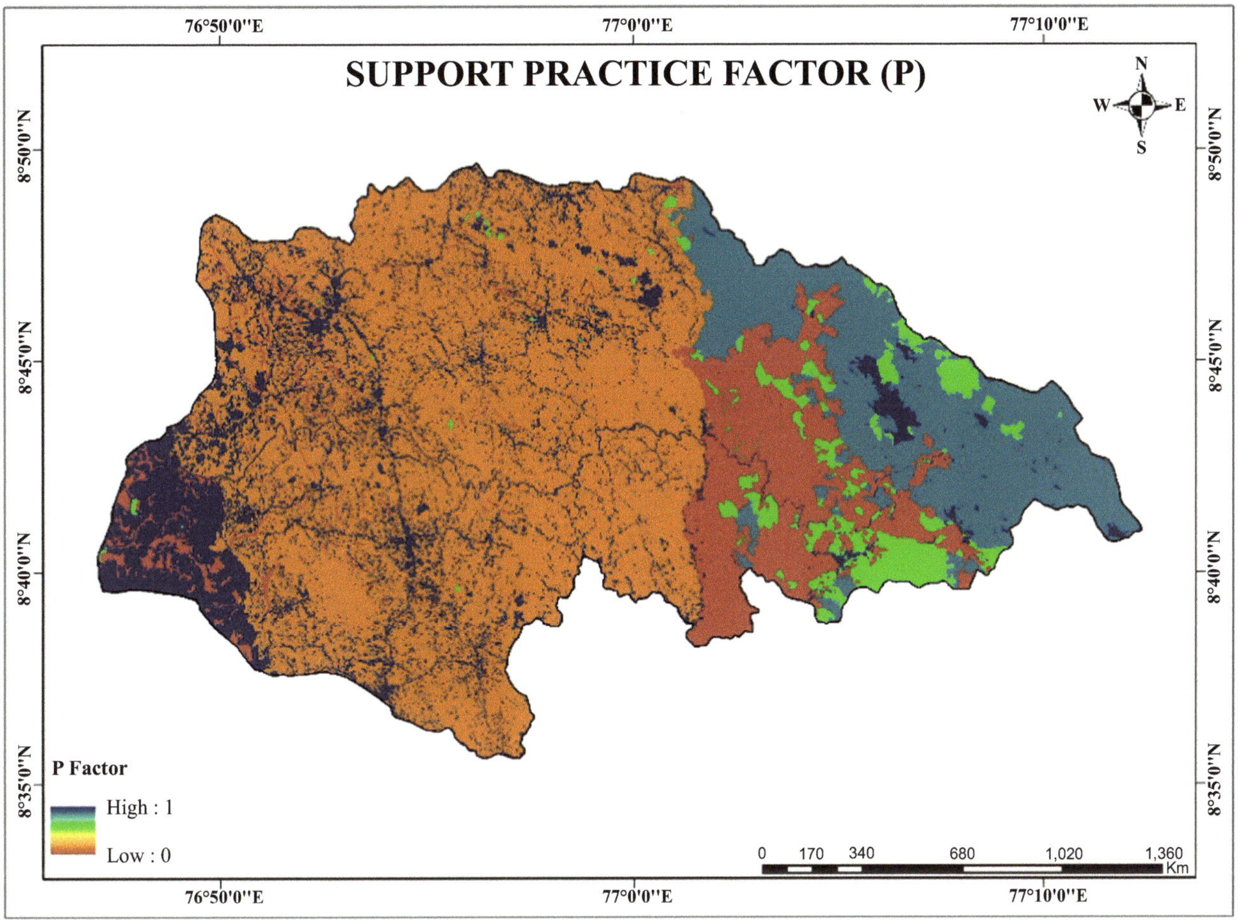

FIGURE 33.6 Support practice factor.

TABLE 33.1 Assigned P factor based on the LULC classes of VRB.

S.no	LULC class	Area (km²)	P factor
1	Built-up	89.62	1
2	Open scrubs	26.95	0.7
3	Forest	99.38	0.9
4	Plantations	82.57	0.2
5	Barren land	7.37	1
6	Agriculture	8.89	0.2
7	Vegetation	361.59	0.3
8	Watershed	3.42	1

where A_w represents catchment area (in km^2); β and α are empirical parameters. The scaling power β provides crucial knowledge on watershed sediment carrying mechanism and is mostly between −0.01 and −0.25 (Walling, 1983; Lu et al., 2006). It is vital to have catchment-specific SDR curves for precise SY estimation; however, it is difficult to

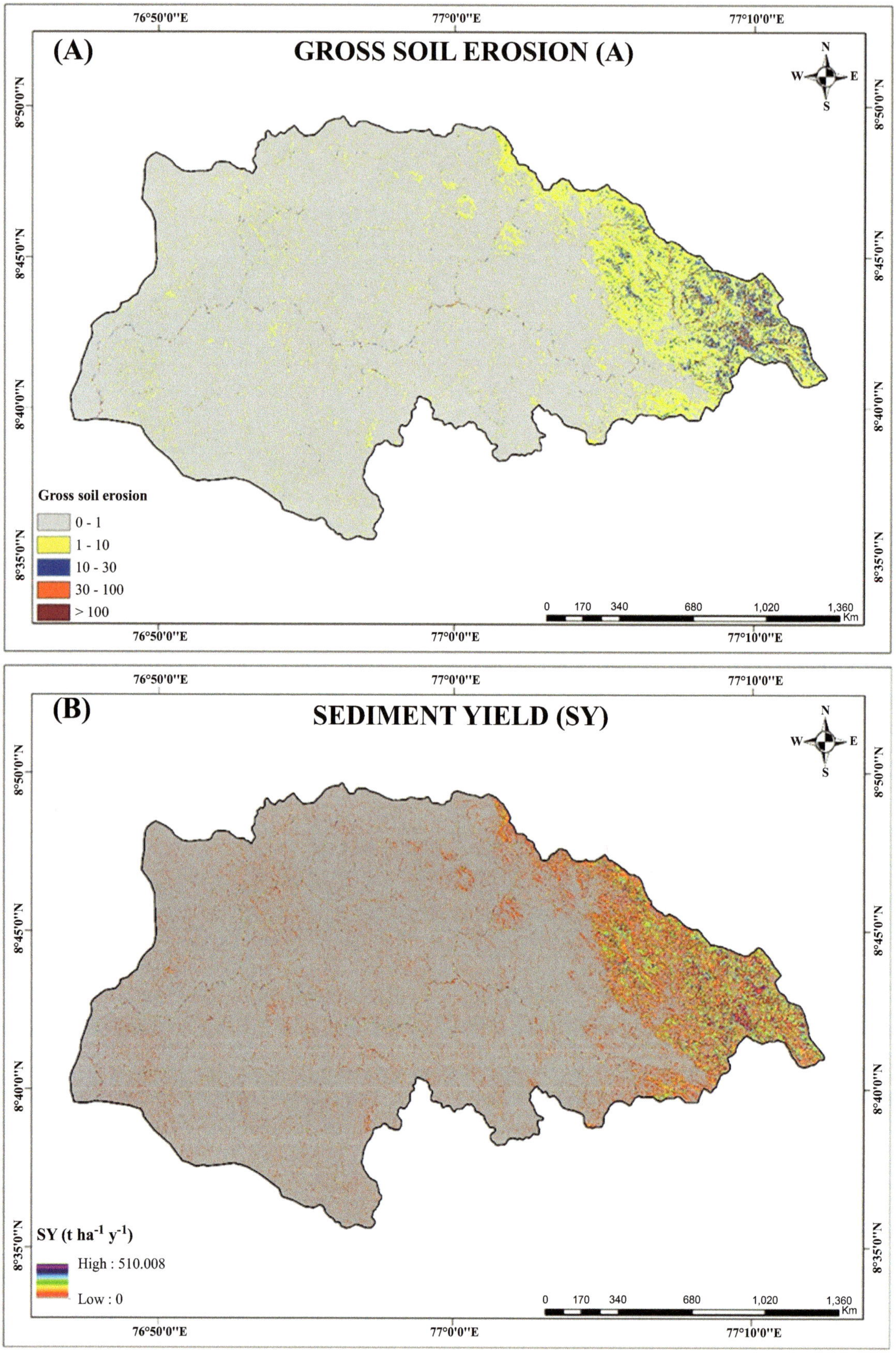

FIGURE 33.7 (A) Classified gross soil erosion, (B) sediment yield.

produce these kinds of curves (Lim et al., 2005). Amidst these diverse conventional SDR curves, the equations proposed by Vanoni (1975), cited in Lim et al. (2005), are used in this study, which is written as follows:

$$\mathbf{SDR} = 0.4724A_w^{-0.125} \tag{33.9}$$

Furthermore, the geographical variation of SY (Fig. 33.7B) within VRB is computed via Eqs. (33.7) and (33.9).

33.3.4 Analytical hierarchic process (AHP)

Estimating vulnerable regions for erosion and mapping them using the soil erosion potential index (SEPI) is employed by the AHP technique (Park et al., 2011; Thomas et al., 2018b; Halefom and Teshome, 2019). The presiding environmental factors for the process are chosen with the help of a literature survey (Park et al., 2011; Pradeep et al., 2015; Thomas et al., 2018b; Halefom and Teshome, 2019) combined with the precise analysis of influential factors in VRB. In the present study, eight parameters are assessed for evaluating risk-prone areas of erosion: slope, soil erodibility (K), rainfall, SPI (stream power index), drainage density, P-factor, geomorphology, and lithology. These parameters are then integrated with the help of AHP in a GIS environment to demarcate the prominent erosion risk zones and to analyze the contribution of different parameters. The AHP is an important multi criteria decision-making process that uses a pairwise comparison of different parameters that control soil erosion. The pairwise comparison was carried out using a parameter matrix and weighting the parameters based on their influence on other parameters represented on a numerical scale, as well as the consistency ratio (Saaty, 1980). The weight can differ on a numerical scale from 1 to 9; the weight 1 represents that both parameters have equal importance, and the weight 9 represents the strong importance of one factor over another. The pairwise comparison matrix and consistency evaluation are carried out with the aid of the Excel AHP template given by Goepel (2013) and using AHP-OS, an online software implemented by Goepel (2017). Each thematic layer of presiding controlling factors of soil erosion is prepared following the methodology depicted in the Table 33.2. Due to the fact that the AHP technique produces quantitative values on erosion-prone areas using a weighted linear sum procedure, it is employed to calculate the soil erosion potential index using following Eq. 33.10:

$$\text{SEPI} = \sum_{i=1}^{n} F_i \times W_i \tag{33.10}$$

TABLE 33.2 Data sources, methods used to create each thematic layer along with their ranks and weights assigned in AHP.

S. no	Theme	Data sources	Methods	Assigned weight	Domain factor	Rank
1	*Slope*	Cartosat, Bhuvan	Slope (in degree) in Spatial Analyst of ArcGIS	25	Flat (< 5)	2
					Gentle (5–11)	2
					Moderate (11–17)	3
					High (17–26)	6
					Steeply high (> 26)	7
2	*K factor*	Soil textural data	Interpolated (IDW) in ArcGIS10.3 after calculating by Eq. (33.3)	22	0.09–0.12	2
					0.12–0.17	2
					0.17–0.24	2
					0.24–0.32	3
					0.32–0.39	4
3	*SPI*	Cartosat, Bhuvan	Calculated flow accumulation and slope and evaluated by eq (Danielson, 2013) in Raster calculator of ArcGIS 10.3	18	Very low (−13.8––5.8)	1
					Low (−5.8––1.4)	1
					Moderate (−1.4–0.6)	1
					High (0.6–3.4)	3
					Very high (3.4–11.9)	5

(Continued)

TABLE 33.2 (Continued)

S. no	Theme	Data sources	Methods	Assigned weight	Domain factor	Rank
4	*Rainfall*	15 years (2001–16) of monthly rainfall dataIMD	Data for 4 stations (Thiruvananthapuram, Thiruvananthapuram airport, Nedumangad and Varkala) interpolated (IDW) in ArcGIS10.3	11	1793–1840 mm	1
					1840–1874 mm	1
					1874–1905 mm	2
					1905–1938 mm	2
					1938–2002 mm	2
5	*Drainage Density*	Cartosat, Bhuvan	Line density tool in ArcGIS10.3	10	Low (1.6–9.1)	1
					Moderate (9.1–12.7)	2
					High (12.7–15.8)	2
					Very High (15.8–19.0)	4
					Extremely high (19.0–27.4)	4
6	*P factor*	LULC from Google Earth	Digitized LULC and assigned values based on Section 33.3.2.5	7	0–0.2	2
					0.2–0.3	2
					0.3–0.7	4
					0.7–1	6
7	*Geomorphology*	Bhukosh	Direct shapefile classified and clipped	4	Waterbody	8
					Valley	3
					Highly dissected hills and valleys	7
					Scarp	4
					Rolling plain	4
					Residual mound	5
					Pediment	5
					Active floodplain	3
					Low-dissected hills and valleys	5
					Active quarry	2
8	*Lithology*	Bhukosh	Direct shapefile classified and clipped	3	Acid to intermediate charnockite	2
					Biotite gneiss	5
					Calc granulite	5
					gabbro	3
					Gar-bio-sil-gneiss + graphite + kyanite	5
					Garnet-biotite gneiss	5
					Garnet gneiss	5
					Laterite	6
					Pyroxene granulite	3
					Quartzite	1
					Sand	4
					Sandstone	4
					Terri sand	4

Where F_i is the themes, rating used to evaluate the soil erosion proneness in the study, *P* factor, *K* factor, geomorphology, lithology, drainage density, mean annual precipitation, SPI, and Slope. The W_i indicates the corresponding weights. The SEPI map developed using AHP and the estimated gross soil erosion rate map generated using the RUSLE model are compared for resemblance, such as high or low soil erosion rates in the A map with high or low index values in the SEPI map.

33.4 Results and discussion

The following sections discuss the outcomes of the soil erosion assessment of VRB using the RUSLE-SDR coupled model, followed by the detailed controlling factors analysis.

33.4.1 Long-term annual average soil loss (A) and sediment yield (SY) estimation

According to Fig. 33.2, the R parameter of VRB is varying in the range 762.96–842.29 MJ mm ha^{-1} h^{-1} $year^{-1}$, and comparatively greater R values than the surrounding area can be found in the eastern and south-eastern regions of the basin. Given that R depends on both monthly and yearly precipitation, the aforementioned interbasin heterogeneity can account for the geographical diversity in monthly rainfall patterns and annual totals. The average annual precipitation of different rain-gauge stations in VRB spans nearly from 1790 mm to greater than 2000 mm. The K factor displays values between 0.009 and 0.3999994 t ha h ha^{-1} MJ^{-1} mm^{-1} (Fig. 33.3), with a 0.1372037 t ha h ha^{-1} MJ^{-1} mm^{-1} average value. The sampled soil fragments from the VRB range in texture from loam to clay. The upstream parts of the VRB, where the majority of the soil sample textures are clayey soil, are spatially dispersed with relatively higher K values. The LS parameter of the VRB is in the range 0 to 14.75, with a 0.03 mean (Fig. 33.4); high values are seen in regions with steep slopes, plateau scarps, terrain slopes with concordant peaks, and escarpments. The comparatively higher LS values in the VRB show that the terrain is more susceptible to soil erosion because the LS factor is referred to as unity at 22.1 m slope length and 9% steepness (Wischmeier and Smith, 1978). The C factor of the basin differs nearly between 0.09256 and 1.25201 (Fig. 33.5). Low C factor values represent the majority of the VRB's area, while high C factor values are very localized and only occur in rocky exposures, open scrub areas, or built-up areas. Since NDVI and C have an inverse relationship [Eq. (33.4)], regions with poor C factor values typically exhibit high NDVI, which suggests the existence of vigorous vegetation. Prasannakumar et al. (2011) state that the C factor can range from almost zero (i.e., for well-secured soils) to 1.5. (i.e., for ridged, finely tilled surfaces). Depending on the types of preservation strategies used in the ground to lessen runoff and soil loss, the basin's P factor (Fig. 33.6) might be in the range 0–1. Since there are no supporting soil and water conservation strategies for the natural vegetation types (such as woods, arid regions, and open scrubs), a *P* value of 1 was assigned.

ArcGIS multiplied the input parameters of the RUSLE model to calculate the geographically distributed A in the VRB (Fig. 33.7A). The range of A in VRB is 0 to 2439.6 t ha^{-1} $year^{-1}$, with a mean value of 1.275 t ha^{-1} $year^{-1}$. Following Singh et al.'s (1992) recommendations for Indian settings, areas with varied soil erosion intensities were identified to comprehend the geographical variability of soil loss. A classification of the VRB area (Fig. 33.7A) based on soil erosion severity suggests that 87.30% (i.e., the major area) experiences only very low erosion (i.e., <1 t ha^{-1} $year^{-1}$), 9.85% experiences low erosion (1–10 t ha^{-1} $year^{-1}$), 2.18% (10^{-30} t ha^{-1} $year^{-1}$) experiences moderate erosion, 0.5% experiences high erosion (30–100 t ha^{-1} $year^{-1}$), and nearly 0.09% experiences severe erosion (> 100 t ha^{-1} $year^{-1}$). The places with the highest rates of soil erosion are typically found near beaches and on steep terrain, with open vegetation covering plateau scarps and slopes with concordant summits. Using the spatially distributed A and SDR [Eq. (33.7)], geographically varying SY in the VRB is computed and is in the range 0 to 510.01 t ha^{-1} $year^{-1}$, with an average number of 0.266 t ha^{-1} $year^{-1}$ (Fig. 33.7B). The spatial discrepancy between A and SY is also comparatively greater in the mountainous landscape with open vegetation, forest, and rocky terrains of LULC type. The geographical variability of SY in VRB is the result of variability in A and heterogeneity in spatial-environmental factors. The majority of VRB (94.5%) yields sediment ≤ 1 t ha^{-1} $year^{-1}$, and almost 5.4% of the basin area acquires sediment above 1 t ha^{-1} $year^{-1}$.

The rate of erosion in VRB is compared with the results of various catchments across the globe. An analysis by Mhangara et al. (2012) predicted that 36.06 t ha^{-1} of soil would be lost per year in the Keiskamma river basin (Eastern Cape Province of South Africa), where soil dispersivity, sparse vegetation, and poor pasture management drive erosion. As stated in Gelagay and Minale (2016), slope and soil erodibility parameters are the primary causes of soil erosion in the Koga river basin (northwestern Ethiopia), where the average annual water erosion is 47.40 t ha^{-1} $year^{-1}$. It is quantified that soil erosion in the Awassa basin (in central Ethiopia) occurs at a rate of between 0 and 201 t ha^{-1} $year^{-1}$, with erosion rates being highest in areas with degraded bare ground, grassland, or shrub cover (Ali and Hagos, 2016). Similarly, rates of soil erosion observed in many Chinese watersheds rank among the highest. According to Pan and Wen (2014), the Caijiamiao watershed in northwest China experiences 70.58 t ha^{-1} $year^{-1}$ of annual soil erosion, while the Yangou watershed in the Chinese Loess Plateau experiences 134.60 t ha^{-1} $year^{-1}$ of yearly soil loss (Tang et al., 2015). The significant difference in A between the mentioned studies (as compared to VRB) may be due to various diverse habitats and significant soil erosion-controlling elements. However, variations in the data and techniques used to derive the model inputs could also be the cause of the variability. Since there are many ways to produce the input parameters (R, C, LS, K, and P) of the model, researchers select the best approach based on the available data.

33.4.2 Soil erosion potential index map (SEPI)

Erosion-vulnerable areas of VRB were delineated by combining the eight parameters through the AHP technique. Normalized principle eigen vector, a comparative pair-wise matrix is given in Table 33.3, and ranks apportioned to each chosen classes as per Saaty's (1977) scale is given in Table 33.2. Since slope affects runoff, erosion potential, etc., and soil type determines erosive characteristics, higher weights are allocated to slope and soil erodibility factors (Moore et al., 1988). The slope of VRB ranges from 0 to 62 degrees and is classified as flat (<5 degrees), low (5–11 degrees), moderate (11–17 degrees), high (17–26 degrees), and very high (>26 degrees) classes. Regions with flat and low slopes receive rank 2, while those with mountainous slopes are ranked 7. The K factor of VRB varies from 0.09–0.39 depending on the texture of soil, as discussed in Section 33.3.2.2. It had been classified into 5, depending upon the erodibility: 0.09–0.12, 0.12–0.17, 0.17–0.24, 0.24–0.32, and 0.32–0.39. A low rank of 2 is assigned to soil texture with less erodibility, that is, 1st 3 classes and high value of 4 is assigned to higher erodible soil which falls in range 0.32–0.39. The stream power index (SPI) is calculated using the equation suggested by Danielson (2013) and classified into 5: −13.8–−5.8, −5.8–−1.4, −1.4–0.6, 0.6–3.4, and 3.4–11.9; a low rank of 1 is assigned for the first 3 classes of low SPI and a high rank of 5 is assigned for last class with high SPI. Similarly, rainfall is also classified into 5 classes: 1793–1840 mm, 1840–1874 mm, 1874–1905 mm, 1905–1938 mm, and 1938–2002 mm; a low value of 1 is assigned to the first 2 classes, increasing to a high value of 2 assigned to the last 3 classes. Drainage density is also classified into 5 classes: 1.6–9.1, 9.1–12.7, 12.7–15.8, 15.8–19.0, and 19.0–27.4; a low value of 1 is assigned to the 1st class with low density, and high value of 4 is assigned to the last 2 classes with high density. The P factor is divided

TABLE 33.3 Pairwise comparison matrix and *Eigen vector for AHP.*

Matrix		Rainfall 1	Soil erodibility 2	P factor 3	Drainage density 4	Geomorphology 5	Lithology 6	Slope 7	SPI 8	Normalized principal Eigenvector
Rainfall	1	1	1/3	2	1	4	5	1/4	1/2	10.56%
Soil erodibility	2	3	1	4	3	4	4	1	1	21.78%
P factor	3	1/2	1/4	1	1/2	3	3	1/4	1/3	6.71%
Drainage density	4	1	1/3	2	1	4	4	1/4	1/2	10.18%
Geomorphology	5	1/4	1/4	1/3	1/4	1	2	1/5	1/4	3.91%
Lithology	6	1/5	1/4	1/3	1/4	1/2	1	1/5	1/4	3.23%
Slope	7	4	1	4	4	5	5	1	1	25.10%
SPI	8	2	1	3	2	4	4	1	1	18.51%

into 4 categories: 0–0.2, 0.2–0.3, 0.3–0.7, and 0.7–1. The higher the P value, the more prone the region to erosion; therefore, a high value of 6 is denoted for the last class, and a lower P value shows less vulnerability, so a lower value of 2 is denoted for first two classes. The geomorphology class is divided into 10 classes, namely water body, valley, highly dissected hills and valleys, scarp, rolling plain, residual mound, pediment, and active floodplain. Low dissected hills and valleys, active quarry. High rank is denoted for a water body since it's more vulnerable area for erosion, while least value is denoted for an active quarry since it is less prone. The lithology class is subdivided into 13 classes: biotite gneiss, acid to intermediate charnockite, gar-bio-sil gneiss + graphite + kyanite, garnet-biotite gneiss, calc granulite, gabbro, garnet gneiss, laterite, pyroxene granulite, quartzite, sand, sandstone, and terri sand. A higher rank of 6 is given to laterite, while a low rank of 2 is given to acidic to intermediate charnockite.

33.4.3 Effects of erosion in different LULC classes of VRB

The data in Table 33.4 displays the statistics of average soil loss detected on different LULC classes of VRB. It shows that a high rate of erosion takes place in forests (mean A = 6.32), followed by barren land (mean A = 5.13). The water body (mean A = 4.57) shows moderate to high erosion. Open shrub (mean A = 1.83) regions and built-up (mean A = 0.45) areas show moderate erosion, while vegetation (mean A = 0.21), plantations (mean A = 0.19), and agricultural (mean A = 0.14) lands show low erosion rates. When comparing plantation classes with anthropogenic structures like built-up areas, it is observed that plantation regions have a low rate of erosion. High observed erosion rates seen in forests in VRB can be compared with the rates of forests in Western Ghats, and this can be due to the topographic effect (Prasannakumar et al., 2012). The land cover types with changeable leaf cover have an impact on gross soil estimates, claim Alexandridis et al. (2015). Areas experiencing moderate to low SER can implement land management measures like afforestation or roper management of vegetative cover. Additionally, to greatly reduce soil erosion in wet tropical systems, Labrière et al. (2015) advise the employment of stringent soil and vegetation management strategies, such as no-till, contour planting, and the use of vegetative buffer strips. Further, to reduce vulnerability, the study recommends avoiding bare soil land use classes by properly managing them or converting them to other useful land use types that can effectively reduce erosion risk. Areas with steep slopes can implement contour cropping or strip cropping strategies to protect the soil from being eroded by runoff.

33.4.4 Effects of rainfall change on soil erosion of VRB

It is possible to calculate the rate of soil erosion in a field under a certain set of circumstances by analyzing the elements in the soil loss equation. By altering the rainfall-runoff erosivity parameters, soil loss modeling can be performed. This part of the study is to understand how rainfall affects soil erosion using GIS software. In this study, varying rainfall is used to find the change in soil erosion in the area (in %). A rainfall erosivity factor map is created for 15 years of mean rainfall with a recent LULC map, and it is compared with mean + standard deviation rainfall and mean −standard deviation rainfall-derived gross soil erosion maps (Table 33.5). A change in the rate of soil erosion has been observed. The gross soil erosion map with 15 years of mean rainfall has a very high rate of soil erosion region of

TABLE 33.4 Mean soil loss with respect LULC classes.

Name	MEAN A (t h^{-1} $year^{-1}$)
Built-up	0.45
Open scrubs	1.84
Forest	6.33
Plantations	0.19
Barren land	5.14
Agriculture	0.15
Vegetation	0.22
Watershed	4.58

TABLE 33.5 Variation in A with respect to rainfall changes.

Class	A with mean annual precipitation (2001–2016)	A with mean + delta rainfall	A with mean – delta rainfall
Very low	87.30	86.99	87.57
Low	9.86	10.03	9.72
Moderate	2.19	2.28	2.11
High	0.55	0.59	0.51
Very high	0.09	0.10	0.09

0.099%, which has increased to 0.105% when mean + standard deviation rainfall is used; likewise, high rate of soil erosion region has also increased from 0.554% to 0.587%, and the moderate region increased from 2.18% to 2.28%. For mean—standard deviation, the very high rate of soil erosion region and the high rate of soil erosion region have an aerial percentage of 0.09% and 0.51%, respectively, and the very low rate of soil erosion region has increased to 87.56% from 87.30%. Thus, it is understood that rainfall has a good influence on the soil erosion of a catchment basin. Natural variations in weather cycles and precipitation patterns throughout time can affect how quickly soil is eroding.

33.5 Conclusions

The present study computes spatially distributed gross soil erosion rates (A) and sediment yield (SY) for a torrid zone mountainous catchment, VRB, southern India, using the coupled RUSLE-SDR model. In the VRB, gross soil erosion rates vary in the range 0–2439.6 t ha^{-1} $year^{-1}$, with 1.275 t ha^{-1} $year^{-1}$ mean soil loss, while the sediment yield (SY) range is 0–510.01 t ha^{-1} $year^{-1}$, with a mean of 0.266 t ha^{-1} $year^{-1}$. The comparison of A and SY indicates that the majority of the degraded soils are redistributed within VRB. According to the geographical heterogeneity of soil loss severity, the regions with intense soil erosion are primarily located in the plateau scarps area and slopy realm, with hills and mountains covering open shrubs. Area-wise, approximately 87.30% of the VRB undergoes tiny erosion ($\leq$1 t ha^{-1} $year^{-1}$), whereas nearly 0.09% exhibit very high erosion rates. Further conservation and management practices to conserve soils are also discussed. Measures like strip cropping and contour cropping are suggested for areas with high erosion-risk potential. Further, analysis of the variation of rainfall on soil erosion rates in VRB is needed considering that climate change is inducing modifications in the hydrological cycle via temperature and rainfall changes, causing extreme events (high rainfall/cyclones) in recent years along the west coast of India. The ensuing soil erosion rates are computed by varying the long-term average rainfall by one standard deviation (δ). This analysis suggests that soil erosion rates are linearly proportional to the amount of rainfall that VRB is receiving; for example, increasing the rainfall by one δ (mean + δ) results in a significant increase in low and medium soil erosion regions by an amount of 9.8% to 10% and 2.18% to 2.28%, respectively, compared with the mean rainfall soil erosion map.

Acknowledgements

Authors sincerely thank the Director of NCESS and the BioGeoChemistry group for their valuable support. The authors also convey gratitude toward the all data providers. Author A.N give his special thanks to faculty members of the department of coastal disaster management, Pondicherry Central University, M.Sc. program. Author B. U especially thanks his doctorial committee of CUSAT-NCESS, the PhD programmers.

Author Contributions

B.U, A.N and M.C contributed to design of the study, data preparation, and manuscript writing. A.N, G.S and K.A contributed to the preparation of maps, figures, and tables, conceiving the ideas, and reviewing the manuscript.

Additional Information

Competing Interests

The authors declare no competing interests.

Data availability statement

The data that support the findings of this study are available within the manuscript, and any additional information regarding the calculation part will be provided on reasonable request.

References

Aju, C.D., Achu, A.L., Raicy, M.C., Reghunath, R., 2021. Identification of suitable sites and structures for artificial groundwater recharge for sustainable water resources management in Vamanapuram River Basin. South India Hydro Res. 4, 24–37.

Akgün, A., Bulut, F., 2007. GIS-based landslide susceptibility for Arsin-Yomra (Trabzon, North Turkey) region. Environ. Geol. 51 (8), 1377–1387. Available from: https://doi.org/10.1007/s00254-006-0435-6.

Alexakis, D.D., Hadjimitsis, D.G., Agapiou, A., 2013. Integrated use of remote sensing, GIS and precipitation data for the assessment of soil erosion rate in the catchment area of "Yialias" in Cyprus. Atmos. Res. 131, 108–124. Available from: https://doi.org/10.1016/j.atmosres.2013.02.013.

Alexandridis, T.K., Sotiropoulou, A.M., Bilas, G., Karapetsas, N., Silleos, N.G., 2015. The effects of seasonality in estimating the C-factor of soil erosion studies. Land. Degrad. Dev. 26 (6), 596–603. Available from: https://doi.org/10.1002/ldr.2223.

Ali, S.A., Hagos, H., 2016. Estimation of soil erosion using USLE and GIS in Awassa Catchment, Rift valley, Central Ethiopia. Geoderma Reg. 7 (2), 159–166. Available from: https://doi.org/10.1016/j.geodrs.2016.03.005.

Ampontuah, E.O., Robinson, J.S., Nortcliff, S., 2006. Assessment of soil particle redistribution on two contrasting cultivated hillslopes. Geoderma 132 (3–4), 324–343.

Atoma, H., 2018. Assessment of Soil Erosion by Rusle Model Using Remote Sensing and GIS Techniques: A Case Study of Huluka Watershed, Central Ethiopia. M.Sc Thesis, Addis Ababa University, Addis Ababa, Ethiopia.

Babu, R., Dhyani, B.L., Kumar, N., 2004. Assessment of erodibility status and refined Iso-Erodent Map of India. Indian. J. Soil. Conserv. 32 (2), 171–177.

Behera, M., Sena, D.R., Mandal, U., Kashyap, P.S., Dash, S.S., 2020. Integrated GIS-based RUSLE approach for quantification of potential soil erosion under future climate change scenarios. Environ. Monit. Assess. 192 (11), 1–18.

Bekele, B., Gemi, Y., 2021. Soil erosion risk and sediment yield assessment with universal soil loss equation and GIS: in Dijo watershed, Rift valley Basin of Ethiopia. Model. Earth Syst. Environ. 7 (1), 273–291.

Boakye, E., Anyemedu, F.O.K., Donkor, E.A., Quaye-Ballard, J.A., 2020a. Spatial distribution of soil erosion and sediment yield in the Pra River Basin. SN Appl. Sci. 2 (3), 1–12.

Boakye, E., Anyemedu, F.O.K., Donkor, E.A., et al., 2020b. Spatial distribution of soil erosion and sediment yield in the Pra River Basin. SN Appl. Sci. 2, 320. Available from: https://doi.org/10.1007/s42452-020-2129-1.

Boggs, G., Devonport, C., Evans, K., Puig, P., 2001. GIS-based rapid assessment of erosion risk in a small catchment in the wet/dry tropics of Australia. Land. Degrad. Dev. 12 (5), 417–434. Available from: https://doi.org/10.1002/ldr.457.

Boyce, R.C., 1975. Sediment routing with sediment delivery ratios. Present and Prospective Technology for Prediction Sediment, Sediment Yields and Sources, pp. 61–65.

Brown, L.R., Wolf, E.C., 1984. Soil erosion: quiet crisis in the world economy. Worldwatch Paper 60.

Chuma, G.B., Bora, F.S., Ndeko, A.B., Mugumaarhahama, Y., Cirezi, N.C., Mondo, J.M., et al., 2022. Estimation of soil erosion using RUSLE modeling and geospatial tools in a tea production watershed (Chisheke in Walungu), eastern Democratic Republic of Congo. Model. Earth Syst. Environ. 8 (1), 1273–1289.

Ciesiolka, C.A., Coughlan, K.J., Rose, C.W., Escalante, M.C., Hashim, G.M., Paningbatan Jr, E.P., et al., 1995. Methodology for a multi-country study of soil erosion management. Soil. Technol. 8 (3), 179–192. Available from: https://doi.org/10.1016/0933-3630(95)00018-6.

Cohen, M.J., Shepherd, K.D., Walsh, M.G., 2005. Empirical reformulation of the universal soil loss equation for erosion risk assessment in a tropical watershed. Geoderma 124 (3–4), 235–252. Available from: https://doi.org/10.1016/j.geoderma.2004.05.003.

Dabral, P.P., Baithuri, N., Pandey, A., 2008. Soil erosion assessment in a hilly catchment of North Eastern India using USLE, GIS and remote sensing. Water Resour. Manag. 22 (12), 1783–1798. Available from: https://doi.org/10.1007/s11269-008-9253-9.

Dai, F.C., Lee, C.F., 2002. Landslide characteristics and slope instability modeling using GIS, Lantau Island, Hong Kong. Geomorphology 42 (3–4), 213–228.

Danielson, T., 2013. Utilizing a high resolution Digital Elevation Model (DEM) to develop a Stream Power Index (SPI) for the Gilmore creek watershed in Winona County, Minnesota. Pap. Resour. Anal 15 (11).

De Vente, J., Poesen, J., 2005. Predicting soil erosion and sediment yield at the basin scale: scale issues and semi-quantitative models. Earth-Sci. Rev. 71 (1–2), 95–125.

Fernández-Raga, M., Palencia, C., Keesstra, S., Jordán, A., Fraile, R., Angulo-Martínez, M., et al., 2017. Splash erosion: a review with unanswered questions. Earth-Sci. Rev. 171, 463–477. Available from: https://doi.org/10.1016/j.earscirev.2017.06.009.

Ganasri, B.P., Ramesh, H., 2016. Assessment of soil erosion by RUSLE model using remote sensing and GIS-a case study of Nethravathi Basin. Geosci. Front. 7 (6), 953–961. Available from: https://doi.org/10.1016/j.gsf.2015.10.007.

Garcia-Ruiz, J.M., Beguería, S., Nadal-Romero, E., Gonzalez-Hidalgo, J.C., Lana-Renault, N., Sanjuán, Y., 2015. A meta-analysis of soil erosion rates across the world. Geomorphology 239, 160–173.

Gelagay, H.S., Minale, A.S., 2016. Soil loss estimation using GIS and Remote sensing techniques: a case of Koga watershed, Northwestern Ethiopia. Int. Soil. Water Conserv. Res. 4 (2), 126–136. Available from: https://doi.org/10.1016/j.iswcr.2016.01.002.

Goepel, K.D., 2013. Implementing the analytic hierarchy process as a standard method for multi-criteria decision making in corporate enterprises–a new AHP excel template with multiple inputs. In: Proceedings of the International Symposium on the Analytic Hierarchy Process (Vol. 2, No. 10, pp. 1–10). Kuala Lumpur, Malaysia: Creative Decisions Foundation Kuala Lumpur.

Goepel, K.D., 2017. Implementation of an online software tool for the analytic hierarchy process—Challenges and practical experiences. Implement. Online Softw. Tool. AHP Chall. Pract. Exp. 10, 1–20. Available from: https://bpmsg.com/ahp/ahp.php.

Guerra, A.J.T., Fullen, M.A., Jorge, M.C.O., Bezerra, J.F.R., Shokr, M.S., 2017. Slope processes, mass movement and soil erosion: a review. Pedosphere 27 (1), 27–41. Available from: https://doi.org/10.1016/S1002-0160(17)60294-7.

Halefom, A., Teshome, A., 2019. Modelling and mapping of erosion potentiality watersheds using AHP and GIS technique: a case study of Alamata Watershed, South Tigray, Ethiopia. Modeling Earth Syst. Environ. 5 (3), 819–831. Available from: https://doi.org/10.1007/s40808-018-00568-6.

Huabin, W., Gangjun, L., Weiya, X., Gonghui, W., 2005. GIS-based landslide hazard assessment: an overview. Prog. Phys. Geogr. 29 (4), 548–567.

Kanwal, S., Atif, S., Shafiq, M., 2017. GIS based landslide susceptibility mapping of northern areas of Pakistan, a case study of Shigar and Shyok Basins. Geomatics, Nat. Hazards Risk 8 (2), 348–366.

Labrière, N., Locatelli, B., Laumonier, Y., Freycon, V., Bernoux, M., 2015. Soil erosion in the humid tropics: a systematic quantitative review. Agric., Ecosyst. Environ. 203, 127–139. Available from: https://doi.org/10.1016/j.agee.2015.01.027.

Lim, K.J., Sagong, M., Engel, B.A., Tang, Z., Choi, J., Kim, K.S., 2005. GIS-based sediment assessment tool. Catena 64 (1), 61–80. Available from: https://doi.org/10.1016/j.catena.2005.06.013.

Lu, D., Li, G., Valladares, G.S., Batistella, M., 2004. Mapping soil erosion risk in Rondonia, Brazilian Amazonia: using RUSLE, remote sensing and GIS. Land. Degrad. Dev. 15 (5), 499–512. Available from: https://doi.org/10.1002/ldr.634.

Lu, H., Moran, C.J., Prosser, I.P., 2006. Modelling sediment delivery ratio over the Murray Darling Basin. Environ. Model. Softw. 21 (9), 1297–1308. Available from: https://doi.org/10.1016/j.envsoft.2005.04.021.

Mancini, F., Ceppi, C., Ritrovato, G., 2010. GIS and statistical analysis for landslide susceptibility mapping in the Daunia area (Italy). Nat. Hazards Earth Syst. Sci. 10, 1851–1864.

Mhangara, P., Kakembo, V., Lim, K.J., 2012. Soil erosion risk assessment of the Keiskamma catchment, South Africa using GIS and remote sensing. Environ. Earth Sci. 65 (7), 2087–2102. Available from: https://doi.org/10.1007/s12665-011-1190-x.

Mitasova, H., 2013. Treatise on geomorphology ‖ 3.9 GIS-Based Soil Erosion Modeling, 228–258. https://doi.org/10.1016/B978-0-12–374739-6.00052-X.

Moore, I.D., Panuska, J.C., Grayson, R.B., & Srivastava, K.P., 1988. Application of digital topographic modeling in hydrology.

Ming, C., Qiangguo, C., Axing, Z., Haoming, F., 2007. Soil erosion along a long slope in the gentle hilly areas of black soil region in Northeast China. J. Geogr. Sci. 17 (3), 375–383. Available from: https://doi.org/10.1007/s11442-007-0375-4.

Morgan, R.P.C., 2005. Soil Erosion and Conservation. Blackwell Publishing, Oxford.

Narayana, D.V., Babu, R., 1983. Estimation of soil erosion in India. J. Irrig. Drain. Eng. 109 (4), 419–434. Available from: https://doi.org/10.1061/(ASCE)0733-9437(1983)109:4(419).

Pan, J., Wen, Y., 2014. Estimation of soil erosion using RUSLE in Caijiamiao watershed, China. Nat. Hazards 71 (3), 2187–2205. Available from: https://doi.org/10.1007/s11069-013-1006-2.

Panagos, P., Borrelli, P., Poesen, J., Ballabio, C., Lugato, E., Meusburger, K., et al., 2015. The new assessment of soil loss by water erosion in Europe. Environ. Sci. Policy 54, 438–447.

Park, S., Oh, C., Jeon, S., Jung, H., Choi, C., 2011. Soil erosion risk in Korean watersheds, assessed using the revised universal soil loss equation. J. Hydrol. 399 (3–4), 263–273. Available from: https://doi.org/10.1016/j.jhydrol.2011.01.004.

Pradeep, G.S., Krishnan, M.V., Vijith, H., 2015. Identification of critical soil erosion prone areas and annual average soil loss in an upland agricultural watershed of Western Ghats, using analytical hierarchy process (AHP) and RUSLE techniques. Arab. J. Geosci. 8 (6), 3697–3711. Available from: https://doi.org/10.1007/s12517-014-1460-5.

Prasannakumar, V., Vijith, H., Abinod, S., Geetha, N.J.G.F., 2012. Estimation of soil erosion risk within a small mountainous sub-watershed in Kerala, India, using Revised Universal Soil Loss Equation (RUSLE) and geo-information technology. Geosci. Front. 3 (2), 209–215. Available from: https://doi.org/10.1016/j.gsf.2011.11.003.

Prasannakumar, V., Vijith, H., Geetha, N., Shiny, R., 2011. Regional scale erosion assessment of a sub-tropical highland segment in the Western Ghats of Kerala, South India. Water Resour. Manag. 25 (14), 3715–3727. Available from: https://doi.org/10.1007/s11269-011-9878-y.

Renard, K.G., 1997. Predicting Soil Erosion by Water: A Guide to Conservation Planning with the Revised Universal Soil Loss Equation (RUSLE). United States Government Printing.

Rouse, J.W.J., Haas, R.H., Schell, J.A., Deering, D.W., 1973. Monitoring vegetation systems in the Great Plains with ERTS. Third ERTS Symposium, NASA SP-351. US Gov. Printing office. Edited by SC Freden, EP Mercanti, and MA Becker. NASA.

Roy, P., Pal, S.C., Chakrabortty, R., Islam, A.R.M.T., Chowdhuri, I., Saha, A., 2022. Impact of ineffective measures on the increasing land degradation in a monsoon-dominated region of India: issues and policy implications. Land. Degrad. Dev. 33 (16), 3174–3185.
Rozos, D., Bathrellos, G.D., Skillodimou, H.D., 2011. Comparison of the implementation of rock engineering system and analytic hierarchy process methods, upon landslide susceptibility mapping, using GIS: a case study from the Eastern Achaia County of Peloponnesus, Greece. Environ. Earth Sci. 63 (1), 49–63. Available from: https://doi.org/10.1007/s12665-010-0687-z.
Saaty, T., 1980. The Analytic Hierarchy Process (AHP) for Decision Making. Kobe, Japan, pp. 1–69.
Saaty, T.L., 1977. A scaling method for priorities in hierarchical structures. J. Math. Psychol. 15 (3), 234–281. Available from: https://doi.org/10.1016/0022-2496(77)90033-5.
Sabins Jr, F.F., 1986. Remote Sensing: Principles and Interpretation. Chevron Oil Field Research Co.
Shamshad, A., Azhari, M.N., Isa, M.A., Hussin, W.W., Parida, B.P., 2008. Development of an appropriate procedure for estimation of RUSLE EI30 index and preparation of erosivity maps for Pulau Penang in Peninsular Malaysia. Catena 72 (3), 423–432. Available from: https://doi.org/10.1016/j.catena.2007.08.002.
Singh, G., Babu, R., Narain, P., Bhushan, L.S., Abrol, I.P., 1992. Soil erosion rates in India. J. Soil. Water Conserv. 47 (1), 97–99.
Tang, Q., Xu, Y., Bennett, S.J., Li, Y., 2015. Assessment of soil erosion using RUSLE and GIS: a case study of the Yangou watershed in the Loess Plateau, China. Environ. Earth Sci. 73, 1715–1724.
Thomas, J., Joseph, S., Thrivikramji, K.P., 2018a. Assessment of soil erosion in a tropical mountain river basin of the southern Western Ghats, India using RUSLE and GIS. Geosci. Front. 9 (3), 893–906. Available from: https://doi.org/10.1016/j.gsf.2017.05.011.
Thomas, J., Joseph, S., Thrivikramji, K.P., 2018b. Assessment of soil erosion in a monsoon-dominated mountain river basin in India using RUSLE-SDR and AHP. Hydrol. Sci. J. 63 (4), 542–560. Available from: https://doi.org/10.1080/02626667.2018.1429614.
Tian, Y.C., Zhou, Y.M., Wu, B.F., Zhou, W.F., 2009. Risk assessment of water soil erosion in upper basin of Miyun Reservoir, Beijing, China. Environ. Geol. 57 (4), 937–942. Available from: https://doi.org/10.1007/s00254-008-1376-z.
TOLIIriSm, M., 1995. Carrying capacity of Himalayan.
Tsegaye, L., Bharti, R., 2021. Soil erosion and sediment yield assessment using RUSLE and GIS-based approach in Anjeb watershed, Northwest Ethiopia. SN Appl. Sci. 3, 582. Available from: https://doi.org/10.1007/s42452-021-04564-x.
Ullah, I., Aslam, B., Shah, S.H.I.A., Tariq, A., Qin, S., Majeed, M., et al., 2022. An integrated approach of machine learning, remote sensing, and GIS data for the landslide susceptibility mapping. Land 11 (8), 1265.
Upendra, B., Ciba, M., Aiswarya, A., Dev, V.V., Sreenivasulu, G., Krishnan, K.A., 2022. Mechanisms controlling the dissolved load, chemical weathering and CO2 consumption rates of Cauvery River, South India: role of secondary soil minerals. Environ. Earth Sci. 81 (3), 1–18. Available from: https://doi.org/10.1007/s12665-022-10222-1.
Van der Knijff, J.M., Jones, R.J.A., Montanarella, L., 2000. Soil erosion risk: assessment in Europe.
Vanoni, V.A., 1975. Closure to "Factors Determining Bed Forms of Alluvial Streams.". J. Hydraul. Div. 101 (11), 1435–1440. Available from: https://doi.org/10.1061/JYCEAJ.0010546.
Walling, D.E., 1983. The sediment delivery problem. J. Hydrol. 65 (1–3), 209–237. Available from: https://doi.org/10.1016/0022-1694(83)90217-2.
Williams, J.R., Berndt, H.D., 1972. Sediment yield computed with universal equation. J. Hydraul. Div. 98 (12), 2087–2098. Available from: https://doi.org/10.1061/JYCEAJ.0003498.
Williams, J.R., Berndt, H.D., 1977. Determining the universal soil loss equation's length-slope factor for watersheds. In: A National Conference on Soil Erosion, pp. 24–26.
Wischmeier, W.H., Smith, D.D., 1978. Predicting rainfall erosion losses: a guide to conservation planning (No. 537). Department of Agriculture, Science and Education Administration.
Wu, Q., Wang, M., 2007. A framework for risk assessment on soil erosion by water using an integrated and systematic approach. J. Hydrol. 337 (1–2), 11–21. Available from: https://doi.org/10.1016/j.jhydrol.2007.01.022.
Yesuph, A.Y., Dagnew, A.B., 2019. Soil erosion mapping and severity analysis based on RUSLE model and local perception in the Beshillo Catchment of the Blue Nile Basin. Ethiopia. Environ. Syst. Res. 8 (1), 1–21.
Yitayew, M., Pokrzywka, S.J., Renard, K.G., 1999. Using GIS for facilitating erosion estimation. Appl. Eng. Agric. 15 (4), 295. Available from: https://doi.org/10.13031/2013.5780.
Yu, B., Rosewell, C.J., 2001. Evaluation of WEPP for runoff and soil loss prediction at Gunnedah, NSW, Australia. Soil. Res. 39 (5), 1131–1145. Available from: https://doi.org/10.1071/SR00091.

Chapter 34

Soil erosion risk assessment and watershed prioritization using remote sensing and GIS techniques: a case study of Tehri Garhwal, Uttarakhand, India

Justin George Kalambukattu, Suresh Kumar, Pujarini Dutta and Anu David Raj
Agriculture and Soils Department, Indian Institute of Remote Sensing (IIRS), Indian Space Research Organization (ISRO), Dehradun, Uttarakhand, India

34.1 Introduction

Soil erosion is one of the major factors of land degradation on the surface of the earth. A complex dynamic process exposes subsurface soil and causes sedimentation in reservoirs as productive surface soils are detached, transported, and deposited in a remote location (Jain et al., 2001). In addition to the natural erosion process, processes brought on by human effects on the ground surface, such as agricultural activities or deforestation, are also included in the term "soil erosion" (Morgan et al., 1998). It can harm the environment, reduce farm business profitability, and annoy the public (Defra, 2005). It has been noted that surface and gully erosion frequently results in the loss of fertile topsoil, and agricultural land has expanded to places with marginal soil cover (Hofer, 1998). Due to land degradation from excessive deforestation, overgrazing, and subsistence farming caused by population pressure, large-scale road development, mining, etc., natural resources in hilly terrain are severely harmed. Due to deforestation and the heavy rainfall impact, fragile terrain with steep slopes is now more vulnerable to catastrophic soil erosion.

As the rate of loss of soil is almost 10–40 times faster than its rate of renewal, soil erosion is one of the most significant environmental and public health issues currently facing human societies (Pimentel, 2006). It poses a serious threat to the long-term sustainability of agricultural productivity and environmental quality. Out of India's 329 Mha overall geographic area, it has been calculated that 167 Mha suffers from severe water and wind erosion. Included in this are 127 Mha of soil erosion and 40 Mha of degradation brought on by gullies, ravines, shifting tillage, waterlogging, salinity, and alkalinity, changing river courses, and desertification (Jain et al., 2001; Das, 1985). According to Narayana and Babu (1983), around 5334 Mt (16.4 t ha^{-1}) of soil is removed from India each year, of which roughly 29%is transported away by rivers into the sea and 10%is deposited in reservoirs, resulting in a significant loss of storage capacity. The entire Himalayan region in northern India is also plagued by a significant problem with soil erosion, and rivers that pass through this area carry substantial amounts of debris. This region contributes over 25% of the dissolved load to the global ocean (Raymo and Ruddiman, 1992). According to Garde and Kothyari (1987), the northern Himalayan region has a high rate of soil erosion, between 2000 and 2500 tons km^{-2} $year^{-1}$.

For the Himalayan mountains' rugged topography to remain productive, soil resources should be maintained sustainably. The Himalayan region's population is heavily reliant on the farming sector, particularly subsistence farming. Utilizing soil and water resources sustainably is essential for mountain development (Ives & Messerli, 1989). The mountainous environment is under tremendous pressure from the ever-increasing population and its dependence on scarce natural resources. In the Himalayan region, landslides, mudslides, collapse of man-made terraces, soil loss from steep slopes, and shrinkage of grazing and forest areas are the primary causes of land resource degradation. In the recent decade in the United States, the loss of valuable land resources and water resources reached $4 billion in a year, not to mention the damage caused by soil erosion, which was calculated up to $ 500 million in a year.

Remote Sensing of Soils. DOI: https://doi.org/10.1016/B978-0-443-18773-5.00036-3

Mitas and Mitasova (1998) reported the use of geographic information systems (GIS) techniques for erosion/deposition modeling with the unit stream power–based erosion and deposition (USPED) soil erosion model. In order to anticipate soil erosion and sediment deposition, Mitasova et al. (2005) reported the validation of a 3D improvement of the universal soil loss equation. In order to evaluate soil erosion using revised universal soil loss equation (RUSLE)-3D and USPED at river basin size in southern Italy, Aiello et al. (2015) used remote sensing (RS) and GIS. To evaluate soil erosion using the RUSLE and USPED models, Bogdan et al. (2012), Rodriguez et al. (2012), and Oliveira et al. (2013) used RS techniques. The soil loss was estimated by using RUSLE model. With the help of RS and GIS techniques, soil erosion risk assessment measurement has become easier, cheaper, and quicker. Using RS and GIS techniques, several approaches are in use for the prioritization of watersheds. Among them, RUSLE and USPED are used in a deterministic erosion prediction model. Using topographic data obtained through RS or field surveys, one can determine stream power, which is the rate of energy consumption along a river (Lea and Legleiter, 2016).

Rahaman et al. (2015) conducted research on the Kallar watershed in Tamil Nadu, prioritizing sub-watersheds based on morphometric parameters utilizing fuzzy analytical hierarchy processes and GIS. Sadeghi et al. (2012) conducted studies on soil erosion evaluation and prioritizing at the plot scale. This study evaluated numerous parameters, including silt content, slope, aspect, and vegetation cover, each of which had four levels of study area in Lar Township.

Strategies for soil conservation are established based on how serious the soil erosion issue is in every given watershed. This severity is demonstrated by the watershed's priority delineation, which takes into account a number of variables, the most significant of which is the annual soil loss from the watershed, which is a quantitative indicator of the loss of top soil that is thought to be fertile for crop development. Strategies for soil conservation are established based on how serious the soil erosion issue is in every given watershed. This severity is demonstrated by the watershed's priority delineation, which takes into account several variables, the most significant of which is the annual soil loss from the watershed, which is a quantitative indicator of the loss of topsoil that is thought to be fertile for crop development. The main objectives of this research were to calculate the amount of soil erosion rate in the Himalayan Mountain region by RUSLE and USPED models, as well as micro watershed prioritization for the utilization of land resources and environmental protection. Therefore, the steepest upper portion of the research region should adopt sustainable soil and water conservation techniques while taking into account the watershed's sensitivity, inhabitants, and potential. These can be integrated with other topographic and ancillary data into a GIS environment to be analyzed and to produce a soil erosion risk map. The present study was undertaken to produce a soil erosion risk map of the study area for soil conservation planning.

34.2 Study area

The study area watershed is located in "Tehri" dam in Tehri Garhwal District, Uttarakhand, and lies between 30°16′30″ N to 30°21′30″ N latitude and 78°24′0″ E to 78°29′30″ E longitude. The elevation of this watershed ranged from 556 to 2147 m above the mean sea level. It is surrounded by the Himalayan Mountains in the northeastern part, and the outlet of this watershed is the Bhagirathi River. Geologically, Tehri Garhwal district is represented by the rocks of the Lesser Himalayas and Central Himalaya. The study area occupies five different age groups of rocks, viz., quartzite, shale, limestone, slate, and phyllite. Even though, quartzite and shale occur as a single unit, the age of these rock types does not belong to the same geological period. All the rocks found in the study area are structurally controlled and exposed on the northern slope. The drainage pattern of the area is mainly paralleled, sub-paralleled, and subdendritic. It is controlled by lithology and structure. The drainage density varies with the lithology. In quartzite, the drainage density is coarse to medium, and the drainage pattern is parallel to subparallel and mostly controlled by joints. In the phyllitic rock, the drainage density is higher, and the pattern of the drainage is subdendritic to subparallel (Fig. 34.1).

The major land use/land cover (LULC) in the study area is the forest with varying densities, viz., dense forest, moderate forest, and open forest, followed by scrubs, fallows, and cropland. The watershed is covered mostly by forest, including oak (*Quercus*), pines (*Pinus*), and deodar (*Cedrus*) or deodar mixed. Shrubs and grasses are mostly covered on the lower slopes. The climate in Tehri Garhwal district varies from cold temperate to tropical and subtropical temperatures. *Saccharum spontanium* and *Cynodon dactylon* are the species of grass found in fallow and scrubland.

34.3 Materials and methods

Different RS techniques and ancillary data were used in this study. Resources at high-resolution imagery of LISS IV data with a resolution of 5.8 m in multispectral were used. Cartosat1, or IRSP5, is a stereoscopic Earth observation satellite in a sun-synchronous orbit and the first one of the Cartosat series of satellites. The watershed boundary was

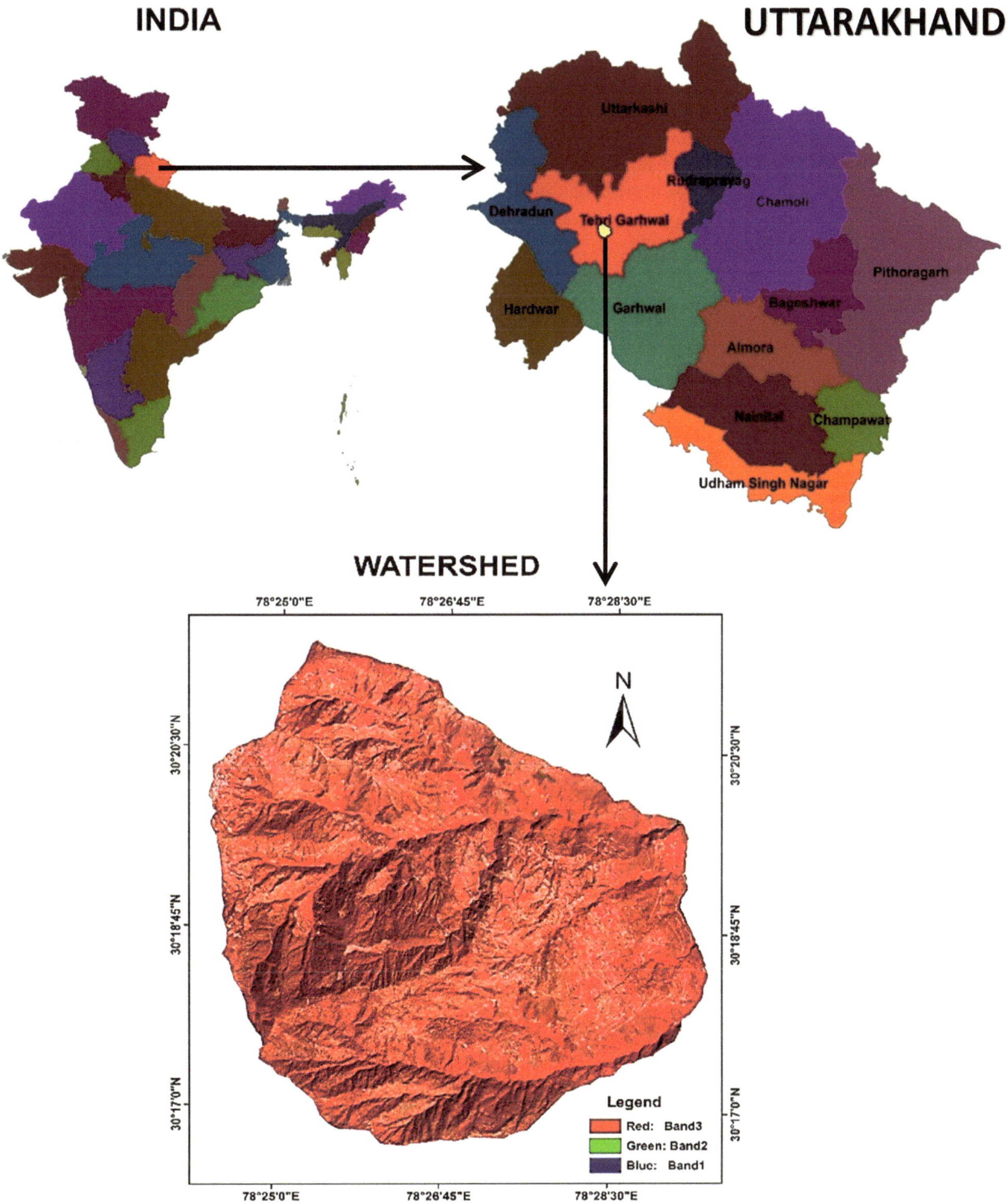

FIGURE 34.1 Location of study area: Tehri Garhwal, Uttarakhand.

marked with the help of a toposheet, the CARTODEM (10 m spatial resolution) data, and GIS tools (ARC-GIS hydrology tool). The slope map was classified into four classes using Cartosat DEM. A total of eight types of land use were delineated by the ground survey. The classes include cropland, valley agriculture, dense forest, moderate forest, open forest, scrub land, fallow land, and settlement. The physiographic map was prepared by the use of the LULC map and slope map, which were prepared by the CARTOSAT DEM (10 m spatial resolution). Two Survey of India topographic maps on 1:50,000 scales of the entire cover were utilized. Global Positioning System receiver, which uses signals from a group of orbiting satellites to locate the position features on the earth's surface, was used to collect ground truth data.

The ground survey was carried out to identify various LULC classes on the field. The ground truth verification was undertaken to cover the area of maximum heterogeneity. Ground truth sites for various LULC types, viz., terraced agriculture, dense forest, moderate forest, open forest, scrub land, and fallow land due to the settlements, were marked on the land use interpretation map with the help of satellite data. The climatic data, such as rainfall, temperature, and humidity, were collected from the Ranichauri University (Automatic Weather Station). The field survey collected 129 surface and subsurface soil samples in different transects. The soil samples were collected during the field visit and then analyzed in the soil lab for pH, EC, texture, and organic carbon. The overall methodology for the study is shown in Fig. 34.2.

34.3.1 RUSLE model

RUSLE is an empirical model of quantitative soil erosion by rain according to the USLE model. RUSLE model updated information of input data and combined with a process of soil erosion. It retained the basic USLE structure but used the algorithm to calculate individual factors that have been changed significantly. The model can be described as (Renard et al., 1997):

$$A = R \times K \times L \times S \times C \times P$$

where A is the average of soil loss by erosion.

34.3.1.1 R-factor (rainfall erosivity factor)

It quantifies the effect of runoff, which is likely to be associated with rain. R is the rainfall erosivity factor. Erosivity is the potential for erosion. It quantifies the impact of raindrops on the amount and rate of runoff likely to be associated with rain. The R factor was estimated using monthly rainfall data for the from a for 1990–2013. The term "rainfall erosivity" refers to the energy required by rain to detach soil particles from one another so that they can be splashed out and then carried away by runoff. Data on precipitation was gathered from a metrological station. The erosion index (EI) for a given storm is the product of the kinetic energy of the falling raindrops and has a maximum intensity of 30 minutes. The relationship between annual rainfall erosivity (R) and rainfall was described by the following linear equation (Babu and Dhyani, CSWCRTI/Annual Report, 2000–2001).

$$R = 81.5 + 0.375 \times A$$

where A is the average annual rainfall.

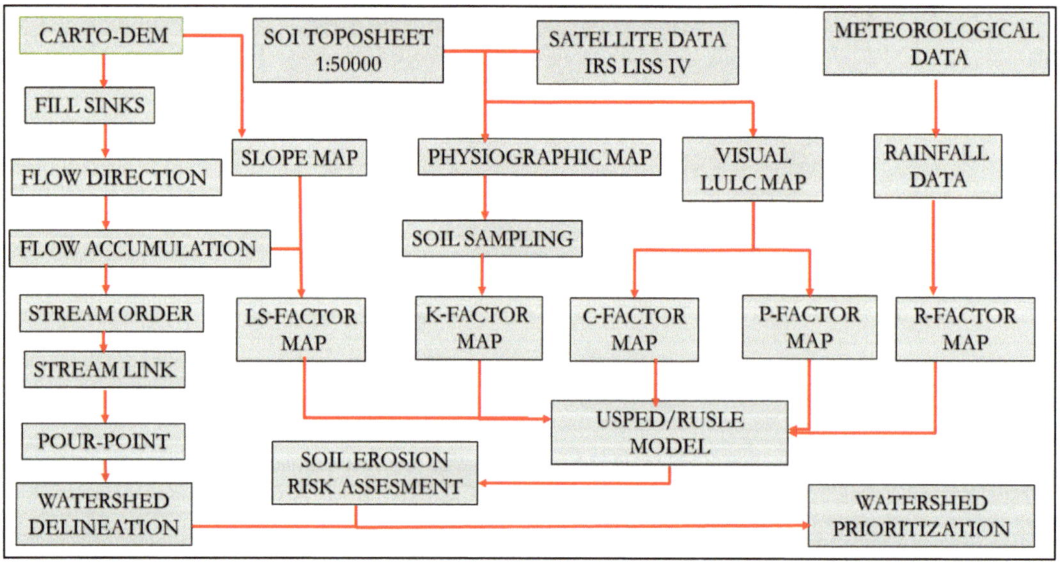

FIGURE 34.2 Methodology flowchart.

34.3.1.2 K-factor (soil erodibility factor)

It measures how susceptible the soil is to erosion. The K-factor represents both the soil's susceptibility to erosion and the volume and flow of runoff, as determined under the conditions of a typical unit plot. Infiltration rate, permeability, water retention forces, porosity, exchangeable ions, primary particle size, and mineralogy, aggregate size and stability, and soil organic matter are just a few of the soil characteristics that have a big impact on how the soil behaves when it comes to soil–water interaction. The impact of soil moisture and temperature on organic matter also affects how easily soil erodes. *K* values for various physiographic soil units were derived based on soil profile research and laboratory investigation, and a soil erodibility K-factor map was created derived from information on soil characteristics, viz., permeability, organic matter, texture, structure, and soil depth for each soil unit. It was computed using the following equation:

$$K = 2.8 \times 10 - 7\mathrm{M}\ 1.14(12 - a) + 4.3 \times 10 - 3(b - 2) + 3.3 \times 10 - 3(c - 3)$$

where K = soil erodibility factor, a = organic matter content b = structure of soil, c = permeability, and M = (% silt + % very fine sand) × (100−% clay). The K values were added to the soil attribute table against each of the 13 physiographic units.

34.3.1.3 LS-factor (length steepness factor)

It represents the erodibility due to the combination of slope length and steepness relative to a standard unit plot. The topological (L and S) factor value was derived by computing slope length (L) and slope steepness (S) factors using map calculation in GIS. To calculate the LS factor, we need DEM, flow accumulation, and slope in the degree map.

34.3.1.3.1 L-factor

The influence of slope length on erosion is represented by this variable. The slope length is the distance an overland flow travels before concentrating or depositing at a specific site. It is determined by dividing the amount of soil lost from a field's slope by the slope of a typical plot, which is 22.13 m. The LS factor takes into account how topography affects erosion. The following equation is used to compute the slope length factor (L).

$$\mathrm{L} = (\text{flow accumulation} \times \text{grid size}/22.13)\hat{}0.6$$

where grid size = 10 m

34.3.1.3.2 S-factor

The S factor represents how steep the slope is. It shows how the steepness of the slope affects erosion; the rate of soil loss increases with slope length. The density of vegetation cover and soil particle size have an impact on how soil loss relates to the gradient. The S-factor is the comparison between soil loss due to a field gradient and soil loss due to a 9% slope. The following equation is used to compute the S-factor:

$$S = (\sin H/0.896)\hat{}1.3$$

where H is the slope map in radians. The LS factor was calculated using the following equation:

$$\mathrm{LS} = \left(\text{Flow accumulation} \times \text{cell size}/22.13\right)0.6 \times \left(\text{Sin } H/0.0896\right)1.3$$

where H = slope map in radian (Mitasova et al., 1996)

34.3.1.4 C-factor (crop management factor)

The expected ratio of soil loss from a farmed area under a particular circumstance to soil loss from a fallow area with a clean surface on a similar slope and rainfall condition is defined as the C-factor. The C-factor is employed to quantify how crops and management techniques affect erosion rates. It is frequently used to examine the relative effects of different management strategies on soil conservation plans' average annual soil loss and how that potential is spread over time throughout various activities like crop rotation or other management strategies. In this study, using satellite data (LISS IV), the LULC map was prepared based on visual interpretation. C-factor values were assigned for each LULC class, based on literature, related to soil erosion studies in similar land cover/land uses and areas of the Himalayas.

34.3.1.5 P-factor (support practice factor)

It is the ratio of soil loss caused by a certain support practice to the equivalent loss caused by uphill and downhill cultivation. It represents the supporting practice of conserving the soil. It depicts how support practices affect the average annual rate of erosion. This study used the FCC's visual interpretation to identify and designate areas that showed various management approaches. The management practice values were then assigned based on the available literature to produce an attribute class reflecting the P value in the region, and ultimately a P-factor map was developed. P-factor describes the effect of those management practices that are conservation-orientated: contouring, ridge planting, strip cropping, and terracing.

34.3.2 USPDE model

A simplistic model called USPED estimates the spatial distribution of erosion and deposition rates for a steady-state overland flow with uniform rainfall excess conditions for the scenario of erosion where transport capacity is constrained. The hypothesis first presented by Moore and Burch (1986) is the foundation of the model, which has undergone multiple improvements. Assuming that the sediment flow rate qs(r) is at the sediment transport capacity $T(r)$, $r = (x,y)$, which is approximately represented by (Julien and Simons, 1985).

$$|qs(r)| = T(r) = Kt(r)|q(r)|m \sin b(r)n$$

where $b(r)$ [deg]is slope, $q(r)$ is water flow rate, $Kt(r)$ is transportability coefficient dependent on soil and cover, and m and n are constants depending on the type of flow and soil properties. For overland flow, the constants are usually set to $m = 1.6$ and $n = 1.3$.

The net erosion/deposition is estimated using the equation, and is expressed as:

$$ED = d(T \cos a)/dx + d(T \sin a)/dy$$

where; a [deg] is the aspect of the terrain surface, $dx = dy$ is the grid resolution, and

$$T = R\ K\ C\ P\ A\ m(\sin b)n$$

where T is the total capacity for transporting sediment, b [deg] is the slope, A is the area that contributes to the upslope, and m and n are constants. Similar to the USLE model, USPED's soil erosion modeling method was developed (Mitasova et al., 1996, Mitas and Mitasova, 1998). Only the LS-factors have changed into the K-, C-, and P-factors maps from the RUSLE model. Where b [deg] is the slope, A is the upslope contributing area, m and n are constants, R is a rainfall–runoff erosivity factor, K is a soil erodibility factor, C is a cover management factor, and P is a support practice factor. The sole essential difference between these two models is that USPED believes that soil erosion depends on both the capacity of surface runoff to carry sediment and rainfall detachment capacity, whereas RUSLE thinks that erosion is primarily dependent on rainfall detachment capacity. Positive numbers in the USPED model indicate net deposition, whereas negative values indicate net erosion.

34.3.3 Prioritization of micro-watershed

The soil loss map using the RUSLE and USPED model was crossed with the micro- watershed map under the GIS environment, and the weighted average soil loss per micro watershed was computed from the attribute table. Based on the annual average soil loss from each micro-watershed, the priority category was assigned as I—High, II—Medium, III—Low, and IV—Very Low for conservation planning. The watershed was delineated into 45 micro watersheds.

34.4 Results and discussions

Using RS and GIS input, spatial erosion modeling by the RUSLE model and USPED model was performed to predict average soil loss. Physiographic soil maps, slope maps, LULC maps, and other theme layers were created as a spatial database for the study. The results obtained in the study are presented below:

34.4.1 Land use/land cover map

The information on land use activities was inferred visually from the toposheet and satellite pictures. It was digitally captured to prepare digital vector coverage and rasterize the map for spatial GIS analysis. Forest, agricultural, scrubland, and settlement were the four major LULC categories used to classify the entire watershed. Agriculture land was further classified into cropland and valley agriculture. The forest was classified into dense forest, moderate forest, and open forest. Scrub land was further subdivided into scrub land and fallow land. Thus, a total of 8 LULC classes were identified in the study area (Fig. 34.3). Forest is the dominant LULC type in this area. About 56.59% of the study area is under forest cover, and 25.4% of the area is under agricultural land. The major crops grown in the area are wheat, vegetables, mustard, fruit trees, etc. Paddy is cultivated in a few areas, where valley and irrigation facilities are present.

34.4.2 Slope map

The slope map was generated using the DEM. The slope in the study area was classified into five classes. From the slope map (Fig. 34.4), we can easily identify that 45.76% of the area falls under very steep slope (>33%), 13.15% of

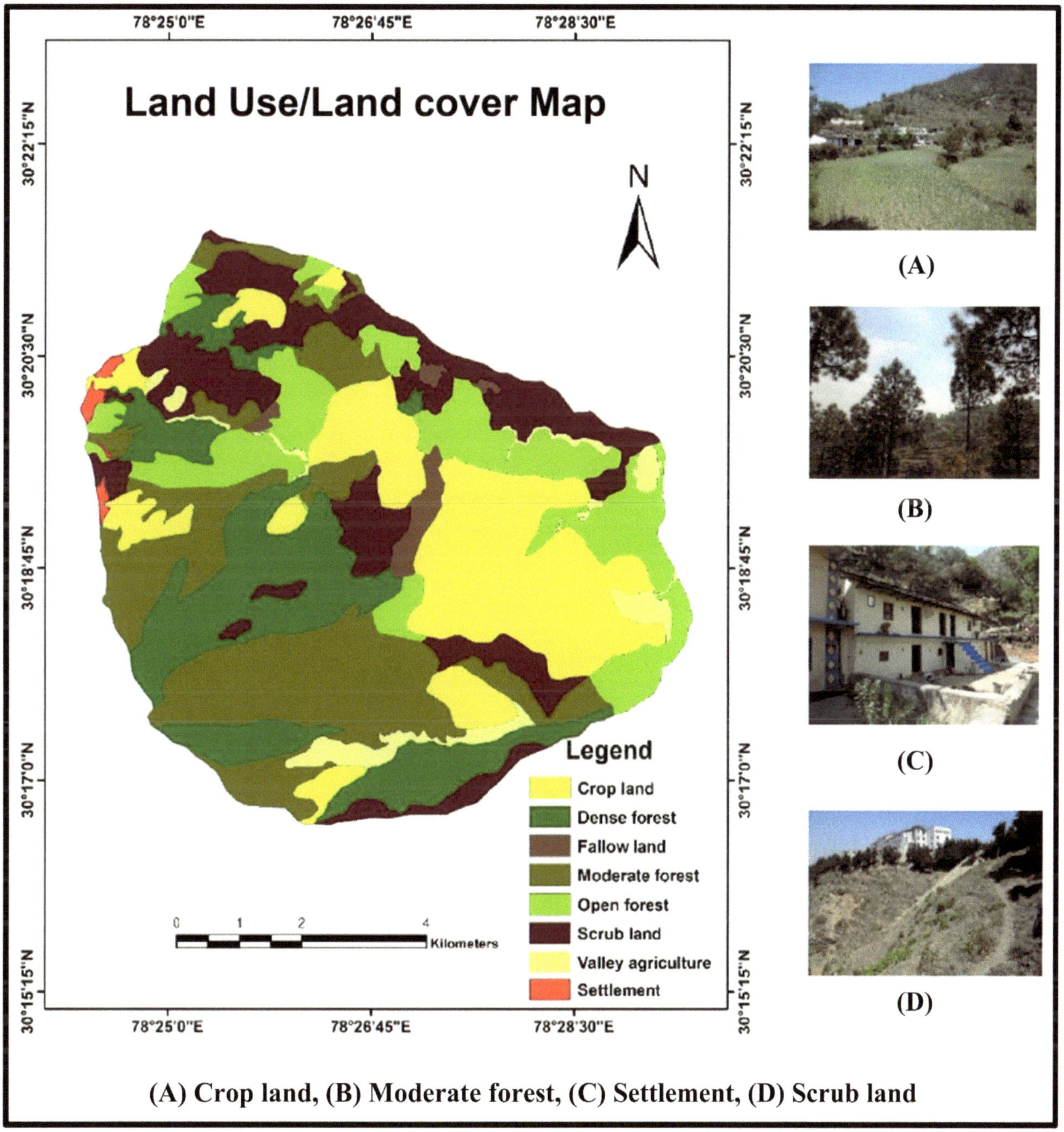

FIGURE 34.3 LULC map.

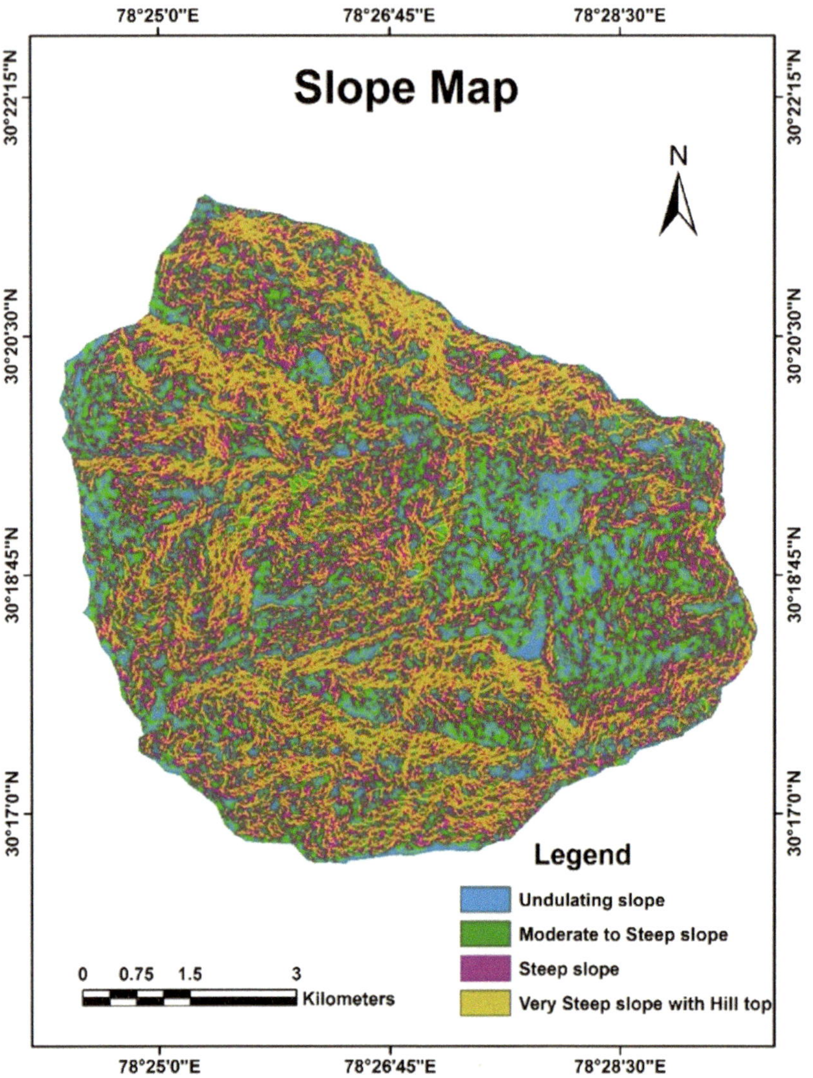

FIGURE 34.4 Slope map of the watershed.

the area falls under steep slope (25%–33%), 16.44% of the area falls under moderate to steep slope (15%–25%), and 24.65% of the area falls under undulating slope (0%–15%) category.

34.4.3 Physiographic soil map

While preparing physiographic soil mapping, landscape factors were taken into account, which denote the details of relief related to the different types of landforms present in the study area, such as hills, rivers, etc. This data was used to distinguish between different physiographic units in satellite imagery, as well as several picture elements related to terrain features such as landforms, topography, slope, erosion condition, relief, drainage network, etc. The aerial distribution of physiographic units is illustrated in Table 34.1 (Fig. 34.5).

34.4.4 Soil erosion prediction using the RUSLE model

Various model parameters required to run the RUSLE model for estimating soil erosion were parameterized and prepared as spatial data layers. The results are given below:

34.4.4.1 R-factor map

To calculate the R-factor, 23-year (1990–2011) rainfall data for the region was obtained from Ranichauri University. Using this data, we calculated the average annual rainfall. As the average annual rainfall was greater than 340 mm and

TABLE 34.1 Areal distribution of physiographic soil map.

Physiographic units	Area (%)
Undulating valley with forest (H11)	1.25
Undulating valley with agriculture (H12)	3.15
Undulating valley with scrub (H13)	0.62
Moderate to steep slope with forest (H21)	2.80
Moderate to steep slope with agriculture (H22)	15.06
Moderate to steep slope with scrub (H23)	1.13
Steep slope with forest (H31)	22.97
Steep slope with agriculture (H32)	6.65
Steep slope with scrub (H33)	5.03
Very Steep slope with forest (H41)	29.36
Very Steep slope with agriculture (H42)	1.06
Very Steep slope with scrub (H43)	10.43
Settlement (S)	0.50

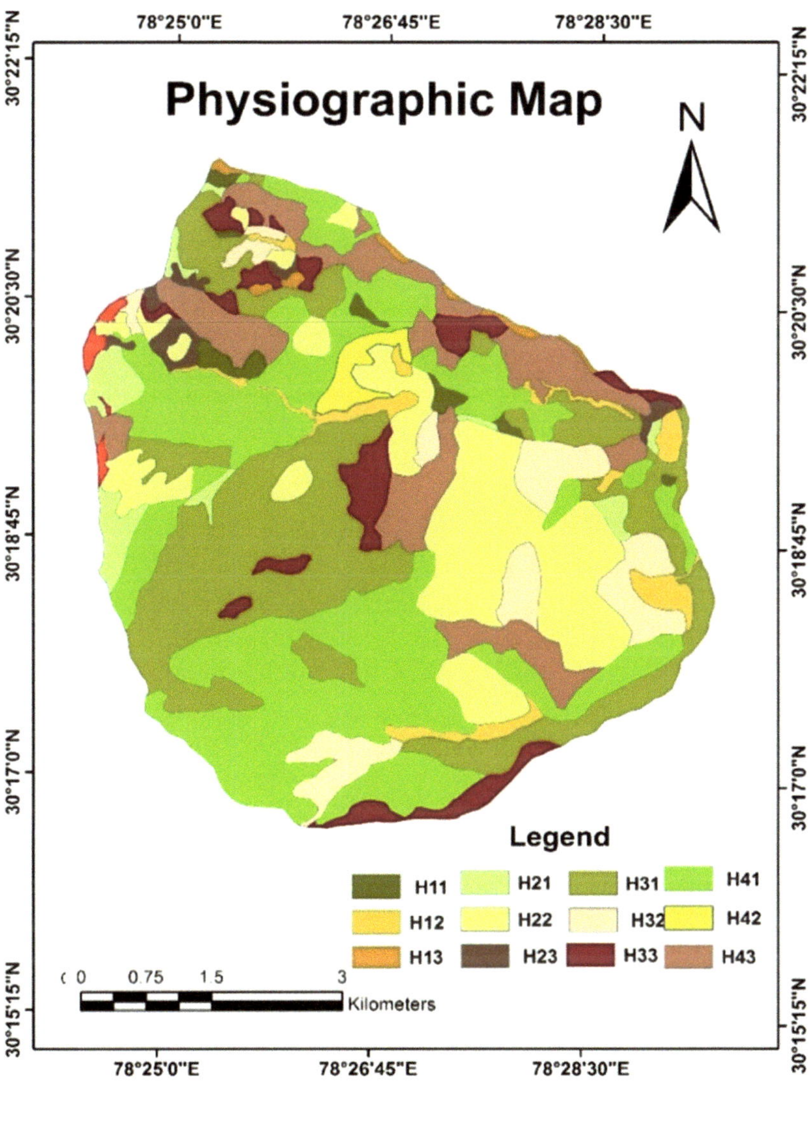

FIGURE 34.5 Physiographic soil map.

less than 3500 mm, we applied Eq. (1). In the watershed area, the average annual rainfall was 1400 mm. The R-factor derived from Eq. (1) is 606.5 mt ha-cm^{-1}.

34.4.4.2 K-factor map

Soil erodibility accounts for the influence or response of soil properties on soil loss during rainfall events. In the soil erodibility factor, the highest value is assigned to silt loam, and the lowest value is assigned to loamy sand. Values are assigned to organic matter, soil structure code, and profile permeability according to the characteristics of soils. K-factor characteristics are given in the table (Table 34.2). The sensitivity of the K-factor to soil texture has been analyzed. Values were assigned based on different textures. Fieldwork was done based on a physiographic unit map, and the soil samples, collected from each physiographic unit were analyzed in the laboratory for pH, EC, organic matter, and texture. The data generated by laboratory analysis was used for the determination of K-values.

Soil erodibility decreases as organic carbon concentration increases. A soil erodibility map was prepared using the data obtained from soil sample analysis in the laboratory. The percentage of sand, silt, and clay in the soil, soil organic matter, soil permeability class, and soil structure class are soil parameters utilized for the K-factor. The soil erodibility values were computed for each of the physiographic units based on the above-mentioned parameters. K map was generated

from the physiographic soil map. The K value in the watershed varied from 0.033 to 0.058. The following map shows the distribution of K values in the watershed.

34.4.4.3 LS-factor

Topographical characteristics play the main role in soil erosion, and the most important is slope. The slope supplies energy for the transport of material down the foot, and it increases the erosion process. The hilly areas have stronger weathered conditions than the others, so the soil loss is higher. The LS-factor was generated by multiplying the L and S factor maps. For this, we used the flow accumulation and slope map of the area, generated using DEM. To calculate the S-factor, we first had a slope map, then converted it from percentage to degree, and then from degree to radian. The map was generated from both the L and S factor (Fig. 34.6) and values of LS varies from 0.036 to 8.37.

34.4.4.4 CP-factor

Land cover represents the effect of plants and soil-disturbing activities. The C-factor was derived from LISS IV and LANDSAT images and based on ground truth data. The land cover factor was prepared from LULC map of the study

TABLE 34.2 Physiographic unit-wise distribution of soil erodibility.

Physiographic units	K-factor
Undulating valley with forest (H11)	0.040
Undulating valley with agriculture (H12)	0.043
Undulating valley with scrub (H13)	0.051
Moderate to steep slope with forest (H21)	0.049
Moderate to steep slope with agriculture (H22)	0.051
Moderate to steep slope with scrub (H23)	0.038
Steep slope with forest (H31)	0.043
Steep slope with agriculture (H32)	0.047
Steep slope with scrub (H33)	0.049
Very steep slope with forest (H41)	0.033
Very steep slope with agriculture (H42)	0.058
Very steep slope with scrub (H43)	0.052
Settlement (S)	0.033

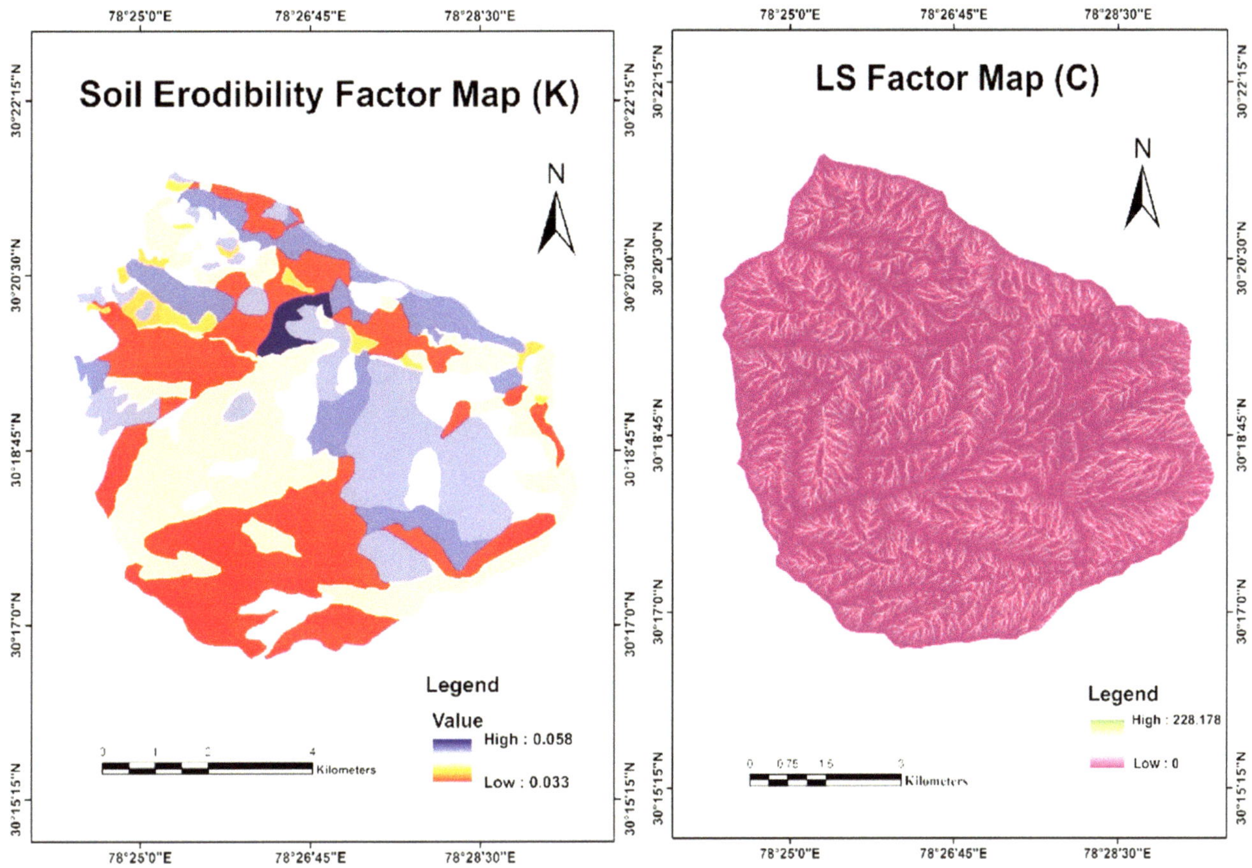

FIGURE 34.6 Soil erodibility and slope length factor.

TABLE 34.3 Land cover–based C- and P-factor of the watershed.

LULC	C factor	P factor
Crop land	0.35	0.5
Valley agriculture	0.25	0.6
Dense forest	0.008	1
Moderate forest	0.08	1
Open forest	0.4	1
Scrub land	0.6	1
Fallow land	0.7	0.8
Settlement	0.1	1

area. The crop management factor was estimated mainly from the literature (Table 34.3). P-factor was calculated based on the LULC map of the study area and literature references. Higher value of P indicates a higher erosion risk. Some fields have good bunds, which can prevent the loss of soil that is included in good management. In the land management factor, the lowest value is assigned to crop land (Fig. 34.7).

Agriculture was assigned the greatest value. In dense forests, the values varied from 0.008 to 0.7; in fallow ground, they were 0.7. The higher values of the C factor imply no cover effect and soil loss similar to that from bare soil, while lower values indicate a very strong cover effect and less soil loss similar to that from bare soil, resulting in reduced or negligible erosion.

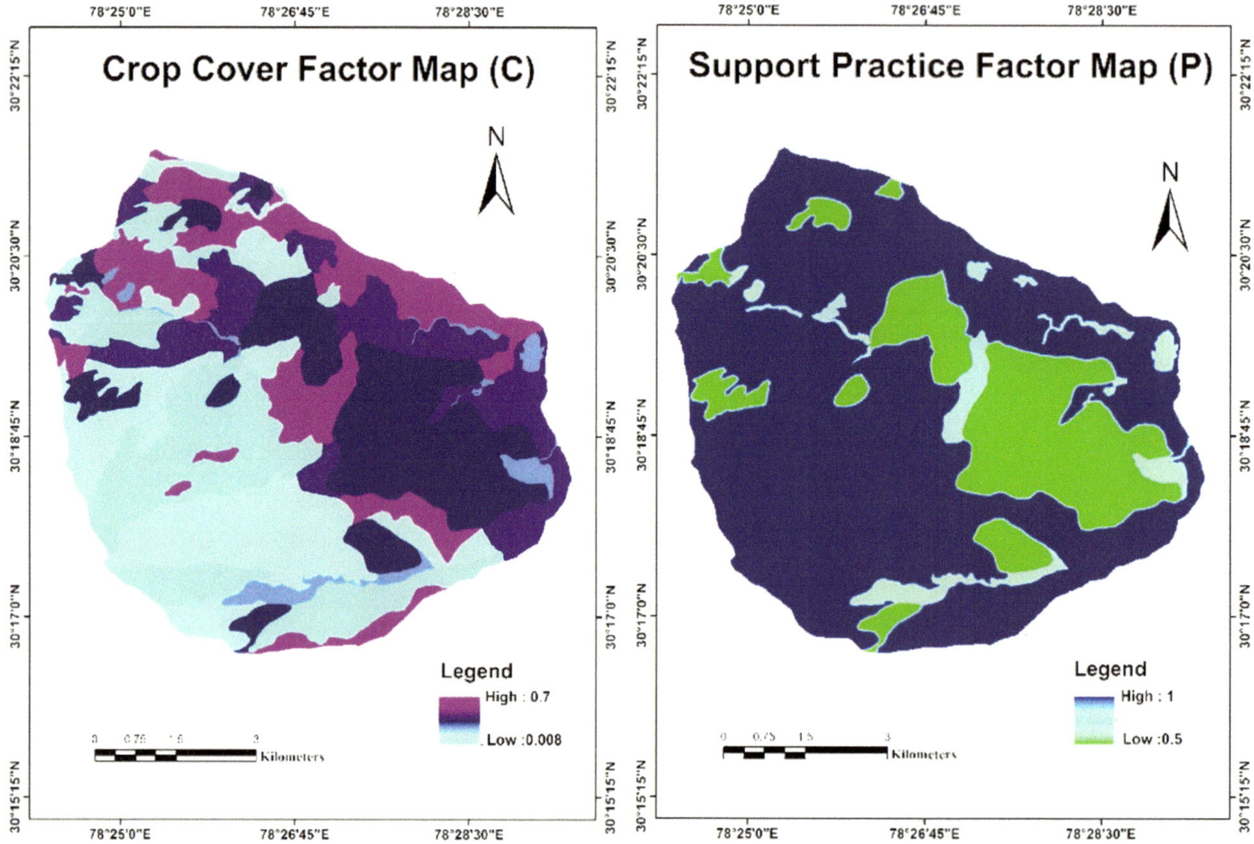

FIGURE 34.7 Crop cover factor and practice factor.

34.4.5 Soil loss—RUSLE

All the maps were integrated to generate an erosion map in a GIS environment. Thus, average annual soil loss ($t\ ha^{-1}\ year^{-1}$) from the watershed was computed (Fig. 34.10). Soil erosion was classified into seven soil erosion risk classes: slight (0–1 $t\ ha^{-1}\ year^{-1}$), slight to moderate (1–4 $t\ ha^{-1}\ year^{-1}$), moderate (4–8 $t\ ha^{-1}\ year^{-1}$), moderate to high (8–20 $t\ ha^{-1}\ year^{-1}$), high (20–40 $t\ ha^{-1}\ year^{-1}$), very high (40–100 $t\ ha^{-1}\ year^{-1}$), and extremely high ($>100\ t\ ha^{-1}\ year^{-1}$). The study revealed that 36.25% of the area of the watershed falls under moderate to high erosion risk class (4–40 $t\ ha^{-1}\ year^{-1}$) and 15.60% of the area falls under slight (0–1 $t\ ha^{-1}\ year^{-1}$) erosion risk class (Table 34.4). LULC is one of the most important factors that affect soil erosion. It shows that the most vulnerable land to soil erosion could be the open scrubs (87.98 $t\ ha^{-1}\ year^{-1}$), and the least erosion could occur in dense mixed forest (3.24 $t\ ha^{-1}\ year^{-1}$) areas. The second-highest soil loss is seen in open forest (48.63 $t\ ha^{-1}\ year^{-1}$), and in crop land areas, the soil loss value is 32.83 $t\ ha^{-1}\ year^{-1}$.

34.4.6 Soil erosion/deposition—USPED model

In USPED model, the R-, K-, C-, and P-factors maps are similar to RUSLE model, with only the LS-factors being changed. A rill erosion map and a sheet erosion map are necessary to generate USPED model. The amount of water erosion caused by sheet erosion increases with slope and distance. Sheet erosion occurs when rainfall intensity exceeds infiltration. Rill erosion occurs when runoff water forms tiny paths as it runs down a slope. Rill erosion increases with slope and distance, and rills may eventually form gullies (Fig. 34.8).

Soil erosion/deposition was estimated by multiplying the USPED factor maps on a GIS platform. In USPED model, positive values represent net deposition, while negative values represent net erosion. All the maps were integrated to generate erosion map in a GIS environment. Thus, the average annual soil loss ($t\ ha^{-1}\ year^{-1}$) from the watershed was computed. Soil erosion was classified into ten soil erosion risk classes of slight erosion/deposition ($\pm$0–15 $t\ ha^{-1}\ year^{-1}$), moderate erosion/deposition ($\pm$15–40 $t\ ha^{-1}\ year^{-1}$), moderate to high erosion/deposition

TABLE 34.4 Areal distribution of soil erosion classes in the watershed (RUSLE).

Classes	Soil loss ($t\ ha^{-1}\ year^{-1}$)	Area(ha)	Area(%)
Slight	0–1	818.61	15.60
Slight to moderate	1–4	279.97	5.33
Moderate	4–8	1039.91	19.81
Moderate to high	8–20	1364.49	26.00
High	20–40	951.55	18.13
Very high	40–100	793.81	15.12
Extremely high	>100	0.44	0.01
Total		5248.78	100.00

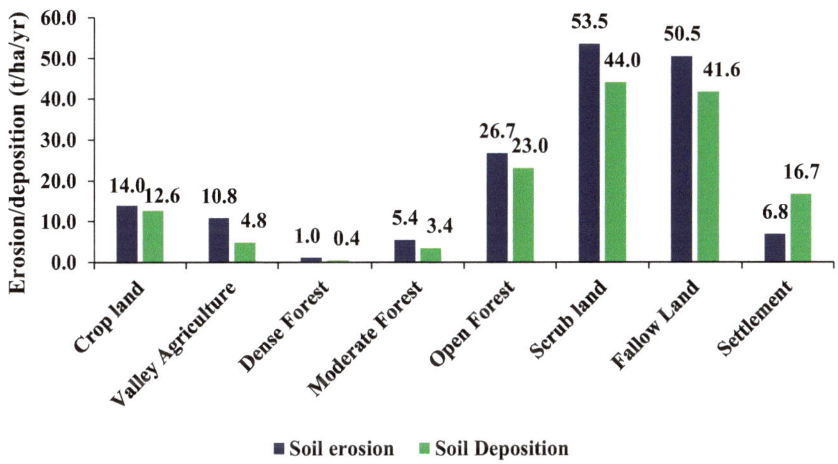

FIGURE 34.8 Soil erosion (RUSLE) and deposition (USPED) estimated.

($\pm$40–80 $t\ ha^{-1}\ year^{-1}$), high erosion/deposition ($\pm$80–100 $t\ ha^{-1}\ year^{-1}$), and very high erosion/deposition ($\pm$ >100 $t\ ha^{-1}\ year^{-1}$). The study revealed that 32.06% of the area of the watershed falls under slight erosion risk class (0– −15 $t\ ha^{-1}\ year^{-1}$), and 15.78% of the area falls under moderate (−15 to −40 $t\ ha^{-1}\ year^{-1}$) erosion risk class (Table 34.5) (Figs. 34.9 and 34.10).

34.4.7 Watershed prioritization

A watershed is an entire area draining through a common point. It can also be described as the terrain that supplies water to a specific location. Imagine this as a line that stretches from the lowest point to the highest point in a region. Small streams or tributaries get runoff from precipitation that falls inside this line, which joins to form rivers. Microwatershed area is from 100 to 1000 hectares in the study area. Microwatershed is created by the catchment area of the stream (Fig. 34.11). The priority class was assigned to the microwatersheds based on the annual soil erosion in each microwatershed (Fig. 34.12). The average annual soil erosion varies from 3.14 tons $ha^{-1}\ year^{-1}$ in TW-29 to 50.49 tons $ha^{-1}\ year^{-1}$ in TW-40 using RUSLE model.

The average annual soil erosion varies from -1.64 tons $ha^{-1}\ year^{-1}$ in TW-29 to −42.75 tons $ha^{-1}\ year^{-1}$ in TW-40 using USPED model. The extent of each microwatershed are given in Table 34.5. Soil erosion of microwatershed was classified into four priority classes, and priority category were assigned as very low, low, medium, and high for conservation planning in 45 microwatersheds (Fig. 34.13).

TABLE 34.5 Areal distribution soil erosion/deposition classes in the watershed (USPED).

Classes	Soil erosion/deposition (tons ha^{-1} $year^{-1}$)	Area(ha)	Area(%)
Very high erosion	−161 to −100	301.36	5.74
High erosion	−100 to −80	162.88	3.10
Moderate to high erosion	−80 to −40	585.53	11.16
Moderate erosion	−40 to −15	828.39	15.78
Slight erosion	−15 to 0	1682.70	32.06
Slight deposition	0 to 15	547.37	10.43
Moderate deposition	15 to 40	310.11	5.91
Moderate to high deposition	40 to 80	241.99	4.61
High deposition	80 to 100	78.20	1.49
Very high deposition	100 to 150	510.24	9.72
Total		5248.78	100.00

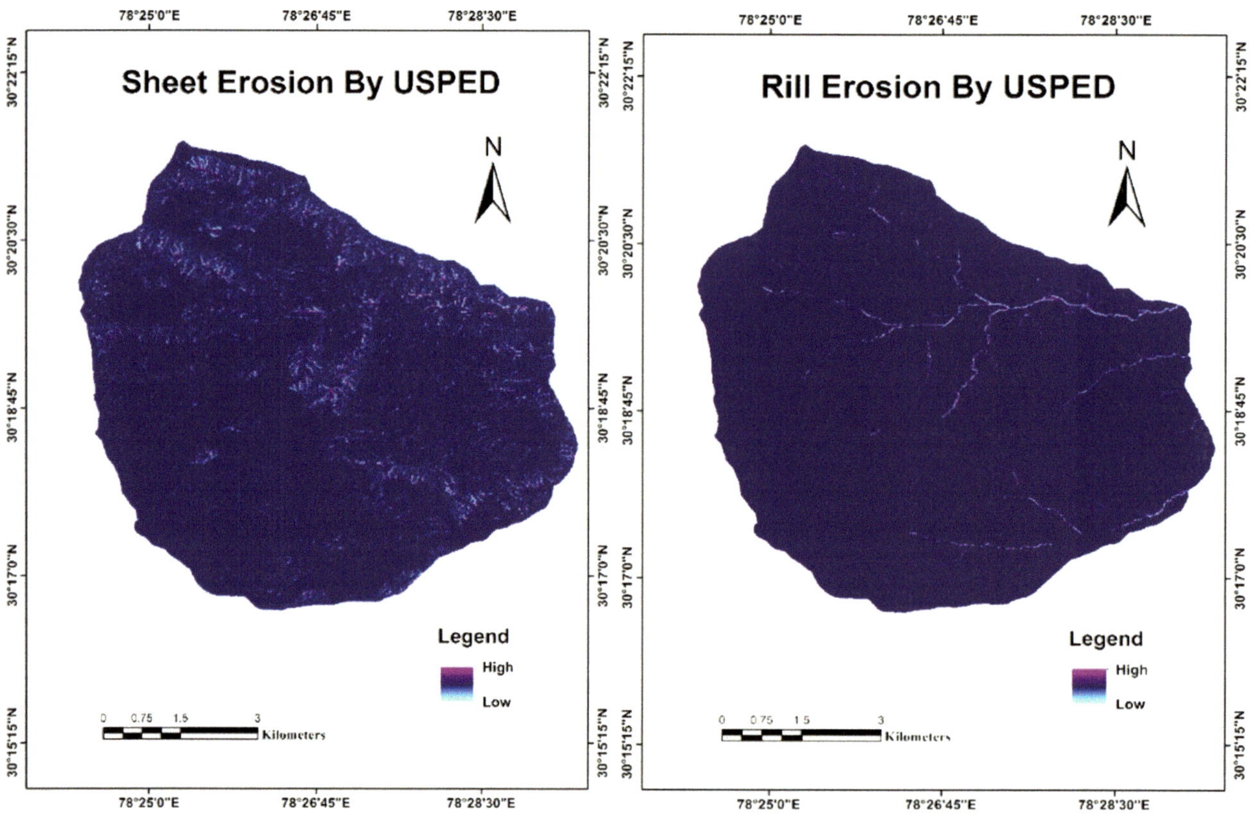

FIGURE 34.9 Sheet erosion and rill erosion by USPED.

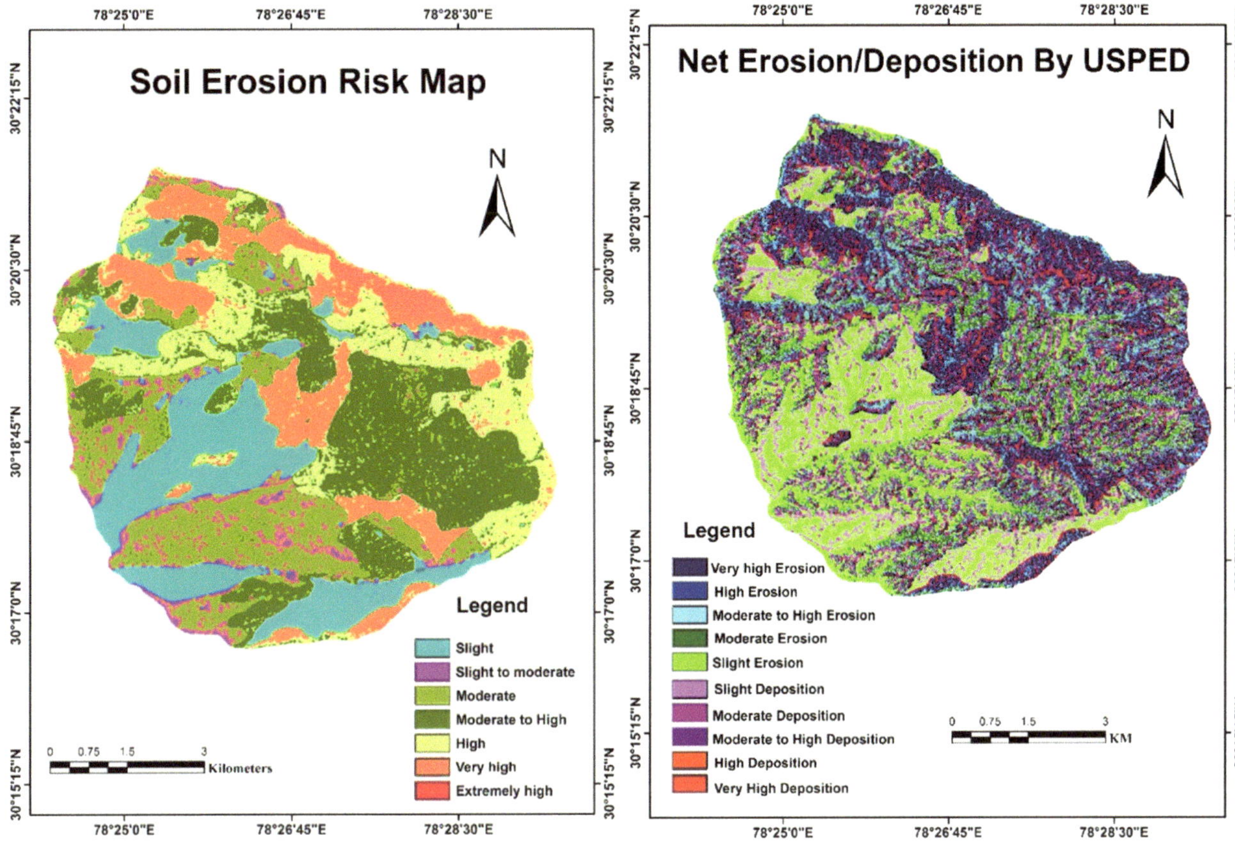

FIGURE 34.10 Soil erosion risk map (RUSLE) and net erosion/deposition map (USPED).

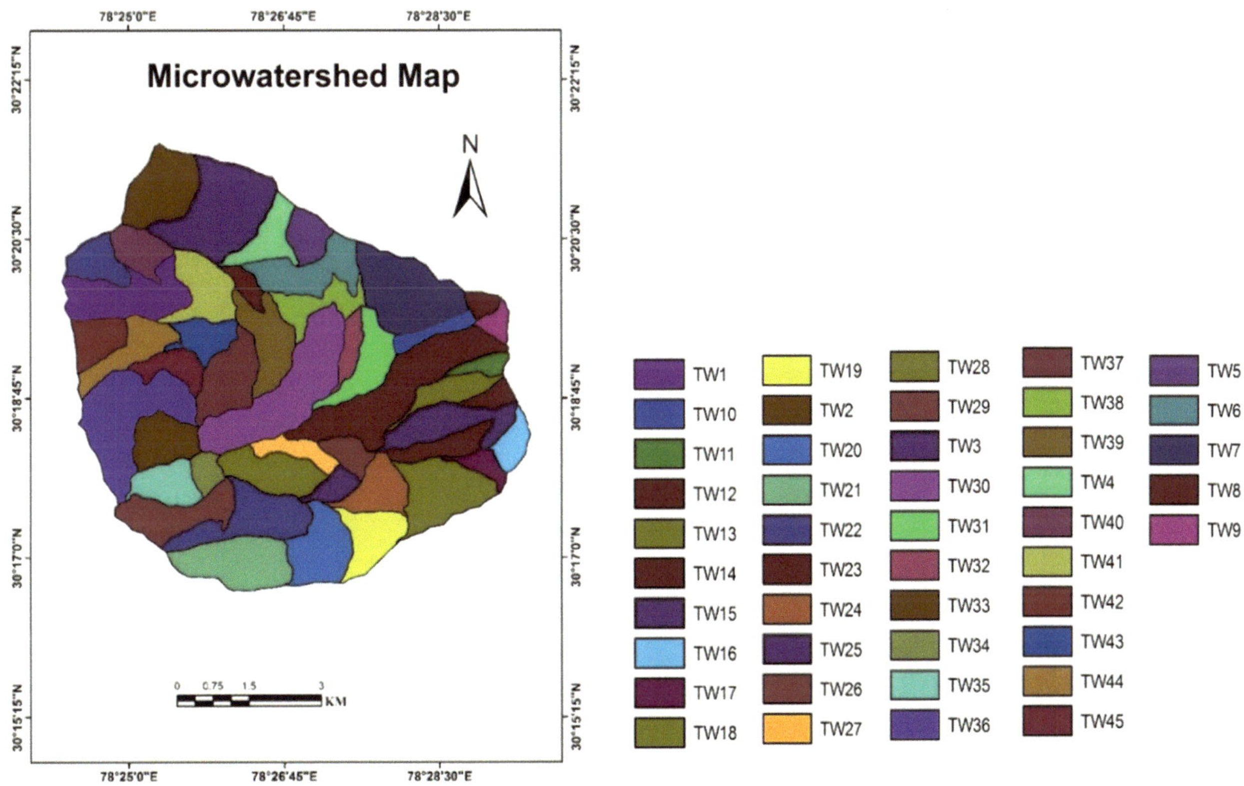

FIGURE 34.11 Microwatershed map of the study area.

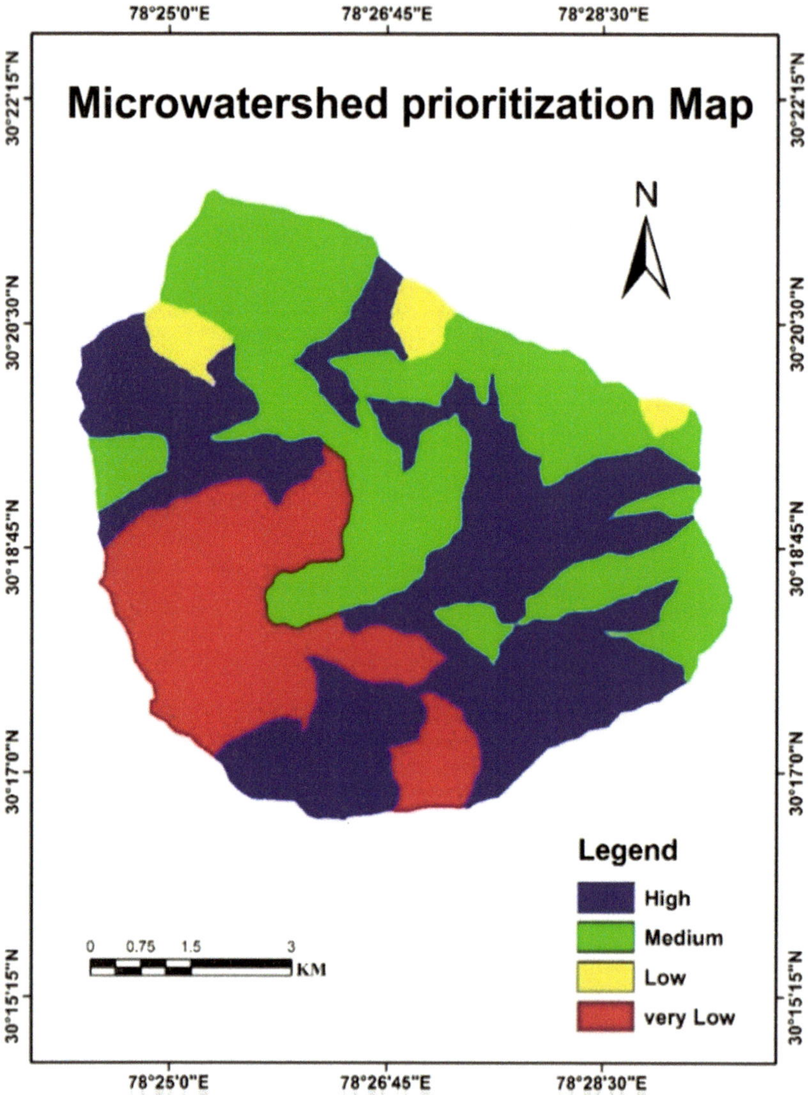

FIGURE 34.12 Microwatershed prioritization map.

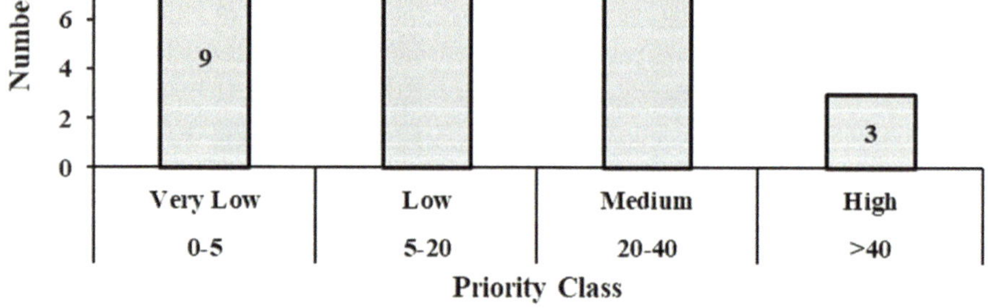

FIGURE 34.13 Distribution of watersheds included in various priority class.

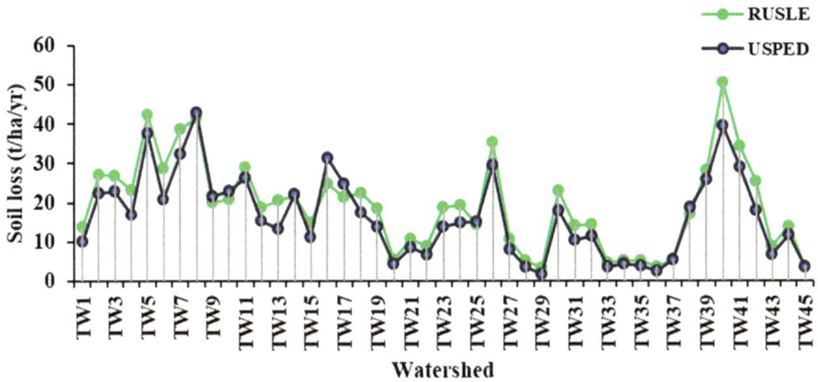

FIGURE 34.14 Soil loss estimated by RUSLE and USPED at various microwatersheds.

The figure shows that Watershed No. TW39, TW37, TW35, TW29, TW27, TW21, TW19, and TW7 are ranked the same in the RUSLE and USPED models. Most of the watersheds are ranked differently due to the USPED model, which also indicates the deposition in this area. The microwatershed TW40, TW8, and TW5 are given the highest priority, respectively, for conservation planning (Fig. 34.14).

34.5 Conclusions

In this study, a watershed database has been created to predict soil erosion risk and map it. This study was largely based on applying RS and GIS techniques along with field surveys. RS and GIS are very useful tools in extracting detailed information about terrains, soils, and LULC, which enabled better spatial prediction of soil erosion. The RUSLE model offered potential utility for soil erosion risk assessment and for identifying areas for conservation measures. It is an automated method to generate slope length values, and its applicability is largely found in mountainous topography. The DEM was generated to visualize and characterize the watershed in an effective manner. The soil textures of this watershed are sandy loam, silt loam, clay loam, sandy clay loam, silty clay loam, and sandy clay. The study area was categorized into eight LULC classes. Moderate forests formed the predominant LULC map. RS data provides us with valuable inputs required for preparing various factor maps used in the computation of soil erosion using the RUSLE and the USPED models. Scrub land (53.51% using RUSLE and 44.00% using USPED) and fallow land (50.45% using the RUSLE and 41.63% using USPED) face more soil erosion in the study area. Nearly 45.76% of the study area falls under the very steep slope category. The majority of the area (44.13%) falls under moderate to high soil erosion risk using the RUSLE model. The majority of the area (32.06%) falls under slight soil erosion risk using the USPED model. The resulting soil erosion map can be used to plan and carry out various management and conservation techniques.

References

Aiello, A., Adamo, M., Canora, F., 2015. Remote sensing and GIS to assess soil erosion with RUSLE3D and USPED at river basin scale in southern Italy. Catena 131, 174–185.

Bogdan et al., 2012. used remote sensing techniques to assess soil erosion with RUSLE and USPED model.

CSWCRTI Annual Report, 2000–2001. P-1: Water erosion Appraisal in different Agro Ecological Regions.

Das, D.C., 1985. September. Problem of soil erosion and land degradation in India. In: National Seminar on Soil Conservation and Watershed Management, pp. 17–18.

Defra, J., 2005. Review of Impacts of Rural Land Use and Management on Flood Generation Impact Study Report.

Garde, R.J., Kothyari, U.C., 1987. Sediment yield estimation. J. Irrig. Power (India) 44 (3), 97–12.

Hofer, T., 1998. Do land use changes in the Himalayas affect downstream flooding? Traditional understanding and new evidences. Mem. Geol. Soc. India 19, 119–141.

Ives, J.D., Messerli, B., 1989. The Himalayan Dilemma: Reconciling Development and Conservation. Routledge and UNU, London and New York, pp. 295–pp.

Jain, S.K., Kumar, S., Varghese, J., 2001. Estimation of soil erosion for a Himalayan watershed using GIS technique. Water Resour. Manag. 15 (1), 41–54.

Julien, P.Y., Simons, D.B., 1985. Sediment transport capacity of overland flow. Trans. ASAE 28 (3), 755–762.

Lea, D.M., Legleiter, C.J., 2016. Mapping spatial patterns of stream power and channel change along a gravel-bed river in northern Yellowstone. Geomorphology 252, 66–79.

Mitas, L., Mitasova, H., 1998. Distributed erosion modeling for effective erosion prevention. Water Resour. Res. 34 (3), 505–516.

Mitasova et al., 2005. Reported the Validation of a 3-D Enhancement of the Universal Soil Loss Equation for Prediction of Soil Erosion and Sediment Deposition.

Mitasova, H., Hofierka, J., Zlocha, M., Iverson, L.R., 1996. Modeling topographic potential for erosion and deposition using GIS. Int. J. Geogr. Inf. Sci. 10 (5), 629–641 (reply to a comment to this paper appears in 1997 in Int. J. Geogr. Inform. Sci., 11).

Moore, I.D., Burch, G.J., 1986. Physical basis of the length-slope factor in the universal soil loss equation. Soil Sci. Soc. Am. J. 50 (5), 1294–1298.

Morgan, R.P.C., Quinton, J.N., Smith, R.E., Govers, G., Poesen, J.W.A., Auerswald, K., et al., 1998. The European Soil Erosion Model (EUROSEM): a dynamic approach for predicting sediment transport from fields and small catchments. Earth Surf. Process. Landforms 23 (6), 527–544.

Narayana, D.V., Babu, R., 1983. Estimation of soil erosion in India. J. Irrig. Drain. Eng. 109 (4), 419–434.

Oliveira et al., 2013. Used Remote Sensing Techniques to Assess Soil Erosion with RUSLE and USPED Model.

Pimentel, D., 2006. Soil erosion: a food and environmental threat. Environ., Dev. Sustain. 8 (1), 119–137.

Rahaman et al., 2015. Carried Out Studies Prioritization of sub Watershed Based on Morphometric Characteristics using Fuzzy Analytical Hierarchy Process and Geographical Information System—A Study of Kallar Watershed, Tamil Nadu.

Raymo, M.E., Ruddiman, W.F., 1992. Tectonic forcing of late Cenozoic climate. Nature 359 (6391), 117–122.

Renard, K.G., Foster, G.A., Weesies, G.A., McCool, D.A., Yoder, D.C., 1997. Predicting soil erosion by water: a guide to conservation planning with the revised universal loss equation (RUSLE). Agri. Handbook703, USDA, WA.

Rodriguez, J.L.G., Suarez, M.C.G., 2012. Methodology for estimating the topographic factor LS of RUSLE3D and USPED using GIS, Sci Verse Science Direct. Geomorphology 175–176, 98–106 (2012).

Sadeghi et al., 2012. Carried Out Studies of Soil Erosion Assessment and Prioritization of Affecting Factors at Plot Scale using Tauguchi Method.

Further reading

Adinarayana, J., Gopal Rao, K., Krishna, N.R., Venkatachalam, P., Suri, J.K., 1999. A rule-based soil erosion model for a hilly catchment. CATENA 37 (3–4), 309–318.

Aiello, A., Canora et al., 2014–2015. Used Remote Sensing and Geographical Information System to Assess Soil Erosion with RUSLE3D and USPED at River Basin Scale in Southern Italy.

AIS & LUS, 2001. Soil and Information for Watershed Development, Technical Bulletin-1/2001 Published by all India Soil and Land Use Survey, New Delhi.

Behera, G., 1986. Integrated Micro-Watershed Management—A Systems Approach, Soil Conservation in India. Gupta, K., Khybri, M.L., (Eds.), Jugal Kishore and Co., Dehradun, pp. 1–8.

Bhattarai, R., Dutta, D., 2007. Watershed Scale hydrologic and non-point source pollution models: review of applications. Trans. ASAE 47 (3), 789–803.

Bhaware et al., 2004–06. Carried Out the Application of Different Soil Erosion Model.

Borah, D.K., Bera, M., 2004. Watershed Scale hydrologic and non-point source pollution models: review of application. Trans. ASAE 47 (3), 789–803.

Burrough, P.A., 1996. Principles of Geographical Information Systems for Land Resources Assessment. Oxford University Press Inc., New York.

Busacca, A.J., Cook, C.A., Mulla, D.J., 1993. Comparing landscape-scale estimation of soil erosion in the palouse using Cs-137 and RUSLE. J. Soil. Water Conserv. 48 (4), 361–367.

Desmet, P.J., Govers, G., 1996a. A GIS-procedure for the automated calculation of the USLE LS-factor on topographically complex landscape units. J. Soil. Water Conserv. 51, 427–433.

Desmet, P.J., Govers, G., 1996b. Comparison of routing systems for DEMs and their implemention for predicting ephemeral gullies. Int. Journal. GIS 10 (3), 311–331.

Efe, E., et al., 2008. Da Ouyang, and Jon Bartholic (2001) Applied RUSLE Soil Erosion Model.

George, R.F., Daniel, C.Y., 2003. RUSLE User's Guide. USDA Agricultural Service, Washington, D.C.

Ghosh, R., 2015. Soil Erosion Risk Assessment Using Remote Sensing and GIS Techniques, A Case Study of Maniyar Watershed, Uttrarakhand.

Hung, Nguyen van, 2013. Integration of Remote Sensing and GIS for Soil Erosion Hazard Modelling—A Case Study in Daknong Province,Vietnam.

Jasrotia, A.S., Dhiman, S.D., Aggarwal, S.P., 2002. Rainfall—runoff and soil erosion modeling using RS and GIS technique—a case study of tons watershed. J. Indian Soc. Remote Sens. 30 (3).

Kumar, R., Sexana, R., Mukherjee, 1999. Application of remote sensing technology for land use/land cover change analysis. J. Indian. Soc. Remote. Sens. 27 (2), 123–127. and.

Lamar, D., 2003. Watershed Prioritization, A Case Study of Part of Ton Watershed.

Li et al., 2004. Applied the Revised Universal Soil Loss Equation (RUSLE), remote sensing, and geographical information system (GIS) to the mapping of soil erosion risk in Brazilian Amazonia.

Maker, M., Angeli, L., Bottai, L., Costantini, R., Ferrari, R., Innocenti, L., et al., 2007. Assessment of land degradation susceptibility by scenario analysis: a case study in Southern Tuscany, Itaily. Geomorphology 93, 120–129.

Millward, A.A., Mersey, J.E., 1999. Adapting the RUSLE to model soil erosion potential in a mountainous tropical watershed. CATENA 38 (2), 109–129.

Mitasova, H., Mitas, L., Brown, B.M., Gerdes, D.P., Kosinovsky, I., 1995. Modeling spatially and temporally distributed phenomena: new methods and tools for GRASS GIS1995 Int. J. GIS 9 (4), special issue on Integration of environmental modeling and GIS, pp. 443–446.

Mitasova et al., 1996. Applied the Modelling Topographic Potential for Erosion and Deposition Using GIS.

Mitasova, H., Mitas, L., 1999. Modeling Soil Detachment with RUSLE 3D Using GIS. University of Illinois at Urbana-Champaign. <http://skagit.meas.ncsu.edu/~helena/gmslab/erosion/usle.html> (accessed July 2010).

Mitasova, H., Mitas, L., Brown, W.M., Johnston, D., 1998. Multidimensional Soil Erosion/Deposition, Modeling and Visualization Using GIS. Final Report for USA CERL. University of Illinois, Urbana-Champaign, IL.

Naipal, R. et al., 2015. Carried Out Studies to Improving the Global Applicability of RUSLE Model-Adjustment of the Topographical and Rainfall Erosivity Factors. They Adjusted the Topographical and Rainfall Erosivity Factors of the RUSLE Model and Compared the Resulting Erosion Rates to Extensive Empirical Databases from the USA.

Nuno de santos Loureiro, Miguel de Azevedo coutinho, 2001. A new procedure to estimate the RUSLE EI30 Index, based on monthly rainfall datd applied to the Algarve region, Portugal. J. Hydrol. 250, 12–18.

Saavedra et al., 2005. Demonstration and Validation of the Unit Stream Power Erosion and Deposition (USPED) Model by Mountain Research Station.

Shrimali, S.S., Agarwal, S.P., et al., 2001. Prioritizing Erosion Prone Areas in Hills Using Remote Sensing and GIS-A Case Study of the Sukhna Lake Catchment, Northern India. JAG 3.

Singh, R.P.D., Bagchi, June, 2011. Ministry of Water Resources, Central Ground Water Board., Ground Water Brochure District Tehri Garhwal, Uttarakhand.

Warren, S. et al., 2005. Validation of a 3D Enhancement of the Universal Soil Loss Equation for Prediction of Soil Erosion and Sediment Deposition.

Wischmeier, W.H., Smith, D.D., 1965. Predicting Rainfall Erosion Losses from Cropland East of the Rocky Mountains: Guide for Practices from Soil and Water Conservation. U.S.Department of Agriculture.

Chapter 35

Mapping of landslide susceptibility using large-scale soil information and GIS-based analytical hierarchy process technique

R. Srinivasan, B.N. Shashikumar, Amar Suputhra, M. Lalitha and S. Dharumarajan

ICAR-National Bureau of Soil Survey and Land Use Planning, Regional Centre, Bangalore, Karnataka, India

35.1 Introduction

Landslides are a key natural disaster in the Himalayan region and cause severe damage to property, road networks and loss of life (Chorley et al., 1985; Pal and Chowdhuri, 2019). Landslides are the result of different geoenvironmental processes; important built-in ingredients are geology (degree of weathering, lithology, and structure), topography (slope gradient, length, relative relief and aspect), soil (depth, structure/texture, water holding capacity, shrink-swell and pore space), land use/land cover (LULC), and hydrological environment. Landslides occur due to innumerable other factors, viz., heavy precipitation, a strong earthquake, detonation, and burst of rainy clouds (Anbalagan, 1992; Sharma et al., 2009; Sengupta et al., 2010). Contrastingly, the human pursuit of intensive farming, construction of buildings, footpaths, and removal of trees tends to steep slopes condition in a poor structure, make more vulnerable to a landslide (Mugagga et al., 2011).

The Geological Survey of India (GSI) reported that 12.6% of geographical areas in India are vulnerable to landslides, among them, 43% fall in the Darjeeling Himalayan hill region, 33% from the northwest Himalayas, 22% in the western Konkan Ghat, and 2% in the eastern, Aruku valley hill. The Eastern Himalayas has criss-cross ridge topography and ravines, which causes frequent migration of soil masses (Rawat et al., 2012). Geological hazards such as earthquakes and volcanic eruptions are rare in these regions. However, landslides are common in several parts of the Himalayan region during the rainy season. Mass movements are recognized as global geomorphic hazards and impact both the economic and social status of the people (Knapen et al., 2006).

Landslide in the eastern Himalayas have a combination of hydro-meteorological, geomorphological, earth-shaking, and geological aspects. Landslide is renowned as mountainous disaster and are responsible for massive economic and social losses (Mandal and Maiti, 2015; Das et al., 2020). Regular monitoring of rainfall, which is indirectly linked with soil infiltration, overflow, and evaporation, provides significant clues for the possibility of landslides in a specific area (Knapen et al., 2006).

Soil characteristics, especially physical attributes such as bulk density, cohesion, and shear force, affect the firmness of soil in steep slope conditions (Mugagga et al., 2011). Shrink-swell behavior of high clay content soils may collapse the slope under excess wet and dry conditions (Baynes, 2008). Soil texture and the arrangement of soil porosities influenced the slope stability and mass movement (Sidle et al., 1985). Large-scale soil mapping at 1:10000 scale accrues detailed information of each parcel of the farm on cadastral number wise soil information, which provides an accurate and scientific inventory of soil types and their nature and extent of distribution so that one can predict their characteristics and potentialities for farm planning of each parcel of land resources (Nisha et al., 2016). It also provides adequate information in terms of landform, terraces, vegetation, and characteristics of different thematic soil layers (viz., texture, depth, organic carbon, stoniness, drainage, acidity, salinity, etc.), which can be used for erosion assessment and other planning and development (Manchanda et al., 2002; Srivastava and Saxena, 2004).

Remote Sensing of Soils. DOI: https://doi.org/10.1016/B978-0-443-18773-5.00020-X

Landslide Hazard Zonation (LHZ) is the identification of homogeneous areas based on the severity of the hazard caused by mass movement (Varnes, 1984). The mapping of land hazards is an important step in natural planning and the mitigation of disasters (Kanwal et al., 2016). Sloped undulated topography caused landslides in the Himalayan region, which are highly inaccessible, and this is a great difficulty for planners. However, geospatial techniques aid in reaching the inaccessible area, identifying the severity of landslides and assessing the landslide-susceptible zones based on different contributive factors (Anbalagan et al., 2014). Geospatial (spatial and temporal) information obtained from remote sensing and ground-based details are required for the finalization of landslide hazard maps (Pandey et al., 2008). The assumption of LHZ is a semblance of future landslides at specific locations, which have the same set of conditions as th those at of past and present landslide locations (Varnes, 1984; Kanungo et al., 2009). Different approaches for landslide hazard assessment using several parameters in the Himalayas region have been carried out by several researchers (Anbalagan, 1992; Arora et al., 2004; Sarkar and Kanungo, 2004; Saha et al., 2005; Dubey et al., 2005; Kanungo et al., 2006; Mathew et al., 2007, 2009; Dahal et al., 2009; Sharma et al., 2012, 2015; Anbalagan et al., 2015; Banerjee et al., 2018; Pal and Chowdhuri, 2019; Bera et al., 2019; Das et al., 2020, Stanley and Kirschbaum, 2017; Hadmoko et al., 2017; Mahdadi et al., 2018).

Landslide hazard zones (LHZs) can be mapped using different techniques; among them, four methods are popular: heuristic, inventory-based probabilistic, deterministic, and statistical techniques. A highly reliable landslide hazard map was prepared using the AHP method. The main advantages of AHP methods are site-specific evaluations of specific causal factors and the avoidance of oversimplification of derived data (Van Westen et al., 2011). Analytic hierarchy process (AHP) (Saaty, 1980) is a known multicriteria decision analysis, which has been applied to several decision-making approaches. In this study, we mapped the LHZ of Mangan block, North Sikkim district, Sikkim state (India) in the eastern Himalayas using GIS tools, characterized by rugged topography with ridges and valleys. The LHZ map was also validated using a field inventory. The LHZ map generated from this study is significant for implementing mitigation measures to stabilize susceptible locations and display warning signs. The map can be used for preliminary analysis of engineering and soil and water conservation projects.

35.2 Materials and methods

35.2.1 Location of the study area

A case study was conducted at Mangan block, North Sikkim district of Sikkim state (Fig. 35.1) covering 8945 ha. The study area extended between 27°33′47″N-27°23′25″N and 88°40′27″E-88°30′42″E Latitude and Longitude. The average elevation of the study area is 956 m. The Teesta valley is characterized by glacial and peri-glacial landforms in the north and northcentral Sikkim and by abundant fluvial terraces in the lower stretches. The geomorphic features observed in the region include the terraces and courses of higher-order streams. Four-tier terrace system is observed in the central and lower stretches of the Teesta River in North Sikkim. Their development is linked with regional uplifts and consequent incisions, and their dispositional irregularities are associated with local tectonics and knick-points.

35.2.2 Field studies

A detailed soil survey at 1:10,000 scale was conducted in the Mangan block from 2017 to 2018. The detailed soil survey was carried out using 4-tier approach, viz., satellite image interpretation, ground truth observation (soil survey fieldwork), laboratory analysis, and mapping using GIS software. The base map prepared used remote sensing satellite data (IRS-P6 LISS IV) along with a survey of India Toposheet (1:50,000 scale), and geological data. The satellite data visually interpreted and delineated different physiographic regions, and based on landforms and LULC, soil profile observations were made. Soil morphological characteristics were studied in all soil profiles (vertical cut and arrangement of soil layers/horizons) (Soil Survey Staff, 1993). The latitude, longitude, and elevation at each sampling site were recorded using a handheld global positioning system. A total of 62 soil profiles were studied in the region.

Based on the soil attributes, the soils were grouped into nine soil series (homogeneous units with the similar response for level management) (Fig. 35.2), and various thematic layers were prepared from the soils maps, viz., soil depth, gravel, texture, and erosion. Land capability class (LCC) was evaluated based on existing resource and soil characteristics (AISLUS, 1970).

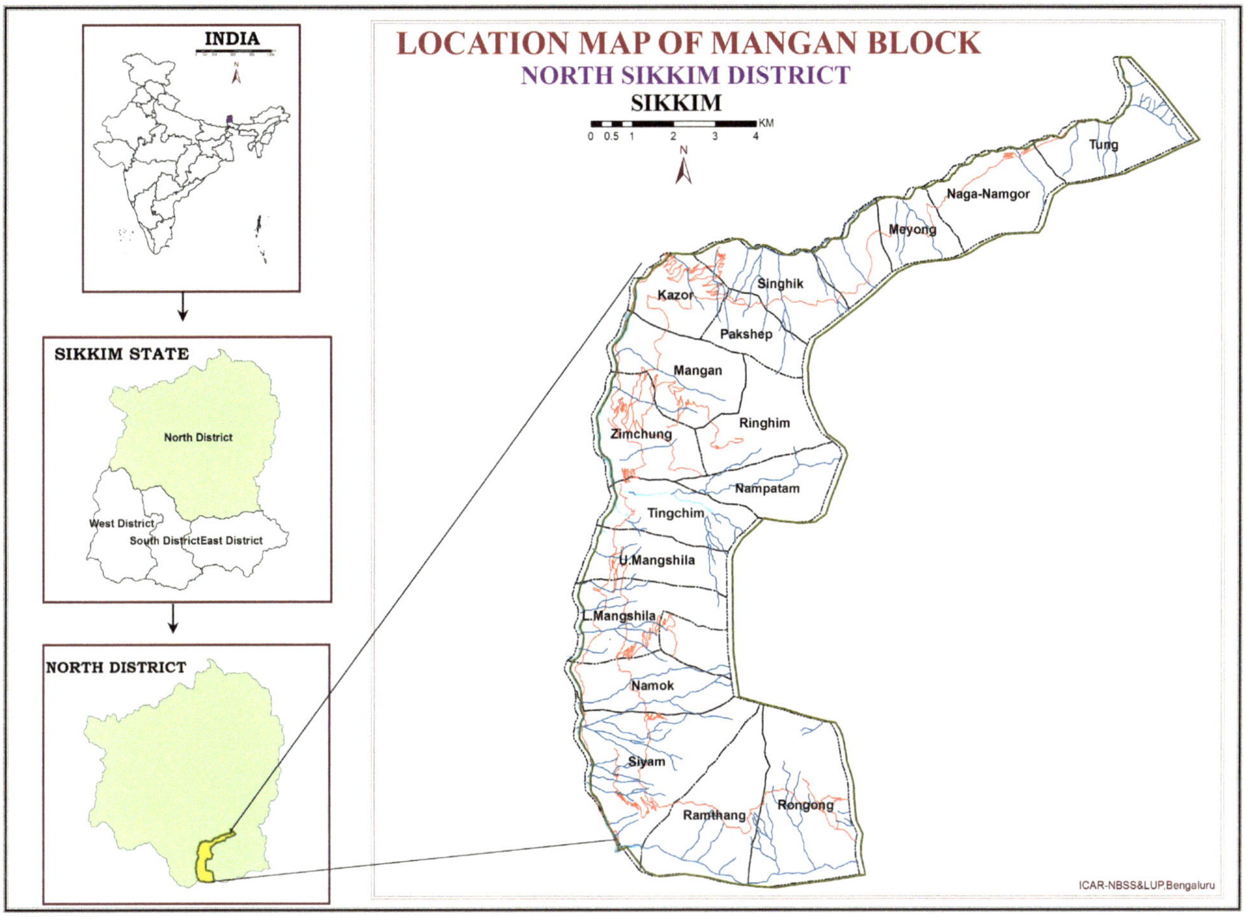

FIGURE 35.1 Location map of the study area.

35.2.3 Laboratory analysis

The soil samples were collected layer-wise, air-dried, powdered, and sieved using a 2 mm sieve and used for the determination of soil physical and chemical characteristics, viz., particle size (soil texture) distribution at different depths, by the international pipette method (Day, 1965). The soils were classified as per the guidelines given in Key to Soil Taxonomy (Soil Survey Staff, 2014).

35.2.4 Other data used

Elevation, slope, drainage density, and slope aspect data were derived from Advanced Land Observing Satellite (ALOS) digital elevation model (DEM). Current land use (single, double, agroforestry, etc.) data were collected during fieldwork. LULC and lineament density data were obtained from satellite data. The geology of the study area were obtained from the GSI and month-wise rainfall distribution data from 2009 to 2018 were obtained from Indian Meteorological Department.

35.2.5 Methodology

The AHP (Saaty, 1980) technique is used to analyze the weight values of each thematic layer. Rating values of each thematic subclass are given based on the severity of landslide activity. Table 35.1 explained calculating weight among three alternatives for a technological advance. Comparisons are made by pairing two factors according to relative preferences of 1–9, as shown in Table 35.2, and the reciprocal condition is satisfied, when the number of alternatives is

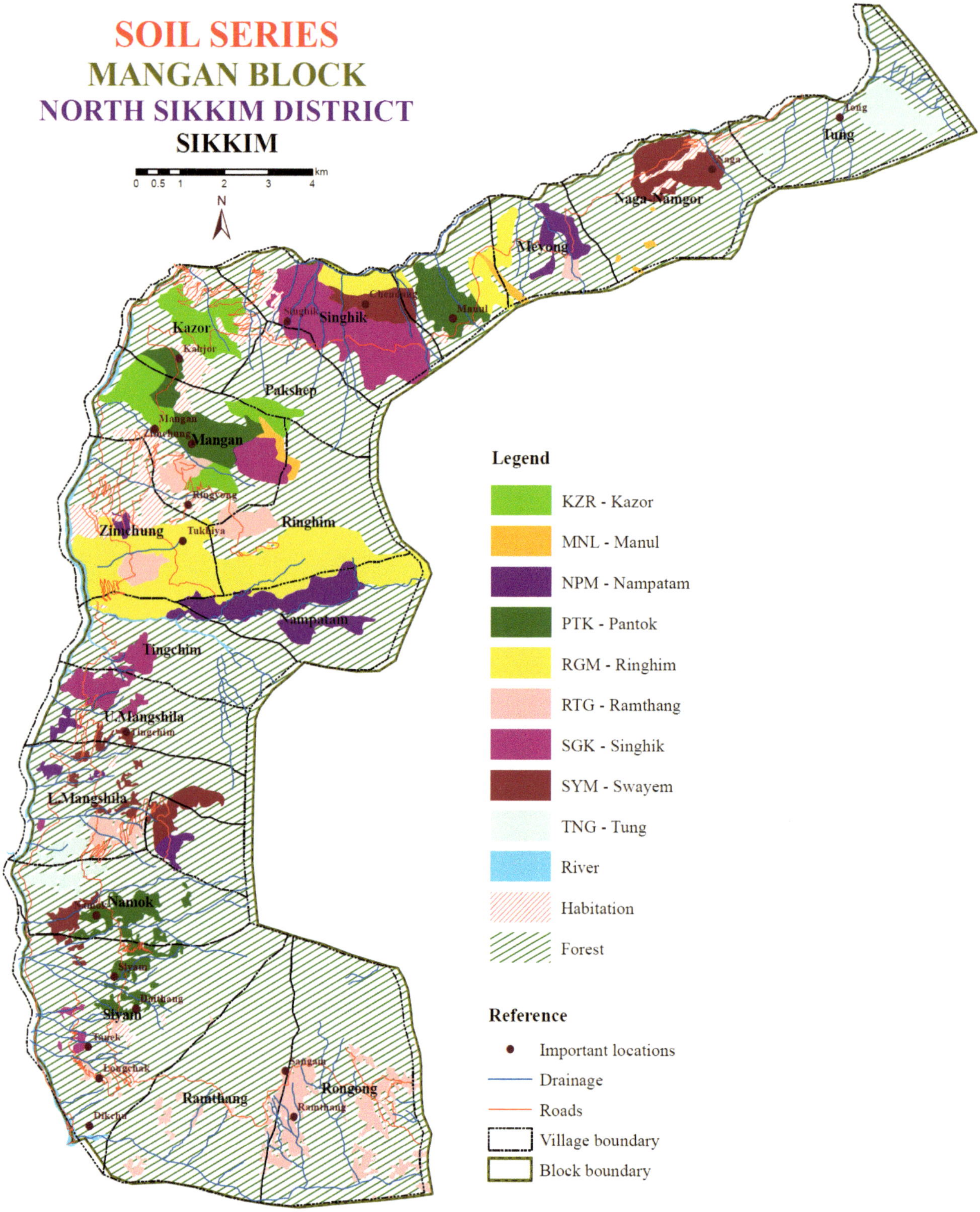

FIGURE 35.2 Soil map of the area.

n, resulting in a total of $n(n-1)/2$ comparisons. For instance, if Alternative 1 is three times as important as Alternative 2, it automatically means that Alternative 2 is 1/3 times as important as Alternative 1, and, subsequently, an additional comparison is omitted.

TABLE 35.1 Example of pair-wise comparison matrix for the criteria.

Technological advance	Alternative 1	Alternative 2	Alternative 3	Weight
Alternative 1	1	3	7	0.643
Alternative 2	1/3	1	5	0.283
Alternative 3	1/7	1/5	1	0.074

λmax = 3.066.
Consistency ratio (CR) = 0.056.

TABLE 35.2 Pair-wise comparison scale for AHP preferences by Al-Subhi Al-Harbi (2001).

Numerical rating	Verbal judgments of preferences
9	Extremely preferred
8	Very strongly to extremely preferred
7	Very strongly preferred
6	Strongly to very strongly preferred
5	Strongly preferred
4	Moderately to strongly preferred
3	Moderately preferred
2	Equally to moderately preferred
1	Equally preferred

According to Al-Subhi Al-Harbi (2001), the consistency ratio (CR) is calculated to verify the consistency of responses. The CR value uses a priority vector, and Saaty (2003) emphasized that the reliability of the responses could be maintained only when the CR is 0.10 or less.

$$\mathrm{CR} = \frac{CI}{RI}$$

where *RI* is the average value of *CI* for random matrices using the Saaty (2003).

Subsequently, the weighted values of each criterion using the AHP pairwise comparison matrix were computed (Table 35.3). Each row of pairwise comparison matrix values describes the relative importance of two criteria. For instance, the first row represents the significance of the slope in comparison with the other fourteen criteria positioned in the column. In this matrix, the rows follow the inverse value of each criterion and its significance with others (e.g., the slope is a little more critical than the rainfall; hence, the value of the slope is 1 and the value of rainfall is 2, thus in the next row rainfall is having a value of ½). CR, order to determine whether pairwise comparisons were consistent or not (Kolat et al., 2012). One of the strengths of the AHP method is that it allows for inconsistent relationships while providing a CR as an indicator of the degree of consistency or inconsistency (Chen et al., 2010; Feizizadeh and Blaschke, 2013). If the CR < 0.10, it indicates a good level of consistency to recognize the class weights. Where λ max is the principal eigenvalue and n is the number of factors λ max $= \Sigma$ of the products between each element of the priority vector and column totals. In this study, the resulting CR was 0.055, which indicates that the comparisons of characteristics were very consistent and that the weights (Table 35.3) were appropriate for the LHZ model. A methodological flowchart is presented in Fig. 35.3.

TABLE 35.3 Pairwise comparison matrix, weights and consistency ratio of the data layers used.

Factors	SL	R	ST	SD	SG	SE	SOC	LCC	LD	EL	DD	G	LULC	SA	Weights
Slope (SL)	1	2	3	3	4	5	5	6	7	7	8	8	9	9	0.2165
Rainfall (R)	1/2	1	2	3	3	4	5	5	6	7	7	8	8	9	0.1761
Soil texture (ST)	1/3	1/2	1	2	3	3	4	5	5	6	7	7	8	8	0.1392
Soil depth (SD)	1/3	1/3	1/2	1	2	3	3	4	5	5	6	7	7	8	0.1109
Soil gravel (SG)	1/4	1/3	1/3	1/2	1	2	3	3	4	5	5	6	7	7	0.0866
Soil erosion (SE)	1/5	1/4	1/3	1/3	1/2	1	2	3	3	4	5	5	6	7	0.0672
Land capability class (LCC)	1/6	1/5	1/5	1/4	1/3	1/3	1/2	1	2	3	3	4	5	5	0.0403
Lineament density (LD)	1/7	1/5	1/5	1/5	1/4	1/3	1/3	1/2	1	2	3	3	4	5	0.0313
Elevation (EL)	1/7	1/7	1/6	1/5	1/5	1/4	1/3	1/3	1/2	1	2	3	3	4	0.0243
Drainage density (DD)	1/8	1/7	1/7	1/6	1/5	1/5	1/4	1/3	1/3	1/2	1	2	3	3	0.0189
Geology (G)	1/8	1/3	1/7	1/7	1/6	1/5	1/5	1/4	1/3	1/3	1/2	1	2	3	0.0151
Land use/land cover (LULC)	1/9	1/3	1/8	1/7	1/7	1/6	1/5	1/5	1/4	1/3	1/3	1/2	1	2	0.0119
Slope aspect (SA)	1/9	1/9	1/8	1/8	1/7	1/7	1/6	1/5	1/5	1/4	1/3	1/3	1/2	1	0.0097

Principal eigenvalues: 15.149562.
Consistency ratio (CR): 0.055967.

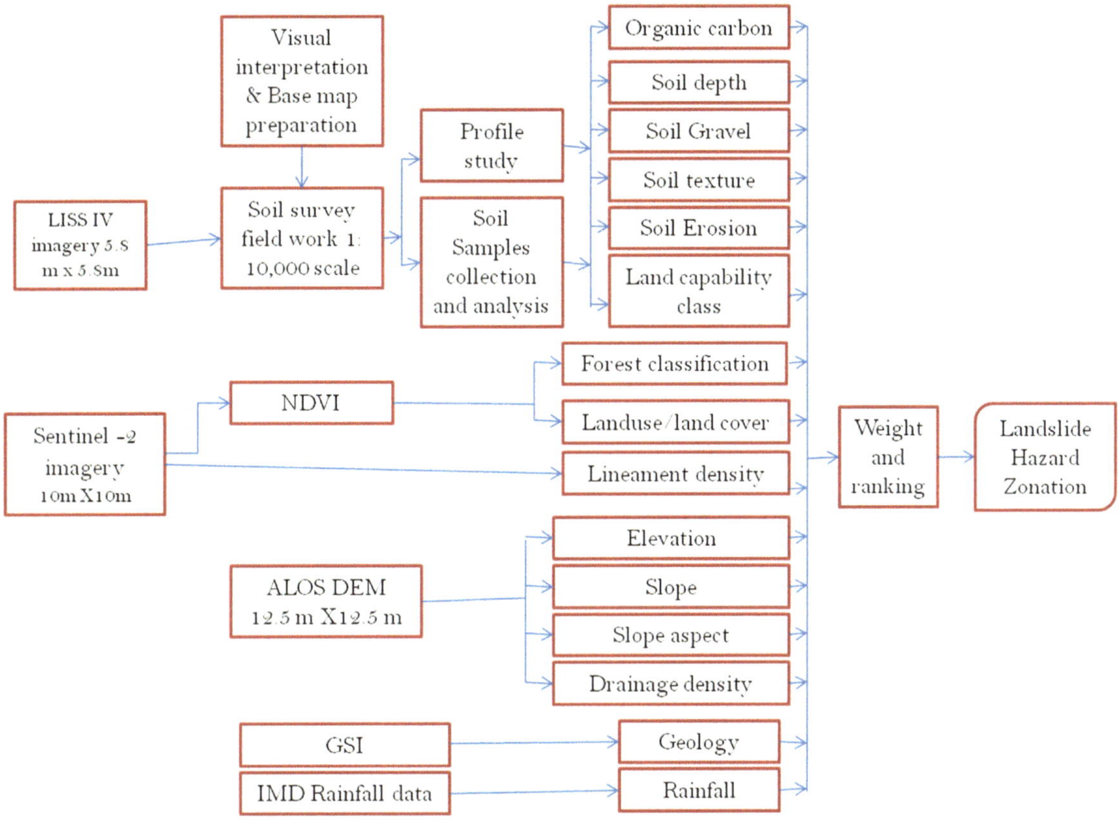

FIGURE 35.3 Methodological flow chart for LHZ mapping.

35.3 Result and discussion

35.3.1 Soil characteristics

Soils of the study area were mapped into nine (9) series representing two soil orders, that is, Inceptisols (Typic Dystrudepts, Humic Dystrudepts, Fluventic Dystrudepts and Humic Pachic Dystrudepts) accounted for 26.6% and mollisols to an extent of 8.24% (Typic Hapludolls, Pachic Hapludolls, Pachic Argiudolls, and Alfic Argiudolls), as presented in Table 35.4 and Fig. 35.4. Forest (67.53%), river (0.93), and habitation (2.65%) occupied the remaining area. Soil series of Tung, Manul, Swayem, and Ringhim are associated with skeletal, excessively drained soil with surface stoniness associated with intense erosion. The Ramthang and Ringhim soil series have acoarse sandy texture and are moderately deep and excessively drained with moderate to intense erosion. Similar results were reported by Sharma et al. (2012) and Bera et al. (2019).

35.3.2 Slope

The slope is a key factor in assessing soil stability and the severity of landslides. The extent of soil erosion is influenced by slope length and degree (Roth, 1983). During the rainy season, the hills' side slope gets saturated with water, resulting in the instability of the soil and movement of the landmass. The increasing slope angle, shear strength of soil, or other unconsolidated materials significantly increases the landslide (Anbalagan, 1992; Dhakal et al., 2000; Lee et al., 2004; Anbalagan et al., 2015). The slopes of the study area were classified by the quantile method into five categories, such as strong slopes (0%–33%), and very strong slopes (33%–44%), moderately steeply (44%–56%), steeply (56%–74%), and very steeply (>74%). A significant portion of the study area is steep to very steep, along with coarse soil texture (sandy), which favors greater susceptibility to landslides. Soils were grouped into various slope classes, and ranking values are presented in Table 35.5 and Fig. 35.5.

TABLE 35.4 Classification of major soils in Mangan block.

S. No	Series	Soil taxonomy				
		Order	Sub order	Great group	Subgroup	Family
P1	Manul	Mollisols	Udolls	Hapludolls	Typic Hapludolls	Coarse loamy
P2	Tung	Inceptisols	Udepts	Dystrudepts	Typic Dystrudepts	Loamy-Skeletal
P3	Kazor	Mollisols	Udolls	Hapludolls	Pachic Hapludolls	Loamy-Skeletal
P4	Swayem	Inceptisols	Udepts	Dystrudepts	Typic Dystrudepts	Fine-loamy
P5	Pantok	Inceptisols	Udepts	Dystrudepts	Humic Dystrudepts	Fine -loamy
P6	Singhik	Mollisols	Udolls	Argiudolls	Pachic Argiudolls	Fine-loamy
P7	Nampatam	Mollisols	Udolls	Argiudolls	Alfic Argiudolls	Fine-loamy
P8	Ramthang	Inceptisols	Udepts	Dystrudepts	Fluventic Dystrudepts	Sandy
P9	Ringhim	Inceptisols	Udepts	Dystrudepts	Humic Pachic Dystrudepts	Sandy

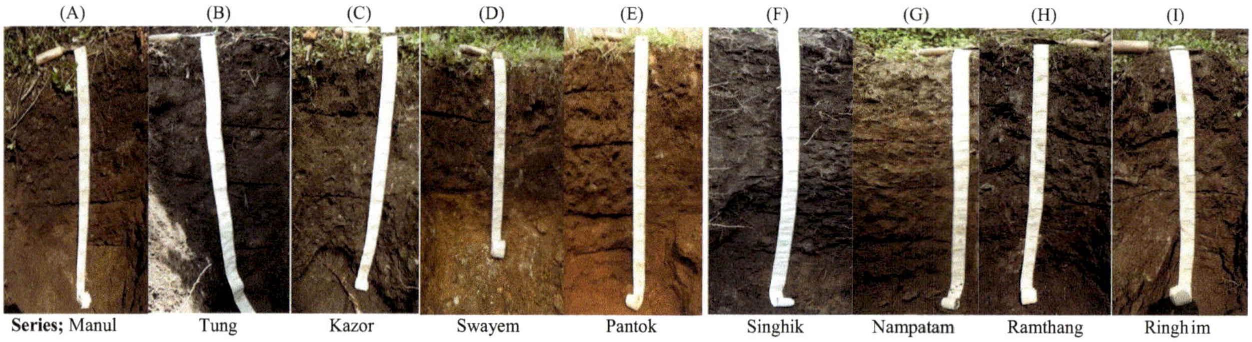

FIGURE 35.4 Soil variability recorded in the study area. That is, Typic Hapludolls (A); Typic Dystrudepts (B, D); Pachic Hapludolls (C); Humic Dystrudepts (E); Pachic Argiudolls (F); Alfic Argiudolls (G); Fluventic Dystrudepts (H); Humic Pachic Dystrudepts (I).

35.3.3 Rainfall

Landslides are activated by aquifer-induced pore pressure during monsoon rains and an escarpment stress environment (Sengupta et al., 2010; Martha et al., 2017b). Month-wise average rainfall was collected from 2009 to 2018 at different rain gauge stations in and around the Mangan area (Fig. 35.6). Maximum rainfall is received from the Southwest monsoon from June to September. The annual average rainfall of the study area is in the range of 2400–3400 mm, with higher intensity of rainfall in the southern part, moderate-intensity rainfall (2800–3200 mm) in the central part, and lowest (2400–2800 mm) in the northern parts. Heavy rainfall during the monsoon prompts high seepage movement in steep slopes and becomes saturated and loosened, finally inducing the slope failure. There is heavy downpour during monsoon, compared to constant low-intensity rainfall in the past. The high volume of runoff flows down the hills during frequent extreme rainfall events, leading to landslides and hence no retention for groundwater recharge. Montgomery et al. (2000) reported that due to heavy rainfall along with rugged topography and sensitive weathering materials caused many new landslides in the Ardesen region.

35.3.4 Soil texture

Soil texture, being an important physical property, is directly linked to other soil properties such as structure, porosity, water content, and consistency. Soil texture refers to the relative proportion of sand, silt, and clay present in the soil. High quantity clay soils favor the very stable aggregates and are resistant to detachment and transportation. Whereas, light texture soils such as sand (coarse texture) and sand mixed with loam containing gravels or stones are easy to detach, resulting in weak stable aggregates (Das and Agarwal, 2002). Consequently, soil with high sand, steep slope,

TABLE 35.5 Landslide hazard weights values and rating system of different thematic layers.

Factor	Class	Weights	Ranking
Slope	Strongly sloping	0.2165	2
	Very Strongly sloping		2
	Moderately Steeply Sloping		3
	Steeply Sloping		4
	Very Steeply Sloping		5
Rainfall (mm)	<2400	0.1761	2
	2400–2800		4
	>3200		5
Soil texture	Loam	0.1392	1
	Sandy clay loam		1
	Sandy loam		2
	Loamy sand		3
Soil depth	Deep (100–150 cm)	0.1109	1
	Moderately deep (75–100 cm)		2
	Moderately shallow (50–75 cm)		3
Soil gravel	Nongravelly	0.0866	1
	Gravelly		2
	Very gravelly		3
Soil erosion	Moderate erosion	0.0672	1
	Severe erosion		2
	Very severe erosion		3
LCC	IVe	0.0403	1
	IVes		2
	VIe		3
	VIes		3
Lineament density (LD)	Very low	0.0313	1
	low		2
	Moderate		3
	High		4
	Very high		5
Elevation (m)	< 1001	0.0243	1
	1001–1326		2
	1326–1642		3
	1642–1983		4
	> 1983		5

(Continued)

TABLE 35.5 (Continued)

Factor	Class	Weights	Ranking
Drainage density (sq. km)	< 1.26	0.0189	1
	1.26–2.49		2
	2.49–3.75		3
	3.75–5.36		4
	> 5.36		5
Geology	Quartzite	0.0151	1
	Granite gneiss (mylonitic)		1
	Quartzite, Gamet kyanite sillimanite biotite schist		1
	Gamet kyanite sillimanite biotite schist		2
	Banded/ streaky migmatile, Bugen bearing (gamet) blottle gneiss with /without kyanite		3
	Inter banded chlorite-sericlie schist/phyllite and quartzite		4
LULC	Forest	0.0119	1
	Single crop		2
	Double crops		2
	Habitation		2
	Agro forestry		3
	Exposed rock		3
	Open field		5
	Open scrub		5
	Water body		0
Slope aspect	Flat	0.0097	0
	North		2
	East		3
	Southern		3
	South		3
	Northeast		4
	Southwest		5
	West		5
	Northwest		5

and heavy rainfall are authorized factors of landslides and damage to different lands (Patanakanog, 2001). Soils of the study area have been grouped into four textural classes. The dominant soil texture is sandy loam (18.75%), followed by sandy clay loam (7.25%), loam (1.81%), and loamy sand (0.39%). The clay texture has shrink-swell properties and shows high expansion characteristics during wet conditions; hence, it is prone to landslides (Mugagga et al., 2011). Based on soil textures, stability ranks were assigned and are presented in Table 35.5 and Fig. 35.7.

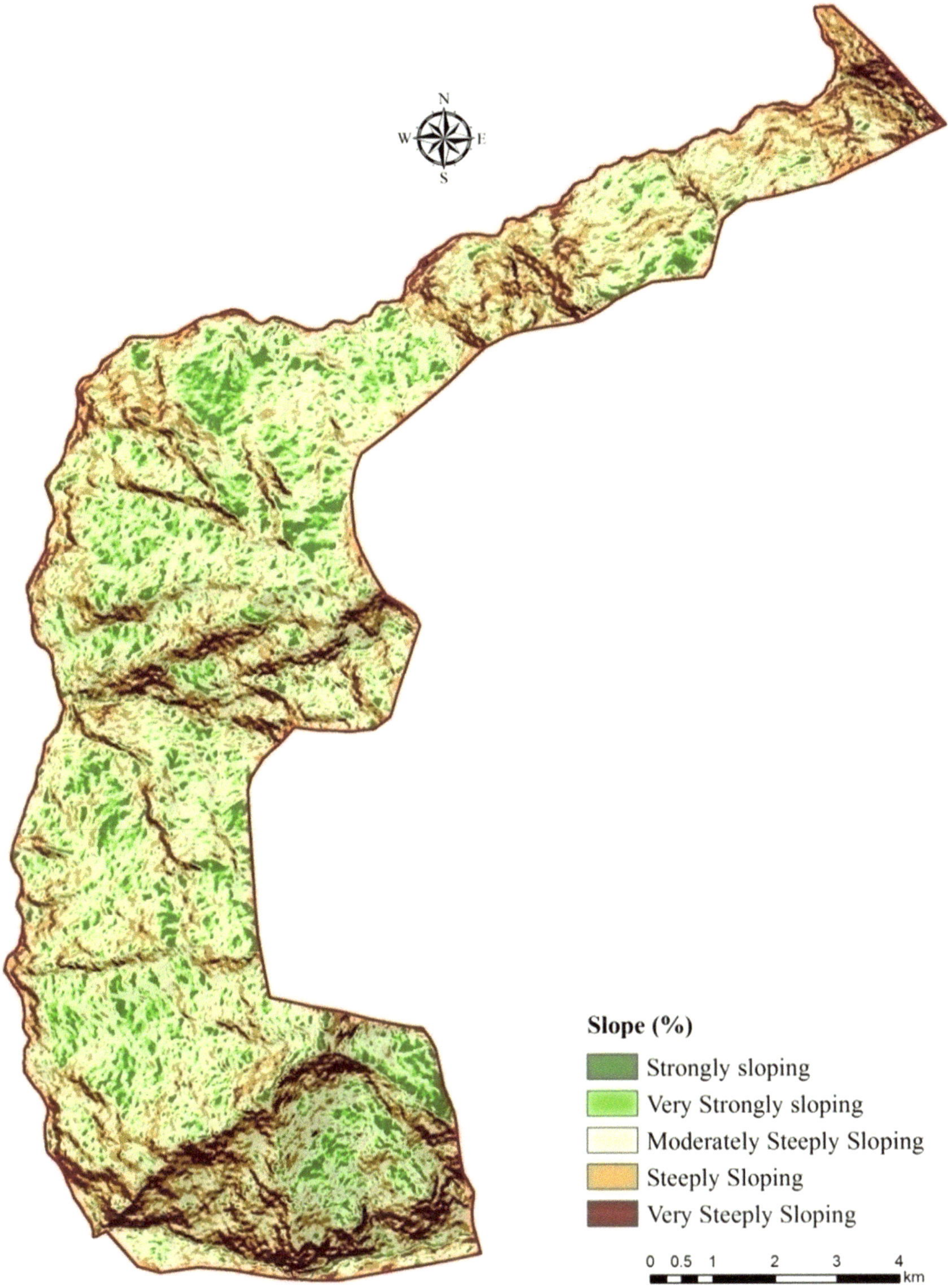

FIGURE 35.5 Slope map of Mangan.

35.3.5 Soil depth

Soil depth indicates the depth of the soil column, which includes different soil layers or horizons, occurring above the parent material or hard rock. Soil depth a crucial factor for assessing slope stability and susceptibility of the landslides. High soil depth increases the moisture absorption of the soil, resulting in reduced runoff. Therefore, shallow soil (low soil depth) is more unstable and highly prone to landslides than deeper soil. Soils are dominantly moderately deep

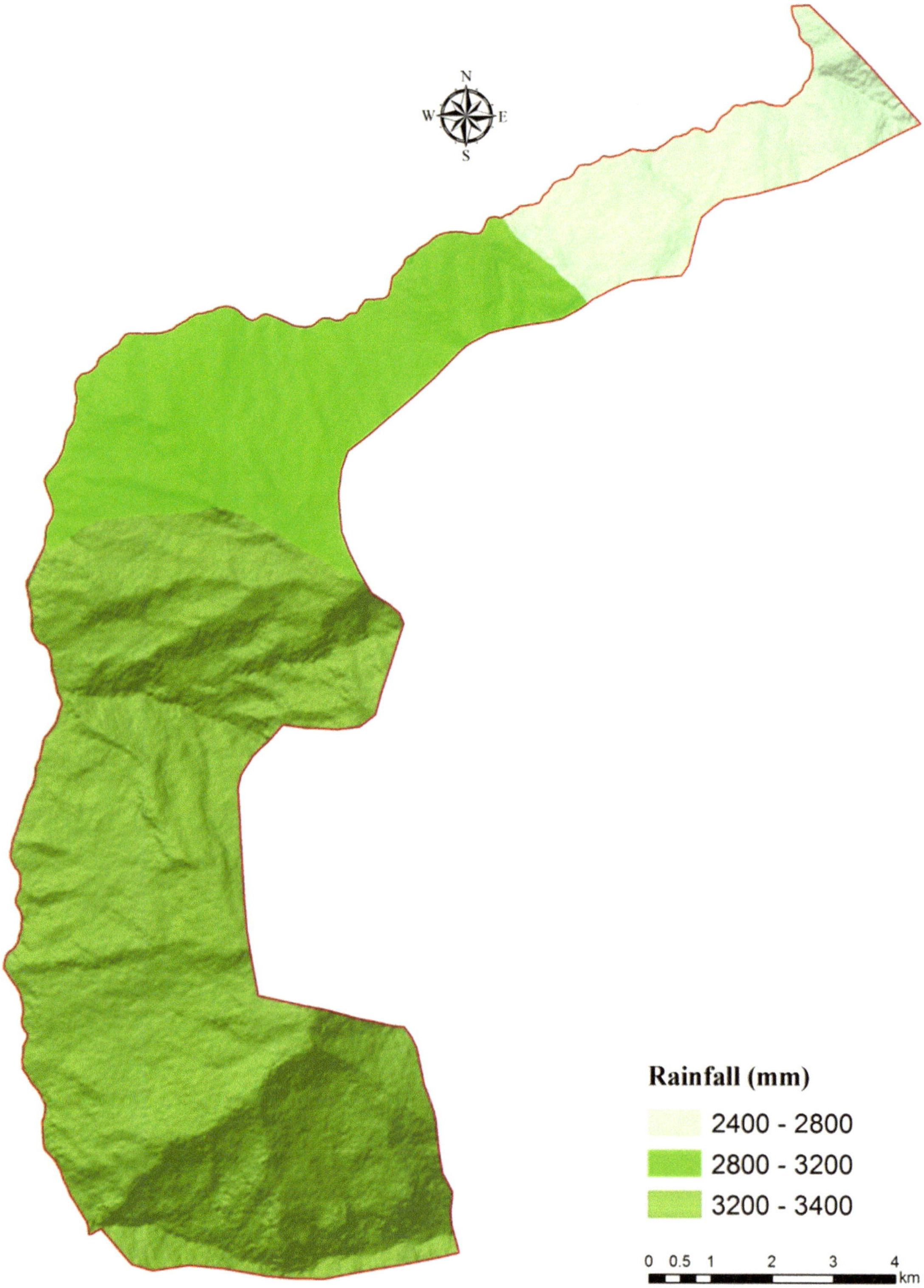

FIGURE 35.6 Rainfall map of Mangan.

(11.50%), followed by moderately shallow (9.35%) and deep (7.36%) (Table 35.5 and Fig. 35.8). Landslide incident probability is low in soil with high thickness and deeper. When heavy rainfall occurs, deeper soils have more space and retain more water (Lee et al., 2004; Sharma et al., 2012).

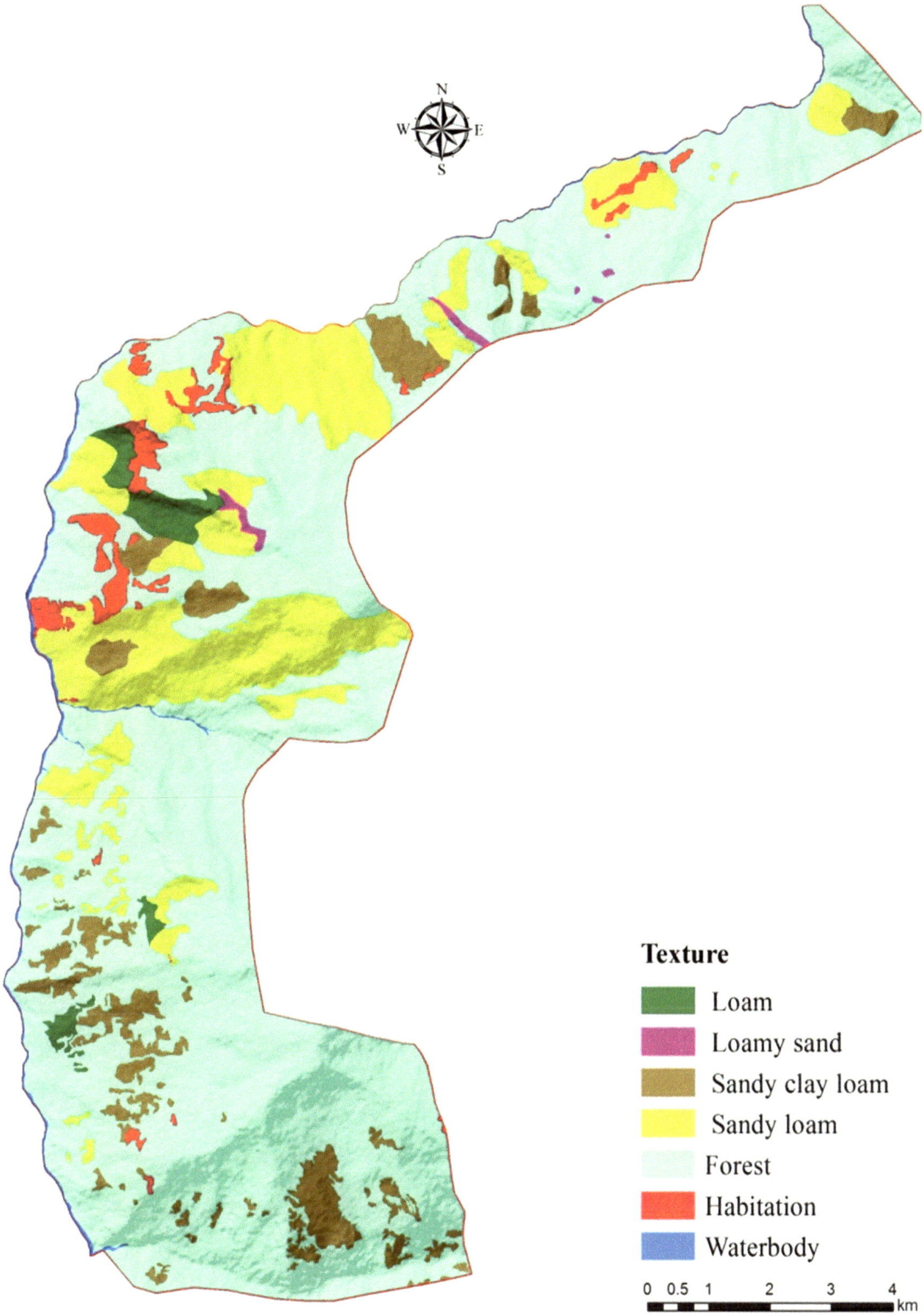

FIGURE 35.7 Soil texture map of Mangan.

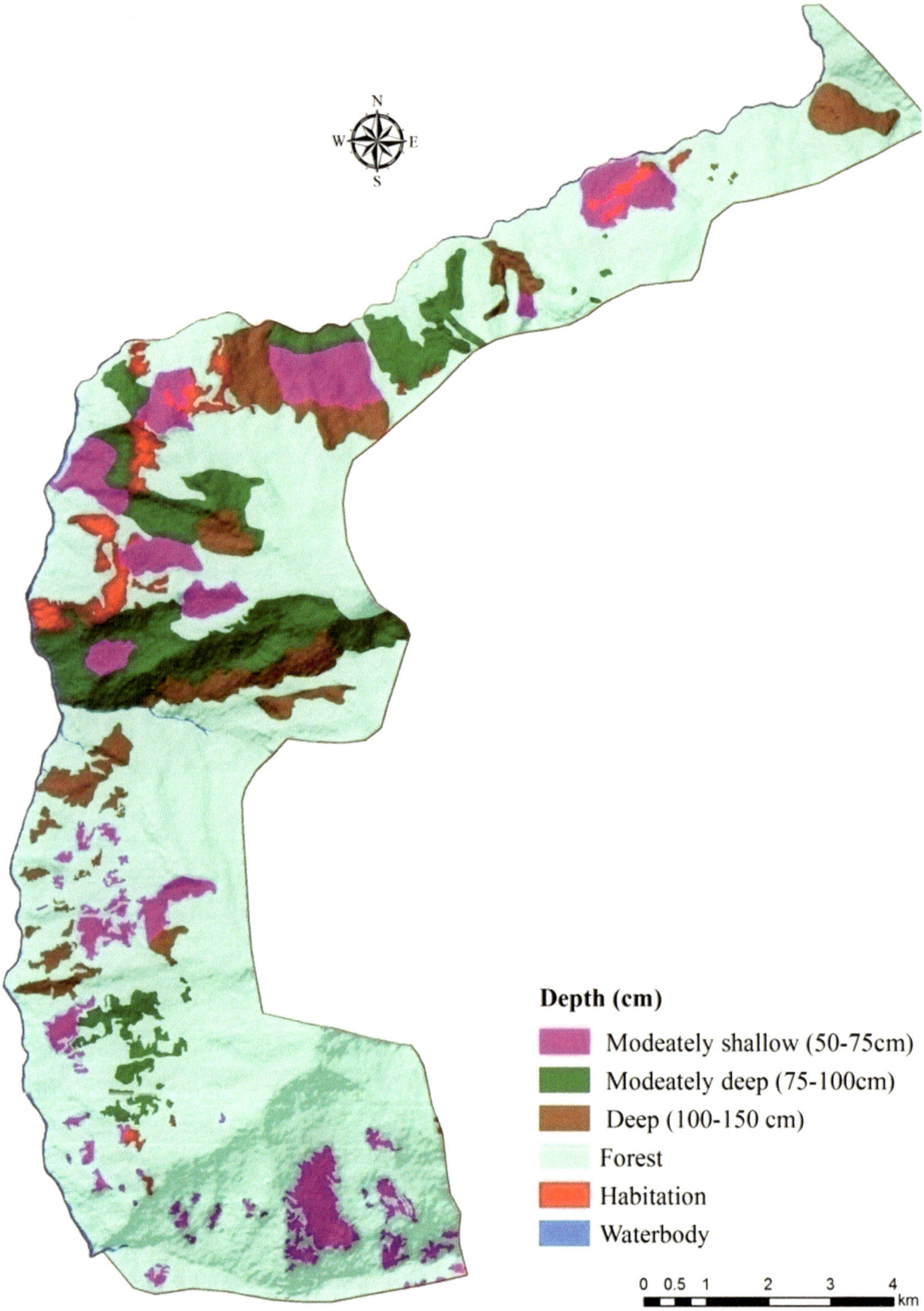

FIGURE 35.8 Soil depth map of Mangan.

35.3.6 Soil gravel

Coarse fragment (gravelliness) refers to discrete unattached pieces of rock, 2 mm in diameter or larger. Different pieces of rock fragment, viz., quartz and sandstone, are observed in the soils of the study area. The presence of a high gravel content in the soil directly reduces the volume of soil and its stability. Soil mixed with gravel and stones may provide easy wash out by heavy rain due to the low soil clay content and binding materials. Therefore, soils with more gravelliness and stones are more susceptible to landslides than soils with no gravelliness. Soils of the study area were grouped (Fig. 35.9) into nongravelly (23.85%), followed by gravelly (3.86%) and very gravelly (0.49%).

35.3.7 Soil erosion

The detachment and removal of soil particles by water or wind are known as soil erosion. It is influenced by slope, rainfall, soil type, vegetative cover, and any interference or disturbance that aids in the removal of particles from the soil surface. The steep slope conditions of different hill and mountain lands are directly linked with severe soil erosion (Samra and Sharma, 2002). The sandy nature of soils, high slope topography and frequency of rainfall in the study area increase the severity of soil erosion and landslides. Soils of the study area were grouped into very severe (14.31%), severe (13.86%), and moderate (0.03%) erosion classes (Fig. 35.10).

35.3.8 Land capability class

Land capability classification is an interpretative grouping made primarily for intensive agricultural and nonagricultural uses. According to the USDA Land Capability Classification, the arable lands in classes I–IV (according to their limitations), grazing and forestry lands in class V–VII, and lands for recreation in class VIII have maximum limitations, wildlife, and quarrying. Further subdivisions of the classes on dominant limiting factors, such as "e" for erosion, "s" for soil, "c" for climate, and "w" for wetness. Soils were grouped into two land capability subclasses, viz., IV and IV, and four subclasses (Fig. 35.11). About, 19.75% of lands are categorized in to be in class IV with the subclasses of IVe and IVes and 8.46% area in class VI with subclasses of VIe and VIes. The major limiting factors in the study area are very steep slopes that lead to severe erosion, strong acidity, and excess gravel caused by poor land quality.

35.3.9 Lineament density

Lineament density helps to understand the cause of landslides through tectonic structures and topographic breaks (Ramli et al., 2010). Regional geological lineaments are major contributing factors to severe landslide phenomena in the Himalayas (Martha et al., 2017a; Mathew et al., 2007; Ramli et al., 2010). The lineament density of the study area was studied using ArcGIS, developed from 10 m resolution, and classified into five density classes, viz., very low, low, moderate, high, and very high (Fig. 35.12). The lineament and drainage patterns are similar behavior in the Mangan region. The results indicate that the high lineament density class is more susceptible to landslides than the lower class. This study was aided by abrupt lithological variations, relief changes, erosional features, and drainage pattern changes. The probability of landslide density along the lineaments is 100% within the buffer zones and is negligible outside (Ayalew and Yamagishi, 2005).

35.3.10 Elevation

The elevation is helpful to group the different reliefs and locate the sites of minimum and maximum heights within the topography. The density of elevation decreases both up and downward from different elevations. Terrain, surface landscape are controlled by flow direction, flow sources, and water holding capacity of the soil, which are essential factors for controlling the density and spatial extent of the landslide. Therefore, DEM is a prime source for deriving different terrain attributes such as slope gradient and density (Moore et al., 1991). Five elevation classes were identified in the study area viz. very low relief (<1001 m), low relief (1001–1326 m), moderate relief (1326–1642 m), high relief (1642–1983 m), and very high relief (> 1983 m). More than 1326 m elevations are highly susceptible to landslides. High elevation along steep slopes, coarse textures, shallow depths, high gravels, etc. were considered high hazard zones for landslides (Fig. 35.13).

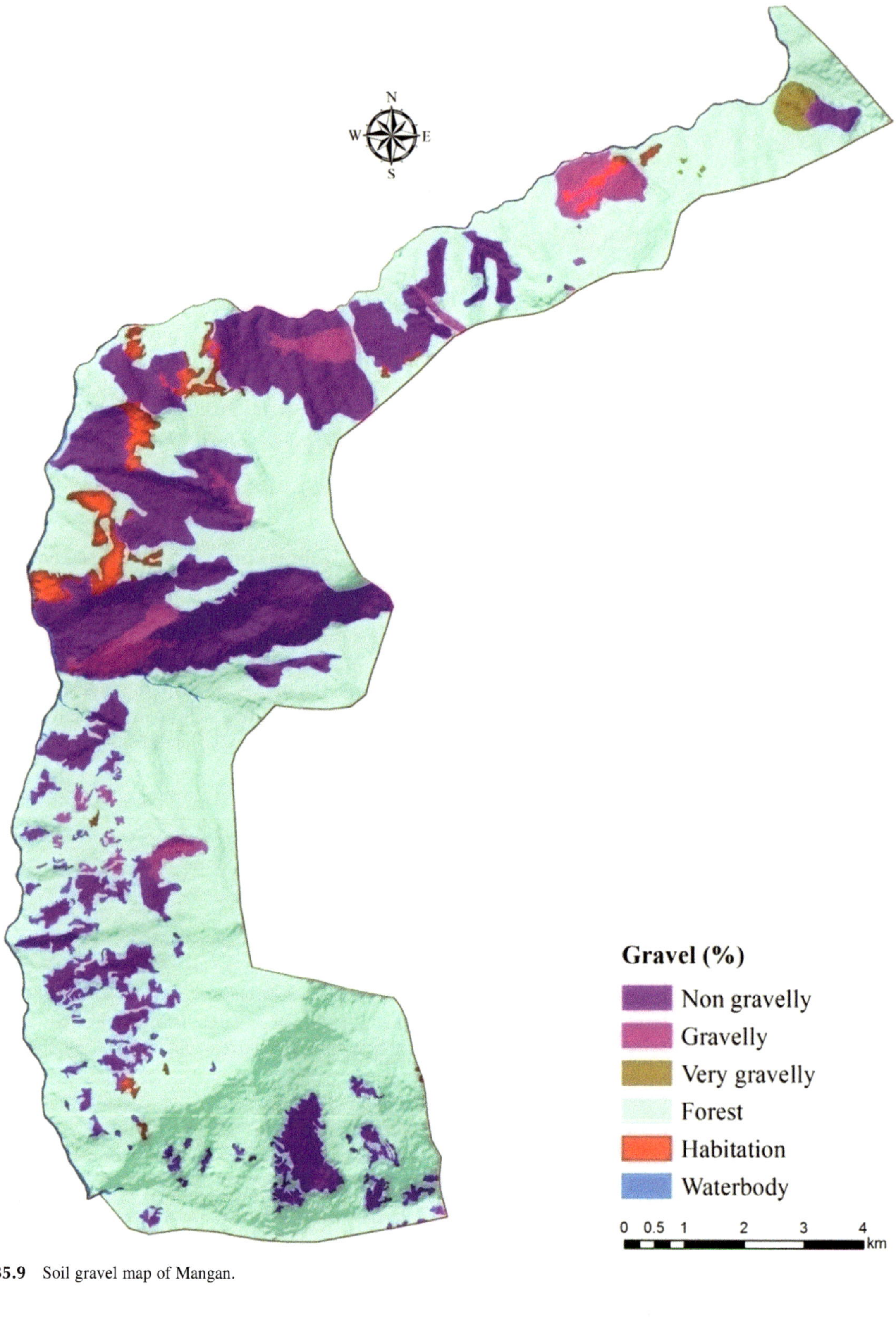

FIGURE 35.9 Soil gravel map of Mangan.

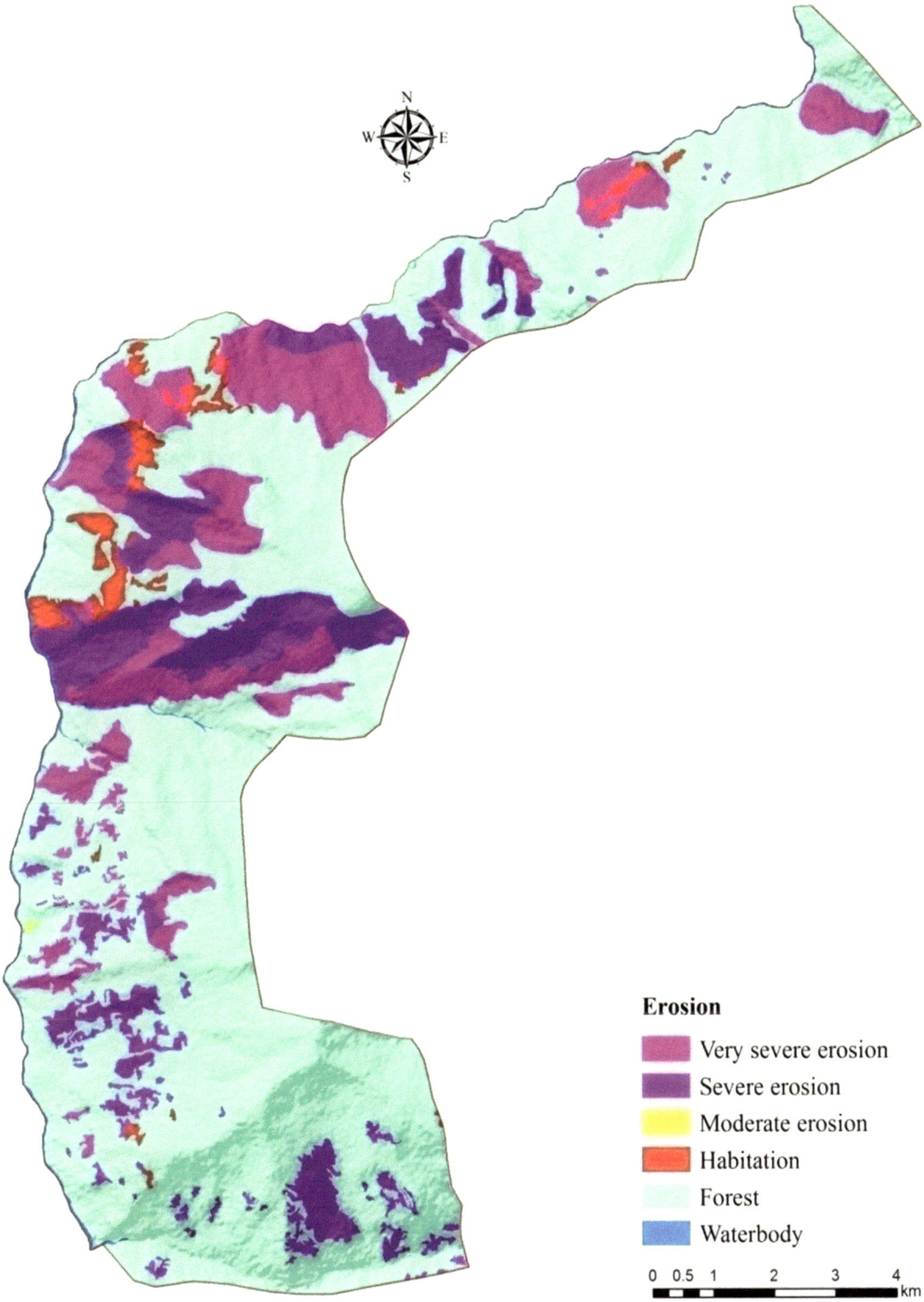

FIGURE 35.10 Soil erosion map of Mangan.

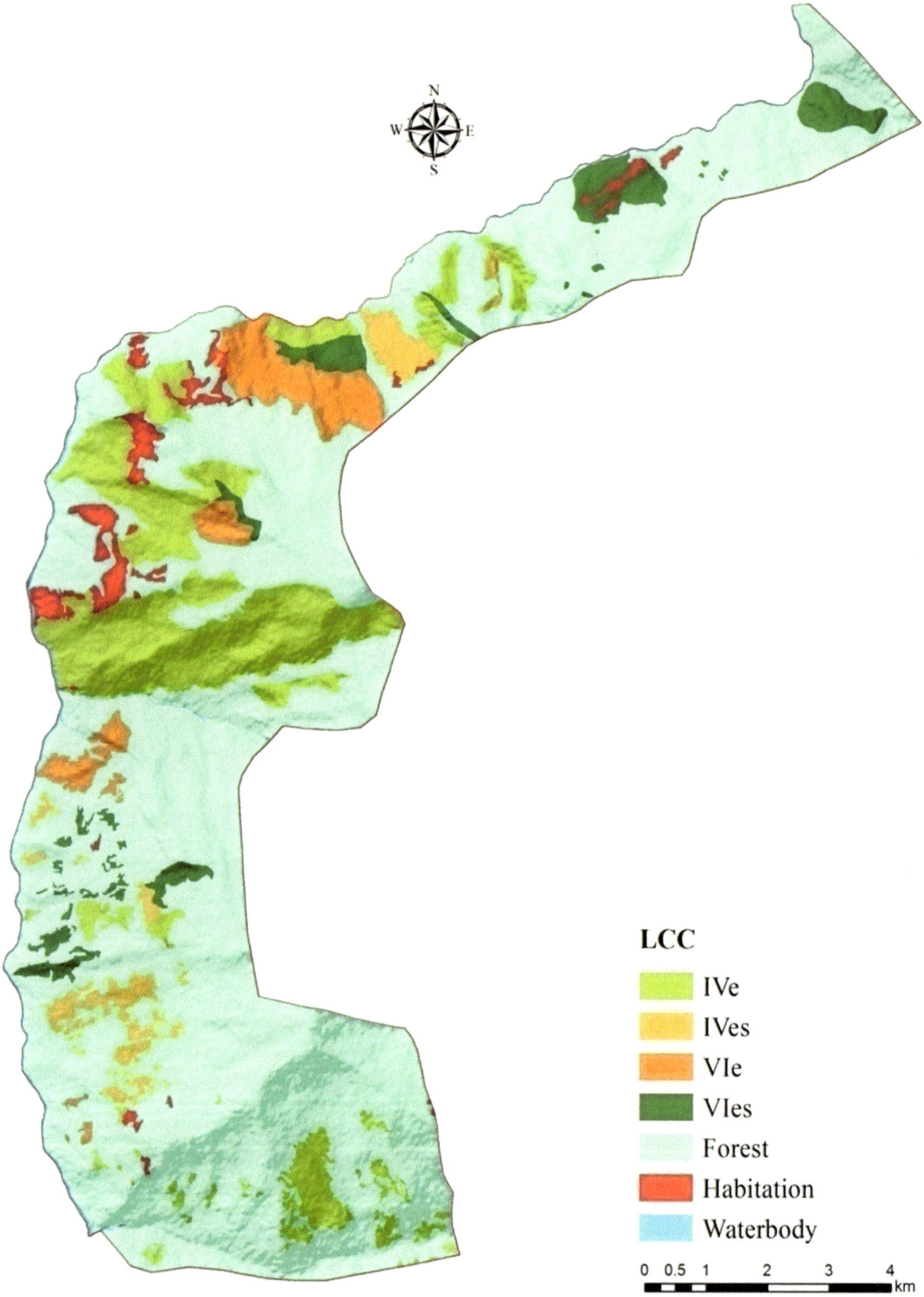

FIGURE 35.11 LCC map of Mangan.

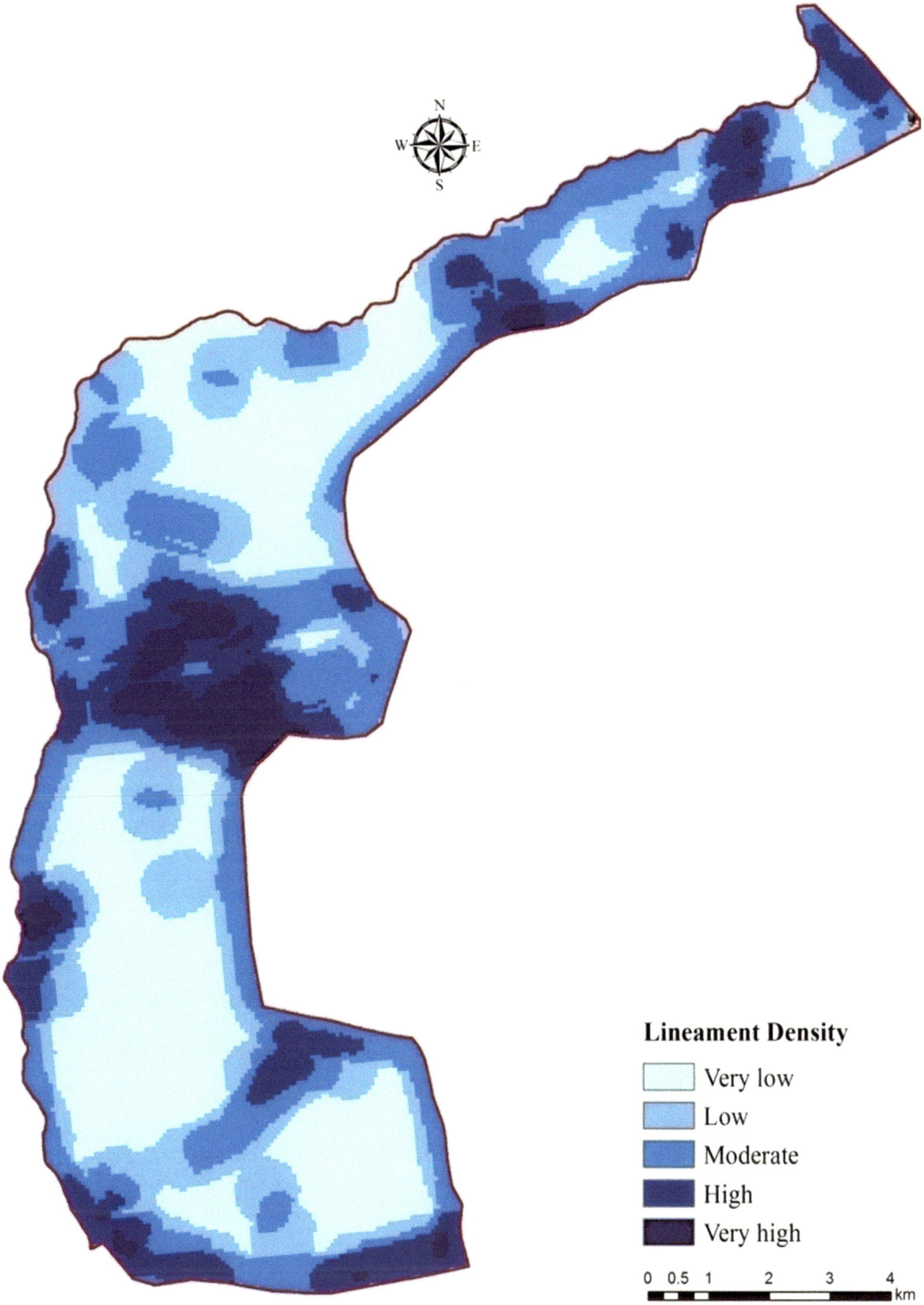

FIGURE 35.12 Lineament density map of Mangan.

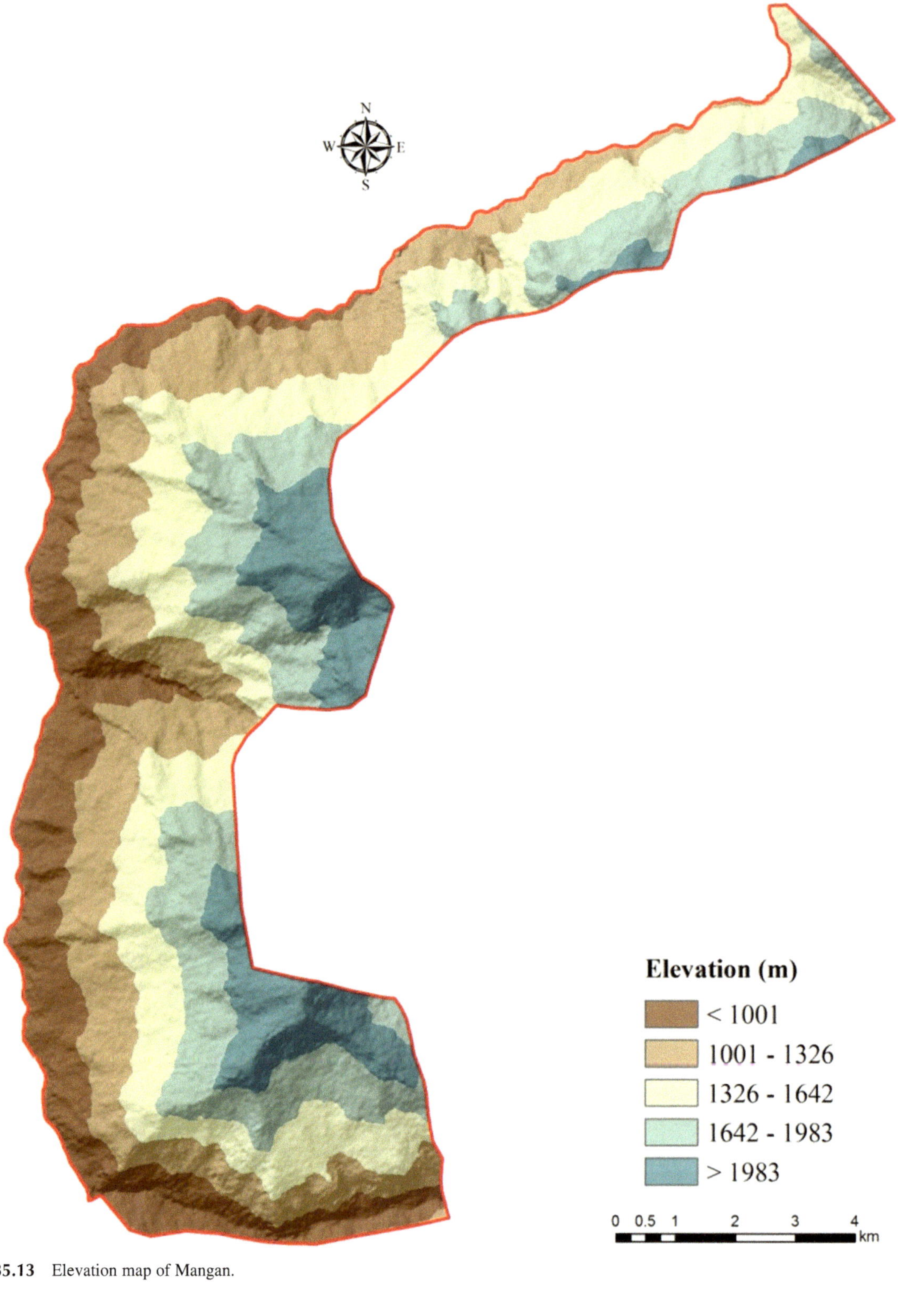

FIGURE 35.13 Elevation map of Mangan.

35.3.11 Drainage density

Drainage density is an important factor accompanying the function of the infiltration rate and control of landslides. Tista is the chief river passing through the lower part of the Mangan region. Most of the tributaries started from the higher hills or ridges and flowed down by cutting deep gorges or ravines and finally joining the main river of Tista. Drainage density is the ratio of the total length of the stream to the area of the drainage basin; the higher the drainage density, the lower the infiltration and the faster the surface runoff (Yalcin, 2005; Shit et al., 2016). The final drainage density map of the study area was prepared by assigning five different textures and ranking values (Table 35.5) to each class and analyzed using ArcGIS software. The high permeable lower soil layer, low relief, and dense vegetation contributed to the low drainage density, whereas poor structure and texture of the subsoil layer and low vegetation on steep slopes favored the high drainage density of the region (Fig. 35.14). Therefore, areas with higher drainage density are accompanied by a higher possibility of landslides (Rawat et al., 2012).

35.3.12 Geology

The Manganregion comprises five litho-stratigraphic units: quartzite, granite gneiss, biotiteschist, migmatile, and chlorite-sericlie (Fig. 35.15). Primary rocks observed in the Himalayan region are the (1) Darjeeling Formation, (2) Daling Formation, and (3) Lingtse Gneiss (Acharyya, 1989). The quartzite and granite gneiss group of rocks are less susceptible to landslides, and interbanded chlorite-sericlie and banded migmatile are highly susceptible to landslide hazards. The critical source of data associated with the geomorphology of an area is the lithologic properties (Dai et al., 2001). Martha et al. (2017b) reported that the "So Bhir" landslide, found that high-grade rocks of proterozoic central crystalline gneissic complex (e.g., granite gneisses, quartzites and schists) are exposed to adjoining area has coarse texture soil, and steep slope caused a valley-blocking landslide in Mantam village, North Sikkim.

35.3.13 Land use/land cover

The surface vegetative cover on any kind of land is considered a dominant parameter to control landslides. Land use patterns are related to different physical, chemical, and biological processes on the surface of the earth (Dai and Lee, 2002). Prominent six categories of LULCin the study area (Fig. 35.16). Single crop, double crop, agroforestry, open scrub, open field, and forest. Within the forest system, land use is further classified into very dense forest, other than forest, open forest, moderately dense forest, and grassland (Fig. 35.17). The land cover with woody vegetation has secured an extensive root system that helps to stabilize the soil structure, improve slope stability, and prevent soil loss (Greenway, 1987). Forestlands area regulates the surface flow and water infiltration. In contrast, farming lands affect slope stability due to the water saturation of the soil. Also, barren and uncultivable lands are extremely vulnerable to soil erosion and mass movement. The study area in general has forest vegetation covering more than 66% of the area. Single crops such as paddy, maize, barley, ginger, cardamom, and are shallow rooting systems that cause poor soil structure formation and are easily prone to landslides.

35.3.14 Slope aspect

The slope aspect is an important criterion for assessing landslide hazards. Exposure to sunlight, rainfall, drying winds, and discontinuities control landslide occurrences (Dai et al., 2001). Different slope degrees into the aspect class as flat, north, west, east, and south. The aspect map of the study area is derived from ALOS DEM (12.5 m resolution) and refers to the direction of mountain slope faces. The slope aspects of the study area are divided into nine categories; those are flat-1, north 0−22.50, northeast 22.5−67.50, east 67.5−112.50, southeast 112.5−157.50, south 157.5−202.50, southwest 202.5−247.50, west 247.5−292.50, and northwest 292.5−337.50. The area aspect map (Fig. 35.18) indicates that the left bank dominant by the west- to the southwest-facing slope and the right bank dominant by northwest- to northeast-facing slopes are along the mainstream. In general, hills facing south and southeast-facing slopes have a lesser vegetation cover than north-facing slopes; which are highly susceptible to severe soil erosion and landslide activity (Dhakal et al., 2000; Singh et al., 2014; Bera et al., 2018).

35.3.15 Landslide hazard zonation

A LHZ map was produced by the weighted overlay index method on the GIS platform with the help of ArcGIS 10.2 software at a 1:10000 scale of the region. Mapping of LHZ of the study area showed five classes such as very low, low, moderate, high, and very high hazard (Fig. 35.19). Table 35.6 clearly shows that 94% (84.4 km^2) of the study area has moderate to high hazard and 4.53% area was very high hazard. The areas that accounted for very low and low hazards were 0.39% and 0.73%, respectively. The results (Fig. 35.20) indicate that different LHZ in different soil series, namely

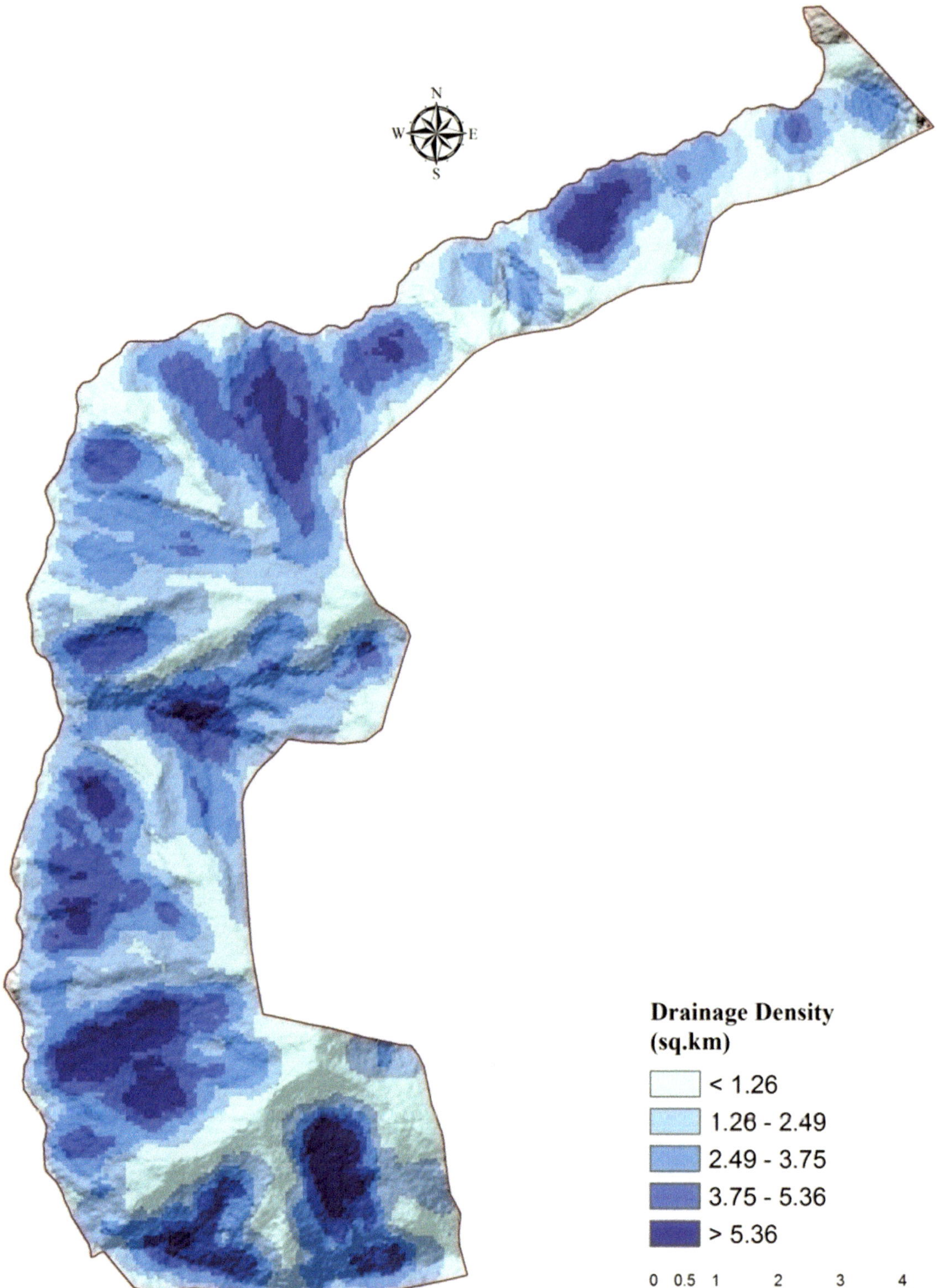

FIGURE 35.14 Drainage density Aspect map of Mangan.

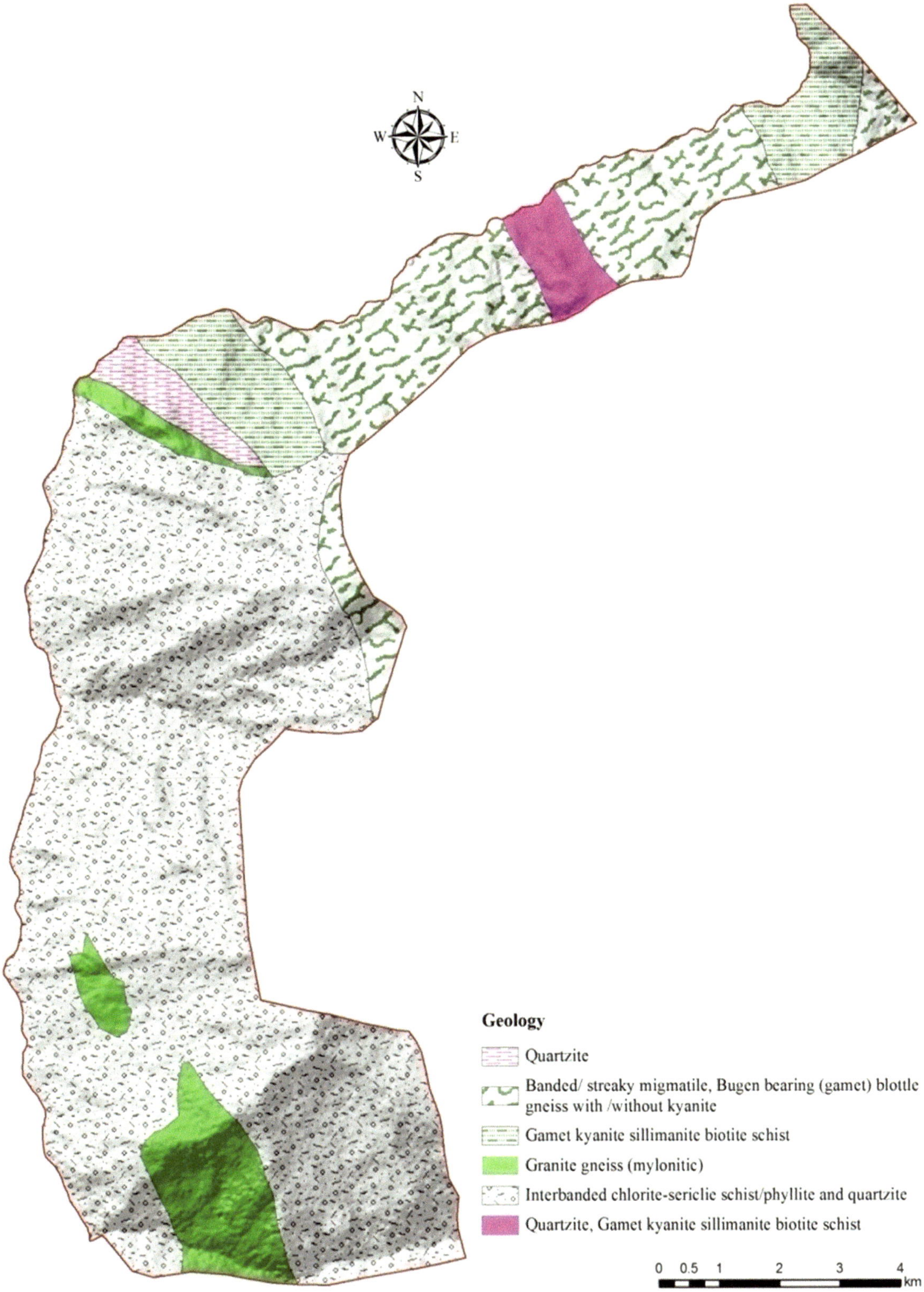

FIGURE 35.15 Geology map of Mangan.

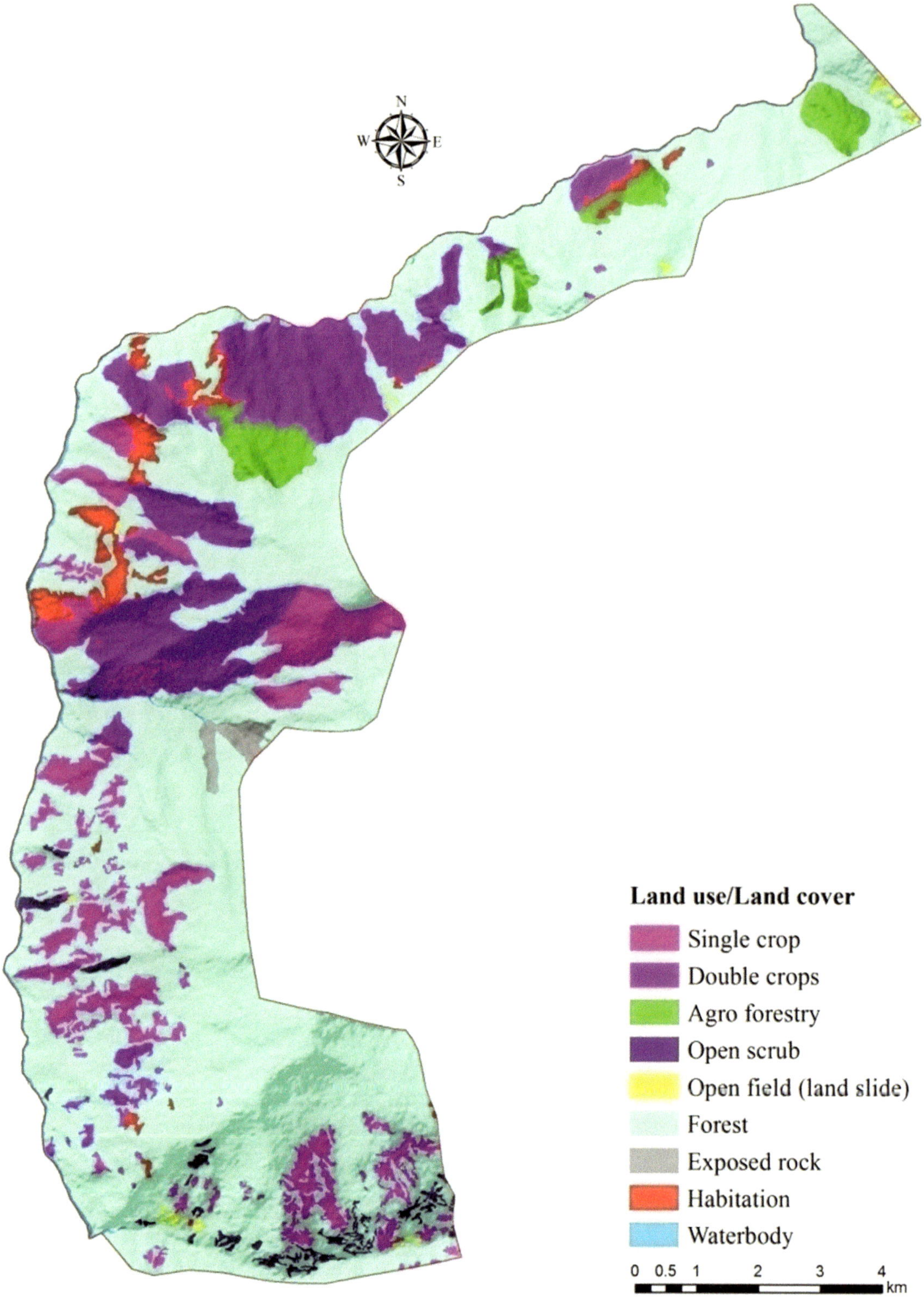

FIGURE 35.16 LULC map of Mangan.

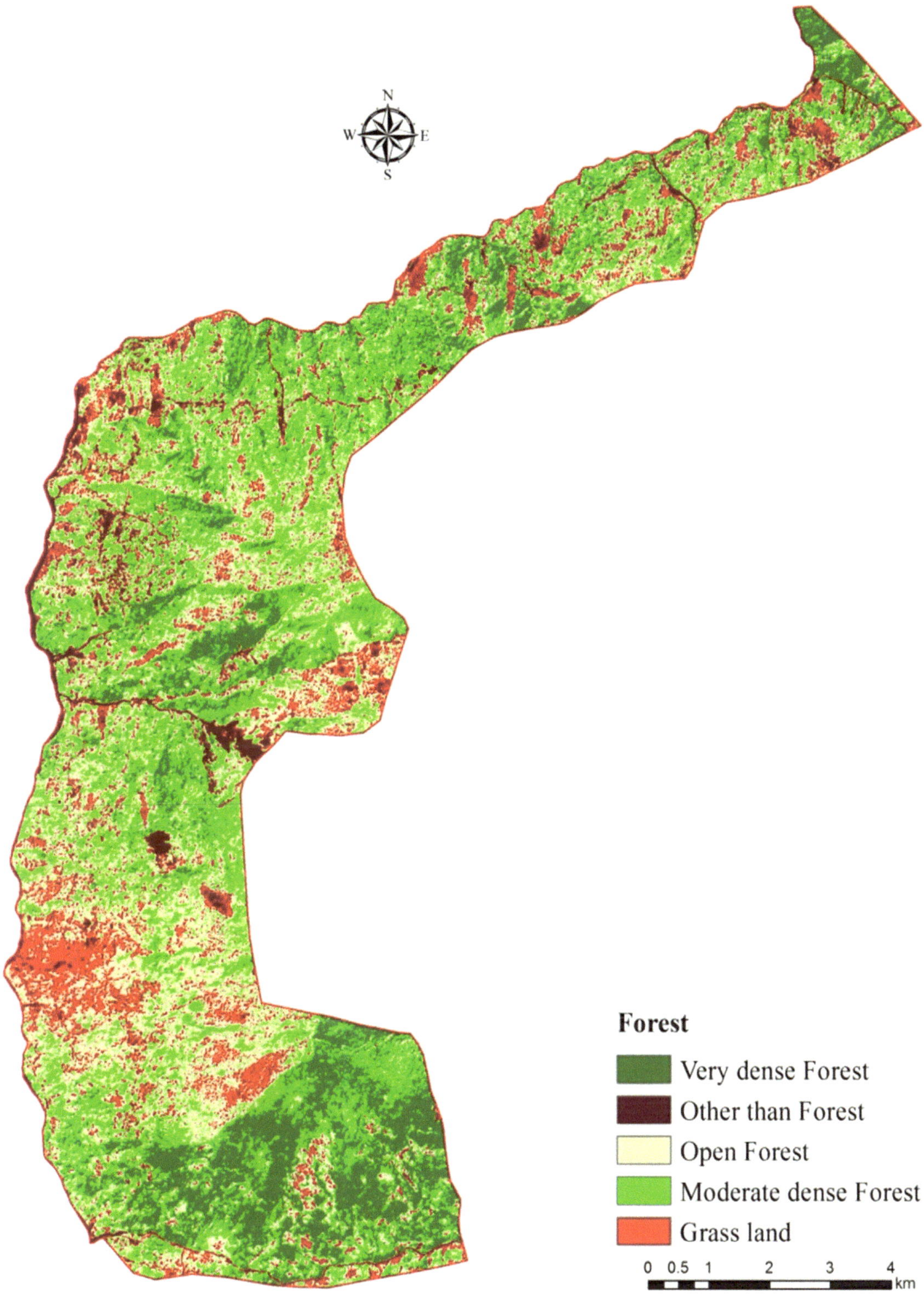

FIGURE 35.17 Forest map of Mangan.

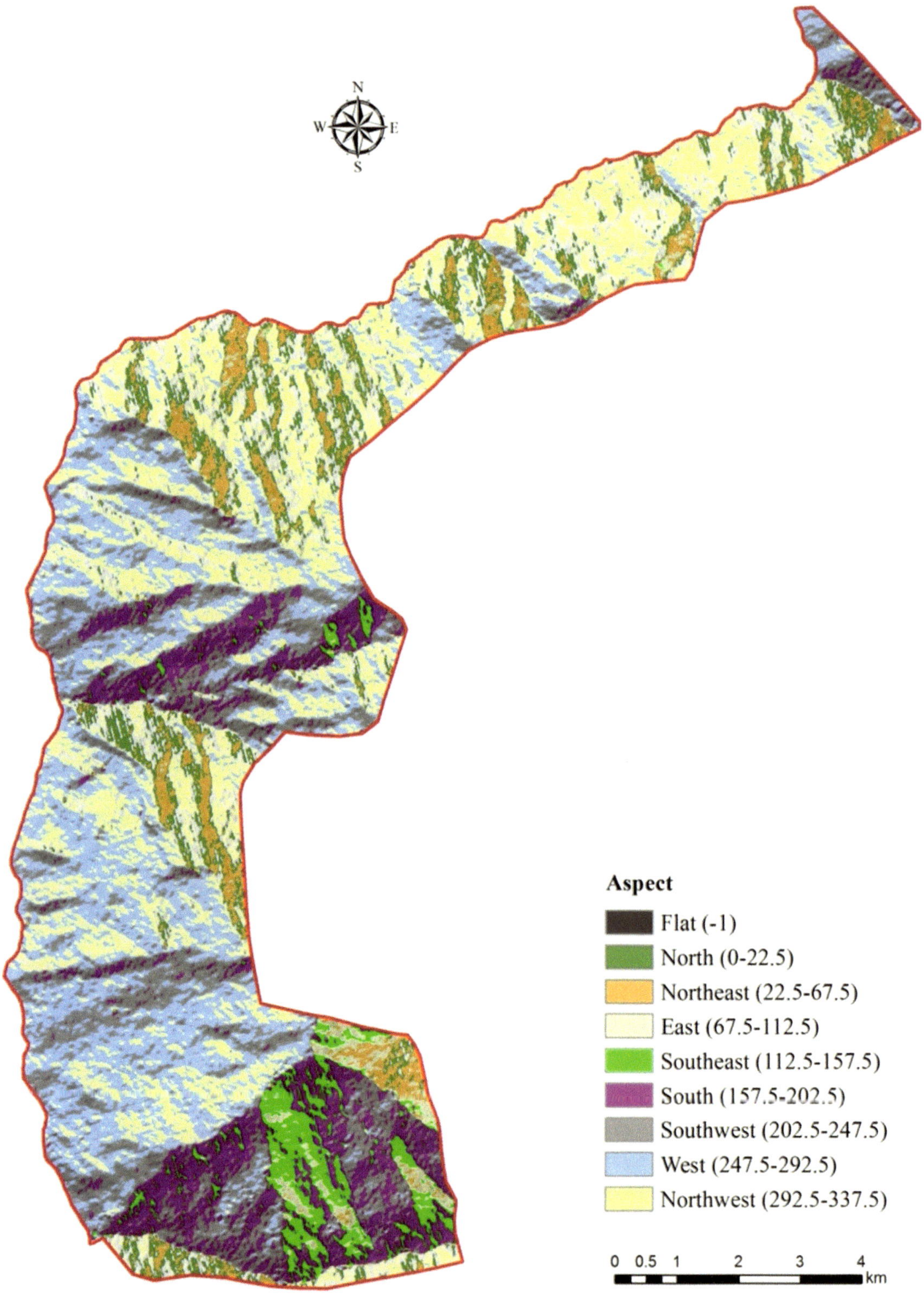

FIGURE 35.18 Slope aspect map of Mangan.

FIGURE 35.19 LHZ map of Mangan.

TABLE 35.6 Percentage area of different LHZ in the Mangan region.

LHZ	Area (sq. km)	Area percentage	Cumulative area percentage
Very low	0.35	0.39	0.39
Low	0.65	0.73	1.12
Moderate	42.8	47.87	48.98
High	41.6	46.48	95.46
Very high	4.05	4.53	100

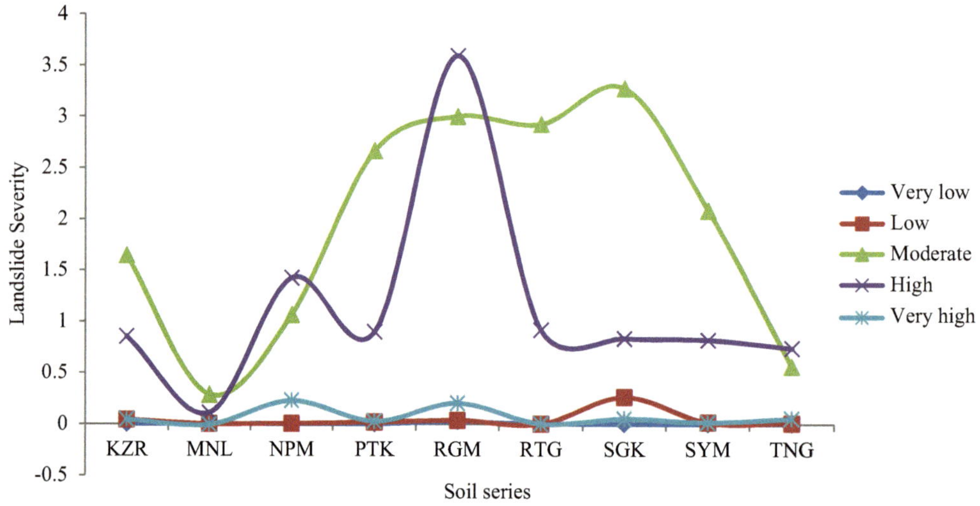

KZR-Kazor MNL- Manul NPM- Nampatam PTK- Pantok RGM- Ringhim
RTG- Ramthang SGK- Singhik SYM- Swayem TNG- Tung

FIGURE 35.20 Soil series wise landslide hazard and severity classes.

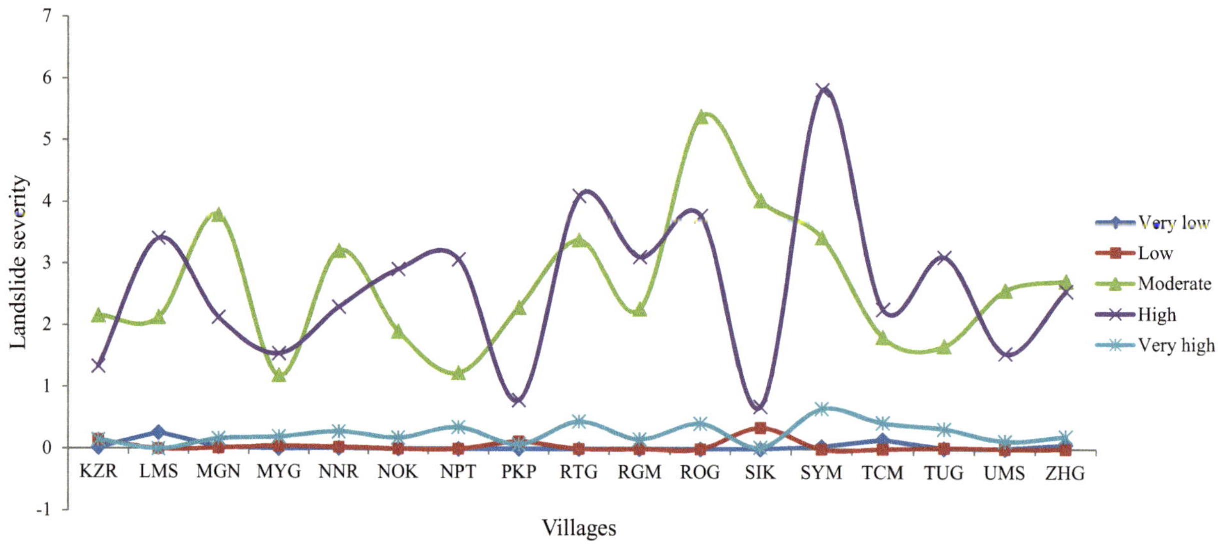

KZR-Kazor LMS-Lower Mangshila MGN- Mangan MYG- Meyong NNR- Naga-Namgor NOK- Namok
NPT- Nampatam PKP- Pakshep RTG- Ramthang RGM- Ringhim ROG- Rongong SIK- Singhik
SYM- Swayem TCM- Tingchim TUG- Tung UMS-Upper Mangshila ZHG- Zimchung

FIGURE 35.21 Village wise landslide hazard and severity classes.

FIGURE 35.22 Landslides observed during the field survey in the study area.

Ringhim, Nampatam, Kazor, Singhik, and Swayem, are higher than others. The moderate to very high LHZ in different villages are given in Fig. 35.21. The majority of the villages are Swayem, Ramthang, Rongong, Nampatam, Tung, Naga-Namgor, Mangan and Tingchim. High-hazard zones in high rainfall, steep slopes, shallow soil depth, high gravel and coarse texture, poor geology, and low LULC falling on the road. Areas with low-hazard zones have moderate slopes and deep soil with fine texture, mostly in the upper and central part of the Mangan area. During ground truth, the observed maximum landslides (Fig. 35.22) were along the steep slope caused by heavy rainfall and weak soil strength.

35.4 Conclusions

Mangan is a segment of the Eastern Himalayas, and Sikkim is regularly prone to landslides. Heavy rainfall and varying soil characteristics are important factors for landslides in this area. Delineated LHZ map using multicriteria analysis (AHP-weighted) of different attributes using remote sensing and GIS techniques. This method is a useful tool to demarcate the susceptible zone of a landslide in anextensive area based on ground truth observations. The LHZ map divides the area into five zones of vulnerability classes. Among them, moderate to high recorded 94% (84.4 km^2) of the study area and 4.53% area was a very high hazard. Hence, such hazard zonation maps should be updated periodically. The resulting LHZ map of the Mangan region can be used as base data to support land-use planning and slope management in this region.

References

Acharyya, S.K., 1989. The Daling group, its nomenclature, tectonostratigraphy and structural grains: with notes on their possible equivalents, Daling Group and Related Rocks, 22. Geological Survey of India, pp. 5–14, Special Publication No.

AISLUS, 1970. Soil Survey Manual. All India Soil and Land Use Survey Organisation, IARI, New Delhi, pp. 1–63.

Al-Subhi Al-Harbi, K.M., 2001. Application of the AHP in project management. Int. J. Constr. Proj. Manag. 19 (1), 19–27.

Anbalagan, R., 1992. Landslide hazard evaluation and zonation mapping in mountainous terrain. Eng. Geol. 32, 269–277.

Anbalagan, R., Kumar, R., Lakshmanan, K., Parida, S., Neethu, S., 2015. Landslide hazard zonation mapping using frequency ratio and fuzzy logic approach, a case study of Lachung Valley, Sikkim. Geoenviron. Disasters 2 (6), 1–17. Available from: https://doi.org/10.1186/s40677-014-0009-y.

Anbalagan, R., Kumar, R., Parida, S., Lakshmanan, K., 2014. GIS-based post-earthquake landslide hazard zonation mapping of Lachung Basin, Sikkim. Int. J. Emerg. Technol. Adv. Eng. 4 (1), 431–441.

Arora, M.K., Das Gupta, A.S., Gupta, R.P., 2004. An artificial neural network approach for landslide hazard zonation in the Bhagirathi (Ganga) Valley, Himalayas. Int. J. Rem. Sens. 25, 559–572.

Ayalew, L., Yamagishi, H., 2005. The application of GIS-based logistic regression for landslide susceptibility mapping in the Kakuda-Yahiko Mountains, Central Japan. Geomorphology 65, 15–31.

Banerjee, P., Ghose, M.K., Pradhan, R., 2018. Analytic hierarchy process and information value method based landslide susceptibility mapping and vehicle vulnerability assessment along a highway in Sikkim Himalaya. Arab. J. Geosci. Available from: https://doi.org/10.1007/s12517-018-3488-4.

Baynes, F.J., 2008. Anticipating problem soils on linear projects. In: Conference Proceedings on Problem Soils in South Africa, 3–4 November, pp. 9–21.

Bera, A., Mukhopadhyay, B.P., Das, D., 2019. Landslide hazard zonation mapping using multiple criteria analysis with the help of GIS techniques: a case study from Eastern Himalayas, Namchi, South Sikkim. Nat. Hazards 96, 935–959. Available from: https://doi.org/10.1007/s11069-019-03580-w.

Bera, A., Mukhopadhyay, B.P., Das, D. 2018. Morphometric analysis of Adula River Basin in Maharashtra, India using GIS and Remote Sensing techniques. In: Geo-spatial Data in Natural Resources, pp 13–35. https://doi.org/10.21523/gcb5.1702.

Chen, Y., Khan, S., Paydar, Z., 2010. To retire or expand? A fuzzy GIS-based spatial multi-criteria evaluation framework for irrigated agriculture. Irrig. Drain. 59 (2), 174–188. Available from: https://doi.org/10.1002/ird.470.

Chorley, R.J., Schumm, S.A., Sugden, D.E., 1985. Geomorphology. Methuen, London.

Dahal, R.K., Hasegawa, S., Yamanaka, M., Dhakal, S., Bhandary, N.P., Yatabe, R., 2009. Comparative analysis of contributing parameters for rainfall-triggered landslides in the Lesser Himalaya of Nepal. Env. Geol. 58 (3), 567–586.

Dai, F., Lee, C., 2002. Landslide characteristics and slope instability modelling using GIS, Lantau Island, Hong Kong. Geomorphology 42 (3–4), 213–228. Available from: https://doi.org/10.1016/s0169-555x(01)00087-3.

Dai, F.C., Lee, C.F., Li, J., Xu, Z.W., 2001. Assessment of landslide susceptibility on the natural terrain of Lantau Island, Hong Kong. Env. Geol. 40, 381–391.

Das, D.K., Agarwal, R.P., 2002. Physical properties of soil. Fundamental of soil science. Indian Society of Soil Science. IARI, New Delhi, India, pp. 75–77.

Das, S., Mallik, J., Dhankhar, S., Suthar, N., Singh, A.K., Dutta, V., et al., 2020. Application of Fracture Induced Electromagnetic Radiation (FEMR) technique to detect landslide-prone slip planes. Nat. Hazards. Available from: https://doi.org/10.1007/s11069-020-03883-3.

Day, P.R., 1965. Particle fractionation and particle size analysis. In: Black, C.A. (Ed.), Methods of Soil Analysis. American Society of Agronomy, Madison, WI, pp. 545–567.

Dhakal, A.S., Amada, T., Aniya, M., 2000. Landslide hazard mapping and its evaluation using GIS: an investigation of sampling schemes for grid-cell based quantitative method. Photogramm. Eng. Rem. S 66 (8), 981–989.

Dubey, C.S., Chaudhry, M., Sharma, B.K., Pandey, A.C., Singh, B., 2005. Visualization of 3-D digital elevation model for landslide assessment and prediction in mountainous terrain: a case study of Chandmari landslide, Sikkim, Eastern Himalayas. Geosci. J. 9 (4), 363–373.

Feizizadeh, B., Blaschke, T., 2013. GIS-multicriteria decision analysis for landslide susceptibility mapping: comparing three methods for the Urmia lake basin, Iran. Nat. Hazards 65 (3), 2105–2128. Available from: https://doi.org/10.1007/s11069-012-0463-3.

Greenway, D.R., 1987. Vegetation and slope stability. In: Anderson, M.G., Richards, K.S. (Eds.), Slope Stability. Wiley, New York, pp. 187–230.

Hadmoko, D.S., Lavigne, F., Samodra, G., 2017. Application of a semi-quantitative and GIS-based statistical model to landslide susceptibility zonation in Kayangan Catchment, Java, Indonesia. Nat. Hazards 87, 437–468. Available from: https://doi.org/10.1007/s11069-017-2772-z.

Kanungo, D.P., Arora, M.K., Sarkar, S., Gupta, R.P., 2006. A comparative study of conventional, ANN black box, fuzzy and combined neural and fuzzy weighting procedures for landslide susceptibility Zonation in Darjeeling Himalayas. Eng. Geol. 85, 347–366.

Kanungo, D.P., Arora, M.K., Sarkar, S., Gupta, R.P., 2009. Landslide susceptibility zonation (LSZ) mapping—a review. J. South. Asia Dis. Stu 2, 81–105.

Kanwal, S., Atif, S., Shafiq, M., 2016. GIS-based landslide susceptibility mapping of northern areas of Pakistan, a case study of Shigar and Shyok Basins. Geom. Nat. Hazards Risk. Available from: https://doi.org/10.1080/19475705.2016.1220023.

Knapen, A., Kitutu, M.G., Poesen, J., Breugelmans, W., Deckers, J., Muwanga, A., 2006. Landslides in a densely populated county at the footsteps of Mount Elgon (Uganda): characteristics and causal factors. Geomorphology 73, 149–165.

Kolat, C., Ulusay, R., Suzen, M.L., 2012. Development of geotechnical micro zonation model for Yenisehir (Bursa, Turkey) located at a seismically active region. Eng. Geol. 127, 36–53. Available from: https://doi.org/10.1016/j.enggeo.2011.12.014.

Lee, S., Ryu, J.H., Won, J.S., Park, H.J., 2004. Determination and application of the weights for landslide susceptibility mapping using an artificial neural network. Eng. Geol. 71, 289–302.

Mahdadi, F., Boumezbeur, A., Hadji, R., Kanungo, D.P., Zahri, F., 2018. GIS-based landslide susceptibility assessment using statistical models: a case study from Souk Ahras province, NE Algeria. Arab. J. Geosci. 11, 476. Available from: https://doi.org/10.1007/s12517-018-3770-5.

Manchanda, M.L., Kudrat, M., Tiwari, A.K., 2002. Soil survey and mapping using remote sensing. Trop. Ecol. 43 (1), 61–74.

Mandal, S., Maiti, R., 2015. Impact assessment of hydrologic attributes and slope instability. Semi-Quantitative Approaches for Landslide Assessment and Prediction. Springer, Singapore, pp. 95–121. Available from: https://doi.org/10.1007/978-981-287-146-6_3.

Martha, T.R., Reddy, S.P., Bhatt, C.M., Govindha Raj, B.K., Nalini, J., Padmanabha, A.E., et al., 2017a. Landslides 14 (1), 373–383.

Martha, T.R., Roy, P., Vinod Kumar, K., 2017b. Assessment of the valley-blocking 'So Bhir' landslide near Mantam village, North Sikkim, India, using satellite images. Curr. Sci. 113 (7), 1228–1229.

Mathew, J., Jha, V.K., Rawat, G.S., 2007. Application of binary logistic regression analysis and its validation for landslide susceptibility mapping in part of Garhwal Himalaya, India. Int. J. Remote. Sens. 28 (10), 2257–2275. Available from: https://doi.org/10.1080/01431160600928583.

Mathew, J., Jha, V.K., Rawat, G.S., 2009. Landslide susceptibility zonation mapping and its validation in part of Garhwal Lesser Himalaya, India, using Binary Logistic Regression analysis and receiver operating characteristic curve method. Landslides 6, 17–26.

Montgomery, D.R., Schmidt, K.M., Dietrich, W.E., Greenberg, H.M., 2000. Forest clearing and regional landsliding in the Pacific Northwest. Geology 28, 311–314.

Moore, I.D., Grayson, R.B., Ladson, A.R., 1991. Digital terrain modeling: a review of hydrological, geomorphological and biological applications. Hydrol. Process. 5, 3–30.

Mugagga, F., Kakembo, V., Buyinza, M., 2011. A characterisation of the physical properties of soil and the implications for landslide occurrence on the slopes of Mount Elgon, Eastern Uganda. Nat. Hazards. Available from: https://doi.org/10.1007/s11069-011-9896-3.

Nisha, S., Singh, S.K., Obi Reddy, G.P., Nirmal, K., Nagaraju, M.S.S., Rajeev, S., 2016. Large-scale soil resource mapping using IRS-P6 LISS-IV and Cartosat-1 DEM in Basaltic Terrain of Central India. J. Indian. Soc. Remote. Sens. 44 (5), 811–819. Available from: https://doi.org/10.1007/s12524-015-0540-7.

Pal, S.C., Chowdhuri, I., 2019. GIS-based spatial prediction of landslide susceptibility using frequency ratio model of Lachung River basin, North Sikkim, India. SN Appl. Sci. 1, 416. Available from: https://doi.org/10.1007/s42452-019-0422-7.

Pandey, A., Dabral, P.P., Chowdary, V.M., Yadav, N.K., 2008. Landslide hazard zonation using remote sensing and GIS: a case study of Dikrong river basin, Arunachal Pradesh, India. Env. Geol. 54 (7), 1517–1529. Available from: https://doi.org/10.1007/s00254-007-0933-1.

Patanakanog, B., 2001. Landslide Hazard Potential Area in 3 Dimension by Remote Sensing and GIS Technique. Land Development Department, Thailand. Available from: http://www.ecy.wa.gov/programs/sea/landslides/help/drainage.html.

Ramli, M.F., Yusof, N., Yusoff, M.K., Juahir, H., Shafri, H.Z.M., 2010. Lineament mapping and its application in landslide hazard assessment: a review. Bull. Eng. Geol. Env. 69, 215–233. Available from: https://doi.org/10.1007/s10064-009-0255-5.

Rawat, M.S., Rawat, B.S., Joshi, V., Kimothi, M.M., 2012. Statistical analysis of landslide in South district, Sikkim, India: using remote sensing and GIS. J. Env. Sci. Toxicol. Food Technol. 2 (3), 47–61.

Roth, R.A., 1983. Factors affecting landslide susceptibility in San Mateo County, California. Bull. Assoc. Eng. Geol. 20 (4), 353–372.

Saaty, T.L., 2003. Decision-making with the AHP: why is the principal eigenvector necessary. Eur. J. Oper. Res. 145 (1), 85–91.

Saaty, T.L., 1980. The Analytic Hierarchy Process. McGraw-Hill, New York.

Saha, A.K., Gupta, R.P., Sarkar, I., Arora, M.K., Csaplovics, E., 2005. An approach for GIS-based statistical landslide susceptibility zonation with a case study in the Himalayas. Landslides 2, 61–69.

Samra, J.S., Sharma, U.C., 2002. Soil erosion and conservation. Fundamental of soil science. Indian Society of Soil Science. IARI, New Delhi, India, p. 162.

Sarkar, S., Kanungo, D.P., 2004. An integrated approach for landslide susceptibility mapping using remote sensing and GIS. Photogramm. Eng. Remote. Sens. 70 (5), 617–625. Available from: https://doi.org/10.14358/pers.70.5.617.

Sengupta, A., Gupta, S., Anbarasu, K., 2010. Rainfall thresholds for the initiation of landslide at Lanta Khola in north Sikkim, India. Nat. Hazards 52 (1), 31–42. Available from: https://doi.org/10.1007/s11069-009-9352-9.

Sharma, L.P., Nilanchal, P., Debnath, P., Ghose, M.K., 2012. Assessing landslide vulnerability from soil characteristics- a GIS-based analysis. Arab. J. Geosci. 5, 789–796. Available from: https://doi.org/10.1007/s12517-010-0272-5.

Sharma, L.P., Patel, N., Ghose, M.K., Debnath, P., 2015. Development and application of Shannon's entropy integrated information value model for landslide susceptibility assessment and zonation in Sikkim Himalayas in India. Nat. Hazards 75, 1555–1576. Available from: https://doi.org/10.1007/s11069-014-1378-y.

Sharma, L.P., Patel, N., Ghosh, M.K., Debnath, P., 2009. Geographic information system-based landslide probabilistic model with trivariate approach—a case study in Sikkim Himalayas. In: 8th United Nations Regional Cartographic Conference for Asia and the Pacific, Bangkok, 26–29 October 2009.

Shit, P.K., Bhunia, G.S., Maiti, R., 2016. Potential landslide susceptibility mapping using weighted overlay model (WOM). Model. Earth Syst. Env. 2 (1), 1–10. Available from: https://doi.org/10.1007/s40808-016-0078-x.

Sidle, R.C., Pearce, A.J., Loughlin, C.L.O., 1985. Hillslope Stability and Land-use. American Geophysical Union, Washington, DC, p. 125.

Singh, P., Gupta, A., Singh, M., 2014. Hydrological inferences from watershed analysis for water resource management using remote sensing and GIS techniques. Egypt. J. Remote. Sens. Space Sci. 17 (2), 111–121. Available from: https://doi.org/10.1016/j.ejrs.2014.09.003.

Soil Survey Staff, 1993. Soil Survey Manual, Agriculture Handbook No. 18, USDA, Washington, DC.

Soil Survey Staff, 2014. Keys to Soil Taxonomy, twelfth ed. USDA-Natural Resources Conservation Service, Washington DC.

Srivastava, R., Saxena, A.K., 2004. Technique of large-scale soil mapping in basaltic terrain using satellite remote sensing data. Int. J. Remote. Sens. 25, 679–688.

Stanley, T., Kirschbaum, D.B., 2017. A heuristic approach to global landslide susceptibility mapping. Nat. Hazards 87 (1), 145–164. Available from: https://doi.org/10.1007/s11069-017-2757-y.

Van Westen, C.J., Ghosh, S., Jaiswal, P., Martha, T.R., Kuriakose, S.L., 2011. From landslide inventories to landslide risk assessment; an attempt to support methodological development in India. In: Proceedings of the Second World Landslide Forum, 3–7 Oct 2011, Rome.

Varnes, D.J., 1984. Landslide Hazard Zonation: A Review of Principles and Practice. UNESCO Press, Paris.

Yalcin, A., 2005. An investigation on Ardesen (Rize) region on the basis of landslide susceptibility. PhD Thesis, Karadeniz Technical University, Trabzon, Turkey. (in Turkish).

Chapter 36

Role of vegetation in soil erosion in the Bhogdoi River Basin, Assam

Mrinal Kumar Dutta, Merina Englengpi and Mrinaljyoti Adhyapok
Civil Engineering Department, Jorhat Engineering College, Jorhat, Assam, India

36.1 Introduction

Soil erosion is one of the most serious problems in recent times, which affects both cultivable and forest lands. Soil erosion depends not only on soil characteristics, climate, and slope (Ruiz Sinoga and Martinez Murillo, 2009), but also on land use and cover plants (Garcia-Ruiz, 2010). Accelerated soil erosion has adverse economic and environmental impacts (Lal, 1998). Many other problems lead to soil erosion, like the deposition of unfertile material on cultivated lands, harmful effects on water supply, fishing, sedimentation of canals and rivers, and, most importantly, the destruction of fertile agricultural land. So, the planning of soil conservation measures has become a prominent agenda in view of river basin management. To plan for measures of soil conservation, information on soil erosion is essential. Saha (2004) described soil erosion as a three-stage process: (1) soil detachment, (2) transport, and (3) deposition of eroded particles. Soil erosion depends not only on soil characteristics, climate, and slope but also on land use and cover plants (Garcia-Ruiz, 2010). Liu et al. (2011) discussed the advances in agronomic, biological, and engineering practices, including several tillage management practices, terraces and strip cultivation, and soil amendments practices to control soil erosion. Blavet et al. (2009) studied the effect of changes in land use on soil water erosion and developed a statistical relationship between runoff, soil losses, and various surface properties and the properties of the 0–5 cm topsoil layer. Greenway (1987) notes that roots reinforce the soil by increasing soil shear strength; they bind the soil particles at the ground surface, reducing their susceptibility to erosion, and they extract moisture from the soil, leading to lower pore–water pressures. Novara et al. (2011) studied the effect of crop cover on erosion control and suggested that planting the appropriate cover crops provides an effective soil and water management system. Liu et al. (2011) discussed various soil erosion control practices in northeast China, which include advances in agronomic, biological, and engineering practices, including several tillage managements, terraces and strip cultivation, and soil amendments. Rapid increase in population and growing demand for agricultural products has generated changes in land use and upland farming systems and resulted in erosion and land degradation over large tracts of northeast India (Paweł Prokop and Grzegorz J. Poręba)

Estimation of soil erosion loss is often difficult due to the complex interplay of many factors, such as climate, land cover, soil, topography, and human activities (Tyagi and Aggarwal, 2004). In addition to the biophysical parameters, social, economic, and political components also influence soil erosion (Ananda and Herath, 2003). Barman et al. (2013) discussed the effect of changes in vegetative cover on soil erosion in the Majuli Island.

The use of remote sensing and geographic information system (GIS) techniques makes soil erosion estimation and its spatial distribution feasible with reasonable costs and better accuracy in larger areas (Wang et al., 2003), wherein GIS tools were used for derivation of the topographic factor from DEM data, data interpolation of sample plots, and calculation of soil erosion loss(Bartsch et al., 2002).

The river Bhogdoi is a small but perennial river that comes down from the foothills of the Assam–Nagaland border into the plains of Assam and finally pours into the Brahmaputra. The river is surrounded by the hills of the Naga-Patkai range on the south and the floodplains of the Brahmaputra on the north. The area of the Bhogdoi basin falls under the very severe soil loss category. As much as 82% of the area of the Bhogdoi basin falls under the very severe soil loss class, where the rate of soil loss is more than 40 tons per hectare per year. The remaining portion of the basin falls

Remote Sensing of Soils. DOI: https://doi.org/10.1016/B978-0-443-18773-5.00037-5

under the moderately slight class, with soil loss ranging from 5 to 10 tons per hectare per year. These results are evident from the measurements made on the map prepared by NBSS & LUP. The basin is considered a heavy rainfall area. Again, the land use/land cover (LULC) of the area is also changing continuously. The LULC changes have been extensive in the past several decades in the Northeastern region (Singh et al., 1992), leading to environmental degradation. Nandy et al. (2006) reported that Jhum cultivation in the eastern region has disrupted the ecological balance of the region as soil erosion resulted from the reduction of Jhum cycle. So, it is obvious that surface erosion takes place due to the impacts of rainfall and LULC change. The Morgan, Morgan, and Finney model has been used in this study to determine the annual soil loss in three different years, viz., 1975, 2008, and 2018. Morgan et al. developed an empirical model to determine the annual soil loss. The MMF model used the concepts proposed by Meyer, Wischmeier, and Kirkby to provide a stronger physical base than the universal soil loss equation, while retaining the advantages of an empirical approach regarding ease of understanding and availability of data. The MMF model is a physically based empirical model (mix model) and needs fewer data than most of the other erosion predictive models. This model divides the soil erosion process into two phases, that is, a water phase and a sediment phase. Another advantage of this model is that it can be easily applied to a raster-based GIS.

The Morgan, Morgan, and Finney model has been applied for the estimation of surface soil erosion using RS and GIS techniques in Bhogdoi river basin, which lies in the geographical territories of both Assam and Nagaland, India.

36.2 Study area

Geographically, the Bhogdoi river basin lies between 26º17′N and 26º50′N latitudes and 93°40′ and 94°30′ E longitudes (Fig. 36.1). The total area of the basin is about 1179.65 km^2, including both plains and hills within it. The part of the basin in Nagaland is formed by low hills not exceeding 1400 m above MSL(mean sea level), while the part in Assam is topographically plain. In Nagaland, the river flows through the hills of the Naga-Patkai ranges (Mukokchung district). In Assam, it flows through the southern Brahmaputra plains of the Jorhat District. The Bhogdoi basin enjoys a subtropical monsoon climate. The average annual rainfall in the area is around 215 cm. Surface soil erosion is a major problem in the area. Shifting cultivation in the hilly area accelerates the surface erosion in the basin, as it depends on

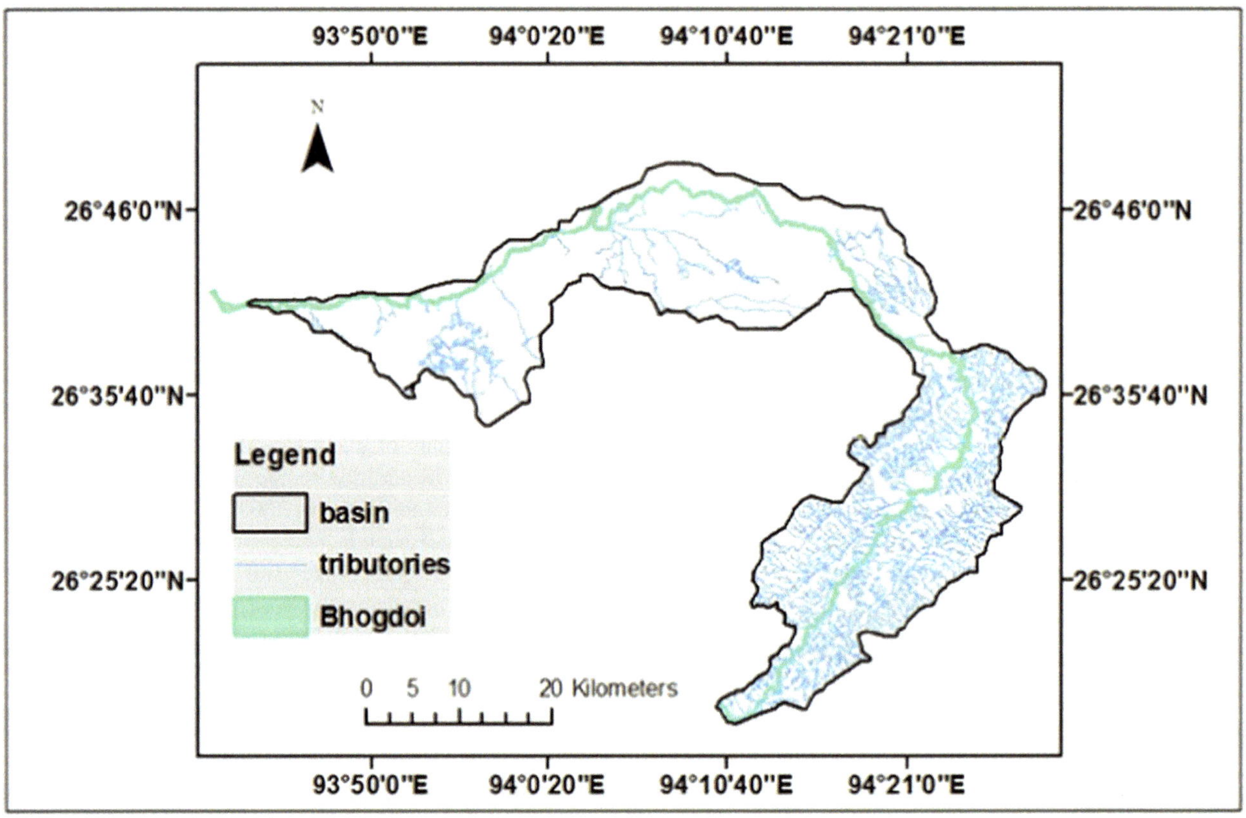

FIGURE 36.1 Location map of the Bhogdoi basin (study area).

the change in vegetation. During the last four decades, the area covered by vegetation has been decreasing continuously, which is causing surface erosion. Heavy rainfall in the area accelerates the process even more.

36.3 Methodologies

36.3.1 Morgan, Morgan, and Finney model

MMF model, which is an empirical model, separates the process of soil erosion into the sediment phase and the water phase. In the water phase, the runoff volume and the rainfall energy are used to determine the soil particle's detachment. The sediment phase uses the function of soil erodibility, energy of rainfall, and the interception of rainfall that is affected by vegetation for the detachment of the soil particles. The model compares the rate of predicted splash detachment with the transport capacity of the runoff or the overland flow, and the lower of these two values is assigned as the annual rate of soil loss. The flowchart of the methodology is shown in the Fig. 36.2.

36.3.2 Water phase

In the water phase, rainfall energy was computed by,

$$E = R \times (11.9 + 8.7 \times \log I) \tag{36.1}$$

Where E is the kinetic energy of rainfall (J m^{-2}), R is the annual rainfall (mm), and I is the intensity of erosive rain (mm hour^{-1}). This is taken as 11 for a temperate climate, 25 for a tropical climate, and 30 for a strongly seasonal climate.

The overland flow was estimated by,

$$Q = R \times \exp(-\mathrm{RC}/\mathrm{Ro}) \tag{36.2}$$

Where Q is the depth of overland flow (mm), R is the annual rain (mm), Rc is the soil moisture storage capacity under land cover (mm), and Ro is the mean rain per day (mm). The soil moisture storage capacity was computed by,

$$\mathrm{Rc} = 1000 \times \mathrm{MS} \times \mathrm{BD} \times \mathrm{RD} \times (\mathrm{Et}/\mathrm{Eo})^{0.5} \tag{36.3}$$

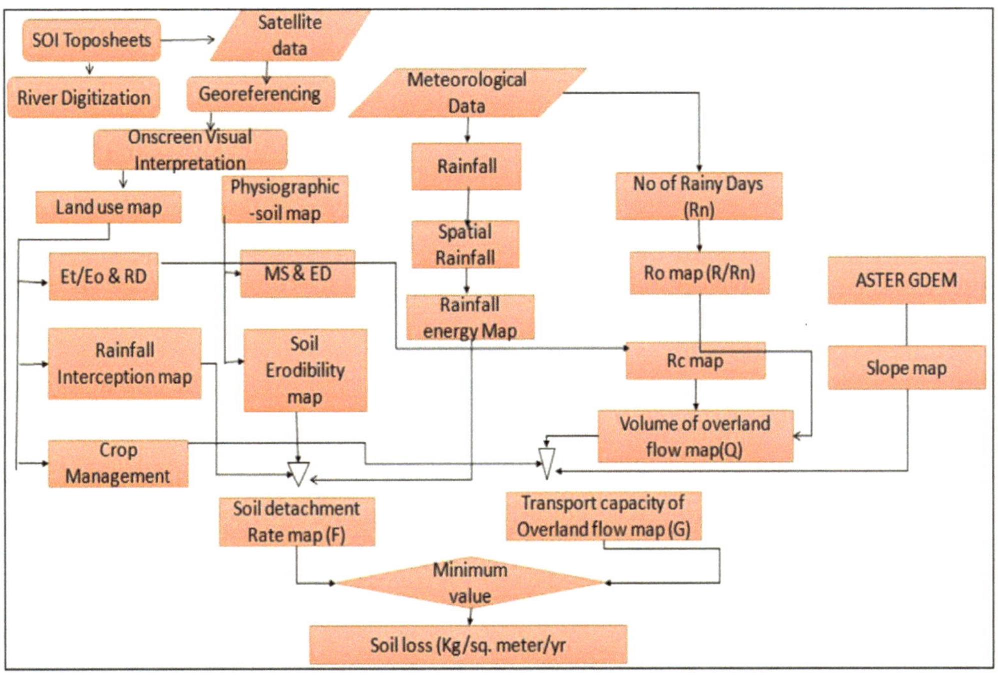

FIGURE 36.2 Flowchart of methodology.

Where MS is the moisture content of soil at field capacity, BD gives the bulk density of soil (mg m^{-3}), RD gives the rooting depth of topsoil (m), *Et* is the actual evaporation (mm day^{-1}), and *Eo* gives the potential evaporation (mm day^{-1})

Mean rain per day (Ro) is calculated by dividing the average annual rain (*R*) by the number of rainy days (Rn) in a year, that is,

$$\text{Ro} = R/\text{Rn} \tag{36.4}$$

36.3.3 Sediment phase

In this phase, soil erosion occurs due to the detachment of soil particles from the mass of soil because of splashes or the impact of raindrops, and the particles are transported by the overland flow. Here, the splash detachment is considered a function of soil erodibility, rainfall energy, and rainfall interception affected by vegetation. The rate of splash detachment is considered as a function of the soil detachment index (K) and is regarded as the weight of the detached soil from the mass of soil per unit of rainfall energy. The detachment of soil by the raindrops was computed by:

$$F = K \times \left(E \times e^{-aA}\right)^{b} \times 10^{-3} \tag{36.5}$$

Where F is the rate of soil detachment by raindrop impact (kg m^{-2}), K is the soil detachability index (g J^{-1}) defined as the weight of soil detached from the soil mass per unit of rainfall energy, E is the rainfall energy, and A is the percentage rainfall contributing to permanent interception and stream flow. The values of the exponents are $a = 0.05$, $b = 1.0$

The distributed transport capacity map *G* was computed by:

$$G = C \times Q^{\mathrm{d}} \times \sin\mathrm{S} \times 10^{-3} \tag{36.6}$$

Where G is the transport capacity of overland flow (kg m^{-2}), C is the crop cover management factor, and S is the steepness of the ground slope (degree). The value of exponent $d = 2.0$.

36.3.4 Preparation of spatial database

The Survey of India toposheets No. 83 F/10, F/14, J/1, J/2, J/3, J/4, J/5, J/6, J/7, J/9, J/10 (Scale 1:50,000) of 1975 were georeferenced, mosaiced, and used for the preparation of the base map using ArcGIS software. From the base map, the Bhogdoi river basin was delineated along with the river Bhogdoi and its tributaries (Fig. 36.1.) by digitizing in a GIS environment. Different attribute data, like area, perimeter of the basin, and length of the river, were generated in GIS and saved in a personal geodatabase.

The base map of 1975 was used to prepare the LULC map of 1975. IRS P6 LISS III digital image of 2008 was used to map the LULC status of the Bhogdoi River basin for the year 2008, and the Google Earth image was used to map the LULC of 2018. The LULC maps of 1975, 2008, and 2018 were prepared following the on-screen visual interpretation method by digitizing in ArcGIS 10.1. The attribute tables were created for the respective years to store information such as the area in square kilometers and type of the various LULC categories. The different types of LULC digitized were settlement, cultivation, tea garden, water body, open scrub, and forest.

36.3.5 Estimation of soil loss using MMF model

To determine the spatial distribution of average annual soil loss, the parameters of MMF models, viz., soil parameters (K, Ms, and BD) and land cover parameters (C, Et/Eo, RD, and A), were stored in an attribute table associated with soil and land cover maps, respectively (Tables 36.1 and 36.2). From the LULC maps, the maps of percentage rainfall contributing to permanent interception and stream flow (A) map, the ratio of actual to potential evapotranspiration (Et/Eo) map, and crop cover management factor (C) map were prepared. The soil map was derived from the National Bureau of Soil Survey and the Land Use Planning Act, and the soil type of the entire region is mainly of sandy loam/ silty loam. From the soil map, the soil moisture content (Ms) at the field capacity map, bulk density (BD) of the top layer map, and soil detachability index (K) map were prepared in the GIS environment. The overland flow (Q) map was generated using Eq. (36.2). Rc map was generated with the integration of parameter maps (Ms, BD, RD, and Et/Eo) using Eq. (36.3). The parameters R and Rn were calculated from the daily rainfall data. Ro was computed using annual rainfall (R) and the number of rainy days (Rn). Finally, the rate of soil detachment by the raindrop impact map

TABLE 36.1 Typical values for soil parameters for the MMF model (Morgan and Duzant, 2008).

Textural category of soil	MS(%w/w)	BD(mg m^{-3})	K
Sand	0.08	1.5	0.3
Sandy loam	0.28	1.2	0.3
loam	0.20	1.3	0.4
Silt loam	0.25	1.3	0.4
Clay loam	0.40	1.3	0.4
Silty clay loam	0.25	1.3	0.4
Clay	0.45	1.1	0.4

TABLE 36.2 Plant parameter values (for the MMF model).

Land cover	A	E_t/E_o	C	RD
Settlement	20	0.10	0.1	0.00
Forest	30	0.9	0.08	0.1
Cultivation	25	0.58	0.3	0.05
Tea garden	25	0.8	0.5	0.1
Open scrub	35	0.8	0.05	0.05
Water body	0	1	0	0

(F) and the transport capacity of the overland flow map (G) were prepared using Eqs. (36.5) and (36.6), respectively, and the minimum values from each of them were used for the preparation of the final soil loss map.

The effect of vegetation was analyzed taking the average rainfalls over the last 43 years, and the same value was applied for the three years (1975, 2008, and 2018).

The derivation of the final soil loss map for the year 2018 is shown below (Figs. 36.3, 36.4, 36.5, 36.6 and 36.7)

In the same way, the soil loss maps for the years 1975 and 2008 were prepared to get the annual soil erosion in kg m^{-2} (Fig. 36.7 and Fig. 36.8). The average soil erosion was then estimated for the three respective years.

36.4 Results and discussions

The change in LULC has a great impact on soil erosion. The LULC maps were derived for the Bhogdoi River basin to represent the present and past LULC conditions and changes that have taken place in the basin during the study period. Five major LULC types have been identified, namely, settlement, cultivation, forest, Tea garden, open scrub, and water body.

Analyzing LULC changes from 1975 to 2018 (Table 36.3), it may be observed that the total forest land has declined by 8.32% in the river basin. A reduction in open scrub (1.42%) is followed by a decrease in areas of tea garden (1.27%) and surface water bodies (0.76%). However, the area covered by cultivation has increased by 0.53%, followed by an increase in settlement by 11.24%. The use of land area for settlement is related to the population. The growth of the population is directly related to the decrease in forest land, open scrub, and tea gardens. So, the increase in settlement area can be attributed to the decrease in the of area of forest land, open scrub, and tea gardens.

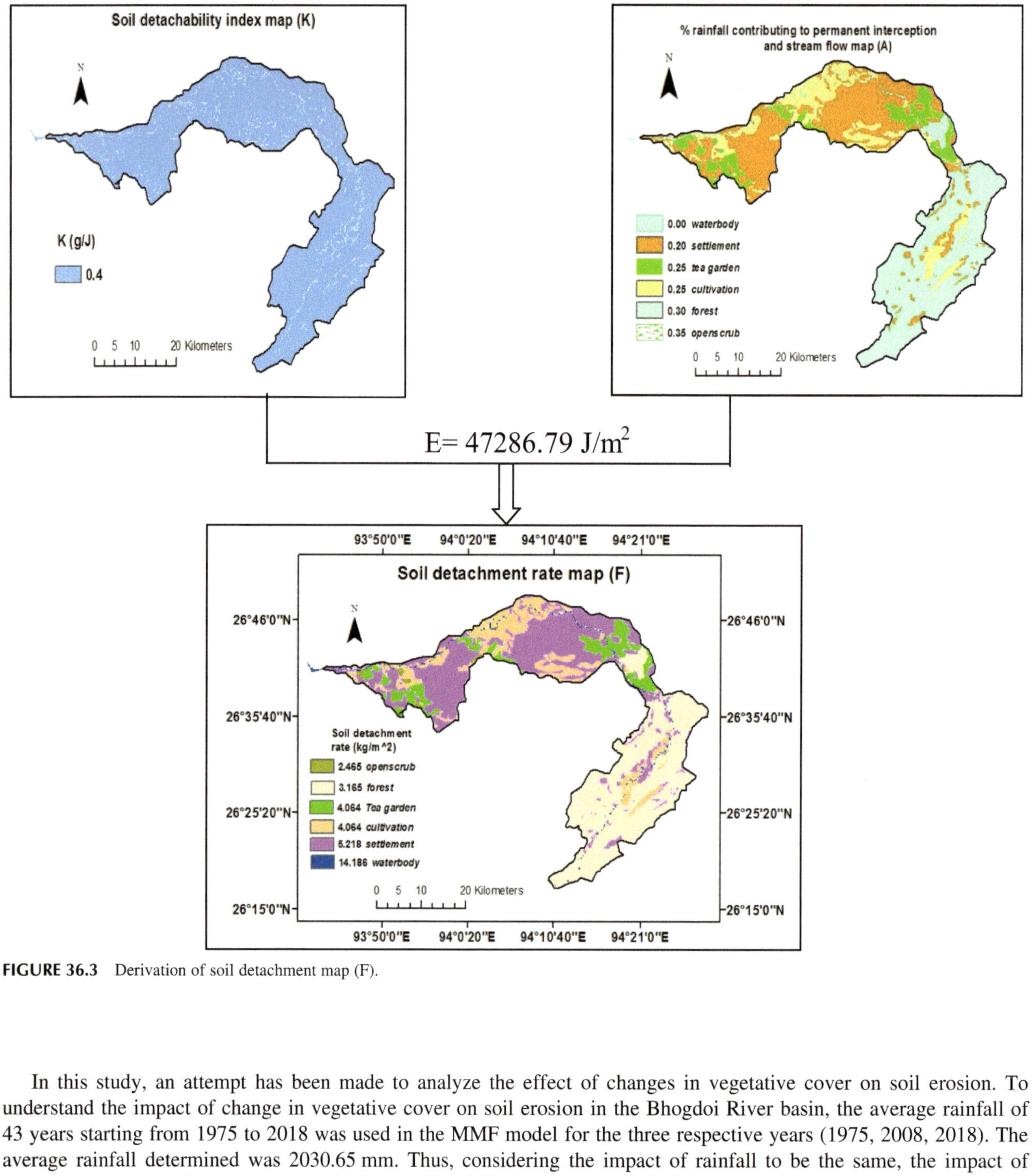

FIGURE 36.3 Derivation of soil detachment map (F).

In this study, an attempt has been made to analyze the effect of changes in vegetative cover on soil erosion. To understand the impact of change in vegetative cover on soil erosion in the Bhogdoi River basin, the average rainfall of 43 years starting from 1975 to 2018 was used in the MMF model for the three respective years (1975, 2008, 2018). The average rainfall determined was 2030.65 mm. Thus, considering the impact of rainfall to be the same, the impact of change in vegetation has been analyzed. The combined effect of forest, cultivation, tea garden, and open scrub has been taken as the effect of vegetation change on annual soil loss for the three years. During the study period from 1975 to 2018, the overall vegetative cover decreased in the basin. Considering the impact of the changed effect of vegetation on soil erosion, it has been observed that the rate of soil erosion increases as the vegetative cover decreases. The average rate of soil erosion was found to be 3.39 kg m^{-2} (Fig. 36.8), 3.43 kg m^{-2} (Fig. 36.9), and 3.64 kg m^{-2} (Fig. 36.10) in 1975, 2008, and 2018, respectively.

It is observed that vegetation plays a role in arresting surface soil erosion. Vegetation acts as a protective layer between the atmosphere and the soil. The above-ground components, such as leaves and stems, absorb some of the

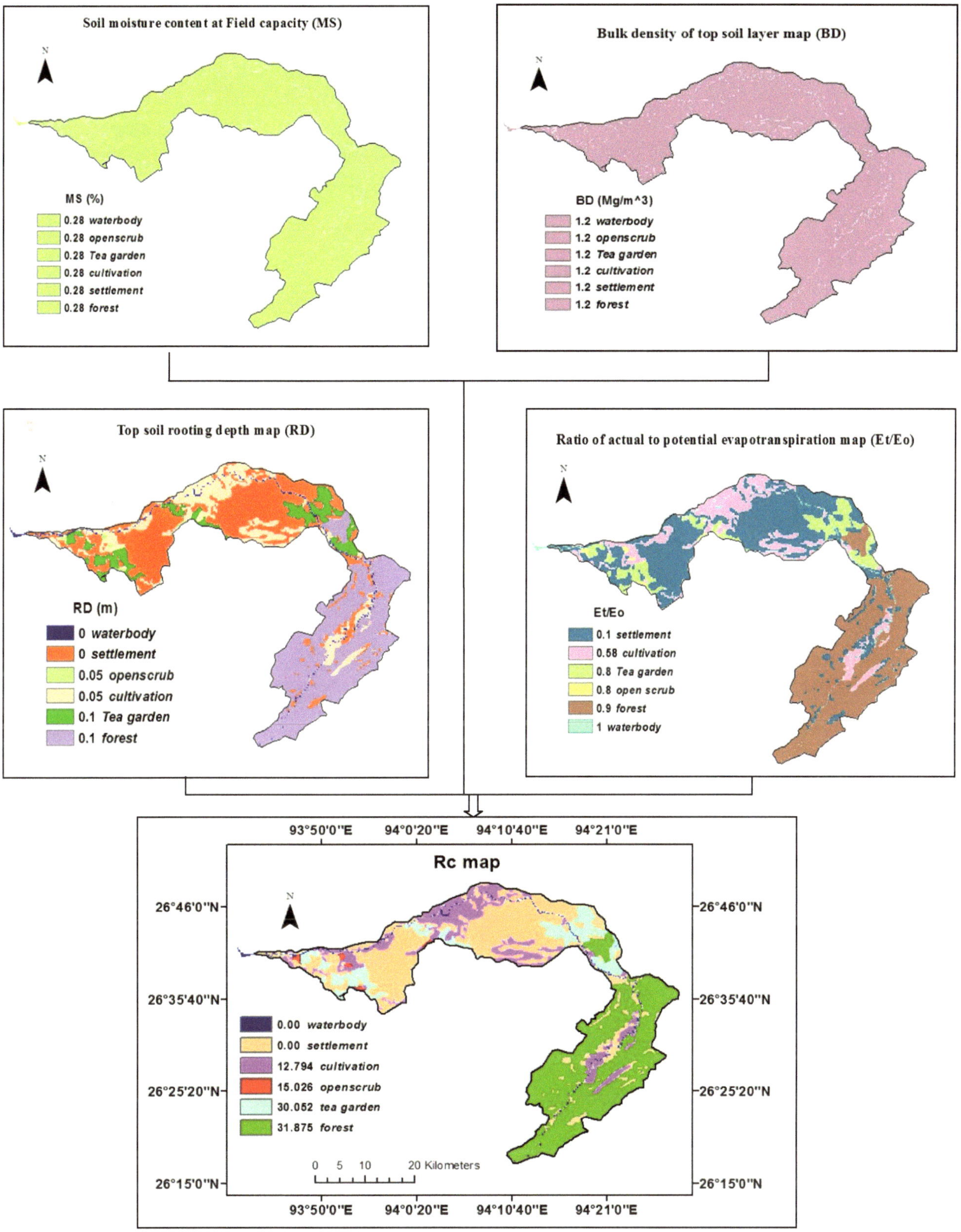

FIGURE 36.4 Derivation of the R_c map.

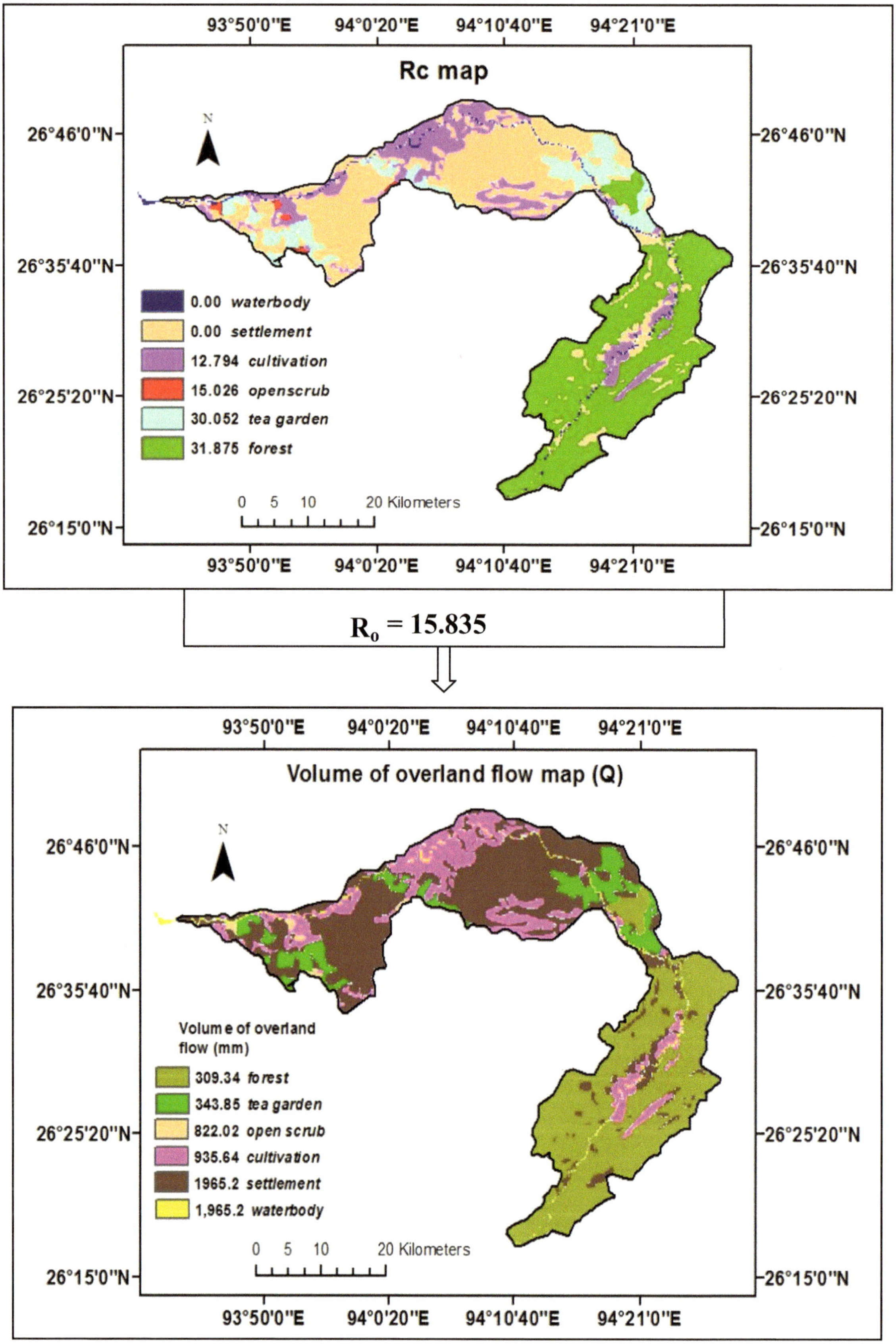

FIGURE 36.5 Derivation of Volume of overland flow map (Q).

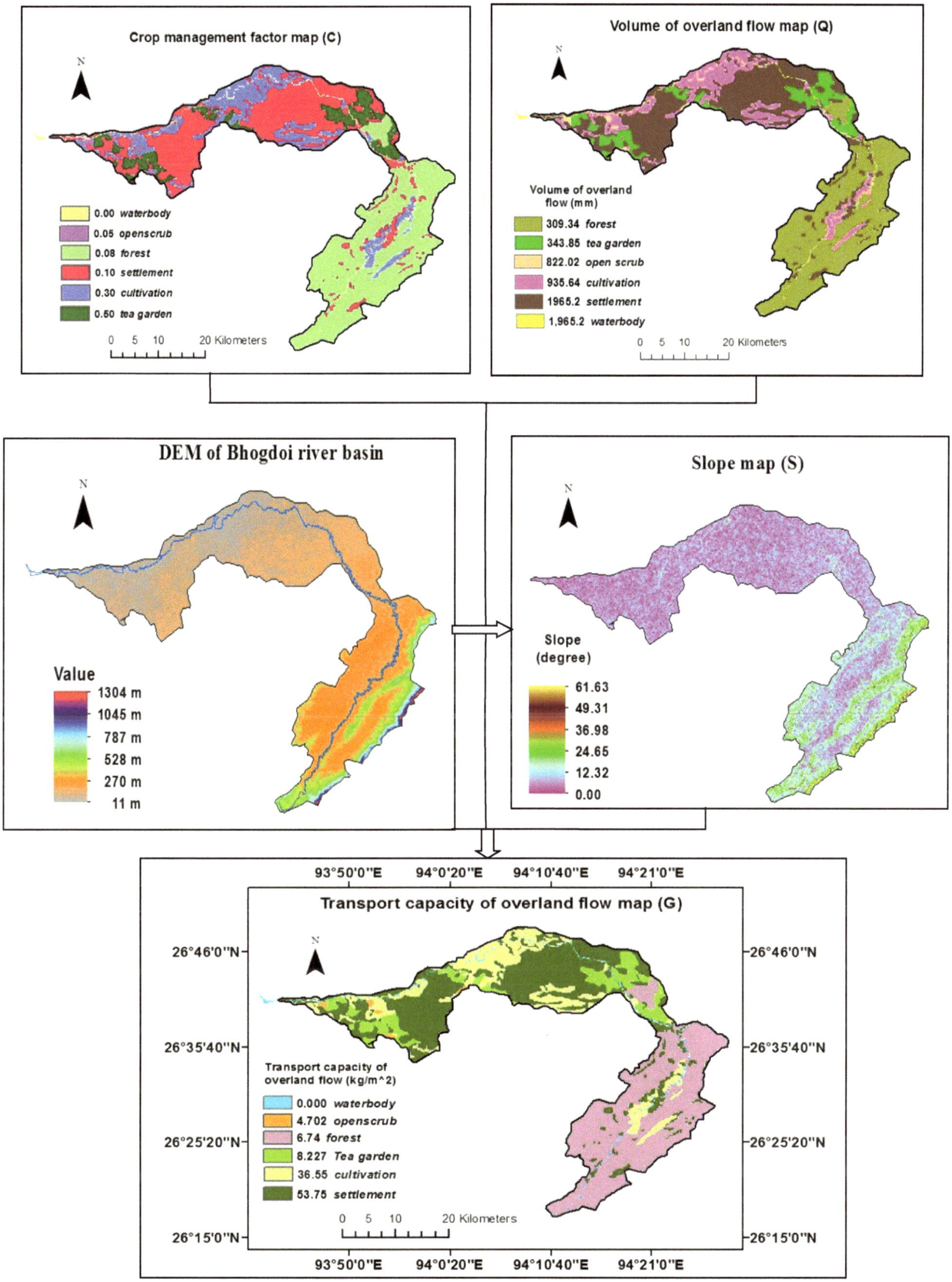

FIGURE 36.6 Derivation of transport capacity of overland flow map (G).

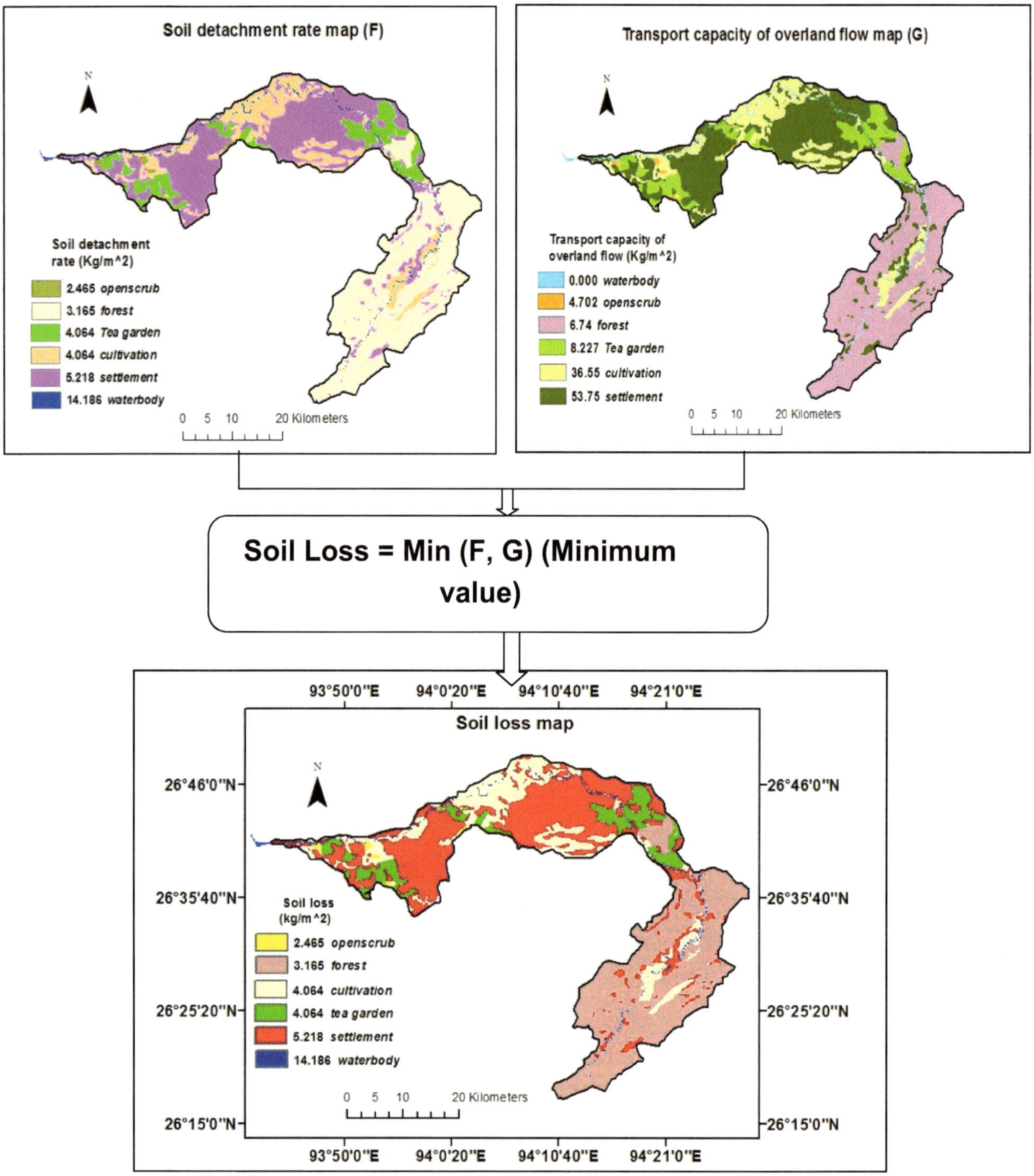

FIGURE 36.7 Final soil loss map.

energy of falling raindrops, running water, and wind, so that less is directed at the soil, while the below-ground components, comprising the root system, contribute to the mechanical strength of the soil. The effectiveness of a plant cover in reducing erosion by raindrop impact depends upon the height and continuity of the canopy and the density of the ground cover. Raindrops intercepted by the canopy may coalesce on the leaves to form larger drops,

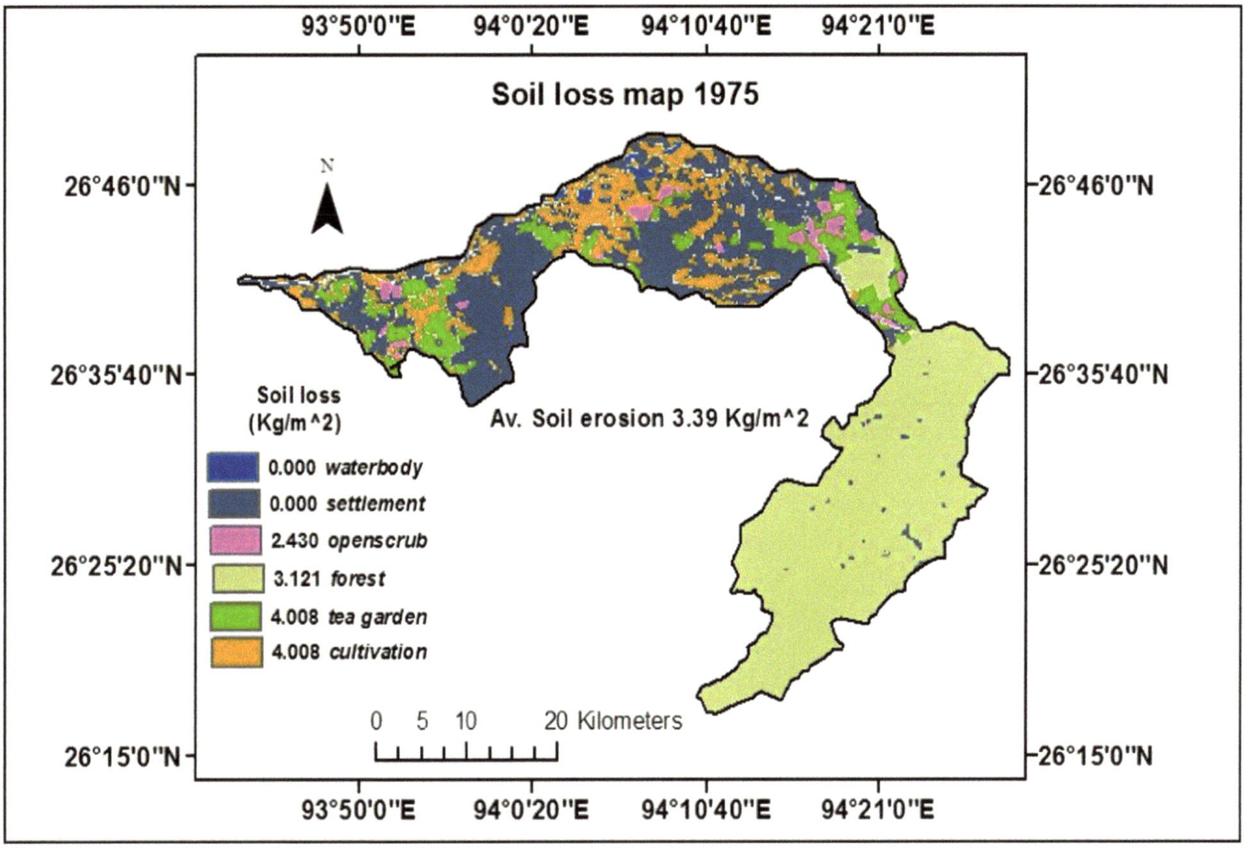

FIGURE 36.8 Annual soil erosion map of 1975, considering the impact of vegetation.

TABLE 36.3 Characteristics of LULC changes in the Bhogdoi River Basin (1975–2018).

Year↓/LULC categories→	Settlement (%)	Forest (%)	Cultivation (%)	Tea garden (%)	Open scrub (%)	Water body (%)
1975	30.52	43.99	12.10	9.56	1.81	2.02
2008	39.42	36.45	11.67	10.97	0.13	1.36
2018	41.76	35.68	12.61	8.29	0.393	1.26

which are more erosive. A plant cover dissipates the energy of running water by imparting roughness to the flow, thereby reducing its velocity. The greatest reductions in velocity occur with dense, spatially uniform, vegetation covers. Due to the rapid decline of vegetation in the basin area during the study period, the rate of soil erosion increased (Table 36.4). From Table 36.4, it has clearly been observed that in 1975, the vegetated area was maximum, that is, 67.48% of the total area of the river basin, which was reduced to 59.22% in 2008 and 56.98% in 2018. Consequently, the average soil erosion in 1975 was the minimum, that is, 3.39 kg m^{-2}, which increased to 3.43 kg m^{-2} in 2008 and to 3.64 kg m^{-2} in 2018.

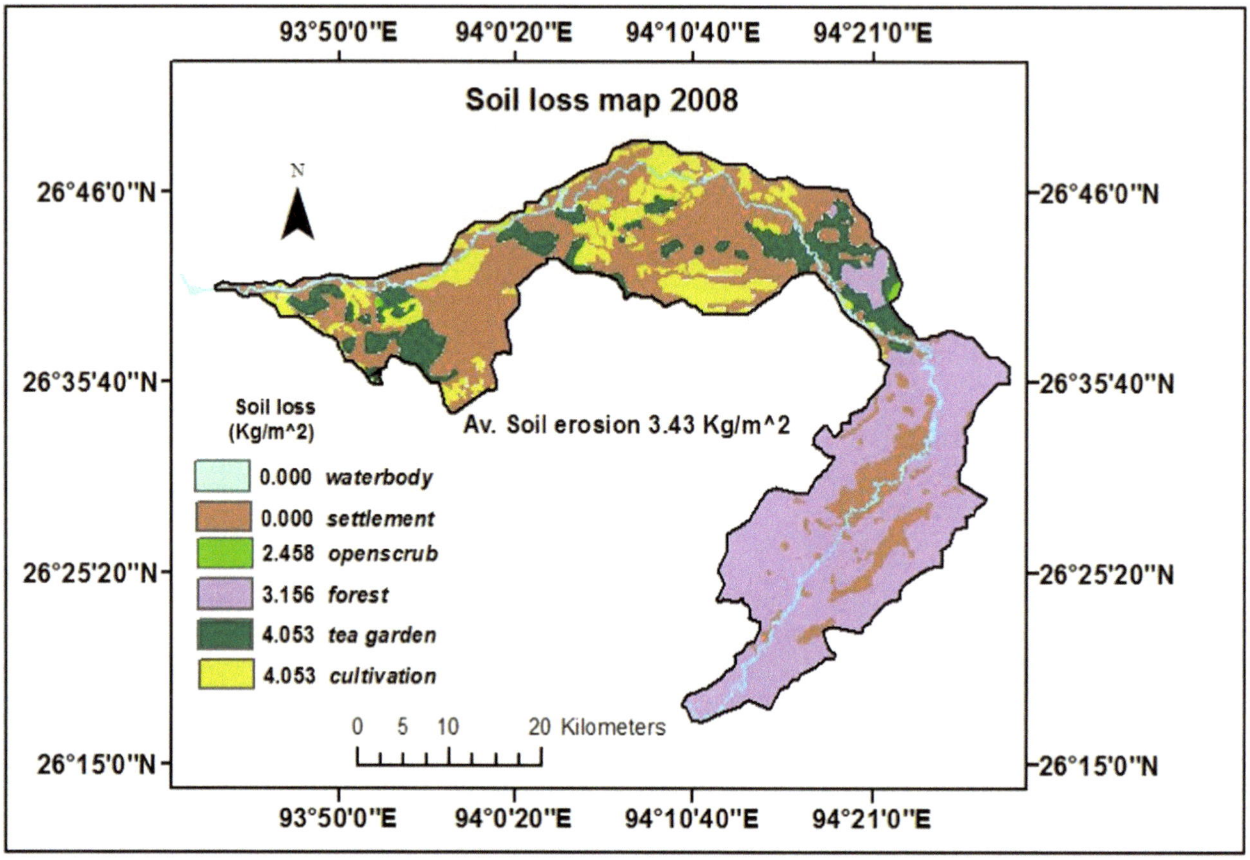

FIGURE 36.9 Annual soil erosion map of 2008, considering the impact of vegetation.

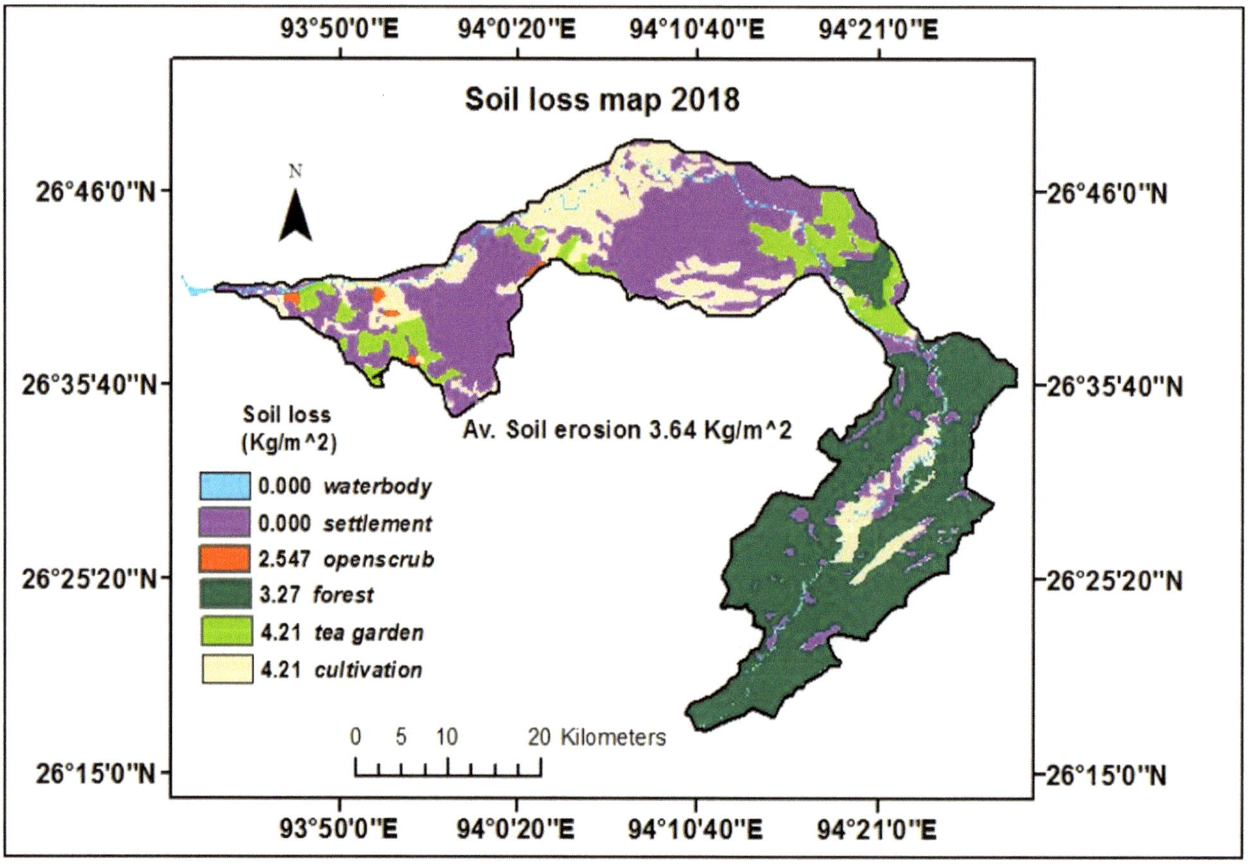

FIGURE 36.10 Annual soil erosion map of 2018, considering the impact of vegetation.

TABLE 36.4 Effect of change in vegetated area on soil erosion.

Class	Area in 1975 (Sq. Km)	% of total area	Area in 2008 (Sq. Km)	% of total area	Area in 2018 (Sq. Km)	% of total area
Vegetation (cultivation + forest + teagarden + open scrub)	796.041	67.48	698.60	59.22	672.22	56.98
Annual soil loss ($kg\ m^{-2}$)	3.39		3.43		3.64	

36.5 Conclusion

The change in area covered by vegetation plays an important role in soil erosion. The result obtained from the Morgan, Morgan, and Finney model shows that surface soil erosion is minimal in forested areas and open scrub areas. Cultivated areas result in more soil loss than forested areas due to the practice of shifting agriculture, locally called "jhum" cultivation, reduces the soil's fertility, preventing any crops from being grown and leaving the land as fallow land. Human built-up areas caused greater soil loss, leaving the soil surface with little or no vegetation, which in turn provided a prime opportunity for soil to be eroded by wind or water moving across the soil surface. Water bodies have resulted in maximum soil loss; they can detach soil particles and transport the detached particles downhill. These particles move from the field and end up in streams and waterways, resulting in an increase in siltation of the river bed. There is an accelerated growth of settlement in the forest area during the study period, which decreases the vegetation cover and results in more soil erosion. Abnormal rainfall patterns because of the changed climate affected vegetation growth in addition to anthropogenic activities.

The study of the effect of vegetation on annual soil loss reveals that, in 1975, the percentage of the area covered by vegetation (67.48%) was the maximum, followed by vegetated areas in 2008 (59.22%) and 2018 (56.98%). Consequently, the average soil erosion in 1975 was the minimum, that is, 3.39 $kg\ m^{-2}$, which was increased to 3.43 $kg\ m^{-2}$ in 2008 and to 3.64 $kg\ m^{-2}$ in 2018. So, it can be concluded that more the vegetation, less is the soil erosion, and vice versa.

References

Ananda, J., Herath, G., 2003. Soil erosion in developing countries: a socio-economic appraisal. J. Environ. Manag. 68, 343–353.

Barman, S., Aggarwal, S., Dutta, M.K., 2013. Soil erosion due to change in vegetated area in the Majuli Island of Assam,". Int. J. Adv. Remote Sens., GIS Geogr. IJARSGG (2013) 1 (1), 8–19.

Bartsch, K.P., Miegroet, H., Boettinger, J., Dobrwolski, J.P., 2002. Using empirical erosion model and GIS to determine erosion risk at Camp Williams. J. Soil. Water Conserv. 57, 29–37.

Blavet, D., Noni, G.D., Bissonnais, Y.L., Leonard, M., Maillo, L., Laurent, J.Y., et al., 2009. Effect of land use and management on the early stages of soil water erosion in French Mediterranean Vineyards. Soil. Tillage Res. 106, 124–136.

Garcia-Ruiz, J.M., 2010. The effects of land uses on soil erosion in Spain: a review. Catena 81, 1–11. Available from: https://doi.org/10.1016/j.catena.2010.01.001.

Greenway, D.R., 1987. Vegetation and slope stability. In: Anderson, M.F., Richards, K.S. (Eds.), Slope Stability. Wiley and Sons, New York.

Lal, R., 1998. Soil erosion impact on agronomic productivity and environment quality. Crit. Rev. Plant. Sci. 17 (4).

Liu, X., Zhang, S., Zhang, X., Ding, G., Cruse, R.M., 2011. Soil erosion control practices in Northeast China: a mini-review. Soil Tillage Res. 117, 44–48.

Morgan, R.P.C., Duzant, J.H., 2008. Modified MMF (Morgan–Morgan–Finney) model for evaluating effects of crops and vegetation cover on soil erosion. Earth SurfaceProcesses Landf. 33, 90–106.

Nandy, S.N., Dhyani, P.P., Samal, P.K. 2006. Resource Information Database of the Indian Himalaya. ENVIS Monograph 3, 108 p.

Novara, A., Gristina, L., Saladino, S.S., Santoro, A., Cerda, A., 2011. Soil erosion assessment on tillage and alternative soil managements in a Sicilian vineyard. Soil Tillage Res. 117, 140–147.

Ruiz Sinoga, J.D., Martinez Murillo, J.F., 2009. Hydrological response of abandoned agricultural soils along a climatological gradient on metamorphic parent material in southern Spain. Earth Surf. Process. Landf. 34, 2047–2056. Available from: https://doi.org/10.1002/esp.1890.

Saha, S.K., 2004. Water and wind induced soil erosion assessment and monitoring using remote sensing and GIS. In: Sivakumar, M.V.K., Roy, P.S. Harsen, K., Saha, S.K. (Eds.), Satellite Remote Sensing and GIS Application in Agricultural Meteorology, Dehra Dun, 2004, pp. 315–330.

Singh, G., Babu, R., Narain, P., Bhusan, L.S., Abrol, I.P., 1992. Soil erosion rates in India. J. Soil. Water Conserv. 47 (1), 97–99.

Tyagi, J.V., Aggarwal, S.P., 2004. Application of ANSWERS model using RS and GIS for simulating runoff and sediment yield. Asian J. Geoinf. 5 (1), 13–20.

Wang, G., Gertner, G., Fang, S., Anderson, A.B., 2003. Mapping multiple variables for predicting soil loss by geostatistical methods with TM images and a slope map. Photogramm. Eng. Remote. Sens. 69, 889–898.

Chapter 37

Assessment of land degradation vulnerability in the semi-arid region of Southern India using GIS-based MEDALUS approach

S. Kaliraj[1], N. Chandrasekar[2], Manish Parmar[3], R.G. Rejith[4], S. Dharumarajan[5], M. Lalitha[5] and K. Chandramohan[6]

[1]*National Centre for Earth Science Studies (NCESS), Ministry of Earth Sciences, Govt. of India, Thiruvananthapuram, Kerala, India,* [2]*Centre for GeoTechnology, Manonmaniam Sundaranar University, Tirunelveli, Tamil Nadu, India,* [3]*Space Applications Centre (SAC-ISRO), Ahmedabad, Gujarat, India,* [4]*Environmental Resources Research Centre (ERRC), Thiruvananthapuram, Kerala, India,* [5]*ICAR-National Bureau of Soil Survey and Land Use Planning, Regional Centre, Bangalore, Karnataka, India,* [6]*ICSSR-Post Doctoral Fellow, Tribal Research Centre, Tamil University, Thanjavur, Tamil Nadu, India*

37.1 Introduction

In the 21st century, land degradation (LD) is one of the major global issues affecting soil, land, and food productivity in arid and semiarid regions (Ferrara et al., 2020; Afzali et al., 2021; Song et al., 2022). UNEP (1992) has stated that LD refers to the loss of ecosystem functions and productivity, focusing on the biological potential of land at the local and regional scales. Understanding the LD processes is essential for sustainable land resource management because it is driven by various factors including soil, land, rainfall, climate, and anthropogenic factors (Kassas, 1976; Kerényi, 1996; Bedoui, 2020; Abuzaid and Abdelatif, 2022). LD causes severe impacts on environmental ecosystems and the socioeconomic development of different countries (Avni et al., 2006; Sepehr et al., 2007; Singh, 2009; Salvati et al., 2015). Globally, 75% of the landscapes fall under LD processes, which will be exceeded by over 90% by 2050 (Uzuner and Dengiz, 2020). LD is defined as the desertification of lands by decreasing or losing their physico-chemical and biological characteristics due to adverse impacts of various factors, including climatic variations and human activities (Conacher and Sala, 1998; Kosmas et al., 1999). Increasing the LD process can reduce soil and land productivity, primarily leads to soil erosion, soil infertility, removal of topsoil, soil nutrients, and organic matter, and increasing soil salinity, alkalinity, calcium carbonate, and gypsum contents (Dutta and Chaudhuri, 2015; Budak et al., 2018; Wijitkosum, 2021). The global spreading of LD is ultimately reducing soil fertility and food productivity, especially in arid, semiarid, and subtropical areas. They are causing adverse impacts like droughts, food insecurity, poverty, violence, emigration, political instability, and social disintegration (Ali and El Baroudy, 2008; Ahmad and Pandey, 2018; Bedoui, 2020).

UNCCD (2014) has reported that 25% of the lands worldwide fall under desertification vulnerability due to climate change and human-induced factors that trigger critical impacts on soil fertility, plant nutrients, and food productivity (UNEP 1992; Ajai et al., 2009; Dasgupta et al., 2013; Kaliraj et al., 2017; Dharumarajan et al., 2018). The LD is a significant issue in many countries like Eastern and Central Asia, mid-parts of South America, Western Australia, North America, Europe, the Mediterranean, and the Sahara region, and drastically creating environmental problems like the expansion of aridity and soil infertility in more extensive areas (Basso et al., 2000; Avni et al., 2006; Ali and El Baroudy, 2008; Ahmad and Pandey, 2018; Afzali et al., 2021; Abuzaid and Abdelatif, 2022). LD occurs in different landscapes mainly due to climate change and human activities (UNEP 1992; Reynolds et al., 2011; SAC 2021), which alter their biophysical settings, restoration capacity, and productivity of environmental ecosystems (Salvati et al., 2015; Ait Lamqadem et al., 2018). In recent decades,

Remote Sensing of Soils. DOI: https://doi.org/10.1016/B978-0-443-18773-5.00038-7

the different landscapes of many countries have experienced impacts of LD due to climate change and increasing anthropogenic pressures from various directions (Kaliraj et al., 2017; Budak et al., 2018; Akbari et al., 2020; Song et al., 2022).

About 20% of the Indian subcontinent covers the black clay soils (also called black cotton soils), and it is enriched with clay contents of weathered igneous rock materials. In the country, about 228 Mha (69%) of the total landmass, especially arid and semiarid landscapes, fall under LD impacts because of dynamics in climate regimes like rainfall, soil moisture, temperature, and other human-induced activities. In the Southern Indian parts, the land use/land cover (LULC) features are found extensively in loamy, clay, and black cotton soils; among them, black cotton soils (black clay and calcareous clay) are chemically deficient by LD processes (Kaliraj et al., 2012). SAC (2021) has reported that the major factors of LD in various parts of the Indian landscapes are erosion by water (11.01%), vegetation degradation (9.15%), and erosion by wind (5.46%), and their cumulative rate increased up to 1.87 Mha during the periods of 2003–05 and 2018–19. In peninsular India, the LULC features of arid, semiarid, and subtropics critically fall under LD vulnerability due to soil infertility, soil salinity, soil alkalinity, surface runoff, soil erosion, and poor rainfall rate (Singh, 2009; Dutta and Chaudhuri, 2015; Dharumarajan et al., 2018; SAC, 2021). The degradation of land resources occurs not only in semiarid areas but also in subtropics and semihumid areas due to the impacts of climatic factors and inappropriate land use practices. In the Southern Indian region, the different landscapes of semiarid and subtropical environments are occurring in LD processes due to climatic factors and improper land management practices (Kaliraj et al., 2016; Dharumarajan et al., 2018). In these areas, LD is a visible symptom of degrading soil fertility matters, that is, deficiency of soil nutrients and biophysical phenomena, and the accelerated condition of the LD process is seriously threatening soil resources and other environmental and socioeconomic structures (Kaliraj et al., 2017). In recent decades, LULC features and soils in arid and semiarid regions are significantly amplified LD due to climatic variability, improper agricultural activities, population growth, and urbanization, etc., which produces adverse impacts on land and soil resources (Masoudi and Zakeri Nejad, 2010; Mohamed et al., 2013; Okin et al., 2018; Ferrara et al., 2020; SAC, 2021; Song et al., 2022). The areas of black cotton soils in the Virudhunagar district are critically experiencing LD processes, that is soil erosion, salinization, alkalinization, and deficiency of soil nutrients. These areas prevail in semiarid to sub-tropic conditions. Their landscapes depend on monsoonal rainfall for agricultural practices, thereby highly prone to droughts when monsoon fails or poor rainfall results from climate change.

Diallo (2008) has stated that the issues of LD and desertification are considered an international priority for agricultural productivity and food security. Therefore, it is essential for combating LD worldwide. Monitoring and assessing LD is very much needed for the sustainable utilization of land resources. Integrated remote sensing (RS) and GIS provide an effective platform for monitoring LD and its driving factors on a spatial and temporal scale. RS is a cost-effective and precise tool for data collection that provides multispectral image products such as Landsat TM, ETM + and operational land imager (OLI) (30 m), IRS—LISS 3 (23.5 m), LISS 4 (5.8 m), Sentinel 2 MSI (10 m), IKONOS (1 m), QuickBird (0.6 m), etc., used for mapping spatiotemporal changes of LULC features (Hanafi and Jauffret, 2008; Mohamed, 2013; Kaliraj et al., 2016; Abuzaid and Abdelatif, 2022). In recent decades, many approaches have been used to measure LD. The GIS techniques provide dynamic tools for storing, analyzing, and interpreting the geodatabase of different layers to demarcate the environmentally sensitive areas (ESAs) to LD and desertification, and the outcome is consequently used for preparing suitable resilient measures to mitigate LD in time (Masoudi and Zakeri Nejad, 2010; Kaliraj et al., 2012; Ait Lamqadem et al., 2018; Kaliraj et al., 2022).

The MEDALUS model (MEditerranean Desertification and Land Use) developed by the European Commission (Kosmas et al., 1999) is used worldwide to assess and map environmental sensitivity to LD processes. The MEDALUS approach is widely used for the assessment of ESAs to LD and desertification by analyzing driving factors including soil, climate, vegetation, and management (Dutta and Chaudhuri, 2015; Ferrara et al., 2020; Abuzaid and Abdelatif, 2022). The GIS-based MEDALUS model is used for mapping risk zones to land and soil degradation at a spatiotemporal scale by analyzing multiple geo-environmental factors, and the outcome ensures potential assessment and mapping with reliable accuracy for decision-making purposes (Budak et al., 2018; Bedoui, 2020). Many researchers used the MEDALUS model for mapping and assessment of LD vulnerability in various countries including Egypt (Ali and El Baroudy, 2008; Abuzaid and Abdelatif, 2022), Morocco (Ait Lamqadem et al., 2018), Italy (Basso et al., 2000; Salvati et al., 2015), Tunisia (Hanafi and Jauffret, 2008; Bedoui, 2020), India (Dutta and Chaudhuri, 2015; Ahmad and Pandey, 2018; Dharumarajan et al., 2018), Israel (Avni et al., 2006), Iran (Sepehr et al., 2007; Masoudi and Zakeri Nejad, 2010), Ethiopia (Song et al., 2022), and Turkey (Budak et al., 2018; Uzuner and Dengiz, 2020).

The MEDALUS model is one of the most widely used tools for assessing and mapping LD vulnerability based on the analysis of soils, vegetation, climate, and management quality indices. Dharumarajan et al. (2018) have studied the desertification status of the Andhra Pradesh, Karnataka, and Telangana regions of Southern India by analyzing multiple parameters like soils, vegetation, climate, and LULC features. The MEDALUS is designed to identify the ESAs to LD based on the geometric average of four indices, namely Soil Quality Index (SQI), climate quality index (CQI), vegetation cover quality index (VQI), and the land Management Quality Index (MQI). These quality indices provide insight into the geo-environmental conditions triggering LD processes, in which, the first three indices represent the environmental influences driving LD processes, while the last one expresses the impacts of anthropogenic activities. Each of these indices is calculated using weightage assigned thematic layers extracted from various data sources including Landsat ETM + and OLI images, SRTM/ASTER DEM data, geology map, soil map, topographical map, rainfall data, and other field survey data (Mohamed, 2013; Bedoui, 2020). The GIS techniques are used for the preparation of thematic layers and quantitative assessment of indices by mapping LD vulnerability. The outcome of this model shows the site-specific LD rate based on the estimated ESA index value, and it is implicitly limited to the assessment of four quality indices and their input parameters (Sepehr et al., 2007; Masoudi and Zakeri Nejad, 2010; Ait Lamqadem et al., 2018; Song et al., 2022).

The present study used the GIS-based MEDALUS model for mapping and assessment of LD vulnerability in the black cotton soil region of Southern India based on available data sources like soil, rainfall, LULC, vegetation, management practice, and other anthropogenic factors. The outcome of this study could be used as a decision-making tool for sustainable land resource management and combating the LD vulnerability on a long-term scale.

37.2 Study area

The study area covers the black cotton soil region of the Virudhunagar district of Tamil Nadu in Southern India. Fig. 37.1 shows the location of the study area situated at the geographical extent of 9° 24′ 28.79″ E to 9° 34′ 58.08″ E latitude and 77° 53′ 47.40″ N to 78° 05′ 25.44″ N longitude and covers the area of 4888 km^2. The Virudhunagar district administratively consists of 8 taluks and 600 villages with a total population of 19.42 lakhs as of Census 2011 report. The study area constitutes a significant plain terrain with a gentle slope toward the east and south-eastern parts and the hilly terrain (ridges and valley complex) of Western Ghats in the western region. The drainage network is formed by nonperennial rivers, namely Vaippar, Arjuna River, Gundar, and Deviar, and their tributaries, and they are seasonal and carry substantial flows during the monsoon periods (October to December). The various geomorphic units identified in multiple parts are upland range, foothills (bajada), rocky outcrops, structural hills, valley fills, shallow and deep buried pediments, pediplains, and flood plains are seen in various parts. Major geological formations are seen in the study area like Archaean Charnockite, igneous rocks, granite, gneisses complex, sandstones, crystalline limestone, limekanker, pleistocene laterite, black clay, and recent alluvium deposits. Moreover, the hydrogeological settings are highly

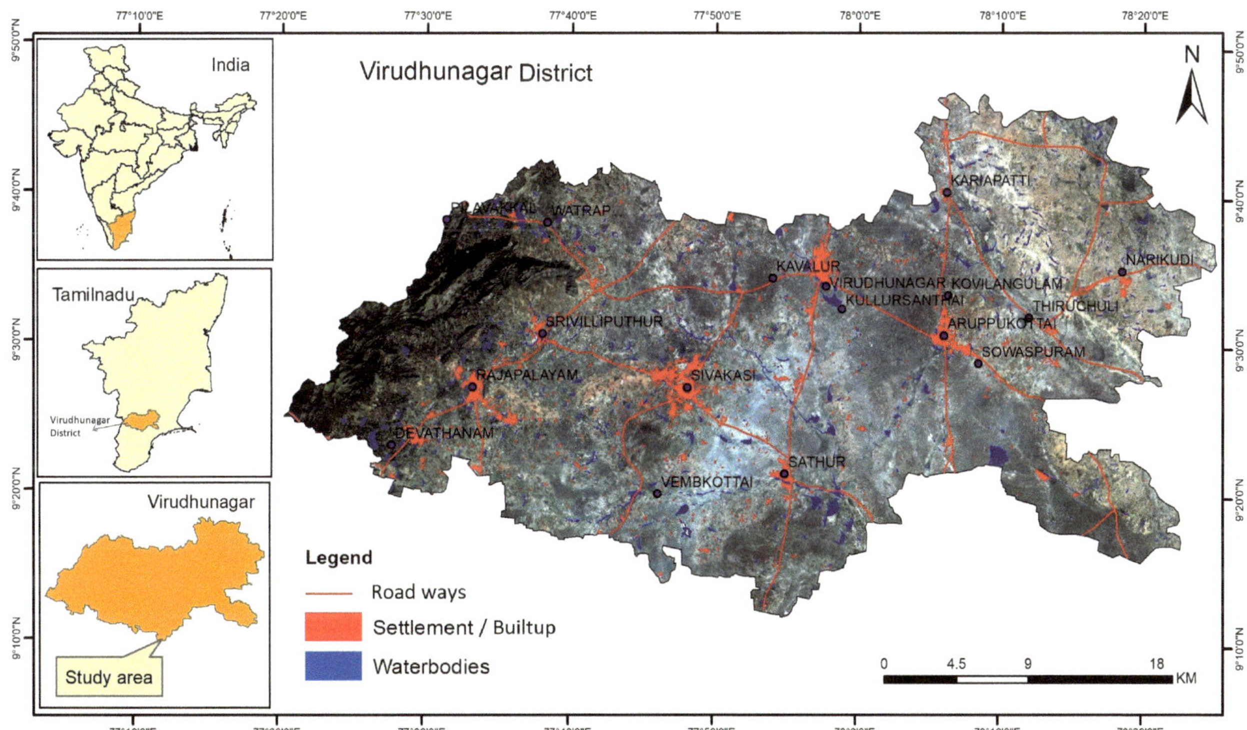

FIGURE 37.1 The geographical location of the study area.

heterogeneous formed by unconsolidated and semiconsolidated quaternary formations and fissured and fractured crystalline rocks, and the thickness of weathered zones (4–15 m) and hard rock terrains (10–15 m) with groundwater table at the range of 25–35 m below ground level. Soil types are noted as red loamy, black soil, lateritic soil, red sandy soil, and sandy alluvium. The majority of the area is covered by black soil (also known as black cotton soil) with enriched clay, lime, iron, magnesium, alumina, and potash but deficient in nitrogen, phosphorus, and organic matter (Muthulakshmi et al., 2010; Sivasankari and Siva Vignesh, 2017; Raja and Neelakantan, 2021). Ferruginous red loamy soil is seen along the foothills of the western part. Red sandy soil is spotted in the middle and eastern parts, and the gravelly loamy is seen along the river courses. LULC features are classified into agricultural lands (irrigated and unirrigated), forest cover, barren, rocky and mining quarries, waterbodies, land with shrubs, and settlements (urban/rural). Wherein agricultural land occupied 33.67% of the total area, forest cover 6.23%, barren land 1.06%, and fallow 0.72%. About 33% of the area falls under irrigation, and the rest of the area depends on rainfed tanks. The major parts of the area are noted as unirrigated croplands, fallows, barrens, and land with shrubs due to unsuitable hydrogeological settings (SAC, 2021). The climate of this region is a prevailing semiarid subtropical type with the optimum temperature ranging from 20°C to 37°C (CGWB report). The study area receives scanty rainfall with an average of 812 mm during both southwest and northeast monsoons. Monsoonal rainfall plays a vital role in the availability of water sources. In contrast, the northeast monsoon (October to December) contributes more rainfall in all parts of the area. The population of the Virudhunagar District has grown from 19.42 to 20.23 during 2011–21, and the growth rate increased to 11.90%. The economic activities associated with the population are textiles, crackers, fireworks, printing, and allied industries. Most of the population depends on agricultural activities, but the irrigation facility is inadequate due to poor rainfall and nonperennial water sources; however, the alluvial deposits along the river courses are significant sources of groundwater supply for domestic and irrigation purposes.

37.3 Materials and methods

The GIS-based MEDALUS model is used to calculate the land degradation vulnerability index (LDVI) for mapping the LD zone by analyzing multiple geo-environmental parameters like soil, rainfall, LULC, vegetation, land management practice, and other climatic factors. Fig. 37.2 shows the workflow of LD assessment using GIS techniques (Masoudi

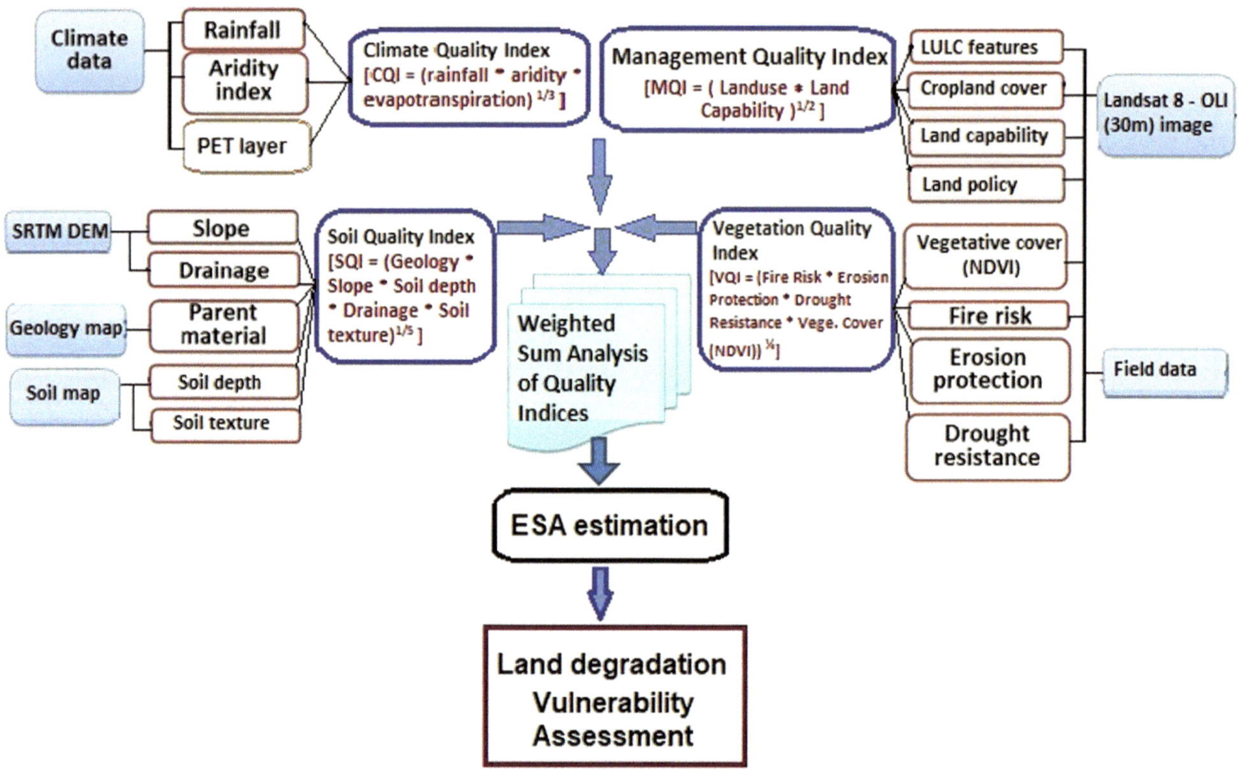

FIGURE 37.2 Methodology workflow of land degradation assessment using GIS techniques.

and Zakeri Nejad, 2010; Ait Lamqadem et al., 2018; Bedoui, 2020). The GIS techniques are systematically executed for storing, analyzing, and interpreting the different layers to demarcate LD vulnerability in the study area. The outcome derived from this model shows the site-specific LD status based on the estimated LDVI value, and it is implicitly limited to the assessment of four quality indices and their input parameters.

37.3.1 Data sources and thematic layer preparation

The MEDALUS parameters are obtained from various geo-environmental data sources including SOI-published topographical maps, Geological Survey of India published geology and geomorphology maps, Landsat OLI images, SRTM DEM (30 m), soil maps, forest and vegetation maps, land management practice, IMD rainfall data, and other climatic factors. Thematic layers of each parameter are prepared with the UTM-WGS 84 projection and coordinate system using GIS functionalities. The thematic layers like administrative boundaries, settlements, and roads were extracted using Survey of India (SOI) published topographical maps with a scale of 1:25K. Landsat 8—OLI/TIRS images acquired from https://glovis.usgs.gov are used for the preparation of thematic layers such as Normalized Difference Vegetation Index (NDVI), LULC, land capability, and forest-vegetative cover, drought resistance, fire-risk zone, and erosion resistance using ENVI 5.3 and ArcGIS 10.6 software techniques. SRTM-DEM (30 m) filled data is used for mapping the slope (topography) and drainage network. The rainfall data of 2000–20 derived from the IMD is used to estimate decadal average rainfall using IDW (inverse distance weighted) interpolation technique in the GIS software environment. The parameters, namely aridity and evapotranspiration, are obtained from the CGIRCSI web portal—Global Aridity Index and Global Evapotranspiration datasets. This data provides high-resolution global raster climate data related to evapotranspiration processes and rainfall deficit for potential vegetative growth. The district resource map published by GIS is used for the preparation of geology and geomorphology layers using ArcGIS software. Soil types with soil depth, texture, and permeability were derived from NBSS-LUP published soil maps and in-situ data (Kaliraj et al., 2017). The feature classes of each thematic layer are added to attributes and weights (value of 1–5) based on the potential magnitude of LD vulnerability. Table 37.1 shows the MEDALUS parameters and its class-wise assigned weights and estimated index score used for assessment of LD in the study area.

37.3.2 Estimation of quality indices

The MEDALUS model is designed to calculate the LDVI using the four quality indicators such as SQI, CQI, VQI, and land MQI based on the assigned weights of the different MEDALUS parameters (Kosmas et al., 1999; Ali and El Baroudy, 2008; Ahmad and Pandey, 2018). The quality index score (QI_s) for each parameter is estimated with values ranging from 1 to 2 based on the parameter's assigned weighted value of 1–5. Wherein the high index value indicates the areas considered to be the most sensitive to LD, while the low index value denotes its least sensitivity to LD; on the other hand, it indicates the land under good productivity (Budak et al., 2018; Bedoui, 2020; Ferrara et al., 2020; Abuzaid and Abdelatif, 2022). The MEDALUS parameter-wise assigned weights and their estimated quality index (QI) value are listed in Table 37.1.

In the first stage, the Quality Index score (QIs) is calculated from the sum of weighted values assigned to subindicators (layers) (P1, P2, P3...Px) of the four quality indicators (SQI, CQI, VQI, and MQI) using the equation (Eq. 37.1), and it is expressed as,

$$QI_S = [(P1) \times (P2) \times (P3) \times ..(Pn)]^{1/n} ---- \tag{37.1}$$

where QI_S represents the Quality Index score, s represents a particular quality index; and n shows the number of subindicators (layers) used to calculate the QIs for each quality indicator.

The MEDALUS—Quality indicators were estimated the Quality Index value (QI_c) by calculating the sum of the QIS of MEDALUS parameters using the different equations as follows,

$$\text{Soil Quality Index(SQI)} = (\text{soil texture} \times \text{parent material} \times \text{soil depth} \times \text{slope} \times \text{soil drainage})^{1/5} ---- \tag{37.2}$$

$$\text{Climate Quality Index (CQI)} = (\text{rainfall} \times \text{aridity index} \times \text{evapotranspiration})^{1/3} ---- \tag{37.3}$$

$$\begin{aligned}&\text{Vegetation Quality Index (VQI)}\\&= (\text{Fire risk resistance} \times \text{soil erosion protection} \times \text{drought resistance} \times \text{vegetation cover})^{1/4} ----\end{aligned} \tag{37.4}$$

$$\text{Land Management Quality Index(MQI)} = (\text{LULC} \times \text{Landuse capability})^{1/2} ---- \tag{37.5}$$

TABLE 37.1 MEDALUS parameters—classification, assigned weights and estimated quality index of each class.

Quality indicator	MEDALUS parameter	Description		Weighted value assigned	Quality Index score (QI_S)
Soil quality index (SQI)	Soil depth (m)	Very deep	> 5	1	1.2
		Deep	3–5	2	1.4
		Moderately deep	2–3	3	1.6
		Moderately shallow	1–2	4	1.8
		Shallow	< 1	5	2
	Soil texture	Rock outcrops/stony waste		1	1.2
		Gravelly clay, Gravelly loamy		2	1.4
		Loamy		3	1.6
		Clayey		4	1.8
		Calcareous clay		5	2
	Soil drainage (percolation capacity)	Well-drained (high percolation)		1	1.2
		Moderately drained		2	1.4
		Low drained (lower percolation)		4	1.8
		Excessive/Poorly drained		5	2
	Geology (parent materials)	Alluvium		1	1.2
		Laterite		2	1.4
		Charnockite		3	1.6
		Granite		4	1.8
		Granite gneiss		5	2
	Slope (in degree)	Plains	< 6	1	1.2
		Gentle slope	6–12	2	1.4
		Slightly steep	13–24	3	1.6
		Steep slope	25–30	4	1.8
		Very-steep slope	> 30	5	2
Climate Quality Index (CQI)	Decadal average rainfall (mm)	> 700		1	1.2
		650–700		2	1.4
		550–650		3	1.6
		500–550		4	1.8
		< 500		5	2
	Aridity Index	< 0.5		1	1.2
		0.5–0.6		2	1.4
		0.6–0.65		4	1.8
		> 0.65		5	2
	Evapotranspiration (mm)	< 1200		1	1.2
		1200–1500		2	1.4
		1500–1700		4	1.8
		> 1700		5	2

(Continued)

TABLE 37.1 (Continued)

Vegetation Quality Index (VQI)	Drought resistance	Very high	1	1.2
		High	2	1.4
		Moderate	3	1.6
		Low	4	1.8
		Very low	5	2
	Fire risk resistance	Very high	1	1.2
		High	2	1.4
		Moderate	3	1.6
		Low	4	1.8
		Very low	5	2
	Erosion protection	Very high	1	1.2
		High	2	1.4
		Moderate	3	1.6
		Low	4	1.8
		Very Low	5	2
	Vegetative cover (NDVI)	> 0.8	1	1.2
		051–0.72	1	1.2
		0.27–0.5	2	1.4
		0.14–0.26	3	1.6
		0.1–0.13	4	1.8
		< 0.1	5	2
Management Quality Index (MQI)	Land use/land cover (LULC)	Settlements/built-ups, water bodies	1	1.2
		Forest cover, plantation, agriculture irrigated	2	1.4
		Land with shrubs, agriculture unirrigated	3	1.6
		Fallows, sandy, barren land	4	1.8
		Rocky outcrops, Stony waste, Mines and Quarries	5	2
	Land capability	Class 1 (forest cover)—high productivity	1	1.2
		Class 2 (plantations/agriculture irrigated)—high productivity	1	1.2
		Class 3 (agriculture—unirrigated)—seasonal productivity	2	1.4
		Class 4 (land with shrubs)—moderate productivity	3	1.6
		Class 5 (barren/fallow lands)—low productivity	4	1.8
		Class 6 (salt-affected/barren/waste/eroded/alkaline)—poor productivity	5	2

TABLE 37.2 MEDALUS—quality indices: classification, quality index value and area extend of each class.

MEDALUS quality indices	Class	QI_c value	Area (%)
SQI	High	<1.13	8.17
	Moderate	1.13–1.45	75.19
	Low	>1.45	10.59
CQI	High	< 1.25	3.81
	Moderate	1.25–1.50	62.83
	Low	>1.50	27.3
VQI	High	<1.15	5.25
	Moderate	1.15–1.40	5.36
	Low	>1.40	83.34
MQI	High	<1.25	5.7
	Moderate	1.25–1.50	60.14
	Low	>1.50	28.1
Area excluded	Settlements		3.66
	Water bodies		2.38

The QI_c for each quality indicator is classified into three classes with a value of 1–2 to represent their potential magnitude to LD processes. Table 37.2 shows the MEDALUS—Quality indicator's classification and QIc in the study area. The QI_c value "1" is considered to be areas of least sensitivity, whereas the QI_c value "2" was considered to be areas with the most sensitivity, and the QI_c value at the range of 1 and 2 reflects relative vulnerability to LD processes.

37.3.3 GIS-based MEDALUS modeling

The GIS-based MEDALUS model developed by the European Commission (Kosmas et al., 1999) is used for assessment and mapping the LD vulnerable zone. In the MEDALUS procedure in the second stage, the LDVI is calculated by the sum of the quality index value (QIA) of four quality indicators (SQI, CQI, VQI, and MQI) using ArcGIS—Raster Calculator tool, and the LDVI equation is expressed as,

$$\mathrm{LDVI} = (\mathrm{SQI} \times \mathrm{CQI} \times \mathrm{VQI} \times \mathrm{MQI})^{1/4} ----. \tag{37.6}$$

In this model, the LDVI represents the LD status in the study area. The estimated LDVI value indicates the cumulative influence of MEDALUS parameters on the LD vulnerability of the study area at the 30 × 30 m scale. The LDVI value is estimated at the range 1–2; thereby, the lower value indicates the land with the least vulnerability to LD, and the high value denotes the land with the most vulnerability to LD. The LDVI class and its estimated index values are given in Table 37.3.

37.4 Results and discussion

The GIS-based MEDALUS model is executed to assess and map LD zone in the study area. The LDVI map is estimated by the sum of four quality indicators which are derived from multiple geo-environmental parameters like soil, climate, rainfall, aridity, erosion, vegetative cover, plantation, land use, land management practice, etc. The MEDALUS—Quality Index classes and their estimated index values are listed in Table 37.2, and it is described as follows,

37.4.1 Soil quality index

The SQI has represented the characteristics of soil depth, soil texture, soil drainage (percolation capacity), soil parental materials (geological settings), and slope (topography) conditions. Fig. 37.3A–E shows the SQI map and its contributing parameters to the study area. Soil depth is defined as the vertical depth of the soil profile between the surface and

TABLE 37.3 Land degradation classes and their LDVI values of the study area.

LDVI class	Sub-class	LDVI value	Area (km^2)	Area (%)	Risk level	Factors inducing land degradation vulnerability
Nonaffected zone	Nonaffected area (N)	< 1.17	22.09	0.51	Lower vulnerability	LULC features retaining soil nutrients, humus, and fertility conditions due to enough rainfall during all seasons.
Potentially resistance zone	Potentially vegetative cover (P)	1.17–1.22	76.07	1.76	Moderately Vulnerability	Seasonal changes in rainfall, LST, soil moisture, and climate depending on monsoonal activities
Fragile zone	F1	1.22–1.26	37.16	0.86	High vulnerability	Dynamic equilibrium of rainfall, LST, soil moisture, and climate resulting from natural and anthropogenic pressures.
	F2	1.26–132	97.52	2.26		
	F3	1.32–1.37	155.54	3.6		
Critical zone	C1	1.37–1.41	421.78	9.78	Severe vulnerability	Degradation of land and soil due to soil erosion, soil salinity, removal of topsoil, lack of rainfall, poor soil moisture, an increase of LST, and aridity conditions
	C2	1.41–1.53	2601.23	60.34		
	C3	> 1.53	638.61	14.81		

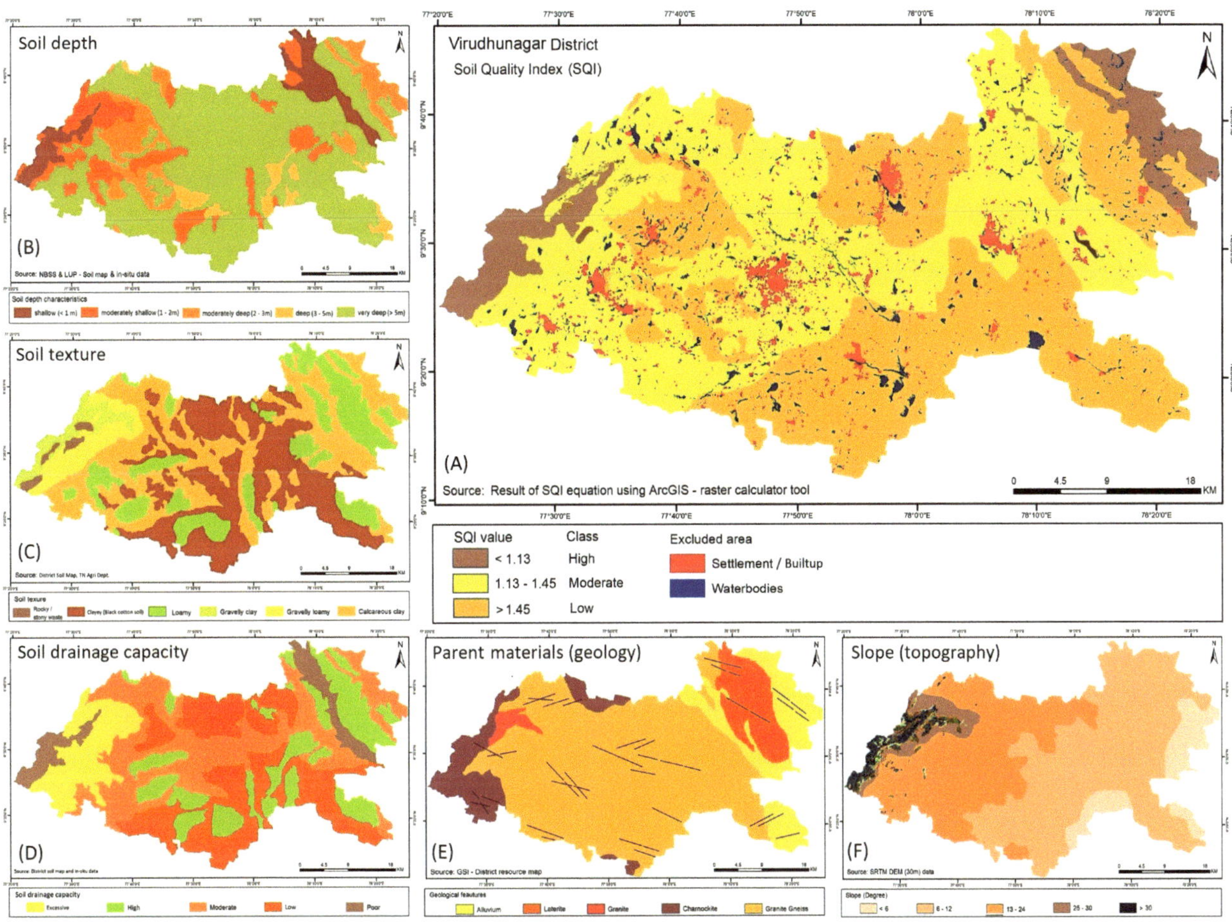

FIGURE 37.3 Spatial characteristics of Soil quality index (SQI) in the study area: (A) SQI map, (B) Soil depth; (C) Soil texture; (D) Soil drainage capacity; (E) Soil parent materials (geological features); (F) Soil slope (surface topography).

the top of the bedrock or weathered parent materials. The deep soil depth transmits water and nutrients to the plant's root zones and is more suitable for agricultural activities. They have a higher water holding capacity that protects soil layers from erosion processes and constitutes proper functioning of soil formation processes. However, soil layers at lower depths indicate poor quality that does not fulfill plant growth requirements. Fig. 37.3B shows the depth of the different soil types based on their vertical depth profiles, and it is classified into five classes such as shallow (< 1 m), moderately shallow (1–2 m), moderately deep (2–3 m), deep (3–5 m), and very deep (> 5 m).

Soil texture refers to the proportion of sand, silt, and clay particles of each soil that make up the mineral compositions that influence the soil-water holding, lateral flow through the soils, and soil fertility. Loamy soil does not hold much water and is low in nutrients, while clay soils generally hold more water and are better at supplying nutrients to the plants. Fig. 37.3C shows the soil texture of the area, which includes clay (black cotton soil), calcareous clay, and loamy were predominantly found in major parts, and other soils like gravelly clay, gravelly loam, and rocky gravels (stony waste) were noted around the uplands in the western parts. The textural composition of sand, silt, and clay determines soil hydraulic properties, that is, sandy textures are highly permeable, have poor water holding capacity, and are highly susceptible to wind erosion, and the humus and nutrient reserves are minimal in them. While clayey textures are deficient in aero-hydric features and have a higher ability to retain excess water, in terms of poor permeability and more susceptibility to salinity and soil–water drainage, it is classified as potential sensitivity to LD and desertification in arid and semiarid regions. Soil drainage refers to the soil's natural ability to allow water to pass through the soil profile at the reaching point of water holding capacity, and it results in the water flow through the soil layers and out of the plant root zone, which causes percolation or soil drainage of an area. Fig. 37.3D shows the spatial distribution of soil drainage capacity such as soil with excessive, high, moderate, and poor drainage based on soil-water trans-flow. The rate of soil drainage causes LD by affecting soil fertility and salinity in the alluvial plains and the pediplains.

Soil parent materials determine soil formations and their compositions that may originate from geological formations of mineral rock and/or organic matter. The parent materials strongly influence soil properties such as texture, grain size, physico-chemical properties, and mineralogical compositions. Fig. 37.3E shows the geological formations of soil parent materials, including alluvium, laterite, granite, charnockite, and granite gneiss. Where clay (black cotton soil) and calcareous clay soil are derived from the weathering of volcanic rocks like gneisses and schists with a mixture of organic and inorganic materials, and the alluvial loams were deposited by work of rivers and streams.

Slope gradient represents the topography of the surface that refers to the angle of inclination of the soil surface concerning horizontal plains. The slope gradient controls soil thickness and soil–water residence time by controlling the lateral transport of soil particles and solutes. The slope gradient is a very important factor affecting the total porosity, runoff rate, and soil erosion intensity of the area. Fig. 37.3F shows the slope gradient of soils that are classified as a very-steep slope (> 30 degrees) in the hilly terrains and a steep slope (25–30 degrees) in the upland forest covers. The gentle slope (< 24 degrees) covers the rest of the other parts including alluvial plains and clayey (black cotton) soils. The slope (topography) of the area undoubtedly influences soil erosion, especially in the semiarid zone, where a slope > 12 degrees causes the removal of topsoil during monsoonal rainfall.

Fig. 37.3A shows the SQI values and their spatial characteristics in the study area. The SQI classes represent the potentiality of LD subject to soil depth, soil texture, soil drainage (percolation capacity), soil parent materials (rocky formations), and surface slope (topography) conditions. Moreover, the SQI indicates the soil state that affects the strength of cohesion between soil particles, soil texture, soil structure, soil aggregation and stability, infiltration rate, and water retention capacity, as well as representing the soil resistance capacity to erosion from water and wind processes (Salunkhe et al., 2018). Table 37.2 shows the SQI estimated value (QI_U) and its spatial distribution in the study area. The result of SQI is classified into three classes such as high (SQI < 1.13), moderate (SQI value of 1.13–1.45), and low (SQI > 1.45). It is noted that the SQI is at a lower value (<1.13), indicating the higher quality of soils that covers about 8.17% of the area, and it is found in vegetative cover, forest, plantations, and irrigated agricultural lands. The SQI is at a moderate level, estimated in 75.19% of the area across the mid-alluvial plains consisting of gravelly loamy, loamy, and calcareous clay soils. However, the higher SQI value (> 1.45) shows lower soil quality, mainly within the unirrigated clayey soils (Black cotton) and eroded lands that extend up to 10.59% of the total area. Upon considering the SQI's constituting parameters, soils are noted to have lower quality in different parts due to processes of erosion and salinization, which causes poor crop productivity. Moreover, soil quality is lower in northeastern parts and across the calcareous clay soils associated with fallows, shrubs, or barren lands. Most loamy and clayey soils along the plains and foothills are noted at moderate soil quality depending on land management practices and monsoonal impacts. Many patches of loamy and black cotton soils (clayey and calcareous clay) are found in good soil quality, mainly around riverbanks, channels, and waterbodies in the middle and south-eastern parts. Similarly, the soils within the forest covers are enriched with humus materials in the western upland range. Overall, the SQI indicates the occurrence of soil degradation in various parts due to erosion and salinity processes.

37.4.2 Climate quality index

The CQI has represented the characteristics of rainfall (precipitation), evapotranspiration (PET), and aridity index (AI) involved in LD processes. Fig. 37.4A–E shows the CQI spatial characteristics and its contributing parameters in the study area. Fig. 37.4C shows the decadal average of rainfall and its spatial variability for the period of 2010–2020. The decadal rainfall average is estimated at the range of 240–830 mm. A higher rainfall rate (> 700 mm) is received in the western uplands and the foothills than in the other parts of the district. The mid-plains have noted a moderate rainfall rate (550–600 mm). It is noted that the rainfall distribution gradually decreased toward the eastern part of the area and progressively increased toward the northern and southern parts. Overall, the study area generally received lower rainfall throughout the year than associated agro-climatic zones.

Fig. 37.4B shows the spatial characteristics of the Aridity index that indicates the ratio between rainfall and evapotranspiration in the area. The Aridity index amplifies water stress on plant growth after intake by evaporation and soil moisture processes. The AI value is noted at a lower rate across the western parts including forest cover (< 0.5), and plantations and irrigated lands (0.5–0.6) which denotes the availability of enough water for plant growth, but the other parts of middle plains are noted higher at AI value (> 0.6) that indicates prevailing semiarid conditions due to water stress and higher PET in there. The increasing aridity in the alluvial plains affects crop productivity due to surface runoff or lack of rainfall, wherein the higher AI value denotes the water stress to plant growth (i.e., crop cultivation and plantation, etc.), and the xerophytic shrubs (plants require less water and moisture) spread over the various parts of the area.

Fig. 37.4D shows the spatial variability of potential evapotranspiration (PET) in the study area. The PET is estimated at a higher rate (> 1500) in the middle alluvial plains and eastern calcareous clay soil areas due to higher land surface temperature (LST), that evaporates water contents from plants and soils. The lower PET rate (< 1200) is noted in forest cover of the western uplands and the moderate PET rate (1200–1500) is found in foothills and associated plantations and irrigated croplands due to the prevailing higher rate of water and soil moisture. The PET represents the amount of water evaporated from the source of vegetation and soil moisture induced by LST and atmospheric temperature, and it affects

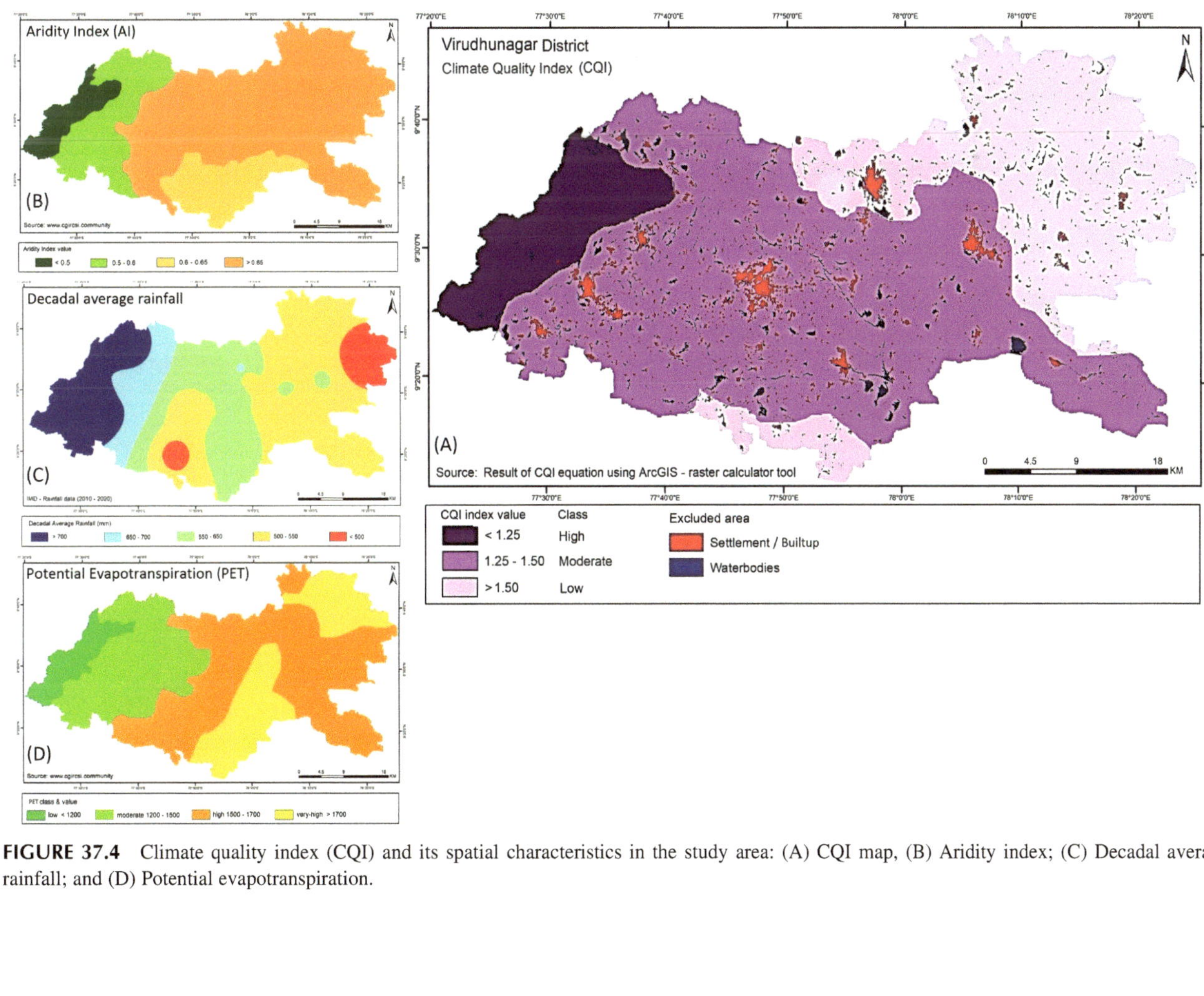

FIGURE 37.4 Climate quality index (CQI) and its spatial characteristics in the study area: (A) CQI map, (B) Aridity index; (C) Decadal average rainfall; and (D) Potential evapotranspiration.

plant growth and land productivity. The higher rate of PET is noted across unirrigated lands and shrubs, which causes lower moisture content on the soil layer. The spatial variability of PET affects soil properties including moisture, fertility, and albedo on a long-term scale.

Fig. 37.4A shows the CQI classes and their spatial characteristics in the study area. The CQI has stated the climatic quality of plant growth is subject to rainfall, soil, and land use of the area. The higher CQI rate (> 1.50) indicates lower climate quality to plant growth, and it is found in the dryland plains (27.3% of the total area) of the eastern and north-eastern parts and small patches in southern parts, and these areas are mainly occupied with calcareous clay and black cotton clay soils. The moderate CQI value (1.25–1.50) has been noted along the larger area of middle-alluvial plains (62.83% of the total area), especially in the irrigated lands of loamy and black clay soils. Significantly, the western uplands and foothills are estimated at a lower CQI rate (<1.25) that shows the higher climate quality to vegetation growth, that is, forest cover, plantations, and irrigated crops, and it constantly supports the maintenance of land and soil fertility in the seasons. The higher CQI of the areas indicates the poor climatic quality of plant growth because of lower rainfall and higher LST, PET, and aridity, affecting the biomass productivity of the area.

37.4.3 Vegetation quality index

The VQI shows the biotic characteristics of the area based on vegetation drought resistance, fire risk, recuperation capacity, soil protection from erosion, and the landscape's vegetative cover. The vegetative cover enriches organic matter in lands and soils, and it reduces soil erosion, fire risk, and runoff while increasing the infiltration rate. Fig. 37.5B shows the area's spatial characteristics of drought-resistant capacity. The higher drought-resistant capacity is found in the upland forest cover range, foothill plantations, and irrigated croplands due to higher rainfall, soil moisture, soil

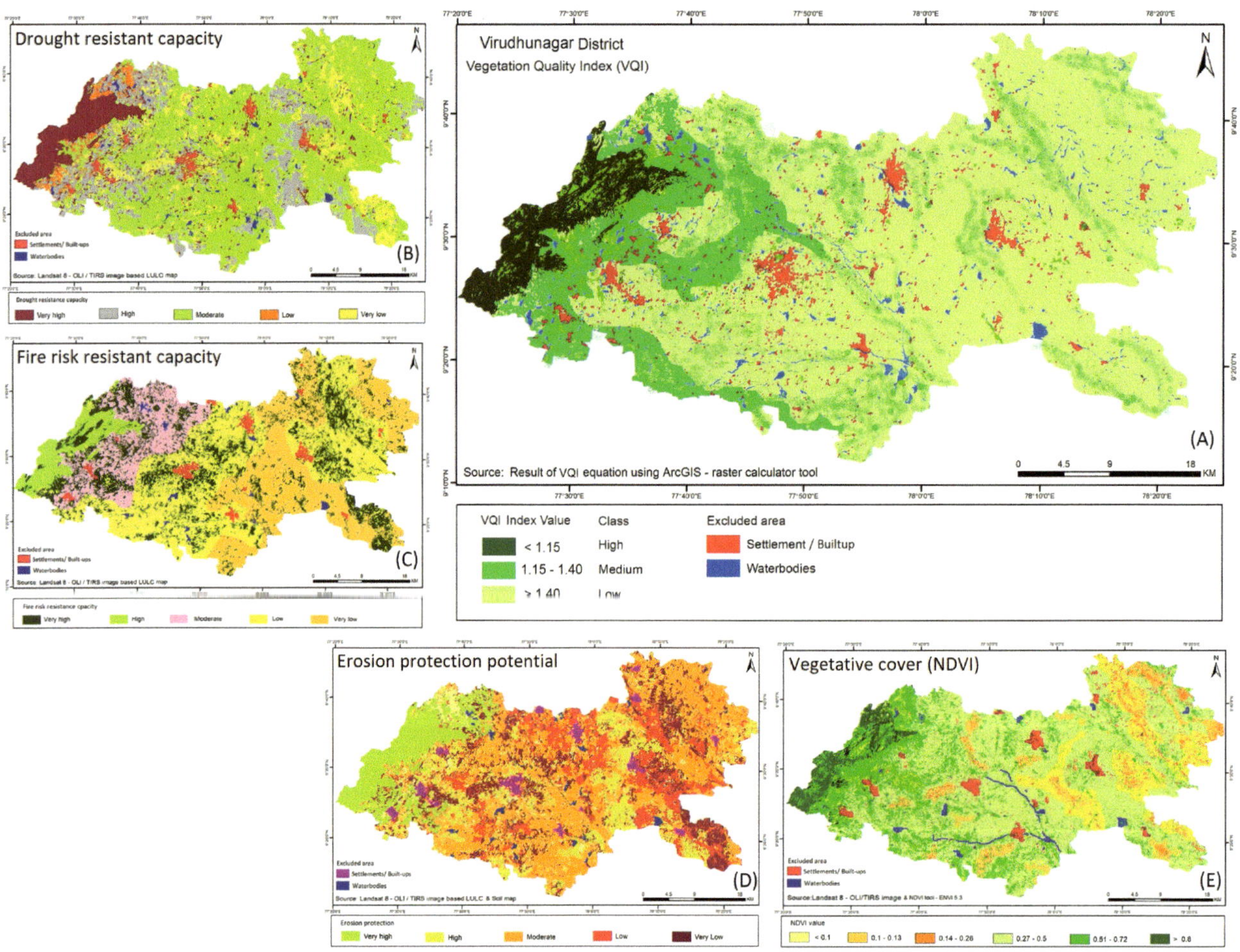

FIGURE 37.5 Spatial characteristics of Vegetation Quality Index (VQI) in the study area: (A) VQI map, (B) Drought-resistant capacity; (C) Fire risk resistant capacity; (D) Erosion protection potentiality; (E) Vegetative cover.

humus, humidity, etc. Major parts including unirrigated croplands and fallows are noted at moderately drought-resistant capacity due to variability of vegetation, LST, and soil moisture based on monsoonal rainfall. The LULC features in many patches are noted at lower drought-resistant capacity, especially land with shrubs and barren lands with grasses and bushes. Severe drought to vegetation occurs in eastern and northeastern parts that reduces biomass productivity and increases erosion and salinity of soils as a reduction of protective vegetative cover.

Fig. 37.5C shows the fire risk resistance and recuperation capacity of the LULC features in the study area. In general, a forest fire is one of the factors affecting biomass and land resources; however, along the western upland range, the forest cover has higher resistance and recuperation capacity to fire risk from the result of frequent rainfall and topographic wetness conditions. The foothill landscapes and associated LULC features have been denoted as moderately fire risk-resistance areas based on prevailing monsoonal impacts in there. Other parts of the middle and eastern plains including unirrigated croplands, land with shrubs, and barren lands have noted a higher sensitivity to fire risk during summer due to an increase of LST and poor rainfall, and it causes a reduction of soil moisture and biomass wetness.

Fig. 37.5D reveals the soil's protection capacity against erosion resulting from water, runoff, and aeolian processes. Throughout the area, soils found with vegetative covers, that is, forests, plantations, croplands, and shrubs, have noted higher protection from erosion during all seasons. However, soil layers within the proxy of slope surfaces, streams, channels, fallows, barrens, and unirrigated croplands have fallen under higher susceptibility to erosion due to runoff and aeolian processes. In the middle and eastern parts, several patches of soils like loamy, clayey, and calcareous clayey are exposed to erosion and removal of topsoil that causes soil infertility and its associated adverse impacts.

Fig. 37.5E shows the spatial distribution of vegetative cover (i.e., forest, plantations, croplands, shrubs, and other biomass) with its health status during April 2020, which is derived from the Landsat 8—OLI image using NDVI analysis. The NDVI detects vegetation and nonvegetation based on the reflected ratio between visible Red and near-infrared (NIR) bands. The NDVI layer is used to monitor the vegetation's greenness and health status and to identify areas that fall under stress or damage in vegetative cover over time. The NDVI value is estimated at the range of -1 and 1 based on the photosynthetic activity (chlorophyll) of vegetation in the area. The NDVI at a higher value (>0.51) shows the occurrence of healthy vegetation, that is, forest cover, plantations, and irrigated croplands, whereas the NDVI value (0.1–0.5) indicates moisture-stressed vegetation with lower biomass in the middle and eastern parts. The NDVI value (<0.1) indicates fallows, barrens, and other eroded landforms in various parts of the study area.

Fig. 37.5A shows the VQI and its spatial characteristics in the study area. The VQI indicates the cumulative characteristics of vegetation properties about soil, land, and climate conditions toward the processes of LD and desertification. The lower VQI value (<1.15) indicates the higher vegetation quality in the area (5.25%) of upland forest covers that comprises enriched vegetation greenness with healthy biomass and deciduous herbaceous layers during all the seasons. Moderate VQI value (1.15–1.40) is noted in the 5.36% of the area, especially in foothills, plantations, and irrigated croplands. However, the higher VQI value (>1.40) indicates, the lower vegetation quality in the LULC features spread over the larger areas (83.34%) in middle and eastern parts, which is found with moisture-stressed vegetations (i.e., unirrigated crops, land with shrubs, and xerophytic plants) due to lower rainfall, and soil moisture, higher LST, and arid climate conditions. The areas showing the higher vegetation quality have occurred with vegetation greenness and herbaceous layers that support maintaining land and soil fertility and protecting the LULC from erosion and degradation activities. The areas of low vegetation quality are mainly found in various parts of middle and eastern plains across fallows, barrens, unirrigated croplands and land with shrubs, and these areas are highly prone to LD from soil erosion and culminating in LD due to climate change and human-induced activities.

37.4.4 Land management quality index

The land MQI is the main driver of LD processes, and this denotes the LULC management practices, especially for croplands, fallows, and other cultivable lands. The MQI is estimated using LULC features and the land capability of the study area. The spatial characteristics of MQI change from place to place depending on natural and anthropogenic factors that affect the land productivity of the area. Fig. 37.6B shows the LULC features include upland forest cover, foothill plantations, and the larger area of irrigated and unirrigated croplands, fallows, barren lands, rocky outcrops, mines, quarries, stony waste, and land with shrubs, waterbodies, and settlements. LULC features in the middle and eastern parts are mainly noted poor land management practices than the other parts. LULC features in some parts include croplands, fallows, barrens, and land with shrubs have inherently exposed more risk to LD due to impacts of erosion and soil salinity. However, LULC features in uplands and irrigated zones are noted at lower sensitivity to LD based on proper management practices and suitable climatic conditions. Fortunately, the LULC features in various parts have been estimated at low sensitivity to degradation processes due to applying proper management practices and less anthropogenic intervention in the area.

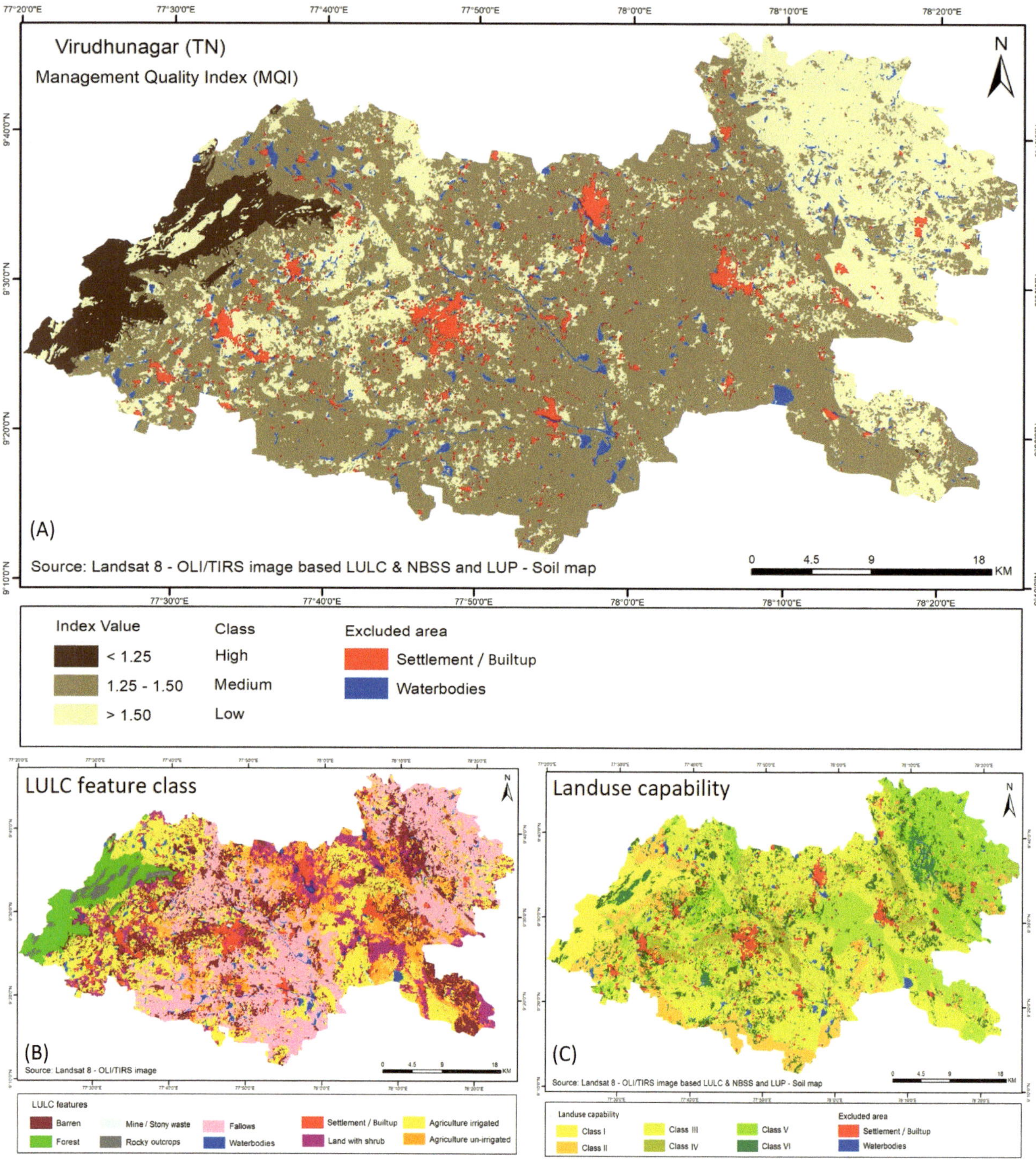

FIGURE 37.6 Spatial characteristics of land management quality index (MQI) in the study area: (A) MQI map, (B) LULC features; (C) Land use capability.

Fig. 37.6C shows the land use capacity of the various LULC features, which is estimated by incorporating soils, rainfall, and climatic factors. This reveals the production capacity of land use under the stress conditions of the different geo-environmental factors. The land use capacity is classified into six classes, in which class I and II indicate the LULC having a higher production capacity, mainly found in forest cover, plantations, and irrigated croplands. The settlements (build-ups) and waterbodies are excluded from these classes, since they play no role in land capability. Likewise, class III and IV shows a moderate production capacity to LULC features based on land management practices and rainfall. Meanwhile, classes V and VI are denoted at lower land use capacity in various parts of fallows, unirrigated croplands, barren lands, and eroded lands, and these areas are highly exposed to degradation processes due to erosion, salinization, and chemical degradation. The areas showing poor land use capacity cause inefficiency in regulating their environmental ecosystem processes.

Fig. 37.6A shows the land MQI and its spatial characteristics in the study area. The MQI varies from place to place, highlighting land productivity subject to land, soil, and climatic factors. The lower MQI value is at a rate (<1.25) revealing, the higher land management quality, which is found in the 5.7% of the area, which covers the upland forest covers, foothills, plantations, and irrigated croplands, and these areas provide good productivity during all seasons. The moderate MQI value (1.25–1.50) is estimated in larger areas (60.14%) in foothills, plantations, irrigated croplands, land use around the urban settlements, and middle plains of unirrigated croplands. The higher MQI value (> 1.50) indicates lower land management quality, and it is estimated that 28.1% of the total area, mainly covers fallows, barrens, land with shrubs, eroded lands, and salt-affected lands in various parts of the study area. Significantly, the majority of agricultural lands fall under moderate land management quality due to intensive land use and poor land management practices. Among these, unirrigated cultivable lands in the east and northeastern parts are highly sensitive to LD due to changes in rainfall, soil moisture, and climatic conditions. Moreover, the LULC features within the urban proxy is noted at a lower quality due to anthropogenic pressure, especially through intensified agricultural activities that cause land and soil degradation.

37.4.5 Assessment of land degradation vulnerability index

LD is commonly referred to as a decrease in land productivity from the adverse impacts of geo-environmental changes like rainfall, LST, aridity, deforestation, soil erosion, soil salinity, and anthropogenic pressures, and these directly reduce the soil fertility, biomass, and net primary productivity. Fig. 37.7 shows the resulting map of LDVI classes with spatial distribution in the study area. The result of LDVI is classified into four classes based on their degree of vulnerability to LD such as (1) nonaffected zone (N) and the estimated LDVI value of < 1.17, (2) potential vegetative cover (P), with the LDVI value of 1.17–1.22, (3) Fragile zones with its three subclasses (F1, F2, and F3), and their LDVI value within the range of 1.22–1.37, and (4) Critical zones with its three subclasses (C1, C2, and C3) and their estimated LDVI value is of

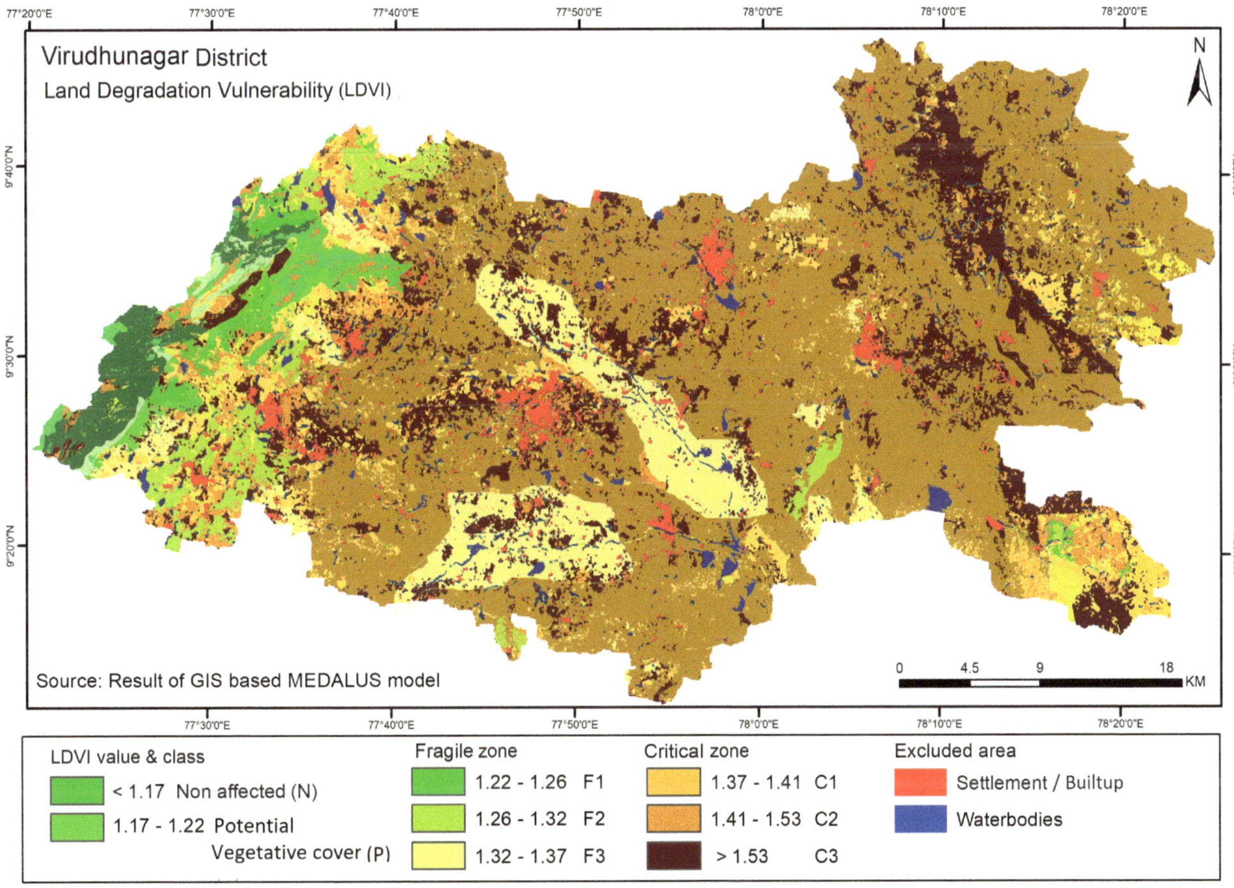

FIGURE 37.7 The Land degradation vulnerability index (LDVI) classes and their spatial characteristics in the study area.

> 1.37. The LDVI map shows that the critical zone and its subclasses of C1, C2, and C3 were predominantly found in the agricultural plains of the middle and eastern parts, indicating severe vulnerability to LD processes. The critical zones are highly sensitive to land and soil degradation due to erosion, removal of topsoil, a decrease in rainfall, soil moisture, biomass increase of aridity, soil salinity, and LST, which affects soil and land productivity in the long-term scenario. The LDVI classes and their estimated index values are given in Table 37.3. About 84.93% of the area falls under critical LD in LULC features due to soil erosion, salinity, and removal of nutrients. The subclasses of critical zones, namely C1, C2, and C3, are spatially estimated at 9.78%, 60.34%, and 14.81%, respectively. Moreover, 6.73% of the area falls under fragile zones, including F1, F2, and F3 subclasses, with areas of 0.86%, 2.26%, and 3.60%, respectively. The LULC features within the proxy of critical zones are noted under soil and LD processes that cause adverse impacts on agriculture productivity. Many patches of land use, like croplands, fallows, barrens, and eroded lands, occur with severe LD processes, and those areas are drastically converted into desertification. The fragile zone with its subclasses (F1, F2, and F3) is distributed in 6.72% of the area. It is mainly found in the western parts of irrigated croplands and land use around the urban proxy. Within the fragile zones, the LULC features are highly sensitive to degradation due to the dynamic equilibrium of rainfall, LST, soil moisture, climate, and various anthropogenic pressures. The area of 1.76% fall under moderate vulnerability to LD, mainly in foothills, pediplains, and various parts of land with shrubs; thereby, the landscapes experience soil and water stress conditions due to rapid runoff and poor infiltration. Significantly, the area of 0.51% is noted as the nonaffected zone in the western parts, which includes upland forest cover, foothill plantations, and irrigated croplands. The upland forest cover and other LULC features like plantations and agriculture-irrigated lands due to retaining soil nutrients, humus, and fertility conditions. The LULC features of these areas are noted at higher biomass productivity due to adequate rainfall, soil moisture, and soil fertility, and also provide a higher resistant capacity to land and soil degradation during all seasons. Similarly, some parts of loamy and black cotton soils having LULC features are noted at lower risk of LD based on their suitability to plant growth under semiarid climatic conditions. However, the LULC features include unirrigated croplands, fallows, and barrens, which are highly prone to LD due to improper land management activities that become converted into eroded lands and salt-affected lands in various parts of the study area. In the eastern parts, the variety of LULC features leads to soil erosion and poor vegetation growth due to lower soil moisture, higher LST, and PET, and these areas are highly prone to LD from erosion, culminating in desertification. Overall observation reveals that the middle and eastern plains are the critical zone for LD, mainly in black cotton clay and calcareous clay soils. Although the western parts are noted at a lower risk of LD, these areas cover upland forest range, dense vegetation, plantations, and irrigated croplands resulting from good quality soils, rainfall, land use, and climatic conditions during all seasons. In the study area, the main driving forces to LD are considered to be (1) spatiotemporal variability of rainfall, soil moisture, climate, aridity, LST, PET, etc., and (2) human-induced factors like encroachments, mining, and quarrying, deforestation, overexploitation, and improper land use management, etc. The LULC features under a variety of anthropogenic pressures that lead to severe LD that affects the fragile environmental ecosystems in the area.

37.5 Conclusion

The MEDALUS approach is widely used for the assessment of LDVI based on the analysis of multiple geo-environmental parameters. MEDALUS approach can be effectively worked for protecting the fragile environment and ecosystems through data inadequacy, lack of in-situ data, and absence of field surveyed data. Furthermore, this approach is a flexible model, which allows users to modify the quality indicators at a site-specific scale. The GIS-based MEDALUS model is a powerful tool for the assessment of LD processes under dynamic environment and climate conditions and its time and cost-effective approach. This study uses the GIS-based MEDALUS model to assess and map LD vulnerability in the semiarid region of Virudhunagar district in Tamil Nadu, South India. The result of LDVI reveals that the area of 84.93% fall under critical zones due to LD in unirrigated croplands, fallows, barrens, and other lands with shrubs, and it becomes a major socioeconomic issue on account of the reduction of biomass and food productivity. However, the uplands forest and land with potential vegetative cover and irrigated croplands are noted in nonaffected zones due to the presence of adequate rainfall and suitable pedological and climatic factors. The LULC features under high risk can be reduced through the implementation of proper management practices on land, soil, and water resources. The GIS-based LDVI results provide a geo-database for planning sustainable land resources management on a long-term scale.

Acknowledgment

The authors are thankful to the Director, National Centre for Earth Science Studies (NCESS), Ministry of Earth Sciences (MoES), Government of India, for providing continuous support and encouragement. We extend our gratitude to IMD, SOI, NBSS & LUP, and USGS—Earth Explorer, for providing necessary data sources.

References

Abuzaid, A.S., Abdelatif, A.D., 2022. Assessment of desertification using modified MEDALUS model in the northern Nile Delta, Egypt. Geoderma 405, 115400. Available from: https://doi.org/10.1016/j.geoderma.2021.115400.

Afzali, S.F., Khanamani, A., Maskooni, E.K., Berndtsson, R., 2021. Quantitative assessment of environmental sensitivity to desertification using the modified MEDALUS model in a semiarid area. Sustainability 13, 7817. Available from: https://doi.org/10.3390/su13147817.

Ahmad, N., Pandey, P., 2018. Assessment and monitoring of land degradation using geospatial technology in Bathinda district, Punjab, India. Solid. Earth 9. Available from: https://doi.org/10.5194/se-9-75-2018.

Ait Lamqadem, A., Pradhan, B., Saber, H., Rahimi, A., 2018. Desertification Sensitivity analysis using MEDALUS model and GIS: a case study of the oases of Middle Draa Valley, Morocco. Sensors 18, 2230. Available from: https://doi.org/10.3390/s18072230.

Ajai, A.A.S., Dhinwa, P.S., Pathan, S.K., Ganesh, R.K., 2009. Desertification/land degradation status mapping of India. Curr. Sci. 97, 1478–1483.

Akbari, M., Shalamzari, M.J., Memarian, H., Gholami, A., 2020. Monitoring desertification processes using ecological indicators and providing management programs in arid regions of Iran.. Ecol. Ind. 111, 106011. Available from: https://doi.org/10.1016/j.ecolind.2019.106011.

Ali, R.R., El Baroudy, A.A., 2008. Use of GIS in mapping the environmental sensitivity to desertification in Wadi El Natrun depression, Egypt. Aust. J. Basic. Appl. Sci. 157–164.

Avni, Y., Porat, N., Plakht, J., Avni, G., 2006. Geomorphic changes leading to natural desertification versus anthropogenic land conservation in an arid environment, the Negev Highlands, Israel. Geomorphology 82. Available from: https://doi.org/10.1016/j.geomorph.2006.05.002.

Basso, F., Bove, E., Dumontet, S., et al., 2000. Evaluating environmental sensitivity at the basin scale through the use of geographic information systems and remotely sensed data: an example covering the Agri basin (Southern Italy). Catena 40 (99), 00062–00064. Available from: https://doi.org/10.1016/S0341-8162.

Bedoui, C., 2020. Study of desertification sensitivity in Talh region (Central Tunisia) using remote sensing, G.I.S., and the M.E.D.A.L.U.S. approach. Geoenviron. Disasters 7, 16. Available from: https://doi.org/10.1186/s40677-020-00148-w.

Budak, M., Günal, H., Çelik, I., et al., 2018. Environmental sensitivity to desertification in northern Mesopotamia; application of modified MEDALUS by using analytical hierarchy process. Arab. J. Geosci. 11, 481. Available from: https://doi.org/10.1007/s12517-018-3813-y.

Conacher, A., Sala, M., 1998. Land degradation in the Mediterranean environments of the world: nature and extent. Causes and Solutions. J. Wiley & Sons, Chichester, p. 491.

Dasgupta, A., Sastry, K.L.N., Dhinwa, P.S., Rathore, V.S., Nathawat, M.S., 2013. Identifying desertification risk areas using fuzzy membership and geospatial technique–a case study, Kota District, Rajasthan. J. Earth Syst. Sci. 122, 1107–1124.

Diallo, H.A., 2008. United Nations convention to combat desertification (UNCCD). In: Lee, C., Schaaf, T. (Eds.), The Future of Drylands. Springer, Dordrecht. Available from: https://doi.org/10.1007/978-1-4020-6970-3_4.

Dharumarajan, S., Lalitha, M., Hegde, R., et al., 2018. Status of desertification in South India - assessment, mapping, and change detection analysis. Curr. Sci. 115. Available from: https://doi.org/10.18520/cs/v115/i2/331-338.

Dutta, S., Chaudhuri, G., 2015. Evaluating environmental sensitivity of arid and semiarid regions in northeastern Rajasthan, India. Geogr. Rev. 105, 441–461.

Ferrara, A., Kosmas, C., Salvati, L., et al., 2020. Updating the MEDALUS-ESA framework for worldwide land degradation and desertification assessment. L Degrad. Dev. 31. Available from: https://doi.org/10.1002/ldr.3559.

Hanafi, A., Jauffret, S., 2008. Are long-term vegetation dynamics useful in monitoring and assessing desertification processes in the arid steppe, southern Tunisia. J. Arid. Environ. 72. Available from: https://doi.org/10.1016/j.jaridEnviron.2007.07.003.

Kaliraj, S., Chandrasekar, N., Ramachandran, K.K., 2016. Mapping of coastal landforms and volumetric change analysis in the south west coast of Kanyakumari, South India using remote sensing and GIS techniques. Egypt. J. Remote. Sens. Space Sci. 20 (2), 265–282. Available from: https://doi.org/10.1016/j.ejrs.2016.12.006.

Kaliraj, S., Chandrasekar, N., Ramachandran, K.K., Srinivas, Y., Saravanan, S., 2017. Coastal landuse and land cover change and transformations of Kanyakumari coast, India using remote sensing and GIS. Egypt. J. Remote. Sens. Space Sci. 20 (2), 169–185. Available from: https://doi.org/10.1016/j.ejrs.2017.04.003.

Kaliraj, S., Meenakshi, M.S., et al., 2012. Application of remote sensing in detection of forest cover changes using geo-statistical change detection matrices—a case study of devanampatti reserve forest, Tamil Nadu, India. Nat. Environ. Pollut. Technol. 11, 261–269.

Kaliraj, S., Parmar, M., Bahuguna, I.M., Rajawat, A.S., 2022. Assessment of desertification and land degradation vulnerability in humid tropics and sub-tropical regions of India using remote sensing and GIS techniques. In: Sajjad, H., Siddiqui, L., Rahman, A., Tahir, M., Siddiqui, M.A. (Eds.), Challenges of disasters in Asia. Springer Natural Hazards. Springer, Singapore. Available from: https://doi.org/10.1007/978-981-19-3567-1_2.

Kassas, M., 1976. Arid and semiarid lands: problems and prospects. Agro-Ecosystems 3. Available from: https://doi.org/10.1016/0304-3746(76)90120-7.

Kerényi, A., 1996. Soil degradation in India: status and impact. Geoderma 72. Available from: https://doi.org/10.1016/0016-7061(96)00025-0.

Kosmas, C., Kirkby, M., Geeson, N., 1999. Manual: key indicators of desertification and mapping environmentally sensitive areas to desertification. European Commission, Energy, Environment and Sustainable Development, EUR 18882, 87p.

Masoudi, M., Zakeri Nejad, R., 2010. Hazard assessment of desertification using MEDALUS model in Mazayjan plain, Fars province, Iran. Eco. Environ. Cons. 16 (3), 425–430.

Mohamed, E.S., 2013. Spatial assessment of desertification in north Sinai using modified MEDLAUS model. Arab. J. Geosci. 6, 4647–4659. Available from: https://doi.org/10.1007/s12517-012-0723-2.

Mohamed, E.S., Belal, A., Saleh, A., 2013. Assessment of land degradation east of the Nile Delta, Egypt using remote sensing and GIS techniques. Arab. J. Geosci. 6, 2843–2853. Available from: https://doi.org/10.1007/s12517-012-0553-2.

Muthulakshmi, L., Ramu, A., Kannan, N., 2010. Assessment of groundwater quality in Virudhunagar district (India): a statistical approach. J. Environ. Sci. Eng. 52, 229–234.

Okin, G.S., D'Odorico, P., Liu, J., 2018. A mechanism of land degradation in Turf-Mantled slopes of the Tibetan plateau. Geophys. Res. Lett. 45, 4041–4048.

Raja, V., Neelakantan, M.A., 2021. Evaluation of groundwater quality with health risk assessment of fluoride and nitrate in Virudhunagar district, Tamil Nadu, India. Arab. J. Geosci. 14 (52), 2021. Available from: https://doi.org/10.1007/s12517-020-06385-5.

Reynolds, J.F., Grainger, A., Stafford Smith, D.M., Bastin, G., Garcia-Barrios, L., Fernandez, R.J., 2011. Scientific concepts for an integrated analysis of desertification. Land Degrad. Dev. 22, 166–183.

Salunkhe, S.S., Bera, A.K., Rao, S.S., Venkataraman, V.R., Raj, U., Krishna Murthy, Y.V.N., 2018. Evaluation of indicators for desertification risk assessment in part of Thar Desert Region of Rajasthan using geospatial techniques. J. Earth Syst. Sci. 127, 116. Available from: https://doi.org/10.1007/s12040-018-1016-2.

Salvati, R., Salvati, L., Corona, P., et al., 2015. Estimating the sensitivity to desertification of Italian forests. Iforest 8. Available from: https://doi.org/10.3832/ifor1111-008.

SAC, 2021. Desertification and Land Degradation Atlas of India (Assessment and Analysis of Changes Over 15 years based on Remote Sensing). Space Applications Centre, ISRO. Ahmedabad, India. 282 pp.

Sepehr, A., Hassanli, A.M., Ekhtesasi, M.R., et al., 2007. Quantitative assessment of desertification in south of Iran using MEDALUS method. Environ. Monit. Assess. 134, 243–254. Available from: https://doi.org/10.1007/s10661-007-9613-6.

Singh, G., 2009. Salinity-related desertification and management strategies: Indian experience. Land. Degrad. Dev. 20, 367–385. Available from: https://doi.org/10.1002/ldr.933.

Sivasankari, R., Siva Vignesh, N., 2017. Study of land use and land cover (LULC) dynamics using remote sensing and geographic information system (GIS) in Virudhunagar district, Tamilnadu, India. Int. J. Civ. Eng. Technol. 8 (12), 744–750.

Song, C., Kim, W., Kim, J., Gebru, B.M., Adane, G.B., Choi, Y.E., et al., 2022. Spatial assessment of land degradation using MEDALUS focusing on potential afforestation and reforestation areas in Ethiopia. Land. Degrad. Dev. 33 (1), 79–93. Available from: https://doi.org/10.1002/ldr.4130.

UNCCD, 2014. Land degradation neutrality: Resilience at local, national and regional levels. Bonn. United Nations. Available from: http://library.unccd.int/Details/fullCatalogue/858

UNEP, 1992. World Atlas of Desertification, Editorial Commentary by N. Middleton and D.S.G. Thomas. Arnold, London.

Uzuner, Ç., Dengiz, O., 2020. Desertification risk assessment in Turkey based on environmentally sensitive areas. Ecol. Indic. 114, 106295.

Wijitkosum, S., 2021. Factor influencing land degradation sensitivity and desertification in a drought prone watershed in Thailand. Int. Soil. Water Conserv. Res. 9 (2), 217–228. Available from: https://doi.org/10.1016/j.iswcr.2020.10.005.

Chapter 38

Impacts of rock quarrying on land use and soil environment of a tropical mountain landscape in south Kerala, India

R.G. Rejith, S. Anirudhan, S. Abinash, Pribin Mathew Thomas and R.S. Ashikraj
Environmental Resources Research Centre (ERRC), Thiruvananthapuram, Kerala, India

38.1 Introduction

Potency, sturdiness, architectural adaptability, and artistic fulfillment make stone or hard rock materials essential for construction (Desai, 1979). Such resources are boundless, and workable deposits of one type or another may be found in Kerala. The invention of sufficient tools and mechanisms for breaking and shaping rock blocks initiated quarrying activities. With the increasing demand and use of rocks, mining or quarrying became a popular industry, and people began to explore more areas for diverse rock types for multiple uses. Thus, the value of rock types depends on their properties, esthetic appearance, and availability. With the growth of quarries and quarrying activities, it was realized that some stones could be extracted and fashioned with better facilities than others, that some stones afforded larger or sounder blocks, and that some stones would afford material of more outstanding architectural and esthetic merit. Even deposits of the same types were found to differ among themselves with respect to these qualities, mainly for geological reasons. Hard crystalline rocks abundant in Kerala, which are matchless construction materials, are Kerala's ubiquitous mineral resources (grouped as minor minerals) (Anand and Reddy, 2015). These rocks are mainly metamorphic types of Precambrian origin (Soman, 1987). They are primarily gneisses of varied types like garnet−sillimanite gneiss or khondalite and garnet-ferrous biotite gneiss. In addition, some igneous intrusions, such as dolerite and granite, are present in some places. Kerala's most widely spread rock type is a brown to gray variety with hypersthene and gray quartz called charnockite.

All these rocks are used as building stones and are extensively quarried, especially in areas where the rock is exposed or found at shallow depths. The dimension stone industry is highly refined and requires state-of-the-art rock cutting machinery. So far, in Kerala, such an industry has yet to pick up because of stiff resistance from people against the greedy excavation process that would spoil the ecology and environment. Nevertheless, a vast quantity of stones is being quarried mainly for the construction of buildings, making coastal protection structures, road pavements, sand-sized particles a substitute for river sand, etc. Therefore, rocks or stone blocks are regularly needed as long as construction works are indispensable. Kerala has witnessed pervasive quarrying activity for the last couple of decades because of the massive demand in the construction sector, which has left the land with several open pits, some of which are as large as a crater. Because the demand is infinite, if freedom is given for incessant quarrying, the whole western ghat region would be flattened or modified into several craters within a few hundred years. The vastness of the problems associated with quarrying activities is manifold and warrants a detailed study of the quarries. This work focuses on the generation of baseline data for quarries abandoned or active in the Thiruvananthapuram, Kollam, and Pathanamthitta districts.

Stone quarrying produces enormous dust emissions, and waste by-products have environmental effects and adverse impacts on soil fertility (Kumar et al., 2017). Most studies reveal the deterioration of soil quality near the quarries by reporting high levels of soil quality indicators, physicochemical parameters, soil nutrients, and metals. Significantly high levels of Cu, Zn, Ni, Pb, Cd, Fe, and Mn were recorded in soil samples collected near quarry and rock-crushing sites (Igwenagu-Ifeanyi et al., 2021). Slightly high pH and increased EC and TDS for soil samples were observed near

Remote Sensing of Soils. DOI: https://doi.org/10.1016/B978-0-443-18773-5.00005-3

sandstone quarrying (Singhal et al., 2020). Due to increased pH values, more acidic soil was observed adjacent to stone quarrying during the rainy season (Kalu and Ogbonna, 2021). Very few studies have been conducted on accounting for stone quarries in Kerala based on different purposes and objectives. Detailed analyses of environmental effects, slope stability, impacts on nearby residents, etc., were carried out on stone quarries in the Muvattupuzha River Basin (Maya et al., 2012), Narippara area, Banasuramala of Wayanad district (Sajinkumar et al., 2014), Netravati–Gurpur and Periyar–Chalakudy river basins (Vandana et al., 2020), and Panniar catchments of the Periyar River Basin (Ramdev et al., 2013). The advent of geospatial tools and remote sensing techniques has led to a breakthrough in studies related to stone quarries. An initial survey of quarries in Kerala was conducted using Google Earth imagery and GIS software in 2017 (Alex and Sajeev, 2017). Later, a few studies were conducted, for instance, on the change in the areal extent of stone quarries and its correlation with landslides in Malappuram District using Google Earth imagery and QGIS desktop software (Biju et al., 2021; George et al., 2022). Furthermore, a GIS-based analysis was carried out on environmental effects due to stone quarrying in the Panniar catchments of the Periyar River Basin (Ramdev et al., 2013). No detailed studies have been conducted in Kerala on the impact analysis of stone quarrying on the soil environment.

The main objective of the present work is to develop a comprehensive geospatial database of granite quarries (building stone) in the Thiruvananthapuram, Kollam, and Pathanamthitta districts of Kerala, South India. A GIS-based assessment of quarrying impacts on adjacent land use and soil was also conducted.

38.2 Study area

Thiruvananthapuram, Kollam, and Pathanamthitta districts form a significant portion of Southern Kerala and are bounded on the west by the Arabian Sea (Lakshadweep Sea) and on the east borders by the Western Ghats (Fig. 38.1). Geographically, Pathanamthitta is located at latitudes 9° 29′ 2.094″N–9° 4′ 19.5708″N and longitudes 76° 28′

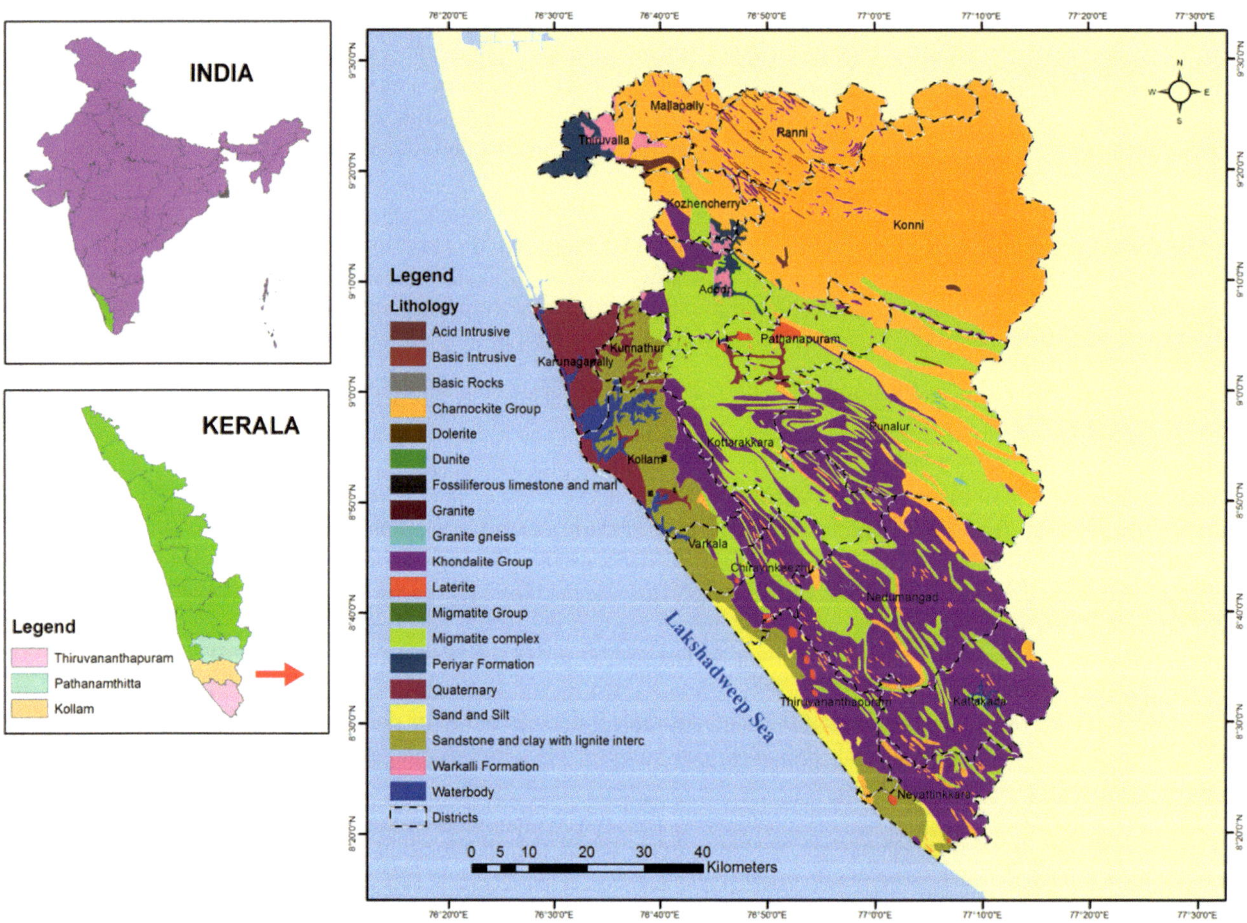

FIGURE 38.1 Study area Map.

24.2832″E−77° 16′ 59.9232″E; Kollam is located at latitudes 9° 10′ 22.2276″N−8° 45′ 41.166″N and longitudes 76° 27′ 49.0716″E−77° 15′ 45.792″E; Thiruvananthapuram is located at latitudes 8° 51′ 44.7372″N−8° 17′ 30.2748″N and longitudes 76° 40′ 22.3068″E−77° 17′ 2.8284″E. The soil formations in these three districts are coastal alluvium, forest soil, grayish onattukara, laterites and hydrogeomorphic association, red loam, and riverine alluvium. Red loam is found only in the south of Thiruvananthapuram district. Physiographically, the Kerala state is divided into highland, midland, and lowland according to their elevations of >75, 7.5−75, and <7.5 m, respectively, above MSL. The Kollam district is divided into four microzones: the Kollam coast, Adoor rolling plain, Kottarakara undulating upland, and Kulathupuzha forest hill (Sheela et al., 2017). The study area covers the Pathanamthitta District in Kerala state, India, divided into 6 taluks, viz., Thiruvalla, Mallapally, Ranni, Kozhenchery, Konni, and Adoor, and 70 villages. Kollam Revenue District is divided into 6 Taluks, viz., Kollam, Kunnathur, Karunagappally, Kottarakkara, Pathanapuram, and Punalur, and 104 villages. Thiruvananthapuram district is divided into 6 taluks—Thiruvananthapuram, Nedumangadu, Chirayinkeezhu, Kattakada, Neyyattinkara, and Varkala—and 124 villages. Geologically, the Pathanamthitta district comprises the (1) charnockite group, khondalite group, and megmatite complex of Archean age, (2) acid intrusive rocks of Proterozoic age, and (3) basic intrusive, Neogene, and quaternary rocks of Cenozoic age. The Kollam district comprises western Quaternary alluvial deposits, late Tertiary sediments, and the easternmost Precambrian metamorphic. The Thiruvananthapuram district is also occupied by eastern Archean crystalline rocks and the western coastal periphery of Tertiary and Quaternary sediments. The average annual rainfall of Pathanamthitta is 3133.9 mm, Kollam is 2428 mm, and Thiruvananthapuram is 2035 mm. About 66% of Thiruvananthapuram is agricultural land, followed by Kollam (56%) and Pathanamthitta (37%). About 8% built-up area per geographic area for Thiruvananthapuram and 7% for Pathanamthitta (Sheela et al., 2017). About 50% of the Pathanamthitta district is covered by forest cover of 1385.27 km^2. The Pathanapuram, Anchal, Kottarakkara, and Chadayamangalam blocks in the Kollam district have large forest areas. The reserve forest area in the Thiruvananthapuram district is 495.145 km^2 (DMG, 2016).

38.3 Materials and method

38.3.1 GPS survey and mapping of quarries

Physical verification of quarries was performed in three districts by a team of experts in the field of geology with the help of a taluk-wise base map prepared using Google Earth imagery and data collected from the website of the Directorate of Mining and Geology, Govt. of Kerala. After a detailed field survey, the quarry locations were plotted in Google Earth Imagery, and the quarry polygon layer was digitized using QGIS Desktop 3.4.1 software. The quarry location and extent will be documented as polygon features, and the area will be calculated.

38.3.2 Mapping of the LULC and soil

The land use/land cover within each quarry's 500 m buffer area was digitized using the Google Earth imagery available in QGIS Desktop 3.4.1 software. The geology of the quarries was identified and explained using a geological map from the Geological Survey of India. The impact of quarrying activities on surrounding land use (within a 500 m buffer zone) was studied by selecting the Veliyam quarry in the Kollam district for 2002 and 2020. The soil types of the study area were mapped using the soil map from the ICAR-National Bureau of Soil Survey and Landuse Planning. The soil fertility status was analyzed using four indicators such as pH, organic carbon (OC), available phosphorous (P), and available potassium (K). Panchayath-level maps indicating the fertility status of soils collected from the Kerala Soil Health Information System, Government of Kerala (https://www.keralasoilfertility.net/) were used for the analyses.

38.4 Results and discussion

38.4.1 GIS-based mapping and estimation of granite quarries

38.4.1.1 Field verification and mapping of quarries

A map showing the spatial distribution of quarries is given in Fig. 38.2. About 684, 327, and 635 quarries, respectively, were identified in Pathanamthitta, Kollam, and Thiruvananthapuram districts. The maximum number of quarries in Pathanamthitta are located in Konni taluk and the least in Ranni. The maximum number of quarries in Kollam occurs in the Kottarakkara taluk and the least in Kunnathur. The maximum number of quarries in Thiruvananthapuram occurs in the Nedumangad taluk and the least in Kattakada taluk. The major quarry types seen in these districts are (1) active,

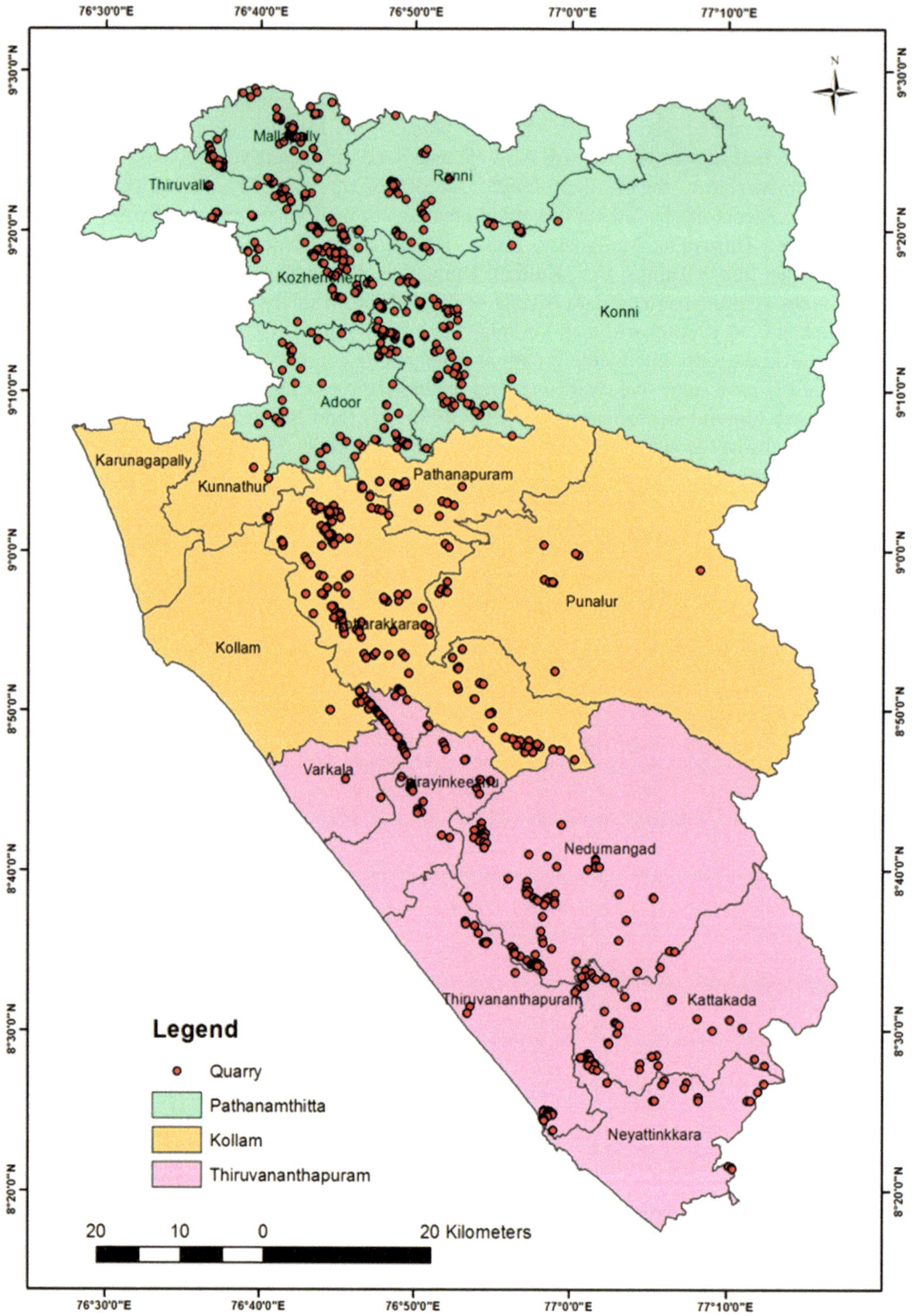

FIGURE 38.2 Spatial distribution of Quarries in each district.

(2) temporarily stopped, (3) abandoned, and (4) special purpose. Since most of the quarries are clustered, more than one type of quarry can be observed on the same site. Detailed statistics on quarry types are given in Table 38.1. From the table, it is clear that most of the quarries in the three districts are abandoned, either water-filled with crushers or spotted with plantation lands. More active quarries are noticed in Thiruvananthapuram, and most abandoned quarries are observed in the Pathanamthitta district. Active quarries are easily identifiable on Google Earth by noticing fresh rocks. Most of the active quarries also have crushers.

TABLE 38.1 Type of quarries that exist in the three districts.

Districts	Active	Temporarily stopped	Abandoned	Special Purpose	Abandoned/ temporarily stopped	Active/ abandoned/ temporarily stopped	Total
Pathanamthitta	70	35	557	3	7	12	684
Kollam	32	2	293	–	–	–	327
Thiruvananthapuram	209	11	416	–	–	–	635

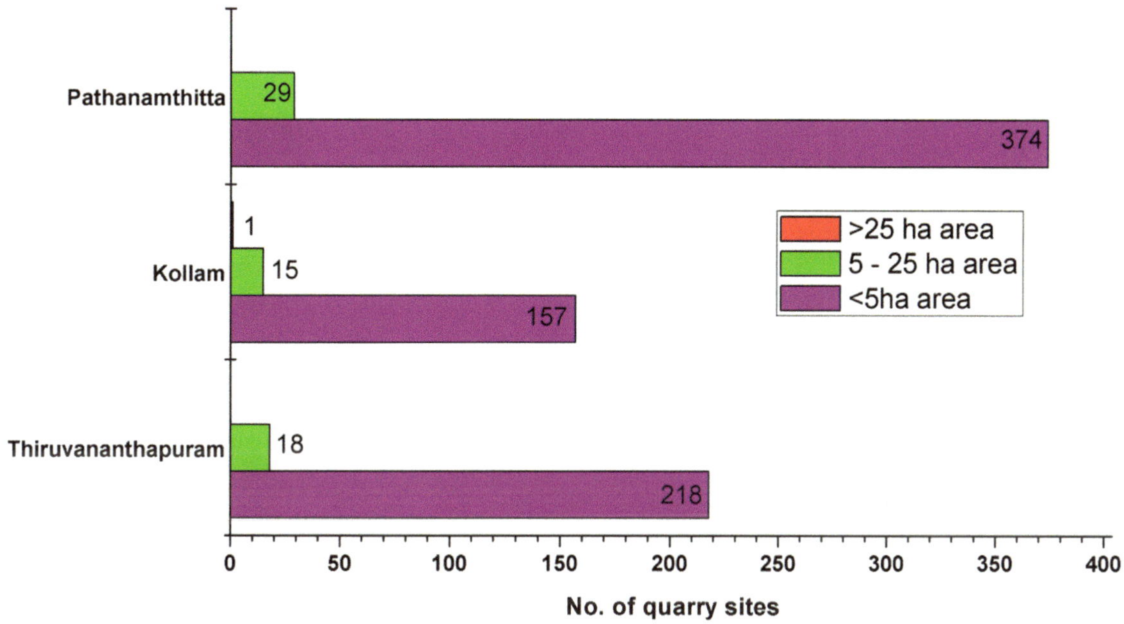

FIGURE 38.3 Area statistics of quarry sites.

Quarry size is indicated by the extent and depth of the quarries. The area of quarries was grouped into three categories, and the number of quarries counted in each category is shown in Fig. 38.3. The majority of quarries have an area less than or equal to 5 ha. There is only one quarry in the Kottarakkara taluk of Kollam with an area >25 ha, which is known as the Veliyam quarry. About 374, 157, and 218 quarry sites in Pathanamthitta, Kollam, and Thiruvananthapuram districts, respectively, have an area less than or equal to 5 ha. Within a range of 5–25 ha, only 29, 15, and 18 quarries, respectively, are noticed in Pathanamthitta, Kollam, and Thiruvananthapuram.

Generally, rock quarrying is performed from the foothill side of a hillock because the topography everywhere in the midland is a rolling type. Although the land of Kerala is geologically underlined by Precambrian crystalline rocks, quarrying is not possible everywhere, but it is preferred in suitable locations determined by factors such as depth of weathered zone, accessibility, and nature of rock types. Usually, denudational hills with a minimum depth of overburden are considered suitable locations. In the low midland portion, such hillocks have a mean altitude of 50 m, but in the higher midland, the hillocks have steep slopes, and the altitude goes up to 75 m. Quarrying starts from the foot slope, and when cutting proceeds toward the mid-slope and crest slope, open pits are usually left with a depth of 5–15 m from the starting point and surrounded by a cliff or quarry wall (circular or linear). Quarrying works for building rock excavation in the plane ground are rare in Kerala. Most of the rock quarries exist in the midland rather than in the highland part of the Trivandrum district because the former topography is more accessible and nearer to the place of demand/construction than the latter.

The spatial distribution of quarries over the geological units is shown in Fig. 38.4. The majority of quarries extract the migmatite complex. Migmatite is a heterogeneous metamorphic layer similar to a rock, with alternate bands of light

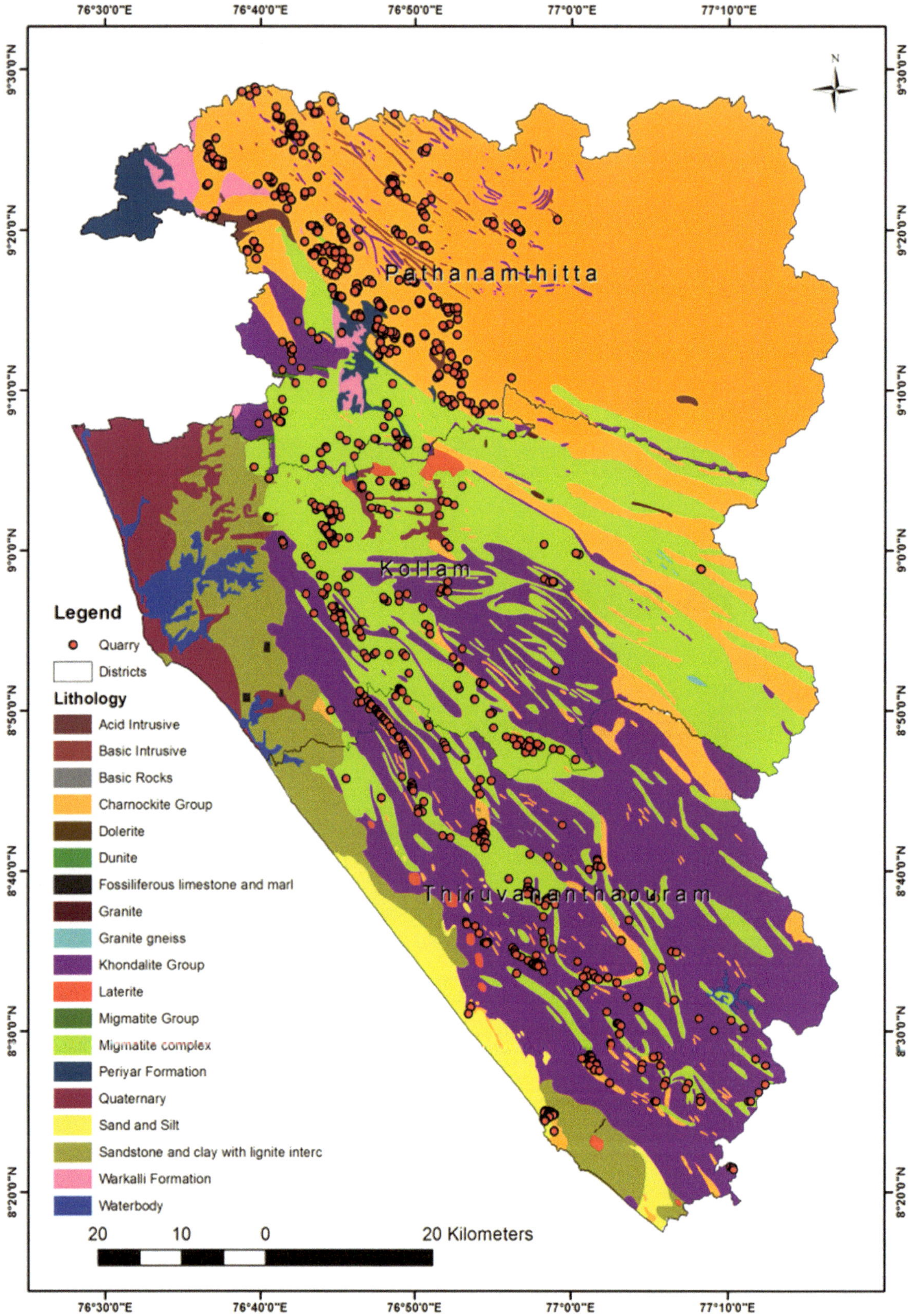

FIGURE 38.4 Geology of the study area with quarry locations.

and dark minerals formed by anatexis that preserve evidence of partial melting. It is a costly rock in the dimensional stone industry and is known as Paradiso. In many places, this rock is interlayered with khondalite. Few quarries also extract charnockite, which is highly valued in the building stone market. In many quarries, there is not much variation in rock types with respect to depth; however, textural changes occur in some parts. Sometimes intrusives of pegmatite or dolerite rocks appear. The rock types encountered in the quarries of the Thiruvananthapuram district are mainly metamorphic rocks of the Precambrian age. These are gneisses and charnockite. The geological map indicates the types of rocks exposed on the surface. In quarries at depth, we can notice the variety of types of gneisses and their extent. In several quarries, the gneisses are intruded by charnockite or dolerite.

38.4.2 Impacts of quarries on land use and land cover

38.4.2.1 LULC within 500 m buffer zone of quarries

The Land use/Land cover (LULC) classes within the 500 m buffer zone of quarries are given in Fig. 38.5. The area corresponding to each class is given in Table 38.2. The Pathanamthitta district of Kerala covers an area of 265765 ha, consisting of 8 blocks, 3 municipalities, and 6 taluks. A vast region of the Pathanamthitta district is covered by forest (1533.79 km^2), which favorably sustains the climate and rainfall of Kerala. The most critical land use classes are built-up land, agricultural land, forest, wastelands, and waterbodies. Achankovil, Manimala, and Pamba are the major rivers in the district. Pathanamthitta is landlocked, with no coastal line or backwaters. About 20166.63 ha of the area was classified into 12 land use/land cover categories and surveyed for further investigation.

The plantation or agricultural land may contain a single crop or multiple crops. This land is mainly used to cultivate and produce food, fiber, and other commercial products (Ganesh et al., 2018). Among the 12 classes, about 12343.42 ha of the area falls under the mixed plantation class, which is 61.21% of the total area. About 13.45% of the surveyed area falls into the habitation category, comprising buildings, transport facilities, and communication infrastructures. About 2713.26 ha area comes under this class. About 1886.08 ha carry natural vegetation grown without human help, covering 9.35% of the total area. About 740.76 ha of the surveyed area is a forest, predominantly consisting of trees, timber, and other produce. It represents 3.67% of the total geographical area. An area where a single crop is cultivated is the plantation mono class, which occupies 607.01 ha and accounts for 3.01% of the total area. Paddy cultivation is carried out on 537.82 ha, 2.67% of the geographical area. About 2.23% of the total surveyed area is in the quarry category. These may be active, abandoned, or water-filled quarries. About 450.52 ha of area fall into this class. About 2.08% contains barren land formed due to a lack of suitable water and soil management practices. This occupies an area of 419.25 ha. Rock exposures are seen in certain parts of the surveyed area, which is about 286.47 ha accounting for 1.42% of the geographical area. Scrub, waterbodies, and waterlogged areas constitute 0.9% of the surveyed area. About 182.03 ha fall under this category. About 83.44 ha of the area includes scrub. The part of the Earth's surface covered with water is known as the waterbody. About 95.07 ha comprise this class in the surveyed area. About 3.52 ha of the area is waterlogged, which occurs when the soil profile or a plant's root zone becomes saturated. Among the 12 land use/land cover categories, mixed plantation comprise the major part, and water-logged regions constitute the most minor portion.

The Kollam district, with an area of 2, 48,788 ha, comprises 11 blocks, 3 municipalities, and 1 urban corporation area. Land use categories include plantations, built-up areas/habitation, waterbodies, etc. Kallada and Ithikkara are the two main rivers that drain through Kollam. Land use/land cover change pattern assessment is essential for planning, utilizing, and managing land resources.

Among the 11 land use/land cover classes surveyed, 60.61% falls under mixed plantation. About 6769.91 ha falls under this category. Plantation mono is about 965.78 ha, 8.63% of the monitored area. It contains a single type of crop. About 11.82% of the area has habitation, which comprises 1322.23 ha. It includes buildings, transport utilities related to communication, water, vegetation, and vacant lands. About 7.96% of the area contains natural vegetation. About 890.29 ha falls under this category. Barren land occupies 354.43 ha, 3.17% of the total area. These are areas devoid of vegetation. Paddy cultivation is moderately observed in the surveyed area. About 364.34 ha comprise paddy land, 3.26% of the total area. About 2.59% of the area under survey falls under the quarry class. Quarries that show the appearance of fresh rocks are active. Some active quarries contain waterbodies in them. Some are abandoned quarries that may or may not include waterbodies. About 290.10 ha of Kollam occupies the quarry region. Rock exposures are present in different parts of the Kollam district. It is 1% of the total area, which occupies 111.36 ha. Rocks of different lithologies are visible here, without any soil or vegetation cover. Scrub, waterbodies, and waterlogged areas constitute 1.07%.

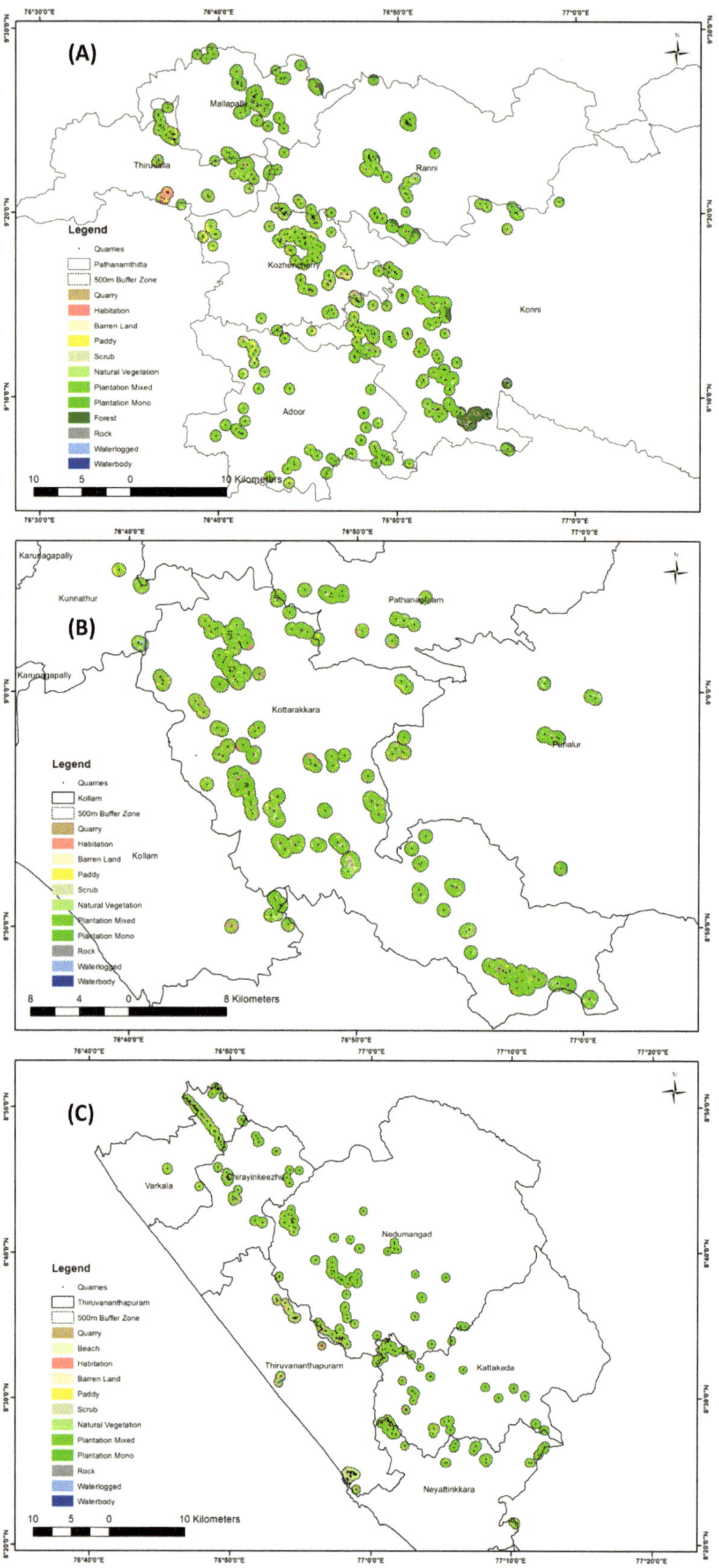

FIGURE 38.5 Spatial distribution of land use classes within 500 m buffer zone of quarries. (A) Pathanamthitta; (B) Kollam; (C) Thiruvananthapuram.

TABLE 38.2 Land use classes and their area within a 500 m buffer zone of quarries.

Sl No.	Land use/land cover	Area (Hectares)				Area (%)
		Pathanamthitta	Kollam	Thiruvananthapuram	Total	
1	Quarry	450.52	290.1	431.46	1172.08	3.44
2	Barren land	419.25	354.43	284.68	1058.36	3.11
3	Beach	–	–	0.97	0.97	0.00
4	Habitation	2713.26	1322.23	2069.88	6105.37	17.93
5	Natural vegetation	1886.08	890.29	1170.86	3947.23	11.59
6	Paddy	537.82	364.34	229.33	1131.49	3.32
7	Plantation mixed	12343.42	6769.91	8843.4	27956.73	82.09
8	Plantation mono	607.01	965.78	504.17	2076.96	6.10
9	Rock	286.47	111.36	164.56	562.39	1.65
10	Scrub	83.44	78.73	73.67	235.84	0.69
11	Waterbody	95.07	29.65	105.19	229.91	0.68
12	Waterlogged	3.52	11.14	9.4	24.06	0.07
13	Forest	740.76	–	–	740.76	2.18
	Total	20166.63	11187.97	13887.57	34054.2	

TABLE 38.3 Land use area statistics correspond to the years 2002 and 2020.

Land use	Area (ha)	
	2002	2020
Barren land	7.08	2.12
Habitation	4.45	29.76
Natural vegetation	196.50	171.84
Paddy	7.67	4.70
Quarry	21.26	30.88
Rock	2.50	0.82
Waterbody	1.94	1.29

Scrub is degraded vegetative cover found in uplands or high grounds. About 78.73 ha area includes scrub. Waterbodies include ponds, lakes, and those waterbodies associated with quarries. These are both natural and man-made features that impound water along a boundary. About 29.65 ha fall under this class. Poor drainage of the area results in the generation of water-logged regions. A deep penetration of water is not possible in such regions. Excessive irrigation can cause waterlogging. About 11.14 ha fall under this category. Among the entire land use land cover classes, mixed plantation occupies a significant portion, while the waterlogged region is the least.

Thiruvananthapuram, the southernmost district of Kerala, has an estimated 2192 km^2, comprising 11 blocks, four municipalities, and one corporation. According to studies based on physical features, Thiruvananthapuram is divided into highlands, midlands, and lowlands (Padmalal et al., 2008). The Attingal–Neyyattinkara undulating upland, Ponmudi–Agasthyar forested hills, and coastal plain are the major physiographic features of the Thiruvananthapuram district. Land use categories include plantations, built-up areas/habitation, waterbodies, etc. Neyyar, Karamana, and Vamanapuram are the three major rivers that drain through Thiruvananthapuram.

About 13887.57 ha of area are divided into 12 land use/land cover classes and are surveyed for the land use/land cover change assessment. It contains multiple crop types. Among the 12 land use/land cover classes, 68.63% come under the mixed plantation category, which is 8843.40 ha in area. About 14.9% of the surveyed area is inhabited. About 2069.88 ha of area fall into this category. About 1170.86 ha contain natural vegetation grown naturally without human aid and are present in 8.43% of the total area. Plantation mono contains a single crop type, accounting for about 3.63% of the surveyed area. About 504.17 ha area comes under this class. About 431.46 ha of Thiruvananthapuram occupy the quarry area, which is 3.11% of the total area. Quarries showing rocks with a fresh appearance are active quarries, and those with a stale appearance are abandoned quarries. About 2.05% of the total area carries 284.68 ha of barren land. It is a soil-dominated area with little to no plant cover. Paddy cultivation is comparatively less in the surveyed areas. About 229.33 ha accommodate paddy cultivation, which is only 1.65%. About 164.56 ha of the area shows rock exposures of changing lithology, free of soil cover and vegetation. It is 1.18% of the observed land. About 0.97 ha of the area comes under the beach. It is present in only 0.01%. Scrub, waterbodies, and waterlogged areas constitute 1.36% of the surveyed area. About 188.26 ha of the area falls under this category. About 73.67 ha include scrubs. Waterbodies may be natural or man-made. Rivers, streams, lakes, canals, etc. are various waterbodies. About 105.19 ha occupy this category. About 9.4 ha of the surveyed area comes from a waterlogged region. It is caused by the soil saturation with water, so it cannot absorb any more water. Among the 12 land use/land cover categories, the buffer zones of the quarries are mainly surrounded by mixed plantations and are least surrounded by beaches.

38.4.2.2 LULC change detection and analysis

The impact of quarrying on adjacent land use was studied by selecting the Veliyam quarry as one of the case studies. The land use classes within the 500 m buffer zone for the Veliyam quarry were mapped for the years 2002 and 2020. The land use area statistics and maps for 2002 and 2020 are shown in Table 38.3 and Fig. 38.6. The results reveal a drastic change in land use over 18 years. A significant decrease occurred in a few land use classes, such as barren land

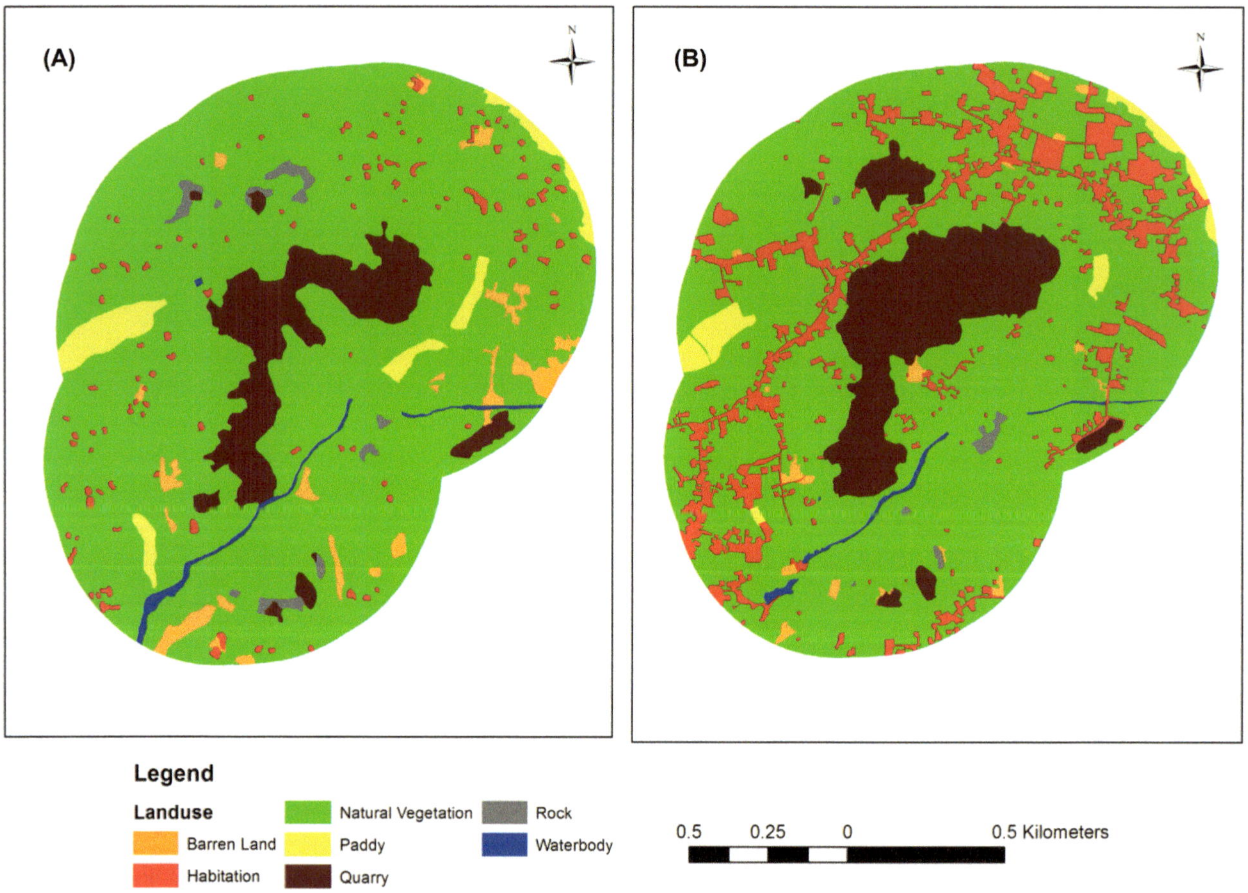

FIGURE 38.6 Land use classes with a 500 m buffer zone of Veliyam quarry for the years (A) 2002 and (B) 2020.

from 7.08 to 2.12 ha, natural vegetation from 196.50 to 171.84 ha, paddy fields from 7.67 to 4.70 ha, and rock from 2.50 to 0.82 ha. On the other hand, there was a remarkable increase in habitation and quarrying. The habitation increased from 4.45 to 29.76 ha, and quarries increased from 21.26 to 30.88 ha. The results conclude that the land once occupied by vegetation, paddy fields, and barren land was used for constructing houses and other built-up. A significant rise in quarrying activities over the last eighteen years. The map, as shown in Fig. 38.6, supports these findings. After 18 years, the quarries became abandoned and water-filled because of continuous quarrying activities. The impacts caused by quarrying activities can be categorized into on-site and off-site impacts, where land use change is one of the significant issues addressed under on-site impacts (Vandana et al., 2020). Vandana et al. and Sajinkumar et al. classified land use under the categories of "extremely high impact" and "appreciably adverse" due to quarrying activities (Sajinkumar et al., 2014; Vandana et al., 2020).

Activities such as road construction to quarries for transportation, construction of built-ups, clearance of land and vegetation, the building of crusher units, and debris dumping are the major modifications done along with quarrying. This may cause severe changes to natural land cover, and vegetation, and deformations such as landscape scars. In the present study, it is well understood that within the 500 m buffer zone of the quarries, the primary land use class is mixed plantation. Mixed plantation refers to mixed groups of different varieties of plantation crops cultivated extensively in a contiguous area. The major crops in Thiruvananthapuram, Kollam, and Pathanamthitta include coconut, rubber, pepper, cassava, banana, cashew, etc. (Guillerme et al., 2011; Srinivasan, 2014; Ashokkumar, 2020). Quarrying activities may adversely affect the growth and development of these crops, thereby reducing crop yield and production. It has already been proven that stone quarries affect the pigment concentration and productivity of paddy (Naik et al., 2006), grassland degradation (Balcha and Oyda, 2021), woody plant species diversity (Belay et al., 2020), etc. About 17.93% of land use within the 500-m buffer zone of the quarries is occupied by habitation. The present analysis was conducted within a buffer zone of 500 m and the results provide baseline information to the governing authorities for sanctioning quarries in the future.

38.4.3 Impacts of quarries on soil

38.4.3.1 Soil formations and their instabilities

As per the records of the ICAR-National Bureau of Soil Survey and Landuse Planning, six soil types are present in these three districts, viz., coastal alluvium, forest soil, grayish onattukara, laterites and brown hydrogeomorphic association, red loam, and riverine alluvium. The distribution of quarries over different soil types is shown in Figs. 38.7 and 38.8. Most quarries in all three districts fall into the Laterites and Brown Hydromorphic association soil type. Laterites are weathered products of hard rocks. Over the years, the hard rocks have undergone several courses of weathering and several mineral transformations under the contributing factors of rainfall and high temperatures. Red soils occur in the catenary sequence sideways on the foot slopes of laterite hills. Brown hydromorphic soil is the washed-down alluvium from adjoining uplands deposited in depressions or bottom areas between hills and hillocks. It can be seen throughout the state.

A few field photographs showing the soil formations and their instabilities near hard rock quarries in this study area are shown in Fig. 38.9. The pictures depict the weathering of rocks and thereby the formation of laterite soil and their instabilities. Quarries located on slopes steeper than 18 degrees have high chances of landslide occurrences, soil erosion, and mass movements (Vandana et al., 2020). Apart from steep slopes, delicate lithology and controlled drainage are other causative factors (Abdallah et al., 2005). The unstable soil layer is a characteristic feature of blasting induced in hard rock quarries. Vandana et al. and Sajinkumar et al. categorized soil cover into "extremely high impact" and "moderately adverse" due to quarrying-related activities and debris dumping. These may cause slumping and loss of topsoil, depletion of macro- and microsoil nutrients, variation in soil characteristics, and deforestation. Loss of soil and vegetation cover due to quarrying results in habitat fragmentation and loss (Vandana et al., 2020).

38.4.3.2 Soil fertility status

The soil fertility indicators used in this study were OC, pH, available P, and available K. Panchayath-wise soil fertility maps with quarry sites are shown in Fig. 38.10. As per the data collected from the Kerala Soil Health Information System, there are three classes such as low, medium, and high for OC, available P, and available K. The pH has four classes defined as extremely acidic, strongly acidic, moderately acid, and neutral and alkaline, assigned to pH ranges of <4.5, 4.5 to 5.5, 5.6 to 6.5, and >6.5. In the case of OC, the maximum quarry sites are seen in the medium region. About 459 quarry sites are seen in the medium, followed by 228 in the high and 126 in the low regions. The reduction

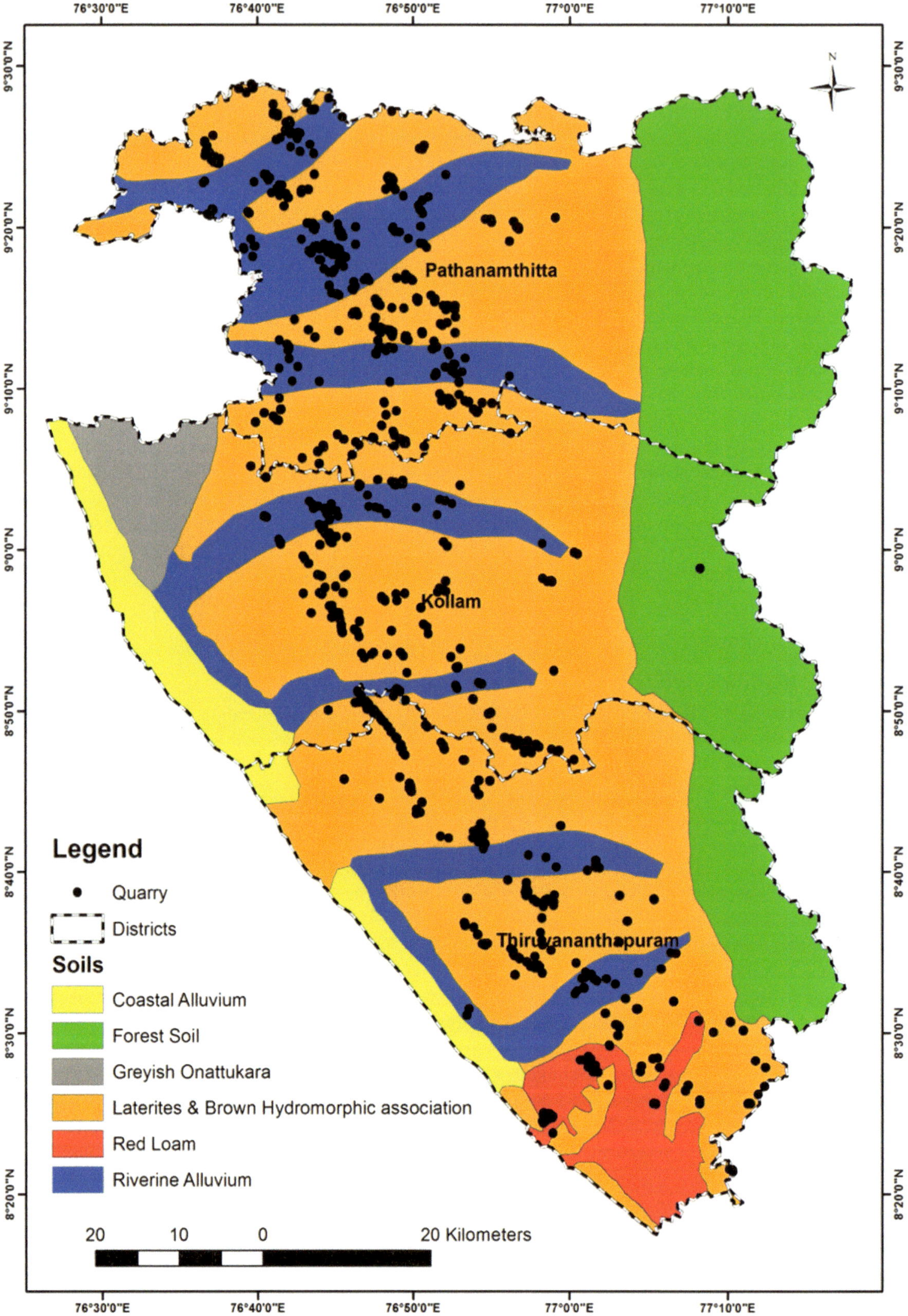

FIGURE 38.7 Soil distribution map.

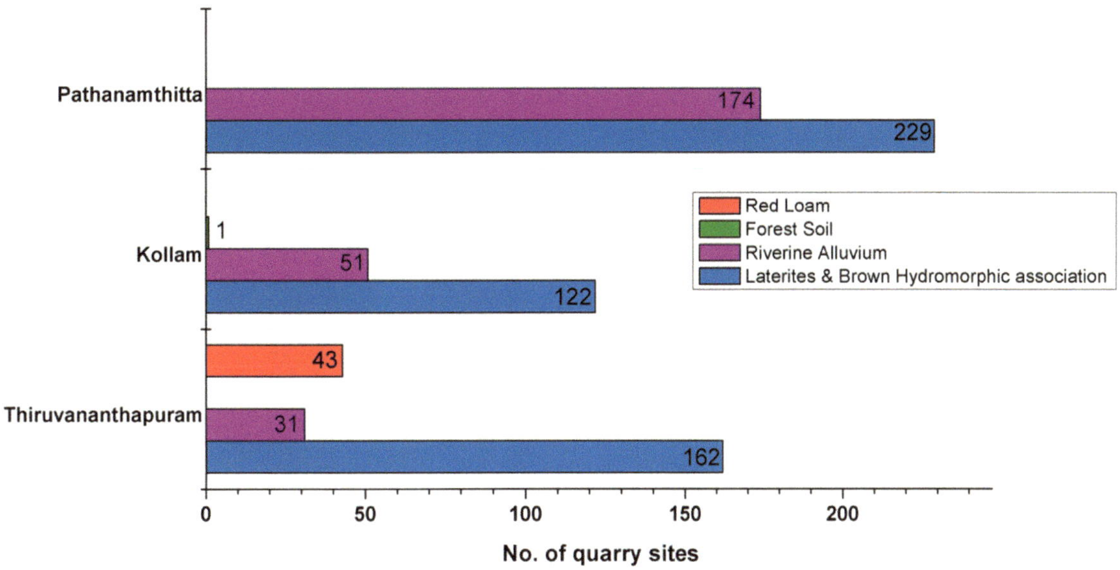

FIGURE 38.8 Soil distribution statistics concerning quarry sites.

FIGURE 38.9 Field photographs showing the soil formation and its instabilities near hard rock quarries.

in OC near quarry sites may be attributed to the addition of quarry dust (Naik and Ushamalini, 2006). In the case of available P, maximum quarries of about 534 are seen in high P content regions, followed by 235 in medium and 44 in low regions. Paramesha et al. confirmed that calcium phosphate formed from rocks in quarries increases the phosphorous content (Naik and Ushamalini, 2006). The available K shows the maximum for medium regions with 408 quarry sites, followed by 286 quarries in high regions and 119 quarries in low regions. Regarding pH, about 458 quarries are observed in strongly acidic regions, followed by 303 quarry sites in moderately acidic regions, and 452 in neutral regions. In addition to this, about ten quarries are located in extremely acidic regions of the Pathanamthitta district. Olatunde et al. stated that continued dust deposition with chemical content from stone quarries might result in severe acidity in nearby soil environments (Olatunde et al., 2021). Quarry wastes, blasted materials, chemicals used for blasting, and dust (rock powder) are significant threats to adjacent soil pollution (Rani et al., 2017). Several studies have shown that stone quarrying activities might adversely affect soil physicochemical properties (Belay et al., 2020; Kalu and Ogbonna, 2021), soil fertility (Naik and Ushamalini, 2006), soil pollution, and metal contamination (Olatunde et al., 2021). Since most of the area surrounding these quarries is occupied by agricultural land, the contaminated soil causes low plant growth and increases the chances of reduced yield.

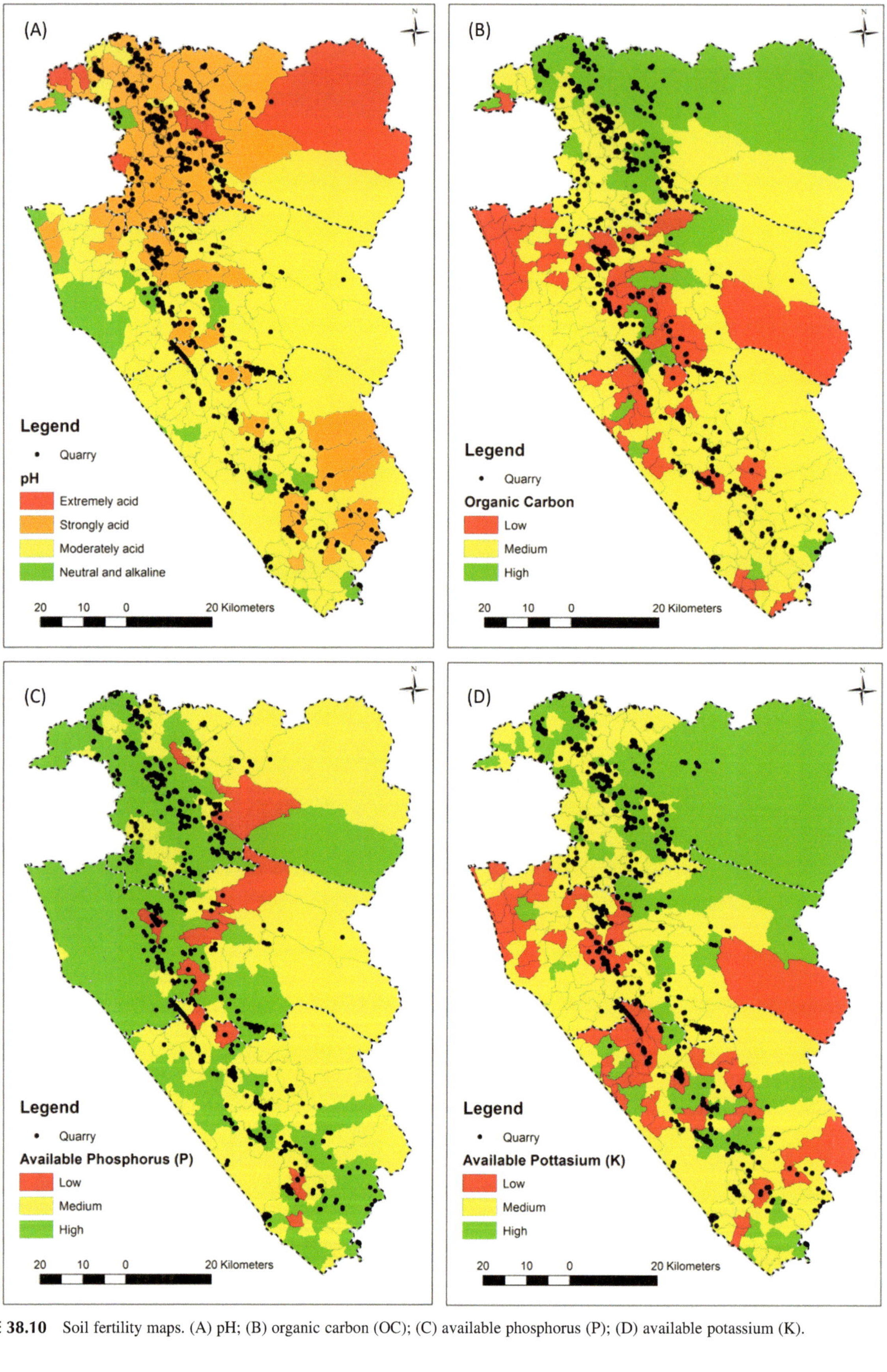

FIGURE 38.10 Soil fertility maps. (A) pH; (B) organic carbon (OC); (C) available phosphorus (P); (D) available potassium (K).

38.5 Conclusion

A geospatial database of stone quarries in three districts, viz., Pathanamthitta, Kollam, and Thiruvananthapuram, was constructed, and their effect on proximate land use and soil environment was analyzed. Not only abandoned quarries but also active and temporarily stopped quarries were mapped and included in the database. The major land use classes and soil types within the quarry's proximity were studied in detail. The impacts of stone quarrying on proximate land use and soil fertility were also analyzed. An increase in habitation was observed within a buffer zone of 500 m of quarries. In the case of soil fertility, the panchayat with the maximum number of stone quarries shows strongly acidic soil, medium OC, high available K, and medium available P. Since rigid conflicts are rising against quarrying in many parts of Kerala, analyses on the number and areal extent of each quarry and its relation to proximate land use and the natural soil environment reveal reliable information to concerned authorities for sanctioning quarries in the future.

Acknowledgments

The authors are grateful to the Directorate of Environment and Climate Change (DoECC), Government of Kerala, for financing the project titled "Assessment, Mapping, and Preparation of Database on Spatial Distribution, Status and Baseline Features of Granite and other Crystalline Rock Quarries in Thiruvananthapuram, Kollam, And Pathanamthitta districts." The authors would like to acknowledge Dr. M. Sundararajan (Principal Scientist, National Institute for Interdisciplinary Science Technology, Council of Scientific & Industrial Research (CSIR)) and Dr. S. Kaliraj (Scientist—C, National Centre for Earth Science Studies (NCESS), Ministry of Earth Sciences (MoES)) for valuable suggestions and GIS technical support.

References

Abdallah, C., Chorowicz, J., Bou Kheir, R., Khawlie, M., 2005. Detecting major terrain parameters relating to mass movements' occurrence using GIS, remote sensing and statistical correlations, case study Lebanon. Remote. Sens. Environ. 99. Available from: https://doi.org/10.1016/j.rse.2005.09.014.

Alex, C.J., Sajeev, T.V., 2017. Mapping of granite quarries in Kerala, India: a critical mapping initiative. In: Erudite Lecture Series of Prof. Madhav Gaadgil.

Anand, R.S., Reddy, D.V., 2015. A geological and geotechnical investigation of some rocks in Trivandrum area, Kerala, India. Int. J. Earth Sci. Eng. 8.

Ashokkumar, K., 2020. Cost and return analysis of natural rubber plantations in Pathanamthitta District in Kerala. MUDRA J. Financ. Acc. 7. Available from: https://doi.org/10.17492/mudra.v7i1.195415.

Balcha, D., Oyda, S., 2021. Assessing the impact of stone quarrying on the landscape, crop and grazing land degradation in Loma Bosa District, Dawuro Zone, Southern Ethiopia. Int. J. Sci. Res. Civ. Eng. 5.

Belay, L., Birhane, E., Zenebe, A., et al., 2020. Effects of stone mining on woody plant species diversity and selected soil properties in northern Ethiopia. Environ. Syst. Res. 9. Available from: https://doi.org/10.1186/s40068-020-00171-8.

Biju, A.K., Mathew, J., George, A.M., et al., 2021. A Study in correlation of landslides and increase in areal extent of quarries in Malappuram District. Int. J. Eng. Res. Technol. 9, 25–30.

Desai, N.D., 1979. Dimension Stones of Gujarat. Maharaja Sayajirao University of Baroda.

DMG, 2016. District Survey Report of Minor Minerals (Except River Sand).

Ganesh, K.M., Jagannadha Rao, M., Jai Sankar, G., Prasad, A.S.S.V., 2018. Evaluation of geomorphological landforms, lineaments and landuse/landcover using Geospatial Technologies in parts of Nalgonda District, Telangana State, India. Int. J. Eng. Technol. 7. Available from: https://doi.org/10.14419/ijet.v7i3.31.18297.

George, A.M., Biju, A.K., Raj, D., et al., 2022. Estimating the increase in areal extent of quarries in malappuram district using open source Software. SSRN Electron. J. . Available from: https://doi.org/10.2139/ssrn.4031956.

Guillerme, S., Kumar, B.M., Menon, A., et al., 2011. Impacts of public policies and farmer preferences on agroforestry practices in Kerala, India. Environ. Manage 48. Available from: https://doi.org/10.1007/s00267-011-9628-1.

Igwenagu-Ifeanyi, V., Belonwu, D.C., Okoro, S.E., 2021. Effects of quarrying activities on soil quality and nutritional composition of selected vegetables in southeastern nigeria. Bionature 24–38.

Kalu, I.E., Ogbonna, N.J., 2021. Investigation of environmental effect of stone quarrying activities on soil and water in Akpoha and Ishiagu communities of Ebonyi state, Nigeria. Int. J. Constr. Manag. 21. Available from: https://doi.org/10.1080/15623599.2019.1604115.

Kumar, S., Gupta, R.C., Shrivastava, S., 2017. Long term studies on the utilisation of quartz sandstone wastes in cement concrete. J. Clean. Prod. 143. Available from: https://doi.org/10.1016/j.jclepro.2016.12.062.

Maya, K., Santhosh, V., Padmalal, D., Kumar, S.A., 2012. Impact of mining and quarrying in Muvattupuzha river basin, Kerala-an overview on its environmental effects. Bonfring Int. J. Ind. Eng. Manag. Sci. 2, 36–40.

Naik, D.P., Ushamalini, S.R., 2006. Impact of stone quarrying on soil quality—a case study in Bangalore Dist., India. J. Ind. Pollut. Control. 21.

Naik, D.P., Ushamalini, Somashekar, R.K., 2006. Impact of stone quarrying on the pigment concentration and productivity of paddy (*Oryza sativa*, L). J. Ind. Pollut. Control. 22.

Olatunde, K.A., Towolawi, T., Kolawole, I., et al., 2021. Soil quality and metal distribution around stone quarries within Abeokuta, Southwest, Nigeria. IFE J. Sci. 23. Available from: https://doi.org/10.4314/ijs.v23i2.6.

Padmalal, D., Maya, K., Sreebha, S., Sreeja, R., 2008. Environmental effects of river sand mining: a case from the river catchments of Vembanad lake, Southwest coast of India. Environ. Geol. 54. Available from: https://doi.org/10.1007/s00254-007-0870-z.

Ramdev, P.R., John, S.E., Maya, K., Padmalal, D., 2013. Environmental effects of hard rock quarrying on land and water resources of small catchments—a case study. In: 23rd Swadeshi Science Congress, pp. 327–331.

Rani, B., Swamy, S., Bharath, et al., 2017. Impact of quarrying and crushing on soil quality: a case study in tumkur district, karnataka. Int. J. Res. Granthaalayah 5. Available from: https://doi.org/10.29121/granthaalayah.v5.i4rasm.2017.3363.

Sajinkumar, K.S., Sankar, G., Rani, V.R., Sundarajan, P., 2014. Effect of quarrying on the slope stability in Banasuramala: an offshoot valley of Western Ghats, Kerala, India. Environ. Earth Sci. 72. Available from: https://doi.org/10.1007/s12665-014-3143-7.

Sheela, A.M., Ghermandi, A., Vineetha, P., et al., 2017. Assessment of relation of land use characteristics with vector-borne diseases in tropical areas. Land. Use Policy 63. Available from: https://doi.org/10.1016/j.landusepol.2017.01.047.

Singhal, A., Goel, S., Sengupta, D., 2020. Physicochemical and elemental analyses of sandstone quarrying wastes to assess their impact on soil properties. J. Environ. Manage 271. Available from: https://doi.org/10.1016/j.jenvman.2020.111011.

Soman, K., 1987. Geology of Kerala. GSI Publications.

Srinivasan, N., 2014. Development of technology and popularization of biological control agents for integrated management of coconut leaf rot disease. Cord 27. Available from: https://doi.org/10.37833/cord.v27i1.123.

Vandana, M., John, S.E., Maya, K., Padmalal, D., 2020. Environmental impact of quarrying of building stones and laterite blocks: a comparative study of two river basins in Southern Western Ghats, India. Environ. Earth Sci. 79. Available from: https://doi.org/10.1007/s12665-020-09104-1.

Chapter 39

Mapping of slope failure hazard zonation subject to soil characteristics: a case study of the Chaliyar River Basin, the Western Ghats, South India

S. Kaliraj[1], S. Bhagyashree[2], B. Mahanta[2], M. Lalitha[3] and N. Chandrasekar[4]

[1]*National Centre for Earth Science Studies (NCESS), Ministry of Earth Sciences, Govt. of India, Thiruvananthapuram, Kerala, India,* [2]*Department of Geology, Central University of Tamil Nadu, Thiruvarur, Tamil Nadu, India,* [3]*ICAR-National Bureau of Soil Survey and Land Use Planning, Regional Centre, Bangalore, Karnataka, India,* [4]*Centre for GeoTechnology, Manonmaniam Sundaranar University, Tirunelveli, Tamil Nadu, India*

39.1 Introduction

Assessment of slope stability conditions in the habitat region plays a vital role in disaster management and is very important for the development of socioeconomic conditions and infrastructure such as buildings, highways, railroads, canals, dams, surface mining, and many other construction and excavation activities. Slope failure is considered the probability of slope instability within specific sites and time due to altering the soil properties due to rainfall and other geoenvironmental factors (Mantovani et al., 1996; Liew, 2004; Duncan and Wright, 2005; Raneesh and Santosh, 2011; Ojo et al., 2016; Kang et al., 2020). Slope failure hazards always result from changes in natural settings due to the activities of the physical, biological, and social systems of the catchments. Slope failure is one of the primary factors inducing various geohazards such as landslides, mass wasting, debris flow, avalanches, flash floods, and other natural disasters (Sarma, 1973; Wu and Siddle, 1995; VanWesten, 2000; Lu and Likos, 2004; Geotechnical Design Manual, 2005; Moayedi, Bui, et al., 2019; Moayedi, Tien Bui, et al., 2019). The slope stability of an area is directly proportional to soil physical characteristics such as texture, depth, porosity, overburden thickness, soil cohesion, soil friction, soil unit weight, soil-water unit weight, moisture capacity, etc. The slope represents the terrain's topography and its relation to soils and rocks; this quantitative information provides insight into landscape features and stability conditions subject to natural and man-made impacts (Marrison and Greenwood, 1989; Mitchel, 1993; Kabir and Taha, 2004; Horpibulsuk et al., 2013; Kaliraj et al., 2021). Slope failure activities in different landscapes mainly depend on changes in soil physical properties that cause various types of slope failure events, such as (1) translational failure—slope fails along a weak zone of the soils; (2) rotational failure—slope fails in homogeneous fine-grained soils; (3) wedge failure—fractured bulk slides down the slope as bricks and wedges; and (4) compound failure—slip surface down the slope from the toe of the hard soil layer. The slope stability of an area is to guarantee that the soil's shear strength is greater than the shear stress; otherwise, this could induce slope failure if soil's shear strength decreases or soil's shear stress increases. The factors affecting soil's shear strength and soil's shear stress directly influence the slope stability of an area. Causes of decline in soil's shear strength are increasing pore−water pressure, lower permeability index of clayey soils, and cracking of the slope's top; and factors increasing soil's shear stress are overloads at the slope top and an increase in soil unit weight due to soil−water contents (Bromhead, 1992; Rahardjo et al., 1995; Faisal, 2000; Huat et al., 2004; Liew, 2004; Duncan and Wright, 2005; Ahmad et al., 2006; Horpibulsuk et al., 2013; Moayedi, Bui, et al., 2019; Moayedi, Tien Bui, et al., 2019). Slope failures are thus influenced by slope geometry, soil type, stratification, groundwater, and seepage. Slope failure is one of the primary factors inducing mass movement by the earth down a slope through the pressure of gravity and causes extreme damage to habitats, the environment, livelihoods, and residences. Slope failure

Remote Sensing of Soils. DOI: https://doi.org/10.1016/B978-0-443-18773-5.00043-0

triggers the process of ground movements such as rocks, debris, top-load materials, and soils, which adversely affect environmental ecosystems (Wu and Siddle, 1995; Abramson et al., 2001; Singhroy, 2004; Rawat et al., 2016; Lalitha et al., 2021). In India, different landscapes such as the Western Ghats, the Himalayan foothills, and Northeast hill ranges are noted as hotspots of slope failure and landslide with the combined processes of debris flowing, land sliding, toppling, down sloping, mass-wasting and falling movements due to intensive rainfall, deforestation, altering the soil properties, changing the groundwater table, encroachments of drainage networks, and unscientific developmental activities (Rahardjo et al., 1995; Ng and Shi, 1998; Mohamed et al., 2006; Ambili and Narayana, 2014; Escobar-Wolf et al., 2021; Ajin et al., 2022; Hao et al., 2022) In the Western Ghats' highlands, many locations are noted under extreme vulnerability to slope failure and landslides during heavy rainfall due to human-induced impacts. During the last decades, many landslides have occurred in the Western Ghats region due to the action of intensive rainfall that altered the soil properties, rapid shifting of water, and transporting debris and rock materials downslope (Wadhawan et al., 2020; Sajinkumar and Oommen, 2021; Yunus et al., 2021; Escobar-Wolf et al., 2021; Ajin et al., 2022; Hao et al., 2022; Ambili and Narayana, 2022). The Chaliyar River Basin comprises various landforms formed by the influence of tectonics and rock uplift on the fluvial and topographic systems using river concavity. In the basin, the knick zones observed at the base of the Western Ghat scarp are highly prone to slope failure and landslide hazards. Geomorphic landforms consist of structural hills, alluvial fans, and unpaired strath terraces due to the tectonic evolution of the basin, whereas erosional and depositional landforms are formed by the high headward erosion capacity of stream networks at various locations (Ambili and Narayana, 2014; Lalitha et al., 2021). The landscapes are noted to have several slope failures and landslide hotspots depending on slope and soil characteristics, and these are unaffected by lithological formations (Wadhawan et al., 2020; Sajinkumar and Oommen,2021; Yunus et al., 2021). The Chaliyar Basin landscapes, especially hilly and mountainous terrains associated with human activities, have frequently experienced various types of mass movements such as landslides, slope failures, and debris flows, mainly from altering soil characteristics by precipitation and other anthropogenic activities such as deforestation, encroachment for settlements, cultivations on sloping terrain, quarrying, toe cutting for roadways, etc. (Ambili and Narayana, 2014, 2022).

Integrated remote sensing and GIS techniques can be effectively used for slope stability analysis from a wider perspective. The advantages of remote sensing and GIS in determining slope stability include (1) estimation of soil properties (i.e., cohesion, internal friction angle, and dry unit weight) using multispectral and hyperspectral data, (2) measurement of slope angle, slope height, topographic conditions, and drainage networks using DEM data, and the GIS techniques used for determining slope stability for a large coverage, emphasizing local conditions. Slope stability analysis is a computational measure used to determine the stability rate of a sloped surface subject to physical and climatological factors. Many researchers have used several approaches, such as numerical, computational, and statistical indices, for mapping and assessment of slope stability or instability in various landscapes worldwide (Mantovani et al., 1996; VanWesten, 2000; Chi et al., 2002; Crozier and Glade, 2004; Kaliraj et al., 2014; Moayedi, Bui, et al., 2019; Moayedi, Tien Bui, et al., 2019). Traditionally, mapping of slope instability conditions is done using a geomorphological approach based on the determination of slope variability in the landscape. This subjective method is applied over large areas for rapid zonation with acceptable accuracy. Integrated remote sensing and GIS techniques provide an effective tool to store, retrieve, update, and analyze multiple-geoenvironmental data for the assessment and mapping of slope failure zones with higher accuracy (Chi et al., 2002; Kaliraj et al., 2015a,b; Yunus et al., 2021). Nowadays, the factor of safety (FoS) model is widely used for slope stability analysis using multiple spatial datasets derived from remote sensing data, ground-based data, and ancillary data sources. GIS-based multivariate statistical analysis was used to demarcate slope stability or instability conditions at slope units or unit river basins. Recent studies on slope stability have been performed using soft computing and big-data analysis such as fuzzy set theories, neural networks, probabilistic analysis, weighted evidential approaches, genetic algorithms, etc., and these models have been successfully applied to the analysis of multisource data (Bromhead, 1992; Wu and Siddle, 1995; Pack et al., 1998; Cho and Lee, 2002; Singhroy, 2004; Ahmad et al., 2006; Glade and Crozier, 2010; Kaliraj et al., 2012; Rawat et al., 2016; Moayedi, Bui, et al., 2019; Moayedi, Tien Bui, et al., 2019; Escobar-Wolf et al., 2021). The FoS is a quantitative analysis that represents the state of strength and the possibility of the slopes failing, the determination of the stability of slopes subject to geological structures, the orientation of the strata, and geomorphological landforms (Pack et al., 1998; Duncan, 2000; Abramson et al., 2001; Liew, 2004; Baharom, 2005; Huat et al., 2006; Bhattarai et al., 2014; Kang et al., 2020). The FoS equation is used to examine the slope stability along the upland hilly regions worldwide by analyzing several parameters such as soil moisture, soil cohesiveness, friction angle, soil unit weight, slope angle, etc., which are produced from Landsat TM, Radarsat SAR data, SRTM DEM, and other ancillary data. The outcome shows hotspots of slope failure and shallow landslides while changing unsaturated residual soil conditions. To assess slope stability in homogeneous terrains with different soil characteristics, the factor of safety model is effectively used to observe slope failure subject to soil properties and slope angle. The finding reveals that the decreasing slope angle can increase the safety factor; thus, higher slopes are seen on the toe slip, whereas lower slopes are seen on the base slide. In addition, the slope slide was dominant on sandy soil at different slope heights and angles (Fredlund and Rahardjo, 1993; Cho and Lee, 2002; Cocka et al., 2004; Huat et al., 2004; Kabir and Taha, 2004; Rawat et al., 2016). GIS-based 3D-FoS (factor of safety) model for calculating slope stability along the Nepal highway stretch based on the input data of soils, geology, hydrology, and topography using ArcGIS software. Many researchers have assessed slope stability using the factor of safety model concerning soil properties under various static conditions. The results show that if the factor of safety value ≥ 1, then the slope is stable, and if it is ≤ 1, the soil slope is noted as unstable (Duncan, 2000; Abramson et al., 2001; Cho and Lee, 2002; Baharom, 2005; Duncan and Wright, 2005; Huat et al., 2006; Moayedi, Bui, et al., 2019; Moayedi, Tien Bui, et al., 2019; Kang et al., 2020).

In the study area, slope failure is one of the most widespread and damaging natural hazards that causes significant damage to properties, infrastructure, and settlements. Therefore, the slope stability index model is executed to map slope failure zonation in the study area using remote sensing and GIS techniques.

39.2 The geographical location of the study area

The Chaliyar River basin lies in the Western Ghats, covering the districts of Malappuram, Kozhikode, and Wayanad in Kerala and a small portion of Tamil Nadu, in Southern India.

Fig. 39.1 shows the location of the study area. The geographical extent of the study area lies between 75′48′E and 76′33′E longitude and 11′06 N and11′36′N latitude, with an area of 2926 km^2. The Chaliyar River, one of the major rivers of Kerala State, originates from the Ilambalari Hills of the Western Ghat at an altitude of 2067 m above MSL.

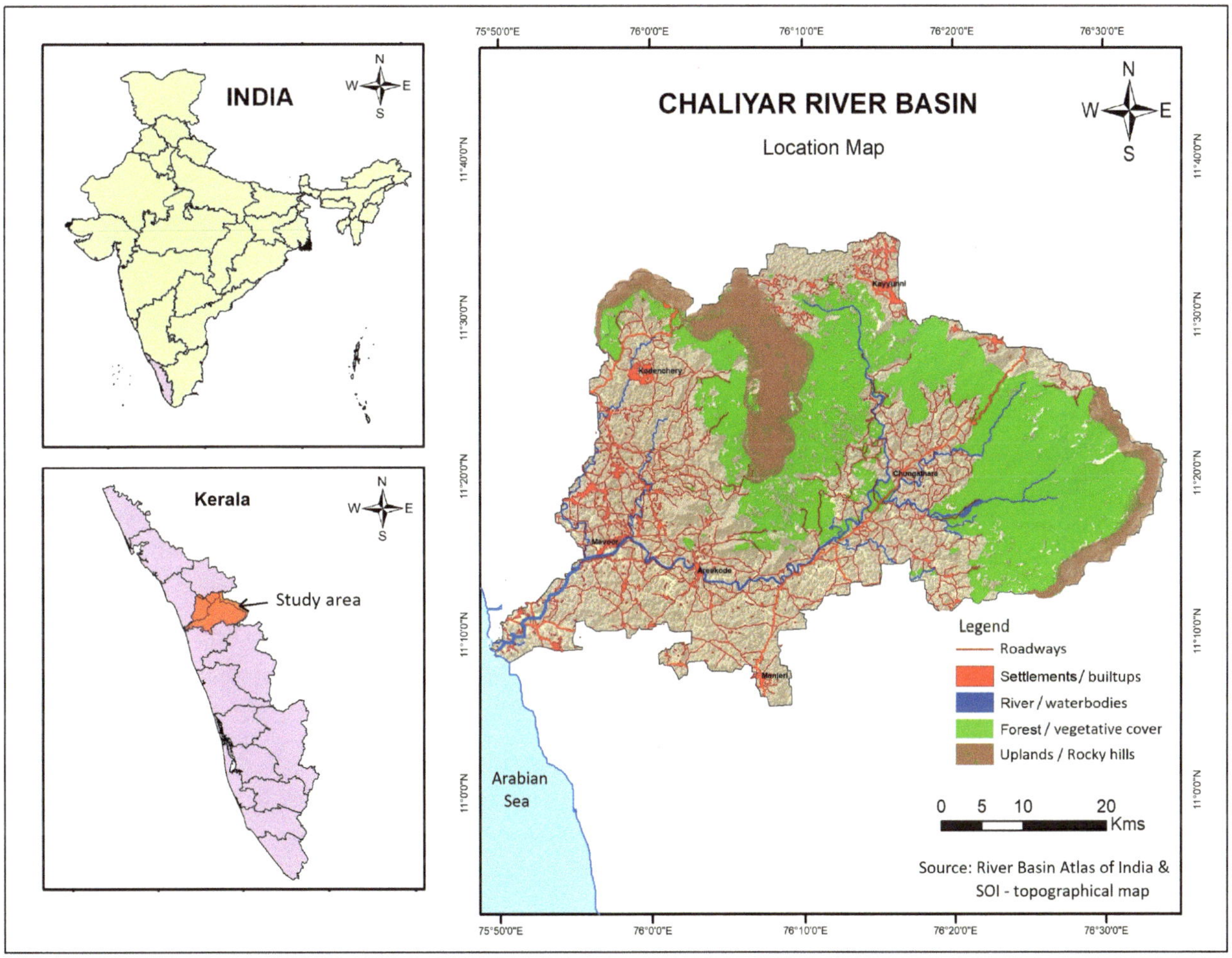

FIGURE 39.1 Geographical location of the study area.

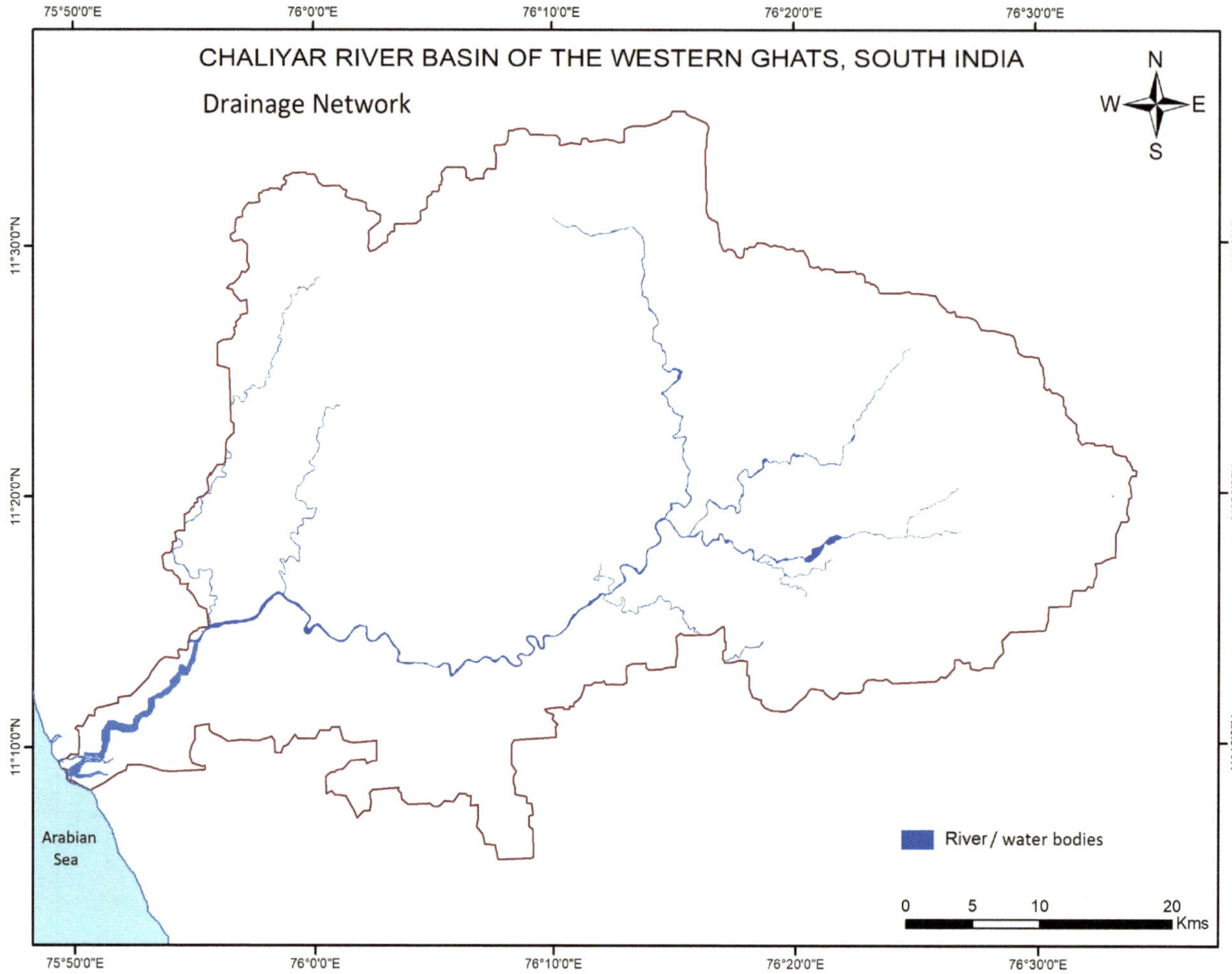

FIGURE 39.2 Drainage network of the study area.

Fig. 39.2 shows the drainage network of the basin. There are six major tributaries, namely Chaliyarpuzha, Punnapuzha, Karimpuzha, Kurumanpuzha, Iruvahnipuzha, and Cherupuzha, are joined to constitute the drainage system of the basin. The basin surface varies in relief pattern and topographic alignment with different physiographic subunits, such as upland of hilly terrains, mid-land of plateaus, and lowlands of alluvial plains and coastal stretches (Kaliraj et al., 2014; Lalitha et al., 2021). The upland consists of hilly terrains formed by undulating topographic structures, abrupt and narrow crests, steep slope valleys, deep canyons, and mountain-fed streams that are potentially susceptible to geo-hazards such as slope failure and landslides while altering soil properties and climatic characteristics (Ambili and Narayana, 2022). The basin covers a variety of soil types, including loam texture (42.74%), clay (28.66%), clay loam (24.18%), and sandy loam (4.42%). Land use/land cover (LULC) have dominated with agriculture (60.04%), forests (38.74%), and other features such as pastures, wastelands, rocky areas, settlements (urban and rural), etc. The basin consists of a total population of about 52.72 lakhs, as per the 2011 census report. Major urban and rural settlements situated along the banks of Chaliyar are Pothukal, Chungathara, Nilambur, Mampad, Nilambur, Kavanoor, Areekode, Vazhakkad, Cheruvadi, Mavoor, Peruvayal, Kalpetta, Mananthavady, Feroke, and Beypore. The basin is characterized by tropical, humid climatic conditions, with an annual average rainfall of 3012 mm during both the southwest monsoon (June–August) and northeast monsoon (September–November). The optimum temperature is recorded in the range of 24°C–34°C and the humidity range of 60%–90% during winter and summer, respectively.

39.2.1 Geological structures

Geological formation is one of the determinant elements in the slope stability analysis of the basin. Fig. 39.3 shows the geological features of the study area, derived from the Geological Survey of India (GSI) published geology map.

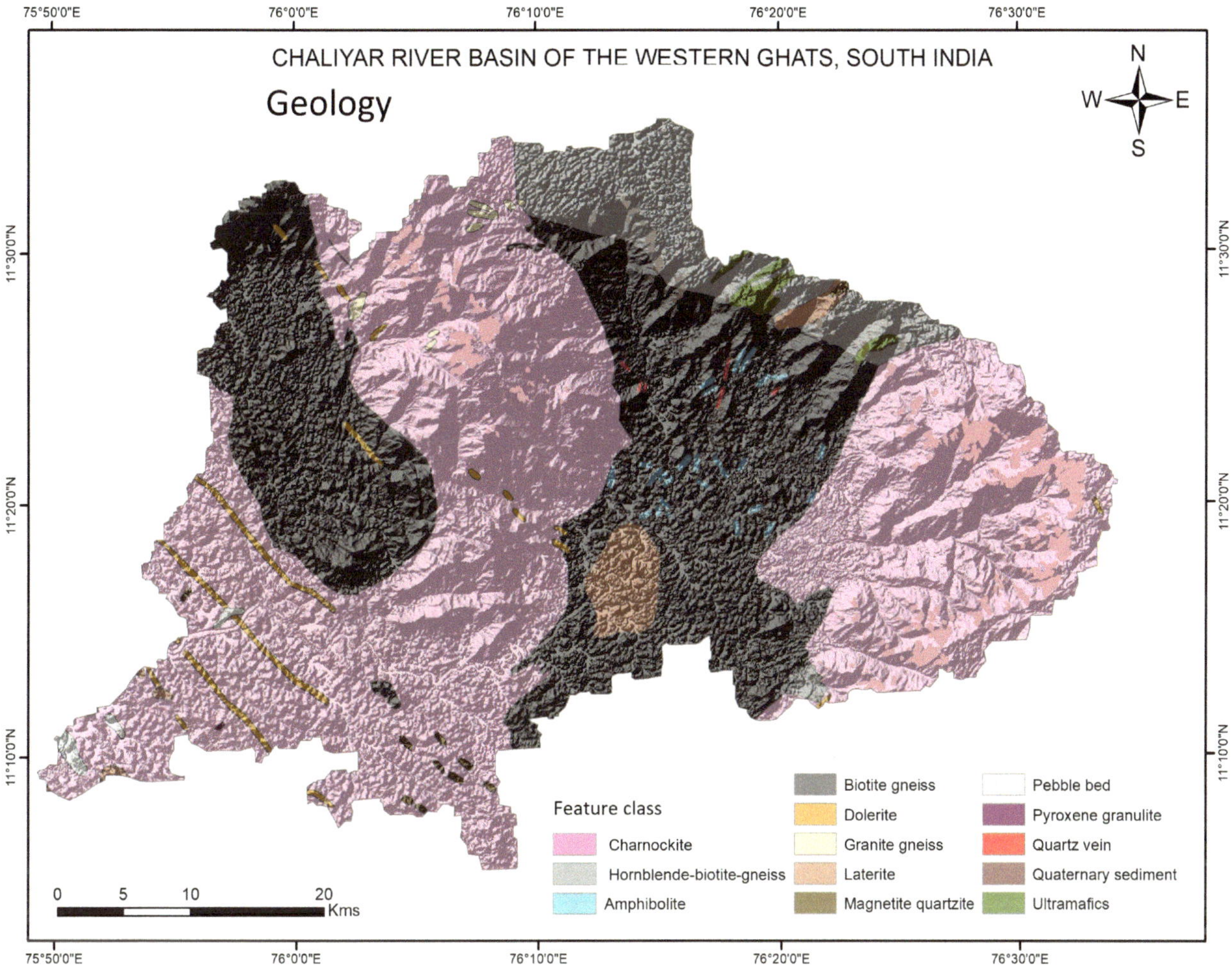

FIGURE 39.3 Geological features of the study area.

The basin consists of the oldest rocky exposures belonging to the Wayanad group, comprising migmatite to banded gneisses, quartz-feldspathic charnockite with interlayered metamorphosed sediments, and minor granites. The Precambrian metamorphic shield comprises a charnockite group and a migmatite complex that were penetrated by younger basic and acid intrusions. Isolated capping of a tertiary deposit (Warkali formation) is formed on the western side. Quaternary sediment formation is found in coastal tracts, river terraces, and valley fills. Charnockite is the most widespread rock, covering approximately 50% of the total area and making up the upland hills, steep slopes, and undulating terrain in the eastern and western parts of the basin. The Migmatite complex consists of hornblende-biotite and granite gneisses with a composition of medium- to coarse-grained rocks with gneissic to granular textures. Laterite formation occurs in the midlands within the charnockite and migmatites with a thickness of 2–8 m. Quaternary sediments consisting of medium- to coarse-grained sand, mainly quartz, are confined to low-land areas such as river terraces, valley fills, and coastal tracts on the western side (Ambili and Narayana, 2014; Sajinkumar and Oommen, 2021). In the Precambrian formation, the structures formed as multiple facies of deformation, metamorphism, and the superimposition of brittle deformation. Moreover, the laterite formations found in various parts are susceptible to slope failure and landslide hazards depending on drainage and climate activities on soil properties and geomorphic landforms.

39.2.2 Geomorphic landforms

The Chaliyar river basin is a major drainage system originating from the Western Ghats, and it is marked by diverse geomorphologic features and a broad valley in the central part due to plate tectonic activities and the work of rivers (Ambili and Narayana, 2014). Fig. 39.4 shows the geomorphic landforms of the study area based on the GSI-published geomorphology map. The geomorphic landforms are divided into four well-defined units: (1) hilly high peaks (altitude

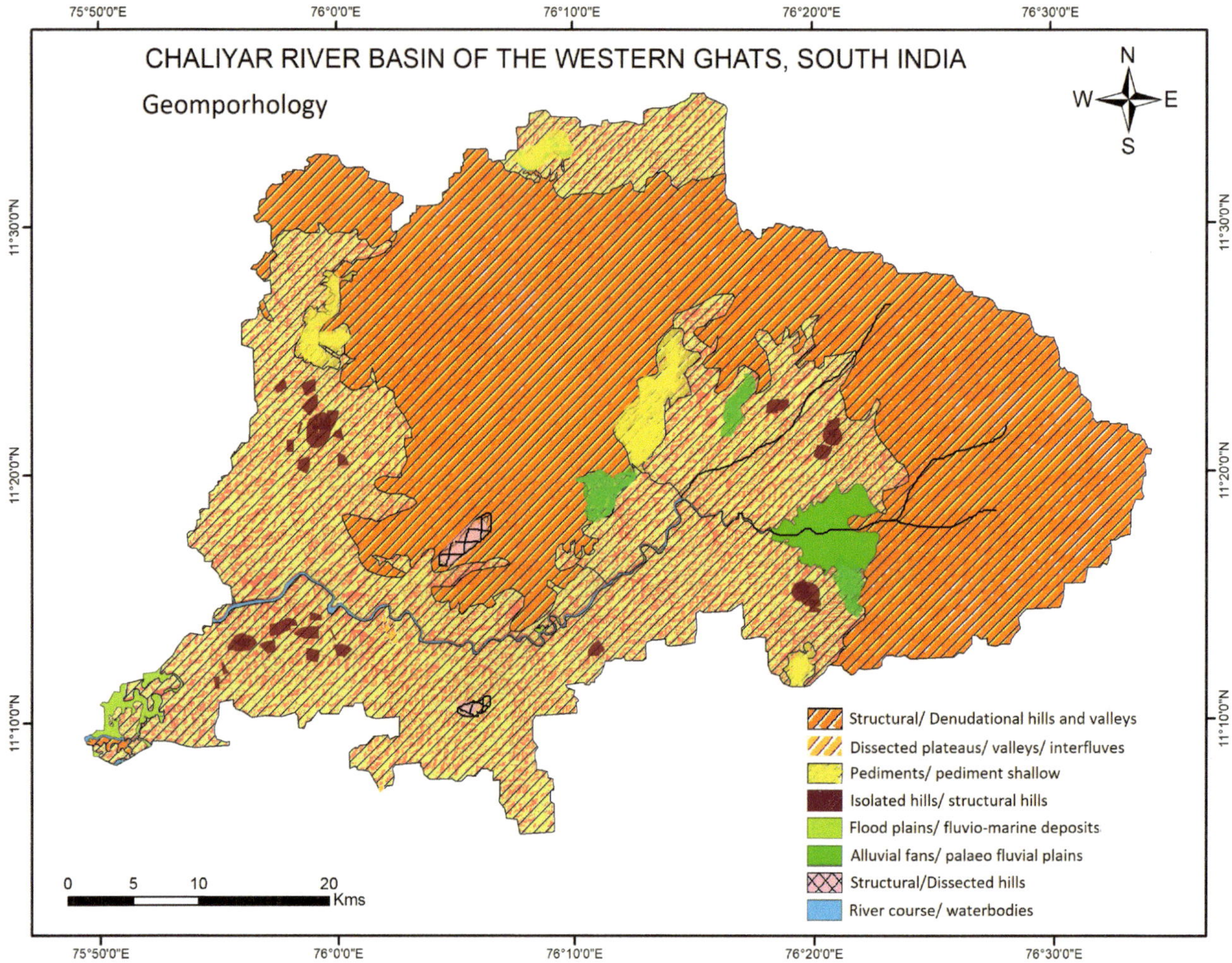

FIGURE 39.4 Geomorphic landforms of the study area.

> 2000 m) formed by abrupt and narrow crests, steep slope canyons, and mountain-fed streams; (2) uplands with ridges, valleys, and isolated hills (altitude 600–2000 m); (3) midlands occupy plateaus with laterite ridges, elongated spurs, and ravine valleys (altitude 20–600 m); and (4) lowland areas (altitude < 20 m) consist of alluvial plains, river terraces, levees, estuaries, paleo-beach ridges, and coastal wetlands. Hill ranges occupied in the eastern parts are characterized by escarpments, gorges, and steeply sloped valleys, wherein the hilltops and hill peaks are generally formed as barren and rocky outcrops and covered by thick and compact laterite. Mid-land ranges of plateaus are formed as flat-topped hillocks with steep valleys and ridges (Wadhawan et al., 2020; Yunus et al., 2021; Ambili and Narayana, 2022). The coastal plains extend in the western part as a narrow stretch with a composition of fluviomarine deposits. Riverine landforms are formed as levees, valley fills, meanders, and floodplains by the work of fluvial systems of rivers and streams. Highlands are formed as erosional, depositional, and structural landforms due to the action of streams and tributaries. They attain a base level of erosion and are prone to slope failure and landslide hazards.

39.2.3 Soil types and characteristics

Soil types in the basin cover loam (42.74%), followed by clay (28.66%), clay loam (24.18%), and sandy loam (4.42%). Fig. 39.5 shows the soil types in the study area, derived from a soil map published by NBSS&LUP. Loamy soils are mainly observed in the lowlands along the alluvial plains, valleys, and coastal plains. Sandy loam occurs in mid-land plateaus, impounded within loamy and clayey soils, and is associated with laterite formations. Clay soil is mainly distributed along the stream flow path in the highlands and along the river course in the middle parts. Clay loam is exclusively found in the scarps and foothills of the Western Ghats and is well-drained and formed by higher-content sand to sandy clay soils (Ambili and Narayana, 2014; Sajinkumar and Oommen, 2021). Clay soil is composed of fine-grained

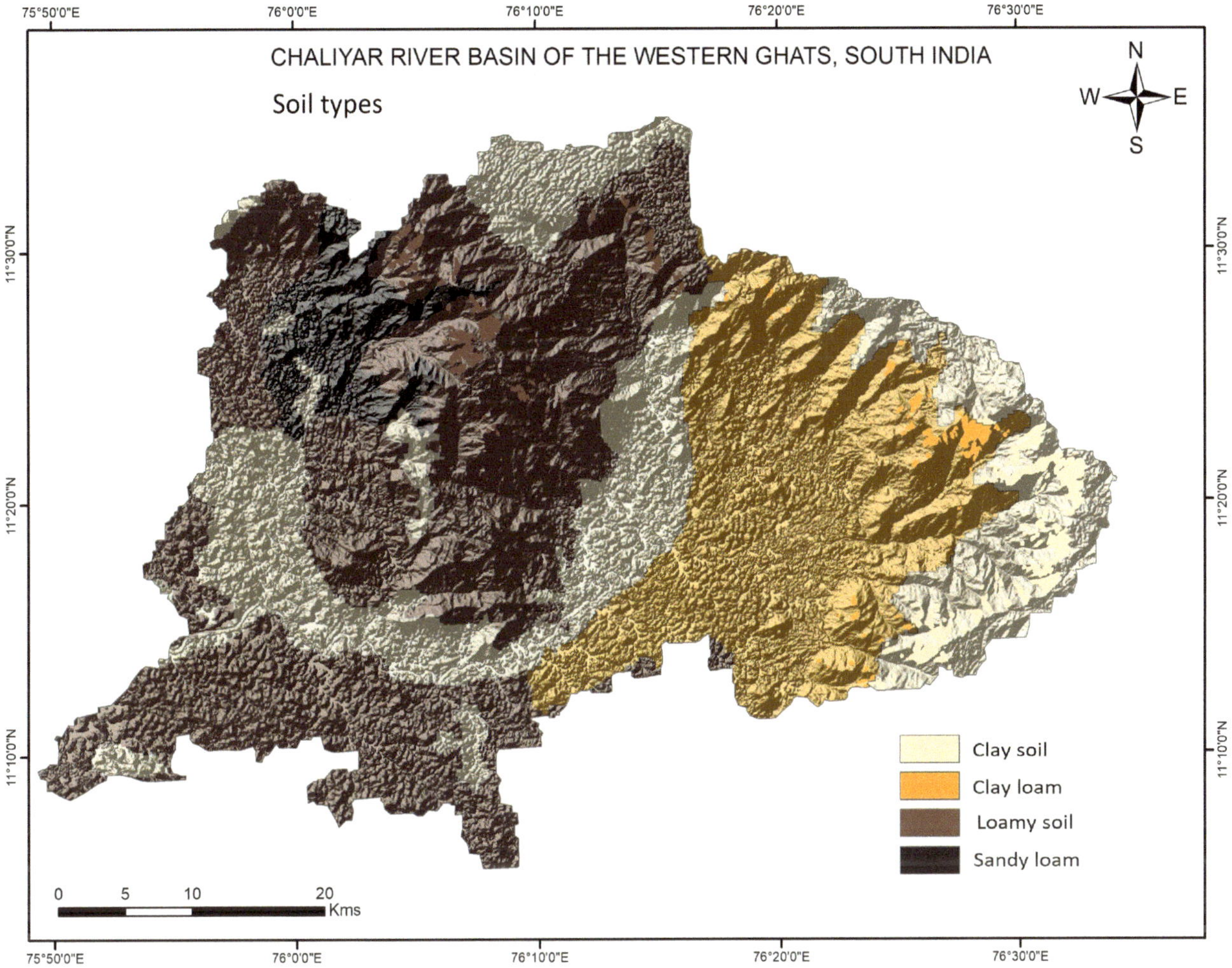

FIGURE 39.5 Soil types in the study area.

natural rock or soil substance made up of clay minerals, metal oxide traces, and organic components. In the basin, clay has an average density of 1.76 g cc^{-1}, with 97 kPa of soil cohesion and 23 degrees of friction angle, whereas loamy soil has an average density of 1.50 g cc^{-1}, with 75 kPa of soil cohesion and 15 degrees of friction angle. Clay and loam soils have the highest frequency ratio, indicating a high risk of slope instability. Thus, clay, clay loam, and loam soils are highly susceptible to slope failure and landslides in association with highlands, steep slope valleys, and laterite formations in various parts of the basin.

39.2.4 Land use/land cover

The LULC features were extracted from Landsat 8—OLI (30 m) images for May 2020 using the maximum likelihood classification technique in ENVI 5.3 software (Kaliraj et al., 2015a,b; Kaliraj et al., 2021). Fig. 39.6 shows the LULC of the study area. The basin comprises a variety of LULC features that include barren, rocky outcrops, deciduous forest, scrubland, plantations, cropland, fallow, grassland, wetlands, settlements (urban and rural), and waterbodies (rivers, channels, reservoirs). Upland hill ranges are covered with deciduous and evergreen forests of the Western Ghats over the ridges, valleys, deep gorges, and mountain-fed streams at different altitudes. Mid-land ranges are occupied with croplands and plantations associated with settlements along the lateritic ridges, plateaus, lateritic spurs, and alluvial valleys. The lowlands consist of urban settlements, croplands, and plantations associated with alluvial plains, floodplains, estuaries, wetlands, and paleo-beach ridges. A high-risk rate is noted for the LULC features such as upland-degraded forest, slope surficial plantations, mixed croplands around settlements, streams and channel flow, mid-land plateaus, undulated lateritic spurs, and valleys, all of which indicate the likelihood of a slope failure and landslide activities due to changes in LULC due to natural and anthropogenic factors (Wadhawan et al., 2020; Ajin et al., 2022).

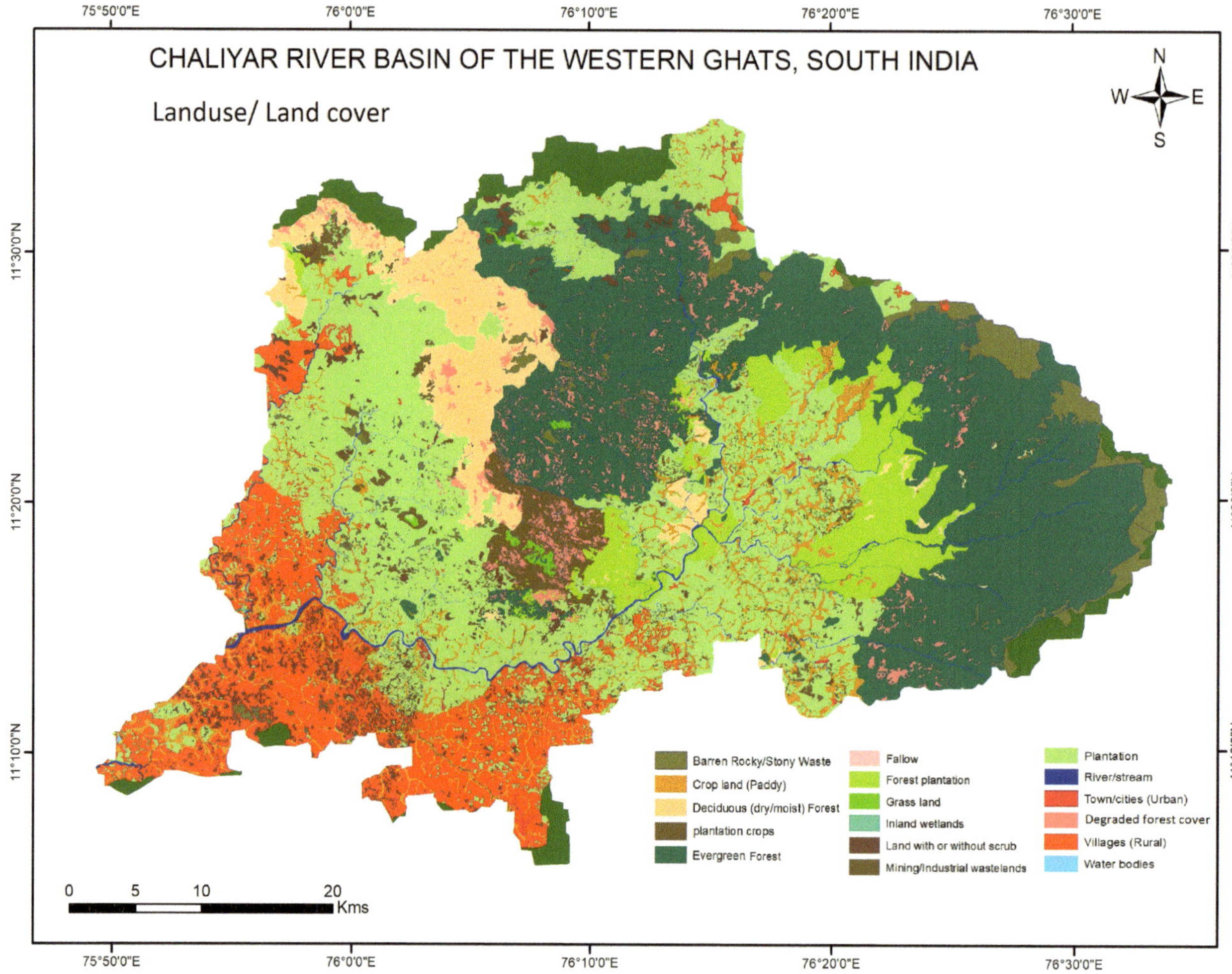

FIGURE 39.6 LULC features in the study area.

39.3 Materials and methods

Fig. 39.7 shows the systematic flowchart of the methodology used in this study. Assessment and mapping of slope failure hazard zones were executed using a GIS-based slope stability index model, which was modified from the standard FoS equation. This model is used to estimate the FoS value to demarcate the zone of slope stability and (or) possible sites to slope failure (slope instability) by analyzing multiple geoenvironmental parameters such as slope gradients, geologic structure, geomorphic landforms, LULC features, fluctuations in groundwater table, and soil physical properties, i.e., soil texture, soil thickness, soil infiltration rate, soil permeability, soil moisture, friction, coefficient, soil unit weight, and soil−water unit weight, etc. These parameters were collected from different spatial sources, such as Landsat 8—OLI images, SRTM—DEM (30 m) data, GSI-published geology and geomorphology maps, CGWB-published groundwater table and borehole data, NBSS-derived soil maps, and in situ data of soil physical properties. Landsat 8—OLI image was used to prepare LULC and landform features. Soil moisture of the different soil types is used to represent the volumetric soil−water content, and it is derived from the brightness temperature of Landsat 8- TIRS data using Walker and Kandel (2004) approach. The output shows a higher correlation ($R^2 = 0.97$) with in situ−measured data. The SRTM - DEM data are used to prepare the layers, namely slope angle (in degrees), topographic conditions, and drainage networks. Thematic layers for each parameter were prepared with their appropriate attributes and quantitative details using ArcGIS 10.6 software.

39.3.1 GIS-based slope stability index modeling

The GIS-based slope stability index is a quantitative analysis that demarcates the status of slope strength and the possibilities of the slope's failure. In this model, the equation is designed to analyze multiple geoenvironmental parameters

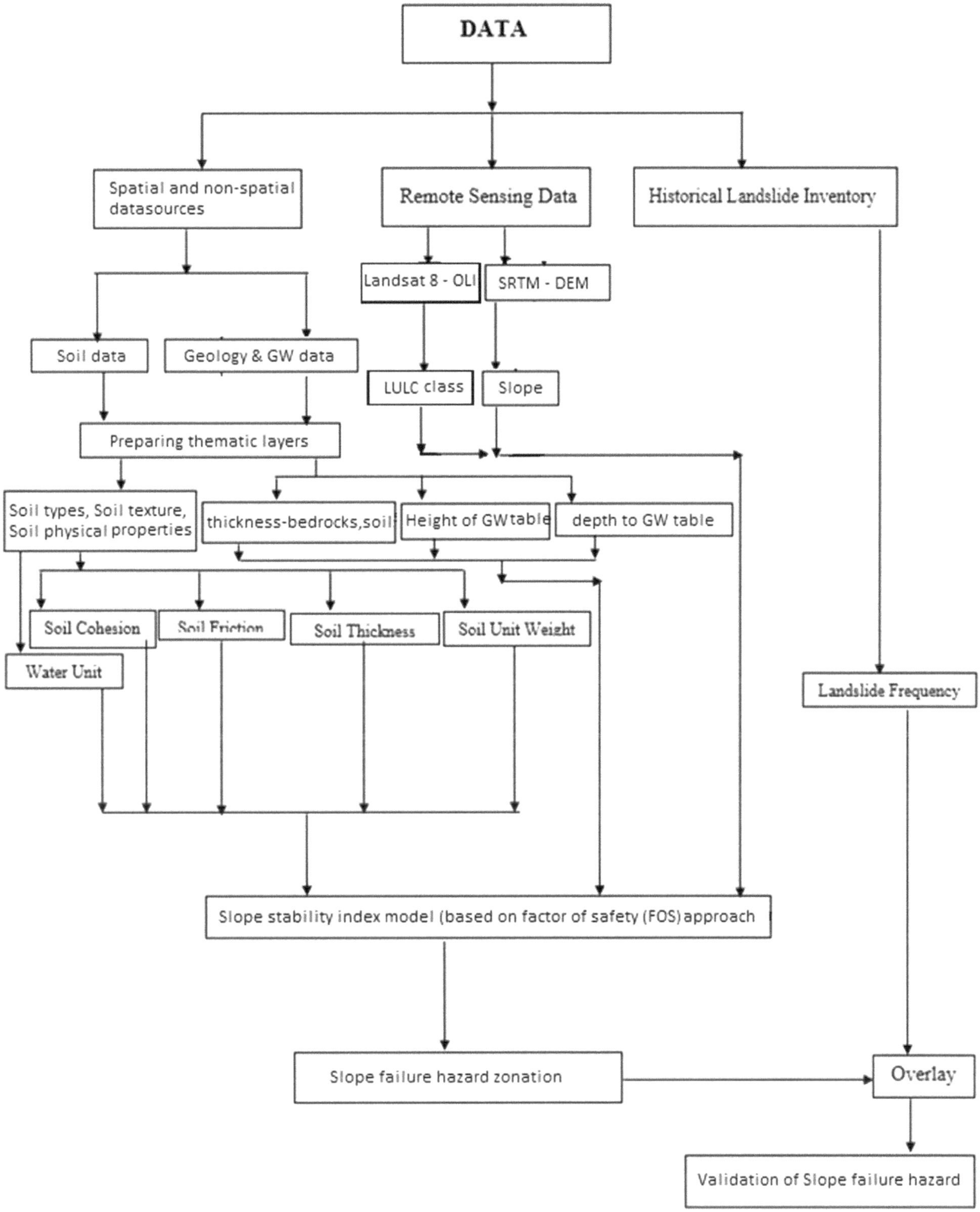

FIGURE 39.7 The methodology flowchart used in this study.

such as soil properties, lithology/geology, slope, relief, landforms, groundwater depth, etc., based on the concept of the safety factor model (Duncan, 2000; Abramson et al., 2001; Baharom, 2005; Moayedi, Bui, et al., 2019; Moayedi, Tien Bui, et al., 2019; Kang et al., 2020). In this study, the slope stability analysis is executed using the revised 1D slope stability model (adopted from Bhattarai et al., 2014). Thus, the equation (Eq. (39.1)) is executed on the raster layer of all parameters to demarcate slope failure hazard zones using the ArcGIS-raster calculator tool, and it is expressed as,

$$FoS = \frac{c + (\gamma - m \times \gamma m) \times z \times \cos 2\beta \times \tan\phi}{\gamma \times z \times \sin\beta \times \cos\beta} \quad (39.1)$$

where, FoS = slope stability index value, γ = unit weight of soil (or) soil density, γ_w = unit weight of soil-water (or) density of soil-water, β = slope angle (terrain surface inclination), ϕ = soil internal friction angle, c = soil cohesion. m = groundwater table depth/soil thickness ratio (z_w/z), in which, z_w = groundwater table depth (or) height of the groundwater table above the failure surface, and z = soil thickness (or) depth of the terrain surface above bedrocks.

As shown in Fig. 39.7, the GIS analysis is systematically executed on the input layers of z, c, φ, β, γ, and m to produce a slope stability map using Arc GIS 10.6 software (VanWesten, 2000; Liew, 2004; Rawat et al., 2016). The output map shows FoS values at a pixel scale (10 × 10 m) that represent the state of slope strength and possibilities of the slopes to failure.

39.4 Results and discussion

Mapping of slope failure hazard zones using a GIS-based slope stability index model is performed by analyzing input parameters such as slope angle, soil properties (i.e., soil moisture, soil cohesiveness, friction angle, soil unit weight, soil-water unit weight), depth to groundwater, geologic structure, geomorphic landforms, LULC features, etc. The details of these parameters are presented in Figs. 39.8–39.14.

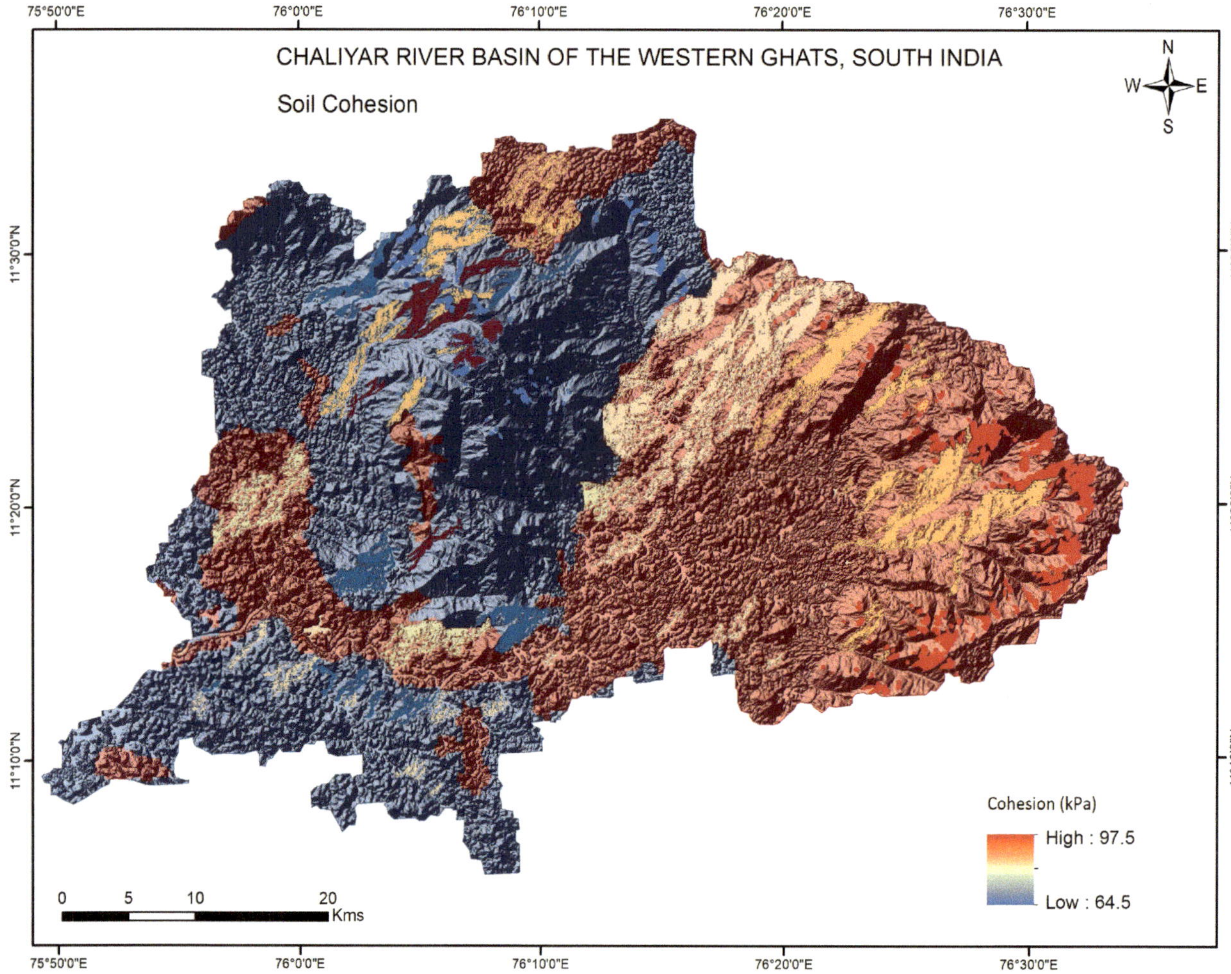

FIGURE 39.8 Spatial characteristics of soil cohesion in the study area.

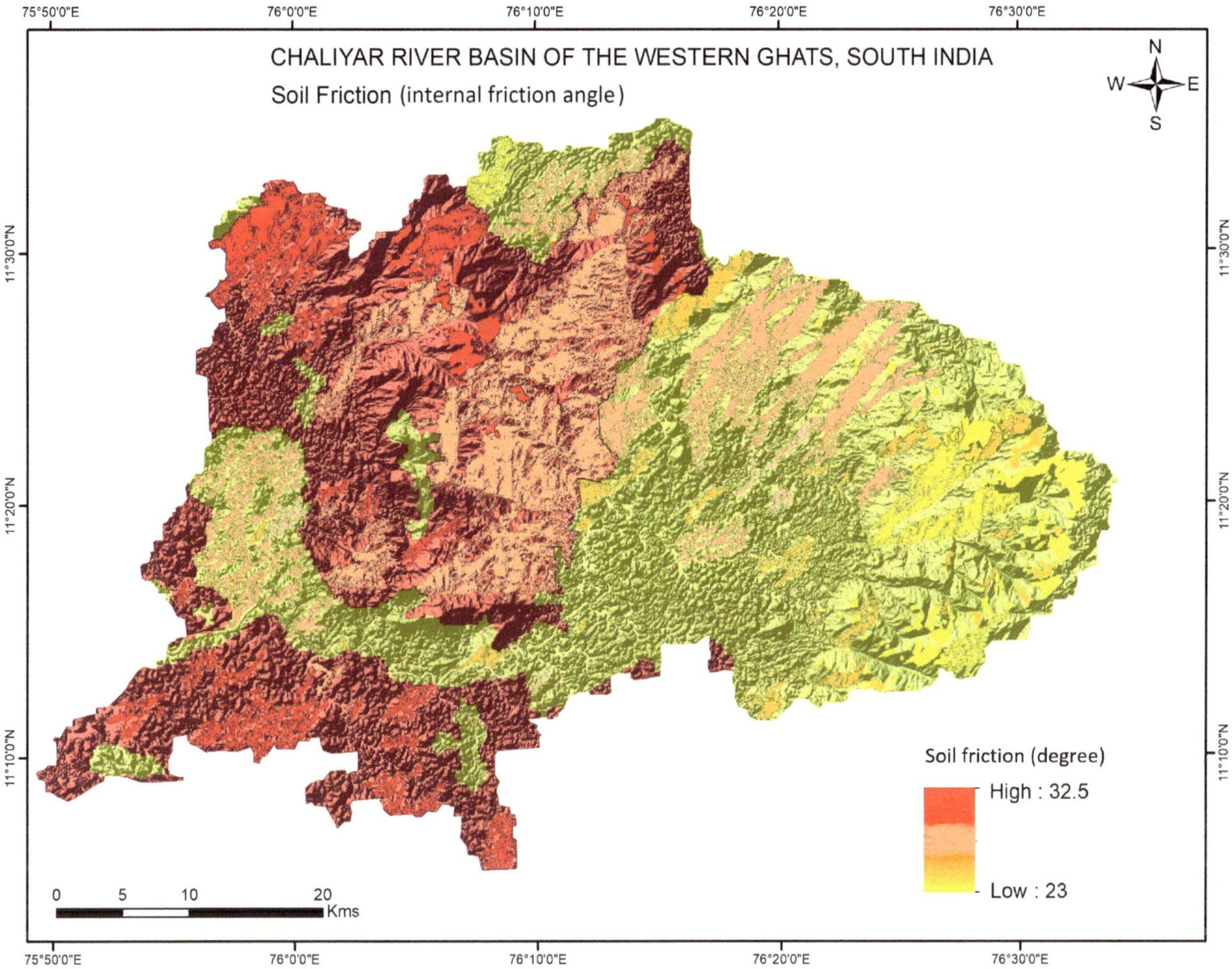

FIGURE 39.9 Spatial characteristics of the soil internal friction angle in the study area.

39.4.1 Soil cohesion

Soil cohesion is a measurement of the shear strength of soils subjects to their texture and particle matter. Soil textures normally comprise fine-grained soil such as silts and clays, and these are commonly referred to as cohesive soil (Fredlund and Rahardjo, 1993; Faisal, 2000; Duncan and Wright, 2005; Huat et al., 2006; Horpibulsuk et al., 2013; Ojo et al., 2016). Table 39.1 shows the quantitative values of soil physical properties in the study area. The amount of moisture present in the particles of cohesive soils typically makes them stick together, which is commonly known as apparent soil cohesion, measured in kPa. Soil cohesiveness depends on consistency, particle size, specific gravity, shear strength, and the saturation state of the soils that occur in the area. Fig. 39.8 shows the soil cohesion value of the basin. The spatial characteristics of soil cohesion vary in different soil types and are estimated as, clay (97.5 kPa), clay loam (82.5 kPa), loam (75 kPa), and sandy loam (64.5 kPa). In addition, clay soil is noted a higher cohesiveness value compared to other soils. The difference in the cohesiveness of soil layers has a stronger effect on slope failures if their physical properties change due to external factors.

39.4.2 Soil internal friction angle (ϕ)

The soil internal friction angle (ϕ) is closely related to the cohesion characteristics of a particular soil, which represents the sliding friction between individual soil grains and within soil particles of a given soil mass (Mitchel, 1993; Ng and Shi, 1998; Lu and Likos, 2004; Lalitha et al., 2021).

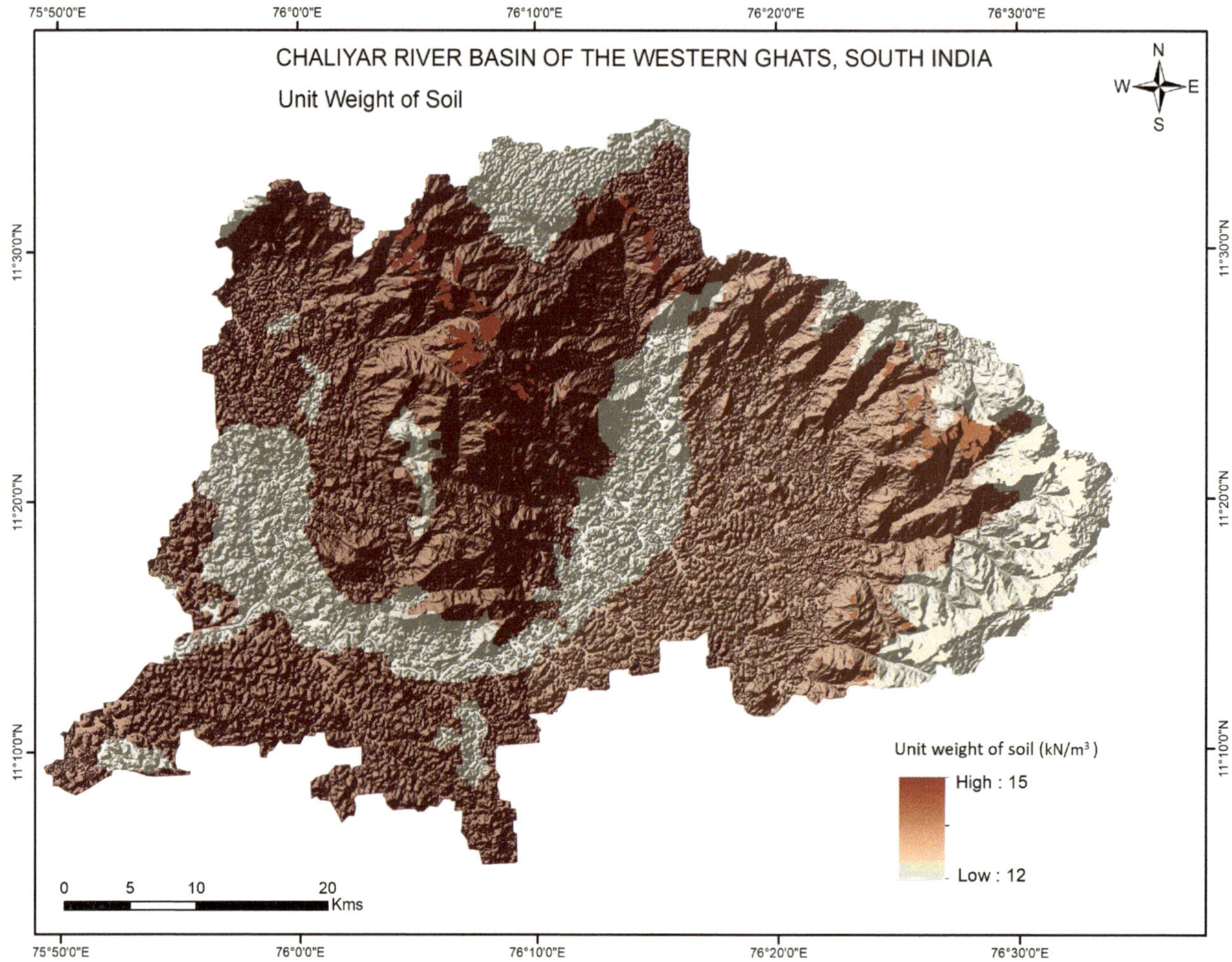

FIGURE 39.10 The quantitative value of the unit weight of soils in the study area.

The angle of internal friction (ϕ) is a function of density, roundness or angularity, and particle size of a particular soil, which represents the sum of sliding friction in slope stability assessment (Rahardjo et al., 1995; Cocka et al., 2004). Fig. 39.9 shows the soil friction and its spatial distribution in the basin. The spatial characteristics of soil friction angle vary depending on grain size (i.e., fine, medium, and coarse) and density of soil, where the higher the density, the larger the friction angle. Table 39.1 shows the statistical values of the internal friction angle of the different soil types in the study area. In the basin, soil friction varies from 23 to 32.5 degrees, with a mean value of 27.5 degrees. Loamy soil consists of a maximum internal friction angle of (32.5 degrees), which is occupied in the middle, northwestern, and southwestern parts along the river course, but clay loam has a lower value at 23 degrees in the eastern parts of the basin. Soil friction value of uplands associated with a steep slope and overburdened materials are highly prone to slope failures that cause mudflow and wedge failure in the basin area.

39.4.3 Unit weight of soil (or) soil density (γ)

The unit weight of soil (also known as unit dry weight) is a physical property of soil that is closely related to the density and aggregate/grain size of the soil. It acts as a controlling factor in the determination of slope stability in association with external forces (Fredlund and Rahardjo, 1993; Duncan and Wright, 2005; Horpibulsuk et al., 2013). The unit weight of soil is defined as the unit soil mass per unit volume. Fig. 39.10 shows the quantitative value of the unit weight of soils in the study area. The unit weight of soil varies from place to place depending on soil type and local topographic conditions, and it is estimated to be in the range of 12–15 kN m^{-3}. The larger area has been noted with a

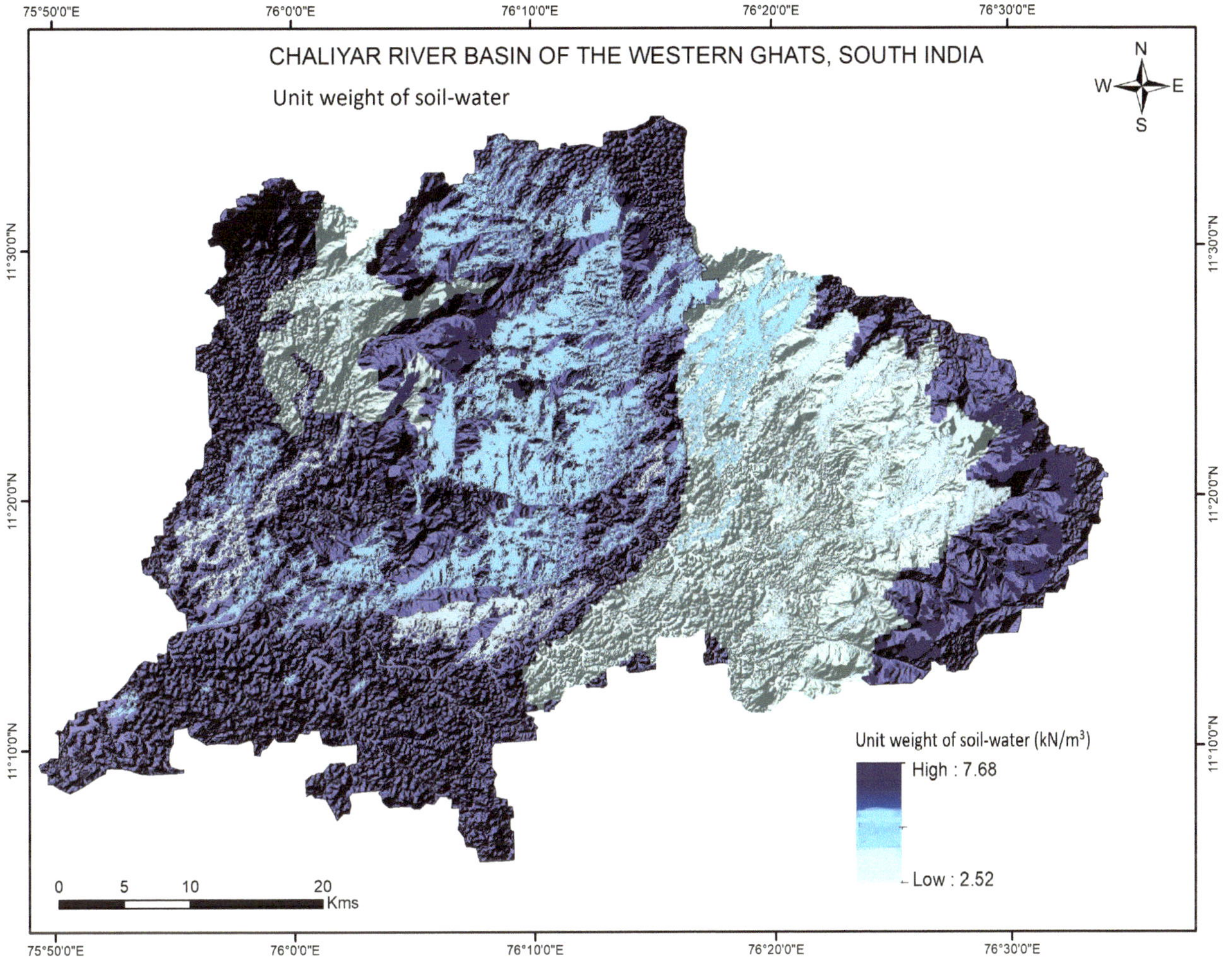

FIGURE 39.11 The quantitative value of the unit weight of soil—water in the study area.

frequent value of 13.5–15 kN m^{-3} in various parts of the basin, including uplands and alluvial plains. The statistical value of the unit weight of soils for different soil types is listed in Table 39.1.

39.4.4 Unit weight of soil—water (or) density of soil—water (γ_w)

The unit weight of soil—water plays a major role in slope stability analysis and influences soil properties such as moisture, cohesiveness, shear strength, and friction angle in the area. Fig. 39.11 shows the quantitative value of the unit weight of soil—water in the study area. The unit weight of soil-water has been noted to be higher in clay (7.68 kN m^{-3}) and clay loam (4.80 kN m^{-3}) found along the river courses, channels, and streams, which helps to retain the soil moisture content of the region throughout all seasons. The 40% of the basin has estimated the unit of weight soil-water as 2.52 kN m^{-3}, and the remaining 60% shows a higher value of 7.68 kN m^{-3}. Table 39.1 shows the statistical value of the unit weight of soil—water in the basin area.

39.4.5 Soil thickness (or) depth of terrain surface (z)

The thickness of the soil represents the average depth of the terrain above the bedrock, which is possible for a sliding zone. Fig. 39.12 shows the soil thickness and its spatial pattern in different soil types of the study area. In the basin, soil thickness is observed to range from 0.12 to 5 m for different soil types. Higher soil thickness in the range of 4–5 m is found in clay and loam soils in the eastern uplands and the southwestern alluvial plains. The terraces along the river course were dominated by clayey and loamy soils composed of gravel, sand, and pebble at different levels.

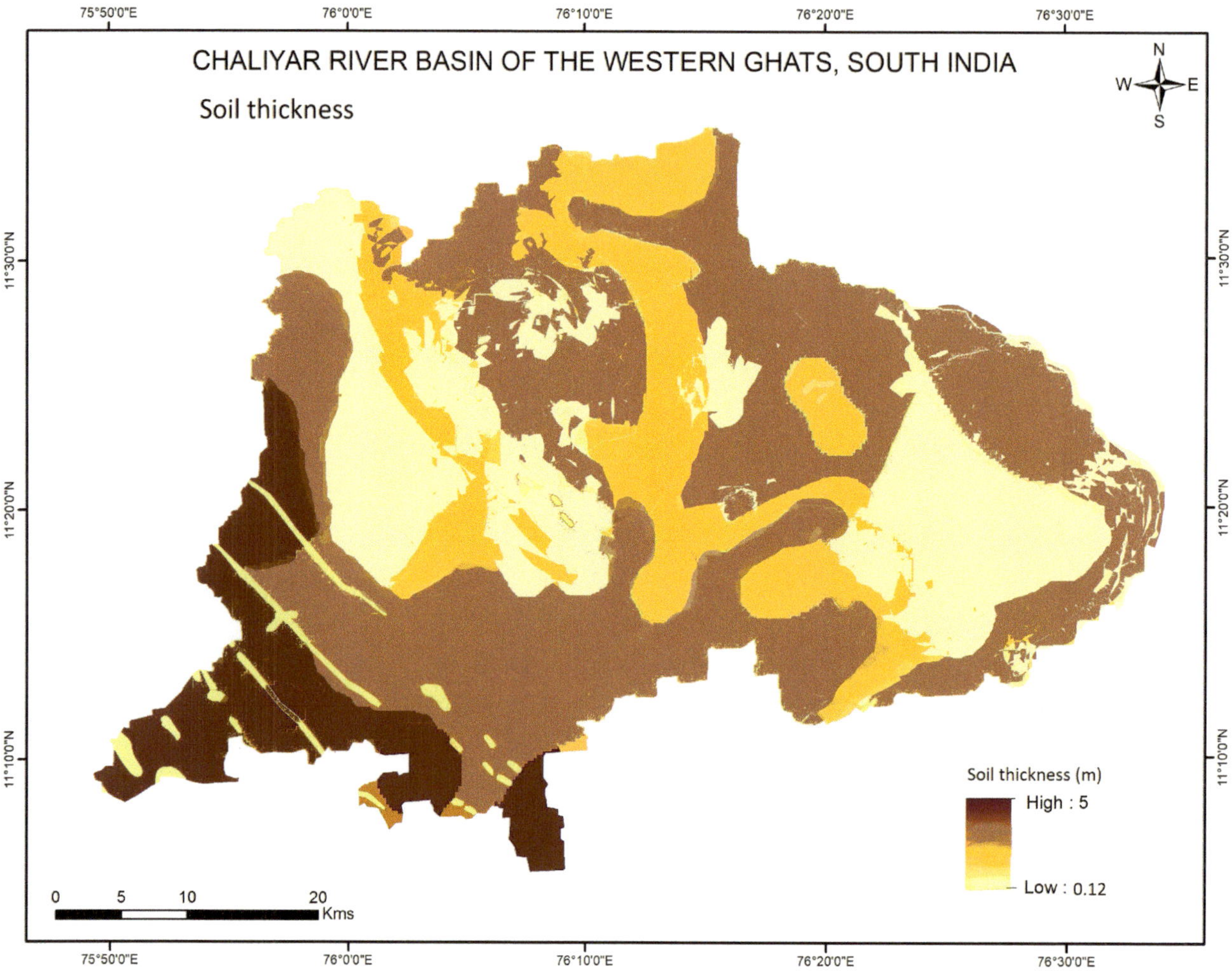

FIGURE 39.12 Spatial pattern of soil thickness in different soil types of the study area.

The mid-land plateau comprises a moderate thickness of clay loam with mixed laterite. Soil thickness is spatially varied along sloped surfaces, valleys, and riverbanks. Spatial variability of soil thickness affects soil cohesiveness and soil friction angle, which influence slope stability characteristics in the area.

39.4.6 Groundwater table depth (or) height of groundwater table (z_w)

In the basin, the majority of the groundwater table has phreatic conditions due to aquifer's formation of weathered and fractured rocks, laterite, and coastal alluvium. Fig. 39.13 shows the groundwater table depth of the study area derived using the inverse distance weighted (IDW) techniques in ArcGIS 10.6 software. The depth of the groundwater table fluctuates in the range of 3.5–5 m above the failure surface and at a depth of < 2.5 m in the rest of the area. Overall, the basin comprises a 1 m depth of groundwater table, but 15% of the area has a 5 m depth. Changing the groundwater table affects the slope stability of an area by altering the soil properties during rainfall seasons.

39.4.7 Groundwater table depth/soil thickness ratio (z_w/z)

The relationship between groundwater table depth and soil thickness plays a significant role in slope failures. Estimating the ratio (m) of groundwater table depth (z_w) and soil thickness (z) was done by dividing their raster layers using ArcGIS—raster calculator tool. In the basin, the soil layer increases with pore water pressure, which decreases the slope stability and leads to potential slope failure in the area. The ratio (m) value is estimated at the range of 0–1

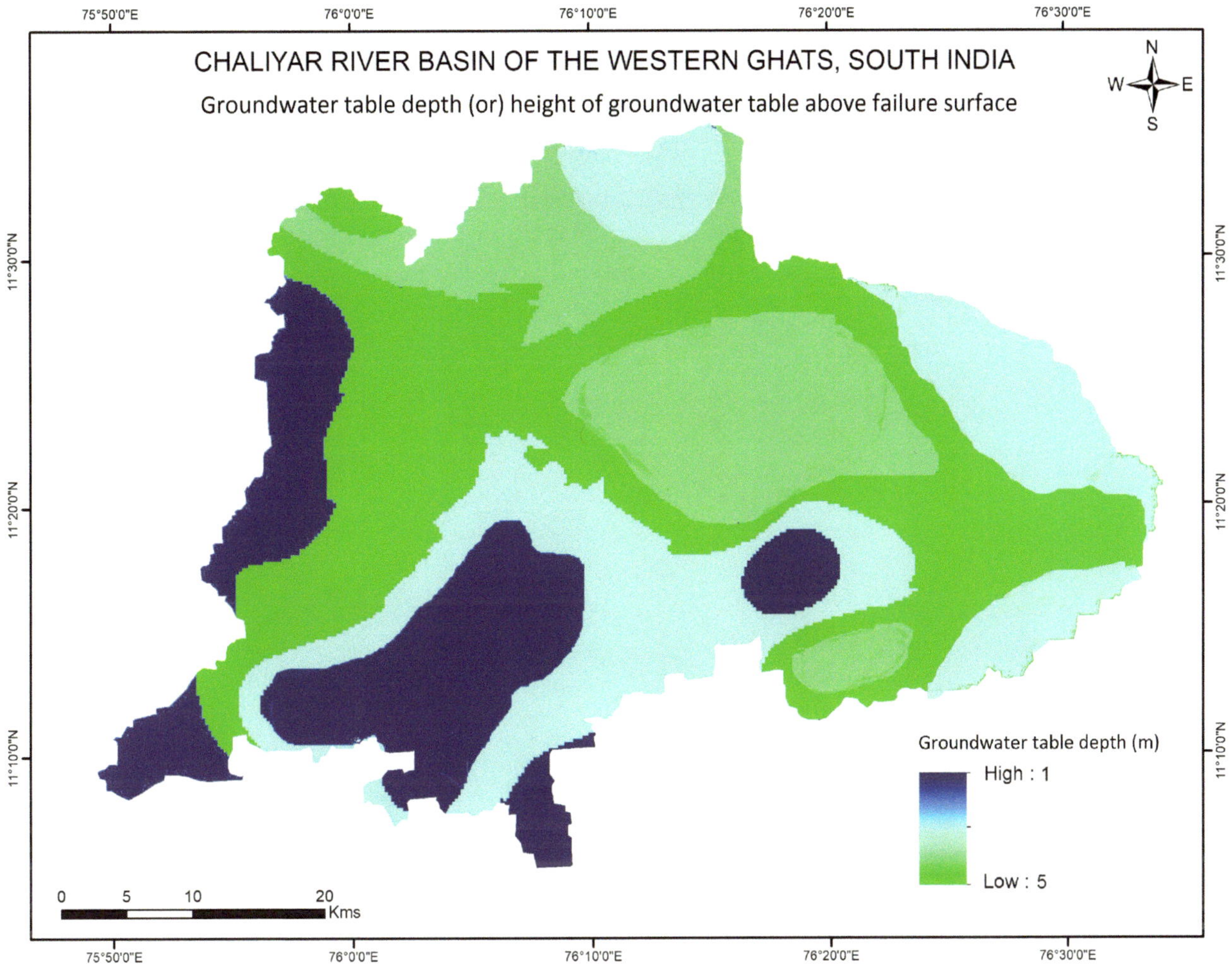

FIGURE 39.13 The depth of the groundwater table below the surface level in the study area.

in various parts of the basin; wherein, the ratio value of 1 is noted along the upland hill ranges and its associated landforms, thereby the groundwater table depth at the range of 2.5–5 m below ground level, within soil layers of deeper thickness, and these zones are more sensitive to slope failure or sliding overburden materials during rainfall seasons.

39.4.8 Slope angle (terrain surface inclination)

Slope angle (also known as slope steepness) is the most important factor for slope stability analyses of an area. The SRTM—DEM (30 m) is used to calculate the slope angle (in degrees) of the basin. Fig. 39.14 shows the slope gradient (topographic) categories of the study area. In the basin, the slope gradient is formed by multidimensional faces with undulated hilly ranges, valleys, laterite plateaus, and coastal plains.

The slope gradient (β) is spatially classified at a range of $>$ 64 degrees in the hilly peaks associated with ridges and steep valleys, and thereby, the slope steepness is involved in the determination of flow accumulation in the basin. The upland ranges have formed at a range of 48–64 degrees, which extends along the interbedded terrain and structural valleys. The midland plateaus have formed a slope gradient in the range of 32–48 degrees, and the coastal plains have noted a slope gradient of $<$ 15 degrees, where the topographic surface is covered by urban and rural settlements, cultivable lands, rivers, and estuaries. In the basin, slopes with greater steepness have a greater risk of slope failure and landslide hazards, whereas surfaces with slope steepness $>$ 60 degrees of slope steepness have been designated as a sensitive zone to slope failure compared to the linear slopes of other areas.

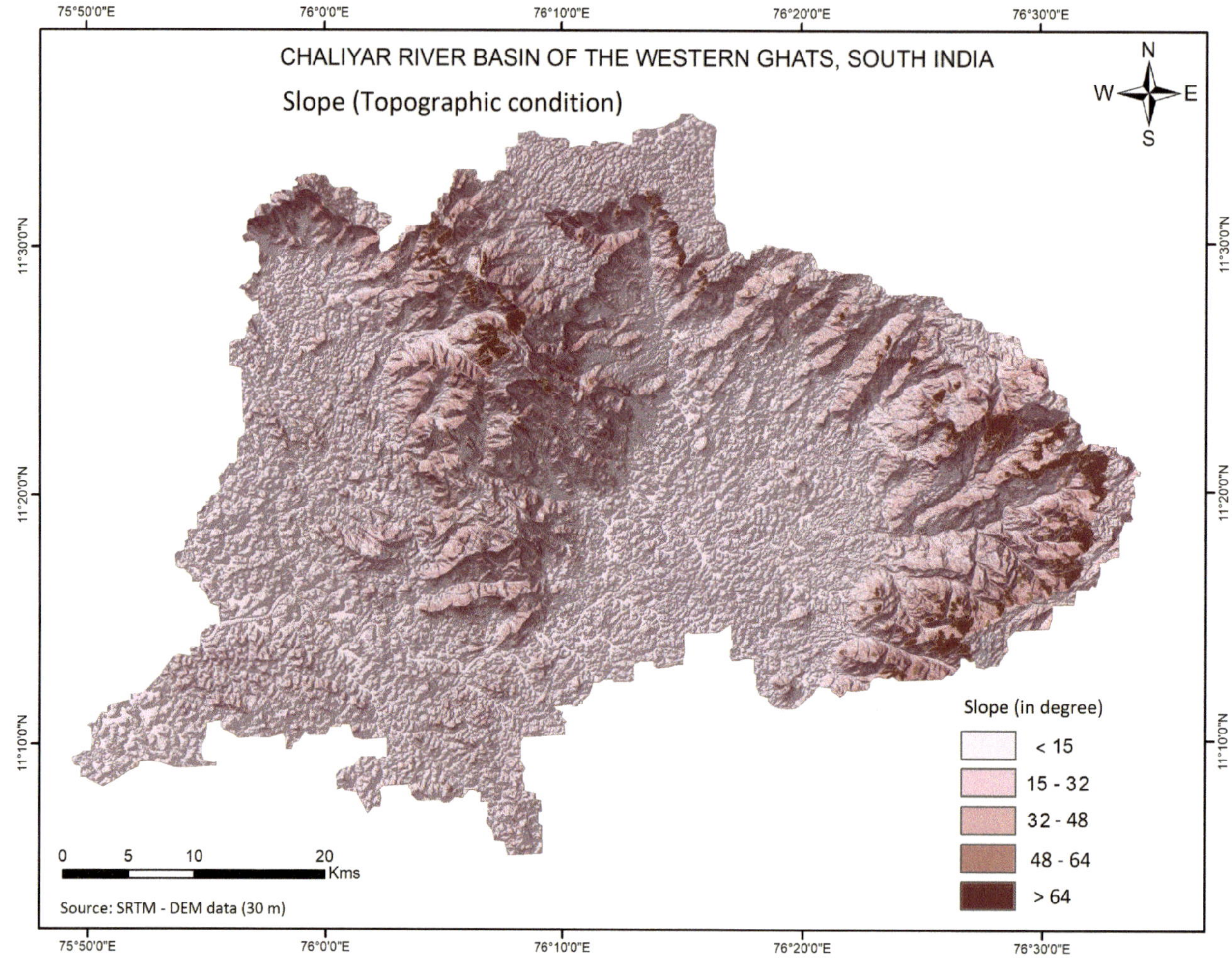

FIGURE 39.14 Slope gradient characteristics (topographic condition) of the study area.

TABLE 39.1 Soil types and their statistical value for physical properties in the study area.

Soil type	Soil physical properties						
	Soil density (g cm^{-3})	Soil moisture (volumetric soil-water content)	Unit weight of soil-water (kN m^{-3})	Soil bulk density (g cm^{-3})	Unit weight of soil (kN m^{-3})	Soil cohesion (c) (kPa)	Soil internal friction angle (ϕ) (degree)
Clay	1.76	60%	7.68	1.45	12.8	97.5	30
Clay loam	1.45	18%	4.80	1.50	12	82.5	23
Loam	1.50	32%	3.45	1.52	14	75	32.5
Sandy loam	1.65	23%	2.52	1.60	15	64.5	25

39.4.9 Slope failure hazard zones: mapping and assessment

The result shows the spatial pattern of FoS in the study area based on the mathematical combination of parametric values of soil properties such as cohesion (c), internal friction angle (ϕ), slope angle (β), unit weight of soil (γ), unit weight of soil–water (γ_w), etc.; thereby, the element of a tangent of internal friction angle (tan ø) and tangent of slope angle (tan α) is directly proportioned to the FoS characteristics and indicates that soil friction angle and slope angle have significant effects on slope stability in the basin area. Fig. 39.15 shows the resulting map of the slope stability index that describes site-specific slope failure vulnerability characteristics of the basin area. The slope stability index value is grouped into five classes: FoS < 1 (severe vulnerability to slope failure), FoS $= 1-2$ (highly vulnerable zone), FoS $= 2-4$ (moderately vulnerable zone), FoS $= 4-5$ (low vulnerable zone), and FoS > 5 (zone of slope stability or very low vulnerable zone). Table 39.2 shows the slope stability index classification and its slope failure vulnerability characteristics for the study area. In the basin, 16% of the area falls under very high vulnerability to slope failure, and this condition mainly occurs in the upland hill ranges formed by hill peaks, ridges, and valleys with greater steepness comprising deeper soil thickness and overburden materials. The area of 22% has been noted as having a high vulnerability to slope failure in the uplands and laterite plateaus associated with sloped valleys and channel flows that accumulate with larger overburden materials. Likewise, 15% of the area covers the moderately vulnerable class in uplands and laterite plateaus that seasonally affect soil properties due to rainfall and other influencing factors. The lowland of coastal, alluvial plains and midland of the plateau, and other gentle sloped landscapes are noted as low-vulnerable zones (18% of the area) and zones of stability (29% of the area), wherein soil properties are constantly stable during all seasons. It is noted that the slope stability of an area is subject to the resistance of soil physical properties under various

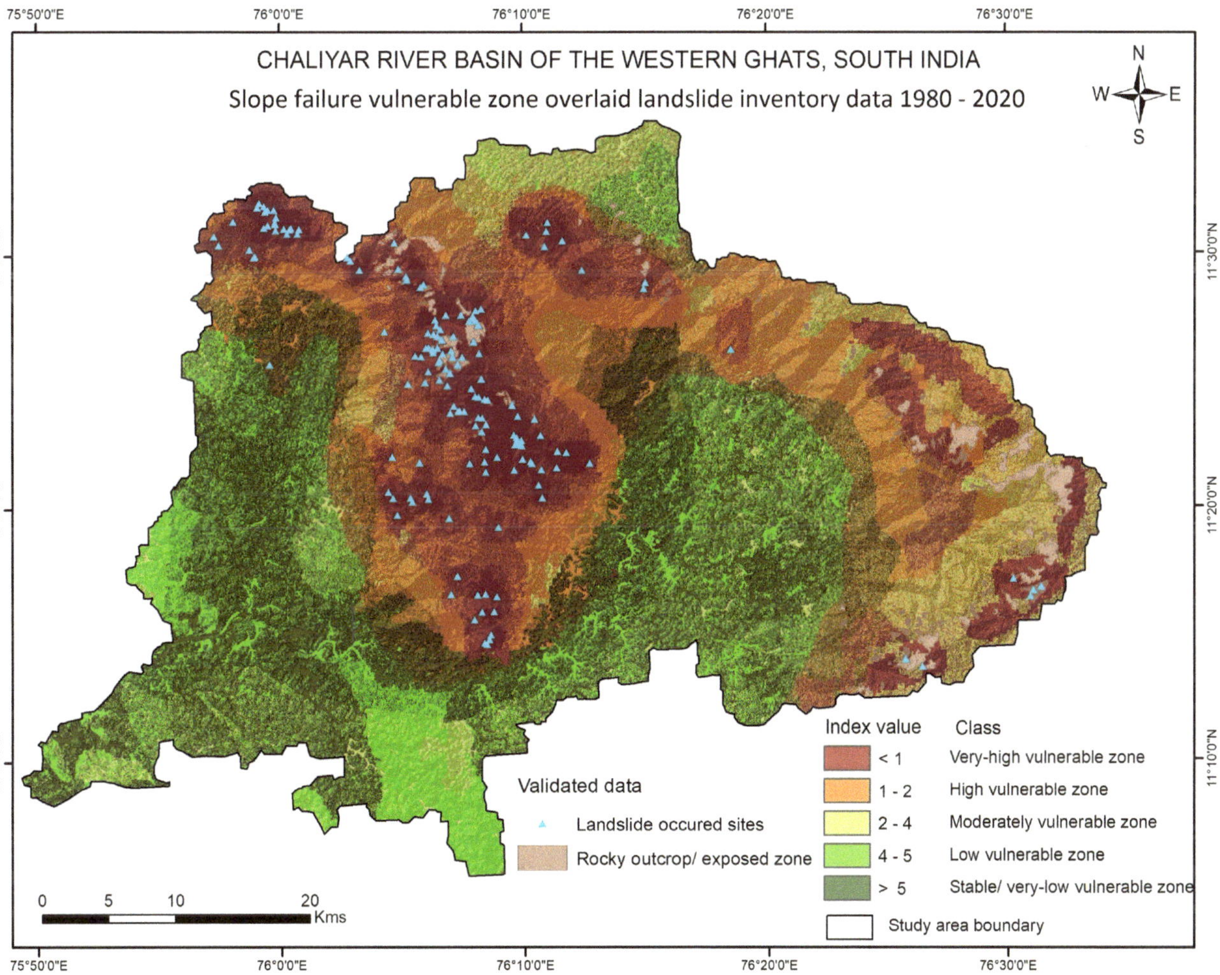

FIGURE 39.15 Slope stability index (FoS) classes and spatial patterns in the study area. *FoS*, Factor of safety.

TABLE 39.2 Classification of the slope stability index and its slope failure vulnerability characteristics in the study area.

Sl. No.	Slope stability index	Vulnerability class	Area (%)	Description
1	> 5	Zone of slope stability (No apparent changes)	29%	Coastal and alluvial plains with higher shear strength of soil properties.
2	4–5	Low-vulnerability zone	18%	Laterite plateau formed as a gentle slope surface with stabilized soil layers
3	2–4	Moderately vulnerable zone	15%	Uplands and laterite plateaus consist of deeper soils that are seasonally affected by rainfall and other influencing factors
4	1–2	High vulnerability to slope failure	22%	Uplands associated with sloped valleys and channel flows accumulated with larger overburdened materials.
5	< 1	Very high vulnerability to slope failure	16%	Upland hilly ranges formed by ridges and valleys with greater steepness that comprise deeper soil thickness and overburdened materials

climatic conditions. From the result, if the area has a slope stability index value < 1, then the soil slope is noted to have unstable conditions and more sensitivity to slope failure, while an index value > 5 denotes that the soil slope is highly stable. Changing soil properties such as shear strength and shear stress due to the adverse impacts of any external factors causes slope failure, followed by landslide activities.

In the basin, the vulnerable zone of slope failure occurs in a widespread area of uplands in the eastern and middle parts and is more susceptible to damage to forest cover, croplands, properties, infrastructure, and settlements. The different parts of the upland hill range have found many hotspots of slope failure, which may be due to the altering of soil layers and overburden materials by stress imposed by gravity or other loading factors; thereby, increasing stress causes a reduction in the strength of the slope (Duncan and Wright, 2005). Within the basin, some locations are categorized as moderately vulnerable (FoS value $2 - 4$); thus, slope strength depends on the topographic (slope gradient) conditions that control slope steepness, soil properties, and flow direction within the area. However, the FoS $= 4-5$ and FoS > 5 were noted in the extensive area, especially covering the plain surface and gently sloped plateaus, and are considered a typical zone of slope stability, indicating many factors contributing to strengthening soil slope characteristics. The different landscapes produce an increase or decrease in soil strength that determines slope stability and/or the possibility of slope failure sites by changing single or combination of multiple influencing factors, including rainfall, soil infiltration rate, soil moisture, fluctuations in the groundwater table, and seepage flow. The resulting map of slope failure vulnerability zones was cross-verified using the historical landslide inventory data of 1980–2020 based on overlay analysis of GIS techniques, and this reveals that a number of shallow landslides were found at several locations within the zones of extremely high and high vulnerability to slope failure. The dynamic change of soil properties, rainfall, and surrounding environmental factors affect slope stability activities in the basin area. Overall, this result shows the slope stability characteristics at a site-specific scale, and it can be used as a primary data source for decision-makers and disaster management authorities.

39.5 Conclusion and limitations

The outcome of the study shows the slope stability characteristics of the basin with its spatial pattern, i.e., the zone of slope stability and (or) the possibility of slope failure sites based on analyzing multiple geoenvironmental parameters using remote sensing and GIS techniques. The resulting map shows that the zone of severe vulnerability to slope failure mainly covers the upland hill ranges associated with hill peaks, ridges, and valleys, which is mainly due to changes in soil properties due to the impacts of any external factors during rainfall. The geo-database of this study can be used for planning and developing effective disaster management activities. The advantage of the resulting map may aid in planning works in the establishment of infrastructure, where cost savings may be made by recognizing, avoiding, or minimizing development works on potentially unstable or unsuitable sites. The findings of this study have certain limitations, i.e., contributing parameters were derived from various secondary data sources; therefore, the predicted result may be found with minimum uncertainty. However, using in-situ samplings and ground-based data, the GIS-based slope stability model can produce results with relatively high accuracy compared to the use of optimum soil properties in this study. Thus, this study has proved that integrated remote sensing and GIS techniques are viable for the assessment and mapping of slope stability in the basin.

Acknowledgment

The corresponding authors thank the Director NCESS for the support and encouragement for the research and are grateful to the Ministry of Earth Sciences (MoES), Government of India, for providing all support. The authors extend their gratitude to USGS—Earth Explorer, IMD, SOI, and GSI for providing the necessary data sources.

References

Abramson, L.W., Lee, T.S., Sharma, S., Boyle, G.M., 2001. Slope Stability and Stabilization Methods, second ed. John Wiley & Sons, Inc.

Ahmad, F., Yahya, A.S., Farooqi, M.A., 2006. Characterization and geotechnical properties of Penang residual soils with emphasis on landslides. Am. J. Env. Sci. 2 (4), 121–128.

Ajin, R.S., Nandakumar, D., Rajaneesh, A., et al., 2022. The tale of three landslides in the Western Ghats, India: lessons to be learnt. Geoenviron. Disasters 9, 16. Available from: https://doi.org/10.1186/s40677-022-00218-1.

Ambili, V., Narayana, A.C., 2014. Tectonic effects on the longitudinal profiles of the Chaliyar river and its tributaries, SW India. Geomorphology 217, 37–47. Available from: https://doi.org/10.1016/j.geomorph.2014.04.013.

Ambili, V., Narayana, A.C., 2022. River drainage response to tectonism: evidence from the Chaliyar river basin, southwestern India. J. Earth Syst. Sci. 131, 96. Available from: https://doi.org/10.1007/s12040-022-01847-8.

Baharom, B., 2005. Stability Analysis of Slopes. Technical Note in Kursus Asas Kejuruteraan Pemulihan Tanah, NASEC, 9–11 May 2005, UiTM, Shah Alam, Malaysia. pp. 23–35.

Bhattarai, P., Tiwari, B., Marui, H., Aoyama, K., 2014. Quantitative slope stability mapping with ArcGIS: prioritize high-way maintenance. Research Gate, https://www.researchgate.net/publication/265119127.

Bromhead, E.N., 1992. Stability of Slopes, second ed. Surrey University Press, London, UK.

Chi, K.H., Park, N.W., Lee, K., 2002. Identification of landslide area using remote sensing data and quantitative assessment of landslide hazard. IEEE Trans. Geosc. Rem. Sens. 36, 2856–2858.

Cho, S.E., Lee, S.R., 2002. Evaluation of surficial stability for homogeneous slopes considering rainfall characteristics. J. Geotech. Geoenv. Eng. 128 (9), 756–763.

Cocka, E., Erol, O., Armangil, F., 2004. Effects of compaction moisture content on the shear strength of an unsaturated clay. Geotech. Geol. Eng. 22, 285–297.

Crozier, M.J., Glade, T., 2004. Landslide hazard and risk: issues, concepts and approach. Landslide Hazard and Risk. John Wiley & Sons, Ltd., pp. 1–40.

Duncan, J.M., 2000. Factors of safety and reliability in geotechnical engineering. Geotech. Geol. Eng. 126 (4), 307–316.

Duncan, J.M., Wright, S.G., 2005. Soil Strength and Slope Stability. John Wiley & Sons, Inc, Hoboken, New Jersey.

Escobar-Wolf, R.V., Sanders, J.D., Oommen, T., Sajinkumar, K.S., Vishnu, C.L., 2021. A GIS tool for infinite slope stability analysis (GIS-TISSA). Geosci. Front. 12 (2), 756–768.

Faisal, H.A., 2000. Unsaturated residual soils and rainfall induced slope failures in Malaysia. In: Rahardjo, Toil & Leong (Eds.), Unsaturated Soils for Asia. Balkema, Rotterdam. pp. 41–52.

Fredlund, D.G., Rahardjo, H., 1993. Soil Mechanics for Unsaturated Soil. John Wiley & Sons, New York.

Geotechnical Design Manual, 2005. Slope Stability Analysis, Chapter 7. Department of Transportation, Washington (WSDOT).

Glade, T., Crozier, M., 2010. Landslide geomorphology in changing environment. Geomorphology 120, 1–90. Available from: https://doi.org/10.1016/j.geomorph.2009.09.018.

Hao, L., van Westen, C., Rajaneesh, A., Sajinkumar, K.S., Martha, T.R., Jaiswal, P., 2022. Evaluating the relation between land use changes and the 2018 landslide disaster in Kerala, India, for land use change analysis. CATENA 216, 106363.

Horpibulsuk, S., Suddeepong, A., Chamket, P., Chinkulkijniwat, A., 2013. Compaction behavior of fine-grained soils, lateritic soils and crushed rocks. Soils Found. 53 (1), 166–172. Available from: https://doi.org/10.1016/j.sandf.2012.12.012.

Huat, B.K., Ali, F., Rajoo, R.S.K., 2006. Stability analysis and stability chart for unsaturated residual soil slope. Am. J. Env. Sci. 2 (4), 154–160.

Huat, B.K., Sew, G.S., Ali, F., 2004. Tropical Residual Soils Engineering. Taylor & Francis Group, London.

Kabir, M.H., Taha, M.R., 2004. Assessment of physical properties of a granite residual soil as an isolation barrier. Electron. J. Geotech. Eng. 9, 0408.

Kaliraj, S., et al., 2012. Application of remote sensing in forest cover change detection using geo statistical change detection matrices—a case study in Devampatti RF, Tamil Nadu. Nat. Env. & Poll. Tech. 11 (2), 261–269.

Kaliraj, S., Chandrasekar, N., Magesh, N.S., 2014. Identification of potential groundwater recharge zones in Vaigai upper basin, Tamil Nadu, using GIS-based analytical hierarchical process (AHP) technique. Arab. J. Geosci. 7, 1385–1401. Available from: https://doi.org/10.1007/s12517-013-0849-x.

Kaliraj, S., Chandrasekar, N., Magesh, N.S., 2015a. Evaluation of multiple environmental factors for site-specific groundwater recharge structures in the Vaigai River upper basin, Tamil Nadu, India, using GIS-based weighted overlay analysis. Env. Earth Sci. 74, 4355–4380. Available from: https://doi.org/10.1007/s12665-015-4384-9.

Kaliraj, S., Chandrasekar, N., Magesh, N.S., 2015b. Morphometric analysis of the River Thamirabarani sub-basin in Kanyakumari District, South west coast of Tamil Nadu, India, using remote sensing and GIS. Env. Earth Sci. 73, 7375–7401. Available from: https://doi.org/10.1007/s12665-014-3914-1.

Kaliraj, S., Ramachandran, K.K., Chandrasekar, N., 2021. Modeling of coastal environmental vulnerability in South India: a multiple parametric approach using remote sensing and GIS. In: Rani, M. et al. (Eds.), Remote Sensing of Ocean and Coastal Environments, 225–249. Available from: https://doi.org/10.1016/B978-0-12-819604-5.00014-7.

Kang, S., Lee, S.R., Cho, S.E., 2020. Slope stability analysis of unsaturated soil slopes based on the site-specific characteristics: a case study of Hwang Yeong Mountain, Busan, Korea. Sustainability 12 (7), 2839. Available from: https://www.mdpi.com/2071-1050/12/7/2839.

Lalitha, M., Kumar, K.S.A., Nair, K.M., Kaliraj, S., et al., 2021. Evaluating pedogenesis and soil Atterberg limits for inducing landslides in the Western Ghats, Idukki District of Kerala, South India. Nat. Hazards 106, 487–507. Available from: https://doi.org/10.1007/s11069-020-04472-0.

Liew, S.S., 2004. Slope Failures in Tropical Residual Soils, Tropical Residual Soils Engineering (TRSE).

Lu, N., Likos, W.J., 2004. Unsaturated Soil Mechanics. John Wiley & Sons, Inc, New Jersey.

Mantovani, F., Soeters, R., VanWesten, C., 1996. Remote sensing techniques for landslide studies and hazard zonation in Europe. Geomorphology 15, 213–225.

Marrison, I.M., Greenwood, J.R., 1989. Assumptions in simplified slope stability analysis by the method of slices. Geotechnique 39 (3), 503. 209.

Mitchel, J.K., 1993. Fundamental of Soil Behaviour, second ed. John Wiley & Sons, Inc, USA.

Moayedi, H., Bui, D.T., Foong, L.K., 2019. Effect of soil texture on drying pattern of soil moisture after saturation. Int. J. Curr. Microbiol. App. Sci. 8 (3), 697–704. Available from: https://doi.org/10.20546/ijcmas.2019.803.086.

Moayedi, H., Tien Bui, D., Kok Foong, L., 2019. Slope stability monitoring using novel remote sensing based fuzzy logic. Sensors 19 (21), 4636. Available from: https://www.mdpi.com/1424-8220/19/21/4636.

Mohamed, T.A., Ali, F., Hashim, S., Huat, B.K., 2006. Relationship between shear strength and soil water characteristic curve of an unsaturated granitic residual soil. Am. J. Env. Sci. 2 (4), 142–145.

Ng, C.W.W., Shi, Q., 1998. A numerical investigation of the stability of unsaturated soil slopes subjected to transient seepage. Comput. Geotech. 22 (1), 1–28.

Ojo, G.P., Igbokwe, U.G., Nwozor, K.K., Egbuachor, C.J., 2016. Geotechnical properties of lateritic overburden materials on the charnockite and gneiss complexes in Ipele-Owo Area, Southwest. Nigeria. Am. J. Eng. Res. (AJER) 5 (9), 53–59. Available from: https://www.ajer.org/papers/v5(09)/I0509053059.pdf.

Pack, R.T., Tarboton, D.G., Goodwin, C.N., 1998. A stability index approach to Terrain stability hazard mapping. SINMAP User's Manual. Forest Renewal British Columbia, Canada.

Rahardjo, H., Lim, T.T., Chang, M.F., Fredlund, D.G., 1995. Shear-strength characteristics of a residual soil. Can. Geotech. J. 32, 60–77.

Raneesh, K.Y., Santosh, G.T., 2011. A study on the impact of climate change on streamflow at the watershed scale in the humid trophics. Hydrol. Sci. J. 56 (6), 946–965.

Rawat, M.S., Joshi, V., Sundriyal, Y.P., 2016. Slope stability analysis in a part of East Sikkim, using Remote Sensing & GIS. In: 2nd International Conference on Next Generation Computing Technologies (NGCT). IEEE, pp. 51–60. Available from: https://ieeexplore.ieee.org/abstract/document/7877389.

Sajinkumar, K.S., Oommen, T., 2021. Landslide atlas of Kerala. Geol. Soc. India 34.

Sarma, S.K., 1973. Stability analysis of embankments and slopes. Geotechnique 23 (3), 423–433.

Singhroy, V., 2004. Remote sensing of landslides. Landslide Hazard and Risk. John Wiley & Sons, Ltd., pp. 469–492.

VanWesten, C.J., 2000. The modeling of landslide hazards using GIS. Surv. Geophys. 21, 241–255.

Wadhawan, S.K., Singh, B., Ramesh, M.V., 2020. Causative factors of landslides 2019: case study in Malappuram and Wayanad districts of Kerala. India Landslides 17, 2689–2697.

Walker, J.P., Kandel, D.D., 2004. Brightness temperature versus surface soil moisture assimilation. American Geophysical Union abstract H12A 01.

Wu, W., Siddle, R.C., 1995. A distributed slope stability model for steep forested Basins. Water Resour. Res. 31 (8), 2097–2110.

Yunus, A.P., Fan, X., Subramanian, S.S., Jie, D., Xu, Q., 2021. Unraveling the drivers of intensified landslide regimes in Western Ghats, India. Sci. Total. Environ. 770, 145357.

Chapter 40

Application of geospatial modeling for assessment of land degradation susceptibility

K.S. Karthika[1], Mohamed A.E. AbdelRahman[2], K.S. Anil Kumar[1] and I. Rashmi[3]

[1]*ICAR-National Bureau of Soil Survey and Land Use Planning, Regional Centre, Bangalore, Karnataka, India,* [2]*Division of Environmental Studies and Land Use, National Authority for Remote Sensing and Space Sciences (NARSS), Cairo, Egypt,* [3]*ICAR-Indian Institute of Soil and Water Conservation, Regional Centre, Kota, Rajasthan, India*

40.1 Introduction

Land degradation is one of the widespread problems that we face globally. It is defined as the set of processes that reduce the capability of land to produce goods or services, thereby decreasing their quality. This could result from various natural or anthropogenic activities. Land degradation could be due to poor management of our natural resources, such as soil, water, and vegetation, resulting in a decline in soil fertility and productivity, loss of soil organic matter, and several other negative impacts. Thus, it could result in loss of biological diversity, ecological integrity, and value to humans (Olsson et al., 2019). The processes that diminish the quality of land could be physical, chemical, or biological, indicating three different aspects of land degradation. Thus, it becomes an environmental threat, as it seriously affects the ecosystem services. Changes in climate, land use, terrain, etc. can result from land degradation. The physical factors that lead to degradation include variations in slope, land use, and soil characteristics, such as erodibility and unsustainable management practices (Dubovyk, 2017).

40.2 Land degradation susceptibility

The susceptibility of land to degradation processes is influenced by several factors. These include soil properties, topographic indicators, climate, geology, and geomorphological processes. These could be influenced mainly by changes in land use and the level of land management. The productivity of land could be reduced mainly by physical, chemical, or biological degradation that occurs due to a decline in soil structure, processes of soil erosion and sedimentation, soil acidity, soil sodicity, etc. The susceptibility of land to degradation is explained hereunder.

40.3 Physical degradation

(i) Decline in soil structure

It is one of the physical degradation processes. When there are changes in soil aggregate stability and porosity, a decline in soil structure naturally occurs, which negatively affects plant growth. The changes in aggregate stability could be due to anthropogenic interventions such as increased cropping, tillage, and excessive grazing, which in turn affect soil organic matter. A decline in soil organic matter, in turn, affects soil porosity. A decline in soil porosity could also result from processes that cause soil compaction. The changes in porosity have effects on soil hydrology, aeration, and development. Thus, soil structure decline is one of the potential reasons for a land to be inherently susceptible to land degradation.

Remote Sensing of Soils. DOI: https://doi.org/10.1016/B978-0-443-18773-5.00042-9

(ii) Soil erosion

Soil erosion is the detachment or breaking away of soil particles from a land surface by some erosive agent, most commonly water or wind, and the subsequent transportation of the detached particles to another location. Erosion is a natural process and is a critical factor in soil formation from rock parent material. However, once productive agricultural soils have been formed over periods of thousands or millions of years, erosion of the soil material is usually very low or negligible because of the impacts of protective natural plants and residual cover. Human activities are responsible for greatly accelerating erosion rates, usually by reducing or eliminating plants and residual cover. This exposes the soil to wind and water erosion forces, weakening the soil cohesive forces by tillage disturbance and increasing the erosive agents, particularly by activities that increase surface runoff. The susceptibility of land to soil erosion depends on many factors, which are interrelated. This could be related to rainfall, soil properties, vegetation cover, and landscape properties. The soil properties could be soil depth, permeability, cohesion, and dispersibility.

(iii) Erosion by water

Water erosion results from the removal of soil material by flowing water. The most common types of soil erosion by water are sheet and rill erosion in upland areas, channel and gully erosion in small watersheds, and stream channel and bank erosion in larger catchments.

(iv) Erosion by wind

Erosion by wind occurs when wind speed exceeds a certain critical or threshold value. Soil particles can be detached and moved through suspension, saltation, or creep. The suspension usually lifts the smallest soil particles (clay, silt, and organic matter) so high into the air mass that they are easily kept in motion and can travel for long distances. Soil particles that move by creep are larger sand grains and aggregates that stay in contact with the soil surface at almost all times; their motion is often through rolling and bouncing. Saltating soil particles are usually moderate in size, and, once detached, move in trajectories up into the air and then back down to the soil surface. The susceptibility of land to wind erosion is governed by the soil properties, climate, and landscape. Topsoil erodibility, soil structure, texture, stoniness, and organic matter are some of the factors that determine the susceptibility of land to wind erosion.

40.4 Soil chemical degradation

Soil chemical degradation can be described as an undesirable change in soil chemical properties such as pH, size, and composition of cation exchange complexes, contents of organic matter, mineral nutrients, and soluble salts. Changes in one or more of these properties often have direct or indirect adverse effects on the chemical fertility of soils, which can lead to a decrease in soil productivity. The most widespread types of chemical degradation in soils are an excessive decrease or increase in pH (acidity or alkalinity), an increase in soluble salt content (salinity), a decrease in organic matter content, and a loss of mineral nutrients through leaching or crop removal.

40.5 Assessment of land degradation susceptibility and vulnerability

The methods to assess land degradation are based on these processes of degradation, such as physical, chemical, or biological. The various approaches to land degradation assessment are summarized in Table 40.1. There are different

TABLE 40.1 Various approaches to assess land degradation.

Sl. No	Approach
1	Subjective approach: direct field observations using diagnostic criteria or simple visual indices
2	Assessing landscape units and subjective rating
3	Parametric semiquantitative method
4	Computer automation (rating system and decision trees)
5	Computer modeling (simulation modeling)

methods of land degradation assessment. These include expert opinion (EO), field monitoring, observations and measurement, and geospatial technologies. Geospatial technology includes remote sensing, the Geographic Information System (GIS), and the Global Positioning System. This involves acquiring data referenced to the earth and using it for analysis, modeling, simulations, and visualization.

Remote sensing has a very significant role in land degradation assessment, mapping, and monitoring at multiple spatial and temporal scales (Dubovyk, 2017). Remote sensing is used in the assessment of land degradation by mapping the extent, type, and severity of land degradation at different spatial scales. Remote sensing also helps in analyzing driving factors of land degradation, spatial decision support for land rehabilitation activities, and spatial assessments of the impacts of either land degradation or land rehabilitation activities (Dubovyk, 2017). The different remote sensing methods developed for land degradation studies are given in Table 40.2.

Assessing land degradation is the primary prerequisite to reversing it. This requires data, which can be obtained through remote sensing and GIS applications. Though there are several methods to assess land degradation, remote sensing is the most cost-effective method, particularly over extensive areas. In the present situation, the only method to monitor land degradation at different spatial and temporal scales in a spatially explicit and continuous manner is remote sensing, which is facilitated through satellite-based assessment (Sivakumar and Stefanski, 2007).

Several steps are involved in the assessment of land degradation using remote sensing and GIS techniques, which include (1) identification of physiographic units; (2) preparation of a base map in order to assess soils, degradation, and conservation; (3) to overlay data layers for different map units; (4) to calculate area; (5) linking spatial data with non-spatial data and to receive detailed attribute data; (6) to make geo-referenced information easily accessible to non-GIS users; (7) to "bridge the scale gap," that is, upgrade experimental results from small plots to larger areas; (8) to present data in map and other graphic format; (9) to map (temporal and spatial changes in) land cover and land use; and (10) to identify areas of degradation (Godert and van Lynden, 2001). Remote sensing techniques play both main and supporting roles in assessing land degradation. The mapping of land degradation, its type, and its severity at different spatial scales are the main roles played by remote sensing. Supporting roles are in the analyses of driving factors of land degradation (Dubovyk et al., 2016), in land rehabilitation activities by providing spatial decision support (Dubovyk et al., 2013a), and in assessing the impacts of land degradation in land rehabilitation activities (Cano et al., 2009).

40.6 Assessment of soil degradation

There are several methodologies to assess soil degradation. The major methods include (1) GLASOD—Global assessment of human-induced soil degradation; (2) ASSOD—Assessment of the status of human-induced soil degradation; (3) WOCAT—World overview of conservation approaches and technologies; (4) classification approach; (5) indicators approach; and (6) modeling of soil erosion and soil loss.

40.6.1 GLASOD

This approach has been applied worldwide based on the GLASOD Survey, which functions based on responses to a questionnaire sent to experts in countries worldwide. This helps gather data on the distribution and intensity of physical and chemical degradation. This method helps in mapping areas of similar severity for erosion risk. Based on the results of the survey, restoration of degraded lands could be planned. However, this method was purely based on EO and did not include remote sensing and field measurements.

TABLE 40.2 Remote sensing methods used for land degradation assessment.

Year	Methods
1970–80	Visual interpretation of aerial photos, photogrammetric methods, manual mapping
1980–90	Image classification, map digitization, expert mapping, photogrammetric methods, manual mapping
1990–2000	Image classification, spectral transformation, change detection, semiquantitative image analysis, expert mapping
2000–10 +	Time series analysis, data fusion, land degradation modeling, image classification, spectral transformation

Source: Modified from Dubovyk O (2017) The role of remote sensing in land degradation assessments: opportunities and challenges. Eur. J. Remote. Sens. 50(1): 601–613.

40.6.2 ASSOD

This is a more accurate study, and it is accurate as it quantifies degradation with its impact on productivity (negligible, light, moderate, strong, or extreme impact). This classification is based on an estimation of the changes in productivity and also takes the level of management into consideration. Changes in productivity are expressed in relative terms, that is, the current average productivity compared to the average productivity in the nondegraded situation (or nonimproved, where applicable) and in relation to inputs.

40.6.3 WOCAT

This is a standardized method for assessing soil degradation and soil conservation. This has developed a set of standardized tools to document, monitor, and evaluate soil and water conservation know-how worldwide. Using this methodology, soil degradation could be mapped, and conservation activities could be planned for better decision-making.

40.6.4 Classification approach

This method to map soil erosion was developed by the Agricultural Research Institute (RALA) and Soil Conservation Service (LR) and is partly based on the methods used to determine the condition of soils in New Zealand and New South Wales in Australia (Arnalds et al., 2001). The method is thus called the RALA/LR classification methods for mapping soil erosion. The methods developed have four primary characteristics: (1) classifying erosion according to erosion forms; (2) applying standard scales to assess the severity of soil erosion; (3) using satellite images as the basic map and as an aid for field mapping; and (4) using GIS.

40.6.5 Indicators approach

Indicators are used to assess soil erosion in this methodology. These could be easily assessable indicators to understand the effects of erosion. The indicators in a study by de Bei (2005) were eroded clods, flow surfaces, pre-rills, and rills. Indicators were expressed in terms of percentage incidence of bare area; the thickness of soil accumulation over solid subsoil was also assessed (de Bei, 2005). The study collected data on soil/terrain, land cover, infrastructure, land management, and the incidence of soil erosion.

40.7 Application of modeling to land degradation susceptibility and vulnerability

40.7.1 Soil erosion and soil loss

Soil erosion is one of the forms of soil degradation. Several factors, either physical or chemical deterioration, can result in accelerated soil erosion. Estimation of soil loss and developing an understanding of the areas degraded due to soil erosion are necessary to carry out conservation planning effectively (Yadav and Sidhu, 2010). Remote sensing and GIS techniques have considerable potential for assessing soil erosion hazards. These are powerful tools for mapping soil erosion risks (Peng et al., 2008). Several studies prove this fact too. Remote sensing and GIS are powerful tools for characterizing soil erosion in large areas as well (Sheikh et al., 2011; Biswas, 2012). The traditional methods yield point-based information; however, remote sensing methods are advantageous over those methods in measuring hydrologic parameters on spatial scales, while GIS integrates the spatial analytical functionality for spatially distributed data.

Integration of remote sensing and GIS could aid in the development of models to assess soil erosion hazard practically by utilizing soil, terrain, and climatic parameters. A flowchart indicating the integration of several factors in modeling soil erosion and mapping land degradation is presented in Fig. 40.1. Soil erosion status was mapped using visual interpretation of the FCCs obtained from LISS III and IV sensors (with 24 and 5.8 m spatial resolution, respectively). These were evaluated for delineation of eroded areas based on soil, slope, and landuse/land cover in Chamarajanagar area of Karnataka (Fig. 40.2).

Remote sensing provides data on soil types, lithological units, and vegetative cover with more accuracy at a relatively reasonable cost. Quantitative assessment of soil erosion can be done by employing GIS techniques in different soil erosion models. Modeling is carried out to assess the land degradation caused by soil erosion and changes in land cover. There are several models that are employed for the same. Some of the known models used in assessing soil erosion are USLE (Universal Soil Loss Equation,Wischmeier and Smith, 1965), EPIC (Erosion/Productivity Impact Calculator, Williams et al., 1990), EUROSEM (European Soil Erosion Model, Morgan et al., 1998), RUSLE (Revised

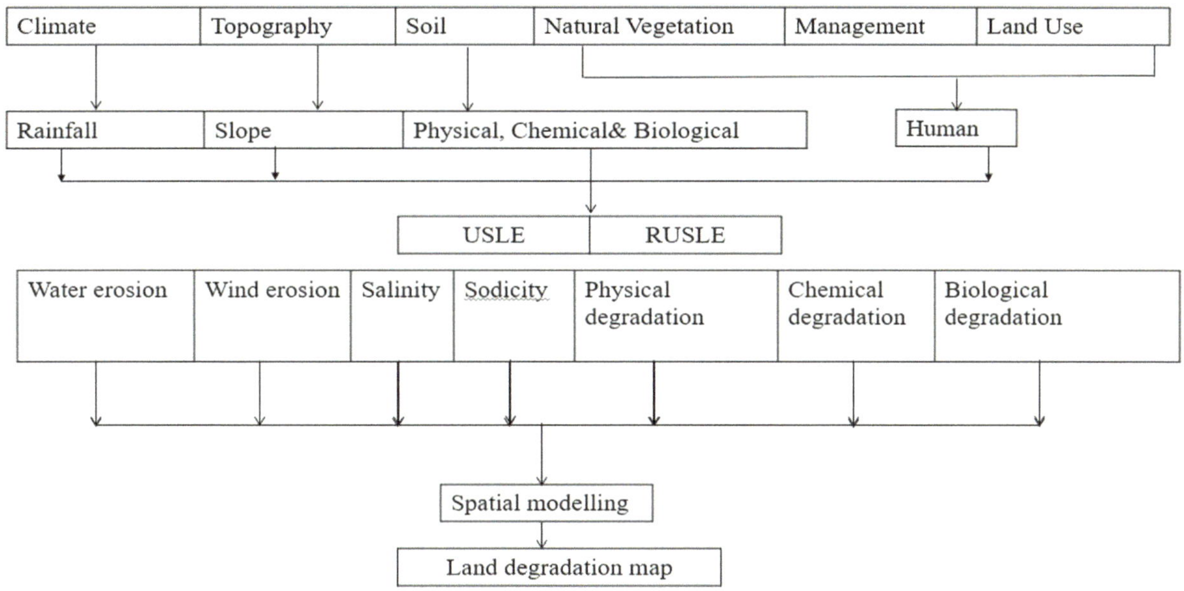

FIGURE 40.1 Flowchart indicating the various processes involved in deriving land degradation map.

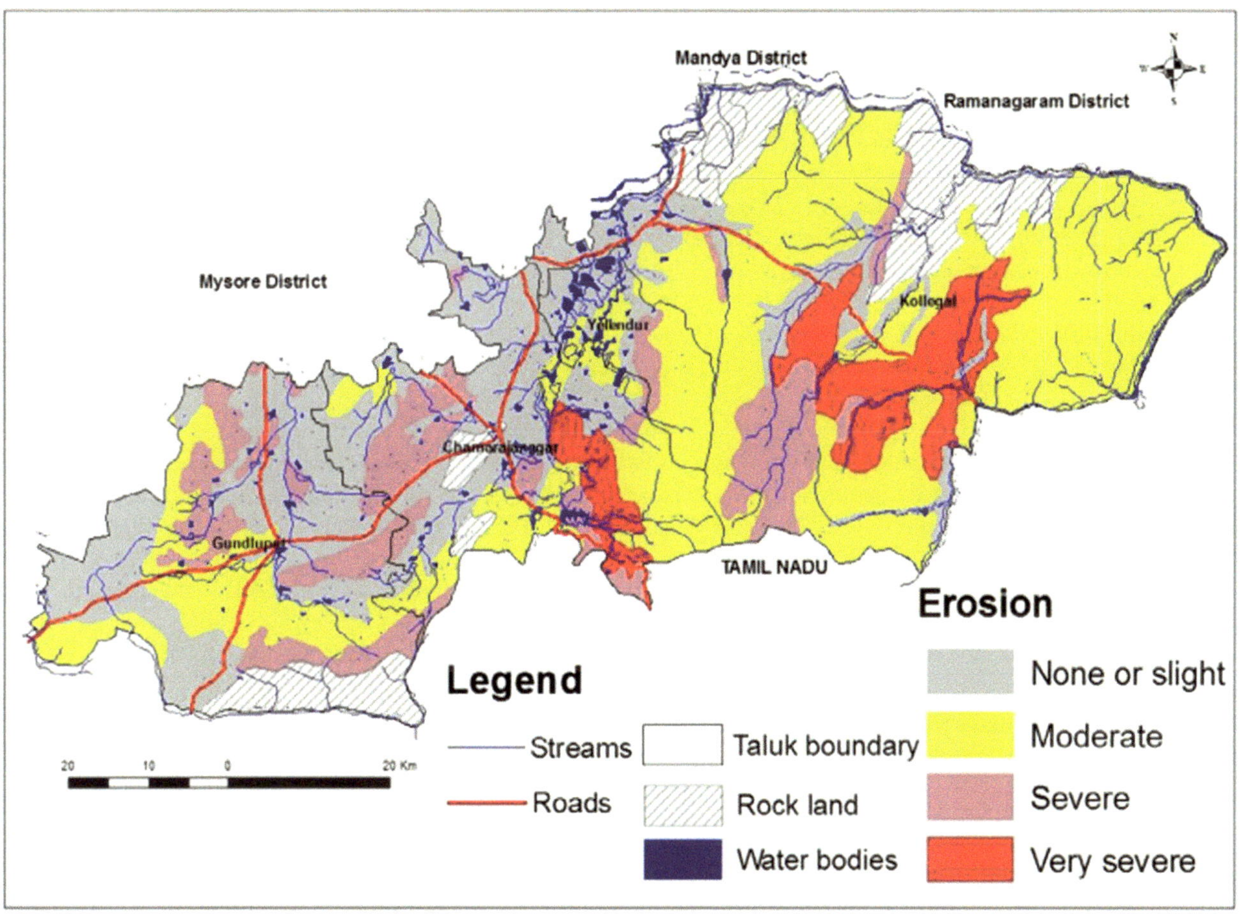

FIGURE 40.2 Mapping soil erosion based on satellite data.

Universal Soil Loss Equation, Renard et al., 1997), Rill Grow (a model for rill initiation and development, Favis-Mortlock et al., 1998), SEMMED (Soil Erosion Model for Mediterranean Regions, de Jong et al., 1999), EGEM (Ephemeral Gully Erosion Model, Woodward, 1999), PESERA (Pan-European Soil Erosion Risk Assessment, Kirkby et al., 2004), and so forth Bahrawi et al., 2016. Remote sensing data provides for the factors in all these models.

40.7.2 USLE model

The USLE model has been used to assess soil loss by water. This is mainly utilized for long-term soil loss. This is a simple empirical model, which is given as

$$A = R \times K \times L \times S \times C \times P$$

where A is the mean annual soil loss (t ha^{-1} year^{-1}), R is the rainfall erosivity factor [(MJ.mm) (ha. h.y)$^{-1}$], K is the soil erodibility factor [(t.ha.h) (ha.MJ.mm)$^{-1}$], L is the slope factor (dimensionless), S is the slope length factor (dimensionless), C is the cover management factor (dimensionless), and P are factors for human practices aimed at erosion control.

The USLE model is considered to be the best model for assessing surface soil loss (Chatterjee et al., 2014; Kourgialas et al., 2016). Considering it as an example, factors such as rainfall erosivity factor (R), soil erodibility factor (K), soil slope length (L), soil slope (S), cover management factor (C), and support practices factor (P) are derived out of the data obtained from remote sensing.

40.7.3 RUSLE model

The RUSLE model uses all factors of USLE, but P-factors for human practices aimed at erosion control were added. This model estimates erosion and sediment yields. The estimation of soil loss is at its best by the revised model of USLE, that is, RUSLE (Napoli et al., 2016; Rawat et al., 2016). An average annual soil loss of 69 t ha^{-1} year^{-1} was estimated using satellite data, GIS, and RUSLE in the Daltonganj watershed of Jharkhand (Tirkey et al., 2013). Soil erosion susceptibility was estimated in the Ikkour watershed in Morocco using a combination of remote sensing, USLE, and spectral indices integrated with GIS techniques (Jazouli et al., 2017). A total soil loss of 2,55,283 t was estimated, and the spatial distribution of soil loss rate in the Koga watershed of northwestern Ethiopia using the RUSLE was simulated by GIS and remote sensing techniques (Gelagay and Minale, 2016). Soil erosion vulnerability in the arid parts of the south Deccan plateau was mapped, and erosion probability zones were delineated using the RUSLE model simulated by remote sensing and GIS (Srinivasan et al., 2021) (Fig. 40.3). Similarly, soil erosion risk mapping was carried out in the Subarnarekha watershed (Samanta et al., 2016). The GIS tools were integrated with the AHP method to map potential erosion watersheds and their spatial patterns in the Chitral district of Pakistan, where the RUSLE model was utilized to quantify soil erosion (Aslam et al., 2021).

40.7.4 CORINE model

CORINE stands for the Coordination of Information on the Environment. This is another model with GIS to estimate erosion and map it by considering the factors of vegetative cover, land slope, meteorological conditions, and soil properties. The CORINE model could be used to delineate soil erosion risk and discriminate the potential soil erosion risk (PSER) areas (Tayebi et al., 2017). This is a semiquantitative model, mainly used by the European community for assessing the risk of soil erosion. However, a study indicated that the CORINE model could also be applicable in the monsoon regions of China (Zhu, 2011). CORINE model could be successful in mapping the PSER based on rainfall erosivity, soil erodibility, and topography and the actual soil erosion risk based on PSER and vegetative cover.

40.7.5 PESERA model

The pan-European Soil Erosion Risk Assessment model is a process-based and spatially distributed model to quantify soil erosion by water and assess the risk mainly in Europe. This model can also be extended to include estimates of wind erosion and tillage. The model is mainly applied at regional levels in Europe and was also found useful for scenario analysis for different land uses and climate changes.

There are several other models for soil erosion assessment. These include SLEMSA: The Soil Loss Estimation Equation for South Africa; MMF: The Morgan-Morgan Finney Model, WEPP: The Water Erosion Prediction Project;

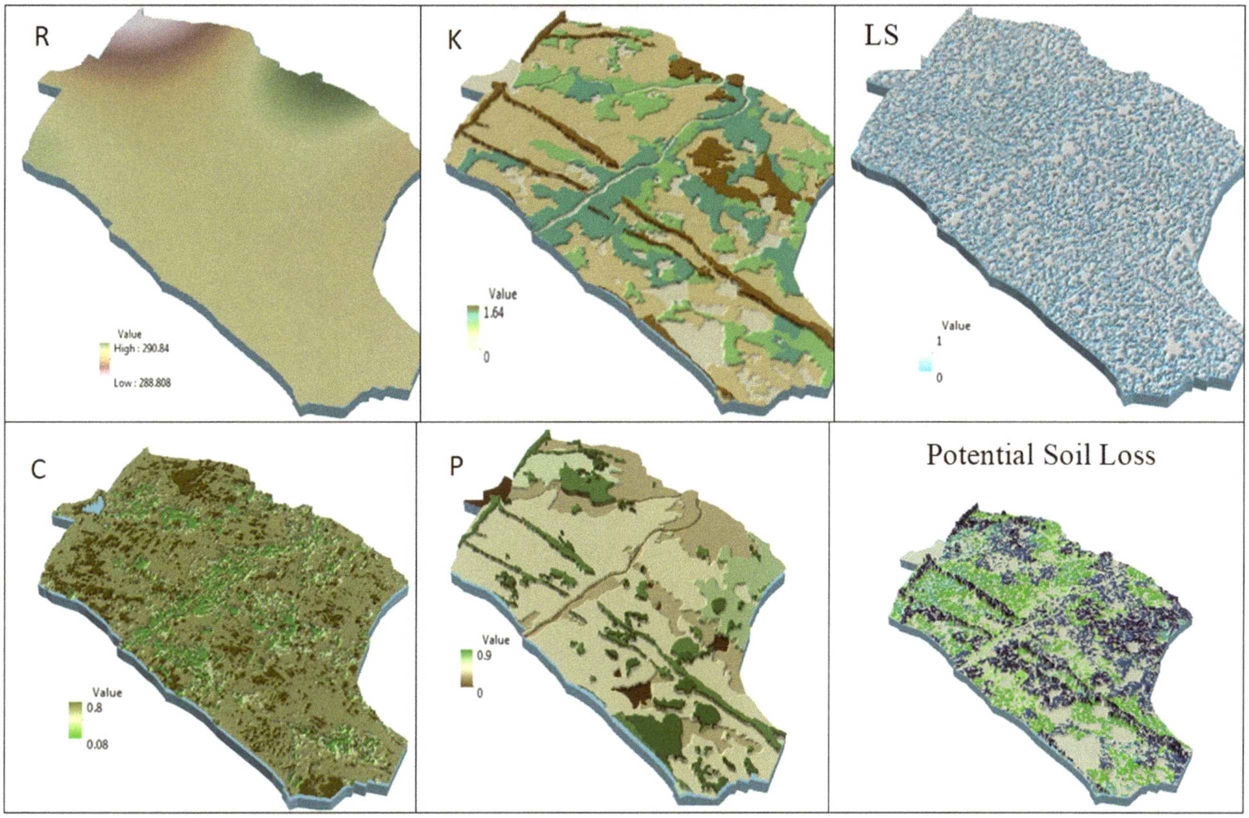

FIGURE 40.3 An example of the RUSLE model in the estimation of potential soil loss. Different soil erosion factors and the potential soil loss map are represented here.

ANSWERS: Areal Non-point Source Watershed Environment Response Simulation; and AGNPS: Agricultural Non-Point Source Pollution Model, which compute soil erosion within a watershed; and the use of IF–THEN–ELSE logic provides an alternative for land degradation assessment.

40.8 Soil salinity/sodicity

Soil salinity and sodicity developed due to the processes of salinization and alkalization are some of the very important chemical degradations happening in soil, resulting in declining soil quality. This chemical deterioration is one of the reasons contributing to land degradation. However, these processes can be prevented or controlled, or at least moderated to an extent (Szabo et al., 1998). These processes of salinization and alkalization can be studied using remote sensing. The possibility of using vegetation as an indicator of soil salinity or sodicity in the interpretation of reflectance data from the field was investigated by Toth et al. (1991).

Soil salinity is a major problem in arid and semiarid areas. Soil salinity may appear in the form of salt crust on soil, by the presence of salt-tolerant crops, or soil moisture conditions. Spectral characteristics of saline soils differ due to the presence of salt encrustation or halophytic vegetation on their surface. The variations in the salt content of soils result in spectral variations (Farifteh and Van der Meer, 2005). Crusted saline soil surfaces are smoother and reflect strongly in visible and near-infrared bands (Singh and Sirohi, 1994; Rao et al., 1995). Halophytes were found to be an indicator to differentiate saline soils from nonsaline areas (Metternicht, 1998).

In India, remote sensing–based studies have been carried out to map and monitor salt-affected soils using data obtained from LANDSAT- MSS and TM, SPOT and IRS - LISS: I, II, III, IRS—LISS IV, and Cartosat (Krishan et al., 2014). Several multispectral satellite sensors are utilized in soil salinity mapping, which includes LANDSAT TM, MSS, ETM + , SPOT, Terra-ASTER, and IKONOS. Narrow bands and their linkages to the salinity status of soils were identified, and based on high-resolution spectra, discrimination among different salinity states was analyzed to understand the soil salinity status (Csillag et al., 1993; Pasztor and Csillag, 1995). In Egypt, the salinity status of agricultural lands was mapped using multitemporal normalized differential vegetation index (NDVI) features (Lenney et al., 1996).

Multispectral and hyperspectral remote sensing have been found to be widely used in mapping salt-affected soils, where the latter is used for quantitative mapping due to the more spectral information available with those. Several studies utilized Landsat TM data for delineating moderately saline-sodic soils from slightly affected soils (Rao et al., 1991; Dwivedi and Rao, 1992). In different studies on soil salinity mapping, Landsat TM (Goossens et al., 1993); Landsat ETM + bands 4, 5, and 7 in combination, with 3 sources of ancillary data (topography, geology, and underground water quality) (Katawatin and Kotrapat, 2005); SPOT XS data (Ahmed and Andrianasolo, 1997); and IKONOS (Dwivedi et al., 2008) were found to have maximum accuracy. The effectiveness of salinity indices and the normalized differential salinity index were analyzed using the LISS III sensor of the IRS 1 B satellite (Khan and Rastoskuev, 2001). The use of various sensors, such as aerial photographs, microwave sensors, multispectral sensors, hyperspectral sensors, and electromagnetic induction meters, and approaches used for remote identification and mapping of salt-affected areas were reviewed by Metternicht and Zinck (2003). The capability of remote sensing data, such as Landsat ETM +, airborne visible/infrared imaging spectrometer (AVIRIS), color infrared aerial photos (CIR), and high-resolution field spectroradiometer (GER 3700), to extract information about soil salinity when studied, revealed the presence of gypsum and halite as the dominant salt crusts in the Rio Grande Valley (Howari, 2003). However, there are certain limitations in the use of multispectral remote sensing in mapping soil salinity due to its coarse spatial and spectral resolution (Furby et al., 1995). These limitations were overcome by the development of airborne and satellite-based hyperspectral sensors for mapping soil salinity at a higher spatial and spectral resolution (Gupta, 2003). Larger areas under soil salinity could be mapped by using airborne hyperspectral imagery (Dutkiewiez, 2006). Remote sensing and GIS techniques were used to assess land degradation by salinity and sodicity in the Bathinda district of Punjab (Ahmad and Pandey, 2018).

NDVI associated with NDSI can be used as an index to map soil salinity. Several studies report the use of spectral vegetation indices in mapping soil salinity. The most widely used vegetation indices in soil salinity studies include the NDVI, soil-adjusted vegetation index, ratio vegetation index (RVI), and the tasseled cap transformation, which consists of the soil brightness index (SBI), the green vegetation Index (GVI), and the wetness Index (WI). Several salinity

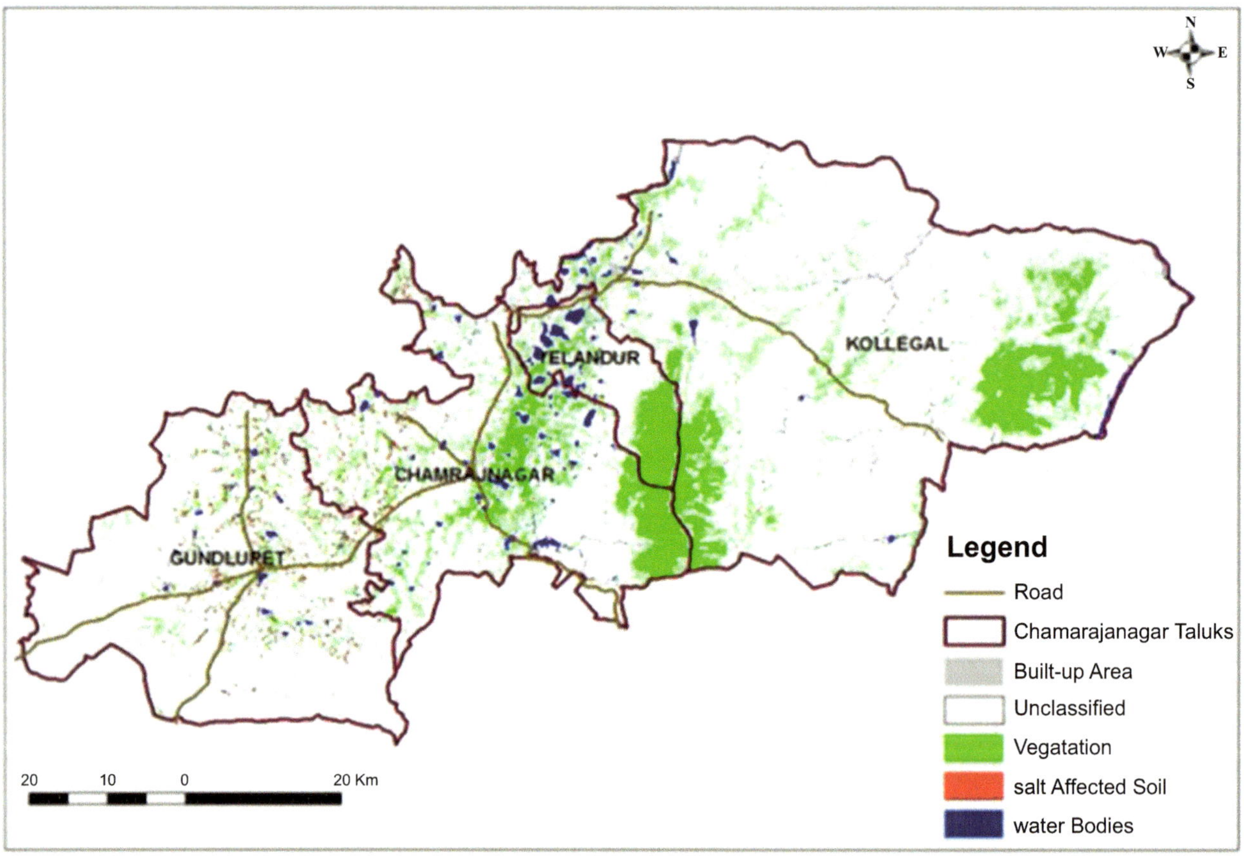

FIGURE 40.4 Salinity index using IRS data.

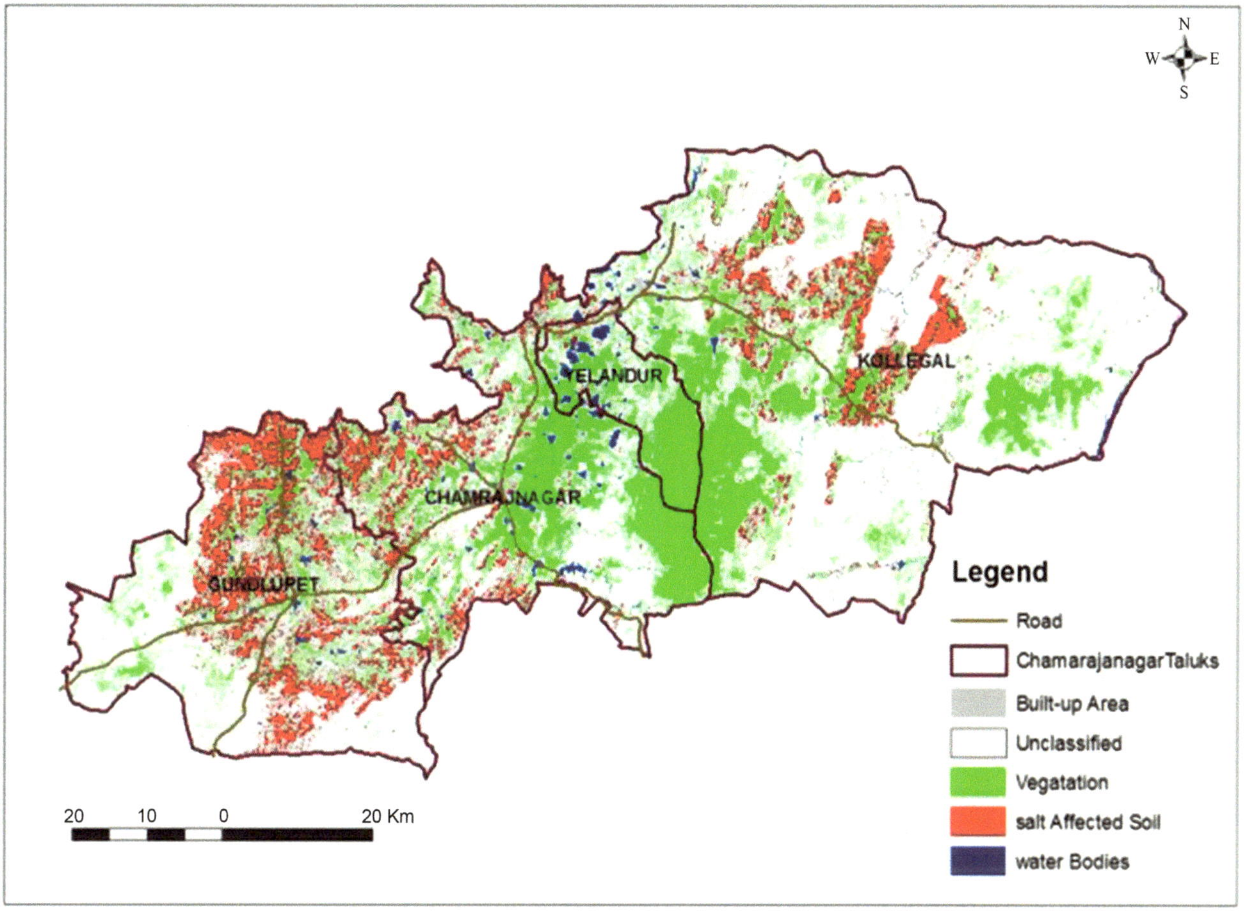

FIGURE 40.5 NDSI using IRS data.

indices are also developed by researchers in order to detect and map soil salinity, including the salinity index (SI), normalized differential salinity index (NDSI), and soil salinity and sodicity index (SASI). Remote sensing data in combination with GIS was employed in developing an SI-based approach to detecting soil salinity by Mohamed et al. (2019). Figs. 40.4 and 40.5 represent the salinity in soils by SI and NDSI, respectively, using IRS data.

Mapping and monitoring of soil salinity and sodicity can be effectively carried out using remote sensing techniques. Multispectral sensors, though have limitations, are widely used in soil salinity mapping due to their cheaper nature and the capacity to map extreme surface expressions of salinity. Thus, to avail better maps in detail with fine spatial and spectral resolutions, hyperspectral remote sensing may be resorted to (Allbed and Kumar, 2013).

40.9 Vegetal degradation

Vegetal degradation is one of the major reasons leading to land degradation. This happens as a result of the decline in vegetative cover, dynamics, and vegetative productivity. Vegetation cover is an indicator of climatic variables such as temperature, rainfall, soil, topography, and anthropogenic interventions. The changes in vegetative cover stem out of a change in land use or land cover. Hence, vegetative cover gives an indication of ecosystem services and sustainability. Climatic factors and anthropogenic interventions to vegetative cover provide a better understanding of ecosystems' feedback to natural stresses (Brown et al., 2006).Thus, an assessment of vegetation cover dynamics and vegetation decline analysis prove useful in understanding the extent of land degradation. This can be carried out using remote sensing–based monitoring, which would enable decisions on soil and land management (Walker et al., 2015). Several methods using satellite images are employed to detect vegetative cover dynamics. Phenometrics, which is based on calculating vegetation-related parameters followed by analyzing the spatiotemporal dynamics, is one of them (Parplies

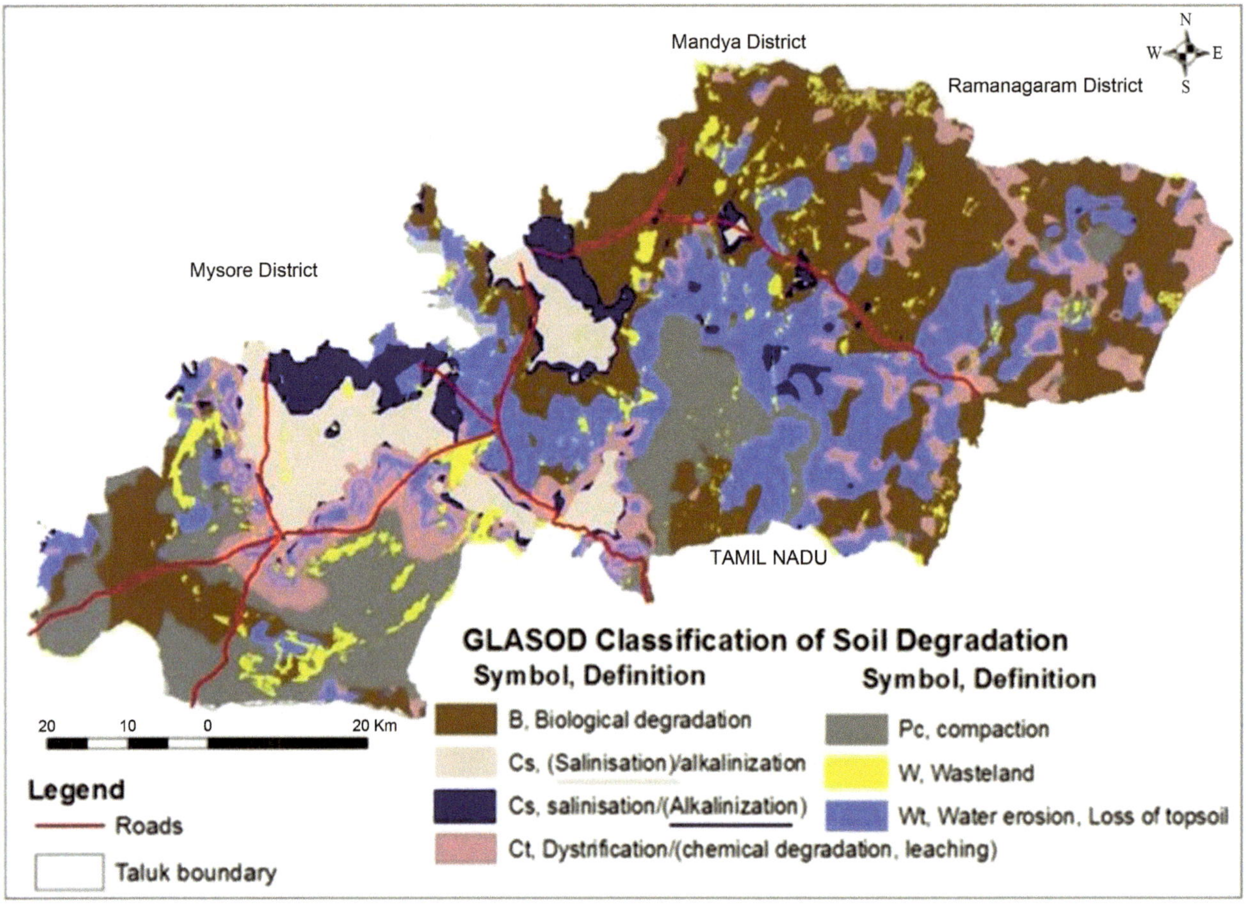

FIGURE 40.6 Land degradation status map of Chamarajnagar district, Karnataka.

et al., 2016; Nagai et al., 2016). These dynamics in vegetative cover behave as a proxy for land degradation when remote sensing is employed for assessing land degradation (Dubovyk, 2017).

In remote sensing, change detection methods are highly employed in analyzing declines in vegetation productivity and cover (Dubovyk et al., 2013b; Higginbottom and Symeonakis, 2014). In a particular land cover class, land degradation is detected by the decline in vegetation cover and productivity parameters between image acquisition dates (Zhao et al., 2004). Image classification is another change detection method by comparing multitemporal land use and land cover maps (Yiran et al., 2011), which is not suitable within a land cover class. Trend analyses are employed for change detection either using multiyear satellite imagery (Le et al., 2016) or coarse and multiscale imagery (Santos et al., 2017). The vegetation index data reader is found to be another remote sensing method employed in assessing land degradation (Higginbottom and Symeonakis, 2014).

Geospatial tools provide reliable, timely, and accurate information on the status of land degradation. The physical, chemical degradation, and biological degradation aspects are geostatistically analyzed to arrive at the land degradation status map for a particular area. Geostatistical analysis of water erosion, and wind erosion, chemical, and vegetal degradation was performed on fertility and land evaluation maps of the Chamarajnagar district in Karnataka by Mohamed et al. (2019), and the land degradation status map is presented in Fig. 40.6.

40.10 Susceptibility of an ecosystem

Sustainability of any ecosystem is defined on the basis of physical, chemical, and biological indicators. Soil properties contribute to the changes in biophysical output of sustainability. The soil properties vary over time due to the processes of erosion, water logging, salinization, compaction, acidity, etc., which are the factors for soil degradation, already discussed in this chapter. The soil quality needs to be good for sustainable productivity. A reduction in soil quality is an indication of an unsustainable ecosystem (Karthika et al., 2020). Hence, through the studies of remote sensing and GIS and by assessing the susceptibility of a land mass toward degradation, hints on the sustainability of that ecosystem can be identified. When a land is degraded by any means, its sustainability also declines. Hence, it becomes imperative that we maintain the qualities of land through proper conservation measures, which could be anthropogenic interventions.

40.11 Conclusion

Geospatial technologies are very significant tools in assessing land degradation susceptibility and vulnerability. These techniques prove to generate spatial and quantitative information on land degradation. Assessment of land degradation susceptibility and vulnerability is essential in understanding the nature of conservation practices to be adopted for the area. Thus, management practices through human interventions could prove essential in improving degraded lands and preventing susceptible lands from degrading.

References

Ahmad, N., Pandey, P., 2018. Assessment and monitoring of land degradation using geospatial technology in Bathinda district, Punjab, India. Solid. Earth 9, 75–90.

Ahmed, I., Andrianasolo, H.H. 1997. Comparative assessment of multisensor data for suitability in study of the soil salinity using remote sensing and GIS in the Fordwah Irrigation Division. In: IEEE International Conference on Geoscience and Remote Sensing, Singapore, 3–8 August 1997, pp. 1627–1629. https://doi.org/10.1109/IGARSS.1997.608989.

Allbed, A., Kumar, L., 2013. Soil salinity mapping and monitoring in arid and semi-arid regions using remote sensing technology: a review. Adv. Remote. Sens. 2013 (2), 373–385. Available from: https://doi.org/10.4236/ars.2013.24040.

Arnalds, O., Þorarinsdottir, E.F., Metusalemsson, S., Jonsson, A., Gretarsson, E., Arnason, A., 2001. Soil Erosion in Iceland. Soil Conservation Services and Agricultural Research Institute, Iceland.

Aslam, B., Maqsoom, A., Alaloul, W.S., Musarat, M.A., Jabbar, T., Zafar, A., 2021. Soil erosion susceptibility mapping using a GIS-based multi-0criteria decision approach: case of district Chitral, Pakistan. Ain Shams Eng. J. 12, 1637–1649.

Bahrawi, A.J., Elhag, M., Aldhebiani, A.Y., Galal, K.H., Hegazy, A.K., Alghailani, E., 2016. Soil erosion estimation using remote sensing techniques in Wadi Yalamalam basin, Saudi Arabia. Advances in Materials Science and Engineering. Hindawi Publishing Corporation, pp. 1–8. Available from: http://doi.org/10.1155/2016/9895962.

Biswas, S., 2012. Estimation of soil erosion using remote sensing and GIS and prioritization of catchments. Int. J. Emerg. Technol. Adv. Eng. 2 (7), 124.

Brown, D.J., Shepherd, K.D., Walsh, M.G., Mays, M.D., Reinsch, T.G., 2006. Global soil characterization with VNIR diffuse reflectance spectroscopy. Geoderma 132, 273–290.

Cano, A., Mermut, A., Arocena, J., Silla, R. (Eds.), 2009. Land Degradation and Rehabilitation. Atena Verlag GMBH, Reiskirchen.

Chatterjee, S., Krishna, A.P., Sharma, A.P., 2014. Geospatial assessment of soil erosion vulnerability at watershed level in some sections of the Upper Subarnarekha river basin. Jharkhand. India Environ. Earth Sci. 71 (1), 357–374.

Csillag, F., Pasztor, L., Biehl, L.L., 1993. Spectral band selection for the characterization of salinity status of soils. Remote. Sens. Environ. 43, 231–242.

de Bei, C.A.J.M., 2005. Assessment of soil erosion indicators for maize-based agro-ecosystems in Kenya. Catena 59 (2005), 231–251.

de Jong, S.M., de Roo, A.P.J., Bertolo, F., Folving, S., Megier, J., 1999. Regional assessment of soil erosion using the distributed model SEMMED and remotely sensed data. Catena 37, 291–308. Available from: https://doi.org/10.1016/s0341-8162(99)00038-7.

Dubovyk, O., 2017. The role of remote sensing in land degradation assessments: opportunities and challenges. Eur. J. Remote. Sens. 50 (1), 601–613.

Dubovyk, O., Menz, G., Conrad, C., Lamers, J., Lee, A., Khamzina, A., 2013a. Spatial targeting of land rehabilitation: a relational analysis of cropland productivity decline in arid Uzbekistan. Erdkunde 67 (2), 167–181. Available from: https://doi.org/10.3112/erdkunde.2013.02.05.

Dubovyk, O., Menz, G., Conrad, C., Thonfeld, F., Khamzina, A., 2013b. Object-based identification of vegetation cover decline in irrigated agro-ecosystems in Uzbekistan. Quat. Int. 311, 163–174. Available from: https://doi.org/10.1016/j.quaint.2013.07.043.

Dubovyk, O., Landmann, T., Dietz, A., Menz, G., 2016. Quantifying the impacts of environmental factors on vegetation dynamics over climatic and management gradients of central Asia. Remote. Sens. 8 (7), 600. Available from: https://doi.org/10.3390/rs8070600.

Dutkiewiez, A., 2006. Evaluating hyper-spectral imagery for mapping the surface symptoms of dryland salinity. School of Earth and Environment Science. University of Adelaide.

Dwivedi, R.S., Rao, B.R.M., 1992. The selection of the best possible Landsat-TM band combinations for delineating salt affected soils. Int. J. Remote. Sens. 13, 2051–2058.

Dwivedi, R., et al., 2008. 5 Generation of farm-level information on salt-affected soils using IKONOS-II multispectral data. In: Metternicht, G., Zinck, J. (Eds.), Remote Sensing of Soil Salinization: Impact on Land Management. CRC Press, Boca Raton, p. 2008. Available from: http://doi.org/10.1201/9781420065039.ch5.

Farifteh, J., Van der Meer F., 2005. Spectral characteristics of salt-affected soils; impact on imaging spectroscopy. In: 4th Workshop on Imaging Spectroscopy, 27–29 April.

Furby, S. et al., 1995. Detecting and Monitoring Salt-Affected Land: A Report from the LWRRDC Project Detecting and Monitoring Changes in Land Condition through Time Using Remotely Sensed Data. CSIRO, Division of Mathematics & Statistics.

Favis-Mortlock, D., Guerra, T., Boardman, J., 1998. A self-organizing dynamic systems approach to hillslope rill initiation and growth: Model development and validation. IAHS-AISH Publication 249.

Gelagay, H.S., Minale, A.S., 2016. Soil loss estimation using GIS and remote sensing techniques: a case study of Koga watershed in Northwestern Ethiopia. Int. Soil. Water Conserv. Res. 4, 126–136.

Godert, W.J., van Lynden, S.Ml, 2001. The role of GIS and remote sensing in land degradation assessment and conservation mapping: some user experiences and expectations. Int. J. Appl. Earth Obs. Geoinf. 3, 61–68. Available from: https://doi.org/10.1016/S0303-2434(01)85022-4.

Goossens, R., el Badawi, M., Ghabour, T., de Dapper, M., 1993. A simulation model to monitor the soil salinity in irrigated arable land in arid areas based upon remote sensing and GIS, EARSeL. Adv. Remote. Sens. 2 (3), 165–171.

Gupta, R.P., 2003. Remote Sensing Geology. Springer Verlag, New York, p. 2003.

Higginbottom, T.P., Symeonakis, E., 2014. Assessing land degradation and desertification using vegetation index data: current frameworks and future directions. [Review]. Remote. Sens. 6 (10), 9552–9575. Available from: https://doi.org/10.3390/rs6109552.

Howari, F.M., 2003. The use of remote sensing data to extract information from agricultural land with emphasis on soil salinity. Aust. J. Soil Res. 41, 1243–1253.

Jazouli, A., Barakaj, A., Ghafiri, A., Moutaki, S.E., Ettagy, A., Khellouk, R., 2017. Soil erosion modeled with USLE,GIS and remote sensing: a case study of Ikkour watershed in middle atlas (Morocco). Geosci. Lett. 4, 25. Available from: https://doi.org/10.1186/s40562-017-0091-6.

Katawatin, R., Kotrapat, W., 2005. Use of LANDSAT-7 ETM + with ancillary data for soil salinity mapping in Northeast Thailand. In: Third International Conference on Experimental Mechanics and Third Conference of the Asian 2004, Sevilla, 3 June 2005, pp. 708–716.

Khan, M.N., Rastoskuev, V., 2001. Mapping salt-affected soils using remote sensing indicators. In: 22nd Asian Conference on Remote Sensing, 5–9 Nov, Singapore.

Kirkby, M.J., Jones, R.J.A., Irvine, B., Gobin, A., Govers, G., Cerdan, O., Van Rompacy, A.J.J., LeBissonnais, Y., Daroussin, J., King, D., Montanarella, L., Grimm, M., Vieillefont, V., Puigdefabregas, J., Boer, M., Kosmas, C., Yassoglou, N., Tsara, M., Mantel, S., Van Lynden, G.J., Huting, J., 2004. Pan-European Soil Erosion Risk Assessment: The PESERA Map, Version 1 October 2003. Explanation of Special Publication Ispra 2004 No. 73 (S.P.I.04.73). European Soil Bureau Research Report No. 16, EUR 21176, 18pp. and 1 map in ISO B1 format. Office for Official Publications of the European Communities, Luxembourg.

Kourgialas, N.N., Koubouris, G.C., Karatzas, G.P., Metzidakis, I., 2016. Assessing water erosion in Mediterranean tree crops using GIS techniques and field measurements: the effect of climate change. Nat. Hazards 83 (1), 65–81.

Krishan, G., Saha, S.K., Kumar, Suresh, Patel, N.R., 2014. Remote sensing in soil fertility evaluation and management. Bioresources for Sustainable Plant Nutrient Management. pp. P 509–P 533.

Le, Q.B., Nkonya, E., Mirzabaev, A., 2016. Biomass productivity-based mapping of global land degradation hotspots. In: Nkonya, E., Mirzabaev, A., von Braun, J. (Eds.), Economics of Land Degradation and Improvement—A Global Assessment for Sustainable Development. Springer International Publishing, Cham, pp. 55–84.

Lenney, M.P., et al., 1996. The status of agricultural lands in Egypt: the use of multitemporal NDVI features derived from Landsat TM. Remote. Sens. Environ. 56, 8–20.

Metternicht, G.I., 1998. Analyzing the relationship between ground based reflectance and environment indicators of salinity processes in the Cochabamba Valley (Bolivia). Int. J. Ecol. Environ. Sci. 24 (4), 359–370.

Metternicht, G.I., Zinck, J.A., 2003. Remote sensing of soil salinity: potentials and constraints. Remote. Sens. Environ. 85, 1–20.

Mohamed, A.E.A., Natarajan, A., Hegde, R., Prakash, S.S., 2019. Assessment of land degradation using comprehensive geostatistical approach and remote sensing data in GIS-model builder. Egypt. J. Remote. Sens. Space Sci. 22, 323–334.

Morgan, R.P.C., Quinton, J.N., Smith, R.E., Grovers, G., Poesen, J.W.A., Auerswald, K., Chisci, G., Torri, D., Styczen, M.E., Folly, A.J.V., 1998. The European Soil Erosion Model (EUROSEM): Documentation and User Guide, . 22. Silsoe College, Cranfield University.

Nagai, S., Nasahara, K.N., Inoue, T., Saitoh, T.M., Suzuki, R., 2016. Review: Advances in in situ and satellite phenological observations in Japan. [Review]. Int. J. Biometeorol. 60 (4), 615–627. Available from: https://doi.org/10.1007/s00484-015-1053-3.

Napoli, M., Cecchi, S., Orlandini, S., Mugnai, G., Zanchi, C.A., 2016. Simulation of field measured soil loss in Mediterranean hilly areas (Chianti, Italy) with RUSLE. Catena 145, 246–256.

Olsson, L., Barbosa, H., Bhadwal, S., Cowie, A., Delusca, K., Flores-Renteria, D., et al., 2019. Land degradation. In: Shukla, P.R., Skea, J., Calvo Buendia, E., Masson-Delmotte, V., Pőrtner, H.O., Roberts, D.C., Zhai, P., Slade, R., Connors, S., RVan Diemen, M., Ferrat, E., Haughey, S., Luz, S., Neogi, M., Pathak, J., Petzold, J.Portugal Pereira, Vyas, P., Huntley, E., Kissick, K., Belkacemi, M., Malley, J. (Eds.), Climate Change and Land: An IPCC Special Report on Climate Change, Desertification, Land Degradation, Sustainable Land Management, Food Security and Green House Gas Fluxes in Terrestrial Ecosystems. In press.

Parplies, A., Dubovyk, O., Tewes, A., Mund, J.P., Schellberg, J., 2016. Phenomapping of rangelands in South Africa using time series of RapidEye data. Int. J. Appl. Earth Obs. Geoinf. 53, 90–102. Available from: https://doi.org/10.1016/j.jag.2016.08.001.

Pasztor, L., Csillag, F., 1995. Reduction of high resolution spectra: application to characterization of salinity status of soils. In: Askne, J. (Ed.), Sensors and Environmental Applications of Remote Sensing. Balkuma Rotterdam, pp. 393–397.

Peng, W., Zhou, J., Yang, C., 2008. Integrated use of remote sensing and GIS for predicting soil erosion process. Int. Arch. Photogram., Remote. Sens. Spat. Inf. Sci. XXXVII (Part B4), 1647–1651. Beijing.

Rao, B.R.M., Dwivedi, L., Venkataratnam, L., 1991. Mapping the magnitude of sodicity in part of the Indo-Gangetic plains of Uttar Pradesh, Northern India using Landsat- TM data. Int. J. Remote. Sens. 12, 419–425.

Rao, B.R.M., Sharma, R.C., Ravisankar, T., Das, S.N., Dwivedi, R.S., Thammappa, S.S., et al., 1995. Spectral behaviour of salt-affected soils. Int. J. Remote. Sens. 16 (12), 2125–2136. Available from: https://doi.org/10.1080/01431169508954546.

Rawat, K.S., Mishra, A.K., Bhattacharyya, R., 2016. Soil erosion risk assessment and spatial mapping using LANDSAT-7 ETM +, RUSLE, and GIS—a case study. Ar. J. Geosci. 9 (4), 1–22.

Renard, K.G., Foster, G.A., Weesies, D.A., Mccool, D.K., Yoder, D.C., 1997. Predicting soil erosion by water: a guide to conservation planning with the revised universal soil loss equation (RUSLE). Agriculture Handbok No. 703. USDA, Washington DC.

Samanta, R.K., Bhunia, G.S., Shit, P.K., 2016. Spatial modelling of soil erosion susceptibility mapping in lower basin of Subarnarekha river (India) based on geospatial techniques. Model. Earth Syst. Environ. 2, 99.

Santos, F., Dubovyk, O., Menz, G., 2017. Monitoring forest dynamics in the andean Amazon: the applicability of breakpoint detection methods using landsat time series and genetic algorithms. Remote. Sens. 9 (1), 68. Available from: https://doi.org/10.3390/rs9010068.

Sheikh, A.H., Palria, S., Alam, A., 2011. Integration of GIS and universal soil loss equation (USLE) for soil loss estimation in a Himalayan watershed. Recent. Res. Sci. Technol. 3 (3), 51–57.

Singh, R.P., Sirohi, A., 1994. Spectral reflectance properties of different types of soil surfaces. ISPRS J. Photogramm. Remote. Sens. 49 (4), 34–40. Available from: https://doi.org/10.1016/0924-2716(94)90045-0.

Sivakumar, M., Stefanski, R., 2007. Climate and Land Degradation. Springer, Berlin.

Srinivasan, R., Karthika, K.S., Amar Suputhra, S., Chandrakala, M., Hegde, R., 2021. Mapping of soil erosion and probability zones using remote sensing and GIS in arid part of south Deccan Plateau, India. J. Indian. Soc. Remote. Sens. 49 (10), 2407–2423.

Szabo, J., Pasztor, L., Suba, Z.S., Varallyay, G.Y., 1998. Integration of remote sensing and GIS techniques in land degradation mapping. Agrokemia ESTalajten Tom 4 (No1–4).

Tayebi, M., Tayebi, M.H., Sameni, A., 2017. Soil erosion risk assessment using GIS and CORINE models: a case study from western Shiraz. Iran. Arch. Agron. Soil. Sci. 63 (8), 1163–1175.

Tirkey, A.S., Pandey, A.C., Nathawar, M.S., 2013. Use of satellite data, GIS and RUSLE for estimation of average annual soil loss in Daltonganj watershed of Jharkhand (India). J. Remote. Sens. Tech. 1 (1), 20–30.

Toth, et al., 1991. Characterization of semi-vegetated salt affected soils by means of field remote sensing. Remote. Sens. Environ. 37, 167–180.

Walker, J.J., de Beurs, K.M., Henebry, G.M., 2015. Land surface phenology along urban to rural gradients in the U.S. Great Plains. Remote. Sens. Environ. 165, 42–52. Available from: https://doi.org/10.1016/j.rse.2015.04.019.

Williams, J.R., Jones, C.A., Dyke, P.T., 1990. The EPIC model documentation. USDA-ARS Technical Bulletin No. 1768. U.S. Department of Agriculture, Washington, DC, pp. 3–92.

Wischmeier, W.H., Smith, D.D., 1965. Predicting Rainfall-Erosion Losses From Cropland East of the Rocky Mountains: A Guide for Selecting Practices for Soil and Water Conservation. USDA Agricultural Handbook No. 282.

Woodward, D.E., 1999. Method to predict cropland ephemeral gully erosion. Catena 37 (3–4), 393–399. Available from: https://doi.org/10.1016/S0341-8162(99)00028-4.

Yadav, R.P., Sidhu, G.S., 2010. Assessment of soil erosion in Himachal Pradesh. J. Indian. Soc. Soil. Sci. 58 (2), 212–220.

Yiran, G.A.B., Kusimi, J.M., Kufogbe, S.K., 2011. A synthesis of remote sensing and local knowledge approaches in land degradation assessment in the Bawku East District, Ghana. Int. J. Appl. Earth Obs. Geoinf. 14 (1), 204–213. Available from: https://doi.org/10.1016/j.jag.2011.09.016.

Zhao, G.X., Lin, G., Warner, T., 2004. Using Thematic Mapper data for change detection and sustainable use of cultivated land: a case study in the Yellow River delta, China. Int. J. Remote. Sens. 25 (13), 2509–2522. Available from: https://doi.org/10.1080/01431160310001619571.

Zhu, M., 2011. Soil erosion risk assessment with CORINE model: a case study in the Danjiangkou reservoir region, China. Stoch. Environ. Res. Risk Assess. . Available from: https://doi.org/10.1007/s00477-011-0511-7.

Chapter 41

Assessment of soil erosion and factor sensitivity analysis using GIS techniques—a case study of Central Region of Meghalaya, India

Amritee Bora[1], B.S. Mipun[1] and Manish Parmar[2]
[1]*Department of Geography, North Eastern Hill University, Shillong, Meghalaya, India,* [2]*Space Applications Centre (SAC-ISRO), Ahmedabad, Gujarat, India*

41.1 Introduction

It is generally acknowledged that soil erosion has become one of the greatest global threats to the human environment (Hu et al., 2019). It has become a serious environmental issue for global ecology, natural resources, and socioeconomic development. It is an environmental threat primarily to agricultural-based developing economies. According to Poerbandono et al. (2006), erosion is the removal of a layer of soil or sediment due to pressure caused by the movement of wind or water on the surface of the soil or water bottom (Herawati, 2010; Khairunnisa et al., 2020). It is a worldwide problem because of its effects on soil productivity and nutrient loss (Benavidez et al., 2018), resulting in the loss of land resources and a reduction in land productivity with adverse economic and environmental impacts (Zhu, 2015; Khairunnisa et al., 2020). As a key consequence of land degradation worldwide, natural erosion is estimated to total some 9.9 billion tons of soil a year, but human-induced accelerated erosion is more than 2.5 times higher, at 26 billion tons^{-1} year^{-1} (Holdgate et al., 1982; O'Keefe, 1983; Brown, 1984a, b; Lal, 1990; Shekinah & Saraswathy, 2005). According to FAO (2021),[1] the drivers of demand for land and water resources are complex, and by 2050, it is estimated that soil erosion could lead to a 10% loss in crop production and remove 75 billion tons of soil. This will put direct pressure on world agricultural land to produce almost 50% more food to satisfy global demand. Despite being a renewable resource, overutilization and exposure of land causes loss and imbalance of microorganisms and other physical properties of topsoil that lead to erosion. It reduces soil moisture and makes the topsoil more vulnerable to erosion. Changes in land use also have an impact on the reduced capacity of soil infiltration, which can lead to increased surface runoff and erosion rates (Desifindiana et al., 2013; Khairunnisa et al., 2020). On the other hand, topography is one of the factors determining the amount of soil erosion, and slope plays an important role in increasing soil loss (Zhang et al., 2019; Hu et al., 2019). It is a complex process that depends on soil properties, ground slope, vegetation cover, rainfall amount and intensity, and wind direction and speed. In general, soil erosion is more severe in mountains than in undulating terrains. Kadomura and Yamamoto (1978) estimated the rate of natural erosion for the world to be on the order of 1.5–7 tons ha^{-1} year^{-1} for mountainous regions and 0.1–7 tons ha^{-1} year^{-1} for undulating uplands. (Rind, 1983; Lal, 1990). In hilly and steeply sloped region, soil erosion may initiate landslides (Michael and Samanta, 2016; Samanta et al., 2016). The ideal soil erosion rate is the equivalent of soil formation. However, due to the variability of the pedologic factors, determining the soil formation rates with precision is not possible (Li et al., 2009; Lense et al., 2019). More often, the process of soil erosion is very slow but continuous, which goes relatively unnoticed. However, in some cases, it can occur at an alarming rate, which can cause serious topsoil loss. The estimation of soil loss depends on three key factors: rainfall–runoff, soil physical properties (viz., sand, silt, clay, and organic carbon), and slope (length and steepness). By understanding the driving forces behind soil erosion, we can more easily identify erosion-prone areas within a landscape to address the problem strategically (Benavidez et al., 2018).

1. Food and Agriculture Organization, United Nations. Founded in 1945.

Remote Sensing of Soils. DOI: https://doi.org/10.1016/B978-0-443-18773-5.00041-7

Regarding Meghalaya, the state is highly susceptible to acute soil erosion problems due to its hilly topography and very high rainfall intensity. Indigenous cultivation practices such as "*Jhum*[2]" and "*Nur Bun*[3]" and unscientific mining are some of the most triggering factors of erosion within the hilly terrain. Water is the most dominant driver of erosion, as it is the single most harmful factor in the degradation of productive land in the state, affecting about 31% of the total geographical area. It has been reported that 14.15 million tons of productive soil are lost every year in the state (Sharma and Prasad, 1995). It has also been estimated that soil loss from unprotected cultivated land is about 120 tons ha^{-1} $year^{-1}$ and may reach as high as 300 tons ha^{-1} $year^{-1}$. Thus, apart from the depletion of fertile soil, erosion results in the loss of plant nutrients and organic matter, siltation of water reservoirs and river beds, and adversely affects the irrigation system of the state (Soil & Water Conservation Department, 2022).

41.2 Study area

The central part of Meghalaya comprises three administrative districts; Southwest Khasi Hills, West Khasi Hills, and Eastern West Khasi Hills. Until the 2011 census, the entire region was called the West Khasi Hills. The total geographical area of the region is 5247 km^2, occupying 23% of the total geographical area of the state. It is extended between 25°10′ east to 25°51′ east longitude and 90°44′ north to 91°49′ north latitude. Physiographically, the region can be divided into two regional divisions; the western Shillong plateau and the Kynhiong (Lynhiong) Uming basin (Konwar, 2010; Bora, 2021). The elevation ranges between 4 and 1919 m, and the slope ranges between 0% and >100% (Aster DEM). The average altitude of the central upland is about 1300 m (CGWB, 2013;[4] Bora, 2021).

Geologically, the region falls within the Shillong or Meghalaya plateau, which is part of the Precambrian rocks of gneissic composition. Hydrologically, this central part of the state can be divided into two major units: consolidated and semiconsolidated. About two-thirds of the region is a consolidated formation (CGWB, 2013). According to the Soil Series of Meghalaya, 2005 (National Bureau of Soil Survey and Land Use Planning), 19 different types of soil have been identified within the region. Red gravelly soil and red loamy soil are the most common soil types in the region. The soil is acidic and rich in organic matter and nitrogen but poor in phosphorous (CGWB, 2013).

According to IMD,[5] the region falls under the subtropical humid climatic zone, which is extremely influenced by the southwest monsoon. Based on weather reports collected between 2005 and 2015, the climatic profile of the region shows that the maximum temperature increases to 27.25°C during July–August and the minimum temperature decreases to 1.39°C during November–December. The maximum rainfall received by the region during July was 985.8 mm. It has been observed that the northeastern corner of the region receives the relatively highest amount of rainfall (an estimated average of 517.89 mm/annum) with 23°C of the annual average temperature.

The drainage system of the region is controlled by the topography. ***Kynshi*** is the major river originating on the southern slope of the Marpna peak. This river is joined by several tributaries on its westerly course, ultimately draining into Bangladesh. Another river is ***Khri***, which rises near Kyllong Rock, following northward into the Assam plains (CGWB, 2013; Bora, 2021). According to FSI (2017),[6] 77.16% of the total geographical area of the region is covered by forest, prominently moderately dense forest and open forest (Meghalaya Forest and Environmental Department). The land use land cover is monitored and estimated with the help of three seasons, LISS III satellite imagery (Karif, Rabi, and Summer) between 2015 and 2016, and ground truth verification, cross-verified with US Army toposheet No: 46-9 and 46-10. It is broadly classified as: tropical deciduous forest (moist), degraded forest, degraded grassland, scrubland, single-crop agricultural land, settlement, and waterbody.

The economy of the region mostly depends on the agriculture and mining sectors. According to the 2011 census, the total number of agricultural laborers is 37,184, and the total number of cultivators is 84,956. It has 366.86 km^2 of total gross cropped area, whereas the total cultivable wasteland is about 1437.38 and 658.73 km^2 is fallow land (CGWB, 2013) (Fig. 41.1).

---

2. Land is cleared by cutting and burning the vegetation cover. It is cultivated for 1–2 years, and then abandoned once the productivity of the soil declines.
3. Land use for potato cultivation. Rows of raised beds, each measuring around 1–1.25 m wide, 20–25 cm in height, and 2–7 m long (depending on the availability of land) are made along the slope (Dubey and Sah, 2009).
4. Central Ground Water Board, India. Established in 1970
5. India Meteorological Department. Established on 15 January, 1875.
6. Forest Survey of India. Established on 1 June, 1981.

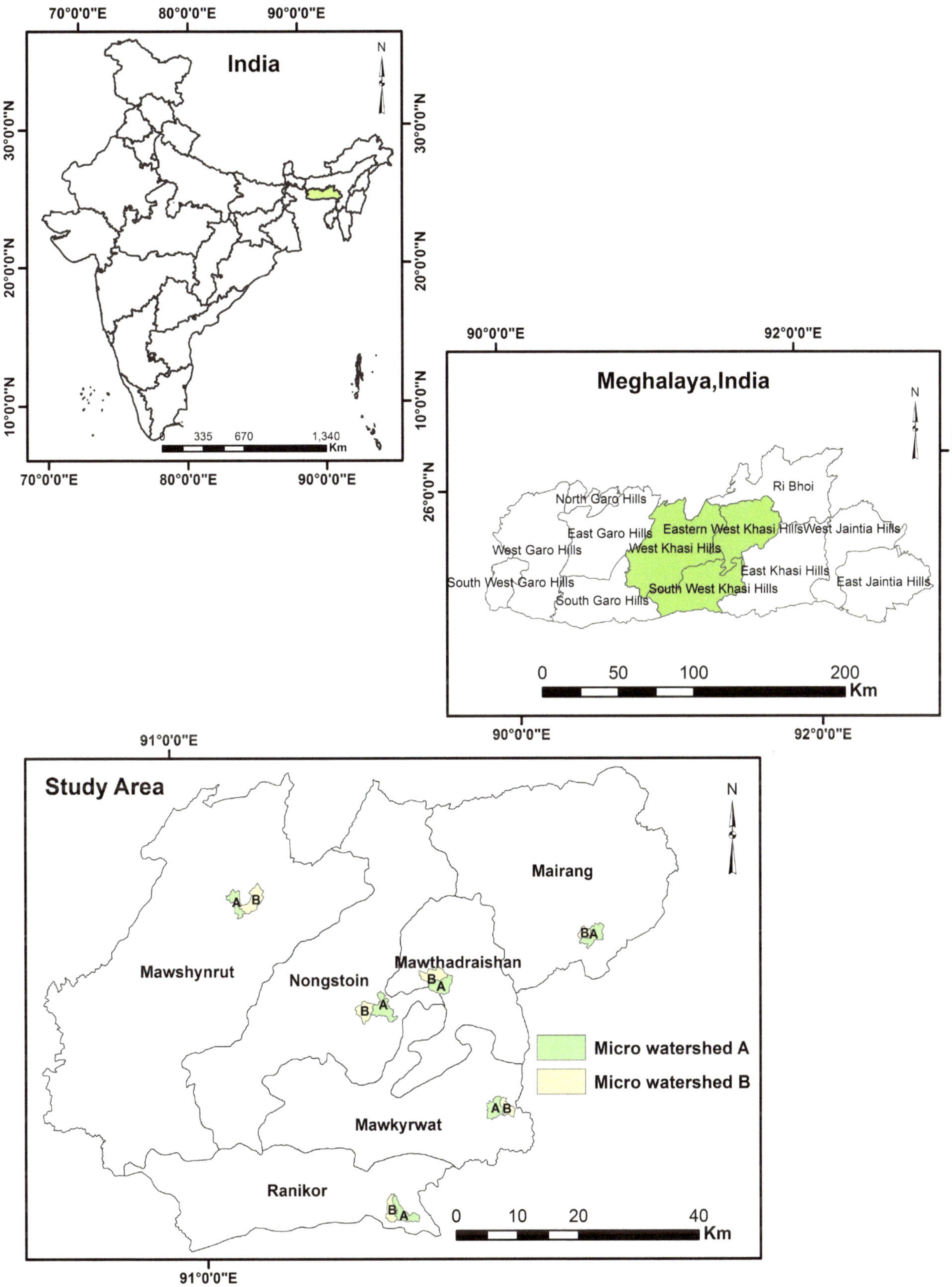

FIGURE 41.1 Study area map.

41.3 Materials and methods

Soil erosion estimation is a multifactor analysis. Quantitative modeling, either physically based or empirically based, has become widely accepted in soil erosion estimation (Bhattarai and Dutta, 2007; Kinnell, 2010; Hu et al., 2019). Many empirical models have been developed for estimating soil erosion (Patowary & Sarma, 2018; Hu et al., 2019), such as the erosion potential method (Gavrilovic, 1988; Lense et al., 2019), the erosion prediction project, the soil and water assessment tool, the universal soil loss equation (USLE), and the revised universal soil loss equation (RUSLE) (Khairunnisa et al., 2020). However, among all of them, the RUSLE is the most commonly accepted and applied quantitative model for soil loss estimation. This study estimated soil erosion using USLE as an estimation of potential soil loss and RUSLE [Eq. (41.1)] as an estimation of actual soil loss. Both equations have been tested over 12 microwatersheds (two adjoining microwatersheds from each administrative CD block) to conduct a comparative analysis. The materials data are primarily based on the period of 2011–2016 due to matriculation of the study (Table 41.1).

The empirical model USLE was first published in 1965 by the US Department of Agriculture, Agricultural Handbook 282. It was calculated using the parameters rainfall erosivity, soil erodibility, and slope length and steepness. However, the model overestimated the soil loss because it did not consider the top layer of vegetation cover. The revised version of USLE was published by the US Department of Agriculture, in 1978 in Agricultural Handbook 537. This revised version is simply termed the revised universal soil loss equation (RUSLE, i.e., Eq. (41.1)), and two more factors have been added; C is the cover management factor and P is the conservation practices factor. In the present day, it is on the digital platform. It is expressed as

$$A = R \times K \times LS \times C \times P \tag{41.1}$$

where, A refers to the rate of soil loss in a unit of area, R is the rainfall erosivity factor, K is the soil erodibility factor, LS is the slope length and steepness factor, C is the cover management factor, and P is the conservation practices factor. The theoretical framework of the study (Fig. 41.2).

41.4 Rainfall erosivity factor (R)

According to USDA agricultural handbook no. 282 and 537, the R factor is the rainfall and runoff factor. It is the number of R units plus a factor for runoff from snowmelt or applied water where such runoff is significant (Wischmeier and Smith, 1965, 1978). It is mentioned in their research that when factors other than rainfall are constant, soil losses from cultivated land are directly proportional to the rainstorm parameter. The total storm energy (E) is considered in the time of a maximum of 30-minute intensity (I3o). The sum of EI3o values of the storm for a given period is the numerical measure of the erosive potential of the rainfall within that period. However, the Modified Fornier Index (MFI) is one of the most widely used methods for calculating the rainfall erosivity factor. In 1960, Fornier developed an index:

$$\mathrm{IF} = \frac{P2m}{P} \tag{41.2}$$

where IF is the Fornier index, Pm is the maximum monthly rainfall depth in mm, and P is the annual rainfall. In 1980, Arnldus modified the Fornier index as the Modified Fornier Index (MFI) to remove the error in monthly rainfall

TABLE 41.1 Data and data sources.

Data	Source	Description
Landsat 8 OLI & TIRS	USGS Earth explorer	Year 2015–16. Spatial resolution is 30 m.
LISS III	Space Applications Centre, Ahmedabad	Year 2015–16. Spatial resolution is 24 m.
Aster DEM	Aster Global digital elevation model	Spatial resolution is 30 m.
Rainfall	Indian Meteorological Department, India.	Year 2012–16. A total of six rain gauge.
Soil data	National bureau of soil survey and land use planning, India.	Soil series of Meghalaya 2005. A total of 19 categories of soil are identified within the districts and a total of 10 categories of soil are identified with the microwatersheds.

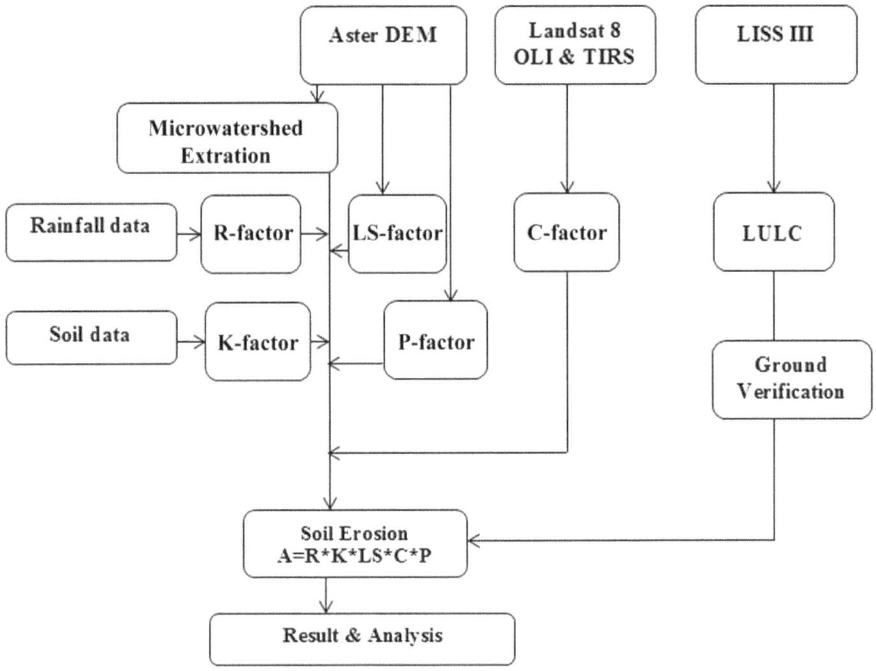

FIGURE 41.2 Flowchart of the study.

distribution. Consequently, IF does not always increase when the number of erosive rainfall events in the year increases. It is expressed as:

$$F = \frac{i = 1Pi2}{P} \tag{41.3}$$

where Pi is the rainfall depth in the month I (mm). For present-day studies, the monthly EI3o value tries to relate to MFI. In India, Babu et al. (1978) and Singh et al. (1981) represented an empirical model for calculating the R factor by relating annual average precipitation with the R factor. The equations are:

$$R = 79 + 0.363 \times \text{AAP (Singh et al., 1981) for the entire India} \tag{41.4}$$

$$R = 22.8 + 0.64 \times \text{AAP (Babu et al., 1978) for Dehradun} \tag{41.5}$$

$$R = 81.5 + 0.375 \times \text{AAP (Babu et al., 1978) for Jharkhand.} \tag{41.6}$$

where AAP is the annual average precipitation.

For this study, Eq. (41.4) is considered. Here, AAP is evaluated from 2012 to 2016 to obtain accuracy for a standard period. Thus, the AAP in this study represents the mean annual average precipitation from 2012 to 2016 (5 years). The R factor for this study was evaluated with the help of rainfall data from six rain gauge stations. Further, it is interpolated to create an R-factor surface (Fig. 41.3 and Table 41.2).

41.5 Soil erodibility factor (K)

The soil erodibility factor is a quantitative value experimentally determined. It is the soil loss rate per erosion index unit for a specific soil as measured on a unit of the plot. However, the meaning of the term "soil erodibility" is distinctly different from that of "soil erosion." Soil erodibility is a function of the complex interactions of a substantial number of its physical and chemical properties and often varies with a standard texture class (Wischmeier and Smith, 1965, 1978). Although it primarily depends on the texture and physical properties of the soil, in some cases, soil organic carbon (SOC), a chemical property of the soil, can also trigger soil erodibility. The input of organic matter into the soil can significantly contribute to a decrease in surface runoff and soil loss (Kadlec et al., 2012). However, the rate of soil erosion in the soil loss equation is more influenced by rainfall erosivity, slope length, and cover management than by soil erodibility. Wischmeier and Smith (1965, 1978) determined the K-factor by the R in a standard plot with a slope of 72.6 ft. in length and 9% steepness. To further simplify the estimation of soil erodibility, several empirical methods

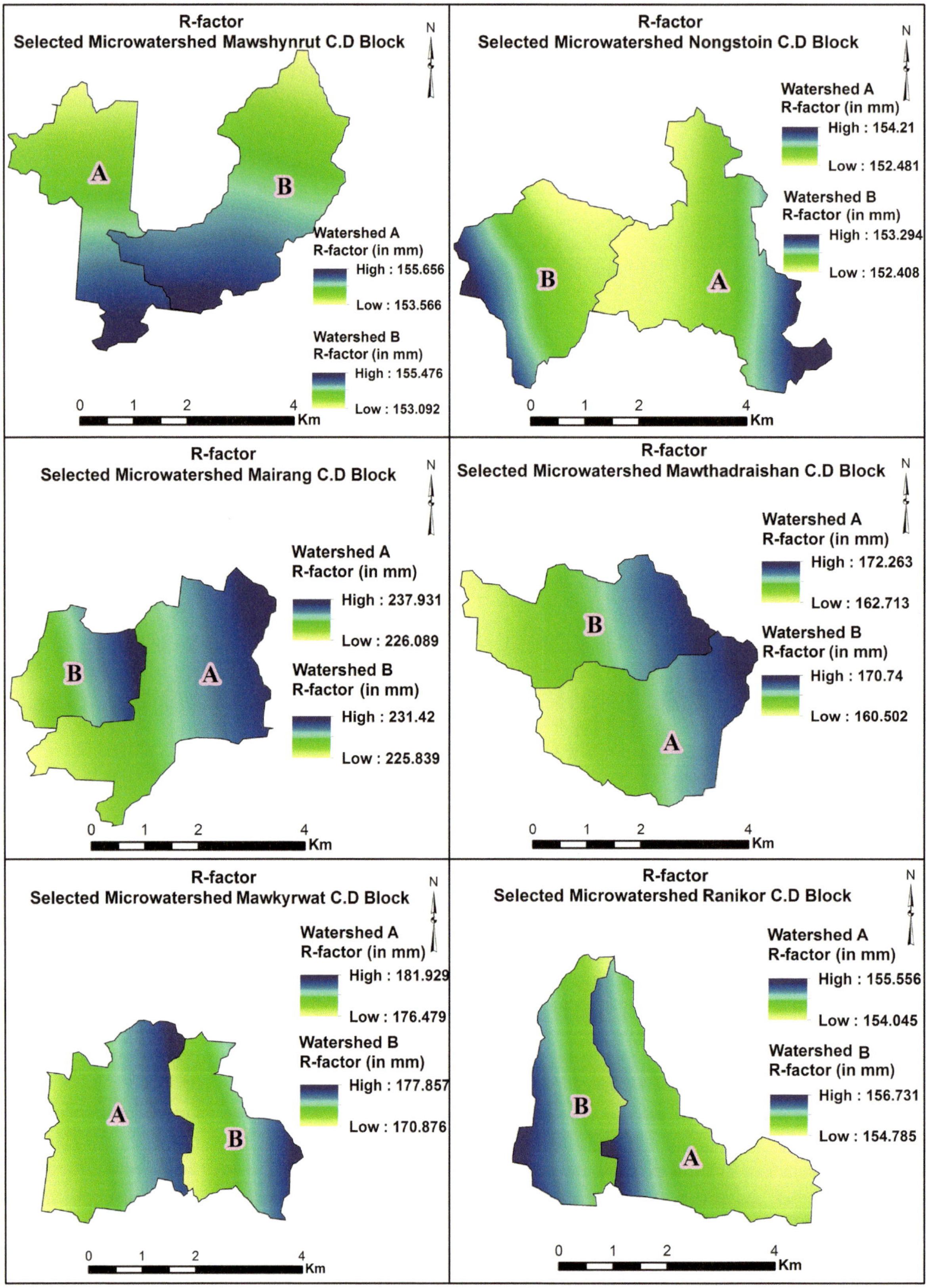

FIGURE 41.3 R factor. *India Meteorological Department, rainfall data for the years 2012, 2013, 2014, 2015 & 2016.*

TABLE 41.2 R-factor for six rain gauge stations.

Station	Longitude	Latitude	AAP in mm (2012–2016)	R factor
Kamrup Rural	91° 14′ 8.52″ E	25° 58′ 42.31″ N	116.75	121.38
East Garo Hills	90° 40′ 38.28″ E	25° 45′ 21.60″ N	279.55	180.48
East Khasi Hills	92° 22′ 26.21″ E	25° 10′ 43.93″ N	539.20	274.73
Jaintia Hills	92° 22′ 26.21′ E	25° 10′ 43.93″ N	234.12	163.98
West Garo Hills	90° 13′ 28.07″ E	25° 34′ 4.57″ N	166.15	139.31
West Khasi Hills	91° 27′ 19.95″ E	25° 21′ 45.12″ N	113.09	120.05

Source: IMD rainfall data; compiled and estimated by the researcher.

TABLE 41.3 K-factor (Source: Soil series of Meghalaya, 2005; compiled and estimated by the researcher).

Soil Code	Horizon	Depth in (Cm)	Sand (%)	Silt (%)	Clay (%)	K factor
1	Ap	0–18	20.40	56.50	23.10	0.23
2	A1	0–23	70.43	8.19	21.38	0.71
3	A1	0–6	16.90	57.60	25.50	0.19
4	A1	0–11	68.10	29.40	2.50	0.80
7	A	0–11	65.79	17.79	16.42	0.67
10	A1	0–11	83.50	7.40	9.10	0.84
12	A1	0–13	62.30	21.20	16.50	0.64
17	A1	0–19	82.60	12.90	4.50	0.85
20	A1	0–22	31.10	43.50	25.40	0.33
23	Ap	0–19	19.00	49.50	31.50	0.21

were developed to determine soil erodibility, considering soil physical properties as input data. For this study, the methodology derived by Denardin (1990) (Eq. (41.7)) is considered, where K is the function of the sand, silt, and clay proportions in soil. It is expressed as:

$$K = \frac{\text{SAN} + \text{SIL}}{\text{CLA}} \times \frac{1}{100} \tag{41.7}$$

where SAN is the percentage of sand, SIL is the percentage of silt, and CLA is the percentage of clay in a particular lasso of soil, respectively (Table 41.3).

The analysis of the K-factor for all 12 microwatersheds in 6 different CD blocks shows the factor is more likely homogeneous and that the fluctuation of the K value within the microwatersheds is minimal. These corresponding K-factor values also indicate a very close composition of sand, silt, and clay within the soil horizons. The texture of the soil is similar to the related proposition of its physical properties. The K-factor value (soil erodibility) of selected microwatersheds, ranges from 0.19 to 0.85, considering all 12 microwatersheds. The highest K factor is recorded in the Mawshynrut CD block in microwatershed B, followed by microwatershed A. of the same. The second-highest K-factor is recorded in the Mairang CD block; here, for both microwatersheds A and B, the highest is 0.70 and the lowest is 0.22 (Fig. 41.4).

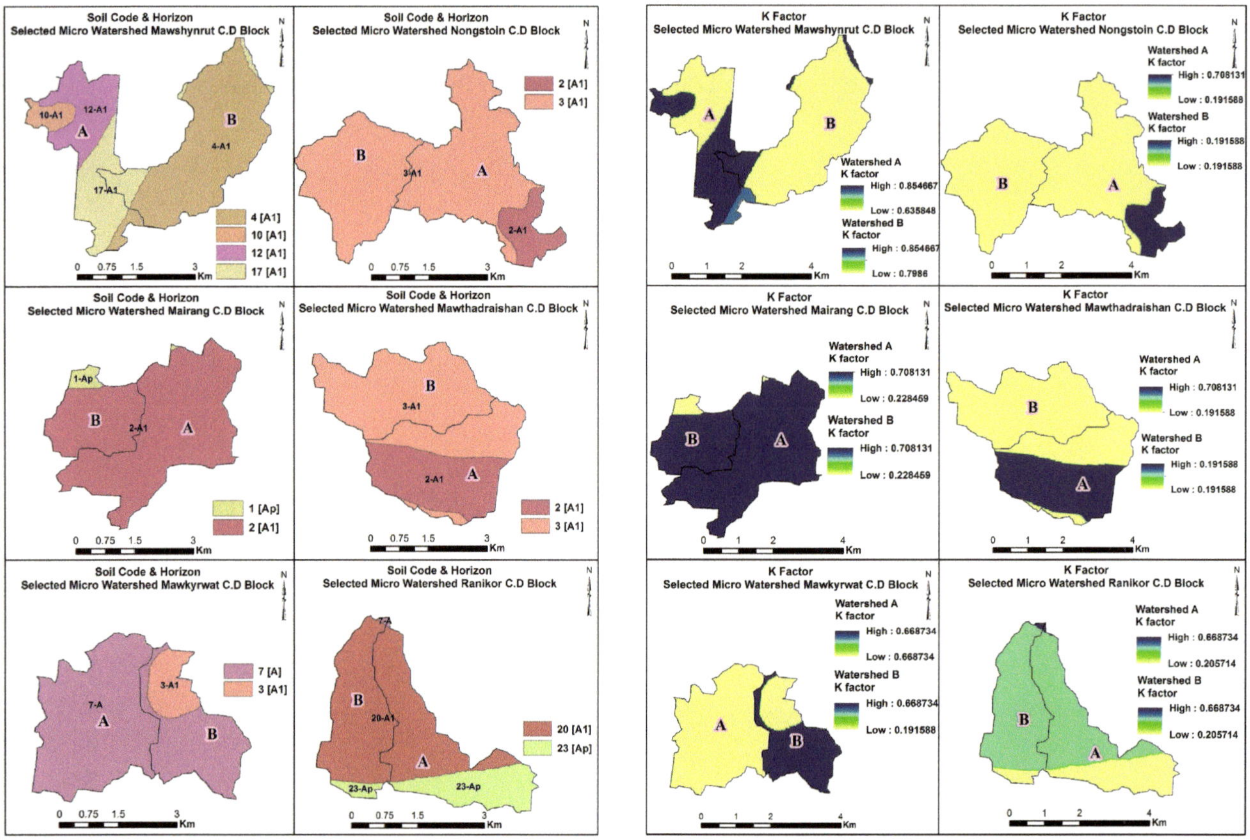

FIGURE 41.4 Soil code, Horizon, and K factor. *National Bureau of soil and land use planning (NBSS & LUP), Soil series of Meghalaya 2005.*

41.6 Slope length and steepness factor (LS)

LS is defined as the slope length and steepness factor. According to USDA Agricultural Handbook No. 282 and 537, LS is the topographical factor consisting of two subfactors:

1. L = slope length factor is the ratio of soil loss to field slope length from a 72.6 ft. or under identical conditions.
2. S = slope steepness factor is the ratio of soil loss from the field with a slope gradient from a 9% slope under otherwise identical conditions.

It is a factor that significantly influences soil erosion by surface runoff. Wischmeier and Smith (1965), Wischmeier and Smith (1978), calculated LS as:

$$\mathrm{LS} = (\lambda/72.6)\mathrm{m} \times (65.4\ \sin^2\theta + 4.56\ \sin\theta + 0.065) \tag{41.8}$$

where λ is slope length in feet; θ angle of the slope; and $m = 0.5$ if the percent slope is 5 or more, 0.4 on slopes of 3.5%–4.5%, 0.3 on slopes of 1%–3%, and 0.2 on uniform gradients of less than 1%.

In 1992, Moore and Wilson modified the Wischmeier and Smith's empirical method using a spatial analysis tool on a digital platform, where slope length is calculated in meters based on flow accumulation and slope steepness in radian value further converted to percentage, where 1 radiant value is equal to 15.91549430919%.

$$\mathrm{LS} = [\text{flow acc} \times \text{cell size}/22.13]0.4 \times [(\sin(\text{slope} \times 3.14/180))/0.0896] \tag{41.9}$$

For the present study, Eq. (41.9) has been used to estimate the LS factor using a digital elevation model (DEM) with 30-m spatial resolution. Further, it has been resampled to 10 m. As a topographical factor, the LS factor denotes slope surface steepness and its length, and the fluctuation or difference in the LS value shows the topographical variations in the study area. The value of the LS factor is always directly proportional to surface runoff or the amount of soil erosion under favorable conditions (Fig. 41.5).

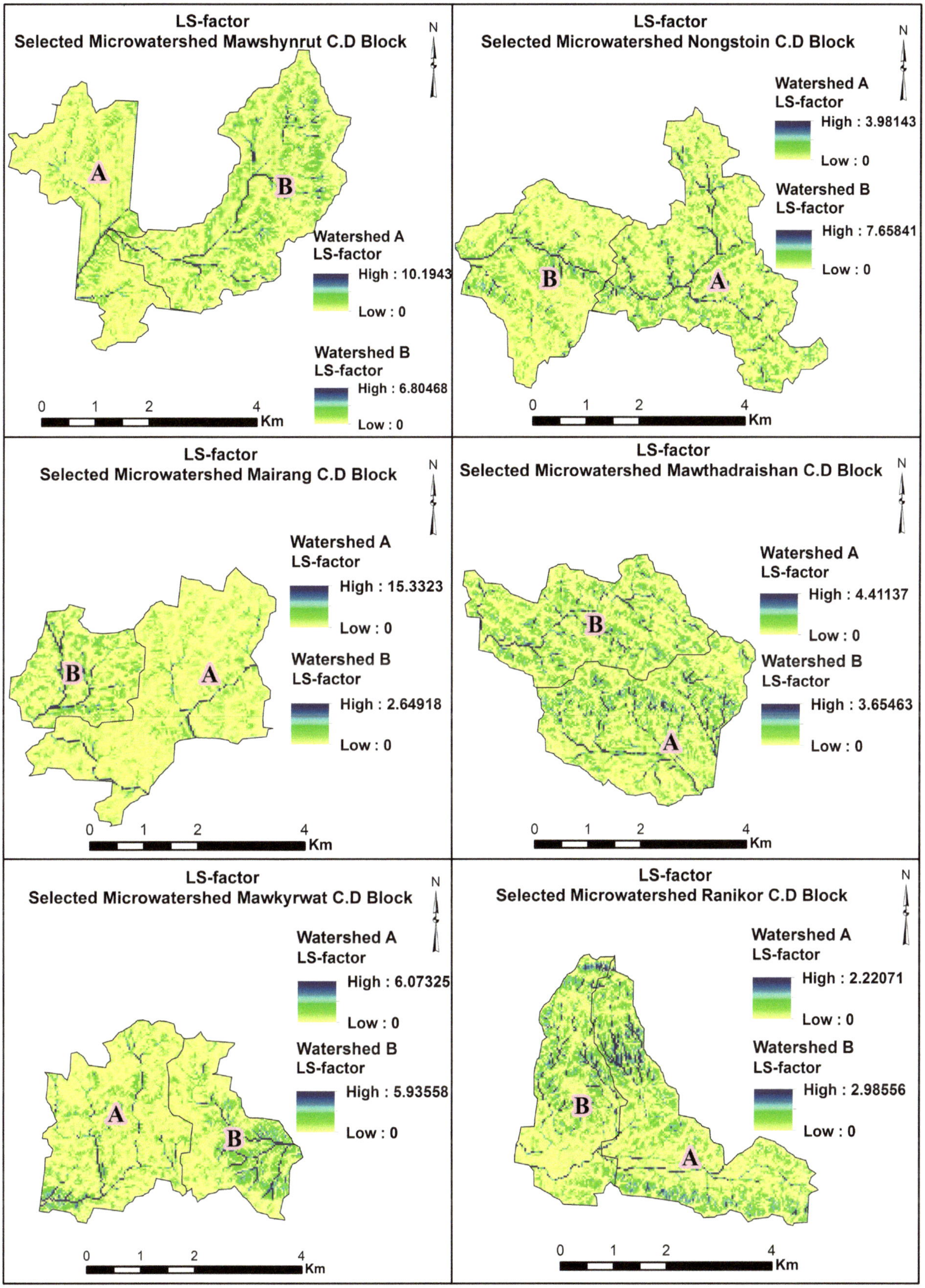

FIGURE 41.5 LS factor. *Aster DEM data (30 m) collected from USGS-Earth Explorer.*

41.7 Cover management factor (C)

The C factor was introduced in USDA Agricultural Handbook No. 537 by Wischmeier and Smith, in 1978. As per its original definition, the cover and management factor is the ratio comparing the relative soil loss between an area with specific cover and management and an identical area that is tilled and continuous fallow. Factor C in the soil loss equation is the ratio of soil loss from land cropped under specified conditions to the corresponding loss from clean-tilled and continuous fallow. This factor measures the combined effects of all interrelated cover and management variables. Wischmeier and Smith, also mentioned that cover and management factors could not be independently evaluated. Their combined effect is influenced by many significant interrelationships. Renard et al. (2011) defined the C factor as a function between prior land use (PLU), canopy cover (CC), surface cover (SC), surface roughness (RS) and soil moisture (SM). Where C = PLU * CC* SC* SR* SM. The C factor in the soil loss equation has also been calculated using NDVI by different scholars, as NDVI represents the vegetation cover and health of a particular area. In 1998, De Jong et al. evaluated the C factor using NDVI:

$$C = 0.431 - 0.805 \times \text{NDVI} \tag{41.10}$$

where NDVI is a geographical indicator that assesses the live green vegetation cover. The spectral reflectance is the ratio of the reflectance of incoming radiation in each spectral band individually. In 1999, 2000, Knijff et al. derived the *C* factor using NDVI:

$$C_{vk} = e\left(-\alpha\frac{\text{NDVI}}{\beta - \text{NDVI}}\right) \tag{41.11}$$

where α and β parameters determine the shape of the curve associated with NDVI with the *C* factor. This method is best for the European climate, where values 2 and 1 are the best representatives of the parameters α and β. However, it has been noticed that in a tropical and subtropical climates with more intense rainfall, the *C* factor tends to be higher than Knijff et al.'s method for the same amount of vegetation cover. In 2014, Durigan et al. proposed a revised method for calculating the *C* factor using NDVI that is best suited for tropical and subtropical regions (Durigon et al., 2014). They derived the *C* factor as:

$$C_r = \left(\frac{-\text{NDVI} + 1}{2}\right) \tag{41.12}$$

where C_r is the *C* factor. This equation represents the *C* factor in 0–1 range. The value 0 corresponds to dense vegetation cover, whereas 1 represents continuous fallow. Here it is noticeable that when NDVI is -1 cover management is 1 and NDVI is $+1$ cover management is 0. The numerical difference between these two ranges varies from ± 1 to ± 2 considering both ends (Bora, 2021). For this study, Eq. (41.12) is used because the study area has a subtropical climate with heavy rainfall and dense vegetation cover. Landsat 8 OLI and TRIS, 2018, with a 30-m spatial resolution are used to estimate the C factor. Furthermore, it is resampled to a 10-m resolution (Fig. 41.6).

41.8 Conservation practices factor (P)

The P factor is known as the practice factor. It is the ratio of soil loss between land with a support practice such as contouring, strip cropping, or terracing and land with straight raw farming up and down the slope. It has been introduced as the C factor in USDA Agricultural Handbook No. 537 by Wischmeier and Smith in 1978. In general, whenever sloping soil is to be cultivated and exposed to erosive rain, the protection offered by sod or close-growing crops in the system needs to be supported by practices that will slow the runoff water and thus reduce the amount of soil it can carry (Wischmeier and Smith, 1978). It is the ratio of soil loss in hilly slope areas, with a specific support practice. Improved tillage practices, such as sod-based rotation of crops, soil fertility treatments, or generating the quantities of the crop concerning the natural slope of the area, can reduce the amount of soil loss and provide major control to the farmer's field.

It is important to note that the benefits derived or extracted from the P factor are incorporated with factor C. The P factor can be evaluated by determining values based on land use concerning soil erosion or can be evaluated based on slope. Singh (1994) estimated the P factor with the help of the slope (Table 41.4).

However, this categorization of the P factor did not provide a clear understanding of practice management regarding the slope of the land. Shin (1999) has estimated the P factor on the basis of existing practice management in terms of the slope. The present study also used Shin's (1999) method to evaluate the P factor (Fig. 41.7 and Table 41.5).

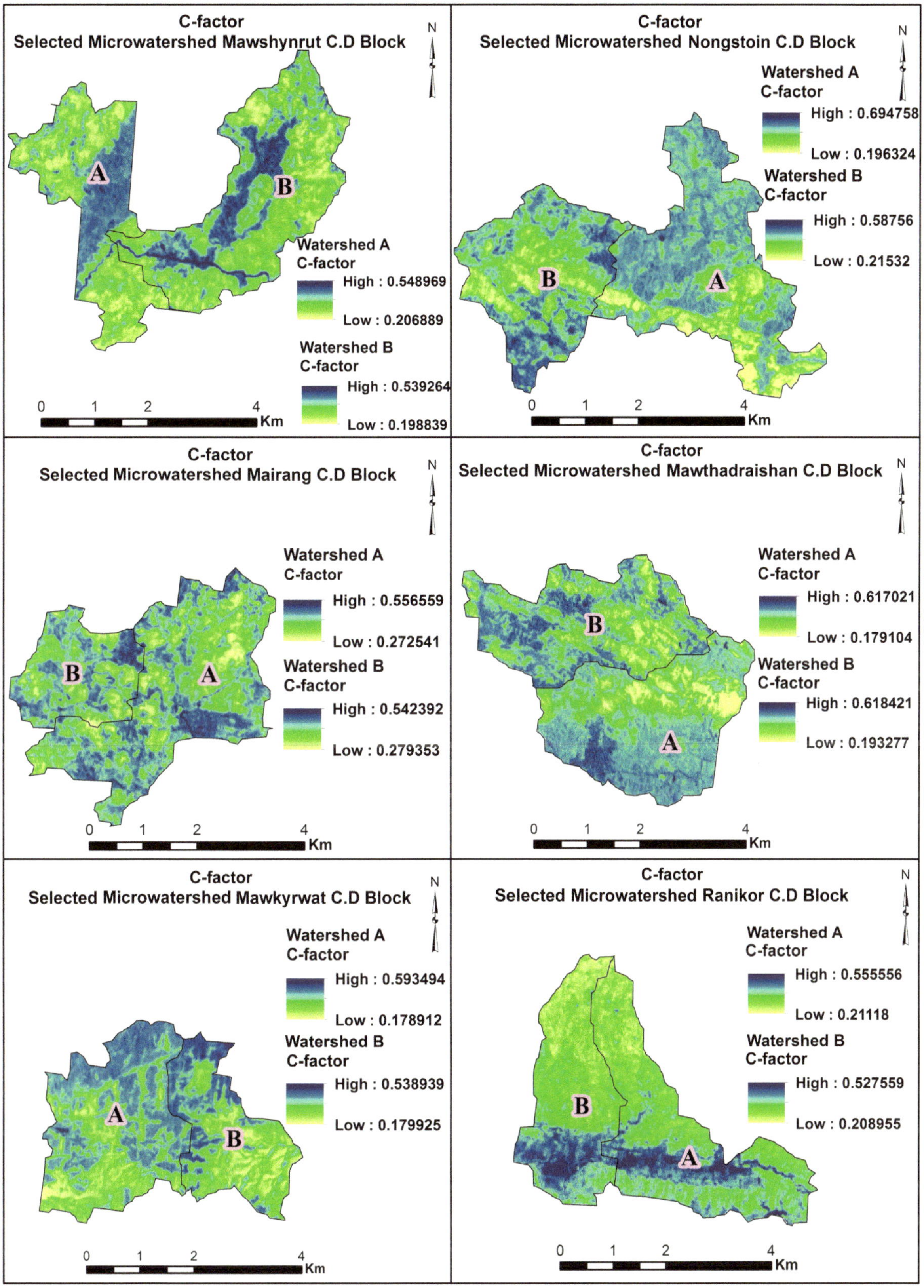

FIGURE 41.6 C factor. *Landsat 8—OLI & TRIS image, USGS Earth Explorer.*

TABLE 41.4 Shing classification of P-factor, 1994 (Source: Singh, V.P. (1994). "Elementary Hydrology").

Land slope in %	P factor
1–2	0.6
3–5	0.5
6–8	0.5
9–12	0.6
13–16	0.7
17–20	0.8
21–25	0.9

In 2010, the Department of Agriculture Malaysia determined the P factor in more specific categories in terms of practice management within the country (Teh, 2011) (Table 41.6).

41.9 Results and discussion

In 2011–13, the region was estimated at 53.01% of the total geographical area under the process of land degradation (Desertification and Land Degradation Atlas of selected districts of India, vol-2, 2018), and during 2015–16, it was 52.42% (Bora, 2021), which is more than half the total geographical area. In SLUSI (2010),[7] estimated that 46.23% of the total land area of the region was under a severe erosion zone, 13.64 was moderate to severe erosion, 36.50% was under moderate erosion, 1.38% was none to slight erosion, and 2.25% was miscellaneous land. The erosion index[8] of the surface soils under paddy, potato, forest, and Jhum land uses varied from 10.42 to 50.94, 30.99 to 73.87, 11.05 to 60.17, and 34.88 to 76.62, respectively (Diengdoh et al., 2016). These facts and figures are alarming. The current study primarily focuses on the intensity of soil erosion within the region and pinpoints the factors controlling erosion by randomly selecting two adjoining microwatersheds from each administrative CD block.

In Mawshynrut C.D Block, microwatershed A shows a comparatively higher amount of soil loss than its adjoining microwatershed B. Within microwatershed **A**, there is a difference of 754.80 tons ha^{-1} $year^{-1}$ of soil between the estimated potential and actual soil erosion in terms of the highest amount of soil loss. For the same microwatershed, B shows a difference of 589.60 tons ha^{-1} $year^{-1}$. Analyzing the factors, it was found that the most dominant factor of soil erosion is slope length and steepness (LS) in these two watersheds. The rainfall erosivity factor (R) is parallel for both watersheds, showing comparatively high erosivity, and the soil erodibility factor (K) is low. In the Nongstoin CD block, both the microwatersheds show nearly equal amounts of actual soil loss. Microwatershed A shows a difference of 93.72 tons ha^{-1} $year^{-1}$ between the estimated potential and actual soil erosion in terms of its highest amount of soil loss. Analyzing the factors, it is discovered that R is moderate but the LS is relatively high. In microwatershed B, the difference between estimated potential and actual soil erosion is 156.05 tons ha^{-1} $year^{-1}$, which is the highest amount of soil loss. Factor analysis shows that the R is parallel to microwatershed A. However, the LS are relatively high within this microwatershed. The K is low for both the microwatersheds.

In the Mairang CD block, the microwatershed A shows a very excessive amount of soil erosion, both in terms of potential and actual soil erosion processes. The difference between estimated potential and actual soil erosion is 1166.68 tons ha^{-1} $year^{-1}$, which is the highest amount of soil loss within the region. Analyzing the factors of soil erosion, it is found that all three primary factors, R, K, and LS, are comparatively high within this microwatershed. The estimated potential and actual soil erosion in microwatershed B are lower than those in adjoining microwatershed A. The difference between the estimated potential and actual soil erosion is 284.53 tons ha^{-1} $year^{-1}$. The R and K are parallel to the corresponding microwatershed A. However, the LS is very low. It is noticeable that LS plays a major role in slowing down the erosion process and reducing the amount of soil loss within this microwatershed (Plate 41.1).

7. Soil and Land Use Survey of India, Ministry of Agriculture and Farmers Welfare. Established in 1958.
8. The erosion index (EI) is estimated by dividing potential erosion by the T value. T is maximum average soil loss.

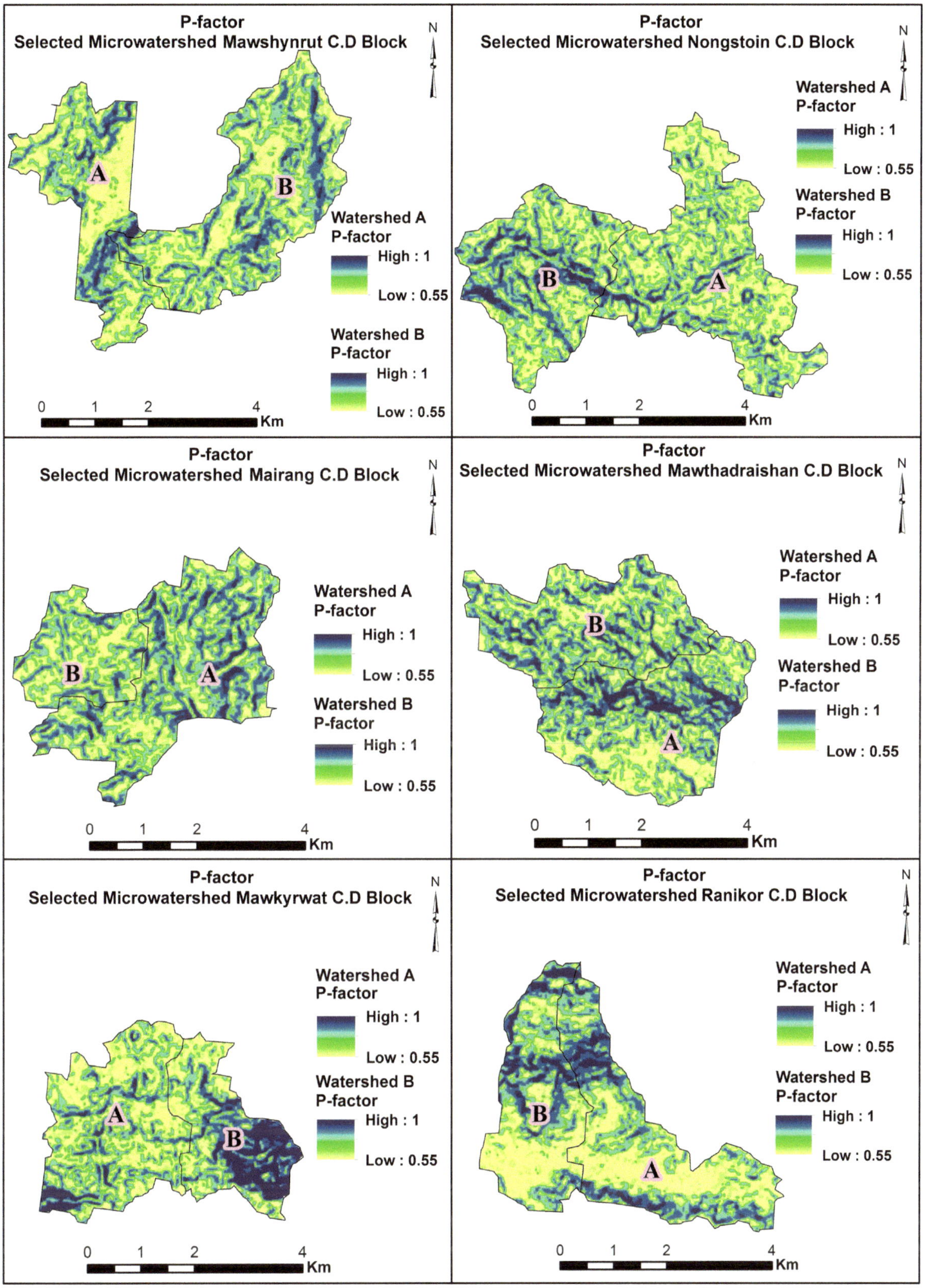

FIGURE 41.7 P factor. *Aster Global Digital Elevation Map & Landsat 8 OLI & TRIS, 2016 -USGS Earth Explorer.*

TABLE 41.5 Shin classification of P factor, Shin (1999).

Slope %	Contouring	Strip cropping	Terracing
0.0–7.0	0.55	0.27	0.10
7.0–11.3	0.60	0.30	0.12
11.3–17.6	0.80	0.40	0.16
17.6–26.8	0.90	0.45	0.18
>26.8	1.00	0.50	0.20

Source: Shin, G.J., 1999. The analysis of soil erosion analysis in watersheds using GIS. Doctoral Thesis, Department of Civil Engineering, Gang-won National University, Gangwon-do, South Korea.

TABLE 41.6 Department of Agriculture, Malaysia; classification of P factor 2010.

Structure/practice	P factor
Planting beds against/perpendicular to contour	0.85
Planting beds along the contour	0.30
Grass strip	0.50
Contour ditches	0.50
Hillside trench (silt trap)	0.60
Rain Shelter	0.10
Contour panning	0.80
Mulching	0.20
Terraces (continuous)	0.20
Terraces (discontinuous)	0.40
Individual basin	0.50
Traditional terraces	0.60
Vertiver contour hedgerow	0.60

Source: Teh, S.H., 2011. Soil erosion modeling using RUSLE and GIS on Cameron highlands, Malaysia for hydropower development. Doctoral Thesis, The School for Renewable Energy Science, University of Iceland, Iceland.

In the Mawthadraishan CD block, microwatershed A shows a fairly moderate amount of soil erosion. The estimated difference between potential and actual soil erosion in terms of the highest amount of soil loss is 276.64 tons ha^{-1} $year^{-1}$. Analyzing the factors of soil erosion, it is found that R is moderate and soil K ranges from high to low. However, LS is considerably low, which prevents soil loss within the microwatershed. In microwatershed B, the estimated potential and actual soil erosion are both lower than those in adjoining microwatershed A. The difference between the estimated potential and actual soil erosion in terms of the highest amount of soil loss is 70.87 tons ha^{-1} $year^{-1}$. The R is parallel to the corresponding microwatershed A. However, the K and LS are very low throughout the microwatershed which obstructs the erosion process and minimizes the amount of soil loss.

In the Mawkyrwat CD block, microwatershed A shows a fairly moderate amount of soil erosion. The estimated difference between potential and actual soil erosion in terms of the highest amount of soil loss is 402.52 tons ha^{-1} $year^{-1}$. Analyzing the factors, it is found that the R and LS are moderate. However, the K is low throughout the microwatershed, which obstructs the erosion process in the watershed. In microwatershed B, the estimated potential and actual soil erosion are both fairly similar to those in adjoining microwatershed A. The difference between estimated potential and actual soil erosion in terms of the highest amount of soil loss is 475.70 tons ha^{-1} $year^{-1}$. Factor analysis shows that

PLATE 41.1 Sever soil erosion; Mairang CD block.

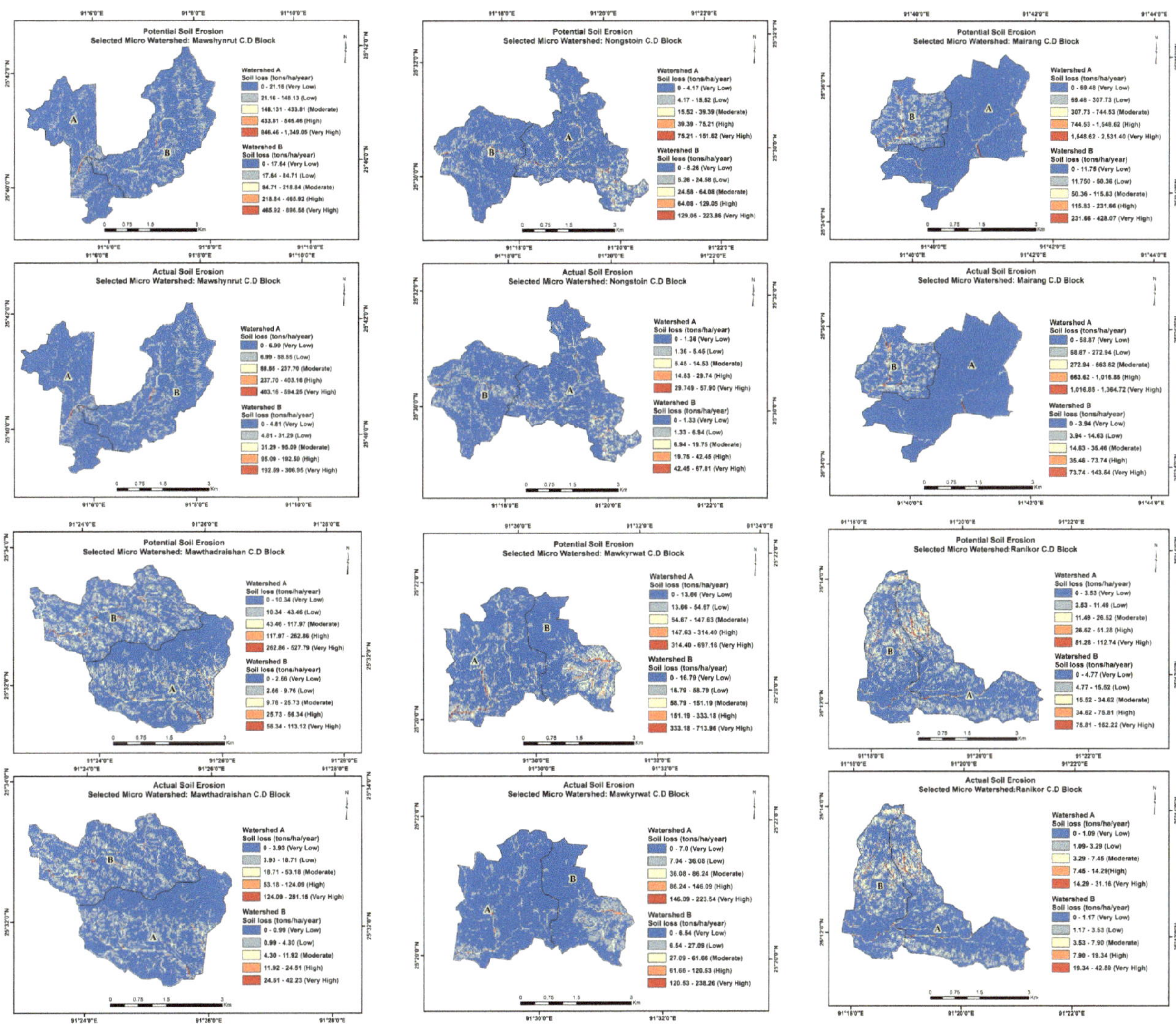

FIGURE 41.8 Potential soil erosion and Actual soil erosion.

the R is parallel to the corresponding microwatershed A. However, the K ranges from low to high, and the LS is moderate in this microwatershed. All three of these factors make the microwatershed very susceptible to soil erosion with moderate intensity.

In the Ranikor CD block, microwatershed A shows a relatively very low amount of soil erosion. The estimated difference between potential and actual soil erosion in terms of the highest amount of soil loss is 81.58 tons ha^{-1} $year^{-1}$. Factor analysis shows the R is moderate in the microwatershed and the K ranges from moderate to low. However, the LS is considerably low, preventing soil erosion in the microwatershed. In microwatershed B, the estimated potential and actual soil erosion are both fairly equal to those in adjoining microwatershed A. The difference between the estimated potential and actual soil erosion in terms of the highest amount of soil loss is 109.33 tons ha^{-1} $year^{-1}$. Analyzing the factors, it is found that the R is moderate throughout the microwatershed and the K ranges from low to high. However, the LS is considerably low, preventing soil erosion in the microwatershed (Fig. 41.8).

In the analysis of the factors of soil erosion, some similarities have been noticed among watersheds with a very high sum of soil loss. Within these, the LS factor (slope length and steepness factor) is comparatively very high. In terms of the very low sum of soil loss, once again similarities have been noticed in the LS factor with fairly low very low LS values. Furthermore, it has also been noticed that the K-factor is also fairly very high within watersheds with a very high sum of soil loss. On the other hand, within microwatersheds with low soil erosion, the K value is also fairly low.

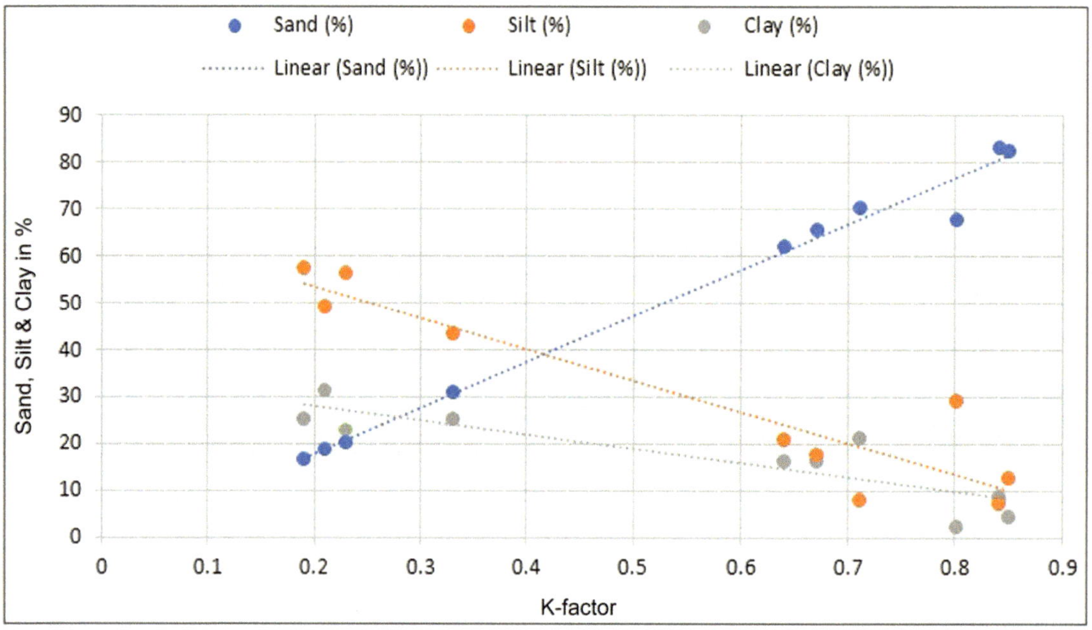

FIGURE 41.9 K-value and soil property graph of the study site.

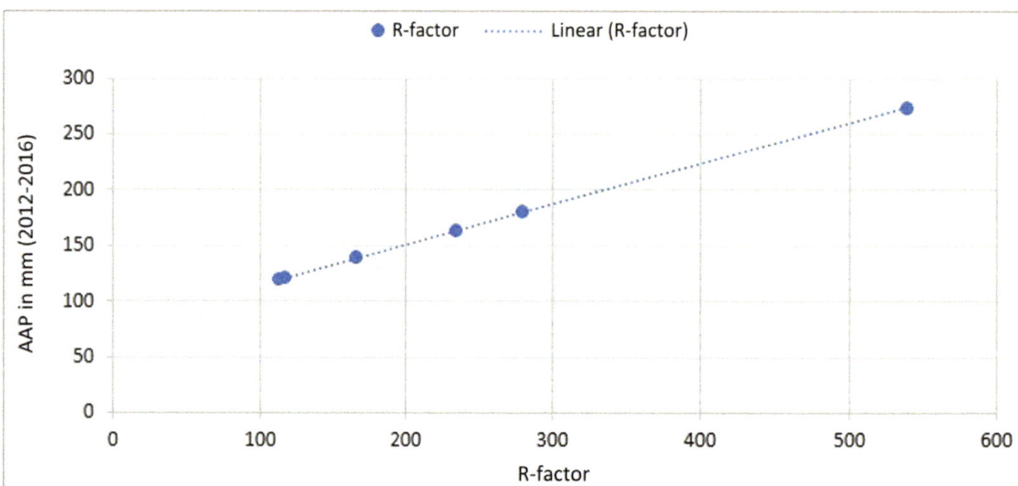

FIGURE 41.10 R-value and annual average precipitation graph of the study site.

The K value and soil property graphs of the study sites show that the presence of sand increases soil erodibility. Similarly, soil with a fairly high proportion of clay reduces soil erodibility. The erodibility of silt also shows decrease in K value for the study region, despite its high erosive character (Fig. 41.9).

The analysis of the R factor (rainfall erosivity) shows that although the factor has a strong impact on the process of soil erosion, it does not function as a decisive factor that controls the amount of soil loss in this region. The R-value and annual average precipitation graph of six rain gauge stations show that the R-value increases as the annual average precipitation increases. As the present study is concentrated in one region, the fluctuation between the amount of rainfall received by different parts of the region is almost negligible, or one could say it is almost similar throughout the region (Fig. 41.10).

41.10 Conclusion

Comparing values of soil loss, both potential and actual soil erosion, it has been observed and found, that microwatershed A of Mairang C.D block recorded the highest amount of soil loss and turned out into a very high zone of actual soil erosion. Most prominently it indicates the uncovered banks of the "***Khri River***" and its substreams. In contrast, microwatershed A, of the Ranikor CD block, recorded the lowest amount of soil loss. However, when all 12 microwatersheds are put together within one single frame and a categorical analysis is performed, the estimated values of soil loss show that, in terms of the very high category, some areas of microwatershed A within the Mirang CD block have the highest potentiality of soil loss, that is, ranging between 1548.62 and 2531.40 tons ha^{-1} $year^{-1}$ and also recorded the highest amount of actual soil loss too, that is, ranging between 1016.85 and 1364.72 tons ha^{-1} $year^{-1}$. On the other hand, in terms of the very low category, some areas of microwatershed A within the Mawthadraishan CD block have the lowest potential of soil loss, that is, ranging between 0 and 2.66 tons ha^{-1} $year^{-1}$, and record the lowest amount of actual soil loss as well, that is, ranging between 0 and 0.99 tons ha^{-1} $year^{-1}$. Regarding the most deriving factors of soil erosion in terms of all twelve (12) microwatersheds the study has established that in the West, South West, and Eastern West Khasi Hills, the most dominant factors of the soil erosion process that cause a considerable amount of soil loss are:

1. LS factor, that is, slope length and steepness, and
2. K-factor, that is, soil erodibility.

The estimated results of actual soil loss show a noticeable difference between the microwatersheds. Considering two adjoining microwatersheds, the amount of actual soil loss is not always equivalent. This also indicates that the pattern of soil erosion in the region varies, regardless of its physical properties. This study does not prove or show any gradual expansion of soil loss based on its topographical and pedological characteristics. Here it is important to mention that the C factor (cover and management factor) and P factor (practices factor) have a strong impact on preventing soil loss. By estimating and analyzing the soil erosion for all the 12 microwatersheds, it was observed that there is an absolute and visible difference between potential and actual soil erosion regarding the amount of soil loss.

Acknowledgment

The authors acknowledge the Department of Geography, North Eastern Hill University, Shillong, for providing laboratory facilities and academic guidance to conduct an effective study. The authors also acknowledge the Space Application Centre, Ahmedabad, ISRO, for providing financial support and assistance throughout the study period.

References

Babu, R., Tejwani, K.G., Agarwal, H.C., Bhusan, L.S., 1978. Distribution of erosion index and Iso—erodent maps of India. Indian J. Soil Conserv. 6 (1), 1–12.

Benavidez, R., Jackson, B., Maxwell, D., Norton, K., 2018. A review of the (revised) Universal Soil Loss Equation ((R) USLE): with a view to increasing its global applicability and improving soil loss estimates. Hydrol. Earth Syst. Sci. 22 (11), 6059–6086.

Bhattarai, R., Dutta, D., 2007. Estimation of soil erosion and sediment yield using GIS at catchment scale. Water Resour. Manag. 21, 1635–1647.

Bora, A., 2021. Land degradation vulnerability assessment in selected micro watersheds of West Khasi Hills district, Meghalaya. Doctoral Thesis, Department of Geography, North Eastern Hill University, Meghalaya, India.

Brown, L.R., 1984a. The global loss of topsoil I. Soil. Water. Conserv. 36 (5), 255–260.

Brown, L.R., 1984b. The global loss of top soil. J. Soil. Water. Conserv. 39, 162–165.

CGWB, 2013. Ground Water Information Booklet West Khasi Hills District, Meghalaya. Central Ground Water Board, Ministry of Water Resources, Govt. of India.
De Jong, S.M., Brouwer, L.C., Riezebos, H.T., 1998. Erosion hazard assessment in the La Peyne catchment, France. *DeMon-2 Project*. Department of Physical Geography, Utrecht University, The Netherlands.
Denardin, J.E., 1990. Erodibilidade de solo estimada por meio de parâmetros físicos e químicos. Piracicaba. *Tese de doutorado*, Escola Superior de Agricultura Luiz de Queiroz. Universidade, de São Paulo.
Desifindiana, M.D., Suharto, B., Wirosoedarmo, R., 2013. Analisa tingkat bahaya erosi pada DAS Bondoyudo Lumajang dengan menggunakan Metode MUSLE (In Press). J. Keteknikan Pertan. Trop. Dan. Biosist. 1 (2), 9–17.
Diengdoh, J., Dutta, M., Ram, S., 2016. Erodibility status of soils under different land uses in West Khasi Hills of Meghalaya. An. Asian J. Soil. Sci. Vol: 11, 217–221.
Dubey, S.K., Sah, U., 2009. Indigenous Nur Bun method of potato cultivation in Meghalaya Hills. Asian Agri–History J. 13 (2), 147–153.
Durigon, V.L., Carvalho, D.F., Antunes, M.A.H., Oliveira, P.T.S., Fernandes, M.M., 2014. NDVI time series for monitoring RUSLE cover management factor in a tropical watershed. Int. J. Remote. Sens. 35 (2), 441–453.
FAO, 2021. The state of the world's land and water resources for food and agriculture. Synthesis Report, Rome.
FSI, 2017. State of forest report 2017. Forest Survey of India. Ministry of Environment & Forests, Govt. of India.
Gavrilovic, Z., 1988. Use of an empirical method(erosion potential method) for calculating sediment production and transportation in unstudied or Torrential streams. In: International Conference on River Regime. Hydraulics Research Limited, Wallingford, Oxon, UK, 1988, pp. 411–422, 5 fig, 4 tab, 8 ref.
Herawati, T., 2010. Analisis Spasial Tingkat Bahaya Erosi Di Wilayah Das Cisadanekabupaten Bogor. J. Penelit. Hutan dan. Konservasi Alam 7 (4), 413–424.
Holdgate, M.W., Kassas, M., White, G.R., 1982. The world environment: soil erosion and land degradation. Tycool Int. Dublin.
Hu, S., Li, L., Chen, L., Cheng, L., Yuan, L., Huang, X., et al., 2019. Estimation of soil erosion in the Chaohu lake basin through modified soil erodibility combined with gravel content in the RUSLE model. Water 11 (9), 1806.
Kadlec, V., Holubík, O., Prochazkova, E., Urbanova, J., Tippl, M., 2012. Soil organic carbon dynamics and its influence on the soil erodibility factor. Soil Water Res. 7 (3), 97–108.
Kadomura, H., Yamamoto, H., 1978. Man-induced erosion: the rate of erosion and an example from Okinawa Island, Southern Japan. J. Geogr. (Chigaku Zasshi) 87 (1), 1–15.
Khairunnisa, F., Tambunan, M.P., Marko, K., 2020. Estimation of soil erosion by USLE model using GIS technique (a case study of upper Citarum Watershed), IOP Conference Series: Earth and Environmental Science, Vol. 561. IOP Publishing, p. 012038, No. 1.
Kinnell, P.I.A., 2010. Event soil loss, runoff and the Universal Soil Loss Equation family of models: a review. J. Hydrol. 385 (1–4), 384–397.
Konwar, M., 2010. Land classification in the hill areas: a study of West Khasi Hills district of Meghalaya. Doctoral Thesis, Department of Geography, North Eastern Hill University, Meghalaya, India.
Knijff, Van der, J.M., Jones, R.J.A., Montanarella, L., 1999. Soil Erosion Risk Assessment in Italy. Ispra: European Commission Directorate General JRC. Joint Research Centre Space Applications Institute. European Soil Bureau.
Knijff, Van der, J.M., Jones, R.J.A., Montanarella, L., 2000. Soil Erosion Risk Assessment in Europe. Ispra: European Soil Bureau. Joint Research Centre. EUR 19044, EN.
Lal, R., 1990. Soil erosion and land degradation: the global risks. Advances in Soil Science. Springer-Verlag Inc., New York.
Lense, G.H.E., Parreiras, T.C., Moreira, R.S., Avanzi, J.C., Mincato, R.L., 2019. Estimates of soil losses by the erosion potential method in tropical latosols. Ciênc. Agrotecnol. 43.
Li, L., Du, S., Wu, L., Liu, G., 2009. An overview of soil loss tolerance. Catena 78 (2), 93–99.
Michael, E.A., Samanta, S., 2016. Landslide vulnerability mapping (LVM) using weighted linear combination (WLC) model through remote sensing and GIS techniques. Model. Earth Syst. Environ. 2, 1–15.
National Bureau of Soil Survey and Land Use Planning India, 2005. Soil series of Meghalaya. Indian Council of Agricultural Research. https://books.google.com/books/about/Soil_Series_of_Meghalaya.html?id=jyjbAAAAMAAJ.
O'Keefe, P., 1983. The causes, consequences and remedies of soil erosion in Kenya. Ambio 12 (6), 302–305.
Patowary, S., Sarma, A.K., 2018. GIS-based estimation of soil loss from hilly urban area incorporating hill cut factor into RUSLE. Water Resour. Manag. 32, 3535–3547.
Poerbandono, A.B., Harto, A.B., Rallyanti, P., 2006. Evaluasi perubahan perilaku erosi daerah aliran sungai Citarum Hulu dengan pemodelan spasial. Jurnal Infrastruktur dan Lingkungan Binaan II (2).
Renard, K.G., Yoder, D.C., Lightle, D.T., Dabney, S.M., 2011. Universal soil loss equation and revised universal soil loss equation. In: Morgan, R.P. C., Nearing, M.A. (Eds.), Handbook of Erosion Modelling, pp. 137–167.
Rind, D., 1983. Global climate in the 21st century. Ambio 13 (3), 148–151.
Samanta, S., Koloa, C., Pal, D.K., Palsamanta, B., 2016. Estimation of potential soil erosion rate using RUSLE and E30 model. Model. Earth Syst. Environ. 2 (3), 149.
Sharma, U.C., Prasad, R.N., 1995. Socio-economic aspects of acid soil management and alternative land use systems for north eastern states of India. In: Plant-Soil Interactions at Low pH: Principles and Management: Proceedings of the Third International Symposium on Plant-Soil Interactions at Low pH, Brisbane, Queensland, Australia, 12–16 September 1993. Springer Netherlands, pp. 689–695.
Shekinah, D.E., Saraswathy, R., 2005. Impacts of soil erosion by water-a review. Agric. Rev.-Agric. Res. Commun. Cent. India 26 (3), 195.

Shin, G.J., 1999. The analysis of soil erosion analysis in watersheds using GIS. Doctoral Thesis, Department of Civil Engineering, Gang-won National University, Gangwon-do, South Korea.

Singh, V.P., 1994. Elementary Hydrology. Prentice Hall.

Singh, G., Ram Babu, V., Chandra, S., 1981. Soil Loss Prediction Research in India. Bulletin Nos.T-12/D-9. Central Soil and Water Conservation Research and Training Institute, Dehradun.

SLUSI, 2010. Soil resource mapping using remote sensing technique, Abstract West Khasi Hills District. Meghalaya. Soil & Land Use Survey of India. Ministry of Agriculture & Farmers Welfare, Govt. of India.

Soil & Water Conservation Department, 2022. Why Soil and Water Conservation. Government of Meghalaya. https://megsoil.gov.in/why_cons.html.

Teh, S.H., 2011. Soil erosion modeling using RUSLE and GIS on Cameron highlands, Malaysia for hydropower development. Doctoral Thesis, The School for Renewable Energy Science, University of Iceland, Iceland.

Wischmeier, W.H., Smith, D.D., 1965. Predicting rainfall-erosion losses from cropland east of the Rocky mountains: guide for selection of practices for soil and water conservation. No. 282. Agric. Res. Serv. Agricultural Research Service.

Wischmeier, W.H., Smith, D.D., 1978. Predicting rainfall erosion losses: a guide to conservation planning. No. 537. Sci. Educ. Adm. U S Dep. Agriculture Cooperation with Purdue Agricultural Experiment Station.

Zhang, P., Yao, W., Liu, G., Xiao, P., 2019. Experimental study on soil erosion prediction model of loess slope based on rill morphology. Catena 173, 424–432.

Zhu, M., 2015. Soil erosion assessment using USLE in the GIS environment: a case study in the Danjiangkou Reservoir Region, China. Environ. Earth Sci. 73, 7899–7908.

Further reading

Cassol, E.A., Silva, T.S.D., Eltz, F.L.F., Levien, R., 2018. Soil erodibility under natural rainfall conditions as the K factor of the universal soil loss equation and application of the nomograph for a subtropical Ultisol. Rev. Bras. Ciênc. Solo 42.

Dabral, P.P., Baithuri, N., Pandey, A., 2008. Soil erosion assessment in a hilly catchment of North Eastern India using USLE, GIS and remote sensing. Water Resour. Manag. 22 (12), 1783–1798.

Das, T., 2017. Development of a regional model to calculate rainfall erosivity factor by using readily available rainfall data. A study report (Post Graduate Research). Department of Civil Engineering Indian Institute of Technology Guwahati.

FAO, 1976. A framework for land evaluation, Soils Bull., 32. FAO and Agriculture Organization of the United Nations, Rome.

Gelagay, H.S., Minale, A.S., 2016. Soil loss estimation using GIS and Remote sensing techniques: a case of Koga watershed, Northwestern Ethiopia, Int. Soil Water Conser. Res., 4, pp. 126–136.

Lal, R., 2001. Soil degradation by erosion. Land. Degrad. Dev. 12, 519–539.

RUSLE—an online soil erosion assessment tool, 2002. C Factor. Institute of Water Research, Michigan State University.

SAC, Ahmedabad, 2018. Desertification and land degradation atlas of selected districts of India (Based on IRS LISS III data of 2011–13 and 2003–05), Vol. 2. Space Application Center, ISRO, Ahmedabad, India, 43–44.

Singh, R.S., Baruah, U., Sarkar, D., Butte, P.S., Gajbhiye, K.S., 2005. Soil of Meghalaya. Tech. Bull., NBSS Publ. 121.

Uddin, K., Abdul Matin, M., Maharjan, S., 2018. Assessment of land cover change and its impact on changes in soil erosion risk in Nepal. Sustainability 10 (12), 4715.

Chapter 42

Spatial assessment of runoff and soil loss of micro watershed using MMF model—a case study of Patapur Micro Watershed, Karnataka, India

Ibrahim Kaleel[1], U. Satish Kumar[2], C.M. Pradeep[3], Mallikarjun Reddy[2], Mallikarjuna[2] and S. Dharumarajan[3]

[1]*Water Technology, Tocklai Tea Research Institute, Jorhat, Assam, India,* [2]*Department of Soil and Water Engineering UAS, Raichur, Karnataka, India,* [3]*ICAR-National Bureau of Soil Survey and Land Use Planning, Regional Centre, Bangalore, Karnataka, India*

42.1 Introduction

The Semiarid Tropic (SAT) regions of the world, which cover parts of 55 developing countries, have over 1.4 billion populations, of whom 350 million are classified as rural poor (Ryan et al., 2002). India comprises a land area of about 53.4% under arid and semiarid regions (First NATCOM, GoI). The United Nations Population Fund projected a problem facing sustainable development in the future for the countries experiencing expeditious population growth (United Nations Population Fund, 2003). One billion of the world's poorest residents who live in SAT regions will be impacted by water scarcity. The impoverished socio-economic condition of the farmers and uncertainty and variability in rainfall is a matter of great concern to invest in today's agriculture. Farmers exploit the groundwater to meet their water needs and to save crops from drought during the rainy season, and postrainy season results in recession of groundwater levels due to insufficient groundwater recharging facilities. Hence, there is a need to develop sustainable and environment-friendly concepts to manage natural resources in this brittle ecosystem for the survival of poor farmers and to increase their productivity and income utterly dependent on natural resources. The way forward to address this complex task is mainly by sustainable management of rainwater and other natural resources in a manageable land unit, which is a watershed.

The annual average rainfall of arid and semiarid regions in India ranges from 300 to 50 mm, and it has been discovered that total water resources in India are of the order of 167 million hectare meters. It has been calculated that only 66 million hectares of water resources in India can be used for irrigation. The total water resources utilizable for the country are assessed as 1086 km^3. Land and water resource management are the basis of the suspendable development of society and economy in arid and semiarid regions (Bekele and Drake, 2008). It is recognized from the present situation that the key issue is the watershed management. If powerful engineering and nonengineering measures are adopted, the problem in arid and semiarid regions in India is really possible to be well solved. In fact, Geographic Information Systems (GIS) and Decision Support Systems (DSS) can play a significant role in land and water resources management, such as surface water, groundwater investigation, dynamic monitoring of ecology, and estimation of water amount necessary for keeping and recovering ecological environment, soil moisture and drought monitoring, monitoring and evaluation of soil loss and runoff, dynamic monitoring of desertification and soil erosion, site selection of conservation measures and its planning, design, construction, and management.

Watershed has been accepted as a basic unit for planning and implementation of protective, curative, and amelioration programs. An accurate understanding of the hydrological behavior of a watershed is most important for its effective management (Neelambika and Satishkumar, 2016). The decisive factors like rainfall, runoff, and soil erosion under a rainfed ecosystem are the main factors for adequate planning to manage land and water resources (Elisabeth and Bas,

Remote Sensing of Soils. DOI: https://doi.org/10.1016/B978-0-443-18773-5.00008-9

2009). Runoff is the portion of precipitation that makes its way toward the stream channels, lakes, or oceans as surface or subsurface flows. The most destructive component of rainfall is runoff, which occurs when the precipitation rate exceeds the infiltration rate. Hence, we need a unique approach to assess runoff and soil loss using spatial tools to conserve the soil and water (Guillermo and Consuelo, 2007). Integrating the soil loss model Morgan, Morgan, and Finney (MMF) model with spatial DSS QGIS leads to qualitative results that can be further utilized for watershed management aspects.

MMF model was considered to predict annual soil loss from field-sized areas on hill slopes, which, while endeavoring to retain the simplicity of the Universal Soil Loss Equation, encompassed some of the recent advances in understanding erosion processes Morgan and Duzant (2008). Further, the versatility of the GIS to perform advanced data analysis through linkages with real-time hydrologic prediction models, development of dynamic interfaces for integrated water resources management, optimal water resources allocation, and multicriteria evaluation support systems, and use of built-in macro programming languages to develop DSS are gaining momentum in these days (Adinarayana, 2001). Also, recent advances in GIS in linking simulation models, remote sensing data, and relational database techniques make it easier for analysis. It is entirely conceivable that within a properly integrated GIS framework, the sustainability profiles of multiple land-use watersheds can be assembled into various scenarios, which may enable the basin managers to identify and avoid possible losses of significant ecological functions while defining management responses and boundaries (Sarangi et al., 2006). The use of spatial decision support systems (SDSS), created within a GIS framework, can enable the decision or appropriate soil and water conservation measures. (Prasanna and Satishkumar, 2014).

42.2 Methodology

42.2.1 Study area

The selected study area was the Patapur micro watershed, which lies between 16°07′35.9″ to 16°08′22.3″N Latitude and 76°51′33.3″ to 76°53′27.7″E Longitudes with an average elevation of 447 m above the mean sea level. This watershed is a typical micro watershed system located under a semiarid agro-climatic condition with an area of about 454.62 ha. It can be located on toposheet No. 56 D/16 (1:50,000) of the Survey of India database.

The study area is a part of the North-Eastern Dry Zone (Zone-2 of Region-1) with subtropical climate. The mean maximum temperature varies from 30.3°C in December to 40.6°C in May, and April and May are the hottest months. The minimum temperature ranges from 15.7°C in December to 25.3°C in May, and December and January are the coldest months. The mean relative humidity fluctuates from 37.5% in March to 69.1% during August. Wind speed exceeds 15 km $hours^{-1}$ during June and July. The recorded maximum annual potential evaporation would be 1950 mm per year (220 mm) in May and the lowest in December (120 mm). The Patapur micro watershed receives an average annual rainfall of 621 mm. The number of rainy days annually ranges between 35 and 60 days per year. Nearly 67% of the rain is received during the southwest monsoon period (June to September), and the remaining 24% gets contributed during northeast monsoon (Fig. 42.1).

The detailed physiographic and soil survey of the Patapur micro watershed was carried out using a cadastral map of the area as a base map and compared with ground truth delineation. The survey was carried out using standard soil survey procedures described in the soil survey manual (IARI, 1970). At first instance, traversing of the study area was carried out to accomplish land resource inventory consisting of physiography, streamlines, geology, soils, vegetation, and land use.

42.2.2 Spatial input maps of study area

Patapur micro watershed is divided into 16 stream-contributing segmental areas based on a stream order map (Fig. 42.2). Stream-contributing areas and soil mapping units were delineated based on Topography (relief, slope), land use/land cover, soil texture, erosion condition, drainage network, etc. Stream junction points or nodes are delineated spatially at the point of streams joining a stream with the same order or with that of a higher order. Soil samples collected from the horizons of various soil mapping units were analyzed for physical characteristics.

42.2.3 Morgan, Morgan, and Finney model as an estimator of runoff and soil loss

MMF model estimates runoff and annual soil loss from field-sized areas on various undulating terrains and includes recent advances compared to the Universal Soil Loss Equation, which helps to study the erosion process

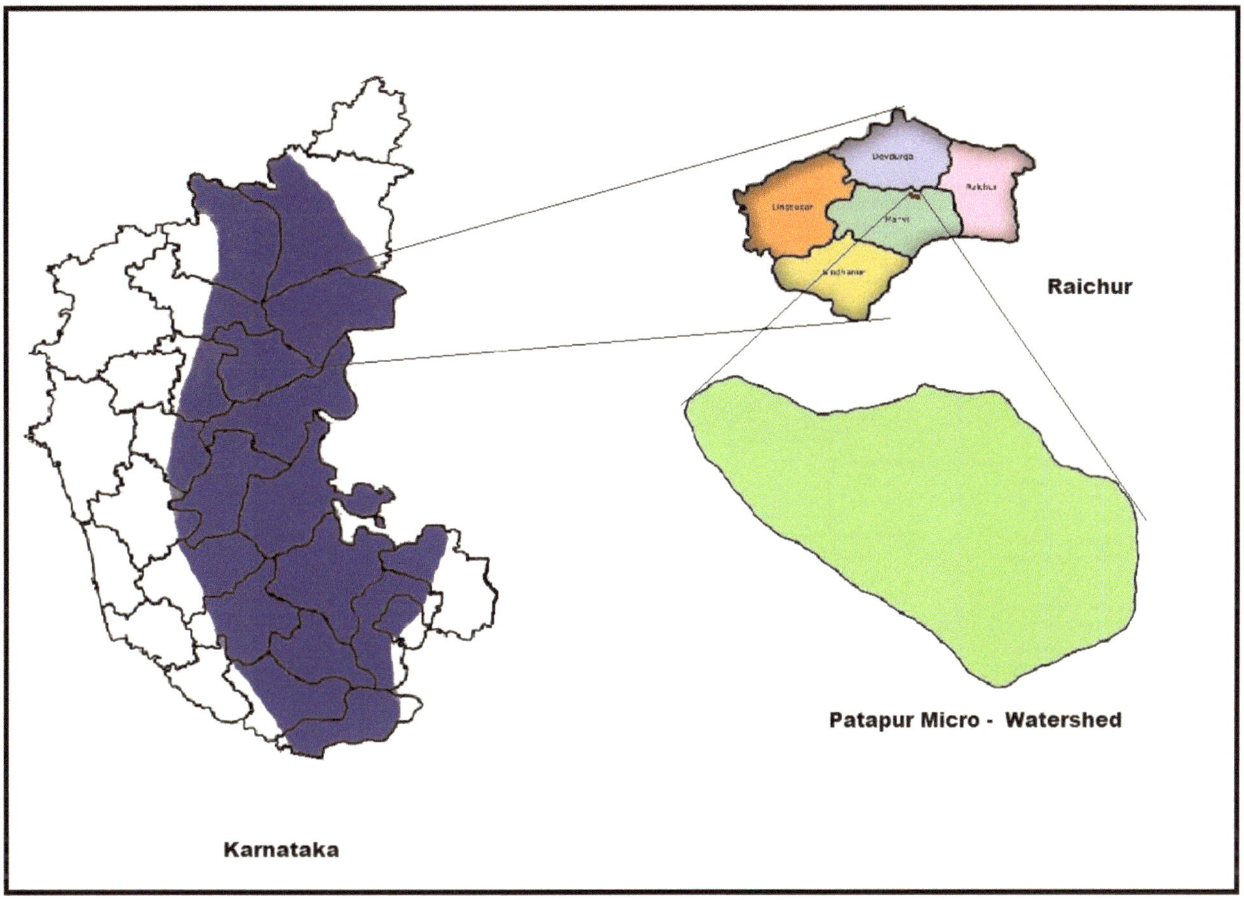

FIGURE 42.1 Location of Patapur micro watershed in Raichur district.

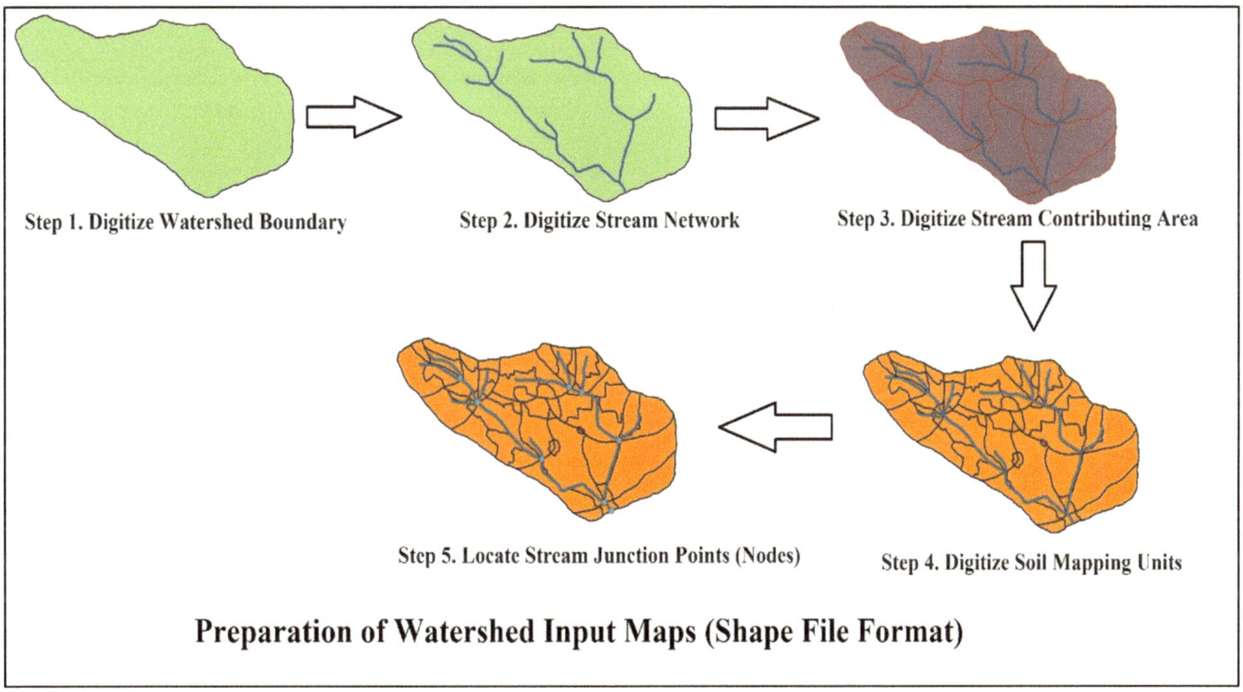

FIGURE 42.2 Digitized stream networks, stream contributing area, soil mapping units & stream junction points of Patapur Micro Watershed using QGIS.

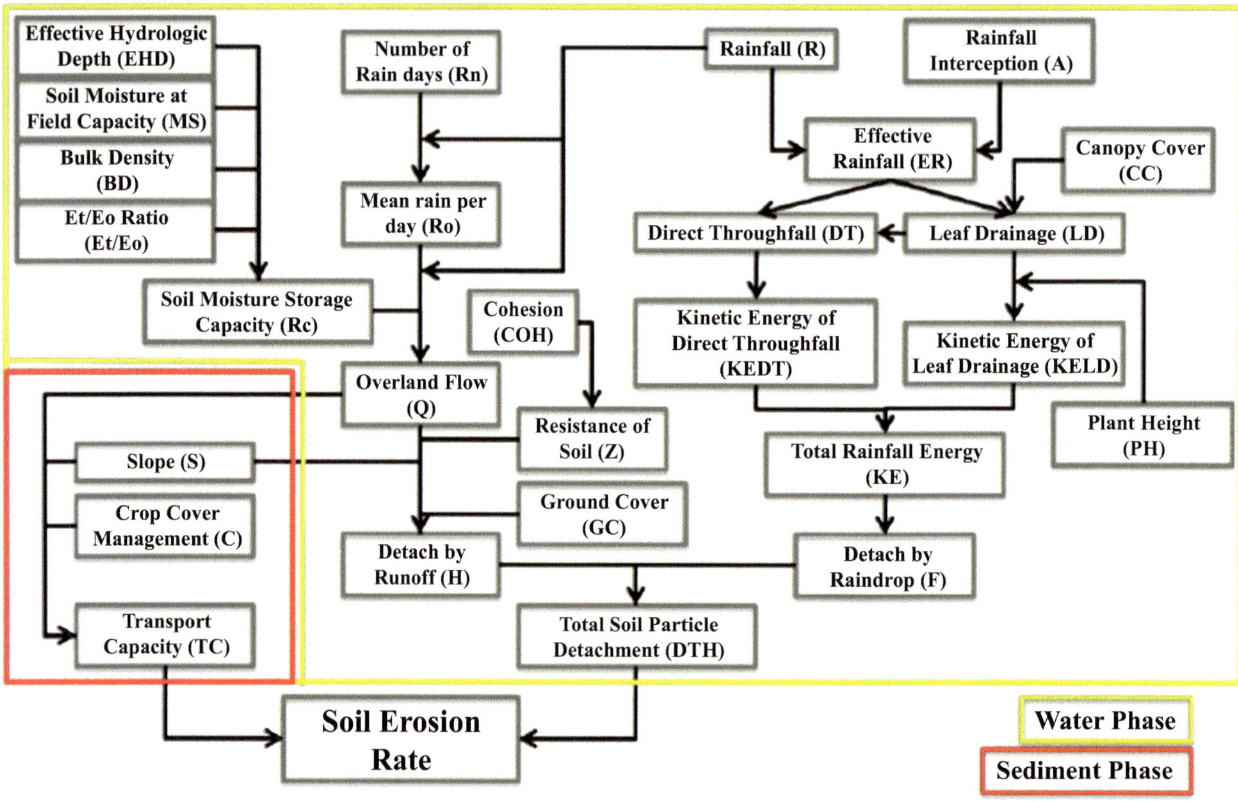

FIGURE 42.3 Flow chart depicting the Morgan–Morgan–Finney method of predicting runoff and soil loss.

Morgan and Duzant (2008). MMF model has two phases: the water phase, which elaborates the process of runoff flow, and the sediment phase, which estimates the soil loss (Fig. 42.3). The sediment phase, which estimates the process of soil loss, is a simplification of the method given by Wischmeier and Smith (1978). The model mainly estimates the impact of a raindrop falling on the soil surface, which detaches the soil particle, and overland flow, which carries the detached soil particles and causes erosion loss. The sediment phase, which estimates the soil loss process, simplifies the method given by Wischmeier and Smith (1978). The sediment phase estimates the soil loss with three empirical equations: firstly, the rate of particle detachment by the impact of rain splash, followed by the rate of particle detachment by runoff, and concluding by estimating the transport capacity of overland flow. The required inputs for sediment phase equations are preceded by the water phase.

42.2.4 Structure of MMF model

The application of the MMF model in the present study was modeled to assess soil erosion by estimating the soil particle detachment by raindrop impact and transportation of detached particles by overland flow. The model comprises two phases, namely a water phase and a sediment phase, which uses six major components and 16 input parameters from rainfall, soil, and land use and land cover.

42.2.5 Water phase

The water phase mainly predicts the detachment of soil particles by rain splash. The data about rainfall such as intensity of rainfall, number of rainy days, and total rainfall is required. The water phase, thus, estimates the kinetic energy of rainfall (Jm^{-2}) and runoff volume of overland flow (mm), and the same is related through the following sequence of relations:

$$\mathrm{ER} = R(1 - A) \tag{42.1}$$

where ER = Effective Rainfall (mm); R = Annual or mean annual rainfall (mm); and A = Proportion of the Rainfall intercepted by the vegetation or crop cover (b/w 0 &1):

$$\text{LD} = \text{ER} \times \text{CC} \tag{42.2}$$

where LD = Leaf Drainage (mm); ER = Effective rainfall (mm); and CC = Percentage of Canopy Cover (b/w 0 & 1):

$$\text{DT} = \text{ER} - \text{LD} \tag{42.3}$$

where DT = Direct through fall (mm); ER = Effective rainfall (mm); and LD = Leaf Drainage (mm):

$$\text{KE(DT)} = \text{DT}(11.9 + 8.7 \log \text{I}) \tag{42.4}$$

where KE (DT) = Kinetic Energy Direct through fall (mm); DT = Direct through fall (mm); and I = Intensity of rainfall (mm):

$$\text{KE(LD)} = \text{LD}\left\{\left(15.8 - \text{PH}^{0.5}\right) - 5.87\right\} \tag{42.5}$$

where KE (LD) = Kinetic Energy of Leaf Drainage (mm); and LD = Leaf Drainage (mm):

$$\text{KE} = \text{KE(DT)} + \text{KE(LD)} \tag{42.6}$$

where KE = Kinetic energy of the rainfall (Jm^{-2}); KE (DT) = Kinetic Energy Direct through fall (mm); and KE (LD) = Kinetic Energy of Leaf Drainage (mm):

$$R_C = 1000 \times \text{MS} \times \text{BD} \times \text{EHD} \times \left(E_t/E_0\right)^{0.5} \tag{42.7}$$

where RC = Soil Moisture Storage Capacity (mm); Ms = Moisture Content at Field Capacity or 1/3 bar tension (wt.%); BD = Bulk Density of the top soil layer (Mgm^{-3}); EHD = Effective Hydrological Depth of Soil (m); value depends on vegetation/crop cover, presence or absence of surface crust, presence of impermeable layer within 0.15 m of the surface; and E_t/E_0 = Ratio of Actual Evapotranspiration (E_t) to Potential Evapotranspiration (E_0):

$$R_0 = R/R_n \tag{42.8}$$

where R_0 = annual rain per rainy days; R = Annual = mean annual rainfall (mm); and R_n = Number of rainy days per year:

$$Q = R \exp\left(-R_C/R_0\right) \tag{42.9}$$

where Q = depth of overland flow (mm); R = Annual or mean annual rainfall (mm); R_C = Soil moisture storage capacity (mm); and R_0 = Annual rain per rain days.

42.2.6 Sediment phase

This is the second phase of the MMF model, which estimates soil loss from the detachment of soil particles by the impact of raindrops, and the transport of detached particles results in overland flow. Thus, the sediment phase comprises two predictive equations, one for the rate of soil detachment by raindrop impact (F kgm^{-2}) and one for the transport capacity of overland flow (G kgm^{-2}). The following relations sequence estimates the annual detachment rate and transport capacity:

$$F = K \times \text{KE} \times 10^{-3} \tag{42.10}$$

where F = Annual rate of soil particle detachment by raindrop impact (kgm^{-2}); K = Soil detachability index (gmJ^{-1}) defined as the weight of soil detached from the soil mass per unit rainfall energy; and KE = Kinetic energy of the rainfall (Jm^{-2}):

$$Z = 1/(0.5\ \text{COH}) \tag{42.11}$$

where Z = constant for runoff detachment depended on soil cohesion; and COH = Cohesion of the Surface Soil (KPa) as measured with a torvane under saturated conditions:

$$H = \text{ZQ}1.5\ \text{Sin}\ S(1 - \text{GC})10^{-3} \tag{42.12}$$

where H = Annual Rate of soil particle detachment by runoff, (kg m^{-2}); Z = Constant for runoff detachment depended on soil cohesion; Q = Depth of overland flow (mm), S = Slope steepness (rad); GC = Percentage of ground cover (b/w 0 & 1):

$$J = F + H \tag{42.13}$$

where J = Annual rate of total soil particle detachment (kg m^{-2}); F = Annual rate of soil particle detachment by raindrop impact (kg m^{-2}); and H = Annual rate of soil particle detachment by runoff (kg m^{-2}):

$$G = CQ2 \text{ Sin } S\ 10 - 3 \tag{42.14}$$

where G = Annual transport capacity of overland flow (kg m^{-2}); and C = Crop cover management factor (Combines the C and P factors of the USLE); and S = Slope steepness (rad).

The model compares and assigns the lower of the two values from the predictions of detachment by rain splash and the transport capacity of the runoff as the annual rate of soil loss, thereby denoting whether detachment or transport is the limiting factor (Morgan and Duzant).

42.2.7 Input parameters of MMF model

The MMF model uses 14 operating functions for which empirical values must be provided for 16 input parameters categorized into three categories: rainfall, soil, and land use factor (Table 42.1).

TABLE 42.1 Input parameters to the Morgan–Morgan–Finney method of predicting soil loss.

Factor	Parameter	Definition
Rainfall	R	Annual or mean annual rainfall (mm)
	R_n	Number of rainy days per year
	I_t	Typical value for intensity of erosive rain (mm hr^{-1}),
Soil	ST	Soil texture
	S	Slope steepness (rad)
	MS	Moisture content at field capacity or 1/3 bar tension (wt.%)
	BD	Bulk density of the top soil layer (Mg m^{-3})
	EHD	Effective hydrological depth of soil (m)
	K	Soil detachability Index (gmJ^{-1}) defined as the weight Of soil detached from the soil mass per unit of rainfall Energy
	COH	Cohesion of the surface soil (KPa) as measured with A torvane under saturated conditions
Land Use & Land Cover Factor	A	Proportion of the rainfall intercepted by the vegetation or crop cover (b/w 0 & 1)
	E_t/E_o	Ratioofactual (E_t) to Potential(E_o) Evapotranspiration
	C	Crop Cover Management factor; combines the C and P factors of the USLE
	CC	Percentage of canopy cover (b/w 0 & 1)
	GC	Percentage of ground cover (b/w 0 & 1)
	PH	Plant height (m), representing the height from which Raindrops fall from the crop or vegetation cover to the Ground surface

42.2.8 QGIS (quantum geographical information system)

It is the main GIS platform used as SDSS, providing free cross-platform and open-source applications for spatial data visualization, editing, and analysis. The MMF model integrated with QGIS plays the role of SDSS and estimates the extent of runoff and soil loss for the Patapur micro watershed. The different layers of functional components of MMF model such as rainfall, soil and land use, and land cover were created using different QGIS plugins and tools, which are open-source to the end users. The generated layers overlaid on the QGIS platform to generate the end maps such as runoff map and annual soil loss map of the Patapur micro watershed. The latest updated version of QGIS (QGIS 3.26 Buenos Aires) was available to download freely from the official website to users worldwide.

42.3 Results and discussion

42.3.1 Estimation of runoff and soil loss using MMF model

The MMF model has two phases' the water phase and the sediment phase. The water phase evaluates the extent of annual runoff across the Patapur watershed by evaluating integral parameters like effective rainfall, which ranges from 573.79 to 691.92 mm across the mapping unit. The total kinetic energy varies across the mapping units, ranging from 12105to 8757 Jm^{-2} and the spatial distribution of soil moisture content at field capacity (Ms) showed a lower value in the hilly part of the catchment as compared to that of flat to gentle slopes under similar management and cover conditions, bulk density was found to be higher (up to 1.33 Mgm^{-3}) in the steep slopes with poor vegetation cover and soil management practices and decreased to about 1.08 Mgm^{-3} around at the foot-slope of the catchment. The effective hydrological top soil depth (EHD) spatial map indicated higher values (0.16 m) in relatively better vegetation-cover areas, stable sites, and flat to gentle slopes. The EHD values were lower, around (0.12 m) in marginalized areas with cultivated and degraded grazing lands. A majority (78%) of the study catchment showed low EHD, indicating that such sites can be the source of higher runoff and soil loss. Similarly, the higher values (0.75–0.85) of $Et/E0$ corresponded to scrubland, whereas lower values (0.62–0.73) corresponded to agricultural areas. The soil moisture storage capacity (RC) calculated as a function of Ms, BD, EHD, and $Et/E0$ varied spatially from 27.3 to 100.68 mm. The estimated runoff (Fig. 42.4) and soil loss (Fig. 42.5) along the soil mapping units and stream junction points (Nodes) is depicted spatially using QGIS. The total average annual runoff depth (Q) estimated for the study area is around 94.9 mm.

The sediment phase calculates the total annual soil loss across the Patapur watershed by estimating soil particle detachment by raindrop impact in the study area, which ranges from 0.34 to 7.91 kgm^{-2}. The sediment phase also estimates the detachment of soil particles by runoff (H), which is a function of the volume of runoff (Q), the detachability of the soil by runoff (Z), the slope angle (S), and the proportion of the soil covered by vegetation (GC). The soil detachability by raindrop impact (F) is also influenced by the soil detachability index (K) and total kinetic energy (KE), which showed higher values (0.79–0.98 $g\,J^{-1}$) on flat areas with coarser soils and poor soil cover and lower values (0.11–0.33 $g\,J^{-1}$) on steep slope with fine soils and good vegetation cover in the study catchment. The spatial variability of the total soil detachment rate (J) due to the summation of F and H for the catchment ranged from 0.34–7.91 kgm^{-2}. The highest rate of F + H (7.91 kgm^{-2}) occurred in soil mapping unit U35, where subsoil-exposed soils had low soil resistance to detaching forces. The lowest rate (0.34 kgm^{-2}) was observed in U14, a protected plantation area and farmlands with high soil quality regardless of the slope steepness.

The spatial distribution of annual soil loss (G) of the Patapur micro watershed showed a higher value (9.74 $t\,ha^{-1}\,y^{-1}$) on steep slopes of the Patapur watershed. A lower value (0.01 $t\,ha^{-1}\,y^{-1}$) was observed on overgrazed lands, sites with bare soils, and intensively cultivated without proper soil management and conservation measures from relatively less disturbed areas (better vegetation and management practices). The obtained results indicated that the extent of soil loss (G) from the study area is affected not only by gradient but also by crop cover, soil conservation practices, and soil erodibility and erosivity conditions. The annual soil loss (G) value was lower than that of the amount of soil detached by raindrop impact (F) + amount of soil separated by the runoff (H), which is attributed to the transport of fewer amounts of detached soils by rainfall drop and runoff impacts. The estimated weighted average soil loss (G) at the outlet of the Patapur micro watershed estimated by the MMF model was 1.6 $t\,ha^{-1}\,y^{-1}$.

42.3.2 Evaluation of MMF model

The SDSS results were evaluated by taking the observed data from the sediment monitoring station of Patapur micro-watershed outlet to verify the applicability of the MMF model in the study area. The total runoff and soil loss observed at the watershed outlet were 108.1 mm and 1.8 $t\,ha^{-1}\,y^{-1}$, respectively, whereas correspondingly estimated values were

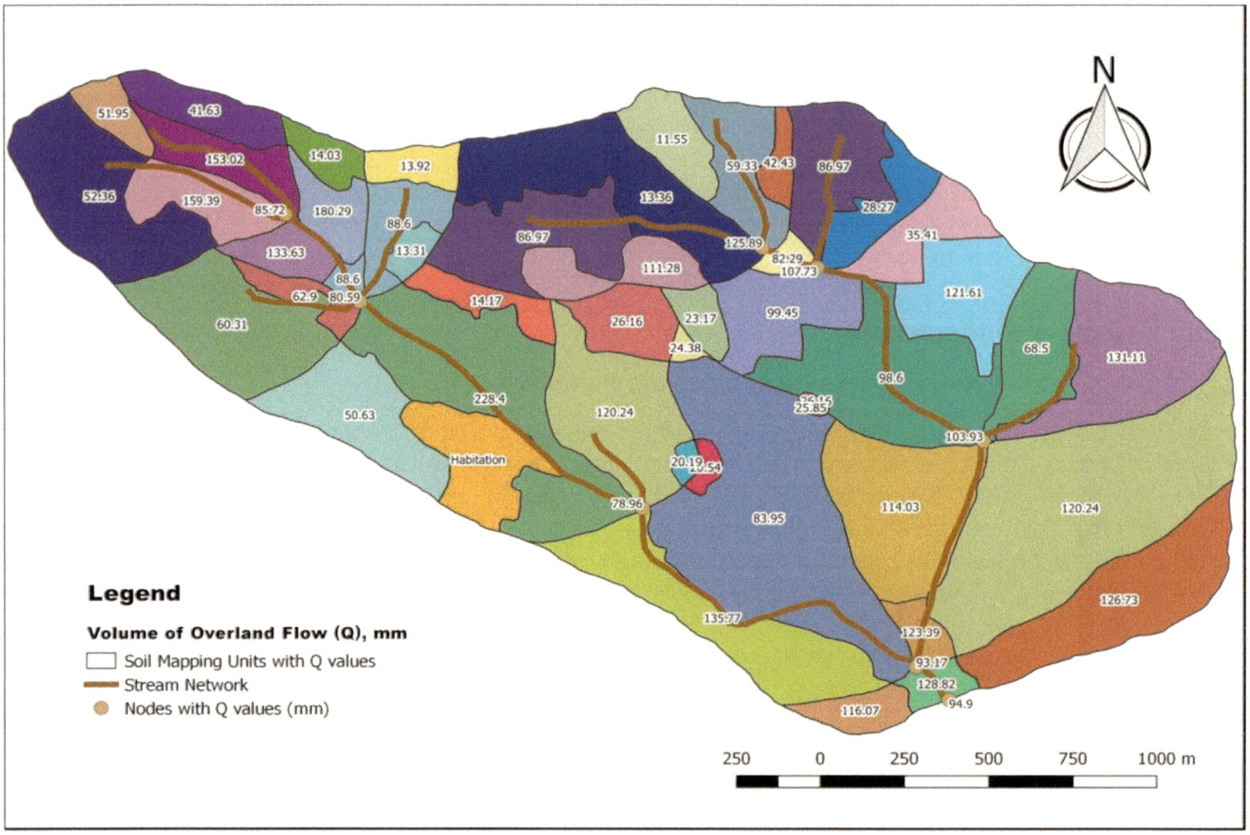

FIGURE 42.4 Soil mapping unit-wise and node-wise depiction of distribution of runoff of Patapur micro watershed.

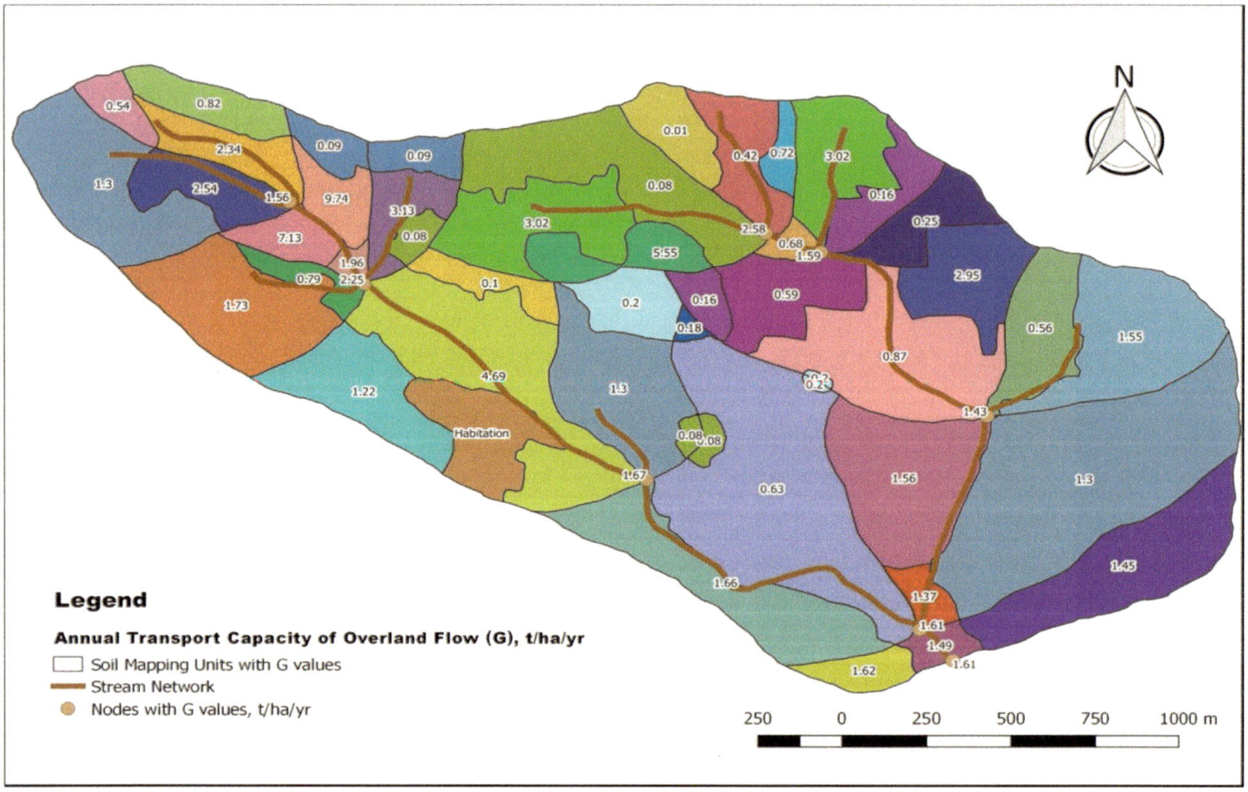

FIGURE 42.5 Soil mapping unit-wise and node-wise annual soil loss in the study area.

94.9 mm and 1.6 t ha^{-1} y^{-1}. These observed runoff and soil loss data were compared with the MMF model's estimated values. The MMF model estimated annual runoff and soil loss rate to an acceptable limit with an accuracy of 88%–89% compared to observed values. Therefore, the spatial pattern of erosion predicted by the MMF model was well correlated with the observed data. Hence, the MMF model can be recommended for widely predicting runoff and soil loss in semiarid climatic regions (Ande et al., 2009; Vaezi, 2014).

42.4 Conclusion

The MMF model used the concepts proposed by Wischmeier and Smith (1978) to provide a more vigorous physical base than the Universal Soil Loss Equation (Wischmeier and Smith, 1978). Yet, it sustains the benefits of the empirical model approach regarding easy availability of data and better understanding. The MMF model has proved simple to use and can give reasonable annual runoff and erosion estimates (Ande et al., 2009).

SDSS is a class of computer-aided spatial data analysis systems in which the technologies of both GIS and Runoff and soil Loss models are applied to various locations, which helps decision-makers resolve the issue with a spatial dimension. The GIS is a general-purpose technology for handling geographic data in digital form, offering a spatial representation of water resource systems. Still, few predictive and related analytical capacities are currently available for solving complex water resource planning and management problems. DSSs are interactive programs that integrate traditional water **resources** simulation and optimization models and adapt new approaches to support users in solving semistructural problems**.**

References

Adinarayana, J., 2001. Spatial decision support system for identifying priority sites for watershed management schemes at district level in India. The Land. 4 (2), 111–130.

Ande, O.T., Alaga, Y., Oluwatosin, G.A., 2009. Soil erosion prediction using MMF model on highly dissected hilly terrain of Ekiti environs in southwestern Nigeria. Inter, J, Phys. Sci. 4 (2), 53–57.

Bekele, W., Drake, L., 2008. Soil and water conservation decision behavior of subsistence farmers in the Eastern Highlands of Ethiopia: a case study of the Hunde-Lafto area. J. Eco. Econ. 46, 437–451.

Elisabeth, F., Bas, V.W., 2009. Predicting runoff from semi-arid hillslopes as source areas for water harvesting in the Sierra de Gador. Southeast. Spain, Catena 79, 83–92.

Guillermo, A.B., Consuelo, C.R., 2007. Assessment of erosion hotspots in a watershed: Integrating the WEPP model and GIS in a case study in the Peruvian Andes. Environ. Model. Soft. 22, 1175–1183.

IARI, 1970. Soil Survey Manual. All India Soil and Land Use Survey. Indian Agricultural Research Institute, New Delhi.

Morgan, R.P.C., Duzant, J.H., 2008. Modified MMF (Morgan–Morgan–Finney) model for evaluating effects of crops and vegetation cover on soil erosion. Earth Surf. Process. Landf. 33 (1), 90–106.

Neelambika, Satishkumar, 2016. Evaluation of hydrologic response of flat land watershed under semi-arid climatic conditions. M.Tech Thesis, UAS, Raichur.

Prasanna, K., Satishkumar, 2014. Development of a decision support system for simulation of runoff and available soil moisture at field scale land holding of watershed. M.Tech Thesis, UAS, Raichur.

Ryan, M.D., Phillip, G., Philip, H., 2002. An Internet-based Spatial Decision Support System for Rangeland Watershed Management .

Sarangi, A., Madramooto, C.A., Enright, P., 2006. Comparison of spatial variability techniques for runoff estimation from a Canadian Watershed. Biosyst. Engg 95 (2), 295–308.

United Nations Population Fund. 2003. UNFPA ANNUAL REPORT: Millennium Development Goals. pp. 30.

Vaezi, A.R., 2014. Modeling runoff from semi-arid agricultural lands in Northwest Iran. Department of Soil Science, Agriculture Faculty, University of Zanjan, Zanjan 45371-38791 (Iran). Soil Science Society of China. Elsevier B.V. and Science Press, ISSN 1002-0160/CN 32–1315/P, 24 (5), 595–604.

Wischmeier, W.H., Smith, D.D., 1978. Predicting rainfall erosion losses. A Guide Conserv. Plan. United States Department of Agriculture, Washington, DC.

Chapter 43

Evaluating the relation of NDVI, NDWI, SMI, and LAI to land and soil degradation processes — a case study of Virudhunagar district, Tamil Nadu, India

K. Chandramohan[1], P. Elayapillai[2], G. Vijayalakshmi[3] and S. Kaliraj[4]

[1]*ICSSR-Post Doctoral Fellow, Tribal Research Centre, Tamil University, Thanjavur, Tamil Nadu, India,* [2]*Department of Literature, Tamil University, Thanjavur, Tamil Nadu, India,* [3]*Bharath Niketan Engineering College, Theni, Tamil Nadu, India,* [4]*National Centre for Earth Science Studies (NCESS), Ministry of Earth Sciences, Govt. of India, Thiruvananthapuram, Kerala, India*

43.1 Introduction

Human land use activities are a major problem across the globe, leading to land degradation and climate change (Klute et al., 1986). However, given that most of the ecosystems' goods and services are not directly measurable by remote sensing datasets (Alistair et al., 2014), satellite imagery is the best source of soil property identification that can be used to prepare digital soil mapping (Ben-Dor et al., 2008; Slaymaker, 2001). Land degradation (LD) is a major issue globally for agriculture and the environment. LD can be studied mostly through two methods, such as field-based observation and remote sensing. They are the best methods when comparing field-based observations, as they have cost- and time-efficient, wherein vast square areas can be studied in a single image to measure the satisfactory extent of degradation (Reddy et al., 2018; El Baroudy, 2011) and soil erosion (Tagore et al., 2012). Remote sensing plays an important role in LD mapping and monitoring at spatial and temporal units. The availability of remote sensing indices and product data on LD is also a prerequisite for land restoration measures (Dubovyk, 2017; Higginbottom and Symeonakis, 2014; Winslow et al., 2011). The current remote sensing product indices assessments vary in their selection of LD classification and spatial datasets used to identify influencing factors (Safriel, 2007). The GIS-based MEDALUS model can used to project the land vulnerability conditions by analyzing multiple geo-environmental parameters.

The normalized difference vegetation index (NDVI) index is used to detect the quantification of a living plant's greenness through the spectral reflectance of visible and NIR bands in the range of $+1$ to -1 (Pradeep Kumar et al., 2022; Kumar et al., 2020; Sekertekin and Bonafoni, 2020; Guha et al., 2018). Positive values of 0 to 1 refer to vegetation availability and health, whereas a negative value of 0 to -1 refers to the occurrence of nonvegetation (other features like rocks, buildups, water bodies, etc.), and no availability of vegetation. This representation of vegetation indices has been highly helpful and quantifies the availability of vegetation to the research community. NDVI has been used to measure the land loss of vegetation and the lack of soil cover offered by vegetation in semiarid environments (Rawat et al., 2021). Prolonged water resources or low rainfall and water demand highly affect vegetation growth in an arid region and lead to land degradation processes (Demetrios et al., 2022; Hadeel. et al., 2011). Normalized difference water index (NDWI) does not remove background soil reflectance effects, NDWI is similar to NDVI, which could be considered an independent vegetation index (Gao, 1996). NDVI has a higher reflectance than NDWI. The combination of the thermal band with empirical value and radiance has been used for soil moisture retrieval. Soil moisture datasets can be obtained through various methods, including in situ measurements, land surface modeling, and remote sensing techniques (Liang and Wang, 2020).

Remote Sensing of Soils. DOI: https://doi.org/10.1016/B978-0-443-18773-5.00040-5

43.2 Study area

Fig. 43.1 shows the location of the study area with sampling points collected from typical sites. The geographical extent of the area is extended between the latitude of 9° 24′ 28.79″ E −9° 34′ 58.08″ E and the longitude of 77° 53′ 47.40″ N − 78° 05′ 25.44″ N, with an area extent of 4888 km^2. LULC features are classified into agricultural lands (irrigated and unirrigated), forest cover, barren, rocky, and mining quarries, water bodies, land with shrubs, settlements (urban and rural), etc. In this study, agricultural land occupied 33.67% of the total area, forest cover occupied 6.23%, barren land occupied 1.06%, and fallow land occupied 0.72%. About 33% of the area is irrigated, and the rest of the area depends on rain tanks. Geographically, the area comprises the major plain with a gentle slope toward the east and south-eastern parts and the hilly terrain (ridges and valley complex) of the Western Ghats in the western part. Soil types consist of red loamy, black soil, lateritic soil, red sandy soil, and sandy alluvium, in which black clay (black cotton) soil is found in most of the area, and the red loamy soils are seen along the foothills of the western part. Red sandy soil is found in the middle and eastern parts, and gravelly loam is seen along the river courses. Most of the area is found as unirrigated croplands, fallows, barrens, and land with shrubs due to poor climate support and lower rainfall rates. However, the majority of the population depends on agricultural activities for their occupations. This study provides primary information for land and soil resource development.

43.3 Materials and methods

This study was conducted for the southern districts of Tamil Nadu; this study investigated the limitations facing the 4291 km^2 area that is geographically referenced with the district administrative boundary. The study area primarily engaged in agricultural practices was more regular, and then the study examined the derived indices that were

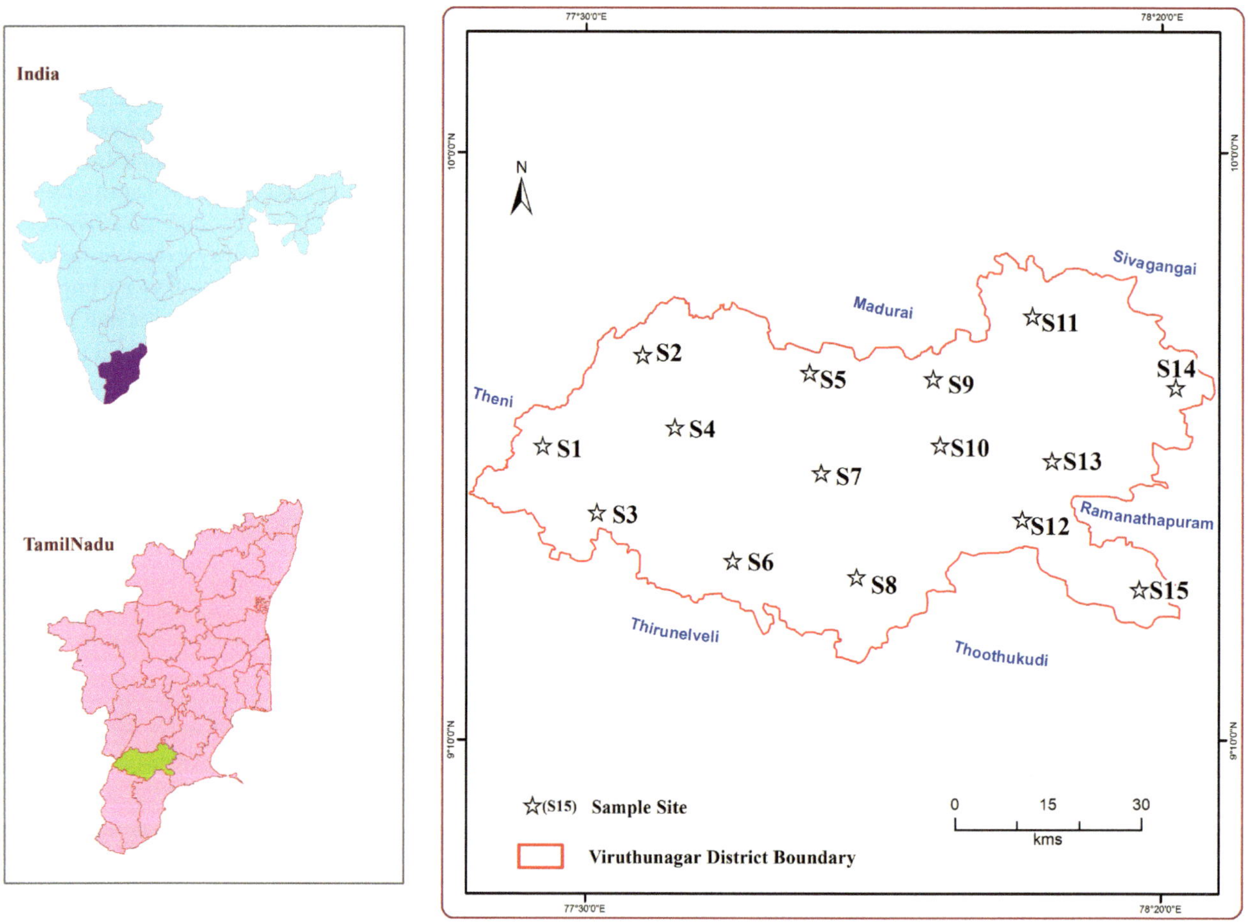

FIGURE 43.1 Location map of the study area—Virudhunagar District, south Tamil Nadu, India.

downloaded from the NASA—USGS Earth Explorer, which have already been derived. The NDVI calculated from the raster in near-infrared (NIR) and red (RED) (Gessesse and Melesse, 2019; Viana, et al., 2019) as follows.

$$\mathrm{NDVI} = \frac{\mathrm{NIR\text{-}RED}}{\mathrm{NIR + RED}} \tag{43.1}$$

NDWI is calculated from the band computation as following

$$\mathrm{NDWI} = \frac{\mathrm{Green\text{-}NIR}}{\mathrm{Green + NIR}} \tag{43.2}$$

Soil moisture index (SMI) can be calculated from the land surface temperature (LST)

$$\mathrm{SMI} = \frac{(\mathrm{LST_{max}} - \mathrm{LST})}{(\mathrm{LST_{max}} - \mathrm{LST_{min}})} \tag{43.3}$$

$$\mathrm{LSTmax} = a1 \times \mathrm{NDVI} + b1 \tag{43.4}$$

$$\mathrm{LSTmin} = a2 \times \mathrm{NDVI} + b2 \tag{43.5}$$

Where the empirical values of a_1, a_2, b_1, and b_2 was derived through linear regression of "a" slope and "b" intercept

$$L = \mathrm{LSTmin} + (((\mathrm{LSTmax} - \mathrm{LSTmin})/(\mathrm{QCALmax} - \mathrm{QCALmin})) \times (\mathrm{DN} - \mathrm{QCALmin})) \tag{43.6}$$

Landsat 8 (Band 10 and 11) radiance maximum and minimum are 0.1003 and 22.0018, respectively (USGS, 2013; Saha et al., 2018).

Leaf area index (LAI) is an important feature of regulatory vegetation and ecological processes like atmospheric exchanges, air moisture, atmospheric gases, the carbon cycle, photosynthesis, evapotranspiration, etc. The study area LAI data was retrieved from Copernicus Global Land Service. Sentinel-3/OLCI Collection 300 m LAI Version 1.1 provides an NRT estimate and three consolidated modes calculated after one (RT1), two (RT2), and six (RT6) decades. One can simply download LAI from https://land.copernicus.eu/global/products/lai. Mapping of land degradation vulnerability is executed using the GIS-based MEDALUS model. The thematic layers of MEDALUS parameters are derived from various data sources like Landsat OLI images (30 m), SOI topographical maps, SRTM DEM (30 m), GSI-published geology maps, and soil maps. Data and a soil analytical method used by the World Soil Information Service (WoSIS) cover 4291 km^2; wherein, the western part of the study area has a deep slope with a hilly region, along with the densely vegetated part of the eastern area toward the Bay of Bengal region of Ramanathapuram and Sivagangai districts, rather than that of a gentle slope. These topographical variations capture the maximum variation in vegetation.

In addition to that, the remote sensing of these derived indices is interrelated to LDVI to identify their spatial relation for the management of land and soil resources. All derived data's pixel information was extracted through spatial analyst tools to extract multivalues from points by random sampling method. Therefore, the image-based analysis was additionally limited to random sampling and data management, with 20 points generated for the extraction of pixel information from each piece of remote sensing data. These points represented the value of pixel identification using the spatial analysis techniques of the "extract multivalue to points" procedure. Finally, the extracted pixel information was assessed to find the influencing or affecting factors and their value.

43.4 Result and discussion

Environmental covariates (Aitkenhead and Coull, 2019) for NDVI, NDWI, and SMI were derived from Landsat-8 multispectral imagery and the LAI download from the Land Copernicus global land survey. Spectral indices allow the acquisition of surface reflectance by satellite for the analysis of multispectral bands. The spectral indices can be used to determine the vegetation's growth and health to improve the accuracy of classification algorithms and increase the quality of the LCLU mapping produced (Karuppaiah et al., 2021; Vignesh et al., 2018; Ustuner et al., 2014). Several indices are available for the spectral measurement of satellite images. Fig. 43.2 depicts the spatial distribution of vegetated and nonvegetated areas. The western part of the study area is fully covered by dense vegetation because of the Western Ghats. The middle part of the study area has low vegetation cover and agricultural activities. The red color is specified for nonvegetated land due to water scarcity. It may lead to land degradation. NDVI ranges from min 0.167 to max 0.254 being in the critical zone and min 0.309 to max 0.404 being in the fragile zone.

The NDWI makes use of reflected NIR and green to improve the presence of such features while eradicating the presence of soil and terrestrial vegetation features. Also, NDWI is highly related to plant water content NDWI (Fig. 43.3) ranges from 0.148 to −0.517 showing the distribution of the water index. The pink color shows a higher

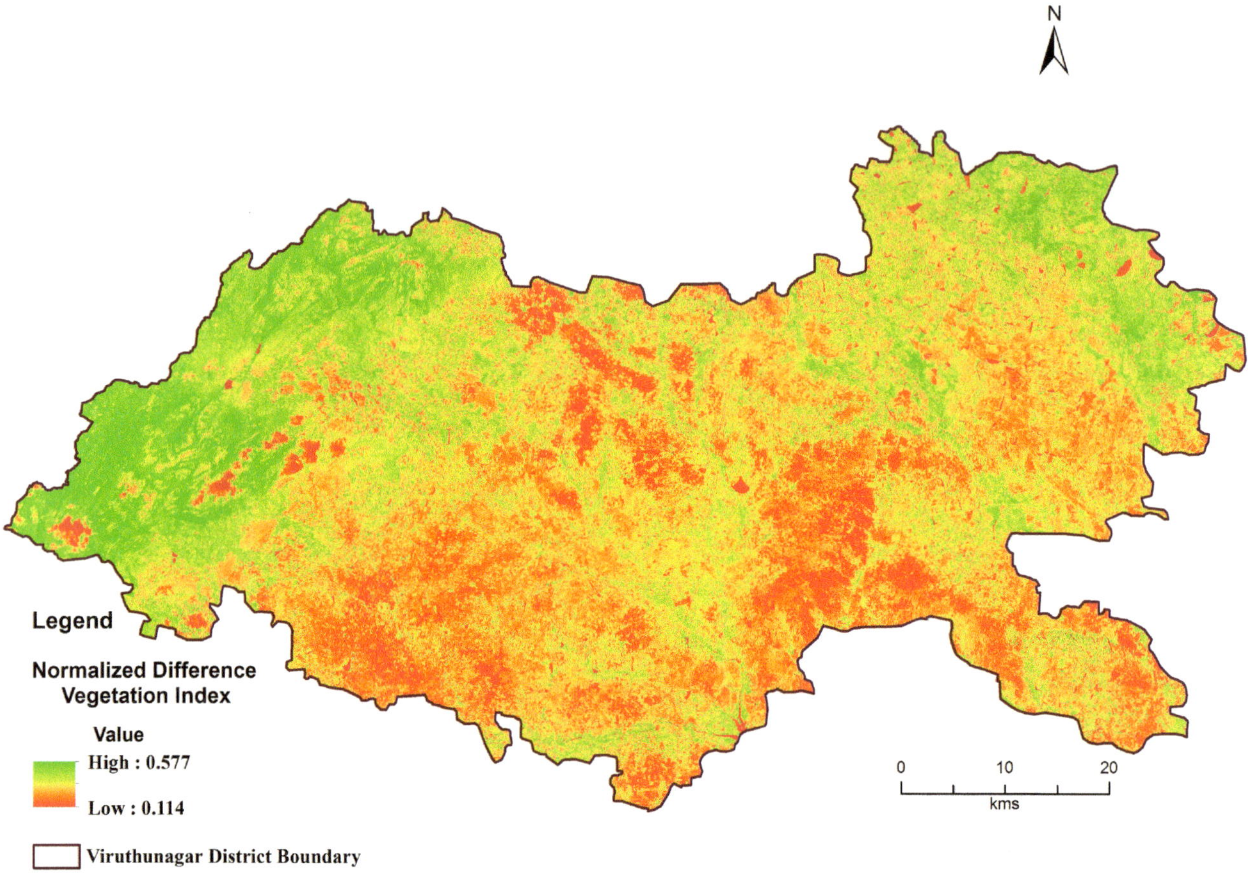

FIGURE 43.2 Normalized difference vegetation index.

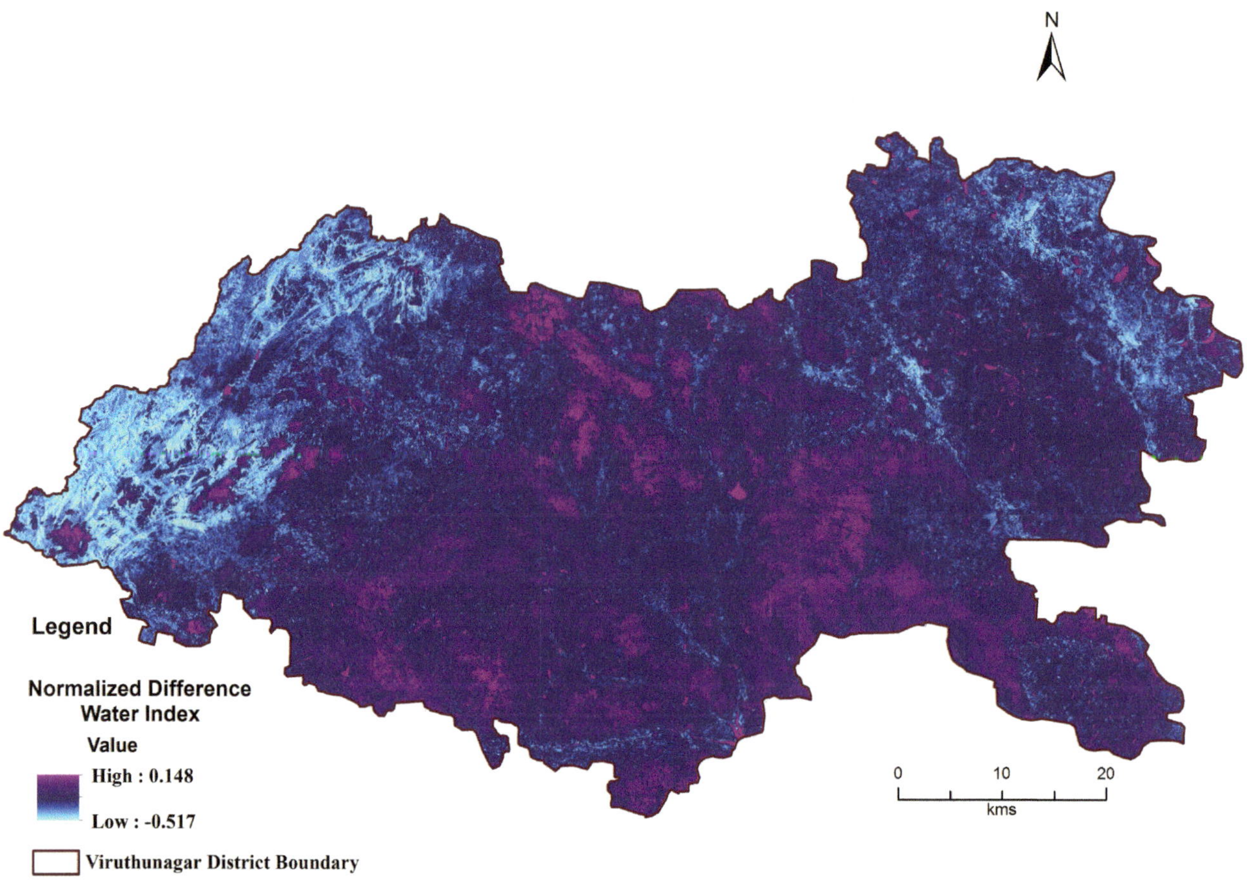

FIGURE 43.3 Normalized difference water index.

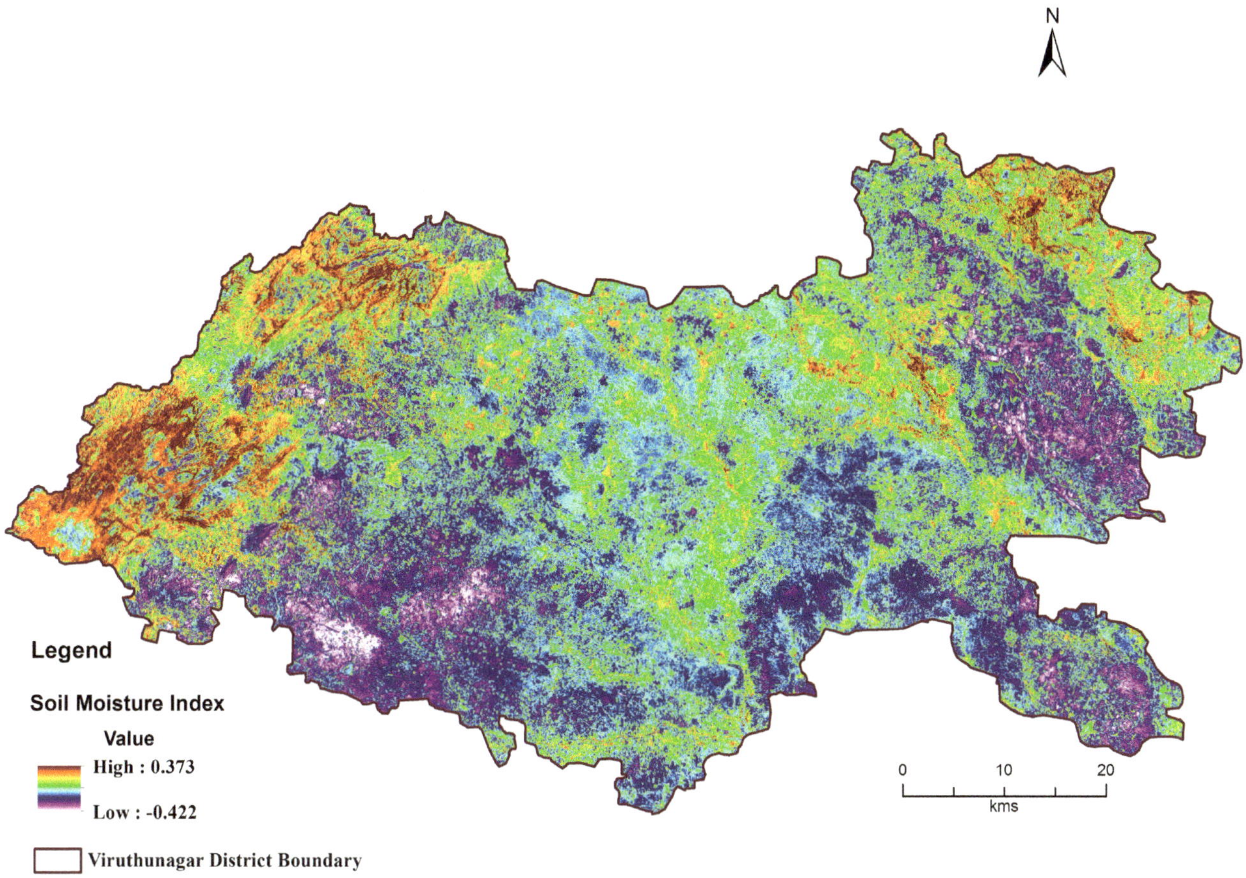

FIGURE 43.4 Soil moisture index.

range than the blue and gray. The normalized difference water index (NDWI) is known to be strongly related to plant water content and plant water stress. Surface soil moisture can spread into water storage, agriculture yield, and vegetation growth and is also an important factor for flood and drought hazard databases. The technique of NIR and green band calculation for NDWI is the more accurate and easiest method for water body information.

The soil moisture index (Fig. 43.4) has been high performing in the areas of the west, east, and river course regions of the study area with a range of 0.373, where white and pink display low moisture in some parts of the study area with the range of −0.422. The green parts of the map show moderately moist soil-indexed areas engaged in agricultural activities.

LAI is one of the most broadly used calculations for labeling plant canopy structure and function because the biosphere and atmosphere exchange energy at the leaf surface. LAI (Fig. 43.5) shows the pixel calculation range of 1.362 to 0.314, which is comparatively equal to all parameters, even in dense to spare LAI conditions.

43.4.1 Land degradation vulnerability assessment

The LDVI (Fig. 43.6) was generated by the GIS-based MEDALUS model for mapping and assessment of land degradation vulnerability in the black cotton soil region of southern India based on the available data sources like soil, rainfall, LULC, vegetation, management practices, and other anthropogenic factors. The result of LDVI is classified into four categories based on their gradation of vulnerability to land degradation: (1) nonaffected zones (N) with an estimated LDVI value of < 1.17; (2) potential vegetative cover (P) with the LDVI value of 1.17−1.22; (3) fragile zones with their three subclasses (F1, F2, and F3), and their LDVI value within the range of 1.22−1.37; and (4) critical zones with their three subclasses (C1, C2, and C3) and their estimated LDVI value is > 1.37.

The results of LDVI C1, C2, and C3 indicate a critical zone found in 84.93% of the total study area, mainly in agricultural and river line areas. In which the subclass C2 covers 60.34% of the area, and subclass C3 is noted under 14.81% of the critical area. The western uplands and forest cover couldn't come under the critical and fragile zone.

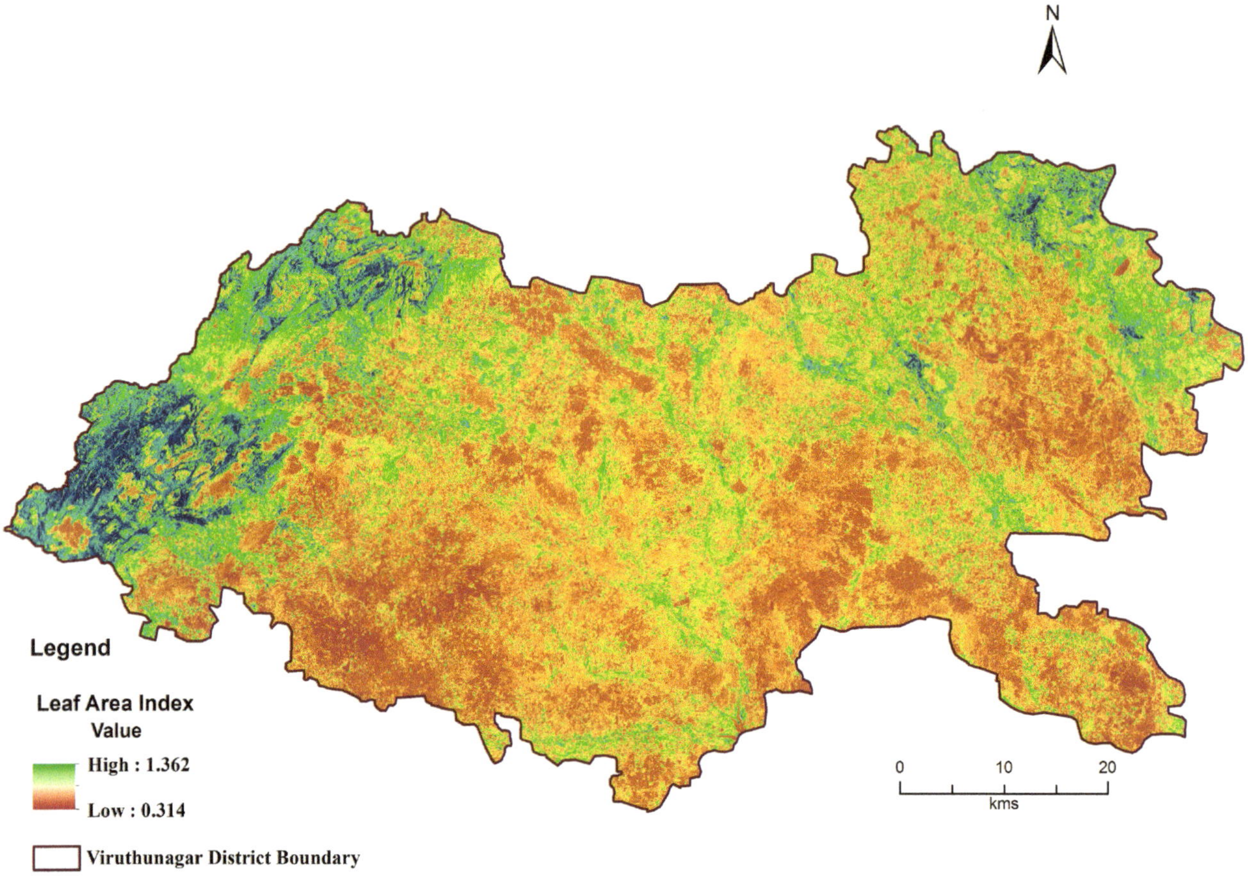

FIGURE 43.5 Leaf area index.

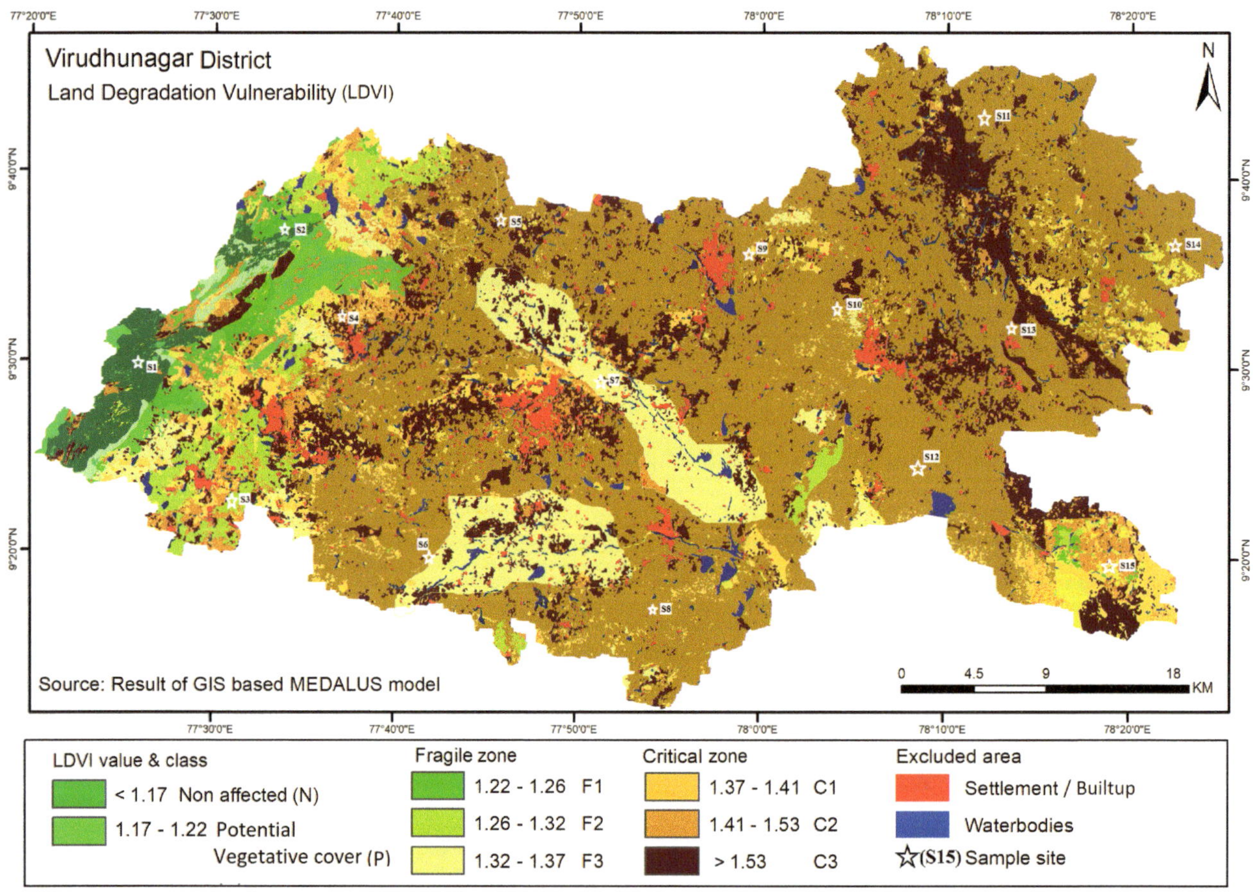

FIGURE 43.6 shows the land degradation vulnerable zones based on the GIS-based MEDALUS model (the LDVI analysis is illustrated by Kaliraj et al. in another chapter).

TABLE 43.1 The spatial relation of NDVI, NDWI, SMI, and LAI with land and soil degradation vulnerable zone in the study area.

Sample ID	LDVI Class	LDVI Value	NDWI	NDVI	SMI	LAI	Factor affecting/influencing
1	Nonaffected (N)	< 1.17	−0.383	0.45	0.275	1.054	LDV is affected by the dynamics of NDVI and LAI; however, it is low (< 1.17)
2	Fragile Zone	1.22 – 1.26 F1	−0.261	0.3	0.097	0.734	All factors influence the fragile zone, and it is estimated to be in the range of 1.22–1.26 except NDWI
3	Critical Zone	> 1.53 C3	−0.174	0.167	0.003	0.586	SMI is very low; it could influence land degradation
4	Critical Zone	1.41 – 1.53 C2	−0.263	0.17	−0.017	0.574	Moderate levels of SMI, NDVI, and LAI cause LDV with moderate vulnerability
5	Excluded area	Settlement/ built-up	−0.237	0.216	−0.028	0.587	Settlements/built-ups are excluded from LDV analysis
6	Critical Zone	1.37 – 1.41 C1	−0.349	0.404	0.206	0.924	SMI, NDVI, and LAI are estimated to a moderate level in LDV critical zone
7	Critical Zone	1.41 – 1.53 C2	−0.215	0.224	0.038	0.641	SMI, NDVI, and LAI are at a moderate level in LDV critical zone
8	Fragile Zone	1.32 – 1.37 F3	−0.285	0.311	0.141	0.783	LAI affects the fragile zone
9	Critical Zone	> 1.53 C3	−0.28	0.309	0.142	0.782	LDVI is at critical zone (C3) affected by NDVI value < 0.3
10	Fragile Zone	1.32 – 1.37 F3	−0.158	0.164	0.042	0.616	NDVI, SMI, and LAI have affected the fragile zone (F3).
11	Fragile Zone	1.22 – 1.26 F1	−0.161	0.173	0.088	0.648	NDVI, SMI, and LAI are at a high rate in the fragile zone
12	Critical Zone	1.41 – 1.53 C2	−0.296	0.335	0.152	0.81	LDVI is observed at the critical zone (C2) with a moderate NDVI < 0.3, SMI < 0.15, and LAI < 0.8, which indicates the adverse anthropogenic impacts affecting LDVI
13	Excluded area	Water bodies	0.126	0.236	0.097	0.694	Water bodies (lakes, ponds, rivers) are excluded from LDVI
14	Fragile Zone	1.32 – 1.37 F3	−0.221	0.245	0.084	0.689	NDVI, SMI, and LAI influence fragile zone (F3) at a higher rate.
15	Critical Zone	1.41 – 1.53 C2	−0.261	0.254	0.044	0.662	SMI, NDVI, and LAI are at a moderate rate that affecting LDVI critical zone

Where the indices derived random samplings data of NDVI, LAI, and SMI are varied by their spectral reflectance but NDWI samplings are coming under the normal as −0.2. The assessment of the Table 43.1 calculation should noticeably be the indices of NDVI, NDWI, SAMI, and LAI directly or indirectly connected with the LDVI. As most of the study area are highly influenced by land degradation, the remote sensing product indices gradually reduce their range. However some parts of the study area moderately affected or did not affect land degradation due to the moderate and high range of remote sensing product indices, respectively.

43.4.2 Spatial relation of NDVI, NDWI, SMI, and LAI with land and soil degradation zones

The LDVI mapping for this area is illustrated in another chapter of this book by Kaliraj et al. (2023). Importantly, this analysis reveals the statistical relation of the above vegetative indices with land degradation zones at 30 m × 30 m pixel level, to provide a primary database for land and soil resource management. The result shows that the NDVI is noted in the range of 0.577 − 0.114, wherein the higher value indicates the occurrence of healthy vegetation, mainly in upland forests and irrigated croplands in the western parts. Likewise, the NDWI is estimated at the range of 0.148− − 0.517, and the areas having a lower value are denoted by higher vegetation−water stress and poor suitability of plant growth and biomass rate, and it is especially found in the middle and eastern parts. Similarly, the SMI is calculated in the range of 0.373− − 0.422, thereby, the high rate is noted as severe soil−water stress for plant growth, while the higher value indicates that the land and soil have enough soil moisture and water content that indicates good quality soils and support for plant growth. The LAI is estimated to be in the range of 1.362 − 0.314, which indicates the growth rate of plants or biomass, wherein the lowest values show poor vegetative cover. Significantly, the areas under the critical LD class have lower NDVI, LAI, and NDWI while the SMI is noted at a higher rate due to poor vegetation, biomass rate, soil moisture, and inadequate rainfall that drastically induce the processes of land and soil degradation on a long-term scale. Importantly, the study area is gradually changing in LULC features that affect land use productivity through the activities of aridity, soil erosion, soil salinity, and other anthropogenic impacts. Major parts of the plains consist of unirrigated lands, fallows, barrens, and land with shrubs experiencing severe drought events happening in tropical regions like the study area. Samples collected from typical sites were analyzed, which indicates that the NDVI is estimated at < 0.450, the NDWI is noted as -0.232, the SMI value < 0.012, and the LAI value is < 0.618 supports and affects the land degradation and all indices coming under the nonaffected area (N) Fig. 43.6. We conclude that the study will perform future developments in machine learning that are acceptable and highly sensible, will initiate the researchers to move to expand their prediction of land degradation, and will lead to analyzing soil erosion, agroeconomic activity, land-use planning, etc.

43.5 Conclusion

The study reveals that the reduced range of NDVI, NDWI, LAI, and SMI influences the processes of land degradation in the area. Land degradation is low in the western parts of the area, where most hills and slopes have slight variations in indices. Significantly, the areas under the critical LD class are related to lower NDVI, LAI, and NDWI, while the SMI is noted at a higher rate due to poor vegetation, biomass rate, soil moisture, and inadequate rainfall that drastically induce the processes of land and soil degradation in long term. Importantly, the study area is gradually changing in LULC features that affect land use productivity through the activities of aridity, soil erosion, soil salinity, and other anthropogenic impacts. Major parts of the plains consist of unirrigated lands, fallows, barrens, and land with shrubs experiencing severe drought events in tropical regions like the study area. Prevention of damage to land degradation related to extreme hazards and measurement based on the regularly updated GIS database. The resulting maps of LDVI, NDVI, NDWI, LAI, and SMI allowed for the uniqueness of areas with higher, medium, and lower vulnerability to land degradation and fragility. This study finds that remote sensing data such as indices have the potential to be powerful supporting variables for improving estimates of land degradation. The dissimilar images derived from the raster calculation of Landsat 8 and Copernicus LAI data support the detection of the number of degraded areas that might be affected by future hazard events. Thus, the combination of the assessment results based on remote sensing index data demonstrates its effectiveness as an input for hazard calculation.

References

Aitkenhead, M., Coull, M., 2019. Mapping soil profile depth, bulk density and carbon stock in Scotland using remote sensing and spatial covariates. Eur. J. Soil. Sci. . Available from: https://doi.org/10.1111/ejss.12916.

Alistair, M.S.S., Kolden, C.A., Tinkham, W.T., Talhelm, A.F., Marshall, J.D., Hudak, A.T., et al., 2014. Remote sensing the vulnerability of vegetation in natural terrestrial ecosystems. Elsevier Inc. Available from: http://doi.org/10.1016/j.rse.2014.03.0380034-4257.

Ben-Dor, E., et al., 2008. Imaging spectrometry for soil applications. Adv. Agron. 97, 321−392.

Demetrios, E.T., Christos, A.K., Kleomenis, K., Efthimios, Z., Constantina, G.V., Nikolaos, A.S., et al., 2022. Evaluating the degradation of natural resources in the Mediterranean environment using the water and land resources degradation index, the case of Crete Island. MDPI, Atmos. 13 (1), 135. Available from: https://doi.org/10.3390/atmos13010135.

Dubovyk, O., 2017. The role of remote sensing in land degradation assessments: opportunities and challenges. Eur. J. Remote. Sens. 50 (1), 601−613. Available from: https://doi.org/10.1080/22797254.2017.1378926.

El Baroudy, A.A., 2011. Monitoring land degradation using remote sensing and GIS techniques in an area of the middle Nile Delta, Egypt. CATENA 87 (2), 201−208. Available from: https://doi.org/10.1016/j.catena.2011.05.023. ISSN 0341-8162.

Gao, B.-c, 1996. NDWI—a normalized difference water index for remote sensing of vegetation liquid water from space. Remote. Sens. Environ. 58 (3), 257−266. Available from: https://doi.org/10.1016/S0034-4257(96)00067-3. ISSN 0034-4257.

Gessesse, A.A., Melesse, A.M., 2019. Chapter 8—Temporal relationships between time series CHIRPS-rainfall estimation and eMODIS-NDVI satellite images in Amhara Region, Ethiopia. In: Melesse, A.M., Abtew, W., Senay, G. (Eds.), Extreme Hydrology and Climate Variability. Elsevier, pp. 81–92. ISBN 9780128159989. Available from: https://doi.org/10.1016/B978-0-12-815998-9.00008-7.

Guha, S., Govil, H., Dey, A., Gill, N., 2018. Analytical study of land surface temperature with NDVI and NDBI using Landsat 8 OLI and TIRS data in Florence and Naples city, Italy Eur. J. Remote. Sens. 51 (1), 667–678.

Hadeel, A., Jabbar, M., Chen, X., 2011. Remote sensing and GIS application in the detection of environmental degradation indicators. Geo-spatial Inf. Sci. 14 (1), 39–47. Available from: https://doi.org/10.1007/s11806-011-0441-z.

Higginbottom, T.P., Symeonakis, E., 2014. Assessing land degradation and desertification using vegetation index data: current frameworks and future directions. Remote. Sens. 6 (10), 9552–9575. Available from: https://doi.org/10.3390/rs6109552.

Karuppaiah, C., Ramachandran, V., Sangu, M., 2021. Role of geographical information system and principal component analysis for identifying vulnerable zone of surface water epidemiology. MDPI Int. J. Environ. Res. Public. Health. Available from: https://doi.org/10.3390/ECERPH-3-08984.

Kumar, B.P., Babu, K.R., Rajasekhar, M., Ramachandra, M., 2020. Identification of land degradation hotspots in semiarid region of Anantapur district, Southern India, using geospatial modeling approaches model. Earth Syst. Environ. 6 (3), 1841–1852.

Pradeep Kumar, B., Raghu Babu, K., Anusha, B.N., Rajasekhar, M., 2022. Geo-environmental monitoring and assessment of land degradation and desertification in the semi-arid regions using Landsat 8 OLI/TIRS, LST, and NDVI approach. Environ. Chall. 8 (2022), 100578. Available from: https://doi.org/10.1016/j.envc.2022.100578. ISSN 2667-0100.

Rawat, M., Karwariya, S., Raushan, R., Kanga, S., Taloor, A.K., Thapliyal, A., 2021. Snow cover and land surface temperature assessment of Mana basin Uttarakhand India using MODIS satellite data, Water, Cryosphere, and Climate Change in the Himalayas, 2021. Springer, Cham, pp. 159–174.

Reddy, G.P.O., Kumar, N., Singh, S.K., 2018. Remote sensing and GIS in mapping and monitoring of land degradation. In: Reddy, G., Singh, S. (Eds.), Geospatial Technologies in Land Resources Mapping, Monitoring and Management. Geotechnologies and the Environment, vol 21. Springer, Cham. Available from: https://doi.org/10.1007/978-3-319-78711-4_20.

Safriel, U., 2007. The assessment of global trends in land degradation. In: Sivakumar, M.K., Ndiang'ui, N. (Eds.), Climate and Land Degradation. Springer, Berlin Heidelberg, pp. 1–38.

Saha, A., Patil, M., Chandra Goyal, V., Singh Rathore, D., 2018. Assessment and impact of soil moisture index in agricultural drought estimation using remote sensing and GIS techniques. MDPI, Proc. . Available from: https://doi.org/10.3390/ECWS-3-05802.

Sekertekin, A., Bonafoni, S., 2020. Land surface temperature retrieval from Landsat 5, 7, and 8 over rural areas: assessment of different retrieval algorithms and emissivity models and toolbox implementation. Remote. Sens. (Basel) 12 (2), 294.

Slaymaker, O., 2001. The role of remote sensing in geomorphology and terrain analysis in the Canadian Cordillera. Int. J. Appl. Earth Obs. Geoinf. 3 (1), 7.

Tagore, Gopal, Bairagi, G., Sharma, N.K., Sharma, R., Verma, S., 2012. Mapping of degraded lands using remote sensing and GIS techniques. J. Agric. Phys. 12, 29–36.

US Geological Survey (USGS) and US Department of the Interior. Normalized Difference Water Index (NDWI). <http://deltas.usgs.gov/fm/data/data_ndwi.aspx>, 2013.

Ustuner, M., et al., 2014. Crop type classification using vegetation indices of RapidEye imagery. ISPRS—Int. Arch. Photogramm., Remote. Sens. Spat. Inf. Sci. XL-7 (2014), 195–198. Available from: https://doi.org/10.5194/isprsarchives-xl-7-195-2014.

Viana, Cláudia M., Oliveira, Sandra, Oliveira, S.érgio C., Rocha, Jorge, 2019. 29—Land use/land cover change detection and urban sprawl analysis. In: Reza Pourghasemi, Hamid, Gokceoglu, Candan (Eds.), Spatial Modeling in GIS and R for Earth and Environmental Sciences. Elsevier, pp. 621–651. ISBN 9780128152263. Available from: https://doi.org/10.1016/B978-0-12-815226-3.00029-6.

Vignesh, M., Chandramohan, K., Santhosh, R., SenthilKumar, M., SharafathNihal, J., 2018. Dynamics of urban sprawl direction of Madurai City, Tamil Nadu using digital image processing of unsupervised classification technique. Int. J. Creat. Res. Thoughts (IJCRT) 6 (2).

Winslow, M.D., Vogt, J.V., Thomas, R.J., Sommer, S., Martius, C., Akhtar-Schuster, M., 2011. Science for improving the monitoring and assessment of dryland degradation. Land. Degrad. Dev. 22 (2), 145–149. Available from: https://doi.org/10.1002/ldr.1044.

Index

Note: Page numbers followed by "*f*" and "*t*" refer to figures and tables, respectively.

B

C

D

E

H

I

J

K

M

Q

R

S

T

U

V

W

X

Y

Z